AEROBIOLOGICAL ENGINEERING HANDBOOK

A Guide to
Airborne Disease Control Technologies

Wladyslaw Jan Kowalski, Ph.D., P.E.
*The Indoor Environment Center
Department of Architectural Engineering
The Pennsylvania State University
University Park, Pennsylvania*

McGRAW-HILL
New York   Chicago   San Francisco   Lisbon   London   Madrid
Mexico City   Milan   New Delhi   San Juan   Seoul
Singapore   Sydney   Toronto

The **McGraw·Hill** Companies

Cataloging-in-Publication Data is on file with the Library of Congress.

Copyright © 2006 by The McGraw-Hill Companies, Inc. All rights reserved. Printed in the United States of America. Except as permitted under the United States Copyright Act of 1976, no part of this publication may be reproduced or distributed in any form or by any means, or stored in a data base or retrieval system, without the prior written permission of the publisher.

1 2 3 4 5 6 7 8 9 0   DOC/DOC   0 1 0 9 8 7 6 5

ISBN 0-07-140245-4

The sponsoring editor for this book was Kenneth P. McCombs and the production supervisor was Pamela A. Pelton. It was set in Times Roman by International Typesetting and Composition. The art director for the cover was Anthony Landi.

Printed and bound by RR Donnelley.

This book was printed on recycled, acid-free paper containing a minimum of 50% recycled, de-inked fiber.

McGraw-Hill books are available at special quantity discounts to use as premiums and sales promotions, or for use in corporate training programs. For more information, please write to the Director of Special Sales, McGraw-Hill Professional, Two Penn Plaza, New York, NY 10121-2298. Or contact your local bookstore.

Information contained in this work has been obtained by The McGraw-Hill Companies, Inc. ("McGraw-Hill") from sources believed to be reliable. However, neither McGraw-Hill nor its authors guarantee the accuracy or completeness of any information published herein, and neither McGraw-Hill nor its authors shall be responsible for any errors, omissions, or damages arising out of use of this information. This work is published with the understanding that McGraw-Hill and its authors are supplying information but are not attempting to render engineering or other professional services. If such services are required, the assistance of an appropriate professional should be sought.

CONTENTS

Preface   xiii
Acknowledgments   xv
List of Symbols   xvii

Section 1   Background and History

Chapter 1. Airborne Disease and the Indoor Environment   3

1.1. Introduction / 3
1.2. Airborne Disease Today / 3
1.3. The History of Airborne Disease / 6
1.4. The Future of Disease Control / 12
     References / 13

Chapter 2. Airborne Pathogens and Allergens   15

2.1. Introduction / 15
2.2. Airborne Pathogens and Allergens / 15
2.3. Viruses / 16
2.4. Bacteria / 18
2.5. Fungal Spores / 19
2.6. Protozoa / 20
2.7. Algae / 20
2.8. Pollen / 21
2.9. Dust Mites and Storage Mites / 23
2.10. Cockroach Allergens / 24
2.11. Animal Dander / 25
2.12. Toxins / 26
2.13. Microbial Volatile Organic Compounds / 27
2.14. Potential and Emerging Airborne Pathogens / 29
2.15. Airborne Biological Weapons / 29
2.16. The Airborne Pathogen and Allergen Database / 30
     References / 32

Chapter 3. The Pathology of Airborne Disease   41

3.1. Introduction / 41
3.2. The Human Respiratory System / 41
3.3. Mechanisms of Lung Infection / 44
3.4. Disease Groups / 46
3.5. Airborne Respiratory Infections / 48

3.6. Upper Respiratory Tract Infections / 48
3.7. Middle Respiratory Tract Infections / 49
3.8. Lower Respiratory Tract Infections / 50
3.9. Noninfectious Respiratory Diseases / 51
3.10. Nosocomial and Opportunistic Respiratory Diseases / 61
3.11. Nonrespiratory Airborne Diseases / 62
3.12. Other Diseases / 63
References / 63

Chapter 4. Epidemiology and Dosimetry   67

4.1. Introduction / 67
4.2. Airborne Disease Statistics / 67
4.3. Routes of Transmission / 69
4.4. The Epidemiology of Airborne Diseases / 82
4.5. Disease Progression Curves / 87
4.6. Dosimetry of Airborne Disease / 90
4.7. Microbial Agent Ingestion Dosimetry / 94
4.8. Toxin Dosimetry / 94
References / 95

Chapter 5. Aerobiology of the Outdoors   99

5.1. Introduction / 99
5.2. Outdoor Pathogens and Allergens / 99
5.3. Outdoor Pathogenic Disease / 104
5.4. Fungal Allergies and Outdoor Air / 106
5.5. Pollen Allergies in Outdoor Air / 108
5.6. Other Allergies in Outdoor Air / 110
5.7. Asthma and Outdoor Air / 110
5.8. Outdoor Airborne Animal Diseases / 111
5.9. Survival of Microorganisms in the Environment / 112
References / 115

Chapter 6. Aerosol Science and Particle Dynamics   119

6.1. Introduction / 119
6.2. Size and Shape of Bioaerosols / 119
6.3. Surface Characteristics / 120
6.4. Fractal Dimension / 122
6.5. Aerodynamic Diameter / 122
6.6. Equivalent Diameter / 123
6.7. Microbial Density / 126
6.8. Size Distribution / 126
6.9. Bioaerosol Dynamics / 130
6.10. Settling and Evaporation of Bioaerosols / 137
6.11. Bioaerosol Viability / 139
References / 140

Chapter 7. Microbial Disinfection Fundamentals   143

7.1. Introduction / 143
7.2. The Classic Exponential Growth Model / 143
7.3. The Classic Exponential Decay Model / 145
7.4. The Classic Two-Stage Decay Model / 148

7.5. The Classic Shoulder Model / *148*
7.6. The Multihit Target Model / *151*
7.7. Recovery and Photoreactivation / *153*
7.8. Specific Disinfection Models / *153*
7.9. Sterilization / *161*
  References / *162*

Chapter 8. Buildings and Enclosures — 165

8.1. Introduction / *165*
8.2. The Evolution of Human Habitats / *166*
8.3. Building Types / *168*
8.4. Building Airtightness / *169*
8.5. Building Dampness / *170*
8.6. Building Materials / *172*
  References / *179*

Section 2 Airborne Disease Control Technologies

Chapter 9. Ventilation Systems and Dilution — 185

9.1. Introduction / *185*
9.2. Ventilation System Types / *185*
9.3. Biocontamination in Ventilation Systems / *189*
9.4. Pressurization Control / *191*
9.5. Dilution Ventilation / *191*
9.6. The Single-Zone Steady-State Model / *192*
9.7. Dilution Modeling with Calculus / *195*
9.8. Single-Zone Transient Modeling / *197*
9.9. Multizone Modeling / *201*
9.10. Computational Fluid Dynamics Modeling / *204*
  References / 204

Chapter 10. Filtration of Airborne Microorganisms — 207

10.1. Introduction / *207*
10.2. Filter Performance Curves / *207*
10.3. MERV Filter Ratings / *210*
10.4. Modeling Filter Performance / *212*
10.5. Mathematical Model of Filtration / *212*
10.6. The Multifiber Filtration Model / *215*
10.7. Fitting the Models to MERV Data / *217*
10.8. Microbial Filtration Test Results / *221*
10.9. Filter Application Test Results / *224*
10.10. Special Filter Types / *227*
  References / *227*

Chapter 11. Ultraviolet Germicidal Irradiation — 231

11.1. Introduction / *231*
11.2. Types of UVGI Systems / *231*
11.3. UVGI Disinfection Theory / *235*
11.4. UVGI Disinfection Modeling / *237*

11.5. Modeling the UV Dose / 240
11.6. UV Lamp Ratings / 243
11.7. The UVGI Rating Value / 243
11.8. UV Irradiance Field Due to Enclosure Reflectivity / 243
11.9. Air Mixing Effects / 248
11.10. Relative Humidity Effects / 249
11.11. Photoreactivation / 250
11.12. Air Temperature Effects / 251
11.13. Performance Optimization / 252
11.14. Cooling Coil Irradiation / 255
11.15. Design Guides / 256
11.16. UVGI Performance in Field Trials / 257
11.17. Combined Performance of UVGI and Filtration / 260
References / 262

Chapter 12. Gas Phase Filtration   267

12.1. Introduction / 267
12.2. Aerobiological Applications of Gas Phase Filtration / 268
12.3. Types of Gas Phase Filters / 269
12.4. Carbon Adsorption / 270
12.5. Impregnated Carbon / 274
12.6. Effect on Microbiological Particles / 275
12.7. Operation and Maintenance / 275
12.8. Applications / 276
References / 279

Chapter 13. Electrostatic Filtration   281

13.1. Introduction / 281
13.2. Electrostatic Precipitation / 281
13.3. Electronic Air Cleaners / 285
13.4. Electrostatic Filters / 286
13.5. Electret Filters / 287
13.6. Electrically Enhanced Filtration / 287
13.7. Corona Wind Air Cleaners / 289
13.8. Aerobiological Performance of ESPs / 291
13.9. Economics of Electrostatic Filters / 292
References / 292

Chapter 14. Photocatalytic Oxidation   295

14.1. Introduction / 295
14.2. PCO Theory and Operation / 295
14.3. Air Disinfection / 297
14.4. Destruction of MVOCs / 301
14.5. Surface Disinfection / 302
14.6. PCO Applications / 302
References / 303

Chapter 15. Pulsed Light   307

15.1. Introduction / 307
15.2. Pulsed White Light and Pulsed UV / 307

15.3. Filtered Pulsed Light / *316*
15.4. Pulsed Electric Fields / *317*
15.5. Pulsed Lasers / *318*
References / *319*

Chapter 16. Ionization — 323

16.1. Introduction / *323*
16.2. Ionization Theory / *323*
16.3. Ionization Equipment / *324*
16.4. Ionization of Airborne Microbes / *326*
16.5. Ionization of MVOCs / *330*
References / *330*

Chapter 17. Ozone — 333

17.1. Introduction / *333*
17.2. Ozone Chemistry / *333*
17.3. Airborne Ozone Disinfection / *336*
17.4. Ozone for Surface Sterilization / *338*
17.5. Ozone Removal / *339*
17.6. Ozone Health Hazards / *340*
17.7. Ozone System Performance / *340*
References / *344*

Chapter 18. Green Technologies — 349

18.1. Introduction / *349*
18.2. Passive Solar Exposure / *350*
18.3. Vegetation Air Cleaning and Biofiltration / *354*
18.4. Material Selectivity / *356*
18.5. Hygienic Protocols / *357*
18.6. Aerobiologically Green Buildings / *360*
References / *363*

Chapter 19. Thermal Disinfection, Cryogenics, and Desiccation — 365

19.1. Introduction / *365*
19.2. Thermal Disinfection / *365*
19.3. Cryogenics / *367*
19.4. Desiccation and Dehumidification / *367*
19.5. Thermal Cycling / *368*
References / *368*

Chapter 20. Antimicrobial Coatings — 371

20.1. Introduction / *371*
20.2. Performance of Materials / *372*
20.3. Health Care and Pharmaceutical Applications / *375*
20.4. Food Industry Applications / *375*
20.5. Residential and Commercial Building Applications / *376*
References / *377*

Chapter 21. Microwaves — 379

21.1. Introduction / 379
21.2. Microwave Thermal Disinfection / 379
21.3. Microwave Athermal Disinfection / 383
References / 386

Chapter 22. Alternative and Developmental Technologies — 389

22.1. Introduction / 389
22.2. Ionizing Radiation / 389
22.3. Ultrasonication / 392
22.4. Plasma Technology / 393
22.5. Corona Discharge / 394
22.6. Chemical Disinfection / 395
22.7. Free Radicals / 397
References / 397

Section 3 Testing and Remediation

Chapter 23. Air and Surface Sampling — 403

23.1. Introduction / 403
23.2. Air and Surface Microflora / 403
23.3. Growing Microbial Cultures / 405
23.4. Surface Sampling / 408
23.5. Air Sampling / 409
23.6. Sampling Procedures / 414
23.7. Sampling Applications / 421
References / 423

Chapter 24. Biodetection and Monitoring — 427

24.1. Introduction / 427
24.2. Particle Detectors / 427
24.3. Mass Spectrometry / 430
24.4. LIDAR / 430
24.5. Polymerase Chain Reaction / 430
24.6. Biosensors / 432
References / 434

Chapter 25. Indoor Limits and Guidelines — 437

25.1. Introduction / 437
25.2. Indoor Bioaerosol Levels / 437
25.3. Indoor/Outdoor Ratios / 441
25.4. Indoor Limits / 443
References / 445

Chapter 26. Remediation — 449

26.1. Introduction / 449
26.2. Air Cleaning / 449

26.3. Surface Cleaning / 450
26.4. Removal of Building Materials / 451
26.5. Whole Building Decontamination / 451
26.6. Time and Weathering / 454
26.7. Mold Remediation Procedures / 454
26.8. Chemical Decontamination Procedures / 456
References / 460

Chapter 27. Testing and Commissioning 463

27.1. Introduction / 463
27.2. Ventilation System Testing / 463
27.3. Building Leak and Pressure Testing / 464
27.4. In-Place Filter Testing / 465
27.5. In-Place UVGI System Testing / 467
27.6. In-Place Testing of Surface Disinfection Systems / 469
27.7. In-Place Carbon Adsorber Testing / 474
27.8. Ventilation System Retrofit Testing / 475
27.9. Commissioning / 475
27.10. Building Protection Factor / 477
References / 480

Section 4 Applications

Chapter 28. Commercial Office Buildings 485

28.1. Introduction / 485
28.2. Office Building Ventilation / 485
28.3. Office Building Epidemiology / 486
28.4. Office Building Aerobiology / 487
28.5. Standards and Guidelines / 491
28.6. Aerobiological Solutions / 491
References / 494

Chapter 29. Educational Facilities 497

29.1. Introduction / 497
29.2. Educational Buildings / 497
29.3. School Building Ventilation Systems / 498
29.4. School Building Epidemiology / 499
29.5. School Building Aerobiology / 504
29.6. School Building Solutions / 506
References / 507

Chapter 30. Residential Housing 511

30.1. Introduction / 511
30.2. Residential Homes / 511
30.3. Apartment Buildings / 516
30.4. Hotels and Dormitories / 516
30.5. Solutions for Homes and Apartments / 517
30.6. Solutions for Hotels / 522
30.7. Homes for the Immunocompromised / 522
References / 524

Chapter 31. Health Care Facilities — 527

31.1. Introduction / 527
31.2. Guidelines, Codes, and Standards / 527
31.3. Isolation Rooms / 530
31.4. Airborne Nosocomial Epidemiology / 531
31.5. Nosocomial Aerobiology / 533
31.6. Nosocomial Control Options / 540
    References / 543

Chapter 32. Biological and Animal Laboratories — 547

32.1. Introduction / 547
32.2. Laboratory Guidelines / 547
32.3. Biological Laboratory Pathogens / 550
32.4. Animal Laboratory Pathogens / 550
32.5. Animal Laboratory Epidemiology / 551
32.6. Laboratory Air Cleaning / 556
32.7. Animal Laboratory Problems and Solutions / 559
    References / 563

Chapter 33. Food Industry Facilities — 567

33.1. Introduction / 567
33.2. Food Industry Codes and Standards / 567
33.3. Foodborne Pathogens and Allergens / 568
33.4. Food Industry Epidemiology / 571
33.5. Food Plant Aerobiology / 574
33.6. Food Plant Solutions / 578
    References / 580

Chapter 34. Agricultural and Animal Facilities — 585

34.1. Introduction / 585
34.2. Agricultural Pathogens and Allergens / 585
34.3. Agricultural Epidemiology / 586
34.4. Agricultural Aerobiology / 590
34.5. Ventilation and Solutions / 592
    References / 597

Chapter 35. Libraries and Museums — 603

35.1. Introduction / 603
35.2. Libraries / 603
35.3. Museums / 610
    References / 615

Chapter 36. Places of Assembly — 619

36.1. Introduction / 619
36.2. Ventilation Systems / 619
36.3. Epidemiology / 622
36.4. Aerobiology / 623
    References / 625

Chapter 37. Aircraft and Transportation — 627

37.1. Introduction / 627
37.2. Aircraft / 627
37.3. Spacecraft / 634
37.4. Ships / 636
37.5. Submarines / 639
37.6. Cars, Buses, and Trains / 640
References / 642

Chapter 38. Industrial Facilities — 647

38.1. Introduction / 647
38.2. Occupational Airborne Diseases / 647
38.3. Wood and Paper Industries / 653
38.4. Textile, Leather, and Fur Industries / 657
38.5. Metal Industries / 658
38.6. Biotechnology Industries / 660
38.7. Chemical and Mineral Industries / 661
38.8. Occupational Legionnaires' Disease / 662
38.9. Other Industries and Hazards / 663
38.10. Industrial Solutions / 664
References / 665

Chapter 39. Sewage and Waste Processing — 671

39.1. Introduction / 671
39.2. Sewage and Waste Aerobiology / 671
39.3. Sewage and Waste Epidemiology / 674
39.4. Air Treatment Solutions / 678
References / 680

Appendix A: Airborne Pathogens and Allergens — 683

Appendix B: Toxins and Associated Fungal Species — 797

Appendix C: Microbial Volatile Organic Compounds — 801

Appendix D: Surface Sampling Test Results Evaluation Form — 805

Appendix E: Settle Plate Test Results Evaluation Form — 807

Appendix F: Air Sampling Test Results Evaluation Form — 809

Glossary 811
Index 821

ABOUT THE AUTHOR

Wladyslaw Jan Kowalski, PhD is widely recognized as an expert in the design and testing of nuclear air cleaning systems, radioactive contamination control systems, building pressurization systems, and HVAC and cooling water systems. Dr. Kowalski has served as a mechanical engineer, project engineer, and senior engineer for various companies in the nuclear power industry, and has been the recipient of many honors and awards. He is currently a researcher/instructor at Pennsylvania State University.

PREFACE

This book is intended to be a general reference to assist and educate engineers, designers, microbiologists, building managers, health care professionals, students, and others in all aspects of the design and application of technologies and methods for controlling aerobiological diseases in indoor environments. The term aerobiological engineering was coined to represent this entire field of research, which is of necessity a fusion of both microbiology and engineering. The information contained herein represents the current state of the art in a developing field and is a compendium of research that began at The Pennsylvania State University in 1995 into methods of controlling respiratory diseases in indoor environments. This book will serve well as a textbook for any course designed to instruct students or professionals in the art and science of protecting buildings against microbial contamination. The chapters have been arranged in as logical an order as would facilitate progressive learning. Every attempt has been made to keep the information in this book as general as necessary so that it may be applicable regardless of new developments, while simultaneously keeping it specific enough to be immediately useful for real-world applications and for continuing research.

Some liberties have been taken with the microbiological aspects out of consideration to engineers who find taxonomic terminology more challenging than beneficial. Genera are often used to represent one or more species, and in many tables traditional italics have been neglected in favor of vivid legibility. From all the microbiologists and taxonomists who may find these matters distressing I beg indulgence.

The attempt to publish quality in quantity results in occasional chance errors, both technical and interpretive. If anyone should find inaccuracies in this text, or any blanks in the Appendices that can be filled, please send them to me at drkowalski@psu.edu and I will post all errata and additions at *http://www.engr.psu.edu/ae/iec/abe/ topics/errata.asp* with credit.

I hope this work inspires and encourages professionals, researchers, and especially students to actively participate into the increasingly urgent war against diseases and realize the full potential they possess to promote human health through the design and construction of healthy buildings and the implementation of air and surface disinfection systems. I also hope this book will foster a new age of awareness regarding how our indoor environments may be practically and economically engineered to protect us against those airborne parasites that have heretofore enjoyed free lodging and even benefited from technology-assisted evolutionary enhancement. The ultimate goal of providing every person on earth with safe and healthy housing by reversing the tendency of our buildings to act as vectors for disease will eventually lead to a future in which many airborne diseases will become extinct. I can think of no nobler goal for the present age than to win the war against diseases—a war that, unlike the perpetually pandemic sort that brings humankind no benefits, has an end instead of just a means, saves lives instead of wasting them, and transcends all ideological, theological, and cultural barriers. *Da audacibus annue cœptis.*

Wladyslaw Jan Kowalski, Ph.D., P.E.

ACKNOWLEDGMENTS

I most warmheartedly acknowledge all those who assisted in the preparation of this book and the review of the manuscript. They include William Bahnfleth, Stanley Mumma, Jim Freihaut, Amy Musser, Richard Mistrick, Eric Burnett, Bill Carey, Brad Hollander, Thomas Whittam, Dave Witham, Chobi Debroy, Cindy Cogil, Doug Benton, Mike Ivanovich, Chuck Dunn, Amy Musser, Jim Kendig, Donna Carey, Bill Carey, Sue Kellerman, Leon Spurrell, Charley Dunn, Rebecca Upham, Sam Speer, Samuel Baron, Ferencz Denes, Joe Ritorto, Kristina Southwell, Julian Laws, David Weber, Tom Beardslee, Bruce Dawson, Raj Jaisinghani, Jelena Srebric, Gretchen Kuldau, Brad Striebig, Rex Coppom, Neil Carlson, Bailey Mitchell, John Beuttner, Thomas Stanifer, Jim Bolton, Tim Pladson, Estellle Levetin, Mary Clancy, Stephen Ells, Helena Boshoff, Yogi Goswami, Qingyan Chen, Louise Fletcher, Philip Mohr, Raymond Schaefer, James Johnson, Brad Hollander, Tim Pladson, Jing Song, Bill Fowler, Sam Guzman, Raimo Vartiainen, Alan Heff, James Woods, Suzanne Blevins, and everyone in the Indoor Environment Center of Penn State.

I also thank my family and friends who gave me unending encouragement and support, especially my father, Stanislaw J. Kowalski and my sister Vicky Chorpenning.

Finally, I thank the hundreds of other academics and professionals who have assisted, promoted, or otherwise supported the last 10 years of research in this field that has culminated in this book. Also, thanks to Art Anderson for the header photos in Chapters 1, 7, 10, 11, 13, 15, and 36. Thanks to Eric Burnett for the header photo in Chapter 18. And thanks to Joe Ritorto for lending me the camera with which I took the rest.

LIST OF SYMBOLS

a	filter media volume packing density (media/total volume), [m³/m³]
ACH	air change rate, [1/h]
a_i	volume i packing fraction
B_r	breathing rate, [m³/min] or [1/min]
$C, C(t)$	concentration, [cfu/m³] for body weight (BW), [µg/m³] for water concentration (CW)
C_a	airborne concentration, [cfu/m³] for BW, [µg/m³] for CW
C_h	Cunningham slip factor
D	duration, [days] or [hours]
D_{37}	37 percent of the duration
D_d	particle diffusion coefficient, [m²/s]
d_f	fiber diameter, [µm]
d_{fi}	fiber i diameter, [mm]
D_L	logmean diameter, [µm]
D_{max}	maximum diameter, [µm]
D_{min}	minimum diameter, [µm]
Dose	dose, [mg·min/m³], [mg] or [cfu]
d_p	particle diameter, [µm]
dP_F	filter pressure loss, [in.w.g.]
E	efficiency of a filter, fractional
I	irradiance, [µW/cm²] or [W/m²]
e	symbol for exponentiation
E_D	diffusion efficiency, fractional
E_R	interception efficiency, fractional
E_s	efficiency of a single filter fiber, fractional
E_{si}	single fiber i efficiency
E_t	exposure time, [s] or [min]
E_{tot}	total UV power, [µW]
E_{uv}	UV power, [µW]
f	resistant population fraction, fractional
F_i	view factor of segment i
F_K	Kuwabara hydrodynamic factor
frac_i	fraction i of fiber diameters
F_{tot}	total view factor
H	height, [cm]
hp	horsepower, [hp]
I_D, I_{Dn}	direct or incident intensity, [µW/cm²]
ID_{50}	mean infectious dose, [cfu]
ID_{99}	99 percent infectious dose, [cfu]
I_R, I_{Rn}	reflected intensity, [µW/cm²]
I_S	UV irradiance, [µW/cm²]

k	Boltzman's constant, 1.3708×10^{-23} J/K
k	UVGI rate constant, [cm²/μW·s]
l	length of lamp, [cm]
L	length, [m] or [cm]
l_g	arclength of lamp, [cm]
l_g, l_b	length of lamp segments, [cm]
MDP	mean disease period, [days] or [hours]
MIP	mean infectious period, [days] or [hours]
$N(t)$	number of microbes at time t
N_r	interception parameter
OA	outside air flowrate, [m³/min]
P	power, [W] or [μW]
Pe	Peclet number
P_v	velocity pressure, [in.w.g.]
Q	airflow, [m³/min]
r	radius of lamp, [cm]
S	filter fiber projected area
$S, S(t)$	source release rate, [cfu·min/m³] for BW, [μg·min/m³] for CW
S_i	fiber i projected area
SSC	steady state concentration, [cfu/m³] or [mg/kg]
T	temperature, [K]
t	time, [s], [min], [h], or [days]
U	media face velocity, [m/s]
V	volume, [m³]
W	width, [cm]
x	distance from lamp axis, [cm]
x	dose, [mg·min/m³], [mg] or [cfu], or arbitrary variable
X	height parameter
y	total number or percentage of cases, casualties, or fatalities
Y	width parameter
Δh	heat increase, [W]
ε	inhomogeneity correction factor
η	absorption efficiency
η	gas absolute viscosity, [N·s/m²]
η_{fan}	efficiency of fan, fractional
η_{motor}	efficiency of motor, fractional
λ	gas molecule mean free path, [μm]
μ	particle mobility, [n·s/m]
π	pi, 3.14159...
ρ	reflectivity, fractional or [%]
σ	standard deviation

S·E·C·T·I·O·N · 1

BACKGROUND AND HISTORY

CHAPTER 1

AIRBORNE DISEASE AND THE INDOOR ENVIRONMENT

1.1 INTRODUCTION

Aerobiological engineering is the art and science of designing buildings and systems for the control of airborne pathogens and allergens in indoor environments, including commercial buildings, hospitals, and residences. Aerobiology is the study of microorganisms in the air that may be detrimental to human health. Included among these organisms are viruses, bacteria, fungi, rickettsias, protozoa, and such microbiological products as endotoxins, mycotoxins, and *microbial volatile organic compounds* (MVOCs).

We live in a world in which every person on earth is constantly exposed to the threat of airborne diseases. Indoor and outdoor air can be full of transient populations of microorganisms, but none actually live in the air. Most microbes die off in the outdoor air as a result of sunlight, temperature extremes, dehydration, oxygen, and pollution. Spores and some environmental bacteria are naturally more resistant and can occur outdoors seasonally in high concentrations. Controlled indoor climates favor the survival and transmission of contagious human pathogens as well as some outdoor fungi and bacteria. Since people spend over 90 percent of their time indoors, the solution to the problem of most airborne infections, therefore, lies in engineering control of the aerobiology of the indoor environment.

1.2 AIRBORNE DISEASE TODAY

Today, we are faced with a world full of microorganisms, many of which have evolved sophisticated mechanisms for enduring the elements or airborne transport. By bringing parasitical microbes along with us on our evolutionary journey from the savannahs into modern cities, we have unwittingly fostered their evolution. The evolution of airborne pathogens is intimately linked to both the evolution of humans and of their habitats.

Given that these microorganisms are taking advantage of our unintentional hospitality and the design of our indoor environments, it is possible now to intentionally design our habitats to eliminate these deadly pests. Knowing what we know about their transmission characteristics and survivability, we can now design our homes and buildings to minimize the possibility of airborne disease transmission.

1.2.1 Building Science

The science of designing buildings implicitly encompasses aspects of both technology and biology. Buckminster Fuller once described a building as "a machine for living in." We design these buildings to meet our needs, and that implies all of our biological needs as well. As we have come to increasingly recognize, paramount among our biological needs is hygiene and the need to prohibit parasites from our living and working environments. By the nature of its purpose, to suit the human need for comfort, warmth, and protection from the elements, the buildings we design are simultaneously comfortable for microorganisms. Most of these microorganisms thrive at temperatures and humidities that we ourselves find necessary for comfort and health.

The source of an airborne pathogen often describes its pathogenic or epidemiological nature. Viruses are almost exclusively communicable (Murray, 1999; Fields and Knipe, 1991). Fungal spores are almost exclusively noncommunicable (Bosche et al., 1993; Burge, 1989). Communicable pathogens like viruses and bacteria come mainly from humans in indoor environments (Kundsin, 1988) while fungal spores come mainly from the outdoor environment (Howard and Howard, 1983; Muilenberg and Burge, 1996). Fungi and bacteria sometimes come from contaminated buildings (Godish, 1995). Almost all fungi of concern form spores but some occur in the yeast form. Some bacteria, particularly some environmental bacteria, also form spores. In general, environmental bacteria are primarily noncommunicable and can survive outdoors. Pathogenic viruses are rarely found in the outdoor environment where they do not survive well (Morey et al., 1990). Certain allergenic fungi show a preference for indoor environments and tend to occur indoors at higher levels than they do outdoors (Flannigan et al., 1991). Dander comes most often from pets and dust mites that thrive in indoor conditions (Esch et al., 2001). Figure 1.1 illustrates some of the

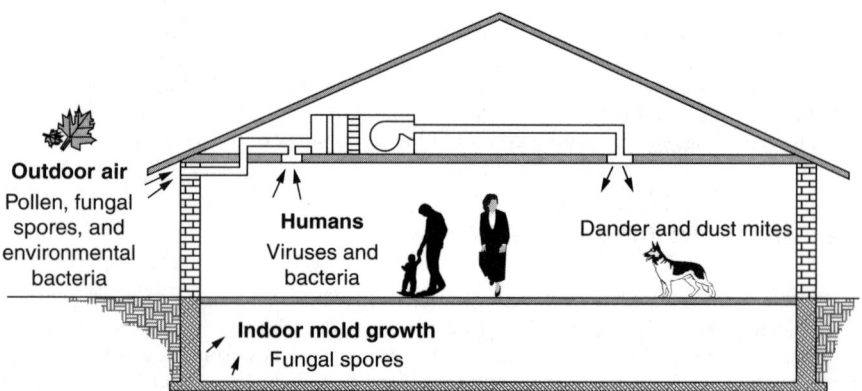

FIGURE 1.1 Typical sources of indoor airborne microbes and allergens.

common sources of airborne pathogens and allergens in relation to their entry into indoor environments.

1.2.2 Indoor Air Quality

The field of *indoor air quality* (IAQ) has seen considerable focus over the past few decades. Much of this study has centered around the simple problem of providing sufficient ventilation for breathing, thermal comfort, and to eliminate odors, be they of biological origin or otherwise.

The same methods used to evaluate the air change rates for removal of CO_2 are adequate to evaluate the removal of airborne pathogens, and again we see the considerable overlap between the problem of removing gases and that of removing microorganisms. The methods used for establishing IAQ can be directly adapted, often without modification, to study, design, or predict the effectiveness of ventilation systems when they are used for the control of airborne disease transmission.

1.2.3 Sick Building Syndrome and Building Related Illness

Sick building syndrome (SBS) and *building related illness* (BRI), jointly known as SBS/BRI, were acknowledged in the 1980s as a category of disease that encompasses a variety of symptoms induced by occupancy of problem buildings. The concept that buildings themselves contribute to or cause diseases highlights the need for a deeper understanding of the various mechanisms that may lead to this condition. In one sense all buildings are subject to BRI since the transmission of common colds and flus is assisted by any building that maintains normal indoor temperatures and humidities. Technically, however, BRI refers to illnesses experienced by a certain fraction of the inhabitants that often consist of varied symptoms and that often cannot be traced to specific causes. A building may be deemed subject to SBS if at least 20 percent of the occupants complain of symptoms (Rostron, 1997). Not everyone suffers the same symptoms and many occupants may suffer no symptoms at all. Symptoms may include headache, rhinorrhea, watery eyes, itchy nose, stuffy nose, itching, loss of concentration, dry eyes, dry skin, lethargy, or irritation of the throat or lungs. The causes of SBS may be multifactorial since the symptoms, and sometimes cause cannot be identified. About 15 to 30 percent of SBS cases are related to airborne microbial contamination (Godish, 1995). Most cases of SBS are related to nonmicrobial sources such as chemical pollutants and smoke. Some cases of SBS are tied to relative humidity problems, ventilation problems, vibration problems, noise problems, or even psychosocial problems. SBS occurs with twice as much frequency in buildings that have air conditioning as opposed to naturally ventilated buildings.

Volatile organic compounds (VOCs) provide one of the highest correlations with SBS. VOCs are primarily due to chemical pollutants but they can also be produced by a variety of bacterial and fungal contaminants. Although pollutants of nonmicrobial origin may exacerbate respiratory infections by lowering immunity, as will be addressed in later chapters, chemical pollutants are not aerobiological contaminants and numerous other publications are available that address this subject in detail.

BRI is a better-defined category in which a specific cause of occupants' symptoms can be identified. The illness may be humidifier fever, occupational asthma, allergic rhinitis, sinusitis, organic dust toxic syndrome, or respiratory irritation from chemicals, but if it can be specifically tied to problems with the building, it is classed as a case of BRI (Hodgson, 2001). Psychosocial symptoms are specifically excluded from BRI. One of the main differences between SBS and BRI is that BRI has a specific etiology and can be diagnosed (Brightman and Moss, 2001).

1.2.4 Engineering Technologies

Many technologies are available for improving the aerobiological quality of our indoor environments, including those in current use—such as dilution ventilation, filtration, *ultraviolet germicidal irradiation* (UVGI), air disinfection, pressurization control, dehumidification—and those that are coming into increasing use—such as *photocatalytic oxidation* (PCO) and antimicrobials—and new technologies such as pulsed light, ozone, ionization, and plasma. To the list of these technologies we could add new methods of building design and air distribution, such as *dedicated outdoor air systems* (DOAS), material selectivity, green building design, passive solar exposure, and other promising new approaches or design methods. Finally we can consider human hygiene as a manageable aspect of our culture that deserves increased focus and holistic integration with our buildings and technology. The engineering design and implementation of these technologies and methods of design are addressed in detail in the various chapters of this book. Although not all aspects of these aerobiological engineering technologies are completely understood at present, this text includes the latest available information for applying these technologies, sizing air disinfection systems, and testing buildings.

1.2.5 Biological Weapons

Of increasing concern today is the potential use of biological weapons against buildings and their occupants. Although such weapons of mass destruction are a relatively recent man-made hazard, their use resembles the natural spread of diseases in buildings in all but scale. Biological weapons may include microbes not normally or previously seen in nature, but the technologies for dealing with biological weapons are the same that would normally be used, and some of the approaches to defend buildings against them need special consideration. These issues have been addressed in detail in a previous work (Kowalski, 2003), and are therefore addressed only incidentally in this book. Readers should be assured, however, that any building aerobiologically engineered to provide a healthy indoor environment will offer considerable protection against intentional releases of any airborne biological agent.

1.3 THE HISTORY OF AIRBORNE DISEASE

Many of the well-established diseases we know today, like tuberculosis, smallpox, plague, and measles, predate history. Although humans carried many diseases with them from before the Stone Age, new diseases arose as animal husbandry and agriculture developed about 15,000 years ago. These industries promoted the exchange of diseases from animals to man, and the close association of people in large communities fostered the further spread of diseases.

Many diseases were widespread in the ancient world including tuberculosis (consumption), smallpox, common colds and influenza, measles, plague, and a variety of protozoal parasites. Then, as now, sneezing, coughing, and direct contact were the most likely means by which respiratory illnesses spread indoors. During cold weather the incidence of respiratory infections increased due to the tendency to stay indoors, a fact that led some of the ancients to associate the seasonal change in winds with the arrival of diseases (Lloyd, 1978). Of course, the outdoor air is essentially self-sterilizing due to the action of sunlight and temperature changes, and apart from outdoor fungal spores, airborne pathogens are unlikely to ever be found in natural outdoor air.

The plague, although typically transmitted by fleas from rats, can also be transmitted by the airborne route, in which form it is known as pneumonic plague. Plague swept through Europe and the Mediterranean on several occasions, in the sixth century, the seventh century, the eighth century, and also from the thirteenth through the eighteenth centuries, approximately coincident with the "Little Ice Age," which may have influenced the spread of this disease (Fagan, 2000). During the Little Ice Age, from about 1450 to 1850, Europeans abandoned failed farms, moved to cities, and learned to tighten up their housing to such a degree that respiratory diseases, especially tuberculosis, became rampant (Imbrie and Imbrie, 1979; Tkachuck, 1983).

The recognition that airborne diseases were being transmitted indoors does not seem to have occurred until the last century. Numerous studies have since shown repeatedly that airborne microorganisms are fragile and incapable of surviving for long while airborne except under the most fortuitous conditions. The myth that outdoor air carries diseases persists to this day and is unwittingly instilled in new generations by parents who tell their children to "bundle up or you'll catch a cold" (Krajick, 1997).

Not more than a century ago, after the discovery of bacteria, the concept of miasma (bad air) held by the ancients to be the cause of diseases was widely rejected. Some decades later it became obvious that a certain few diseases, like TB, were indeed being transmitted in the air and yet some sources fervently disputed this theory. Various researchers eventually demonstrated that a number of diseases had the capacity to transmit by the airborne route. By the early 1930s a dozen or so diseases had been verified as airborne and the disbelievers gradually abated. By the 1950s a dozen more airborne diseases had been identified and the recognition of indoor allergens and pathogenic fungi greatly increased the list (Langmuir, 1961; Pady, 1957). In this volume are presented over a hundred airborne pathogens and allergens that may cause diseases by the airborne route. This number is sure to grow, as even at the time of this writing a new airborne virus (SARS) entered the aerobiological milieu. The ancient Greek theory of miasma wasn't so far off, but it was mainly indoor air, not outdoor air that caused seasonal epidemics. How it came to pass that airborne infectious disease agents thrived throughout history and proliferated in our modern world can be explained through the evolution of pathogens, humans, and our habitats, and it is informative to explore these processes as a basis for understanding how the principles of aerobiological engineering can reverse the evolutionary trend and lead to the eradication of many of the airborne pathogens that continue to plague mankind.

1.3.1 The Evolution of Airborne Pathogens

In order to understand the complicated relationship between humans, their pathogens, and the indoor environments that foster their spread, it is most beneficial to examine the evolution of pathogenic bacteria and viruses, human immune system evolution, and the evolution of our habitats. Figure 1.2 illustrates this interacting triad of evolutionary factors. The present state of affairs in which we are under constant threat from airborne diseases is the result of a complex and ongoing interplay between these three main factors—and also to some degree our technological evolution—and the hygienic practices, or lack thereof, that form a part of human culture. First we will examine pathogen evolution, then the coevolution of pathogens and the human immune system, and finally the impact of human habitats on pathogen adaptation.

Metazoan evolution is relatively well understood, primarily due to the morphological similarities among the species and differences between the various phyla. Less well understood is the mode of bacterial evolution, but certain patterns can be inferred, partly based on morphology but mostly based on recent genetic research. Prokaryotes are single-celled organisms, and are distinct from eukaryotes, or multicellular organisms like ourselves.

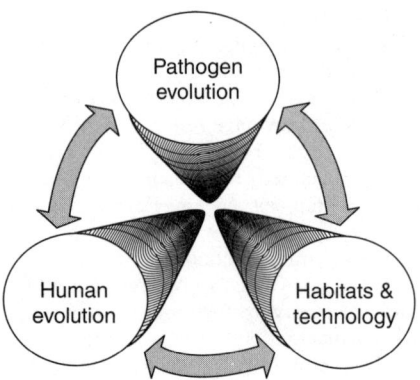

FIGURE 1.2 The coevolution of pathogens, humans, and their habitats and technology have brought about the current state of airborne diseases in the world.

Prokaryotes evolved from common progenitors called progenotes about 3.5 billion years ago (Schleifer and Stackebrandt, 1985). It wasn't until the development of an oxygen rich atmosphere some 1.5 billion years ago that eukaryotic cells appeared. Bacteria at this point and this period represented the beginnings of both symbiotic and parasitic relationships between prokaryotes and eukaryotes. The divergence of multicellular life forms appears to have carried many dependent pathogens along the evolutionary path to the present.

Although the ultimate origin of most pathogens, especially viruses, is obscure, certain inferences can be made due to various genetic and morphological similarities. Smallpox, for example, is considered a mutant form of cowpox (Davey and Halliday, 1994). Measles resembles canine distemper, and likely jumped species soon after the dog was domesticated by hunter-gatherers at least 14,000 years ago, and probably much longer (Morey, 1994). Diphtheria, caused by *Corynebacterium diphtheria*, is transmitted from cattle, which were domesticated in about 7000 BC, but is relatively new, being not more than 2000 years old. Cattle also suffer from a form of TB caused by *Mycobacterium bovis*, and evidence of TB can be found in neolithic skeletons from 5000 BC. *Mycobacterium tuberculosis* is at least 15,000 years old, based on genetic studies, which would correlate well with domestication of cattle (Kapur et al., 1994). Horses are the only other animal species to harbor rhinoviruses, and although they may not have been domesticated until about 4000 BC, they were likely husbanded much earlier since they were hunted for food ages before this. It is not certain that these disease agents jumped from animals to man, as they could just as well have jumped from man to animals, but these examples highlight the fact that there was close contact between animal species and exchange of pathogens, followed by adaptation to new hosts and pathogenic species divergence.

Chlamydia pneumonitis is an example of an airborne pathogen that may have jumped species not from domesticated animals to man but from rodents, which, at least in former times often lived in close association with man in his habitats. A comparison of *C. pneumonitis* with *C. trachomatis*, which infects rodents, shows a 69 to 70 percent similarity in sequence of the MOMP gene, which codes for outer membrane proteins (Zhang et al., 1993). The study also shows that both bacteria derived from a common ancestor, and the evidence furthermore indicates evolutionary pressure to conserve regulation of MOMP mRNA transcription.

Figure 1.3 shows the hypothetical first appearance of airborne pathogens. This chart represents the approximate earliest limits estimated based on actual or hypothetical reasons. The airborne viruses, for example, are placed in a recent context due to their remarkable sophistication, streamlined genomes, rapid evolutionary rates, the likelihood that they could not have existed independently before any host bacteria, and the fact that they possibly represent fragments of DNA shed from bacteria that developed reproducibility. All of these concepts are subject to great debate and theories abound as to whether viruses actually preceded bacteria or not (Holland and Domingo, 1998; Domingo et al., 2000). To be fair, the view that viruses coexisted with vertebrates from the beginning is tenable and has

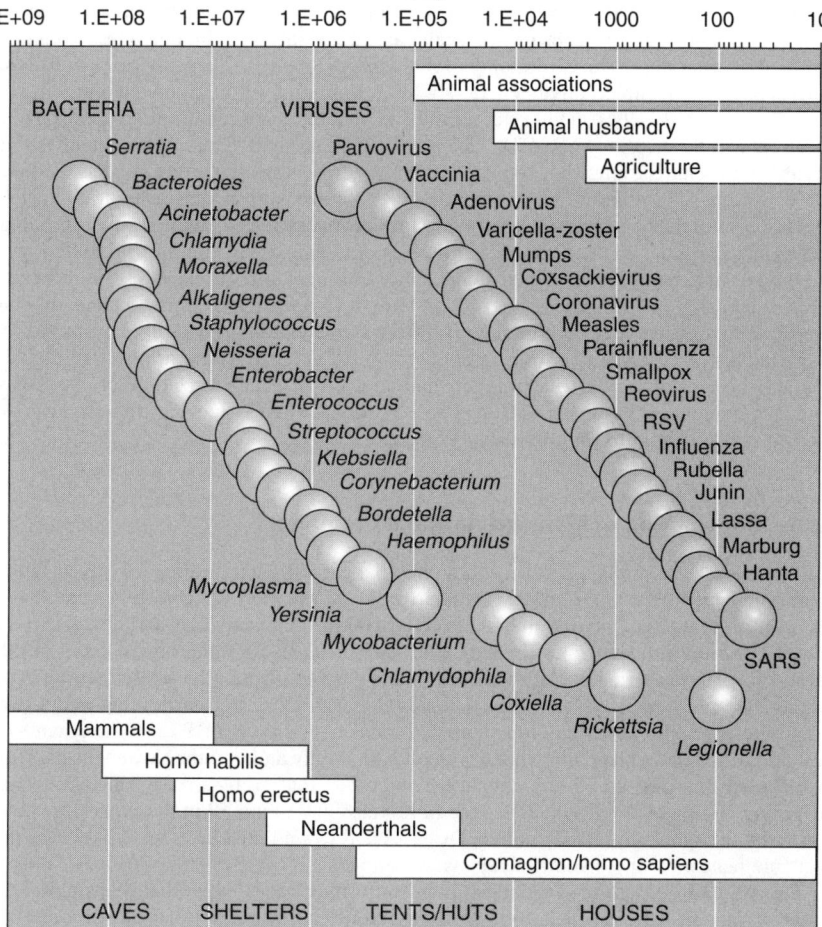

FIGURE 1.3 Hypothetical first appearance of airborne pathogens.

support in the fact that some mammalian viruses, like adenovirus, have counterparts in reptilian species (Davison et al., 2003). The postulated framework in Fig. 1.3 only represents the hypothetical first appearance of these pathogens in humans, and the first human viral diseases may have been preceded by a billion years of virus evolution. In this figurative chart those pathogens that seem to have jumped species are placed alongside the earliest estimates for the beginnings of animal husbandry. The more benign pathogens are assumed to have been around longer, and commensals to have been around the longest of all, in accordance with the hypothesis of obligate evolution toward commensalism considered by some to be an implicit process in natural selection (Ewald, 1994). The majority of the chart, however, represents a great deal of speculation but it illustrates the fact that airborne pathogens have recent links in our evolutionary and technological (i.e., agricultural) history, and that pathogens are likely to continue to emerge and evolve in the future.

Influenza A virus occurs in several species besides humans, including chickens, ducks, seals, swine, horses, and other birds. These animals have lived in close association with humans for some time, except for the seals, an oddity that may be due to coincidentally analogous physiology. The evolutionary relationships among influenza viruses of different species have been studied phylogenetically by examination of the ribonucleoprotein complex (Gorman et al., 1990). These studies indicate that human virus strains evolve at a much faster rate, 1.82×10^{-3} nucleotide changes per year, than their avian cousin strains. Phylogenetic charts show that human influenza virus is most closely related to that of swine, and that the ultimate source of influenza A virus was avian (Prescott, 1996). The implied evolutionary scenario is that avian influenza viruses existed long before humans domesticated swine, chickens, ducks, or horses, and that it was first transmitted to swine and then to man from domesticated swine. Once ensconced in man as a reservoir the virus could then take full advantage of indoor human-to-human transmission and diversify rapidly. The time at which the virus jumped from swine to humans would correspond to the first attempts to domesticate swine, or about 7000 BC at least. *Avian influenza virus* (AIV) is a newly emerging human pathogen and molecular characterization of the ssRNA sequence indicates it has evolved from a new genotype of AIV that was present during the 2003 to 2004 Asian bird flu outbreaks (Wan et al., 2005).

1.3.2 Coevolution of Man and Airborne Pathogens

Although bacteria surely predated mammals, parasitical bacteria either evolved later or were transferred from previously existing arthropods. The first vertebrates appeared about 600 million years ago, the first mammals about 200 million years ago, and bipedalism first occurred 7.5 million years ago (Leakey and Lewin, 1977). As sheltering in caves or other natural formations was already a natural tendency by this time, it is likely that the evolution of airborne pathogens was already underway at the time our earliest ancestors began to use fire approximately 4 million years ago (Leakey and Lewin, 1992). When man's earliest ancestors first walked out from the Great Rift Valley and into the savannahs of Africa, they already carried numerous parasitical microbes that were already on their way to becoming commensals. The average age of humans was about 19 in the Stone Age, and it was rare for anyone to live beyond 35. Diseases and wound infections were likely the most probable reason for these short life spans.

Theoretically, microbes first transmitted only by direct contact between humans, or between humans and animals, but as caves, tents, and huts were made progressively warmer and cozier these microbes could survive briefly on surfaces or while airborne from coughing, sneezing, or talking. Some microbes gradually evolved the ability to survive exposure or airborne transport, and every successful transmission of an infection ensured the proliferation of new variants of pathogenic microbial species.

Hunter-gatherer societies may have been the first to exist in close association with animals, perhaps 100,000 years ago. Wolves may have served as hunting associates to groups of men long before any were actually domesticated. Some dog diseases may have first infected and adapted to man at about this time. When hunter-gatherer groups turned to the more reliable methods of animal husbandry and agriculture, a new round of microbial evolution was fostered. They built sturdier, more permanent, and air-tight homes. A combination of these warm indoor environments coupled with close proximity to husbanded animals caused a large number of new diseases to jump species, both to humans and from humans. Roughly half the diseases we endure today seem to have developed within the past 15,000 years. The explosion of the earth's population following the success of agriculture led to the development of large cities beginning about 12,000 BC. The close living conditions invariably spawned epidemics and fostered the further evolution of airborne diseases.

The human immune system evolved in concert with the pathogens to which they were exposed. All vertebrates possess antibody responses, but in jawed vertebrates, antibody diversity is specified by the same heterodimeric immunoglobulin molecule. The diversity of antibodies is generated both by inherited characteristics as well as by somatic rearrangement within the host. High degrees of nucleotide similarity occur in these immunoglobulin gene loci, but marked differences in organization and recombination mechanisms exist between phylogenetically divergent species (Litman et al., 1993). This evidence suggests a very ancient divergence of the types of bacterial pathogens afflicting the various vertebrate species, possibly in direct accordance with the divergence of the species themselves, but also indicates a recurring similarity of the mechanisms that these pathogens use to infect their hosts. The gene locus of the bacterial DNA that codes for the M proteins has been shown to be important for generating antigenic heterogeneity, which is a prime virulence determinant. Extreme sequence diversity is present in portions of the alleles that code for surface-exposed parts of the M proteins. The diversity of this region, as opposed to the lack of diversity in some of the other regions is an indicator of coevolution with the immune system.

A good example of the coevolution of bacterial pathogens and the human immune system can be found in *Neisseria meningitidis*, the causative agent of meningitis, and member of a most ancient bacterial genera. The outer membrane proteins of *N. meningitidis* are encoded by a single locus. Any given strain of this bacterium will express one or the other of a class of homology groups coded by this locus, but never both. These regions determine amino acid composition in the outer protein loops of the surface membrane, which are exposed antigenic surfaces and therefore subject to immune system responses. Comparisons of the synonymous and nonsynonymous regions of these alleles indicate they have accumulated significantly more nonsynonymous substitutions per site than synonymous substitutions (Smith et al., 1995). The high ratio of nonsynonymous to synonymous changes is most adequately explained by positive Darwinian selection for diversity in response to selection by the human immune system.

Streptococcus pyogenes is the etiological agent of streptococcal pharyngitis. This pathogen is capable of defending itself against phagocytic attack from lung defense mechanisms through the use of M proteins. This pathogen is an example of a very ancient bacterium, based on the slow evolutionary rate, diversity of strains, and high rates of nonsynonymous substitutions (Hollingshead et al., 1994).

Parvoviruses comprise a family that includes the human B19 virus and also canine, feline, bovine, and porcine viruses. The parvoviruses themselves seem to have coevolved along with vertebrates and also exist in nondomesticated species such as mink and mice. The relationships between these viruses have been studied in terms of the genomic homologies (Fisher and Mayor, 1991), and these indicate that human parvovirus B19 is most closely related to the bovine parvovirus, and then to cats, dogs, and mice before swine. The mink parvoviruses represent the most distant phylogenetic branch from all these species, as one would expect.

Coronaviruses provide another example of a virus that may have evolved independently from another virus when a strain developed the capability of airborne transport. Berne virus in horses and Breda virus in cattle represent a morphologically distinct torovirus family. These viruses do not infect humans but were recently found to have some very specific similarities in their genomes to human coronaviruses (Snijder et al., 1991). Both of these RNA viruses express their genetic information from a nested set of mRNAs with identical coterminals, are of similar length (25 to 30 kilobases), and display the same basic gene order— the coding for the spike protein followed by the coding for the membrane protein and then the coding for the nucleocapsid protein. These and other similarities are strong indicators of common ancestry. In this particular case the evidence indicates that a recombination event, not uncommon with RNA viruses, caused the sudden acquisition of morphological characteristics that enabled the virus to transition from one species to another.

A classic example of coevolution between parasite and host is found in the phylogenetic study of *Mycobacterium tuberculosis*, the causative agent of TB. Certain proteins that serve as targets for the host immune system have been shown to evolve much more rapidly under this selective pressure (Hughes, 1993). The purpose of this rapid evolution is to evade immune system defenses. These proteins, called chaperonins, are common to another bacteria, *Mycobacterium leprae*, and also to various other prokaryotes and eukaryotes. In *Streptomyces albus*, the two chaperonin genes have evolved at the same rate, whereas in both *M. tuberculosis* and *M. leprae* one of the chaperonin genes has evolved much more rapidly at nonsynonymous nucleotides sites, indicating a response to selective pressures, specifically those of the human immune system. Phylogenetic trees constructed from analysis of these chaperonin genes also provide insight into the relationship between bacteria and humans. In particular it is interesting to note that the common ancestor of *M. tuberculosis* and *M. Leprae* is older than mammals, and that *Chlamydia pneumoniae* is older than humans.

Within the realm of airborne pathogens the initial diversity of the precursor pathogens resulted in the emergence of those characteristics that facilitate respiratory infection, namely the ability to survive outside the host and during airborne transport. Regardless of the specific origin of airborne pathogens, their analogous characteristics entail the ability to survive outside of hosts almost exclusively in indoor environments.

1.3.3 Coevolution of Airborne Pathogens and Habitats

Since airborne pathogens do not remain viable for long in outdoor air, and are difficult to transmit outdoors, their genesis must have been coeval with the first use of enclosed shelters. The earliest shelters were found, caves in particular, which offered protection against the weather and predators. Huddling together for warmth, as other primates are known to do, would have fostered viral and bacterial exchange even if the pathogens were not very viable in air. The time period of these first exchanges in natural shelters probably predates the appearance of bipedalism about 7.5 million BC. Pathogens that first transmitted by direct contact would have eventually evolved the ability to survive first on surfaces, and then for short periods in air. The advent of modern technology has brought forced ventilation into buildings, and although this provided fresh, tempered air and helped reduce diseases over much of the past century, the trend has recently reversed, partly due to attempts to save energy by reducing outdoor airflow. The evolutionary adaptation of airborne microbes to indoor environments would have transpired in conjunction with the evolution of human habitats. This subject will be dealt with in more detail in Chap. 8.

1.4 THE FUTURE OF DISEASE CONTROL

The information and tools presented in these chapters form the first comprehensive attempt to address the problem of airborne diseases from an engineering standpoint. The engineering solutions to the disease control problem may be sufficient to reduce or even eliminate many of the airborne diseases known to man today. But new problems may emerge, and new solutions can always be sought that are more effective, or cost less to implement.

It has been suggested by some that exposure to diseases protects us against diseases and that a disease-free environment may not be a desirable eventuality (Dowling, 1966). However, there is no such thing as general immunity—immunity to measles does not protect us against influenza, and immunity to influenza is only temporary. A large portion of the population has some inherent immunity to smallpox, but as smallpox no longer exists in nature this immunity, bought at the cost of countless millions of lives, is of no real value today. It would be far better to rid the world of pathogenic diseases than to go on with halfway measures that merely prolong the problem. This is why the long-term focus of the war on diseases must be total eradication rather than what is essentially triage for those in developed nations while dangerous pathogens continue to evolve and spread among the poor and in Third World nations. Tuberculosis was almost an anachronism in the West a few decades ago but today it is a reemerging worldwide threat. The halfway measure of protecting the few has led to new strains of TB so well adapted to our antibiotics that we are rapidly running out of effective antibiotics to use against them.

With the growing limitations of our vaccines and antibiotics, aerobiological engineering of our indoor environments may be the only alternative in the long run. Indeed, the design of our very cities may need rethinking since the population density itself puts us at risk (Aicher, 1998). What is required today is the design and construction of healthy building designs and implementation of effective disease control technologies on a worldwide scale. Standardization of healthy buildings and cities, combined with a renewed focus on the culture and technologies of hygiene, will be capable of stopping epidemics and will ultimately drive many species of airborne pathogenic microbes to extinction. This is the paradigm of aerobiological engineering.

REFERENCES

Aicher, J. (1998). *Designing Healthy Cities: Prescriptions, Principles, and Practice.* Krieger Publishing, Malabar, FL.

Bosche, H. V., Odds, F., and Kerridge, D. (1993). *Dimorphic Fungi in Biology and Medicine.* Plenum Publishers, New York.

Brightman, H. S., and Moss, N. (2001). Chapter 3: Sick building syndrome studies and the compilation of normative and comparative values. *Indoor Air Quality Handbook*, J. D. Spengler, J. M. Samet, and J. F. McCarthy, eds., McGraw-Hill, New York. 3.1–3.32.

Burge, H. A. (1989). "Airborne allergenic fungi: Classification, nomenclature, and distribution." *Immunol Allergy Clinics North Amer* 9(2):307–309.

Davey, B., and Halliday, T. (1994). *Human Biology and Health: An Evolutionary Approach.* Open University Press, London.

Davison, A. J., M. Benk, and Harrach, B. (2003). "Genetic content and evolution of adenoviruses." *J Gen Virol* 84:2895–2908.

Domingo, E., Webster, R., and Holland, J. (2000). *"Origin and Evolution of Viruses."* Academic Press. San Diego, CA.

Dowling, H. F. (1966). "Airborne infections—the past and the future." *Bact Rev* 30(3):485–487.

Esch, R. E., Hartsell, C. J., Crenshaw, R., and Jacobson, R. S. (2001). "Common allergenic pollens, fungi, animals, and arthropods." *Clin Rev Allerg Immunol* 21:261–279.

Ewald, P. W. (1994). *Evolution of Infectious Disease.* Oxford University Press, Oxford, NY.

Fagan, B. (2000). *The Little Ice Age: How Climate Made History 1300–1850.* Basic Books, New York.

Fields, B. N., and Knipe, D. M. (1991). *Fundamental Virology.* Raven Press, New York.

Fisher, R. E., and Mayor, H. D. (1991). "The evolution of defective and autonomous Parvoviruses." *J Theoret Biol* 149:429–439.

Flannigan, B., McCabe, E. M., and McGarry, F. (1991). Allergenic and toxigenic microorganisms in houses. *Pathogens in the Environment*, B. Austin, ed., Blackwell Scientific Publications, Oxford, UK.

Godish, T. (1995). *Sick Buildings: Definition, Diagnosis and Mitigation.* CRC/Lewis Publishers, Boca Raton, FL.

Gorman, O. T., Donis, R. O., Kawaoka, Y., and Webster, R. G. (1990). "Evolution of Influenza A virus PB2 genes: implications for evolution of the ribonucleoprotein complex and origin of human Influenza A virus." *J Virol* 64:4893–4902.

Hodgson, M. (2001). Chapter 54: Building-related diseases. *Indoor Air Quality Handbook*, J. D. Spengler, J. M. Samet, and J. F. McCarthy, eds., McGraw-Hill, New York. 3.1–3.32.

Holland, J., and Domingo, E. (1998). "Origin and evolution of viruses." *Virus Genes* 16(1):13–21.

Hollingshead, S. K., Arnold, J., Readdy, T. L., and Bessen, D. E. (1994). "Molecular evolution of a multigene family in Group A Streptococci." *Mol Biol Evol* 11:208–219.

Howard, D. H., and Howard, L. F. (1983). *Fungi Pathogenic for Humans and Animals*. Marcel Dekker, New York.

Hughes, A. L. (1993). "Contrasting evolutionary rates in the duplicate chaperonin genes of Mycobacterium tuberculosis and M. leprae." *Mol Biol Evol* 10:1343–1359.

Imbrie, J., and Imbrie, K. P. (1979). *Ice Ages: Solving the Mystery*. Harvard University Press, Cambridge, MA.

Kapur, V., Whittam, T. S., and Musser, J. M. (1994). "Is Mycobacterium tuberculosis 15,000 years old?" *J Infect Dis* 170:1348–1349.

Kowalski, W. J. (2003). *Immune Building Systems Technology*. McGraw-Hill, New York.

Krajick, K. (1997). "Floating Zoo." *Discover* 18(2):67–73.

Kundsin, R. B. (1988). *Architectural Design and Indoor Microbial Pollution*. Oxford Press, NY.

Langmuir, A. D. (1961). "Epidemiology of airborne infection." *Bacteriol Rev* 25:173–181.

Leakey, R. E., and Lewin, R. (1977). *Origins*. E. P. Dutton, New York.

Leakey, R., and Lewin, R. (1992). *Origins Reconsidered: In Search of What Makes Us Human*. Doubleday, New York.

Litman, G. W., Rast, J. P., Shamblott, M. J., Haire, R. N., Hulst, M., Roess, W., Litman, R. T., Hinds-Frey, K. R., Zilch, A., and Amemiya, C. T. (1993). "Phylogenetic diversification of immunoglobulin genes and the antibody repertoire." *Mol Biol Evol* 10:60–72.

Lloyd, G. E. R. (1978). *Hippocratic Writings*. Penguin, New York.

Morey, P. R., Feeley, J. C., and Otten, J. A. (1990). *Biological Contaminants in Indoor Environments*. ASTM, Philadelphia, PA.

Morey, D. F. (1994). "The early evolution of the domestic dog." *Amer Scientist* 82:336–347.

Muilenberg, M., and Burge, H. (1996). *Aerobiology*. CRC Press, Boca Raton, FL.

Murray, P. R. (1999). *Manual of Clinical Microbiology*. ASM Press, Washington, DC.

Pady, S. M. (1957). "Quantitative studies of fungus spores in the air." *Mycologia* 49:339–353.

Prescott, L. M., Harley, J. P., and Klein, D. A. (1996). *"Microbiology."* Wm. C. Brown Publishers. Dubuque, IA.

Rostron, J. (1997). *Sick Building Syndrome: Concepts, Issues, and Practice*. E & FN Spon, London.

Schleifer, K. H., and Stackebrandt, E. (1985). *Evolution of prokaryotes*. Academic Press, Orlando.

Smith, N. H., Smith, J. M., and Sprat, B. G. (1995). "Sequence of evolution of Neisseria gonorrhea and Neisseria meningitis: evidence of positive Darwinian selection." *Mol Biol Evol* 12:363–370.

Snijder, E. J., den Boon, J. A., Horzinek, M. Z., and Spaan, W. J. M. (1991). "Comparison of the genome organization or Toro and Corona viruses: Evidence for two nonhomologous RNA recombination events during Berne virus evolution." *Virology* 180:448–452.

Tkachuck, R. D. (1983). "The Little Ice Age." *Origins* 10(2):51–65.

Wan, X. F., Ren, T., Luo, K. J., Liao, M., Zhang, G. H., Chen, J. D., Cao, W. S., Li, Y., Jin, N. Y., Xu, D., and Xin, C. A. (2005). "Genetic characterization of H5N1 avian influenza viruses isolated in southern China during the 2003-2004 avian influenza outbreaks." *Arch Virol* 150(6):1257–1266.

Zhang, Y., Fox, J. G., Ho, Y., Zhang, L., Stills, H. F., and Smith, T. H. (1993). "Comparison of the major outer membrane protein gene of mouse pneumonitis and hamster SFPD strains of Chlamydia trachomatis with other Chlamydia strains." *Mol Biol Evol* 10:1327–1342.

CHAPTER 2

AIRBORNE PATHOGENS AND ALLERGENS

2.1 INTRODUCTION

The classification of airborne pathogens and allergens is broadly defined here to include all microbes that can transmit diseases by the airborne route, all allergenic airborne microbes, and all organisms or microbial products that cause respiratory disease or cause respiratory irritation. Pathogens are parasitical disease-causing infectious microorganisms. Allergens are microbes or materials from microbes and other organisms that induce allergies or allergic reactions. Respiratory irritants are a loosely defined class of microbes or agents that cause temporary symptoms and are considered here to be included under a broader definition of allergens. Although most airborne pathogens and allergens cause respiratory diseases, some may cause other types of infections like skin diseases, eye and ear infections, and even some gastrointestinal infections. The single defining characteristic of these agents is that they are transported in whole or in part by the airborne route, either by natural or man-made mechanisms. In the following sections the categories of airborne pathogens and allergens that afflict mankind are identified and discussed.

2.2 AIRBORNE PATHOGENS AND ALLERGENS

Airborne microorganisms consist of viruses, bacteria, fungi, pollen, and sometimes protozoa. Bacteria can be subdivided into bacterial spores and nonsporulating bacteria. Bacterial spores include an important class of bacteria called actinomycetes. The remaining allergens and respiratory irritants are not microbes but consist of material or parts of organisms that include dust mites, dander, insect allergens, toxins, mycotoxins, endotoxins, and *microbial volatile organic compounds* (MVOCs). These categories are addressed individually in the

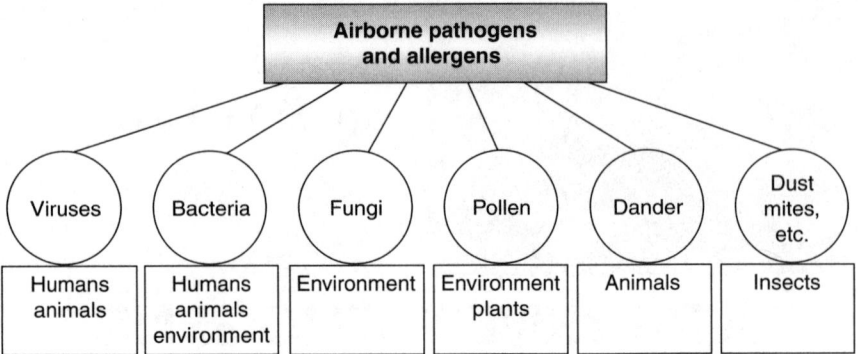

FIGURE 2.1 The major groups of airborne pathogens and allergens and their primary sources.

following sections. Prions are not addressed since they have never yet been known to transmit by the airborne route, although the possibility for such transmission exists.

Three microbial groups, viruses, bacteria, and fungi, include all the airborne pathogens and many of the most common airborne allergens. No protozoa have been identified as being a major airborne hazard. Pollen, dust mites, and dander form a separate group of allergens and respiratory irritants. Figure 2.1 shows a breakdown of these groups and their primary sources.

A complete summary of airborne microorganisms is provided in App. A, the airborne pathogen and allergen database. A summary of toxins from airborne fungal spores is provided in App. B, and App. C provides a summary of MVOCs that may be produced from airborne microbes. Summaries of allergenic algae, pollen, animal allergens, insect allergens, and toxigenic fungi are provided later in this chapter.

2.3 VIRUSES

Viruses are the smallest microbes and span a size range from about 0.02 to 0.22 μm. They include many of the most lethal, highly infectious pathogens (Fraenkel-Conrat, 1985; Fields and Knipe, 1991). Human pathogenic viruses can only reproduce in a host and as such their only reservoirs are humans or animals (Freeman, 1985; Murray, 1999). On occasion human viruses have been isolated in sewage but their ability to reproduce in sewage is limited at best (Austin, 1991; Mitscherlich and Marth,1984). Another exception is in laboratories where viruses are deliberately cultured and maintained (Collins and Kennedy, 1993; Fleming et al., 1995).

Two basic types of viruses exist—DNA viruses and RNA viruses. They may be enveloped or naked, single stranded or double stranded, and of several basic shapes such as icosahedral or helical (see Fig. 2.2).

Figure 2.3 illustrates the subdivisions of virus morphology and taxonomy for airborne contagious respiratory viruses. The primary characteristics of interest are their size, which affects their filtration rates, their pathogenesis, their contagiousness and lethality, and their susceptibility to disinfection methods. Appendix A and the references may be consulted for such information.

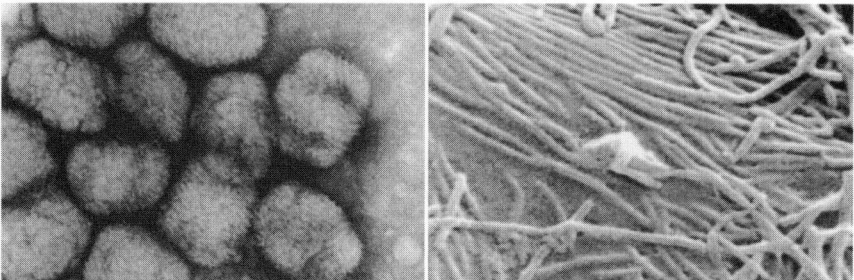

FIGURE 2.2 Most viruses tend to be icosahedral or round like the smallpox virions at left (PHIL#2292), or extended helices like the Ebola virions at right (PHIL#1836). (*Images reprinted courtesy of the CDC Public Health Image Library.*)

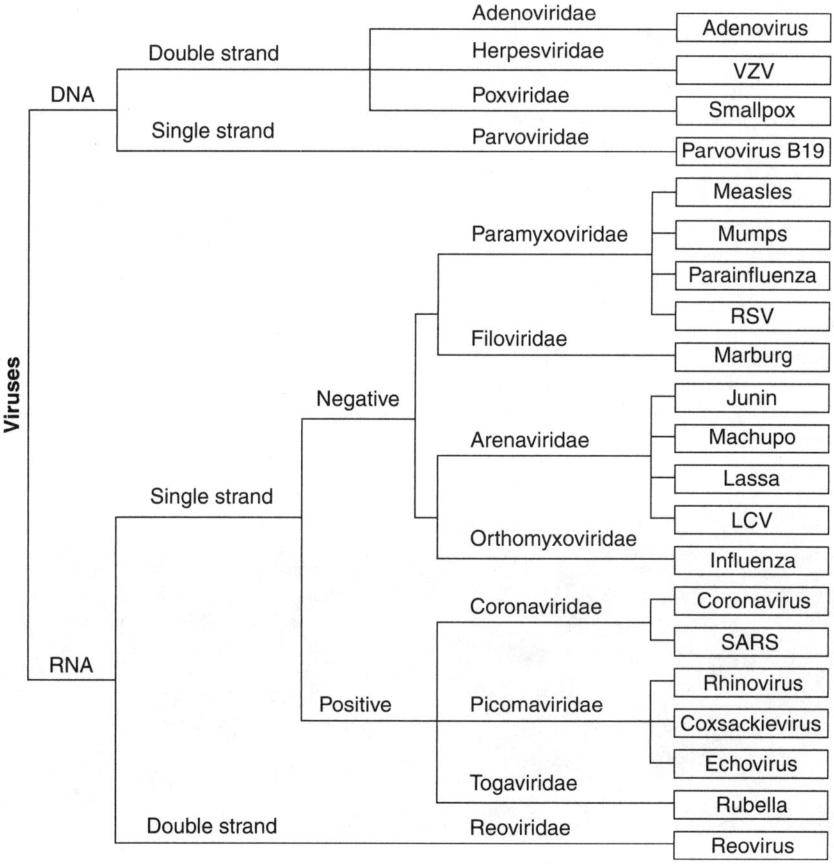

FIGURE 2.3 Breakdown of contagious respiratory viruses by type and taxonomy.

2.4 BACTERIA

The only common characteristic of bacteria is their diversity. They cause so many different types of diseases and follow so many infectious pathways that generalizations for this group are difficult. They span a size range of about 0.2 to 5.0 μm that renders some of them highly filterable and other highly penetrating (Kowalski et al., 1999). They vary in shape from spherical to elongated (see Fig. 2.4). Some bacteria hail from environmental sources but most pathogenic bacteria come from humans or animals. Man-made equipment can sometimes foster the spread of diseases, as in the case of *Legionella pneumophila*.

In the previous chapter the evolution of bacterial pathogens was examined. Part of the basis for the presumed first appearances of airborne pathogens (see Fig. 1.3) was the taxonomy of bacteria. Taxonomy is the phylogenetic classification of organisms and is based on similarities that include morphology and genetics. Figure 2.5 presents one version of the phylogeny of airborne bacterial pathogens based on traditional phylogeny and taxonomy, which no doubt will be subject to revision as genomic research progresses.

Most bacteria exist as cells, either singularly or in groups, but some bacteria form spores. There are several spore-producing bacteria that may transmit infections by the airborne route. These include *Bacillus anthracis, Clostridium botulinum, Clostridium perfringens*, and the actinomycetes. The actinomycetes are an important class of bacteria that produce fungilike mycelia and often produce spores. They exist environmentally but tend to be amplified in agricultural environments. Sporulating bacteria that cause respiratory infections behave epidemiologically like fungal spores (Sikes and Skinner, 1973; Slack and Gerencser, 1975). They cause noncontagious infections of the lungs and sometimes other locations and grow in a mycelia-like fashion in the presence of moisture and nutrients (Ortiz-Ortiz et al., 1984; Smith, 1989). Most of the spore-forming bacteria in the database are members of actinomycetes and are also called sporoactinomycetes.

Although actinomycete spores are common in the environment, the pathogenic species are predominant in agricultural environments and agro-industrial facilities (Lacey and Crook, 1988; Woods et al., 1997). Farmer's lung, for example, describes a category of similar infections that can be caused by a variety of actinomycete species.

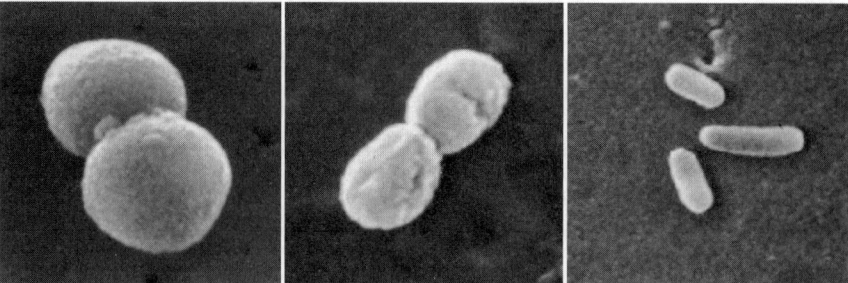

FIGURE 2.4 Most bacteria tend to be round, ovoid, or rodlike as in these images of *Streptococcus* (left, PHIL#263), *Acinetobacter* (center, PHIL#185), and *Pseudomonas* (right, PHIL#230). (*Images reprinted courtesy of the CDC Public Health Image Library.*)

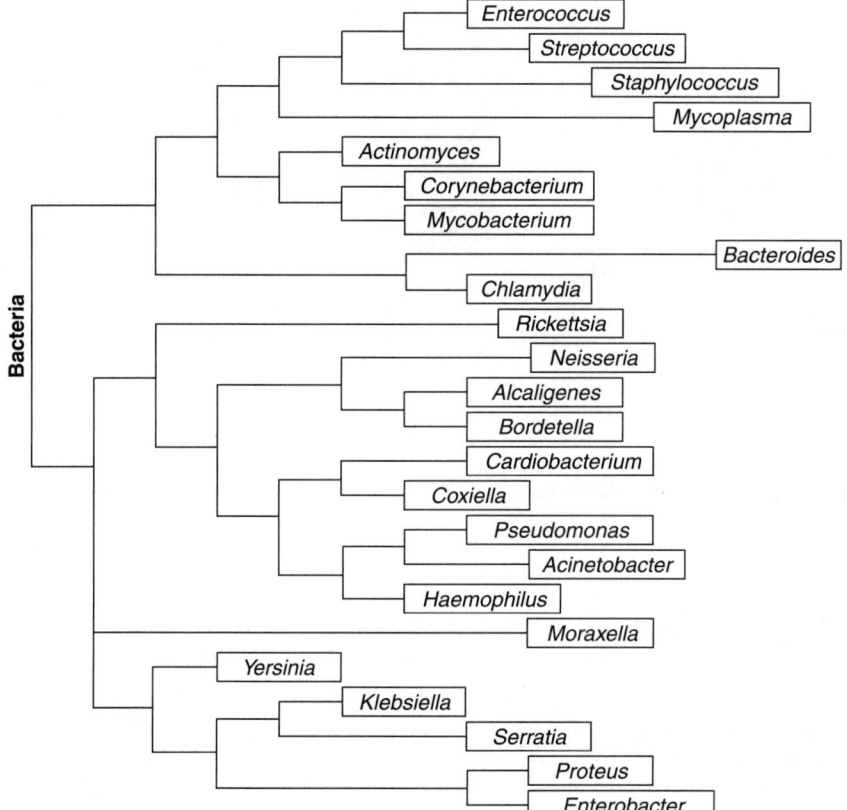

FIGURE 2.5 The phylogeny of airborne pathogenic bacteria. Based on Prescott et al. (1996) and standard taxonomic classifications.

2.5 FUNGAL SPORES

Fungal spores originate predominantly from the environment, especially where soils become dry and windblown (Austin, 1991; Mitscherlich and Marth, 1984). In general, it is the fungal spore that is of most concern as an indoor airborne hazard, but in a few species (i.e., *Candida*) it is the yeast form that is a hazard. Germinated spores typically form mycelia and sometimes fragments of mycelia may produce allergenic hazards. Fungal spores vary in shape from spherical to barrel (see Fig. 2.6). Certain fungal spores produce toxins that can be deadly even in nonsusceptible individuals (Pope et al., 1993). In all reported cases the problem appears to have been indoor amplification in the presence of moisture and certain building materials like gypsum (Woods et al., 1997).

The colonization of indoor environments by fungal spores, and their subsequent growth in the presence of moisture and nutrients, can lead to indoor levels exceeding outdoor levels (Godish, 1995; Rao and Burge, 1996). Fungal spores can cause respiratory infection, allergies, and toxic reactions, but not contagious diseases (Howard and Howard, 1983).

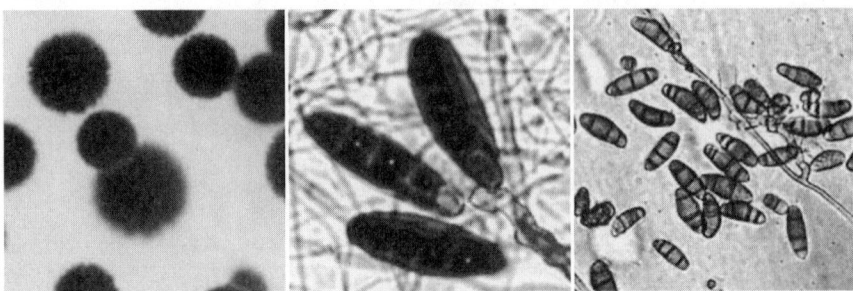

FIGURE 2.6 Fungal spores tend to be spherical or ovoid as in these examples of *Aspergillus* (left), *Bipolaris* (center), and *Curvularia* (right). *(Images provided courtesy of Neil Carlson, University of Minnesota.)*

They are overwhelmingly opportunistic and so the casual exposure to spores in the outdoors rarely causes diseases in healthy persons (Murray, 1999). Millions of Americans, for example, have had histoplasmosis without realizing it (Ryan, 1994). Susceptible individuals, however, can suffer fatal consequences.

2.6 PROTOZOA

Protozoa are single-celled microorganisms that resemble yeasts morphologically. They can range in size from 2 to 100 µm. Protozoan classes include rhizopods (amoebas), ciliates, flagellates, and sporozoa. A number of protozoa are important parasites like *Giardia lamblia* and trypanosomes, but none have been identified as being airborne. They are predominantly waterborne, foodborne, sexually transmitted, and blood-borne or vector-borne parasites. The only protozoa ever identified as an airborne pathogen, *Pneumocystis carinii*, has recently been reclassified as a fungus, albeit an unusual one (Ryan, 1994). Therefore, no protozoa transmit by the airborne route. It is possible that by-products or materials from protozoa may act as allergens but no conclusive information has been developed on this matter.

2.7 ALGAE

Algae are chlorophyll-bearing plants that are widely distributed in salt waters and fresh waters around the world. They include seaweed, kelp, and blue-green algae (cyanobacteria). Algae were first identified as possible airborne allergens by McElhenney et al. (1962). Investigators have since identified a number of algae that may be responsible for allergies and these have been listed in Table 2.1, per Flannigan et al. (2001). Particles from the algae may become aerosolized on fragmentation. They may also be ingested by those exposed to them.

Some algae can produce or facilitate the production of toxins that may be involved in causing respiratory problems or toxic reactions. Anatoxin A, brevetoxin B, and microcystin are toxins produced in algal blooms. Brevetoxin B is associated with the "red tide" catastrophes that killed massive numbers of fish and resulted in a number of human poisonings.

TABLE 2.1 Potentially Allergenic Algae

Algae	Class	Algae	Class
Anabaena	Blue-green algae	*Microcystis*	Blue-green algae
Ankistrodesmus	Green algae	*Myrmecia*	Green algae
Aphanothece	Blue-green algae	*Myxosarcina*	Green algae
Bracteacoccus	Green algae	*Nostoc*	Blue-green algae
Chlamydomonas	Green algae	*Oocystis*	Green algae
Chlorella	Green algae	*Oscillatoria*	Blue-green algae
Chlorococcum	Green algae	*Phormidium*	Blue-green algae
Coccomyxa	Green algae	*Porphyridium*	Red algae
Ecklonia	Brown algae	*Scytonema*	Blue-green algae
Fucus	Brown algae	*Spygrogia*	—
Hormidium	Blue-green algae	*Stichococcus*	Green algae
Laminaria	Brown algae	*Tetracystis*	Green algae
Lessonia	Brown algae	*Trebouxia*	Green algae
Lyngbya	Blue-green algae	*Ulva*	Green algae
Mesotaenium	Green algae	*Westiellopsis*	Blue-green algae

Microcystin is associated with blue-green algae while anatoxin comes from the cyanobacteria *Anabaena flosquae*. Algae are generally single-celled species that grow mostly in water. They provide food for aquatic animals and play a part in natural water purification, but under favorable conditions algal blooms may spread widely and generate high levels of toxins. Algal toxins can have varying effects, like central nervous system damage, neurotoxic effects, and paralysis. Algal toxins have, however, not been known to spread by the airborne route.

2.8 POLLEN

Pollen spores are plant seedlings that can cause allergic reactions in atopic individuals (Pope et al., 1993). Their mean size ranges from about 6 μm to about 120 μm and their shape may vary considerably (see Fig. 2.7). They are large enough to be almost completely removed by moderate levels of filtration (Kowalski et al., 1999). However, they tend to

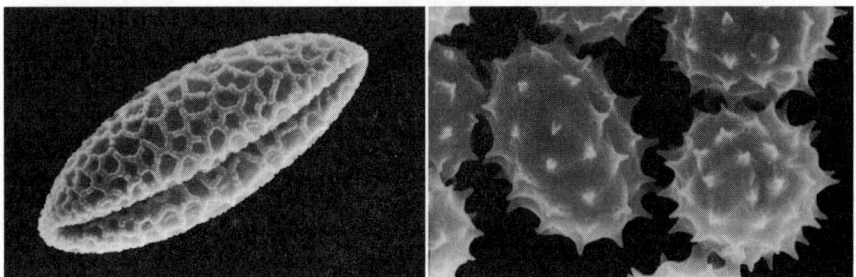

FIGURE 2.7 Scanning electron micrograph of Lilium (left) and Rudbeckia (right) pollen. (*Image provided courtesy of the Israeli Botanical Society and Dr. Yoav Waisel.*)

TABLE 2.2 Most Common Allergenic Pollen

Latin name or genera/family	Common name	Minimum size, μm	Maximum size, μm	Logmean size, μm
Acacia	Wattle, mimosa	38	140	73
Acer	Maple, box elder	22	51	33
Alnus	Alder	18	35	25
Amaranthaceae	Pigweed, amaranth, cottonweed	14	65	30
Ambrosia	Ragweed	22	32	27
Artemisia	Sage, wormwood, mugwort	19	26	22
Asteracea	Goldenrod, aster, cocklebur, and the like	15	40	24
Betulaceae	Birch, hazelnut, hop hornbeam	18	35	25
Calliandria	Calliandria	38	140	73
Cannabinaceae	Wax myrtle	14	28	20
Carpinus	Hornbeam	30	40	35
Carya	Hickory, pecan	10	68	26
Chenopodiaceae	Goosefoot, saltbush, Russian pigweed, Lamb's quarters	20	30	24
Chrysanthemum	Ox-eye	32	37	34
Corylus	Hazel	18	28	22
Cryptomeria	Evergreen, Japanese cedar	25	35	30
Cupressaceae	Juniper family, cedar, conifers	20	36	27
Cyperaceae	Sedge family	21	65	37
Fagus	Beech	31	54	41
Fraxinus	Ash, black ash, green ash	15	33	22
Juglans	Walnut, butternut	10	68	26
Liquidambar	Sweet gum	4.8	8	6
Morus	Mulberry	11	25	17
Olea	Olive	15	33	22
Parietaria	Pellitory	12	16	14
Pinaceae	Pine family	85	160	117
Pinus radiata D	New Zealand pine	60	85	71
Plantago	English plantain	26	31	28
Platanus	Sycamore, plane tree	17	20	18
Poaceae	Grass family, Gramineae	22	122	52
Polygonaceae	Buckwheat, Rumex, sorrel	18	56	32
Populus	Poplar, cottonwood, aspen	25	40	32
Quercus	Oak	19	39	27
Salix	Willow	16	40	25
Taraxacum	Dandelion	32	37	34
Tilia	Basswood, linden	20	47	31
Typha	Cattail	20	26	23
Ulmus	Elm	16	50	28
Urticaceae	Nettle family, pellitory	12	15	13

cause seasonal allergies in those exposed to outside air or in naturally ventilated buildings, or in buildings with unfiltered air or excessive infiltration.

Pollen spores were among the first allergens ever identified and they continue to be a leading cause of morbidity among atopic individuals. The term hay fever describes the symptoms of what is known as seasonal allergic rhinitis, and symptoms include runny nose, sneezing, watery eyes, and other reactions. A wide variety of tree pollen, flower pollen, grass pollen, and weed pollen may induce allergic reactions. Table 2.2 summarizes the most common or suspected allergenic pollen that have been identified based on a review of the literature (Cox and Wathes, 1995; AAAAI, 2000; Lewis et al., 1983; Burge, 2002; Middleton et al., 1983).

The airborne pollen largely responsible for allergies are wind-pollinated, or anemophilious, plants. These include common trees such as elm, oak, mulberries, and hackberries, which can release massive quantities of pollen. Many flowers, grasses, and weeds may also produce huge quantities of pollen during flowering periods. Humans become incidental or accidental receivers of the pollen grains. The pollen is the male gametophyte that, on contact with surfaces, will release proteins that serve as species-specific recognition compounds (Lewis et al., 1983). If these surfaces happen to be human mucosa, the immune system responds, or overresponds, in atopic individuals to produce allergic symptoms. A period of sensitization is needed before an allergic reaction can develop, but this probably happens early in life. Some evidence suggests that the time of year a person is born, if it is the flowering season for certain pollen, is a predictor for atopy (Middleton et al., 1983).

2.9 DUST MITES AND STORAGE MITES

Dust mites are often identified as allergens, but generally only where there is considerable dust disturbance (Pope et al., 1993). Dust mites are acarids that thrive in warm and damp environments. The primary species responsible for most allergies are the dust mites *Dermatophagoides farinae* and *D. pteronyssinus* (see Fig. 2.8). Table 2.3 lists all mites that have been identified as allergens, including storage mites and spider mites (O'Rourke et al., 1993, Platts-Mills, 2001).

Allergy to dust mites is far more prevalent due to their constant presence in indoor environments and the inevitable exposure of occupants. Dust mites can be found in high concentrations in upholstered furniture, mattresses, pillows, and carpets. Because of their size—some 300 µm—dust mites are not truly capable of airborne transport for any but the shortest distances. Allergic reactions to dust mites are usually due to much smaller particulate matter that cannot be well characterized in terms of size or shape. These particles may become aerosolized through disturbances caused by walking over rugs, changing bed sheets, or working with stored food or grain products. Studies show that most dust mite allergens tend to be greater than 10 µm in size and that 96 percent will settle out of the air after 15 to 35 min following any disturbance that aerosolizes them (Platts-Mills et al., 1986). Dust mites are one of the most common causes of asthma throughout the world. The majority of asthmatic children are sensitized to dust mite allergens (Platts-Mills et al., 1987).

Storage mites pose problems for individuals only in certain environments, such as where cereal-based foods or grains are stored. Storage mites that have been implicated in allergies include *Lepidoglyphus destructor*, *Acarus siro*, and *Tyrophagous putrescentior*, and also species in the genera *Glycyphagus* (O'Rourke et al., 1993; Platts-Mills, 2001). There are seven groups of specific allergenic proteins that are associated with allergies and these are labeled Group 1 through Group 7. In addition, there are several additional compounds from mites that have been implicated in allergies. See Flannigan et al. (2001) for

24 BACKGROUND AND HISTORY

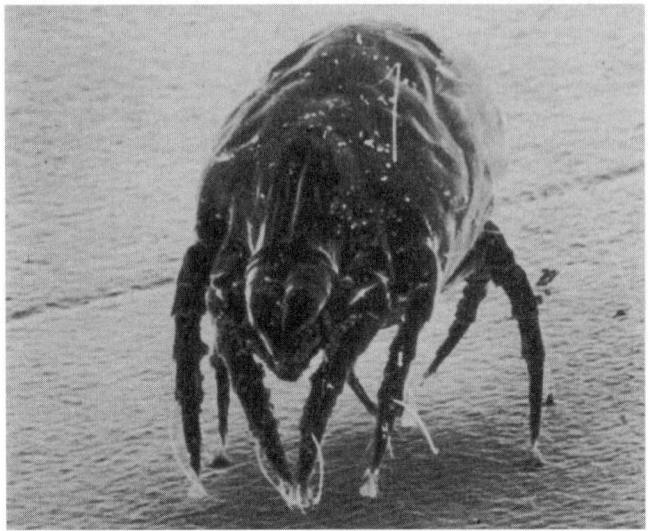

FIGURE 2.8 The most common dust mite, *Dermatophagoides pteronyssinus*. (*Photo courtesy of Sally Wirtz and American Academy of Allergy, Asthma, & Immunology.*)

biochemical details on these allergens. These allergens are mainly products found in dust and storage mite excreta or body parts. Spider mites are primarily an occupational hazard among fruit growers and greenhouse workers.

2.10 COCKROACH ALLERGENS

The most significant insect allergens besides dust mites in indoor environments are cockroach allergens. Cockroach allergens were only recently recognized as being important causative agents of asthma among inner city children (Rosenstreich et al., 1997). Sensitivity to cockroach allergens was more prevalent among inner city children than sensitivity

TABLE 2.3 Allergenic Mites

Dust mites	*Dermatophagoides farinae*
	Dermatophagoides pteronyssinus
	Dermatophagoides microceras
	Euroglyphus maynei
	Blomia tropicalis
Storage mites	*Lepidoglyphus destructor*
	Acarus siro
	Tyrophagous putrescentior
	Glycyphagus spp.
Spider mites	*Tetranychus urticae*
	Panonychus citri

to dust mite allergens. Approximately 85 percent of inner city bedrooms had detectable cockroach allergen. Like dust mite allergens, cockroach allergens exist in the excreta or body parts of cockroaches and may become aerosolized on disturbance and inhaled. At the size typical of such matter, about 10 µm or larger, they should be easily removable by air filters.

2.11 ANIMAL DANDER

Dander refers to particles of dead skin (squames) or hair that are normally shed by animals. Dander can come from furred or feathered pets of animals, including cats, dogs, birds, mice, hamsters, gerbils, horses, cows, and other animals or pets (Esch et al., 2001; Platts-Mills, 2001). Table 2.4 lists the various animal and bird species that have been identified as allergenic.

Cat dander is the most prevalent allergen and shows up everywhere from schools to airplanes. When these particles are microscopic in size they may become airborne and end up being inhaled. To atopic individuals, these particles are allergens. To others they may be respiratory irritants or simply annoying odors. Dander will tend to accumulate in upholstered furniture, carpets, and other furnishings, and will become aerosolized on disturbance. Dander may occur in various sizes and shapes (see Fig. 2.9) but studies suggest they

TABLE 2.4 Allergenic Animals and Birds

Class	Common name	Species
Mammals	Cat	*Felis domesticus*
	Cow	*Bos domesticus*
	Dog	*Canis familiaris*
	Eastern mole	*Scalopus aquaticus*
	Gray fox	*Urocyon cinereoargenteus*
	Horse	*Equus caballus*
	Minks and ferrets	*Mustela* spp.
	Mouse	*Mus musculus*
	Opossum	*Didelphis marsupialis*
	Pig	*Sus scrofa*
	Raccoon	*Procyon lotor*
	Rat	*Rattus norvegicus*
	Sheep	*Ovis* spp.
	Shrews	*Sorex* spp.
	White-tailed deer	*Odocoileus virginianus*
Birds	Chicken	*Gallus domesticus*
	Common canary	*Serinus canaria*
	Common hamster	*Cricetus cricetus*
	Eastern fox squirrel	*Sciurus niger*
	Gerbil	*Gerbillus campestris*
	Guinea pig	*Cavis porcellus*
	Parakeet	*Melopsittacus undulatus*
	Pigeon	*Columba livia*
	Rabbit	*Oryctologus cuniculus*
	Turkey	*Meleagris gallopavo*

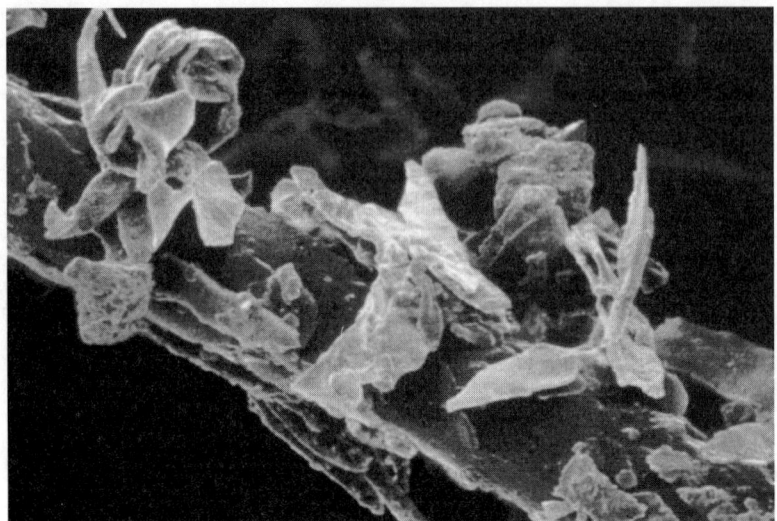

FIGURE 2.9 Image of animal dander. [*Photo provided courtesy of Jerome Schultz and American Academy of Allergy, Asthma and Immunology (AAAAI)*].

will often tend to be particles of 2 μm or larger. Particles this size should be removed at high efficiencies by moderate efficiency filters (i.e., MERV 6 to MERV 8).

2.12 TOXINS

Toxins are types of organic poisons that induce diseases or can cause toxic reactions (Sorenson, 1990; Smith and Moss, 1985). Some toxins are known to cause cancer and to decrease the effectiveness of the immune system (CAST, 1989). Endotoxins and exotoxins are produced by certain bacteria. Mycotoxins are produced by fungi. Some toxins are produced by algae, plants, insects, and animals. The toxins of most concern in airborne bacterial diseases are endotoxins, exotoxins, and mycotoxins.

2.12.1 Endotoxins and Exotoxins

Endotoxins are toxic lipopolysaccharides present in or released from the outer membrane of certain gram-negative bacteria (Ryan, 1994). They can be an important component of disease. Gram-positive bacteria produce no endotoxins. Exotoxins are toxic proteins liberated from certain bacterial cells, usually gram-positive bacteria. They impact the host immune system and may cause severe reactions such as toxic shock.

In general, the quantities of bacteria that may exist in indoor air are normally insufficient to pose a toxic hazard even if these bacteria are toxin producers. The bacteria typically must multiply under the right conditions in order to produce a sufficient quantity of toxin to create a health risk, but this is what occurs during some infections and in cases of food poisoning. However, in environments where gram-negative bacteria proliferate the airborne levels of endotoxins may reach unnatural levels and pose an inhalation hazard.

Endotoxins are produced by a variety of gram-negative bacteria, especially the *Enterobacteriaceae*, including the potentially airborne pathogens *Klebsiella pneumoniae, Proteus, Yersinia, Serratia,* and *Enterobacter.*

2.12.2 Mycotoxins

Mycotoxins are toxins produced by toxigenic fungi when environmental conditions are appropriate. The toxins are secreted by fungal hyphae directly into the growth substrate. Mycotoxins are products of the secondary metabolism of fungi. Products of the primary metabolism include RNA, DNA, and proteins. The primary metabolism involves the consumption of organic compounds to produce essential components or to produce the energy necessary to drive the process of biomass construction. Secondary metabolism occurs after the growth phase. The products of secondary metabolism are often species specific and diversity of products tends to be the rule (Hendry and Cole, 1993; Husman, 1996).

Mycotoxin production can vary with the substrate and environmental conditions (Kemp et al., 1997; Scott and Yang, 1997). Certain substrates favor the growth of certain mycotoxic fungi (Nielsen et al., 1999). Adverse environmental conditions, such as insufficient nutrients and water, or temperature extremes, stimulate toxin production (Smith and Moss, 1985). One school of thought holds that in the competition for scarce resources, fungi release mycotoxins into the local environment as a form of biological warfare against their neighbors, presumably other fungi or bacteria (CAST, 1989; Amman, 2001). Humans become incidental casualties of this struggle for survival. Table 2.5 lists the main fungi that can produce toxins and their associated or potential toxins.

The most commonly occurring fungi in indoor air, *Penicillium, Cladosporium, Aspergillus, and Alternaria,* are all potential producers of toxins (Etzel et al., 1998; Li and Kendrick, 1995). Most of the fungi that commonly grow on building materials, *Penicillium, Aspergillus, Cladosporium, Chaetomium, Fusarium, and Stachybotrys,* are also potential toxin producers (Johanning et al., 1999; Flannigan et al., 1991; Miller et al., 1988).

Mycotoxins are nonvolatile, however, airborne spores containing toxins are potentially respirable, and the risk due to inhalation of toxins has been estimated as greater than that due to ingestion (Sorensen et al., 1999). Not all species that grow indoors necessarily produce toxins. Not all toxin-producing fungi will produce toxins. The presence of potentially mycotoxic fungi does not necessarily imply a problem, although it may indicate a potential hazard. Conversely, the absence of fungi may not imply an acceptable environment since the mycotoxins may persist after the fungi is gone. A list of mycotoxins and their fungal sources is provided in App. B.

2.13 MICROBIAL VOLATILE ORGANIC COMPOUNDS

Volatile organic compounds (VOCs) are vapors given off by microbes and other sources that may impact human health. Most VOCs are produced by man-made products such as tobacco, paint, adhesives, insulation, building materials, and cleaning fluids and these are not the subject of this book. *Microbial volatile organic compounds* (MVOCs) are produced by bacteria and fungi that may grow on building materials or wherever condensation occurs (Pasanen et al., 1997). MVOCs may be odorous or toxic secondary metabolites and may affect human health, causing lethargy, headache, and irritation of the eyes and mucous membranes of the nose and throat (Fischer and Dott, 2003). MVOCs may be a contributing factor in cases of SBS/BRI. Some MVOCs may be toxigenic or cytotoxic but little data are available to

TABLE 2.5 Mycotoxins Produced by Airborne Indoor Fungi

Fungal spore genera	Species	Toxins
Acremonium		Crotocin
Alternaria	A. alternata	Alternariol, altenuisol, tenuazonic acid Alternariol monomethyl ether, tenuazonic acid, altertoxin I, altertoxin II, altenuene, altenusin
Aspergillus	A. flavus	Aflatoxin B1, aflatoxin B2, aflatoxin M1, cyclopiazonic acid, kojic acid, aflatrem
	A. fumigatus	Aflatoxin, fumigaclavines, gliotoxin, fumigatoxin, fumagillin, fumitremorgens, helvolic acid, tryptoquivaline tremorgens, verruculogen, ergot alkaloids, viriditoxin, endotoxin
	A. niger	Malformin C, oxalic acid
	A. ustus	Austocystins, austamide, austdiol, brevianimide
	A. versicolor	Aspercolorin, averufin, cyclopiazonic acid, sterigmatocystin, versicolorin
	A. ochraceus	Ochratoxins, penicillic acid, destruxin B
	A. chevalieri	Xanthocillin
	A. clavatus	Patulin, cytochalasin E, tryptoquivaline
	A. parasiticus	Aflatoxins (B1, B2, G1, G2, M1)
Chaetomium	C. globosum	Chaetoglobosins Chaetomin
Cladosporium		Cladosporin, emodin, epicladosporic acid
Epicoccum		Flavipin, epicorazine A, epicorazine B, indole-3-acetonitrile
Eurotium	E. chevalieri	Xanthocillin
Fusarium		Trichothecenes (Type B), T-2 toxin, zearalenone (F-2 toxin), vomitoxin, deoxynivalenol, fumonisin
	F. nivale	Trichothecenes, butenolide
	F. graminearum	Trichothecenes, zearalenone
	F. solani	Trichothecenes
	F. moniliforme	Fusarin
	F. sporotrichioides	T-2 toxin
Paecilomyces		Paecilotoxins, byssochlamic acid, variotin, ferrirubin, viriditoxin, indole-3-acetic acid, fusigen, patulin
Penicillium	P. citrinum	Citrinin
	P. expansum	Citrinin, patulin
	P. viridicatum	Citrinin, ochratoxin, viridicatin, xanthomegnin, viomellein
	P. cyclopium	Ochratoxin A, penitrem A, cyclopiazonic acid, penicillic acid
	P. islandicum	Luteoskyrin, islanditoxin, cyclochlorotine
	P. purpurogenum	Rubratoxin
	P. roquefortii	P.R. toxin, roquefortine

TABLE 2.5 Mycotoxins Produced by Airborne Indoor Fungi (*Continued*)

Fungal spore genera	Species	Toxins
	P. crustosum	Penitrem A
		Penicillic acid, peptide nephrotoxin, viomellein, xanthomegin, xanthocillin X, mycophenolic acid, rocquefortine C, roquefortine D, penicillin, cyclopiazonic acid, isofumigaclavine A, penitrem A, decumbin, citreoviridin, griseofulvin, verruculogen, chrysog
Phoma		Phomenone
Rhizopus		Rhizonin A
Stachybotrys	*S. chartarum*	Trichothecenes, verrucarin J, roridin E, satratoxin F, satratoxin G, satratoxin H, sporidesmin G, trichoverrol, cyclosporins, stachybotryolactone, trichoverrins, verrucarol
Trichoderma	*T. viridae*	Trichodermin, isocyanides

specifically link any microbial VOCs to diseases other than SBS/BRI. MVOCs are often used as indicators of the presence of mold contamination in buildings (Cochrane, 2001). However, MVOCs do not derive solely from active microbial growth, but may be released into the air from materials that have become wet. A list of MVOCs is provided in App. C.

2.14 POTENTIAL AND EMERGING AIRBORNE PATHOGENS

A number of microbes that are rare, or that are not directly connected to infections but merely suspect by association have been excluded from the main database of airborne pathogens and allergens in App. A, but have been included in a list of potential pathogens and allergens at the end of App. A. Many of these have loose or doubtful associations with allergic or infectious diseases, or else have been associated with diseases in immunocompromised hosts. In addition, it is likely that pathogens will continue to emerge and evolve, especially viruses, and so the database cannot be considered exclusive. A recent virus, avian influenza, shows every sign of being able to jump species to humans, but so far there is no evidence of it being airborne, so it has not been included.

2.15 AIRBORNE BIOLOGICAL WEAPONS

The deliberate creation of biological weapons has added a number of potential airborne pathogens to the array of natural human hazards. Toward the development of disease agents of unusual infectiousness and virulence, the technologies of weaponization are surpassed only by the deviousness of bioengineering. From unnatural selection of the most virulent variants of existing pathogens to genetically engineered mutants of Ebola, designed through recombinant DNA techniques to be highly toxic and aerosolizable, there is increasing potential that such research may produce new plagues for which neither the human immune system nor nature has any adequate defense. Immune buildings may offer the only feasible protection against agents with the potential to wipe out the entire human race. The topic of biological weapons is not specifically addressed in this book since they have been adequately treated in a previous work (Kowalski, 2003). However, the methods and technologies of aerobiological engineering may be equally well applied regardless of the airborne disease agent and any building designed to protect occupants against common airborne pathogens will protect against future threats as well.

2.16 THE AIRBORNE PATHOGEN AND ALLERGEN DATABASE

Appendix A is a database of airborne pathogens and microbial allergens that contains a summary of all known and suspected airborne microbes that are pathogens, allergens, or respiratory irritants. It has been designed for ease of reference with the most relevant parameters and descriptions necessary to determine what form of protection may be required and how effective an engineered system may be in protecting building occupants. It includes over 110 microbial agents that have been well established as potential health hazards, and lists several dozen more for which the hazards are limited, uncertain, or doubtful. In addition to respiratory hazards, it includes a number of nonrespiratory airborne microbes, some foodborne pathogens that have been known to be airborne, and many agents that are common nosocomial threats or that are hazardous to the immunocompromised. This list is by no means exclusive, since as new pathogens arise, and existing pathogens can evolve new infectious mechanisms. Following are explanations of the standardized groupings of information in App. A.

In most cases the microbial species is identified at the top of each page in App. A, but in some cases only a genus is identified either because there are several species in the genera or because the exact species is uncertain. At the top right is a block providing some basic information identifying the type of microbe, including taxonomic information, communicability, and biosafety level. The Group block typically describes the agent as a virus, a bacterium, or a fungal spore. The taxonomic identification information may variously include Phylum, Class, Order, Family, Genus, and (for mitosporic fungi) the Type (not a formal designation). The Type is used for special designations including mitosporic fungi and the virus DNA or RNA type. DNA viruses may be double stranded (ds) or single stranded (ss), and the RNA viruses may have positive or negative strands. As the taxonomic classifications of microbes are currently in constant, but necessary, flux due to recent advances in genetic research, and as species names are in seemingly endless flux, some of these classifications and names may even be obsolete by the time of publication. Every effort has been made to keep the species names up to date and older obsolete names are provided where space allows.

The Disease Group describes whether the microbe is communicable (contagious), noncommunicable, or an endogenous (or commensal) member of human microflora (i.e., skin, intestines) and therefore not specifically definable as contagious or noncontagious. The Biosafety Level gives the Risk Group, 1 through 4, to which the microbe has been assigned by the Centers for Disease Control (CDC). Unassigned fungi that are minor allergens have been listed here as Risk Group 1 pending any formal assignment by the CDC.

Below the identification block is a summary block of epidemiological information, including Incubation Period, Peak Infection, Annual Cases, and Annual Fatalities. In many cases this information is limited or not known with certainty, and for the allergens this information is often not applicable. The Infectious Dose, is given as the ID_{50} or the quantity of agent that will result in 50 percent infections in an exposed population. The Lethal Dose is similarly given as the ID_{50}. Both these values, if known, are based on the best available information and are typically from laboratory studies on animals. The Infection Rate, Annual Cases, and other epidemiological data are provided based on the best estimates or published data for recent years.

Below the image is a block providing an abbreviated write-up describing the microbe, its epidemiology and pathogenesis, and other relevant information. Below the write-up are summarized the Disease or Infection, the Natural Source of the microbe, a list of any Toxins produced, the primary Point of Infection, common Symptoms, and an abbreviated description of the typical infection Treatment. In cases of allergens that have infectious potential, only the infection is addressed by the Treatment—for nontoxic allergens there often is no specific treatment other than avoidance. Next, the Untreated Fatality Rate is provided if known, along with notes on whether any Prophylaxis or Vaccine exists.

The Shape describes the physical appearance of the microbe and for the fungi it almost always refers to the spores, not the yeast form. The mean diameter generally represents the logarithmic mean diameter estimated based on the microbe's dimensions and, for non-spherical microbes, is computed from the algorithms described in Chap. 10. The Size Range on which the Mean Diameter is based is also summarized. Estimates of the size range can vary in the literature and the most representative dimensions have been used.

The Growth Temperature represents either the optimum growth temperature (i.e., in laboratory cultures) of the growth range (i.e., for indoor fungal growth). A note about the ability of the organism's Survival Outside Host (or survival outdoors for nonpathogenic microbes) is provided.

The Inactivation block describes the thermal disinfection requirements for the particular organism. Where available both the moist heat (i.e., steam sterilization) conditions are provided along with dry heat sterilization information.

The Disinfectants describe the various common disinfectants that are considered effective against the particular microbe, although this list is never exclusive and there are numerous other disinfectants available that may be equally well used to disinfect surfaces of the microbe in question.

The next box consists of a table giving the Filter Nominal Rating of five typical filters, and the associated percent Removal Efficiency. These removal rates are based on filter models and the mean diameter of the microbe. Details of the filter models can be found in Chap. 10. The particular filter models are shown graphically in Fig. 2.10 and are based on MERV test results for several actual manufacturers filters (Kowalski and Bahnfleth, 2002). The filters used are representative only and other filters of the same rating may outperform the ones shown in Fig. 2.10, especially for the MERV 6–11 filters in the higher size range.

The UVGI Rate Constant shows a representative value, if any is available, the Media (water, air, or plates) for which the rate constant was determined per the indicated reference (Ref.). Also provided is the Dose for 90 percent Inactivation (or fluence for 90 percent inactivation). Most rate constants are unknown and a substitute rate constant for a similar or closely related species is provided where available. The surrogate microbe will be indicated in parentheses alongside the reference.

Next is provided a Suggested Indoor Limit. Few limits have actually been established for most pathogens and allergens. In general, any dangerous pathogen will have a zero limit indoors, as unachievable as this may be. For most allergenic fungi, the limit stated is simply the limit of total fungi, typically 150 to 500 cfu/m^3 of which the fungi in question may be a part of the natural mixture of fungi.

The Genome Size (bp) states the established or estimated size of the microbial genome in base pairs. Many of these have recently been determined for bacteria and viruses but few are known with certainty for fungi. Alongside the genome size are given the percent Guanine plus Cytosine (G+C) and the complementary percent Thymine and Adenine (T+A).

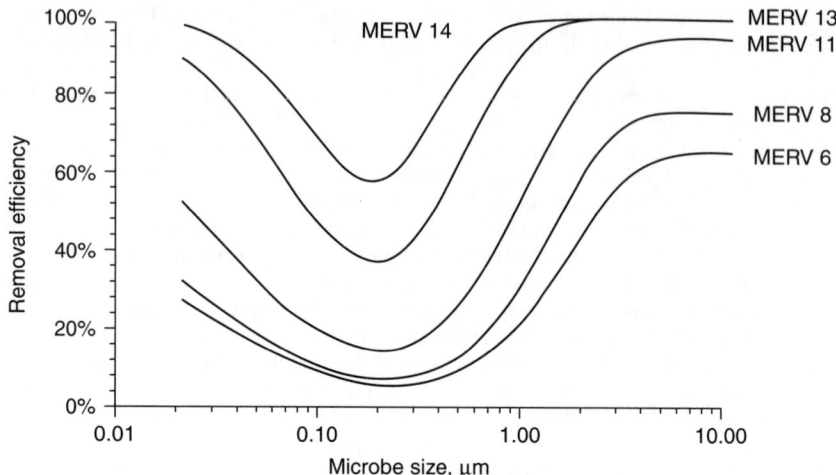

FIGURE 2.10 Filter models used for removal efficiencies in App. A.

Since UVGI inactivates microbes primarily by creating thymine-thymine dimers, there exists some yet-to-be-determined relationship between thymine content and the UVGI rate constant, and these values may be of some use in estimating unknown UVGI rate constants.

Related Species identifies closely related or similar species that may or may not pose similar airborne hazards or that may identify some or all of the species of a stated genus. The type species (TS) may also be provided in this block.

The Notes provides additional information that may not suit other blocks, and usually identifies whether an infectious disease due to the microbe is reportable or not. The Photo Credit is then given for the image at the top. The References at the bottom of each page in App. A are mostly abbreviated and can be found following.

REFERENCES*

AAAAI (2000). *Pollen & Spore Report.* American Academy of Allergy, Asthma and Immunology, Milwaukee, WI.

Al-Doory, Y., and Ramsey, S. (1987). *Moulds and Health: Who is at Risk?* Charles C. Thomas, Springfield, IL.

Allen, E. G., Bovarnick, M. R., and Snyder, J. C. (1954). "The effect of irradiation with ultraviolet light on various properties of typhus rickettsiae." *J Bacteriol* 67:718–723.

Amman, H. M. (2001). "Is indoor mold contamination a threat to health?" Washington State Department of Health, Olympia, WA.

Antopol, S. C., and Ellner, P. D. (1979). "Susceptibility of *Legionella pneumophila* to ultraviolet radiation." *Appl Environ Microbiol* 38(2):347–348.

*Including references for App. A.

Arnow, P. M., Sadigh, M., Costas, C., Weil, D., and Chudy, R. (1991). "Endemic and epidemic Aspergillosis associated with in-hospital replication of *Aspergillus* organisms." *J Infect Dis* 164:998–1002.

Ashford, D. A., Hajjeh, R. A., Kelley, M. F., Kaufman, L., Hutwagner, L., and McNeil, M. M. (1999). "Outbreak of histoplasmosis among cavers attending the National Speleological Society Annual Convention, Texas, 1994." *Am J Trop Med Hyg* 60(6):899–903.

Asthana, A., and Tuveson, R. W. (1992). "Effects of UV and phototoxins on selected fungal pathogens of citrus." *Int J Plant Sci* 153(3):442–452.

Austin, B. (1991). *Pathogens in the Environment*. Blackwell Scientific Publications, Oxford, England.

Beebe, J. M. (1959). "Stability of disseminated aerosols of *Pasteurella tularensis* subjected to simulated solar radiations at various humidities." *J Bacteriol* 78:18–24.

Bell, C., and Kyriakides, A. (1998). *Listeria: A Practical Approach to the Organism and Its Control in Foods*. Blackie Academic & Professional, London.

Bendinelli, M., and Friedman, H. (1988). *Coxsackieviruses: A General Update*. Plenum Press, New York.

Berendt, R. F., Young, H. W., Allen, R. G., and Knutsen, G. L. (1980). "Dose-response of guinea pigs experimentally infected with aerosols of *Legionella pneumophila*." *J Infect Dis* 141(2):186–192.

Brachman, P. S., Kaufmann, A. F., and Dalldorf, F. G. (1966). "Industrial Inhalation Anthrax." *Bacteriol Rev* 30(3):646–657.

Braude, A. I., Davis, C. E., and Fierer, J. (1981). *"Infectious Diseases and Medical Microbiology,* 2d ed." W. B. Saunders, Philadelphia, PA.

Burge, H. A. (1996). Health effects of biological contaminants. *Indoor Air and Human Health*. CRC Press, Boca Raton, FL.

Burge, H. A. (2002). "An update on pollen and fungal spore aerobiology." *J Allergy Clin Immunol* 110(4):544–552.

Canada (2001). "Office of Laboratory Security Material Safety Data Sheets." Canada Population and Public Health Branch. http://www.hc-sc.gc.ca/pphb-dgspsp/msds-ftss/index.html.

CAST (1989). "Mycotoxins: Economic and health risks." *Council for Agricultural Science and Technology*. Ames. IA.

Castle, M., and Ajemian, E. (1987). *Hospital Infection Control*. John Wiley & Sons, New York.

Chick, E. W., Hudnell, J. A. B., and Sharp, D. G. (1963). "Ultraviolet sensitivity of fungi associated with mycotic keratitis and other mycoses." *Sabouviad* 2(4):195–200.

Cochrane, P. (2001). "Microbial volatile organic compounds." *Aerotech Monitor* 4(1):2.

Collier, L. H., McClean, D., and Vallet, L. (1955). "The antigenicity of ultra-violet irradiated vaccinia virus." *J Hyg* 53(4):513–534.

Collins, F. M. (1971). "Relative susceptibility of acid-fast and nonacid fast bacteria to ultraviolet light." *Appl Microbiol* 21:411–413.

Collins, C. H., and Kennedy, D. A. (1993). *Laboratory-Acquired Infections*. Butterworth Heineman, Oxford, UK.

Cox, C. S., and Wathes, C. M. (1995). *Bioaerosols Handbook*. CRC/Lewis Publishers, Boca Raton, FL.

Dalton, A. J., and Haguenau, F. (1973). *Ultrastructure of Animal Viruses and Bacteriophages: An Atlas*. Academic Press, New York.

David, H. L. (1973). "Response of mycobacteria to ultraviolet radiation." *Am Rev Resp Dis* 108:1175–1184.

del Rey Calero, J. (2004). "Epidemiological perspectives on SARS and avian influenza." *An R Acad Nac Med (Madr)* 121(2):289–304.

DiSalvo, A. F. (1983). *Occupational Mycoses*. Lea & Febiger, Philadelphia, PA.

Dolman, P. J., and Dobrogowski, M. J. (1989). "Contact lens disinfection by ultraviolet light." *Am J Ophthalmol* 108(6):665–669.

Elke, K., Begerow, J., Oppermann, H., Kramer, U., Jermann, E., and Dunemann, L. (1999). "Determination of selected microbial volatile organic compounds by diffusive sampling and dual-column capillary GC-FID—a new feasible approach for the detection of an exposure to indoor mould fungi?" *J Environ Monit* 1(5):445–452.

Ellison, D. H. (2000). *Handbook of Chemical and Biological Warfare Agents*. CRC Press, Boca Raton, FL.

Esch, R. E., Hartsell, C. J., Crenshaw, R., and Jacobson, R. S. (2001). "Common allergenic pollens, fungi, animals, and arthropods." *Clin Rev Allerg Immunol* 21:261–279.

Etzel, R. A., Balk, S. J., Bearer, C. F., Miller, M. D., Shannon, M. W., and Shea, K. M. (1998). "Toxic effects of indoor molds." *Pediatrics* 101(4):712–714.

Farmer, J. J., Asbury, M. A., Hickman, F. W., Brenner, D. J., and Group, T. E. S. (1980). "*Enterobacter sakazakii*: A new species of *"Enterobacteriaceae"* isolated from clinical specimens." *Intl J System Bacteriol* 30(3):569–584.

Fields, B. N., and Knipe, D. M. (1991). *Fundamental Virology.* Raven Press, New York.

Fischer, G., and Dott, W. (2003). "Relevance of airborne fungi and their secondary metabolites for environmental, occupational and indoor hygiene." *Arch Microbiol* 179(2):75–82.

Flannigan, B., McCabe, E. M., and McGarry, F. (1991). Allergenic and toxigenic micro-organisms in houses. *Pathogens in the Environment,* B. Austin, ed., Blackwell Scientific Publications, Oxford, UK.

Flannigan, B., Samson, R. A., and Miller, J. D. (2001). *"Microorganisms in Home and Indoor Work Environments."* Taylor & Francis. Andover, Hants, UK.

Fleming, D. O., Richardson, J. H., Tulis, J. J., and Vesley, D. (1995). *"Laboratory Safety Principles and Practices,* 2d ed." ASM Press, Washington, DC.

Fraenkel-Conrat, H. (1985). *The Viruses: Catalogue, Characterization, and Classification.* Plenum Press, New York.

Freeman, B. A. (1985). *"Burrows Textbook of Microbiology."* W. B. Saunders, Philadelphia, PA.

Fuortes, L., and Hayes, T. (1988). "An outbreak of acute histoplasmosis in a family." *Am Fam Physician* 37(5):128–132.

Gilpin, R. W. (1984). Laboratory and Field Applications of UV Light Disinfection on Six Species of Legionella and Other Bacteria in Water. *Legionella: Proceedings of the Second International Symposium,* C. Thornsberry, ed., American Society for Microbiology, Washington, DC.

Godish, T. (1995). *Sick Buildings:Definition, Diagnosis and Mitigation.* CRC/Lewis Publishers, Boca Raton, FL.

Goodman, R. A., Beuhler, J. W., Greenberg, H. B., McKinley, T. W., and Smith, J. D. (1982). "Norwalk gastroenteritis associated with a water system in a rural Georgia community." *Arch Environ Health* 37(6):258–260.

Goodman, N. L. (1983). Sporotrichosis. *Occupational Mycoses,* A. F. DiSalvo, ed., Lea & Febiger, Philadelphia, PA.

Gordon, M. A. (1983). Cryptococcosis. *Occupational Mycoses,* A. F. DiSalvo, ed., Lea & Febiger, Philadelphia, PA.

Grigoriu, D., J. Delacretaz, and Borelli, D. (1987). *Medical Mycology.* Hans Huber Publishers, Toronto, Ontario, Canada.

Gunn, R. A., Terranova, W. A., Greenberg, H. B., Yashuk, J., Gary, G. W., Wells, J. G., Taylor, P. R., and Feldman, R. A. (1980). "Norwalk virus gastroenteritis aboard a cruise ship: An outbreak on five consecutive cruises." *Am J Epidemiol* 112(6):820–827.

Harris, G. D., Adams, V. D., Sorenson, D. L., and Curtis, M. S. (1987). "Ultraviolet inactivation of selected bacteria and viruses with photoreactivation of the bacteria." *Water Res* 21(6):687–692.

Harris, M. G., Fluss, L., Lem, A., and Leong, H. (1993). "Ultraviolet disinfection of contact lenses." *Optom Vis Sci* 70(10):839–842.

Hawksworth, D. L., Sutton, B. C., and Ainsworth, G. C. (1983). *Ainsworth & Bisby's Dictionary of the Fungi.* Commonwealth Mycological Institute, Kew, Surrey.

Heijden, K. v. d., Younes, M., Fishbein, L., and Miller, S. (1999). *"International Food Safety Handbook: Science, International Regulation, and Control."* Marcel Dekker, New York.

Henderson, D. A. (1999). "Smallpox: Clinical and epidemiological features." *Emerg Infect Dis* 5(4):537–539.

Hendry, K. M., and Cole, E. C. (1993). "A review of mycotoxins in indoor air." *J Toxicol and Environ Health* 38:183–198.

Henis, Y. (1987). *Survival and Dormancy of Microorganisms.* Wiley-Interscience, New York.

Higgins, I. J., and Burns, R. G. (1975). *The Chemistry and Microbiology of Pollution.* Academic Press, London.

Hill, W. F., Hamblet, F. E., Benton, W. H., and Akin, E. W. (1970). "Ultraviolet devitalization of eight selected enteric viruses in estuarine water." *Appl Microbiol* 19(5):805–812.

Howard, D. H., and Howard, L. F. (1983). *Fungi Pathogenic for Humans and Animals.* Marcel Dekker, New York.

Husman, T. (1996). "Health effects of indoor-air microorganisms." *Scand J Work Environ Health* 22:5–13.

Hyllseth, B., and Banrud, H. (1998). "Literature on UVC (J/m2) microbe killing/inactivation (%)." *Enclosure 01/Enclosure 02.* Norwegian College of Veterinary Medicine, Oslo, Norway.

Inglesby, T. V., Henderson, D. A., Bartlett, J. G., Ascher, M. S., Eitzen, E., Friedlander, A. M., Hauer, J., McDade, J., Osterholm, M. T., O'Toole, T., Parker, G., Perl, T. M., Russell, P. K., and Tonat, K. (1999). "Anthrax as a biological weapon: Medical and public health management." *JAMA* 281(18):1735–1745. http://jama.ama-assn.org/issues/v281n18/ffull/jst80027.html.

Jensen, M. M. (1964). "Inactivation of airborne viruses by ultraviolet irradiation." *Appl Microbiol* 12(5):418–420.

Jepson, J. D. (1973). "Disinfection of water supplies by ultraviolet radiation." *Wat Treat Exam* 22:175–193.

Jiang, Y., Li, X. -F., Zhao, B., Zhang, Z. -Q., and Zhang, Y. F. (2003). "SARS and Ventilation." *The Fourth International Symposium on HVAC,* Beijing, China. 27–36.

Johanning, E., Landsbergis, P., Gareis, M., Yang, C. S., and Olmsted, E. (1999). "Clinical Experience and Results of a Sentinel Health Investigation Related to Indoor Fungal Exposure." *Environ Health Perspect* 107(Suppl. 3):489–494.

Johnson, E., Jaax, N., White, J., and Jahrling, P. (1995). "Lethal experimental infections of Rhesus monkeys by aerosolized Ebola virus." *Intl J Experim Path* 76:227–236.

Joseph, J. M. (1983). Aspergillosis. *Occupational Mycoses,* A. F. DiSalvo, ed., Lea & Febiger, Philadelphia, PA.

Kashino, S. S., Calich, V. L. G., Burger, E., and Singer-Vermes, L. M. (1985). "In vivo and in vitro characteristics of six *Paracoccidioides brasiliensis* strains." *Mycopathologia* 92:173–178.

Katila, M. L., and Mantjarvi, R. A. (1978). "The diagnostic value of antibodies to the traditional antigens of farmer's lung in Finland." *Clin Allergy* 8(6):581–587.

Keller, L. C., Thompson, T. L., and Macy, R. B. (1982). "UV light-induced survival response in a highly radiation-resistant isolate of the *Moraxella-Acinetobacter* group." *Appl Environ Microbiol* 43(2):424–429.

Kemp, P. C., Neumeister, H. G., Kircheis, U., Schleibinger, H., Franklin, P., and Ruden, H. (1997). "Fungal genera in an office building with a central HVAC system in an Australian mediterranean climate." *Healthy Buildings/IAQ '97.* ASHRAE, Bethesda, MD. 257–260.

Kenyon, R. H., Green, D. E., Maiztegui, J. I., and Peters, C. J. (1988). "Viral strain dependent differences in experimental Argentine Hemorrhagic Fever (Junin virus) infection of guinea pigs." *Intervirology* 29:133–143.

Knowles, T. (2002). *Food Safety in the Hospitality Industry.* Butterworth Heineman, Oxford, UK.

Knudson, G. B. (1986). "Photoreactivation of ultraviolet-irradiated, plasmid-bearing, and plasmid-free strains of *Bacillus anthracis*." *Appl Environ Microbiol* 52(3):444–449.

Kotimaa, M. H. (1990). "Occupational exposure to fungal and actinomycete spores during the handling of wood chips." *Grana* 29:153–156.

Kowalski, W. J., Bahnfleth, W. P., Whittam, T. S. (1999). "Filtration of airborne microorganisms: Modeling and prediction." *ASHRAE Trans* 105(2):4–17. http://www.engr.psu.edu/ae/wjk/fom.html.

Kowalski, W. J., and Bahnfleth, W. P. (2002). "MERV filter models for aerobiological applications." *Air Media* Summer:13–17.

Kowalski, W. J. (2003). *Immune Building Systems Technology.* McGraw-Hill, New York.

Kreja, L., and Seidel, H. -J. (2002). "Evaluation of genotoxic potential of some microbial volatile organic compounds (MVOC) with the comet assay, the micronucleus assay, and the HPRT gene mutation assay." *Mut Res* 513:143–150.

Kundsin, R. B. (1968). "Aerosols of Mycoplasmas, L forms, and bacteria: Comparison of particle size, viability, and lethality of ultraviolet radiation." *Appl Microbiol* 16(1):143–146.

Lacey, J., and Crook, B. (1988). "Fungal and actinomycete spores as pollutants of the workplace and occupational illness." *Ann Occup Hyg* 32:515-533 32:515–533.

Larsh, H. W. (1983). Histoplasmosis. *Occupational Mycoses*, A. F. DiSalvo, ed., Lea & Febiger, Philadelphia, PA.

Lewis, W. H., Vinay, P., and Zenger, V. E. (1983). *Airborne and Allergenic Pollen of North America.* Johns Hopkins University Press, Baltimore, MD.

Li, D. -W., and Kendrick, B. (1995). "A year-round comparison of fungal spores in indoor and outdoor air." *Mycologia* 87(2):190–195.

Lidwell, O. M., and Lowbury, E. J. (1950). "The survival of bacteria in dust." *Annu Rev Microbiol* 14:38–43.

Linton, A. H. (1982). *Microbes, Man, and Animals: The Natural History of Microbial Interactions.* Wiley & Sons, New York.

Little, J. S., Kishimoto, R. A., and Canonico, P. G. (1980). "In vitro studies of interaction of rickettsia and macrophages: Effect of ultraviolet light on *Coxiella burnetti* inactivation and macrophage enzymes." *Infect Immun* 27(3):837–841.

Luckiesh, M. (1946). *Applications of Germicidal, Erythemal and Infrared Energy.* D. Van Nostrand, New York.

Lytle, C. D. (1971). "Host-cell reactivation in mammalian cells. 1. Survival of ultra-violet-irradiated herpes virus in different cell-lines." *Int J Radiat Biol Relat Stud Phys Chem Med* 19(4):329–337.

Madoff, S. (1971). *Mycoplasma and the L Forms of Bacteria.* Gordon & Breach Science Publishers, New York.

Mahy, B. W. J., and Barry, R. D. (1975). *Negative Strand Viruses.* Academic Press, London.

Malherbe, H. H., and Strickland-Cholmley, M. (1980). *Viral Cytopathology.* CRC Press, Boca Raton, FL.

Mandell, G. L., Gerald, L., Bennett, J. E., and Dolin, R. (2000). *Principles and Practice of Infectious Diseases.* Churchill Livingstone, Philadelphia, PA.

Maniloff, J., McElhaney, R. N., Finch, L. R., and Baseman, J. B. (1992). *Mycoplasmas: Molecular Biology and Pathogenesis.* ASM Press, Washington, DC.

Marks, P. J., Vipond, I. B., Carlisle, D., Deakin, D., Fey, R. E., and Caul, E. O. (2000). "Evidence for airborne transmission of Norwalk-like virus (NLV) in a hotel restaurant." *Epidemiol Infect* 124(3):481–487.

Martini, G. A., and Siegert, R. (1971). *Marburg Virus Disease.* Springer-Verlag, New York.

McCaul, T. F., and Williams, J. C. (1981). "Developmental cycle of *Coxiella burnetii*: Structure and morphogenesis of vegetative and sporogenic differentiations." *J Bacteriol* 147(3):1063–1076.

McElhenney, T. R., Bold, H. C., Brown, R. M., and McGovern, J. P. (1962). "Algae: A cause of inhalant allergy in children." *Ann Allergy* 20:739–743.

Middleton, E., Reed, C. E., and Ellis, E. F. (1983). *Allergy: Principles and Practice,* Vol. 2. C. V. Mosby, St. Louis, MO.

Miller, J. D., Laflamme, A. M., Sobol, Y., Lafontaine, P., and Greenlaugh, R. (1988). "Fungi and fungal products in some Canadian houses." *Intl Biodeterior* 24:103–120.

Mitscherlich, E., and Marth, E. H. (1984). *Microbial Survival in the Environment.* Springer-Verlag, Berlin, Germany.

Miyaji, M. (1987). *Animal Models in Medical Mycology.* CRC Press, Boca Raton, FL.

Mongold, J. (1992). "DNA repair and the evolution of transformation in *Haemophilus influenzae*." *Genetics* 132:893–898.

Montana, E., Etzel, R., Sorenson, W., Kullman, G., Allan, T., and Dearborn, D. (1998). "Acute pulmonary hemorrhage is infants associated with exposure to *Stachybotrys atra* and other fungi." *Arch Pediatr Adolesc Med* 152:757–762.

Morey, P. R., Feeley, J. C., and Otten, J. A. (1990). *Biological Contaminants in Indoor Environments.* ASTM, Philadelphia, PA.

Murray, P. R. (1999). *Manual of Clinical Microbiology.* ASM Press, Washington, DC.

Myint, S. H. (1995). Human Coronavirus Infections. *The Coronaviridae,* S. G. Siddell, ed., Plenum Press, New York.

Nagy, R. (1964). "Application and measurement of ultraviolet radiation." *AIHA J* 25:274–281.

NATO (1996). *Handbook on the medical aspects of NBC defensive operations FM 8-9.* Dept. of the Army, Washington, DC.

Nielsen, K. F., Gravesen, S., Nielsen, P. A., Andersen, B., Thrane, U., and Frisvad, J. C. (1999). "Production of mycotoxins on artificially and naturally infested building materials." *Mycopathologia* 145:43–56.

Nikulin, M., Reijula, K., Jarvis, B. B., and Hintikka, E. (1996). "Experimental lung mycotoxicosis in mice induced by *Stachybotrys atra*." *Int J Exp Path* 77:213–218.

O'Rourke, M. K., Fiorentino, L., Clark, D., Ladd, M., Rogan, S., Carpenter, J., Gray, D., McKinley, L., and Sorenson, E. (1993). "Building characteristics and importance of house dust mite exposure in the Sonoran Desert, USA." *Proceedings of the Sixth International Conference on Indoor Air Quality and Climate,* Helsinki, Finland, 155–160.

Oldstone, M. B. A. (1987). "Arenaviruses." *Current Topics in Microbiology and Immunology.* Springer-Verlag, New York.

Ortiz-Ortiz, L., Bojalil, L. F., and Yakoleff, V. (1984). *Biological, Biochemical, and Biomedical Aspects of Actinomycetes.* Academic Press, Orlando, FL.

Pappagianis, D. (1983). Coccidioidomycosis. *Occupational Mycoses,* A. F. DiSalvo, ed., Lea & Febiger, Philadelphia, PA.

Pasanen, P., Korpi, A., Kalliokoski, P., and Pasanen, A. -L. (1997). "Growth and volatile metabolite production of *Aspergillus versicolor* in house dust." *Environ Int* 23(4):425–432.

Pattison, J. R. (1988). *Parvoviruses and Human Disease.* CRC Press, Boca Raton, FL.

Peters, C. J., Jahrling, P. B., Liu, C. T., Kenyon, R. H., Jr., K. T. M., and Oro, J. G. B. (1987). Experimental studies of Arenaviral hemorrhagic fevers. *Arenaviruses,* M. B. A. Oldstone, ed., Springer-Verlag, New York.

Platts-Mills, T. A. E., Heymann, P. W., Longbottom, J. L., and Wilkins, S. R. (1986). "Airborne allergens associated with asthma: Particles sizes carrying dust mite and rat allergens measured with a cascade impactor." *J Allergy Clin Immunol* 77:850–857.

Platts-Mills, T. A. E., and Chapman, M. D. (1987). "Dust mites: Immunology, allergic disease, and environmental control." *J Allergy Clin Immunol* 80:755–775.

Platts-Mills, T. A. E. (2001). Chapter 34: Allergens derived from arthropods and domestic animals. *Indoor Air Quality Handbook,* J. D. Spengler, J. M. Samet, and J. F. McCarthy, eds., McGraw-Hill, New York. 3.1–3.32.

Pope, A. M., Patterson, R., and Burge, H. (1993). *"Indoor Allergens,"* Inst. of Medicine, ed., National Academy Press. Washington, DC.

Prescott, L. M., Harley, J. P., and Klein, D. A. (1996). *"Microbiology."* Wm. C. Brown. Dubuque, IA.

Rao, C. Y., and Burge, H. A. (1996). "Review of quantitative standards and guidelines for fungi in indoor air." *J Air Waste Manage Assoc* 46(Sep):899–908.

Ray, B. (1996). *Fundamental Food Microbiology.* CRC Press, Boca Raton, FL.

Razum, O., Becher, H., Kapaun, A., and Junghanss, T. (2003). "SARS, lay epidemiology, and fear." *Lancet* 361(9370):1739–1740.

Restrepo-Moreno, A. (1983). Paracoccidioidomycosis. *Occupational Mycoses,* A. F. DiSalvo, ed., Lea & Febiger, Philadelphia, PA.

Rosenstreich, D. L., Eggleston, P., Kattan, M., Baker, D., Slavin, R. G., Gergen, P., Mitchell, H., McNiff-Mortimer, K., Lynn, H., Ownby, D., and Malveaux, F. (1997). "The role of cockroach allergy and exposure to cockroach allergen in causing morbidity among inner-city children with asthma." *N Engl J Med* 336:1356–1384.

Ryan, K. J. (1994). *"Sherris Medical Microbiology."* Appleton & Lange, Norwalk, CT.

Salvato, M. S. (1993). *The Arenaviridae.* Plenum Press, New York.

Sandstrîm, M. (2003). "Microbial volatile organic compounds (MVOC:s) emitted from building materials affected by microorganisms—their suitability as indicators of growth of microorganisms." Department of Biology and Environmental Sciences, Umea University, Sweden. http://www.bmg.umu.se/samarbeta/D20/MH02-22.htm.

Sarosi, G. A., Parker, J. D., Doto, I. L., and Tosh, F. E. (1970). "Chronic pulmonary Coccidioidomycosis." *N Engl J Med* 283(7):325–329.

Schaal, K. P., and Pulverer, G. (1981). *Actinomycetes.* Gustav Fischer Verlag, Cologne, MN.

Schleibinger, H. W., Wurm, D., Moritz, M., Bock, R., and Ruden, H. (1997). "Sick building syndrome and HVAC system: MVOC from air filters." *Zentralbl Hyg Umweltmed* 200(2–3):137–151.

Scott, R., and Yang, C. (1997). "Comparison of successful and unsuccessful *Stachybotrys chartarum* remediation projects." *Healthy Buildings/IAQ '97,* ASHRAE, Bethesda, MD.

Shapton, D. A., and Board, R. G. (1972). "Safety in Microbiology." Academic Press, London.

Sharp, G. (1939). "The lethal action of short ultraviolet rays on several common pathogenic bacteria." *J Bacteriol* 37:447–459.

Sharp, G. (1940). "The effects of ultraviolet light on bacteria suspended in air." *J Bacteriol* 38: 535–547.

Sikes, G., and Skinner, F. A. (1973). *Actinomycetes: Characteristics and PracticalImportance.* Academic Press, London.

Slack, J. M., and Gerencser, M. A. (1975). *Actinomycetes, Filamentous Bacteria: Biology and Pathogenicity.* Burgess Publishing, Minneapolis, MN.

Smith, J. E., and Moss, M. O. (1985). *Mycotoxins: Formation, Analysis and Significance.* John Wiley & Sons, Chichester, UK.

Smith, J. M. B. (1989). *Opportunistic Mycoses of Man and other Animals.* BPCC Wheatons, Exeter, RI.

Sorensen, K. N., Clemons, K. V., and Stevens, D. A. (1999). "Murine models of blastomycosis, coccidioidomycosis, and histoplasmosis." *Mycopathologia* 146:53–65.

Sorenson, W. G. (1990). "Mycotoxins as potential occupational hazards." *Develop Indust Microb* 31:205–211.

Storz, J. (1971). *Chlamydia and Chlamydia-Induced Diseases.* Charles C. Thomas, Springfield, IL.

Su, H. J., Burge, H. A., and Spengler, J. D. (1992). "Association of airborne fungi and wheeze/asthma symptoms in school-age children." *J Allergy Clin Immunol* 89(Pt. 2):251.

Sussman, A. F., and Halvorson, H. O. (1966). *Spores: Their Dormancy and Germination.* Harper & Row, New York.

Sutton, D. A., Fothergill, A. W., and Rinaldi, M. G. (1998). *Guide to Clinically Significant Fungi.* Williams & Wilkins, Baltimore, MD.

Tuder, R. M., Ibrahim, E., Godoy, C. E., and Brito, T. D. (1985). "Pathology of the human paracoccidioidomycosis." *Mycopathologia* 92:179–188.

von Bodrotti, H. S., and Mahnel, H. (1982). "Comparative studies on susceptibility of viruses to ultraviolet rays." *Zbl Vet Med B* 29:129–136.

Wagner, F. S., Eddy, G. A., and Brand, O. M. (1977). "The African green monkey as an alternate primate host for studying Machupo virus infection." *Am J Trop Med Hyg* 26(1):159–162.

Walker, D. H. (1988). *Biology of Rickettsial Diseases,* Vol. 1, 2. CRC Press, Boca Raton, FL.

Wang, Y., and Casadevall, A. (1994). "Decreased susceptibility of melanized *Cryptococcus neoformans* to UV light." *Appl Microbiol* 60(10):3864–3866.

Weinstein, R. A. (1991). "Epidemiology and control of nosocomial infections in adult intensive care units." *Am J Med* 91(Suppl 3B):179S–184S.

Welshimer, H. J. (1960). "Survival of *Listeria monocytogenes* in soil." *J Bacteriol* 80:316–320.

Wilson, R., Anderson, L. J., Holman, R. C., Gary, G. W., and Greeberg, H. B. (1982). "Waterborne gastroenteritis due to the Norwalk agent: Clinical and epidemiologic investigation." *AJPH* 72(1):72–74.

Woods, J. E., Grimsrud, D. T., and Boschi, N. (1997). *Healthy Buildings/IAQ'97.* ASHRAE, Washington, DC.

Yoshida, K., Ando, M., Sakata, T., and Araki, S. (1989). "Prevention of summer-type hypersensitivity pneumonitis: Effect of elimination of *Trichosporon cutaneum* from the patients' homes." *Arch Environ Health* 44(5):317–322.

Youmans, G. P. (1979). *Tuberculosis.* W. B. Saunders, Philadelphia, PA.

Zemke, V., Podgorsek, L., and Schoenen, D. (1990). "Ultraviolet disinfection of drinking water. 1. Communication: Inactivation of *E. coli* and coliform bacteria." *Zentralbl Hyg Umweltmed* 190(1–2):51–61.

Zerbini, M., Musiani, M., Gentilomi, G., Venturoli, S., Gallinella, G., and Morandi, R. (1995). "Comparative evaluation of virological and serological methods in prenatal diagnosis of *Parvovirus B19* fetal hydrops." *J Clin Microbiol* 34(1):603–607.

Zhang, Y., Li, X., Zhu, Y., and Jiang, Y. (2003). "Research on infectious concentration of airborne SARS virus." *The Fourth International Symposium on HVAC,* Beijing, China.

CHAPTER 3

THE PATHOLOGY OF AIRBORNE DISEASE

3.1 INTRODUCTION

Most airborne pathogens and allergens impact human health via the respiratory system. Throughout millions of years of coevolution with pathogens, human physiology has evolved a variety of protective mechanisms that include the primary and general defenses of lung clearance and the final and specific defenses of antibodies generated by the immune system. Sometimes it is the immune system reaction that causes the problem, as in the case of allergies. A basic understanding of respiratory physiology, immune system response, and the mechanisms by which pathogens attempt to overcome these defenses will contribute to the overall understanding of how these microorganisms cause diseases, and how such diseases might be protected against. This chapter provides a general overview respiratory infections and airborne diseases with a number of examples. For detailed information on specific diseases the reader should consult the various references provided here or the references cited in App. A that are provided in Chap. 2.

3.2 THE HUMAN RESPIRATORY SYSTEM

The human lungs are intended to extract oxygen from the air we breathe and transfer it into the bloodstream by diffusion. Other gases and vapors may also be absorbed by this route. In addition, liquids and solids may be transported into the blood via the lymphatic system. Finally, foreign bacteria and indigestible solids are removed by physical clearance mechanisms and immune system defenses. Problems occur when the clearance mechanisms and lung defenses are defeated, or when toxins are transferred into the blood stream.

The respiratory system is divided into three basic regions—nasopharyngeal, tracheobronchial, and pulmonary. The nasopharyngeal region extends from the nasal cavity to the larynx in the throat. It includes the oral cavity, the sinuses, and the pharynx. The lining of the nasopharynx consists primarily of cellular tissue and mucous membranes. Most of the largest particles inhaled will be removed in the nasopharynx, and this is also the area where most of the heat and moisture in inspired air is exchanged.

The tracheobronchial region includes the trachea, the bronchi, and the bronchiole. These components represent the conducting airways between the trachea and the bronchi in the lungs. They are lined with cilia and coated with mucous secreted by the mucous membrane. Cilia are critical for physical clearance of foreign matter from the airways. Cilia move in a whiplike fashion and move mucous out of the airways and the lungs. Cilia in the nose beat downward and clear mucous and other matter toward the pharynx. Cilia in the respiratory tract beat upward and also move matter toward the pharynx. In the pharynx, any foreign matter is transferred to the digestive tract where it is typically destroyed or broken down by digestive juices. Microbes that may be dangerous in the lungs are often harmless in the stomach for this reason.

Coughing and sneezing are also clearance mechanisms that force material out from the lungs and tracheobronchial regions. The bronchi and trachea are highly sensitive and may react to any foreign particles by producing coughs or sneezes. During infections, the coughing and sneezing reaction may be triggered both as a lung clearance mechanism and because the infecting microorganism depends on this reaction for airborne transport to new hosts.

The pulmonary region is the respiratory airspace of the lungs and consists of several hundred million alveoli. Alveoli are tiny sacs approximately 300 μm in diameter that provide airspace for the exchange of oxygen, carbon dioxide, and other volatiles between the inspired air and the blood in the capillaries. Inside the alveoli will typically be found mobile white blood cells known as macrophages that provide a second line of defense against foreign microorganisms. Macrophages are approximately 7 to 10 μm in diameter and will attempt to engulf and digest foreign bacteria, spores, viruses, and other particulate matter. The process of metabolizing such particles is known as phagocytosis.

Phagocytosis and lung clearance together provide effective protection against most pathogens and allergens on a daily basis provided the inhaled dose is not beyond normal or ambient levels and the health of the host remains unimpaired. Some infectious agents, like *M. tuberculosis* and *Legionella*, are able to resist or circumvent phagocytosis.

The left and right lungs are not symmetrical. The right lung is larger (in most people) and has a lower airflow resistance. As a result, most inspired air flows into the right lung. Air is then exchanged between the right lung and the left lung by diffusion.

The lungs are never completely evacuated. A typical breath contains about 500 mL or air, while the alveolar volume of the lungs is typically about 3000 mL (Heinsohn, 1991). Therefore, each breath exchanges only about 16 percent of the total air contained within the lungs. The airways may contain an additional 1500 mL of air. At a normal breathing rate of about 12 to 15 breaths per minute, it may take several minutes to completely exchange all the air in the lungs, considering that air mixing in the lungs approaches perfect mixing.

Breathing rates can vary greatly depending on the level of activity, and can also vary from person to person. Healthy people tend to breathe more slowly and have lower heart rates, since their systems tend to be more efficient. Table 3.1 shows typical breathing rates

TABLE 3.1 Breathing Rate Variation with Activity Level

Level of activity	Rest	Light	Moderate	Heavy
Breathing rate, L/min	11.6	32.2	50	80.4
Breathing rate, m^3/h	0.70	1.93	3	4.82
Breathing rate, cfm	0.41	1.14	1.77	2.84

and how they vary under moderate to heavy exercise. These breathing rates can be used to estimate inhaled doses of pathogens, allergens, or toxins.

The human lungs act much like filters in that they remove particles from inspired air or particular sizes. The three different regions of the respiratory system possess different particle removal characteristics. The largest particles tend to be removed in the nasopharyngeal region while the smaller particles tend to penetrate to deeper parts of the lungs before becoming attached to surfaces (Lippmann, 1998; Scheuch and Stahlhofen, 1992). Figure 3.1 illustrates the removal efficiencies of the three regions of the lungs as a function of particle size (based on data from Perera and Ahmed, 1979). Note the rough similarity of the overall removal efficiency curve to the performance curves of air filters.

The particle removal efficiencies predicted by Fig. 3.1 are basically corroborated by empirical data from Seinfeld (1986) as shown in Fig. 3.2. In this figure, data from multiple studies have been fitted to a curve. Again, the resemblance to a filter performance curve is remarkable. Additional data and studies from Darquenne (2002), Zhang et al., (2002) add corroboration to this model of deposition efficiency.

The U.S. Environmental Protection Agency (USEPA) defines inhalable particles as those having an aerodynamic diameter of less than 10 μm. The American Conference of Government Industrial Hygienists (ACGIH) defines respirable particles as those with an aerodynamic diameter of less than 2 μm. The aerodynamic diameter of any particle is the diameter of a sphere of unit density (water density) that would have the same settling rate in still air as the actual particle. The aerodynamic diameter is defined mathematically as

$$D_a = D_p \sqrt{\frac{\rho_p}{\rho_w}}$$

where D_a = aerodynamic diameter, μm
D_p = particle diameter, μm
ρ_p = density of particle, kg/m^3
ρ_w = density of water, kg/m^3 (1000 kg/m^3)

For bacteria and spores, the density is typically about 1030 to 1100 kg/m^3 (Bakken and Olsen, 1983; Bratbak and Dundas, 1984).

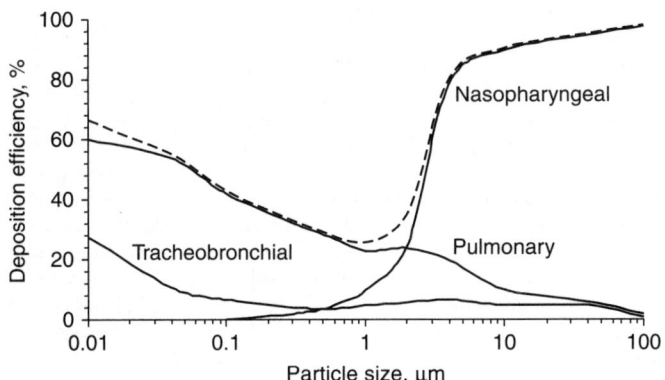

FIGURE 3.1 Predicted deposition efficiencies for the three regions of the respiratory system. Net performance is shown by the dotted line.

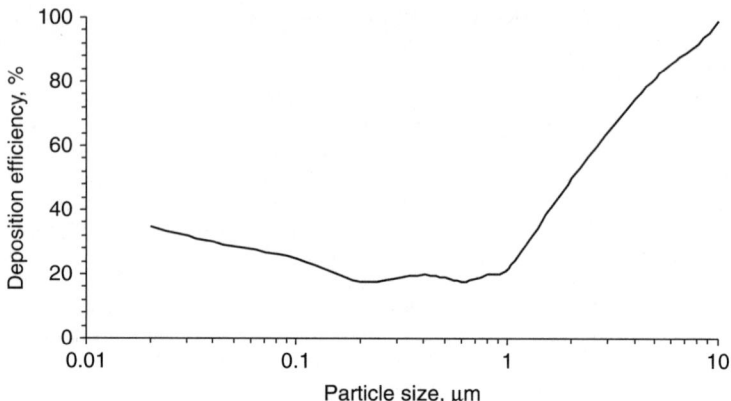

FIGURE 3.2 Deposition efficiency of the respiratory system. [*Based on combined data from Seinfeld (1986)*].

3.3 MECHANISMS OF LUNG INFECTION

Most inhaled microorganisms, whether they are viruses, bacteria, fungal spores, or other organisms, will be cleared from the lungs, dissolved and absorbed, or phagocytized. In the event that none of the primary defensive mechanisms are successful, the microorganism may take hold in the lungs and begin either multiplying or reactions. The results of such attachment or growth may be physical irritation of the lungs, allergic reaction, toxic reaction, or infection.

The entry of any fumes or foreign particles, including pollen and fungi, into the lungs may cause irritation due to the sensitivity of the bronchi and trachea (Chan-Yeung, 1994). Certain *microbial volatile organic compounds* (MVOCs) produced by bacteria and fungi may cause respiratory irritation, or sometimes just unpleasant odors. Asthma may be exacerbated by dust and respirable particles. Often, the difference between ordinary respiratory irritation and actual allergic or toxic reactions to low levels of allergens and toxins may not be distinguishable. Allergic hypersensitivity and hay fever may be triggered by airborne pollen, spores, or atmospheric pollutants (Sherman, 1968). These latter categories of respiratory irritation are not the same as allergies or allergic responses, which are full-blown reactions to specific allergens.

Figure 3.3 depicts the possible outcomes of the inhalation of an airborne microorganism. The first three, lung clearance, phagocytosis, and dissolution, are normal daily occurrences without consequences. If an infection develops, it will resolve naturally, resolve with treatment, or persist indefinitely. It is possible for certain infections to progress to toxic reactions. Allergic reactions may not resolve without removal from the source. Allergic reactions may also result in toxic reactions if the inhaled dose is sufficiently high.

The deposition of airborne particles in the human respiratory tract is governed primarily by the size of the particle as shown by the approximate breakdown in Table 3.2.

None of the values in Table 3.2 represent an absolute cutoff, merely a high probability limit. Cilia in the upper part of the respiratory tract normally expel the particles that impact, except that motile bacteria, such as *Legionella* and *Mycobacteria*, are capable of resisting clearance. The importance of the size limits given in Table 3.2 is that deeper penetration of the lungs will favor infection, therefore extreme smallness will tend to favor airborne pathogens. Entry into the alveolar region can facilitate infection by avoiding the mucociliary clearance mechanism.

THE PATHOLOGY OF AIRBORNE DISEASE 45

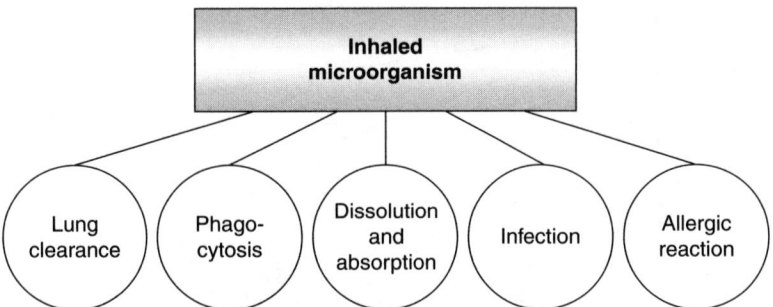

FIGURE 3.3 Possible outcomes from inhalation of an airborne microorganism.

The ciliated epithelia of the respiratory tract have their own vulnerabilities. The cilia have nine fibers surrounded by a membrane and extending into the cell's cytoplasm. The cells are covered with nonmotile microvilli. As a primary deposition site, some airborne pathogens have developed a high affinity for this site. In particular, two viruses, influenza and parainfluenza, and two bacteria, *Mycoplasma pneumoniae* and *Bordetella pertussis*, seem to have a rate of attachment to this site which exceeds the ability of the cilia to clear them (Davey and Halliday, 1994). In contrast, *Haemophilus influenzae* produces a ciliostatic substance, but only after it has established itself. *Mycoplasma pneumoniae* also reduces ciliary motility, but through cell damage.

An additional factor that can impact infectivity is the clearance time. Particles that deposit on the epithelia have clearance half-times of minutes. Phagocytes plus ciliary mucus transport clear the alveolar region with a half-life of about 24 to 48 hours. Figure 3.4 graphs the half-lives of micron-sized particles in the lung based on data from Hatch and Gross (1964) for three sizes and low lung burden. Although these curves are based on a limited data set, they illustrate two important points. One is that the largest and smallest particles are cleared more quickly than midsized particles, and the other is that clearance is unlikely to occur sufficiently fast to avoid acquiring an infection. Since most infections have incubation periods of about 1 to 3 days, it would appear that lung clearance may not be able to play a significant role in protecting against airborne disease unless, of course, the initial inhaled dose was already quite low.

Airborne pathogens must first survive on mucous surfaces, then adhere to the surface and then be unaffected by pH and the bactericidal or bacteriostatic action of host secretions. The invaders must also successfully compete with the indigenous microflora in the lungs to survive. A second line of pulmonary defense is phagocytosis, whereby the phagocytes attempt to digest any foreign invaders. Some pathogens are resistant to phagocytosis, such as *Francisella tularensis* and *Bacillus anthracis* spores. Other pathogens, such as *Legionella pneumophila*, have the ability to parasitize the phagocytes. A few days after infection the

TABLE 3.2 Particle Size and Deposition Zone

Particle size, μm	Deposition zone
<2 μm	Alveolar region (lower respiratory tract)
2–5 μm	Lungs and conducting airways
5–10 μm	Upper and lower respiratory tract
>10 μm	Nasopharyngeal region (upper respiratory tract)

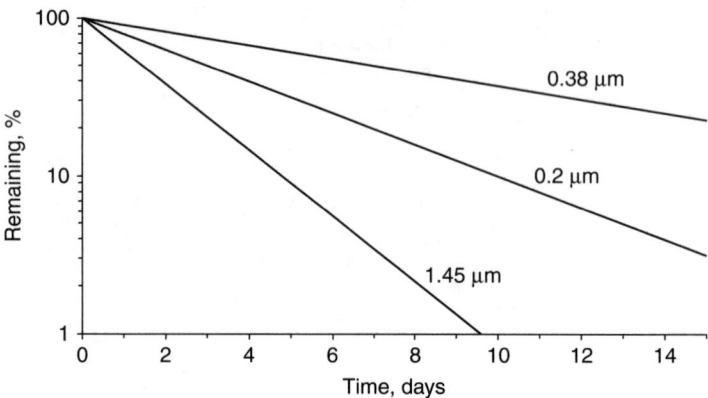

FIGURE 3.4 Lung clearance of micron-sized particles over time. [*Based on data from Hatch and Gross (1964)*].

host immune system releases antibodies which will bind to specific antigens and target the pathogens for destruction by phagocytes. These antibodies can also neutralize bacterial aggressins (compounds which inhibit host defenses). In addition to the main antibodies present in blood, IgA, IgG, and IgM, respiratory surfaces are covered with mucus containing a wide spectrum of immunoglobulins, the main one being a dimer form of IgA. All of these plus interferon help increase host resistance.

3.4 DISEASE GROUPS

Three disease groups, communicable, noncommunicable, and endogenous, are used to define the epidemiological hazards associated with airborne pathogens. An alternative disease group, those that are primarily nosocomial and opportunistic, includes many members of each category and is treated as an overlapping group. Communicable respiratory diseases are contagious infections that transmit between humans. Transmission may also occur by other than airborne routes, including direct contact and contact with fomites left on surfaces (Braude et al., 1981; Ryan, 1994). Sometimes a disease may transmit from animals to humans, as with plague, after which it may transmit to other humans (Murray, 1999).

All airborne human respiratory viruses except Hantavirus are communicable (Fleming et al., 1995). No spores are known to be communicable although some studies suggest the possibility that transmission of fungal infections may occur between the immunodeficient (Weinstein, 1991). Certain viruses occur primarily as nosocomial infections, and these are identified in App. A.

Noncommunicable airborne infections do not transmit between humans by either airborne or other routes. Some infections, like cowpox or brucellosis, may transmit from animals to humans but do not cause secondary infections among humans (Murray, 1999). Such zoonotic diseases may be transmissible to humans from the animals but are classified here as noncommunicable. Figure 3.5 illustrates the general relationship between communicable, noncommunicable, and opportunistic airborne respiratory infections. This figure highlights the fact that essentially all communicable diseases come from humans and essentially all noncontagious disease agents come from the environment or from animals. The less definable place of the opportunistic or nosocomial infections shows up as the middle ground between

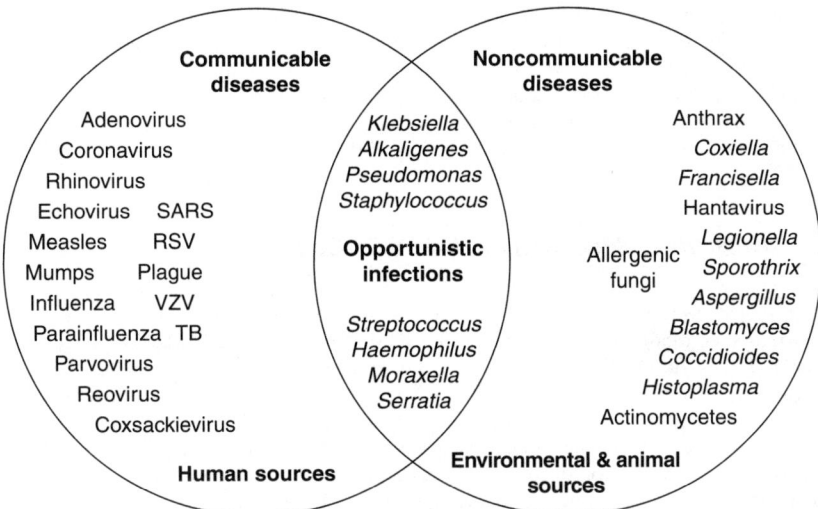

FIGURE 3.5 Figurative illustration of the differences between contagious and noncontagious respiratory diseases and the place of opportunistic infections.

contagious and noncontagious diseases. There are a few exceptions to this general rule since the arenaviruses may come from rodents and still transmit between humans, and plague also comes from rodents. Furthermore, a number of primarily nosocomial agents also cause diseases in healthy people, like *Corynebacteria, Bordetella, Mycoplasma,* and *Neisseria.*

Essentially all fungal spores are noncommunicable under normal conditions, as are environmental bacteria (Howard and Howard, 1983; Mitscherlich and Marth, 1984). Hantavirus, the only noncommunicable respiratory virus, is found in the feces of many rodents and can be rapidly fatal if inhaled (Murray, 1999). Noncommunicable opportunistic infections can occur in those who have increased susceptibility due to impaired immune systems, burns, or open wounds (Castle and Ajemian, 1987). Certain noncommunicable infections are predominantly nosocomial, and these are identified in the database.

Certain bacteria and fungi that exist commensally in humans may cause respiratory or other infections opportunistically when health is compromised (Ryan, 1994). Endogenous bacteria would seem to have originally been pathogens that evolved toward commensalism (see Chap. 1) and, on the decline of coevolved host immunity, their pathogenicity seems to reassert itself and poses a threat. When endogenous bacteria transmit from a healthy individual to a susceptible person, the distinction between a communicable microbe and a noncommunicable microbe becomes blurred. Microbes listed as endogenous in App. A are all potentially opportunistic. Endogenous microbes may contaminate water or medical equipment and result in subsequent nosocomial infections (Weinstein, 1991).

Nosocomial, or hospital-acquired, infections are a unique hazard in the health care field. Several reasons can be cited for the predominance of certain nosocomial infections including (1) the susceptibility of patients with compromised health, burns, or wounds, (2) the preponderance or concentration of pathogens within the hospital environment, and (3) the use of equipment or invasive procedures that provide an opportunity for contamination (Castle and Ajemian, 1987).

Laboratories performing work with any of the identified respiratory pathogens should always be aware of the possible hazards. Perhaps the most dangerous hazard is the sudden or accidental aerosolization of high numbers of microorganisms due to mechanical agitation or spills. Technicians should be ever alert to the possibility of cultures of airborne pathogens being transmitted by contact with the mucous membranes of the mouth, nose, eyes, or ears (Fleming et al., 1995; Shapton and Board, 1972). Inhalation of large concentrations of any microorganism could result in previously unknown problems. The relevant or requisite guidelines for laboratory safety should be followed at all times, with particular awareness of the hazards of handling infectious airborne species.

3.5 AIRBORNE RESPIRATORY INFECTIONS

The most serious type of respiratory disease is an infection. Pathogenic microorganisms are those that specifically cause diseases, other than allergies or syndromes. Pathogens may include viruses, fungi, and bacteria. Some protozoa and amoebae may cause infections but little, if any, evidence exists that these cause airborne or inhalation hazards. In an infection, one or more microbes may succeed in attaching to the surface of the respiratory system or otherwise gaining entry. Multiplication and growth then begins, and the infection progresses, either to resolution or to fatality. In some cases the infection may remain asymptomatic (i.e., *Histoplasma*) or may persist almost indefinitely (i.e., tuberculosis).

Respiratory infections may occur in the upper, middle, or lower respiratory tract. These sites tend to be favored by specific pathogens and the size of the inhaled microbe (i.e., respirable, inhalable, and the like) may play an important role in allowing it to transport to its preferred location. Figure 3.6 shows a breakdown of the respiratory tracts and the infections that may affect these sites.

3.6 UPPER RESPIRATORY TRACT INFECTIONS

Upper respiratory tract infections tend to be centered on the nasal cavity and pharynx, together known as the nasopharynx. Over 80 percent of the infections of the nasopharynx are due to viruses, with the remainder being mainly bacteria and certain fungal or yeast infections (Ryan, 1994). The diseases or symptoms that affect the nasopharynx include

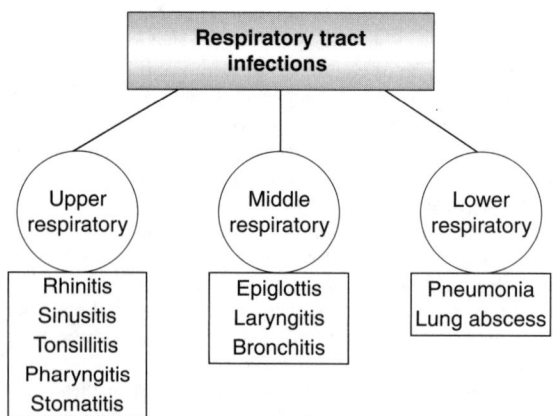

FIGURE 3.6 Breakdown of respiratory tract infections.

rhinitis, sinusitis, pharyngitis, and tonsillitis. Stomatitis is an infection localized in the oral cavity but it is not commonly the result of airborne infection. Only the general categories of upper respiratory infections are described in the following sections. For specific information on particular infections like tuberculosis or influenza, refer to App. and the associated references.

Rhinitis. Rhinitis is an infection of the nasal mucosa and is characterized by variable fever, inflammation of the nasal mucosa, and an increase in mucosal secretions, or rhinorrhea. Nasal obstruction is normally the result, with the mucosal secretions becoming thicker as the disease progresses. The primary agents of rhinitis include rhinoviruses, adenoviruses, coronaviruses, parainfluenza, respiratory syncytial virus, and coxsackieviruses. Bacterial and fungal infections resulting in rhinitis are rare. Rhinitis may be the result of allergy or it may be nonallergic in nature (Nelson and Lockey, 1998). Occupational allergens that may induce rhinitis include enzymes from *Bacillus subtilis*, red cedar, ash wood dust, laboratory animals, psyllium, guar gum, and wheat flour.

Sinusitis. Sinusitis is a mucosal inflammation of the sinuses. It is often caused by an infectious microorganism and is characterized by symptoms of nasal obstruction, hyposmia, purulent nasal secretions, posterior pharyngeal irritation, fetid breath, fatigue, malaise, and headache (Nelson and Lockey, 1998). Sinusitis often occurs together with rhinitis. Viral upper respiratory infections cause acute viral-induced sinusitis. Acute sinusitis is considered to be sinusitis that lasts less than 6 weeks to 3 months. It is typically the result of infection from bacteria such as *Streptococcus pneumoniae, Haemophilus influenza,* and *Moraxella catarrhalis*. Chronic sinusitis may have similar but les severe symptoms and may also include hoarseness and sore throat. It may be caused by bacteria such as *Bacteroides*.

Pharyngitis. Pharyngitis is a localized infection in the pharynx or throat. It often involves the tonsils also. Sore throat is the most prominent manifestation, along with inflammation and erythema. There may be exudates and ulcers or hemorrhaging, depending on the specific pathogen causing the infection. The primary causative airborne agents of tonsillitis include adenoviruses, parainfluenza, influenza, rhinoviruses, coxsackieviruses, *Streptococcus pyogenes,* and *Corynebacterium diphtheriae.*

Tonsillitis. Tonsillitis is a localized infection of the tonsils. It often accompanies pharyngitis. Soreness and inflammation are the prevailing characteristics but, like pharyngitis, the development of ulcers or hemorrhages depends on the specific pathogen causing the infection. The main causative airborne agents of tonsillitis include adenoviruses, parainfluenza, influenza, rhinoviruses, coxsackieviruses, *Streptococcus pyogenes,* and *Corynebacterium diphtheriae.*

Stomatitis. Stomatitis is an inflammation of the oral cavity. Soreness and inflammation are the common manifestations and the progression of the disease may result in ulcers or hemorrhage. Stomatitis is not a common result of airborne infectious agents but it may be caused by coxsackieviruses and *Candida*.

3.7 MIDDLE RESPIRATORY TRACT INFECTIONS

The middle and lower respiratory tract consists essentially of the epiglottis, larynx, trachea, and associated tissues. These comprise the areas between the throat and the trachea. The diseases associated with the middle respiratory tract include epiglottitis, laryngitis, and bronchitis.

Epiglottitis. Epiglottitis is an inflammation of the epiglottis, the region below the throat and above the larynx. It is typically characterized by soreness, fever, and breathing difficulty. It may also result in difficulty in speaking and swallowing. Epiglottitis can become serious if the airways become obstructed to the point of preventing breathing. Viruses rarely cause infections of the epiglottis. Over 90 percent of epiglottitis is due to bacteria, and the most common airborne causes include *Haemophilus influenzae, Streptococcus pyogenes, Streptococcus pneumoniae, Corynebacterium diphtheria*, and *Neisseria meningitidis*.

Laryngitis. Laryngitis is an infection of the larynx, which includes the vocal chords. It is an inflammation of the larynx and is characterized by variable fever, hoarseness, and cough. It may cause the loss of the ability to speak, or aphonia. In its most severe form it is known as croup. The infection sometimes extends to the trachea and bronchi where it may produce severe coughing with chest pain. Bacterial causes of laryngitis are rare but the airborne bacteria most commonly involved include parainfluenza, influenza, adenoviruses, respiratory syncytial virus, rhinoviruses, coronaviruses, and echoviruses. Laryngitis that also involves the trachea and bronchi may be due to the bacteria *Haemophilus influenzae* or *Staphylococcus aureus*.

Bronchitis. Bronchitis is an infection of the bronchi in the lungs that affects up to 15 percent of the general population (Barnhart, 1994). It may result from a primary infection or from infectious spread from the upper respiratory tract. It is characterized by inflammation of the bronchi, coughing, variable fever, and sputum production. Viruses that most commonly cause bronchitis include parainfluenza, influenza, respiratory syncytial virus, adenoviruses, and measles virus. The airborne bacteria and fungi that most commonly cause bronchitis include *Bordetella pertussis, Haemophilus influenzae, Mycoplasma pneumoniae,* and *Chlamydia pneumoniae*. Chronic bronchitis is typically the result of long-term damage to the bronchial epithelium from factors like smoking, pollution, or occupational exposure to airborne hazards. Although it may be due to persistent infection like tuberculosis, it is more commonly due to noninfectious agents and is treated separately in a later section.

3.8 LOWER RESPIRATORY TRACT INFECTIONS

Lower respiratory tract infections involve the alveoli, the interstitium, and the terminal bronchioles. Infection may be the result of airborne pathogens penetrating into the lower lungs or by spread of infection from the upper or middle respiratory tract. When lower respiratory tract infection occurs, it is often the result of some compromise of the upper airway mechanisms for filtering or clearing inhaled infectious agents (Ryan, 1994). Some small microbes may be able to penetrate the passages of the upper and middle airways and reach the lower respiratory tract to produce a primary infection. The most common infection of the lower airways is pneumonia, which may be caused by a variety of infectious microbes.

Pneumonia. Pneumonia is an infection of the lower respiratory tract. It may be acute pneumonia or chronic pneumonia. Acute pneumonia may develop over hours or days and persist for days or weeks if untreated. It is characterized by fever, malaise, cough, and sputum production. If the disease progresses, dyspnea, or difficulty breathing, may occur. Chills, purulent sputum, or bloody sputum (hemoptysis) may result. If the exchange of oxygen in the lungs becomes substantially reduced by filling of the lungs with fluid, cyanosis and death may occur. When the bronchi become covered with exudate, it is called broncho pneumonia. If the lungs become filled with exudate, it is called lobar pneumonia. Chronic pneumonia may develop over weeks or months after a gradual onset. The symptoms are the same as for pneumonia but they develop more slowly. Loss of appetite, weight loss, insomnia, and night sweats may develop. Chronic pneumonia may develop from a microbial infection or from various noninfectious causes, such as neoplasms, allergic conditions, or toxic injury. The etiologic agents of pneumonia are numerous. Table 3.3 lists the most common potentially airborne causes of pneumonia. Viruses cause over 80 percent of pneumonia in infants and children while causing only 10 to 20 percent of adult pneumonia. Viral pneumonia can predispose patients to bacterial pneumonia. *Streptococcus pneumoniae* is the

TABLE 3.3 Potentially Airborne Etiologic Agents of Pneumonia

Microbe	Group	Type	Notes
Adenovirus	Virus	Acute	
Aspergillus species	Fungi	Acute	Primarily immunocompromised
Blastomyces dermatitidis	Fungi	Chronic	North central and eastern United States
Candida albicans	Fungi	Acute	Primarily immunocompromised
Chlamydia pneumoniae	Fungi	Acute	Often nosocomial
Chlamydia trachomatis	Fungi	Acute	Infants
Coccidioides immitis	Fungi	Chronic	South America
Cryptococcus neoformans	Fungi	Chronic	
Enterobacter species	Bacteria	Acute	Rare
Haemophilus influenzae	Bacteria	Acute	Often nosocomial
Histoplasma capsulatum	Fungi	Chronic	Southeast United States
Influenza	Virus	Acute	Seasonal
Klebsiella pneumoniae	Bacteria	Acute	Uncommon
Legionella	Bacteria	Acute	
Mycobacterium tuberculosis	Bacteria	Chronic	
Mycoplasma pneumoniae	Bacteria	Acute	
Nocardia species	Bacteria	Chronic	Often nosocomial
Parainfluenza	Virus	Acute	Often nosocomial
Pneumocystis carinii	Fungi	Acute	Primarily immunocompromised
Proteus species	Bacteria	Acute	Nosocomial
Pseudomonas aeruginosa	Bacteria	Acute	Primarily immunocompromised
Respiratory syncytial virus	Virus	Acute	Primarily infants
Serratia marcescens	Bacteria	Acute	Nosocomial
Staphylococcus aureus	Bacteria	Acute	Nosocomial
Streptococcus pneumoniae	Bacteria	Acute	Most common bacterial cause

most common cause of acute bacterial pneumonia. Nosocomial pneumonia is the second most common infection in hospital settings and the most common infection in *intensive care units* (ICUs) (Jarvis, 2000). Nosocomial bacterial pneumonias are frequently polymicrobial and gram-negative bacilli are the predominant causative agents (Tablan et al., 1994).

Lung abscess. Lung abscess is usually a complication that results from chronic or acute pneumonia. It may result from organisms that cause localized damage to the lungs, or from nonmicrobial causes. It may develop as the result of a blood-borne infection. The symptoms are not specific but may resemble those of pneumonia that has failed to resolve.

3.9 NONINFECTIOUS RESPIRATORY DISEASES

Noninfectious respiratory diseases may involve airborne microorganisms that induce a variety of conditions, most notably asthma, allergies, hypersensitivity pneumonitis, chronic bronchitis, and respiratory irritation. There are many noninfectious respiratory diseases that are not microbial in origin but that can exacerbate microbial infections and so the distinction between microbial and nonmicrobial respiratory diseases becomes blurred. The following sections address each of the major noninfectious respiratory diseases that are potentially caused by airborne microbes either in whole or in part.

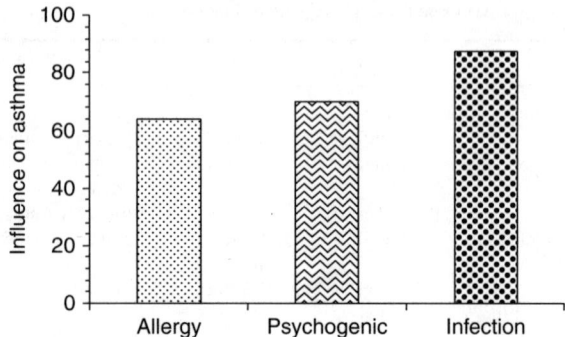

FIGURE 3.7 Relative influence of factors on asthma. [*Based on data from Lane (1979).*]

Asthma. Asthma is a condition that renders a person susceptible to respiratory irritation in the form of asthma attacks. Asthma attacks may be caused by allergic reactions, by virus infections, and by long-term exposure to pollutants. Asthma attacks typically involve constriction of the airways and wheezing or shortness of breath and may be triggered by a variety of microbial or chemical exposures or even by stress. Figure 3.7 illustrates the relative influence some factors have in triggering asthma attacks. In a study reported by Lane (1979) infection was the sole cause in only 11 percent of asthma cases, allergy was the sole cause in 3 percent, and psychological stress caused only 1 percent of attacks by itself.

It was once believed that individuals were born with a predisposition to asthma but the incidence of asthma has been steadily increasing for decades, suggesting the possibility that asthma may be an acquired condition resulting from exposure to allergens, pollutants, or other respiratory irritants (Pope et al., 1993). Asthma was recognized as a disease in the ancient world but it is not much better understood today than it was then. The condition is easy to characterize but a definition of the exact cause has eluded researchers. Many asthmatics suffer frequent attacks as children but attacks often diminish in frequency and severity as they grow older. Figure 3.8 illustrates the decrease in asthma with age.

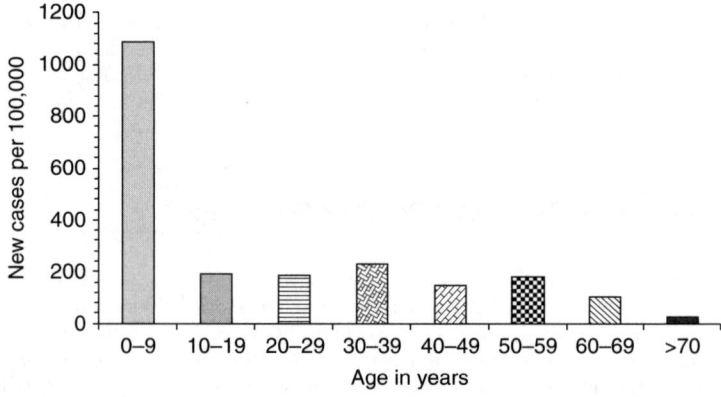

FIGURE 3.8 Decrease in asthma incidence with age. [*Based on data from Derrick (1971) per Lane (1979).*]

During an asthma attack there is a narrowing or blockage of the bronchi which causes wheezing (Lane, 1979). The blockage may be caused in part by excess production of mucus, by widespread narrowing of the airways, swelling of the lining of the airways, spasms of the muscle in the walls of the airways, or a combination of these factors. The condition resembles bronchitis, which is an inflammation of the bronchial passages, but is distinct from it. Acute bronchitis may occur in asthmatics, but this will usually trigger an asthma attack. Asthma has a predilection for boys, with twice as many cases of asthma in males as opposed to females. Asthma is the most common single chronic disease in children, as shown in Fig. 3.9.

Occupational Asthma. Occupational asthma is an inflammatory disorder of the airways with episodic restriction of breathing (Brooks, 1998). It can arise from exposure to microbial or other pollutants in the workplace, of which over 200 have been identified that can induce sensitization. It may occur in those who already have asthma or it may develop through workplace exposure to allergens or irritants. It is usually characterized by asthma symptoms during the workweek or during working hours, but clears after working hours or on the weekends. When asthma develops due to some sensitizing agent in atopic individuals it is often known as atopic asthma. The prevalence of occupational asthma varies with profession. Overall, some 2 percent of all cases of asthma in the United States are occupationally related (Brooks, 1998). Some materials that may induce occupational asthma include pollutants from soldering, painting, printing, preparing chemicals, or working with food products like raw coffee beans or eggs. Table 3.4 lists some of the pollutants in the workplace of several professions that may induce occupational asthma, reproduced from information per Lane (1979) and Brooks (1998).

Reactive Airways Dysfunction Syndrome. Reactive airways dysfunction syndrome (RADS) is a term that describes the symptoms of coughing, wheezing, and shortness of breath that can develop within a few hours or minutes of acute exposure to high levels of irritant vapor, fumes, or smoke (Chan-Yeung, 1994). The symptoms tended to persist for several years in most patients. In addition to a wide variety of chemicals, organic compounds, insect, and plant matter that may induce asthma there are a handful of microbes. *Alternaria, Aspergillus, Cladosporium, Verticillium,* and *Paecilomyces* have all been associated with occupational asthma (Brooks, 1998).

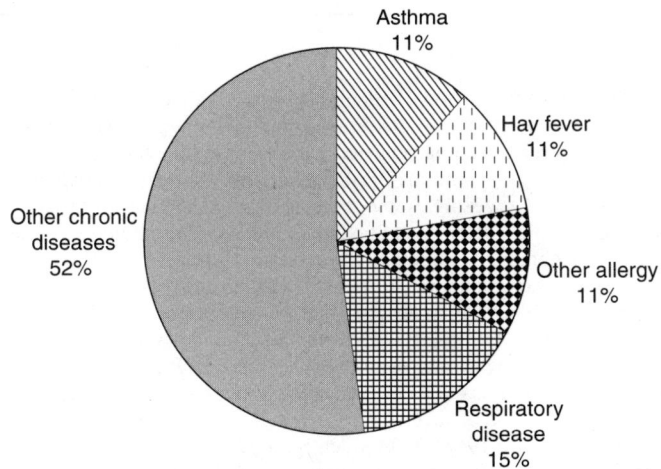

FIGURE 3.9 Breakdown of chronic diseases in children.

TABLE 3.4 Some Causative Agents of Occupational Asthma

Occupation	Asthma sensitizer
Animal handlers	Hair, dander, dust mites, storage mites, insects
Aquatic work	Amebae, crabs, oysters, pearl shells
Bakers	Flours, dust, moulds, weevils
Biochemical workers	Sulphanilomides, penicillin, other antibiotics
Bird breeders	Bird feathers, bird droppings
Chemical workers	Formalin, piperazine, isocyanates, epoxy resins
Coffee industry	Raw coffee beans, green coffee dust
Cotton workers	Cotton, flax, hemp
Detergent industry	Proteolytic enzymes or proteases
Domestic workers	House dusts, dander
Electricians	Soldering fluxes
Farm workers	Animal dander, vegetable dusts
Food industry	Tartrazine
Grain handler	Grain dust, insect debris
Hairdressers	Sodium and potassium persulfate
Leather workers	Formalin, chromium salts
Metal workers	Nickel, vanadium, chromium
Millers	Flour, insects, storage mites
Oil extractors	Castor bean oil, linseed oil, cotton seed oil
Outdoor workers	Crickets, flies, various insects
Painters	Toluene diisocyanates
Paper product workers	Natural glues
Pharmaceutical workers	Penicillin, ampicillin
Photographers	Complex salts of platinum
Plastics workers	Diisocyanates, anhydrides, diethylene
Poultry workers	Feathers
Printers	Gum arabic, tragacanth
Rubber industry	Ethylene diamine, paraphenylene diamine
Veterinarians	Animal dander
Wood workers	Dust from red cedar and other woods
Textile workers	Wool, cotton

Allergy. Allergy is the reaction of allergic or atopic individuals who are susceptible to developing reactions to specific allergens as a result of their having been previously exposed to those allergens. Allergic reactions may occur in response to pollen, mold, animal dander, food, and insect bites. The symptoms of allergy are characterized by what is commonly called hay fever—rhinitis, sneezing, watery eyes, and itching. When allergens are inhaled by atopic individuals, the mast cells, or antibodies, in the lungs send automatic signals to the immune system. The immune system typically overreacts in such cases and floods the body with chemical compounds, including histamines, which provoke physiological reactions that may include those of hay fever. Severe allergic reactions may result in death.

Allergic reactions are a by-product of the human immune system in its attempt to protect against foreign substances. The immune system recognizes and responds to foreign compounds, especially proteins that may enter the body. On the first exposure to foreign proteins an incubation period of several days is required for the development of protective countermeasures (Sherman, 1968). Subsequently, the immune system is sensitized to the particular agent to which it is exposed to the point that it responds immediately and automatically. Once this process has run its course, the substance becomes an antigen, a specific recognizable agent that provokes an immune system response that usually results in the production of protective antibodies. In some cases, where the specificity of the antigen may become altered, there is a delayed response to the antigen. In other cases, if an antigen appears when antibodies from the immune system are already circulating throughout the body, the result may be an immediate anaphylactic reaction. It is the physiological disturbance produced by these immune system reactions that cause the problem of allergies or allergic reactions.

Atopy. Atopy is a hereditary susceptibility in which some 5 to 10 percent of the population may develop sensitization spontaneously as the result of casual exposure to pollen or other allergens. However, the antibodies of atopic individuals differ from other antibodies both in their physicochemical properties and in their being essentially devoid of protective properties. They generally do not facilitate elimination of the antigens from the body. Figure 3.10 illustrates the types of allergic reactions that can result from airborne exposures to allergens.

The primary respiratory allergies of concern include hay fever, nonseasonal allergic rhinitis, bronchial asthma, infiltrative lung diseases, and inhalation fever. Infiltrative lung diseases include farmer's lung, bagassosis, and pigeon breeder's disease. Inhalation fever diseases include byssinosis, mill fever, organic dust toxic syndrome, humidifier fever, Pontiac fever, sump fever, heckling fever, grain fever, swine fever, silo-unloader's disease, and wood-trimmer's fever.

Hay Fever. Hay fever is an allergic reaction of the nasal mucosa due to sensitization to a specific pollen that proliferates seasonally. It is also known as pollinosis. Hay fever affects about 5 to 10 percent of the population although many cases are not severe and are simply tolerated by individuals. Many of the pollen to which atopic individuals are allergic are not only seasonal but geographic in origin. The incidence of ragweed allergy, for example, is high in areas where ragweed grows but much lower in areas where it does not. Often, the onset of pollinosis peaks among the age bracket of 20 to 30. All races and sexes suffer an approximately equal incidence of pollinosis except Native Americans, who are reportedly immune to hay fever (Sherman, 1968). The most common allergenic pollen are identified in Chap. 2. These include grass pollen, flower pollen, weed pollen, and tree pollen. Pollen occur seasonally but the seasons may vary according to species. Various trees pollinate in April, May, or early June. Weeds pollinate from approximately the second week in August to the first frost. Grasses pollinate from late May through July. Figure 3.11 shows the percentage of hay fever occurring in each of these seasons.

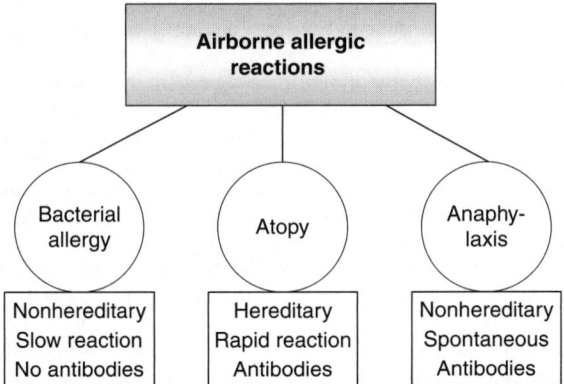

FIGURE 3.10 Breakdown of the types of airborne allergic reactions. Antibodies refer to the presence or absence of circulating antibodies.

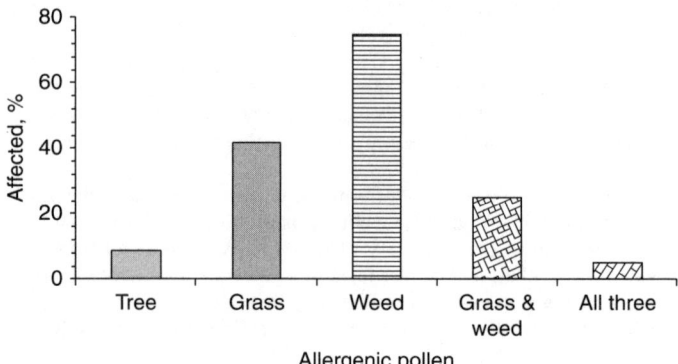

FIGURE 3.11 Percentage of atopic individuals affected by hay fever during the associated pollen season. [*Based on data from Sherman (1968)*].

Nonseasonal Allergic Rhinitis. Nonseasonal allergic rhinitis is an atopic reaction of the nasal mucosa due to causes other than pollen, and a condition that may occur throughout the year either continuously or intermittently. It is also called perennial allergic rhinitis and is due to sensitization to a specific antigen. Many patients with hay fever also suffer from nonseasonal allergic rhinitis. Hay fever and nonseasonal allergic rhinitis together represent about two-thirds of cases involving the respiratory tract (Sherman, 1968). The most important causative agents of allergic rhinitis other than pollen include house dust, animal dander, feathers, silk, wool, and vegetable matter. Other allergens may become causative agents as the result of occupational exposure, including flour, grain dust, castor beans, and certain chemicals. Once these materials become airborne and inhaled, they are generally trapped in the nasal mucosa, from where the allergy usually begins. The nasal passages are also susceptible to irritation due to smoke, fumes, and other pollutants, which may exacerbate the allergy or render an individual more susceptible to other allergens. Once nasal inflammation has developed due to allergy, the normal resistance to bacteria and viruses may be diminished and render the subject susceptible to colds and flus. The predisposition to sinus infection may lead to further complications and sinusitis. As many as one-third of patients with allergic rhinitis may eventually develop asthma.

Farmer's Lung. Farmer's lung is a condition that results from occupational exposure to moldy hay. Moldy hay may contain a variety of fungal and actinomycetes spores suspected of being causative agents, including *Aspergillus, Cladosporium, Mucor, Actinomyces, Micromonospora faeni, Saccharopolyspora rectivirgula, Thermoactinomyces vulgaris, Thermomonospora viridis, Saccharomonospora viridis, Thermopolyspora polyspora,* and *Thermoactinomyces candidus* (Sherman, 1968; Ortiz-Ortiz et al., 1984; Waksman, 1967; Reijula, 1993). Farmer's lung is believed to be a form of hypersensitivity to actinomycetes. The symptoms consist of shortness of breath, cough, fever, chills, weight loss, and hemoptysis. Farmer's lung is distinct from silo-filler's disease, which is an acute bronchiolitis due to inhalation of nitrogen oxides remaining from fertilizer use.

Bagassosis. Bagassosis is an occupational disease in workers handling bagasse, the dry fiber of sugarcane that remains after the sugar is extracted. It is similar to farmer's lung and has been reported all over the world. The onset of bagassosis is gradual or insidious and is characterized by dyspnea, fever, and occasionally cyanosis. Bagasse consists of cellulose, silica, and protein, but when stored in the outdoors it may become moist and develop mold growth. One gram of dust from bagasse may contain 240 million spores consisting of 20 different species (Sherman, 1968). Some of the mold spores on bagasse that may contribute to inhalation hazards include *Thermoactinomyces sacchari, Micromonospora vulgaris, Thermoactinomyces candidus, Saccharopolyspora hirsuta, Thermoactinomyces vulgaris,* and *Thermoactinomyces sacchari* (Ortiz-Ortiz et al., 1982; Lacey, 1981).

Pigeon Breeder's Disease. Pigeon breeder's disease is a recurrent acute febrile illness that occurs in some patients who handle or breed pigeons. Symptoms include fever, dyspnea, and breathing difficulty. The etiologic agent is pigeon feathers and other pigeon matter, and there is no evidence of any microbial involvement (Sherman, 1968).

Byssinosis. Byssinosis is a respiratory disease that occurs in patients with occupational exposure to cotton dust (Sherman, 1968). Most cases occur in workers of cotton mills but it is considered distinct and separate from mill fever (Rose and Blanc, 1998). Symptoms develop slowly after prolonged exposure of several years or more and include nasal irritation, sneezing, dry cough, and wheezing. Symptoms often abate on weekends and resume during the workweek. The disease may progress to pulmonary emphysema and permanent respiratory damage. It is unclear whether byssinosis is the result of the cotton fibers or antigens contained in cotton dust that may include microbes.

Mill Fever. Mill fever, mattress-maker's fever, heckling fever, and card fever are syndromes characterized by fever, chills, nausea, malaise, cough, and rhinitis that may occur in workers newly exposed to plant fibers like cotton, hemp, flax, and kapok (Rose and Blanc, 1998). The acute symptoms of these fevers occur within 1 to 6 hours of initial exposure and resolve within a few hours or days. Tolerance develops to the fibers and symptoms tend to disappear in subsequent exposures. The precise causes of mill fever are unclear but evidence suggests that endotoxins from gram-negative bacteria contaminating the dust may be at fault. In mattress-maker's fever it was found that mattresses stained due to microbial growth caused high rates of inhalation fever that was linked to endotoxin production in the gram-negative bacteria.

Organic Dust Toxic Syndrome. *Organic dust toxic syndrome* (ODTS) describes a form of inhalation fever that develops following the inhalation of organic dust from moldy hay, damp silage, contaminated wood chips for mulch, or other agricultural dusts (Rose and Blanc, 1998). Agricultural dusts may contain bacteria, bacterial endotoxins, fungal spores and hyphae, animal dander, pollen, and insects, among other things. ODTS has also been called silo-unloader's disease. ODTS has also been called pulmonary mycotoxicosis due to the large component of mycotoxin-producing fungal spores often found in such dust. ODTS tends to occur seasonally in the summer and the fall, when farmers are most likely to encounter very moldy grain and silage. ODTS is considered to be different from hypersensitivity pneumonitis, although the distinctions are not entirely clear. Symptoms of ODTS include an onset of fever about 4 to 12 hours after exposure, with chills, cough, chest tightness, malaise, nausea, and headache. Some of the microbes suspected of involvement in ODTS include *Penicillium, Aspergillus fumigatus, Aspergillus terreus*, and gram-negative bacteria that produce endotoxins. Endotoxins have the strongest and most consistent relationship to the symptoms of ODTS.

Humidifier Fever. Humidifier fever results from contaminated water in humidifiers. Humidifier fever is an influenza-like illness with symptoms that include fever, chills, malaise, headache, and some respiratory tract problems (Rose and Blanc, 1998). It typically occurs 4 to 12 hours after exposure to aerosols of contaminated water generated by humidifiers or air-conditioning systems. It can be an occupational disease when the workplace is air conditioned or humidified. This disease is considered similar to but distinct from humidifier lung, which is a form of hypersensitivity pneumonitis. Microorganisms play an uncertain role in humidifier fever but those that have been linked to the disease include *Naegleria gruberi*, species of *Pseudomonas*, and other gram-negative bacteria. Amoebae have also been found as one possible cause.

Pontiac Fever. Pontiac fever is a severe influenza-like illness that may result from exposure to aerosols of water contaminated with *Legionella pneumophila*. Pontiac fever is clinically and epidemiologically distinct from Legionnaires' disease but the causative difference between the two is unclear and may be primarily related to host immunity (Rose and Blanc, 1998). The attack rate for Pontiac fever is usually 95 to 100 percent and the incubation period is between 5 hours and 3 days. Symptoms include fever, chills, myalgia, headache, and malaise. Cough, diarrhea, vomiting, and sore throat may also occur.

Sump Fever. Sump fever or sump bay fever is an inhalation fever linked to contaminated water supplies that may become aerosolized from water pumps. Symptoms include fever, chills, myalgia, and cough beginning at least 4 hours after exposure. Occurrence has been rare but in one case it has been linked to species of *Pseudomonads* (Anderson et al., 1996). Other cases have been associated with water contaminated with gram-negative bacteria and may be associated with endotoxins (Rose and Blanc, 1998).

Grain Fever. Grain fever is an acute febrile illness that can occur due to exposure to high concentrations or airborne grain dust (Rose and Blanc, 1998). It is an inhalation fever with symptoms that include fever, chills, myalgia, headache, and cough. Symptoms usually subside after a few hours but may persist for several days. The illness is most often associated with damp or moldy grain, suggesting a microbial cause. Grain dust may contain bacterial endotoxins, fungi, and dust or storage mites. Grain may also contain aflatoxin (Ghosh et al., 1997).

Swine Fever. Swine fever is an illness that may be inhalation fever, or one or more of the occupational diseases typical in animal confinement facilities. Workers in swine confinement facilities are subject to occupational diseases that include acute and chronic bronchitis, airways hyperreactivity, hypersensitivity pneumonitis, and ODTS. Animal confinement units that are used to hold swine, poultry, cattle, sheep, and other animals often have high levels of airborne microbes and organic materials that may include *Staphylococcus, Micrococcus, Pseudomonas, Bacillus, Listeria, Enterococcus, Nocardia,* and *Penicillium* (Crook et al., 1991).

Wood-Trimmer's Disease. Wood-trimmer's disease describes an inhalation fever that may occur in sawmill workers who are exposed to high levels of dust from contaminated wood (Rose and Blanc, 1998). Workers in sawmills, pulp and paper mills, and workers handling wood chips are at risk for both inhalation fever and hypersensitivity pneumonitis. Up to 22 percent of sawmill workers may have experienced wood-trimmer's disease. Some of the microbes that have been found in the air of lumber processing facilities include *Penicillium, Aspergillus, Rhodotorula, Cladosporium, Aureobasidium, Paecilomyces, Mucor, Trichoderma, Rhizopus, Alternaria, Botrytis, Phialophora,* and *Ulocladium* (Sarantila et al., 2001).

Hypersensitivity Pneumonitis. Hypersensitivity pneumonitis, also called allergic alveolitis, is a type of disease that is similar to an allergic reaction, but it results from intense exposures to fungal or bacterial spores by (presumably) nonallergic individuals or in cases where irreversible damage to the lungs has occurred (Cormier, 1998). Individuals may become sensitized to spores due to chronic or high-level exposure, which can occur in agricultural industries. Farmer's lung is one common example of such diseases. Difficulties exist in both defining and diagnosing hypersensitivity pneumonitis due to the wide range of symptoms and possible etiologic agents. Symptoms typically occur about 3 to 8 hours after exposure. Table 3.5 summarizes many of the types of hypersensitivity pneumonitis and some of the identified airborne microbes that have been associated with these diseases (Cormier, 1998; Sarantila et al., 2001; Sherman, 1968; Rose, 1994).

Toxic Reactions. Toxic reactions result from ingesting or inhaling toxins, which are biological poisons. Endotoxins produced by gram-negative bacteria and mycotoxins produced by fungi are of primary concern. The inhalation or ingestion of toxins may produce toxicosis, or acute toxic poisoning. Toxins are produced by a variety of bacteria and fungi (see Chap. 2 and App. C). In general, the amount of toxin present in airborne bacteria or spores is insufficient, by itself, to present a health hazard. However, if the same microbes succeed in causing an infection in the host, the quantity of toxin produced may become dangerous. In other words, an infection must usually progress before the quantity of toxin present in the microbes becomes a health hazard. Exceptions exist, however, where toxic

TABLE 3.5 Airborne Microbial Causes of Hypersensitivity Pneumonitis

Disease	Source of antigens	Probable antigen
Farmer's lung	Moldy hay	*Aspergillus, Cladosporium, Mucor, Actinomyces, Micromonospora faeni, Saccharopolyspora rectivirgula, Thermoactinomyces vulgaris, Thermomonospora viridis,* Saccharomonospora *viridis, Thermopolyspora polyspora,* and *Thermoactinomyces candidus*
Bagassosis	Bagasse	*Thermoactinomyces sacchari, Micromonospora vulgaris, Thermoactinomyces candidus, Saccharopolyspora hirsuta, Thermoactinomyces vulgaris,* and *Thermoactinomyces sacchari*
Mushroom worker's disease	Moldy compost	*Thermoactinomyces vulgaris, Saccharopolyspora rectivirgula*
Malt-worker's lung	Contaminated barley	*Aspergillus clavatus*
Maple bark stripper's disease	Contaminated maple logs	*Cryptostroma corticale*
Sequoiosis	Contaminated wood dust	*Graphium, Pullularia*
Wood pulp worker's disease	Contaminated wood pulp	*Alternaria*
Humidifier lung	Contaminated water	*Thermoactinomyces candidus, Thermoactinomyces vulgaris, Penicillium,* Amebae
Compost lung	Compost	*Aspergillus*
Cheese washer's disease	Cheese casings	*Penicillium*
Wood-trimmer's disease	Contaminated wood trimmings	*Penicillium, Aspergillus, Rhodotorula, Cladosporium, Aureobasidium, Paecilomyces, Mucor, Trichoderma, Rhizopus, Alternaria, Botrytis, Phialophora,* and *Ulocladium*
Thatched roof disease	Dried grasses and leaves	*Saccharomonospora viridis*
Paprika splitter's lung	Paprika dust	*Mucor stolonifer*
Japanese summer house HP	House dust or bird droppings	*Trichosporon cutaneum*
Detergent worker's disease	Detergent	*Bacillus subtilis* enzymes
Potato-riddler's lung	Moldy straw	*Thermoactinomyces vulgaris, Saccharopolyspora rectivirgula, Aspergillus*
Tobacco-worker's disease	Mold on tobacco	*Aspergillus*
Hot tub lung	Mold on walls, ceiling	*Cladosporium*
Wine-grower's lung	Mold on grapes	*Botrytis cinerea*
Suberosis	Cork dust	*Penicillium*
Woodman's disease	Mold on bark	*Penicillium*
Grain-worker's disease	Grain dust	Fungi, bacterial endotoxins, mycotoxins
Sauna-taker's disease	Contaminated sauna water	*Pullularia*

microbes occur in unusually high airborne concentrations, such as in certain agricultural environments. High levels of endotoxins have been identified in the air of some agricultural and industrial settings where organic dust is present (Heederik et al., 1994). Facilities where wastewater becomes aerosolized have also been found to have high airborne endotoxin levels (Sarantila et al., 2001).

Occupational Mycoses. Occupational mycoses of the lung are infections caused by agents inhaled in workplace settings (DiSalvo, 1983). Mycoses are diseases induced by inhalation of fungal spores. In general, very few serious respiratory infections are caused in the natural environment by passive inhalation of fungal spores, but in some occupational environments the airborne levels can be excessive and lung infections can develop. Some of the more important occupational mycoses include coccidioidomycosis, histoplasmosis, paracoccidioidomycosis, blastomycosis, cryptococcosis, aspergillosis, and pulmonary mycotoxicoses.

Pulmonary Mycotoxicosis. Pulmonary mycotoxicosis is a toxic reaction to inhaled mycotoxins produced by certain fungi (Emanuel, 1983). It can produce an atypical form of farmer's lung disease. It usually occurs among agricultural workers handling moldy hay or silage. Symptoms commonly include chills, fever, cough, and dyspnea that appear about 2 to 4 hours after exposure. Symptoms may last several weeks after the initial exposure. Some of the toxins that may be responsible include aflatoxin, ochratoxin, T-2 toxin, zearalenone, citrinin, penicillic acid, and patulin.

Legionnaires' Disease. Legionnaires' disease, or legionellosis, is an important disease that requires special consideration because of its close association with engineered systems. It causes a pneumonia that attacks 2 to 5 percent of those exposed, and these represent those who are susceptible. It is distinct from the less severe version called Pontiac fever but is caused by the same bacteria—*Legionella pneumophila* (Ryan, 1994). Between 5 and 15 percent of those who contract legionellosis die from it. Factors influencing susceptibility include the elderly and those with suppressed immune systems, heavy smokers, and alcoholics, and others with weak lungs or constitutions. Males are over twice as susceptible as females. It incubates in human hosts within 2 to 10 days and will not abate without medication. Pneumonia caused by *Legionella* is indistinguishable on the basis of symptoms, and therefore many cases may go unreported. *Legionella* is also considered to be a nosocomial disease (Hart and Makin, 1991).

Legionella is ubiquitous in the natural environment and has colonized plumbing systems and cooling towers throughout the world (Thornsberry et al., 1984). In the natural aquatic environment, *Legionella* feeds on various nutrients but is most adept in the role of an intracellular parasite on other bacteria. Once it is uptaken by, or insinuates itself in, a larger bacteria it resists bacterial defenses and then multiplies. It avoids ciliary clearance through the use of its polar flagella. It resists immune system defenses in the lungs by cleavage of immunoglobulins, a characteristic of most bacteria. The primary lung defense at the cellular level includes phagocytes and alveolar macrophages, which attempt to ingest the invader. After surrounding and engulfing the *Legionella*, the phagocyte or macrophage fails to digest it. Instead, the *Legionella* parasitizes the phagocyte and begins multiplying. After doubling its numbers every 2 hours the phagocyte eventually ruptures. Cilia in the upper part of the respiratory tract normally expel the particles, except among smokers and alcoholics, where the process is muted. Conditions that may cause *Legionella* to amplify are summarized in Table 3.6.

The two most frequent sources of *Legionella* are neglected cooling towers and domestic hot water systems. Most equipment becomes contaminated through potable water supplies. The bacteria die off quickly on drying out in an aerosol and also from UV exposure on sunny days, but *Legionella* survives well in aerosols in comparison with other bacteria. Airborne *Legionellae* survive over greater distances when the relative humidity is 65 percent or greater. Amplifiers and disseminators of *Legionella* include cooling towers,

TABLE 3.6 Conditions Favoring the Growth of *Legionella*

Water temperatures between 20 and 50°C (68 and 122°F)
Optimal growth occurs at temperatures between 35 and 46°C (95 and 115°F)
Stagnant water, Absence of sunlight
A pH range of 2.0 to 8.5
Sediment in water which supports the growth of supporting microbiota
Microbiota including algae, protozoa, and others
The presence of L-cysteine-HCL and iron salts

---

evaporative condensers, domestic hot water systems, spas and whirlpools, humidifiers, decorative fountains, reservoir misters in supermarkets, portable cooling units with stagnant water, faucets, and showerheads.

3.10 NOSOCOMIAL AND OPPORTUNISTIC RESPIRATORY DISEASES

Nosocomial infections are those that are acquired in hospital and health care facilities as by-products of health care activities, especially surgery. Exposure may be through inhalation, via the settling of airborne pathogens on wounds or burns, or as the result of intrusive procedures with contaminated instruments. Most nosocomial infections are believed to transmit by direct contact, either between personnel and the patients, or via contaminated medical equipment like catheters. Perhaps the most intractable type of nosocomial infections are those that are caused by the patient's own microflora, or what are known as endogenous infections. Nosocomial infections are often opportunistic infections, or infections due to microbes that do not normally cause infection in healthy individuals. Opportunistic infections do not necessarily occur in health care settings, since they may also be anywhere outside of hospitals. Appendix A provides information on which agents may cause nosocomial infections, and further information is provided in Chap. 31.

Figure 3.12 shows those airborne nosocomial pathogens that predominate in ICUs (Weinstein, 1991). Again, these include pathogens that may settle on surfaces after skin shedding or be transmitted through direct contact, processes that leave them vulnerable during periods of exposure.

Nosocomial infections affect over 2 million patients annually in the United States (CDC, 2000). At least 5 percent of patients hospitalized in acute care institutions acquire an infection that was not present on admission. In ICUs, patients have nosocomial rates that are 5 to 10 times higher than those in general wards (Fagon et al., 1992).

Pneumonia is the second most common nosocomial infection in the United States. Pneumonias may be caused by intubation but *Legionella, Aspergillus,* and influenza virus most often cause pneumonia by inhalation of contaminated aerosols. Pneumonias and surgical wound infections account for approximately 15 percent of all hospital-acquired infections. Nosocomial pneumonia affected 16.6 percent of patients in ICUs. The mortality rate of these patients was 52.4 percent compared with 22.4 percent for patients without ICU-acquired pneumonia (Tablan, et al., 1994).

Per Ryan (1994), quoting CDC studies, the average nosocomial infection rate is 18.5 per 10,000 for bacteremia, 75 per 10,000 for surgical wound infections, and 53.5 per 10,000 for lower respiratory tract infections. Assuming that half of surgical wound and respiratory

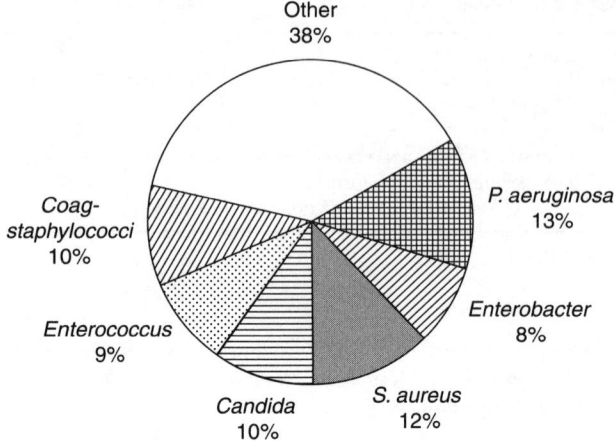

FIGURE 3.12 Predominant pathogens in ICUs. [*Based on data from Weinstein (1991)*.]

infections may be airborne, an airborne nosocomial infection rate of perhaps 50 per 10,000 patients is anticipated.

More detailed data are available from the National Nosocomial Infections Surveillance System (CDC, 2000), which identifies infection rates for hospital-wide infections, ICUs, high-risk nursery, and surgical patient units. Many of the types of infections and pathogens represent airborne or direct contact pathways such as pneumonia or catheter-associated *urinary tract infections* (UTI). See also Chap. 31 for additional information on nosocomial diseases.

3.11 NONRESPIRATORY AIRBORNE DISEASES

Airborne microbes can cause a variety of nonrespiratory infections including infections of the eyes, ears, and skin. Examples of such infections include otitis media, keratitis, mycetoma, and candidamycosis. Some respiratory infections may also transmit to the blood and result in infections like meningitis, endocarditis, and bacteremia. Some foodborne and waterborne diseases can result from the settling of airborne microbes onto exposed food or water.

Food poisoning may occur when *Clostridium botulinum* spores settle on food (Ryan, 1994). *Staphylococcus aureus* may also be transmitted to foods via handling or by shedding, resulting in food poisoning with enterotoxin (Cliver, 1990). Some evidence exists to suggest that *Listeria* may enter food processing systems via the airborne route (Bell and Kyriakides, 1998). Although the airborne transport of such microbes may be brief, they are still technically airborne and may be amenable to the same control technologies used for air and surface disinfection.

Although a number of fungi are capable of causing skin infections, these are almost entirely limited to those with compromised immune systems. Burn victims in recovery, AIDS patients, transplant patients, and others with impaired immune systems may be susceptible to infection from levels of fungal spores that pose no threat to healthy individuals.

3.12 OTHER DISEASES

The previous examples of diseases have addressed many of the most common or representative airborne diseases and general categories of respiratory disease. There are many more airborne diseases that are so common, so rare, or so complex in etiology that they cannot all be satisfactorily addressed here. Some of these, like meningitis, diphtheria, measles, mumps, chicken pox, hemorrhagic fever, plague, anthrax, Q fever, scarlet fever, nocardiosis, zygomycosis, and others are referred to in this book but not described in detail. The reader may find more specific information on these diseases both in App. A and in the associated references cited in Chap. 2.

REFERENCES

Anderson, K., McSharry, C. P., Clark, C., Clark, C. J., Barclay, G. R., and Morris, G. P. (1996). "Sump bay fever: inhalational fever associated with a biologically contaminated water aerosol." *Occup Environ Med* 53(2):106–111.

Bakken, L. R., and Olsen, R. A. (1983). "Buoyant densities and dry-matter contents of microorganisms: Conversion of a measured biovolume into biomass." *Appl Environ Microbiol* 45:1188–1195.

Barnhart, S. (1994). 11.3 Irritant Bronchitis. *Textbook of Clinical Occupational and Environmental Medicine*, L. Rosenstock and M. R. Cullen, eds., W. B. Saunders, Philadelphia, PA. 224–232.

Bell, C., and Kyriakides, A. (1998). *Listeria: A Practical Approach to the Organism and Its Control in Foods*. Blackie Academic & Professional, London.

Bratbak, G., and Dundas, I. (1984). "Bacterial dry matter content and biomass estimations." *Appl Environ Microbiol* 48:755–757.

Braude, A. I., Davis, C. E., and Fierer, J. (1981). "*Infectious Diseases and Medical Microbiology*," 2d ed.," W. B. Saunders. Philadelphia, PA.

Brooks, S. M. (1998). Chapter 33: Occupational and environmental asthma. *Environmental & Occupational Medicine*, W. H. Rom, ed., Lippincott-Raven Publishers, Philadelphia, PA. 481–524.

Castle, M., and Ajemian, E. (1987). *Hospital infection control*. John Wiley & Sons, New York.

CDC (2000). "National Nosocomial Infections Surveillance (NNIS) System Report, Data Summary from." *AJIC* 28(6):429–448.

Chan-Yeung, M. (1994). 11.1 Asthma. *Textbook of Clinical Occupational and Environmental Medicine*, L. Rosenstock and M. R. Cullen, eds., W. B. Saunders , Philadelphia, PA. 197–209.

Cliver, D. O. (1990). *Foodborne Diseases*. Academic Press, San Diego, CA.

Cormier, Y. (1998). Chapter 31: Hypersensitivity pneumonitis. *Environmental & Occupational Medicine*, W. H. Rom, ed., Lippincott-Raven Publishers, Philadelphia, PA. 457–465.

Darquenne, C. (2002). "Heterogeneity of aerosol deposition in a two-dimensional model of human alveolated ducts." *J Aerosol Sci* 33:1261–1278.

Davey, B., and Halliday, T. (1994). *Human Biology and Health: An Evolutionary Approach*. Open University Press, London.

DiSalvo, A. F. (1983). *Occupational Mycoses*. Lea & Febiger, Philadelphia, PA.

Emanuel, D. A. (1983). Histoplasmosis. *Toxic Reactions (Pulmonary Mycotoxicosis)*, A. F. DiSalvo, ed., Lea & Febiger, Philadelphia, PA.

Fagon, J., Chastre, J., Vuagnat, A., Trouillet, J., A. N., and Gibert, C. (1996). "Nosocomial pneumonia and mortality among patients in intensive care units." *JAMA* 275:866–869.

Fleming, D. O., Richardson, J. H., Tulis, J. J., and Vesley, D. (1995). "*Laboratory Safety Principles and Practices*, 2d ed.," ASM Press, Washington, DC.

Ghosh, S. K., Desai, M. R., Pandya, G. L., and Venkaiah, K. (1997). "Airborne aflatoxin in the grain processing industries in India." *Am Ind Hyg Assoc J* 58(8):583–588.

Hart, C. A., and Makin, T. (1991). "*Legionella* in hospitals: A review." *J Hosp Infect* 18(June Suppl. A):481–489.

Hatch, T. F., and Gross, P. (1964). *Pulmonary Deposition and Retention of Inhaled Aerosols.* Academic Press, New York.

Heederik, D., Smid, T., Houba, R., and Quanjer, P. H. (1994). "Dust-related decline in lung function among animal feed workers." *Am J Ind Med* 25(1):117–119.

Heinsohn, R. J. (1991). *Industrial Ventilation: Principles and Practice.* John Wiley & Sons, New York.

Howard, D. H., and Howard, L. F. (1983). *Fungi pathogenic for Humans and Animals.* Marcel Dekker, New York.

Jarvis, W. R. (2000). *Nosocomial Pneumonia.* Marcel Dekker, New York.

Lacey, J. (1981). Airborne actinomycete spores as respiratory allergens. *Actinomycetes,* K. P. Schaal and G. Pulverer, eds., Gustav Fischer Verlag, New York.

Lane, D. J. (1979). *Asthma: The Facts.* Oxford University Press, New York.

Lippmann, M. (1998). Chapter 19: Particle deposition and pulmonary defense mechanisms. *Environmental & Occupational Medicine,* W. H. Rom, ed., Lippincott-Raven Publishers, Philadelphia, PA.

Mitscherlich, E., and Marth, E. H. (1984). *Microbial Survival in the Environment.* Springer-Verlag, Berlin, Germany.

Murray, P. R. (1999). *Manual of Clinical Microbiology.* ASM Press, Washington, DC.

Nelson, R. P., and Lockey, R. F. (1998). Chapter 45: Rhinitis and sinusitis. *Environmental & Occupational Medicine,* W. H. Rom, ed., Lippincott-Raven Publishers, Philadelphia, PA. 667–673.

Ortiz-Ortiz, L., Bojalil, L. F., and Yakoleff, V. (1984). *Biological, biochemical, and biomedical aspects of actinomycetes.* Academic Press, Orlando, FL.

Perera, F. P., and Ahmed, A. K. (1979). *Respirable Particles: Impact of Airborne Fine Particulates on Health and the Environment.* Ballinger Publishing, Cambridge, MA.

Pope, A. M., Patterson, R., and Burge, H. (1993). *Indoor Allergens.,* I. O. Medicine, ed. National Academy Press. Washington, DC.

Reijula, K. E. (1993). "Two bacteria causing farmer's lung: fine structure of *Thermoactinomyces vulgaris* and *Saccharopolyspora rectivirgula.*" *Mycopathologia* 121(3):143–147.

Rose, C. (1994). 11.5 Hypersensitivity pneumonitis. *Textbook of Clinical Occupational and Environmental Medicine,* L. Rosenstock and M. R. Cullen, eds., W.B. Saunders, Philadelphia, PA. 242–248.

Rose, C. S., and Blanc, P. D. (1998). Chapter 32: Inhalation fever. *Environmental & Occupational Medicine,* W. H. Rom, ed., Lippincott-Raven Publishers, Philadelphia, PA. 467–480.

Ryan, K. J. (1994). "*Sherris Medical Microbiology.*" Appleton & Lange, Norwalk, CT.

Sarantila, R., Reiman, M., Kangas, J., Husman, K., and Savolainen, H. (2001). "Exposure to endotoxins and microbes in the treatment of wastewater and in the industrial debarking of wood." *Bull Environ Contam Toxicol* 67:171–178.

Scheuch, G., and Stahlhofen, W. (1992). "Deposition and dispersion of aerosols in the airways of the human respiratory tract: The effect of particle size." *Exp Lung Res* 18:343–358.

Seinfeld, J. H. (1986). *Atmospheric Chemistry and Physics of Air Pollution.* Wiley Interscience Publications, New York.

Shapton, D. A., and Board, R. G. (1972). *Safety in Microbiology.* Academic Press, London.

Sherman, W. B. (1968). *Hypersensitivity: Mechanisms and Management.* W. B. Saunders, Philadelphia, PA.

Tablan, O. C., Anderson, L. J., Arden, N. H., Beiman, R. F., Butler, J. C., MacNeil, M. M., and HIC-PAC (1994). "Guideline for the prevention of nosocomial pneumonia." *Am J Infect Control* 22:247–292.

Thornsberry, C., Balows, A., Feeley, J., and Jakubowski, W. (1984). *Legionella: Proceedings of the Second International Symposium.,* Atlanta, GA.

Waksman, S. A. (1967). *The Actinomycetes: A Summary of Current Knowledge.* Ronald Press, New York.

Weinstein, R. A. (1991). "Epidemiology and control of nosocomial infections in adult intensive care units." *Am J Med* 91(Suppl. 3B):179S–184S.

Zhang, Z., Kleinstreuer, C., and Kim, C. S. (2002). "Aerosol deposition efficiencies and upstream release positions for different inhalation modes in an upper bronchial airway model." *Aerosol Sci Technol* 36:828–844.

CHAPTER 4

EPIDEMIOLOGY AND DOSIMETRY

4.1 INTRODUCTION

Epidemiology is the study of disease transmission through the use of statistics. Statistical data about diseases can elucidate the sources and mechanisms of transmission, quantify risks, highlight good health practices and the effectiveness of remedial measures, and assist the economic evaluation of risks/benefits. It often tells a more complete story than clinical or biological analysis of a disease. Both contagious and noncontagious diseases can be studied with epidemiology. A basic understanding of epidemiology is essential to understanding the spread of disease and can assist engineers and medical personnel in seeking solutions to disease problems.

In this chapter the focus is on the epidemiology and dosimetry of airborne respiratory diseases, both contagious and noncontagious, but some aspects of nonrespiratory disease epidemiology are addressed. Some statistics on airborne diseases are presented and the routes of disease transmission are described and summarized to provide perspective on the various transmission modes and pathways of airborne diseases in indoor environments. Next, the mathematics of disease progression and epidemiology is summarized, including an epidemiological model of buildings as airborne disease vectors. Finally, the dosimetry of airborne diseases is presented.

4.2 AIRBORNE DISEASE STATISTICS

The statistics of airborne diseases and the costs to society are worth some preliminary consideration. Figure 4.1 shows a breakdown of per capita medical costs in the United States, based on data from Smyth (1987). The per capita cost of 10 percent provides an indication of average cost to the individual of respiratory infections. It is estimated that the average adult experiences at least two colds, flus, or other respiratory infections per year, although some of these may not be symptomatic.

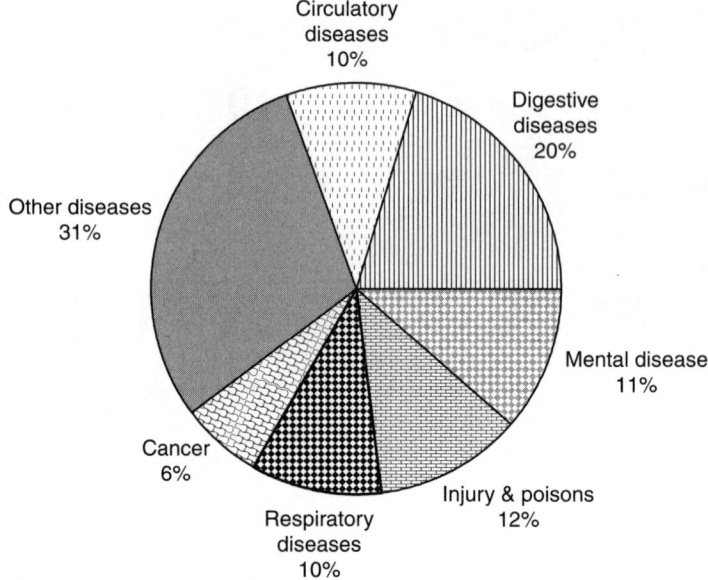

FIGURE 4.1 Per capita medical costs of diseases. [*Based on data from Smyth (1987)*.]

Respiratory diseases account for one-fourth of all medical costs in the United States today, as shown in Fig. 4.2. Only cardiovascular diseases outweigh respiratory diseases in total costs. Not all respiratory diseases are the result of infections by microorganisms, since some may have nonmicrobial causes, such as asbestosis, but a majority of them are caused by viruses, bacteria, and fungi. Annual health care costs for acute respiratory infections are estimated to total $30 billion per Fisk and Rosenfeld (1997).

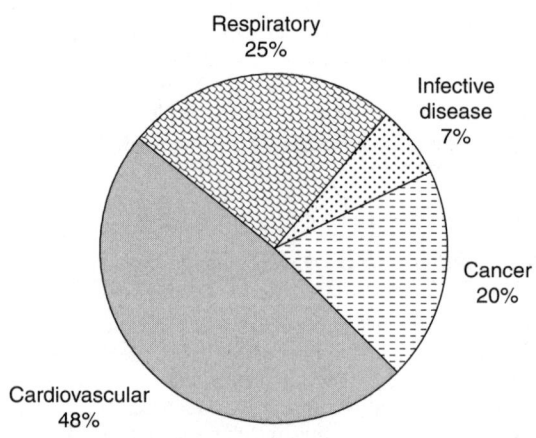

FIGURE 4.2 Medical costs of diseases in the United States. [*Based on data from Tolley et al. (1994)*.]

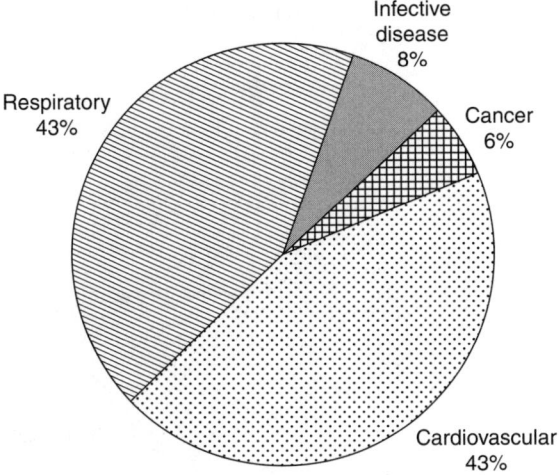

FIGURE 4.3 Lost earnings from diseases. [*Based on data from Tolley et al. (1994).*]

If the actual lost earnings due to diseases are compared, it can be seen that respiratory infections account for a much greater proportion of the national wealth than the medical costs alone indicate. Figure 4.3 illustrates the lost earnings from the same diseases shown in Fig. 4.2, and it shows that respiratory disease accounts for some 43 percent of total lost earnings. Annual losses in work time and productivity for acute respiratory infections are estimated to total $35 billion per Fisk and Rosenfeld (1997).

Figure 4.4 comes from a detailed study of the etiological causes of respiratory infections. It shows that the general category of *upper respiratory infections* (URI), which includes the common cold, can account for over half of all respiratory infections. Colds comprise the largest single respiratory infection, although during flu epidemics, influenza will predominate. Although not all of these respiratory infections necessarily transport via the airborne route, there is clearly much to be saved in terms of life, health, and costs if technologies could be implemented to disinfect indoor air and reduce the overall incidence of respiratory infections.

4.3 ROUTES OF TRANSMISSION

Respiratory diseases can be transmitted by at least three routes—inhalation of airborne microbes, direct contact, and ingestion (Thompson, 1962). The airborne route might be the most direct route by which an organism may cause a lung infection but it is not always the predominant route (Langmuir, 1961). Direct contact with microbes may occur in person-to-person contact or in contact with a contaminated surface. In most cases where a respiratory disease is acquired by direct contact the hands become contaminated and the pathogens are then transferred to the eyes, nose, or mouth. Contact with other areas of the body (i.e., feet, genitalia, and the like) is not considered a pathway for respiratory infections. Direct contact between lips, however, is a possible pathway for respiratory diseases. Ingestion of respiratory pathogens is an uncommon mechanism but it is possible for pathogens on food or in water to contaminate the mouth and subsequently spread to other areas of the

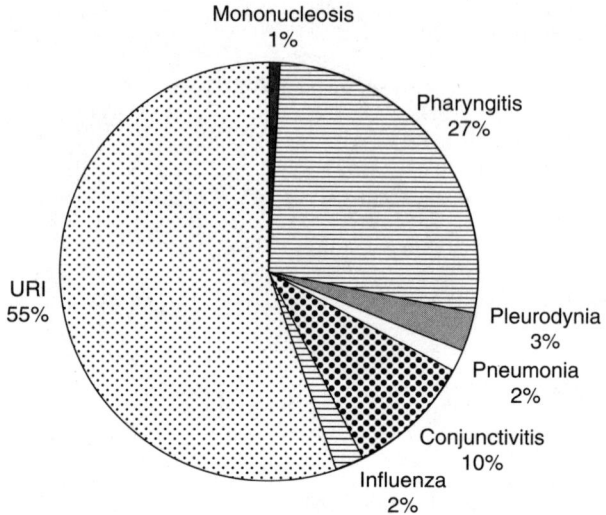

URI: Upper respiratory infection, including colds

FIGURE 4.4 Breakdown of respiratory infections. [*Based on data from Robinson et al. (1960)*.]

respiratory system. Airborne microbes that settle on surfaces or skin are referred to as fomites. Fomites may be droplets or dried droplet nuclei containing one or more pathogenic microbes. Hand contact with fomites is considered an important pathway to infectious diseases. Fomites on surfaces may also become reaerosolized due to disturbance and then become inhaled. Figure 4.5 illustrates the major transmission mechanisms for airborne respiratory diseases.

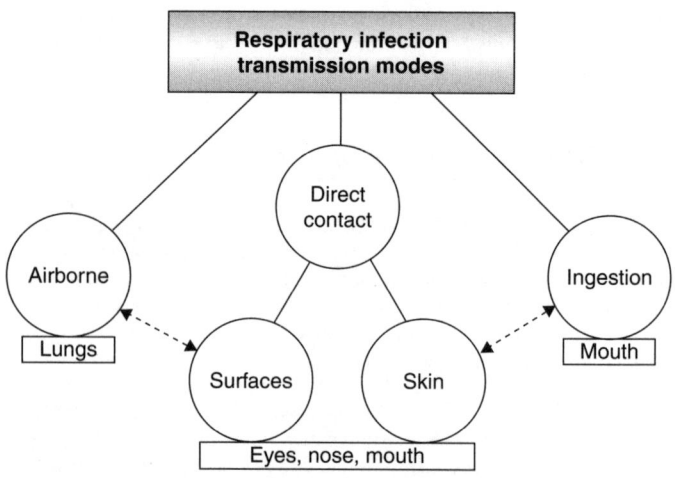

FIGURE 4.5 Transmission mechanisms for airborne respiratory diseases.

The question of whether infections transmit primarily by the airborne route or through direct contact has occupied the interest of researchers for the past century. While definitive evidence exists to prove both routes of infections for respiratory diseases, little data are available that would allow a specific breakdown of the importance of each route for any of the known respiratory diseases. As Langmuir explained in 1956, if transmission is by direct contact, then control is to be based on personal measures, but if the spread is airborne, the indicated measures are to be drawn from engineering (Gordon and Ingalls, 1957).

The transmission mechanisms for nonrespiratory diseases are similar except that the target may be different. Food- and waterborne pathogens target the intestines or gastrointestinal region and therefore ingestion is the primary mode of transmission. However, microbes that reach the hands may also end up transmitting to the mouth and ultimately the stomach. Food- or waterborne pathogens may also become aerosolized and settle on food, as in staphylococcal food poisoning or botulism.

Respiratory diseases are most often transported from an infected individual by sneezing, coughing, talking, and rhinorrhea or runny nose. Sneezing, coughing, and talking all generate droplets of various sizes, as shown in Fig. 4.6. Obviously, sneezing produces a considerably larger quantity of droplets than coughing or talking. The quantity of aerosolized microbes produced by talking might appear to be negligible, but instances have occurred in which arguing or singing has resulted in diseases being transmitted to others.

Additional data from Rubbo and Benjamin (1953) seems to corroborate the idea that sneezing produces more aerosolized particles than coughing, at least on a per unit basis. Figure 4.7 shows the results of viable microorganisms produced by coughing and sneezing in tests on two common commensal microbes, *Staphylococcus* and *Streptococcus*. Sneezing ejects saliva mostly from the anterior part of the mouth and tends to contain only those microbes resident in that area (Williams, 1960). Sneezing may contain very few viable microbes but it atomizes them, perhaps to single microbes in many cases, which allows for extended residency in air and increased likelihood of deep inhalation. Coughing produces far fewer droplets than sneezing and ejects microbes that are primarily resident in the pharynx.

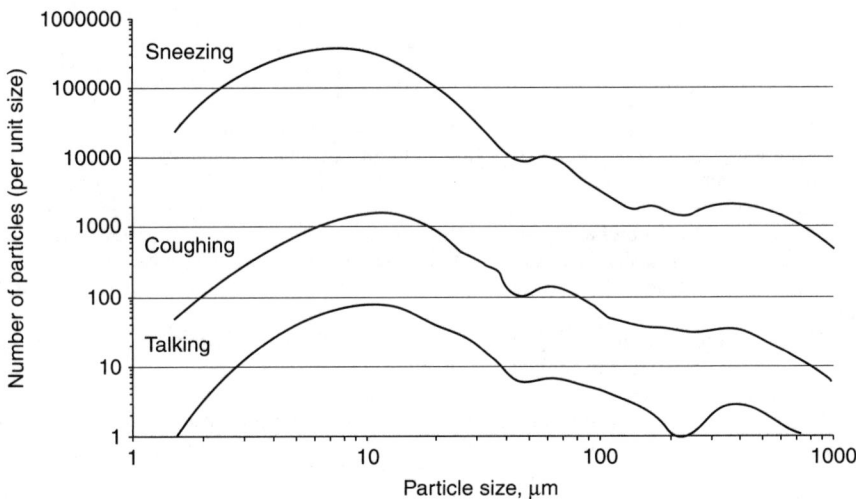

FIGURE 4.6 Size distribution of particles produced by various expulsion modes. [*Based on data from Duguid (1945).*]

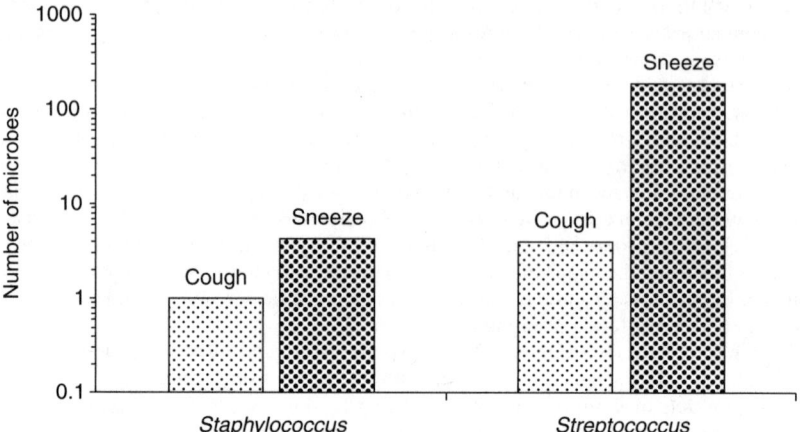

FIGURE 4.7 Comparative production of microbes from coughing and sneezing. [*Based on data from Rubbo and Benjamin (1953).*]

The ultimate fate of aerosolized particles is to settle out of the air. Larger droplets settle first and the smallest droplets may remain suspended for many hours before they settle. Particles that are inhaled may cause infections. Particles that settle out may remain viable as fomites for hours or even days if conditions are right. If hand contact is made with fomites and the eyes, nose, or mouth is rubbed or touched, the infectious agent may find its way into the respiratory tract and induce an infection.

The actual number of airborne microbes generated by an infectious individual is not known with any certainty for any species. Nardell et al. (1991) describe the TB release rate from an infected individual as about 1 to 250 quanta per hour, where the quanta could be as little as a single TB bacilli. Wheeler (1993) refers to a case in which measles in a school were produced at the rate of 5480 quanta per hour. In the case of a virus like measles, it may be that each quantum represents thousands of virions. Remington et al. (1985) report on a case in which an index patient was producing 8640 quanta per hour of measles virus.

The size of droplet nuclei may impact their infectiousness due to both the settling rate and the point in the respiratory system they are likely to settle in. Studies have shown that the larger droplets are less likely to cause infections (Druett et al., 1953). The larger the droplet nuclei, the more likely it is to settle out in the nasopharynx, and even though it may contain a larger number of microbes, it still requires more microbes in this location to induce an infection. Clouds of single microorganisms are more likely to penetrate to the lower respiratory tract and are more infectious in this form (Druett et al., 1956). Particles less than 3 to 5 µm in size have the potential to reach the alveoli and remain there, while particles smaller than 0.25 µm may actually be exhaled part of the time (Gordon and Ingalls, 1957).

Noncommunicable disease agents often hail from the environment or from animals. Fungal spores grow in the soil or on plant material and are predisposed to become aerosolized on maturity. They easily become airborne at the slightest wind or disturbance of the soil and plant material. Animal diseases, called zoonoses, that can transmit to humans are primarily noncommunicable. Exceptions include some arenaviruses (hemorrhagic fever) and pneumonic plague. Infection with zoonotic diseases generally requires direct contact or close inhalation near the animals. An exception is Hantavirus, which becomes aerosolized from rodent feces. The sequence of events that are required to produce a respiratory disease will normally include (1) aerosolization and transport or direct contact,

(2) inhalation or direct contact, and (3) successful infection. The latter factor requires overcoming the primary lung defenses and immune system. Since everyone's level of health or immunity may be different at any given time, not all exposures will result in infection. Figure 4.8 illustrates the basic pathways and mechanisms that may result in a respiratory infection.

The percentage of infections that result from inhalation as opposed to direct contact between the hands and the eyes, nose, or mouth has been a subject of several investigations. It is likely that the fraction of infections that transmits purely by the airborne route, by direct contact, or via fomites is species dependent. Some microbes survive well in air while others do not. Some microbes survive well on surfaces while others do not. Rhinovirus can survive 3 to 4 hours on hands while parainfluenza and *respiratory syncytial virus* (RSV) last less than 1 hour on hands (Hurst, 1996). In a pair of studies by Gwaltney et al. (1967, 1978) on rhinoviruses, one of the causes of the common cold, it was found that the hands contained the highest concentrations of rhinovirus while objects in the home contained the lowest. These results are illustrated in Fig. 4.9 and suggest that more rhinovirus infections are probably produced by direct person-to-person contact than are produced through contact with fomites on objects.

Figure 4.10 shows the route of cold infections for coxsackievirus, based on data from Buckland et al. (1965). In this test subjects were inoculated at the various sites with different doses of the virus. The results indicate that the nose and the eyes are by far the most vulnerable routes of virus invasion. This study also tested the ability of contaminated hands to transmit the infection to the mouth. The hand-to-mouth route could not be demonstrated to play any part in cold infections, although several other investigators have suggested this is an important route. The route of hand-to-eyes was not studied but is believed to play a major role in causing infections from contaminated hands.

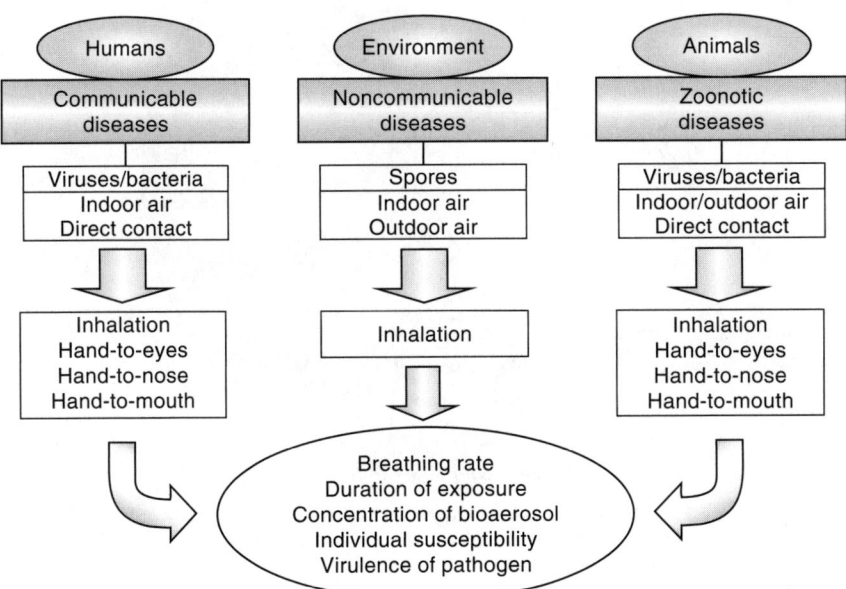

FIGURE 4.8 Pathways of airborne infections and determining factors of disease for communicable and noncommunicable respiratory diseases.

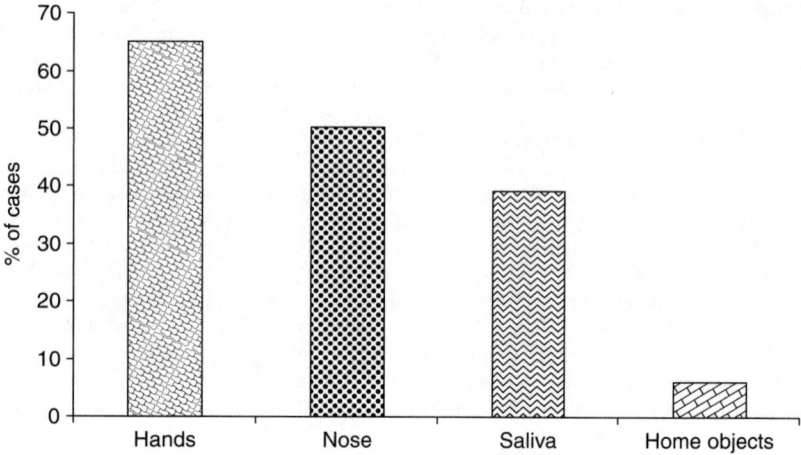

FIGURE 4.9 Location of rhinovirus in people with colds. [*Based on data from Gwaltney et al. (1978).*]

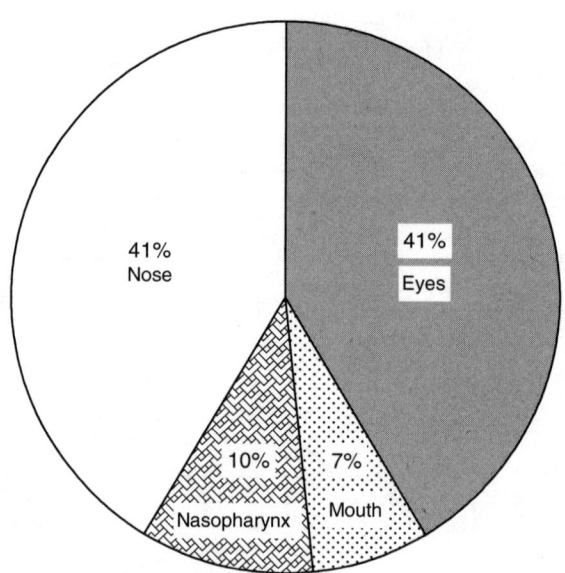

Hand-to-nose route had 0% infections

FIGURE 4.10 Route of infection for colds in a clinical study of coxsackievirus at various doses. [*Based on data from Buckland et al. (1965).*]

EPIDEMIOLOGY AND DOSIMETRY

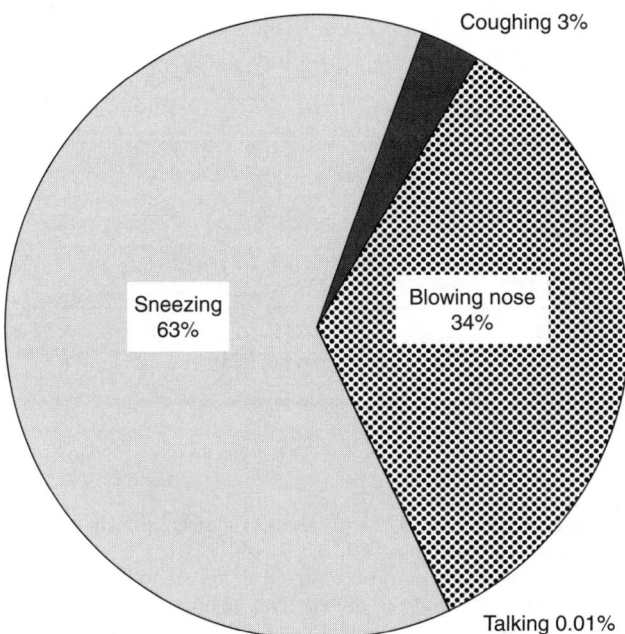

FIGURE 4.11 Source of cold virus dispersion. [*Based on coxsackievirus data from Buckland et al. (1965).*]

Figure 4.11 provides some perspective on the source of airborne viruses based on clinical data from Buckland et al. (1965). Sneezing and nose blowing generate and disperse the largest quantities of aerosolized viruses. Coughing accounts for a much lower proportion of aerosolized viruses; however, coughing generally occurs much more frequently for most types of colds. Since there seems to be insufficient data on the comparative frequency of coughing and sneezing, it is difficult to estimate what poses the greater risk. Contradicting these results are those of Couch et al. (1966) who found that the virus concentrations produced by coughing were much the same as those produced by sneezing. Talking accounted for only 0.01 percent of the aerosolized cold virus dispersion and would seem to play a very minor role in causing infections. However, talking may be continuous, and may even be agitated. In one outbreak of influenza, a singer in a rock band caused numerous infections in the audience.

Airborne droplets containing viable viruses or bacteria tend to evaporate rapidly in air and condense to droplet nuclei (Nardell, 1990). Droplet nuclei are the evaporated residues of infected respiratory secretions and are of such a small diameter, being typically less than 5 μm, that they can remain airborne for extended periods of time. Particles in this size range and smaller can be inhaled and deposited in the lower respiratory tract. Data showing the settling time for droplets in the size range of about 1 to 20 μm were taken by Duguid (1945) and are shown plotted in Fig. 4.12. Obviously the smaller particles in the 1- to 2-μm size range can remain airborne for extended periods of time—long enough to diffuse throughout a room or be recirculated by ventilation systems.

In a clinical study by Couch et al. (1966), the specific site of infection, sinus versus lungs, was studied by inoculating these sites with infectious doses of coxsackievirus and

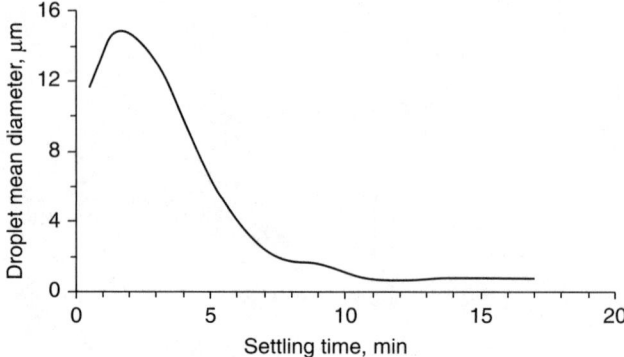

FIGURE 4.12 Setting time of droplets in air vs. size. [*Based on data from Duguid (1945).*]

rhinovirus. The results, shown in Fig. 4.13, seem to indicate the sinus is somewhat more vulnerable to infection than the lungs.

Since direct contact and person-to-person contact appear to be such important vectors for respiratory infections, the study of the survival of microbes on surfaces has attracted some attention over the years. In a study by Reed (1975) skin and surfaces were inoculated with rhinovirus to determine how long the virus would remain viable at these locations. The back of the hands were used for inoculation, as were ball point pens, tabletops, and spoons. The results of these studies are summarized graphically in Fig. 4.14. It would appear, based on these results, that the viability of microbes drops off after a few hours on the hands but that viruses may remain viable on inanimate surfaces for several days. Obviously the risk

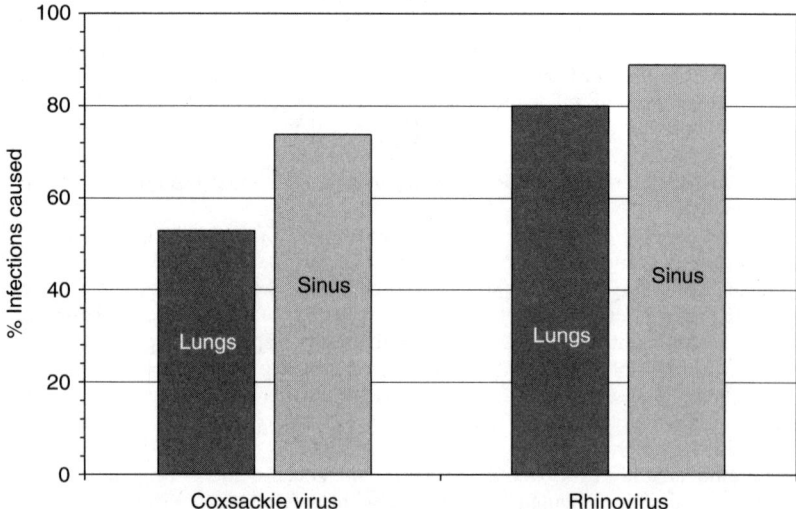

FIGURE 4.13 Route of infection of cold viruses. [*Based on data for coxsackievirus from Couch et al. (1966).*]

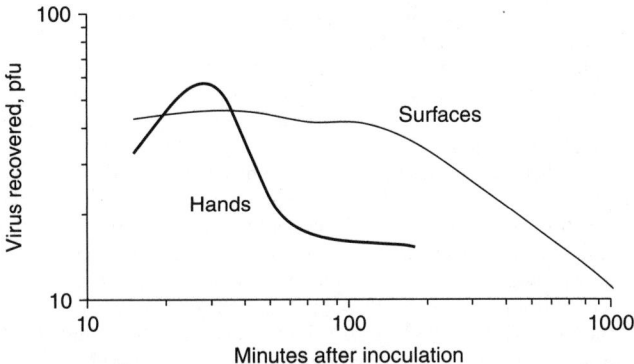

FIGURE 4.14   Rhinoviruses remaining on skin and surfaces over time. [*Based on data from Reed (1975).*]

of contamination by fomites in either location is not one to be ignored. The same studies also determined that acute contamination of the fingers with viruses is common in subjects with colds, and that the rate of recovery of viruses from the fingers was 42 percent. Furthermore, the transfer rate between contaminated fingers and noncontaminated fingers was found to be 17 percent between dry hands but 34 percent when the hands were moist. The transfer rate between dry surfaces and hands was found to be 15 percent but this figure rose to 21 percent when the surfaces were moist.

Studies on the transfer of *Klebsiella* and *Staphylococcus* between soiled cloth and hands showed that not only could moist cloth preserve and transfer microbes to the hands, but that these microbes could even grow and multiply on cloth, and can remain viable on surfaces for many hours (Scott and Bloomfield, 1990). Figure 4.15 shows the results for several of these tests on three species on laminate surfaces. Whether the surfaces or cloth were clean or soiled had little effect on the survival but moisture played a role in prolonging their

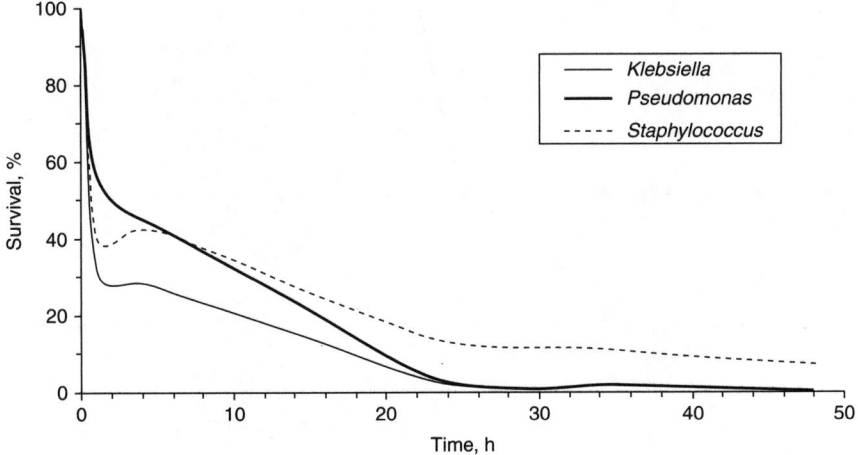

FIGURE 4.15   Survival times for airborne microbes on laminate surfaces. [*Based on data from Scott and Bloomfield (1990).*]

survival or even promoting growth. Clearly, there is a threat from fomites on hands, surfaces, and cloth that is amplified when they are moist.

The type of surface may sometimes impact the viability of microbes copper and silver, for example, can generate ions that can be lethal to microbes (Thurman and Gerba, 1989). Microbial survival times on surfaces or in the environment have been provided in App. A where information is available.

If direct contact between persons (i.e., hand to hand) or with fomites on surfaces (i.e., tabletops, telephones, keyboards, pens, doorknobs, faucets, towels, and the like) is a major factor in the transmission of respiratory diseases, then intervention programs focused on hygiene should have a significant impact on reducing disease incidence. In a study of the effects of a handwashing program in an elementary school, it was found that girls washed their hands after bathroom use about 58 percent of the time while boys only washed about 48 percent of the time (Guinan et al., 2002). Soap usage was 28 percent among girls and a mere 8 percent among boys. Hand drying tended to be neglected about 75 percent of the time. When a comprehensive handwashing program was implemented at the school, absenteeism dropped by over 50 percent. Both respiratory and gastrointestinal illnesses would have been included in these absenteeism reports, but the results are still significant. Perhaps one-third to one-half of absenteeism is due to acute respiratory diseases, and although there are insufficient data here to draw any quantitative conclusions, it could be conjectured that if about half of illnesses could be reduced by handwashing programs, then the other half may be airborne infections.

Whether most infections are transmitted by the airborne route or through direct contact with hands or fomites on surfaces, the net effect of proximity to an infectious individual for extended time periods can be a risk factor for infection. This is exemplified in Fig. 4.16, which shows the risk of acquiring an infection as a function of distance between persons during 8 hours of exposure, as would be the case in a typical office environment. This figure suggests that a distance of about 6 feet represents a cutoff below which the risk of a acquiring a coworker's infection increases considerably. It also suggests that the risk never

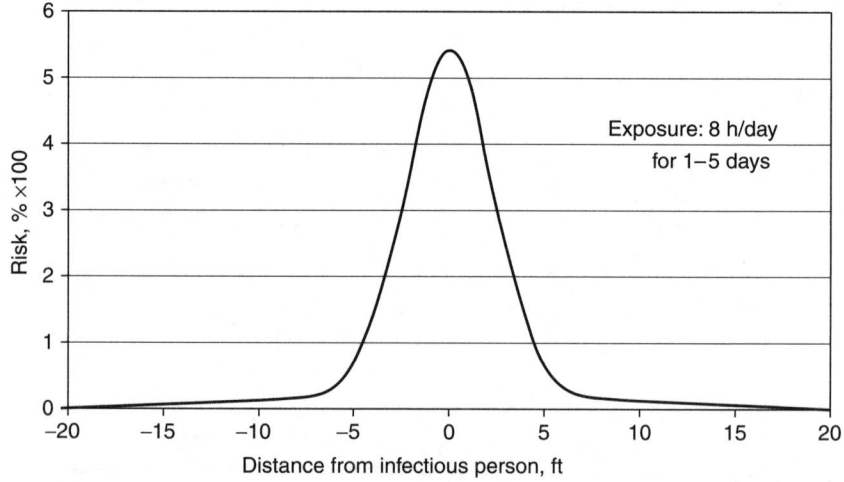

FIGURE 4.16 Proximity to an individual as a risk factor for infection. [*Based on data from Lidwell and Williams (1961).*]

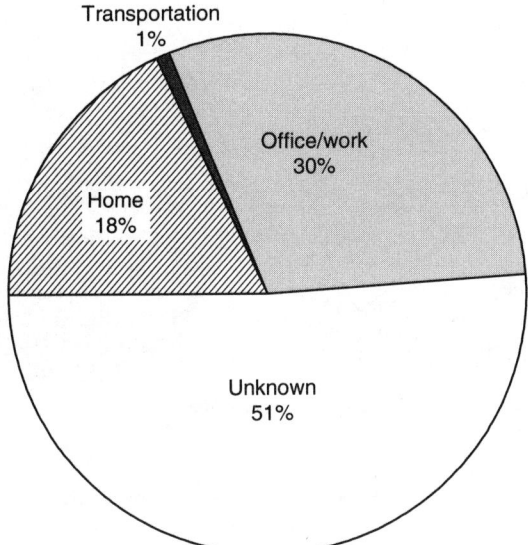

FIGURE 4.17 Source of cold infections in adults. [*Based on data from Lidwell and Williams (1961)*.]

truly drops to zero no matter what the distance, if people are inside the same building, and being served by a common ventilation system.

The limited data available indicate that most cold infections are acquired in the office or work environment, home, and in schools. Figure 4.17 shows that the home environment provides the second largest source of infections after the work environment. It also shows that some 51 percent come from unknown sources, which could also be due to home and office infections, or any enclosed locations where prolonged exposure to infective individuals could occur, such as bars, dance halls, restaurants, churches, other people's home, indoor sporting events, or theaters.

Children are particularly nonhygienic in their behavior and the rates of disease among children in schools, day care centers, and homes make this clear. In a study on chicken pox and measles among children summarized in Wells (1955) it was found that the school and home together account for most of the disease transmission. Figure 4.18 illustrates these results.

The fact that children are both highly susceptible to diseases and act as vectors for diseases is exemplified by the results of a study shown in Fig. 4.19, which compares the risk of acquiring a cold at home. Clearly, infants are highly susceptible, perhaps due to insufficiently developed immunity or perhaps due to hygiene. Figure 4.19 indicates that children bring home most of the colds from school. The wife is twice as likely to catch a cold as her husband perhaps because she is the one who cares for the children at home. This situation may change as more husbands become homemakers but there are no data available yet.

In a study of intrafamily spread of respiratory syncytial virus during an outbreak, Hall et al. (1976) found that the secondary attack rate among family members was 27 percent. The virus had infected some 44 percent of families and 22 percent of all members. An infant's older sibling appeared to be most likely to introduce the virus into the family. This pattern may be true for many other types of infections. Figure 4.20 illustrates the infection rates that occur among members in a family at home. It can again be seen that children and

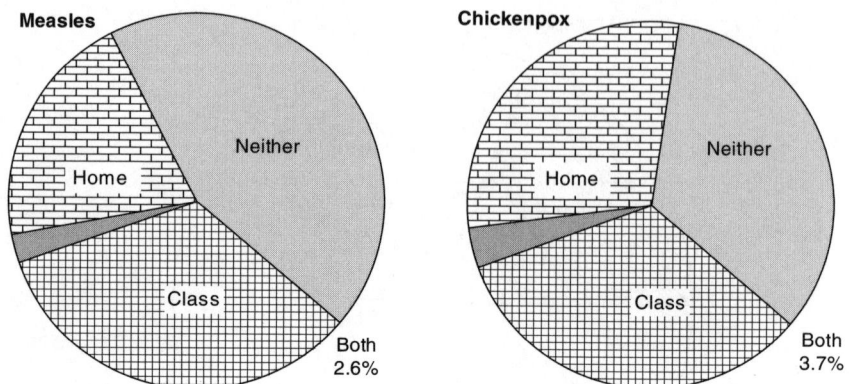

FIGURE 4.18 Breakdown of disease source in schoolchildren. [*Based on data from Wells (1955)*.]

infants are most susceptible to infections, and that the housewife who cares for them is likely to catch the same colds and flus.

Figure 4.21 shows the results of a study on transmission rates of rhinovirus colds between members of a household under various conditions. There is a significant difference between transmission rates of someone in the infectious stage, and when they are not in the infectious stage, of a cold. The presence of rhinovirus on the hands and saliva suggests that transmission might be by direct contact instead of airborne, but it may also be reflective of cold symptoms only. These results suggest that some sort of quarantine of infected members of a household, when they are in the peak infectious period, might reduce transmissions by about half.

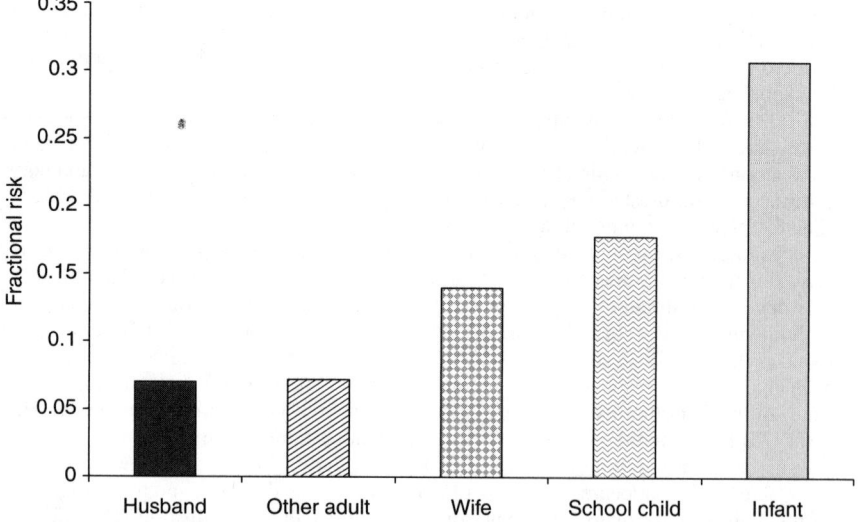

FIGURE 4.19 Risk of acquiring a cold at home. [*Based on data from Lidwell and Williams (1961)*.].

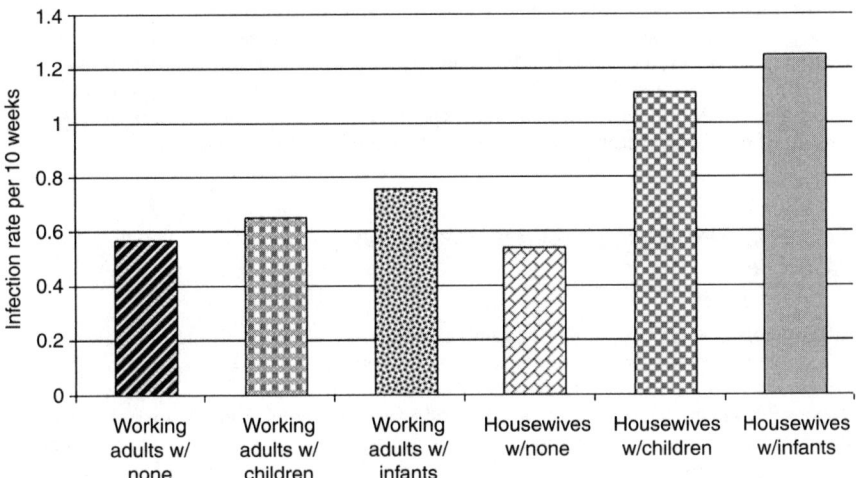

FIGURE 4.20 Adult infection rates at home with and without children. [*Based on data from Hendley et al. (1969).*]

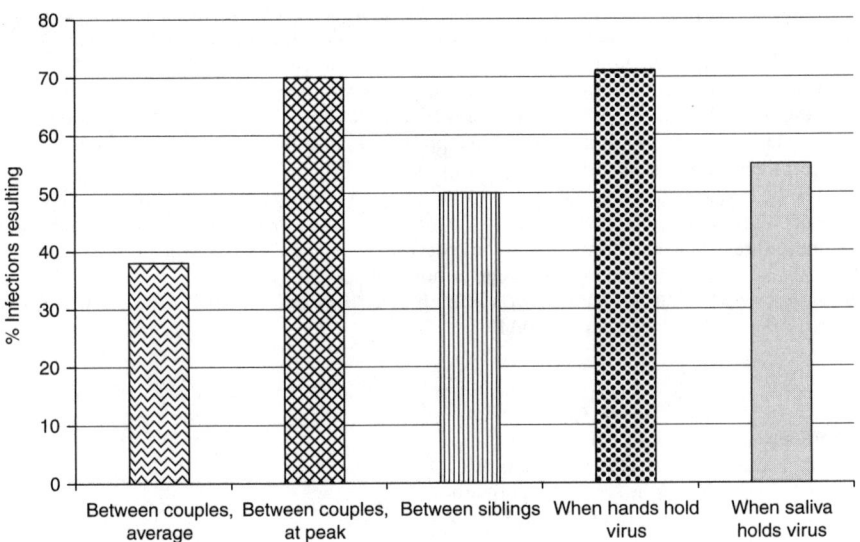

FIGURE 4.21 Rhinovirus transmission rates in households. [*Based on data from D'Alessio et al. (1976)*].

The fact that several infectious diseases may be passing through a population at any given time has been noted in previous studies. It is not uncommon for several virus infections to exist within the same family simultaneously, resulting in members thinking they have the same infection. A study by Dick et al. (1967) identified two strains of rhinovirus, two strains of parainfluenza, one strain of RSV and one strain of influenza passing through a group of families in three buildings over a period of less than 3 months.

4.4 THE EPIDEMIOLOGY OF AIRBORNE DISEASES

Airborne diseases can be subdivided into communicable and noncommunicable diseases. Communicable diseases are contagious between persons. Most viruses are contagious and will spread from person to person in epidemics. Influenza, for example, spreads around the globe approximately every 2 years. Most noncommunicable diseases are either environmental microbes or come from other animals. Inhalation or exposure to noncommunicable agents might cause epidemics when a large number of people are involved but there will be no secondary infections and the epidemic will not be sustained in the absence of a source. Epidemics of noncommunicable diseases are often referred to as outbreaks. Outbreaks of Legionnaires' disease, for example, may affect dozens of people, but once the source is identified and remediated the outbreak ceases.

The susceptibility to diseases plays the same role in the spread of diseases as dry kindling does in fire. One detailed case study performed in an isolated Alaskan community provides insight into the spread of diseases in a susceptible population (Clark et al., 1970). Five individuals in the town of Klawock had attended a conference in Ketchikan where they had been exposed to influenza B. Shortly thereafter the flu spread through 33 of the 37 households in the town. Of the 181 residents, 150 contracted the illness and one person died in the 4-week course of the epidemic. The attack rate was 83 percent, which is much higher than the normal attack rates for influenza, which are about 20 to 40 percent. The long periodic isolation of the community may have protected them from diseases in the interim but had left them highly susceptible to the introduction of a new flu.

The degree to which a population is affected by an outbreak of a noncommunicable disease depends mainly on the virulence of the microbes, the dose received by individuals, and the natural resistance or immunity of the individuals. All of these factors tend to be probabilistic and are describable by a gaussian or normally distributed curve. The basic equation, sometimes called the Soper equation, which describes the statistics of epidemics is as follows (Wilson and Worcester, 1944):

$$C = rIS \qquad (4.1)$$

where C = number of new infections
r = average contact rate, fractional
I = number of infected disseminators
S = number of susceptible individuals

Figure 4.22 shows a graph of this standard epidemiological model in terms of the new cases and the susceptibles. Note that the new cases result in a normally distributed curve.

The production of new infections in a susceptible population is termed a generation. Several generations typically comprise an epidemic, as the infection is constantly retransmitted to new individuals. Once the supply of susceptible individuals has been exhausted the epidemic must end. Normally, the epidemic will end before everyone is infected

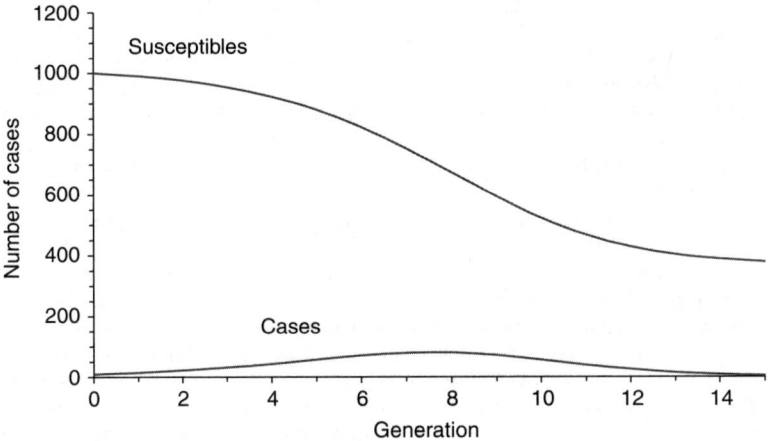

FIGURE 4.22 Standard epidemiological model, nonepidemic conditions.

because there will be insufficient population to propagate the epidemic. That is, the density of susceptibles becomes so low that no new transmissions occur. Mathematically, if $C/I > 1.0$, it denotes a propagating epidemic. If $C/I < 1.0$, the epidemic will rapidly fizzle out.

If the cases of infection are summed up, the result will be total cases. Redefining the cases as new cases to distinguish it from the total cases, Fig. 4.23 is a redrawn version of Fig. 4.22.

In a rearrangement of Eq. (4.1):

$$\frac{C}{I} = rS \qquad (4.2)$$

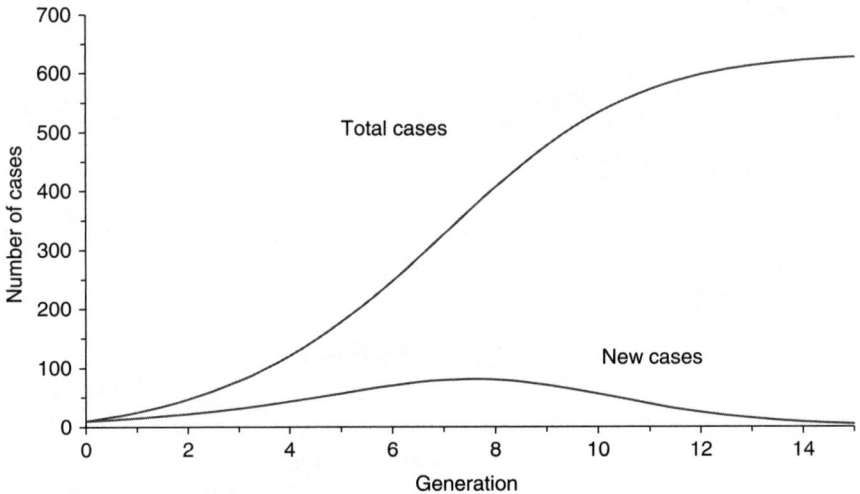

FIGURE 4.23 The standard epidemiological model shown with total cases.

The value rS determines the value of C/I and is called the contagious potential. The rate r can only be determined by epidemiological data, but a value of between 0.1 and 0.2 would be typical and conservative for most respiratory infections. Other computational methods are available to model epidemics but they produce results virtually identical to Fig. 4.23 (Alocilja, 2001; Ackerman, 1984; Boccara and Cheong, 1993; Frauenthal, 1980; Daley and Gani, 1999). Although these various models agree in general form with data from outbreaks, they also have the same basic deficiencies. Figure 4.24 graphs the incidence of acute respiratory diseases in a naval training station during a 10-week training period, based on data from Miller et al. (1948). This study included all respiratory infections even though they may have been caused by different microorganisms. Since all recruits arrived on the same day, and were more or less isolated during the training period, the graph represents a summation of the various infections that were transmitted from person to person. Note that the general form is an excellent representation of the curves in the previous epidemiological models. Not all epidemic data sets lend themselves so easily to bell curves, as will be seen later.

Equation (4.2) does not always provide an accurate picture of the epidemic spread of respiratory infections since they can have complex transmission factors that include inhalation, direct contact, and fomite spread. A proposed modification to the Soper equation has been presented by Riley (1980), who introduced a new definition of the contact rate in terms of the release rate of the agent, the breathing rate of individuals, the exposure time, and the dilution rate of the room air. Using these concepts, the previous equations can be adapted to model infections spread by ventilation systems by defining these terms:

d = quanta of infection produced by each infective individual

Q = volumetric flow of fresh ventilation air (m³/min)

p = volume of air breathed by each susceptible individual (m³/min)

x = building characteristic constant

The ventilation system model equation can then be written as

$$C = \left(\frac{xpd}{Q}\right) IS \quad (4.3)$$

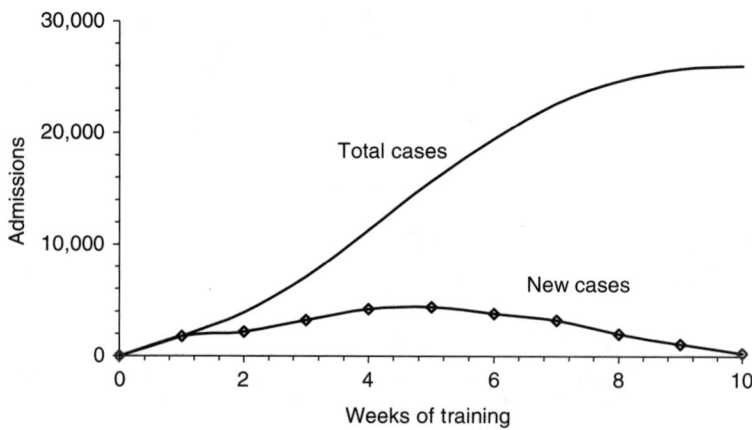

FIGURE 4.24 Acute respiratory infections in recruits at a naval training station. [*Based on data from Miller et al. (1948).*]

Comparison with Eq. (4.1) makes it clear that the contact rate will be

$$r = \frac{xpd}{Q} \quad (4.4)$$

For breathing at rest the value of p is approximately 0.015 m^3/min. The value of d depends on the particular virus or bacteria. For measles the value has been found from epidemiological studies to be about 9.1. The factor x will account for the effect of the air-exchange rate of each building on causing new infections. Figure 4.25 graphs the ventilation model showing the susceptible occupants, total cases, and new cases. It is, of course, virtually identical to the standard epidemiological model.

All buildings act to some degree as incubators, sustaining microbes in the air or on surfaces for periods ranging from minutes to weeks. Buildings are vectors in the epidemic transmission cycle and it should be possible to develop an epidemiological model of multiple buildings in a community. Although the epidemiological information on disease transmission in particular buildings is scant, we can develop a first-order model of an epidemic based on actual data for a large outbreak and effectively average out the building differences. Figure 4.26 shows the data for the Hong Kong outbreak of SARS in 2003, which involved several buildings including apartment buildings and a hospital (HWFB, 2003a, 2003b). The curve shown in the graph is the ventilation model adjusted to match the total number of infections. It has been assumed for the sake of this model that the initial number of susceptibles in the affected buildings was 10,000, and although this number is not certain, any other reasonable value would produce similar results. In this particular outbreak almost 57 percent of occupants of one apartment building contracted the infection and the airborne route was implicated in at least some of the infections.

Having fit the ventilation model to the data, the effective contact rate in Eq. (4.4) proves to be approximately 0.000109, which might allow us to compute the value of x if we knew the infectious quanta and the building airflows. However, it is clear that airborne spread accounts for only part of the transmission of this disease, and it would be necessary to

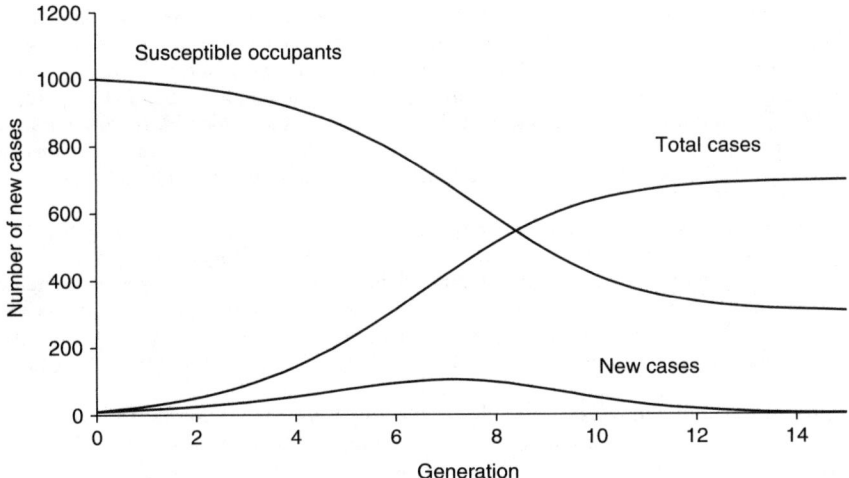

FIGURE 4.25 The ventilation model, nonepidemic conditions.

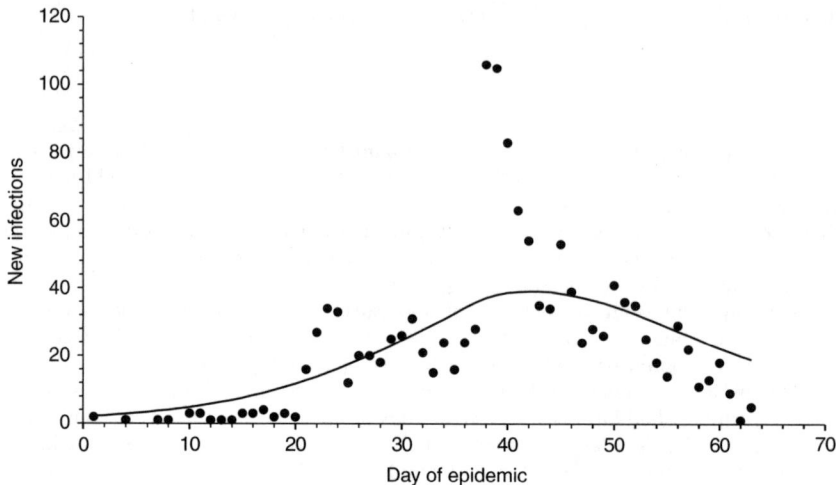

FIGURE 4.26 Ventilation model of SARS epidemic compared to data from 2003 Hong Kong outbreak.

model every building and also to separately model the contact rate from nonairborne spread. Furthermore, the interaction between the airborne infections and the direct contact interactions would need to be defined. Given a sufficient database, such a model could be constructed, but it will be left for future research. A more valuable lesson can be extracted from the crude model developed in Fig. 4.26, and that is an estimate of how many buildings might have to be immunized, or equipped with air treatment systems, in order to prevent the epidemic.

Let us assume that 25 percent of infections in our model SARS epidemic are transmitted by the airborne route, and that 30 percent transmit in the office environment (see Fig. 4.17). Now let us use the ventilation model to test the impact of immunizing some percentage of the office buildings and observe the effect on the course of the epidemic. Figure 4.27 shows the results of this analytical experiment. The uppermost curve in Fig. 4.27 is the same as the curve shown in Fig. 4.26 but it has been normalized on a percentage scale. The bottom curves show the effect of immunizing 25, 50, and 75 percent of the office buildings that are assumed to play a part in the epidemic. It would appear that with 50 percent of the buildings immunized the epidemic is considerably attenuated, and that with 75 percent of the office buildings immunized the epidemic practically ceases to be an epidemic at all.

The previous example shows how sensitive the epidemic is to interruption of the infection transmission process. Just as with vaccination programs, it is not necessary to vaccinate everyone, since herd immunity is acquired at some lower percentage (Jordan and Burrows, 1946). The same effect will occur with immunized buildings and although the SARS epidemic model is simplistic, the potential clearly exists to make great strides against airborne diseases through engineering technology and building science. This example highlights one fact about contagious pathogens—their evolved ability to transmit and cause epidemics is most tenuous and easily interrupted. If the epidemic cycle could be broken, it is entirely possible that engineering buildings to inhibit airborne transmission could drive some contagious airborne pathogens to extinction.

EPIDEMIOLOGY AND DOSIMETRY

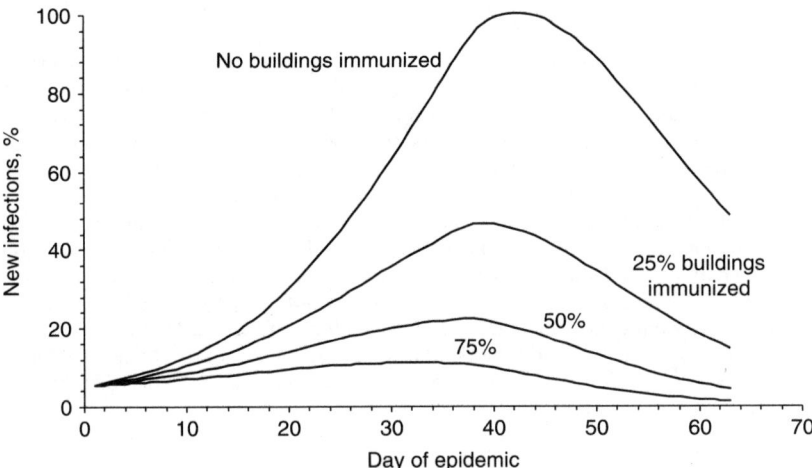

FIGURE 4.27 Effect of immunizing office buildings on the course of model SARS epidemic. Upper curve is a normalized graph of the SARS epidemic model in Fig. 4.26.

4.5 DISEASE PROGRESSION CURVES

The dose response of pathogens operates in a characteristic fashion and generally follows a typical disease progression curve such as that shown in Fig. 4.28. Initially the symptoms may be subclinical but will usually become manifest within a few days, depending on the incubation period. Following the incubation period will be a period of potentially severe symptoms and during which the host is infectious if the disease is contagious. A long period of recovery may follow during which infectiousness may diminish, or else infectiousness may persist even after symptoms become subclinical. This curve applies to many, but not

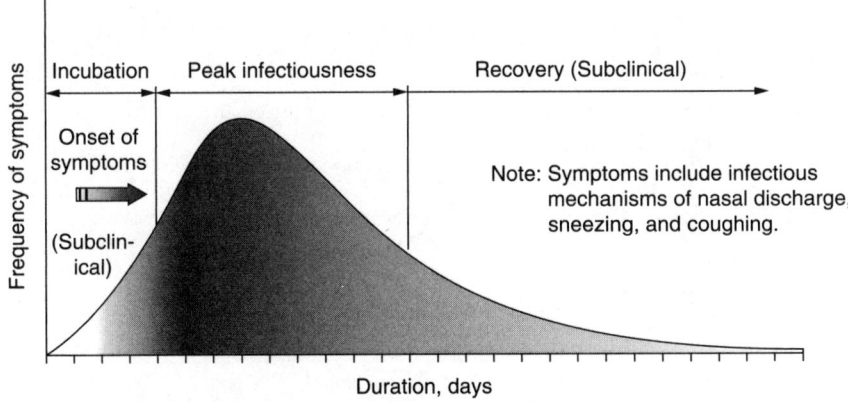

FIGURE 4.28 Generic representation of the progression of a disease.

all, of the pathogens identified in App. A. Tuberculosis, for example, persists indefinitely without treatment (Ryan, 1994).

The basic progression of the disease symptoms in Fig. 4.28 can be modeled in terms of the percentage of the symptoms or severity at the peak of the infection. Diseases generally incubate rapidly and in an exponentially manner, and then they decrease in severity over time as the immune system gets in gear. This progression could be easily modeled as two competing processes—one exponentially increasing function and one overlapping exponentially decreasing function. However, the curve resembles a single lognormal curve, which provides a simpler mathematical model overall and will facilitate prediction of the average duration and the peak infection period. The peak infection period is one of significance to epidemiological studies since it normally defines the period of contagiousness for communicable diseases. Figure 4.29 shows an example of symptom data for the progression of a rhinovirus infection, in which the idealized form of Fig. 4.28 is manifest.

The standard lognormal distribution curve can be adapted to model the progression of any disease with some simple modifications. The equation for a lognormal distribution is as follows:

$$y = \frac{1}{\sigma\sqrt{2\pi}} e^{-0.5 \frac{(x-\mu)^2}{\sigma^2}} \quad (4.5)$$

In Eq. (4.5), x represents the time in days, μ is the mean, and σ is the standard deviation. Deleting the multipliers of the exponent to normalize the equation to a maximum value of 1 (or 100 percent), produces the following:

$$y = e^{-0.5 \frac{(x-\mu)^2}{\sigma^2}} \quad (4.6)$$

The next step in lognormalizing this curve is to replace x with $\ln x$, the natural log of x, and replace the mean with the natural logarithm of the peak. The peak of the illness corresponds approximately to the peak of the lognormal curve, and so this conversion should be easy to grasp. This produces the following intermediate form:

$$y = e^{-0.5 \left(\frac{\ln x - \ln \text{peak}}{\sigma} \right)^2} \quad (4.7)$$

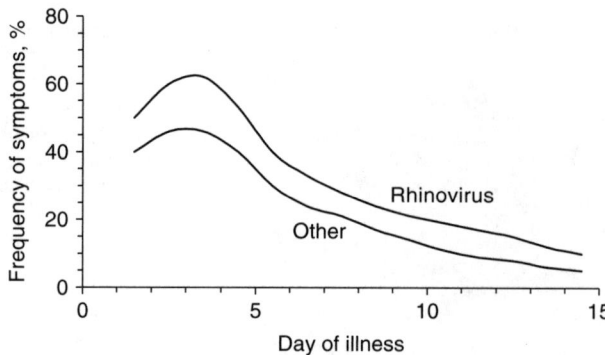

FIGURE 4.29 Progression of common colds as a function of symptom frequency. [*Based on data for nasal discharge from Gwaltney et al. (1967).*]

Next, the standard deviation is defined as the natural logarithm of one-half of the duration D divided by the peak. This normalizes the duration in terms of the peak and the definition, after simplifying, is

$$\sigma = \ln\left(\frac{D}{\text{peak}}\right) \quad (4.8)$$

However, the equation in this form does not decay fast enough and it is necessary to multiply the exponent 0.5 by a factor of 10, which produces the correct type of rapid decay that characterizes most diseases. Of course, this is a simplistic generalized model and individual diseases may not conform precisely, but it serves the purposes of engineering design and helps overcome the limited knowledge of actual disease progression for many of the microbes in the database. Simplifying the terms then produces the following equation for defining the progression of the disease in terms of y, the fractional percent of the symptoms:

$$y = e^{-5\left(\frac{\ln(x/\text{peak})}{\sigma}\right)^2} \quad (4.9)$$

Of some importance in the epidemiology of contagious diseases is the period of infectiousness. In general the infectious period is coincident with apparent symptoms. This may not always be the case since asymptomatic diseases can be transmitted in some cases. But as a general rule of thumb, the infectious period could loosely be defined as any symptoms at about 37 percent or higher, based on the well-known D_{37} criteria used in radiation biology as a cutoff or dose limit (Casarett, 1968; Harm, 1980). The significance of the value of 37 is that it is the inverse of the natural exponent e, and that it has been found through experience in radiation biology to be a safe measure of the maximum dose to use in radiation therapy. Other than that it has no significance but serves as an approximation in lieu of more specific data on infectious diseases. Therefore, for purposes of estimating the infectious period, we can define it as the time when any value of y from Eq. (4.9) is greater than or equal to 37 percent:

Infectious period: whenever

$$y \geq 37\% \quad (4.10)$$

To determine the infectious period based on the model, we can invert Eq. (4.9) as follows:

$$x = \text{peak} \cdot e^{-(\sigma/\sqrt{5})} \quad (4.11)$$

Equation (4.11) will produce two possible values since the square root in the exponent could be positive or negative. The smaller number, when rounded off, will be the first day of infectiousness, while the larger will be the last day of infectiousness. These days, the first and last days of infectiousness, are defined as D_{37} and D'_{37}, respectively. Furthermore, the period between these days can be defined as the *mean disease period* (MDP) for noncontagious diseases. This term will be synonymous with *mean infectious period* (MIP) for contagious diseases. Equation (4.11) can be written as the following two equations by defining the first with a negative exponent:

$$D_{37} = \text{peak} \cdot e^{-(\sigma/\sqrt{5})} \quad (4.12)$$

$$D'_{37} = \text{peak} \cdot e^{(\sigma/\sqrt{5})} \quad (4.13)$$

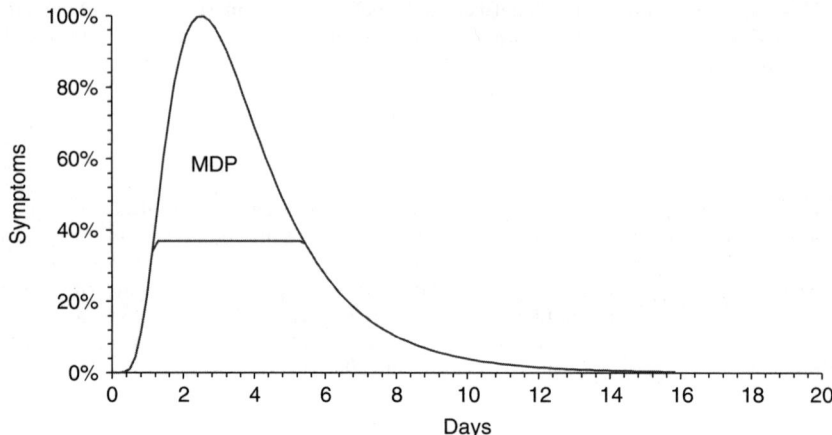

FIGURE 4.30 Mean disease progression curve for *Bacillus anthracis*.

Equations (4.12) and (4.13) may be more useful than merely computing the mean infectious period for contagious diseases. They may also be used to estimate the mean time until treatment measures could still be effective. That is, some treatments, like vaccines or antibiotics, may only be effective against a disease if they are administered in the early stages. Equation (4.11) provides an estimate of how advanced the disease is, and may be a useful indicator of the limits within which treatment may be effective. The parameters necessary to define the progression of most microbial agents are provided in App. A where they are known. The peak for each infection can be taken as the mean of the peak or as the mean of the incubation period given in App. A. The duration can be estimated likewise.

Figure 4.30 shows an example of the hypothetical disease progression for anthrax. The estimated MDP is identified as the cutoff line above 37 percent of the disease symptoms. The disease may be asymptomatic for the first day or two, depending on the dose inhaled. After the D_{37} point, estimated to be 1.2 days on the average, the symptoms progress like those of a cold or flu, but will exceed them in severity within a day or two.

4.6 DOSIMETRY OF AIRBORNE DISEASE

Dosimetry is the science of determining infectious or lethal doses of microorganisms. It is important in the aerobiological engineering of any system or building to be able to estimate the infections or fatalities that may occur from exposure to disease agents. For contagious diseases there is the additional consideration of how secondary infections might spread.

The dose of any disease agent depends on the route of exposure. There are three possible routes of exposure—inhalation, skin exposure, and ingestion. The two main factors relevant to the study of dosimetry are defined as follows:

ID$_{50}$ (Mean Infectious Dose):   The dose or number of microorganisms that will cause infections in 50 percent of an exposed population. Units are in terms of the number of colony forming units per cubic meter (cfu/m^3). Similarly, the proper term used to describe viable viruses in culture is pfu or plaque forming units, but for simplicity the term cfu is often used in place of pfu in this text.

EPIDEMIOLOGY AND DOSIMETRY   91

LD$_{50}$ (Mean Lethal Dose): The dose or number of microorganisms that will cause fatalities in 50 percent of an exposed population. Applies to both microorganisms and toxins. For microorganisms, the unit is cfu/m^3. For toxins, the unit is mg/kg, which represents an absorbed, inhaled, injected, or ingested dose per bodyweight.

The dose from exposure to airborne pathogens applies to the inhaled dose. The dose from foodborne pathogens applies to the ingested dose. An infectious dose can cause infections in individuals but not necessarily any fatalities. The lethal dose will always be higher than the infectious dose, although not necessarily by much. Many of the infectious doses and lethal doses for microorganisms are not known with certainty, or are based on animal studies. Sometimes the doses that will produce 50 percent infections or fatalities are defined by a range, which can be broad.

When exposed to a range of doses, the number of infections produced in a population will typically result in a normal curve or bell curve such as shown in Fig. 4.31. Some members will acquire infections at very low doses while others require a large dose to become infected. This is because some people may be susceptible, such as the elderly or the ill, while others may be in a healthy state or have strong immune systems. In Fig. 4.31 it can be observed that the ID$_{50}$ is 10 and that it produces approximately 10 percent new infections. If the curve were integrated to the ID$_{50}$ point of 10, it would produce a value for total infections of 50 percent. If the entire curve were integrated, it would produce 100 percent total infections.

If the concentration of airborne microorganisms is approximately constant, then the acquired dose is a linear function of exposure time. Defining E_t = exposure time and C_a = airborne concentration, we can write this equation as follows:

$$\text{Dose} = E_t C_a \qquad (4.14)$$

If the airborne concentration is such that the LD$_{50}$ is achieved at 4 hours, then 50 percent of the exposed population will have been infected by that time. Figure 4.32 shows a plot of both new infections and total infections over an 8 hours exposure period. In this case 100 percent infections are reached after 8 hours of exposure. It can be observed that the graph of infections is virtually identical to the epidemiological curve presented in Fig. 4.31.

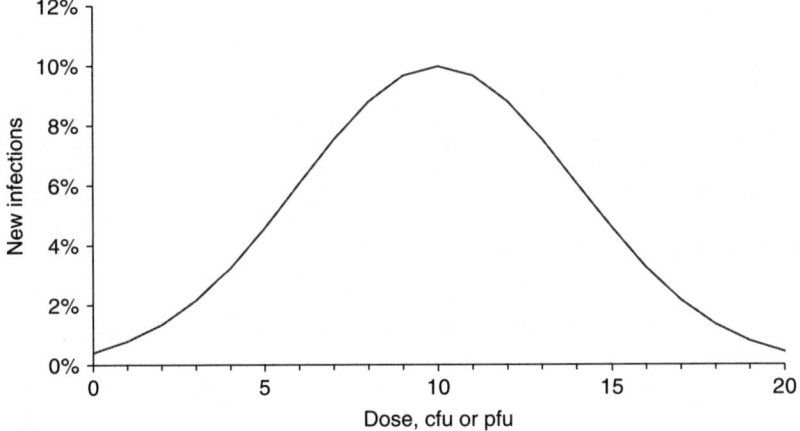

FIGURE 4.31 Normal distribution of new infections in a population receiving the same dose. The infectious dose has an ID$_{50}$ of 10.

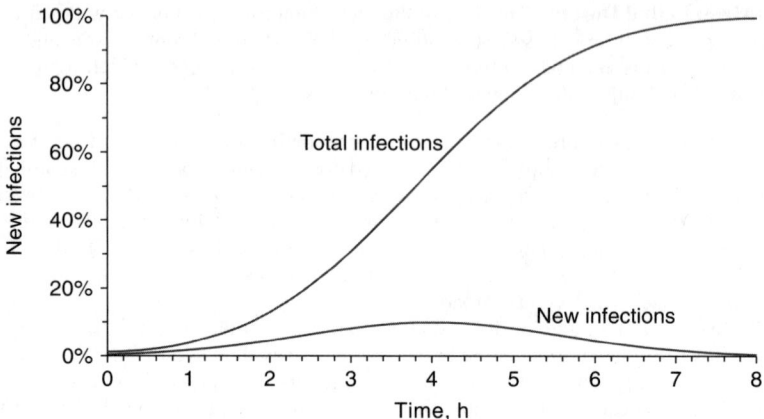

FIGURE 4.32 New infections and total infections over time under exposure to a constant concentration of airborne pathogens.

The total infections in Fig. 4.32 represent the sum of the new infections. The mathematical relation for predicting the total infections can be developed by beginning with the statistical definition of a normal bell curve. If y represents the number of new cases, the normal distribution curve is the same as shown previously in Eq. (4.5):

$$y = \frac{1}{\sigma\sqrt{2\pi}} e^{-0.5\frac{(x-\mu)^2}{\sigma^2}} \tag{4.15}$$

In Eq. (4.15), x represents the dose, with units depending on the units for the ID_{50} value, normally cfu for bacteria and fungi and pfu for viruses. The mean μ represents the ID_{50}, and the equation can be rewritten as follows:

$$y = \frac{1}{\sigma\sqrt{2\pi}} e^{-0.5\left(\frac{x-ID_{50}}{\sigma}\right)^2} \tag{4.16}$$

It is now necessary to define the standard deviation in order to be able to use Eq. (4.16) to predict infections. The standard deviation may not be the same for all pathogens, yet it would be convenient to have some representative value that would produce reasonable results for all pathogens. The standard deviation could be any value between a small fraction of the mean and some multiple thereof. If it were too small, all the infections would occur in a narrow range of time, and this would be unnatural. If it were too large, the same number of infections would occur at time zero as at the peak, and this also would be contrary to the observations of disease epidemiology.

Comparison with epidemiological data suggests the standard deviation must be between about 0.25 and 0.5 of the mean to provide reasonable and elastic results. Some data are available from published sources that have tabulated dose-response data for several viruses and bacteria (Haas, 1983; Haas et al., 1999). The standard deviations ranged from 0.028 of the mean to as high as two times the mean. Some of the data sets were both limited and erratic, and if the extremes are discounted, there are some reasonable values, such as 0.527

EPIDEMIOLOGY AND DOSIMETRY   93

for *Shigella dysenteriae* and 0.431 for echovirus. Based on this review, the above discussion, and some trial and error, the value of 0.5 is adopted as representative in lieu of any future data for specific species. Given a standard deviation equal to 0.5 of the mean or ID_{50} value, Eq. (4.16) is rewritten as follows:

$$y = \frac{2}{ID_{50}\sqrt{2\pi}} e^{-2\left(\frac{x-ID_{50}}{ID_{50}}\right)^2} \quad (4.17)$$

Equation (4.17) can be used to predict either new infections or new fatalities due to exposure to any dose of airborne pathogens. In order to predict the total infections, however, it is necessary to sum the results of Eq. (4.17). It could be integrated mathematically, but the resulting equation is not simple to resolve. Numerical integration is also possible, but it would be far more convenient to have a closed form of the equation that will directly predict the total infections for any given dose. The simplest form of equation that will satisfy this purpose is known as a Gompertz curve (Whiting, 1993; Boyce and DiPrima, 1997). It has the following general form, in which y is a function of x, and the parameters a and b are arbitrary constants:

$$y = a^{b^x} \quad (4.18)$$

Equation (4.18) is capable of producing a curve almost identical to that shown in Fig. 4.32 for total infections. It remains only to find the appropriate constants a and b to match the integrated form of Eq. (4.18). The constants a and b must be fractional to produce the Gompertz curve. Equation (4.18) must also be normalized such that it equals 0.5 (or 50 percent) at the normalized ID_{50} value. Therefore, the constant a must be 0.5, and the exponent x must be normalized around the ID_{50}. Inserting these parameters, the following equation is obtained:

$$y = 0.5^{b^{\left(\frac{x-ID_{50}}{ID_{50}}\right)}} \quad (4.19)$$

In order to establish a value for the constant b, it is necessary to consider the same criteria for an acceptable dose-response curve as discussed previously. That is, the infections must be normally distributed over the range of doses such that low dose infections are nonzero and the slope in the midrange is not severe. This can be accomplished by fitting Eq. (4.19) to the graphs previously produced, since they were based on these criteria. Figure 4.33 shows a graph of Eq. (4.19) plotted against the dose-response curve from Fig. 4.32, and in which the constant b has been adjusted to a value of 0.1. Although the curve is not a perfect fit, the range of poorer fit is an area of uncertainty. The only area which is reasonably certain is the range near the ID_{50} value, and for this the curve is an excellent match.

Having found an appropriate constant the final equation for prediction of total infections is as follows:

$$y = 0.5^{0.1\left(\frac{x-ID_{50}}{ID_{50}}\right)} \quad (4.20)$$

Equation (4.20) is not really intended to supplant the classic epidemiological model but it serves as a convenient closed-form means of estimating infections. Although it may have some inherent inaccuracy, it compares reasonably well with dose-response data on inhalation anthrax from other sources (Haas, 2002; Druett et al., 1953). The actual application of

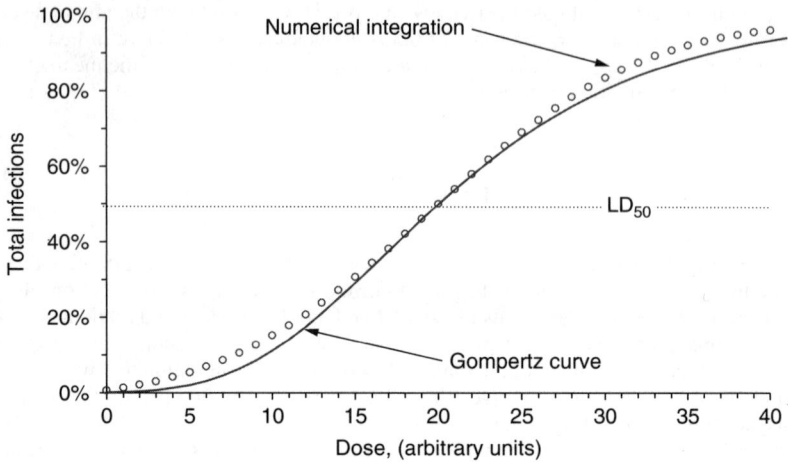

FIGURE 4.33 Comparison of fitted Gompertz curve with total infections computed by numerical integration.

this equation will become clear in later applications in which buildings are analyzed to determine the effectiveness of air cleaning equipment. The ID_{50} and LD_{50} values given in App. A represent the best estimates in the available literature, although many are based on animal experiments and actual human doses are often not known with certainty.

4.7 MICROBIAL AGENT INGESTION DOSIMETRY

Doses for ingestion apply mainly to toxins and pathogens. No equations or charts are necessary for computing dosages of ingested agents. The information provided in App. A for pathogens can be used directly for estimating infections or fatalities using Eq. (4.20).

Ingestion of biological agents is primarily a problem for foodborne or waterborne pathogens, and the references can be consulted for additional detailed information on their epidemiological characteristics (Ray, 1996; Hunter, 1997). The stomach has natural protection against many pathogens that are infectious by the inhalation route. Most respiratory viruses and bacteria are unlikely to cause any infections as a result of being swallowed, since stomach acids will destroy them, but to reach the stomach they may have to pass through the mouth and there is a potential for any agent present in the mouth to reach the nasal mucosa or the lungs. The ingestion of toxins is a hazard that is treated in the following section.

4.8 TOXIN DOSIMETRY

Dose-response curves operate the same for toxins as for microbes. Toxins can be inhaled, ingested, or absorbed through the skin, but their toxicities are invariably specified as LD_{50} in units of mg/kg of bodyweight. Equation (4.20) can be used to determine the dose-response curve for any microbial toxins. These curves have been generated for all toxins of

TABLE 4.1 Lethal Doses for Microbial Toxins

Toxin	Type	Lethal dose, LD_{50} mg/kg bodyweight
Aflatoxin	Mycotoxin	0.3
a-latrotoxin	Neurotoxin	0.01
Anatoxin A	Neurotoxin	0.05
b-bungarotoxin	Neurotoxin	0.014
Botulinum	Neurotoxin	0.000001
Citrinin	Mycotoxin	35
C. perfringens toxin	Enterotoxin	0.0003
Diptheria toxin	Exotoxin	0.0001
Microcystin	Peptide	0.05
Palytoxin	Neurotoxin	0.00015
Saxitoxin	Neurotoxin	0.002
Shiga toxin	Exotoxin	0.000002
Staphylococcal enterotoxin A	Enterotoxin	0.00005
Staphylococcal enterotoxin B	Enterotoxin	0.027
T-2 toxin	Mycotoxin	1.21
Tetanus toxin	Neurotoxin	0.000002
Tetrodoxin (TTX)	Neurotoxin	0.008
Textilotoxin	Neurotoxin	0.0006

interest for which dose information is available. Table 4.1 provides a summary of all toxins for which dose information is available (Kowalski, 2003).

REFERENCES

Ackerman, E., Elveback, L. R., and Fox, J. P. (1984). *Simulation of Infectious Disease Epidemics.* Charles C. Thomas, Springfield, IL.

Alocilja, E. (2001). Chapter 4: Growth and feedback in population biology. Michigan State University. 79–108. http://www.egr.msu.edu/classes/be230/chapter4.pdf.

Boccara, N., and Cheong, K. (1993). "Critical behavior of a probabilistic automata network SIS model for the spread of an infectious disease in a population of moving individuals." *J Phys A-Math Gen* 26(5):3707–3717.

Boyce, W. E., and DiPrima, R. C. (1997). *Elementary Differential Equations and Boundary Value Problems.* John Wiley & Sons, New York.

Buckland, F. E., Bynoe, M. L., and Tyrrell, D. A. J. (1965). "Experiments on the spread of colds." *J Hyg* 63(3):327–343.

Casarett, A. P. (1968). *Radiation Biology.* Prentice-Hall, Englewood, CO.

Clark, P. S., Feltz, E. T., List-Young, B., Ritter, D. G., and Noble, G. R. (1970). "An Influenza B epidemic within a remote Alaska community." *JAMA* 214(3):507–512.

Couch, R. B., Cate, T. R., Douglas, R. G., Gerone, P. J., and Knight, V. (1966). "Effect of route of inoculation on experimental respiratory viral disease in volunteers and evidence for airborne transmission." *Bact Rev* 30:517–529.

D'Alessio, D. J., Peterson, J. A., Dick, C. R., and Dick, E. C. (1976). "Transmission of experimental rhinovirus colds in volunteer married couples." *J Inf Dis* 133(1):28–36.

Daley, D. J., and Gani, J. (1999). *Epidemic Modelling: An Introduction.* Cambridge University Press, New York.

Dick, E. C., Blumer, C. R., and Evans, A. S. (1967). "Epidemiology of infections with rhinovirus types 43 and 55 in a group of University of Wisconsin student families." *Am J Epid* 86:386–400.

Druett, H. A., Henderson, D. W., Packman, L., and Peacock, S. (1953). "The influence of particle size on respiratory infection with anthrax spores." *J Hyg* 51:359.

Druett, H. A., Robinson, J. M., Henderson, D. W., Packman, L., and Peacock, S. (1956). "Studies on respiratory infection, II & III." *J Hyg* 54:37–57.

Duguid, J. P. (1945). "The size and the duration of air-carriage of respiratory droplets and droplet-nuclei." *J Hyg* 54:471–479.

Fisk, W., and Rosenfeld, A. (1997). "Improved productivity and health from better indoor environments." *Center for Building Science Newsletter* 15:5.

Frauenthal, J. C. (1980). *Mathematical Modeling in Epidemiology.* Springer-Verlag, New York.

Gordon, J. E., and Ingalls, T. H. (1957). "Preventive medicine and epidemiology." *Prog Med Sci* 233:334–357.

Guinan, M., McGuckin, M., and Ali, Y. (2002). "The effect of a comprehensive handwashing program on absenteeism in elementary schools." *Am J Infect Control* 30(4):217–220.

Gwaltney, J. M., Hendley, J. O., Simon, G., and Jordan, W. S. (1967). "Rhinovirus infections in an industrial population." *JAMA* 202(6):158–164.

Gwaltney, J. M., Moskalski, P. B., and Hendley, J. O. (1978). "Hand-to-hand transmission of rhinovirus colds." *Ann Int Med* 88:463–467.

Haas, C. N. (1983). "Estimation of risk due to low doses of microorganisms." *Am J Epidem* 118(4):573–582.

Haas, C. N., Rose, J. B., and Gerba, C. P. (1999). *Quantitative Microbial Risk Assessment.* John Wiley & Sons, New York.

Haas, C. N. (2002). "On the risk of mortality to primates exposed to anthrax spores." *Risk Anal* 22(6):1035–1036.

Hall, C. B., Geiman, J. M., Biggar, R., Kotok, D. I., Hogan, P. M., and Douglas, G. R. (1976). "Respiratory syncytial virus infections within families." *N Engl J Med* 294(8):414–419.

Harm, W. (1980). *Biological Effects of Ultraviolet Radiation.* Cambridge University Press, New York.

Hendley, J. O., Gwaltney, J. M., and Jordan, W. S. (1969). "Rhinovirus infections in an industrial population. IV. Infections within families of employees during two fall peaks of respiratory illness." *Am J Epid* 89(2):184–196.

Hunter, P. R. (1997). *Waterborne Disease: Epidemiology and Ecology.* John Wiley & Sons, Chichester, UK.

Hurst, C. J. (1996). *Modeling Disease Transmission and Its Prevention by Disinfection.* Cambridge University Press, Cambridge, MA.

HWFB (2003a). "SARS Bulletin (17 April 2003)." Health, Welfare, and Food Bureau, Government of the Hong Kong Special Administrative Region. Hong Kong. http://www.emergency-management.net/pdf/sars_bulletin/bulletin0417.pdf.

HWFB (2003b). "SARS Bulletin (24 April 2003)." Health, Welfare, and Food Bureau, Government of the Hong Kong Special Administrative Region. Hong Kong. http://www.emergency-management.net/pdf/sars_bulletin/bulletin0417.pdf.

Jordan, E. O., and Burrows, W. (1946). *Textbook of Bacteriology.* W. B. Saunders, Philadelphia, PA.

Kowalski, W. J. (2003). *Immune Building Systems Technology.* McGraw-Hill, New York.

Langmuir, A. D. (1961). "Epidemiology of airborne infection." *Bacteriol Rev* 25:173–181.

Lidwell, O. M., and Williams, R. E. O. (1961). "The epidemiology of the common cold." *J Hyg* 59:309–334.

Miller, W. R., Jarrett, E. T., Willmon, T. L., Hollaender, A., Brown, E. W., Lewandowski, T., and Stone, R. S. (1948). "Evaluation of ultra-violet radiation and dust control measures in control of respiratory disease at a naval training center." *J Infect Dis* 82:86–100.

Nardell, E. A. (1990). "Dodging droplet nuclei." *Am Rev Respir Dis* 142:501–503.

Nardell, E. A., Keegan, J., Cheney, S. A., and Etkind, S. C. (1991). "Airborne infection: Theoretical limits of protection achievable by building ventilation." *Am Rev Respir Dis* 144:302–306.

Ray, B. (1996). *Fundamental Food Microbiology.* CRC Press, Boca Raton, FL.

Reed, S. E. (1975). "An investigation of the possible transmission of rhinovirus colds through indirect contact." *J Hyg* 75:249–258.

Remington, P. L., Hall, W. N., Davis, I. H., Herald, A., and Gunn, R. A. (1985). "Airborne transmission of measles in a physician's office." *JAMA* 253(11):1574–1577.

Riley, R. L. (1980). "The role of ventilation in the spread of measles in an elementary school." *Airborne Contagion, Ann N Y Acad Sci* 353:25–34.

Robinson, R. Q., Hoshiwara, I., Schaeffer, M., Gorrie, R. H., and Kaye, H. S. (1960). "A survey of respiratory illnesses in a population." *Am J Hyg* 75:18–27.

Rubbo, S. D., and Benjamin, M. (1953). "Transmission of haemolytic streptococci." *J Hyg* 51:278–292.

Ryan, K. J. (1994). *Sherris Medical Microbiology.* Appleton & Lange. Norwalk, CT.

Scott, E., and Bloomfield, S. F. (1990). "The survival and transfer of microbial contamination via cloths, hands, and utensils." *J Appl Bacteriol* 68(3):271–278.

Smyth, W. (1987). *Respiratory and Infectious Disease.* Facts on File Publications, New York.

Thompson, L. R. (1962). *Microbiology and epidemiology.* W. B. Saunders, Philadelphia, PA.

Thurman, R., and Gerba, C. (1989). "The molecular mechanisms of copper and silver ion disinfection of bacteria and viruses." *CRC Crit Rev Environ Control* 18:295–315.

Tolley, G., Kenkel, D., and Fabian, R. (1994). *Valuing health for policy.* University of Chicago Press, Chicago, IL.

Wells, W. F. (1955). *Airborne Contagion and Air Hygiene.* Harvard University Press, Cambridge, MA.

Wheeler, A. E. (1993). Better filtration: A prescription for healthier buildings. *IAQ 93: Operating and Maintaining Buildings for Health, Comfort, and Productivity,* K. Y. Teichman, ed., American Society of Heating, Refrigerating and Air-Conditioning Engineers, Atlanta, GA. 201–207.

Whiting, R. C. (1993). "Modeling bacterial survival in unfavorable environments." *J Ind Microbiol* 12:240–246.

Williams, R. E. O. (1960). "Intramural spread of bacteria and viruses in human populations." *Annu Rev Microbiol* 14:43–64.

Wilson, E. B., and Worcester, J. (1944). "The law of mass action in epidemiology." *Proc Nat Acad Sci* 31:24–34.

CHAPTER 5

AEROBIOLOGY OF THE OUTDOORS

5.1 INTRODUCTION

Respiratory disease transmission is generally regarded as an indoor phenomenon but there are a number of pathogenic microbes that can occur and cause diseases in the outdoor air. Pollen and fungal allergens are ubiquitous in the outdoor air, but due to the fact that people spend over 90 percent of their time indoors it is difficult to estimate what proportion of allergies are induced in the outdoors. Complicating the whole question is the fact that many homes and buildings use unfiltered outdoor air and so the microbial composition of indoor air is largely similar to outdoor air. Although there is little if anything that can be done about microbes in the outdoor air, a detailed review of outdoor aerobiology can form a basis for understanding indoor aerobiology and the kinds of hazards that may come from outdoor air. This chapter addresses the pathogens and allergens that can occur in outdoor air, the diseases that can be transmitted in outdoor air, and the environmental fate of airborne microorganisms. Some examples of diseases that have transmitted in the outdoor air are provided, including some airborne diseases transmitted among animals.

5.2 OUTDOOR PATHOGENS AND ALLERGENS

The primary bioaerosols encountered in the outdoor air are pollen, fungal spores, and environmental bacteria. Environmental bacteria can include bacteria and bacterial spores. Actinomycetes are bacterial spores most often encountered in agricultural settings. Apart from pollen, fungal allergens are common in the outdoor air. Some of the fungi are potentially causative agents of infections and therefore considered pathogenic. Other allergens like dander, dust mites, and cockroach allergen are uncommon in outdoor air. Appendix A contains detailed information on all known allergenic fungi, which are virtually all environmental and may hail from soil, plants, or water. Some fungi come primarily from animal sources.

Airborne viruses and bacteria that are known to exist environmentally or that have been found in outdoor sources are summarized in Table 5.1 (Bitton, 1980; Hurst, 1991; Jenkins, 1991; Austin, 1991; McDade et al., 1964; Wright et al., 1969; Henis, 1987; Mitscherlich and Marth, 1984).

The microbes in Table 5.1 are mostly pathogenic, or disease causing. The four viruses shown have been found in sewage or environmental waters but have apparently never been isolated in outdoor air (Abbaszadegan et al., 2003; Hunter, 1997; Austin, 1991). Many of the bacteria that can exist environmentally often come from animal or agricultural sources, or from sewage. These microorganisms may have originally existed naturally at some level, but they proliferate today due to agricultural technology. Many bacterial spores, especially actinomycetes, primarily hail from agricultural sources and are, to a large degree, man-made occupational hazards.

The primary outdoor allergens include pollen, fern spores, soy dust, and fungal spores or hyphae, while algae and arthropods also contribute small numbers of allergenic particles to outdoor air (Burge and Rogers, 2000). Particles are released from sources into the air by wind, rain, disturbance by mechanical or human means, or natural discharge mechanisms.

Many studies have been performed on the fungal composition of outdoor air. Results can vary geographically but in general there are several common species, like *Cladosporium* and *Penicillium*, and the rest are often a varied mix. Figure 5.1 shows the results of a study by Richards (1954) of the fungi in the outdoor air.

In the largest study of airborne indoor and outdoor fungal species and concentrations conducted with a standardized protocol to date Shelton et al. (2002) examined over 12,000 fungal air samples indoors and outdoors from 1717 buildings located across the United States from 1996 to 1998 (see Fig. 25.2 in Chap. 25). The culturable airborne fungal concentrations in indoor air were mainly lower than those in outdoor air. The fungal levels were highest in the fall and summer and lowest in the winter and spring. Geographically, the highest fungal levels were found in the Southwest, Far West, and Southeast. The fungi most commonly found in both indoor and outdoor air in all seasons are *Cladosporium, Penicillium, Aspergillus*, and nonsporulating fungi. *Stachybotrys chartarum*, one of the most hazardous fungi, was identified in the indoor air in 6 percent of the buildings studied and in the outdoor air of 1 percent of the samples. Figure 5.2 illustrates the annual median concentrations of airborne fungi by region. Totals are shown to exemplify the kinds of typical outdoor levels that could be expected. It can be observed that although there are some regional variations, the general trends are the same in all parts of the country, and the most predominant species of fungi show up heavily in all these areas.

Table 5.2 summarizes the median seasonal outdoor airborne concentrations of the most common allergenic fungi, based on data from Shelton et al. (2002). Totals are shown to illustrate the new seasonal variations.

Outdoor levels of fungi vary with the seasons. Figure 5.3 shows the seasonal variations that occur with overalls airborne levels of all species, based on data for Sardinia, Italy (Palmas and Consentino, 1990).

The most common pollen, in terms of allergic reactions in North America, have been listed in Table 2.2 in Chap. 2. Other allergens that may occur in outdoor air, like algae and animal dander, are also listed in Chap. 2.

Outdoor studies of pollen and fungal spores indicate significant seasonal variations. In an Illinois study of outdoor air, levels of pollen were measured along with the fungal spores *Alternaria, Aspergillus/Penicillium, Cladosporium, Curvularia, Drechslera, Epicoccum,* and *Rhodotorula* (Chung et al., 1996). The pollen types counted were birch, maple, oak, ragweed, and grass, and the range of total pollen concentrations was 8.4 to 4105 grains/m^3. Birch and oak pollens were found from early April to early June, maple pollens from late March to early May, grass pollens from late April to early September, and ragweed pollens from late July to late October, periods that are similar with those throughout America. The

TABLE 5.1 Viruses and Bacteria That Exist or Survive Environmentally

Pathogen	Group	Natural source	Survival outside host
Adenovirus	Virus	Sewage	Survives in sewage for weeks
Coxsackievirus	Virus	Feces, sewage	Survives in stool for weeks
Norwalk virus	Virus	Environmental waters	Survives in water
Vaccinia virus	Virus	Agricultural, cattle	Limited
Acinetobacter	Bacteria	Environmental, soil, sewage	Survives outdoors
Actinomyces israelii	Bacteria	Cattle	Survives outdoors
Aeromonas	Bacteria	Environmental, water, soil	Survives outdoors
Alcaligenes	Bacteria	Soil, water	Survives outdoors
Brucella	Bacteria	Goats, cattle, swine, dogs, sheep, caribou, coyotes, camels	32–135 days
Burkholderia cepacia	Bacteria	Environmental	—
Burkholderia mallei	Bacteria	Environmental, horses, mules	30 days in water
Burkholderia pseudomallei	Bacteria	Environmental, rodents, soil, water	Years in soil and water
Chlamydophila psittaci	Bacteria	Birds, fowl	2–20 days
Clostridium botulinum	Bacteria	Environmental	Indefinitely in soil, water
Clostridium perfringens	Bacteria	Environmental, animals, soil	Years
Enterobacter cloacae	Bacteria	Environmental, soil, and water	7–21 days in food
Enterococcus faecalis	Bacteria	Feces	—
Francisella tularensis	Bacteria	Wild animals, natural waters	31–133 days
Klebsiella pneumoniae	Bacteria	Environmental, soil	4 hours to several days
Legionella pneumophila	Bacteria	Environmental	Months in water
Mycobacterium avium	Bacteria	Environmental, water, dust, plants	Survives outdoors
Mycobacterium kansasii	Bacteria	Water, cattle, swine	Survives outdoors
Pseudomonas aeruginosa	Bacteria	Environmental, sewage	Survives outdoors
Serratia marcescens	Bacteria	Environmental	35 days or more
Staphylococcus aureus	Bacteria	Sewage	7–60 days
Staphylococcus epidermis	Bacteria	Sewage	—
Yersinia pestis	Bacteria	Rodents, wild animals.	Limited
Coxiella burnetii	Bacteria	Cattle, sheep, goats	Years
Bacillus anthracis	Bacteria	Cattle, sheep, other animals, soil	Years
Micromonospora faeni	Bacteria	Agricultural, moldy hay	Survives outdoors
Nocardia asteroides	Bacteria	Environmental, soils, sewage	Indefinitely in soil, water
Nocardia brasiliensis	Bacteria	Environmental, soils, sewage	Survives outside host
Saccharopolyspora rectivirgula	Bacteria	Agricultural	Survives outdoors
Thermoactinomyces sacchari	Bacteria	Agricultural, bagasse	Survives outdoors
Thermoactinomyces vulgaris	Bacteria	Agricultural	Survives outdoors
Thermomonospora viridis	Bacteria	Agricultural	Survives outdoors

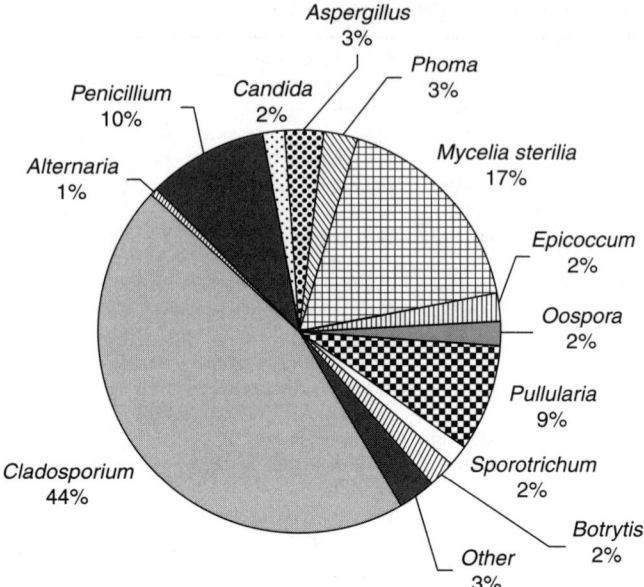

FIGURE 5.1 Outdoor fungi as measured by Richards (1954) in the United States.

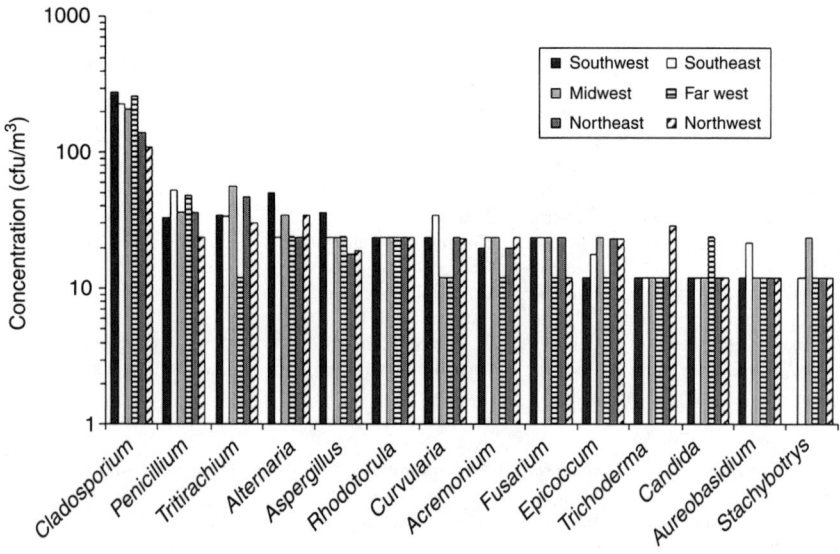

FIGURE 5.2 Outdoor levels of the most common airborne fungal allergens. [*Based on data from Shelton et al. (2002).*]

AEROBIOLOGY OF THE OUTDOORS   103

TABLE 5.2 Outdoor Concentrations of Allergenic Fungi (cfu/m^3)

Fungal spore	Fall	Spring	Summer	Winter	Average
Cladosporium	290	130	290	94	201
Penicillium	71	24	60	35	48
Tritirachium	47	12	47	24	33
Curvularia	24	30	35	35	31
Rhodotorula	35	24	24	24	27
Alternaria	36	12	25	13	22
Aspergillus	24	18	24	18	21
Fusarium	24	18	24	18	21
Acremonium	30	12	24	12	20
Epicoccum	24	18	17	12	18
Stachybotrys	24	12	12	8	14
Aureobasidium	17	12	12	12	13
Candida	12	12	12	12	12
Trichoderma	12	12	12	12	12
Total	670	346	618	329	491

range of total fungal spore concentrations was 27 to 41,901 spores/m^3. Total spore concentrations increased from April to September and began to decrease in October. Spores of *Cladosporium* were the most common with a median concentration 1451 spores/m^3, followed by *Alternaria* at 98 spores/m^3. Table 5.3 lists the results of various studies on bioaerosol levels in outdoor air.

Fungal spores and bacteria have been found at altitudes near 10,250 ft (3127 m) although they tend to predominate at lower altitudes (Fulton et al., 1966). At or above 1600 ft approximately 90 percent of the organisms found were bacteria, and these were primarily *Bacillus* and *Micrococcus* species. Predominant fungi genera were *Alternaria, Hormodendron, Penicillium,* and *Aspergillus*, with *Rhizopus, Helminthosporium, Trichoderma, Fusarium,* and *Scopulariopsis* also occasionally found. Figure 5.4 shows the results of this study in which the measured concentrations were averaged over a 30-hour period.

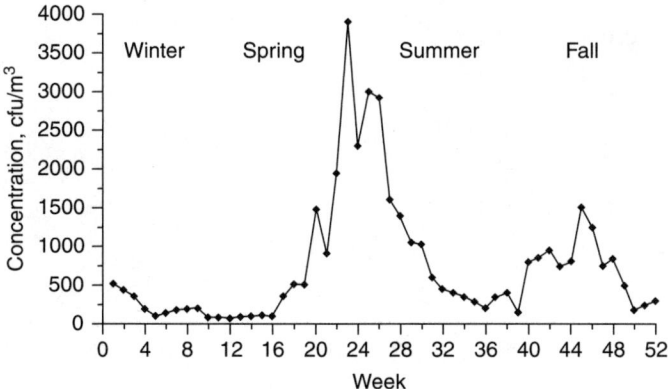

FIGURE 5.3 Seasonal variations of airborne fungi outdoors in an urban environment. [*Based on data from Palmas and Consentino (1990).*]

104 BACKGROUND AND HISTORY

TABLE 5.3 Typical Bioaerosal Levels in Outdoor Air

Bioaerosol	Concentration	Location	Season	Reference
Bacteria cfu/m^3	327	Agricultural area	—	Mullins, 2001
	81	Residential area	—	Mullins, 2001
	146	Business area	—	Mullins, 2001
	880–590,000	Oregon	Summer	Mullins, 2001
	220–400	Hong Kong	—	Lee et al., 2002
Fungal spores cfu/m^3	85,000	England	—	Mullins, 2001
	449–547	Taiwan	—	Mullins, 2001
	20,000	Belgium	Summer	Nolard et al., 2001
	27–41,901	Illinois	All	Chung et al., 1996
Pollen g/m^3	8.4–4,105	Illinois	All	Chung et al., 1996
	26–700	Illinois	Summer	Nelson et al., 1933
	313–2,094	Spain	Spring	Carinanos et al., 2004
	71–196	Texas	All	Sterling and Lewis, 1998
	110–5,000	Philadelphia	All	Spiegelman and Friedman, 1968

5.3 OUTDOOR PATHOGENIC DISEASE

In general, outdoor air is relatively free of pathogens and rarely results in any form of respiratory disease. Exceptions can include *Legionella* infections from hot springs, exposure to bird droppings, and inhalation of certain pathogenic fungi during soil disturbances. The wearing of facemasks outdoors practiced on occasion in Far Eastern cities results from misconceptions about respiratory disease etiology. In spite of the fact that most of the microorganisms previously listed in Table 5.1 can occur in outdoor air, the probability of contracting a respiratory or other infection from inhaling outdoor air is infinitesimally small. Airborne diseases are acquired almost exclusively indoors from other infected individuals. Contagious airborne bacteria and viruses have evolved the ability to survive for only short periods in the air and are highly susceptible to dehydration, oxygenation, and exposure to sunlight. Concentrations of pathogens will always tend to be much higher in

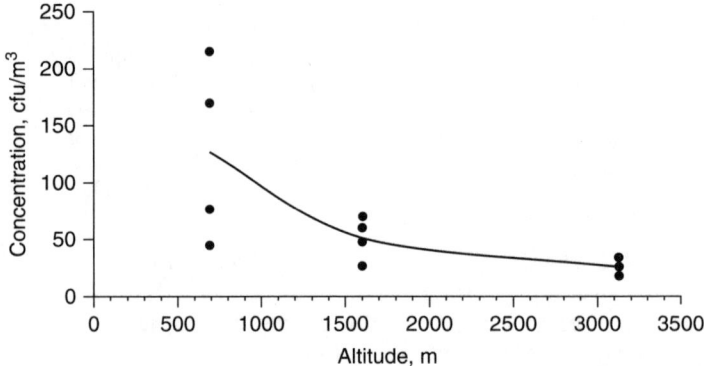

FIGURE 5.4 Airborne fungal and bacterial concentrations as a function of altitude. Line shows mean values. [*Based on data from Fulton et al. (1966).*]

the confines of a building whereas in outdoor air releases from human sources will be dispersed to harmless concentrations rather quickly. The rare cases in which noncontagious respiratory infections have been acquired in outdoor air exposure are reviewed next.

Histoplasma capsulatum is a naturally occurring fungus that has the distinction of having caused respiratory infections in the outdoor air. In 1980 an outbreak of acute pulmonary histoplasmosis occurred among participants in a wagon train as it traveled through eastern Tennessee (Gustafson et al., 1981). Of the 85 people on the train 81 percent became infected with *Histoplasma capsulatum*, although only 64 percent of cases were symptomatic. The source of infection was traced to the site of a former winter blackbird roost that had been partially cleared 5 years earlier to make a park. Most of the soil samples from this site were found to contain *H. capsulatum*. *Histoplasma* exposure in the outdoors normally occurs after excavation, construction, or tree-cutting, making this particular case unusual.

In another unusual case of disease acquired in the outdoors, two individuals contracted blastomycosis while working outdoors in Colorado relocating prairie dogs (Lenaway et al., 1999). Blastomycosis is caused by *Blastomyces dermatitidis* and this was the first reported case of the infection in Colorado. The men had been using hand trowels and gas-powered equipment to excavate abandoned prairie dog burrows, an activity which stirred up a lot of dust. It was believed that unusually heavy rainfall combined with fecal matter in the burrows contributed to causing the infections.

Man-made outdoor sources of airborne microbes can be a source of exposure and respiratory diseases. In a study by Herr et al. (2003) the prevalence of symptoms in people living in the vicinity of composting plants was evaluated. Microorganisms were measured in the air of residential areas located closest to the plants at the same time an epidemiological investigation was performed in the neighborhood within 150 to 1500 m of three plants. The study sample living closest to the site, who were exposed to bioaerosols and odors, had the highest rates of complaints. The type of symptoms reported most often was not strongly influenced by environmental odors and bioaerosol concentrations, except for nausea in context with annoying residential odors.

Legionella pneumophila is a naturally occurring bacteria that can become airborne in both the indoors and outdoors due to man-made technology. In 1978 an outbreak of Legionnaires' disease characterized by high fever, prostration, and pneumonia occurred at an Atlanta, Georgia, country club in which all eight cases involved club members who had been golfing (Cordes et al., 1980). The amount of time spent golfing was a risk factor for acquiring the illness. *Legionella pneumophila* was isolated from the evaporative condenser within the clubhouse where a stream of air blowing from the exhaust duct of the evaporative condenser was directed toward the golf course. Although most outbreaks of Legionnaires' disease have occurred in indoor environments, this was a fairly clear case of outdoor airborne dissemination.

In another example of man-made aerosolization of *Legionella*, samples of tap water used for sprinkling of plants grown in outdoor gardens, and in greenhouses, in Poland were examined for the presence of *Legionella*, along with the samples of soil, artificial medium, and air collected in modern greenhouses (Stojek and Dutkiewicz, 2002). The strains of *Legionella* were isolated 22 percent of samples collected from outdoor taps used for sprinkling plants cultivated in outdoor gardens, and from 25 percent of samples of water collected from indoor taps used for sprinkling of plants cultivated in greenhouses. In both samples collected from outdoor and indoor taps *Legionella pneumophila* was more common than other *Legionella* species. The results of this preliminary study suggest that water aerosolized at sprinkling of plants represents a potential source of *Legionella* infection among gardeners. Cooling tower water has frequently been cited as a source of infection in outbreaks of Legionnaires' disease but few studies have identified legionellae in aerosols from cooling towers. Ishimatsu et al. (2001) demonstrated that *L. pneumophila* was present in the air around a contaminated cooling tower using an air sampler. Levels in the outdoor air were found to be 90 cfu/m^3.

There are some remediation options for outdoor areas that develop a contamination problem. Bird roosts may develop excessive levels of dangerous fungi in the soil below, and these can be remediated. In one bird roosting site that had been in use for 10 years, 8 acres of land was decontaminated by applying 1 gal (3.78 L) of 3 percent formalin solution per square foot of soil (Bartlett et al., 1982). Bird droppings had accumulated to a depth of 2 to 5 cm in some areas. *Histoplasma capsulatum* was isolated from 83 percent of soil samples. In the previous years, at least five cases of histoplasmosis had been reported among residents living adjacent to the property. After application of the solution, no soil samples tested positive for *H. capsulatum*.

Another attempt to decontaminate environmental reservoirs of pathogens was undertaken in 1960 on Gruinard Island, which had been used for testing anthrax as a biological weapon. Decades after the testing was complete, the soil still contained high levels of anthrax. Attempts to decontaminate the soil with formaldehyde, glutaraldehyde, and peracetic acid (Titball et al., 1991) were made. An attempt was made to use dodecylamine to cause the spores to germinate so that they would become susceptible to soil acids but it failed. Formaldehyde was ultimately selected, mixed in a 5 percent solution with seawater, and used to irrigate a 20-mi-wide zone. After 1 month of decontamination the zone was deemed free of anthrax.

Air pollution is a nonaerobiological hazard of outdoor air but it increases the incidence of upper- and lower-respiratory infections in children. Studies have shown an increase in infant mortality in relation to outdoor air pollution. Acute respiratory infections are the most common cause of illness and death in children in the developing world (Romieu et al., 2002). Children may be at greater risk, given the poor environmental and nutritional conditions prevalent in developing countries.

5.4 FUNGAL ALLERGIES AND OUTDOOR AIR

Allergic reactions from outdoor air may be due to fungal spores, pollen, or other allergens. This section focuses on the relationship between outdoor fungi and allergies and the other allergens are addressed in the following section. Some sources suggest that outdoor exposure is a primary source of allergenic reactions and asthma since even short-term peak outdoor exposures can elicit acute symptoms (Burge and Rogers, 2000). Although this may be true for pollen during allergy season, it is not clear that it necessarily applies to fungal allergens, since people spend so much of their time indoors, and indoor air contains much more fungi than pollen. Indoor pollen levels, especially in air-conditioned spaces can be about 6 percent of outdoor levels, while typical fungal spore levels are 15 percent of outdoor levels (Solomon et al., 1980; Shelton et al., 2002). The question here is—are more fungal spores inhaled during extended indoor occupation at low levels or from short-term outdoor exposure to high levels? This question can be evaluated based on available data on outdoor levels and inhalation exposure.

Mitakakis et al. (2000) used two types of special samplers to measure the quantities of microbes inhaled by adults during passive and active behavior in indoor and outdoor air. The particles studied consisted primarily of *Cladosporium, Alternaria*, grass pollen, and nongrass pollen. The mean quantity of particles inhaled was 12.1 particles per hour for indoor air, and 27.15 for outdoor air. If we assume that people spend, on the average, 90 percent of their time indoors, then the quantity of particles inhaled in the indoor air, Q_i, is

$$Q_i = 12.1 \times 24 \times 0.90 = 261 \tag{5.1}$$

The quantity of articles inhaled in the outdoor air, Q_o, is

$$Q_o = 27.15 \times 24 \times 0.10 = 65 \tag{5.2}$$

Based on this estimate, only about 20 percent of inhaled particles come from outdoor air exposure. Using these same inhalation rates, Fig. 5.5 is constructed to determine the influence of time spent indoors on particle inhalation. Obviously, since indoor levels are generally lower than outdoor air, the more time spent indoors, the less the total number of particles inhaled. Also, there is a cutoff at about 70 percent below which more particles will be inhaled from outdoor air as opposed to indoor air.

The above exercise could be repeated if indoor and outdoor levels of fungi, or any allergen, were known for any given location. Unfortunately, the range of outdoor fungal spores is so broad (see Table 5.3) that the choice of levels to use will change the calculated results dramatically. The most extensive study on indoor and outdoor fungal spores indicates the mean ratio of indoor to outdoor fungal spores is 0.15 (Shelton et al., 2002). The percent of inhaled spores due to outdoor air, assuming 10 percent of time is spent outdoors, can then be simply computed as follows:

$$P_o = 0.15 \times 0.10 = 0.09 \tag{5.3}$$

In other words, 9 percent of inhaled fungal spores, on the average, are due to inhalation of outdoor air. This is, admittedly, a simplistic exercise and ignores the fact that there may be a threshold concentration or short-term inhalation dose that will initiate allergic reactions in outdoor air during allergy season, but it does suggest that most allergic reactions to fungi are likely due to indoor air. This is an area that could use further research, and perhaps the best test would be a study of whether atopic individuals spending most (i.e., 90 percent) of their time living in allergen-free houses still suffer reactions during allergy season.

One of the complicating factors in allergies and respiratory disease etiology is the fact that outdoor air pollution exacerbates the effects of exposure to microbial agents. Both the prevalence and severity of respiratory allergic diseases have increased in recent years and both indoor and outdoor airborne pollutants have been cited as contributing factors (D'Amato et al., 2002a). High levels of vehicle emissions in urban environments tend to

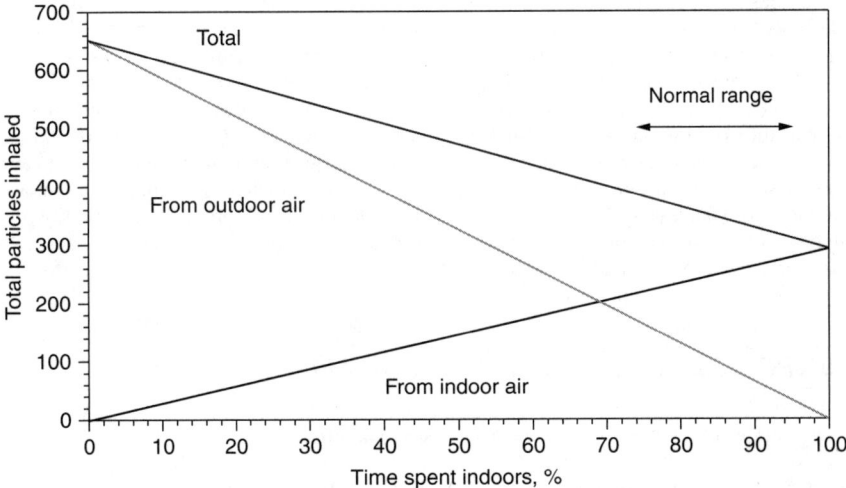

FIGURE 5.5 Total particles inhaled as a function of time spent indoors. In the normal range of occupancy, most particles are inhaled from indoor air.

increase respiratory allergies and infections in most industrialized countries, and people who live in urban areas tend to be more affected by such disease than those of rural areas. In atopic subjects, exposure to air pollution increases airway responsiveness to aeroallergens. Biological aerosols carrying antigenic proteins, such as pollen grains, can produce allergic symptoms and the simultaneous exposure to airborne pollutants can affect the degree of airway sensitization and hyperresponsiveness of exposed subjects. Airway mucosal damage and impaired mucociliary clearance induced by air pollution may also facilitate the penetration and the impact of inhaled allergens on the immune system.

A wealth of evidence suggests that allergic respiratory diseases such as rhinitis, sinusitis, and bronchial asthma have become more common worldwide in recent years. There is also some evidence that increased atmospheric concentrations of pollutants such as ozone, oxides of nitrogen, respirable particulate, and volatile organic compounds, which result from increased use of liquid petroleum gas or kerosene, may be linked to the increased prevalence of allergic diseases (D'Amato et al., 2002b). Experimental studies have also shown that diesel exhaust causes respiratory symptoms and is able also to modulate the immune response in predisposed animals and humans. There is also some evidence that air pollutants can interact with aeroallergens in the atmosphere and/or on human airways, potentiating their effects. Air pollution and climatic changes may also have an indirect effect on the allergic response by influencing quantitatively and qualitatively the pollen production of allergenic plants.

Respiratory allergy affects all age groups but children are the most susceptible single group (Singh and Kumar, 2002). Pollen grains, fungal spores, insect, and other materials of biological origin form the most important allergens in indoor and outdoor air. Among the fungi, *Alternaria, Candida, Aspergillus, Cladosporium, Fusarium, Helminthosporium*, and *Ustilago* are the most important aeroallergens. Outdoor airborne fungal spore levels can vary in the total number of spores and the frequency of species according to weather conditions (Nolard et al., 2001).

Allergic sensitization to the fungi *Alternaria* and *Cladosporium* and to pollen are common and important risk factors for asthma. A study in Australia by Mitakakis et al. (2000) found that inhaled levels of pollen grains and fungal spores were higher during periods of activity than during rest, and higher while subjects were outdoors than indoors. During the active outdoor period, the number of *Alternaria* spores averaged 11 spores per hour (range 4 to 794), *Cladosporium* averaged 4 spores per hour (range 0 to 396), grass pollen averaged 1 grain per hour (range 0 to 81), and nongrass pollen averaged 5 grains per hour (range 0 to 72). Exposure was found to be highly variable between individuals and the amount of particles inhaled depended both on location of the individual and the activity being performed.

In a study of outdoor airborne spores in Uganda by Ismail et al. (1999), a total of 39 genera and over 52 species were trapped from outdoor exposures at outdoor and 35 genera and over 49 species from indoor environments. The total fungal catches of outdoor air spores obtained from all exposures were more than twice that of the indoors. The most highly polluted sites were parks, forests, or riverbanks for outdoor exposures, or teaching laboratories, libraries, latrines, or bathrooms for indoor exposures.

5.5 POLLEN ALLERGIES IN OUTDOOR AIR

Pollen represent the most common cause of seasonal allergic reactions. Allergic reactions to outdoor pollen and spore levels tend to occur in the summer and fall when allergy sufferers may experience the most respiratory distress (Middleton et al., 1983). Atopic individuals who spend most of their time indoors may be symptom free until they go outside. Pollen are often preferentially removed by ordinary dust filters and air conditioners but may exist in high levels outdoors seasonally. Seriously atopic individuals tend to avoid the outdoors during allergy season. The question of whether allergic pollen reactions occur more often due to short-term exposure to high outdoor levels or from long-term indoor exposure to low levels is similar to the fungal allergy question examined previously. If we assume that atopic individuals spend 90 percent of their time indoors, we can estimate the inhalation exposure. According to one study by Carinanos et al. (2004), indoor pollen levels are less than 5 percent of outdoor levels. The percent of inhaled pollen due to outdoor air is computed as follows:

$$P_o = 0.05 \times 0.10 = 0.005 \qquad (5.4)$$

In other words, 0.5 percent of inhaled pollen, on the average, are due to exposure in outdoor air. Again, this ignores the fact that extreme levels in outdoor air may result in brief but heavy inhaled doses, which is probably fairly common. Mitakakis et al. (2000) measured inhalation exposure in indoor and outdoor settings and speculated that greater total exposure to outdoor allergens, including pollen, occurred indoors rather than outdoors. If generally true, this is good news for allergy sufferers since it is fairly easy to completely control indoor airborne pollen with even a moderate efficiency air filter.

Changes in climate have altered pollen distribution in the environment (Burge, 2002). Predictive modeling of outdoor pollen and spore dispersion is a method that can be used to forecast long- and short-term changes in pollen concentrations (Isard and Gage, 2001).

Pollen counts vary not only seasonally but diurnally as well. Levels of pollen tend to increase during the hours of the morning and midafternoon during pollen season. Figure 5.6 shows data for the hourly variation of pollen levels during late summer. Avoidance of the outdoors during these particular hours may be a way for allergy sufferers to circumvent outdoor exposure.

Atmospheric surveys carried out in different parts of India reveal that *Alnus nitida, Amaranthus spinosus, Argemone mexicana, Cocos nucifera, Betula utilis, Borassus flabellifer, Carica papaya, Cedrus deodara, Cassia fistula, Parthenium, Chenopodium album, Dodonaea viscosa, Mallotus phillipensis, Plantago ovata, Prosopis juliflora, Ricinus communis,* and *Holoptelea integrifolia* are the allergenically important pollens of the country (Singh and Kumar, 2002).

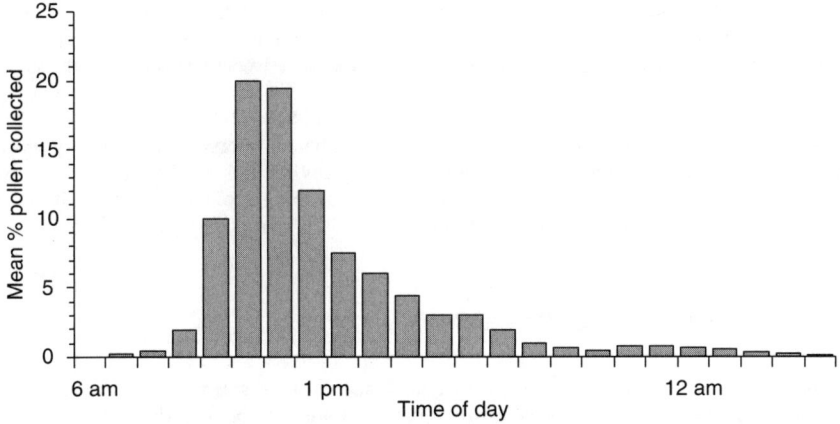

FIGURE 5.6 Diurnal variation of airborne ragweed pollen in England during late summer. [*Based on data from Smith and Rooks (1954).*]

Polluted urban outdoor air may be enriched with large amounts of submicron-sized respirable pollen allergen particles that penetrate into street-level shops and businesses. In a study by Holmquist et al. (2001) concentrations of birch and grass pollen allergens were mapped in indoor air of street-level shops and the effects of electrostatic air cleaning under authentic working conditions were evaluated. The air cleaner reduced all allergen concentrations by an average of 27 percent.

Allergic sensitization to pollen is a common and important risk factor for asthma. Inhaled levels of pollen grains and fungal spores can be higher during periods of activity than during rest, and higher while subjects were outdoors than indoors. Mitakakis et al. (2000) found that during the active outdoor period grass pollen averaged 1 grain per hour (range 0 to 81) and nongrass pollen averaged 5 grains per hour (range 0 to 72).

5.6 OTHER ALLERGIES IN OUTDOOR AIR

Aside from fungi and pollen, the remaining allergens are primarily those found in indoor environments, such as cat dander, dog dander, cockroach allergens, and dust mite allergens. There have been some cases of occupational hypersensitivity due to chronic exposure to allergens in certain industrial and agricultural settings, but by and large the outdoor air outside these man-made environments is relatively free of any other allergens.

Differences in sensitization rates to outdoor aeroallergens have been found between children living in urban and rural areas. A recent investigation by Lee et al. (2001) has suggested that the citrus red mite *Panonychus citri* is the most important allergen in citrus-cultivating Korean farmers with asthma and allergic rhinitis. The prevalence of wheezing and chronic rhinitis symptoms was 8.3 and 35.7 percent in the rural children and 10.5 and 22.4 percent in the control group. The most common sensitizing allergens in order of decreasing frequency were *Dermatophagoides pteronyssinus* (26.6 percent), *D. farinae* (22.7 percent), citrus red mites (14.2 percent), cockroach allergens (11.3 percent), and Japanese cedar pollen (9.7 percent) among the rural children, but the sensitization rates to the citrus red mite and Japanese cedar pollen were only 1.3 and 0.2 percent among the urban children, respectively.

Various airborne allergens may be found in agricultural settings. In a study by Emenius et al. (2001) air samples were collected in the ambient air surrounding a horse stable. Airborne levels of horse allergens were over 500-fold higher inside the stable than just outside the stable and over 3000-fold higher than at a residential building located only 12 m from the stable. Similarly, an inverse correlation was found between the distance to the stable and levels of horse allergens that had settled in the dust.

Atmospheric surveys carried out in India by Singh and Kumar (2002) identified the dust mites *D. farinae* and *D. pteronyssinus* as important inhalant allergens, particularly in the coastal areas of the country. Cockroaches, beetles, weevils, mosquitoes, house flies also contribute toward the aeroallergen load and have been implicated as allergens. It is not clear, however, whether the allergenic exposures resulted from indoor or outdoor air.

5.7 ASTHMA AND OUTDOOR AIR

Some studies have produced evidence that suggests a link between outdoor spore counts and severity of asthma. Sensitivity to some fungal spores can be a predictor for the existence of asthma (Burge and Rogers, 2000). In a study relating outdoor exposure to spores with incidence of attacks of asthma, hospital admissions for asthma for children and adults in England were compared with daily counts of spores from volumetric traps (Newson et al., 2000). Although results were not entirely conclusive, there was some evidence that high rates of admission for asthma tended to correlate with high total mould spore counts. No specific fungi were implicated. Although high outdoor spore levels may correlate with some respiratory problems like asthma, they will also correlate with indoor spore levels. Asthma has been on the increase in recent years and both indoor and outdoor airborne pollutants have been cited as contributing factors (D'Amato et al., 2002a). Various studies have demonstrated that inhalation of air pollutants such as ozone, nitrogen dioxide and sulfur dioxide, either individually or in combination, can enhance the airway response to inhaled allergens in atopic subjects inducing and exacerbating asthma.

Acute effects on humans due to outdoor and indoor exposures to certain pollutant classes, such as ozone, sulfur dioxide, acid rain, airborne toxics, and diesel exhaust, have been demonstrated in epidemiological studies but the effects of these environmental factors on susceptible individuals are not known conclusively (Lebowitz, 1996). These pollutants act as irritants that increase airway hyperreactivity and are thought to be causal factors that act to modulate the immune response. Respiratory allergic diseases appear to be increasing in most countries and asthma in particular has been reported to be increasing (D'Amato et al., 1994). It has been also observed that subjects living in urban and industrialized areas are more likely to have respiratory allergic symptoms than those living in rural areas. This increase has been linked to air pollution. Further research is needed to determine the contributions of the time-related activities of individuals in different environments including the outdoors, in homes, offices, and in transportation, and the interactive effects of pollutants.

An observed increase in asthma admissions in Norway during the 1980s prompted a study to examine whether indoor and outdoor air pollution was associated with asthma development in young children (Loedrup Carlsen, 2002). It was found that home dampness and low ventilation, but not outdoor air pollution, increased the risk. Home dampness is generally associated with increased levels of indoor allergens. Pollen have been associated with asthma epidemics, especially after thunderstorms (Levetin and Van de Water, 2001).

Besides air pollution, respiratory viral infections have also been recognized to be important triggers of asthmatic reactions. In a study by Tarlo et al. (2001) a total of 57 adults and children with asthma were monitored under various levels of outdoor air pollutants, pollen, and fungal spores. Results showed that 47 percent of asthmatic reactions were associated with cold symptoms and that they were also associated with higher levels of sulfur dioxide and nitric oxide in the outside air.

5.8 OUTDOOR AIRBORNE ANIMAL DISEASES

Animals that live in the outdoors, including in agricultural settings, are subject to a number of diseases, both natural and otherwise, some that may transmit by the airborne route. Animal diseases are generally called zoonoses. Most zoonotic pathogens have never been permitted the luxury of adapting to and proliferating inside protective indoor environments. As a result, some animal pathogens have evolved the ability to transmit in outdoor air under conditions that would kill most human pathogens. Even so, it would appear the outdoor conditions must be just right, meaning warm, moist air, and cloudy conditions, for such pathogens to transmit effectively. Exceptions include some diseases of poultry and other farm animals that are kept in indoor barns during winter.

It has long been thought by many in agriculture that outdoor windborne spread of diseases could not occur (Alexander, 1993). Studies on enzootic pneumonia among pig herds led to the conclusion that this disease, at least, was probably spread by wind. More definitive evidence came with the foot-and-mouth disease epidemic of 1967 to 1968 in England, where a collaboration of meteorologists and veterinarians realized that the epizootic was following wind patterns over land and water. The disease traveled over water longer distances than it traveled over land. The most critical factor influencing airborne survival was relative humidity, with higher relative humidity favoring virus survival. Foot-and-mouth disease is a common and highly contagious viral infection among livestock, especially young cattle. This virus spread from French Brittany 300 km to the Isle of Wight along the wind patterns, and a little later it reached the British coast.

Subsequent to the foot-and-mouth outbreak it was realized that several other zoonoses were transmitting via the outdoor air. Porcine respiratory coronavirus, similar to some human cold viruses, was first recognized to be windborne about 1984 (Alexander, 1993). Aujeszky's disease, also called pseudorabies, was also identified as a windborne virus, having carried at least 9 km in England (Christensen et al., 1990). It may also travel farther over water than over land. Outbreaks that started in Denmark and Germany spread to islands off the coast, covering distances of up to 40 km. Another disease known as porcine reproductive and respiratory syndrome, which was first recognized in the United States in 1987, also spreads via the wind. It reportedly has traveled up to 3 km in outdoor air. A bacterium, *Mycoplasma hypopneumoniae*, that causes enzootic pneumonia, is also thought to travel outdoors on wind currents.

It is thought that foot-and-mouth disease can travel up to 150 km when the weather conditions are right (Hugh-Jones and Wright, 1970; Rumney, 1986). Overcast skies may play a part since clouds screen out sunlight and UV rays. Rain, overcast skies, high relative humidity, calm air, and warm temperatures were common during the 1967 to 1968 foot-and-mouth disease epizootic in Europe. To some degree, these conditions simulate warm, moist indoor environments, raising the possibility that some human pathogens might, under the right circumstances, also transmit in the outdoors, but little, if any, evidence exists to suggest this has ever actually happened.

5.9 SURVIVAL OF MICROORGANISMS IN THE ENVIRONMENT

Microorganisms that are airborne in the environment are subject to a variety of biocidal factors including sunlight, dehydration, thermal heating, freezing, oxygenation, and the effects of man-made pollution. Environmental microorganisms that may become airborne but are resident in soil or water may succumb to other factors such as soil acidity (or pH), water availability, and naturally occurring toxins. Microorganisms will also tend to die out from natural decay unless they are preserved by freezing. Not all microorganisms can survive freezing, but spores are more likely to survive freezing in the outdoor environment than bacterial cells or viruses.

Thermal death of microbes can depend on whether exposure is to moist heat or dry heat. Experiments on vegetative cells and bacterial spores have shown that microorganisms can resist dry heat more successfully than moist heat. Saturated steam kills microbes more effectively than superheated steam, which acts as dry heat. The reason moist heat is more lethal than dry heat is apparently related to the rate of heat conduction to the bacterial cell or spore.

The ability of any microbe to resist heat destruction depends on various conditions, including the initial water content of the cell, fat content, and the nature of the medium in which it is exposed. The ability of a microorganism to resist heating is considered to be genetically determined (Mitscherlich and Marth, 1984). The mathematics of thermal disinfection is treated in detail in Chap. 7. Low temperatures can cause cell death at a rate that is temperature dependent except at freezing. Once frozen, no additional decrease in temperature has any significant effect. Although freezing may preserve many microbes, especially spores, many may also die either from cellular damage or from related factors like desiccation or increased susceptibility to sunlight or UV. Death below the freezing point may be due to water crystallization (Henis, 1987).

Sunlight possesses some wavelengths of biocidal UV light but visible light also has some biocidal effects (Futter and Richardson, 1967). Visible light is defined as wavelengths between 400 and 759 nm while UV includes wavelengths between 100 and 400 nm. Natural ionizing radiation (i.e., cosmic rays, gamma rays, and the like) also plays a small part in the death of microbes in the environment. The mathematics of UV disinfection and ionizing radiation is treated in detail in Chap. 7.

The death of microorganisms due to dehydration or desiccation results when sufficient moisture is removed from the cell (Benbough and Hood, 1971; de Mik and de Groot, 1977). Microbes tolerate desiccation to various degrees, with bacteria being the most vulnerable and spores being relatively hardy in this regard. The effect of relative humidity (RH) on the survival of airborne microbes is a complex species-dependent function, apparently due to internal changes of state the microbes may undergo (Goldberg and Ford, 1973). Figure 5.7 shows one example in which airborne microbe survival in air is seen to be a function of RH. See also the section "Desiccation" in Chap. 7 for more information.

Viruses may not behave the same as bacteria in terms of their susceptibility to variations in relative humidity. Low humidities appear to have a limited effect on virus viability while high humidities appear to decrease viability. Figure 5.8 shows a graph of the survival of influenza versus relative humidity at several sample times. The pattern is clearly a decrease in viability at increasing relative humidity. Similar results were obtained for vaccinia but results for poliomyelitis and Venezuelan equine encephalitis showed the opposite tendency, with higher survival at higher humidities (Harper, 1961). However, these latter microbes are not naturally airborne viruses and may not have the same resistance to desiccation.

The death of microbes due to desiccation is impacted by the initial moisture content, growth phase, rate of drying, relative humidity or composition of the surrounding medium, temperature, the coincident presence of sunlight, and other factors (Henis, 1987). Dehydration

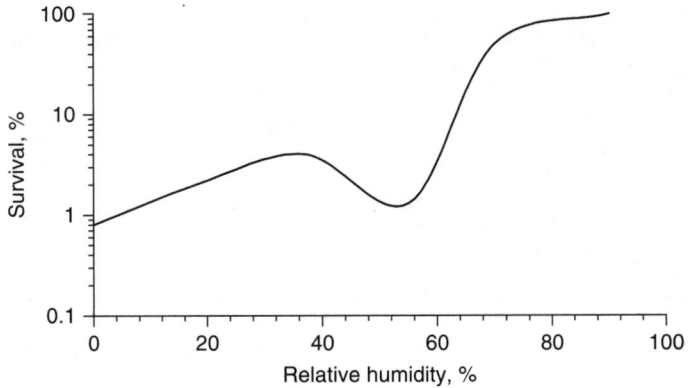

FIGURE 5.7 Effect of relative humidity on survival of airborne *Klebsiella pneumoniae*. [*Based on data from Goldberg and Ford (1973).*]

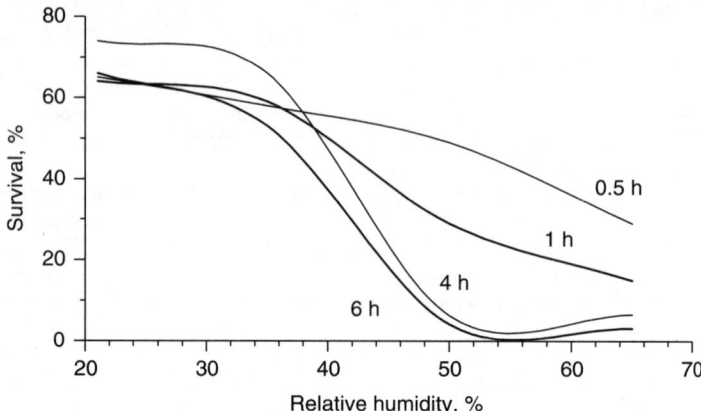

FIGURE 5.8 Viability of influenza virus over time at various relative humidities. [Based on data from Harper (1961).]

renders microbes more susceptible to the effects of oxygenation and the contents of the cell will oxidize over time (Cox 1967, 1973; May et al., 1969). Dried cultures, especially frozen cultures, will tend to be preserved in the absence of oxygen. Further details on the death of microbes due to desiccation are provided in Chap. 7.

Microbes are classified based on their growth capacity at various temperatures. They have been broadly classified as shown in Table 5.4. Cryophiles are also known as psychrophiles. There do not seem to be any cryophiles of concern as pathogens or allergens except for blue-green algae, which may be allergenic. Most pathogens are mesophiles and prefer temperatures between ambient and body temperatures. Some microbes, like the actinomycetes prefer much higher temperatures.

In the absence of biocidal factors microbes die a natural death over time, like any other living thing, except that although individual cells may die the population may remain constant provided some minimum level of nutrients is available. This is not the case for airborne clouds of single microbes since they no longer have access to nutrients while airborne. Airborne microbes in the outdoors will often die from various factors like sunlight and desiccation long before they reach their natural age of death and so this factor can generally be neglected in aerobiology.

More information will be presented on the survival of microbes under various exposures in later chapters. See Chap. 7 for disinfection fundamentals and Chap. 20 for more information on solar exposure.

TABLE 5.4 Growth Temperature Classification of Microbes

Type	Minimum °C (°F)	Optimum °C (°F)	Maximum °C (°F)	Examples
Cryophiles	−7 (19)	<20 (<68)	25 (77)	Blue-green algae
Mesophiles	10 (50)	20–30 (68–86)	45 (113)	All pathogens
Thermophiles	30 (86)	55–60 (131–140)	70–80 (158–176)	Actinomycetes

REFERENCES

Abbaszadegan, M., LeChevallier, M., and Gerba, C. (2003). "Occurrence of viruses in U.S. ground waters." *J Am Water Works Assoc* 95:107–120.

Alexander, T. J. L. (1993). "A winter of windborne spread." *Br Vet J* 149:507–509.

Austin, B. (1991). *Pathogens in the Environment*. Blackwell Scientific Publications, Oxford, UK.

Bartlett, P. C., Weeks, R. J., and Ajello, L. (1982). "Decontamination of a *Histoplasma capsulatum*-infested bird roost in Illinois." *Arch Environ Health* 37(4):221–223.

Benbough, J. E., Hood A. M. (1971). "Viricidal activity of open air." *J Hyg* 69:619–626.

Bitton, G. (1980). *Introduction to Environmental Virology*. John Wiley & Sons, New York.

Burge, H. A., and Rogers, C. A. (2000). "Outdoor allergens." *Environ Health Perspect* 108(Suppl. 4):653–659.

Burge, H. A. (2002). "An update on pollen and fungal spore aerobiology." *J Allergy Clin Immunol* 110(4):544–552.

Carinanos, P., Alcazar, P., Galan, C., Navrro, R., and Diminguez, E. (2004). "Aerobiology as a tool to help in episodes of occupational allergy in work places." *J Invest Allergol Clin Immunol* 14(4):300–308.

Christensen, L. S., Mousing, J., Mortensen, J., Soerensen, K. J., Strandbygard, B. S., Henriksen, C. A., and Andersen, J. B. (1990). "Evidence of long distance airborne transmission of Aujeszky's disease (pdeudorabies) virus." *Vet Rec* 119:282–283.

Chung, J., Wadden, R. A., and Scheff, P. A. (1996). "Outdoor aeroallergen sampling and identification in the Iowa/Illinois Quad Cities area." *Air & Waste Management Association's 89th Annual Meeting & Exhibition*, Nashville, TN

Cordes, L. G., Fraser, D. W., Skaliy, P., Perlino, C. A., Elsea, W. R., Mallison, G. F., and Hayes, P. S. (1980). "Legionnaires' disease outbreak at an Atlanta, Georgia, Country Club: evidence for spread from an evaporative condenser." *Am J Epidemiol* 111(4):425–431.

Cox, C. S., Baldwin F. (1967). "The toxic effect of oxygen upon the aerosol survival of *Escherichia coli*." *J Gen Microbiol* 49:115–117.

Cox, C. S., Baxter, J., Maidment B. J. (1973). "A mathematical expression for oxygen-induced death in dehydrated bacteria." *J Gen Microbiol* 75:179–185.

D'Amato, G., Liccardi, G., and Cazzola, M. (1994). "Environment and development of respiratory allergy: I. Outdoors." *Monaldi Arch Chest Dis* 49(5):406–411.

D'Amato, G., Liccardi, G., D'Amato, M., and Cazzola, M. (2002a). "Outdoor air pollution, climatic changes and allergic bronchial asthma." *Eur Respir J* 20(3):763–776.

D'Amato, G., Liccardi, G., D'Amato, M., and Cazzola, M. (2002b). "Respiratory allergic diseases induced by outdoor air pollution in urban areas." *Monaldi Arch Chest Dis* 57(3–4):161–163.

de Mik, G., de Groot I. (1977). "The germicidal effect of the open air in different parts of the Netherlands." *J Hyg* 78:175–187.

Emenius, G., Larsson, P. H., Wickman, M., and Harfast, B. (2001). "Dispersion of horse allergen in the ambient air, detected with sandwich ELISA." *Allergy* 56(8):771–774.

Fulton, J. D. (1966). "Microorganisms of the upper atmosphere, I - V." *Appl Microbiol* 14(2):232.

Futter, B. V., and Richardson, G. (1967). "Inactivation of bacterial spores by visible radiation." *J Appl Bacteriol* 30(2):347–353.

Goldberg, L. J., and Ford, I. (1973). "Airborne transmission and airborne infection." *VIth International Symposium on Aerobiology*, J. F. P. Hers and K. C. Winkler, eds., Oosthoek Publishing, Utrecht, The Netherlands.

Gustafson, T. L., Kaufman, L., Weeks, R., Ajello, L., Hutcheson, R. H. Jr., Wiener, S. L., Lambe, D. W. Jr, Sayvetz, T. A., and Schaffner, W. (1981). "Outbreak of acute pulmonary histoplasmosis in members of a wagon train." *Am J Med* 71(5):759–765.

Harper, G. J. (1961). "Airborne microorganisms: Survival tests with four viruses." *J Hyg* 59:479–486.

Henis, Y. (1987). *Survival and Dormancy of Microorganisms*. Wiley-Interscience, New York.

Herr, C. E., Nieden, A. Z., Bodeker, R. H., Gieler, U., and Eikmann, T. F. (2003). "Ranking and frequency of somatic symptoms in residents near composting sites with odor annoyance." *Int J Hyg Environ Health* 206(1):61–64.

Holmquist, L., Weiner, J., and Vesterberg, O. (2001). "Airborne birch and grass pollen allergens in street-level shops." *Indoor Air* 11(4):241–245.

Hugh-Jones, M. E., and Wright, P. B. (1970). "Studies on the 1967-1968 foot-and-mouth disease epidemic: The relation of weather to the spread of disease." *J Hyg* 68:253–271.

Hunter, P. R. (1997). *Waterborne Disease: Epidemiology and Ecology.* John Wiley & Sons, Chichester, UK.

Hurst, C. J. (1991). *Modelling the Environmental Fate of Microorganisms.* American Society for Microbiology, Washington, DC.

Isard, S. A., and Gage, S. H. (2001). *Flow of Life in the Atmosphere.* Michigan State University Press, East Lansing, MI.

Ishimatsu, S., Miyamoto, H., Hori, H., Tanaka, I., and Yoshida, S. (2001). "Sampling and detection of *Legionella pneumophila* aerosols generated from an industrial cooling tower." *Ann Occup Hyg* 45(6):421–427.

Ismail, M. A., Chebon, S. K., and Nakamya, R. (1999). "Preliminary surveys of outdoor and indoor aeromycobiota in Uganda." *Mycopathologia* 148(1):41–51.

Jenkins, P. A. (1991). Mycobacteria in the environment. *Pathogens in the Environment,* B. Austin, ed., Blackwell Scientific Publications, Oxford, UK.

Lebowitz, M. D. (1996). "Epidemiological studies of the respiratory effects of air pollution." *Eur Respir J* 9(5):1029–1054.

Lee, M. H., Kim, Y. K., Min, K. U., Lee, B. J., Bahn, J. W., Son, J. W., Cho, S. H., Park, H. S., Koh, Y. Y., and Kim, Y. Y. (2001). "Differences in sensitization rates to outdoor aeroallergens, especially citrus red mite (Panonychus citri), between urban and rural children." *Ann Allergy Asthma Immunol* 86(6):691–695.

Lee, S. C., Li, W.-M., and Ao, C.-H. (2002). "Investigation of indoor air quality at residential homes in Hong Kong—case study." *Atmos Environ* 36:225–237.

Lenaway, D. D., Bailey, A. M., Smith, H., Gershman, K., and Hoffman, R. E. (1999). "Blastomycosis acquired occupationally during prairie dog relocation—Colorado, 1998." *JAMA* 282(1):21–22.

Levetin, E., and VandeWater, P. (2001). "Environmental contributions to allergic disease." *Curr Allergy Asthma Rep* 1(6):506–514.

Loedrup Carlsen, K. C. (2002). "The environment and childhood asthma (ECA) study in Oslo: ECA-1 and ECA-2." *Pediatr Allergy Immunol* 13(Suppl. 15):29–31.

May, K. R., Druett, H. A., Packman, L. P. (1969). "Toxicity of open air to a variety of microorganisms." *Nature* 221:1146–1147.

McDade, J. J., Hall, L. B., and Street, A. R. (1964). "Survival of gram-negative bacteria in the environment." *Am J Hyg* 80:192–204.

Middleton, E., Reed, C. E., and Ellis, E. F. (1983). *Allergy: Principles and Practice,* Vol. 2. C.V. Mosby, St. Louis, MO.

Mitakakis, T. Z., Tovey, E. R., Xuan, W., and Marks, G. B. (2000). "Personal exposure to allergenic pollen and mould spores in inland New South Wales, Australia." *Clin Exp Allergy* 30(12):1733–1739.

Mitscherlich, E., and Marth, E. H. (1984). *Microbial Survival in the Environment.* Springer-Verlag, Berlin.

Mullins, J. (2001). Fungal spores in outdoor air. *Microorganisms in Home and Work Environments,* B. Flannigan, R. A. Samson, and J. D. Miller, eds., Taylor & Francis, London. 3–16.

Nelson, T., Rappaport, B. Z., and Welker, W. H. (1933). "The effect of air filtration in hay fever and pollen asthma." *JAMA* 100(18)

Newson, R., Strachan, D., Corden, J., and Millington, W. (2000). "Fungal and other spore counts as predictors of admissions for asthma in the Trent region." *Occup Environ Med* 57(11):786–792.

Nolard, N., Beguin, H., and Chasseur, C. (2001). "Mold allergy: 25 years of indoor and outdoor studies in Belgium." *Allerg Immunol* 33(2):101–102.

Palmas, F., and Cosentino, S. (1990). "Comparison between fungal airspore concentration at two different sites in the south of Sardinia." *Grana* 29:87–95.

Richards, M. (1954). "Atmospheric mold spores in and out of doors." *J Allergy* 25:429–439.

Romieu, I., Samet, J. M., Smith, K. R., and Bruce, N. (2002). "Outdoor air pollution and acute respiratory infections among children in developing countries." *J Occup Environ Med* 44(7):640–649.

Rumney, R. P. (1986). "Meteorological influences on the spread of foot-and-mouth disease." *J Appl Bacteriol Symp.* 15(Suppl.):105S–114S.

Shelton, B. G., Kirkland, K. H., Flanders, W. D., and Morris, G. K. (2002). "Profiles of airborne fungi in buildings and outdoor environments in the United States." *Appl Environ Microbiol* 68(4):1743–1753.

Singh, A. B., and Kumar, P. (2002). "Common environmental allergens causing respiratory allergy in India." *Indian J Pediatr* 69(3):245–250.

Smith, R. D., and Rooks, R. R. (1954). "The diurnal variation of airborne ragweed pollen as determined by a continuous recording particle sampler and implications of this study." *J Allergy* 25:36–45.

Solomon, W. R., Burge, H. A., and Boise, J. R. (1980). "Exclusion of particulate allergens by window air conditioners." *J Allergy Clin Immunol* 65(4):305–308.

Spiegelman, J., and Friedman, H. (1968). "The effect of air filtration and air conditioning on pollen and microbial contamination." *J Allergy* 42(4):193–202.

Sterling, D. A., and Lewis, R. D. (1998). "Pollen and fungal spores indoor and outdoor of mobile homes." *Ann Allergy Asthma Immunol* 80:279–285.

Stojek, N. M., and Dutkiewicz, J. (2002). "*Legionella* in sprinkling water as a potential occupational risk factor for gardeners." *Ann Agric Environ Med* 9(2):261–264.

Tarlo, S. M., Broder, I., Corey, P., Chan-Yeung, M., Ferguson, A., Becker, A., Rogers, C., Okada, M., and Manfreda, J. (2001). "The role of symptomatic colds in asthma exacerbations: Influence of outdoor allergens and air pollutants." *J Allergy Clin Immunol* 108(1):52–58.

Titball, R. W., Turnbull, P. C. B., and Hutson, R. A. (1991). The monitoring and detection of Bacillus anthracis in the environment. *Pathogens in the Environment*, B. Austin, ed., Blackwell Scientific Publications, Oxford, UK.

Wright, T. J., Greene, V. W., and Paulus, H. J. (1969). "Viable microorganism in an urban atmosphere." *J Air Pollution Contr Assoc* 19:337–341.

CHAPTER 6

AEROSOL SCIENCE AND PARTICLE DYNAMICS

6.1 INTRODUCTION

Airborne microbes are particles and obey the laws that govern particle dynamics. Whether airborne microbes exist as individual cells, clumps of cells, or as components of droplets or droplet nuclei they still behave as particles. The ACGIH defines the general category of bioaerosols as airborne particles, large molecules, or volatile compounds that are living, contain living organisms, or were released from living organisms. This chapter reviews the fundamentals of aerosol science and bioaerosol dynamics as they relate to airborne microbes. These subjects relate to important aspects of aerobiological engineering including airborne transmission, filtration, detection methods, plate-out and distribution in buildings, inhalation characteristics, and some aspects of epidemiology and pathogenesis.

The primary determinants of bioaerosol behavior and their physical removal or detection are size, shape, and size distribution. Characteristics that may have a minor or potential impact on airborne behavior and filterability include density, adherence or surface characteristics, electrical charge, and motility. These characteristics are addressed individually in the following sections.

6.2 SIZE AND SHAPE OF BIOAEROSOLS

The size of a bioaerosol particle may vary from about 0.01 to over 100 μm. Figure 6.1 illustrates the size of viruses, bacteria, spores, yeast, and pollen in an approximately proportional scale. Sizes shown are representative only since each of these types of microbes may span a range much smaller or larger than indicated. Dust mites are not shown because they are too large, being about 300 μm, to truly become airborne particles. Dander is difficult to classify with any mean size or definitive size range but can be any size from about 2 μm to larger than the size of pollen. Particles much larger than pollen will tend to rapidly settle out of the air should they become airborne and do not pose a significant inhalation hazard. The database in App. A provides complete information on the mean sizes and size ranges spanned by microbes.

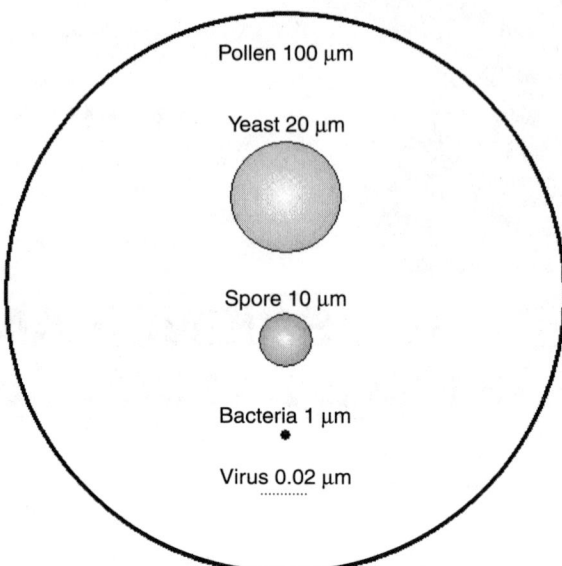

FIGURE 6.1 Size of microbes in approximate proportion.

Most microbes are spherical or ovoid and can therefore be reasonably well described with a simple mean diameter. Airborne microbes that are rod shaped or elongated will still behave according to some mean aerodynamic diameter. However, these are not known for microbes since they can only be determined by testing and so the mean diameters used to represent elongated microbes must be calculated by methods that will be detailed herein.

Airborne microorganisms differ from particulate matter in several regards, including their individually definable sizes, shapes, size distributions, surface characteristics, and density. Modeling of these pathogen characteristics on an individual basis is necessary to assure predictive accuracy. Microbes also differ from particulate matter in certain respects, such as their density and the presence of hydrophobic capsules or slime layers. Some microbes have flagella that enable motility, although this may not be a factor in their aerosolization or filtration.

Since not all microbes are spherical, shape becomes a secondary determinant of size. The types of shapes that pathogens may have are identified graphically in Table 6.1.

6.3 SURFACE CHARACTERISTICS

Most bacteria have a tough outer cell wall made from a tightly woven mesh of polymer-like molecules. The cell wall encloses a balloon-like inner cell containing plasma and cellular components. The internal pressure of a bacterial cell can be on the order of 50 to 60 psia (Koch, 1995; Ryan, 1994), or about the pressure in a truck tire. Bacteria are therefore similar to solid particles except that some species may have detachable capsules or an outer surface covered with a slime layer. Such gelatinous surfaces could conceivably decrease microbial filterability.

Surface stickiness, or adherence, may affect filterability. Bacteria have a strong affinity for adherence to glass surfaces (Ellwood et al., 1979) and fiberglass filter media may

AEROSOL SCIENCE AND PARTICLE DYNAMICS

TABLE 6.1 Shape and Aspect Ratios of Microorganisms

Shape	Type	Description	AR
	Icosahedral / Helical	All respiratory viruses, whether icosahedral or helical, are so small that they can be considered spherical for purposes like filtration and settling calculations.	1
	Spherical	Most bacteria and spores are approximately spherical.	1
	Ovoid	Some bacteria and spores are ovoid.	1–3
	Rods	Bacteria classed as bacilli are rod-shaped.	1–10
	Diplo-cocci	Certain bacteria normally occur in pairs.	1–3
	Strepto-cocci	Some bacteria occur in strings (i.e., streptococcus).	NA
	Staphylo-cocci	Some bacteria occur in bunches (i.e., staphylococcus).	NA
	Flagella	Some bacteria have flagella, enabling motility.	NA
	Capsule	Some bacteria have hydrophobic capsules.	1–3
	Slime layer	Some microbes produce external slime layers.	1–3
	Droplets	Aerosolized droplets, typically 20–100 microns, may contain numerous microbes and other particles. These evaporate to droplet nuclei.	1–3
	Droplet nuclei	Droplet nuclei may contain one or more viable microbes and residue that remains from condensation. Shape may be arbitrary.	NA
	Dander	Dander may consist of squames, hair, or other particles of irregular shape.	NA

already be the ideal filter material. Polyester filter media are also in use and a comparison of their relative ability to filter airborne microbes might provide interesting conclusions on the significance of material-dependent adherence.

Most microorganisms have a negative surface charge (Ellwood et al., 1979). If this charge were significant, then a fiber with a natural positive charge would be likely to provide enhanced microbial filtration. Artificial means of inducing electrostatic charges, either on the particles or on the filter media, have been promising in theory but empirical data indicate electrostatic efficiency improvements are marginal (Offerman et al., 1991).

Droplet nuclei are defined as droplets that have evaporated to form clumps of microbes and condensable matter. Theoretically, a droplet several microns in diameter should evaporate within seconds even in humid air (Reist, 1993). Empirical data, however, suggest that some small fraction of micron-sized droplets can remain airborne for hours (Duguid, 1945).

Since droplets and droplet nuclei are held together by surface tension or other minor forces, they will break up on impact with filter fibers and reduce to smaller constituents. The breakup of water droplets enhances evaporation, as it does in cooling towers and evaporative coolers. The end result is that most microbes contained in a droplet will be subject to filtration at their actual diameters.

6.4 FRACTAL DIMENSION

Particle shape is a fundamental property and is important in assessing health hazards and interpreting data from some sampling methods. The shape frequency distribution can be developed using Fourier analysis of the signature waveform, of which the first five harmonics represent the basic shape, and the higher harmonics describe the texture of the surface.

The surface texture is characterized by the fractal dimension, which is defined as 1.0 plus the absolute value of the slope of the profile perimeter estimate based on a specified increment (Mandelbrot, 1983). Typical fractal dimensions for two-dimensional profiles are shown in Table 6.2.

The usefulness of fractal dimensions in aerobiological engineering is presently theoretical only. At a microscopic level, bacteria, viruses, and spores often have highly distinctive patterns. This is true at a molecular level also. It is possible that a more perfect definition of the fractal dimension of the exterior surfaces of microbes may provide a new, more accurate means of predicting their filterability, adsorption, and behavior. This, in turn, may lead to new technologies for their control and is a matter for future research.

6.5 AERODYNAMIC DIAMETER

If all bioaerosols were perfectly spherical, their defining sizes would be unambiguous. Since few particles are perfectly spherical, and since many have oblong or rodlike shapes, the most useful size definition is their aerodynamic diameter. This is because the aerodynamic diameter acts as the primary determinant of their behavior in essentially all airborne processes or physical removal processes of interest in aerobiology. This may not be true in some particle detection processes but this matter will be addressed in Chap. 24.

TABLE 6.2 Fractal Dimensions for Surfaces

Type of surface	Fractal dimension
Smooth surfaces	1
Mildly convoluted surfaces	1.15
Severely convoluted surfaces	1.3

The aerodynamic diameter is the diameter of a spherical particle of water (a unit density sphere) with which the bioaerosol or microbe has the same settling velocity in air (Heinsohn, 1991). It can be computed empirically from the following relationship:

$$D_a = \left[\frac{18 v_t \mu}{\rho_w g C(D_a)} \right]^{\frac{1}{2}} \tag{6.1}$$

where D_a = aerodynamic diameter, μm
  v_t = settling velocity, cm/s
  μ = dynamic viscosity, kg/m s
  ρ_w = density of water, 1000 kg/m³
  g = gravitational constant, 980 cm/s²
  $C(D_a)$ = Cunningham slip factor based on D_a (see Chap. 10)

Unfortunately, Eq. (6.1) requires measured values of the settling velocity of airborne microbes, data of a type which is practically nonexistent for most of the microbes of concern. Therefore, the definition of aerodynamic diameter most often used simply assumes the microbes are spherical and corrects for the difference in density only, as follows:

$$D_a = D_p \left[\frac{\rho_p}{\rho_w} \right]^{\frac{1}{2}} \tag{6.2}$$

where ρ_p is the density of bioaerosol, kg/m³.

6.6 EQUIVALENT DIAMETER

Although no empirical data on settling velocities are available for most pathogens and allergens, some empirical relationships have been published that allow equivalent spherical diameters to be computed for rods and elongated microbes. First, it is necessary to define the aspect ratio of microbes as follows:

$$A_r = \frac{L}{W} \tag{6.3}$$

where A_r = aspect ratio
  L = length (or longest dimension), μm
  W = width or diameter, μm

The aspect ratio provides a convenient means of identifying microbes that deviate significantly from spherical or ovoid shapes. For microbes with an aspect ratio of greater than about 3.5, the effective diameter can be computed based on an empirical study by Benarie (Matteson and Orr, 1987). This study determined statistically that particles with a large aspect ratio that arrive at a filter surface unoriented, the effective diameter D_e is

$$D_e = 0.285 L \quad \text{for Re} < 2.0 \tag{6.4}$$

where L is the diameter, μm, and Re is the Reynolds number for the particle.

Equation 6.4 was applied to the pathogens whenever $A_r > 3.5$ to determine the equivalent diameters shown in App. A. The value calculated from Eq. (6.4) was selected for use only when it resulted in a larger equivalent diameter. The Reynolds number was based on a particle velocity of 0.66 m/s.

An alternative definition of the equivalent or aerodynamic diameter comes from the study of fibers in the micron size range. Schneider (1987) provides the following relationship for aerodynamic diameter of elongated particles whose aspect ratio is 3 or greater:

$$D_a = 1.29 \left[\frac{\rho_p}{\rho_w} \right]^{\frac{1}{2}} WA_r^{0.13} \quad (6.5)$$

The width W in Eq. (6.5) can be the diameter, or smaller of the two microbial dimensions, the same as in Eq. (6.3). Equation (6.5) was defined for use with fibers below 5 µm in length and for diameters no greater than 3 µm. Since dimensions of microbes with large aspect ratios may typically range from 1 to 10 µm, and Eq. (6.5) overlaps this range, it is of interest to compare these two equations. Figure 6.2 shows the comparison for a particle with a diameter or width of 3 µm. Obviously, considerable differences can exist under certain combinations of diameter and aspect ratio. In lieu of further data corroborating either model, the Benarie model, Eq. (6.4), has been arbitrarily selected for use in correcting the diameters of those microbes in App. A for which the aspect ratio exceeds 3.5.

A third method for adjusting the shape of nonspherical objects to their equivalent spherical diameters was suggested by Fuchs (1964) who introduced the dynamic shape factor X, defined as the ratio of the drag forces between a nonspherical particle and a sphere having the same volume and velocity. An approximate relation for X is provided by Leith (1987) as follows:

$$X = \left[0.33 + 0.67 \frac{D_{s,p}}{D_{p,p}} \right] \left[\frac{D_{e,p}}{D_{p,p}} \right] \quad (6.6)$$

where $D_{s,p}$ = diameter of a sphere with the same surface area of the actual particle
$D_{p,p}$ = diameter of a sphere with the same projected area, or the same cross-sectional area of the actual particle normal to the direction of flow
$D_{e,p}$ = volume equivalent diameter

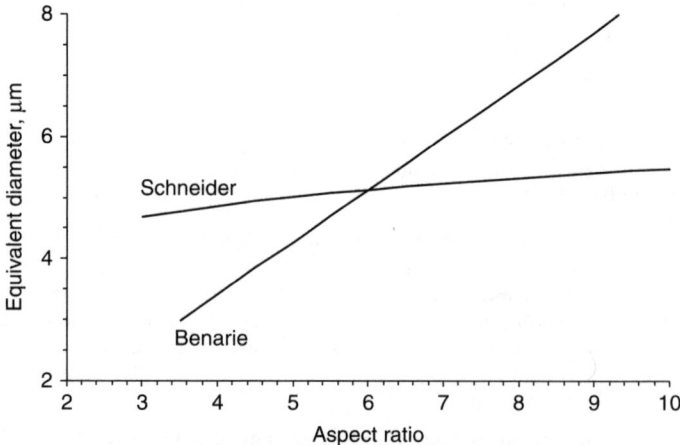

FIGURE 6.2 Comparison of two models for estimating equivalent diameters of nonspherical particles. Models apply when the width/diameter ratio is 3 or greater.

The volume equivalent diameter $D_{e,p}$ is defined as follows:

$$D_{e,p} = \left[\frac{6V_p}{\pi}\right]^{0.33} \tag{6.7}$$

The shape factor is used as a multiplier on the drag force of a sphere moving through the air. For a spherical particle settling in still air, the drag force is given by

$$F = -c_D \left(\frac{\rho}{2}\right)\left[\frac{\pi D_p^2}{4}\right] v^2 \tag{6.8}$$

where c_D = drag coefficient
ρ = density, g/cm^2
D_p = diameter of sphere, cm
v = velocity, cm/s

The drag coefficient of a sphere is a function of the Reynolds number. The relations for the drag coefficient and the Reynolds number for a sphere settling in still air are as follows (Heinsohn, 1991):

$$c_D = 0.4 + \frac{24}{\text{Re}} + \frac{6}{(1+\text{Re}^{0.5})} \tag{6.9}$$

$$\text{Re} = \frac{\rho D_p v}{\mu} \tag{6.10}$$

where μ is the viscosity of air, 1.818×10^{-6} g/cm·s at 21.1°C (70°F).

Table 6.3 provides values for some typical microbial shapes averaged over all orientations, abstracted from the references (Fuchs, 1964).

Clusters of microbes, streptococci, staphylococci, and droplet nuclei, are held together by very weak natural forces. They may settle as clumps but are likely to break up on impact with surfaces like filter fibers. The end result of filtration is that microbes will be reduced to singular forms during the process of filtration.

TABLE 6.3 Dynamic Shape Factors for Microbial Shapes

Shape	Shape factor X
Sphere	1
Cylinder with $L/D = 4$	
Horizontal axis	1.32
Vertical axis	1.07
Ellipsoid, across polar axis, $H/W = 4$	1.2
Chain of 2 spheres	1.12
Chain of 3 spheres	1.27
Chain of 4 spheres	1.15
Cluster of 3 spheres	1.32
Cluster of 4 spheres	1.17

6.7 MICROBIAL DENSITY

The density of an airborne microbe has an effect on its settling rate in air and on filterability. Information on actual virus, bacteria, or fungal spore densities is not readily available for any given species. Bacteria consist mainly of water and proteins. Spores have less water content. Viruses are so small that their density may not be a factor, since they tend to move or be filtered by diffusion, which is not highly dependent on the mass or density of the particle (Davies, 1973; Raber, 1986). The density of viruses is assumed to be the same as the density of water.

The density of bacteria and spores plays a role in both settling and filtration. The specific gravity of water is 1.0 and the density is 1.0 g/cm^3. The actual buoyant density of bacteria has been found to be 1.03 to 1.10 times that of water (Bakken and Olsen, 1983; Bratbak and Dundas, 1984). According to Porter (1946), the specific gravity of bacteria averages 1.1 with 80 percent water on a per weight basis and the waterfree components have a specific gravity equal to 1.83. However, van Veen and Paul (1979) estimated the biomass content of a variety of bacteria and suggested the actual water content is about 51 percent, and the density of bacteria is 1.3 times that of water.

Bacterial spores, being mostly devoid of water, could be expected to have a specific gravity approaching 1.83. Lindsay et al. (1985) found the dry density of bacterial spores to be 1.46 times that of water, and that the water content varied from 26 to 55 percent.

Limited information is available on the density of fungal spores, but it is likely to be very similar to that of bacterial spores since they are composed of basically similar proteins. McConnell and Cone (1992) determined the density of *Myxosporea* spores to be 1.06 times that of water. Dormant spores are reported to have a specific gravity of 1.28 and a water content of 35 percent per Sussman and Halvorson (1966) who cite earlier sources. They also state the specific gravity of germinated spores to be 1.11, but the spores are generally of more interest since they are airborne.

Pollen are often assumed to have a specific gravity of between 1.0 and 1.5. Data for 16 species of pollen are summarized in Stanley and Linskens (1974) from earlier sources that show an average specific gravity of 0.5, but the same source notes that many spores sink in water. The water content of pollen is estimated to be between 15 and 35 percent per Dumas et al. (1983), which would produce a specific gravity of 1.5 to 1.7 based on a dry weight specific gravity equal to that of bacterial dry weight. Per Stanley and Linskens (1974) the water content is between 20 and 50 percent, yielding an estimated specific gravity of 1.4 to 1.7.

6.8 SIZE DISTRIBUTION

The distribution of the size of any microbe tends to follow a lognormal distribution curve, or a curve in which the logarithm of size is distributed normally. This is typical of bioaerosols and even most particulates near the micron size range (Reist, 1993; Painter and Marr, 1968; Duguid, 1945; Koch, 1966a). Studies on cell size distribution indicate the lognormal distribution of cell size predominates both during division, growth phases, and for full-grown microbes (Koch, 1966b, 1995; Koppes et al., 1978; Zaritsky, 1975). Other mathematical expressions can be used to describe these characteristic distributions, such as Pearson curves, but the lognormal curve is adequate.

Figure 6.3 illustrates the theoretical size distribution of *Histoplasma capsulatum*, based on a standard deviation of 0.2. *H. capsulatum* has a size range of 1 to 5 µm. The size distribution in Fig. 6.3 is called lognormal because it forms a normal curve (a bell curve) on a logarithmic scale. In Fig. 6.4 the same size distribution is graphed on a logarithmic

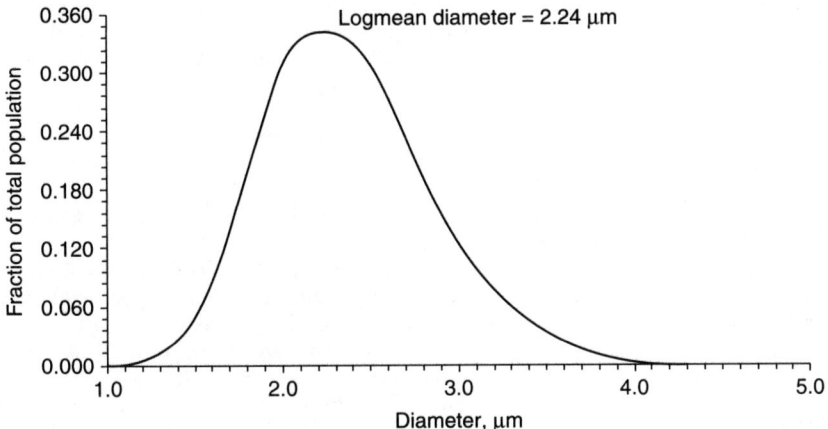

FIGURE 6.3 Size distribution of *Histoplasma capsulatum*.

horizontal axis. It can be observed to be a normal curve. This curve has a normal distribution about the mean diameter, which is defined by the standard deviation. The logmean diameter is 2.24 μm and it can be seen in Fig. 6.4 to be representative of the mean size.

The lognormal distribution can be computed based on the probability density of the normal distribution (Bliss, 1967; Reist, 1993) of a set of diameters

$$Z = \frac{1}{\sigma\sqrt{2\pi}} e^{-\frac{1}{2}((y-\psi)/\sigma)^2} \qquad (6.11)$$

where Z = probability density
σ = standard deviation of ln(diameter)
ψ = mean of ln(diameter)
y = ln(diameter)

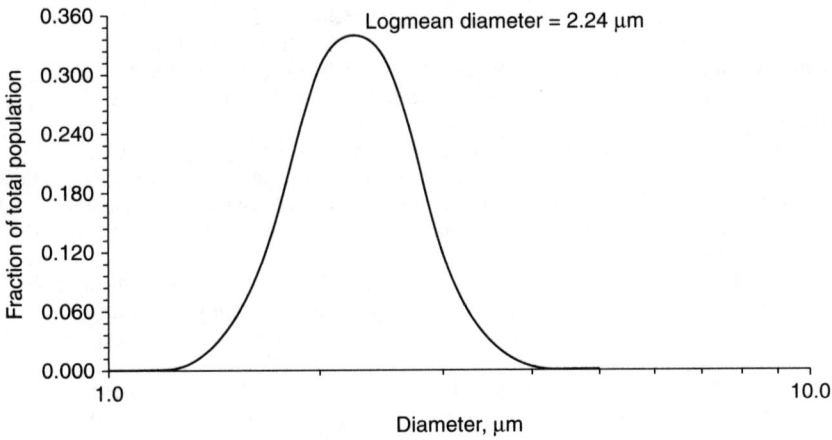

FIGURE 6.4 Size distribution of *Histoplasma capsulatum* shown on a logarithmic scale.

The only parameters necessary to develop a lognormal distribution curve are the range of diameters, the standard deviation, and the logmean diameter. In the absence of additional distribution data, the logmean diameter is approximately equal to the average of the logarithms of the minimum and maximum diameters, as follows:

$$\psi = \frac{\ln(D_{p\min}) + \ln(D_{p\max})}{2} \quad (6.12)$$

Equation (6.12) is only meant to be used as an approximation where exact data for the size distribution are unavailable.

The standard deviation for lognormal size distributions of bacteria is typically from 10 to 30 percent of the size range, based on inspection of published data (Koppes et al., 1978; Tyson, 1986; Harvey and Marr, 1966). However, since the size distribution is lognormal, the standard deviation is actually a function of the mean size and the coefficient of variation. The coefficient of variation is the standard deviation divided by the mean (Bliss, 1967; Powell and Errington, 1963). For lognormal curves the coefficient of variation is the standard deviation divided by the natural log of the mean. The standard deviation can be computed from the *coefficient of variation* (COV) as follows:

$$\sigma = \psi \cdot \text{COV} \quad (6.13)$$

The coefficient of variation typically lies between 0.10 and 0.20 for most microbes (Koppes et al., 1978). Table 6.4 summarizes a number of studies on the COV for microbes. The COV of microbial size is exactly one-half of the COV of the microbial age at division. The consistency of this range of COVs suggests it may be a reasonable estimate to use for other microbes for which no data are available. In fact, the ranges of diameters for all the subject pathogens and allergens are so narrow that the results are not sensitive to large changes in the coefficient of variation. In summary, if the size distribution is known from empirical data, the standard deviation can be computed, but if no size distribution data are available, the standard deviation can be estimated by assuming some reasonable coefficient of variation and using Eq. (6.12).

The size distribution curve is generated by establishing 10 population blocks between the minimum and maximum logarithmic diameters as shown in Fig. 6.5. Each block is represented by a mean and spans ±5 percent of the total range. The population fraction of each block was determined by numerically integrating Eq. (6.11) to get the population fractions. It was found, postanalysis, that no significant difference resulted from simply using the logmean diameters as opposed to performing size-distribution calculations. This is the method recommended because of its simplicity and the logmean diameters, incorporating aspect ratios and determined by the size-distribution calculations, are provided in App. A for use.

The results of a size distribution calculation for *Mycobacterium tuberculosis* are illustrated graphically in Fig. 6.5, in which the continuous probability density curve has been converted into the blocks of population fractions at each of the indicated mean diameters. A similar procedure has been performed to establish the population fractions (blocks) versus size for each microbe in App. A.

Figure 6.6 shows actual data for *Legionella* taken by Kowalski and Bahnfleth (2002). In this graph the data only extend down to 0.5 μm due to measurement limitations. The model is extended down to about 0.2 μm. The coefficient of variation used for this model was 0.175.

Figure 6.7 shows the size ranges and mean diameters for the viruses listed in App. A. Figure 6.8 shows the size ranges and mean diameters for bacteria, and Fig. 6.9 shows the same for fungal spores.

AEROSOL SCIENCE AND PARTICLE DYNAMICS 129

TABLE 6.4 Summary of Microbial Coefficient of Variation Studies

Microbe	Type	Size COV	COV Age at division	Reference
Enteric bacteria		0.1	0.2	Painter and Marr, 1968; Powell, 1956
Bacilli		0.175	0.35	Painter and Marr, 1968; Powell, 1956
Bacteria—all		0.1	0.2	Tyson, 1986
Gram-positive average	GP	0.120		
Gram-negative average	GN	0.096		
Enterobacter cloacae	GN	0.0903	0.1806	Powell, 1958
Bacillus cereus	GP	0.1		Collins and Richmond, 1962
Bacillus megaterium	GP	0.17395	0.3479	Powell, 1956
Bacillus mycoides	GP	0.248	0.496	Powell, 1956
Average	GP	0.1740		
Chromobacterium prodigiosum	GN	0.1055	0.211	Powell, 1956
E. coli	GN	0.100725	0.20145	Powell and Errington, 1963
	GN	0.0792	0.1584	Errington et al., 1965
	GN	0.0849	0.1698	Schaechter et al., 1962
Average	GN	0.0883		
Proteus morganii	GN	0.1054	0.2108	Errington et al., 1965
Proteus vulgaris	GN	0.0978	0.1956	Powell and Errington, 1963
Proteus vulgaris	GN	0.103	0.206	Schaechter et al., 1962
Average		0.1021		
Pseudomonas aeruginosa	GN	0.09125	0.1825	Powell and Errington, 1963
	GN	0.0687	0.1374	Powell, 1958
	GN	0.0705	0.141	Powell, 1956
	GN	0.08905	0.1781	Errington et al., 1965
Average		0.0799		
S. typhimurium	GN	0.0902	0.1804	Schaechter et al., 1962
Serratia marcescens	GN	0.0833	0.1666	Powell, 1958
	GN	0.1456	0.2912	Errington et al., 1965
Average	GN	0.1145		
Streptococcus faecalis	GP	0.0665	0.133	Powell, 1956

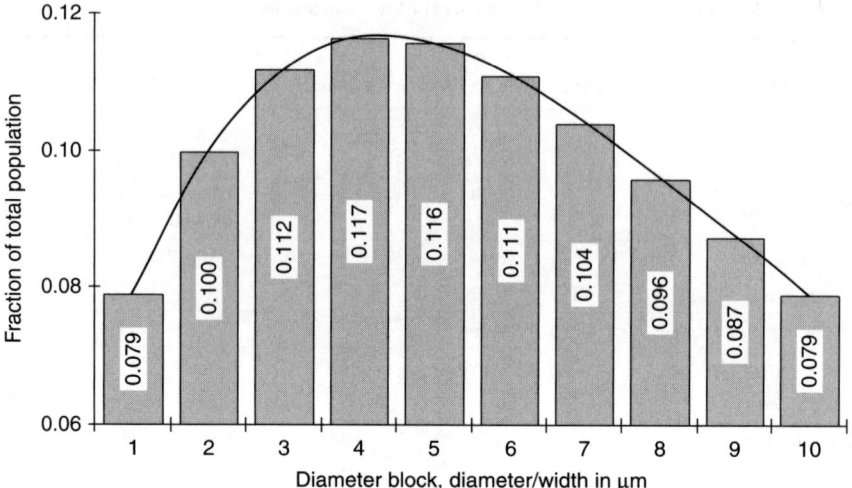

FIGURE 6.5 Size distribution of *Mycobacterium tuberculosis*. Modeled with a coefficient of variation of 0.175.

6.9 BIOAEROSOL DYNAMICS

Bioaerosols are subject to aerodynamic and other forces which define the way they behave in air, how they distribute or diffuse, and their behavior when removed by physical methods like filtration. These principles of dynamic behavior are introduced here although they will be partly readdressed in Chap. 11.

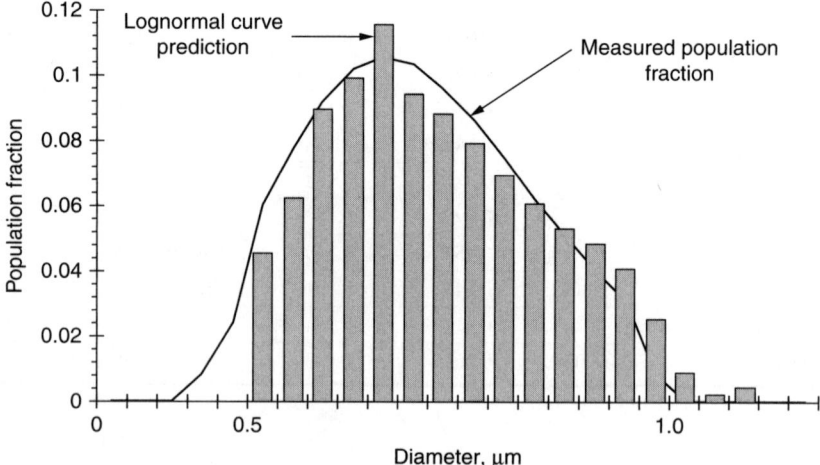

FIGURE 6.6 Lognormal size distribution data and modeled curve for *Legionella pneumophila*. [Based on data from Kowalski and Bahnfleth (2002).]

AEROSOL SCIENCE AND PARTICLE DYNAMICS

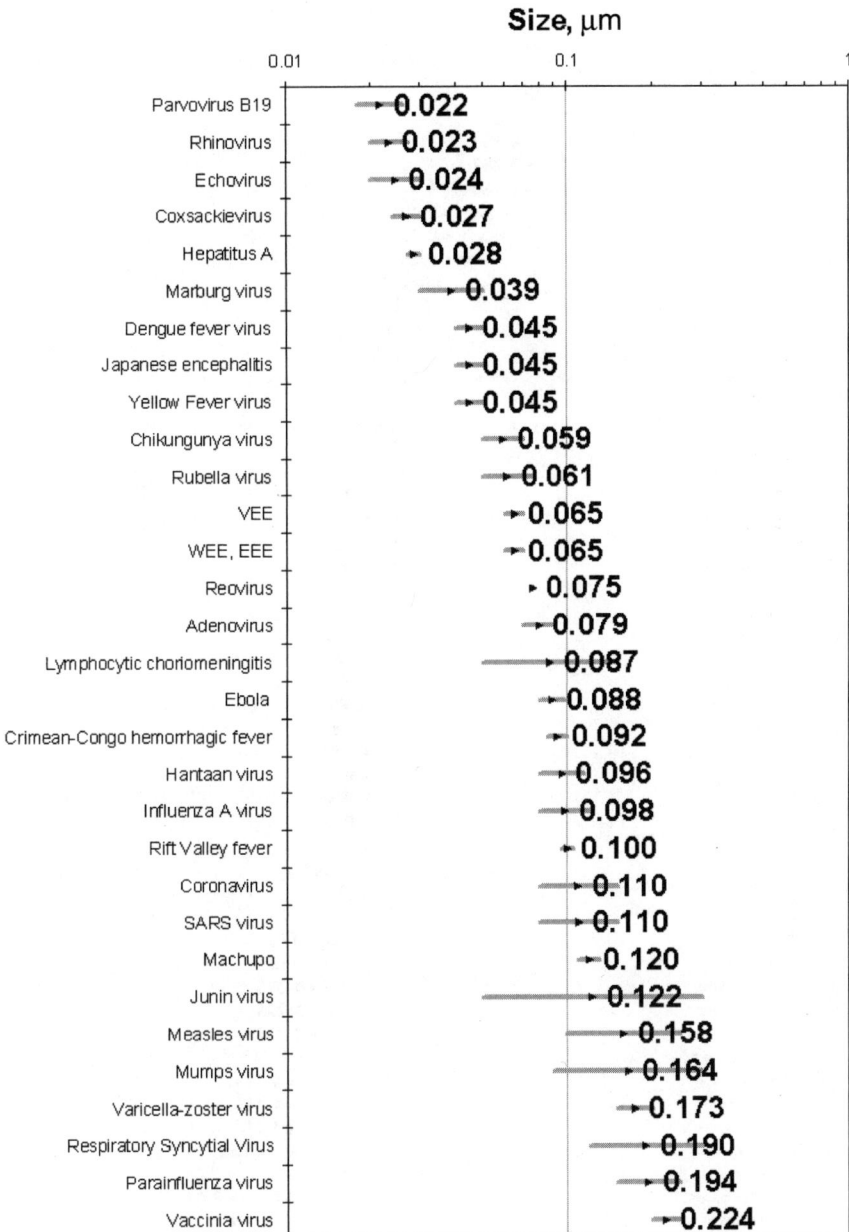

FIGURE 6.7 Size ranges and mean diameters of viruses.

132 BACKGROUND AND HISTORY

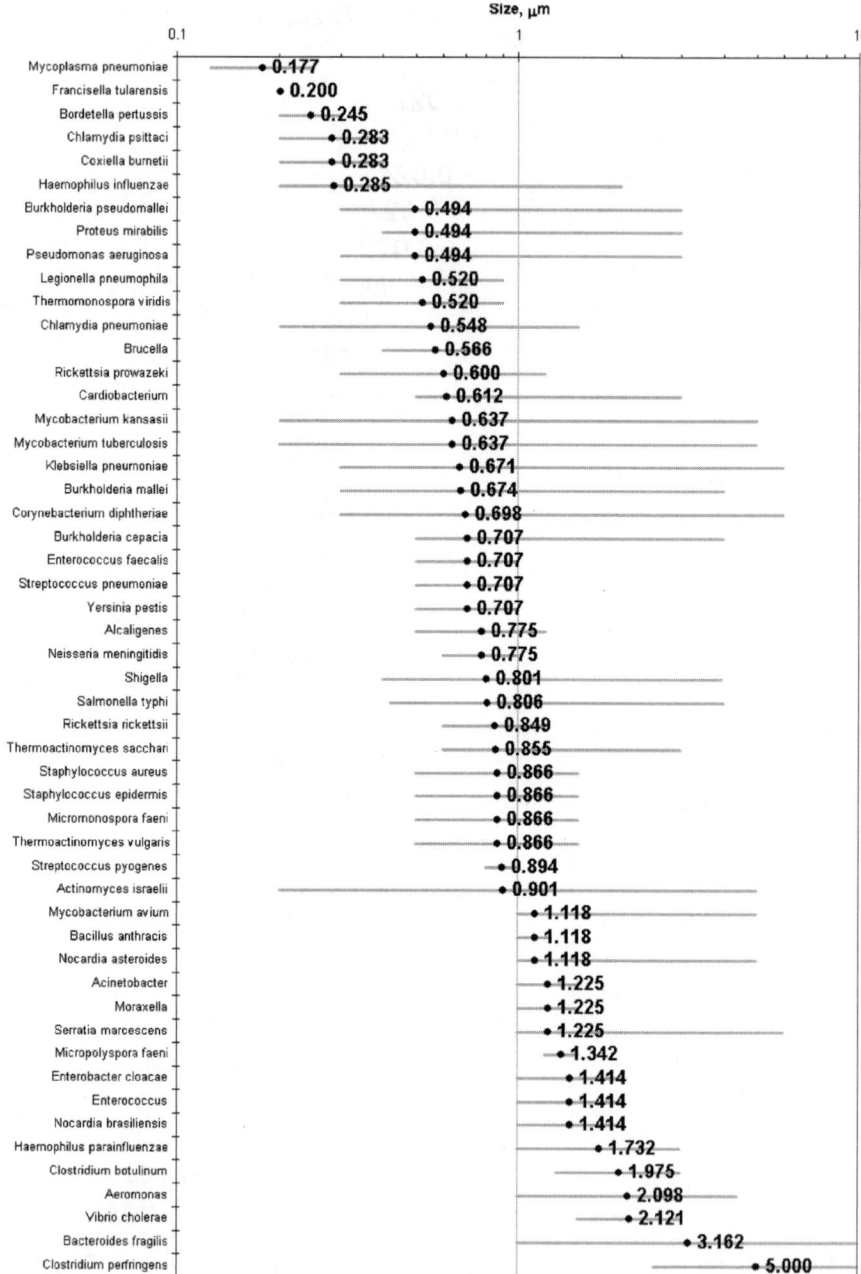

FIGURE 6.8 Size ranges and mean diameters of bacteria.

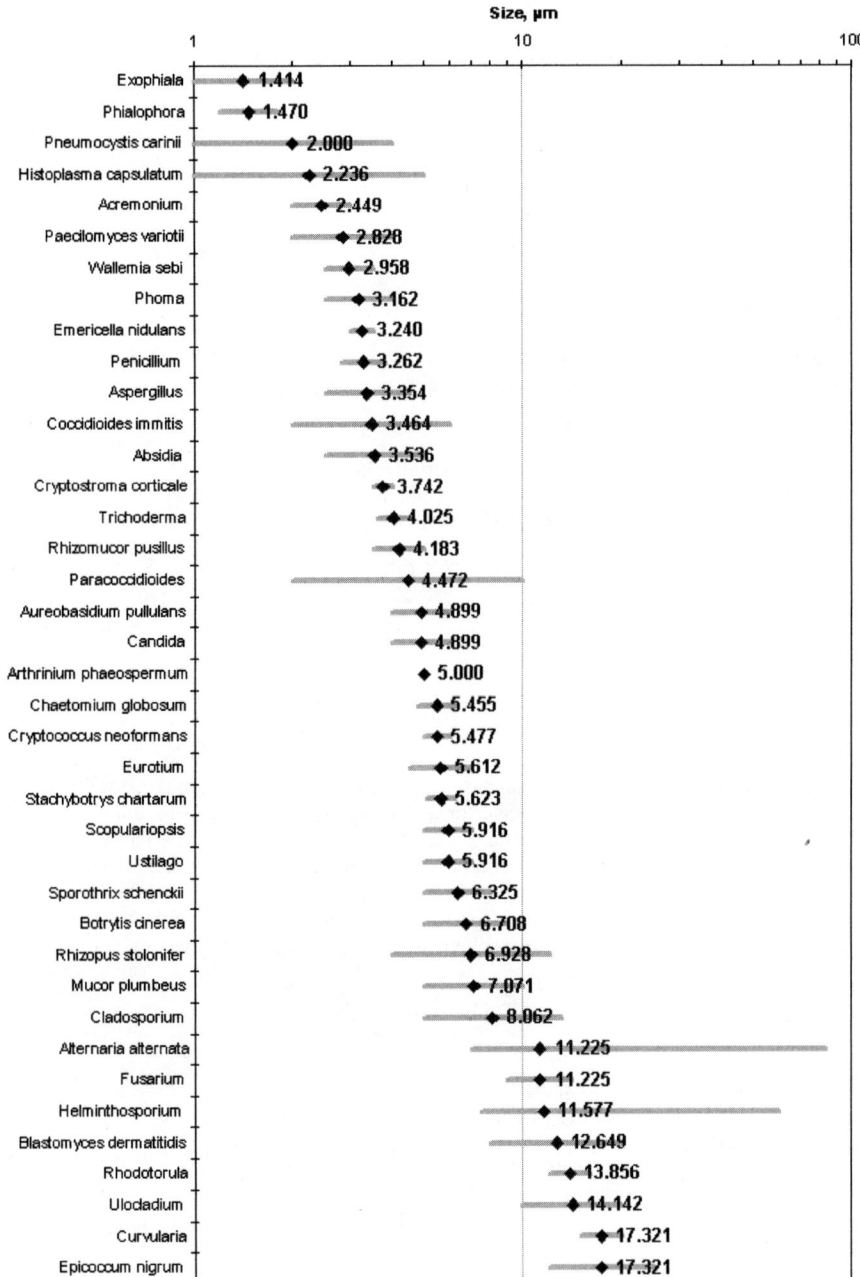

FIGURE 6.9 Size ranges and mean diameters of fungal spores.

6.9.1 Brownian Motion

Bioaerosol particles are subject to Brownian motion according to Einstein's equation

$$\overline{X} = 5 \cdot 10^{-6} \sqrt{\frac{t}{r}} \qquad (6.14)$$

where X = root mean square particle displacement, cm
t = time, s
r = particle radius, cm

6.9.2 Gravitation

For particles greater than 1 μm, diffusion due to gravitational settling is dominant. For spherical particles the terminal velocity (in calm air) due to gravity is:

$$V = 32\rho d^2 \qquad (6.15)$$

where ρ is the particle density and d is the particle diameter, cm.
For particles less than 0.5 μm the terminal velocity is approximately 0. Physical decay (not biological decay) due to gravitational settling is approximated by the first-order decay process

$$N_t = N_0 e^{-kt} \qquad (6.16)$$

where N_0 = number of particles at time $t = 0$
N_t = number of particles at time t
k = first-order decay rate constant

6.9.3 Electrical Forces

Even though the overall bioaerosol may have a neutral charge, the bioaerosol particles themselves are invariably charged to a greater or lesser degree as the result of natural electrical charging due to random collisions that take place between ions and particles and is called *diffusion charging*. Charging occurs in accordance with Boltzmann's distribution and charged particles will have charges that vary over time and depend on statistical distributions of the random particle and ion velocities, and the local fields that develop among the charged particles. The charge (q_d) that develops on a spherical particle at time t is given by (Vincent, 1995)

$$q_d = \left\{\frac{\pi \varepsilon_0 dkT}{2e}\right\} \ln\left\{\frac{\pi d c_i N_i t}{2kt}\right\} \qquad (6.17)$$

where q_d = charge on a spherical particle
ε_0 = absolute permittivity of vacuum, 8.85×10^{-12} A·s/V·m
d = particle diameter
$k = 1.38 \times 10^{-23}$ J/K, Boltzmann's constant
$c_i = 2 \times 10^2$ m/s for air ions at STP
N_i = ion density
t = time

6.9.4 Thermal Gradients and Electromagnetic Radiation

Thermal gradients can induce aerosol movements. Aerosol particles interact with electromagnetic radiation primarily through reflection, refraction, absorption, and scattering. In both cases transparent particles move toward heat sources because they act as a lens thereby focusing energy on the distal side. Opaque particles move in the opposite direction. This phenomenon is known as thermophoresis (for thermal gradients) and photophoresis (for radiation). Evidence of thermophoresis exists in the soiling of walls near radiators. Particles in a thermal field where there is a temperature gradient are buffeted by gas molecules and will move with a thermophoretic velocity. For very fine particles, which are small compared to the mean free path between gas molecules, the thermophoretic velocity has the following form:

$$v_T = -k_1 \partial T \tag{6.18}$$

where v_T = thermophoretic velocity
k_1 = constant depending on the local temperature
∂T = temperature gradient

For larger particles where the gas is essentially a continuum, the thermal conductivity of the particle and of the surrounding air becomes a factor. In this region the thermophoretic velocity is defined by

$$v_T = -k_2 \partial T \tag{6.19}$$

where k_2 = function of σ_p and σ_a
σ_p = thermal conductivity of the particle
σ_a = thermal conductivity of the air

Figure 6.10 shows a plot of empirical data for thermophoretic velocity as a function of particle size for aerosols typical of those found in the workplace. It is based on data from Vincent (1995) for hardwood dust with a ratio of particle thermal conductivity to air conductivity of approximately 5.

6.9.5 Turbulent Diffusion

In laminar flow particles are carried along the airstream with the air molecules, but on a change in direction the heavier bioaerosol particles will break the streamlines. As a result,

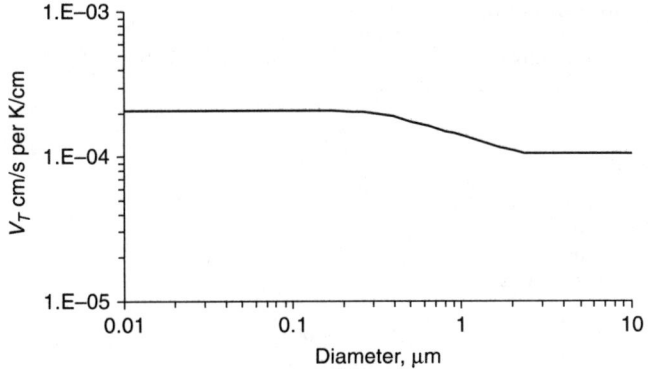

FIGURE 6.10 Thermophoretic velocity as a function of particle size. Velocity is in units of cm/s of thermophoretic velocity per degrees K/cm of thermal gradient. [*Based on data from Vincent (1995)*.]

the particles may deposit on curved or angled surfaces. Consider flow in a curved pipe, where the linear air velocity is given by

$$V = \frac{Q}{\pi(D/2)^2} \qquad (6.20)$$

where Q is the flow, cm³/s, and D is the pipe diameter, cm.
For the airflow, the Reynolds number is given by

$$Re = \frac{\rho VD}{\mu} \qquad (6.21)$$

where D = pipe diameter, cm
ρ = air density, g/cm³
μ = air viscosity, g/cm-s

Although this may be laminar, the Reynolds number for the particle, which is experiencing higher inertial forces, is

$$Re_p = \frac{\rho V_p r}{\mu} \qquad (6.22)$$

where r is the particle radius, cm, and V_p is the particle velocity, cm/s.
The particle velocity is given by Stoke's law. For spherical particles this is

$$V_p = \frac{\rho d^2 V^2}{18\mu R} \qquad (6.23)$$

where ρ = particle density, g/cm³
d = particle diameter, cm
R = radius of pipe bend, cm

For laminar flow in which the particle is turbulent, deposition is more likely to occur. For turbulent flow, deposition is less likely. This factor is important in the design of inertial samplers, or impactors, as described in the following section.

6.9.6 Inertial Impaction

The ideal impactor consists of a laminar airstream turning before a flat surface. The distance traveled by a particle is defined as

$$D = V_p t \qquad (6.24)$$

where $t = L/V$
V_p = particle velocity, cm/s
t = travel time, s
L = length of curved trajectory, cm
V = air velocity, cm/s

Substituting and rearranging we have

$$V_p = \frac{DV}{L} \qquad (6.25)$$

combining with Stoke's law [Eq. (6.23)] we have

$$\frac{DV}{L} = \frac{\rho d^2 V^2}{18\mu R} \qquad (6.26)$$

and

$$D_m = \frac{d^2 V^2 L}{18\mu R} \qquad (6.27)$$

which is called the Sinclair stopping distance. This defines the probability of a given size particle impacting the plate under a given airflow. The collection efficiency is affected by the adhesion of the surface, particle bounce, and particle shape.

6.10 SETTLING AND EVAPORATION OF BIOAEROSOLS

Airborne microbes and droplet nuclei will settle in air over time. The smaller the microbe, the longer it is likely to remain airborne. The terminal velocity has been given by Eq. (6.15). Droplets will simultaneously evaporate as they settle, which alters the terminal velocity. Droplets can be characterized on the basis of size as shown in Table 6.5. Droplets in the size range of bacteria may be best represented by fog, at 1 to 10 μm, which tends to remain airborne for extended periods of time.

Figure 6.11 shows the results of theoretical computations of the evaporation rate of droplets in still, dry air. Based on these theoretical results, a 100-μm droplet would evaporate in a fraction of a second. Of course, this is an idealized case. If the air is saturated, the droplet will evaporate much slower, and if the droplet contains microbes, it will not evaporate past some point.

Microbes invariably tumble when airborne due to natural buffeting and diffusive forces in the air, which further complicates estimates of evaporation and settling rates. Consider aerosolized microbes from a sneeze that evaporate to droplet nuclei, as shown in Fig. 6.12. The droplet nuclei may contain anywhere from one to tens of thousands of microbes. The heavier droplet nuclei will settle out in short order while the lighter ones will remain airborne until they are inhaled, or perhaps removed by filtration. The droplet nuclei in the figure are shown with arbitrary shapes that are assumed to be composed of some remaining water and organic matter that remains after evaporation. It is not known with any certainty what form droplet nuclei take (i.e., spherical or irregular) but it is known that some water

TABLE 6.5 Droplet Names and Sizes

Name	Droplet diameter, μm	
	Minimum	Maximum
Fog	1	10
Mist	10	100
Drizzle	100	1000
Rain	1000	-

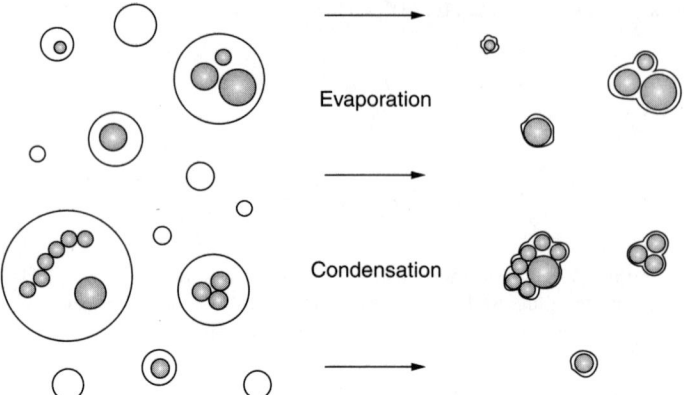

FIGURE 6.11 Theoretical time for particles to evaporate in air (based on the Fuch-Langmuir equation).

will remain bound to the surface of microbes to a depth of several molecules and will not evaporate due to bonding forces.

Figure 6.13 plots empirical data for droplets that simultaneously settle and evaporate. It shows the percentage of droplets of specific size ranges that remain airborne after measured time periods. The result of the complex process of settling and evaporation produces a remarkably linear curve on a logarithmic scale. In contrast to Fig. 6.11, these sneeze droplets do not disappear in fractions of a second due at least partly to the fact that they contain microbes.

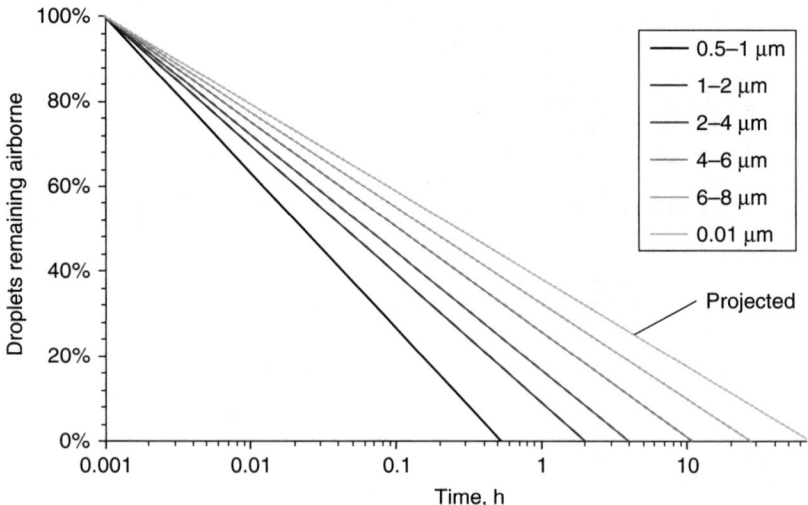

FIGURE 6.12 Aerosolized droplets from a sneeze or cough will evaporate or condense down to droplet nuclei containing one or more microbes.

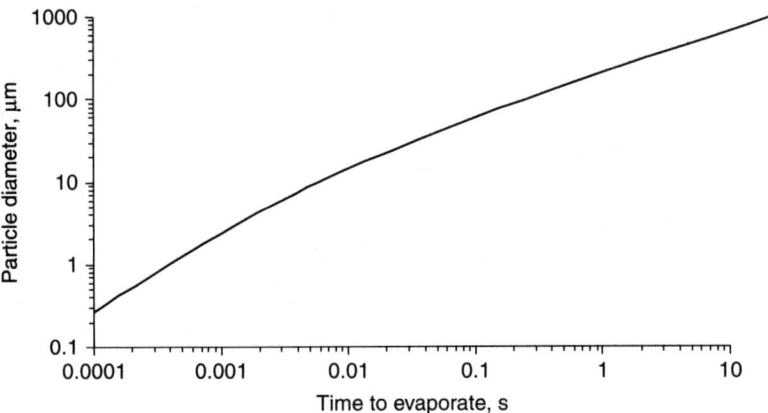

FIGURE 6.13 Settling of airborne sneeze droplets over time, by particle size. [*Based on data from Duguid (1945)*.]

6.11 BIOAEROSOL VIABILITY

One factor that impacts bioaerosols almost as much as their residence time in the air is the time they remain viable, or infectious, while airborne. Microbes that become airborne, even those that depend on this route of transmission, tend to die off in the air. Sometimes this effect is referred to as *open air factor* (OAF) (Cox and Baldwin, 1967; Cox, 1995), but die-off in the outside air generally involves solar radiation or pollution, or temperature extremes (Mitscherlich and Marth, 1984). The rate at which microbes die off in indoor air is relatively slow, but in order for the model to be realistic, the die-off rate for each microbe

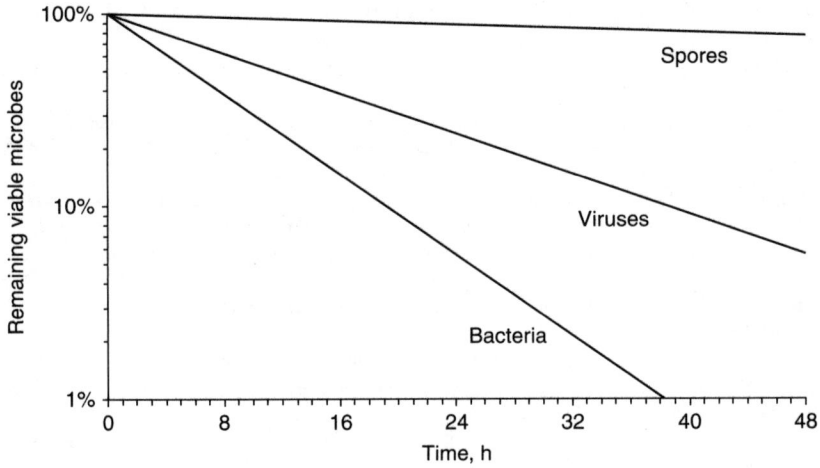

FIGURE 6.14 Viability of airborne microbes indoors in absence of sunlight (based on averages for each microbial group).

group, spores, bacteria, and viruses, has been determined based on a literature review (May et al., 1969; Crowe and Clegg, 1978; El-Adhami et al., 1994; Futter and Richardson, 1967; Benbough and Hood, 1971; Cox et al., 1973). The values for the decay rate of microbes in indoor environments are not significant in comparison with their removal rate by mechanical means. They also surely vary with species. The estimated average values for the natural indoor decay rate for the three microbe groups are as follows:

- Spores—0.009 percent per minute
- Bacteria—0.20 percent per minute
- Viruses—0.10 percent per minute

These decay rates in air have been graphed in Fig. 6.14. Note that bacteria survive less well in air than viruses, apparently due to their increased susceptibility to dehydration.

REFERENCES

Bakken, L. R., and Olsen, R. A. (1983). "Buoyant densities and dry-matter contents of microorganisms: Conversion of a measured biovolume into biomass." *Appl Environ Microbiol* 45:1188–1195.

Benbough, J. E., and Hood, A. M. (1971). "Viricidal activity of open air." *J Hyg* 69:619–626.

Bliss, C. I. (1967). *Statistics in Biology.* McGraw-Hill, New York.

Bratbak, G., and Dundas, I. (1984). "Bacterial dry matter content and biomass estimations." *Appli Environ Microbiol* 48:755–757.

Collins, J. F., and Richmond, M. H. (1962). "Rate of growth of *Bacillus cereus* between divisions." *J Gen Microbiol* 28:15–33.

Cox, C. S., F. Baldwin (1967). "The toxic effect of oxygen upon the aerosol survival of *Escherichia coli.*" *J Gen Microbiol* 49:115–117.

Cox, C. S., Baxter, J., Maidment, B. J. (1973). "A mathematical expression for oxygen-induced death in dehydrated bacteria." *J Gen Microbiol* 75:179–185.

Cox, C. S. (1995). Chapter 6: Stability of airborne microbes and allergens. *Bioaerosols Handbook,* C. S. Cox and C. M. Wathes, eds., CRC-Lewis Publishers, Boca Raton, FL.

Crowe, J. H., and Clegg, J. S. (1978). *Dry Biological Systems.* Academic Press, New York.

Davies, C. N. (1973). *Air Filtration.* Academic Press, London.

Duguid, J. P. (1945). "The size and the duration of air-carriage of respiratory droplets and droplet-nuclei." *J Hyg* 54:471–479.

Dumas, C., Duplan, J.-C., Said, C., and Soulier, J. P. (1983). 1H Nuclear magnetic resonance to correlate water content and pollen viability. *Pollen: Biology and Implications for Plant Breeding,* D. L. Mulcahy and E. Ottaviano, eds., Elsevier Biomedical, New York. 15–27.

El-Adhami, W., Daly, S., and Stewart, P. R. (1994). "Biochemical studies on the lethal effects of solar and artificial ultraviolet radiation on *Staphylococcus aureus.*" *Arch Microbiol* 161:82–87.

Ellwood, D. C., Melling, J., and Rutter, P. (1979). *Adhesion of Microorganisms to Surfaces.* Academic Press, London.

Errington, F. P., Powell, E. O., and Thompson, N. (1965). "Growth characteristics of some gram-negative bacteria." *J Gen Microbiol* 39:109–123.

Fuchs, N. A. (1964). *The Mechanics of Aerosols.* Pergamon Press, Oxford, UK.

Futter, B. V., and Richardson, G. (1967). "Inactivation of bacterial spores by visible radiation." *J Appl Bacteriol* 30(2):347–353.

Harvey, R. J., and Marr, A. G. (1966). "Measurement of size distributions of bacterial cells." *J Bacteriol* 92(4):805–811.

Heinsohn, R. J. (1991). *Industrial Ventilation: Principles and Practice.* John Wiley & Sons, New York.

Koch, A. L. (1966a). "The logarithm in biology: Mechanisms generating the lognormal distribution exactly." *J Theoret Biol* 12:276–290.

Koch, A. L. (1966b). "Distribution of cell size in growing cultures of bacteria and the applicability of the Collins-Richmond principle." *J Gen Microbiol* 45:409–417.

Koch, A. L. (1995). *Bacterial Growth and Form.* Chapman & Hall, New York.

Koppes, L. J. H., Woldringh, C. J., and Nanninga, N. (1978). "Size variations and correlation of different cell cycle events in slow-growing *Escherichia coli.*" *J Bacteriol* 134:423–433.

Kowalski, W. J., and Bahnfleth, W. P. (2002). "Airborne-microbe filtration in indoor environments." *HPAC Eng* 74(1):57–69. http://www.bio.psu.edu/people/faculty/whittam/research/amf.pdf.

Leith, D. (1987). "Drag on nospherical objects." *Aerosol Sci Technol* 6:153–161.

Lindsay, J. A., Beaman, T. C., and Gerhardt, P. (1985). "Protoplast water content of bacterial spores determined by buoyant density sedimentation." *J Bacteriol* 163(2):735–737.

Mandelbrot, B. B. (1983). *The Fractal Geometry of Nature.* W. H. Freeman and Company, New York.

Matteson, M. J., and Orr, C. (1987). "Filtration: principles and practices." C. Industries, ed., Marcel Dekker, New York.

May, K. R., Druett, H. A., Packman, L. P. (1969). "Toxicity of open air to a variety of microorganisms." *Nature* 221:1146–1147.

McConnell, C. J., and Cone, D. K. (1992). "Settling rates and density of spores of *Henneguya doori* (Myxosporea) in water." *J Parasitol* 78(3):427–429.

Mitscherlich, E., and Marth, E. H. (1984). *Microbial Survival in the Environment.* Springer-Verlag, Berlin.

Offerman, F. J., Loiselle, S. A., and Sextro, R. G. (1991). Performance comparison of six different air cleaners installed in a residential forced-air ventilation system. *Healthy Buildings/IAQ '91,* Washington, DC.

Painter, P. R., and Marr, A. G. (1968). "Mathematics of microbial populations." *Annu Rev Microbiol* 22:519–549.

Porter, J. R. (1946). *Bacterial Chemistry and Physiology.* John Wiley & Sons, New York.

Powell, E. O. (1956). "Growth rate and generation time of bacteria, with special reference to continuous culture." *J Gen Microbiol* 15:492–511.

Powell, E. O. (1958). "An outline of the pattern of bacterial generation times." *J Gen Microbiol* 18:382–417.

Powell, E. O., and Erington, F. P. (1963). "Generation times of individual bacteria: Some corroborative measurements." *J Gen Microbiol* 31:315–327.

Raber, R. R. (1986). "Fluid filtration: Gas." Symposium on Gas and Liquid Filtration, Philadelphia, PA.

Reist, P. C. (1993). *Aerosol Science and Technology.* McGraw-Hill, New York.

Ryan, K. J. (1994). *Sherris Medical Microbiology.* Appleton & Lange. Norwalk, CT.

Schaechter, M., Williamson, J. P., Hood, J. R., and Koch, A. L. (1962). "Growth, cell, and nuclear divisions in some bacteria." *J Gen Microbiol* 29:421–434.

Schneider, T. (1987). "Mass concentration of airborne man-made mineral fibers." *Ann Occup Hyg* 31(2):211–217.

Stanley, R. G., and Linskens, H. F. (1974). *Pollen: Biology, Biochemistry, Management.* Springer-Verlag, New York.

Sussman, A. F., and Halvorson, H. O. (1966). *Spores: Their Dormancy and Germination.* Harper & Row, New York.

Tyson, J. J. (1986). "Sloppy size control of the cell division cycle." *J Theor Biol* 118:405–426.

van Veen, J. A., and Paul, E. A. (1979). "Conversion of biovolume measurements of soil organisms, grown under various moisture tensions, to biomass and their nutrient content." *Appl Environ Microbiol* 37:686–692.

Vincent, J. H. (1995). *Aerosol Science for Industrial Hygienists.* Pergamon, New York.

Zaritsky, A. (1975). "On dimensional determination of rod-shaped bacteria." *J Theoret Biol* 54:243–248.

CHAPTER 7

MICROBIAL DISINFECTION FUNDAMENTALS

7.1 INTRODUCTION

All populations of microorganisms grow and die according to physical laws that can be modeled in basic mathematical terms. Microorganisms that are exposed to any biocidal factor, whether heat, dehydration, chemical biocides, or radiation, will die off according to logarithmic relationships. In the presence of growth factors, such as nutrients, microorganisms will grow at an exponential rate. The logarithmic function most often describes both the growth and death of microbial populations for various reasons that have been studied at length (Koch, 1995). Because logarithmic and exponential functions are fairly well defined in most empirical data sets on microbial growth and decay, the prediction of growth and decay rates of microbial populations is often accomplished with reasonable precision.

The various mathematical models described here are not the only models that will produce accurate predictions, but they are the ones most commonly used. These models are presented without derivation, as the weight of empirical evidence has fairly well established their reliability. These models work because they provide the best fit to the data, and whether or not they are exact descriptors of microbiological processes is a matter for ongoing research. It is sufficient that these models are simple and flexible enough to use in evaluating disinfection processes to degrees of accuracy that are acceptable for engineering purposes. This chapter presents the mathematical models that describe microbial population death and shows how they are applied to predict microbial disinfection. It is informative to begin by reviewing the mathematics of microbial growth in order to better understand the basis of the exponential decay models.

7.2 THE CLASSIC EXPONENTIAL GROWTH MODEL

Growth may be defined as a growth in the size of a microorganism or growth in a population of microorganisms. The mechanisms that determine the course of the former are related to the course of the latter. Bacteria provide the best example of cell growth and are loosely representative of the growth of viruses and fungi, although the latter may have a considerably more complex life cycle. Bacteria draw nutrients from their surroundings through their cell walls, which are selectively permeable membranes for the appropriate growth factors. There are three major classes of growth factors—(1) amino acids, (2) purines and pyrimidines, and (3) vitamins. Given these growth factors, the internal machinery of the bacterial cell will manufacture the additional components it needs to grow. The bacterial cell will expand in size until it reaches a limiting size at which point it will begin to divide. The limiting size is not exact, but is a range within which the probability of cell division increases. The critical size at which a microbe divides into two may also be viewed as a critical age at which the cell divides, since growth is essentially a function of time provided nutrients are available and ambient conditions are suitable. The age of division is characteristic of each species and represents a mean age at which the probability of cell division will occur under ideal conditions.

A cell will grow continuously if conditions are right and then it will divide. As the process repeats itself the number of microbes will double for each time period known as the mean doubling time. Figure 7.1 shows a species with a doubling time of 20 minutes. Beginning from a single cell the population will exceed 4000 in 4 hours.

The period of growth is called an exponential or log phase. In microbiological cultures it is often preceded by a lag phase during which adaptive processes occur, and is followed by a stationary phase during which the population remains constant due to its having reached a critical density for the available space and nutrients. Ultimately the nutrients may become depleted and the population may go into a death phase.

Since beyond the first doubling time period the population is a function of powers of two, the growth phase can be modeled in terms of the number of generations as follows:

$$N_t = N_0 2^n \tag{7.1}$$

where N_t = population at time t
N_0 = initial population
n = generation number

Equation (7.1) is equivalent to the Malthus law of geometric growth (Malthus, 1798). Although base 2 seems a logical choice with which to define growth rates, the natural

Time, min	Population
0	1
20	2
40	4
60	8
80	16
100	32
120	64
140	128
160	256
180	512
200	1024
220	2048
240	4096

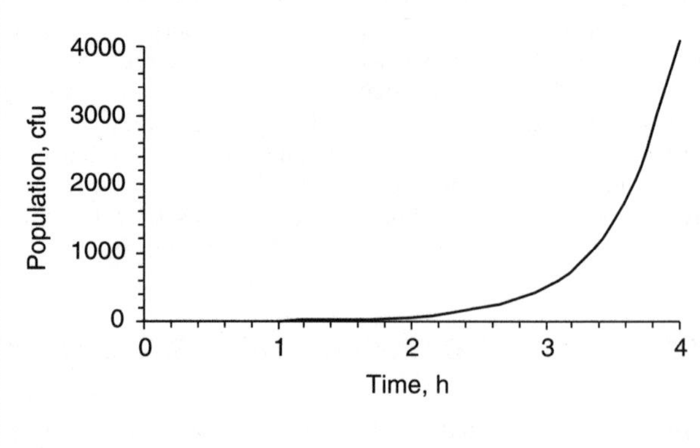

FIGURE 7.1 Population increase for a doubling time of 20 min.

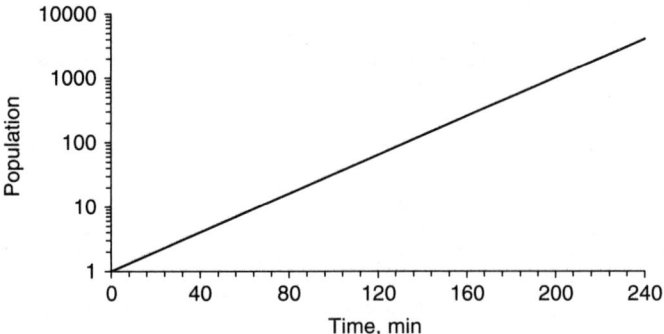

FIGURE 7.2 Population increase for a doubling time of 20 min, logarithmic scale.

base e has come to be commonly used. In exponential terms the growth rate can be written as

$$N_t = N_0 e^{kt} \quad (7.2)$$

where k is the growth rate constant, units of inverse time (i.e., s^{-1} or min^{-1}).

Plotting Eq. (7.2) on a logarithmic scale for the same example in Fig. 7.1 produces the graph in Fig. 7.2. In this example the growth rate constant can be computed by converting from base 2 to base e by taking the natural log of 2, and then dividing by the doubling time as follows:

$$\frac{\ln 2}{20} = 0.03466 \quad (7.3)$$

The growth rate constant is then inserted into Eq. (7.2) to yield the following equation for the population growth shown in Fig. 7.2, with an initial population of $N_0 = 1$:

$$N_t = e^{0.03466t} \quad (7.4)$$

It can be seen that Eq. (7.4) will provide the same results as Eq. (7.1) for any given mean doubling time. The fact that microbial growth is logarithmic has been demonstrated here. The fact that microbial death by almost any means is also logarithmic is not so easy to prove, but it is an observable fact. The logarithm and the related exponential function are ubiquitous in microbiology and the various reasons for this can be further explored in Koch (1966, 1969) and Painter and Marr (1968).

7.3 THE CLASSIC EXPONENTIAL DECAY MODEL

Any microbial population subject to a biocidal factor will tend to decay exponentially over time. The survival of the population is generally measured as a fraction, or as a percentage, of the original population. The survival fraction is defined as

$$S = \frac{N_t}{N_0} \quad (7.5)$$

In Eq. (7.5), N_t represents the number of cells or microbes at some time after exposure to the biocidal factor, while N_0 represents the original population. The survival fraction at any time t after exposure can be defined by the following simple exponential decay equation:

$$S = e^{-t} \tag{7.6}$$

Figure 7.3 illustrates a classic exponential decay curve. Because the survival function is essentially always exponential, it is far more convenient to represent the survival on a logarithmic scale. Figure 7.4 shows the same curve in Fig. 7.3, but on a logarithmic scale. Since the survival curve will always be a straight line on a logarithmic scale, it is easier to perceive the microbial response.

Alternatively, the logarithm of the survival could be plotted on a linear scale, but this approach is rarely used anymore. In such a case Eq. (7.6) is more easily written as

$$\ln S = -t \tag{7.7}$$

The slope of the logarithmic decay curve (the slope of the line in Fig. 7.4) is governed by some factor called a rate constant and which is designated as k. The microbial decay equation is then written as follows:

$$S = e^{-kt} \tag{7.8}$$

In Eq. (7.8), the constant k is called the rate constant. In Fig. 7.4 the rate constant k is 1. The rate constant will determine how fast the population decreases under exposure. The value of the rate constant depends on both the species and the intensity of the biocidal factor. Figure 7.5 illustrates how the rate constant affects the survival curve.

Obviously, the rate constant determines how fast the microbial population decays under the influence of some biocidal factor. This assumes the biocidal factor is constant for all exposures. Often, the biocidal factor, such as heat or radiation, may vary in intensity. The intensity may be the concentration of a biocide, the temperature, or the irradiance of radiation. The variation of intensity is accounted for by a multiplier designated I. The classic exponential decay equation is then written as

$$S = e^{-kIt} \tag{7.9}$$

In the form shown in Eq. (7.9), the rate constant k is known as the standard rate constant and it represents the susceptibility of the species for unit intensity only. In general, k is a

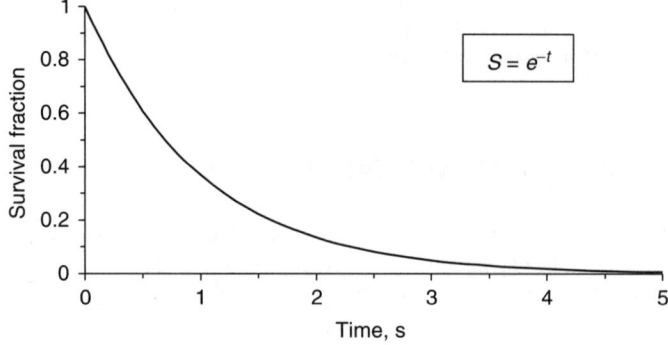

FIGURE 7.3 Basic single-stage survival curve.

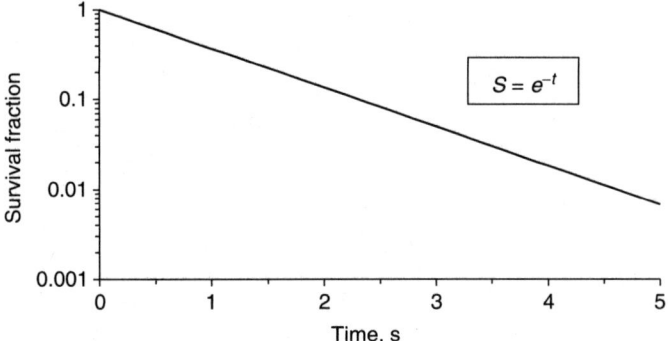

FIGURE 7.4 Single-stage survival curve on a log scale.

true constant—it is unique to the species and is usually always independent of the biocidal factor. Often the quantity "It" is combined into a single term called the dose. The dose can therefore be defined as

$$D = It \tag{7.10}$$

When the dose is defined as in Eq. (7.9), the exponential decay equation is simply written as

$$S = e^{-kD} \tag{7.11}$$

When the dose is defined as a distinct quantity, the graphs are often drawn with the dose as a parameter in place of the time. Figure 7.6 illustrates this style of graph generically. Since the dose is simply the time multiplied by some constant, there is no effective difference in these graphs. On occasion, however, the intensity may vary (during an experiment, for example) and plotting the survival curve versus the dose becomes a convenience. The units of dose depend on the biocidal factor, which may be heat, radiation, or chemical disinfectant concentration. Specific applications will be given later in this chapter.

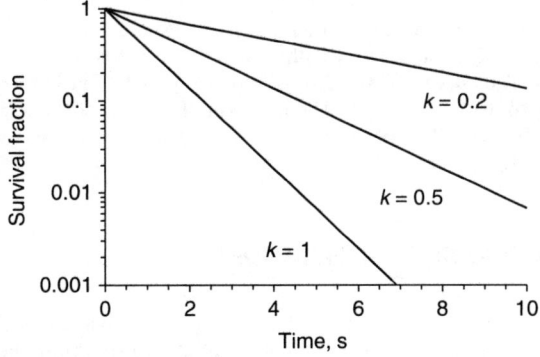

FIGURE 7.5 Survival curves for various rate constants.

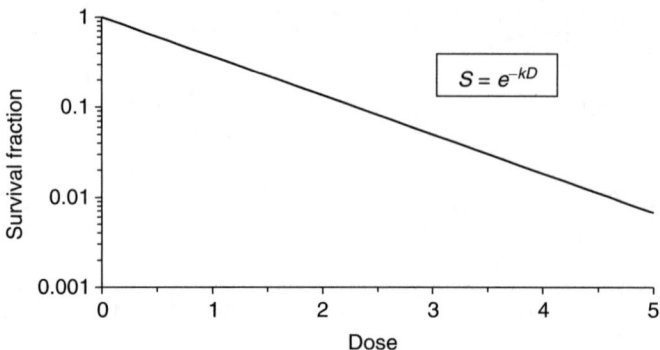

FIGURE 7.6 Exponential decay as a function of dose, in arbitrary units.

In the preceding equations, the units depend on the biocidal factor. Obviously the units for the rate constant and the intensity will be different for heating, radiation, chemical exposure, or any other factor. The appropriate units will be defined in the subsequent chapters where they are applicable.

7.4 THE CLASSIC TWO-STAGE DECAY MODEL

Sometimes a microbial population under exposure to some biocidal factor behaves as if it is two separate populations—one that succumbs rapidly to the factor and another that resists the factor. This effect has often been referred to as tailing or as nonlogarithmic survivor curves (Fujikawa and Itoh, 1996; Moats et al., 1971). Under these conditions the result is a two-stage decay curve. The two-stage curve is treated mathematically as if it were two distinct and separate populations that are simply added together.

Equation (7.12) illustrates the mathematical addition of these populations. Each population has a unique rate constant, denoted by k_1 and k_2. The fraction of the population that is resistant is denoted by f, while the complementary fraction is denoted by $(1-f)$.

$$S = (1-f)e^{-k_1 I t} + f e^{-k_2 I t} \tag{7.12}$$

Figure 7.7 shows a survival curve fitted to Eq. (7.12) based on thermal decay data for a rhinovirus held at 55°C in a solution with a pH of 6.6 to 6.8. Data were selected from Dimmock (1967). The curve was fitted by splitting Eq. (7.12) into two halves and fitting them individually to the split data set. The intercept of the second stage provided the population fraction. The two halves of the equation were then combined and adjustments made to provide the best fit.

7.5 THE CLASSIC SHOULDER MODEL

The exponential decay of a microbial population in response to any biocidal factor is often subject to a slight delay (Cerf, 1977; Munakata et al., 1991; Pruitt and Kamau, 1993). Figure 7.8 shows the survival curve for *Aspergillus niger*, where a shoulder is evident.

MICROBIAL DISINFECTION FUNDAMENTALS 149

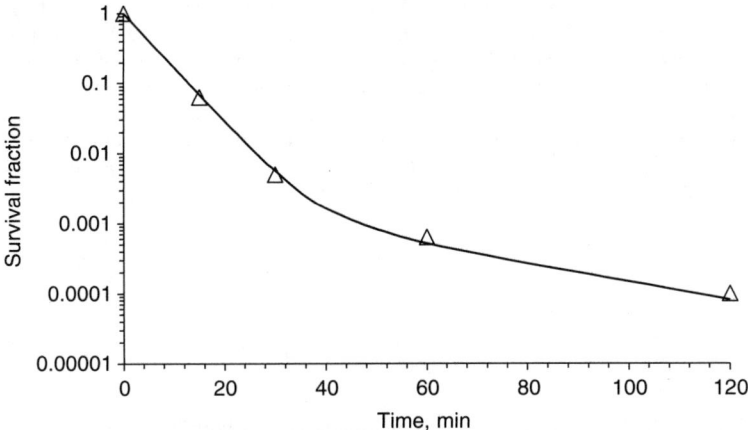

FIGURE 7.7 Survival of rhinovirus at 55°C, at pH 6.6 to 6.8. Resistant population fraction is 0.997. [*Two-stage curve fitted to data from Dimmock (1967)*.]

Shoulder curves start out horizontally before developing into full exponential decay. This will always be true if the initial dose is zero, at time $t = 0$.

The lag in response to the stimulus implies that either a threshold dose is necessary before measurable effects occur or that repair mechanisms actively deal with low-level damage (Casarett, 1968). The threshold is the point at which the exponential decay curve is fully developed. The effect is species dependent and intensity dependent. In many cases it can be neglected, especially for susceptible microbes or for high intensities. However, for some species and sometimes for low intensity exposure, the shoulder can be significant and prolonged.

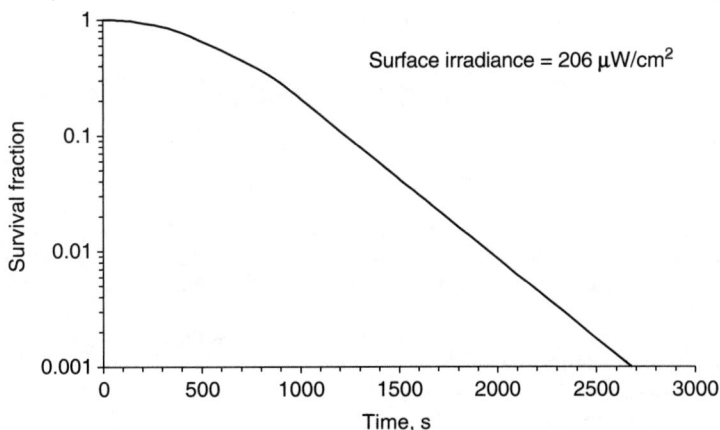

FIGURE 7.8 Survival curve for *Aspergillus niger* under UVGI exposure. [*Based on data from UVDI (2001)*.]

In general, any data set describing single-stage microbial decay can be easily fit to a single-stage exponential decay curve. Normally, the y-intercept is fixed at $S = 1$ when fitting data to a curve. If a shoulder is suspected, the constraint on the y-intercept should be removed and the coefficient of the exponential will then have some value greater than 1. This assumes, of course, that the shoulder is real and not a result of measurement uncertainty.

In two-stage curves there is a separate shoulder for both stages, although the contribution due to the second stage (the resistant fraction) is typically small. The classical model of the complete single-stage survival curve can be defined as the piecewise continuous function (Kowalski et al., 2000):

$$\ln S(t) = \begin{cases} -\dfrac{kI}{4t_c}t^2 & t \leq 2t_c \\ -kI(t-t_c) & t \geq 2t_c \end{cases} \tag{7.13}$$

The time delay (the threshold t_c) may approach zero at high intensities and it may be infinitely long for low intensities. For $t_c = 0$, Eq. (7.13) reduces to Eq. (7.9).

A theoretical basis can be found for the same relationship in the Arrhenius rate equation, which describes the influence of temperature or radiation on process rates as being that of simple exponential decay (Rohsenow and Hartnett, 1973). Assuming, therefore, that the threshold t_c is an exponential function of the intensity I, we can write

$$t_c = Ae^{-BI} \tag{7.14}$$

where A is a constant defining the intercept at $I = 0$ and B is a constant defining the slope of the plotted line of $\ln(t_c)$ versus I.

Given any two sets of data for t_c and I, Eq. (7.14) can be used to determine the values of A and B. Prediction of t_c for any arbitrary value of intensity I then becomes possible. The complete equation can be defined by combining Eqs. (7.12) and (7.14), where a shoulder is considered to be present in both stages:

$$S(t) = fe^{-k_1 I t'} + (1-f)e^{-k_2 I t'} \tag{7.15}$$

where

$$t' = \begin{cases} \dfrac{t^2}{4t_c} & t \leq 2t_c \\ (t-t_c) & t \geq 2t_c \end{cases}$$

Figure 7.9 shows an application of the classical shoulder model to the UV exposure of *Rhizopus nigricans* at different intensities, based on data from UVDI (2001). As the irradiance is increased, the shoulder becomes shorter and will ultimately become negligible at very high values.

Sometimes the first stage intercept may prove to be less than 1, which can be due to experimental error, limited data sets, or even reactivation effects. Any time a data set is evaluated, an error analysis should be performed to verify that the results defining the shoulder and second stage are meaningful.

Although the previous model may be useful for elucidating the effects of intensity on shoulder curves, it is otherwise cumbersome. Somewhat easier to use is the multihit model, which incorporates shoulders and is easily adapted to two stages.

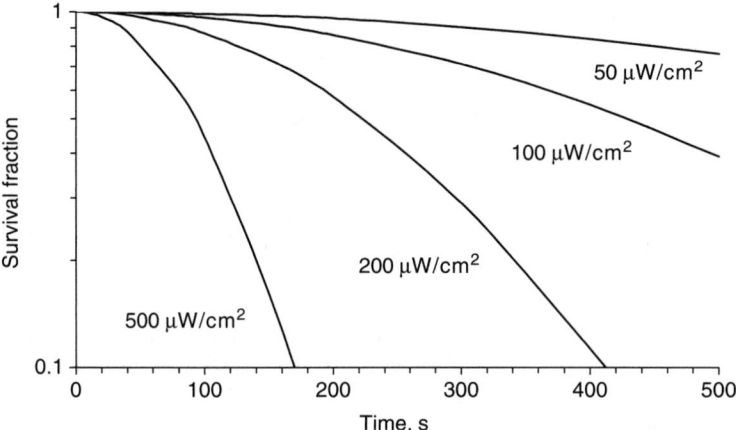

FIGURE 7.9 Exposure of *Rhizopus nigricans* to various levels of UVGI irradiance. [*Based on data from UVDI (2001)*.]

7.6 THE MULTIHIT TARGET MODEL

Alternate mathematical models have been proposed to account for the shoulder including the multihit model or multitarget model, recovery models, split-dose recovery models, and empirical models (Russell, 1982; Harm, 1980; Casarett, 1968). The use of the multihit target model, for example, to determine shoulder characteristics is similar in form to the methods for the classical model (Anellis et al., 1965), and is addressed here for comparison purposes.

The multihit target model (Severin et al., 1983) can be written as follows:

$$S(t) = 1 - (1 - e^{-kIt})^n \tag{7.16}$$

The parameter n represents the number of discrete critical sites that must be hit to inactivate the microorganism, and is unique for each species. In theory n is an integer, but in practice this is not always the case. In Eq. (7.16) the number of targets n must be unique to each population fraction in a two-stage curve, since these behave as though they were independent. Figure 7.10 shows an example of the multihit model being used to fit to a curve with a shoulder. In this case the exponent $n = 2$ and the rate constant $k = 0.4$, for an iodine concentration of 0.194 meq/L (units per Berg et al., 1964). Both the rate constant and the exponent were found through simple trial and error.

By analogy to Eq. (7.12), we can write the complete two-stage equation for the multihit model as follows:

$$S(t) = (1-f)[1-(1-e^{-k_1 It})^{n_1}] + f[1-(1-e^{-k_2 It})^{n_2}] \tag{7.17}$$

In Eq. (7.17), n_1 represents the number of targets for the species in population 1, the fast decay population, while n_2 represents the number of targets in the resistant fraction. Figure 7.11 shows a comparison of shoulder curves generated by the classical model and the multihit model compared against test data on *Staphylococcus aureus* irradiated on petri dishes. The curves do not exactly coincide, but the question of which model is a more accurate predictor is indeterminate due to experimental error. It remains for future research to determine

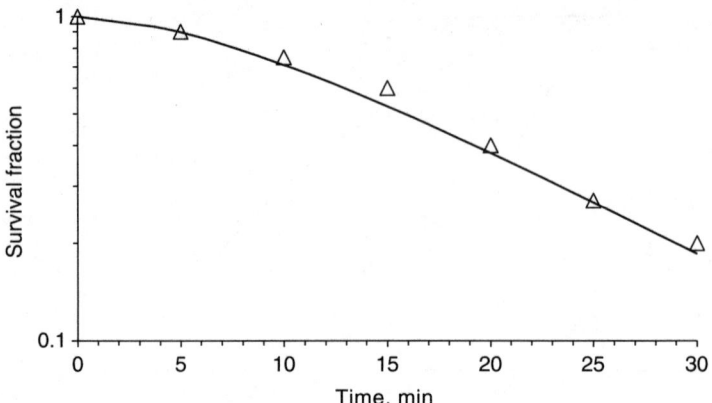

FIGURE 7.10 Population decay of poliovirus exposed to iodine at a concentration of 0.194 meq/L. Line shows fitted multihit model with $n = 2$. [*Poliovirus data from Berg et al. (1964)*.]

which mode is a more accurate predictor of shoulder curves. Either model should suffice for basic analysis and design purposes.

Figure 7.11 shows an example of Eq. (7.17) fit to data for the thermal death of *Listeria* under low pH, based on a select data set from Cole et al. (1993). The curve was fitted through trial and error. Fitting Eq. (7.17) can be challenging and is best facilitated by splitting the data into the first and second stage and fitting the parts independently. The population fraction can then be estimated by the intercept of the second stage. The two halves of the curve can then be combined and any final adjustments made for a best fit (i.e., least squares fit).

Several additional models have been proposed, including variations of the classic model (Fujikawa and Itoh, 1996), variations of the multihit model (Casarett, 1968), multitarget models (McNally, 1982), a logistic model (Whiting, 1993), a sigmoid curve model (Whiting, 1991),

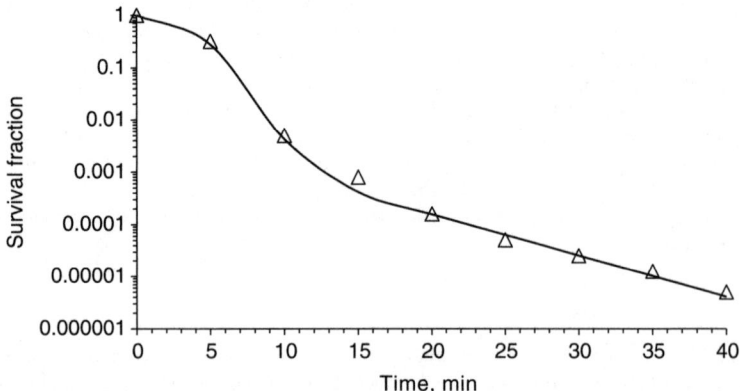

FIGURE 7.11 Death of *Listeria monocytogenes* under heat exposure between 56 and 62°C and pH of 4.53 to 4.83. Line shows a fitted two-stage multihit curve with $n_1 = 30$, $n_2 = 2$, $f = .997$. [*Listeria data from Cole et al. (1993)*.]

a series event model (Severin et al., 1983), and others. Most of these models are functional, if not easy to use, but provide no additional benefits over the four fundamental models presented here. These four models are summarized for easy reference in Table 7.1, which indicates the conditions under which they are applicable. It should be noted that if a model is applied inappropriately, it may grossly overpredict or underpredict inactivation rates for any disinfection process. For example, if a simple single-stage model is used to predict inactivation rates for a microbe subjected to extended exposure, perhaps in the attempt to achieve sterilization, the microbe may end up in the resistant second stage. The result is that the predicted exposure time will be far too short to obtain sterilization, which is defined as six logs of reduction.

7.7 RECOVERY AND PHOTOREACTIVATION

Microbes may recover from a biocidal process for various reasons. If the biocidal factor is not severe enough to destroy the entire population, only the susceptible portion may succumb, and the resistant portion may continue to survive or even grow. Some of the factors that may contribute to recovery include relative humidity, visible light, and nutrient availability. Recovery during UV exposure is a phenomenon called photoreactivation. Visible light can actually repair thymine dimers and other DNA or RNA damage done by UV.

In most cases, recovery due to growth during irradiation can be assumed to be negligible or accounted for by the model if the data for the UV rate constant were taken under conditions in which photoreactivation occurs—this should be at least partly true if the parameters are based on a broad range of empirical data. A variety of studies and models have been developed to address photoreactivation and these can be consulted for more detailed information (Peccia and Hernandez, 2001; Setlow, 1966). This issue will be addressed further in Chap. 11.

Recovery of spores, although not well understood, is recognized as a process associated with germination (Russell, 1982). The recovery of spores is a self-limiting factor under UV or ionizing radiation since a germinated spore invariably becomes less resistant to UVGI irradiation (Harm, 1980).

7.8 SPECIFIC DISINFECTION MODELS

Examples of the previous disinfection models will demonstrate the use of these models under a variety of common biocidal factors. The factors of most interest include thermal disinfection, desiccation, ionizing radiation, nonionizing radiation, nutrient deprivation, freezing, chemical disinfection, pressure, aerosolization, oxygenation, and hydrogen ion concentration. Not all of these factors are relevant to aerobiological engineering but they may arise as questions in related areas like foodborne pathogens and laboratory analysis. Some of these factors will be revisited in later chapters that address specific air disinfection technology applications.

7.8.1 Thermal Disinfection

The death rate of any microbe under heating will depend on the temperature to which the microbe is exposed. Under temperatures that are ideal for growth, the rate constant would be zero. Under higher temperatures, the rate constant would have some positive value. The value under dry heat differs from that under moist heat. Many species have varying degrees of heat resistance and some can withstand prolonged exposure to high temperatures

TABLE 7.1 Summary of Exponential Decay Models

Single-stage model

$S = e^{-kD}$

Simple log-linear decay
Valid in the first-stage region only
Not valid in the shoulder region (if any)
Not valid in the second-stage region

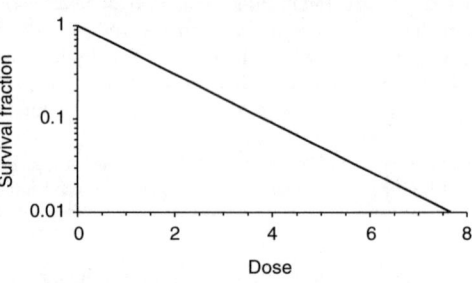

Single-stage shoulder model

$S(t) = 1 - (1 - e^{-kD})^n$

Log-linear decay with shoulder
Not valid in the second-stage region
Typical values for $n = 1\text{–}1000$

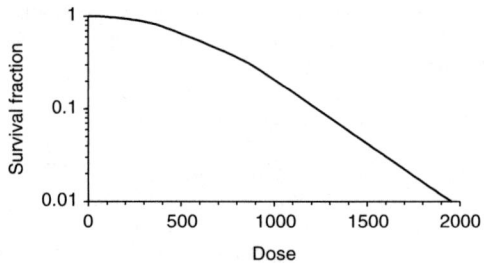

Two-stage model

$S = (1-f)e^{-k_1 D} + fe^{-k_2 D}$

Two stages of log-linear decay
Not valid in the shoulder region (if any)
Typical resistant fraction = 0.05–0.5

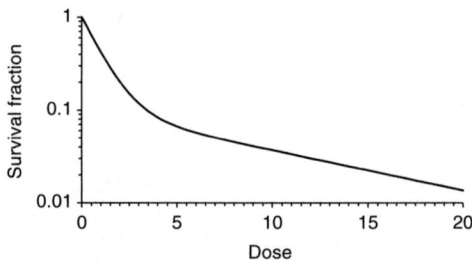

Two-stage shoulder model

$S(t) = (1-f)[1-(1-e^{k_1 D})^{n_1}] + f[1-(1-e^{k_2 D})^{n_2}]$

Valid everywhere
Typical resistant fraction = 0.05–0.5
Typical values for $n = 1\text{–}1000\text{-}$
Two stages of log-linear decay
Shoulder in both stages

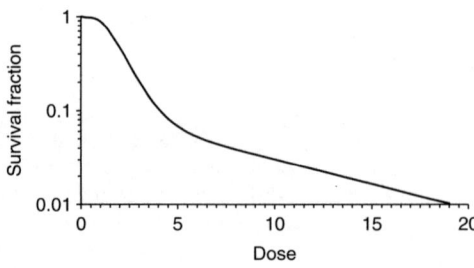

(Sapru and Labuza, 1993). Most empirical results come from tests done in liquid solution, where it is necessary to account for the pH of the liquid since pH affects the decay rate. The thermal rate constant is a function of the temperature, but it is not a linear function as it tends to have lower values at lower temperatures (Mitscherlich and Marth, 1984). The food industry has used thermal decay computations to establish disinfection rates with good success (Whiting, 1991). The common method is to use versions of Eq. (7.1) with specific parameters called F values, z values, and D values defined as follows (Mitscherlich and Marth, 1984; Ray, 1996):

F value. Number of minutes required to sterilize specific bacteria at 121.1°C (250°F). F values at other temperatures may be designated with a subscript as in F_{60} for F value at 60°C (140°F).

D value. Number of minutes at 121.1°C (250°F) to disinfect the bacterial species by 90 percent (10 percent survival). D values at other temperatures may be designated with a subscript T as in D_{60} for D value at $T = 60°C$ (140°F). It is typically determined as follows, with all parameters as previously defined:

$$D_T = \frac{t}{\log_{10} N_0 - \log_{10} N_t} \tag{7.18}$$

k value. Natural logarithm of the slope of the survivor curve. This is the same as the rate constant defined in the previous exponential decay equations.

K value. Base 10 logarithm of the survivor curve. The K value is converted to k by multiplying by the natural log base $e = 2.3026$, or $K = ek = 2.3026k$.

z values. The slope of the thermal death time curve over one log. A z value of 10 (in °C units $z = 10°C$) implies that if the D value is 50 min at 100°C, the D value will be 5 min at 110°C.

The thermal death time curve is a useful tool in the food industry and is a plot of the D values versus temperatures. Figure 7.12 shows an example of a thermal death time curve in

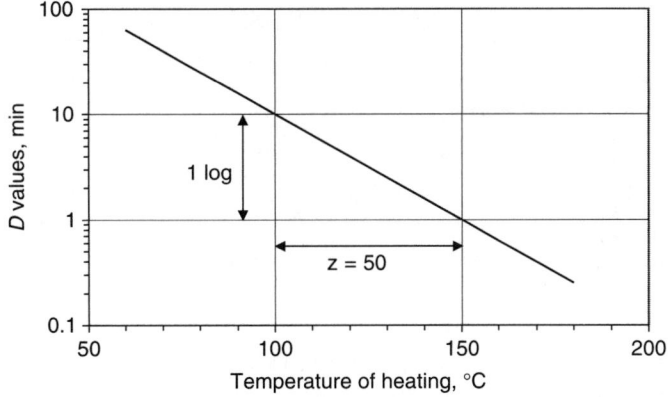

FIGURE 7.12 Hypothetical thermal death time curve with a z value of 50.

which the z value is 50°C, implying that a 50°C temperature rise (above some baseline temperature) would be required to obtain a one-log reduction in microbial population. The thermal death time curve is assumed to be linear, but this would only be true within some limited range since the value of z often tends to be lower at lower temperatures. Equation (7.18) represents, in effect, a single-stage model.

The single-stage equations are sufficient for ordinary thermal disinfection applications but they are merely simplistic models. The actual survival curves of microbes under heating, or almost any biocidal factor, are subject to tailing (existence of a second stage in the decay curve), and sometimes shoulder effects. The previously detailed two-stage multihit model, Eq. (7.16), can be used for more predictive accuracy in food or water sterilization applications. Temperature may impact the susceptibility of microorganisms to other factors like UV irradiation but the impact of the normal range of indoor temperatures is insignificant and can be ignored (Rentschler et al., 1941).

7.8.2 Desiccation

The death of bacteria may occur during or after desiccation. This is caused primarily due to the removal of water, which is essential for normal bacterial cellular functions, but the actual death is due to a complex interplay of factors (Mitscherlich and Marth, 1984). Dried bacteria die from simultaneous stresses that mainly relate to oxidation processes. Airborne microorganisms invariably lose water that is replaced upon host infection or when they reach moist environments. The decay rate of airborne microbes under various relative humidities defies analysis with simple rate constants. Due to the effects of relative humidity on conformational stability in component macromolecules, microbes may go through a phase transition in which survival is lower in midrange humidities (Cox, 1987). Figure 7.13 illustrates the relationship between survival and relative humidity for airborne *Francisella tularensis*. Survival was highest for high humidities, but midrange humidities produced the lowest survival.

Insufficient data are available at present to derive any general mathematical models for the relative humidity effect. Relative humidity has an effect on UV rate constants that can be quite significant and resembles Fig. 7.13 in form (Peccia et al., 2001).

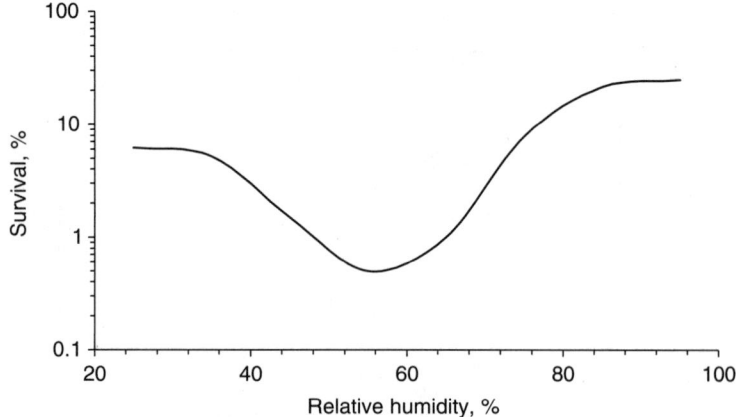

FIGURE 7.13 Survival of airborne *Francisella tularensis*. [*Based on data from Hers and Winkler (1973).*]

7.8.3 Oxygenation or Open Air Factor

Oxygen at normal levels does not harm most airborne microorganisms but in combination with desiccation it can destroy them. Oxygen toxicity is observed only at relative humidities below about 70 percent and the toxic effect of higher concentration reaches a maximum at about 30 percent (Cox and Wathes, 1995; Cox et al., 1973). Oxygen susceptibility usually increases with the degree of desiccation (Cox, 1987). However, the effects of relative humidity on airborne microbes are complex and involve phase changes at the molecular level, as shown in the previous section. Oxygen toxicity will increase with increased oxygen concentration, as shown in Fig. 7.14, for different values of oxygen concentration.

In addition to oxygen, the outdoor air (the open air) often contains a variety of pollutants including ozone which can act as biocides (de Mik and de Groot, 1977; Benbough and Hood, 1971). If a cubic meter of air contains 1000 cfu of bacteria and a pollutant in the air is at 1 ppm, then the mass of the pollutant is about four million times greater than the mass of the bacteria (May et al., 1969). The survival of many species of airborne pathogens is much lower in the outdoor air than in indoor air (Cox and Wathes, 1995). In general, the effects of oxygenation and open air under normal conditions are so minor that they can be ignored in comparison with other disinfection mechanisms like UV irradiation.

7.8.4 Ionizing Radiation

Ionizing radiation includes gamma rays, alpha rays, beta rays, x-rays, electron beams, emitted protons, and emitted neutrons. Exposure of microorganisms to ionizing radiation produces exponential decay in populations and can be described by the previous equations (Mitscherlich and Marth, 1984; Bertani, 1960). Much research has been done on the effects of ionizing radiation on mammalian cells and the results parallel those for bacterial cells (Coggle, 1971). The repair of injury to mammalian cells exposed to lethal radiation is similar to the recovery of bacterial cells but is a topic of intense interest in radiation biology which is often applied to destroy tumor cells while minimizing damage to the patient (Alpen, 1990; Pizzarello and Witcofski, 1982). This aspect is of somewhat less interest in microbial disinfection since the object is normally to maximize the dose and achieve

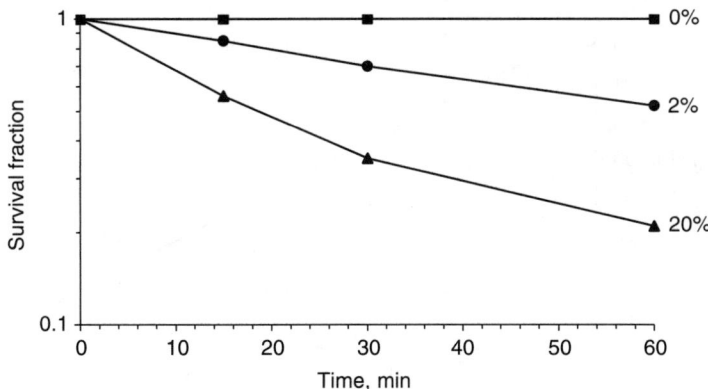

FIGURE 7.14 Survival of airborne *Serratia marcescens* at 25°C, 50 percent relative humidity, and the oxygen concentration in percent as indicated. [*Based on data from Cox (1987).*]

sterilization. Sterilization with ionizing radiation is a common technique in the food industry where it has proven highly successful due to its penetrating ability (WHO, 1988; Loaharanu and Thomas, 1999). At high energy levels ionizing radiation can make certain organic constituents radioactive, but below a certain threshold these kinds of reactions do not occur.

Several factors may influence the susceptibility of a microorganism to radiation including growth phase, temperature, water activity, the oxygen concentration, and the presence of sensitizing agents. There is considerable evidence that genetic material, DNA or RNA, may be the primary site of lethal radiation damage in an exposed virus or bacteria (Casarett, 1968). Cell inactivation may be caused by double-strand breaks in DNA or RNA. However, under some conditions self-repair of chromosome damage may take place and it is possible that nonchromosomal damage may also contribute to irreversible cell death (Grecz et al., 1977). The formation of oxidizing compounds like hydroxyl radicals may also contribute to DNA double-strand breakage (McNally, 1982). There is evidence that DNA base composition correlates with the susceptibility or resistance of bacteria to radiation (Kaplan and Zavarine, 1962).

The multihit and multitarget models are widely applied in radiation biology. Various units of measurement are used to define ionizing radiation including radiation absorbed dose (rad), which is defined as 0.01 J/kg, the gray (Gy), which is defined as 1000 rad, the Roentgen, which is defined as 2.58×10^{-4} C/kg, and the electron volt, among others. Regardless of what units are used, the survival curve produced will have the same exponential decay characteristics. Figure 7.15 shows the survival curve of *Serratia marcescens* under x-ray exposure. Data have been fit to the single-stage multihit model of Eq. (7.16) with a rate constant of 0.0032 krad^{-1} and an exponent of $n = 4$:

$$S = 1 - (1 - e^{-0.0032D})^4 \qquad (7.19)$$

7.8.5 Nonionizing Radiation

Both visible light and ultraviolet light are considered to be nonionizing radiation. *Ultraviolet light* (UV) in the range of 100 to 400 nm wavelengths has strong biocidal effects. Visible light or daylight in the range 400 to 759.4 nm wavelengths contains trace

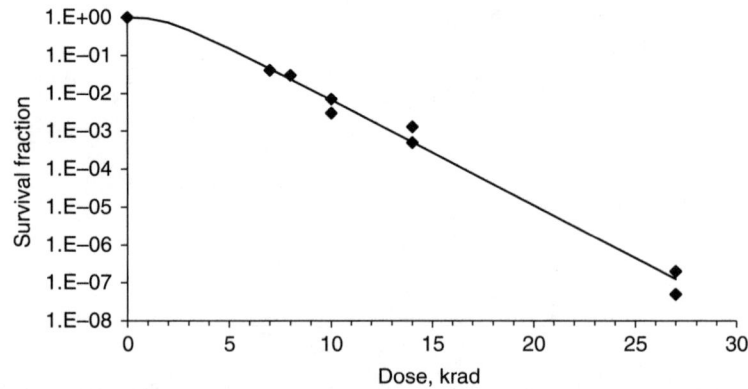

FIGURE 7.15 Survival of *Serratia marcescens* under exposure to x-rays in oxygen. Curve is a fitted multihit model. [*Data from Casarett (1968).*]

levels of UV and will have the effect of disinfecting exposed microbial populations even without the presence of UV wavelengths (Futter and Richardson, 1967; Webb, 1961). Irradiation with UV produces characteristic exponential decay of microbial populations and often produces both a shoulder and a second stage. Maximum bactericidal effectiveness occurs at a wavelength of about 265 nm as shown in Fig. 7.16. UV lamps put out about 95 percent of their light at a wavelength of 253.4 nm, which has a relative bactericidal effectiveness, or the degree of germicidal activity, of about 84 percent (Rea, 1990). It is the close coincidence of the UV light spectrum with the germicidal peak of microbes that gives UV its disinfection capability. The literature on the germicidal effects of UV is extensive and the reader should see Chap. 11 for more detailed treatment of the subject. See also Fig. 7.8 for an example of a UV survival curve.

7.8.6 Nutrient Deprivation

Reducing nutrient availability will literally starve bacteria. Bacteria grown in culture that reach a critical density in a finite medium will first enter the stationary phase and then the death phase. The death phase represents exponential decay of the population. The death phase is a result of both nutrient deprivation and buildup of toxic wastes. After some major reduction in the population a second stage may be entered reflecting extended survival of a resistant fraction (Prescott et al., 1996). The result is virtually identical to a two-stage decay curve. Limited data are available that would allow fitting a two-stage curve but Postgate and Hunter (1962) report that deprivation of certain essential nutrients will cause death to occur linearly over time at rates of about 6 to 12 percent per hour. They also report the occasional occurrence of two stages in the death curve and of limited recovery after extended incubation.

7.8.7 Freezing

Low temperatures can cause cell death depending on whether the microbes are frozen or just maintained at low temperatures. Death from freezing may be due to more than one factor. Freezing may cause water crystallization and induce cellular damage. Freezing generally causes rapid death while death from prolonged cold temperatures is known as

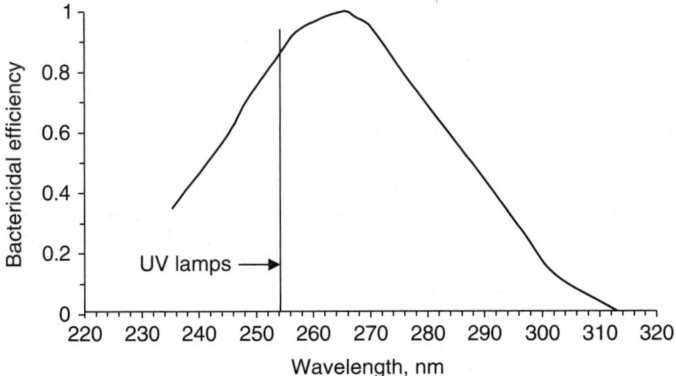

FIGURE 7.16 Bactericidal efficiency of UV on *E. coli* as a function of wavelength. [*Based on data from Rea (1990).*]

storage death. Storage death is a function of time and temperature similar to death by heating. Death by freezing is independent of the freezing temperature, although the rate of storage death is much greater at 0 to –10°C than at –17 to 30°C and below. Storage at temperatures from –70 to –195°C results in preservation or low death rates. Factors that influence cell damage due to freezing include the cooling rate to the freezing point, the freezing temperature, population density, nutritional status, composition of the chilling medium, and growth phase. The extent of cell damage due to freezing can depend on the chemical composition of the chilling medium, salt concentration, pH, and other factors. Loss of internal solute can also influence death of cells due to freezing.

The more rapidly the freezing and thawing process, the more likely bacteria are to remain viable. Rapid freezing may prevent crystallization. Bacterial cultures may be preserved by dehydrating them at low temperatures through sublimation at low pressure, and various additives may assist preservation (Henis, 1987). Cryogenic freezing is used to rapidly destroy bacteria in the food industry (Andrews et al., 2000, Ray, 1996). Typically, freezing from –2 to –20°C will destroy bacteria through water crystallization but insufficient data are available to quantify the survival curves or characterize rate constants at various temperatures.

7.8.8 Disinfectants

A wide variety of chemicals have biocidal properties and are commonly used as disinfectants. Some of the most commonly used disinfectants include alcohol, formaldehyde, glutaraldehyde, formalin, sodium hypochlorite, phenolics, ethanol, iodine, chlorine, and sodium hydroxide. These disinfectants are generally mixed in solutions with water for use. Most disinfectants are effective against a broad range of microorganisms but there are varying levels of resistance among species. Most disinfectants are used for disinfection and not for sterilization (Wilson, 2001). The survival curves of microbes exposed to disinfectants are adequately modeled by the exponential decay equations previously summarized. See Fig. 7.6 for an example of microbial population decay under iodine exposure. Antibiotics have a similar effect on microbial populations but operate by different mechanisms. There are several classes of antibiotics and they generally work by inhibiting microbial growth or by interfering with normal cellular processes and causing cell death directly or indirectly (Ryan, 1994). The difference between antibiotics and disinfectants is primarily that the antibiotics pose no hazard to hosts but selectively inhibit infectious microbial agents. Antibiotics have selective toxicity for microbes and must be chosen based on the specific pathogen to be targeted (Prescott et al., 1996).

7.8.9 Pressure

High pressure can kill microbes and sudden variations in pressure can rupture microbes. Pressures as high as 200 atm (2×10^6 bar) can be lethal for many bacteria (Atlas, 1995). Temperature can impact the hydrostatic inactivation pressure (McClements et al., 2001). Osmotic pressure that exists across cell walls in solution can cause cells to lyse if it becomes extreme. Spores tend to be more resistant to pressure than bacterial cells. The sudden heating of bacteria, such as may occur with pulsed light, can induce internal pressure spikes that rupture cell walls or lyse the bacterial cell (Wekhof et al., 2001).

Low pressure can also damage cells. Some microbes, especially bacterial and fungal spores, have extraordinary resistance to low pressures and can survive near-vacuum conditions, as well as survive in space (Sussman and Halvorson, 1966; La Duc et al., 2004). Insufficient data are available to determine the characteristic survival curves of bacteria exposed to high or low pressure or to sudden pressure changes.

7.8.10 Aerosolization

The very act of aerosolization can reduce microbial populations. When bacteria are aerosolized via a sneeze or through aerosolization equipment, they are subject to die off as the result of physical stresses (Willeke et al., 1995). These stresses may include sudden pressure changes or simply the friction involved in high velocity airstreams. A secondary effect occurs purely as the result of being airborne—microorganisms will die off in the air due to several factors that may include desiccation, oxygenation, and exposure to visible light or pollutants. Due to the complex nature of death due to aerosolization, no modeling is likely to succeed in conforming data, if it could be acquired, to any survival curves.

7.9 STERILIZATION

By definition, sterilization is the complete destruction of bacteria or inactivation of viruses in a sample. However, for purposes of modeling inactivation it is common to use six logs of reduction as a mathematical definition of sterilization. If indoor air was disinfected to six logs of reduction, it is improbable that surviving microbes, if any, would be capable of inducing respiratory infections and so this approach is reasonable for these design purposes. Table 7.2 illustrates this mathematical definition with an initial microbial population of 1 million cfu. Six logs of reduction would leave 1 cfu, which would not be sterilization in a laboratory setting, but if an air cleaner achieved a six log reduction, the air could be considered sterilized, since airborne concentrations of 1 million cfu (i.e., cfu/m^3) would never be seen in a normal indoor environment. The modeling problem can be seen in Table 7.2 at seven logs of reduction—the zeros don't compute. Instead of targeting zero survival (i.e., when sizing an air cleaning system) it is more practical to target six logs of reduction.

Table 7.2 also illustrates a point that is sometimes confusing—whether six logs of reduction is the same in natural logs or not. Six base 10 logs are not the same as six natural logs. It can be seen that six base 10 logs equal 13.8155 natural logs, as follows:

$$\ln(S) = 2.3026 \log(S) \quad (7.20)$$

Related terms include cleaning, disinfection, decontamination, and pasteurization. Pasteurization is the elimination of pathogens, but not necessarily the elimination of all microorganisms (Atlas, 1995). Cleaning, disinfection, and decontamination refer to reductions in microbial populations but do not imply any specific level of reduction.

TABLE 7.2 Comparison of Sterilization in Base 10 log and Natural log

Time	P	$\log_{10}(P)$	$\ln(P)$	S %	$\log_{10}(S)$	$\ln(S)$	log Reduction
0	1,000,000	6	13.81551	100	2	4.60517	0
1	100,000	5	11.51293	10	1	2.302585	1
2	10,000	4	9.21034	1	0	0	2
3	1,000	3	6.907755	0.1	−1	−2.30259	3
4	100	2	4.60517	0.01	−2	−4.60517	4
5	10	1	2.302585	0.001	−3	−6.90776	5
6	1	0	0	0.0001	−4	−9.21034	6
7	0	—	—	0	—	—	7

REFERENCES

Alpen, E. L. (1990). *Radiation Biophysics.* Prentice-Hall, Englewood, CO.

Andrews, L. S., Park, D. L., and Chen, Y. P. (2000). "Low temperature pasteurization to reduce the risk of vibrio infections from raw shell-stock oysters." *Food Addit Contam* 17(9):787–791.

Anellis, A., Grecz, N., and Berkowitz, D. (1965). "Survival of *Clostridium botulinum* spores." *Appl Microbiol* 13(3):397–401.

Atlas, R. M. (1995). *Microorganisms in Our World.* Mosby, St. Louis, MO.

Benbough, J. E., and Hood A. M. (1971). "Viricidal activity of open air." *J Hyg* 69:619–626.

Berg, G., Chang, S. L., and Harris, E. K. (1964). "Devitalization of microorganisms by iodine." *Virology* 22:469–481.

Bertani, G. (1960). "Sensitivities of different bacteriophage species to ionizing radiations." *J Bacteriol* 79:387–393.

Casarett, A. P. (1968). *Radiation Biology.* Prentice-Hall, Englewood, CO.

Cerf, O. (1977). "A review: Tailing of survival curves of bacterial spores." *J Appl Bacteriol* 42:1–19.

Coggle, J. E. (1971). *Biological Effects of Radiation.* Wykeham, London.

Cole, M. B., Davies, K. W., Munro, G., Holyoak, C. D., and Kilsby, D. C. (1993). "A vitalistic model to describe the thermal inactivation of *Listeria monocytogenes*." *J Ind Microbiol* 12:232–239.

Cox, C. S., Baxter, J., Maidment B. J. (1973). "A mathematical expression for oxygen-induced death in dehydrated bacteria." *J Gen Microbiol* 75:179–185.

Cox, C. S. (1987). *The Aerobiological Pathway of Microorganisms.* John Wiley & Sons, New York.

Cox, C. S., and Wathes, C. M. (1995). *Bioaerosols Handbook.* CRC-Lewis Publishers, Boca Raton, FL.

de Mik, G., and de Groot I. (1977). "The germicidal effect of the open air in different parts of the Netherlands." *J Hyg* 78:175–187.

Dimmock, N. (1967). "Differences between the thermal inactivation of Picornaviruses at high and low temperatures." *Virology* 31:338–353.

Fujikawa, H., and Itoh, T. (1996). "Tailing of thermal inactivation curve of *Aspergillus niger* spores." *Appl Microbiol* 62(10):3745–3749.

Futter, B. V., and Richardson, G. (1967). "Inactivation of bacterial spores by visible radiation." *J Appl Bacteriol* 30(2):347–353.

Grecz, N., Lo, H., Kang, T. W., and Farkas, J. (1977). Characteristics of radiation survival curves of spores of *Clostridium botulinum* strains. *Spore Research 1976*, Vol. 2, A. N. Barker, J. Wolf, D. J. Ellar, G. J. Dring, and G. W. Gould, eds., Academic Press, London.

Harm, W. (1980). *Biological Effects of Ultraviolet Radiation.* Cambridge University Press, New York.

Henis, Y. (1987). *Survival and Dormancy of Microorganisms.* Wiley-Interscience, New York.

Hers, J. F. P., and Winkler, K. C. (1973). *Airborne Transmission and Airborne Infection.* Technical University, Enschede Oosthoek Publishing, The Netherlands.

Kaplan, H. S., and Zavarine, R. (1962). "Correlation of bacterial radiosensitivity and deoxyribonucleic acid base composition." *Biochem Biophys Res Commun* 8:432.

Koch, A. L. (1966). "The logarithm in biology: Mechanisms generating the lognormal distribution exactly." *J Theoret Biol* 12:276–290.

Koch, A. L. (1969). "The logarithm in biology: Distributions simulating the lognormal." *J Theoret Biol* 23:251–268.

Koch, A. L. (1995). *Bacterial Growth and Form.* Chapman & Hall, New York.

Kowalski, W. J., Bahnfleth, W. P., Witham, D. L., Severin, B. F., and Whittam, T. S. (2000). "Mathematical modeling of UVGI for air disinfection." *Quant Microbiol* 2(3):249–270. http://www.kluweronline.com/issn/1388-3593.

La Duc, M. T., Satomi, M., and Venkateswaran, K. (2004). "*Bacillus odysseyi* sp. nov., a round-spore-forming bacillus isolated from the Mars Odyssey spacecraft." *Int J Syst Evol Microbiol* 54(2004):195–201.

Loaharanu, P., and Thomas, P. (1999). *Irradiation for Food Safety and Quality.* Technomic Publishing Co., Lancaster, PA.

Malthus, T. R. (1798). *An Essay on the Principle of Population.* J. Johnson, London.

May, K. R., Druett, H. A., Packman L. P. (1969). "Toxicity of open air to a variety of microorganisms." *Nature* 221:1146–1147.

McClements, J. M. J., Patterson, M. F., and Linton, M. (2001). "The effect of growth stage and growth temperature on high hydrostatic pressure inactivation of some psychotrophic bacteria in milk." *J Food Prot* 64(4):514–522.

McNally, N. J. (1982). Chapter 2: Cell Survival. *Radiation Biology,* D. J. Pizzarello, ed., CRC Press, Boca Raton, FL.

Mitscherlich, E., and Marth, E. H. (1984). *Microbial Survival in the Environment.* Springer-Verlag, Berlin, Germany.

Moats, W. A., Dabbah, R., and Edwards, V. M. (1971). "Interpretation of nonlogarithmic survivor curves of heated bacteria." *J Food Sci* 36:523–526.

Munakata, N., Saito, M., and Hieda, K. (1991). "Inactivation action spectra of *Bacillus subtilis* spores in extended ultraviolet wavelengths (50-300 nm) obtained with synchrotron radiation." *Photochem Photobiol* 54(5):761–768.

Painter, P. R., and Marr, A. G. (1968). "Mathematics of microbial populations." *Annu Rev Microbiol* 22:519–549.

Peccia, J., Werth, H. M., Miller, S., and Hernandez, M. (2001). "Effects of relative humidity on the ultraviolet induced inactivation of airborne bacteria." *Aerosol Sci Technol* 35:728–740.

Peccia, J., and Hernandez, M. (2001). "Photoreactivation in Airborne *Mycobacterium parafortuitum*." *Appl and Environ Microbiol* 67:2001.

Pizzarello, D. J., and Witcofski, R. L. (1982). *Medical Radiation Biology.* Lea & Febiger, Philadelphia, PA.

Postgate, J. R., and Hunter, J. R. (1962). "The survival of starved bacteria." *J Gen Microbiol* 29:233–263.

Prescott, L. M., Harley, J. P., and Klein, D. A. (1996). *Microbiology.* Wm. C. Brown Publishers. Dubuque, IA.

Pruitt, K. M., and Kamau, D. N. (1993). "Mathematical models of bacterial growth, inhibition and death under combined stress conditions." *J Ind Microbiol* 12:221–231.

Ray, B. (1996). *Fundamental Food Microbiology.* CRC Press, Boca Raton, FL.

Rea, M. S. (1990). *Lighting Handbook Reference & Application,* 8th ed. Illuminating Engineering Society of North America, New York.

Rentschler, H. C., Nagy, R., and Mouromseff, G. (1941). "Bactericidal effect of ultraviolet radiation." *J Bacteriol* 42:745–774.

Rohsenow, W. M., and Hartnett, J. P. (1973). *Handbook of Heat Transfer.* McGraw-Hill, New York.

Russell, A. D. (1982). *The Destruction of Bacterial Spores.* Academic Press, New York.

Ryan, K. J. (1994). *"Sherris Medical Microbiology."* Appleton & Lange. Norwalk, CT.

Sapru, V., and Labuza, T. P. (1993). "Temperature dependence of thermal inactivation rate constants of bacterial spores in a glassy state." *J Ind Microbiol* 12:247–250.

Setlow, J. K. (1966). "Photoreactivation." *Radiat Res Suppl* 6:141–155.

Severin, B. F., Suidan, M. T., and Englebrecht, R. S. (1983). "Kinetic modeling of U.V. disinfection of water." *Water Res* 17(11):1669–1678.

Sussman, A. F., and Halvorson, H. O. (1966). *Spores: Their Dormancy and Germination.* Harper & Row, New York.

UVDI (2001). "Report on survival data for *A. niger* and *R. nigricans* under UVGI exposure." Valencia, CA.

Webb, S. J. (1961). "Factors affecting the viability of airborne bacteria: V. The effect of desiccation on some metabolic systems of *Escherichia coli*." *Can J Microbiol* 7:621–632.

Wekhof, A., Trompeter, I.-J., and Franken, O. (2001). "Pulsed UV-Disintegration, a new sterilization mechanism for broad packaging and medical-hospital applications." *Proceedings of the First International Congress on UV-Technologies,* Washington, DC. http://www.wektec.com/.

Whiting, R. C. (1991). Predictive Modeling. *Food Microbiology,* M. P. Doyle, ed., ASM Press, Washington, DC. 728–739.

Whiting, R. C. (1993). "Modeling bacterial survival in unfavorable environments." *J Ind Microbiol* 12:240–246.

WHO (1988). *Food Irradiation.* WHO, Geneva.

Willeke, K., Grinshpun, S. A., Ulevicius, V., Terzieva, S., Donnelly, J., Stewart, S. and Jouzaitis, A. (1995). "Microbial stress, bounce and re-aerosolization in bioaerosol samples." *J Aerosol Sci* 26(S1):s883–s884.

Wilson, J. (2001). *Infection Control in Clinical Practice.* Balliere Tindall, Edinburgh.

CHAPTER 8

BUILDINGS AND ENCLOSURES

8.1 INTRODUCTION

Buildings provide a barrier to the outdoor air and the weather, and create enclosed environments that are comfortable for living and suitable for working or storage. Since environments that are comfortable for human habitation have the unintended effect of fostering the transmission of microbial parasites as well as providing for the growth of certain fungi and bacteria, it can be useful to examine the types of buildings and their vulnerabilities to aerobiological contamination. Figure 8.1 shows a proposed breakdown of buildings according to how well they protect against airborne or other diseases. Although the focus here is on the aerobiology of indoor environments, it is sometimes difficult to separate the problems caused strictly by microbes and those caused by man-made nonmicrobial sources. *Sick building syndrome* (SBS), for example, may be due to microbial causes in only 15 to 30 percent of cases (Godish, 1995; Lippmann, 1998).

The proposed breakdown of buildings in Fig. 8.1 is somewhat arbitrary and loosely defined but it illustrates the fact that there are significant differences in the relative healthiness of buildings. The problem building classification in Fig. 8.1 can be considered to include buildings that are definitively unhealthy and in need of immediate remediation. Normal buildings do not specifically protect against airborne diseases and passively allow airborne diseases to transmit, but are otherwise not problem buildings. Healthy buildings are often defined as those that promote good *indoor air quality* (IAQ) or have reduced indoor levels of airborne microbes relative to outdoors (Woods et al., 1997). Immune buildings are defined as those which actively or passively suppress the transmission of airborne disease and remove aerobiological contaminants (Kowalski, 2003).

This chapter reviews how buildings contribute to or detract from human health in the context of respiratory diseases. Ventilation systems and disinfection technologies are not specifically addressed here since they are addressed in later chapters. This chapter covers types of buildings, building materials, microbial growth, building airtightness, building dampness, and a discussion of sick building syndrome and building related illness. Also addressed is the informative topic of the evolution of human habitation and what part it has played in fostering the evolution of airborne diseases.

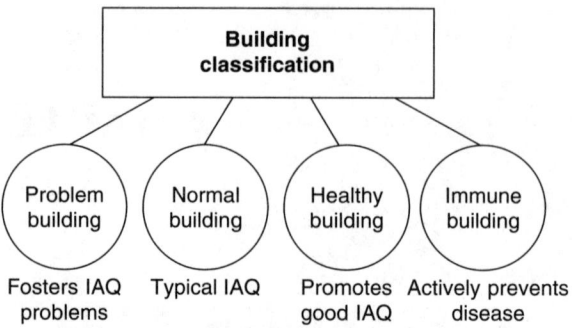

FIGURE 8.1 Building aerobiological classifications.

8.2 THE EVOLUTION OF HUMAN HABITATS

The evolution of human habitats has provided for the coevolution of airborne diseases (see Chap. 1). Indoor environments act as a vector facilitating the transmission of respiratory pathogens that have not evolved survivability in the external environment (Ewald, 1994). Since before the Stone Age, humans and their ancestors sought refuge from the elements by huddling together within caves and found shelters. The common protection they obtained also fostered the exchange and proliferation of various pathogens, most of which probably transmitted originally by direct contact. When man's ancestors first ventured out of the Rift Valley and onto the plains and the savannahs of Africa, they naturally sought shelter as protection against the elements and predators (Leakey and Lewin, 1977). Where no caves or shelter could be found, they built them up from branches, bones, and stones. Archeological evidence indicates Homo habilis was constructing crude forms of shelter about 1.8 million years ago, probably wood/branch/leaf-enclosed structures secured with stones at the base (Jelinek, 1975). The airtightness of such structures is minimal, but the close quarters are likely to have preserved the already existing forms of direct-contact pathogens.

Beginning in the Aechulean period, around 1.5 to 0.5 million years ago, humans developed improved weapons for hunting and learned to make tents from animal skins. The oldest tent found in this period was actually constructed inside a cave and made of animal skins draped over a wooden structure, and about 12 ft. × 35 ft. in size and divided into two rooms, sufficient for sleeping an extended family. As shelters became progressively cozier and more airtight, the pathogens they brought with them became more adapted to survive indoors and less likely to survive outdoors. At this point some microbes probably developed the ability to survive for short periods in air long enough to be inhaled and infect a new host.

The sudden improvement in tool technology about 500,000 years ago, during the Mousterian Period, corresponded with a refinement in the quality of the shelters that could be made. Larger tents and huts with increasingly sophisticated structures and designs for indoor heating appear right down to the Upper Paleolithic, from about 50,000 BC (Binford, 1983). Both Neanderthals and Cro-Magnons coexisted with similar Neolithic technologies at this time, indicating some degree of trade and interaction within the limited overlap of their preferred climates (Leakey and Lewin, 1992). It is possible that these two species of humans were also exchanging pathogens. During the last Ice Age, about 30,000 to 15,000 BC, when Neanderthals were already extinct and Cro-Magnons had adapted to cold climates, the practice of living inside tents and wooden huts was surely the norm. The mammoth hunters of Eastern Europe spent millennia hunting mammoths to extinction, and lived in tents made exclusively of mammoth hides and bones (Jelinek, 1975). About the same time in the warmer climates to the south, in the Middle East, Africa, Asia, Australia and South America, people were already building wooden huts, herding and possibly husbanding animals.

Permanent structures would have provided the first ideal environment for airborne pathogens to develop more sophisticated means of respiratory infections. Although permanent houses are thought to correspond with the beginnings of agriculture about 12,000 to 9000 BCE, the existence of trade in hunter-gatherer societies could place the first permanent housing much earlier. Evidence suggests that late in the Stone Age every habitable cave in Europe was occupied not just seasonally but permanently (Wymer, 1982).

Vitruvius, in his book "On Architecture" in about 27 BCE, details the construction of the first shelters from branches and leaves, the digging of caves under hills, and the development of wattle-work or mud-coated branch-covered shelters which led to the first huts. The first houses are described as being built from logs and had clay, oak shingles, or thatched roofs. The logs were stacked with gaps between them that were filled with wood, sod, and clay. The logs were stacked in a pyramidal shape to create the roof, which was covered with leafage and clay. Where wood was not plentiful, trenches were dug through mounds and roofs built over them, then covered with soil. These, Vitruvius said, were quite warm in the winter and quite cool in the summer. In some countries, like England, houses could be built up with flat stones instead of logs for the roofs. The relative airtightness of these structures and the long months spent inside would have provided airborne microbes extensive opportunities for exchange and adaptation.

In ancient mild climates buildings were naturally ventilated, but in cold climates, especially in Europe during the Little Ice Age, AD 1350 to 1850, and in America during the latter half of this period, people would have been forced to make their homes more airtight (Imbrie and Imbrie, 1979). This period corresponds with the explosive spread of a number of airborne pathogens, including TB and also the plague. Figure 8.2 shows a figurative

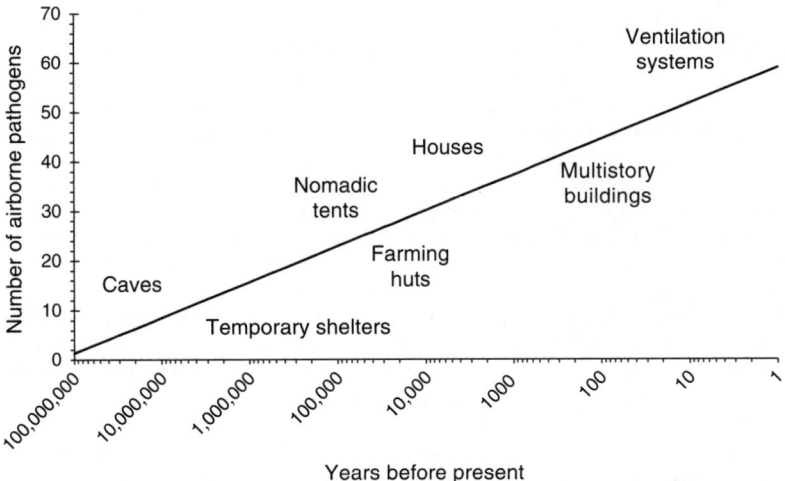

FIGURE 8.2 Development of housing and the parallel increase in airborne diseases (*based on estimates of first appearances of pathogens in Chap. 2.*)

estimate of the development of housing and the associated first appearance of the 50 or so airborne pathogens that we have today (see Chap. 1).

The advent of modern technology brought forced ventilation into buildings, and although this provided fresh, tempered air and helped reduced disease over much of the past century, the trend has been partly reversed due to attempts to save energy by reducing outdoor airflow. We have seen a number of outbreaks directly traceable to buildings and their ventilation systems in recent decades. More attention is now being paid to building design, maintenance, and microbial disinfection than has been the case in the past, but the problem is likely to continue, and furthermore, selective pressures are now at work.

The selective pressures in indoor environments are mostly favorable to airborne pathogens because living and working spaces are comfortable by design. The protection from direct sunlight, the humidity and moisture, the available host reservoir, and ventilation systems aiding in dispersion represent selective pressures that are favorable to airborne pathogens. Pathogens adapted in these environments should be capable of evolving increased diversity and virulence, with the building playing the part of a disease vector. With humans further vectoring the microbes from building to building, they can periodically establish themselves in new buildings long enough to cause additional infections.

There are some factors that could produce selective pressures on well-established indoor pathogens. Microbes that can pass through low efficiency filters, or tolerate momentary chilling to 12°C or heating to 50°C as they pass through the cooling and heating coils may be selected over time. Also the occupancy schedule of the buildings, being often unoccupied over night for 8 to 10 hours or over a weekend for 48 hours, could select for longer surface survival or even short-term dormancy of pathogens. *M. tuberculosis*, for example, is already capable of surviving this length of time on surfaces. Any method of disinfecting air, if not highly effective, could play a part in selecting for future pathogens in exactly the same way that inadequate dosages of antibiotics have produced drug-resistant strains of various bacterial pathogens.

8.3 BUILDING TYPES

The modern world provides an almost endless variety of buildings and structures with new types being developed all the time. Table 8.1 lists buildings designed for daily human occupation, grouped and categorized along with estimated floor space and occupancy ranges (EIA, 1989; Bell, 2000). These are based on EIA (1989) categories and do not correspond exactly to the distinct aerobiological categories of buildings and environments addressed individually in Chaps. 28 through 39.

Commercial buildings include some of the largest and most diverse structures in the world today, and are the type of building in which most people spend much of their day. They include office buildings, manufacturing facilities, and businesses of every variety. Airports are considered large commercial buildings although they more closely resemble assembly buildings due to their large enclosed volumes.

Most government buildings are similar to commercial buildings in form and operation. Health care and research facilities are a distinctly separate type of building due to their special ventilation requirements. Lodging facilities include residential homes, apartment buildings, dormitories, and hotels and are also a special category since they have a wide range of occupancies and often have no specific ventilation requirements. This latter category is of great interest because of the large fraction of diseases that may be transmitted in the home environment or in school dormitories. Food and entertainment facilities, mercantile facilities, education facilities, and assembly buildings have transient occupancies that can be very high.

TABLE 8.1 Building Types, Sizes, and Occupancies

Category	Building type	Median size, ft^2		Occupancy per 1000 ft^2	
		Min	Max	Min	Max
Commercial industrial	Commercial office	2,500	500,000	7	13
	Banks	2,500	25,000	7	20
	Manufacturing	2,500	500,000	3	10
	Airports	10,000	500,000	10	20
Government	Courthouses	2,500	100,000	7	20
	Municipal	2,500	200,000	7	20
	Police station	2,500	100,000	2	10
	Post office	2,500	200,000	2	10
	Prisons	5,000	250,000	3	20
	Military facilities	5,000	250,000	2	50
Food and entertainment	Food processing	2,500	200,000	3	10
	Restaurants	2,500	25,000	20	67
	Nightclubs	2,500	25,000	20	67
Health care	Hospitals	5,000	250,000	7	20
	Clinics	2,500	10,000	7	20
	Laboratories	2,500	10,000	2	20
	Dental offices	2,500	10,000	2	20
	Nursing homes	5,000	100,000	3	10
Lodging	Residential	1,000	10,000	2	5
	Dormitories	5,000	25,000	10	30
	Apartments	5,000	100,000	3	10
	Hotels	5,000	200,000	5	10
Education	Schools	5,000	250,000	33	50
	Day care	2,500	10,000	33	50
	Libraries	5,000	250,000	10	33
	Museums	5,000	250,000	10	33
Mercantile	Department stores	5,000	250,000	13	67
	Supermarkets	5,000	100,000	10	20
	Malls	10,000	500,000	10	20
Assembly	Auditoriums	10,000	100,000	50	200
	Stadiums	10,000	200,000	50	200
	Churches	2,500	10,000	50	200

8.4 BUILDING AIRTIGHTNESS

Building airtightness plays a major role in the energy consumption of buildings and simultaneously impacts indoor air quality. This may not be a factor for naturally ventilated buildings but for any building with forced air ventilation, it is essential that the buildings be reasonably airtight so as to maintain control over the indoor air quality. Although outdoor air is generally clean and healthy for most people, it is still necessary to provide filtration of dust, pollen, and spores in order to provide the highest level of protection against microbial hazards for indoor occupants, especially for those with allergies. Health care facilities require extra levels of air

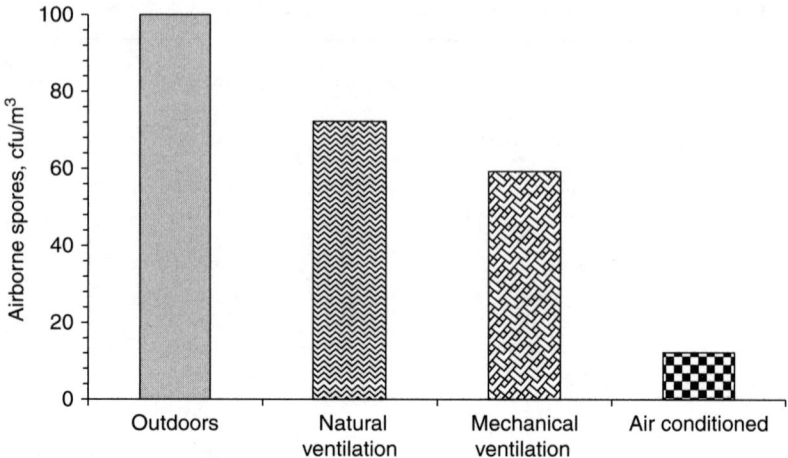

FIGURE 8.3 Comparison of spore levels in buildings with ventilation. [*Based on data from the California Healthy Buildings Study (Woods et al., 1997).*]

cleaning, especially where immunodeficient patients are treated who are vulnerable to outdoor airborne spores and bacteria that pose no threat to healthy people.

Figure 8.3 illustrates the differences in indoor airborne contamination levels between naturally ventilated buildings and buildings with forced air systems. This chart highlights the fact that even naturally ventilated buildings tend to have lower indoor airborne levels of spores, due in large part to settling or plate-out of microbes on building materials. However, the fact that spores may settle into building materials and furnishings may cause secondary problems like mold growth under moist conditions.

During the energy crisis of the 1970s, increased attention was paid to building airtightness at the expense of indoor air quality. The result was a steady increase in the number of IAQ complaints by building occupants. Building airtightness can cause problems when the amount of fresh outside air delivered was less than adequate or inadequately distributed. Airtightness is simultaneously important because it can aid in preventing moist outdoor air from infiltrating through walls and condensing on building materials, a process that can lead to microbial growth and decreased air quality.

Some older houses that are less airtight often have more problems due to the fact that they may have accumulated microbes and moisture damage over time. The prevalence of upper respiratory symptoms was found to be about 26 percent in new houses and almost 48 percent in old houses, according to one survey of 60 houses (Ren et al., 1993). Airborne levels of bacteria in old houses were almost twice that of both new houses and outdoor air.

8.5 BUILDING DAMPNESS

Building dampness has long been correlated with indoor air quality and respiratory diseases among occupants, especially children (Dales et al., 1991; Platt et al., 1989). Condensation can occur in a variety of places, including on walls, inside walls, in basements, under-floor crawlspaces, roof space, windows, kitchen and bathroom areas, on pipes, and on ductwork. Whenever a temperature gradient exists across a porous wall, condensation can occur in the interstices. Warm moist air may infiltrate the wall and condense out moisture when it reaches the dew point. If this should occur where there are building materials such as cellulose or insulation that provide a nutrient or substrate, fungal growth may occur. Vapor barriers or vapor retarders are used to control migration of moisture from the outside into building materials. Vapor retardant materials or coatings are often used inside walls to prevent the migration of moisture. High moisture levels in a foundation can rapidly degrade the performance of most foundation insulation and mold and mildew can grow where the relative humidity at a surface is greater than 70 percent (ASHRAE, 1999).

Standing water may also provide a medium in which certain microbes can proliferate. A number of bacteria multiply in or on aquatic environments, such as *Legionella pneumophila* and *Flavobacterium, Staphylococcus* spp., and *Bacillus*. Some fungi that may grow in aquatic environments include *Alternaria, Aspergillus, Mucor, Penicillium,* and *Exophiala* (Flannigan et al., 2001). Algae and amoeba have also been known to grow in indoor water. Although not all of these microbes are harmful, the ones that are not can provide nutrients for the growth of more hazardous pathogens.

Three main types of dampness can occur in buildings—(1) penetrating dampness, (3) rising dampness, and (3) condensation. Penetrating dampness arises as the result of defects in the external building envelope, including the leakage of rain through the roof and walls. Internal penetrating dampness can occur as the result of failed plumbing and defects in appliances using water, including central heating and cooling systems (Oliver, 1988).

Rising dampness is similar to but distinct from penetrating dampness. It occurs when moisture in the ground rises through capillary action through any porous materials, including concrete, mortar, brickwork, wood, and stonework. Cracks that develop in building walls and foundations create capillaries by which moisture can rise and enter the building. Figure 8.4 illustrates two possible pathways by which rising moisture through cracks can penetrate walls and foundations to enter the building. Cycles of freezing and thawing can exacerbate the problem. Remediation of this problem generally involves the installation of a damp proof course that may either be a physical barrier or a chemical treatment.

The third mechanism by which moisture may penetrate building materials is via condensation from moist air. This is perhaps the most common form of dampness in buildings. Moisture from outside air may condense in building walls that allow infiltration during hot, humid weather. Moisture from internal sources, especially the kitchen and bathrooms, may condense on surfaces and exfiltrate through the walls during cold weather. Various other mechanisms, such as diffusion, capillary action, air movement through temporary openings, wind, and air pressure, may bring moisture into the indoors. An airtight building

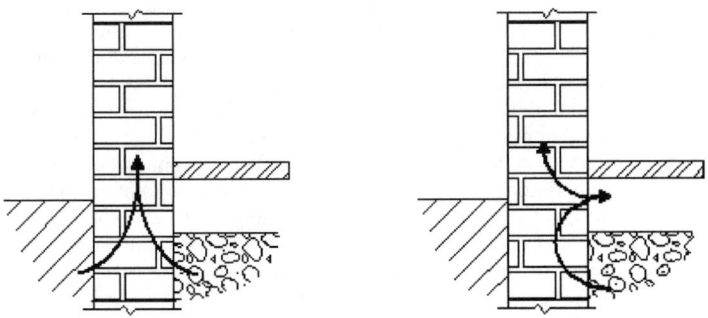

FIGURE 8.4 Two possible pathways of rising damp penetration of moisture through capillary cracks in foundations, walls, and floors.

TABLE 8.2 Condensation Remediation Approaches

1. Inspect or adjust heating system.
2. Unblock or increase ventilation.
3. Clean duct and open air registers.
4. Clear combustion air inlets on equipment.
5. Retrofit windows with double-pane glass.
6. Vent laundry driers to outside.
7. Install dehumidifiers.

envelope with moisture barriers is one way to help control such problems. Another approach may be to pressurize the building positively or negatively depending on conditions. Some common methods for remediating condensation are summarized in Table 8.2. See ASHRAE (1999) for details on designing and constructing walls, roofs, and foundations to control moisture.

8.6 BUILDING MATERIALS

Some building materials support the growth of bacteria and fungi either by providing a nutrient source or by providing a substrate that may become moist. Table 8.3 identifies a number of building materials and components that may foster the growth of fungi.

Table 8.4 identifies some bacteria and the building materials or locations that may support their growth indoors, with references as indicated. In general, the growth of bacteria in indoor settings is uncommon and generally involves environmental bacteria of limited pathogenicity, the one exception being *Legionella*.

A building can act like a large air filter for microbial contaminants, in which adsorption to surfaces and deposition on building materials and furnishings tend to remove and hold contaminants, at least temporarily, and the building envelope itself may act as a filter for infiltrating or exfiltrating air (Taylor et al., 1999; Jamriska and Morawska, 2003). Approximately 20 percent of airborne particles may be removed by the building itself. Carpeted environments provide a reservoir for biocontamination that may impact air quality (Cole et al., 1993). Routine carpet cleaning tends to reduce indoor levels of fungal spores, bacteria, and endotoxins. Carpet cleaning itself can increase airborne levels of microbes, but levels after cleaning are generally reduced. In buildings with regular carpet maintenance, airborne fungal spore levels of 380 cfu/m^3 and bacteria levels of 109 cfu/m^3 have been measured. These are considered to be normal and not hazardous. Carpet dust levels in cleaned carpets containing about 8.5×10^5 cfu/g of fungi, 1.2×10^7 cfu/g of bacteria, and 4.5×10^3 cfu/g of actinomycetes are also considered to be normal and not major contributors to airborne biocontamination. Water damage to rugs will sometimes result in rapid mold growth due to the fact that mold spores have settled or been tracked into the rug over time. Cleaning rugs periodically and exposing them to direct sunlight can help.

Dust can provide a nutrient base on which fungi can grow (Kalliokoski et al., 1996). In HVAC systems, dust that collects on surfaces or in crevices is sufficient to support fungal growth in the presence of moisture from condensation (Sugawara, 1997). The most frequently isolated species of fungi in house dust include *Alternaria, Penicillium, Aspergillus, Cladosporium*, and *Aureobasidium*.

The growth of bacteria and fungi on building materials can lead not only to diseases, allergies, and respiratory irritation, but also to degradation of those same building materials.

BUILDINGS AND ENCLOSURES   173

TABLE 8.3 Fungi That May Grow Indoors

Airborne fungal spore	Location of growth	Reference
Acremonium spp.	Humidifier water, HVAC fiberglass insulation	Heinemann et al., 1994; Samson, 1994
Alternaria spp.	Cooling systems, paint mildew, carpet dust, floor dust, refrigerator coils, filters, dust in ductwork	Neumeister et al., 1997; Gravesen et al., 1999; Godish, 1995
Aspergillus spp.	Carpet and floor dust, evaporative air cooler, HVAC fiberglass insulation, cooling coils, filters, dust in ductwork	Neumeister et al., 1997; Hyvarinen et al., 1993; Macher and Girman, 1990; Price et al., 1994; Samson, 1994
Aureobasidium pullulans	Moist building materials, latex paint, filters	Pasanen et al., 1992; Godish, 1995
Chaetomium spp.	HVAC fiberglass insulation, filters, dust in ductwork	Heinemann et al., 1994; Samson, 1994; Price et al., 1994
Cladosporium spp.	Wet carpet, wet walls, moist building materials, latex paint, floor and carpet dust, filters, evaporative coolers, fans, dust in ductwork, fiberglass insulation	Pasanen et al., 1992; Heinemann et al., 1994; Gravesen et al., 1999; Sugawara, 1996; Godish, 1995
Cryptococcus spp.	Floor dust	Samson, 1994
Epicoccum spp.	Fiberglass insulation	Morey et al., 1990; Kemp et al., 1997
Eurotium herbariorum	Gypsum-based finishes	Kujanpaa et al., 1999; Godish, 1995
Exophiala spp.	Humidifier water	Heinemann et al., 1994; Samson, 1994
Fusarium spp.	Filters, floor dust, humidifier water	Neumeister et al., 1997; Fouad et al., 1999; Heinemann et al., 1994
Mucor spp.	Floor dust, fans, filters, dust in ductwork	Heinemann et al., 1994; Kemp et al., 1997
Paecilomyces spp.	Humidifier water	Heinemann et al., 1994
Penicillium spp.	Latex painted surfaces, carpet dust, air conditioners, evaporative air coolers, HVAC ducts, filters, humidifier water	Pasanen et al., 1992; Gravesen et al., 1999; Burge 1990; Hyvarinen et al., 1993; Macher and Girman, 1990; Godish, 1995; Chang et al., 1996
Phialophora spp.	Humidifier water	Heinemann et al., 1994
Phoma spp.	Paint mildew, floor dust, filters, humidifier	Heinemann et al., 1994; Neumeister et al., 1997
Rhizopus spp.	Floor dust, ductwork dust, fans, filters	Heinemann et al., 1994; Neumeister et al., 1997
Rhodotorula spp.	Wet carpet, wet walls	Kemp et al., 1997
Scopulariopsis spp.	Floor dust, filters	Heinemann et al., 1994

(*Continued*)

TABLE 8.3 Fungi That May Grow Indoors (*Continued*)

Airborne fungal spore	Location of growth	Reference
Stachybotrys spp.	Moist building materials, fans, humidifier water	Heinemann et al., 1994; Scott and Yang, 1997; Pasanen et al., 1992
Trichoderma spp.	Moist building materials, fans, filters, ductwork dust	Heinemann et al., 1994; Neumeister et al., 1997; Sugawara, 1996; Pasanen et al., 1992
Ulocladium spp.	Floor dust, filters, humidifier water	Heinemann et al., 1994; Neumeister et al., 1997
Wallemia sebi	Floor dust, filters	Godish, 1995

Table 8.5 identifies materials that are known to support growth of the indicated mold and can experience biodeterioration by the same (Goynes et al., 1995; Klens and Yoho, 1984; Zyska, 1997; Montegut et al., 1991; Oppermann, 1984; Nigam et al., 1994; Zabel and Terracina, 1980).

Different molds will favor different levels of ambient moisture for their growth. Hydrophilic molds prefer wet conditions while xerophilic molds prefer dry conditions. The range of temperatures and relative humidities over which different mold growth is fostered have been reviewed by Clarke et al. (1998) and Pasanen et al. (1992). Figure 8.5 provides a triangular chart illustrating the favored growing conditions of molds, and in which the nutrient axis is merely figurative and not based on any data. The normal indoor range of conditions shows typical interior temperatures. Although indoor humidities are preferably kept between 25 and 60 percent, the actual humidities near the surfaces where condensation may occur and cause mold growth typically exceed interior air humidities. The indoor relative humidity itself is less an indicator of mold growth than is the *water activity* (Aw) of the building materials. Fungal growth is likely if the water activity exceeds 0.76 to 0.96,

TABLE 8.4 Bacteria That May Grow Indoors

Airborne pathogen	Location of growth	Reference
Acinetobacter	Potable water	Kundsin, 1988
Klebsiella pneumoniae	Potable water	Kundsin, 1988
Legionella pneumophila	Potable water Cooling towers	Ryan, 1994 Kundsin, 1988
Micropolyspora faeni	Home humidifiers	Flannigan et al., 1991
Pseudomonas aeruginosa	Indoors Indoor dust Potable water Evaporative air cooler Humidifiers	Godish, 1995 Mitscherlich and Marth, 1984 Kundsin, 1988 Macher and Girman, 1990 Flannigan et al., 1991
Pseudomonas spp.	Filters	Godish, 1995
Serratia marcescens	Potable water	Kundsin, 1988
Thermoactinomycetes vulgaris	Air conditioners Humidifier water	Flannigan et al., 1991 Samson, 1994

TABLE 8.5 Biodeterioration of Materials by Molds

Fungal genera	Leather	Textiles	Wood	Paper	Paints
Alternaria					X
Aspergillus	X	X		X	X
Aureobasidium		X	X		X
Botryodiplodia		X			
Ceratostomella			X		
Chaetomium		X		X	
Cladosporium		X	X	X	X
Curvularia		X			X
Dendryphiella					X
Fusarium		X	X		X
Gliomastix		X			
Graphium			X		
Memnoniella		X			
Monascus	X				
Mucor	X				
Myrothecium		X			
Oidiodendron			X		
Paecilomyces	X	X			X
Penicillium		X	X	X	X
Pestalotia					X
Phialophora			X		
Phoma			X		X
Scopulariopsis	X				
Sporendonema	X				
Stachybotrys		X		X	X
Stemphylium				X	
Torula		X			
Trichoderma		X	X	X	X

depending on fungal species, temperature, time, and composition of the material (Pasanen et al., 1992). Water activities for fungi, where they are known, are summarized in App. A.

Different fungi favor certain materials and their growth rates reflect both the suitability of the nutrients as well as the conditions. Figure 8.6 illustrates the growth of various common fungal allergens on building materials. Materials normally present in most buildings provide nutrients for fungal growth. These include building materials like wood- or cellulose-based materials, and organic textile materials found in rugs and curtains. Gypsum, or gypsum board, can support many types of fungi and is often subject to dampness in modern buildings, in which it is widely used (Flannigan and Miller, 2001). The cardboard and starch glue, which are used to bind the gypsum plaster, are highly biodegradable sources of nutrients for molds. Because the cardboard is hygroscopic, it readily absorbs moisture. *Stachybotrys*, as well as numerous other species of fungi, have been found growing on gypsum, as well as on wallpaper. Both latex paint and cellulose-based paints can support microbial growth (Smith, 1980; Flannigan and Miller, 2001). Evidence exists to suggest that many of the building materials like gypsum and wallpaper that favor the growth of fungi also influence the generation of potentially hazardous mycotoxins (Nielsen et al., 1999). Bacteria can influence the growth of fungi. Environmental bacteria can grow biofilms, and thereby provide fungal spores a nutrient base. Biofilms generally grow where

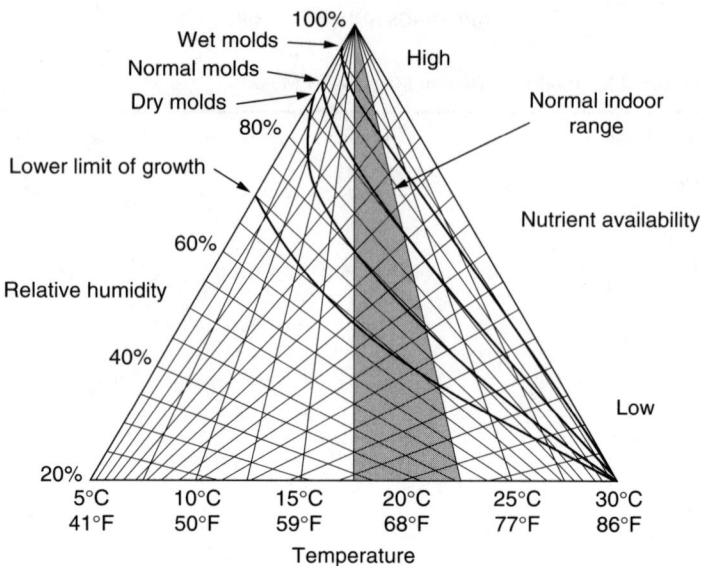

FIGURE 8.5 Range of growth conditions for various fungi. [*Adapted from Clarke et al. (1998).*]

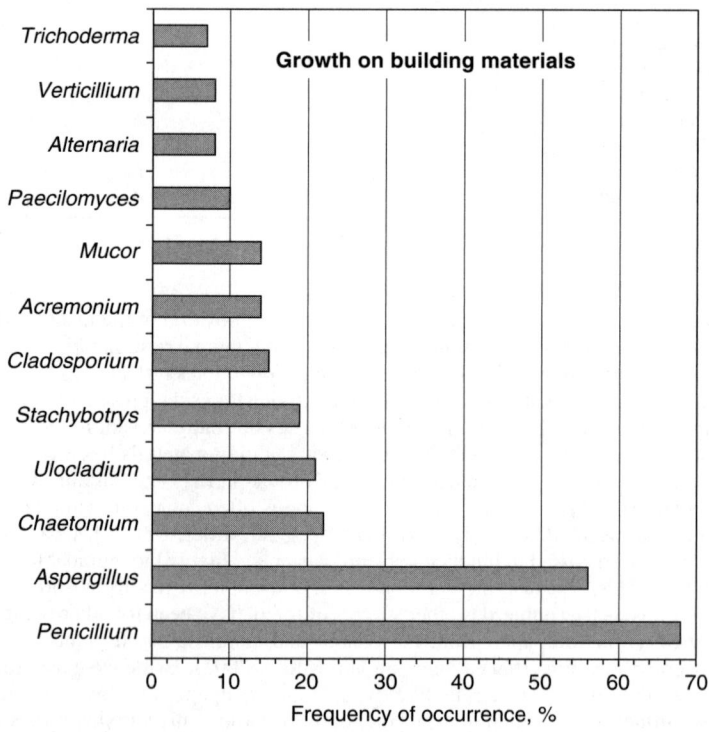

FIGURE 8.6 Growth of fungal allergens on building materials. [*Based on data from Gravesen et al., (1999).*]

there is excessive moisture or where water collects, such as drain pains, dehumidification cooling coils, and sump type humidifiers (Morey, 2001).

Figure 8.7 shows the most frequently isolated species of fungi that have been found to grow on walls in homes. Most of these fungi are recognized allergens. Most walls tend to be normally sheathed with gypsum board and are often coated with either cellulose-based wallpaper or paint.

Insulation can provide a substrate for microbial growth, especially if it is subject to condensation. Figure 8.8 shows the types of fungi and their frequency of occurrence from growth on foam insulation. Fiberglass or mineral wool insulation can also support microbial growth but the degree to which fungi can colonize fiberglass insulation is highly dependent on the local relative humidity (Ezenou et al., 1994; Price et al., 1994; Kujanpaa et al., 1999).

Fabric office panels also accumulate fungal spores and may produce mold growth under high humidity or when they become moist (Hung et al., 1993). Thorough vacuum cleaning of fabric panels can effectively reduce the fungal concentrations to low levels. Mold that grows on textiles, leather, paper, and materials is often referred to as mildew. Cotton- or cellulose-based textile materials can support the growth of a variety of fungi, including a number of allergenic species (Montegut et al., 1991). Figure 8.9 graphically summarizes

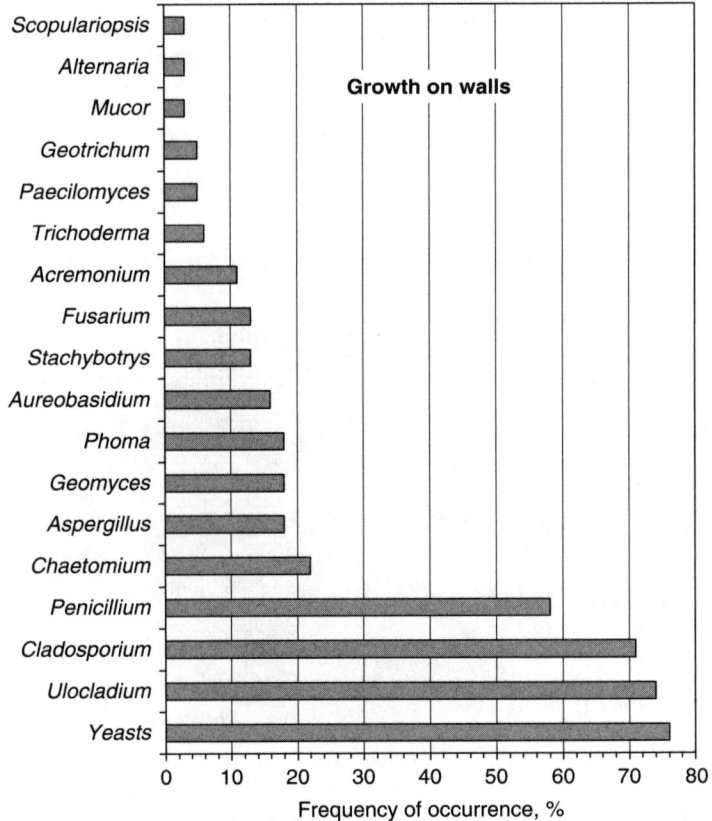

FIGURE 8.7 Growth of fungi on walls. [*Based on data from Hunter and Bravery (1989).*]

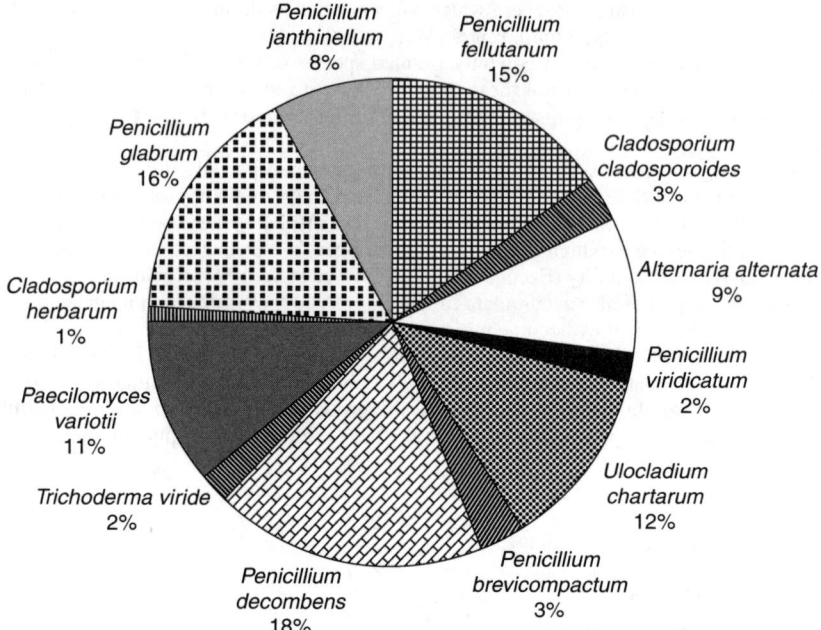

FIGURE 8.8 Growth of fungi in foam insulation shown as frequency of isolates. [*Based on data from Flannigan et al. (2001).*]

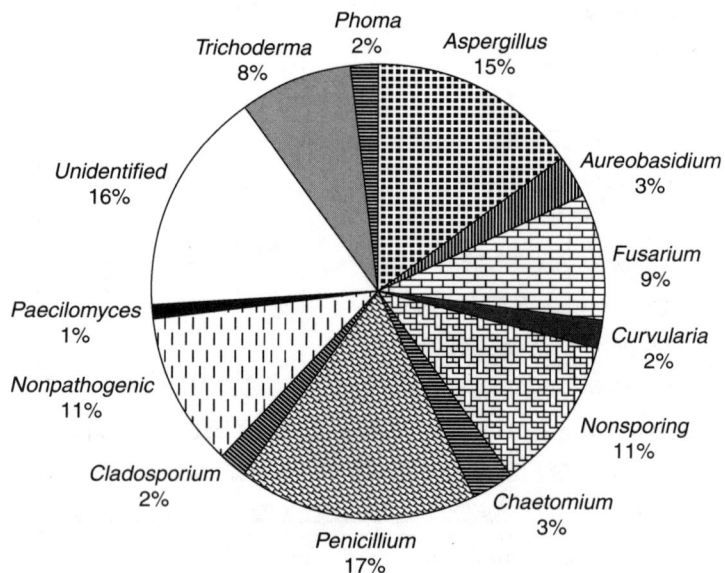

FIGURE 8.9 Occurrence of fungal species that grow on cotton textiles. [*Based on data from Siu (1951).*]

data from Siu (1951) on the frequency of occurrence of fungal species that grow on cotton textiles. The fungi may not only cause airborne health hazards but will also cause biodeterioration of fabrics (Goynes et al., 1995).

It seems clear that the materials commonly used in buildings and for building construction can, under moist conditions, provide a nutrient and/or substrate on which a wide variety of fungi can thrive, including a great many fungi that are potentially allergenic, toxigenic, or that may even cause serious respiratory disorders like *Stachybotrys*. Coupled with the availability of nutrients is the ever-present potential for moisture to be produced by condensation or by leakage. In order to forestall microbial growth in buildings, building materials should be restricted to those that will not support fungal growth, buildings should be designed in a fashion to prevent condensation and leakage problems, and the use of proper ventilation, humidification, and air cleaning should be considered where necessary (Bower, 1997; Oliver, 1988; Teichman, 1993).

It would seem improved methods of designing and constructing buildings are needed and that a departure from what has become the norm is necessary to ensure the health of building occupants. The definition of green building design includes not only the use of sustainable materials but also their application to the construction and design of healthy buildings (ASHE, 2002). The various problems with existing buildings point directly to solutions. Vapor barriers and waterproof courses should perhaps be standard in new construction. Preventive maintenance may be one option, along with regular cleaning of interior surfaces and building design should facilitate disinfection practices. Another option may be the use of antimicrobials coating or combined with the cellulose-based building materials that are unlikely to be abandoned due to their cheapness and widespread use today. Antimicrobial fabrics may also be in order for furnishings and textiles used in the home. New types of carpets that either resist dust accumulation or contain antimicrobials may provide some benefit, although abandoning the concept of carpets in favor of under-floor heating systems (i.e., radiant heating) may prove to be a healthy alternative for the future. See Chap. 18 for more information on the subject of green building technologies.

REFERENCES

ASHE (2002). "Green healthcare construction design guidance statement." American Society for Healthcare Engineering. Chicago, IL. http://www.healthybuildings.net/healthcare/ASHE_Green_Healthcare_2002.pdf.

ASHRAE (1999). Chapter 42: Building Envelopes. *HVAC Applications*. American Society of Heating, Refrigerating and Air-Conditioning Engineers, Atlanta, GA.

Bell, A. A., Jr. (2000). *HVAC Equations, Data, and Rules of Thumb*. McGraw-Hill, New York.

Binford, L. R. (1983). *In Pursuit of the Past: Decoding the Archaeological Record*. Thames and Hudson, New York.

Bower, J. (1997). *The Healthy House: How to Buy One, How to Build One, How to Cure a Sick One*. The Healthy House Institute, Bloomington, IN.

Chang, J. C. S., Foarde, K. K., and VanOsdell, D. W. (1996). "Assessment of fungal (*Penicillium chrysogenum*) growth on three HVAC duct materials." *Environ Int* 22(4):425.

Clarke, J. A., Johnstone, C. M., Kelly, N. J., McLean, R. C., Anderson, J. A., Rowan, N. J., and Smith, J. E. (1998). "A technique for the prediction of the conditions leading to mould growth in buildings." *Build Environ* 34:515–521.

Cole, E. C., Foarde, K. K., Leese, K. E., Franke, D. L., and Berry, M. A. (1993). "Biocontaminants in carpeted environments." *Indoor Air 93,* Helsinki, Finland

Dales, R. E., Zwanenburg, H., Burnett, R., and Franklin, C. A. (1991). "Respiratory health effects of home dampness and molds among Canadian children." *Am J Epidemiol* 134(2):196–203.

EIA (1989). *Commercial Buildings Characteristics.* U.S. Department of Energy, Energy Information Administration, Washington, DC.

Ewald, P. W. (1994). *Evolution of Infectious Disease.* Oxford University Press, New York.

Ezenou, I. M., Noble, J. A., Simmons, R. B., Price, D. L., Crow, S. A., and Ahearn, D. G. (1994). "Effect of relative humidity on fungal colonization of fiberglass insulation." *Appl Environ Microbiol* 60(6):2149–2151.

Flannigan, B., McCabe, E. M., and McGarry, F. (1991). Allergenic and toxigenic microorganisms in houses. *Pathogens in the Environment,* B. Austin, ed., Blackwell Scientific Publications, Oxford, UK.

Flannigan, B., Samson, R. A., and Miller, J. D. (2001). *Microorganisms in home and indoor work environments.* Taylor & Francis. Andover, Hants, UK.

Flannigan, B., and Miller, J. D. (2001). Microbial growth in indoor environments. *Microorganisms in Home and Indoor Work Environments,* B. Flannigan, R. A. Samson, and J. D. Miller, eds., Taylor & Francis, New York.

Fouad, H. G., Donn, M. R., Isaacs, N. P., and Baird, G. (1999). "Results of an analysis of airborne bacterial and fungal levels in fully sealed New Zealand offices." *Indoor Air 99 : Proceedings of the 8th International Conference on Indoor Air Quality and Climate,* Edinburgh, Scotland, UK. 246–251.

Godish, T. (1995). *Sick Buildings: Definition, Diagnosis and Mitigation.* CRC- Lewis Publishers, Boca Raton, FL.

Goynes, W. R., Moreau, J. P., DeLucca, A. J., and Ingber, B. F. (1995). "Biodeterioration of nonwoven fabrics." *Textile Res J* 65(8):489–494.

Gravesen, S., Nielsen, P. A., Iversen, R., and Nielsen, K. F. (1999). "Microfungal contamination of damp buildings—examples of risk constructions and risk materials." *Environ Health Perspect* 107(Suppl. 3):505–508.

Heinemann, S., Beguin, H., and Nolard, N. (1994). Biocontamination in air-conditioning. *Health Implications of Fungi in Indoor Environments,* R. A. Samson, ed., Elsevier, Amsterdam, The Netherlands. 179.

Hung, L.-L., Yang, C. S., Lewis, F. A., and Zampiello, F. A. (1993). "Accumulation of fungal spores in fabric modular office panels." *Indoor Air 93,* Helsinki, Finland.

Hunter, C. A., and Bravery, A. F. (1989). Requirements for growth and control of surface moulds in dwellings. *Airborne Deteriogens and Pathogens,* B. Flannigan, ed., The Biodeterioration Society, Surrey, UK.

Hyvarinen, A., Reponen, T., Husman, T., Ruuskanen, J., and Nevalainen, A. (1993). "Composition of fungal flora in mold problem houses determined with four different methods." *IAQ '93,* Helsinki, Finland. 273–278.

Imbrie, J., and Imbrie, K. P. (1979). *Ice Ages: Solving the Mystery.* Harvard University Press, Cambridge, MA.

Jamriska, M., and Morawska, L. (2003). "Quantitative assessment of the effect of surface deposition and coagulation on the dynamics of submicrometer particles indoors." *Aerosol Sci Technol* 37:425–436.

Jelinek, J. (1975). *The Pictorial Encyclopedia of the Evolution of Man.* Hamlyn, London.

Kalliokoski, P., Pasanen, A.-L., Korpi, A., and Pasanen, P. (1996). "House dust as a growth medium for microorganisms." *The Seventh International Conference on IAQ and Climate,* Nagoya, Japan. 131–135.

Kemp, P. C., Neumeister, H. G., Kircheis, U., Schleibinger, H., Franklin, P., and Ruden, H. (1997). "Fungal genera in an office building with a central HVAC system in an Australian mediterranean climate." *Healthy Buildings/IAQ '97,* Bethesda, MD. 257–260.

Klens, P. F., and Yoho, J. R. (1984). "Occurrence of alternaria species on latex paint." *The Sixth International Biodeterioration Symposium,* Washington, DC.

Kowalski, W. J. (2003). *Immune Building Systems Technology.* McGraw-Hill, New York.

Kujanpaa, L., Haatainen, S., Kujanpaa, R., Vilkki, R., and Reiman, M. (1999). "Microbes in material samples taken from base boardings, gypsum boards and mineral wool insulation." *Indoor Air 99:*

Proceedings of the Eighth International Conference on Indoor Air Quality and Climate, Edinburgh, Scotland, UK. 892–896.

Kundsin, R. B. (1988). *Architectural Design and Indoor Microbial Pollution.* Oxford University Press, New York.

Leakey, R. E., and Lewin, R. (1977). *Origins.* E. P. Dutton, New York.

Leakey, R., and Lewin, R. (1992). *Origins Reconsidered: In Search of What Makes Us Human.* Doubleday, New York.

Lippmann, M. (1998). Chapter 112: Sick Building Syndrome and Building-Related Illness. *Environmental & Occupational Medicine,* W. H. Rom, ed., Lippincott-Raven, Philadelphia, PA. 1471–1477.

Macher, J. M., and Girman, J. R. (1990). "Multiplication of microorganisms in an evaporative air cooler and possible indoor air contamination." *Environ Int* 16:203–211.

Mitscherlich, E., and Marth, E. H. (1984). *Microbial Survival in the Environment.* Springer-Verlag, Berlin, Germany.

Montegut, D., Indictor, N., and Koestler, R. J. (1991). "Fungal deterioration of cellulosic textiles: a review." *Int Biodeterior* 28(1–4):209–226.

Morey, P. R., Feeley, J. C., and Otten, J. A. (1990). *Biological Contaminants in Indoor Environments.* ASTM, Philadelphia, PA.

Morey, P. R. (2001). *Control and remediation of microbial growth in problem buildings.* Microorganisms in Home and Work Environments, B. Flannigan, R. A. Samson, and J. D. Miller, eds., Taylor & Francis, London. 83–100.

Neumeister, H. G., Kemp, P. C., Kircheis, U., Schleibinger, H. W., and Ruden, H. (1997). "Fungal growth on air filtration media in heating ventilation and air conditioning systems." *Healthy Buildings/IAQ '97,* Bethesda, MD 569–574.

Nielsen, K. F., Gravesen, S., Nielsen, P. A., Andersen, B., Thrane, U., and Frisvad, J. C. (1999). "Production of mycotoxins on artificially and naturally infested building materials." *Mycopathologia* 145:43–56.

Nigam, N., Dhawan, S., and Nair, M. V. (1994). "Deterioration of feather and leather objects of some Indian museums by keratinophilic and nonkeratinophilic fungi." *Int Biodeter Biodegr* 33(2): 145–152.

Oliver, A. C. (1988). *Dampness in Buildings.* Nichols Publishing, New York.

Oppermann, R. A. (1984). "The Anaerobic Biodeterioration of Paint." *The Sixth International Biodeterioration Symposium,* Washington, DC

Pasanen, A.-L., Juutinen, T., Jantunen, M. J., and Kalliokoski, P. (1992). "Occurrence and moisture requirements of microbial growth in building materials." *Int Biodeter Biodegr* 30:273–283.

Platt, S. D., Martin, C. J., Hunt, S. M., and Lewis, C. W. (1989). "Damp housing, mould growth, and symptomatic health state." *Brit Med J* 298:1673–1678.

Price, D. L., Simmons, R. B., Ezeonu, I. M., Crow, S. A., and Ahearn, D. G. (1994). "Colonization of fiberglass insulation used in heating, ventilation and air conditioning systems." *J Ind Microbiol* 13:154–158.

Ren, J., Guo, R. R., and Wang, X. (1993). "Indoor airborne bacteria concentrations and respiratory disease in old and new rural dwelling." *Proceedings of the Sixth International Conference on Indoor Air Quality and Climate,* Helsinki, Finland 207–212.

Ryan, K. J. (1994). *"Sherris Medical Microbiology."* Appleton & Lange. Norwalk, CT.

Samson, R. A., ed. (1994). *Health Implications of Fungi in Indoor Environments.* Elsevier, Amsterdam, The Netherlands.

Scott, R., and Yang, C. (1997). "Comparison of successful and unsuccessful *Stachybotrys chartarum* remediation projects." *Healthy Buildings/IAQ '97,* Bethesda, MD., ASHRAE. Atlanta, GA. 269.

Siu, R. G. H. (1951). *Microbial Decomposition of Cellulose.* Reinhold Publishing, New York.

Smith, R. A. (1980). Chapter 46: Fungi surviving in a latex paint film after surface disinfection. *Developments in Industrial Microbiology,* R. A. Zabel, ed., Society for Industrial Microbiology, Fairfax, VA. 565—569.

Sugawara, F. (1996). "Microbial contamination in ducts of air conditioning systems." *The Seventh International Conference on IAQ and Climate,* Nagoya, Japan 161–166.

Sugawara, F. (1997). "Components of dust and microbial proliferation in ducts of air conditioning systems." *Healthy Buildings/IAQ '97,* Bethesda, MD. ASHRAE, Atlanta, GA. 563—566.

Taylor, B. J., Webster, R., and Imbabi, M. S. (1999). "Building envelope as an air filter." *Build Environ* 34(3):353–361.

Teichman, K. Y. (1993). *IAQ 93: Operating and Maintaining Buildings for Health, Comfort, and Productivity.* American Society of Heating, Refrigerating and Air-Conditioning Engineers. Atlanta, GA.

Woods, J. E., Grimsrud, D. T., and Boschi, N. (1997). *Healthy Buildings / IAQ '97.* ASHRAE, Washington, DC.

Wymer, J. J. (1982). *The Paleolithic Age.* Croon Helm, London.

Zabel, R. A., and Terracina, F. (1980). "The role of Aureobasidium pullulans in the disfigurement of latex paint films." *Developments in Industrial Microbiology,* Pittsburgh, PA. 179–189.

Zyska, B. (1997). "Fungi isolated from library materials: A review of the literature." *Int Biodeter Biodegr* 40(1):43–51.

SECTION·2

AIRBORNE DISEASE CONTROL TECHNOLOGIES

CHAPTER 9

VENTILATION SYSTEMS AND DILUTION

9.1 INTRODUCTION

Ventilation systems provide breathing air inside buildings, and also typically provide heating, cooling, and humidity control. Ventilation systems are intended to create a safe and comfortable indoor environment but they may also contribute to the spread of diseases by disseminating pathogens and allergens. They can, furthermore, even promote the growth of microorganisms by creating condensation, standing water, and accumulating dust. Although ventilation systems often contribute to the problem of indoor diseases, they can form an integral component of an immune building system designed to prevent the spread of diseases by airborne routes. At a minimum, ventilation systems provide dilution and removal of airborne contaminants. They can also be used for pressurization of areas within buildings to provide contamination control. Air cleaning technologies can form an integral part of any ventilation system. This chapter reviews the types and functions of ventilation systems that are common today along with the aerobiological problems they may cause. Specialized types of ventilation systems, which include pressurization or isolation systems and laboratory hood systems, are also addressed here. Three methods of modeling dilution ventilation are presented in this chapter, and they are the steady state model, the single-zone model, and the multizone model. *Computational fluid dynamics* (CFD) modeling is also discussed although it is too broad a topic to be treated in complete detail.

9.2 VENTILATION SYSTEM TYPES

There are six basic types of ventilation systems in use today—natural ventilation, constant volume, variable air volume, 100 percent outside air, *dedicated outside air systems* (DOAS), and displacement ventilation. Figure 9.1 shows a breakdown of the forced ventilation system types.

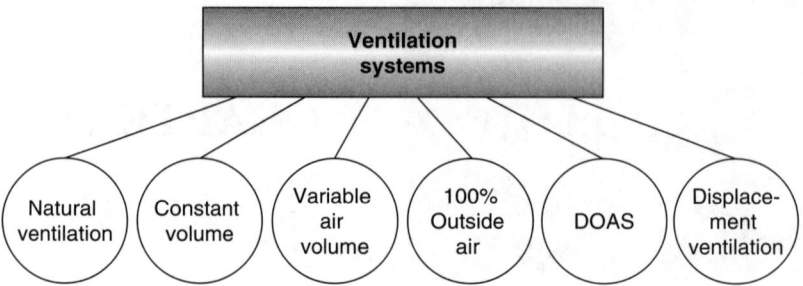

FIGURE 9.1 Basic types of ventilation systems.

The nature of each of these ventilation systems provides various advantages and disadvantages in terms of both air cleaning ability and economics. These properties are individually discussed in the following sections.

9.2.1 Natural Ventilation

The most ancient and most common ventilation system for buildings is natural ventilation, in which no mechanical ventilation is used. Many countries in the world still depend on natural ventilation in their office buildings, but in the United States these are now the exception rather than the rule. Climate is a limiting factor in the construction of naturally ventilated buildings—a mild climate is required. Naturally ventilated buildings in cold climates invariably have separate heating systems. In naturally ventilated buildings leakage into and out of the building envelope provides the air exchange necessary for human occupancy, and also takes care of the cooling loads. Older buildings are often simply designed with windows the occupants can adjust to obtain the amount of air they desire. In such older buildings little, if any, attention was paid to internal airflow and such buildings can contain dead spots with little or no air exchange.

The actual *air change per hour* (ACH) obtainable by natural ventilation is difficult to predict and is heavily dependent on how many windows are open, outdoor air temperature, and local wind conditions. The air-exchange rate in naturally ventilated buildings can vary from 1 to 6 ACH and is usually lower, on the average, than any building with forced ventilation.

New, naturally ventilated green building designs that avoid use of energy consuming systems are possible today through the use of sophisticated CFD software (Flomerics, 2003). Detailed analysis of air currents through a building can lead to designs that enhance airflow and obtain adequate air-exchange rates as well as providing airflow to every corner of the building. Wind-driven natural ventilation for cooling can also be modeled in wind tunnels. Such designs can take good advantage of both external wind effects and internal stack effects to produce acceptable levels of air exchange and cooling. Even with such advanced techniques it may not be possible to entirely do away with forced ventilation since such designs may be subject to seasonal variations and some amount of forced air may be required to meet building needs and occupant comfort.

The main problem with natural ventilation is that there is no control over outdoor microbial contaminants entering the building. Pollen, fungal spores, and environmental bacteria will enter the building and may accumulate. In a healthy building the indoor levels of airborne microbes will tend to be lower than in the outdoor air due to plate-out and other factors, as shown in Fig. 9.2.

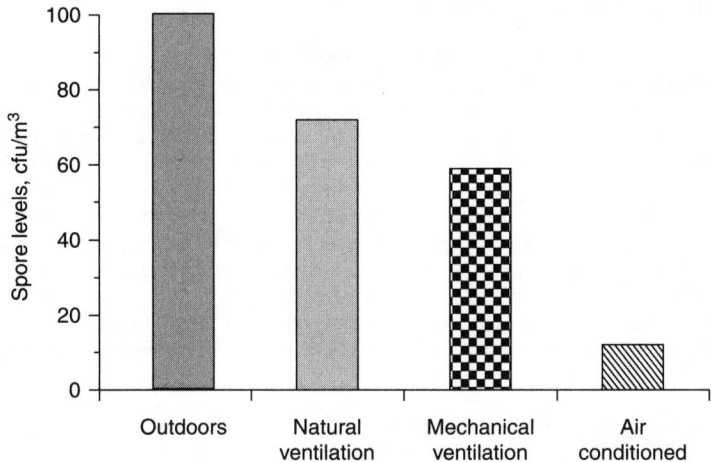

FIGURE 9.2 Comparison of airborne spore levels in buildings with different ventilation systems. [*Based on the California Healthy Buildings Study (Fisk, 1994).*]

A study by Parat et al. (1997) determined that an office building's *heating, ventilation, and air-conditioning* (HVAC) system may be responsible for the production and spread of airborne microorganisms. Eight sets of measurements were made over a 1-year period comparing airborne microbiological flora in air-conditioned building with that in a naturally ventilated building. An Andersen single-stage sampler was used to culture fungi, staphylococci, and mesophilic bacteria. One group of offices was found to be more contaminated than others, and there was a marked seasonal variation in airborne fungal spore concentrations. A comparison of the two buildings showed that the mean airborne levels of microbes were significantly higher and more variable in the naturally ventilated building than in the air-conditioned building. Also, the interior fungal content of the naturally ventilated building was strongly dependent on the outdoor content, while in the air-conditioned building fungal concentrations remained constant in spite of outdoor variations. The effect of the HVAC system, which included high efficiency filters, was to prevent the intake of outdoor particles and to dilute the indoor concentrations.

9.2.2 Constant Volume Ventilation

Constant volume (CV) systems comprise the majority of forced ventilation system types and are the most common type of system in commercial buildings today. Constant volume systems are forced ventilation systems driven by a constant volume fan. They may or may not include cooling and heating equipment. They provide a constant flow rate of both outside and return air to the building. The design and operation of such systems is simple and, if well designed, can be comfortable and cost-effective. These systems normally employ dust filters and therefore the indoor levels of airborne microbes will tend to be lower than outdoors, as shown in Fig. 9.2. When cooling coils are present in a mechanically ventilated building, indoor spore levels may be lower than in any other systems due to the filtration effect of the cooling coils when moisture condenses on them (Seigel and Walker, 2001). In general, contemporary well-designed, air-conditioned buildings provide a healthier environment for office workers than naturally ventilated ones (Parat et al., 1997).

9.2.3 Variable Air Volume Systems

Variable air volume (VAV) systems alter the amount of outside air in response to either outside air temperature or outside air enthalpy. When conditions are right, the amount of outside air drawn in is maximized. The purpose of these systems is to save energy by taking advantage of mild air conditions. Normally these buildings have some fixed amount of minimum outside air that may be between 15 and 25 percent. Under minimum outside air conditions these systems operate almost identically to constant volume systems, while under maximum outside air, if it is 100 percent, they will operate like 100 percent outside air systems. However, there is sometimes a limit on maximum outside air, which may be as low as 50 percent, and therefore not all VAV systems will provide the same benefits as 100 percent outside air systems even during the best outside air conditions. Because of the fact that VAV operating conditions may be unpredictable, these systems are not specifically analyzed here but they can be considered to lie between the extremes of constant volume systems and 100 percent outside air systems.

9.2.4 100 percent Outside Air Systems

Systems that use 100 percent outside air are often found in some hospitals, health care facilities, clinics, and laboratories. They are often used in situations where the risk of internal contamination outweighs the costs associated with using large volumes of outside air in winter or summer. Because of the energy costs usually associated with this type of system, they are not common in commercial office buildings except for some older buildings in mild climates. However, when used in conjunction with air-to-air heat exchangers or enthalpy wheels such high air-exchange systems have potential as a means of improving IAQ (Tamblyn, 1995; Nardell et al., 1991). Even outside air, though virtually free of pathogens, still requires filtration for allergens, and at the high flow rates associated with 100 percent outside air system, the energy question remains a critical one. Perhaps the principle of using 100 percent outside air to improve IAQ is better applied in the more sophisticated and more energy efficient systems known as DOAS.

9.2.5 Dedicated Outside Air Systems

Dedicated outside air systems (DOAS) are distinct from 100 percent outside air systems in that they only deliver the minimum outside air needed for ventilation, or about 15 to 25 percent as much airflow as that of an all-air system (Mumma, 2001a). Typically, the thermal loads are accommodated with hydronic systems, rather than air systems. DOAS are relatively new in the United States but have a variety of features, including dehumidification, that give them some inherent protection against dispersion of contaminants and indoor microbial growth (Mumma, 2001b). Since they use 100 percent outside air, there is no recirculation of contaminants, and since they operate at lower total airflows, the total air that is filtered is only about 20 percent of conventional all-air systems, resulting in significant savings in first costs and operating costs over an all-air system (Mumma, 2002).

9.2.6 Displacement Ventilation

The air-exchange rate is often considered to be the primary parameter that determines how well biological contaminants are removed from indoor air (Morey et al., 1990; Nardell et al., 1991). Another factor that is perhaps at least as important as the air change rate is the actual efficiency with which the interior air of any room is displaced and removed. Displacement ventilation systems are those which attempt to control the pattern of supply and exhaust air such that they approach the ideal of plug flow or piston type flow through a room or zone, and thereby remove contaminants more efficiently than normal systems in which considerable air mixing occurs (Chen and Glicksman, 2003; Skistad, 1994). In theory, displacement ventilation systems supply air at floor level and depend on thermal currents to lift VOCs and CO_2 to the upper levels of a room where they are removed by the exhaust outlets. The problem with this approach, when applied to aerobiological contaminants rather than gases, is that most biocontaminants tend to settle over time and therefore more efficient removal would theoretically be accomplished with exhausting the air at floor level rather than at the ceiling. The solution to both aerobiological and VOC problems may be to exhaust air at both the ceiling and the floor, and to supply fresh air at breathing levels (i.e., sitting or standing head height).

9.3 BIOCONTAMINATION IN VENTILATION SYSTEMS

Ventilation systems contribute to microbial loading of the indoor environment by drawing in microbes from the outdoor air and by creating conditions for growth. They may also recirculate internally generated fungi and pathogens from human sources. If the outside or supply air is filtered or there are cooling coils present, most of the spores will be intercepted inside the ventilation system. If fungal spores and microbes enter a building by infiltration or by being physically carried inside on clothes or objects and find a suitable growth environment, the microbial loading of the building may be amplified above that of the outdoor air. The specific locations in and around ventilation systems where microbial growth most commonly occurs include drain pans, filters, cooling coils, evaporative coolers, and sump-type dehumidifiers (Morey, 2001). Microbial growth may also occur anywhere where humidity or moisture requirements are met, including inside walls and on water-damaged building materials.

Cooling coils are a common source of microbial growth due to the fact that they provide moisture and tend to intercept spores much like a low efficiency filter. Figure 9.3 illustrates a model of cooling coil deposition rates as a function of particle size based on theoretical

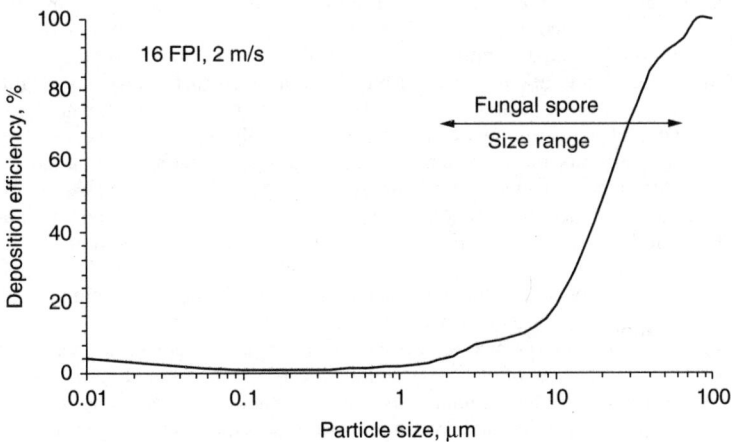

FIGURE 9.3 Deposition of particles on a model cooling coil. [*Based on results from Seigel and Walker (2001)*.]

and experimental results from Seigel and Walker (2001). In this example the velocity is 2 m/s (394 fpm) and the coil has 16 *fins per inch* (FPI). The deposition efficiency in the fungal spore size range will clearly result in the removal of large fractions of spores from the airstream. These results, furthermore, represent dry coils. During cooling in humid weather condensation will occur on the fins and tubes, which will enhance the deposition rate, as it does in air conditioners (Solomon et al., 1980; Spiegelman et al., 1963).

Spores that deposited on cooling coils may grow in the presence of moisture and nutrients like dust particles and may be dispersed to other parts of the building (Banaszak et al., 1970; Schicht, 1972). Although the temperature of cooling coils during operation, about 13°C (55°F) may be too cool for optimum growth of most fungi, about 20 to 40°C (68 to 104°F), cooling coils do not usually operate continuously and temperatures will approach those of ambient air, about 23°C (73°F), during periods when they cycle off. As a result there is plenty of opportunity for fungal spores on cooling coils to germinate and begin growing. Bacterial spores and vegetative bacteria can similarly be intercepted and grow on cooling coils under the right conditions.

Dust filters are often present to protect the cooling coils but these are rarely sufficient to completely prevent spores from entering the ventilation system. With moderate efficiency filters used for dust control (i.e., MERV 6–8 or less), some spore penetration will occur (Kowalski and Bahnfleth, 2002). Even high efficiency filters in the MERV 10–13 will not prevent complete entry of fungal spores unless there is no filter bypass. It only takes a few spores to initiate growth in the presence of moisture and nutrients. Once growth begins the development of larger numbers of spores becomes inevitable. Filters used for controlling dust, like MERV 1–6 filters (~20 percent DSP filters) or higher, will accumulate fungal spores, bacterial spores, and bacteria from outdoor air or from internal recirculated air. Often, microbes intercepted in this fashion will tend to die from dehydration and natural causes. However, if moisture is present, microbial growth may occur (Kemp et al., 1995; Maus, 1996). It is even possible for a microbial mass to grow right through the filter and appear on the downstream side, at which point the microbes may become aerosolized (Foarde et al., 2000; Jankowska et al., 2000).

The presence of dust inside ductwork or anywhere in the ventilation system may provide a growth medium for microbes (Flannigan et al., 1991; Sugawara, 1996). Often the dust itself carries microbes and only moisture is needed to facilitate the growth cycle. Since condensation may occur inside of ventilation systems, the potential for amplification of microbial contamination exists. Condensation may occur inside or outside of ventilation ductwork, and on any components subject to temperature and humidity differentials. Ductwork supplying cooled or chilled air may develop condensation on the exterior surfaces. Often, such ductwork is insulated but this may lead to the problem of microbial growth on insulation since insulation may provide a substrate or nutrient source for fungi. It is essential that insulation either be of a type (i.e., noncellulolytic or nonorganic) that does not provide any nutrients, or that it be installed in an airtight fashion such that moisture will not enter and produce condensate. Periodic maintenance and cleaning of ductwork is one option, as is maintenance of filtration systems. Additional options include antimicrobial duct coatings and *ultraviolet germicidal irradiation* (UVGI) systems, both of which will be addressed in later chapters.

Condensation may occur on other components like water tanks, pumps, piping, and refrigeration systems. Standing water provides an aquatic environment that may favor certain species of fungi and bacteria such as *Alternaria, Aspergillus, Penicillium,* and *Legionella* (Flannigan et al., 2001). One of the most common locations where standing water may occur is in the drain pains under the cooling coils inside air handling units. If the water does not drain properly by gravity, microbial growth may occur. It is possible for algae and amoeba to grow in such pools of water, and although these are unlikely health hazards, they may provide nutrients for more dangerous microbes like *Legionella* (Thornsberry et al., 1984).

Humidifiers also often contain standing water and these are particularly susceptible to microbial growth during periods when they are turned off and stagnant (Heinemann et al., 1994; Flannigan et al., 1991). Evaporative coolers have been found to contain microbes (Macher and Girman, 1990) but have never been found to contain *Legionella* (Millar et al., 1997).

Any potable water supply may contain microbial contamination although it is typically at such low levels as not to be a direct hazard (Kundsin, 1988; Ryan, 1994). Still, the use of potable water should be considered in terms of possible amplification if the water is left standing in the presence of a growth medium or nutrient source. More specific details on the types of microbes that occur in buildings and ventilation systems can be found in Chap. 8 (see Tables 8.3 and 8.4).

9.4 PRESSURIZATION CONTROL

Maintaining a building or zone under positive or negative pressure can control the flow of biocontaminants between areas. The use of pressurization to control contaminants is common in certain industries, especially in health care, biological research facilities, animal laboratories, nuclear facilities, and even computer chip manufacturing. Since air will tend to flow from areas of high pressure to areas of low pressure, contaminants can be prevented from entering designated areas as long as the air velocity through the gaps or doorways is high enough. This technique is most commonly applied in hospital isolation rooms and laboratories and is treated in more depth in Chaps. 31 and 32.

The use of pressurization to control contaminants in commercial buildings is not uncommon and lobbies are often pressurized to some degree to control the influx of outside air. In general this involves supplying more airflow than is returned or made up by outside air, but is sometimes implemented on a zonal basis. Pressurization is also often used as a fire protection scheme in stairwells used as escape routes.

9.5 DILUTION VENTILATION

Any ventilation system that provides any amount of outside air is purging indoor air in a process that is referred to as dilution ventilation. If the outdoor air perfectly displaced indoor air, this would be called piston flow, but such a process would be an idealization and is not practically achievable in real buildings. Instead of perfect displacement of indoor air with outdoor air, we have a gradual mixing and purging of indoor air that occurs at a rate that depends on the building volume, the outside airflow rate, and the degree of mixing.

Dilution ventilation removes all airborne microbes at approximately equal rates, but it will also add microbes from the outdoor air. Dilution ventilation can provide a considerable degree of control over the aerobiology of the indoor air but it cannot, by itself, provide a complete solution because of the outdoor microbes that are brought into the system. At the very least, some level of filtration is needed to control the number of ambient environmental microbes that enter the indoor environment. Figure 9.4 illustrates the effect of these ventilation systems on indoor contaminants assuming complete air mixing. Displacement ventilation performance cannot be compared here since it is a function of the building volume.

The only condition in which purging with outside air is not a solution to an indoor microbial contamination problem is when microbial growth has occurred inside the air handling unit, because this may increase respiratory distress throughout the building.

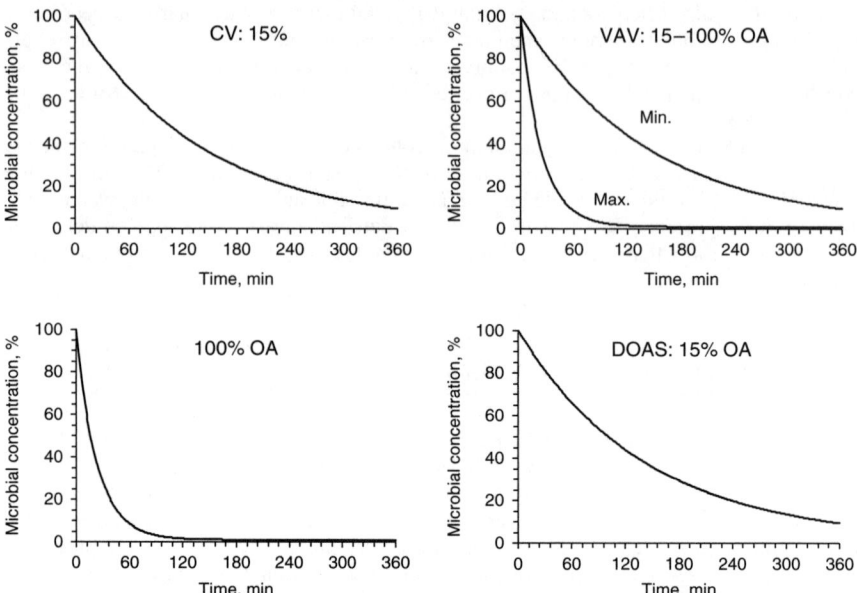

FIGURE 9.4 Comparison of purge rates of ventilation systems (based on typical parameters for a large office building and an initial concentration of 10,000 cfu/m^3.)

Therefore, under normal conditions, purging a building with outside air is an acceptable way of removing airborne pathogens, especially contagious human pathogens.

A variety of tools and methods are available to model the effect of dilution ventilation on indoor levels of airborne microbes, including steady state models, transient models, single-zone models, multizone models, and CFD. These will be addressed in the following sections as a prelude to modeling dilution ventilation in combination with other air cleaning technologies such as filtration and UVGI.

9.6 THE SINGLE-ZONE STEADY-STATE MODEL

In a single-zone model the entire building volume is represented as one zone whether it is divided up into rooms or not. This is a simplistic model since most buildings have multiple rooms, but it can provide insight into the basic dilution and purging process.

A steady-state model simply represents the condition inside the single zone after the system has been in operation for some time. Under steady-state conditions, the quantity of airborne microbes being brought into the zone equals the quantity of microbes being purged from the zone by ventilation air, as follows:

$$N_i = N_o \qquad (9.1)$$

where N_i is the number of microbes entering zone and N_o is the number of microbes exiting zone.

Assuming the infiltration or exfiltration through the walls is not significant, we can write Eq. (9.1) in terms of the airflow concentrations in the *outside air* (OA) provided at the inlet, and the outlet or exhaust air

$$Q_i C_a = Q_o C \qquad (9.2)$$

where Q_i = inlet airflow (OA), m³/min
Q_o = outlet airflow (exhaust), m³/min
C_a = OA concentration of microbes, cfu/m³
C = concentration of airborne microbes indoors, cfu/m³

The concentrations of microbes in air are stated in units of cfu/m³, which are *colony forming units* (cfu) of bacteria or fungi. Technically, the term *plaque forming units* (pfu) should be used for viruses, but this distinction will not be made here for purposes of simplicity, and the units cfu/m³ will be used to refer to all microbes except where stated otherwise.

Assuming exfiltration through the walls is significant, Eq. (9.2) is rewritten as

$$Q_i C_a = (Q_o + Q_e) C \qquad (9.3)$$

where Q_e is the exfiltration through walls, m³/min.
In the event leakage is from the outside to the inside, Eq. (9.3) must be written as

$$(Q_i + Q_x) C_a = Q_o C \qquad (9.4)$$

where Q_x is the infiltration, m³/min.
If it is assumed that infiltration and exfiltration occur simultaneously, which is not an unreasonable assumption in modern buildings, the equation can be written as follows:

$$(Q_i + Q_x) C_a = (Q_o + Q_e) C \qquad (9.5)$$

Figure 9.5 depicts the balance of airflows in the single-zone model. Note that in the steady-state model the zone volume does not figure into the calculations.

Some quantity of the aerosolized microbes will tend to settle out or adsorb to internal building surfaces in a process known as plate-out. The removal rate of microbes due to plate-out is given in terms of a rate constant K_d and the unit area A_d. In effect, this makes

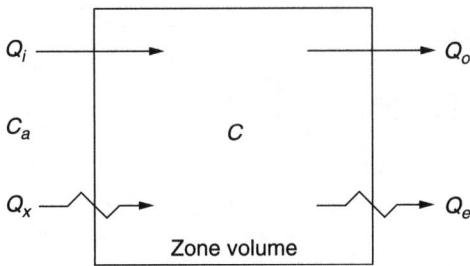

FIGURE 9.5 Single-zone model showing entering airflows and exhaust airflows.

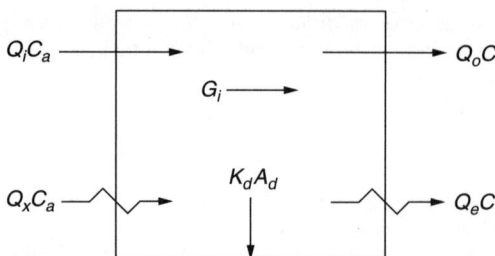

FIGURE 9.6 Flow diagram for microbes entering and exiting the single-zone model.

the building act like a giant, low efficiency filter. Some of these microbes may reaerosolize, such as spores in carpets, but if we ignore this tertiary effect can write Eq. (9.5) to account for plate-out as follows:

$$(Q_i + Q_x)C_a = (Q_o + Q_e)C + K_d A_d \tag{9.6}$$

where K_d is the deposition rate constant, cfu/min·m², and A_d is the deposition area, m².

Furthermore, no model would be complete if we didn't include internal generation of microbes. Typically the source of most indoor pathogens like viruses and bacteria will be humans, or sometimes pets. In some buildings that act as amplifiers, the building itself may be generating fungal spores or bacteria. The generation rate can be included in Eq. (9.6) as follows:

$$(Q_i + Q_x)C_a + G_i = (Q_o + Q_e)C + K_d A_d \tag{9.7}$$

where G_i microbes released internally to indoor air, cfu.

Figure 9.6 depicts the total balance in terms of the entering and exiting microbes.

Assuming for the sake of a simple example that the exfiltration, infiltration, and plate-out are all zero, we can compare the predicted indoor concentrations of Eq. (9.6) with increasing levels of outside airflow. Figure 9.7 shows the results of this model.

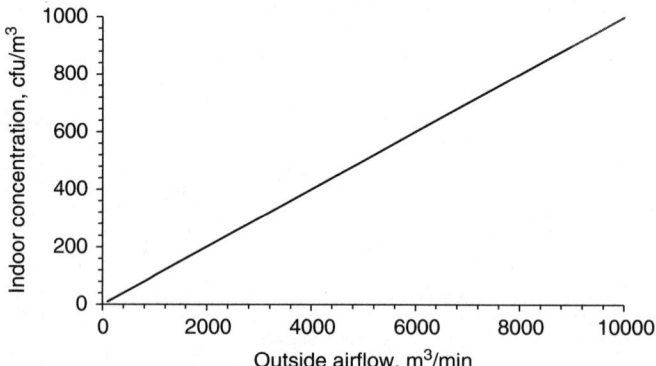

FIGURE 9.7 Effect of increasing outside airflow with constant outside air concentration and constant internal generation.

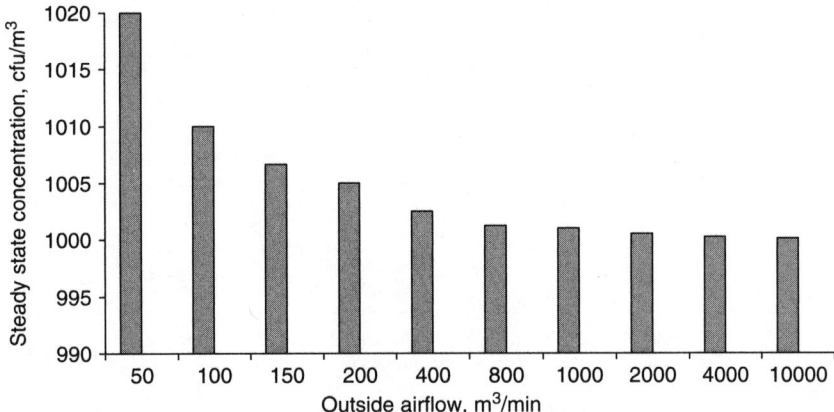

FIGURE 9.8 Effect of increasing outside airflow on indoor concentrations.

As an example of single-zone steady-state model prediction of microbial concentrations in a building, the system model presented in Grimm and Rosaler (1990) is restated here in aerobiological terms and assumes only one filter and a UVGI system is present:

$$C_i = \frac{G_i + 0.01(P_i Q_i + 0.01_u Q_m)C_0}{Q_{v2} + K_d A_d} \tag{9.8}$$

where C_i = steady-state indoor concentration, cfu/m^3
C_0 = outdoor concentration, cfu/m^3
G_i = indoor generation or release rate, cfu/min

Figure 9.8 provides an example of the results of Eq. (9.8) applied to a zone with constant internal contaminant generation and increasing levels of outside air. It can be observed that the steady-state condition will level off at some point, and that increasing outside air beyond this point brings diminishing returns.

9.7 DILUTION MODELING WITH CALCULUS

Calculus can be used to estimate indoor contaminant concentrations in a perfectly mixed single-zone model. The classic model for mixing is a linear first-order differential equation (Boyce and DiPrima, 1997; Heinsohn, 1991). Figure 9.9 shows a single building volume into which outside air flows at a rate of Q and which has a contaminant source discharging into the room at a rate of S.

The change in concentration over time can be found by first writing the mass flow balance for the concentration. The change in concentration over time will equal the amount of contaminants entering from the outside air minus the amount of contaminants exiting through the exhaust air plus the release rate of the source.

$$\frac{d(VC)}{dt} = QC_a - QC + S \tag{9.9}$$

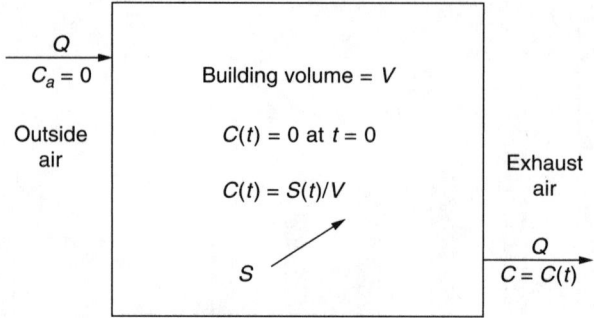

FIGURE 9.9 Single volume model for a building with a constant source of contaminants and an initial concentration of $C = 0$.

In Eq. (9.9), V is the volume, C is the concentration, C_a is the concentration of contaminants in the outside air, Q is the flow rate, and S is the source or release rate of the contaminant. Taking the outside air concentration to be zero, or $C_a = 0$, produces the following:

$$V\frac{dC}{dt} = -QC + S \qquad (9.10)$$

Rearranging Eq. (9.10) for integration over time t results in

$$\int_0^{C(t)} \frac{dC}{S - QC} = \int_0^t \frac{1}{V} dt \qquad (9.11)$$

The solution obtained for the concentration as a function of time $C(t)$ is as follows:

$$C(t) = \frac{S}{Q}\left[1 - \exp\left(-\frac{Q}{V}t\right)\right] \qquad (9.12)$$

In Eq. (9.12), S/Q represents the steady-state concentration. It can be rewritten in a slightly more convenient form by defining the number of room-air changes per hour, N, as the airflow Q divided by the room volume V as follows:

$$N = \frac{Q}{V} \qquad (9.13)$$

Rewriting the equation in its simplified form produces

$$C(t) = \frac{S}{Q}[1 - \exp(-Nt)] \qquad (9.14)$$

Figure 9.10 illustrates the generic response of the room concentration over time for a constant source and for various air change rates.

If the initial concentration is some value C_0, as in the case of a sudden release of an agent, the decay of the building concentration can be derived in a similar fashion and

VENTILATION SYSTEMS AND DILUTION 197

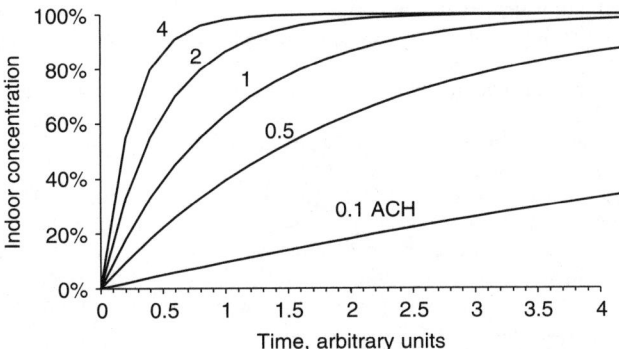

FIGURE 9.10 Effect of various air change rates (ACH) on building concentrations from a constant contaminant source.

written as the following:

$$C(t) = C_0[\exp(-Nt)] \qquad (9.15)$$

Figure 9.11 shows the results of Eq. (9.15) for the same generic conditions as Fig. 9.10.

9.8 SINGLE-ZONE TRANSIENT MODELING

The concentrations of airborne contaminants inside a building vary continuously over time due to variations in outdoor airborne levels and changing internal inputs due to everything from sneezing to moving furniture. The previous steady-state model can only tell us the final concentrations of airborne microbes after conditions have stabilized. How long this will take depends primarily on the building volume, the airflow rates, and the degree of mixing. A single-zone transient model can predict how the indoor concentrations vary over

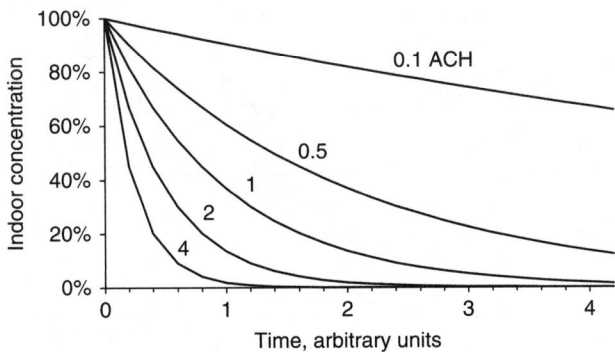

FIGURE 9.11 Decay of building concentrations after a sudden release of a biocontaminant for different air change rates.

time, which will allow determination of how effective dilution ventilation is at controlling indoor aerobiology. It will also allow us to estimate the doses of microorganisms inhaled by building occupants.

The single-zone transient model presented here uses finite time steps to estimate indoor concentrations for each minute of building occupation. This computational approach can easily be performed on a spreadsheet, as in the examples shown here, or via a programming language such as C++, Basic, or Fortran.

In this single-zone transient model several assumptions must be made, the first of which is complete mixing. All the air in the single-zone model of the building is assumed to mix completely on a minute-by-minute basis. Complete air mixing will slow the removal of airborne pathogens in an exponential manner, as opposed to plug flow, in which the removal rate is a linear function of time. Complete mixing represents the limiting case for normal buildings, and is a reasonable and simple model to use for evaluating the removal rate of airborne pathogens. Given this assumption, the primary factor determining the removal of airborne pathogens is the air change rate. Figure 9.12 shows a schematic of a typical constant volume ventilation system with recirculation and outside air.

For each time step a finite volume of fresh air replaces an equal volume of mixed room air. The choice of the time step can influence the concentration, but a value of 1 min provides a close approximation of continuous flow without requiring excessive computations.

The removal rate of microorganisms is given by the indoor concentration $C(t)$ at any given time t multiplied by the outside airflow rate Q_o, which is the volume of air displaced into and out of the building volume. For each minute of the analysis, the number of microorganisms removed is determined as

$$N_{out} = C(t) \cdot Q_o \tag{9.16}$$

Likewise, the number of microorganisms added will be the outside airflow rate Q_o multiplied by the concentration of microbes (spores) $C_a(t)$ in the outdoor air. For each minute of analysis, the number of spores added is given by

$$N_{in} = C_a(t) \cdot Q_o \tag{9.17}$$

The rates of bacteria and viruses generated internally are specified per minute, and are designated B_{in} and V_{in}, respectively. The total population of microbes $N(t)$ that will exist in the building, for any given minute t, will then be the previous minute's population plus the

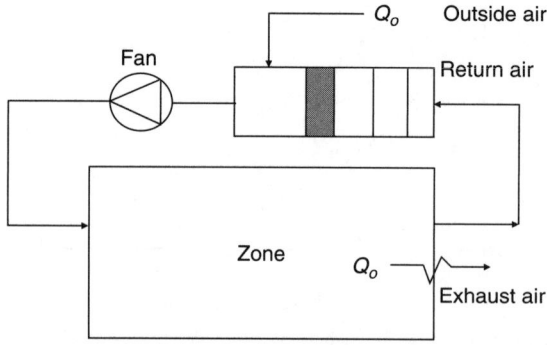

FIGURE 9.12 Schematic of single-zone building model.

VENTILATION SYSTEMS AND DILUTION 199

current minute's additions minus the number exhausted to the outside air:

$$N(t) = N(t-1) + N_{in} - N_{out} + B_{in} + V_{in} \qquad (9.18)$$

In the model, complete mixing is assumed, and therefore the building microbial concentration will be defined as the building microbial concentration divided by the building volume, for any given minute, or

$$C(t) = \frac{N(t)}{V} \qquad (9.19)$$

In Eq. 9.12 the term V refers to the building volume, in cubic meters, which is given, per Table 9.1, as 3401 m³.

Normal conditions, in which initial concentrations would be lower, would not provide for a drastic enough distinction between low ACH and high ACH. In order to provide a useful means of comparing results, it is necessary to begin the analysis with a building already saturated with a high concentration of airborne pathogens. The scenario could be assumed to represent the condition in which the system has been turned off for several hours while indoor airborne concentrations have risen to a high level, and at which point the ventilation system is suddenly turned on.

This transient analysis compares several outside air conditions as shown in Table 9.2. The minimum outside air rate of 15 percent is typical for most constant volume systems. The 100 percent outside air case is normally only seen in laboratory and hospital settings. The initial concentration of microbial contamination assumed for this analysis is 20,000 cfu/m³.

Airborne pathogen hazards are dependent on the species of microbe. Some microbes are extremely lethal at very low doses. Some guidelines exist for levels of airborne fungi, and these are used as general indicators. The ACGIH and the AIHA define 1000 cfu/m³ as an upper limit for concentrations in indoor environments, while the CEC defines 2000 cfu/m³ as a very high level (Rao and Burge, 1996). A value of 20,000 cfu/m³ of nondescript airborne microbes could therefore be considered a hazardous level for indoor environments.

The calculated value of the building concentration, per Eq. 9.12, is then used to compute the values for the next minute. This incremental calculation process, carried out for 300 minutes, provides a close approximation of continuous flow, with the result being an exponential decrease in the building microbial concentration over time. Table 9.3 shows a sample of the first few minutes of computations for the 15 percent outside air case.

Results for all four cases are shown graphically in Fig. 9.13. It can be seen that zone concentrations are drawn down rapidly at first and then more slowly, in a classic exponential

TABLE 9.1 Building Design Parameters

Total people	1,000	
Number of floors	10	
Floor area	929 m²	10,000 ft²
Floor height	3.6576 m	12 ft
Floor volume	3,401 m³	120,000 ft³
Total floor area	9,290 m²	100,000 ft²
Unit area per person	9.29 m²	100 ft²
Total volume	33,980 m³	1E+06 ft³
Initial concentration	20,000 cfu/m³	566.34 cfu/ft³
OA concentration	1,000 cfu	

TABLE 9.2 Parameters for OA Airflow Cases

Outside air, %	15	25	50	100
OA Air change rate per hour (ACH)	1	2	3	5
OA per person, m³/min	0.56634	0.707925	1.41585	2.8317
OA Total, m³/min	425	708	1416	2832
Total zone m³/min	2832	2832	2832	2832
Total OA ACH	0.75	1.25	2.5	5
Total ACH	5	5	5	5

TABLE 9.3 Transient Zone Concentrations for 15 Percent OA

Time, min	Total zone, cfu	Concentration, cfu/m³	cfu In (OA air)	cfu Out (Exhaust air)
0	679,608,000	20,000	424,755	8,495,100
1	671,537,655	19,763	424,755	8,394,221
2	663,568,189	19,528	424,755	8,294,602
3	655,698,342	19,296	424,755	8,196,229
4	647,926,868	19,068	424,755	8,099,086
5	640,252,537	18,842	424,755	8,003,157
6	632,674,135	18,619	424,755	7,908,427
7	625,190,463	18,399	424,755	7,814,881
8	617,800,338	18,181	424,755	7,722,504
9	610,502,588	17,966	424,755	7,631,282
10	603,296,061	17,754	424,755	7,541,201

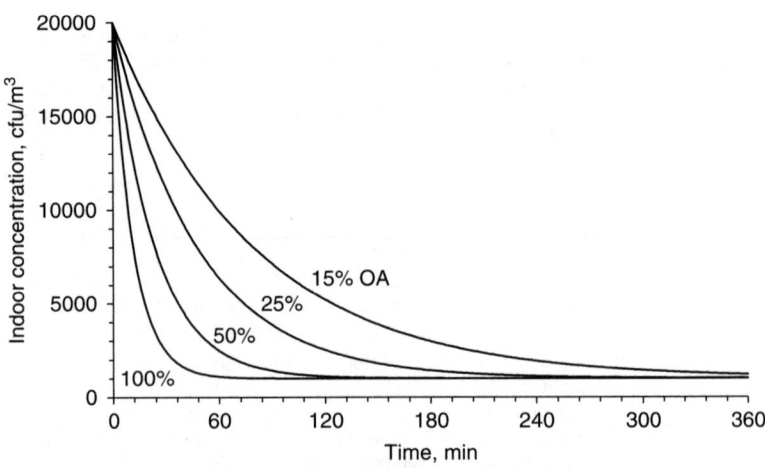

FIGURE 9.13 Transient concentrations for single-zone model.

decay curve, as would be expected. The final concentration in all four cases reaches a steady state of approximately 1000 cfu/m^3 due to the fact that the outside air has a constant concentration of 1000 cfu/m^3 and the indoor air cannot be brought down any lower without filters or air cleaning devices.

A good method for checking the results of any spreadsheet or program method such as the previous example is to verify the steady-state condition with a simple calculation. In this example, for the 15 percent outside air case, the steady-state concentration can be directly computed by dividing the rate of input of contaminants by the outside airflow. In the example presented, where the input microbes come only from the outdoor air, it can be logically deduced that the inside air concentration will approach the outdoor concentration over time. If there were any internal generation, at a rate of G_i cfu/min, the steady state would be computed as follows:

$$C_{ss} = \frac{G_i + Q_o C_a}{Q_o} \qquad (9.20)$$

where C_{ss} is the steady-state concentration, cfu/m^3.

The question of whether to use outdoor air for dilution and purging of indoor airborne contaminants hinges on several factors that all involve economic and energy questions. If the climate is favorable, it can be economical to use large quantities of outdoor air. If not, the economics of heat exchangers needs to be investigated to determine what percentage of outdoor air it might be feasible to use. Since filtration of outdoor air will almost always be appropriate for controlling indoor aerobiology, the cost of filtration should be accounted for also.

The previous example did not include infiltration, exfiltration, or plate-out. These factors can be included by simply adding extra columns to the spreadsheet in Table 9.3. Additional zones can be added to this model but this would become akin to a multizone model. Multizone models of buildings are better handled by existing software packages.

9.9 MULTIZONE MODELING

Single-zone modeling can be a simple and convenient way of estimating the impact of dilution ventilation on indoor airborne contaminants but may give a misleading picture if the building is, in fact, composed of distinct and separate zones with separate airflows. In order to more accurately account for separate zones with separately mixed air volumes, it is necessary to model the building with multiple zones. In general, large office buildings or other commercial facilities are more realistically modeled with multiple zones, especially when they have multiple ventilation systems.

It is possible to use the previously developed spreadsheet to create a multizone model. It would require the addition of a new set of columns for each zone with lots of cross-linked cells and can become quite complicated. Fortunately, some very sophisticated software tools have been developed over the previous decades that perform multizone modeling and which incorporate a variety of factors such as interior leakage, exterior leakage, wind pressure effects, stack effects, and plate-out effects. Several multizone modeling packages are currently available including CONTAMW, DOE-2, ESP-r, Risk V1.0, and others (Dols et al., 2000; Sparks, 1995; Axley, 1987; Demokritou, 2001).

The previous methods for modeling buildings are single-zone models that treat the entire building as one large area. The single-zone approach should be adequate for modeling most buildings for basic release scenarios like an air intake release, or a release in the supply or return duct. For improved modeling realism, or where multiple internal zones may have different airflows and volumes, the best approach is to use a multizone model such as the CONTAMW program. Other programs are also available for multizone modeling, but the CONTAMW program is public domain and has been used in a number of published studies on contaminant dispersion inside buildings (NIST, 2002).

The CONTAMW program simulates the release and distribution of contaminants in a building with a ventilation system in much the same way as the previously described numerical integration or spreadsheet model. It has a variety of advantages including the ability to model multiple zones, varying infiltration and exfiltration rates, and modeling of wind pressure or buoyancy effects. CONTAMW has the ability to model complex building systems and the contaminant source is easy to move around or manipulate.

Modeling with CONTAMW begins with a definition of the floors and zone volumes, as shown in Fig. 9.14. To each zone are added icons that denote the supply air volume and path, the exhaust air volume and path, the leakage pathways through doors, walls, windows, and floors, air handling units, and contaminant sources. Additional data input involves defining the scheduled air supply, minimum outside air, time steps, and the like. Options are available in CONTAMW to model wind effects, stack effects, plate-out on surfaces, and air filtration equipment.

Output data from CONTAMW can be exported to a text file from which it can be easily read into a spreadsheet for postprocessing. Table 9.4 shows some typical output from a CONTAMW analysis of a multizone, multistory building with a recirculated contaminant. The date and times are shown in the first three columns and the contaminant concentration is shown in the final column. The units given in this case are kilograms of contaminant per kilograms of dry air, but any units are possible including ppm, or ng/m^3. In fact, it is convenient to ignore the output units (i.e., kg/kg) and simply assign units of cfu/m^3 to the output (and input) values, since the results are generic regardless of units. Postprocessing of data is also useful for determining inhaled dosages of contaminants, and the inhalation dosimetry methods presented in Chap. 4 can be used along with the epidemiological models (see also Kowalski, 2003).

Figure 9.15 shows the graphical output of the transient modeling results from the CONTAMW program for a multizone building with a biocontaminant released in one zone. In this case the zone with the release, a hallway, is seen to have much higher levels than some of the adjacent areas.

Extensive information is available in the literature on multizone modeling methods from NIST and other sources (Musser et al., 2002; Musser, 2000, 2001; Yaghoubi et al., 1995;

FIGURE 9.14 CONTAMW model of a mall with icons denoting air supply, exhaust, and leakage paths. Labels have been added to identify major zones.

TABLE 9.4 CONTAMW Output for a Multizone Building

Level: date	First time of day	Zone: time, s	Stairwell1 Agent-1, kg/kg
1-Jan	0:00:00	0	0.000
1-Jan	1:00:00	3,600	0.001
1-Jan	2:00:00	7,200	0.002
1-Jan	3:00:00	10,800	0.003
1-Jan	4:00:00	14,400	0.004
1-Jan	5:00:00	18,000	0.005
1-Jan	6:00:00	21,600	0.006
1-Jan	7:00:00	25,200	0.007
1-Jan	8:00:00	28,800	0.008
1-Jan	9:00:00	32,400	0.009
1-Jan	10:00:00	36,000	0.010
1-Jan	11:00:00	39,600	0.011
1-Jan	12:00:00	43,200	0.012
1-Jan	13:00:00	46,800	0.013
1-Jan	14:00:00	50,400	0.014
1-Jan	15:00:00	54,000	0.015
1-Jan	16:00:00	57,600	0.016

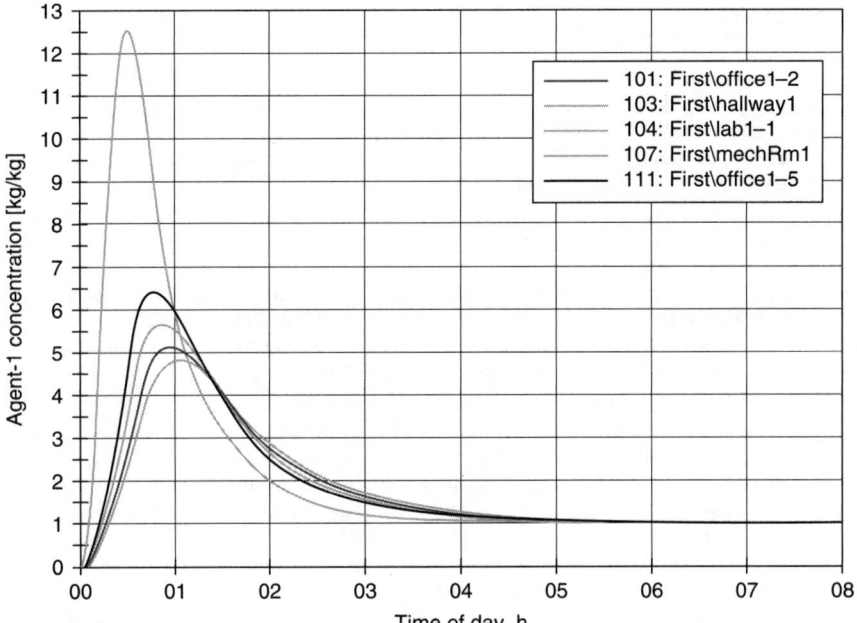

FIGURE 9.15 Example of CONTAMW graphical output showing airborne concentrations in several zones over an 8-hour period due to a contaminant release in the first floor hallway.

Walton, 1989), especially through the NIST website (NIST, 2002). Operation of the CONTAMW program is not simple and it is best to refer to the cited documents or consult an expert.

9.10 COMPUTATIONAL FLUID DYNAMICS MODELING

Computational fluid dynamics (CFD) software uses numerical solutions of the Navier-Stokes equations governing conservation of mass, conservation of energy, and conservation of momentum to create a realistic picture of the behavior of air currents in defined volumes. CFD methods divide a room volume into finite three-dimensional cells in which the equations are individually balanced and rebalanced iteratively until all the cells are in agreement. The result is that CFD modeling can be used to predict the movement of air currents through rooms and around objects.

CFD is a computationally intensive approach that can be time-consuming but will provide more accurate predictions of air movement than are possible with the previously discussed methods. CFD can be very useful in certain situations that require a high degree of detail in the definition of the airflow movement. CFD methods can provide assistance in the location of supply and exhaust registers to facilitate air mixing in rooms and to facilitate the removal of contaminants from rooms (FDI, 1988; Kundu, 1990). Programs that are currently available include FLUENT, FIDAP, FLOW-3D, and others.

REFERENCES

Axley, J. W. (1987). "Indoor Air Quality Modeling." *NBSIR 87-3661* NBS. Gaithersburg, MD.

Banaszak, E. F., Thiede, W. H., and Fink, J. N. (1970). "Hypersensitivity pneumonitis due to contamination of an air conditioner." *New Engl J Med* 283(6):271–276.

Boyce, W. E., and DiPrima, R. C. (1997). *Elementary Differential Equations and Boundary Value Problems*. John Wiley & Sons, New York.

Chen, Q., and Glicksman, L. (2003). *System Performance Evaluation and Design Guidelines for Displacement Ventilation*. American Society of Heating, Refrigerating and Air-Conditioning Engineers, Atlanta, GA.

Demokritou, P. (2001). Modeling IAQ and Building Dynamics. *Indoor Air Quality Handbook*. J. D. Spengler, J. M. Samet, and J. F. McCarthy, eds., McGraw-Hill New York.

Dols, W. S., Walton, G. N., and Denton, K. R. (2000). "CONTAMW 1.0 users manual." NTIS. Springfield, VA. http://fire.nist.gov/bfrlpubs/build00/art041.html.

FDI (1998). *FIDAP 8*. Fluid Dynamics International, Lebanon, NH.

Fisk, W. (1994). "The California healthy buildings study." *Center for Building Science News* 12:7, 13.

Flannigan, B., McCabe, E. M., and McGarry, F. (1991). Allergenic and toxigenic microorganisms in houses. *Pathogens in the Environment*, B. Austin, ed., Blackwell Scientific Publications Oxford, UK.

Flannigan, B., Samson, R. A., and Miller, J. D. (2001). *Microorganisms in home and indoor work environments*. Taylor & Francis. Andover, Hants, UK.

Flomerics (2003). "Application of CFD to naturally ventilated buildings: A guide for practitioners." http://www.flovent.com/applications/hga/hgapdf1.pdf.

Foarde, K. K., Hanley, J. T., and Veeck, A. C. (2000). "Efficacy of antimicrobial filter treatments." *ASHRAE J* 42(12):52–58.

Grimm, N. R., and Rosaler, R. C. (1990). *Handbook of HVAC Design*. McGraw-Hill, New York.

Heinemann, S., Beguin, H., and Nolard, N. (1994). Biocontamination in air conditioning. *Health Implications of Fungi in Indoor Environments*, R. A. Samson, ed., Elsevier, Amsterdam, The Netherlands. 179.

Heinsohn, R. J. (1991). *Industrial Ventilation: Principles and Practice*. John Wiley & Sons, New York.

Jankowska, E., Reponen, T., Willeke, K., Grinshpun, S. A., and Choi, K.-J. (2000). "Collection of fungal spores on air filters and spore reentrainment from filters into air." *J Aerosol Sci* 31(8):969–978.

Kemp, S. J., Kuehn, T. H., Pui, D. Y. H., Vesley, D., and Streifel, A. J. (1995). "Filter collection efficiency and growth of microorganisms on filters loaded with outdoor air." *ASHRAE Trans* 101(1):228.

Kowalski, W. J., and Bahnfleth, W. P. (2002). "MERV filter models for aerobiological applications." *Air Media*, Summer:13–17.

Kowalski, W. J. (2003). *Immune Building Systems Technology*. McGraw-Hill, New York.

Kundsin, R. B. (1988). *Architectural Design and Indoor Microbial Pollution*. Oxford University Press, New York.

Kundu, P. K. (1990). *Fluid Mechanics*. Academic Press, San Diego, CA.

Macher, J. M., and Girman, J. R. (1990). "Multiplication of microorganisms in an evaporative air cooler and possible indoor air contamination." *Environ Int* 16:203–211.

Maus, R. (1996). "Viability of microorganisms in fibrous air filters." *The Seventh International Conference on IAQ and Climate*, Nagoya, Japan. 137–142.

Millar, J. D., Morris, G. K., and Shelton, B. G. (1997). "Legionnaires' disease : Seeking effective prevention." *ASHRAE J* 39(1):22–29.

Morey, P. R., Feeley, J. C., and Otten, J. A. (1990). *Biological Contaminants in Indoor Environments*. ASTM, Philadelphia, PA.

Morey, P. R. (2001). Control and remediation of microbial growth in problem buildings. *Microorganisms in Home and Work Environments*. B. Flannigan, R. A. Samson, and J. D. Miller, eds., Taylor & Francis, London. 83–100.

Mumma, S. A. (2001a). "Dedicated outside air systems." *ASHRAE IAQ Applications* 2(1). http://doas-radiant.psu /IAQ_Winter2001pgs20-22.pdf.

Mumma, S. A. (2001b). "Designing dedicated outdoor air systems." *ASHRAE J* 43(5):28–31. http://doas-radiant.psu /journal_01_doas.pdf.

Mumma, S. A. (2002). "Safety and comfort using DOAS: Radiant cooling panel systems." *IAQ Applications* Winter:20–21. http://doas-radiant.psu /IAQ_winter_02.pdf.

Musser, A. (2000). "Multizone modeling as an indoor air quality design tool." *Healthy Buildings 2000*. Espoo, Finland. ASHRAE, Atlanta, GA.

Musser, A. (2001). "An analysis of combined CFD and multizone IAQ model assembly issues." *ASHRAE Trans* 106(1):371–382.

Musser, A., Kowalski, W., and Bahnfleth, W. (2002). "Stack and mechanical system effects on dispersion of biological agents in a tall building." *Ninth Symposium on Measurement and Modeling of Environmental Flows, International Mechanical Engineering Congress and Exposition*. New Orleans, LA.

Nardell, E. A., Keegan, J., Cheney, S. A., and Etkind, S. C. (1991). "Airborne infection: Theoretical limits of protection achievable by building ventilation." *Am Rev Resp Dis* 144:302–306.

NIST (2002). *CONTAMW: Multizone Airflow and Contaminant Transport Analysis Software*. National Institute of Standards and Technology, Gaithersburg, MD. http://www.bfrl.nist.gov/IAQanalysis/CONTAMWdownload1.htm.

Parat, S., Perdrix, A., Fricker-Hidalgo, H., Saude, I., Grillot, R., and Baconnier, P. (1997). "Multivariate analysis comparing microbial air content of an air-conditioned building and a naturally ventilated building over one year." *Atmos Environ* 31(3):441–449.

Rao, C. Y., and Burge, H. A. (1996). "Review of quantitative standards and guidelines for fungi in indoor air." *J Air Waste Mgt Assoc* 46(Sep.):899–908.

Ryan, K. J. (1994). *Sherris Medical Microbiology*. Appleton & Lange. Norwalk, CT.

Schicht, H. H. (1972). "The diffusion of microorganisms by air conditioning installations." *The Steam and Heating Engineer* October:6–13.

Seigel, J. A., and Walker, I. S. (2001). "Deposition of biological aerosols on HVAC heat exchangers." Lawrence Berkely National Laboratory. Berkely, CA. http://eetd.lbl.gov/ied/viaq/pubs/LBNL-47660.pdf.

Skistad, H. (1994). *Displacement Ventilation.* John Wiley & Sons, New York.

Solomon, W. R., Burge, H. A., and Boise, J. R. (1980). "Exclusion of particulate allergens by window air conditioners." *J Allergy Clin Immunol* 65(4):305–308.

Sparks, L. E. (1995). *IAQ Model for Windows: Risk Version 1.0 User Manual.* EPA-600/R-96-037 EPA.

Spiegelman, J., Friedman, H., and Blumstein, G. I. (1963). "The effects of central air conditioning on pollen, mold, and bacterial concentrations." *J Allergy* 34(5):426–431.

Sugawara, F. (1996). "Microbial contamination in ducts of air conditioning systems." *The Seventh International Conference on IAQ and Climate.* Nagoya, Japan. 161–166.

Tamblyn, R. T. (1995). "Toward zero complaints for office air conditioning." *HPAC Eng* 67(3):67–72.

Thornsberry, C., Balows, A., Feeley, J., and Jakubowski, W. (1984). *Legionella: Proceedings of the Second International Symposium.* Atlanta, GA.

Walton, G. N. (1989). "Airflow network models for element-based building airflow modeling." *ASHRAE Trans* 1989:611–620.

Yaghoubi, M. A., Knappmiller, K., and Kirkpatrick, A. (1995). "Numerical prediction of contaminant transport and indoor air quality in a ventilated office space." *Particul Sci Technol* 13:117–131.

CHAPTER 10

FILTRATION OF AIRBORNE MICROORGANISMS

10.1 INTRODUCTION

Filtration can be an effective means of controlling the aerobiology of indoor air. This chapter details the performance of filters in the size range of airborne microbes. A classic mathematical model of filtration is presented to predict filter performance below the limits of typical filter performance curves and at operating velocities other than design velocities. Standards for rating filters are reviewed and test results are summarized that demonstrate filter performance against airborne microbes. Results from field trials are summarized to show both the problems and benefits in actual installations. The models presented apply to all filters, including cartridge filters like those shown in Fig. 10.1, bag filters, and prefilters or dust filters. The model may not necessarily apply to all electrostatic filters, which are detailed in Chap. 13, and it does not apply to carbon filters, which are addressed in Chap. 12.

10.2 FILTER PERFORMANCE CURVES

Filters performance is generally described by filter performance curves such as those in Fig. 10.2. These curves depict the filtration efficiency for a range of particle sizes, typically from about 0.3 to 0.4 μm up to about 10 μm, and at a design velocity of 400 to 500 fpm (ASHRAE, 1992). The main problem with using catalog performance curves to predict microbial filtration is that they do not extend far enough down to accommodate predictions for microbes that span the range of 0.02 to 0.3 μm. Another problem is that filters are often operated beyond their design velocity.

There are a number of rating methods and standards for classifying filters. Until recently the most common rating standard for high-efficiency filters was ANSI/ASHRAE Standard 52.1-1992 but the current standard is ANSI/ASHRAE Standard 52.2-1999 (ASHRAE, 1999). Dust spot efficiency is a measure of the ability of the filter to remove dust from a test airstream. It primarily applies to medium to high-efficiency filters. Dust spot efficiency

FIGURE 10.1 Cartridge type high-efficiency filters. (*Photos provided courtesy of the Donaldson Company, Inc., Minneapolis, MN.*)

testing does not apply to *high efficiency particulate air* (HEPA) filters, which are typically rated with a thermal DOP penetration test (per US MIL-STD 282). HEPA filters must remove particles of 0.3 μm diameter with at least a 99.97 percent efficiency (ASHRAE, 1992). Figure 10.3 shows a representative performance curve for a HEPA filter with the performance point of 99.97 percent at 0.3 μm indicated. The reason for this particular definition of HEPA filter performance relates to its original use in the nuclear industry of filtering out radioactive isotopes in the 0.3-μm size range. HEPA filters have seen widespread use outside the nuclear industry, including computer industry clean rooms, the pharmaceutical industry, and hospital operating rooms. Many clean rooms actually use *ultra-low penetration air* (ULPA) filters, which may have ratings as high as 99.999 percent on particles in the range 0.1 to 0.2 μm.

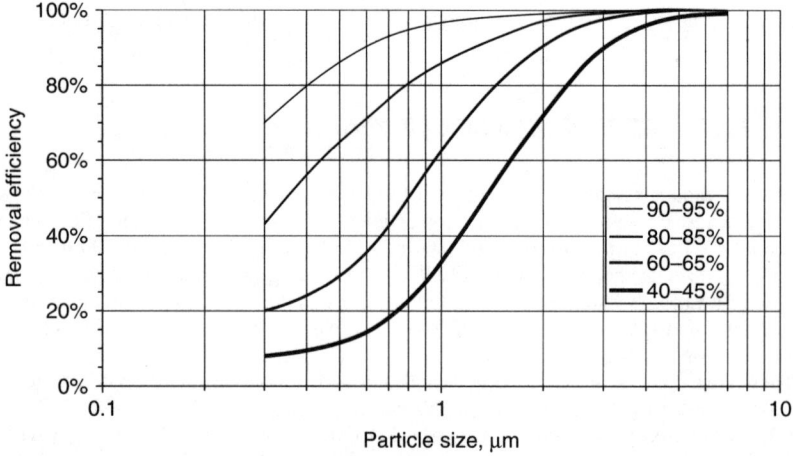

FIGURE 10.2 Typical filter performance curves for four dust spot (DSP)-rated filters from a vendor catalog. The minimum size limit of most performance curves is usually 0.3–0.4 μm.

FILTRATION OF AIRBORNE MICROORGANISMS 209

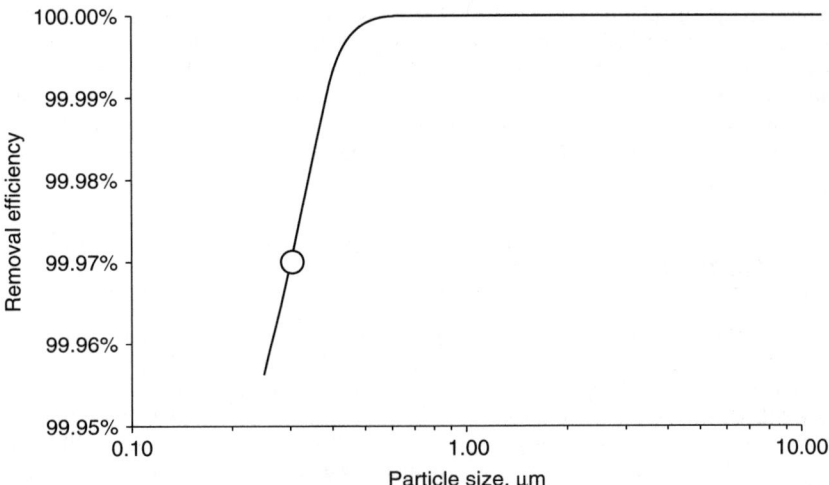

FIGURE 10.3 Performance curve for a HEPA filter showing rating point at 0.3 μm.

Variations in performance occur between HEPAs from different manufacturers, but performance often exceeds requirements (Matteson and Orr, 1987). Differences between HEPAs may also be reflected in terms of the most penetrating particle size, or the inflection point (where the removal efficiency begins to decrease) that exists at or below the design point of 0.3 μm.

Figure 10.4 provides one set of performance curves for DSP-rated filters based on laboratory measurements (Ensor et al., 1988). These data indicate what both empirical data and theory predict—the removal efficiency of particles below about 0.2 μm increases,

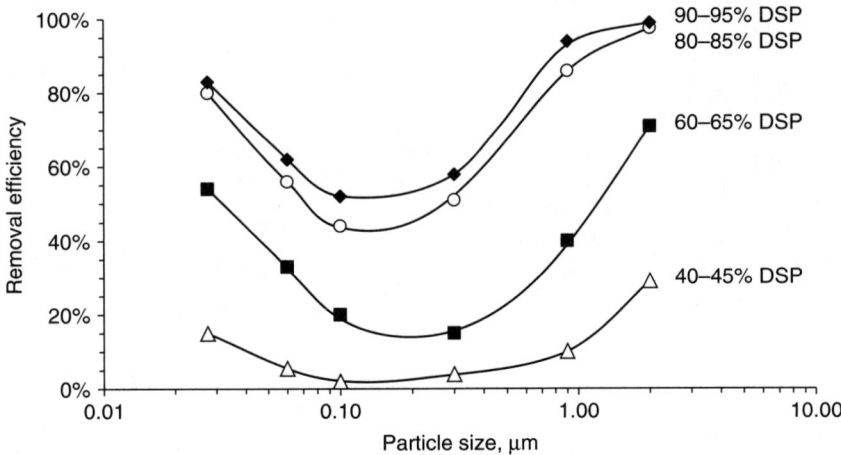

FIGURE 10.4 Performance results for four DSP-rated filters. [*Based on data from Ensor et al. (1988).*]

resulting in higher removal rates for virus-sized particles. The DSP rating of each of the four filters in Fig. 10.4 is indicated. The efficiency at any point can fall far below or above the arrestance. Vendors generally specify arrestance as a range (i.e., 90 to 95 percent)—the lower limit is used here as a matter of convenience. A review of various vendor catalog curves and a variety of empirical data suggests that filter performance variations at any point in the performance curve of about ±15 percent are normal for any grade of high-efficiency filter.

10.3 MERV FILTER RATINGS

The ASHRAE performance scale known as *minimum efficiency reporting value* (MERV) is used to define the performance of medium to high-efficiency filters. Unlike the dust spot efficiency method, the MERV standard evaluates filter performance over a broad range of sizes. MERV ratings are determined using ASHRAE Standard 52.2-1999 to test filter performance in the 0.3- to 10.0-μm size range. Table 10.1 lists the main types of filters and compares MERV ratings with their approximate equivalent DSP ratings per ASHRAE (1999). Although approximate MERV ratings are shown for HEPA and ULPA filters, these filters would not normally be tested or rated on the MERV scale since the DOP penetration test remains an adequate measure of their performance.

Determination of filter MERV under ASHRAE Standard 52.2 involves measuring the penetration at each of a series of 12 size ranges between 0.3 and 10 μm. Several tests are performed in succession during which the filter may load up and increase (or sometimes decrease) in efficiency. A composite of the results is used to assign the final filter rating. Table 10.2 shows the results for one filter test that produced a MERV 10 rating.

Although MERV ratings are an improvement over the previous DSP standard, they still do not extend down into the smaller size range of interest in aerobiological engineering.

TABLE 10.1 Filter Types and Approximate Ratings

Filter type	Applicable size range, μm	Dust spot efficiency, %	Total arrestance, %	MERV rating (estimated)
Dust filters	>10	<20	<65	1
		<20	65–70	2
		<20	70–75	3
		<20	75–80	4
High efficiency	3–10	<20	80–85	5
		<20	85–90	6
		25–35	>90	7–8
	1–3	40–45	>90	9
		50–55	>95	10
		60–65	>95	11
		70–75	>98	12
	0.3–1	80–90	>98	13
		90–95	NA	14
		>95	NA	15–16
HEPA	<0.3	NA	NA	17–18
ULPA	<0.3	NA	NA	19–20

FILTRATION OF AIRBORNE MICROORGANISMS   211

TABLE 10.2 ASHRAE 52.2 Test Results for a MERV 10 Rated Filter

Particle diameter, μm			Removal efficiency, %				
Minimum	Maximum	Mean	Initial	Conditioning	1st	2nd	Composite
0.3	0.4	0.35	6.4	11.7	0	0	0
0.4	0.55	0.47	6.9	14.9	0	0	0
0.55	0.7	0.62	18.4	30.8	16.5	0	0
0.7	1	0.84	19	37.5	27.7	7	7
1	1.3	1.14	35.7	57.9	47.9	19.2	19.2
1.3	1.6	1.44	57.9	74.4	65.2	43.3	43.3
1.6	2.2	1.88	65	84.1	78.9	63.6	63.6
2.2	3	2.57	76.8	94.4	92.5	90.4	76.8
3	4	3.46	87.1	97.8	96.4	96.4	87.1
4	5.5	4.69	91.2	98.3	98	95.3	91.2
5.5	7	6.2	92.4	97.8	97.6	95.4	92.4
7	10	8.37	93	99	97.8	95.3	93

They do, however, provide an improved data set from which to predict performance in the lower size range using filter modeling.

The European Union uses a different standard to rate filters that is based on the same concepts and testing methods as ASHRAE Standard 52.2 but is differentiated by filter types. As a result the ratings do not exactly correspond to the more homogeneous U.S. standards. Table 10.3 summarizes the European filter rating standards (both old and new) and suggests the approximate MERV equivalent.

TABLE 10.3 European Filter Classes Comparison

Type	New class	Eurovent class	Efficiency, %	MERV (approximate)	Testing
Coarse dust filter	G1	EU1	<65	1	Dust arrestance (weight)
	G2	EU2	65–80	2	
	G3	EU3	80–90	—	
	G4	EU4	>90	—	
Fine dust filter	F5	EU5	40–60	1–2	Dust spot (DSP)
	F6	EU6	60–80	2–4	
	F7	EU7	80–90	5–6	
	F8	EU8	90–95	7–10	
	F9	EU9	>95	11–13	
High efficiency	H10	EU10	85	5–9	NaCl or aerosol
	H11	EU11	95	10–13	
	H12	EU12	99.5	15	
	H13	EU13	99.95	16	
HEPA	H14	EU14	99.995	17–18	
ULPA	U15	EU15	99.9995	19	Aerosol
	U16	EU16	99.99995	19–20	
	U17	EU17	99.999995	20	

10.4 MODELING FILTER PERFORMANCE

The performance of high-efficiency ASHRAE filters can be modeled using single-fiber filtration theory. A variety of filtration models have been developed in the past and most of them provide results that are in good agreement with actual laboratory measurements of filter efficiency across the entire size range of interest, or about 0.2 to 10 µm.

Several mathematical models for filters exist and were considered for possible use. The basic single-fiber models of Liu and Rubow (Raber, 1986), Liu (Matteson and Orr, 1987), Davies (1973), Brown (1993), and Lee and Liu (Rivers, 1988) were compared under real-world design parameters for a 90 percent filter in Kowalski et al. (1999). It was found that none of the models gave perfect fits for all grades of filters and that all models could be adjusted to yield close fits by adjusting air velocities within the normal range (i.e., 250 to 800 fpm) for the respective nominal filter or by adjusting fiber diameters within normal ranges. The Rubow model (Liu and Rubow, 1986) was selected for development because of the closeness of the fit and the fact that the model's lack of an impaction component simplified the computations. Exhaustive testing showed that all of these filter models could be adapted to yield comparable predictive accuracy based on the data and therefore this choice was essentially arbitrary.

Figure 10.5 shows a photomicrograph of glass filter fibers. Shown for comparison is a spore of 5 µm diameter. Fiber diameters range from about 1 to 20 µm. It should be clear from this comparison that filters do not act as sieves for microbes, but that they are removed by impaction, interception, and diffusion processes.

10.5 MATHEMATICAL MODEL OF FILTRATION

The efficiency of a single fiber forms the theoretical foundation for overall filter efficiency. The single-fiber efficiency depends on particle size, air velocity, and fiber properties. The equation defining overall filter efficiency (E) for any particle size and set of conditions

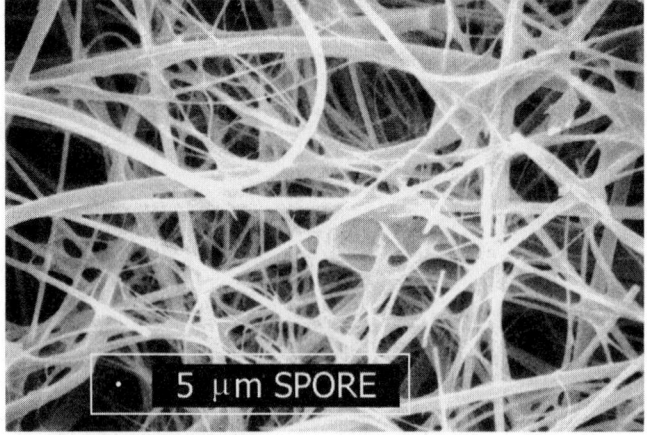

FIGURE 10.5 Photomicrograph of glass fiber filter shown with a 5-mm spore for comparison. (*Photo provided courtesy of Filterite, Timonium, MD.*)

(Davies, 1973) is as follows:

$$E = 1 - e^{-E_s S} \tag{10.1}$$

where S is the fiber projected area, dimensionless, and E_s is the single-fiber efficiency, fractional.

The fiber projected area (S) is a dimensionless constant combining the three main determinants of filter efficiency—filter thickness (length normal to airflow), filter packing density (packing fraction or volume fraction), and fiber diameter (Davies, 1973; Brown, 1993). The fiber projected area can be represented in forms that are more physically intuitive, but the mathematical definition simplifies to

$$S = \frac{4 \times 10^6 L a}{\pi d_f} \tag{10.2}$$

where L = length of filter media in direction of airflow, m
d_f = fiber diameter, μm
a = filter media volume fraction, (m³/m³)

Three primary mechanisms operate in filtration—impaction, interception, and diffusion. Impaction occurs when the particle inertia is so high that it breaks the air streamlines and impacts the fiber. This process is not significant for normal filter velocities and microbial sizes, and is neglected in most filter models since interception satisfactorily accounts for it (Brown and Wake, 1991; Brown, 1993; Matteson and Orr, 1987; Stafford and Ettinger, 1972; VanOsdell, 1994; Wake et al., 1995).

Interception occurs when a particle following a normal airflow streamline carries a particle within contact range of a fiber, at which point it will become attached by natural forces. An airstream passes through so many fibers that the probabilities are high that any particle 1 μm or larger will be intercepted in a typical high-efficiency filter.

Diffusion is a removal process that dominates for particles smaller than about 0.1 μm (Brown, 1993; Davies, 1973). Since these particles are subject to the effects of Brownian motion, they randomly traverse areas much wider than their diameters. This phenomenon causes attachment whether airflow streamlines bring a particle within a single radius of a fiber or not. Lower air velocities increase the removal of small particles by diffusion since they spend more time in the vicinity of a fiber.

Figure 10.6 illustrates the contribution to total filter efficiency from the components of diffusion and interception. These are computed from Eq. (10.1) as if each component were a separate filter. Since interception efficiency increases with increasing particle size while diffusion efficiency decreases, a minimum occurs near where the separate efficiency curves of diffusion and interception cross. This minimum efficiency defines the most penetrating particle size. It should be noted that the theoretical curves in Fig. 10.6 show the removal efficiency of the smallest particles reaching 100 percent, contrary to the empirical data in Fig. 10.4.

The single-fiber efficiencies for diffusion and interception are summed to obtain the total single-fiber efficiency:

$$E_s = E_R + E_D \tag{10.3}$$

where E_R is the interception efficiency, fractional, and E_D = diffusion efficiency, fractional.

Lee and Liu (Matteson and Orr, 1987) define the single-fiber diffusion efficiency as

$$E_D = 1.6125 \left(\frac{1-a}{F_K}\right)^{1/3} Pe^{-2/3} \tag{10.4}$$

where F_K is the Kuwabara hydrodynamic factor and Pe is the Peclet number, dimensionless.

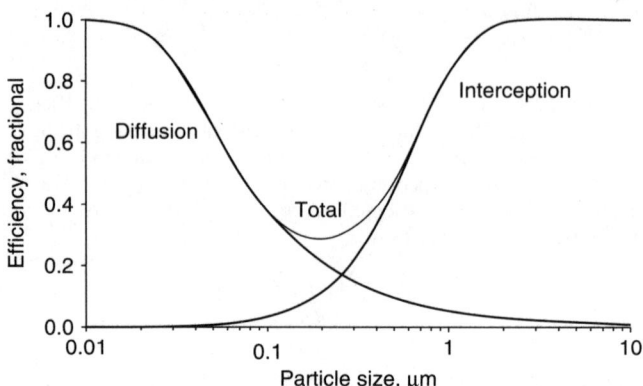

FIGURE 10.6 The two major components of filtration used in the model.

The Peclet number characterizes the intensity of diffusional deposition, and an increase in the Peclet number will decrease the single-fiber diffusion efficiency. The Peclet number is defined as

$$Pe = \frac{1 \times 10^{-6} U d_f}{D_d} \qquad (10.5)$$

where U is the media face velocity, (m/s), and D_d is the particle diffusion coefficient, (m²/s).

The Kuwabara hydrodynamic factor (Liu and Rubow, 1986; Brown, 1993; Matteson and Orr, 1987) is defined as

$$F_K = a - \frac{(a^2 + 2\ln a + 3)}{4} \qquad (10.6)$$

where a is the volume packing density, (m³/m³).

The particle diffusion coefficient measures the degree of diffusional motion. It is closely approximated by the same relation that defines molecular diffusion, the Einstein equation:

$$D_d = \mu k T \qquad (10.7)$$

where μ = particle mobility, (N·s/m)
k = Boltzmann's constant, 1.3708×10^{-23} J/K
T = temperature, K

The particle mobility is defined as

$$\mu = \frac{C_h}{3 \times 10^{-6} \pi \eta d_p} \qquad (10.8)$$

where η = gas absolute viscosity, N·s/m²
C_h = Cunningham slip factor, dimensionless
d_p = particle diameter, μm

FILTRATION OF AIRBORNE MICROORGANISMS   215

The Cunningham slip factor accounts for the aerodynamic slip that occurs at the particle surface, and is defined as

$$C_h = 1 + \left(\frac{\lambda}{d_p}\right)\left(2.492 + 0.84 e^{\left(-0.435 d_p/\lambda\right)}\right) \quad (10.9)$$

where λ is the gas molecule mean free path, 0.067 μm (Reist, 1993).
Liu and Rubow (1986) defined the single-fiber interception efficiency E_R as

$$E_R = \frac{1}{\varepsilon}\left(\frac{1-a}{F_K}\right)\left(\frac{N_r^2}{1+N_r}\right) \quad (10.10)$$

where N_r = Interception parameter, dimensionless
a = volume fraction (m³/m³)
F_K = Kuwabara hydrodynamic factor, dimensionless
ε = correction factor for inhomogeneity, dimensionless

The interception parameter is

$$N_r = \frac{d_p}{d_f} \quad (10.11)$$

In Eq. (10.10), the correction factor $1/\varepsilon$ accounts for filter media inhomogeneity. Yeh and Liu (Raber, 1986) have determined experimentally that the value of ε is approximately 1.6 for polyester filters. The same value provides acceptable agreement with empirical data for glass fiber filters as well. Equations (10.1) through (10.11) can define the performance of any filter based on a single-fiber diameter. This model can now be extended to form a multifiber model.

10.6 THE MULTIFIBER FILTRATION MODEL

Modern high-efficiency filter media are composed of fibers of more than one diameter (Vaughan and Brown, 1996). Both high-efficiency and HEPA filters consist of fiber diameters ranging from 0.65 to 6.5 μm, usually in three nominal diameter groups, according to vendor consensus.
For each nominal fiber diameter d_i, the fiber projected area is given by the following:

$$S_i = \frac{4 \times 10^6 L a_i}{\pi d_{f i}} \quad (10.12)$$

where $d_{f i}$ is the diameter of fiber i and a_i is the volume fraction of fiber i, (m³/m³).
The sum of the volume fractions for the fiber diameters must equal the total volume fraction. For computation purposes, each collection of discrete fiber diameters can be visualized as separate filters arranged in series. For a multifiber filter, the exponent in Eq. (10.1) is the sum of the products of E_s and S for each of the discrete fiber diameters:

$$E_s S = \sum E_{si} S_i \quad (10.13)$$

where E_{si} is the single-fiber efficiency for fiber i.

The single-fiber efficiency E_{si} is calculated individually for each fiber diameter using Eqs. (10.3) through (10.11) as described before, and the results are combined in Eq. (10.13). The total efficiency is then computed per Eq. (10.1).

Two problems are evident with this theoretical filter model when compared with actual data on filter performance. As noted previously, the diffusion efficiency approaches 100 percent efficiency near 0.01 μm and will actually reach 100 percent efficiency for smaller particles. Not only is it unlikely that particles smaller than 0.01 μm will be completely removed, but published data demonstrate that removal rates for submicron-sized particles never reach 100 percent, even for HEPA filters (Ensor et al., 1988).

The second problem is that many filters, especially in the MERV 6–10 range, often never reach 100 percent removal efficiency on the high end, contrary to theoretical filter model predictions. This may not be true for every manufacturer's filter. For some filters, the upper limit of efficiency often plateaus at less than 100 percent in the 5- to 10-μm size range.

The modified classical model presented here corrects both these deficiencies in a way that facilitates the modeling of filters based on MERV data. The diffusion component E_D in Eq. (10.3), is corrected with a factor that reduces the removal efficiency as a function of particle mean diameter. This diffusion efficiency correction factor, called D_f, is based on a Gompertz curve with constants set by least squares curve fitting of data from Ensor et al. (1988,1991). These data and the fitted models for four DSP-rated filters are shown in Fig. 10.7. This curve can be observed to provide an improved fit compared with the curve fits in the source that were obtained without a diffusion correction factor (Kowalski and Bahnfleth, 2002).

The diffusion efficiency correction factor, called a counter-diffusion factor is as follows:

$$D_f = 1 - z^{(d_p - g_m)^\beta} \quad (10.14)$$

where d_p = particle diameter, μm
z = a constant, 1×10^{-10}
g_m = gas molecule size, 0.003 μm
β = exponent, 0.75 for high-efficiency filters, 1.25 for HEPA filters

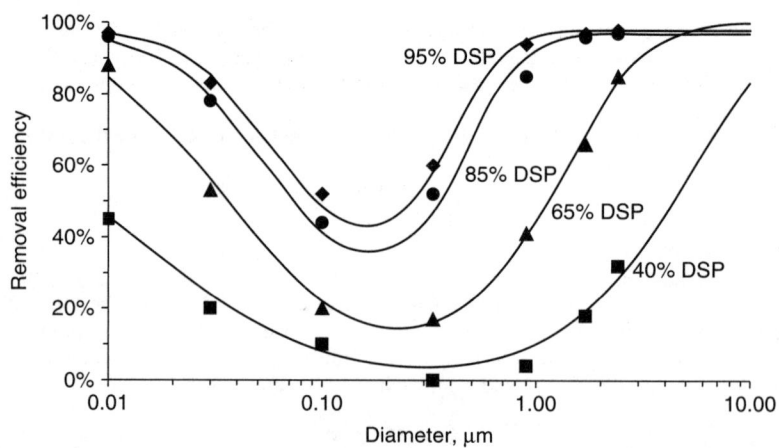

FIGURE 10.7 Comparison of four DSP-rated filters modeled with a diffusion correction factor and compared with data from Ensor et al. (1988).

Equation (10.14) is strictly an empirical correction fitted to a limited data set and is presented here without derivation. It mathematically defines the fact that diffusional efficiency decreases toward zero as a particle approaches the size of a gas molecule. It predicts removal efficiency accurately for high-efficiency filters when the exponent $\beta = 0.75$. However, for HEPA filters it fails unless the exponent is set to a value of 1.25.

The second problem with the existing filter model is that high-efficiency filters, especially those with lower MERV ratings, do not all approach 100 percent efficiency for large particles of about 10 μm. An upper limit correction factor for the interception parameter, called LU, has been added to define the upper limit of the curve efficiency in the interception range. Admittedly this is an artificial correction without theoretical basis, but it is an expedient that allows the creation of filter models to match any manufacturer's MERV testing results. Both correction factors are now applied to Eq. (10.14) to produce the following modified filter model:

$$E = L_U(1 - e^{-(D_f E_D + f E_R)S})$$ (10.15)

where D_f is the counter diffusion factor and L_U is the upper limit factor.

The corrected filter model has been applied to the data previously shown in Fig. 10.4, and the results are shown in Fig. 10.7. This corrected filter model can be compared with a similar figure shown in Kowalski et al. (1999) to see the improvement in the fit of the performance curve to the data.

10.7 FITTING THE MODELS TO MERV DATA

Equation (10.15) can now be used to fit a performance curve to any filter based on a single mean fiber diameter. The two correction factors L_U and f provide considerable flexibility in matching the model to MERV data. The factor L_U can be set equal to the highest efficiency in the MERV test results. The factor f can then be used to make further curve-fitting adjustments if necessary. It should be noted that the variation in filter efficiency at any point on the performance curve for any given MERV rating is probably on the order of at least ±20 percent, and therefore it may not be critical that a tight curve fit is obtained. Manual adjustment of parameters can be used to fit a curve although the most accurate approach would be to use a least squares curve fit.

Manufacturers of fibrous filter media use varying proportions of fibers at each diameter in order to obtain the desired grade or performance. This method of proportioning fiber diameters is duplicated mathematically to fit the filter models to empirical data. The fractions a_i for each diameter are adjusted while holding all other parameters constant, including the total packing fraction a and the filter length L. For HEPA filters, however, an increase in length L is necessary to achieve the order-of-magnitude increase in efficiency.

The actual diameters of fibers and proportions at each fiber diameter are considered proprietary information by vendors and were not available. The absence of such data leaves only the recourse of fitting the multifiber model to the empirical data in Fig 10.7. For the purposes of this analysis, the assumption is made that there are three discrete fiber diameters only—a high mean, a low mean, and an arbitrary value in-between. Table 10.4 summarizes the design parameters that act as constraints on filter efficiency, such as media length and filter face velocity. All of these parameters were chosen to represent typical filter design and operating parameters based on catalog data from various vendors.

Fitting of the model equations to the data was accomplished by adjusting the fiber diameters and volume fractions to minimize the r^2 values in relation to the data from Ensor

TABLE 10.4 Multifiber Filter Model Design Parameters

Parameter	40–45%	60–65%	80–85%	90–95%
Total volume fraction a	0.002039	0.002039	0.002039	0.002039
Media length, m	0.015	0.015	0.015	0.015
Face area, ft^2	3.8	3.8	3.8	3.8
Media area/face area	15.28	15.28	15.28	15.28
Nominal face velocity, fpm	500	500	500	500
Media velocity, fpm	34.48	34.48	34.48	34.48
Media velocity, m/s	0.175	0.175	0.175	0.175
Mass of filter media per unit area, kg/m^2	0.078	0.078	0.078	0.078
Fiber density, kg/m^3	2550	2550	2550	2550
Fiber 1				
d_{f1}, μm	1	0.7	0.65	0.65
fraction of a	0.01	0.05	0.1	0.05
Fiber 2				
d_{f2}, μm	6	5	2.2	1.2
fraction of a	0.1	0.25	0.3	0.3
Fiber 3				
d_{f3}, μm	20	8	10	7.5
fraction of a	0.89	0.7	0.6	0.65

et al. (1988). The resulting filter models are shown in Figs. 10.8 and 10.9 along with the data points for the MERV filters that were used in developing the models. Filters that had no data were derived by interpolating other data sets or estimating filter parameters.

The HEPA filter model was fit to a single data point, 99.97 percent efficiency at 0.3 μm. Not surprisingly, the overprediction of diffusional efficiency, as mentioned previously, causes a deviation from empirical data that is not insignificant. The counter diffusion correction factor does not work well with the extreme efficiency of the HEPA filter. To adjust the HEPA filter performance, the efficiency was limited based on empirical data from Matteson and Orr (1987) and Sinclair (1976) for sizes below 0.2 μm. The Sinclair (1976) study used the comparatively high velocity of 0.7 m/s and found a most penetrating particle size at a lower range of about 0.03 to 0.05 μm.

All of the filter models compare well with available data from other sources (Whitby, 1965; Thorne and Burrows, 1960; Harstad and Filler, 1969). The models were generally within ±10 percent of the empirical data after adjusting to match all known study parameters such as velocity and fiber diameter.

Figure 10.10 graphically summarizes the various MERV filter models shown in the previous figures. Figure 10.11 illustrates how the filter performance will vary according to the air velocity, using the MERV 12 filter as an example.

The distribution of the size of any microbial species follows a lognormal distribution curve, or a curve in which the logarithm of size is distributed normally (see Chap. 6). This is typical of bioaerosols and most particulates near the micron size range (Reist, 1993; Painter and Marr, 1968; Duguid, 1945). Ideally, the mean size of a microbial species would be determined empirically and this value would be used to predict removal rates for any given filter. However, very few microbes have had their size distribution evaluated and in place of the mean diameter it has been found that the logmean diameter of the size range will serve the same purpose without significant errors (Kowalski et al., 1999).

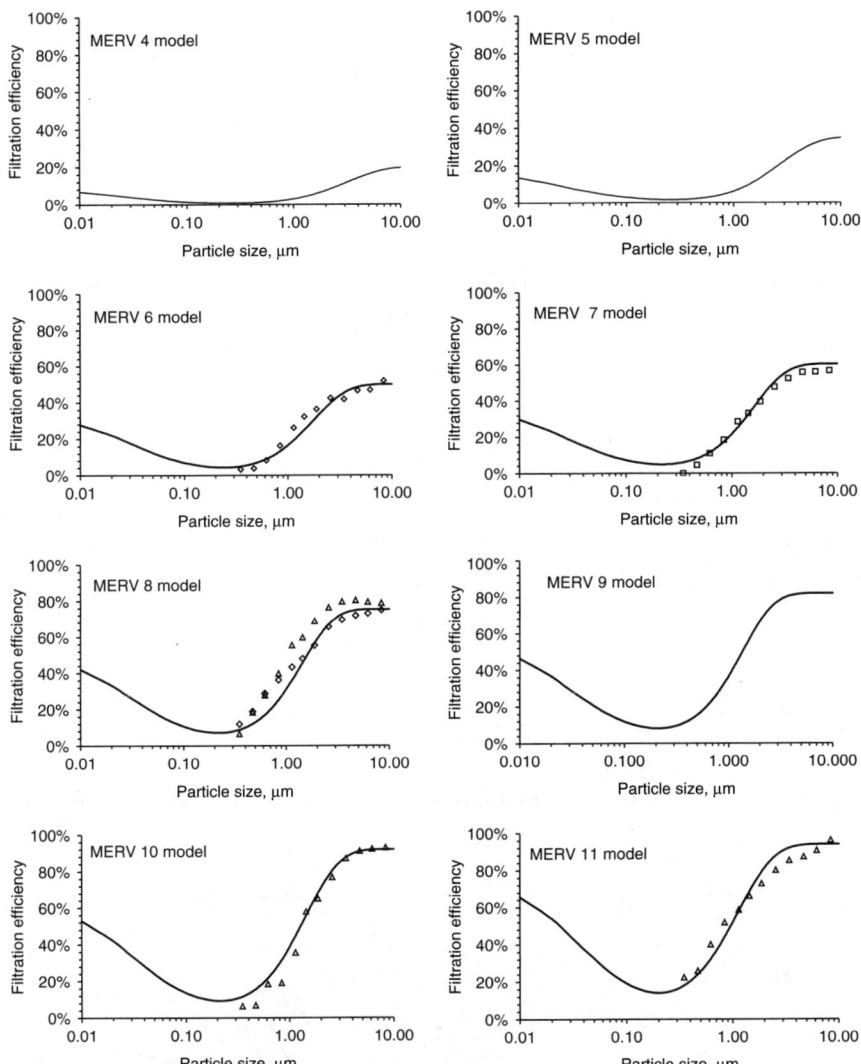

FIGURE 10.8 MERV filter models fitted to MERV test data results, except for MERV 4, 5, and 9 filters, which were interpolated. Gray lines show the ranges of the MERV filter definition.

Prediction of microbial filtration using the above models is a matter of determining the mean diameter of the microbes. Since microbe sizes are distributed in a lognormal fashion, this means using the logmean diameter (see Chap. 6). Corrections must also be made for nonspherical microbes (Kowalski et al., 1999). The mean diameters shown in App. A have incorporated these corrections and can be used to determine filtration rates, estimates of which are also provided. Figure 10.12 illustrates the performance of a nominal MERV 11 filter against 100 of the allergens and pathogens shown in App. A.

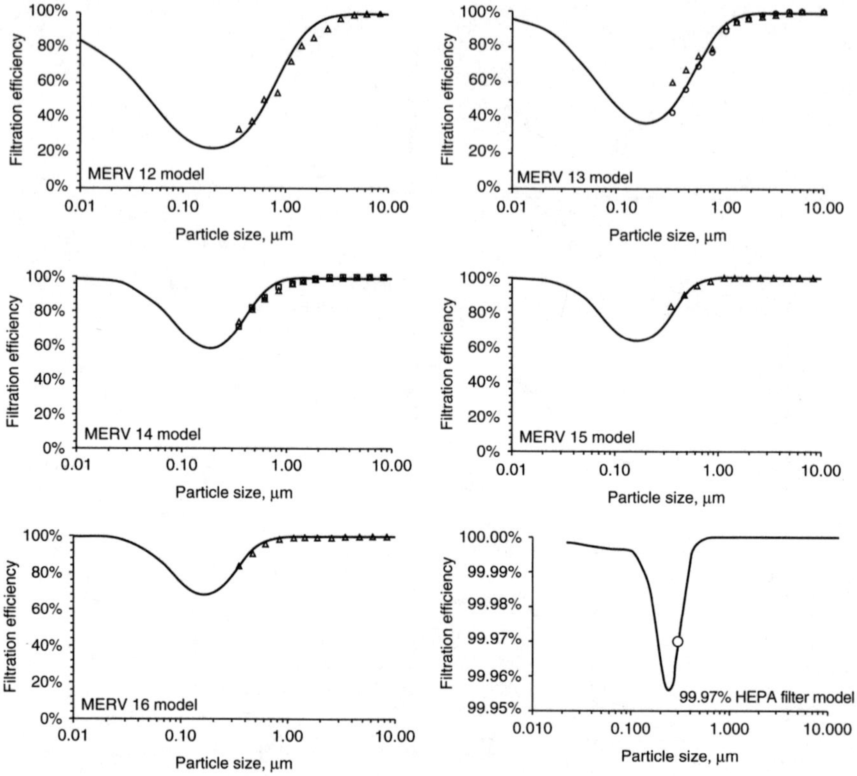

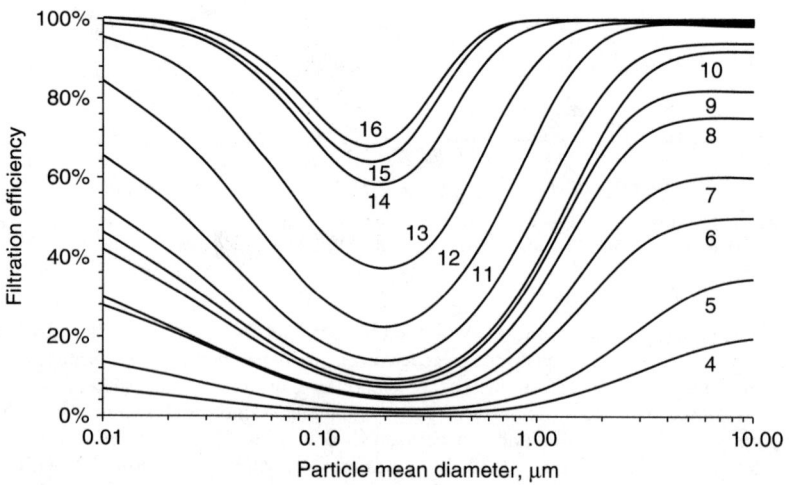

FIGURE 10.9 MERV 12–16 filter models fitted to MERV test data results. HEPA filter model shown for reference.

FIGURE 10.10 Estimated performance curves for MERV 4–16 filters.

FILTRATION OF AIRBORNE MICROORGANISMS

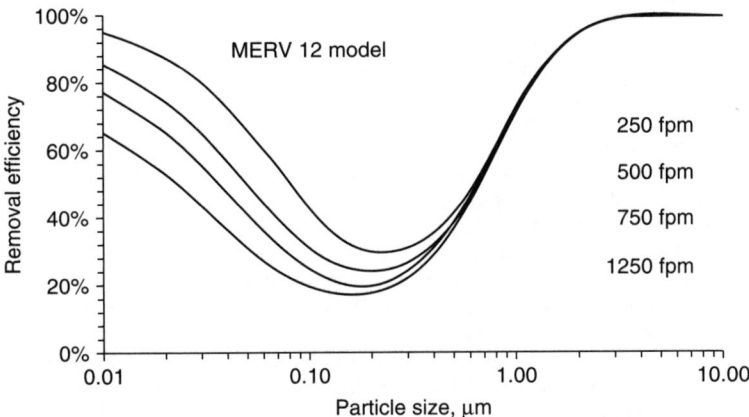

FIGURE 10.11 Estimated performance of a MERV 12 filter at different air velocities.

10.8 MICROBIAL FILTRATION TEST RESULTS

In theory, filters will remove airborne microbes at efficiencies based on their particle size (i.e., their aerodynamic mean diameter). In actuality there has been limited testing done to corroborate filter removal rates of microbes. The available test results in the literature are reviewed here to provide some confidence that microbes do, in fact, behave like airborne

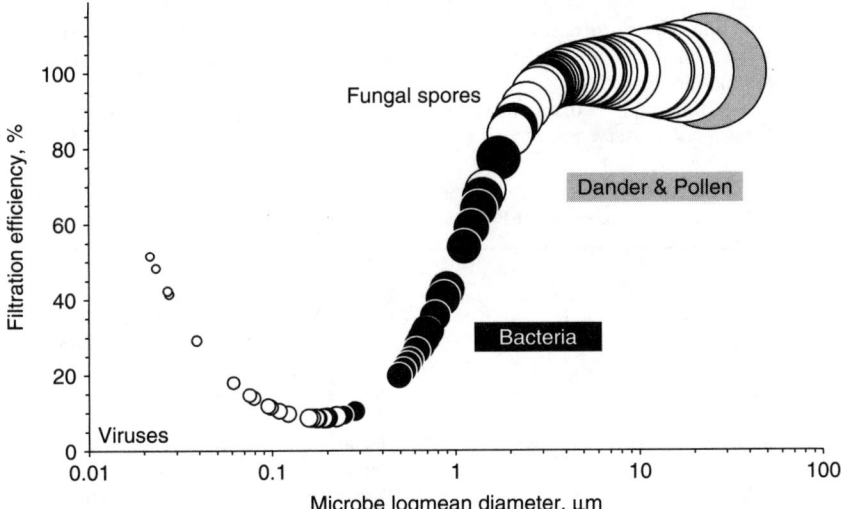

FIGURE 10.12 Performance of a nominal MERV 11 filter against 100 allergens and pathogens from App. A. Circles represent relative size of microbes—area of circles is equivalent in proportion to the microbe mean diameters.

particles and are removable at predictable rates based on their mean diameter or aerodynamic diameter (for nonspherical microbes). The results summarized here are primarily laboratory tests, and highlight both the effectiveness of microbial filtration and some of the problems that can be encountered.

Testing has generally corroborated the concept that microorganisms are filtered based strictly on their mean size. Some tests have found that the filtration of microorganisms nearly approximates the filtration of particles of identical size but that it is not exactly the same (Ginestet et al., 1996). Other tests have shown that microbes are removed at rates slightly less than particles of equivalent diameters (Sinclair, 1976).

In the earliest test that measured the actual removal rate of airborne pollen by a filter, ordinary commercial glass wool filters were found to remove an average of 96.5 percent of pollen (Criep et al., 1936). Other studies have shown that air filtration reduced pollen levels by up to 98 percent and relieved most symptoms in allergy sufferers (Wheeler, 1993; Nelson et al., 1933). In some of the earliest tests on the filtration of microorganisms by glass fiber filters, Decker et al. (1954) demonstrated a 99 percent removal rate for *Serratia indica*, *E. coli*, and the virus T-3 bacteriophage. The operating velocity was a mere 20 fpm and the filter efficiency was not known but was described as having a depth of 0.5 in. and a mean fiber diameter of 1.28 µm.

A number of tests have been performed on the filtration of viruses, and these results are summarized in Table 10.5 and graphed in Fig. 10.13, along with some data on submicron particle filtration. These tests all produced results that are essentially in agreement with expectations for the indicated filters. The viruses that were filtered were removed at rates that suggest they existed as single particles in the test aerosols and offer corroboration for both the filter models and the size distribution models for viruses.

Studies on the ability of bacteria to survive while attached to filters have generally shown that the bacteria die rapidly from dehydration, even under high relative humidity. A study by Ruden and Botzenhart (1974) found that HEPA filters do not act as a growth medium for microorganisms. Filters were inoculated with test bacteria and fungi and maintained at 70 and 90 percent RH for 3 weeks. Microbes tested included *Staphylococcus*, *Streptococcus*, *Pseudomonas*, *Corynebacterium*, *Bacillus*, *Clostridium*, *Candida*, *Penicillium*, and *Aspergillus*. All species populations decreased at 70 percent RH and all species decreased under 90 percent RH except *Bacillus*, which remained constant, and *Candida*, which increased.

In a test of the viability of microbes on unused filter media, Maus et al. (1997) found that *Mycobacterium luteus* and *E. coli* collected in air filter media and exposed to low air

TABLE 10.5 Summary of HEPA Filter Test Results

Test agent	Mean diameter, µm	Removal, %	Reference
Actinophage *S. virginiae* S1	0.05	99.997	Roelants et al., 1968
T phage *E. coli* B	0.12	99.915	Washam, 1966
T phage *E. coli* B	0.12	99.99	Jensen, 1967
T phage *E. coli* B	0.1	99.997	Harstad et al., 1967
Foot-and-mouth disease virus	0.027	99.998	Thorne and Burrows, 1960
Ag aerosol	0.06	99.996	Sinclair, 1976
Ag aerosol	0.06	99.997	Sinclair, 1976
Ag aerosol	0.08	99.997	Sinclair, 1976

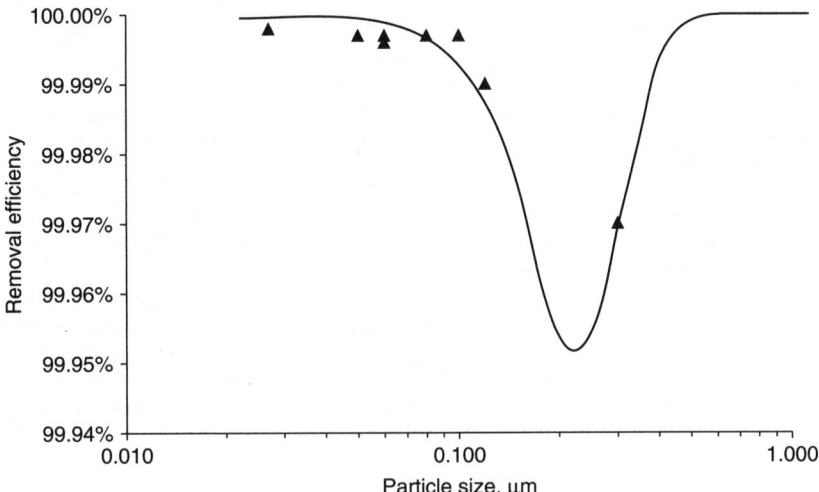

FIGURE 10.13 HEPA filter model showing test data from Table 10.5 for viruses and submicron particles.

humidity in the range of 30 to 60 percent lost their viability in less than 1 h. The survival of bacterial and mold spores in air filter media was studied by Maus et al. (2001) who found that spores of *Bacillus subtilis* and *Aspergillus niger* experienced no growth regardless of relative humidity in most filter media. However, some samples of filter media produced mold growth when exposed to 98 percent RH. In general, the lack of available nutrients on clean filters precludes most mold growth even under high relative humidity, but this may not be the case for filters that have been in operation and accumulated dust and organic debris.

Kemp et al. (1995) report that microorganisms can grow on untreated air filters and even grow through the filters and release spores downstream. Other studies have reported or shown that microbes can survive or grow on filters, including Godish (1995), Samson (1994), Pasanen et al. (1992), Fuoad et al. (1997), Chang et al. (1996), and Neumeister et al. (1997). Coincident with the problem of growth of fungi on filters is the possible generation of VOCs. Laboratory experiments proved that species of *Aspergillus, Cladosporium, Acremonium*, and possibly other microbes may produce acetone, ethanol, formaldehyde, acetaldehyde, and other compounds on fiberglass and cellulose filter media (Schleibinger and Ruden, 1999).

The collection of fungal spores on ventilation filters and their potential for subsequent re-entrainment were compared with the collection and re-entrainment of standard test particles by Jankowska et al. (2000). Test microbes *Penicillium brevicompactum* and *Penicillium melinii* were used to represent common fungal spores. Two filters were tested, a medium-efficiency prefilter and a higher efficiency fine filter. Filters were operated at design velocity. The collection efficiency was found to be slightly lower for fungal spores than for test particles of the same aerodynamic size. When the re-entrainment velocity was the same as the loading velocity through the filter medium the re-entrainment rate was less than 0.4 percent. When the re-entrainment velocity was increased to 3.00 m/s, the re-entrainment of fungal spores was higher than that of the test particles. Re-entrainment rates were 2 to 6 percent for *P. brevicompactum*, 5 to 12 percent for *P. melinii*, and 0.2 to 0.6 percent for the test particles. The differences in re-entrainment between fungal spores and the test particles were attributed to aggregation and de-aggregation of the fungal spores. The results suggest that during the start-up of a ventilation system or when the air velocity may suddenly increase, re-entrainment of fungal spores can be significant. Furthermore, if fungal spores grow on the ventilation filters, the re-entrainment rate may become even higher than measured in this study. Other studies have shown that the shedding of microorganisms was generally low and independent of the loading of the filters, air velocity, or humidity (Ginestet et al., 1996).

Antimicrobial filters are available that inhibit the growth of microorganisms but they are not always successful (Foarde et al., 2000). At relative humidities below 90 percent none of the antimicrobial filters tested permitted mold growth, but dust loading enabled fungi to grow even on filters with antimicrobial treatment.

Filters contaminated with microorganisms may pose a risk to maintenance workers who remove the filters. Many facilities now have procedures for handling used filters and some types of filter installations are designed to facilitate filter removal without posing hazards to workers. One option is the use of UVGI during normal operation, which will sterilize filters as well as the air and other duct surfaces, a subject that will be addressed in Chap. 11.

10.9 FILTER APPLICATION TEST RESULTS

Filters may be individually effective at removing microbes in laboratories but the real test of their effectiveness is how well they do in actual applications. Performance in real world settings is often limited by how they are applied, actual operating velocities, and local airflow conditions (McDonald and Ouyang, 2001; Miller-Leiden et al., 1996). The success of filter applications is generally measured in terms of the indoor airborne concentrations before and after installation. Following are some representative studies that show varying degrees of effectiveness and some of the associated problems that can be encountered in aerobiological applications of filters.

Some limited data are available on the effectiveness of high-efficiency filters used in recirculating units from tests done in agricultural animal houses (Carpenter et al., 1986a, 1986b). In the first test levels were reduced to 31 percent, compared against the unfiltered condition, and in the second test they were reduced to 18 percent of the unfiltered condition. The filter ratings were not specified in these tests but appear to have been 60 to 90 percent DSP filters (MERV 11–14). Although the airborne concentrations were significantly reduced, the performance is somewhat less than might be expected for high-efficiency filters. The reason for the reduced performance was apparently due to poor airflow distribution in the areas where these units were placed. This highlights the fact that airflow distribution is an important component of an efficient filtration system and attention must be given to placement of recirculation units in order to achieve optimum performance.

Recirculating air filtration systems were tested to determine reductions in dust and bacteria levels inside the feeder barns of a hog farm (Lau et al., 1996). Multiple fabric filters and a high-voltage, plate-type electrostatic precipitator were used as dust collection devices for the grower and finisher barns, respectively. Recirculation airflow rate was set at 20 air changes per hour which lies between the summer and winter ventilation rates. The air filtration systems effectively reduced dust levels and aerial bacteria counts inside the pig barns although their efficiency was dependent on air recirculation flow rate, location, size, and number of inlets and outlets, and humidity. Air filtration resulted in reduced prevalence of enzootic pneumonia and atrophic rhinitis among the pigs. In terms of bacteria reduction, the electrostatic precipitator was consistently more effective than the fabric filters.

Tests performed by Cheng et al. (1998) evaluated the efficiency of a commercial air cleaner in removing indoor pollen and fungal spores, as well as the effect of the air ventilation rate. At high ventilation rates, the cleaner helped to maintain a low concentration of pollen and spores. At a moderate ventilation rate, the filtration helped to remove 80 percent of the particles. At a low air-exchange rate, pollen and fungal spore concentrations decreased rapidly, indicating sedimentation of large particles. The air cleaner was useful in filtering out pollen and spores and was more effective when doors and windows are closed.

The results of studies on typical offices by Law et al. (2001) revealed that the air change rate inside the office environment had less significant effects than filtration on airborne bioaerosols. The primary objectives of these tests were to determine the temporal concentration profile of bioaerosol inside office environments during office hours, and the effects of air change rate on the concentration profile. The highest bacteria concentrations were recorded to be 2912 cfu/m^3 at the early morning hours during the start-up of the HVAC systems. The highest fungi concentrations were recorded to be 3852 cfu/m^3 during the weekend mornings. The background fungal concentration was found to have a strong correlation with the indoor relative humidity. Of the sampled bacteria 80 percent were found to be gram-positive, while the dominating genera of fungi was found to be *Cladosporium* and *Penicillium*.

Analysis by Fisk et al. (2002) found that increasing filter efficiencies above approximately ASHRAE DSP 65 percent in both in-duct and recirculation filter units would produce only moderate reductions of indoor concentrations of dust mite and cat allergens. Filters with a wide efficiency range were considered. Predicted reductions in cat and dust-mite allergen concentrations range from 20 to 80 percent. Results suggest that to obtain substantial reductions in indoor concentrations of these allergens, the rate of airflow through the filter must be at least a few air changes per hour. Increasing the filter efficiency above ASHRAE 85 percent results in only modest predicted incremental decreases in indoor concentration.

The capability of medium efficiency air filters to retain airborne outdoor microorganisms was examined in field experiments in two HVAC systems by Moritz et al. (2001). At the beginning of the 15-month investigation period, the first filter stages of both HVAC systems were equipped with new unused air filters. The number of airborne bacteria and molds entering and leaving the filters were determined simultaneously in 14 days intervals. Under relative humidities below 80 percent the air filters led to a marked reduction of airborne microorganism concentrations. Bacteria levels were reduced by approximately 70 percent and mold spores by over 80 percent. However, during periods of high relative humidity, above 80 percent, a proliferation of bacteria on air filters resulted in a subsequent release into the filtered air. These microorganisms were mainly smaller than 1.1 μm. The results showed furthermore that one way to avoid microbial proliferation may be to limit the relative humidity in the area of the air filters to 80 percent or less by using preheaters in front of the filters.

Air filtration units incorporating a HEPA filter are theoretically able to remove almost all potential airborne allergens. This may have implications for subjects with allergic lower respiratory diseases. In a field test by Warburton et al. (1994) filter units were placed in the living room of 12 atopic asthmatics, and the internal filters were inserted and removed in a double-blind fashion. No difference in subjective symptoms, spirometry, or bronchial reactivity, was demonstrated. *Peak expiratory flow rate* (PEFR) variability was significantly improved from baseline readings, and there was a trend toward higher mean PEFRs when the filters were present in the AFU. Trends toward lower levels of airborne microorganisms were also demonstrated when the filters were present, however, no effect on total airborne dust and airborne allergens could be demonstrated.

In an early test on the ability of filters to decrease hay fever symptoms, commercial filters were installed outside of rooms and used to provide filtered air to atopic individuals (Nelson et al., 1933). With an average outdoor pollen count of 139, about 73 percent of patients were relieved of their symptoms within hours. In an early hospital field trial in which filtered air was used in a dressing station for burns, isolates of *Staphylococcus* experienced a reduction of approximately 58 percent after filtration was implemented

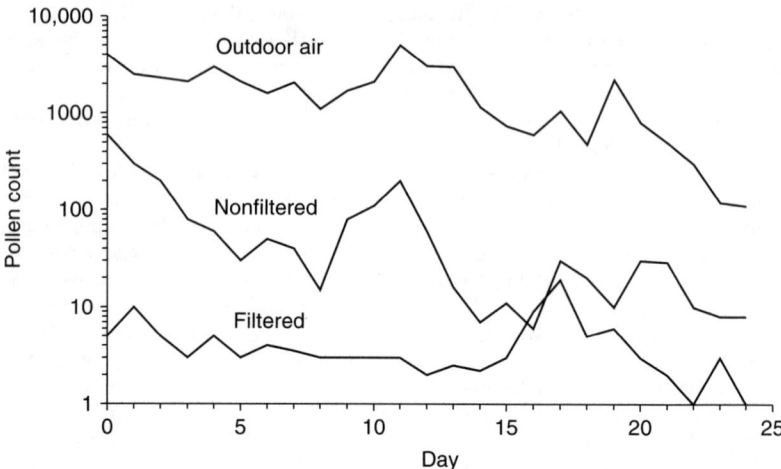

FIGURE 10.14 Pollen levels in the outdoors, in an air-conditioned house, and in an air-conditioned house with an 80–90 percent DSP filter installed. [*Based on data from Spiegelman and Friedman (1968).*]

(Lowbury, 1954). Isolates of other bacteria experienced similar reductions from the use of air filtration.

In a test of a filter installed in an air-conditioning system of a house, Spiegelman and Friedman (1968) demonstrated that indoor levels of pollen and bacteria were significantly lowered. The air conditioner was itself responsible for a significant reduction of pollen, as shown in Fig. 10.14, but the addition of the filter reduced pollen levels even more. The filter in this test was described as an 80 to 90 percent DSP (i.e., MERV 13).

Figure 10.15 summarizes averages for 4 months of the same testing from Fig. 10.14. It can be observed that there is little difference between operating the air conditioner with and

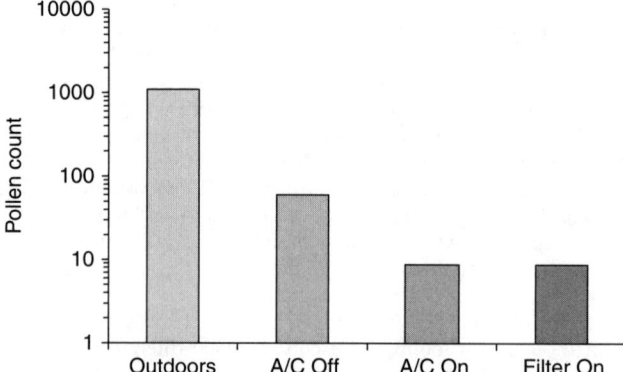

FIGURE 10.15 Comparison of average pollen counts in test house. Filter on condition includes the A/C system operating. [*Based on data from Spiegelman and Friedman (1968).*]

without filtration. For pollen spores, the wet coils of the air conditioner will have a major filtration effect. Testing also showed that levels of indoor bacteria were approximately 52 percent less in the house with filtered air, on the average.

In another study on the potential for microbes to penetrate HEPA filters, 888 samples were collected from in-place filters over a 26-month period and no bacteria or fungi were detected in any samples (Rechzeh and Dontenwill, 1974). After extensive testing of the dust-laden upstream side of the HEPA filters, a number of microbes were found. Some 59 percent of these were gram-positive spore-forming bacteria, 20 percent were fungi, 13 percent were gram-positive bacteria, and the remainder were cocci or gram-negative rods.

10.10 SPECIAL FILTER TYPES

In addition to standard filters using fiberglass or other media there are several other types of filters available today, including electret filters, charged filters, chemically impregnated filters, membrane filters, and carbon media filters. Electret filters retain a natural electrical charge that may be induced or enhanced by passing air. In theory these filters would have improved filtration efficiencies against smoke or dust particles in the range of 0.1 to 5 μm but no data are yet available that they result in improved filtration of microorganisms. Results of studies indicate that the efficiency and reliability of electret filters depend to a large degree on the stability of the electrostatic charge in the filter material and that the efficiency can decrease significantly under exposure to aerosols or high relative humidity (Lehtimaki and Heinonen, 1994; Tennal et al., 1991).

Electrically charged filters with external power sources are more properly considered electrostatic filters and these are dealt with separately in Chap. 13. Carbon-based filters are more properly considered gas phase filters and these are treated in Chap. 12. Chemically treated filters are uncommon and basically a developmental technology. Membrane filters are composed of fiber media with such a fine mesh that they essentially act as sieves for particulates. Some types of membrane filters do not rely on a sieve effect but use materials that have a high affinity for bacterial cells (Kawabata and Kawato, 1998). The biggest problem with membrane filters is their extraordinary pressure drops, which make them impractical for most high velocity ventilation applications. Another type of filter under development is the ePTFE filter, which is made from a Teflonlike fiber with remarkable meshlike regularity on a microscopic scale, as opposed to the irregular mesh of glass fiber filters (see Fig. 10.5). Theoretically, the ability to control the fine mesh of fiber media may result in filters with superior chemical resistance, lower volume, and potentially lower cost, but at present such filters have limited applications due to the high pressure losses associated with them (Folmsbee and Ganatra, 1996). Another method of enhancing filtration efficiency is the use of a corona precharger or ion generator to ionize the incoming filter air (Lee et al., 2001). This type of electrically enhanced filtration may increase filtration efficiency but will be addressed further in Chap. 13.

REFERENCES

ASHRAE (1992). Chapter 25: Air cleaners for particulate contaminants., *HVAC Systems and Equipment*, American Society of Heating, Refrigerating and Air-Conditioning Engineers, Atlanta, GA.

ASHRAE (1999). Standard 52.2-1999: Method of testing general ventilation air-cleaning devices for removal efficiency by particle size. American Society of Heating, Refrigerating and Air-Conditioning Engineers. Atlanta, GA.

Brown, R. C., and Wake, D. (1991). "Air filtration by interception—Theory and experiment." *J of Aerosol Sci* 22(2):181–186.

Brown, R. C. (1993). *Air Filtration.* Pergamon Press, Oxford, UK.

Carpenter, G. A., Smith, W. K., MacLaren, A. P. C., and Spackman, D. (1986a). "Effect of internal air filtration on the performance of broilers and the aerial concentrations of dust and bacteria." *Brit Poultry Sci* 27:471–480.

Carpenter, G. A., Cooper, A. W., and Wheeler, G. E. (1986b). "The effect of air filtration on air hygiene and pig performance in early-weaner accommodation." *Anim Prod* 43:505–515.

Chang, J. C. S., Foarde, K. K., and VanOsdell, D. W. (1996). "Assessment of fungal (*Penicillium chrysogenum*) growth on three HVAC duct materials." *Environ Int* 22(4):425.

Cheng, Y. S., Lu, J. C., and Chen, T. R. (1998). "Efficiency of a portable air cleaner in removing pollen and fungal spores." *Aerosol Sci Technol* 29(2):93–101.

Criep, L. H., Green, M. D., and Green, M. A. (1936). "Air cleaning as an aid in the treatment of hay fever and bronchial asthma." *J Allergy* 7:120–131.

Davies, C. N. (1973). *Air Filtration.* Academic Press, London.

Decker, H. M., Harstad, J. B., Piper, F. J., and Wilson, M. E. (1954). "Filtration of microorganisms from air by glass fiber media." *HPAC Eng* 26(5):155–158.

Duguid, J. P. (1945). "The size and the duration of air-carriage of respiratory droplets and droplet-nuclei." *J Hyg* 54:471–479.

Ensor, D. S., Viner, A. S., Hanley, J. T., Lawless, P. A., Ramanathan, K., Owen, M. K., Yamamoto, T., and Sparks, L. E. (1988). "Engineering solutions to indoor air problems." *IAQ 88 / Engineering Solutions to Indoor Air Problems,* Atlanta, GA.

Ensor, D. S., Hanley, J. T., and Sparks, L. E. (1991). *Particle-Size-Dependent Efficiency of Air Cleaners.* Healthy Buildings/IAQ '91, Washington, DC.

Fisk, W. J., Faulkner, D., Palonen, J., and Seppanen, O. (2002). "Performance and costs of particle air filtration technologies." *Indoor Air* 12(4):223–234.

Foarde, K. K., Hanley, J. T., and Veeck, A. C. (2000). "Efficacy of antimicrobial filter treatments." *ASHRAE J* 42(12):52–58.

Folmsbee, T. W., and Ganatra, C. P. (1996). "Benefits of membrane surface filtration." *World Cement* 27(10):59–61.

Fuoad, H., Baird, G., Donn, M., and Isaacs, N. (1997). "Indoor airborne bacteria and fungi in New Zealand office buildings." *Healthy Buildings/IAQ '97,* Bethesda, MD. 233.

Ginestet, A., Mann, S., Parat, S., Laplanche, S., Salazar, J. H., Pugnet, D., Ehrler, S., and Perdrix, A. (1996). "Bioaerosol filtration efficiency of clean HVAC filters and shedding of microorganisms from filters loaded with outdoor air." *J Aerosol Sci* 27(Suppl. 1):S619–S620.

Godish, T. (1995). *Sick Buildings : Definition, Diagnosis and Mitigation.* CRC-Press Lewis Publishers, Boca Raton, FL.

Harstad, J. B., Decker, H. M., Buchanan, L. M., and Filler, M. E. (1967). "Air filtration of submicron virus aerosols." *Am J Public Health* 57(12):2186–2193.

Harstad, J. B., and Filler, M. E. (1969). "Evaluation of air filters with submicron viral aerosols and bacterial aerosols." *Am Ind Hyg Assoc J* 30:280–290.

Jankowska, E., Reponen, T., Willeke, K., Grinshpun, S. A., and Choi, K.-J. (2000). "Collection of fungal spores on air filters and spore reentrainment from filters into air." *J Aerosol Sci* 31(8):969–978.

Jensen, M. (1967). "Bacteriophage aerosol challenge of installed air contamination control systems." *Appl Microbiol* 15(6):1447–1449.

Kawabata, N., and Kawato, S. (1998). "Removal of airborne bacteria by filtration using a composite microporous membrane made of a pyridinium-type polymer showing strong affinity with microbial cells." *Epidemiol Infect* 121:349–356.

Kemp, S. J., Kuehn, T. H., Pui, D. Y. H., Vesley, D. and Streifel, A. J. (1995). "Filter collection efficiency and growth of microorganisms on filters loaded with outdoor air." *ASHRAE Trans* 101(1):228.

Kowalski, W. J., Bahnfleth, W. P., Whittam, T. S. (1999). "Filtration of airborne microorganisms: Modeling and prediction." *ASHRAE Trans* 105(2):4–17. http://www.engr.psu.edu/ae/wjk/fom.html.

Kowalski, W. J., and Bahnfleth, W. P. (2002). "MERV filter models for aerobiological applications." *Air Media* Summer:13–17.

Lau, A. K., Vizcarra, A. T., Lo, K. V., and Luymes, J. (1996). "Recirculation of filtered air in pig barns." *Canadian Agric Eng* 38(4):297–304.

Law, A. K. Y., Chau, C. K., and Chan, G. Y. S. (2001). "Characteristics of bioaerosol profile in office buildings in Hong Kong." *Build Environ* 36(4):527–541.

Lee, J.-K., Kim, S. C., Shin, J. H., Lee, J. E., Ku, J. H., and Shin, H. S. (2001). "Performance evaluation of electrostatically augmented air filters coupled with a corona precharger." *Aerosol Sci Technol* 35:785–791.

Lehtimaki, M., and Heinonen, K. (1994). "Reliability of electret filters." *Build Environ* 29(3):353–355.

Liu, B. Y. H., and Rubow, K. L. (1986). "Air filtration by fibrous media." *Fluid Filtration: Gas.* Philadelphia, PA.

Lowbury, E. J. L. (1954). "Air-conditioning with filtered air for dressing burns." *Lancet* 1:293–295.

Matteson, M. J., and Orr, C. (1987). *Filtration: Principles and Practices.* C. Industries, ed. Marcel Dekker, New York.

Maus, R., Goppelsroder, A., and Umhauer, H. (1997). "Viability of bacteria in unused air filter media." *Atmos Environ* 31(15):2305–2310.

Maus, R., Goppelsroder, A., and Umhauer, H. (2001). "Survival of bacterial and mold spores in air filter media." *Atmos Environ* 35:105–113.

McDonald, B., and Ouyang, M. (2001). Air cleaning—particles. *Indoor Air Quality Handbook.* J. D. Spengler, J. M. Samet, and J. F. McCarthy, eds., McGraw-Hill, New York. 9.1–9.28.

Miller-Leiden, S., C. Lobascio, and Nazaroff W. W. (1996). "Effectiveness of in-room air filtration and dilution ventilation for tuberculosis infection control." *J Air Waste Manage Assoc* 46(9):869.

Moritz, M., Peters, H., Nipko, B., and Ruden, H. (2001). "Capability of air filters to retain airborne bacteria and molds in heating, ventilating and air-conditioning (HVAC) systems." *Int J Hyg Environ Health* 203(5–6):401–409.

Nelson, T., Rappaport, B. Z., and Welker, W. H. (1933). "The effect of air filtration in hay fever and pollen asthma." *JAMA* 100(18):1385–1392.

Neumeister, H. G., Kemp, P. C., Kircheis, U., Schleibinger, H. W., and Ruden, H. (1997). "Fungal growth on air filtration media in heating ventilation and air conditioning systems." *Healthy Buildings/IAQ '97*, Bethesda, MD. 569–574.

Painter, P. R., and Marr, A. G. (1968). "Mathematics of microbial populations." *Ann Rev Microbiol* 22:519–549.

Pasanen, A.-L., Juutinen, T., Jantunen, M. J., and Kalliokoski, P. (1992). "Occurrence and moisture requirements of microbial growth in building materials." *Intl Biodeterior Biodegradation* 30:273–283.

Raber, R. R. (1986). "Fluid filtration: Gas." *Symposium on Gas and Liquid Filtration*, Philadelphia, PA.

Rechzeh, G., and Dontenwill, W. (1974). "Contribution to the question of the contamination of suspended-substances filters by germs." *Zbl Bakt Hyg* 159:272–283.

Reist, P. C. (1993). *Aerosol Sci Technol.* McGraw-Hill, New York.

Rivers, R. D. (1988). "Interpretation and use of air filter particle-size-efficiency data for general-ventilation applications." *ASHRAE Trans* 88(2):1835–1849.

Roelants, P., Boon, B., and Lhoest, W. (1968). "Evaluation of a commercial air filter for removal of viruses from the air." *Appl Microbiol* 16(10):1465–1467.

Ruden, H., and Botzenhart, K. (1974). "Experimental studies on the capacity of glass-fibre HEPA-filters to retain microorganisms." *Zbl Bakt Hyg* 159:284–290.

Samson, R. A., ed. (1994). *Health Implications of Fungi in Indoor Environments.* Elsevier, Amsterdam, The Netherlands.

Schleibinger, H., and Ruden, H. (1999). "Air filters from HVAC systems as possible source of volatile organic compounds (VOC)—laboratory and field assays." *Atmos Environ* 33:4571–4577.

Sinclair, D. (1976). "Penetration of HEPA filters by submicron aerosols." *J Aerosol Sci* 7:175–179.

Spiegelman, J., and Friedman, H. (1968). "The effect of air filtration and air conditioning on pollen and microbial contamination." *J Allergy* 42(4):193–202.

Stafford, R. G., and Ettinger, H. J. (1972). "Filter efficiency as a function of particle size and velocity." *Atmos Environ* 6:353–362.

Tennal, K. B., Mazumder, M. K., Siag, A., and Reddy, R. N. (1991). "Effect of loading with an oil aerosol on the collection efficiency of an electret filter." *Particul Sci Technol* (9):19–29.

Thorne, H. V., and Burrows, T. M. (1960). "Aerosol sampling methods for the virus of foot-and-mouth disease and the measurement of virus penetration through aerosol filters." *J Hyg* 58:409–417.

VanOsdell, D. W. (1994). "Evaluation of test methods for determining the effectiveness and capacity of gas-phase air filtration equipment for indoor air applications." *ASHRAE Trans* 100(2):511.

Vaughan, N. P., and Brown, R. C. (1996). "Observations of the microscopic structure of fibrous filters." *Filtr Separat* 33(9):741–748.

Wake, D., Redmayne, A. C., Thorpe, A., Gould, J. R., Brown, R. C., and Crook, B. (1995). "Sizing and filtration of microbiological aerosols." *J Aerosol Sci* 26(S1):s529–s530.

Warburton, C. J., Niven, R. M., Pickering, C. A., Fletcher, A. M., Hepworth, J., and Francis, H. C. (1994). "Domiciliary air filtration units, symptoms and lung function in atopic asthmatics." *Respir Med* 88(10):771–776.

Washam, C. J. (1966). "Evaluation of filters for removal of bacteriophages from air." *Appl Microbiol* 14(4):497–505.

Wheeler, A. E. (1993). Better Filtration: A Prescription for Healthier Buildings. *IAQ 93: Operating and Maintaining Buildings for Health, Comfort, and Productivity,* K. Y. Teichman, ed., American Society of Heating, Refrigerating and Air-Conditioning Engineers, Atlanta, GA. 201–207.

Whitby, K. T. (1965). "Calculation of the clean fractional efficiency of low media density filters." *ASHRAE J* 7:56.

CHAPTER 11

ULTRAVIOLET GERMICIDAL IRRADIATION

11.1 INTRODUCTION

Ultraviolet germicidal irradiation (UVGI) is defined as electromagnetic radiation in the range of about 200 to 320 nm that is used to destroy microorganisms. UVGI systems for water and surface disinfection have demonstrated reliability and effectiveness for the past century (AWWA, 1971). Airstream disinfection systems have been installed for the past 50 years but have had a checkered performance history with varying and unpredictable performance. Recent research, however, has advanced the science of UVGI air disinfection to the point that reliable and predictable performance is now possible. This chapter reviews the varied applications of UVGI for controlling indoor airborne microbial contamination and summarizes methods for accurate and precise design of such systems.

11.2 TYPES OF UVGI SYSTEMS

An increasing variety of UVGI systems are available today for the control of microbial contamination in buildings and ventilation systems, with new specialized applications being developed all the time. The following sections describe the main types of UVGI systems used for air and surface disinfection.

Figure 11.1 shows the applications and types of UVGI systems that are sold today based on estimates from a number of major manufacturers. The use of systems for disinfecting air and controlling microbial growth is growing in the United States and Europe, according to manufacturers. In the Third World, however, demand for upper air disinfection systems is high because of the TB pandemic, strained economics, and the common use of natural ventilation. These proportions are in flux today due to increased interest in UVGI from both the commercial and residential sectors.

The types of air disinfection systems installed can be subdivided into three categories—in-duct air disinfection, recirculation units, and upper air disinfection systems. In addition, there are at least four types of surface disinfection systems—microbial growth control

232 AIRBORNE DISEASE CONTROL TECHNOLOGIES

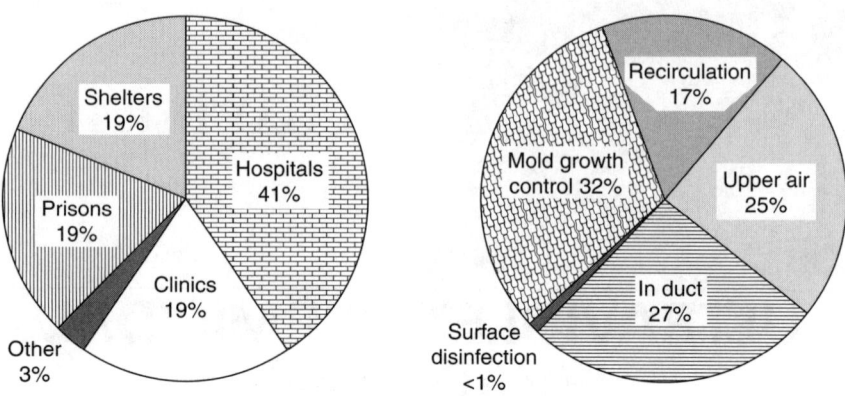

FIGURE 11.1 Breakdown of UVGI systems by application (left) and by type (right).

systems, laboratory disinfection systems, portable area disinfection systems, and mail decontamination systems.

11.2.1 In-duct Air Disinfection Systems

The first step in the design of an airstream or surface disinfection system is to characterize the application. This includes describing the airstream, identifying the specific surface, and, sometimes, targeting specific microbes, such as TB.

UVGI units are commonly located in an *air handling unit* (AHU) downstream from the mixing box. Figure 11.2 shows a typical airstream disinfection system installed downstream from the filter bank and upstream from the cooling coils.

FIGURE 11.2 UVGI air disinfection system installed upstream of cooling coils with reflective aluminum diamond plate installed. (*Photo provided courtesy of Immune Building Systems, Inc., New York.*)

Although UVGI systems can also be placed in the return-air duct to deal with recirculated, contagious pathogens, they are rarely placed in outside air supply duct. Spores, which hail from the outdoors, are more efficiently removed by filtration alone.

The characteristics of an airstream that can impact UVGI design are *relative humidity* (RH), temperature, and air velocity. Increased RH tends to decrease decay rates under *ultraviolet* (UV) exposure. Air temperature has a negligible impact on microbial susceptibility to UVGI (Rentschler et al., 1941). However, it can greatly impact the power output of UVGI lamps if it exceeds design values. Operating a UVGI system at air velocities above design will reduce UV output because of the cooling effect of the air on the lamp, a factor that will be addressed later in this chapter.

11.2.2 Upper Room UVGI

Upper air disinfection systems (aka upper room UVGI systems) create a beam of UV rays parallel to the ceiling that are confined to the upper portion of a room volume (Dumyahn and First, 1999). Air that gravitates into this field is disinfected. Since the levels of UV below this area are kept below safe limits, 0.2 $\mu W/cm^2$ for 8 hour, these systems operate continuously and are considered to be safe for human occupied areas (ACGIH, 2005; NIOSH, 1972). Some studies have demonstrated the effectiveness of upper room UVGI systems in reducing airborne microbial levels (Xu et al., 2003). However, because room air currents are unpredictable, it is difficult to predict performance of these systems in advance, except by empirical data or perhaps by CFD analysis of room airflows. Some methods, however, exist to estimate the effectiveness of upper air units (Beggs and Sleigh, 2002). Figure 11.3 shows one type of upper room air disinfection system.

11.2.3 Microbial Growth Control

UVGI for microbial-growth control has been undergoing much study recently and has enjoyed success in field applications, especially for decontaminating cooling coils (Shaughnessy et al., 1999; Scheir and Fencl, 1996). UVGI was first recommended for controlling mold growth in the late 1940s and early 1950s, when mold had begun to be recognized as an indoor health hazard (Luckiesh et al., 1949; GE, 1950; Richards, 1954). Placement of UV lamps in air

FIGURE 11.3 Upper air disinfection system. (*Photo courtesy of Lumalier/Commercial Lighting Design, Inc., Memphis.*)

conditioners and between filters and cooling coils had been suggested by several sources (Buttolph and Haynes, 1950; Harstad et al., 1954; GE, 1950). Luciano (1977) recommended placing UVGI between the filters and the cooling coils for hospital applications. The first UVGI system designed specifically for disinfecting the surfaces of air handling equipment, including humidifier water and filters, was detailed by Grun and Pitz (1974). In Europe, microbial-growth control on cooling coils had been practiced in breweries since at least 1975 and one manufacturer recommended placing a 15-W lamp 1 m from the surface of cooling coils or walls where condensation could occur (Philips, 1985). In spite of decades of successful applications in Europe, it wasn't until 1996 that the first UVGI system for controlling microbial growth on cooling coils in the United States was installed, by Public Service of Omaha in Tulsa, based on a recommendation by a European expert (McKain, 2003).

Direct UVGI exposure can sterilize any surface given enough time. Theoretically, low-irradiance UVGI could be used for microbial growth because the exposure time is extended. In practical applications, however, microbial growth can occur in crevices, shadowed areas like insulation, and stagnant water where UVGI may not completely penetrate.

UVGI can control microbial growth on filters subject to moisture or high humidity. Figure 11.4 shows an unirradiated and irradiated filter bank. The unirradiated filters show natural contamination from various fungal species, including *Aspergillus* and *Penicillium*, while the irradiated filters show no evidence of microbial growth. Microbial growth control systems have a proven record of maintaining cooling coil cleanliness, to the point that short payback periods can be established in terms of energy saved.

11.2.4 Air Recirculation Units

UVGI systems that recirculate room air or that are placed in a return-air duct or mixing-air plenum deliver multiple doses to airborne microorganisms. Although the effect is partially dependent on the air change rate, the result is an effective increase in removal rate in

FIGURE 11.4 Photos showing mold growth on filters without UV (left) and with UV (right). (*Photos provided courtesy of Airguard Industries, Inc., Louisville, KY.*)

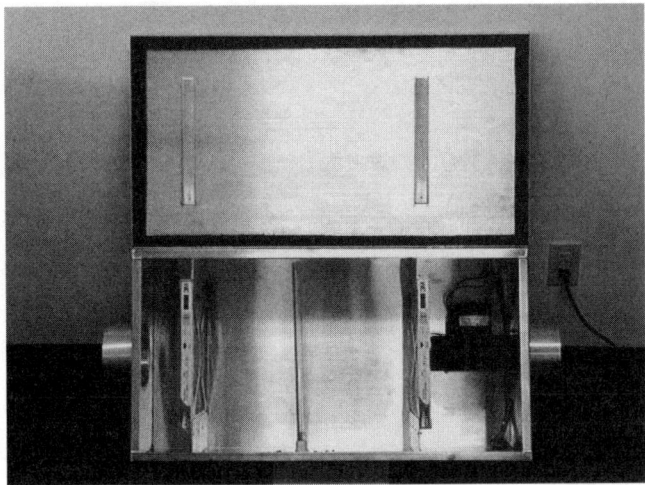

FIGURE 11.5 UVGI system designed for retrofitting residential central air units. (*Photo courtesy of Air Clean Assurance Corporation, Houston, TX.*)

comparison with a single-pass system. Room recirculation units and upper air systems can be installed to augment in-duct systems or where in-duct installation is not feasible. Figure 11.5 shows an example of a UVGI unit that can be retrofit in ductwork.

11.2.5 Surface Disinfection

Area disinfection or decontamination systems can be used to eliminate surface contamination in open areas, either for remediation or for prevention of potential hazards. Portable UVGI systems are available to decontaminate open areas and these are, of course, operated only in unoccupied areas. After-hours UVGI systems are permanently mounted on walls or ceilings and are engaged during times when no occupants are present. The latter type of system is engaged either by timers or disengaged by motion detectors, and provides options for mailrooms or laboratories where contamination potential exists. Figure 11.6 shows an after-hours UVGI unit that has been used in food industry applications.

Mail disinfection systems are similar to medical disinfection equipment except that they can be as large as rooms. Typically these require about 20 to 30 minutes to sterilize the surface of envelopes and packages, similar to medical equipment systems.

11.3 UVGI DISINFECTION THEORY

The ultraviolet spectrum of light extends from wavelengths of about 100 to 400 nm, with subdivisions of UVA (315 to 400 nm), UVB (280 to 315 nm), UVC (200 to 280), and VUV (100 to 200 nm). Although all UV wavelengths cause some photochemical effects, UVC wavelengths are especially damaging to cells because they are absorbed by proteins, RNA, and DNA (Bolton, 2001; Rauth, 1965). The germicidal effectiveness of UVC is illustrated in Fig. 11.7, where it can be observed that germicidal efficiency peaks at about 260 to 265 nm. This peak corresponds to the peak of UV absorption by bacterial DNA. The germicidal

236 AIRBORNE DISEASE CONTROL TECHNOLOGIES

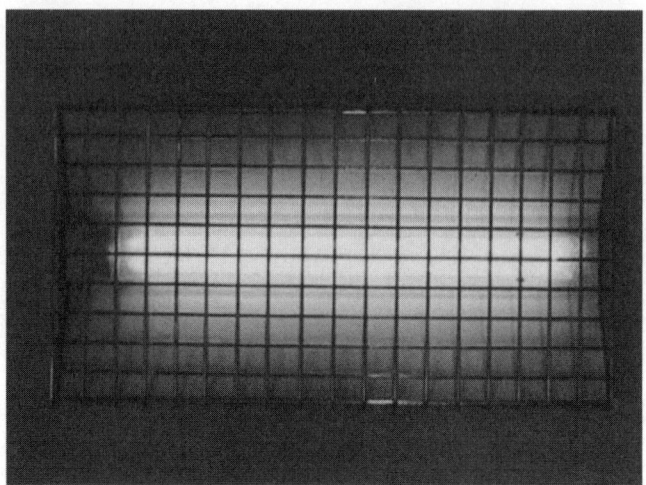

FIGURE 11.6 After-hours UVGI system, for area surface disinfection. (*Photo provided courtesy of Lumalier/Commercial Lighting Design, Inc., Memphis.*)

effectiveness can vary between species and the broader range wavelengths also make a small contribution to inactivation (Webb and Tuveson, 1982). Low-pressure mercury vapor lamps radiate about 95 percent of their energy at a wavelength of 253.4 nm, which is coincidentally so close to the DNA absorption peak that it has a high germicidal effectiveness (IESNA, 2000).

UV wavelengths inactivation microorganisms by cross-linking and breaking bonds between nucleic acids. The absorption of UV can result in the formation of intrastrand cyclobutyl-pyrimidine dimers in DNA, which can lead to mutations or cell death (Harm, 1980; Koller, 1952; Kuluncsics et al., 1999). Pyrimidines are molecular components in the biosynthesis process and include thymine and cytosine. Thymine and cytosine are components

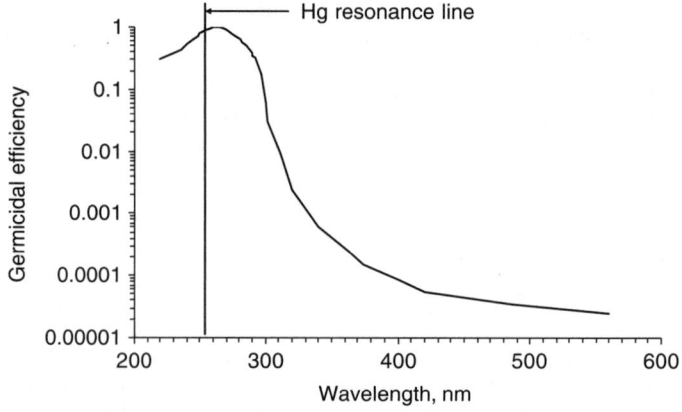

FIGURE 11.7 Germicidal efficiency of UV wavelengths. [*Based on data from Luckiesh (1946) and IESNA (2000).*]

of DNA. The dimers formed are also known as thymine dimers. The lethal effect of UV radiation is primarily due to the structural defects caused when the thymine dimers form (David, 1973).

Since the disruption of normal DNA processes occurs as the result of the formation of thymine dimers, there may be a relationship between the thymine content and the UV susceptibility or rate constant. Thymine (T) combines with adenine (A) in DNA and cytosine (C) combines with guanine (G), and the percentage of thymine will vary from species to species. David (1973) inferred that for a constant G+C content (or T+A content), the sensitivity to UV radiation is a reciprocal function of the molecular weight of the genome, suggesting that the smaller the DNA molecule, the higher the probability that a hit would be lethal. Some data on mycobacteria strongly suggest a relationship between molecular weight of the genome, the G+C content, and UV rate constants, at least within specific genera (David, 1973). In theory, thymine dimers could occur not only between adjacent amino acids but also across different strands of DNA or RNA, making the problem a complex matter of T+C content, probability, geometry, and conformational state of DNA (i.e., state A or B), which may be a function of RH. Further research in this area could be most fruitful, since laboratory studies on UV rate constants have yet to address the majority of airborne pathogens and allergens.

11.4 UVGI DISINFECTION MODELING

The classical exponential decay model addressed in Chap. 7 is adequate for most UVGI design purposes. The single-stage decay equation for microbes exposed to UV irradiation is repeated here:

$$S = e^{-kIt} \qquad (11.1)$$

The variables are as previously defined in Chap. 7. This equation applies as long as shoulder effects can be ignored and the inactivation rate is not well into the second stage. Figure 11.8 illustrates the difference between the simple model of Eq. (11.1) and the

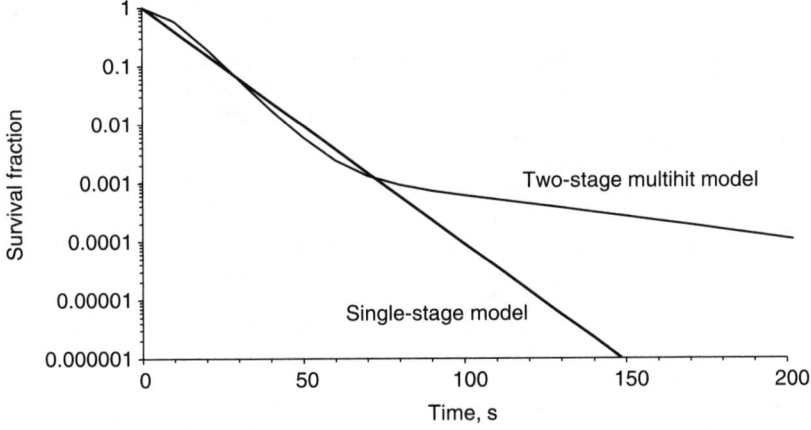

FIGURE 11.8 Comparison of single-stage exponential decay model with two-stage multihit model for anthrax exposed to 3000 µW/cm².

complete model of Eq. (7.17) for anthrax spores exposed to 3000 µW/cm² (Kowalski, 2003). Equation (11.1) is seen to provide a reasonable estimate of the inactivation rate provided no more than three logs of reduction are required. As a general rule, if sterilization is required, which is considered to be six logs of reduction for the purpose of modeling, it will probably be necessary to use the complete model to obtain an accurate prediction of the necessary UV dose.

The UV rate constant defines the sensitivity of a microorganism to UV exposure and is unique to each microbial species. If the average irradiance is constant, or can be calculated, then the standard rate constant can be computed as

$$k = \frac{-\ln S}{It} \qquad (11.2)$$

The parameter I is the irradiance of UV exposure. It can be used to refer to either irradiance, which is the radiative flux through a flat surface, or the fluence rate, which is the radiative flux through an external surface (i.e., a spherical microbe). It has the same units in both cases and the choice of which term to use depends on the context. In most cases the data used to define the rate constant are measured as irradiance, but in cases where spherical actinometry is used the rate constant is defined in terms of the fluence (Rahn et al., 1999). As a matter of convenience and convention the term irradiance will be used throughout.

The Z value is an alternative term occasionally used in the literature as a substitute for the rate constant k (Lai et al., 2004; Ko et al., 2000). Riley et al. (1976) credit Kethley with the definition $Z = \ln(N_0/N)/\text{Dose}$, but Peccia and Hernandez (2001) define Z in terms of the first-order rate coefficient and the spherical irradiance. There is certainly a need for more specific terminology and definitions of terms related to UVGI inactivation of microbes, as well as further development of new sensors to measure actual fluence or spherical irradiance, but to avoid confusion the traditional terminology is maintained here. In general, the terms are defined by the context in which they are used and there is little chance for confusion.

UV rate constants are often expressed in units of m²/J or cm²/µW·s. Appendix A provides selected UV rate constants for many pathogens and allergens along with the associated reference, media, and a computed dose for 90 percent inactivation rate. Only a limited number of these rate constants come from studies of airborne exposure. A comparison of the UV rate constants from multiple studies of the same microbe in different media suggests that water-based rate constants are higher than those performed on plates or in air, as shown in Fig. 11.9. It could be presumed that water-based rate constants would be similar to those of airborne microbes in 100 percent RH, although this is not certain.

Figure 11.10 compares the average rate constants of all three microbial groups, viruses, bacteria, and fungi, in terms of their average survival under exposure to an irradiance of 100 µW/cm². This chart highlights the fact that fungi are relatively resistant to UV and viruses are highly susceptible, on the average.

The complete model given by Eq. (7.17) in Chap. 7, which includes the shoulder and the second stage, requires two rate constants, two multihit exponents, and a resistant population fraction. Limited data are available that would enable extracting these values for any but a handful of microbes, and these are summarized in Table 11.1. The resistant population fraction (Res. Pop. = f) defines the fraction of the population that resists UV exposure, and has the rate constant k_2. The first-stage rate constant k_1 applies to the susceptible population $(1 - f)$. The multihit exponents are not known for some of these microbes and those that are, are assumed to be the same for both stages. In general, the multihit exponent for the second-stage rate constant has little or no effect.

The two-stage rate constants computed from data can differ somewhat from the single-stage rate constants due to the fact that the second stage and the shoulder, if any, will affect the first-stage rate constant. This is exemplified by the two-stage curve data in Table 11.1

ULTRAVIOLET GERMICIDAL IRRADIATION 239

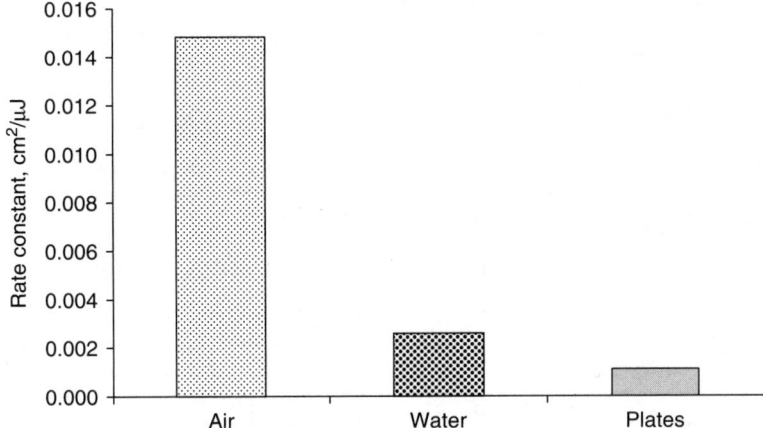

FIGURE 11.9 Comparison of average rate constants for bacteria in different media.

and the graph of the data in Fig. 11.11. In Table 11.1, the rate constant for *Mycobacterium tuberculosis* based on fitting all the data to a single-stage curve without a shoulder proved to be 0.035 m²/J. When a two-stage curve with a shoulder (Eq. 7.11 in Chap. 7) is fitted to the same data, the first-stage rate constant proves to be approximately 0.055 m²/J. The reason for the difference is that when the shoulder and second stage are modeled separately the first stage will have a steeper slope.

Often, it is more convenient to measure the effectiveness of a UVGI system in terms of the inactivation rate IR which is the complement of the survival, computed as follows:

$$S = e^{-klt} \tag{11.3}$$

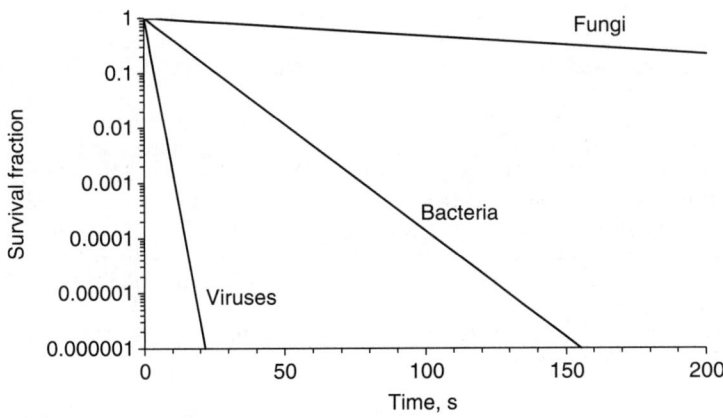

FIGURE 11.10 Comparative decay rates for the three microbial groups (based on averages of rate constants for microbes shown in App. A.)

TABLE 11.1 Two-Stage and Multihit Parameters

Airborne microorganism	k_1 m²/J	k_2 m²/J	Res. pop. (f)	Multihit exponent		Reference
				n_1	n_2	
Adenovirus Type 2	0.0048	0.7784	0.0001	1.29	1.29	Rainbow and Mak, 1973
Coxsackievirus B-1	18.5500	14.5500	0.2622	1	1	Hill et al., 1970
Coxsackievirus A-9	28.8200	10.2200	0.0193	1	1	Hill et al., 1970
Staphylococcus aureus	1.7020	0.9100	0.0860	4.9	4.9	Sharp, 1939
Streptococcus pyogenes	0.2870	0.0167	0.1484	1	1	Lidwell and Lowbury, 1950
Serratia marcescens	64.6500	43.6100	0.6246	1.71	1.71	Riley and Kaufman, 1972
Bacillus anthracis spores	0.0042	0.0006	0.0016	2.6	2.6	Knudson, 1986
Mycobacterium smegmatis	0.0500	0.0108	0.00062	6	1	Boshoff et al., 2003
Mycobacterium tuberculosis	0.0550	0.0244	0.012	30	1	Boshoff et al., 2003

11.5 MODELING THE UV DOSE

The complete three-dimensional (3D) irradiance field for any enclosed UVGI system needs to be resolved in order to establish the dose received by any airborne microbe passing through the field. In theory, the UV dose is the radiative energy absorbed by the subject microbial population but in fact the actual absorbed dose is unknown and only the fluence (the incident irradiance) can be measured or calculated, but in keeping with common usage, the term dose is used here. Both the irradiance field of the lamp and the irradiance field due to a reflective enclosure can be determined through the use of thermal radiation view factors (Kowalski and Bahnfleth, 2000a).

Various models of the irradiance field due to UV lamps have been proposed in the past, including point source, line source, integrated line source, and other models (Jacob and Dranoff, 1970; Qualls and Johnson, 1983; Beggs et al., 2000; Krasnochub, 2005). The model used here is based on thermal radiation view factors, which define the amount of diffuse radiation transmitted from one surface to another (Modest, 1993). Figure 11.12 illustrates a lamp modeled as a cylinder where the planar area at which the UV irradiance is to be determined is perpendicular to the axis and is at the edge of the cylinder.

The fraction of radiative irradiance that leaves the cylindrical body and arrives at a differential area (Modest, 1993) is

$$F = \frac{L}{\pi H}\left[\frac{1}{L}\text{ATAN}\left(\frac{L}{\sqrt{H^2-1}}\right) - \text{ATAN}(M) + \frac{X-2H}{\sqrt{XY}}\text{ATAN}\left(M\sqrt{\frac{X}{Y}}\right)\right] \quad (11.4)$$

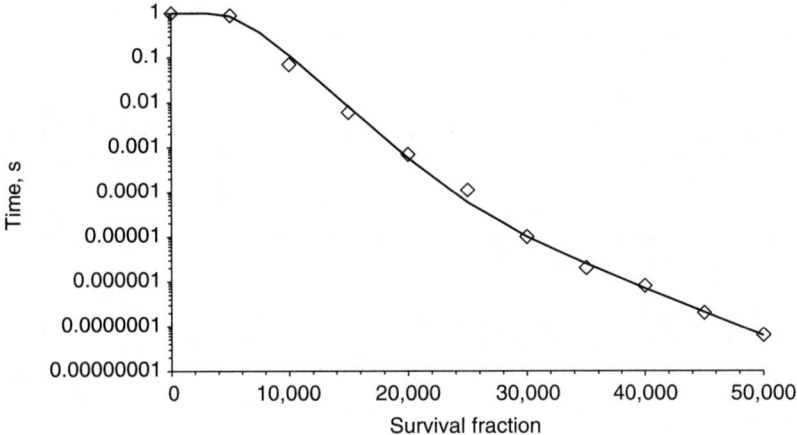

FIGURE 11.11 Survival curve for *Mycobacterium tuberculosis*. Data fitted to a two-stage multihit model. [*Based on data from Boshoff et al. (2003).*]

The parameters in Eq. (11.4) are defined as follows:

$$H = \frac{x}{r}$$

$$L = \frac{l}{r}$$

$$X = (1 + H)^2 + L^2$$

$$Y = (1 - H)^2 + L^2$$

$$M = \sqrt{\frac{H-1}{H+1}}$$

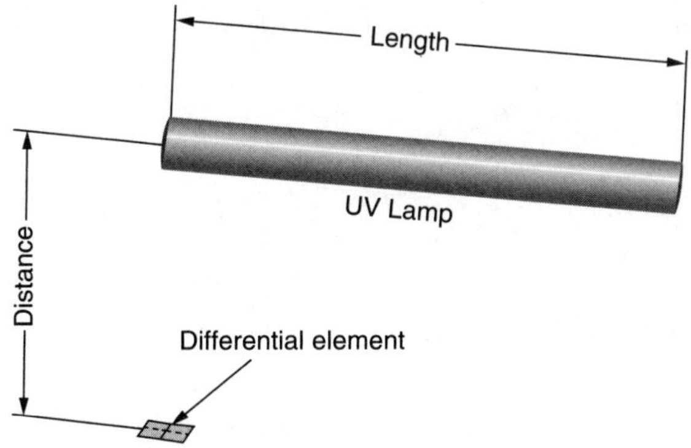

FIGURE 11.12 UV lamp modeled as a cylinder some finite distance from a differential element at which the irradiance is to be computed.

where l = length of the lamp segment (arclength), cm
 x = distance from the lamp, cm
 r = radius of the lamp, cm

This equation applies to a differential element located at the edge of the lamp segment. In order to compute the view factor at any point along a lamp it must be divided into two segments. Figure 11.13 shows the predicted results of the irradiance at a series of points that extend from the lamp midpoint. Comparison of the view factor model with lamp photosensor data generally gives excellent agreement. No other models that have been used, including point source, line source, and integrated line source models, approach the accuracy of the view factor model (Kowalski and Bahnfleth, 2000a; Kowalski et al., 2000).

Equation (11.4) can be used to compute the irradiance at any point beyond the ends of the lamp by applying it twice—once to compute the view factor for an imaginary lamp of the total length (distance between some point and the far end of the lamp) and then subtracting the view factor of the nonexistent portion, or ghost portion. This method, known as view factor algebra, is detailed in Kowalski and Bahnfleth (2000a), Kowalski (2001), and elsewhere (Modest, 1993).

Implicit in the use of this view factor is the assumption that microbes are spherical and receive UV rays that pass right through the cell. In this model the cross-sectional area of a sphere is a flat disc that remains perpendicular to a line passing through the lamp axis. Although not a perfect model, since the flat disc only faces the lamp axis and not the lamp length, the possible error appears to be minor or insignificant in most cases (Kowalski et al., 2000).

The irradiance field as a function of distance from the lamp axis is simply the product of the surface irradiance and the view factor, where the surface irradiance is computed by dividing the UV power output by the surface area of the lamp:

$$I = \frac{E_{uv}}{2\pi r l} F_{total} \tag{11.5}$$

where E_{uv} is the UV power output of lamp, µW.

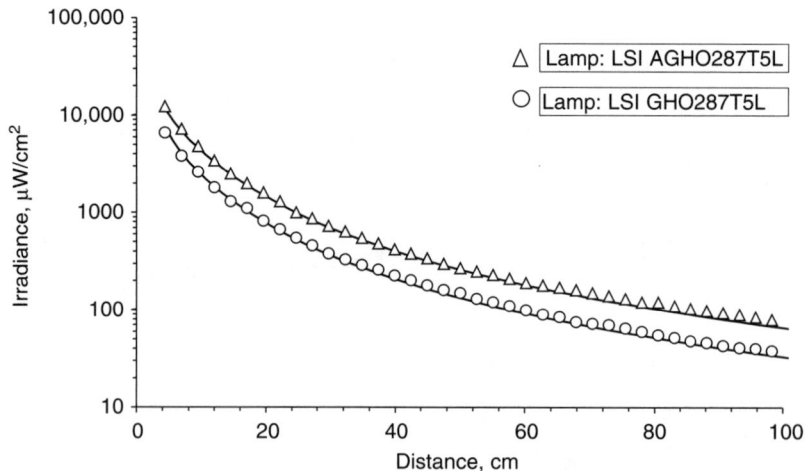

FIGURE 11.13 Comparison of view factor model predictions of irradiance at the midpoint for two lamps with photosensor data.

Several sources may contribute errors to the predictions of the view factor model, including the round-off error in the lamp wattage and variations of surface irradiance along the lamp. Manufacturers typically state the error in the nominal UV wattage is ±1 percent but this may be optimistic. Blatchley (1997) measured wattage variations in the same model lamp of +3.7/−2.2 percent. The lamp rating for any tubular lamp can be computed by using Eq. (11.5) and modeling the lamp as two equal lamp segments. A comparison of predicted ratings using the view factor model for over 90 different UVGI lamps indicates predictive errors within ±9 percent (Kowalski and Bahnfleth, 2000a). More detailed information on the 3D model of the UV irradiance field, including source code,* is provided in Kowalski (2003, 2001).

11.6 UV LAMP RATINGS

UV lamps are rated by taking photosensor measurements at 1 m from the midpoint of the lamp axis in still air (IESNA, 2000). Lamp UV output typically decreases after initial operation and then remains fairly steady for most of its useful life. Many manufacturers will burn-in lamps prior to use or sale for about 10 to 100 h.

11.7 THE UVGI RATING VALUE

The dose is the average irradiance of a UVGI system multiplied by the exposure time and can be used as a sizing criterion. Typically, exposure times are on the order of fractions of a second, and this is defined by the available space in a duct or air handling unit and the air velocity. The *UVGI rating value* (URV) has been introduced as a simplified means of rating UVGI systems (Kowalski and Dunn, 2002; Kowalski and Bahnfleth, 2004). Table 11.2 summarizes the URV scale and the minimum dose each URV represents, along with examples of the typical inactivation rates for some representative microbes.

The URV system has been designed to span the range of typical UVGI system sizes and to complement the MERV filter rating system such that a UVGI system of one URV can be combined with a filter with the same MERV to obtain approximately equal reductions across the entire spectrum of airborne microbes (Kowalski et al., 2003b). Systems like those used for microbial growth control, or upper air systems, may use much lower values than 100 $\mu W/cm^2$, and URV ratings do not apply to these systems.

11.8 UV IRRADIANCE FIELD DUE TO ENCLOSURE REFLECTIVITY

Reflectivity of the UVGI enclosure can greatly enhance the irradiance field depending on the enclosed surface area, the reflectivity, and the type of reflectivity. Reflectivity may be diffusive like the reflectivity of clouds or white paper, or it may be specular like the reflectivity of mirrors.

---

*Author's Note: In the source code provided in Kowalski (2003), the following correction is needed in the subroutine DirectIntField(): the code line
  paxis = fabs(Position(i,j,k,l)); should be changed to paxis = Position(i,j,k,l);
  The original subroutine in Kowalski (2001) was intended to model lamps with one end at $x = 0$ only. This modification allows the lamp end to be at any position. Without this change predictive errors can occur for lamps with one end located at any position other than $x = 0$.

TABLE 11.2 UVGI Rating Values and Typical Inactivation Rates

	Dose					
URV	µJ/cm²	J/m²	Anthrax %	Influenza %	Smallpox %	TB %
1	1	0.01	0	0	0	0
2	10	0.1	0	1	2	2
3	20	0.2	0	2	3	4
4	30	0.3	0	3	4	6
5	50	0.5	1	6	7	10
6	75	0.75	1	9	11	15
7	100	1	2	11	14	19
8	150	1.5	2	16	20	27
9	250	2.5	4	26	32	41
10	500	5	8	45	53	66
11	1,000	10	15	69	78	88
12	1,500	15	22	83	90	96
13	2,000	20	28	91	95	99
14	3,000	30	39	97	99	100
15	4,000	40	49	99	100	100
16	5,000	50	57	100	100	100
17	6,000	60	63	100	100	100
18	8,000	80	74	100	100	100
19	10,000	100	81	100	100	100
20	20,000	200	96	100	100	100
Rate constant k, cm²/µJ			0.000167	0.001187	0.001528	0.002132

11.8.1 Diffuse Reflectivity

The reflective surfaces of rectangular ducts produce a 3D irradiance field that depends on the field of the lamp. Assuming that the surfaces are diffuse, with the same value of reflectivity, provides a simple means of approximating the reflected irradiance field.

First determine the direct irradiance at each surface with Eq. (11.5). The resulting matrix, a surface contour really, of the intensities for each wall can be averaged to simplify computations without introducing much error. The average reflected irradiance for each surface is

$$I_{vl} = \bar{I}_{Dvl}\rho \quad I_{vr} = \bar{I}_{Dvr}\rho \quad I_{ht} = \bar{I}_{Dht}\rho \quad I_{hb} = \bar{I}_{Dhb}\rho \tag{11.6}$$

where $\bar{I}_D$ is the average direct irradiance for surface vl, vr, ht, hb.

The view factor F_h represents a horizontal element exposed to the diffuse irradiance of a facing parallel wall (view factor 10 per Modest, 1993):

$$F_h = \frac{1}{2\pi}\left[\frac{X}{\sqrt{1+X^2}}\text{ATAN}\left(\frac{Y}{\sqrt{1+X^2}}\right) + \frac{Y}{\sqrt{1+Y^2}}\text{ATAN}\left(\frac{X}{\sqrt{1+Y^2}}\right)\right] \tag{11.7}$$

where $X = \text{height}/x$
$Y = \text{length}/x$
$x = \text{distance to the vertical wall.}$

The view factor F_v represents an element perpendicular to the bottom side of the duct (view factor 11 per Modest, 1993):

$$F_v = \frac{1}{2\pi}\left[\text{ATAN}\left(\frac{1}{Q}\right) - \frac{Q}{\sqrt{P^2+Q^2}}\text{ATAN}\left(\frac{1}{\sqrt{P^2+Q^2}}\right)\right] \quad (11.8)$$

where P = height/width
$Q = Y$/width
Y = distance to the horizontal wall

For a rectangular duct use Eq. (11.7) for each of the two vertical surfaces and Eq. (11.8) for each of the two horizontal faces to compute the 3D irradiance field due to the first reflections. The reflected irradiance I_R at any x, y, z point will be

$$I_R = I_{vl}F_{vl} + I_{vr}F_{vr} + I_{ht}F_{ht} + I_{hb}F_{hb} \quad (11.9)$$

where F_{vl}, F_{vr} are the view factors to vertical left and right wall, respectively, Eq. (11.7), and F_{ht}, F_{hb} are the view factors to horizontal top and bottom wall, respectively, Eq. (11.8).

Equation (11.9) will provide an approximation of the irradiance field due to the first reflection and this should prove adequate for most design purposes. For information on computing multiple diffusive reflections see Kowalski (2001, 2003).

11.8.2 Specular Reflectivity

A separate model has been developed for specularly reflective, or mirror-like, surfaces (Kowalski et al., 2005). The specular model uses a single view factor to define the lamp irradiance field as described previously, but uses multiple virtual images of the lamp to compute the reflected irradiance field. The virtual images are treated as separate lamps with their UV power output reduced by the reflectivity of each of the reflective surfaces through which the image passes. Figure 11.14 shows a photograph in which six to eight virtual images of the UV lamp can be seen reflected. The reflectivity is approximately 85 percent in the visual field.

Figure 11.15 illustrates the model in terms of the real lamp in the center and the virtual image of the first reflection. The irradiance contribution of the first specular reflection can be computed by modeling a lamp at the equivalent distance and position of its virtual image. Since reflectivity is always less than 100 percent, the lamp irradiance must be multiplied by the fractional reflectivity. For example, if the surface reflectivity is 75 percent, then the irradiance of the first virtual lamp image must be multiplied by 0.75, which is the same as multiplying the UV power (E_{uv} in Eq. 11.5) by 0.75. As a result, the entire array of virtual images can be modeled as separate lamps, with the UV power of each lamp reduced by the reflectivity of each virtual reflective surface it passes through.

Figure 11.16 illustrates the array of virtual images that are created in a four-sided specular reflective UVGI enclosure. Note that the position of the lamp is mirrored at each reflection. Note also that the first image in the corner is actually a second reflection since the image must pass through two virtual reflective surfaces. The first specular reflection creates four virtual images, the second creates eight images, the third creates 12, and so on. The number of images in each reflection progresses by the following series:

$$(4, 8, 12, 16, 20, 24, \ldots) = 4\sum_{n=1}^{\infty} n \quad (11.10)$$

where n is the number of reflections.

246 AIRBORNE DISEASE CONTROL TECHNOLOGIES

FIGURE 11.14 UVGI system made from one-way mirrors. At least six virtual images can be seen in this photograph. (*Photo provided courtesy of Lumalier, Memphis.*)

Since the maximum specular reflectivity likely to be encountered is no more than about 90 percent, the number of reflections that need be considered in a normal UVGI system is about five before the contribution diminishes to insignificant levels. Although 90 percent reflectivity through five virtual surfaces will only reduce the fifth lamp virtual image by about 60 percent of its irradiance (i.e., $0.90^5 = 0.59$), the drop-off in irradiance due to the

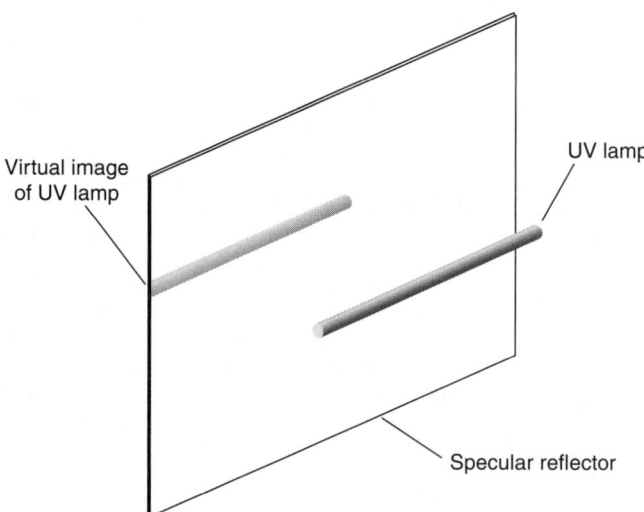

FIGURE 11.15 A specular reflector (mirror) will show a virtual image of the real lamp an equivalent distance behind the reflector surface.

ULTRAVIOLET GERMICIDAL IRRADIATION 247

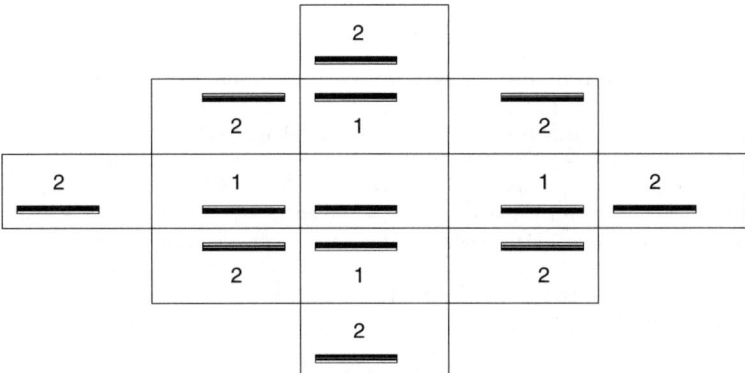

FIGURE 11.16 Virtual images of UV lamps in an enclosure with four specular surfaces. Only the first two reflected virtual lamp images are shown.

distance to the virtual image is far greater. In the following specular model example, the first six reflections are represented by 80 virtual lamps. Analysis of reflections beyond the first six is best handled with a geometric series, especially since the distant specular images will become diffused, since even the most specular surface has a diffuse component.

The specular model produces results that are similar in quantity to the diffuse reflective model previously presented but differ in certain qualities. Since specular reflections are more focused there can be differences in the predicted inactivation rates due to the fact that concentrated irradiance fields may be less efficient. The specular model also requires considerably more computation time (IBSI, 2003). Figure 11.17 shows the combined results of the irradiance field contributions from the example in Fig. 11.16. Each of the successive specular reflections 1 to 6 represents the virtual image from four or more walls of the rectangular enclosure.

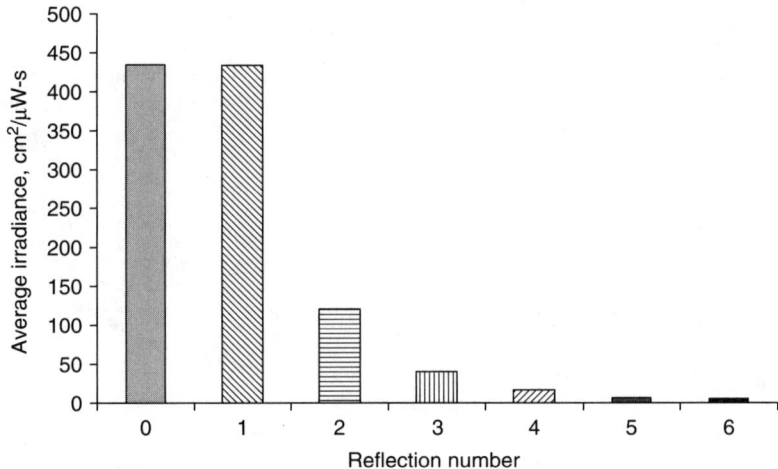

FIGURE 11.17 Irradiance contribution of six specular reflections inside a rectangular UVGI enclosure with a reflectivity of 90 percent. Reflection 0 is the direct irradiance contribution from the UV lamp.

11.9 AIR MIXING EFFECTS

In long ducts the velocity profile of a laminar airstream will approach a parabolic shape, with the velocity higher toward the center. However, fully developed laminar velocity profiles are unlikely to be achieved in laboratory or real-world installations. The design velocity of a typical UVGI system is about 2.54 m/s (500 fpm), producing a Reynolds number of approximately 150,000. Turbulent mixing is therefore more likely to be the norm. Even laminar flow involves mixing by diffusion and so real world operating conditions will lie somewhere between complete mixing and the idealized condition of completely unmixed flow, as shown by Severin et al. (1984) for water-based systems. These bounding conditions, complete mixing and unmixed flow, assume a flat velocity profile. Rate constants for complete mixing can be computed using the average irradiance, and the survival is computed as follows:

$$S = e^{-kI_{avg}t} \qquad (11.11)$$

where I_{avg} is the average irradiance in the irradiation chamber, $\mu W/cm^2$.

The average irradiance can be computed for any irradiation chamber using Eq. (11.5) for each lamp and lamp virtual image and using a 3D matrix of sufficient resolution. If the airflow through a UVGI chamber followed perfectly parallel streamlines without mixing, then each streamline would be subject to a dose that depends on the distance from the lamp. The survival rate in this case must be calculated for each streamline segment and summed or integrated to obtain the net survival. The survival S_i for each streamline segment is computed by Eq. (11.11) and the total survival would be the sum of all streamline segments as follows:

$$S = \sum_{j=1}^{l}\sum_{i=1}^{m}\sum_{k=1}^{n} e^{-kI_{ijk}t} \qquad (11.12)$$

where I_{ijk} = Irradiance at point i, j, k
i = point defining the x coordinate (width)
j = point defining the y coordinate (height)
k = point defining the z coordinate (length)
t = exposure time for each segment defined by point ijk

Equation (11.12) permits the development of a contour map of the inactivation zones of the unmixed condition. Consider two typical systems as shown in Fig. 11.18a, which is

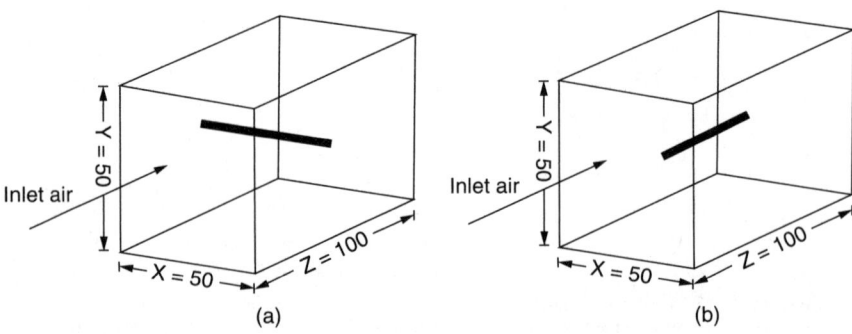

FIGURE 11.18 Schematic of crossflow (a) and axial flow (b) configurations.

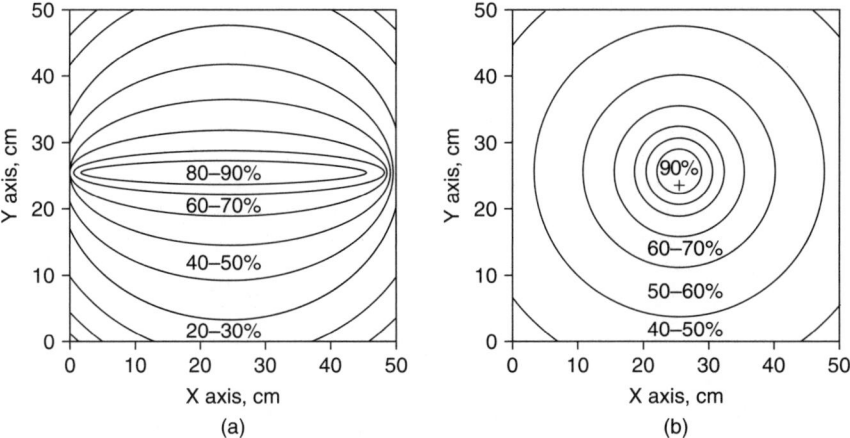

FIGURE 11.19 Inactivation zones determined for the crossflow (*a*) and axial flow (*b*) configurations, shown as percentage of population inactivated in unmixed air.

commonly known as crossflow, and Fig. 11.18*b*, which is called axial flow. The inactivation zones developed from the above computational methodology are shown in Fig. 11.18*a* and 11.18*b* for a single lamp system in a reflective chamber and *S. marcescens* as the test microbe.

In the crossflow condition, the inactivation rate was predicted to be 59 to 64 percent (unmixed-mixed). In Fig. 11.18*b*, using the same 5.5-watt lamp in the axial flow case, the range of inactivation rates was predicted to be 53 to 56 percent. Figure 11.19*a* and 11.19*b* illustrates the predicted inactivation zones inside the units shown in Fig. 11.18*a* and 11.18*b*. The crossflow configuration could be said to be more efficient due to the wider spread of the highest inactivation zone.

11.10 RELATIVE HUMIDITY EFFECTS

Various sources state that increased relative humidity (RH) decreases the decay rate under UVGI exposure (Riley and Kaufman, 1972). Lidwell and Lowbury (1950) showed the rate constant for *Serratia marcescens* decreasing with increasing RH. Rentschler and Nagy (1942) showed the rate constant for *Streptococcus pyogenes* increasing with higher RH. Fletcher (2004) showed the rate constant for *Burkholderia cepacia* decreases at higher RH. One recent study on three bacteria species indicates that the decay rate decreases with higher RH (Peccia et al., 2001). Figure 11.20 shows the effect of relative humidity on the rate constant of *Serratia marcescens*.

Lai et al. (2004) obtained a similar response to high RH for *S. marcescens* as Peccia et al. (2001) and others but demonstrated that the suspending solution used for aerosolization solution had a significant effect on the rate constant. They recommended that a synthetic saliva would more accurately account for real-world conditions.

It is known that DNA undergoes a conformation change, from the A state to the B state, at higher RH (Fletcher, 2004; Peccia and Hernandez, 2001). It has also been shown in Chap. 7 that airborne microbial survival varies with RH (Cox, 1987). The influence of RH on UV susceptibility cannot be quantified at present but remains a matter for ongoing research.

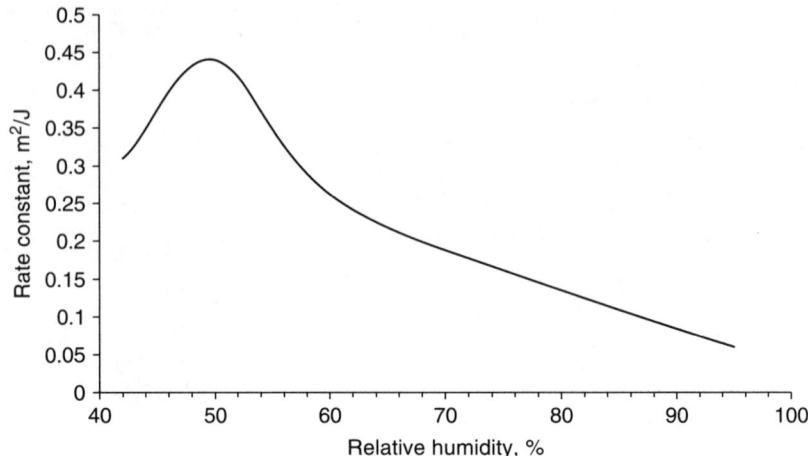

FIGURE 11.20 Effect of relative humidity on the rate constant of *Serratia marcescens*. Based on data from Peccia et al. (2001).

In indoor environments, ASHRAE (1999) defines comfort zones as having an RH below 60 percent and this would be the operating RH range of any UVGI system that recirculated room air or disinfected return air. In an air handling unit, however, the RH could vary greatly depending on where the UVGI system was located. Obviously, the nature of the RH effect on UV susceptibility may dictate the preferred location of any installed UVGI system, and may even allow a means of boosting UVGI efficiency through RH control, but without further quantitative results on RH effects, no detailed modeling of RH effects is possible here.

11.11 PHOTOREACTIVATION

Another factor that cannot be adequately addressed at present is the phenomenon of photoreactivation. Photoreactivation occurs when microorganisms are exposed to visible light during or after UV irradiation (Setlow, 1966; Fletcher et al., 2003). This process can result in self-repair of damaged microbes and can cause a significant percentage of the population to recover from UV inactivation. Photoreactivation has been studied at length in water-based UV experiments but the data for photoreactivation of airborne microbes are limited at present (Linden and Darby, 1994; Masschelein, 2002). One study indicates the decay rate of *Mycobacterium parafortuitum* under UV exposure in liquid suspension is effectively decreased by simultaneous exposure to visible light (Peccia and Hernandez, 2001). The same study suggests that airborne microbial populations can recover significantly if allowed sufficient time. The photoreactivation effect may be dependent on RH, with the effect absent when RH is less than approximately 65 percent. Evidence suggests that a conformal DNA change occurs above some discrete range of RH that may allow microbes to experience photoreactivation (Rahn and Hosszu, 1969; Munakata and Rupert, 1974). Some microbes do not seem to experience photoreactivation, including viruses, *H. influenzae*, *B. subtilis*, and others (Masschelein, 2002).

Although no general quantitative model for photoreactivation of airborne microbes exists at present, some considerations for the design of UVGI systems can be given based on what is known about the photoreactivation effect. Ideally, the RH would be kept below approximately 65 percent in any UVGI system, and the UVGI enclosure should admit little or no internal visible light. Studies on photoreactivation generally involve narrow band UVC lamps, whereas the evidence suggests that broad band UV lamps produce no photoreactivation (Masschelein, 2002). Apparently the broader range UV bands damage the enzymes needed for photoreactivation.

11.12 AIR TEMPERATURE EFFECTS

The air temperature has a negligible impact on microbial survival during UVGI irradiation provided that neither freezing nor heat damage occurs (Rentschler et al., 1941). This would be true for most constant volume ventilation systems since the air temperature tends to remain in a narrow range between approximately 13 and 27°C (55 and 80°F), depending on location and other factors.

Air temperature can also impact UV lamp output by overcooling the lamp, especially when air velocity is beyond design limits. Most UV lamps are designed to operate, and are rated, at an air temperature of approximately 21.5°C (70°F) and an air velocity of 2 to 2.54 m/s (400 to 500 fpm). Lamp UV output may decrease or increase outside this range. Some lamps can lose 25 percent or more of their UV output when the air temperature drops from 27°C (80°F) to 16°C (60°F) (Westinghouse, 1982). Figure 11.21 shows a typical example of the cooling effects on the UV power output of a UV lamp. When lamps are placed into service under conditions where the cooling effects are significant, they should be sized for the output at those conditions. Lamp manufacturers should be consulted regarding lamp performance outside design conditions as they can vary between lamp models.

Certain UV lamps have controls that will boost UV output in response to the cooling effects of airflow on the plasma temperature (Kowalski and Bahnfleth, 2000b). Still other lamps may include an infrared blocking UV-transparent shield to reduce cooling effects. There are well over 100 different types of UV lamps and their individual performance cannot be generalized. Data should be obtained from the manufacturer of any UV lamp to

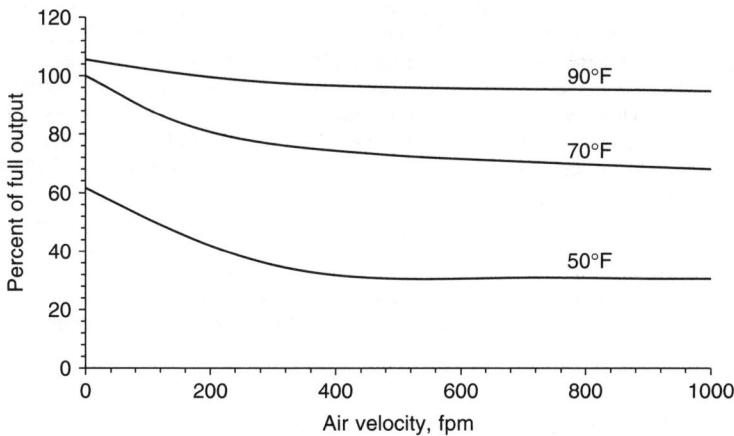

FIGURE 11.21 Effects of air temperature and air velocity on a model TUV36W-PLL lamp.

determine the cooling effects or the limiting design air velocities and temperatures within which the lamps can be efficiently operated.

11.13 PERFORMANCE OPTIMIZATION

The performance of UVGI systems can be optimized to produce maximum inactivation rates while minimizing energy costs if attention is paid to certain aspects of design. Both reflectivity and duct length, for example, can be used to reduce required UV lamp power. There are several variables that define the design and operation of any UVGI system, and these include the dimensions ($W \times H \times L$), the lamp coordinates (x, y, z), airflow (Q), reflectivity (ρ), lamp UV wattage (P), lamp length (l), and lamp radius (r). Some of the variables have optimum values that are fairly well known. The optimum air velocity for most lamps is considered to be the design velocity range of 400 to 500 fpm. The optimum operating air temperature is considered to be 20 to 22°C (68–72°F). The critical design parameters of UVGI systems and their impact on performance has been evaluated using dimensional analysis (Kowalski et al., 2003a). There are eight dimensionless parameters, excluding RH, that define UVGI system performance and these are defined in Table 11.3 in terms of the basic variables.

Figure 11.22 compares the effects of diffuse reflectivity and specific dose. Increasing reflectivity produces an approximately linear increase in inactivation rates but the gains level off at inactivation rates near 100 percent. It is not too difficult to see from this chart that if the same inactivation rates can be achieved by increasing reflectivity, then the expense of increasing lamp power may be unnecessary, but this is ultimately a matter of economics (i.e., lamp power versus materials). The specular model produces very similar results for these two parameters (Kowalski et al., 2005).

Figure 11.23 shows the response surface for the X ratio versus the Y ratio. This figure shows that the inactivation rate increases by almost 15 percent as the lamp is moved toward the diffusive reflective surface. Because there must be a limit to how close a lamp can be placed to a reflective surface before interference with the airflow occurs, it can be hypothesized that an optimum Y ratio must exist for any given UVGI diffusely reflective system configuration. The airflow between the lamp and the surface may be reduced if it is placed too close to a wall, based on studies of the flow fields around cylinders (Kundu, 1990; Gordon, 1978). In Figure 11.23 the inactivation rate is also boosted by about 15 percent as the lamp length is shortened relative to the duct width. The combined minimization of both the Y ratio and the X ratio results in a maximum 25 percent boost in the inactivation rate. This effect does not occur in specularly reflective systems (Kowalski et al., 2005).

For the Z ratio the maximum inactivation rate clearly occurs at a value of 0.5, or exactly centered along the length of the duct, which matches common industry practice. For additional information on dimensionless analysis of UVGI systems, see Kowalski (2001) and Kowalski et al., (2001, 2005).

TABLE 11.3 Dimensionless Parameters of UVGI Systems

Parameter	Definition	Parameter	Definition
Aspect ratio	W/H	Y ratio	y/H
Lamp aspect ratio	r/l	Z ratio	z/L
Specific dose	kPL/Q	Reflectivity	ρ
X ratio	x/W	Height ratio	H/L

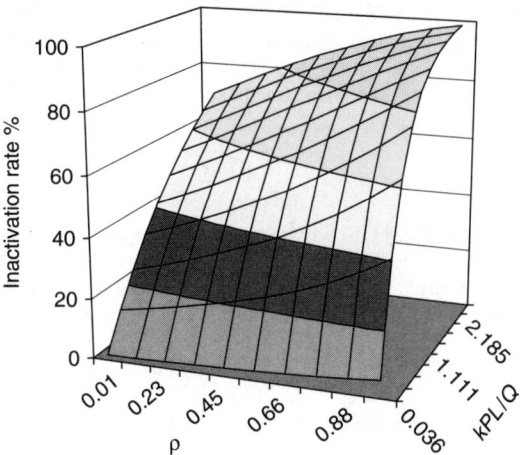

FIGURE 11.22 Inactivation rate for specific dose versus reflectivity, diffusive reflectivity model.

The previous dimensionless parameters can be used to estimate the system wattage for any given set of mass flow rate, duct length, and reflectivity. For example, the specific dose can be multiplied by the reflectivity to obtain a dimensionless parameter that encompasses all the defining parameters of a UVGI system. We can call this resulting dimensionless parameter an F value as follows:

$$F = \rho\left(\frac{kPL}{Q}\right) \quad (11.13)$$

For any given set of design parameters, F will be a constant. We can further simplify this relation by ignoring the units and assuming that the mass flow rate Q is essentially equivalent

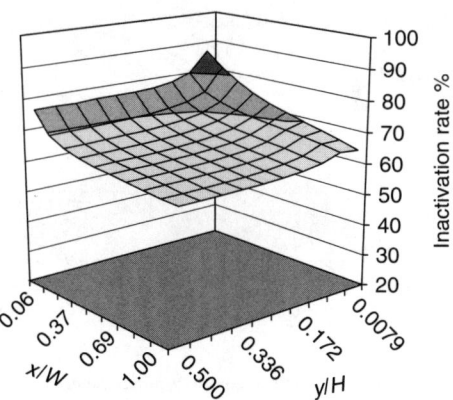

FIGURE 11.23 Inactivation rate for X ratio versus Y ratio.

to the airflow in *cubic feet per minute* (CFM). This will be approximately true for all UVGI systems operated under a normal range of air temperatures. The simplified relation is now dimensional and we have no need of the rate constant k, which acted as a dimensionless conversion factor. The result is a dimensional F value equal to some constant, or

$$F_d = \rho\left(\frac{PL}{\text{CFM}}\right) \tag{11.14}$$

The value of F_d given in Eq. 11.14 will be approximately constant for all systems of comparable performance or specific dose, and the same reflectivity. Since the dose produced by the system is also essentially a function of the URV, we can define an F constant for each URV. Using the term F_u to represent the constant F as a function of the URV, we can write

$$F_u = f(\text{URV}) = \text{constant} \tag{11.15}$$

This allows the writing of a convenient relation between the power and the airflow per unit length of duct as follows:

$$P = F_u \rho\left(\frac{\text{CFM}}{L}\right) \tag{11.16}$$

Equation (11.16) implies that the quantity of airflow per unit length is a prime determinant of UVGI system effectiveness, not unlike similar relations used for sizing pumps and fans. Values of F_u may be estimated based on well-designed UVGI systems of known performance parameters, effectively scaling new systems from functional designs. Although this simplified relationship may provide estimates of UV wattage for any given URV, reflectivity, and airflow per unit duct length, it ignores the remaining dimensionless parameters and errors could well exceed ±30 percent. Equation (11.16) may be useful for

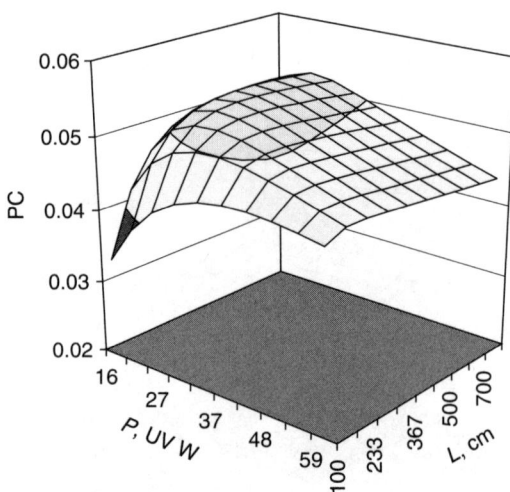

FIGURE 11.24 Optimization of UV power versus duct length for a constant velocity system in terms of the performance cost (inactivation rate/annual cost).

estimating total lamp wattage for any proposed application, but it should be corroborated and/or adjusted with a detailed analysis of the UV dose. Equation (11.16) will not apply when the dose is so low that the shoulder curve dominates or when the inactivation rate is so high that the second stage of the decay curve dominates.

Economic optimization of UVGI systems is possible by assigning a cost to any system in terms of the defining parameters. Since the quantity of most interest is usually the performance in terms of the inactivation rate, we can define a *performance cost* (PC) for any system as the inactivation rate (of some design basis microbe) divided by the annualized cost of the system. This quantity, PC, can be developed for any system parameters taken two at a time to determine the optimum UVGI system for any application. Two of the most critical parameters, UV lamp power P and duct length L, are compared in Fig. 11.24 in terms of the performance cost, as an example of optimization. The system examined here has a constant face area and operates at a design velocity of 500 fpm. This figure suggests there is an optimum for a power of about 27 W and a length of about 367 cm. The implication is that there is an ideal power level that would correspond to any given length, as is also suggested by Eq. (11.16).

11.14 COOLING COIL IRRADIATION

The use of UV to disinfect and clean cooling coils of biocontamination has proven to be one of its most effective and economic applications (Levetin et al., 2001; Shaughnessy et al., 1999). For cooling coils the irradiance on the coil surface need be only a fraction of the average irradiance used in air disinfection applications since the exposure is typically continuous. Figure 11.25 shows the irradiance levels needed on a surface to sterilize it of *Cladosporium* spores. These decay curves were modeled with a resistant population fraction of 0.001 and a second-stage rate constant assumed to be 1/10 that of the first stage. If sterilization is assumed to be six logs of reduction, it can be seen that even at 10 $\mu W/cm^2$ the surface will be sterilized within about 100 hours.

Ideally, both sides of a cooling coil would be irradiated, but this is not always possible due to space constraints. When only the upstream or downstream side of a coil is irradiated,

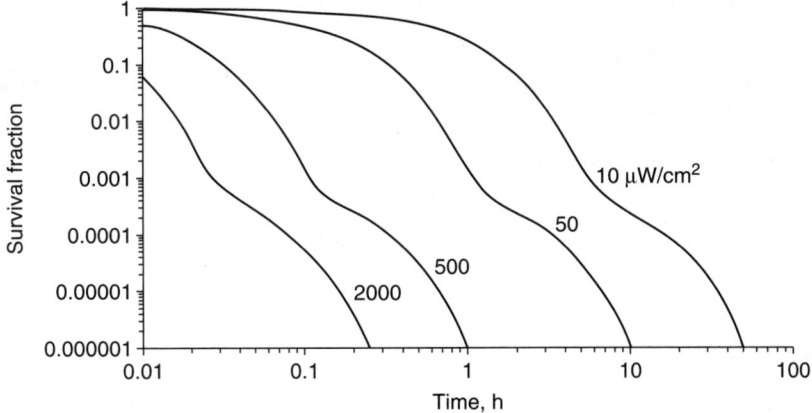

FIGURE 11.25 Time to sterilize cooling coil surfaces of *Cladosporium* spores at different levels of surface irradiance. Numbers indicate the irradiance in $\mu W/cm^2$.

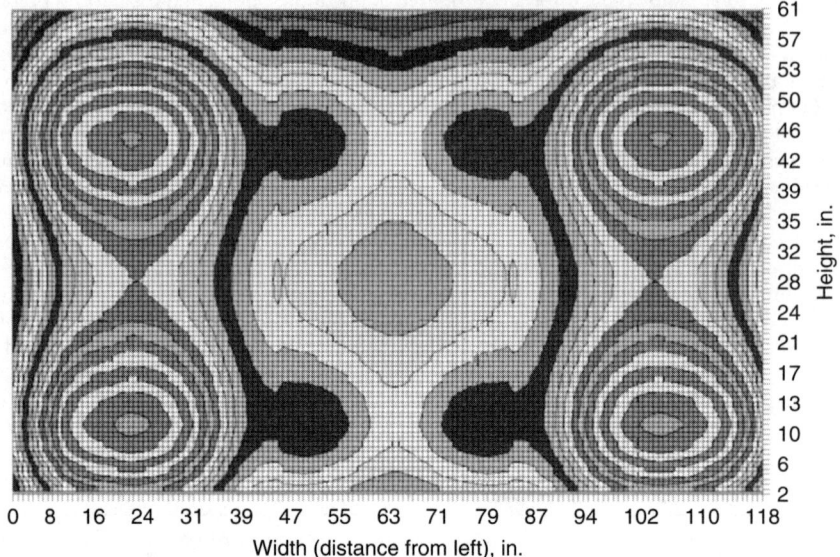

FIGURE 11.26 Irradiance contours on the surface of a cooling coil. Eight lamps of two different wattages have been positioned in two horizontal rows 12 inches from the face of the coil. (*Image generated by the Vmod program, provided courtesy of UVDI, Valencia, CA.*)

UV may not necessarily penetrate cooling coils sufficiently to sterilize the opposite side in the same amount of time. Therefore, it may be prudent to oversize the upstream side to assure the downstream side will also be sterilized. Also, the minimum irradiance on the surface of the coils should be used as a design basis, since the corners may have much lower irradiance levels than the center. This is illustrated in Fig. 11.26, which shows the irradiance contours on the face of a 60 inches by 120 inches cooling coil in which eight lamps with a total of 56 W of UV power has been located 12 inches away from the face of the coil. The minimum irradiance actually occurs along the edge of the coil in the shadow zone of the lamp, and is 186 $\mu W/cm^2$. Using reflective panels on two or more sides of the enclosing duct will raise the minimum or corner irradiance.

In general, a minimum irradiance of approximately 50 $\mu W/cm^2$ on the upstream side should provide eventual sterilization on both sides of the cooling coil. Placement of air disinfection systems in air handling units often requires the lamps to be located around the coiling coils and the coils become irradiated as an unavoidable by-product. An interesting artifact of coil irradiation systems placed on biofouled cooling coils is that as the microbiological contamination is destroyed the reflectivity of the fins tends to increase over time and UV levels on the opposite side of the coils increase accordingly.

11.15 DESIGN GUIDES

A limited number of guidelines, catalog methods, and articles are available for assisting designers with various aspects of implementing various types of UVGI systems (Westinghouse, 1982; Sylvania, 1981; Philips, 1985; Luciano, 1977; Luckiesh and Holladay,

1942; Luckiesh, 1945, 1946; GE, 1950; Buttolph and Haynes, 1950; VanOsdell and Foarde, 2002). Many of the older guidelines are based on a limited set of test results and may not always provide an effective design, or may oversize systems. The results of these various methods should be used with caution, as there is no substitute for a detailed analysis of a UVGI system. New guidelines currently under development may provide a more comprehensive approach to designing effective UVGI systems (IUVA, 2005). At present, the detailed analysis of UVGI systems is only available from a limited number of software programs and published information (UVDI, 2001; IBSI, 2003; Kowalski, 2001). See the appendix in Kowalski (2003) for a table of typical UVGI system operating parameters and inactivation rates.

11.16 UVGI PERFORMANCE IN FIELD TRIALS

Due to mixed opinions on whether UVGI is effective or not at reducing disease incidence, it is worthwhile to review historical successes and failures, and the degree to which UVGI may control disease outbreaks. The first laboratory studies on UVGI disinfection of air in the 1920s showed such promise that the elimination of airborne diseases seemed a possibility. In 1936 Hart used UVGI to sterilize air in a surgical operating room (Hart and Sanger, 1939). In 1937 the first application of UVGI to a school ventilation system dramatically reduced the incidence of measles, with subsequent applications enjoying similar success (Wells, 1955). A number of other epidemiological field trials were run, which focused on upper room UVGI systems. Problems with both the design of UVGI systems and problems with the studies created a mixture of successes and failures. Just as the industry was poised to make major strides a 1954 study on the use of upper room UVGI in London schools was reported to have no effect on reducing diseases (MRC, 1958). In fact, this study did reduce respiratory disease incidence but the researchers reported UVGI as failing in one-half of the cases, opting for a pessimistic interpretation due to statistical significance problems. Although dozens of studies had obtained definitive results, this single study became widely quoted, bringing a halt to expansion in the UVGI air disinfection industry, and leaving misconceptions about UVGI effectiveness that remain even today. A painstakingly thorough experiment by Riley and O'Grady (1961) demonstrated the complete elimination of *tuberculosis* (TB) bacilli from hospital ward exhaust air. Unfortunately, UVGI had already been seemingly discredited by the MRC study and apart from limited applications in the health care industry, UVGI air disinfection remained out of the limelight. It has only been in the past few years that new research in response to increased airborne diseases has begun to again advance this science, but mostly there are only older studies on which to base estimates of disease reduction. Table 11.4 summarizes all those past studies from which useful or conclusive data could be extracted. The net decrease indicates the change in the stated infection rates. The "%" decrease indicates the overall percentage decrease of the infection rates.

The overall results in Table 11.4 would seem to suggest that by and large UVGI has proven effective in reducing airborne diseases. Although two tests reported 0 percent reduction in infection, the review of these results by Wells (1955) suggests extenuating circumstances. In the National Training School for Boys study, the irradiated classes were not isolated from the control group and the students mixed freely outside of the classrooms. In the New York study of three schools, an epidemic of measles swept through the juvenile population, and although the in-classroom infections were lowered by the UVGI systems, the extra-classroom infections equalized the number of cases, leading to no discernable difference other than that the epidemic occurred more slowly in the irradiated classrooms. It is true that many of these studies had similar deficiencies but overall the results suggest

TABLE 11.4 Results of Epidemiological Field Tests of UVGI Systems

Location	Infection	Infection cases		Decrease		Reference
		Before	UVGI	Net	%	
Duke University Hospital	SSI	5%	1%	4%	80%	Kraissl et al., 1940
NE Deaconess Hospital	SSI	15%	6.53%	8.5%	56%	Overholt and Betts, 1940
Infant & Children's Hospital, Boston	SSI	12.5%	2.7%	9.8%	78%	Del Mundo and McKhann, 1941
The Cradle, Evanston	URI	14.5%	4.6%	9.9%	68%	Sauer et al., 1942
St. Luke's Hospital, NY	URI	—	—	—	33%	Higgons and Hyde, 1947
Home for Hebrew Infants, NY	Varicella epidemic	97%	0%	97%	100%	Wells, 1955
National Training School for Boys, DC	URI	—	—	—	0%	Schneiter et al., 1944
Camp Sampson Naval Training Station, NY	URI	—	—	—	20%	Wheeler et al., 1945
Great Lakes Naval Training Station, IL	URI	—	—	—	19%	Miller et al., 1948
Germantown Friends School	Mumps	11	2	—	82%	Wells, 1938
Germantown Friends School	Measles	—	—	—	20%	Wells et al., 1942
Combined results for 4 PA schools	Measles	75%	43%	32%	43%	Wells et al., 1942
Germantown Friends School	Cold viruses	2122	1738	—	18%	Wells et al., 1942
New York state 3-school study	Measles epidemic	—	—	—	0%	Perkins et al., 1947
Mexico and Cato-Meridian schools	Mumps epidemic	235	59	—	75%	Wells, 1955
Port Byron School	Mumps epidemic	49%	45.90%	3.1%	6%	Wells, 1955

TABLE 11.4 Results of Epidemiological Field Tests of UVGI Systems (*Continued*)

Location	Infection	Infection cases		Decrease		Reference
		Before	UVGI	Net	%	
Mexico, Cato-Meridian & Port Byron schools	Chicken pox	38.5%	28.9%	9.6%	25%	Bahlke et al., 1949
Pleasantville & Mt. Kisco	Measles	227	217	—	4%	Wells, 1955
Pleasantville & Mt. Kisco	Chicken pox	297	104	—	65%	Wells, 1955
Southall Elementary schools, England	Measles	12.97%	11.10%	1.9%	14%	MRC, 1958
Southall Elementary schools, England	Mumps	9.97%	3.92%	6.1%	61%	MRC, 1958
Southall Elementary schools, England	Chicken pox	7.49%	6.24%	1.3%	17%	MRC, 1958
Duke University Hospital	Hip arthroplasty infection	5%	0.5%	4.5%	90%	Lowell et al., 1980
Veterans Administration Hospital/ Duke Univ. Hosp.	SSI	1.7%	0.34%	1.4%	80%	Goldner and Allen, 1973
Boston Homeless Shelter	Tuberculosis	—	—	—	78%	Nardell, 1988
Average				15%	45%	

UVGI is potentially effective and it would appear that the widely reported failure of UVGI to control infections was exaggerated.

One of the most ambitious studies involved the attempt to install UVGI systems on a community-wide basis, recognizing that airborne diseases are transmitted not just in schools but in all buildings in a community. Figure 11.27 shows the results of a community-wide study in which numerous upper room UVGI systems were installed throughout buildings in the community to decrease the incidence of respiratory infections in school and at home. Infection in a community with UVGI systems installed, Pleasantville, was compared

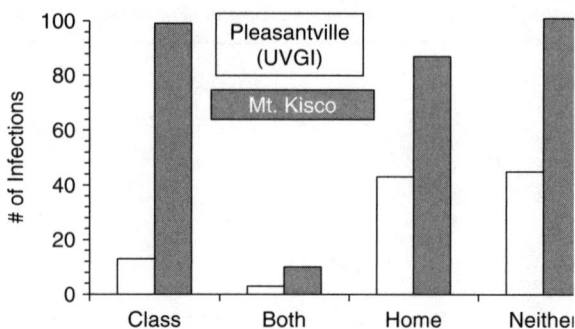

FIGURE 11.27 Results of a study on the use of UVGI to decrease the incidence of respiratory infections in school and at home. Bars show the infections acquired in each of the respective locations. [*Based on data summarized in Wells (1955).*]

to a similar community (Mt. Kisco) without UVGI. Results appeared to be better than would be expected for infections that perhaps transmit more often by direct contact than by air. It is possible that the synergy of reducing airborne microbes reduced the incidence of direct contact, a fact that would render the promising predictions of the epidemiological model in Chap. 4 even more conservative.

It should be noted that most of the UVGI systems listed in Table 11.4 were upper air UVGI systems. The use of in-duct or recirculation UVGI systems could be expected to produce greater effects, but only one recent study, Menzies et al. (2003), is available on the effectiveness of UVGI in forced air ventilation system. In this study respiratory symptoms in commercial office buildings were reduced from 3.9 to 2.9 percent after the installation of an in-duct UVGI system, representing a net decrease of 27 percent in respiratory symptoms. If this value could be taken as an indicator of the reduction of actual respiratory infections, it becomes possible to compute payback periods for the installation of UVGI systems in commercial office buildings by accounting for savings in sick time and lost work due to morbidity.

11.17 COMBINED PERFORMANCE OF UVGI AND FILTRATION

Filtration and UVGI are mutually complementary technologies. Filtration removes most of the microbes that tend to be resistant to UVGI, and vice versa. This is because the larger microbes, like spores, tend to be hardier. Figure 11.28 shows the passage of a select group of microbes through a prefilter, an 80 percent filter, and a UVGI system. Notice how the microbes in the most penetrating particle size range of the filters tend to be susceptible to UVGI. Likewise, UVGI destroys small microbes like viruses that may be difficult to filter out, while filtration easily removes spores, which tend to be resistant to UVGI (Kowalski and Bahnfleth, 2000a).

Not all the microbes in App. A can be used in this example because most UVGI rate constants for microbes remain unknown. It can be reasonably assumed, however, that most viruses will succumb to UVGI exposure and most spores will be taken out by the filters, hence combination systems offer an ideal solution.

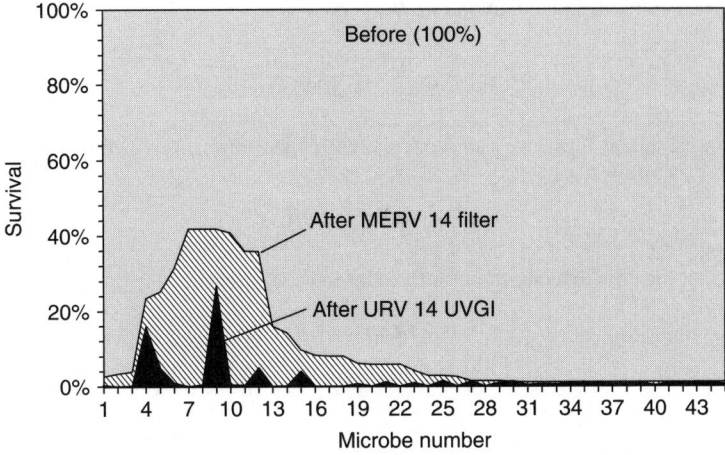

FIGURE 11.28 Microbial populations before and after MERV 14 filter and an URV 14 UVGI system. Only microbes with known UVGI rate constants were included here, ordered in size from smallest (1) to largest (45) diameters.

Such a combined filtration and UVGI system can be "tuned" to target the microbes of concern for any particular facility, and to achieve any desired level of disinfection, up to and including sterilization. In this example, tuning the system to remove all microbes completely could be accomplished by (1) decreasing the airflow rate, (2) increasing UV power, or (3) changing the filter to a more efficient model. Economics would likely dictate any such choice, and basic techniques of economic optimization can be used to seek the perfect solution.

Combining the filter removal rate with the UVGI inactivation rate is a simple algebraic process. The population that penetrates the filter is subject to the inactivation rate of the UVGI system. Figure 11.29 illustrates the process, in which the removal rate or filter efficiency is termed an *inactivation rate* to conveniently identify it as the same type of process as in the UVGI system. The term S_0 defines the initial population, while S_n denotes the survivors after each step.

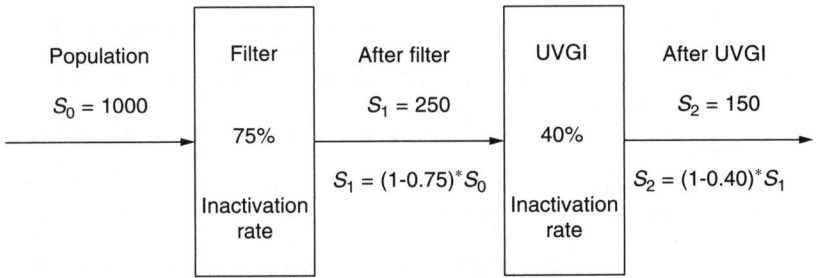

FIGURE 11.29 Schematic for computing total combined inactivation rate when a filter is placed in series with a UVGI system.

The total combined inactivation rate IR_T is then

$$IR_T = \frac{S_0 - S_2}{S_0} = \frac{1000 - 150}{1000} = 85\% \qquad (11.17)$$

In mathematical form, we can write the combined total survival for the example in Fig. 11.17 as follows:

$$S_2 = (1 - IR_1)(1 - IR_2)S_0 \qquad (11.18)$$

If $S_0 = 1$, then the total survival for three systems in series is

$$S_3 = (1 - IR_1)(1 - IR_2)(1 - IR_3) \qquad (11.19)$$

This pattern simply repeats itself for as many systems as there may be, and the total inactivation rate is

$$IR_T = (1 - S_T) \qquad (11.20)$$

REFERENCES

ACGIH (2005). "2005 Threshold limit values and biological exposure indices." American Conference of Governmental Industrial Hygienists. Cincinnati, OH.

ASHRAE (1999). *Handbook of Applications.* American Society of Heating, Refrigerating and Air-Conditioning Engineers, Atlanta, GA.

AWWA (1971). *Water Quality and Treatment.* The American Water Works Association, McGraw-Hill, New York.

Bahlke, A. M., Silverman, H. F., and Ingraham, H. S. (1949). "Effect of ultra-violet irradiation of classrooms on spread of mumps and chicken pox in large rural central schools." *Am J Pub Health* 41:1321–1330.

Beggs, C. B., Kerr, K. G., Donelly, J. K., Sleigh, P. A., Mara, D. D., and Cairns, G. (2000). "An engineering approach to the control of Mycobacterium tuberculosis and other airborne pathogens: a UK hospital based pilot study." *Trans R Soc Trop Med Hyg* 94:141–146.

Beggs, C. B., and Sleigh, P. A. (2002). "A quantitative method for evaluating the germicidal effects of upper room UV fields." *J Aerosol Sci* 33:1681–1699.

Blatchley, E. F. (1997). "Numerical modelling of UV intensity: Application to collimated-beam reactors and continuous-flow systems." *Wat Res* 31(9):2205–2218.

Bolton, J. R. (2001). *Ultraviolet Applications Handbook.* Bolton Photosciences, Ayr, Ontario, Canada.

Boshoff, H. I. M., Reed, M. B., Barry, C. E., and Mizrahi, V. (2003). "DnaE2 polymerase contributes to in vivo survival and the emergence of drug resistance in *Mycobacterium tuberculosis.*" *Cell* 113:183–193.

Buttolph, L. J., and Haynes, H. (1950). "Ultraviolet Air Sanitation." *LD-11.* General Electric, Cleveland, OH.

Cox, C. S. (1987). *The Aerobiological Pathway of Microorganisms.* John Wiley & Sons, New York.

David, H. L. (1973). "Response of mycobacteria to ultraviolet radiation." *Am Rev Resp Dis* 108:1175–1184.

Del Mundo, F., and McKhann, C. F. (1941). "Effect of ultra-violet irradiation of air on incidence of infections in an infant's hospital." *Am J Dis Child* 61:213–225.

Dumyahn, T., and First, M. (1999). "Characterization of ultraviolet upper room air disinfection devices." *Am Ind Hyg Assoc J* 60(2):219–227.

Fletcher, L. A., Noakes, C. J., Beggs, C. B., Sleigh, P. A., and Kerr, K. (2003). "The ultraviolet susceptibility of aerosolised microorganisms and the role of photoreactivation." *Proceedings of the 2d International Congress on Ultraviolet Technologies,* Vienna, Austria.

Fletcher, L. (2004). "The influence of relative humidity on the UV susceptibility of airborne gram negative bacteria." *IUVA News* 6(1):12–19.

GE (1950). "Germicidal lamps and applications." *SMS TAB: VIII-B.* General Electric. Cleveland, OH. 14.

Goldner, J. L., and Allen, B. L. (1973). "Ultraviolet light in orthopedic operating rooms at Duke University." *Clin Ortho* 96:195–205.

Gordon, D. (1978). "Numerical calculations on viscous flows fields through cylindrical arrays." *Computers & Fluids* 6(1):1–13.

Grun, L., and Pitz, N. (1974). "U.V. radiators in humidifying units and air channels of air conditioning systems in hospitals." *Zbl Bakt Hyg* B159:50–60.

Harm, W. (1980). *Biological Effects of Ultraviolet Radiation.* Cambridge University Press, New York.

Harstad, J. B., Decker, H. M., and Wedum, A. G. (1954). "Use of ultraviolet irradiation in a room air conditioner for removal of bacteria." *Am Ind Hyg Assoc J* 2:148–151.

Hart, D., and Sanger, P. W. (1939). "Effect on wound healing of bactericidal ultraviolet radiation from a special unit: Experimental study." *Arch Surg* 38(5):797–815.

Higgons, R. A., and Hyde, G. M. (1947). "Effect of ultra-violet air sterilization upon incidence of respiratory infections in a children's institution." *N Y State J Med* 47(7).

Hill, W. F., Hamblet, F. E., Benton, W. H., and Akin, E. W. (1970). "Ultraviolet devitalization of eight selected enteric viruses in estuarine water." *Appl Microbiol* 19(5):805–812.

IBSI (2003). *UVS: Specular Reflectivity UVGI Analysis Program.* Immune Building Systems, New York.

IESNA (2000). *Lighting Handbook 9th Edition IESNA HB-9-2000.* Illumination Engineering Society of North America, New York.

IUVA (2005). "Guideline for Design and Installation of UVGI In-Duct Air Disinfection Systems." *IUVA-G03A-2005.* International Ultraviolet Association, Ayr, Ontario, Canada. www.iuva.org.

Jacob, S. M., and Dranoff, J. S. (1970). "Light intensity profiles in a perfectly mixed photoreactor." *AIChE J* 16(3):359–363.

Knudson, G. B. (1986). "Photoreactivation of ultraviolet-irradiated, plasmid-bearing, and plasmid-free strains of *Bacillus anthracis.*" *Appl Environ Microbiol* 52(3):444–449.

Ko, G., First, M. W., and Burge, H. A. (2000). "Influence of relative humidity on particle size and UV sensitivity of *Serratia marcescens* and *Mycobacterium tuberculosis.*" *Tuber Lung Dis* 80:217–228.

Koller, L. R. (1952). *Ultraviolet Radiation.* John Wiley & Sons, New York.

Kowalski, W. J., and Bahnfleth, W. P. (2000a). "Effective UVGI system design through improved modeling." *ASHRAE Trans* 106(2):4–15. http://www.engr.psu.edu/ae/wjk/uvmodel.html.

Kowalski, W. J., and Bahnfleth, W. P. (2000b). "UVGI design basics for air and surface disinfection." *HPAC Eng* 72(1):100–110. http://www.engr.psu.edu/ae/wjk/uvhpac.html.

Kowalski, W. J., Bahnfleth, W. P., Witham, D. L., Severin, B. F., and Whittam, T. S. (2000). "Mathematical modeling of UVGI for air disinfection." *Quant Microbiol* 2(3):249–270. http://www.kluweronline.com/issn/1388-3593.

Kowalski, W. J. (2001). Design and optimization of UVGI air disinfection systems. PhD Thesis. The Pennsylvania State University State College. http://etda.libraries.psu.edu/theses/available/etd-0622101-204046/.

Kowalski, W. J., and Dunn, C. E. (2002). "Current trends in UVGI air and surface disinfection." *INvironment Professional* 8(6):4–6.

Kowalski, W. J. (2003). *Immune Building Systems Technology.* McGraw-Hill, New York.

Kowalski, W. J., Bahnfleth, W. P., and Rosenberger, J. L. (2003a). "Dimensional Analysis of UVGI Air Disinfection Systems." *Intl J HVAC&R Res* 9(3):17.

Kowalski, W. J., Bahnfleth, W. P., and Musser, A. (2003b). "Modeling immune building systems for bioterrorism defense." *J Arch Eng* 9(2):86–96.

Kowalski, W. J., and Bahnfleth, W. P. (2004). "Proposed standards and guidelines for UVGI air disinfection." *IUVA News* 6(1):20–25.

Kowalski, W. J., Bahnfleth, W. P., and Mistrick, R. G. (2005). "A specular model for UVGI air disinfection systems." *IUVA News* 7(1):19–26.

Kraissl, C. J., Cimiotti, J. G., and Meleney, F. L. (1940). "Considerations in the use of ultra-violet radiation in operating rooms." *Ann Surg* 111:161–185.

Krasnochub, A. V. (2005). "UV disinfection of air: Some remarks." *IUVA News* 7(2):9–13.

Kuluncsics, Z., Perdiz, D., Brulay, E., Muel, B., and Sage, E. (1999). "Wavelength dependence of ultraviolet-induced DNA damage distribution: Involvement of direct or indirect mechanisms and possible artifacts." *J Photochem Photobiol* 49(1):71–80.

Kundu, P. K. (1990). *Fluid Mechanics*. Academic Press, San Diego, CA.

Lai, K. M., Burge, H., and First, M. W. (2004). "Size and UV germicidal irradiation susceptibility of *Serratia marcescens* when aerosolized from different suspending media." *Appl Environ Microbiol* 70(4):2021–2027.

Levetin, E., Shaughnessy, R., Rogers, C. A., and Scheir, R. (2001). "Effectiveness of germicidal UV radiation for reducing fungal contamination within air-handling units." *Appl Environ Microbiol* 67(8):3712–3715.

Lidwell, O. M., and Lowbury, E. J. (1950). "The survival of bacteria in dust." *Annu Rev Microbiol* 14:38–43.

Linden, K. G., and Darby, J. L. (1994). "Ultraviolet disinfection of wastewater: effect of dose on subsequent reactivation." *Water Res* 28:805–817.

Lowell, J. D., Kundsin, R. B., Schwartz, C. M., and Pozin, D. (1980). "Documentation of airborne infection during surgery." *Airborne Contagion, Ann N Y Acad Sci,* R. B. Kundsin, ed., NYAS, New York. 285–293.

Luciano, J. R. (1977). *Air Contamination Control in Hospitals*. Plenum Press, New York.

Luckiesh, M., and Holladay, L. L. (1942). "Designing installations of germicidal lamps for occupied rooms." *General Electric Review* 45(6):343–349.

Luckiesh, M. (1945). "Disinfection with germicidal lamps: Air—II." *Electr World* 13(Oct.):109–111.

Luckiesh, M. (1946). *Applications of Germicidal, Erythemal and Infrared Energy*. D. Van Nostrand, New York.

Luckiesh, M., Taylor, A. H., Knowles, T., and Leppelmeier, E. T. (1949). "Inactivation of molds by germicidal ultraviolet energy." *J Franklin I* 248(4):311–325.

Masschelein, W. J. (2002). *Ultraviolet Light in Water and Wastewater Sanitation*. Lewis Publishers, Boca Raton, FL.

McKain, T. (2003). Public service of Omaha, Tulsa, OK. *Private communication with W. J. Kowalski* 4/29/03.

Menzies, D., Popa, J., Hanley, J. A., Rand, T., and Milton, D. K. (2003). "Effect of ultraviolet germicidal lights installed in office ventilation systems on workers' health and well-being: double-blind multiple crossover trials." *Lancet* 362(Nov. 29):1785–1791.

Miller, W. R., Jarrett, E. T., Willmon, T. L., Hollaender, A., Brown, E. W., Lewandowski, T., and Stone, R. S. (1948). "Evaluation of ultra-violet radiation and dust control measures in control of respiratory disease at a naval training center." *J Infect Dis* 82:86–100.

Modest, M. F. (1993). *Radiative Heat Transfer*. McGraw-Hill, New York.

MRC (1958). "Air disinfection with ultra-violet irradiation; its effect on illness among school-children by the air hygiene committee." *283* Medical Research Council, Her Majesty's Stationary Office, London.

Munakata, N., and Rupert, C. S. (1974). "Dark Repair of DNA Containing "Spore Photoproduct" in *Bacillus subtilis.*" *Molec Gen Genet* 130:239–250.

Nardell, E. A. (1988). Chapter 12: Ultraviolet air disinfection to control tuberculosis. *Architectural Design and Indoor Microbial Pollution,* R. B. Kundsin, ed., Oxford University Press, New York. 296–308.

NIOSH (1972). "Occupational exposure to ultraviolet radiation." *HSM 73-110009*. National Institute for Occupational Safety and Health, Cincinnati, OH.

Overholt, R. H., and Betts, R. H. (1940). "A comparative report on infection of thoracoplasty wounds." *J Thoracic Surg* 9:520–529.

Peccia, J., Werth, H. M., Miller, S., and Hernandez, M. (2001). "Effects of relative humidity on the ultraviolet induced inactivation of airborne bacteria." *Aerosol Sci Technol* 35:728–740.

Peccia, J., and Hernandez, M. (2001). "Photoreactivation in Airborne *Mycobacterium parafortuitum*." *Appl Environ Microbiol* 67:2001.

Perkins, J. E., Bahlke, A. M., and Silverman, H. F. (1947). "Effect of ultra-violet irradiation of classrooms on the spread of measles in large rural central schools." *Am J Pub Health* 37:529–537.

Philips (1985). *UVGI Catalog and Design Guide*. Catalog No. U.D.C. 628.9, The Netherlands.

Qualls, R. G., and Johnson, J. D. (1983). "Bioassay and dose measurement in UV disinfection." *Appl Microbiol* 45(3):872–877.

Rahn, R. O., and Hosszu, J. L. (1969). "Influence of relative humidity on the photochemistry of DNA films." *Biochim Biophys Acta* 190:126–131.

Rahn, R. O., Xu, P., and Miller, S. L. (1999). "Dosimetry of room-air germicidal (254 nm) radiation using spherical actinometry." *Photochem Photobiol* 70(3):314–318.

Rainbow, A. J., and Mak, S. (1973). "DNA damage and biological function of human adenovirus after U.V. irradiation." *Int J Radiat Biol* 24(1):59–72.

Rauth, A. M. (1965). "The physical state of viral nucleic acid and the sensitivity of viruses to ultraviolet light." *Biophys J* 5:257–273.

Rentschler, H. C., Nagy, R., and Mouromseff, G. (1941). "Bactericidal effect of ultraviolet radiation." *J Bacteriol* 42:745–774.

Rentschler, H. C., and Nagy, R. (1942). "Bactericidal action of ultraviolet radiation on air-borne microorganisms." *J Bacteriol* 44:85–94.

Richards, M. (1954). "Atmospheric mold spores in and out of doors." *J Allergy* 25:429–439.

Riley, R. L., and O'Grady, F. (1961). *Airborne Infection*. The Macmillan Company, New York.

Riley, R. L., and Kaufman, J. E. (1972). "Effect of relative humidity on the inactivation of airborne *Serratia marcescens* by ultraviolet radiation." *App Microbiol* 23(6):1113–1120.

Riley, R. L., Knight, M., and Middlebrook, G. (1976). "Ultraviolet susceptibility of BCG and virulent tubercle bacilli." *Am Rev Resp Dis* 113:413–418.

Sauer, L. W., Minsk, L. D., and Rosenstern, I. (1942). "Control of cross infections of respiratory tract in nursery for young infants." *JAMA* 118:1271–1274.

Scheir, R., and Fencl, F. B. (1996). "Using UVC Technology to Enhance IAQ." *HPAC Eng* 68(2):28.

Schneiter, R., Hollaender, A., Caminita, B. H., Kolb, R. W., Fraser, H. F., duBuy, H. G., Neal, P. A., and Rosenblum, H. G. (1944). "Effectiveness of ultra-violet irradiation of upper air for the control of bacterial air contamination in sleeping quarters." *Am J Hyg* 40:136.

Setlow, J. K. (1966). "Photoreactivation." *Radiat Res Suppl* 6:141–155.

Severin, B. F., Suidan, M. T., and Englebrecht, R. S. (1984). "Mixing effects in UV disinfection." *J Water Pollut Control Fed* 56(7):881–888.

Sharp, G. (1939). "The lethal action of short ultraviolet rays on several common pathogenic bacteria." *J Bacteriol* 37:447–459.

Shaughnessy, R., Levetin, E., and Rogers, C. (1999). "The effects of UV-C on biological contamination of AHUs in a commercial office building: Preliminary results." *Indoor Environ '99*:195–202.

Sylvania (1981). "Sylvania Engineering Bulletin 0-342, Germicidal and Short-Wave Ultraviolet Radiation." GTE Products Corp, Stamford, CT.

UVDI (2001). *UVD: Ultraviolet Air Disinfection Design Program*. Ultraviolet Devices, Valencia, CA.

VanOsdell, D., and Foarde, K. (2002). "Defining the effectiveness of UV lamps installed in circulating air ductwork." *ARTI-21CR/610-40030-01* ARTI. Arlington, VA.

Webb, R. B., and Tuveson, R. W. (1982). "Differential sensitivity to inactivation of NUR and NUR+ strains of *Escherichia coli* at six selected wavelengths in the UVA, UVB and UVC ranges." *Photochem Photobiol* 36:525–530.

Wells, W. F. (1938). "Air-borne infections." *Mod Hosp* 51:66–69.

Wells, W. F., Wells, M. W., and Wilder, T. S. (1942). "The environmental control of epidemic contagion; I—An epidemiologic study of radiant disinfection of air in day schools." *Am J Hyg* 35:97–121.

Wells, W. F. (1955). *Airborne Contagion and Air Hygiene*. Harvard University Press, Cambridge, MA.

Westinghouse (1982). *Booklet A-8968, Westinghouse Lighting Handbook*. Westinghouse Electric Corp., Lamp Div., Fairfield, CT.

Wheeler, S. M., Ingraham, H. S., Hollaender, A., Lill, N. D., Gershon-Cohen, J., and Brown, E. W. (1945). "Ultra-violet light control of airborne infections in a naval training center." *Am J Pub Health* 35:457–468.

Xu, P., Peccia, J., Fabian, P., Martyny, J. W., Fennelly, K. P., Hernandez, M., and Miller, S. L. (2003). "Efficacy of ultraviolet germicidal irradiation of upper-room air in inactivating airborne bacterial spores and mycobacteria in full-scale studies." *Atmos Environ* 37:405–419.

CHAPTER 12
GAS PHASE FILTRATION

12.1 INTRODUCTION

The removal of gases or vapors from an airstream is known as gas phase filtration. Carbon adsorption is the most common type of gas phase filtration. Carbon adsorption systems involve the use of activated carbon or activated carbon impregnated with various compounds. Carbon is impregnated with certain compounds to enhance the adsorption of gases and vapors, or to catalytically destroy them. Although forms of carbon adsorption have been successfully used for decades, a number of gas phase removal systems have recently been developed that provide new options. One of these new technologies, *photocatalytic oxidation* (PCO), is dealt with in Chap. 14 since it is not merely a gas phase filter but also a particle filter and an air disinfection system. Another technology with potential for destroying gases is plasma destruction, but this technology is developmental and is treated in Chap. 22. This chapter deals primarily with passive gas adsorption systems like activated carbon that are used to remove volatile organic compounds created by microbial contamination.

The units used to measure gaseous concentration are typically variations of ppm or mg/m^3 related to air volumes. The most common units are defined as follows:

ppmv   parts per million of contaminant by volume; units of contaminant per million units of air

ppbv   parts per billion of contaminant by volume; units of contaminant per billion parts of air

mg/m^3 milligrams of contaminant per cubic meter of air

$\mu g/m^3$ micrograms of contaminant per cubic meter of air

Based on the ideal gas law, the conversions between ppmv and mg/m^3 are as follows:

$$\text{ppmv} = \frac{8.314(\text{mg/m}^3)(273.15+t)}{Mp} \quad (12.1)$$

$$\text{mg/m}^3 = \frac{\text{ppmv} \times Mp}{8.314(273.15+t)} \quad (12.2)$$

where 8.314 = gas constant, J/(k·gmol)
M = gram molecular mass of contaminant, g/gmol
p = mixture absolute pressure, kPa
t = mixture temperature, °C

12.2 AEROBIOLOGICAL APPLICATIONS OF GAS PHASE FILTRATION

Perhaps the most common application of carbon adsorbers in the HVAC industry is to control odors and *volatile organic compounds* (VOCs). A large body of information is available on removal rates of various pollutants by activated carbon but most of these pollutants are not microbiological in origin. The VOCs of concern in aerobiological engineering are *microbial volatile organic compounds* (MVOCs). It is estimated that there are hundreds or thousands of MVOCs but only a few dozen have been studied in detail. Appendix C provides a list of MVOCs that have been associated with indoor microbial contamination. The references for App. C are provided at the end of this chapter. Many of these MVOCs have been proposed as indicator compounds that can be used to identify the presence of microbial contamination in buildings; however, some of these MVOCs may be generated by other sources such as humans or wet building materials.

Both bacteria and fungi may produce MVOCs. Some of the fungi often cited as producers of MVOCs include *Aspergillus, Chaetomium, Fusarium, Paecilomyces, Penicillium*, and *Trichoderma* (Macher, 1999; Gao et al., 2002; Fischer et al., 1999; Sandström, 2003). Viruses do not produce MVOCs since they are intracellular parasites. Figure 12.1 illustrates the primary biological sources of MVOCs in indoor environments. Moist building materials may produce MVOCs but this is not specifically an aerobiological problem, although it may be a cause of microbial growth.

MVOCs may be cytotoxic, mutagenic, respiratory irritants, or just annoying odors (Fischer and Dott, 2003; Kreja and Seidel, 2002). Due to limited study, little information is available on the threshold limits of most MVOCs and less is known about their ability to be removed by gas phase filtration. Table 12.1 summarizes some of the MVOCs for which limits have been established, and for which the retentivity in carbon adsorbers is known (ASHRAE, 1999). The TWA8 represents the time-weighted average that is not to be exceeded in any 8-hour shift of a 40-hour week. The retentivity represents the amount of the agent that can be removed from the air by activated carbon, based on an inlet concentration of 1000 ppm.

Table 12.2 shows some typical levels of several MVOCs that were measured in schools and buildings suspected of having mold contamination problems (AQS, 2003). The average indoor level of total MVOCs was 19 µg/m³.

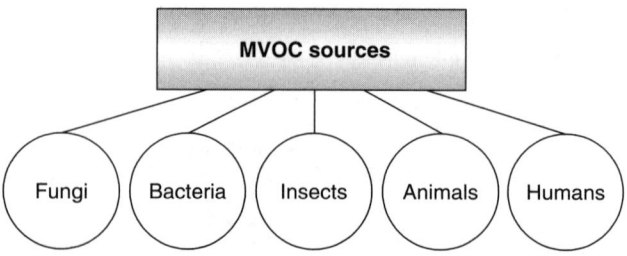

FIGURE 12.1 Breakdown of common MVOCs sources.

TABLE 12.1 Characteristics of Select MVOCs

MVOC	Type	TWA8, mg/m^3	Odor threshold, mg/m^3	Retentivity, %
Acetyldehyde	Aldehyde	360	1.2	8
Acetone	Ketone	2400	47	16
2-Butanone (MEK)	Ketone	590	30	12
Formaldehyde	Aldehyde	4	1.2	0.4

12.3 TYPES OF GAS PHASE FILTERS

The most common gas phase filtration technologies are illustrated in Fig. 12.2. They are subdivided into carbon adsorber-type equipment and noncarbon equipment. This is not necessarily an inclusive breakdown since a number of technologies are currently under research and development that have potential for removing gaseous contaminants, but limited or insufficient information is available on these (see Chaps. 13 and 22).

The several types of gas phase air filters available today include carbon adsorbers or *granulated activated carbon* (GAC), activated alumina impregnated with potassium permanganate, activated silica, and carbon impregnated with other compounds to improve performance against specific contaminants (VanOsdell and Sparks, 1995; ASHRAE, 1999). When carbon filters are impregnated with chemical compounds the process is referred to as chemisorption. Table 12.3 summarizes the types of adsorption and chemisorption media available today. Polar organic compounds include alcohols, phenols, and amines.

There are also fibrous filters impregnated with carbon or related compounds to give them the ability to remove gaseous contaminants. There are, in addition, types of carbon adsorbers that utilize electrical currents for both destruction of adsorbed compounds, and for regeneration purposes (Pathogenus, 2002). Some new types of adsorber materials have been developed recently, like fullerenes and heterofullerenes, microporous glasses, and nanoporous materials, but insufficient data are available to quantify their use or performance (Dabrowski, 2001).

TABLE 12.2 Typical Levels of MVOCs in Buildings

MVOC	Concentration, μg/m^3
1-Butanol	17
2-Beptanone	1.9
2-Hexanone	1.2
2-Methyl-1-propanol (Isobutanol)	20
2-Methyl-2-butanol	1.4
2-Methyl-isoborneol	0.49
2-Pentylfuran	1.2
3-Methylfuran	1.4
Terpineol	2.8

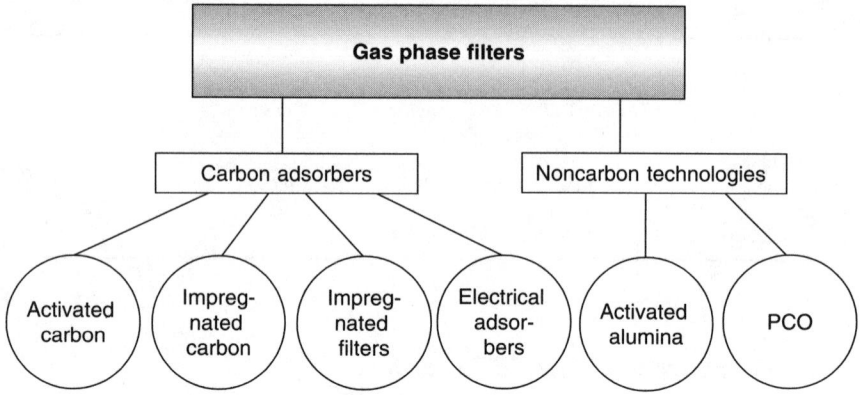

FIGURE 12.2 Breakdown of some of the more common gas phase filtration technologies.

12.4 CARBON ADSORPTION

Carbon adsorption is effective against VOCs and other airborne pollutants but is not used for control of airborne microorganisms. Where dust may be present, filtration is invariably needed upstream of the carbon adsorber to prevent clogging of the adsorbent bed. Activated carbon forms the single main component of most gas phase filters. Activated carbon possesses an enormous amount of surface area per unit mass. This huge surface area increases the probability of a gas molecule adhering to the surface through natural attractive forces (Strauss, 1975; Cheremisinoff and Ellerbusch, 1978; Walker and Thrower, 1975). The molecular attractive forces that attract and hold gas molecules to the surface of solid carbon are primarily van der Waals forces (Kinoshita, 1988). The van der Waals force may be vanishingly small, but when the total surface area available for adsorption is large, high removal rates are possible. The biggest problem with carbon adsorber design and applications is that with large surface areas the pressure loss through the adsorber unit can be high. One way to compensate for excessive pressure losses is to expanding the total face area, but this can require considerable space.

TABLE 12.3 Types of Gas Phase Filter Media

Material	Type	Impregnant	Contaminant captured
Activated carbon	Adsorption	None	Organic vapors
Activated alumina	Adsorption	None	Polar organic compounds
Silica gel	Adsorption	None	Polar organic compounds
Molecular sieves (Zeolites)	Adsorption	None	CO_2, iodine
Activated alumina	Chemisorption	$KMnO_4$	H_2S, SO_2
Activated carbon	Chemisorption	I_2, Ag, S	Mercury vapor
Activated carbon	Chemisorption	KI_3, amines	Iodine
Activated carbon	Chemisorption	$NaHCO_3$	NO_2
$NaOH + Ca(OH)_2$	Chemisorption	None	Acid gases
Activated alumina	Catalysis	Metal salts	Ozone

The bed depths of commercial carbon filters range from 1 to 8 cm with nominal residence times of 0.025 to 0.1 second. Carbon adsorbers are normally tested with contaminants with concentrations from 400 to 4000 mg/m^3. Such levels are comparatively high for normal indoor contaminants.

The adsorption of a gaseous agent onto carbon is a complex function of pore size, gas molecule size, relative humidity, temperature, and residence time. Various analytical and computer models may be available for sizing carbon adsorbers (Lordgooei et al., 2001; Cheremisinoff and Ellerbusch, 1978; Lodewyckx and Vansant, 2000; Balieu and Bjarnov, 1990; Grubner and Burgess, 1979), but catalog sizing methods are commonly used to select units for specific applications.

The basic procedure for sizing a carbon adsorber is to first identify the gases to be removed and then define the operating parameters. Air velocities will typically be predefined by the system. The exact type of carbon adsorber to be selected, whether impregnated or not and what type of impregnation, is generally dependent on the gases to be removed. Table 12.4 lists some common indoor odor-causing agents and the relative retentive capacity of activated carbon or impregnated carbon for these agents.

The time a carbon adsorber can operate before it becomes overloaded and has to be regenerated or replaced is defined by the breakthrough time. No general computing method can be provided for the efficiency of various types of carbon adsorbers, but the basic procedures for determination of breakthrough times for standard GAC systems are treated in the *ASHRAE Applications Handbook* (ASHRAE, 1999). A simplified method is to use the semiempirical Wheeler equation (Underhill, 2001):

$$t_b = \frac{v_e}{C_{in}Q}\left\{M_c - \left[\frac{\rho_c Q}{k_v}\ln\left(\frac{C_{in}}{C_{out}}\right)\right]\right\} \quad (12.3)$$

where t_b = breakthrough time, min
v_e = equilibrium mass adsorbed, g per g of activated carbon
C_{in} = inlet concentration, g/cm^3
C_{out} = outlet concentration, g/cm^3
M_c = mass of carbon in bed, g
ρ_c = bulk density of carbon, g/cm^3 (typically about 0.3 g/cm^3)
k_v = adsorption rate constant, min^{-1}

TABLE 12.4 Retentive Capacity for Odorous Gases and MVOCs (ASHRAE 1999)

Odorous gas or vapor	Retentive capacity of carbon	Media type
Acetone	Adsorbs 10–25% of agent	Activated carbon
Ammonia	Relatively low adsorption capacity	Activated carbon
Animal odors	Adsorbs 10–25% of agent	Activated carbon
Decaying substances	Adsorbs 20–40% of agent	Activated carbon
Dimethyl sulfide	Adsorbs 20–40% of agent	Activated carbon
Formaldehyde	Relatively low adsorption capacity	Impregnated carbon
Methane	Negligible capacity for removal	Activated carbon
Mildew odor	Adsorbs 10–25% of agent	Activated carbon
Paper deteriorations	Adsorbs 20–40% of agent	Activated carbon
Pet odors	Adsorbs 20–40% of agent	Activated carbon

The adsorption rate constant is determined by

$$k_v = 14.41 d_p^{1.5} \sqrt{U} \qquad (12.4)$$

where U is the inlet air velocity, cm/min, and d_p is the carbon granule diameter, cm (typically about 0.1 cm).

The adsorber's capacity will decrease over time from adsorbing pollutants in the air, after which breakthrough will occur and the penetration rate will increase. Figure 12.3 shows an example of breakthrough times for a carbon adsorber challenged with different concentrations of toluene (Van Osdell and Sparks, 1995). Air temperature was 25°C (77°F) with 50 percent relative humidity. The bed depth of the carbon was 1 inches and had a face velocity of 0.23 m/s (45 fpm). The breakthrough time for the higher concentration was about 11 hours, while the breakthrough time for the lower concentration was about 750 hours.

Carbon adsorbers are normally regenerated by passing a stream of hot air through them to remove moisture or destroy any adsorbed contaminants. The adsorber material is typically replaced periodically depending on the amount of use. Unfortunately, there is no simple way to determine how much gas has been adsorbed by the carbon without testing a sample, which is the primary method for determining adsorber performance (VanOsdell, 1994). One rule of thumb is to replace 4.5 lb of carbon for each 1000 ft^3 of space served per year, but this may be a conservative estimate (Underhill, 2001).

Carbon adsorbers are usually rated in terms of the minimum activity, which expresses the degree of pore activation. The most common type of activated carbon used is 60 percent minimum activity level, which has an internal surface area of approximately 1000 to 1100 m^2/g. This form of carbon is either 6/8 mesh (3 mm) pellets or 6/12 mesh in granular form. Metal frames or screens are typically filled with these pellets or granules and either inserted or permanently fixed inside ductwork. Some adsorbers are available in which the pellets are processed into a solid medium, such as that shown in Fig. 12.4.

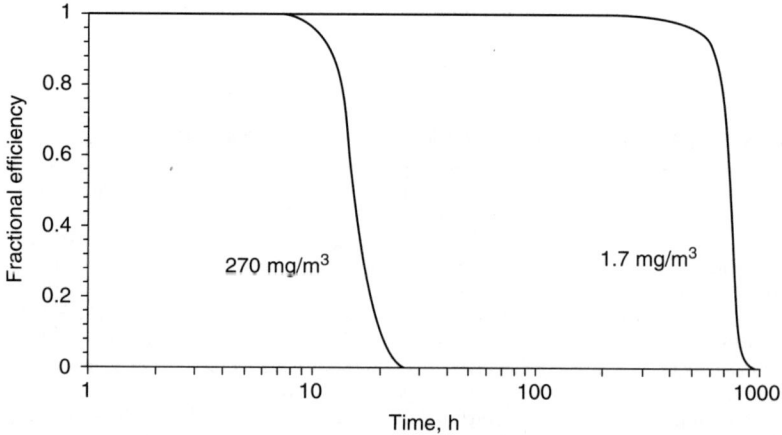

FIGURE 12.3 Efficiency of carbon adsorber showing breakthrough times for different concentrations of toluene. [*Based on data from VanOsdell and Sparks (1995).*]

GAS PHASE FILTRATION

FIGURE 12.4 Cutaway model showing carbon media located behind filter media. (*Model provided courtesy of D-Mark, Inc., Chesterfield, MI.*)

Carbon adsorbers usually last about 1 year before needing replacement or regeneration, depending on the level of service. As a carbon adsorber loads up with contaminants, its efficiency will decrease. Figure 12.5 shows a test of a packed bed carbon adsorber that was continuously challenged with m-xylene at a concentration of 10 ppm (436 µg/m^3). Although not an MVOC, m-xylene is a test gas that provides an example of VOC removal rates.

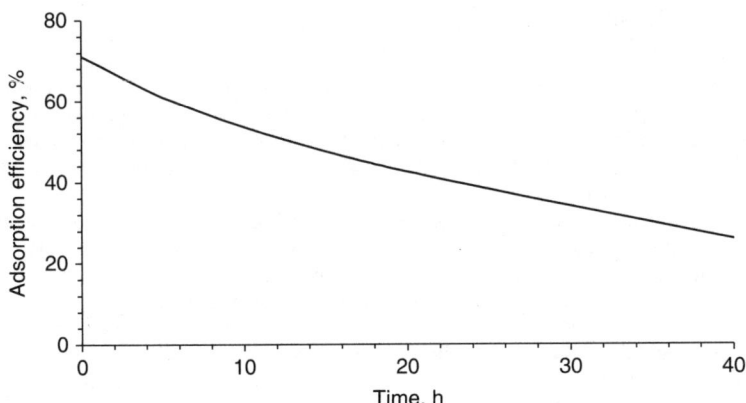

FIGURE 12.5 Adsorption efficiency as a function of time for a packed bed carbon filter with a challenge of 10 ppm m-xylene. (*Holmberg et al., 1993*)

The data shown in Figure 12.5 can be effectively modeled as exponential decay, using the following relation:

$$E = Ce^{-kIt} \qquad (12.5)$$

where E = efficiency, %
C = a constant
k = rate constant for process
I = intensity in terms of concentration, ppm
t = time, hour

A curve fit of the data in Fig. 12.5 yields the following:

$$E = 71e^{-0.0253t} \qquad (12.6)$$

Dividing the exponent by the concentration produces a normalized or standard rate constant as follows:

$$k_s = \frac{0.0253}{10} = 0.00253 \qquad (12.7)$$

where k_s is the standard rate constant, ppm/h.

Using Eq. (12.5) with the standardized rate constant, the life of the carbon adsorber can be estimated for normal levels of indoor MVOCs, as shown in Table 12.2. Assuming normal levels were approximately 19 μg/m³, which converts by Eq. (12.1) to 0.436 ppm (molecular weight of m-xylene = 106.17), we rewrite Eq. (12.5) as follows:

$$E = 71e^{-0.00253(0.436)t} \qquad (12.8)$$

Assuming that 0.1 percent efficiency represents the end of the carbon adsorber's life cycle, we can invert Eq. (12.8) and solve for t, which proves to be 5952 hours, or about 8 months. Carbon adsorbers that see light duty have been reported to last for several years (Weschler et al., 1994).

12.5 IMPREGNATED CARBON

Impregnated carbon adsorbers are used to remove low molecular weight compounds like ammonia, hydrogen sulfide, and formaldehyde (Underhill, 2001). These compounds do not adhere well to ordinary activated carbon. Activated carbon may be impregnated with a variety of compounds that include chemical compounds that react with pollutants or catalytic agents that break down pollutants.

When chemical compounds are used for impregnation the process is called chemisorption. The chemical reacts with the pollutants and is consumed in the process. The chemical impregnant may ultimately have to be replaced or regenerated. The carbon serves as a substrate in such applications and provides the immense surface area necessary to create contact between the gaseous pollutant and the chemical impregnant. The contact time between the chemical and the pollutant must be sufficiently long for the desired reaction to occur. Typically the contact time will be between 0.01 and 0.4 seconds (Underhill, 2001).

Chemically impregnated carbon adsorbers are available for the specific removal of ammonia, formaldehyde, nitrogen dioxide, sulfur dioxide, and other gases. Activated carbon is not always essential as the substrate. Chemical impregnation of alumina with potassium permanganate can be effective against sulfur dioxide, formaldehyde, toluene, and other VOCs. Carbon adsorbers have some degree of effectiveness in removing ozone (Takeuchi and Itoh, 1993; Deitz and Bitner, 1972, 1973; Rakitskaya et al., 1996). However, catalytic removal of ozone with metal salt impregnated alumina is far more effective than activated carbon (Lee and Davidson, 1999; Kowalski et al., 2003). The use of metallic salts as catalysts to remove ozone has been shown to be highly effective (Singh et al., 1997). Carulite 200, which is an activated carbon-like substrate impregnated with magnesium dioxide, copper oxide, and aluminum oxide, has a density of 0.93 g/cm^3 and a surface area of at least 175 m^2/g (Carus, 2000). The catalyst has no upper limit of ozone concentration but in order to achieve a high rate of ozone removal it has an optimal residence time of 0.36 seconds and an optimal face velocity of 0.671 m/s (2.2 fps) in dry air. Under these conditions, performance is stated to be predictable and independent of relative humidity.

12.6 EFFECT ON MICROBIOLOGICAL PARTICLES

Carbon adsorption has some reported ability to remove botulinum toxin (Gomez, 1995). It can also be processed to manipulate the pore size and remove large molecules (Tamai et al., 1996). Carbon adsorbers might have some effect on small viruses near 0.01 μm in diameter but the activated carbon pore size is typically 0.001 μm, or about 10 times too small to accommodate the smallest virus. Since diffusion is the primary mechanism by which viruses are removed (see Chap. 10) it may be possible to use activated carbon with a pore size on the order of 0.05 μm, to remove airborne viruses, but this possibility has not been studied.

12.7 OPERATION AND MAINTENANCE

All adsorption media are subject to accumulations of dust, which will increase the pressure losses and degrade the capacity for gaseous filtration. A particulate dust filter of approximately MERV 6–8 (DSP 25 to 30 percent) should be installed upstream of any carbon adsorber.

Higher relative humidity in the airstream lowers the gaseous filtration efficiency due to adsorption of water molecules. In most cases, performance of carbon adsorbers is stable up to about 40 to 50 percent RH, but may degrade above this. The factors that affect the efficiency of gas phase filters under variations in relative humidity are complex and dependent on the type of media and the loading. Manufacturers should be consulted regarding the limits of relative humidity under which their product will operate satisfactorily.

Media will typically be changed out according to some manufacturer's schedule. However, loading or deterioration of the media may require changing the filters prior to the end of a change-out cycle. Occupant complaints may be one indicator that the useful life of the adsorber has been reached. Testing is another option. Samples of the media may be regularly tested or special coupons may be inserted alongside the media to monitor performance. Some systems may allow carbon regeneration in place, but most do not due to the high regeneration temperatures. In one type of carbon adsorber, an electrical charge is passed through the carbon to regenerate it (Pathogenus, 2002). In the latter system, the carbon is packaged as solid media and during regeneration the airflow is reversed as it is charged.

FIGURE 12.6 Modular carbon adsorber unit with metal framing. Round aluminum sections are filled with activated carbon and perforated to allow air passage. Opposite (exhaust) side has an integral air filter. (*Unit provided courtesy of Donaldson Company, Inc., Minneapolis, MN.*)

Cartridge-type carbon adsorbers, like that shown in Figure 12.6, are fairly easy to change out, being handled similarly to cartridge filters. Packed bed media can be somewhat more difficult to empty and fill due to spilling and aerosolization of carbon particles.

12.8 APPLICATIONS

Carbon adsorbers are used in a variety of applications to control odors and gaseous or vaporous contaminants, including hospitals, laboratories, restaurants, animal facilities, libraries, airports, commercial buildings, and respiratory protective equipment (Dabrowski, 2001; Brown, 1995; D-Mark, 1995). The type and size of a carbon adsorber often depends on the nature of the contaminant, as discussed previously, but simple activated carbon or activated alumina are the most common types used. One manufacturer recommends using 30 to 45 lb of activated carbon per 1000 cfm of air (D-Mark, 1995). Panel-type carbon filters contain approximately 1.75 to 5 lb of carbon per square foot. Woven fabric carbon filters may have 100 to 600 g (0.22 to 1.32 lb) of carbon per square foot.

Carbon adsorbers are typically designed to operate at 300 to 500 fpm. The typical pressure drop of a carbon adsorber at 500 fpm is about 0.3 inches of water but increases with bed depth. Figure 12.7 shows some typical values for pressure drop versus bed depth

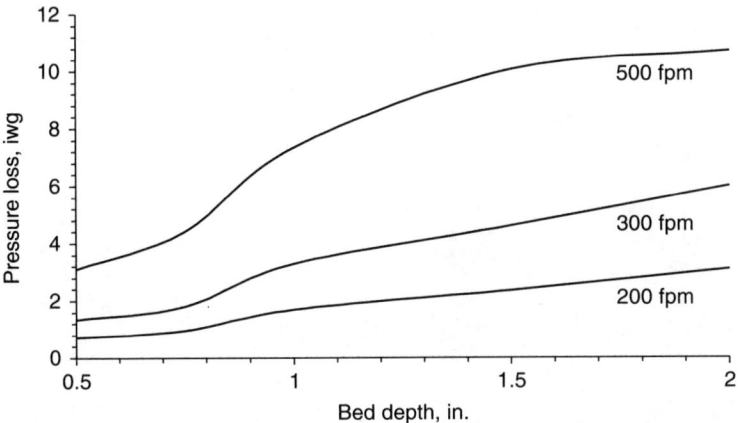

FIGURE 12.7 Pressure losses vs. bed depth for activated carbon filters. [*Based on data from D-Mark (1995).*]

of activated carbon based on manufacturer's catalog information (D-Mark, 1995). Obviously, pressure losses from a deep bed carbon filter can be high, and it may be best to expand the face area in any system that is deigned for activated carbon so as to lower the energy costs. Thinner panel filters will have much lower pressure losses but their ability to remove contaminants will also be reduced.

Carbon adsorbers that operate in recirculation will purge contaminants from a room in the same way as filters remove particulates, for relatively constant adsorption efficiency. Figure 12.8 shows the results of a carbon adsorber used to remove formaldehyde from a room with a volume of 868 ft^3 (Shaughnessy et al., 1993). The adsorber contained 13 lb of

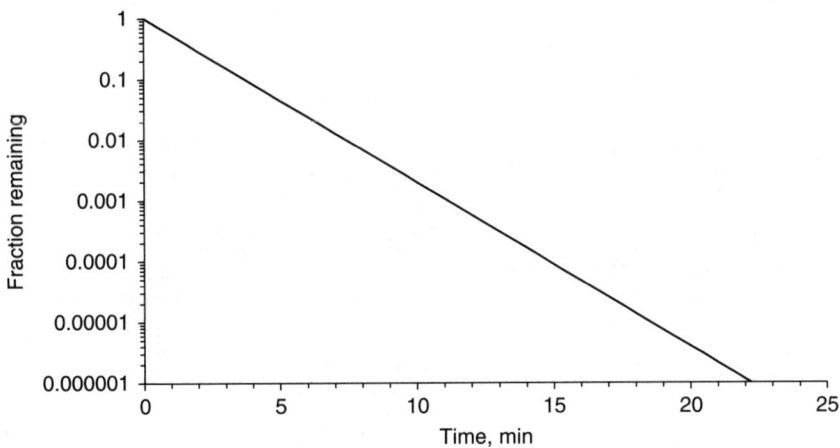

FIGURE 12.8 Removal of formaldehyde from a room by a recirculating carbon adsorber. [*Based on data from Shaughnessy et al. (1993).*]

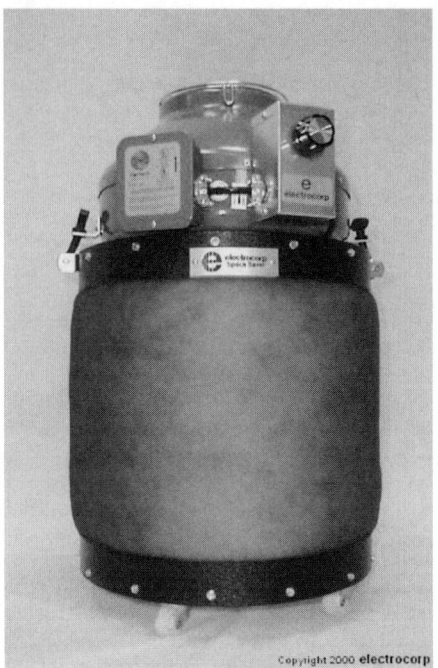

FIGURE 12.9 A carbon adsorber recirculation unit. (*Image courtesy of Electrocorp, California.*)

activated carbon and operated at a flowrate of 180 cfm for an air change rate of over 12 per hour. The carbon adsorber efficiency was reported as 30 percent and the initial formaldehyde concentration was 4 ppm.

Stand-alone recirculation units with carbon adsorbers generally include filters and serve a dual purpose for cleaning air of particulates and gaseous contaminants. Figure 12.9 shows a portable recirculating carbon adsorber unit with integral filters.

Table 12.5 provides some estimated usage rates of activated carbon for office buildings, based on toluene and 4000 operating hours (Liu 1993a, 1993b). Usage was specified in terms of a unit floor area of 100 m^2 and was converted to square feet and extrapolated for lower MVOC values (see Table 12.2). This chart is based on generation rates for VOCs, and not specifically for MVOCs, but should be conservative for MVOC applications.

Finally, it should be noted that if there are high MVOC levels in any building, the problem may actually be mold contamination or possibly moisture problems in building materials. In either case, the use of carbon may only be an expedient method of reducing levels that will not, by itself, solve the underlying problem. For information on remediating mold problems, see Chap. 26.

TABLE 12.5 Retentive Capacity for Odorous Gases and MVOCs

MVOC generation rate, mg/h rate, kg/year per 1000 ft^2	Carbon usage rate, kg/year per 1000 ft^2		
	85% GAC efficiency	50% GAC efficiency	10% GAC efficiency
17	1.1	1.0	0.6
33	1.4	1.2	0.9
134	4.4	4.1	3.1
268	8	7.5	—
535	15.1	14	—

REFERENCES

AQS (2003). *Detecting Mold, MVOCs in Problem Buildings.*" Air Quality Sciences, Marietta, GA. http://www.snipsmag.com/CDA/ArticleInformation/features/BNP__Features__Item/0,3374,69666.

ASHRAE (1999). *Handbook of Applications.* American Society of Heating, Refrigerating and Air-Conditioning Engineers, Atlanta, GA.

Balieu, E., and Bjarnov, E. (1990). "Activated carbon filters in air cleaning processes—II. Prediction of breakthrough times and capacities from laboratory studies of model filters." *Ann Occup Hyg* 34(1):1–11.

Brown, R. C. (1995). "Review: Activated carbon filters in respiratory protective equipment." *Int J Occup Saf Ergon* 1(4):330–373.

Carus (2000). "Carulite 200 Catalyst Material Safety Data Sheet (MSDS)." Carus Chemical Corporation, LaSalle, IL.

Cheremisinoff, P. N., and Ellerbusch, F. (1978). *Carbon Adsorption Handbook.* Ann Arbor Science, Ann Arbor, MI.

D-Mark, I. (1995). "Concept of molecular screening in micropores." D-Mark, Chesterfield, MI.

Dabrowski, A. (2001). " Adsorption—from theory to practice." *Adv Colloid Interface Sci* 93(1–3):135–224.

Deitz, V. R., and Bitner, J. L. (1972). "The reaction of ozone with adsorbent charcoal." *Carbon* 10:145–154.

Deitz, V. R., and Bitner, J. L. (1973). "Interaction of ozone with adsorbent charcoals." *Carbon* 11:393–401.

Fischer, G., Schwalbe, R., Moeller, M., Ostrowski, R., and Dott, W. (1999). "Species-specific production of microbial volatile organic compounds (MVOC) by airborne fungi from a compost facility." *Chemosphere* 39(5):796–810.

Fischer, G., and Dott, W. (2003). "Relevance of airborne fungi and their secondary metabolites for environmental, occupational and indoor hygiene." *Arch Microbiol* 179(2):75–82.

Gao, P., Korley, F., Martin, J., and Chen, B. T. (2002). "Determination of unique microbial volatile organic compounds produced by five Aspergillus species commonly found in problem buildings." *AIHA J* 63:135–140.

Gomez, A. F. (1995). "Adsorption of Botulinum Toxin to activated charcoal with a mouse bioassay." *Ann Emerg Med* 25:818.

Grubner, O., and Burgess, W. A. (1979). "Simplified description of adsorption breakthrough curves in air cleaning and sampling devices." *Am Ind Hyg Assoc J* 40(3):169–179.

Holmberg, R., Torkelsson, S., and Strindehag, O. (1993). "Suitability of activated carbon for air handling units.", *Proceedings of the 6th International Conference on Indoor Air Quality and Climate. Indoor Air '93,* Helsinki, Finland.

Kinoshita, K. (1988). *Carbon: Electrochemical and Physical Properties.* John Wiley & Sons, New York.

Kowalski, W. J., Bahnfleth, W. P., Striebig, B. A., and Whittam, T. S. (2003). "Demonstration of a hermetic airborne ozone disinfection system: Studies on E. coli." *AIHA J* 64:222–227.

Kreja, L., and Seidel, H.-J. (2002). "Evaluation of genotoxic potential of some microbial volatile organic compounds (MVOC) with the comet assay, the micronucleus assay and the HPRT gene mutation assay." *Mut Res* 513:143–150.

Lee, P., and Davidson, J. (1999). "Evaluation of activated carbon filters for removal of ozone at the PPB level." *Am Ind Hyg Assoc J* 60:589–600.

Liu, R.-T. (1993a). "Model simulation of the performance of activated carbon adsorbers for the control of indoor VOCs." *Indoor Air '93,* Helsinki, Finland.

Liu, R. T. (1993b). Use of Activated Carbon Adsorbers in HVAC Applications. *IAQ 93: Operating and Maintaining Buildings for Health, Comfort, and Productivity,* K. Y. Teichman, ed., American Society of Heating, Refrigerating and Air-Conditioning Engineers, Atlanta, GA. 209–215.

Lodewyckx, P., and Vansant, E. F. (2000). "The influence of humidity on the overall mass transfer coefficient of the Wheeler-Jonas equation." *AIHA J* 61(4):461–468.

Lordgooei, M., Rood, M. J., and Rostam-Abadi, M. (2001). "Modeling effective diffusivity of volatile organic compounds in activated carbon fiber." *Environ Sci Technol* 35(3):613–619.

Macher, J. (1999). *Bioaerosols: Assessment and Control.* The American Conference of Government Industrial Hygienists, Cincinnati, OH.

Pathogenus (2002). *Personal communication with W.J. Kowalski.* Pathogenus, Inc., Etobicoke, Ontario, Canada.

Rakitskaya, T. L., Bandurko, A. Y., and Boginskaya, O. V. (1996). "Low-temperature decomposition of ozone trace concentrations by fibrous carbon materials." *Russ J Appl Chem* 69(1):148–150.

Sandström, M. (2003). "Microbial volatile organic compounds (MVOC:s) emitted from building materials affected by microorganisms—their suitability as indicators of growth of microorganisms." Department of Biology and Environmental Sciences, Umea University, Sweden. http://www.bmg.umu.se/samarbeta/D20/MH02-22.htm.

Shaughnessy, R. J., Levetin, E., and Sublette, K. (1993). "Effectiveness of portable indoor air cleaners in particulate and gaseous contaminant removal." *Indoor Air '93*, Helsinki, Finland.

Singh, N., Pisarczyk, K. S., and Sigmund, J. J. (1997). "Catalytic destruction of ozone at room temperature." *90th Annual Meeting & Exhibition*, Toronto, Ontario, CA.

Strauss, W. (1975). *Industrial Gas Cleaning.* Pergamon Press, New York.

Takeuchi, Y., and Itoh, T. (1993). "Removal of ozone from air by activated carbon treatment." *Sep Technol* 3:168–175.

Tamai, H., Kakii, T., Hirota, Y., and Kumamoto, T. (1996). "Synthesis of extremely large mesoporous activated carbon and its unique adsorption for giant molecules." *Chem Mater* 8(2):454.

Underhill, D. (2001). Removal of Gases and Vapors. *Indoor Air Quality Handbook*, J. D. Spengler, J. M. Samet, and J. F. McCarthy, eds., McGraw-Hill, New York. 10.1–10.19.

VanOsdell, D. W. (1994). "Evaluation of test methods for determining the effectiveness and capacity of gas-phase air filtration equipment for indoor air applications." *ASHRAE Trans* 100(2):511.

VanOsdell, D. W., and Sparks, L. E. (1995). "Carbon adsorption for indoor air cleaning." *ASHRAE J* 37(2):34–40.

Walker, P. L., and Thrower, P. A. (1975). *Chemistry and Physics of Carbon*, Vol. 12. Marcel Dekker, New York.

Weschler, C. J., Shields, H. C., and Naik, D. V. (1994). "Ozone-removal efficiencies of activated carbon filters after more than three years of continuous service." *ASHRAE Trans* 100:1121–1129.

CHAPTER 13

ELECTROSTATIC FILTRATION

13.1 INTRODUCTION

A variety of filtration systems exist which use electrostatic effects to enhance or produce filtration of microbial-sized particles. Electrostatic effects result when either the airborne particles or the filtering media are charged with electricity. Opposite charges on the particles or filter fibers will result in attractive forces that increase filter efficiency above what would be predicted by the traditional filter models previously addressed in Chap. 10. Some electrostatic filters are charged with an electrical power supply while others require no power supply. All of these systems are subject to relative humidity effects and their use in aerobiological applications depends on conditions as well as economics. The usefulness and feasibility of using electrostatic filters is reviewed here along with a summary of existing data and a summary of basic modeling techniques.

Figure 13.1 shows a breakdown of the main types of electrostatic or electronic filters. This is by no means an exclusive breakdown as additional types of electrical filters are under development, like electrical plasma or corona discharge systems (see Chap. 22), or may be developed in the future.

The first electrostatic filtration systems were electrostatic precipitators designed to remove large dust particles. Although electrostatic precipitators are generally industrial-purpose dust removal systems without aerobiological applications, they are addressed here because they exemplify fundamental principles and, furthermore, some new types of miniaturized electrostatic precipitators, often called electronic air cleaners, may be useful for removing airborne allergens. One type of electronic air cleaner, electronic ionizers, which induce their own airflow, are addressed here but ionization units, which cause no airflow, are only addressed in Chap. 16. Other electrically based air cleaning systems, such as electrical plasma, corona destruction, and charged activated carbon systems, are specifically not covered here but in other chapters.

13.2 ELECTROSTATIC PRECIPITATION

Electrostatic precipitators are commonly used to remove particles from airstreams having large steady flow rates. Typical applications include coal-burning plants and cement kilns. A typical two-stage *electrostatic precipitator* (ESP) has a stage of corona

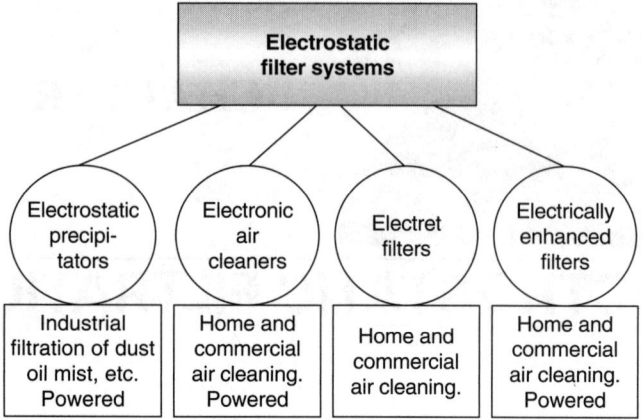

FIGURE 13.1 Breakdown of the most common electrical or electronic filter systems.

wires (sometimes called the ionizing unit) and a stage of collecting plates, as illustrated in Fig. 13.2.

The corona wires are typically positively charged at several thousand volts, often as high as 35 kV, which produces a corona that strips electrons from molecules, leaving them positively ionized. The electrons quickly attach to dust particles and give them a net negative charge. Charged particles are deflected into the interspersed grounded collection plates. Gas molecules may also become ionized in the corona (i.e., oxygen may ionize and form ozone molecules). The collecting plates are periodically rapped by mechanical rappers to dislodge the collected dust, which then drop into hoppers below. In many industrial units the air or gas may flow upward or downward through the collecting plates. The air velocity between the plates needs to be sufficiently low to allow the dust to fall and not to be re-entrained in the airstream (Parker, 1997).

It takes between 0.01 and 0.1 second for dust particles to acquire a charge in the corona region. Industrial systems are normally designed with more than 1 second residence time in the first stage to assure the charging of dust particles. The limiting charge acquired by a particle in an electric field is proportional to the surface area of the particle (Rose and Wood, 1956).

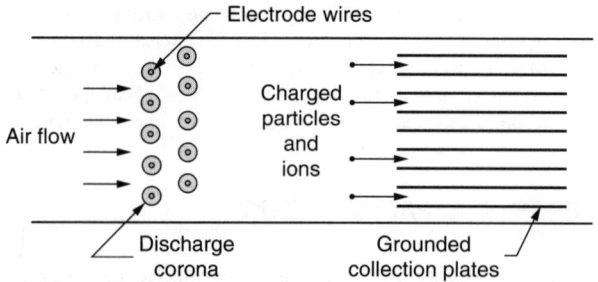

FIGURE 13.2 The electrostatic precipitation process. Particles become charged in the corona and molecules may become ionized.

Particles below about 5 μm may be more influenced by diffusion than by electrical forces. Within the range 5 to 40 mm, the efficiency of precipitation varies from 92 to 96 percent. Industrial systems are capable of removing particles 10 μm and less, and can achieve efficiencies in the neighborhood of 95 percent, but the voltages and power consumption required to remove particles in the submicron range render electrostatic precipitation somewhat less economic than ordinary high-efficiency filters for the removal of microbes. Small units may be able to remove pollen and some fungal spores, and has been noted in the past, the ozone produced as a by-product in electrostatic precipitators may also have a biocidal effect (Award and Castle, 1975).

A number of models of electrostatic precipitation and electrical filtration have been developed and applied (Parker, 1997; Khare and Sinha, 1996; Henry and Ariman, 1983; Zhibin and Guoquan, 1992). Most of these models are based on two conceptual driving forces, electrostatic deposition and diffusional deposition (Alonso and Alguacil, 2002). These forces act simultaneously with electrostatic deposition being the dominant force for large particles (approximately >1μm) and diffusional deposition being dominant for smaller particles (approximately <1μm). These mechanisms are primarily dependent on field strength and ionic current. Application of these models often involves defining the flow field as being either laminar or turbulent.

All of the models begin with defining the charge q_d on a particle of diameter d_p that has passed through the ionizer stage of an ESP, as shown in Fig. 13.3. The charge it acquires can be determined through the following equation, which represents the saturation charge on the particle:

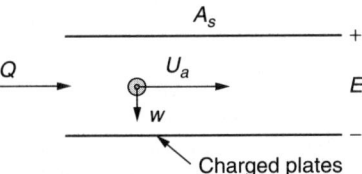

FIGURE 13.3 Electrical forces acting on a charged particle in an electrical field.

$$q_d = \left\{\left(1+\frac{2\lambda}{d_p^2}\right)+\left(\frac{2}{1+2\lambda/d_p}\right)\left(\frac{\varepsilon_r-1}{\varepsilon_r+2}\right)\right\}\pi\varepsilon_0 d_p^2 E \tag{13.1}$$

where ε_r = electrical permittivity of the particle, assumed = 10
ε_0 = electrical permittivity of a vacuum
λ = mean free path of the molecules

As the particle travels with velocity U_a through the electrical field E, it experiences a transverse electrical force of magnitude F_e as follows:

$$F_e = q_d E \tag{13.2}$$

The electrical field imparts a drift velocity w (m/min) to the particle that acts perpendicular to the airflow direction. The velocity is limited by the drag force of the air and the equilibrium condition is defined as

$$q_d E = c_D \frac{\pi d_p^2}{4C_h}\left(\frac{\rho w^2}{2}\right) \tag{13.3}$$

where c_D = drag coefficient
ρ = air density, kg/m^3
C_h = Cunningham slip factor

The Cunningham slip factor has been previously defined in Chap. 10 by Eq. (10.9). The drag coefficient can be written in terms of the Reynolds number under Stokes flow:

$$c_D = \frac{24}{Re} = \frac{24\mu}{\rho d_p w} \tag{13.4}$$

where μ is the dynamic viscosity.

The particle mobility has been previously defined in Chap. 10 by Eq. (10.8). Solving for the drift velocity produces the following relation:

$$w = \frac{q_d E C_h}{3\pi \mu d_p} \tag{13.5}$$

Given the velocity w, the grade efficiency of the ESP can be determined by the Deutsch equation as follows:

$$\eta = 1 - \exp\left(-\frac{w A_s}{Q}\right) \tag{13.6}$$

where A_s is the total surface area of collecting electrode, m², and Q is the volumetric flow rate of air, m³/min.

Figure 13.4 shows an example of the prediction of the Deutsch equation for an air velocity of 2.54 m/s, and 2 m² of surface area for each m³/min of airflow.

The Deutsch equation falls short of accurately predicting the filtration efficiency of most electrostatic precipitators, tending to overpredict efficiencies at small particle diameters and underpredict at large particle diameters (Riehle, 1997). This is evident in the comparison of the model and data from Huang and Chen (2001) shown in Fig. 13.5. Several improvements to the Deutsch equation have been proposed, including the Matts-Ohnfeld equation, the F. L. Smidth equation, and the extended Deutsch equation (Paulson, 1997). One of the problems accounted for in these latter models is the fact that ESPs will filter

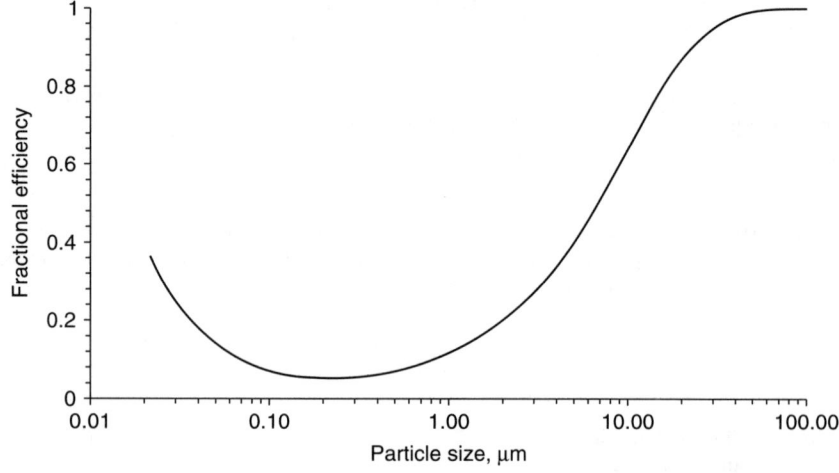

FIGURE 13.4 Typical ESP performance curve predicted by the Deutsch equation.

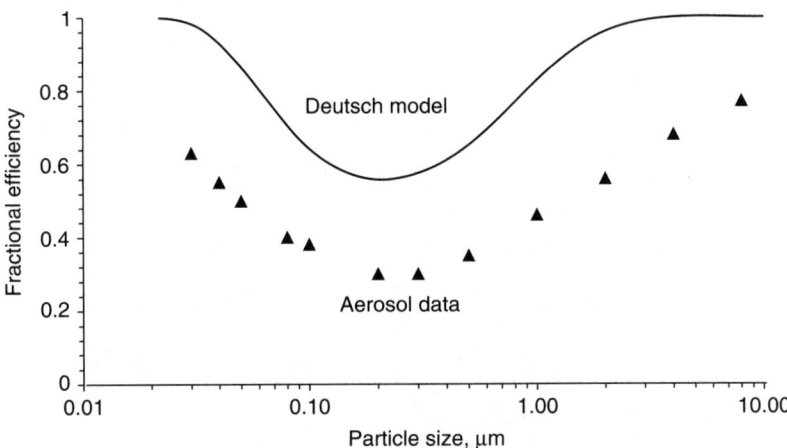

FIGURE 13.5 Comparison of Deutsch model prediction with aerosol data from Huang and Chen (2001) at a flow rate of 120 L/min and a voltage of 8 kV.

large particles even when turned off. Another problem is that the Deutsch model predicts efficiencies that approach 100 percent for submicron particles, a problem that also affects classical filtration models (see Chap. 10).

Electrostatic precipitators also remove droplets or mists of oil that are large enough to receive an electrical charge, depending on the voltage. However, any type of liquid droplet may evaporate and reaerosolize and so the removal efficiency of oil droplets can be difficult to quantify.

Small electrostatic precipitators designed for home or other nonindustrial applications are known as electronic air cleaners. These do not have rappers, but must be taken apart and cleaned periodically. Also, these devices are often inserted into airstreams without regard to residence time or air velocities, and hence efficiencies can be much lower than those used in industrial applications. A well-designed electronic air cleaner for home or office building applications might not only be larger and have a higher energy demand than an air filter, but it might also generate ozone at potentially hazardous levels.

Even a well-sized, efficiently operating electronic air cleaner cannot achieve the efficiency necessary to guarantee complete interception of airborne pathogens and allergens. However, as a means of simply improving air quality and decreasing levels of dust, smoke, and airborne microbes, electronic air cleaners do indeed have some value in home and office building environments (Offerman et al., 1992).

13.3 ELECTRONIC AIR CLEANERS

Electronic air cleaners (EACs) are a class of electrically powered filters that are basically similar to electrostatic precipitators. These systems use variations of electrostatic precipitation and are sized and designed to remove smaller particles in the micron size range, such as indoor dust, smoke, pollen, allergens, and spores. Most of them operate identically to the electrostatic precipitators described in the previous section. Electrostatic filters use an electrical charge maintained by an external source to attract particles, or else to precipitate them toward an oppositely charged filter fibers (McLean, 1988). Such filters have lately become

FIGURE 13.6 Electrically powered electrostatic filter for in-duct applications. (*Image courtesy of LakeAir International, Inc., Racine, WI.*)

popular in residential markets. One notable exception is the corona wind generator, which is addressed later in this chapter.

The performance of EACs has been studied to a limited extent and results do not always suggest any improvement over standard filters. Furthermore, they are subject to losing their charge under high relative humidity. Little is known about their ability to remove airborne pathogens although studies abound on particulate removal (Huang and Chen, 2001; Cheng et al., 1981). These units are reported to generate some ozone although the levels may not be hazardous unless they accumulate (Boelter and Davidson, 1997; Hautanen et al., 1986). Figure 13.6 shows one example of an electrostatic air filter for in-duct applications.

The main operational problem with EACs is arcing between the positively charged plates and the ground plates when collected particles bridge the small gap between these two sets of plates. This arcing can occur even after cleaning if a large particle collects and narrows the gap between plates, or a hair/fiber bridges the two sets of plates. This arcing is noticeable as a popping sound in residential EACs and voltage is automatically reduced for safety concerns. This arcing process may also produce ozone.

13.4 ELECTROSTATIC FILTERS

Filters called electrostatic are not electrically powered or enhanced and generally rely on air movement across the filter fibers to generate a small electric charge on the passing airborne particles. The small charge is supposed to enhance the removal efficiency of the filter although the effect may not be significant. Filter efficiency can be reduced by particle loading, temperature, and relative humidity (Blackford et al., 1986; Ackley, 1985).

13.5 ELECTRET FILTERS

Electret filters refer to filters that have a permanent electrostatic charge which is imparted to the fibers during the manufacturing process (Tennal et al., 1991). It refers to filters whose fibers have individual electrostatic charges while the media as a whole remains electrically neutral (West, 1997). The types of media include tribocharged polypropylene/acrylic, corona-charged polypropylene, fibrillated electret film, and other media. Table 13.1 lists some types of electret media and their typical properties, based on Barrett and Rousseau (1998).

The electrical charge imparted to the filter fibers in electret media enhances the filtration efficiency and permits the same performance as ordinary filters with lower pressure drops (Barrett and Rousseau, 1998). However, the efficiency of electret filters can be reduced by relative humidity and exposure to certain aerosols (Blackford et al., 1986). Liquid aerosols with a relatively high dielectric constant can decrease the filtration efficiency of electret filters. Liquid droplets that collect on electret filters cause degradation of the electrostatic enhancement of efficiency due to charge neutralization and the formation of a dielectric coating over the charged fibers (Tennal et al., 1991). Electret media can also experience slight decreases in their efficiency due to long-term storage and heating (Tsai and Wadsworth, 1994).

Figure 13.7 shows a comparison of an electret filter performance curve with and without charging of the airborne particles (Lee et al., 2002). The air velocity was 5 cm/s (152 fpm). Performance of this particular electret filter, called a high-performance electret filter, approximates that of a *high efficiency particulate air* (HEPA) filter.

13.6 ELECTRICALLY ENHANCED FILTRATION

Some types of electrostatic filters use variations like the ionization of the air entering an electrostatic filter. Such systems are called *electrically enhanced filters* (EEFs) or electrostatically augmented filters. Typically, a precharger is used to generate a high-intensity ionizing field. Airborne bacteria that survive the ionizing field are trapped by a grounded filter and become subject to a continuous flow of ions and are ultimately destroyed on the filter media (Jaisinghani, 1999). There are at least two types of EEFs available today. The first type of EEF uses a purely static field with a separate ionizer and a separate electrostatic collection field is applied to increase capture. The second type of EEF maintains the filter under an ionizing field in which no direct voltage is applied to the filter media. The technology of EEFs is more complex than can be adequately addressed here, and interested

TABLE 13.1 Electret Filter Media Properties

Media	Basis weight, g/m^2	Thickness, mm	dP at 7.8 cm/s in.w.g.	Fiber Dia., μm
Fiberglass	60–95	0.5–0.7	0.24–2.16	1–3
Tribocharged polypropylene/acrylic	150	3.2	0.03	31
Fibrillated polypropylene film	100	2.4	0.03	22
Charged polypropylene BMF	60	0.9	0.13	8
N-type advanced electret	60	0.9	0.12	8
R-type advanced electret	60	0.9	0.11	8
P-type advanced electret	60	1	0.15	7

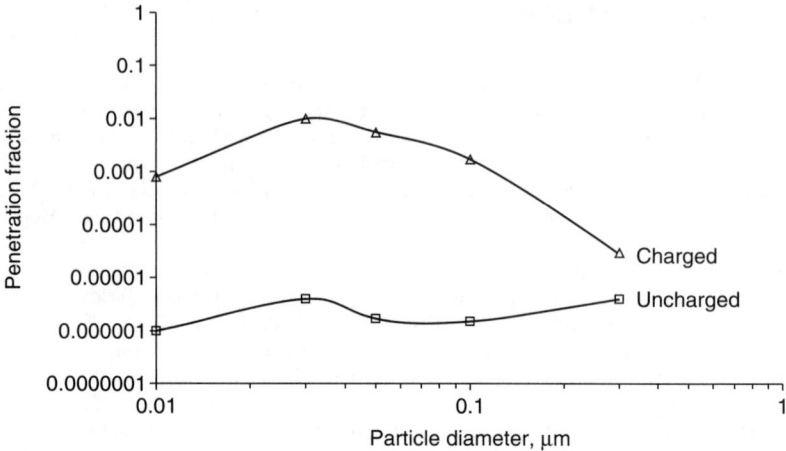

FIGURE 13.7 Penetration through a high-performance electret filter. [*Based on data from Lee et al. (2002).*]

readers are encouraged to refer to the various cited references, including Mainelis et al. (2002a) and Jaisinghani et al. (1998). The arcing problem is reportedly controlled by alternate designs that included insulating the precharging electrode (Bergman et al., 1983). Figure 13.8 shows examples of EEF units currently available from different manufacturers.

A typical precharger consists of an array of wires with a distance between the electrodes of about 30 mm, and an applied voltage of about 7 kV. In one test the efficiency of an EAA was measured at 92.9 percent for particles of 1.96 μm diameter while a comparable conventional filter was measured to 70 percent efficiency for the same particles (Lee et al., 2001). In a test of bacterial filtration efficiency with an EEF, Jaisinghani and Bugli (1991) reported that almost 100 percent of aerosolized *Staphylococcus epidermis* were eliminated by the filter using a field strength of 4.5 kV/cm. In a comparison test performed in a tissue culture laboratory, the EEF reduced concentrations in a test room 60 to 75 percent below

FIGURE 13.8 Electrically enhanced filter (EEF) models. (*Left photo courtesy of Technovation Systems, Inc., Midlothian VA; Right photo courtesy of StrionAir, Inc., Louisville, CO.*)

the levels achieved using HEPA filters (Jaisinghani 1999). In a test on a different EEF system, inoculations of *Serratia marcescens* and *Staphylococcus aureus* were destroyed after extended exposure to the flow of ions (LMS, 2004). EEF units have performed well in clean room applications (Jaisinghani et al., 2000).

Performance claims of EEF units vary considerably. Both the units in Fig. 13.8 have achieved ASHRAE ratings of MERV 15, a performance level that guarantees high removal efficiencies in the bacteria size range (see Fig. 10.9 in Chap. 10). These types of filters can achieve better than 90 percent removal efficiencies against diesel fume and smoke particles in the virus size range (i.e., 0.03 to 0.3 mm). However, such fume and smoke particles are preferentially adsorbed on electrostatic surfaces due to their initial charged state and high charge to mass ratios, and they are not truly representative of bioaerosols. Although viruses may carry a net negative charge in nature, the electrical energy required to induce electrostatic precipitation is high. It remains for further research to demonstrate the ability of EEFs to remove viruses.

One novel type of precharging suggests that irradiation with UV can improve the deposition rates of the smaller particles in the most penetrating particle size range between 0.1 and 1 μm (Mohr et al., 1993). The reason is that ESP imparts the least charge to particles in this size range while UV is not size specific, and the absorption of UV photons will result in some charging of the particles through photoemission. Estimates indicate this effect, called aerosol photoelectric charging, can boost the removal efficiency of a 0.2-μm particle from 95.4 to 99.98 percent. The UV photoemission effect was also used to produce electrostatic precipitation by Seto et al. (1995). A similar phenomenon has been demonstrated using x-rays (Forster, 2004). In the event EEF units prove to be less effective against viruses than they are against bacteria, they could be augmented with UVGI lamps to enhance deposition rates. The curious paradox is that UV will, by itself, destroy viruses and the enhanced deposition seemingly becomes redundant. However, there may be an energy optimization problem to be solved that would result in lower requirements for UV power.

As with electrostatic filters, relative humidity can degrade performance and increase power consumption (Jaisinghani and Hamade, 1987; Ackley, 1985). MERV test results on EEF units have been run in the 44 to 50 percent RH range, but no results seem to be available for higher levels of RH. High humidities that may be encountered in outside or mixed return air (i.e., 90 percent) may conceivably cause high amperage and automatic cutout of EEF systems, thereby reducing the MERV 15 filter performance (on loss of power) to perhaps an equivalent of MERV 11. One option is to place the EEF in a controlled humidity environment or to augment the system, for example, with a desiccant dehumidifier. There is clearly a need for further research in some areas before this promising technology can be applied with predictable performance against aerobiological agents comparable to that of nonelectrical filters combined with UVGI.

13.7 CORONA WIND AIR CLEANERS

One type of popular ionizing air cleaner available today takes advantage of an electrically generated air current to produce airflow through the electrodes, and thus operates without fans. These novel units are capable of removing smoke and other particulates in local environments or small rooms, but their ability to remove pathogens and allergens is unknown.

The phenomenon of electrical wind, also called ionic wind or corona wind, creates air movement by generating ions in a corona between electrodes (Loreth and Torok, 1993). The ions impart a motive force to air molecules they collide with, thus creating a light breeze. The driving electrical field converts electrical energy to kinetic energy but is limited in the amount of force it can generate. The ion velocity, in m/s, is given by

$$v = kE \qquad (13.7)$$

where k is the ion mobility, typically 2.5×10^{-4} m^2/Vs, and E is the electrical field, V/m.

Typically the electrical field between the corona electrode and the target electrode is about 10^5 V/m for most of the space, which produces an air velocity of about 25 m/s (4920 fpm). The force, in N/m^3, acting on a volume of air is the force on the light air ions within the electrical field:

$$F = \rho E \qquad (13.8)$$

where ρ is the space charge density, As/m^3.

The current density, in A/m^2, represented by the moving air ions is given by

$$i = \rho v \qquad (13.9)$$

The volume force, in N/m^3, is thus proportional to the ionic current density as follows:

$$F = \frac{i}{k} \qquad (13.10)$$

The total force acting on a straight filament through which a current runs is proportional to the length L of that filament and independent of the cross-sectional area.

$$\Delta F = \frac{\Delta I}{k} L \qquad (13.11)$$

where ΔI is the the current density in the filament, A.

The total ionic force acting on the air between the corona electrode and the target electrode is proportional to the product of the total corona current and the distance measured along the line of action. For a device with a given geometry the airflow is proportional to the square root of the corona current. Typical air velocities of 0.3 to 0.6 m/s are attained (Loreth and Torok, 1993). The larger the distance between the electrodes, the larger the voltage needed to stimulate airflow. Figure 13.9 illustrates one arrangement of electrodes that will drive air through an enclosure as ions (shown with dotted lines) are driven toward the electrode plates. The actual system has numerous electrodes and plates. Any particles caught in the field will be ionized and drawn to the negative plates, where they will remain attached by van der Waals forces until they are forcibly removed (i.e., by cleaning).

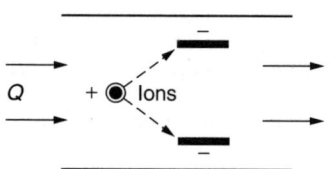

FIGURE 13.9 Corona wind air cleaning principle of operation. The ions impart a motive force to the air molecules, while the electrodes attract dust and particles.

One problem with corona wind air cleaners, besides the low flow rate, is that they generate ozone (Award and Castle, 1975). The ozone may have some bactericidal effects, and may be responsible for the clean air smell reported by these devices, but is not a desirable pollutant in indoor environments. No data are available on the performance of the consumer versions of corona wind air cleaners against any aerobiological agents or particles, although they certainly would tend to remove smoke and microscopic dust from the air at the least. Considering the low flow rates of these types of air cleaners, it is unlikely they would provide more than a fraction of the air cleaning efficiency of conventional filters that cost far less. However, the novelty of silent air cleaning and transparent operation has made these units quite popular.

13.8 AEROBIOLOGICAL PERFORMANCE OF ESPS

A very limited number of studies have evaluated the performance of electrostatic filters against airborne microbial contaminants (Zhibin and Guoquan, 1992). Most electronic air filters have pollen removal efficiencies in excess of 99 percent (Sutton et al., 1964). The use of electrostatic air cleaners to control allergens and reduce allergy symptoms has been demonstrated (Silverman and Dennis, 1956). Electrostatic precipitation has also been successfully used for the capture of microorganisms in air samplers (Alonso and Alguacil, 2002).

An early study on the ability of ESP units to remove pollen from the air and reduce hay fever symptoms in hospital rooms was conducted by Criep et al. (1936). A unit supplying 27,000 cfm was used to supply air to 61 patients. In once-through tests, the electrostatic precipitator effectively removed approximately 99.45 percent of pollen. After operating the ESP unit for 24 to 36 hours, many atopic patients experienced relief from their symptoms.

In one study on the electrical effects of a positive corona and a negative corona of an electrostatic precipitator, four bacterial species were aerosolized and passed through the electrodes of a charging chamber (Varekhov and Smirnov, 1993). The species included *Candida albicans* and *Staphylococcus aureus*, which were removed at efficiencies of 98.46 percent and 93 percent, respectively. Part of the removal was due to precipitation and part due to electrical damage to the cells.

In a test of a number of air cleaners that included three electrostatic precipitator units, Shaughnessy et al. (1993) found that these units removed spores and pollen from rooms at rates that compared with removal rates of HEPA filters. Table 13.2 summarizes these results. The ESP units included prefilters and carbon adsorbers, and the effect of the ESP was not isolated for comparison. It is likely the prefilters played a major part in the removal of the pollen and spores.

A new type of electrostatic precipitator developed for use in airborne sampling has been found to be effective in capturing various microorganisms (Mainelis et al., 2002b). Using a precharger and a precipitation voltage of 4 kV, over 80 to 90 percent of bacteria and fungi were removed from the air and onto agar plates. Microbes captured by this method included *Pseudomonas fluorescens*, *Bacillus subtilis*, and *Penicillium brevicompactum*. Microorganisms sent through the precharger were found to maintain their charge for as long as 1 hour.

TABLE 13.2 ESP Removal Rates in a Room Environment

ESP unit	Airflow, cfm	ACH	Pollen removal, %	Spore removal, %	Notes
EP1	350	24	95	95	Included prefilter and GAC
EP2	320	22	83	82	Included prefilter and GAC
EP3	330	23	39	64	Included prefilter and GAC

13.9 ECONOMICS OF ELECTROSTATIC FILTERS

Under ideal conditions, electronic air cleaners can outperform ordinary high-efficiency filters. The question of whether the improved performance of electronic cleaners justifies their initial cost or operating cost, especially when they are electrically powered, remains to be demonstrated. Jaisinghani and Bugli (1991) report that electrically enhanced filters are cost-effective for indoor applications, especially where space constraints are a factor. A review of the relative costs of various filtration systems indicates that electrostatic precipitation is economically competitive in comparison with high-efficiency and HEPA filtration, but that electret and passive electrostatic filters do not provide competitive performance for the cost (Offerman, 1992). This study estimates that cost-effective air cleaners should remove 75 to 80 percent of particles and cost $0.28 to $0.56 per cfm, while poorly performing air cleaners will cost $4 to $6 per cfm.

REFERENCES

Ackley, M. W. (1985). "Degradation of electrostatic filters at elevated temperature and humidity." *Filtr Separat* 22(4):239–242.
Alonso, M., and Alguacil, F. J. (2002). "Electrostatic precipitation of ultrafine particles enhanced by simultaneous diffusional deposition on wire screens." *J Air Waste Manage* 52:1342–1347.
Award, M. B., and Castle, G. S. P. (1975). "Ozone generation in an electrostatic precipitator with a heated corona wire." *J Air Pollut Control Assoc* 25(4):369–374.
Barrett, L. W., and Rousseau, A. D. (1998). "Aerosol loading performance of electret filter media." *AIHA J* 59:532–539.
Bergman, W., Bierman, A., Kuhl, W., Lum, B., Bogdanoff, A., Hebard, H., Hall, M., Banks, D., Mazunder, M., and Johnson, J. (1983). "Electric air filtration: Theory, laboratory studies, hardware development, and field evaluations." *UCID-19952*. Lawrence Livermore Laboratory, Berkeley, CA.
Blackford, D. B., Bostock, G. J., Brown, R. C., Loxley, R., and Wake, D. (1986). "Alteration in the performance of electrostatic filters caused by exposure to aerosols." *Fourth World Filtration Conference*, Belgium.
Boelter, K. J., and Davidson, J. H. (1997). "Ozone generation by indoor electrostatic air cleaners." *Aerosol Sci Technol* 27(6):689–708.
Cheng, Y. S., Yeh, H. C., and Kanapilly, G. M. (1981). "Collection efficiencies of a point-on-plane electrostatic precipitator." *Am Ind Hyg Assoc J* 42:605–610.
Criep, L. H., Green, M. D., and Green, M. A. (1936). "Air cleaning as an aid in the treatment of hay fever and bronchial asthma." *J Allergy* 7:120–131.
Forster, S. (2004). *Inactivation of Bacterial Aerosols Using an X-ray Enhanced Electrostatic Precipitator*. Washington University, St. Louis, MO.
Hautanen, J. H., Janka, K., Lehtimaki, M., and Kivisto, T. (1986). "Optimization of filtration efficiency and ozone production of the electrostatic precipitator." *J Aerosol Sci Technol* 17(3):622–626.
Henry, F., and Ariman, T. (1983). "A staggered array model of a fibrous filter with electrical enhancement." *Particul Sci Technol* 1:139–154.
Huang, S.-H., and Chen, C. C. (2001). "Filtration characteristics of a miniature electrostatic precipitator." *Aerosol Sci Technol* 35:792–804.
Jaisinghani, R. A., and Hamade, T. A. (1987). "Effect of relative humidity on electrically stimulated filter performance." *JAPCA* 37(7):823–828.
Jaisinghani, R. A., and Bugli, N. J. (1991). "Conventional air cleaners versus electrostatic air cleaners in HEPA and indoor applications." *Particul Sci Technol* 9:1–18.
Jaisinghani, R. A., Inzana, T., Glindemann G. (1998). "New Bactericidal Electrically Enhanced Filtration System for Cleanrooms." *IEST 44th Annual Technical Meeting and Exposition*, Phoenix, AZ. http://www.cleanroomsys.com/downloads.htm.

Jaisinghani, R. (1999). "Bactericidal properties of electrically enhanced HEPA filtration and a bioburden case study." *InterPhex Conference,* New York. http://www.cleanroomsys.com/downloads.htm.

Jaisinghani, R. A., Smith, G., Macedo G. (2000). "Control and monitoring of bioburden in biotech/pharmaceutical cleanrooms." *J Validation Technol* 6(4):686.

Khare, M., and Sinha, M. (1996). "Computer aided simulation of efficiency of an electrostatic precipitator." *Environ Int* 22(4):451–462.

Lee, J.-K., Kim, S. C., Shin, J. H., Lee, J. E., Ku, J. H., and Shin, H. S. (2001). "Performance evaluation of electrostatically augmented air filters coupled with a corona precharger." *Aerosol Sci Technol* 35:785–791.

Lee, M., Ohtani, Y., Namiki, N., and Emi, H. (2002). "Prediction of collection efficiency of high-performance electret filters." *J Chem Eng Jpn* 35(1):57–62.

LMS (2004). *Test of the Germicidal Effect of the StrionAir GC Unit on Bacteria.* LMS Technologies, Edina, MN.

Loreth, A., and Torok, V. (1993). "Transport and cleaning of indoor air using the corona wind." *Indoor Air '93,* Helsinki, Finland

Mainelis, G., A, A., Willeke, K., Lee, S.-A., Reponen, T., and Grinshpun, S. A. (2002a). "Collection of airborne microorganisms by a new electrostatic precipitator." *J Aerosol Sci* 33:1417–1432.

Mainelis, G., Górny, R. L., Reponen, T., Trunov, M., Grinshpun, S. A., Yadav, J., Baron, P. A., and Willeke, K. (2002b). "Effect of electrical charges and fields on injury and viability of airborne bacteria." *Biotech Bioeng* 79:229–241.

McLean, K. J. (1988). "Electrostatic precipitators." *IEE Proceedings* 135(6):347–362.

Mohr, M., Kwetkus, B. A., and Burtscher, H. (1993). "Improvement of electrostatic precipitation by UV-charging of submicron particles." *J Aerosol Sci* 24(S1):s247–s248.

Nelson, T., Rappaport, B. Z., and Welker, W. H. (1933). "The effect of air filtration in hay fever and pollen asthma." *JAMA* 100(18):1385–1392.

Offerman, F. J., Loiselle, S. A., and Sextro, R. G. (1992). "Performance of air cleaners in a residential forced air system." *ASHRAE J* 34(7):51–57.

Parker, K. R. (1997). *Applied Electrostatic Precipitation.* Blackie Academic and Professional, London.

Paulson, C. (1997). Chapter 9A: Precipitator sizing methods. *Applied Electrostatic Precipitation,* K. R. Parker, ed., Blackie Academic & Professional, London.

Riehle, C. (1997). Chapter 3: Basic and theoretical operation of ESPs. *Applied Electrostatic Precipitation,* K. R. Parker, ed., Blackie Academic & Professional, London.

Rose, H. E., and Wood, A. J. (1956). *An Introduction to Electrostatic Precipitation in Theory and Practice.* Constable & Company, London.

Seto, K., Okuyama, K., and Inuoe Y. (1995). "Electrostatic precipitation of fine particulate contaminants by UV/photoelectron method under low pressure condition." *J Aerosol Sci* 26(S1):s17–s18.

Shaughnessy, R. J., Levetin, E., and Sublette, K. (1993). "Effectiveness of portable indoor air cleaners in particulate and gaseous contaminant removal." *Indoor Air '93,* Helsinki, Finland.

Silverman, L., and Dennis, R. (1956). "Removal of air-borne particulates and allergens by a portable electrostatic precipitator." *Air Cond Heat Vent,* December.

Sutton, D. J., Cloud, H. A., McNall, P. E., Nodolf, K. M., and McIver, S. H. (1964). "Performance and application of electronic air cleaners in occupied spaces." *ASHRAE J* 6(6):55–62.

Tennal, K. B., Mazumder, M. K., Siag, A., and Reddy, R. N. (1991). "Effect of loading with an oil aerosol on the collection efficiency of an electret filter." *Particul Sci Technol* 9:19–29.

Tsai, P. P., and Wadsworth, L. C. (1994). "Air filtration improved by electrostatically charging fibrous materials." *Particul Sci Technol* 12:323–332.

Varekhov, A. G., and Smirnov, O. W. (1993). "The electrical effects of the bioaerosols." *Indoor Air '93,* Helsinki, Finland.

West, R. D. (1997). "Performance enhanced electret composites." *Fluid/Part Separat J* 10(2):148–153.

Zhibin, Z., and Guoquan, Z. (1992). "New model of electrostatic precipitation efficiency accounting for turbulent mixing." *J Aerosol Sci* 23(2):115–121.

CHAPTER 14

PHOTOCATALYTIC OXIDATION

14.1 INTRODUCTION

Photocatalytic oxidation (PCO) is a recently developed air and surface cleaning technology that has the unique capacity to destroy both microorganisms and *microbial volatile organic compounds* (MVOCs). It has spawned a great deal of interest and some PCO-based systems are currently available on the market. Although the use of PCO has yet to be demonstrated to be superior to either filtration or UVGI for air disinfection, and has yet to be proven more economical than carbon adsorption for most gas-phase filtration applications, the use of PCO to create self-sterilizing surfaces on building components like doorknobs makes it a most promising technology for use in an integrated building design employing both air and surface disinfection systems. Since the catalytic material is excited by sunlight and by UV and is not consumed in the process, PCO is essentially a self-cleaning process that requires no regeneration. This chapter reviews the performance and application of PCO systems for air disinfection, destruction of MVOCs, and surface disinfection applications for buildings.

14.2 PCO THEORY AND OPERATION

PCO is a process in which surfaces coated with titanium dioxide (TiO_2) become chemically reactive to organic compounds when exposed to ultraviolet or visible light and (Lyons, 1995). The photoreactive properties of titanium dioxide were first described by Renz (1921). It wasn't until the early 1970s that several researchers began exploring the photocatalytic potential of TiO_2 (Gravelle et al., 1971; Formenti et al., 1971; Bickley and Stone, 1973). When a TiO_2 coating is irradiated, electrons are released for a brief period during which they may participate in localized chemical reactions that may produce hydroxyl radicals and other ions (Jacoby et al., 1998; Goswami et al., 1997). Hydroxyl radicals are highly reactive species that will oxidize *volatile organic compounds* (VOCs) and many chemical pollutants (Jacoby et al., 1996).

Titanium dioxide is a semiconductor photocatalyst with a band gap energy of 3.2 eV. When this material is irradiated with photons of less than 385 nm, the band gap energy is exceeded and an electron is promoted from the valence band to the conduction band. With holes (h^+) and hydroxyl radicals (OH) generated in the valence band, and electrons and superoxide ions (O_2^-) generated in the conduction band, illuminated TiO_2 photocatalysts can decompose and mineralize organic compounds by participating in a series of oxidation reactions that culminate in the end product carbon dioxide (Huang et al., 2000). The electron-hole pairs produced have a lifetime in the space-charge region sufficiently long enough to enable participation in chemical reactions that allow the formation of hydroxyl radicals and superoxide ions. The various reactive oxygen species produced can be highly biocidal to microorganisms.

The simplest and most widely postulated reactions are shown here.

$$OH^- + h^+ \rightarrow OH \qquad (14.1)$$

$$O_2 + e^- \rightarrow O_2^- \qquad (14.2)$$

An alternate postulated reaction is as follows:

$$H_2O + h^+ \rightarrow OH + H^+ \qquad (14.3)$$

The actual chemical reactions that may occur may be considerably more complex and involve numerous stages, especially when pollutants and organic compounds are involved. The process is also affected by relative humidity. In the absence of water vapor, the photocatalytic oxidation of organics is seriously retarded (Peral et al., 1997).

Hydroxyl radicals and superoxide ions are highly reactive species that will oxidize VOCs that are either adsorbed on the catalyst surface or come in close enough contact to react chemically. They will inactivate and decompose adsorbed viruses, bacteria, and fungi (Blake, 1994). The microbiocidal effects were first demonstrated by Matsunaga (1985).

Photocatalytic oxidation can be used for wastewater treatment and can convert contaminants into environmentally acceptable products such as CO_2 and H_2O (Vohra and Davis, 1993; Romero et al., 1999). Some trace level production of acids, methane, and other compounds can occur as by-products of the process (Lee et al., 2001). For maximum removal of organics, there is an optimum value of photocatalyst-specific surface area (Vohra and Davis, 1993). The PCO reaction rate increases with the incident irradiance. The TiO_2 photocatalyst remains generally robust under extended exposure to a wide variety of compounds but some compounds may partially deactivate TiO_2 and it may be rapidly deactivated by hexamethyldisilazane (Peral et al., 1997; Turchi et al., 1995).

Although most PCO systems depend on UV irradiation, some systems have been designed that use only solar radiation or visible light from fluorescent lamps (Cardona et al., 1999; DOE, 1995; Chapuis et al., 2002). In this context the process is sometimes referred to as solar detoxification (Lyons, 1995). It has been shown that PCO effectiveness can be enhanced under visible light by using TiO_2 doped with nitrogen (Asahi et al., 2001).

The decomposition of VOCs under photocatalysis can be modeled as exponential decay (Sattler and Liljestrand, 2003). The limited data for destruction of microorganisms in air and in water suggest that a single stage of exponential decay will be effective for modeling purposes but is insufficient to establish whether a second stage exists (Block and Goswami, 1995; Nagame et al., 1989; Wolfrum et al., 2002). Since PCO generally involves UV and filtration, these effects may need to be separated from the PCO effect. Furthermore, since the effect of PCO depends on contact or proximity with the media or substrate carrying the TiO_2 catalyst, a filtration model (see Chap. 10) combined with both a residence time model and an irradiation intensity model (see Chap. 11) might prove to be the ultimate predictor of performance.

Various methods have been explored to coat or impregnate materials with concentrations of TiO_2. Solutions used to spray on coatings of TiO_2 typically contain particles of 10 to 30 nm in diameter. The latter represents a surface area of 50 m^2/g. TiO_2 is used as the primary catalyst because it is inert, inexpensive, available in many forms, nontoxic, and its energy gap is 3.2 eV, which extends to the oxidation and reduction values for water- and carbon-based constituents (Speer, 2005).

14.3 AIR DISINFECTION

PCO systems are currently available as self-contained recirculation units and generally consist of tubing, filter material, or packing material coated with TiO_2 that is exposed to a UV lamp. Early systems often consisted of irradiated plates coated with TiO_2 over which air or water flowed. Other systems consisted of UV lamps inside glass tubes that were coated on the inside with TiO_2. More recent designs involve coating a packing material, such as glass beads, with TiO_2 and surrounding a UV lamp with a packed bed through which air or water would flow. Attempts to coat filter media or fibrous meshes with TiO_2 have also proven successful (Ernesto et al., 2002). Figure 14.1 illustrates a typical PCO system in which filters coated with TiO_2 sit upstream and downstream of a UV lamp. In some systems coating is only upstream, while in other variations all six enclosure walls are made of coated filter material.

In all such systems, designed for air disinfection, there is a combined effect of the filter, the UV irradiation, and the photocatalysis. There has been limited quantitative study that allows differentiation of these effects. Figure 14.2 shows an example of a PCO system disinfecting air in a recirculating system (Goswami et al., 1997). The exposed microorganism, *Serratia marcescens*, received multiple doses during transit and therefore this example is not the same as a constant exposure system, but it provides clear evidence of the air disinfection properties of PCO. In this figure the disinfection due to UV alone is graphed for comparison. Based on comparison of the effective rate constants, the effect due to UV appears to account for 63 percent of the disinfection, while PCO accounts for 37 percent. The effect of the filter material, described as an air-conditioning filter, could not be isolated from these data.

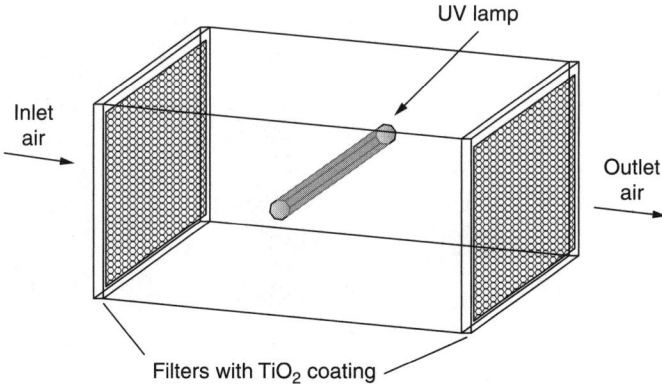

FIGURE 14.1 Typical arrangement of a PCO unit showing TiO_2-coated filters on the inlet and outlet.

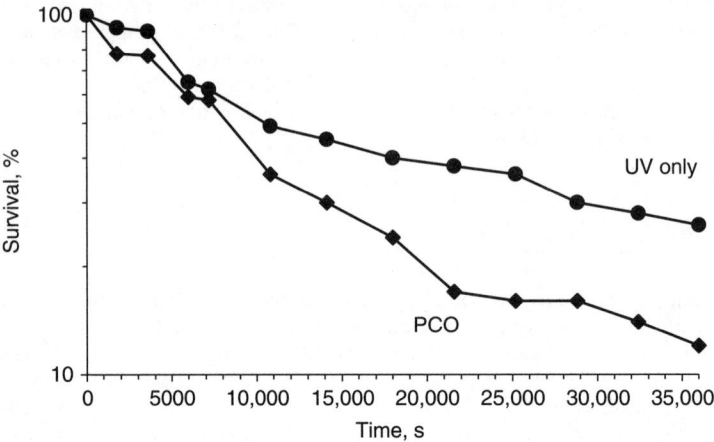

FIGURE 14.2 PCO disinfection of airborne *Serratia marcescens* in a recirculation system, compared with UV disinfection. [*Based on data from Goswami et al. (1997).*]

Huang et al. (2000) report that the bactericidal effect of PCO begins with oxidation of the external cell wall of bacteria where contact with TiO_2 occurs. After elimination of the protective cell wall, oxidative damage takes place on the underlying cytoplasmic membrane. Photocatalytic action progressively increases cell permeability, leading to cell lysis and death.

The photocatalytic effect is dependent on the wavelength and the irradiance of the UV (Raupp and Junio, 1993; Nozawa et al., 2001). Some data suggest the photocatalytic effect peaks at about the 400-nm wavelength, but that it is also active between about 300 and 500 nm (Formenti et al., 1971). Higher UV irradiance increase the photocatalytic effect, as shown in Fig. 14.3 (Goswami et al., 1997). The presence of sunlight can also add to the photocatalytic

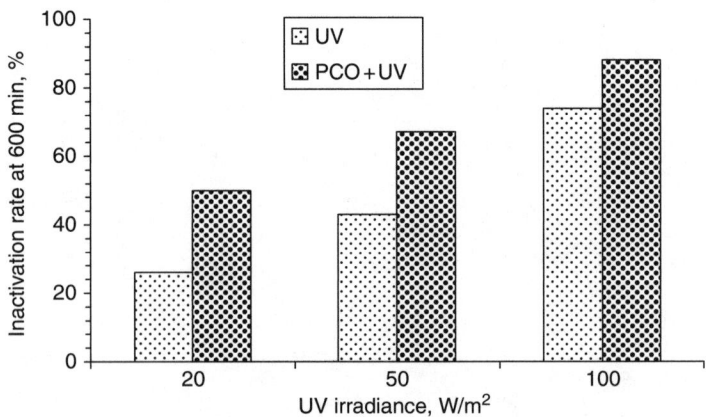

FIGURE 14.3 PCO inactivation rates vs. UV inactivation rates after 600 min in reactor. [*Based on data from Goswami et al. (1997).*]

effect (Block and Goswami, 1995). PCO reaction rates are also sensitive to temperature (Magrini et al., 1996). It has been suggested that only the first 2 μm of TiO_2 actually participate in the photocatalytic reaction, suggesting that maximizing surface area would be the best approach for increasing the effectiveness of PCO systems (Jacoby et al., 1995).

PCO systems promise low power consumption, potentially long service life, and low maintenance requirements (Block and Goswami, 1995). One study, however, suggests these systems are least cost-effective for gas phase filtration than is carbon adsorption (Henschel, 1998).

Figure 14.4 shows an example of water disinfection used to destroy the bacteria *Serratia marcescens* (Block and Goswami, 1995). In this test a phosphate buffer was added to the test runs and clearly reduced the disinfection rate compared to the case with no buffer. The buffered case is included here as it is strongly suggestive of a two-stage decay rate, as occurs in most, if not all, other disinfection processes. Once again, the fact that these data are from a water-based studies probably precludes its usefulness in air-based disinfection predictions due to the dominance of the hydroxyl radical reaction in water. Additional tests in solution have been performed showing the effectiveness of PCO against *E. coli, Streptococcus mutans, Streptococcus cricetus, Streptococcus rattus, Actinomyces viscosus, Aspergillus niger, Bacillus subtilis, Micrococcus luteus,* and *Candida albicans* (Ireland et al., 1993; Nagame et al., 1989, Onoda et al., 1988; Saito et al., 1991; Wolfrum et al., 2002).

Results for the inactivation of several microorganisms were reported by Lin and Li (2003), who used a TiO_2-coated filter irradiated with 8- and 36-W UV lamps to inactivate *E. coli, B. subtilis* spores, *Candida famata* yeast cells, and *Penicillium citrinum* spores. The PCO system measured 25 cm × 15 cm × 5.45 cm and included a TiO_2-coated filter. Pore size of the filter material was measured at 20 to 500 μm. With an 8-W UV lamp the irradiance on the filter surface was measured to be 7400 μW/cm². Results for the survival of *Penicillium* spores are shown in Fig. 14.5 in terms of the residence time within the unit. Differences in microbial survival with and without the UV lamp operating are shown in the figure. In many such UV-PCO systems, the UV dose imparted seems to be responsible for most of the microbial disinfection.

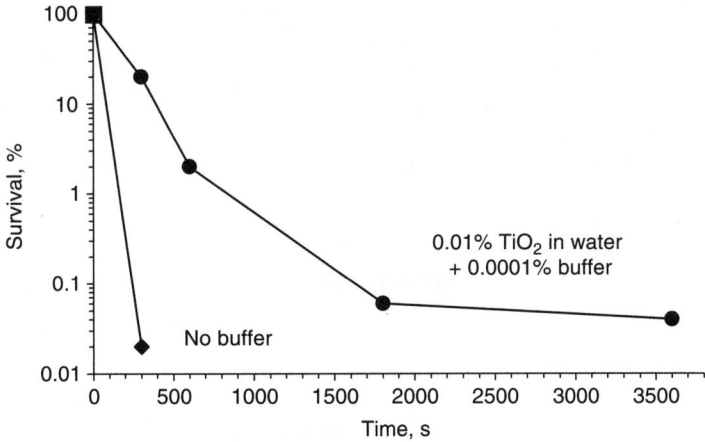

FIGURE 14.4 Destruction of *Serratia marcescens* in buffered water. [*Based on data from Block and Goswami (1995).*]

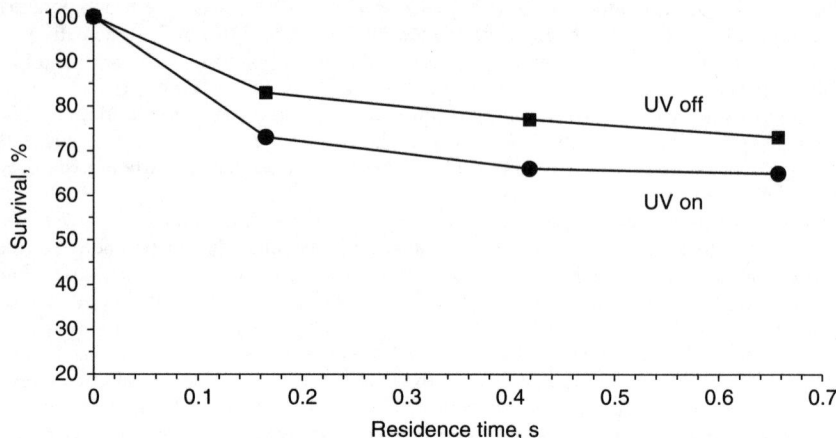

FIGURE 14.5 Survival of airborne *Penicillium citrinum* spores after passing through a PCO system with an 8-W lamp. [*Based on data from Lin and Li (2003).*]

Photocatalysis of microorganisms results in the oxidation of their organic components. The end products of the oxidation process will be mainly CO_2 for carbon and inorganic nitrates, sulfates, and phosphates for the components of biological materials. Experiments have shown that a mass balance can be established in which the destruction of bacterial cells and spores can be measured in terms of the total CO_2 released over time (Wolfrum et al., 2002).

Photocatalytic reactors may be integrated into new and existing *heating, ventilation, and air-conditioning* (HVAC) systems due to their modular design, room temperature operation, and negligible pressure drop. PCO reactors also feature low power consumption, potentially long service life, and low maintenance requirements. These attributes contribute to the potential of PCO technology to be an effective process for removing and destroying low-level pollutants in indoor air, including bacteria, viruses, and fungi.

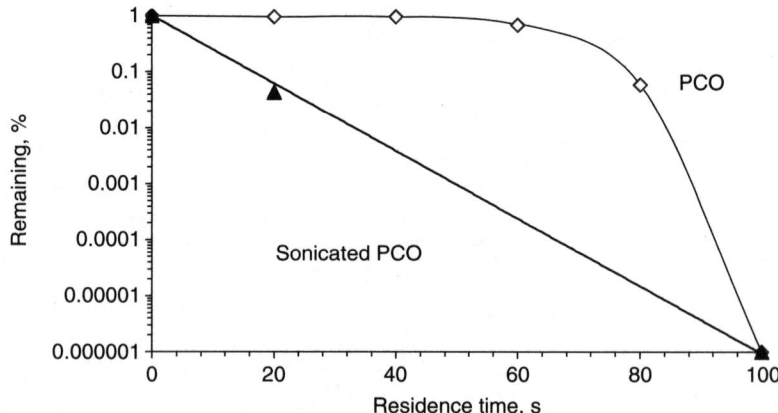

FIGURE 14.6 Comparison of sonicated and nonsonicated samples of *Streptococcus mutans* in a liquid photocatalytic reactor. [*Based on data from Morioka et al. (1988).*]

It is reported that sonication can enhance the biocidal effectiveness of TiO_2 in liquid slurries (Huang et al., 2000; Morioka et al., 1988). Figure 14.6 illustrates one example of the interaction of sonication and photocatalysis. It is unlikely that this process would work in air unless some means of aerosolizing and recovering TiO_2 particles could be developed, but it is possible that ultrasonication in air may increase contact between airborne microbes and TiO_2.

PCO filters are created by binding TiO_2 to the filter media of fibrous mesh. This can be accomplished by soaking the media in a solution of TiO_2 and a binding compound. A metal microfibrous mesh can also be coated with titanium dioxide by spraying TiO_2 on the surface of the mesh using an airbrush (Ortiz et al., 2002). Such coatings may lead to an increase in air pressure drop through the mesh.

14.4 DESTRUCTION OF MVOCS

Pollutants, particularly VOCs, are preferentially adsorbed on the surface and oxidized to (primarily) carbon dioxide (CO_2). Rather than simply changing the phase and concentrating the contaminant, the absolute toxicity of the treated airstream is reduced, allowing the photocatalytic reactor to operate as a self-cleaning filter relative to organic material on the catalyst surface. Photocatalytic oxidation can convert VOCs like heptane, trichloroethylene, ethanol, methanol, acetone, isopropanol, toluene, formaldehyde, acetaldehyde, acetone into mainly harmless products like CO_2, O_2, and H_2O (Romero et al., 1999; Stevens et al., 1998; Turchi et al., 1995; Muradov et al., 1996). Photocatalytic oxidation under solar exposure can completely mineralize organic contaminants in water and air although the reactivation rates are slow compared to the use of UV as a light source (Romero et al., 1999). Destruction rates of 95 to 99 percent or better are possible depending on the specific VOCs, their levels in air, and other factors (Turchi et al., 1995; Cardona et al., 1999).

Models are under development that define the radiation field and flow field in order to predict the rate of conversion of airborne VOCs (Hossain et al., 1999; Raupp et al., 1997). These models involve predicting the UV irradiance field on the TiO_2-coated surfaces and can achieve reasonably accurate predictions.

TiO_2 is a photostable and relatively inexpensive catalyst (Cardona et al., 1999). However, PCO system first costs and operating costs remain to be demonstrated as competitive in comparison with other air cleaning technologies like filtration, UVGI, and carbon adsorption (Henschel, 1998). In spite of high potential costs, PCO may be uniquely suitable for certain applications where VOCs exist in high concentrations, where maintenance is a problem, or where space is a constraint (Turchi et al., 1995; Block and Goswami, 1995).

Some of the (MVOCs) that PCO has been shown to be effective against include formaldehyde, acetaldehyde, and acetone (Stevens et al., 1998; Obee and Brown, 1995; Blount et al., 2001; Raupp and Junio, 1993; Jacoby et al., 1996). Figure 14.7 shows the reduction of acetone in a fixed-bed photocatalytic reactor. The rate constant for this particular system is 0.218 s^{-1}.

In a comparative study of the performance and economics of PCO systems versus carbon adsorption for the removal of various VOCs, Henschel (1998) estimated that the cost of a full-size PCO system would be several times more than the cost of a carbon adsorption system for cleaning the same quantity of air. The bulk of the cost differences come from the initial cost of the photocatalytic material. The calculations, however, focused strictly on VOCs and did not account for the benefits of air disinfection. Also, some of the performance characteristics and material costs are likely to improve as PCO systems undergo further research and development. Ollis and El-Akabi (1992) suggest that PCO is competitive with catalytic incineration.

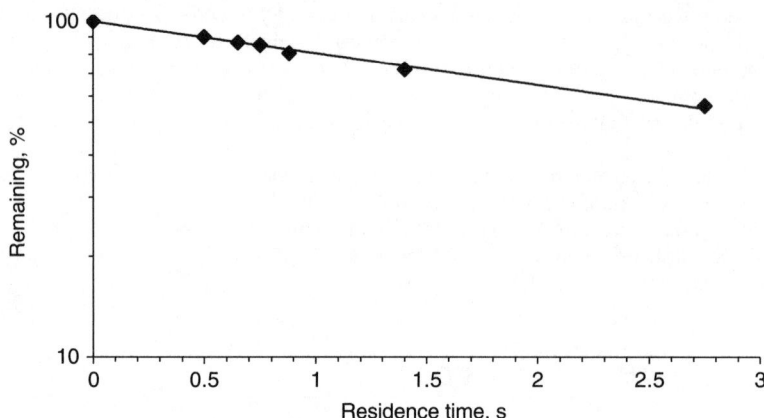

FIGURE 14.7 Reduction of acetone through a packed-bed photocatalytic reactor. [*Based on data from Raupp et al. (1997).*]

14.5 SURFACE DISINFECTION

Transparent PCO films have been developed that have reaction rates comparable to standard PCO surface coatings (Blount et al., 2001). Transparent coatings may be useful for indoor surfaces that are subject to contamination with fomites (i.e., infectious particles on doorknobs, handles, telephones, keyboards, and the like). Glass coated with TiO_2 can be self-cleaning under exposure to sunlight (Romeas et al., 1999). PCO coatings have the potential to revolutionize building materials, providing for self-sterilization of surfaces properties under extended exposure to visible light. PCO films may be used in applications where antimicrobial coatings may be impractical to apply.

In a study in which plate glass was coated with a transparent film of TiO_2, the removal rates of the atmospheric pollutants—palmitic acid and fluoranthene—were reduced, with about 20 percent of the organic carbon contained in the compounds being transformed into volatiles (Romeas et al., 1999). Grease stains typically contain palmitic acid. The reduction of organics on indoor surfaces may reduce the potential for microbial growth, as well as destroying mold, although no data are currently available to demonstrate the latter.

Specialized application may develop as research continues, especially in relation to VOCs that are not removed by GAC filters, like anesthesia gases. In cold room storage, for example, many fruits are susceptible to low levels of ethylene, a VOC easily removed by PCO. Another promising application for TiO_2 is surface disinfection of wood structures. Experimental spraying of PCO solutions onto mold-contaminated wood in the presence of sunlight has significantly reduced visible mold growth (Speer, 2005).

14.6 PCO APPLICATIONS

No studies appear to be currently available on actual implementations of PCO systems in buildings. Such systems are widely marketed in the Far East but reports of their effectiveness have not been forthcoming. Technical issues that remain to be addressed before PCO air disinfection units become practical for use in air disinfection applications include the formation of products of incomplete oxidation, reaction rate inhibition due to humidity, mass transport issues associated with high-flow rate systems, catalyst deactivation, the effects of inorganic contamination (dust and soil), the improvement of contact time without undue increase in pressure drop, and the performance as a function of cost in comparison with other air cleaning technologies.

REFERENCES

Asahi, R., Morikawa, T., Ohwaki, T., Aoki, K., and Taga, Y. (2001). "Visible-light photocatalysis in nitrogen-doped titanium oxides." *Science* 293:269–271.

Bickley, R. I., and Stone, F. S. (1973). "Photoadsorption and photocatalysis at rutile surfaces I. Photoadsorption of oxygen." *J Catal* 31:389.

Blake, D. M. (1994). "Bibliography of work on the photocatalytic removal of hazardous compounds." *NREL/TP-430-6084*. National Renewal Energy Laboratory, Golden, CO.

Block, S. S., and Goswami, D. Y. (1995). "Chemically enhanced sunlight for killing bacteria." *Solar Engineering 1995, Proceedings of the 1995 ASME International Solar Energy Conference*, W. B. Tine, T. Tanaka, and D. E. Claridge, eds., ASME, New York. 431–438.

Blount, M. C., Dong, H. K., and Falconer, J. L. (2001). "Transparent thin-film TiO_2 photocatalysts with high activity." *Environ Sci Technol* 35(14):2988–2994.

Cardona, A. I., Sanchez, B., Romero, M., Fabrellas, B., Garcia, E., and Blanco, J. (1999). "Solarized photoreactors for degradation of chlorinated organics in air." *J Phys—Paris* 9(Pr3):271-273.

Chapuis, Y., Klvana, D., Guy, C., and Kirchnerova, J. (2002). "Photocatalytic oxidation of volatile organic compounds using fluorescent visible light." *J Air Waste Manage* 52:845–854.

DOE (1995). "Solar Industrial Program 1994 Review." *DOE/GO-10095-125*. Department of Energy. Golden, CO.

Ernesto, J., Lopez., O., and Jacoby, W. A. (2002). "Microfibrous mesh coated with titanium dioxide: A self-sterilizing, self-cleaning filter." *J Air Waste Manage* 52:1206–1213.

Formenti, M., Juillet, F., Meriaudeau, P., and Teichner, S. J. (1971). "Heterogeneous photocatalysis for partial oxidation of paraffins." *Chem Technol* 1:680.

Goswami, D. Y., Trivedi, D. M., and Block, S. S. (1997). "Photocatalytic disinfection of indoor air." *J Sol Energ Eng* 119:92–96.

Gravelle, P. C., Julliet, F., Meriadeau, P., and Teichner, S. J. (1971). "Surface reactivity of reduced titanium dioxide." *Faraday Discuss Chem Soc* 52:140.

Henschel, D. B. (1998). "Cost analysis of activated carbon versus photocatalytic oxidation for removing organic compounds from indoor air." *J Air Waste Manage* 48(10):985–994.

Hossain, M. M., Raupp, G. B., Hay, S. O., and Obee, T. N. (1999). "Three-dimensional developing flow model for photocatalytic monolith reactors." *AIChE J* 45(6):1309–1321.

Huang, Z., Maness, P. C., Blake, D., Wolfrum, E. J., Smolinski, S. L., and Jacoby, W. A. (2000). "Bactericidal mode of titanium dioxide photocatalysis." *J Photochem Photobiol A* 130(2–3):163–172.

Ireland, J. C., Klosterman, P., Rice, E. W., and Clark, R. M. (1993). "Inactivation of Escherichia coli by titanium dioxide photocatalytic oxidation." *Appl Environ Biol* 59(5):1668–1670.

Jacoby, W. A., Blake, D. M., Nobel, R. D., and Koval, C. A. (1995). "Kinetics of the Oxidation of Trichloroethylene in Air via Heterogeneous Photocatalysis." *J Catal* 157:87–96.

Jacoby, W. A., Blake, D. M., Fennell, J. A., Boulter, J. E., and Vargo, L. M. (1996). "Heterogeneous photocatalysis for control of volatile organic compounds in indoor air." *J Air Waste Manage* 46:891–898.

Jacoby, W. A., Maness, P. C., and Wolfrum, E. J. (1998). "Mineralization of bacterial cell mass on a photocatalytic surface in air." *Environ Sci Technol* 32(17):2650–2653.

Lee, G. D., Tuan, V. A., and Falconer, J. L. (2001). "Photocatalytic oxidation and decomposition of acetic acid on titanium silicalite." *Environ Sci Technol* 35(6):1252–1258.

Lin, C.-Y., and Li, C.-S. (2003). "Effectiveness of titanium dioxide photocatalyst filters for controlling bioaerosols." *Aerosol Sci Technol* 37:162–170.

Lyons, C. (1995). "Photocatalytic oxidation, an effective and cost competitive process for destroying volatile organic chemical pollutants is now being commercialized." *Mater Technol* 10(11–12): 236–238.

Magrini, K. A., Rabogo, R., Larson, S. A., Turchi, C. S. (1996). "Control of gaseous solvent emissions using photocatalytic oxidation." *Proceedings of the Air & Waste Management Association's Annual Meeting & Exhibition,* Nashville, TN.

Matsunaga, T. (1985). "Sterilization with particulate photosemiconductor." *J Antibact Antifung Agents* 13:211–220.

Morioka, T., Saito, T., Nara, Y., and Onoda, K. (1988). "Antibacterial action of powdered semiconductor on a serotype of *Streptococcus mutans*." *Caries Res* 22:230–231.

Muradov, N. Z., T-Raissi, A., and Muzzey, D. (1996). "Selective photocatalytic oxidation destruction of airborne VOCs." *Sol Energy* 56:445–453.

Nagame, S., Oku, T., Kambara, M., and Konishi, K. (1989). "Antibacterial effect of the powdered semiconductor TiO_2 on the viability of oral microorganism." *J Dental Res* 68:1696–1697.

Nozawa, M., Tanigawa, K., Hosomi, M., Chikusa, T., and Kawada, E. (2001). "Removal and decomposition of malodorants by using titanium dioxide photocatalyst supported on fiber activated carbon." *Water Sci Technol* 44(9):127–133.

Obee, T. N., and Brown, R. T. (1995). "TiO_2 photocatalysis for indoor air applications: Effects of humidity and trace contaminant levels on the oxidation rates of formaldehyde, toluene, and 1,3 butadiene." *Environ Sci Technol* 29(5):1223–1231.

Ollis, D. F., and Al-Ekabi, H. (1992). Photocatalytic Purification and Treatment of Water and Air. *Proceedings of the 1st International Conference on TiO_2*, Elsevier, London, Ontario, Canada.

Onoda, K., Watanabe, J., Nakagawa, Y., and Izumi, I. (1988). "Photocatalytic bactericidal effect of powdered TiO_2 on *Streptococcus mutans*." *Denki Kagaku* 56(12):1108–1109.

Ortiz, L., Jose, E., and Jacoby, W. A. (2002). "Microfibrous mesh coated with titanium dioxide: A self-sterilizing, self-cleaning filter." *J Air Waste Manage Assoc* 52(10):1206–1213.

Peral, J., Domenech, X., and Ollis, D. F. (1997). "Heterogeneous photocatalysis for purification, decontamination, and deodorization of air." *J Chem Technol Biotechnol* 70:117–140.

Raupp, G. B., and Junio, C. T. (1993). "Photocatalytic oxidation of oxygenated air toxics." *Appl Surf Sci* 72:321–327.

Raupp, G. B., Nico, J. A., Annangi, S., Changrani, R., and Annapragada, R. (1997). "Two-flux radiation-field model for an annular packed-bed photocatalytic oxidation reactor." *AIChE J* 43(3): 792–801.

Renz, C. (1921). "Photo-reactions of the oxides of titanium, cerium, and earth-acids." *Helv Chim Acta* 4:961.

Romeas, V., Pichat, P., Guillard, C., Chopin, T., and Lehaut, C. (1999). "Self-cleaning properties of TiO_2-coated glass: Degradation, under simulated solar light, of palmitic (hexadecanoic) acid and fluorathene layers deposited on the glass surface." *J Phys—Paris* 9(Pr3): 247–252.

Romero, M., Blanco, J., Sanchez, D., Vidal, A., Malato, S., Cardona, A. I., and Garcia, E. (1999). "Solar photocatalytic degradation of water and air pollutants: challenges and perspectives." *Sol Energy* 66(2):169–182.

Saito, T., Iwase, T., Horie, J., and Morioka, T. (1991). "Mode of photocatalytic bactericidal action of powdered semiconductor TiO_2 on *Mutans Streptococci*." *J Photochem Photobiol* 14:369–379.

Sattler, M. L., and Liljestrand, H. M. (2003). "Method for predicting photocatalytic oxidation rates of organic compounds." *J Air Waste Manage Assoc* 53:3–12.

Speer, S. (2005). *Personal communication with W. Kowalski.* Catalyx Technologies, Media, PA.

Stevens, L., Lanning, J. A., Anderson, L. G., Jacoby, W. A., and Chornet, N. (1998). "Investigation of the photocatalytic oxidation of low-level carbonyl compounds." *J Air Waste Manage* 48(10): 979–984.

Turchi, C. S., Ranago, R., and Jassal, A. (1995). "Destruction of Volatile Organic Compound (VOC) Emissions by Photocatalytic Oxidation (PCO): Benchscale test Results and Cost Analysis." *Technology Transfer 95082935A-ENG.* Sematech, Austin, TX.

Vohra, M. S., and Davis, A. P. (1993). "Photocatalytic oxidation: the process and its practical applications." *Hazardous and Industrial Wastes—Proceedings of the 25th Mid-Atlantic Industrial Waste Conference,* College Park, MD. 275–282.

Wolfrum, E. J., Huang, J., Blake, D. M., Maness, P.-C., Huang, Z., Fiest, J., and Jacoby, W. A. (2002). "Photocatalytic oxidation of bacteria, bacterial and fungal spores, and model biofilm components to carbon dioxide on titanium dioxide-coated surfaces." *Environ Sci Technol* 36(15):3412–3419.

CHAPTER 15

PULSED LIGHT

15.1 INTRODUCTION

Pulsed light is a relatively new technology that has had remarkable success in the sterilization of surfaces. It is closely related to the technology of pulsed electric fields that has been highly successful in disinfecting foods and liquids. Several variations and applications of pulsed light technology have been developed or are under research. All of these technologies are actually subsets of pulsed electromagnetic fields, as illustrated by Fig. 15.1. Pulsed electric fields are included because the technology is rather similar to pulsed light even though they are primarily used for liquid disinfection. Specifically omitted from this list of pulsed electromagnetic fields is pulsed microwave irradiation because microwaves are separately addressed in Chap. 21. Also absent from Fig. 15.1 is pulsed infrared light (spectrum 780 to 1100 nm), which has not yet been applied as a disinfection technology.

Figure 15.2 illustrates the wavelength span of the various forms of electromagnetic radiation, of which pulsed light systems occupy a narrow range covering mainly visible light and UV.

15.2 PULSED WHITE LIGHT AND PULSED UV

Pulsed white light (PWL) is also sometimes called pulsed light or *pulsed UV* (PUV) light. Because PWL generally contains UV light and UV is often the primary factor involved in disinfection, the terms PWL and PUV are almost synonymous. However, some systems may produce only UV light, and these could be termed PUV systems. The distinction may not be significant and therefore these two types of systems are treated here together.

PWL involves the pulsing of a high-power xenon lamp for about 0.1 to 3 ms for some types of systems (Johnson, 1982), or about 100 µs to 10 ms for other types (Wekhof, 2000). The spectrum of light produced resembles the spectrum of sunlight but is momentarily about 20,000 times as intense (Bushnell et al., 1997). The spectrum of PWL includes a large component of ultraviolet light, which may be responsible for most of the biocidal effects. Figure 15.3 shows a comparison of the spectrum of one pulsed lamp with a germicidal effectiveness curve.

308 AIRBORNE DISEASE CONTROL TECHNOLOGIES

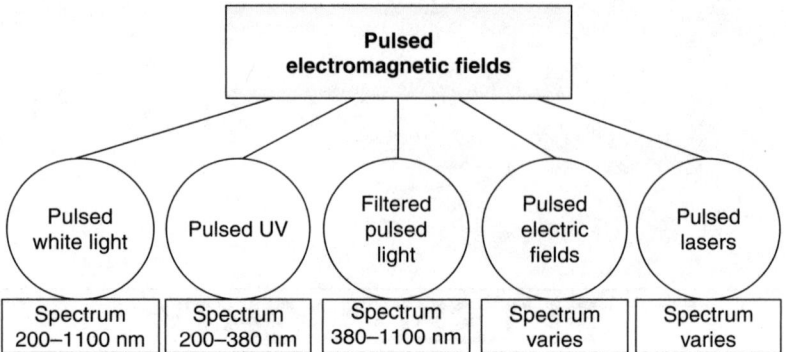

FIGURE 15.1 Breakdown of pulsed electromagnetic disinfection technologies.

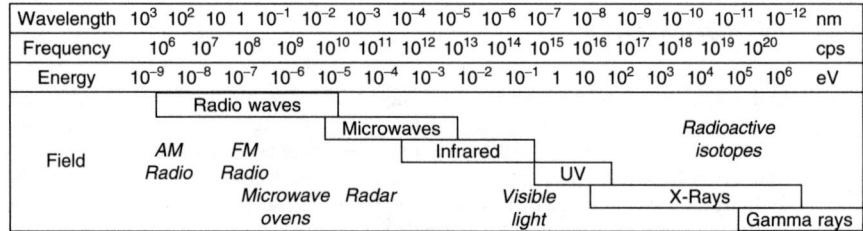

FIGURE 15.2 Illustration of the wavelengths of various types and sources of electromagnetic energy.

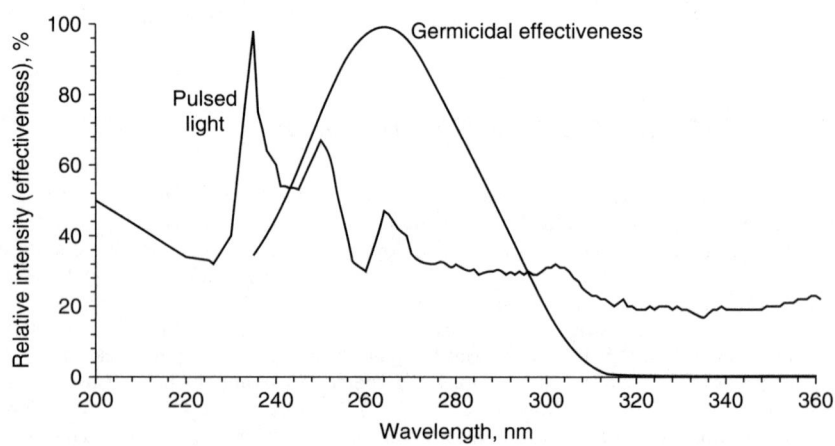

FIGURE 15.3 Relative irradiance of pulsed light compared against a germicidal effectiveness curve. [*Based on data from Panico (2002) and IESNA (2000).*]

The ability of pulsed light to penetrate more deeply than UVGI allows its use in various applications where material translucence would limit the effectiveness of UVGI. PWL is currently being applied for medical equipment sterilization and in the pharmaceutical packaging industry where translucent aseptically manufactured bottles and containers are sterilized in a once-through light treatment chamber (Wallen et al., 2001; Bushnell et al., 1998). The chamber generates a light irradiance at the surface of the exposed containers of about 1.7 J/cm^2, or 1.7 W s/cm^2. Pulsed light is sometimes called flash pasteurization in the food industry where it has been used to extend the shelf life of foods (Deng and Cliver, 2001; Sharma and Demirci, 2003). Hillegas and Demirci (2003) have used pulsed light to inactivate *Clostridium* in clover honey. Jun et al. (2003) have demonstrated the inactivation of *Aspergillus* spores after treatment of corn meal.

Figure 15.4 shows a survival curve for *Aspergillus niger* spores exposed to pulsed UV light, based on data from Dunn et al. (1997). The pulsed light decay curve in this case can be fitted to a single-stage exponential decay equation (Eq. (7.11) in Chap. 7).

Pulsed light systems vary from large to desktop size, as shown in Fig. 15.5. Some systems generate so much heat they require an external cooling unit or connections for fluid cooling. A new type of pulsed lamp, called a surface discharge lamp, has been shown to be effective in disinfecting water (Schaefer and Linden, 2001)

In some cases, only two or three pulses are sufficient to completely eradicate bacteria and fungal spores. Two pulses at 0.75 J/cm^2 each (1 J = 1 Ws) were sufficient to reduce plate cultures of *Staphylococcus aureus* by more than seven logs of cfu (Dunn et al., 1997). Spores of *Bacillus subtilis, Bacillus pumilus, Bacillus stearothermophilus,* and *Aspergillus niger* were completely inactivated from an initial six to eight logs of cfu with one to three pulses (Bushnell et al., 1998).

The pulses generated may last several microseconds and several pulses may be emitted per second. The pulses can be approximated as square waves, as shown in Fig. 15.6. Assuming the pulse is perfectly rectangular, the energy in each pulse can be computed by multiplying the amplitude in watts by the duration in seconds as follows:

$$E = A \cdot t_d \tag{15.1}$$

where E = energy, Ws
A = amplitude of pulse, W
t_d = duration of pulse, s

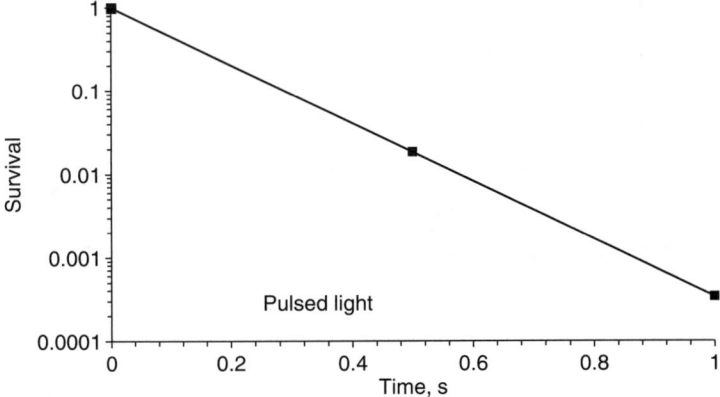

FIGURE 15.4 Pulsed light inactivation decay curve for *Aspergillus niger*. [*Based on data from Dunn et al. (1997).*]

FIGURE 15.5 Pulsed Light sterilization system showing hoses for cooling water. (*Image courtesy of PurePulse of California.*)

It should be obvious that as the number of pulses is increased and the amplitude decreased, it will approach, in the limit, a continuous wave with the same total power. Therefore, the equivalent power of a continuous wave can be computed by multiplying the energy of each pulse by the number of pulses per second as follows:

$$P = E \cdot n \qquad (15.2)$$

where P is the power, W, and n is the number of pulses per second.

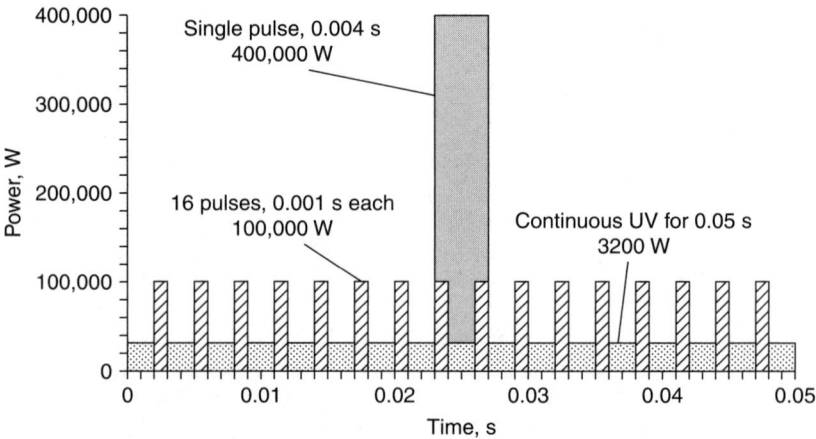

FIGURE 15.6 Comparison of pulses at different power levels. The single pulse shown is equivalent to the 16 pulses in terms of total energy.

In Fig. 15.6, the energy of the single pulse is

$$E = 400,000(\text{W}) \times 0.004(\text{s}) = 1600(\text{W} \cdot \text{s}) \qquad (15.3)$$

The energy in the 16 smaller pulses is seen to be equivalent as follows:

$$E = 100,000(\text{W}) \times 0.001(\text{s}) \times 16 = 1600(\text{W} \cdot \text{s}) \qquad (15.4)$$

Both the single pulse and the 16 smaller pulses can be equated to the energy of continuous wave at 100 W for 16 second as follows:

$$E = 100(\text{W}) \times 16(\text{s}) = 1600(\text{W} \cdot \text{s}) \qquad (15.5)$$

The actual number of pulses in a pulsed light system is typically on the order of 1 to 10 per second. This is known as the pulse frequency and is normally specified in units of Hz, which is cycles per second (or pulses per second). Often, the power input to a pulsed light system will be specified in unit of Ws, which allows the equivalent or effective wattage to be computed. For example, if a pulsed power input of 500 Ws is used to generate three pulses per second for 1 second, the effective wattage is

$$P_e = 500(\text{W} \cdot \text{s}) \times 3(\text{Hz}) = 1500(\text{W}) \qquad (15.6)$$

where P_e is the effective wattage, W.

Equations (15.1) through (15.6) implicitly include the entire pulsed light spectrum. For purposes of computing a PWL rate constant it may be necessary to separate the UV spectrum, assuming that the UV portion is responsible for most of the biocidal effect. This may or may not be true, as will be seen later, but for most purposes it is reasonable to assume that the bulk of the disinfection is due to UV. In this case Eq. (15.2) can be written as follows:

$$P_{\text{uv}} = f_{\text{uv}} \cdot E \cdot n \qquad (15.7)$$

where P_{uv} is the UV power output and f_{uv} is the fraction of pulsed light output in the UV range (i.e., UVC). The UV power output can be used with the UV lamp model in Chap. 11 to compute the irradiance.

The conversions from Eqs. (15.1) through (15.7) are used in the following data sets, in which the pulsed light rate constant is estimated. In these examples, which involved pulsed UV light, it was necessary to first determine the equivalent lamp wattage in order to apply the lamp irradiance equations from Chap. 11. Once the irradiance at the surface of the plates in which the microorganisms were being exposed was determined, the pulsed light rate constant can be determined in the same way as for UVGI exposure. Table 15.1 shows the estimated rate constants for several microbial species exposed to various frequencies and pulsed power levels.

The doses in Table 15.1 were provided in the source documents. Most of the values for irradiance were measured with calorimetric methods, but the irradiance in UVDI (2002) was computed using the effective lamp wattage, the lamp dimensions, and the enclosure characteristics using the algorithms in Chap. 11. A comparison of the PWL rate constants with UVGI rate constants is summarized in Table 15.2. In each case the average irradiance was computed using the methods of Chap. 11 and the resulting PWL rate constant is compared against UVGI rate constants for identical or similar microorganisms. In most cases it is clear that PWL

TABLE 15.1 Summary of PWL Studies and Estimated Rate Constants

Microbe	Medium	Dose, J/m^2	Est., % UV	UV Dose, J/m^2	Survival, frac	PWL k, m^2/J	Reference
Staphylococcus aureus	Plates	15,000	25	3,750	1.0E–07	0.001075	Dunn, 1996
	Solution	3,500	25	875	1.0E–08	0.005263	Dunn, 2000
	Plates	100,000	40	40,000	1.0E–12	0.000691	Wekhof, 2000
	Plates	168,000	25	42,000	1.3E–05	0.000269	Krishnamurthy et al., 2003
Staphylococcus epidermis	Air	1,558	100	1,558	4.8E–04	0.000480	UVDI, 2002
Bacillus subtilis spores	Plates	40,000	25	10,000	1.0E–07	0.000403	Dunn, 1996
	Glass	11,000	30	3,300	1.0E–04	0.000837	Wekhof, 2001
	Plates	100,000	40	40,000	1.00E–09	0.00052	Wekhof, 2000
	Solution	15,000	25	3750	4.0E–07	0.00393	Wallen et al., 2001
	Air	8,237	100	8,237	3.0E–03	0.000810	UVDI, 2002
	Glass	1000	40	400	1.0E–06	0.03454	Schaefer and Linden, 2001
B. pumilis spores	Plastic	45,000	56	25,200	1.0E–06	0.000307	Bushnell et al., 1998
Aspergillus niger spores	Solution	53,000	25	13,250	1.0E–06	0.000261	Dunn, 2000
	Plastic	45,000	56	25,200	1.6E–01	0.000167	Bushnell et al., 1998
	Plastic	45,000	65	29,250	1.0E–01	0.000079	Bushnell et al., 1998
	Plastic	45,000	79	35,550	2.6E–03	0.000072	Bushnell et al., 1998
	Glass	11,000	30	3,300	0.12589	0.000628	Wekhof et al., 2001
Aspergillus versicolor spores	Air	1,781	100	1,781	2.2E–03	0.004690	UVDI, 2002
Candida albicans	Solution	9,500	25	2,375	1.0E–07	0.001697	Dunn, 2000
E. coli	Solution	10,000	25	2500	1.0E–12	0.00069	Wekhof, 1991
	Glass	624	25	156	1.0E–04	0.05904	Schaefer and Linden, 2001
Salmonella enteritidis	Eggshells	40,000	25	10,000	8.1E–07	0.000351	Dunn, 1996
Cryptosporidium parvum	Water	120,000	25	30,000	2.2E–02	0.00013	Arrowood et al., 1996
MS2 coliphage	Glass	400	40	160	1.0E–04	0.05756	Schaefer and Linden, 2001

requires considerably more energy than UVGI to achieve the same disinfection rate, but this effect seems also to be dose dependent.

Since the effect of PWL is considered by many to be mostly due to the UV component, then theoretically the same rate constants should be obtained at equal power levels. Figure 15.7 plots the rate constants of Table 15.2 versus the PWL dose. The results seem to suggest that at similar doses both PWL and UVGI will have similar rate constants. The

TABLE 15.2 Ratio of UV Rate Constants to Estimated PWL Rate Constants

Microbe	PWL k, m²/J	UVGI k, m²/J	UVGI Reference	UVGI/PWL k ratio
Staphylococcus aureus	0.001075	0.05760	Chang et al., 1985	54
	0.005263			11
	0.000691			83
	0.000269			214
Staphylococcus epidermis	0.000480	0.01430	Harris et al., 1993	30
Bacillus subtilis spores	0.000403	0.01920	Nagy, 1964	48
	0.000837			23
	0.000520			37
	0.00393			12
	0.000810			24
Aspergillus niger spores	0.000261	0.00174	Nagy, 1964	6.7
	0.000167			10
	0.000079			22
	0.000072			24
	0.000628			3
Aspergillus versicolor spores	0.004690	0.00344	Luckiesh, 1946	0.7
Candida albicans	0.001697	0.92100	Ishida et al., 1991	543
E. coli	0.00069	0.05120	Zemke et al., 1990	74
Salmonella enteritidis	0.000351	0.22300	Nagy, 1964	635
Cryptosporidium parvum	0.00013	0.00014	Lorenzo et al., 1993	1.08

corollary is that the PWL process is progressively less efficient as the dosage (and the associated speed of disinfection) is increased. This is a familiar phenomenon, since most physical processes become less efficient as the speed of the process is increased. Although PWL may be more effective than UVGI as is often reported (Marshall, 1999; McDonald et al., 2000), it would appear to be only in terms of speed of inactivation and not in terms of efficiency. However, it is not certain why the UV and PWL rate constants are so divergent at high doses and this may be an artifact of the experimental method or an irradiance measurement problem. It remains for future research to resolve these issues.

It has often been reported that there is no tailing, or second-stage curve, in PWL disinfection, but this is not necessarily true. In most cases it appears complete sterilization is achieved so rapidly that no evidence of a second stage remains. In lower dose studies of PWL and PEF, however, there is evidence of a second stage. Figure 15.8 shows the combined results for three tests of the pulsed light inactivation of spores of *Aspergillus niger* (Sonenshein, 2001). The spore samples were placed at various positions from the pulsed UV lamp as indicated in the figure. The lamp power levels were not reported and so it is not possible to model the inactivation rates, but it is clear the decay is approximately exponential. The highest irradiance exposure (on lamp axis) produced a second stage of inactivation in all three tests.

Figure 15.9 shows an example of a decay curve of *Bacillus subtilis* under pulsed light exposure. This curve suggests both a shoulder and a second stage, although the last data point may be below the detection limit.

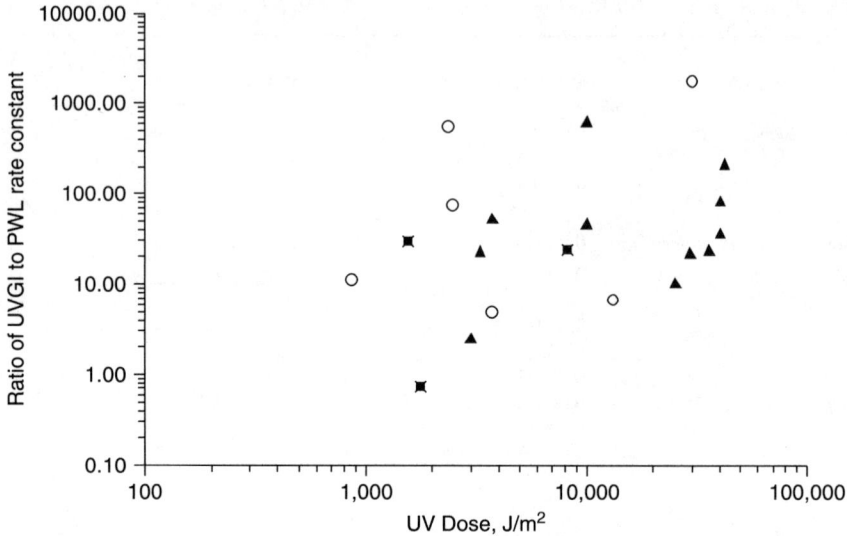

FIGURE 15.7 Ratio of UVGI to PWL rate constants. Triangles represent PWL test results for surfaces, circles for liquids, and squares for air.

The data in Fig. 15.9 were fitted to a multihit model two-stage curve with a shoulder as defined by Eq. (7.17) in Chap. 17 as

$$S = (1-f)[1-(1-e^{-k_1 D})^{n_1}] + f[1-(1-e^{-k_2 D})^{n_2}] \quad (15.8)$$

Since the second stage is such a tiny fraction f is approximately unity, as is n_2, and the equation can be simplified to

$$S = [1-(1-e^{-k_1 D})^n] + e^{-k_2 D} \quad (15.9)$$

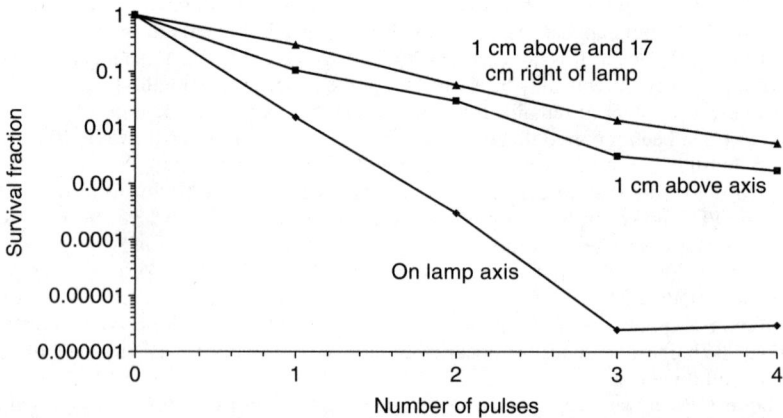

FIGURE 15.8 Pulsed light inactivation curves for *Aspergillus niger* at three different locations from the lamp. [*Based on data from Sonenshein (2001).*]

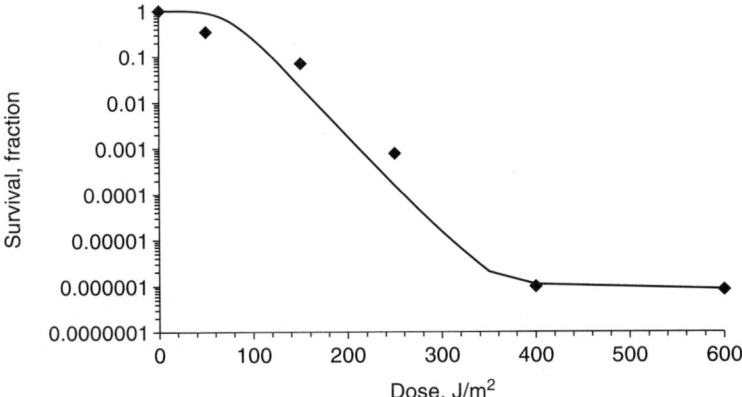

FIGURE 15.9 Pulsed light inactivation curves for *Bacillus subtilis*. [*Based on data from Schaefer and Linden (2001).*]

The first- and second-stage rate constants were approximated by separately fitting the data within the well-defined first and second stages, respectively. The value of n was then found by trial and error with minor adjustments made to k_1 and k_2. The resulting fitted curve is as follows:

$$S = [1 - (1 - e^{-0.05D})^{40}] + e^{-0.00081D} \tag{15.10}$$

The kind of damage done to bacterial cells by PWL can be distinctly different from most other forms of disinfection. Figure 15.10 shows an example of spores that have been subjected to pulsed light. The cavitation of the cells is likely due to dehydration or destruction of internal cell constituents. Although pulsed UV is capable of causing thermal damage, it would appear that a secondary disinfection mechanism is at work in higher power PWL systems that operates independently of the UV effects.

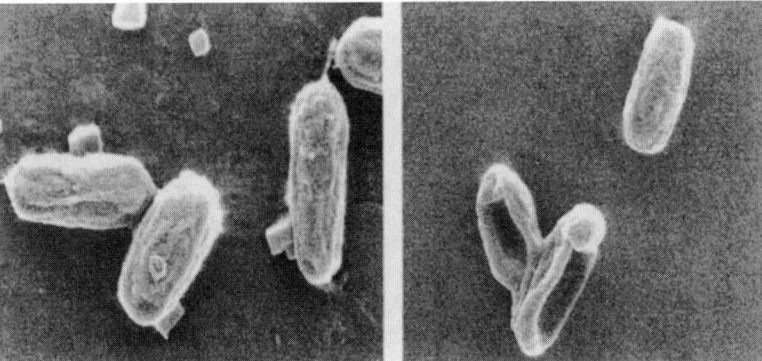

FIGURE 15.10 Effects of pulsed light on *Bacillus subtilis* spores. Left shows spores before, and the right image shows spores after exposure. (*Images courtesy of Alex Wekhof, Wektec Inc., Germany.*)

15.3 FILTERED PULSED LIGHT

A secondary effect, here called pulsed light disintegration, can occur when power levels are increased above those normally used. This effect is not dependent on the UV component of white light and can be achieved even after filtering out the UV spectrum. Filtering out the UV component of PWL and boosting total energy produces pulsed light that is not necessarily hazardous to humans and yet maintains biocidal properties (Wekhof et al., 2001). Bacterial cell wall rupture can result from sudden overheating with or without the presence of the UVC band. The overheating results primarily from the UVA content, and under high doses the disintegration effect can be dominant.

Figure 15.11 show *Aspergillus niger* spores before and after exposure to two different levels of filtered pulsed light. In the middle image, the spore has been collapsed by five pulses of 5 kW/cm^2 each. The spore on the right, which is clearly ruptured, was subjected to two pulses of 33 kW/cm^2 each. It is not yet known at what irradiance levels the disintegration effect becomes dominant, but Dunn (2000) suggested that at 175 J/cm^2 the inactivation slope steepens inexplicably for Bacillus spores.

Pulsed light disintegration may be able to destroy bacterial cells on skin surfaces without necessarily harming skin cells due to the fact that skin cells are packed in a matrix, giving them protection against sudden overpressure. Even if skin cells were damaged, they may eventually recover from natural regeneration processes. The ability of this process to produce surface disinfection has promising potential applications in the medical industry where it could be used in operating rooms to control surgical site infections. More research is needed on this technology to assure that any new hazards posed will at least outweigh the benefits. If this technology proves viable, it may also have potential applications in personnel decontamination where it might be used to disinfect contaminated skin surfaces.

If the pulsed light disintegration effect becomes manifest at some as-yet-unknown power level, then there may be a threshold at which the majority of the disinfection is due to disintegration rather than normal UV disinfection. Figure 15.12 shows a figurative graph of how the transition from UV decay to disintegration might look if it could be separated and plotted against power levels. Presumably the power level required for disintegration is more a function of the rapid delivery of a dose than it is of the dose itself, offering the possibility that doses delivered in extremely short periods might even approach or possibly surpass UV efficiencies.

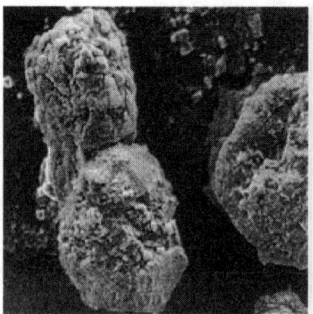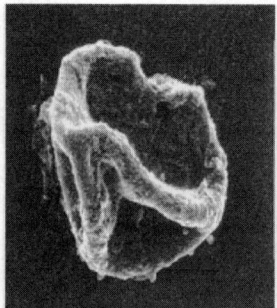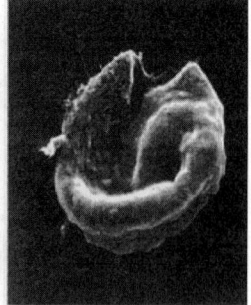

FIGURE 15.11 Pulsed light effects on *Aspergillus niger* spores. The leftmost image shows spores before pulsing, the middle image shows normal levels of pulsed light, while the image on the right shows the disintegrating effects of increased pulsed light irradiance. (*Photos courtesy of Alex Wekhof, Wektec Inc., Germany.*)

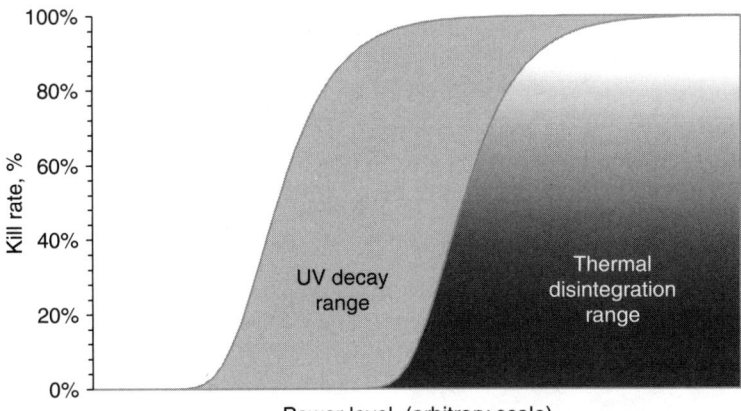

FIGURE 15.12 A figurative graph illustrating the transition from UV effects to disintegration effects as the pulsed power level is increased.

15.4 PULSED ELECTRIC FIELDS

Pulsed light is actually a variation of the more general field of pulsed electric field technology. Electric fields and light are both electromagnetic radiation, although the mechanisms of inactivation are distinctly different. The use of *pulsed electric fields* (PEF) for processing foods was pioneered by H. Doevenspeck in Germany, whose PEF system was patented in 1960 (Barbosa-Canovas and Zhang, 2001). Pulsed electric fields can be used to disinfect fluids like milk and juices, and are currently in use in the food industry for such applications. PEF technology has also been applied to water systems, such as for the eradication of *Cryptosporidium*, and systems are currently available for such applications (Clark et al., 1997; Bendicho et al., 2002; Liang et al., 2002). Although effective in liquids, it is unlikely that a PEF system could be developed that would be effective in air due to the immense electrical resistance of air. PEF may, however, have some applications in surface disinfection.

PEF disinfection involves the pulsing of an electric field of about 4 to 14 kV/cm through a liquid medium. PEF sterilization typically requires an electric field of no less than 8 kV/cm (Peleg, 1995). The result of this momentary field is a membrane potential across the bacterial cell wall of more than 1.0 V, which is sufficient to lyse or damage the cell irreparably (Bruhn, 1997; Qin et al., 1996; Lado and Yousef, 2002). The inactivation of various microbes, including *Escherichia coli, Lactobacillus brevis, Listeria, Pseudomonas fluorescens, Clostridium* spores, *Staphylococcus, Bacillus cereus* spores, Rotavirus, and *Salmonella*, has been found to be dependent on field strength and treatment times that are unique to each species (Khadre and Yousef, 2002; Wuytack et al., 2003; Bolton et al., 2002; Lado and Yousef, 2002; Grahl and Markl, 1996; MacGregor et al., 1998; Rowan et al., 1999). Since this method has little effect on proteins, enzymes, or vitamins, it is perfectly suited for food processing.

PEF disinfection conforms to the standard exponential decay equation but often exhibits two-stage behavior. The temperature of fluids will normally increase from electrical pulsing, and there may be some degree of heat inactivation concurrent with PEF inactivation. Figure 15.13 shows a typical survival curve for spores exposed to PEF. Three different pulse rates were used in this test but they all produced the same total energy, being 3 µs at

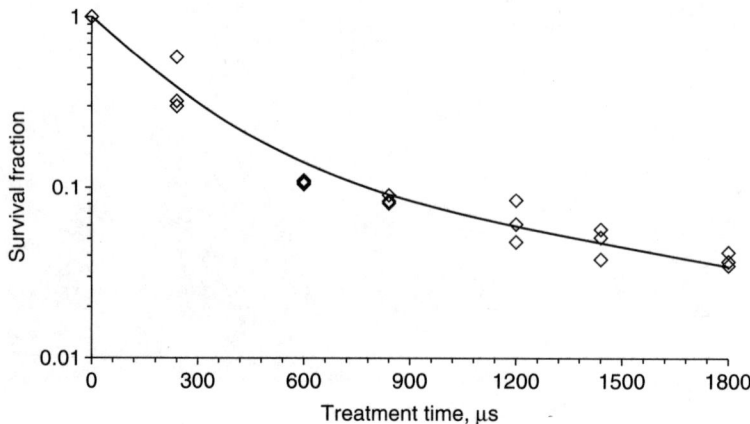

FIGURE 15.13 Inactivation of *Bacillus subtilis* spores exposure to PEF. Line shows a two-stage curve fitted to the test results. [*Based on data from Jin et al. (2001).*]

2 kHz, 6 μs at 1 kHz, and 12 μs at 0.5 kHz (Jin et al., 2001). Two stages were evident in the decay curve and the data were fitted to a two-stage exponential decay curve without a shoulder (Eq. (7.12) in Chap. 7), since there were insufficient data to resolve any shoulder.

15.5 PULSED LASERS

Lasers are by nature a form of pulsed light, but the pulsing is so rapid that they approximate continuous waves. The frequencies of laser pulses are typically 5 to 35 Hz and energy densities of 0.1 to 26 mJ/cm^2 are effective at destroying bacterial cells (Tuszynski et al., 1986). Figure 15.14 shows an example of a UV laser inactivation curve for *Saccharomyces* yeast

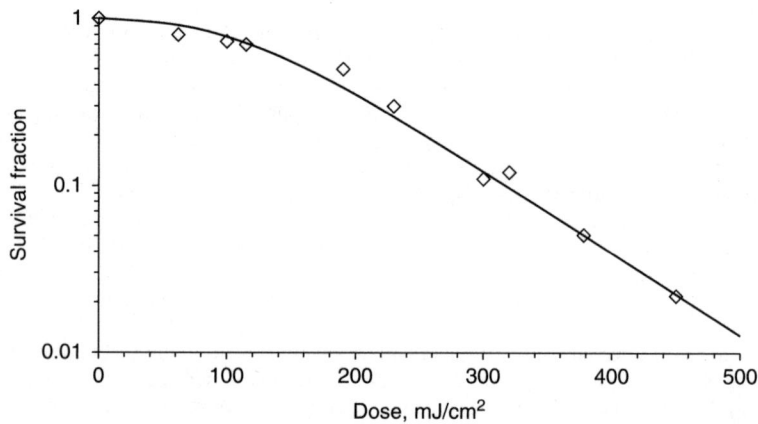

FIGURE 15.14 Survival curve for *Saccharomyces cerevisiae* yeast cells exposed to 308 nm laser pulses at 1.5 mJ/cm^2 per pulse. [*Based on data from Tuszynski et al. (1986).*]

cells. Data were fitted to a single-stage multihit model (Eq. (7.16) in Chap. 7) with an exponent of $n = 4$.

The biocidal effects of lasers have been demonstrated in a handful of experiments on other microbial species (Macmillan et al., 1966; Takahashi et al., 1975). In spite of their effectiveness, lasers suffer from the disadvantage of being an energy-intensive and expensive method in comparison with UVGI. However, some special applications in surface disinfection may exist for the use of lasers (Sadoudi et al., 1997).

REFERENCES

Arrowood, M. J., Xie, L.-T., Rieger, K., and Dunn, J. (1996). "Disinfection of *Cryptosporidium parvum* oocysts by pulsed light treatment evaluated in an in vitro cultivation model." *J Euk Microbiol* 43(5):88S.

Barbosa-Canovas, G. V., and Zhang, Q. H. (2001). *Pulsed Electric Fields in Food Processing.* Technomic Publishing Co., Lancaster, PA.

Bendicho, S., Espachs, A., Arantegui, J., and Martin, O. (2002). "Effect of high intensity pulsed electric fields and heat treatments on vitamins of milk." *J Dairy Res* 69(1):113–123.

Bolton, D. J., Catarame, T., Byrne, C., Sheridan, J. J., and McDowell, D. A. (2002). "The ineffectiveness of organic acids, freezing and pulsed electric fields to control *Escherichia coli* O157:H7 in beef burgers." *Lett Appl Microb* 34:139–143.

Bruhn, R. E. (1997). "Electrical environment surrounding microbes exposed to pulsed electric fields." *IEEE T: Dielect El In* 4(6):806.

Bushnell, A., Clark, W., Dunn, J., and Salisbury, K. (1997). "Pulsed light sterilization of products packaged by blow-fill-seal techniques." *Pharm Eng* 17(5):74–84.

Bushnell, A., Cooper, J. R., Dunn, J., Leo, F., and May, R. (1998). "Pulsed light sterilization tunnels and sterile-pass-throughs." *Pharm Eng* 18:48–58.

Chang, J. C. H., Ossoff, S. F., Lobe, D. C., Dorfman, M. H., Dumais, C. M., Qualls, R. G., and Johnson, J. D. (1985). "UV inactivation of pathogenic and indicator microorganisms." *Appl Environ Microbiol* 49(6):1361–1365.

Clark, W., Bushnell, A., Dunn, J., and Ott, T. (1997). "Pulsed light and pulsed electric fields for food preservation, Paper 65f." *AIChE Annual Meeting*, Los Angeles, CA.

Deng, M. Q., and Cliver, D. O. (2001). "Inactivation of *Cryptosporidium parvum* oocysts in cider by flash pasteurization." *J Food Prot* 64(4):523–527.

Dunn, J. (1996). "Pulsed light and pulsed electric field for food and eggs." *Poult Sci* 75(9):1133–1136.

Dunn, J., Burgess, D., and Leo, F. (1997). "Investigation of pulsed light for terminal sterilization of WFI filled blow/fill/seal polyethylene containers." *Parenterl Drug Assoc J Pharm Sci Tech* 51(3):111–115.

Dunn, J. (2000). "Pulsed light disinfection of water and sterilization of blow/fill/seal manufactured aseptic pharmaceutical products." Report by Automatic Liquid Packaging. Woodstock, IL.

Grahl, T., and Markl, H. (1996). "Killing of microorganisms by pulsed electric fields." *Appl Microbiol Biotechnol* 45:148–157.

Harris, M. G., Fluss, L., Lem, A., and Leong, H. (1993). "Ultraviolet disinfection of contact lenses." *Optom Vis Sci* 70(10):839–842.

Hillegas, S. L., and Demirci, A. (2003). "Inactivation of *Clostridium sporogenes* in clover honey by pulsed UV-light treatment." *ASAE 2003 International Meeting*, Las Vegas, NV

IESNA (2000). *Lighting Handbook 9th Edition IESNA HB-9-2000.* Illumination Engineering Society of North America, New York.

Ishida, H., Nahara, Y., Tamamoto, M., and Hamada, T. (1991). "The fungicidal effect of ultraviolet light on impression materials." *J Prosthet Dent* 65(4):532–535.

Jin, Z. T., Su, Y., Tuhela, L., Zhang, Q. H., Sastry, S. K., and Yousef, A. E. (2001). Chapter 11: Inactivation of Bacillus subtilis spores using high voltage pulsed electric fields. *Pulsed Electric*

Fields in Food Processing, G. V. Barbosa-Canovas and Q. H. Zhang, eds., Technomic Publishing Co., Lancaster, PA.

Johnson, T. (1982). "Flashblast: the light that cleans." *Pop Sci* 233(July):82–84.

Jun, S., Irudayaraj, J., and Geisner, D. (2003). "Pulsed UV-light treatment of corn meal for inactivation of *Aspergillus niger*." *Int J Food Sci Tech*, 38:883–888.

Khadre, M. A., and Yousef, A. E. (2002). "Susceptibility of human rotavirus to ozone, high pressure, and pulsed electric field." *J Food Prot* 65(9):1441–1446.

Krishnamurthy, K., Demirci, A., and Irudayaraj, J. (2003). "Paper # 03-037: Inactivation of *Staphylococcus aureus* using pulsed UV treatment." *NABEC 2003 Northeast Agricultural Biological Engineering Conference,* Storrs, CT.

Lado, B. H., and Yousef, A. E. (2002). "Alternative food-preservation technologies: efficacy and mechanisms." *Microbes and Infect* 4:433–440.

Liang, Z., Mittal, G. S., and Griffiths, M. W. (2002). "Inactivation of Salmonella Typhimurium in orange juice containing antimicrobial agents by pulsed electric field." *J Food Prot* 65(7):1081–1087.

Lorenzo-Lorenzo, M. J., Ares-Mazas, M. E., de Maturana, I. V.-M., and Durn-Oreiro, D. (1993). "Effect of ultraviolet disinfection of drinking water on the viability of Cryptosporidium parvum oocysts." *J Parasitol* 79(1):67–70.

Luckiesh, M. (1946). *Applications of Germicidal, Erythemal and Infrared Energy.* D. Van Nostrand Co., New York.

MacGregor, S. J., Rowan, N. J., McIlvaney, L., Anderson, J. G., Fouracre, R. A., and Farish, O. (1998). "Light inactivation of food-related pathogenic bacteria using a pulsed power source." *Lett Appl Microbiol* 27:67–70.

MacMillan, J. D., Maxwell, W. A., and Chichester, C. O. (1966). "Lethal photosensitization of microorganisms with light from a continuous-wave gas laser." *Photochem Photobiol* 5:555–565.

Marshall, T. (1999). "Deadly pulses." *Water Environ Technol* 11(2):37–41.

McDonald, K., Curry, R. D., Clevenger, T. E., Unklesbay, K., Eisenstark, A., Golden, J., and Morgan, R. D. (2000). "Comparison of pulsed and continuous ultraviolet light sources for the decontamination of surfaces." *IEEE T Plasma Sci* 28(5):1581–1587.

Nagy, R. (1964). "Application and measurement of ultraviolet radiation." *AIHA J* 25:274–281.

Panico, L. R. (2002). "Instantaneous surface sanitization with pulsed UV." *Hygienic Coatings Global Conference,* Brussels, Belgium. http://www.xenon-corp.com/press/images/SurfaceSanitization.pdf.

Peleg, M. (1995). "A model of microbial survival after exposure to pulsed electric fields." *J Sci Food Agric* 67:93–99.

Qin, B., Pothakamury, U. R., Barbosa-Canovas, G. V., and Swanson, B. G. (1996). "Nonthermal pasteurization of liquid foods using high-intensity pulsed electric fields." *Crit Rev Food Sci Nutr* 36(6):603–627.

Rowan, N. J., MacGregor, S. J., Anderson, J. G., Fouracre, R. A., McIlvaney, L., and Farish, O. (1999). "Pulsed-light inactivation of food-related microorganisms." *Appl Environ Microbiol* 65(3):1312–1315.

Sadoudi, A. K., Herry, J. M., and Cerf, O. (1997). "Elimination of adhering bacteria from surfaces by pulsed laser beams." *Lett Appl Microbiol* 24(3):177–179.

Schaefer, R. B., and Linden, K. (2001). "Innovative ultraviolet light source for disinfection of drinking water." *First International Congress on Ultraviolet Technologies,* Washington, DC.

Sharma, R. R., and Demirci, A. (2003). "Inactivation of *Escherichia coli* O157:H7 on inoculated alfalfa seeds with pulsed ultraviolet light and response surface modeling." *J Food Sci* 68(4):1448–1453.

Sonenshein, A. L. (2001). "Killing of *Bacillus* spores by high-intensity ultraviolet light." Tufts University School of Medicine/Xenon Corp. Boston, MA. http://www.xenon-corp.com/products/images/Final_Report.pdf.

Takahashi, P. K., Toups, H. J., Greenberg, D. B., and Dimopoullos, G. T. (1975). "Irradiation of *Escherichia coli* in the visible spectrum with a tunable organic-dye laser energy source." *Appl Microbiol* 29(1):63–67.

Tuszynski, W., Schaarschmidt, B., and Lamprecht, I. (1986). "Inactivation of *Saccharomyces* cells by 8-methoxypsoralen plus pulsed laser irradiation in the wavelength range 308 nm-380 nm." *Radiat Environ Biophys* 25:55–63.

UVDI (2002). "Report on pulsed light disinfection of microorganisms prepared by K. Foarde and Research Triangle Institute." Ultraviolet Devices Incorporated, Valencia, CA.

Wallen, R. D., May, R., Reiger, K., Holoway, J. M., and Cover, W. H. (2001). "Sterilization of a new medical device using broad-spectrum pulsed light." *Biomed Instrum Technol* 35(5):323–330.

Wekhof, A. (1991). "Treatment of contaminated water, air, and soil with UV flashlamps." *Environ Prog* 10(4):241–247.

Wekhof, A. (2000). "Disinfection with flashlamps." *PDA J of Pharm Sci Technol* 54(3):264–267. http://www.wektec.com/.

Wekhof, A., Trompeter, I. -J., and Franken, O. (2001). "Pulsed UV-Disintegration, a new sterilization mechanism for broad packaging and medical-hospital applications." *Proceedings of the First International Congress on UV-Technologies*, Washington, DC. http://www.wektec.com/.

Wuytack, E. Y., Phuong, L. D., Aertsen, A., Reyns, K. M., Marquenie, D., deKetelaere, B., Masschalck, B., vanOpstal, I., Diels, A. M., and Michiels, C. W. (2003). "Comparison of sublethal injury induced in *Salmonella enterica serovar Typhimurium* by heat and by different nonthermal treatments." *J Food Prot* 66(1):31–37.

Zemke, V., Podgorsek, L., and Schoenen, D. (1990). "Ultraviolet disinfection of drinking water. 1. Communication: Inactivation of *E. coli* and coliform bacteria." *Zentralbl Hyg Umweltmed* 190(1–2):51–61.

CHAPTER 16

IONIZATION

16.1 INTRODUCTION

Ions are produced when electrons are stripped from or added to atoms, leaving a temporary charge imbalance. These charged atoms can cause particles like dust to clump together. Negative air ionization has an effect on reducing the incidence of respiratory infection transmission, but it is somewhat species dependent and can be affected by relative humidity (Estola et al., 1979; Happ et al., 1966). The research has been limited, but several studies have determined that ionization has the potential to reduce the concentration of bioaerosols (Makela et al., 1979; Phillips et al., 1964). Some reports indicate that ions can kill bacteria as well as precipitate them from air (Seo et al., 2001; Krueger, 1985).

Specific sources of ions include cosmic radiation, ionizing radiation from radioactive earth sources, UV light, frictional charging by wind, water droplet breakup (waterfalls, showers), electrical discharges, lightning, combustion, and strong electrical fields (corona). Ions can be scavenged from indoor air environments by combustion processes, smoke, processed ventilation air can deplete ions, and video displays deplete local charges (Daniels, 2002; Kroling, 1985).

16.2 IONIZATION THEORY

Ionization occurs when an electrically neutral atom or molecule acquires a positive or negative electrical charge. An atom is ionized when it absorbs energy in excess of its electron potential. This process yields a free electron and a positively charged atom. The specific reactions that occur in an atmospheric environment depend on the composition of the air and on the physical properties of the atoms and molecules including ionization potential, electron affinity, proton affinity, dipole moment, polarizability, and chemical reactivity (Daniels, 2000). The primary positive ions, N_2^+, O_2^+, N+, and O^+, are converted within microseconds to protonated hydrates, $H^+ (H_2O)^n$ ($n < 10$), while the free electrons rapidly attach to oxygen to form the superoxide radical anion, $^3O_2^-$, which can also form hydrates. These intermediate species are known as cluster ions.

TABLE 16.1 Typical Ionization Levels

Location	Ion levels, ions/cm^3
Smoky room	0–100
Commercial office building	0–250
Airplane cabin	20–250
Naturally ventilated building	250–500
Outdoor urban area	250–750
Country air	1,000–2,000
Mountain air	1,000–5,000
Inside caves	5,000–20,000
Waterfalls, thunderstorms	25,000–100,000

Cluster ions may exist for approximately 1 minute, during which time they can react with airborne gases and particulates. The chain of chemical reactions that can take place are myriad. Protonated hydrates may be about 0.001 mm in diameter and can have electrical mobilities of 1 to 2 cm^2/V·s. Ion clusters may be about 0.01 to 0.1 mm and have electrical mobilities of about 0.3 to 1×10^{-6} m^2/V·s. The typical lifetime of a naturally generated small air ion in clean air is about 100 to 1000 second (Daniels, 2002). The lifetime of ions is strongly dependent on the relative humidity of the air and the presence of trace contaminants. Dolezalek (1985) differentiates between fast (small) ions that last 50 to 250 second and slow (large) ions that can last for days. When a neutral atom loses an electron the result is a positive and negative (fast) ion pair that usually recombine, but in an electric field they may separate. The electron reattaches itself to a neutral molecule within about 10^{-8} second and becomes a negative ion. A slow ion is a charged aerosol particle.

The presence of air ions and free electrons results in a space charge. In fair weather, the natural level of both positive and negative ions at sea level is approximately 200 to 3000 ions/cm^3. During rainfall and thunderstorms, the level of negative ions in the air may increase to 14,000 ions/cm^3 and the positive ions to 7000 ions/cm^3. Smoke, from cigarettes or other sources, may decrease the ion concentration in room air to approximately 10 to 100 ions/cm^3. Table 16.1 lists some examples of ionization levels that may occur in various environments.

The complex chain of reactions that may result from the presence of ions in air may produce temporary species of highly reactive species like superoxide anion radicals, hydroxyl radicals, peroxyl radicals, hydrogen peroxide, and ozone. These species may have biocidal effects on airborne microorganisms and may break down some MVOCs. The ionization of airborne particles, including microbes, may cause them to attach to interior surfaces, including walls, floors, and filters. It will also cause them to cluster or attach to dust particles, a process known as agglomeration, which may subsequently cause them to precipitate out of the air.

16.3 IONIZATION EQUIPMENT

Air ionizers are distinct from both electrostatic precipitators and ozone generators. Recent developments in large ion generator design and operation have led to the commercial availability of energy-efficient units which can now produce controlled outputs of specific ions on demand, while minimizing the formation of undesirable byproducts, such as ozone (Daniels, 2000). Ion generators have been used to control surface static charges, especially in industrial facilities where static charges must be controlled. Air ionizers, or ion generators, are becoming more common for cleaning air in some indoor

TABLE 16.2 Typical Specifications for Ion Generators

Ion generation method	Pulse ionization field
Power supply	9–15 kV
Wattage	0.75–2.7 W
Ozone production	<0.02 ppm

environments where dust is a problem, and are becoming more popular among consumers for their presumed health benefits.

Ionization can also be used to enhance the efficiency of air filters in dry air by causing agglomeration of airborne particulate matter and imparting a static charge to the particles. Such techniques are similar to those used with electrostatic filters (Lee et al. 2001; Chen and Huang, 1998).

Many types of equipment can produce ionization, including electrostatic precipitators, UV lamps, and ozone generators, but ionization is not always the intent of these systems. The electronic ionizer, also called a corona-discharge ionizer, uses a high voltage applied to sharp emitter points or grids to induce a strong local electric field. Gas molecules that enter the field will be ionized and acquire a charge of the same polarity of the electric field. Table 16.2 lists the typical specifications for an ion generator.

Ionizers are classified according to the type of electrical current that they employ, whether pulsed *direct current* (DC), steady-state DC, or *alternating current* (AC). DC devices produce positive or negative ionization depending on their charge. AC devices produce bipolar ionization, generating alternate clouds of positive and negative ions with each cycle.

Many commonly available ion generators are stand-alone devices that are suitable for in-room use, like the one shown in Fig. 16.1. Negative ionization is far more common as a means of cleaning air than positive ionization, and many commercial units are referred to as negative ion generators.

FIGURE 16.1 Negative ion generator for indoor use. (*Unit provided courtesy of Electrocorp, Cotati, CA.*)

Ionization units are also available for installation inside air handling units. They may be located anywhere, including upstream or downstream of the filters. They will typically be sized to produce natural levels of ionization in indoor air, or about 1000 to 10,000 ions/cm^3 or more. Ionization levels can be monitored with charged plate monitors or electrostatic field meters to ensure performance or to control output (ANSI, 1991). Higher levels of ionization may produce undesirable by-products, such as ozone, and an ozone sensor may be installed to keep levels below safe limits.

16.4 IONIZATION OF AIRBORNE MICROBES

In some situations, as where dust may carry microorganisms, negative air ionization can be used to reduce infections by precipitating the dust out of the air, such as in poultry houses (Gast et al., 1998; Petkov et al. 1987). Some success has also been reported in burn wards, classrooms, and dental offices (Gabbay, 1990; Makela et al., 1979).

Negative air ionization has the potential to reduce the concentration of airborne microorganisms. The effect appears to result from the ionization of bioaerosols and dust particles that may carry microorganisms, causing them to settle out more rapidly. Settling tends to occur on horizontal surfaces, especially metallic surfaces, and generally in the area near the ionization unit. Ionization may enhance agglomeration, creating larger particles out of smaller particles, thereby increasing the settling rate. Ionization may also cause attraction between ionized particles and grounded surfaces.

In situations where dust may carry microorganisms, negative air ionization can be economical to use to reduce infections. It has been used economically to reduce the incidence of Newcastle disease virus in poultry houses (Mitchell, 1994). Poultry houses can be notoriously dusty.

Figure 16.2 shows the *colony forming units* (cfu) measured with and without ionization in a dental clinic by Gabbay (1990). Airborne microbial levels were reduced by 32 to 52 percent with ionization. He also found that horizontal plates picked up considerably more cultures than vertical plates, strongly suggesting that settling out of ionized particles was the primary mode of removal.

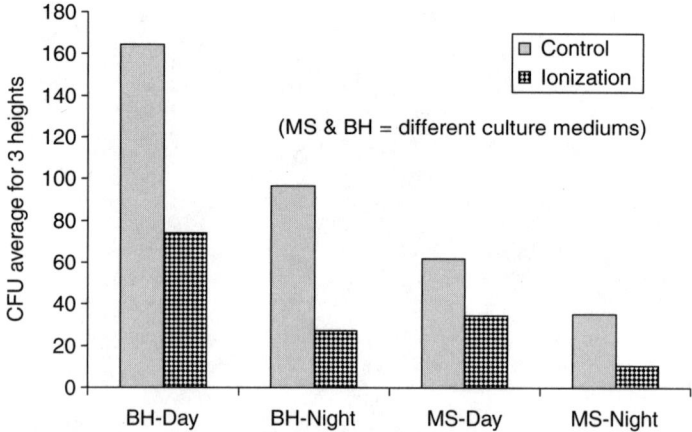

FIGURE 16.2 Reduction of microbial air pollution in a dental clinic. [*Based on data from Gabbay (1990).*]

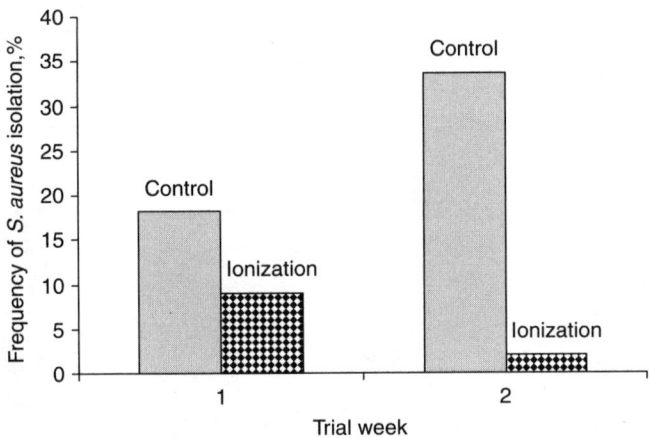

FIGURE 16.3 Reduction in mean cfu by ionization in a patient room. [*Based on data from Makela et al. (1979).*]

Figure 16.3 summarizes the results of studies by Makela et al. (1979), who found that bacterial aerosols in patient rooms of a burns and plastic surgery unit could be reduced with air ionization. Variations in the bacterial levels were associated with bed changing and other room activities. The humidity in the rooms was low, which may have enhanced the effect.

In Fig. 16.4, also based on results from Makela et al. (1979), specifically identified *Staphylococcus aureus* levels in a room with and without ionization. The average for 2 days

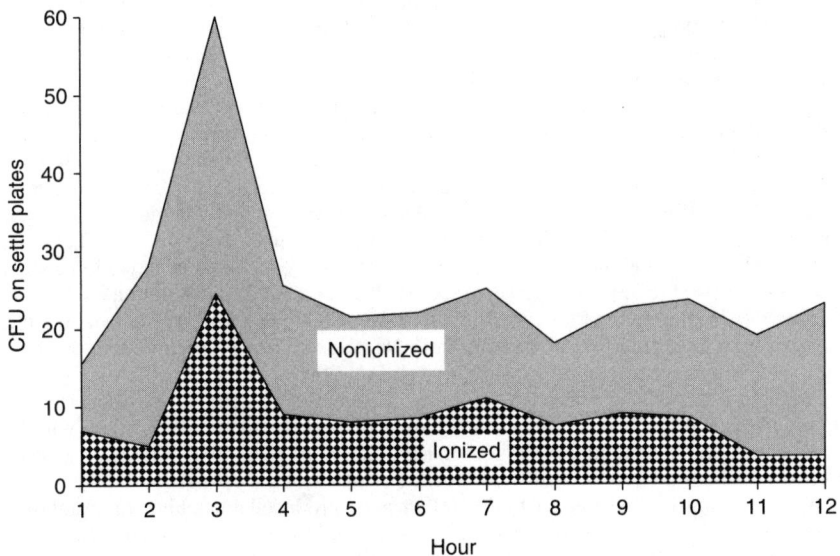

FIGURE 16.4 Reduction of airborne *Staphylococcus aureus* in a patient waiting room with ionization. [*Based on data from Makela et al. (1979).*]

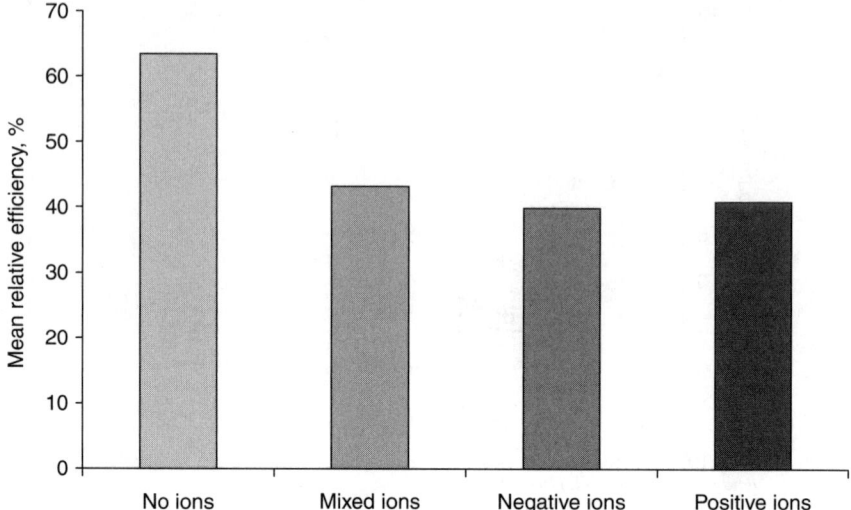

FIGURE 16.5 Effect of ions on T1 phage recovery from aerosols. [*Based on data from Happ et al. (1966).*]

of monitoring indicated a definitive reduction in airborne levels. *S. aureus* is a potential nosocomial infectious agent of wounds and burns.

Figure 16.5 summarizes some results from Happ et al. (1966), who found that levels of aerosolized virus T1 bacteriophage were reduced under various types of ionization, which included mixed ions, negative ions, and positive ions. All three types of ionization had comparable results in terms of reducing airborne levels. The method used involved measuring the filtration efficiency, and in which lower filter efficiencies demonstrated lower recoveries from the air. These lower recoveries suggested either that the phage was not present in the air or had perhaps been inactivated.

Most of the effects of negative air ionization can be explained in terms of agglomeration and the ensuing precipitation of the heavier particles out of the air (Lehtimaki and Graeffe, 1976; Mitchell, 1998). Some evidence suggests that ionization may inhibit microbial growth or that it may have some biocidal effects (Krueger et al., 1957; Krueger and Reed, 1976; Phillips et al., 1963). The biocidal effects may actually be due to the generation of radicals but insufficient information is available to quantify this effect. The growth of cultured bacteria and the viability of bioaerosols were reduced by negative ion concentrations of 50,000 to 5,000,000 ions/cm^3 per Phillips et al. (1964). Tanimura et al. (1997) demonstrated that growth of *E. coli* was inhibited by 1,000,000 ions/cm^3 but that no effects occurred with levels below 100,000 ions/cm^3. Negative air ionization may also be able to reduce microbial contamination on surfaces (USDA, 2000).

Ionization may enhance the biocidal effects of ozonation, possibly due to increased generation of radical species. As shown in Fig. 16.6, the combination of ionization with low levels of airborne ozone provides higher inactivation rates or removal rates than ozone by itself. According to Fan et al. (2002), ozone levels of about 0.1 ppm (100 nL/L) combined with negative ionization levels of 1,000,000 ions/cm^3 produced much higher rates of disinfection for several bacterial species than ozone alone. Ozone concentrations of 0.03 ppm combined with 3,000,000 ions/cm^3 was effective in killing 98 percent of *S. aureus* on plates after 72 hours of exposure (Li et al., 1989). Tanimura et al. (1997) also report that negative

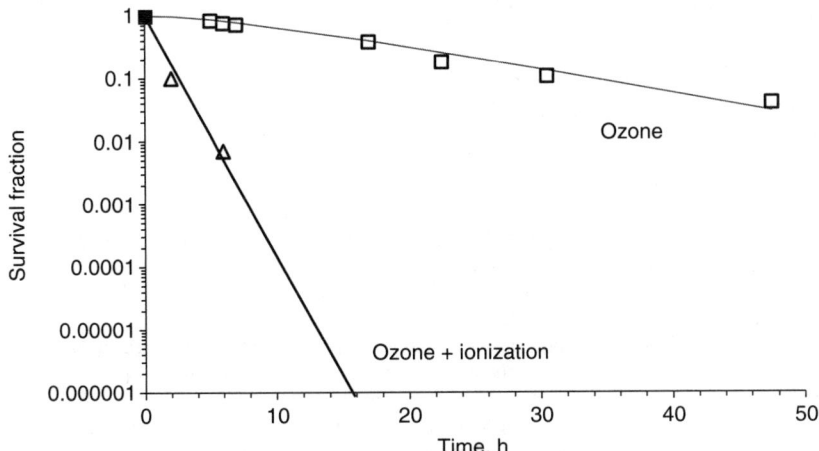

FIGURE 16.6 Enhancement of ozone disinfection with ionization. Lines show a single-stage multi-hit model curve fit to the data. [*Based on data from Fan et al. (2002).*]

air ionization enhanced the effectiveness of ozone at inhibiting microbial growth. The attempt to use ozone in combination with negative air ionization to control surface mold in fruits has met with limited success (Fan et al., 2002; Hildebrand et al., 2001).

The ability of high levels of negative ions to reduce air and surface levels of *Salmonella enteritidis* in a laboratory test was demonstrated by Seo et al. (2001). Tests indicated better than a 90 percent reduction on all surfaces except the top, which had a 72 percent reduction. Researchers report that most of the effect is through direct killing of the bacteria. Arnold and Mitchell (2002) report on a laboratory test in which negative air ionization effectively decreased the survival levels of bacteria on stainless steel, with a reduction efficiency of 99.8 percent.

Mitchell and Waltman (2002) studied the use of an electrostatic space charge system in a commercial broiler hatchery to reduce airborne pathogens that may lead to disease transmission during the hatch. The system transfers a strong negative electrostatic charge to dust and microorganisms that are aerosolized during the hatch and collects the charged particles on grounded surfaces. In studies with three poultry companies, the system achieved 93 to 96 percent reductions of airborne *Enterobacteria*, and reduced airborne *Salmonella* by 33 to 83 percent. Davies and Breslin (2003) reported that an ionized air generator cabinet had negligible effect in reducing surface contamination of *Salmonella* on eggs.

Lee et al. (2004) studied the ability of negative and positive ions to remove fine particulates from an indoor environment. They found that a density of 10^5 to 10^6 electrons/cm^3 of either positive or negative ions would reduce airborne particulates in the 0.1- to 1-μm size range by about 95 to 97 percent after 30 minute. They also noted that high ion densities could produce static charges in the room.

An epidemiological study is currently being conducted by Clive Beggs of the University of Leeds on the use of ionization in hospital wards (Chard, 2005; McDowell, 2003). Air and surface sampling, and patient records are being collected to determine if airborne levels and infections are being reduced. Initial reports indicate that after the first year the nosocomial agent *Acinetobacter* was completely eliminated from the ward air and new infections fell to zero.

16.5 IONIZATION OF MVOCS

Ionization can dissociate molecules of VOCs and reduce them to less harmful compounds. The degree to which ionization reduces VOCs depends on their individual ionization energy. The *ionization energy* (IE) is the amount of energy required to remove an electron from a molecule or an atom and which leads to the formation of an ion. The IE is measured in units of *electron volts* (eV). Many VOCs may be amenable to control by ionization, based on a review of their IE values, but limited information is available on specific MVOCs (Daniels, 2002). The MVOCs acetaldehyde and formaldehyde have IE values of 10.23 and 10.88, respectively, which are less than that of oxygen (IE = 12.07), and which should enable them to be controlled with air ionization.

Air ionization systems have been used for removal of odors, VOCs, and airborne microbes in a variety of installations, including commercial buildings, agricultural facilities, shopping centers, restaurants, meat processing facilities, anatomy laboratories, pathology laboratories, stadiums, and animal handling facilities (Daniels, 2000; Soyka and Edmonds, 1991). Negative ion generators have been used to rid a geriatric lounge of unpleasant orders and clear an out-patient program's hall of tobacco smoke in a psychiatric center (Miller, 1984). Ceiling mounted ion generators proved to be more cost-effective than other air purifiers.

REFERENCES

ANSI (1991). "ESD Association Ionization Standard." *ANSI EOS/ESD S3.1-1991*. American National Standards Institute, New York.

Arnold, J. W., and Mitchell, B. W. (2002). "Use of negative air ionization for reducing microbial contamination on stainless steel surfaces." *J Appl Poultry Res* 11:179–186.

Chard, A. (2005). *New Weapon to Fight Hospital Infections*. University of Leeds. Leeds, UK.

Chen, C. C., and Huang, S. H. (1998). "The effects of particle charge on the performance of a filtering facepiece." *Am Ind Hyg Assoc J* 59:227–233.

Daniels, S. L. (2000). "Applications of air ionization for control of VOCs and PNx." *Air & Waste Management Association 94th Annual Conference & Exhibition,* Orlando, FL. http://www.precisionair.com/news/iaq.pdf.

Daniels, S. L. (2002). "On the ionization of air for removal of noxious effluvia (Air ionization of indoor environments for control of volatile and particulate contaminants with nonthermal plasmas generated by dielectric-barrier discharge)." *IEEE T Plasma Sci* 30(4):1471–1481.

Davies, R. H., and Breslin, M. (2003). "Investigations into possible alternative decontamination methods for *Salmonella enteritidis* on the surface of table eggs." *J Vet Med B Infect Dis Vet Public Health* 50(1):38–41.

Dolezalek, H. (1985). "Remarks on the physics of atmospheric ions (natural and artificial)." *Int J Biometeorol* 29(3):211–221.

Estola, T., Makela, P., and Hovi, T. (1979). "The effect of air ionization on the airborne transmission of experimental Newcastle disease virus infections in chickens." *J Hyg* 83:59–67.

Fan, L., Song, J., Hildebrand, P. D., and Forney, C. F. (2002). "Interaction of ozone and negative air ions to control microorganisms." *J Appl Microbiol* 93:144–148.

Gabbay, J. (1990). "Effect of ionization on microbial air pollution in the dental clinic." *Environ Res* 52(1):99.

Gast, R. K., Mitchell, B. W., and Holt, P. S. (1998). "Application of negative air ionization for reducing experimental airborne transmission of salmonella enteritidis to chicks." USDA. http://www.nal.usda.gov/ttic/tektran/data/000009/20/0000092027.html.

Happ, J. W., Harstad, J. B., and Buchanan, L. M. (1966). "Effect of air ions on submicron T1 bacteriophage aerosols." *Appl Microbiol* 14:888–891.

Hildebrand, P. D., Song, J., Forney, C. F., Renderos, W. E., and Ryan, D. A. J. (2001). "Effects of corona discharge on decay of fresh fruits and vegetables. Proceedings of the Fourth International Conference on Postharvest Science." *Acta Hortic* 553:425–426.

Kroling, P. (1985). "Natural and artificially produced air ions—a biologically relevant climate factor?" *Int J Biometeorol* 29(3):233–242.

Krueger, A. P., Smith, R. F., and Go, I. G. (1957). "The action of air ions on bacteria." *J Gen Physiol* 41:359–381.

Krueger, A. P., and Reed, E. J. (1976). "Biological Impact of Small Air Ions." *Science* 193(Sep.):1209–1213.

Krueger, A. P. (1985). "The biological effect of air ions." *Int J Biometeor* 29(3):205–206.

Lee, J.-K., Kim, S. C., Shin, J. H., Lee, J. E., Ku, J. H., and Shin, H. S. (2001). "Performance evaluation of electrostatically augmented air filters coupled with a corona precharger." *Aerosol Sci Technol* 35:785–791.

Lee, B. U., Yermakov, M., and Grinshpun, S. A. (2004). "Removal of fine and ultrafine particles from indoor air environments by the unipolar ion emission." *Atmos Environ* 38(29):4815–4823.

Lehtimaki, M., and Graeffe, G. (1976). "The effect of the ionization of air on aerosols in closed spaces." *Proceedings of the 3d International Symposium on Contamination Control.* Copenhagen, Denmark.

Li, J., Wang, X., Yao, H., Yao, Z., Wang, J., and Luo, Y. (1989). "Influence of discharge products on post-harvest physiology of fruit." *Proceedings of the International Symposium on High Voltage Engineering,* New Orleans, LA, 6:1–4.

Makela, P., Ojajarvi, J., Graeffe, G., and Lehtimaki, M. (1979). "Studies on the effects of ionization on bacterial aerosols in a burns and plastic surgery unit." *J Hyg* 83:199–206.

McDowell, N. (2003). "Air ionizers wipe out hospital infections." NewScientist.com news service. http://www.newscientist.com/article.ns?id=dn3228.

Miller, N. J. (1984). "Ions purify patient's air." *J Ment Health Adm* 11(1):36–37.

Mitchell, B. W. (1994). "Effect of negative air ionization on airborne transmission of Newcastle Disease Virus." *Avian Dis* 38(4):725.

Mitchell, B. W. (1998). "Effect of negative air ionization on ambient particulates in a hatching cabinet." *Appl Eng Agric* 14(5):551–555.

Mitchell, B. W., and Waltman, W. D. (2002). "Reducing airborne pathogens and dust in commercial hatching cabinets with an electrostatic space charge system." *Avian Dis* 47(2):247–253.

Petkov, G., Baikov, B. D., and Russak, G. (1987). "Possibilities for decontaminating the air in commercial poultry breeding." *Vet Med Nauki* 24(3):67–72.

Phillips, G., Harris, G. J., and Jones, M. W. (1963). "The effect of ions on microorganisms." *Int J Biometerol* 8:27–37.

Phillips, G., Harris, G. J., and Jones, M. V. (1964). "Effect of air ions on bacterial aerosols." *Int J Biometeorol* 8:27–37.

Seo, K. H., Mitchell, B. W., Holt, P. S., and Gast, R. K. (2001). "Bactericidal effects of negative air ions on airborne and surface *Salmonella enteritidis* from an artificially generated aerosol." *J Food Prot* 64(1):113–116.

Soyka, F., and Edmonds, A. (1991). *The Ion Effect.* Bantam Books, New York.

Tanimura, Y., Nakatsugawa, N., Ota, K., and Hirotsuji, J. (1997). "Inhibition of microbial growth using negative air ions." *Antibacterial and Antifungal Agents* 25(11):625–631.

USDA (2000). "Use of negative air ionization for reducing microbial contamination on stainless steel surfaces." *J Appl Poultry Res.* http://www.nps.ars.usda.gov/publications/publications.htm?SEQ_NO_115=116660.

CHAPTER 17

OZONE

17.1 INTRODUCTION

Ozone has been successfully used for the disinfection of water since the early 1900s, although it had been used experimentally as early as 1892 (McCarthy and Smith, 1974). The possibility of using ozone for the disinfection of air and surfaces is implicit in the body of test results in the literature, but no systems are currently available for air disinfection that remove the ozone to safe levels. Ozone systems that produce low levels of ozone in indoor air for air quality control are currently being marketed to consumers but the usefulness of such approaches has not been conclusively demonstrated to outweigh the potential hazards. Ozone has promise as a means of disinfecting surfaces and may therefore be useful for remediation. It has also been developed as a practical method of deodorizing buildings. This chapter explores the current state of ozonation technology and the possible aerobiological applications. Figure 17.1 shows the four major applications in which ozone is or could be applied. Water disinfection is not addressed here except that it applies to airborne microbes, but extensive information on this subject is available elsewhere (Rice, 1997; Kowalski et al., 1998).

17.2 OZONE CHEMISTRY

Ozone is an allotrope of oxygen; it differs from oxygen in chemical properties but not in composition (Mortimer, 1975). Table 17.1 summarizes the properties of ozone as compared with oxygen conversion factors for airborne ozone.

Ozone is created by the addition of one electron to an oxygen molecule, as shown in Figure 17.2, which results in three oxygen molecules combining in a temporary bond. Ozone is 50 percent denser than oxygen and over 10 times more soluble in water. It can only be produced in air in quantities of about 2 percent by weight. In pure oxygen the quantity produced is approximately doubled. The basic chemical equation for the creation of ozone is as follows, but is left unbalanced for simplicity's sake:

$$O_2 + e^- \rightarrow O_3 \qquad (17.1)$$

TABLE 17.1 Properties of Oxygen and Ozone

Characteristic	Oxygen	Ozone
Symbol	O_2	O_3
Molecular weight	32	48
Specific gravity (air = 1)	1.105	1.658
Density (STP)	1.429 g/L	2.141 g/L
Solubility in water (0°C)	.06981 g/L	1.09 g/L
Melting point, °C	−218.4	−250
Boiling point, °C	−182.9	−112
Critical temperature, °C	118	−12.2
Critical pressure, atm	49.7	55
Oxidizing potential	−1.23 V	−2.07 V
Absorption wavelength	760/688 nm	253.7 nm
Odor	None	Pungent
Conversion factors	Ozone equivalence	
By weight in air	1 ppm	1.28 mg/m^3
	1 g/m^3	782 ppm
	1 mg/L	782 ppm
	1 mg/m^3	0.782 ppm
	1%	12.8 g/m^3
By volume in air	1 ppm	1 g/m^3
	1 ppm	1 mg/L

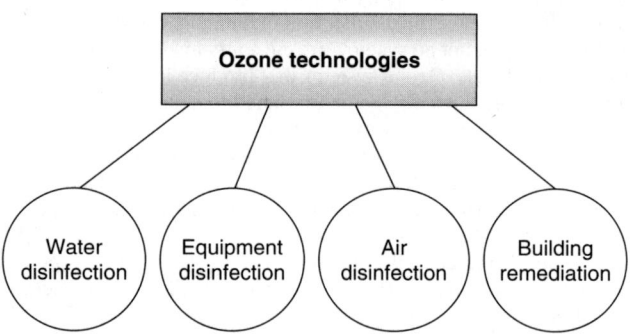

FIGURE 17.1 Breakdown of ozone disinfection applications.

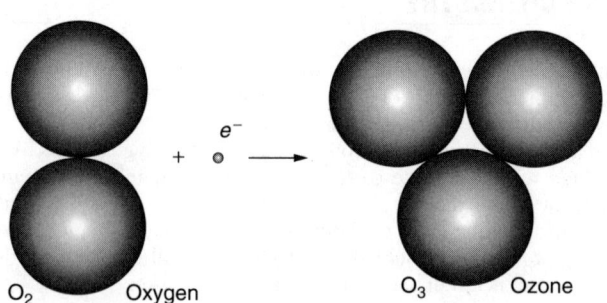

FIGURE 17.2 Generation of ozone by the addition of an electron to oxygen.

Free electrons for the ozone generation process can be generated in air through the use of electrical discharges. Electrons liberated by an electrical discharge, such as that in the spark between electrodes, may impact oxygen molecules, splitting them apart. The ionized single oxygen molecules then react with O_2 to form ozone. Typically, an alternating current is generated between electrodes to send free electrons back and forth. The glow produced by the incomplete breakdown of oxygen molecules is called a corona, and the process is often referred to as corona discharge.

Ozone can also be produced by ultraviolet lamps. Ultraviolet radiation in the range 175 to 210 nm can produce ozone in air (Scott and Lesher, 1962). Many ultraviolet lamps produce a range of wavelengths that includes the ozone-producing range and these are often used to generate ozone. Since ozone also absorbs and is decomposed by UV at the wavelength of 253.7 nm (see Table 17.1), these lamps will simultaneously create and destroy ozone.

Ozone is far more reactive than oxygen and can produce a variety of by-products in humid air, including various hydroxyl radicals. These products are usually unstable and ultimately break down into water and oxygen. Figure 17.3 graphically illustrates one of the most important products, the OH radical, which is thought to be responsible for much of the oxidizing or disinfecting effects of ozone in both air and water. Additional short-lived radicals generated may include HO_2 and HO_3 but these are thought to be less significant in the ozone disinfection process.

A variety of possible chemical pathways exist in which hydroxyl radicals may be generated (Rice, 1997), but the simplified or unbalanced chemical equation for the process depicted in Fig. 17.3 is as follows:

$$O_3 + H_2O \rightarrow OH^- \tag{17.2}$$

Ozone is unstable and will tend to break down per the following chemical equation:

$$O_3 \rightarrow O_2 + O \tag{17.3}$$

Obviously a variety of intermediate reactions can occur between the oxygen molecules and the products, rendering the actual process somewhat more complicated than has been described earlier, but most of these processes and products are not critical to the disinfection process. They may, however, play a part on reacting with volatile organic compounds in the air to produce secondary products. Some of these secondary products may be hazardous although they likely exist in minute quantities under normal circumstances. Bacterial cell constituents are typically high molecular weight compounds with a variety of low-energy bonds that are susceptible to reactions with ozone (Langlais et al., 1991). Although the outer cell wall of bacteria is thought to be somewhat resistant to ozone,

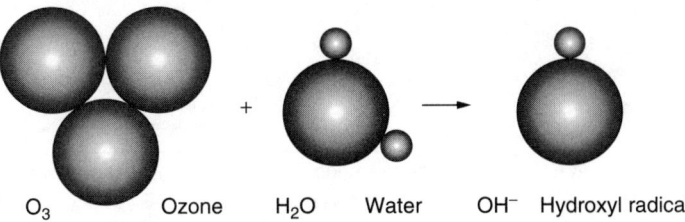

FIGURE 17.3 Generation of hydroxyl radicals from ozone. Only the hydroxyl product is shown.

penetration of ozone into the cytoplasmic membrane, and especially into the nucleic acids, offers opportunities for ozone degradation (Ishizaki et al., 1986).

17.3 AIRBORNE OZONE DISINFECTION

Ozone can be used as an aerial disinfectant for both airstreams and surfaces. The use of ozone generators in indoor environments is not recommended as they may pose a health hazard (Steiber, 1995). The only applications addressed here are those in which ozone is used and then removed from the airstream or those in which ozone is used for remediation in unoccupied areas or equipment decontamination.

Airborne ozone disinfection produces microbial decay rates that conform to classical exponential decay models. The appearance of shoulders at low doses and a second stage of decay under prolonged exposure are common (Kowalski et al., 2003). Figure 17.4 shows a graph of the decay rates of *Bacillus cereus* spores under exposure to ozone at various *relative humidities* (RH). *Bacillus cereus* is morphologically identical to *Bacillus anthracis*, and genetically identical also except for two plasmids, making this spore a good model for anthrax spores. The ozone dose given in Fig. 17.4 is specified in units of ppm-hour, which is the concentration in ppm multiplied by the hours of exposure. The original test was performed at 3 mg/L in air, or 2346 ppm (Ishizaki et al., 1986). However, concentrations of ozone in this range are likely well above the threshold value, meaning that lower doses of ozone might produce the same result.

The ozone dose is computed as follows:

$$\text{Dose} = C_a E_t \tag{17.4}$$

In Eq. (17.4) C_a is the concentration of ozone in air and E_t is the exposure time. If the airborne concentration in ppm is known, then the time to achieve sterilization of anthrax spores can be computed using Fig. 17.4. Sterilization is often assumed as a six-log (base 10) reduction. For example, if the ozone concentration was 200 ppm, and the RH was 90 percent, then the sterilization dose is approximately 9600 ppm-hour. The exposure

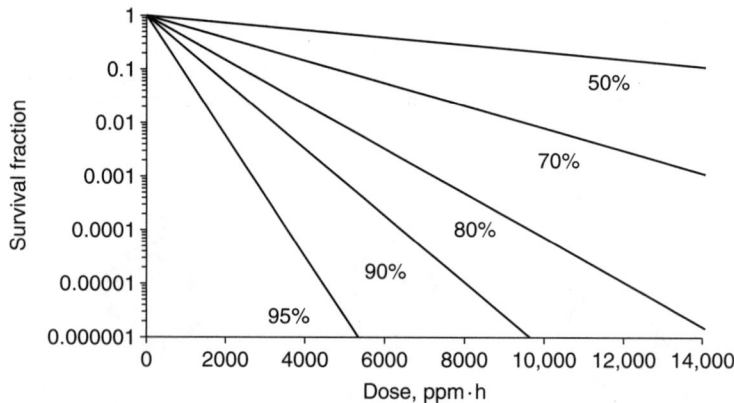

FIGURE 17.4 Survival of *Bacillus cereus* spores under ozone exposure at different relative humidities. [*Based on data from Ishizaki et al. (1986).*]

time can now be computed based on Eq. (17.4) as follows:

$$E_t = \frac{\text{Dose}}{C_a} = \frac{9600}{200} = 48 \text{ hours} \qquad (17.5)$$

It should be noted that the decay rate of *Bacillus cereus* spores in ozone increases with increasing relative humidity. The relative humidity can therefore be used to manipulate the decay rate. That is, if the building ventilation system had humidity control, then it would be desirable to maximize RH in conjunction with ozonation. This would minimize the damage to building materials as the result of high ozone levels.

Although the data on which Fig. 17.4 is based (Ishizaki et al., 1986) suggest only a single stage of decay, this may not reflect the complete decay curve. Many spores show evidence of two stages under ozone exposure (Dyas et al., 1983) just as they do under UVGI. Therefore, the estimated sterilization dose should be used with caution until data become available on the presence or absence of a second stage.

Ozone should not be used in breathing air spaces, but the development of efficient ozone filters (Reiger et al., 1995; Takeuchi and Itoh, 1993) and ozone-destruction catalysts (Rodberg et al., 1991) offers some means of removing residual ozone and producing sterilized, breathable air. In addition, the low doses that have proven effective in water (Katzenelson and Shuval, 1973; Hart et al., 1995; Beltran, 1995; Chang et al., 1996) coupled with the rapid decomposition rates that have been observed (Horvath et al., 1985; McCarthy and Smith, 1974; Harakeh and Butler 1985), indicate that engineered alternatives to zone filters may be feasible. These alternatives include ultraviolet irradiation and heating to enhance decomposition, extended residence time in mixing plenums, and the catalytic effect of glass and silica (Oeuderni et al., 1996).

Ozone has been previously investigated as an aerial disinfectant, but with inconclusive results (Elford and van den Eude, 1942; Rodberg et al., 1991). The disinfection of entire rooms with ozone shows some promise (Masaoka et al., 1982), but data on ozone disinfection of air remain limited. A summary of results obtained by previous investigators is shown in Table 17.2 for viruses and bacteria. Various researchers have studied the biocidal

TABLE 17.2 Summary of Previous Airborne Ozone Test Results

Test microbe	Ozone, ppm	RH%	Time, min	Survival, %	Researchers
S. aureus	0.3–0.9	—	240	0.5	Dyas et al., 1983
P. aeruginosa	0.3–0.9	—	240	31	Dyas et al., 1983
Serratia spp.	0.3–0.9	—	240	3.2	Dyas et al., 1983
Proteus	0.3–0.9	—	240	0.9	Dyas et al., 1983
A. fumigatus	0.3–0.9	—	240	8	Dyas et al., 1983
S. salivarius	0.6	60–75	10	2	Elford and van de Eude, 1942
F. oxysporum	0.1	35–75	240	2	Hibben and Stotzky, 1969
A. niger	0.1	35–76	240	84	Hibben and Stotzky, 1969
R. stolonifer	0.1	35–77	240	43	Hibben and Stotzky, 1969
S. epidermis	0.47	60–75	60	1	Heindel et al., 1993
P. chrysogenum	3–9	90	1380	0.1	Foarde et al., 1997
E. coli	300	18–21	0.67	0.001	Kowalski et al., 1998
S. aureus	300	18–21	1.5	0.001	Kowalski et al., 1998
B. cereus	2346	95	60	0.013	Ishizaki et al., 1986
B. subtilis	2346	95	60	0.03	Ishizaki et al., 1986

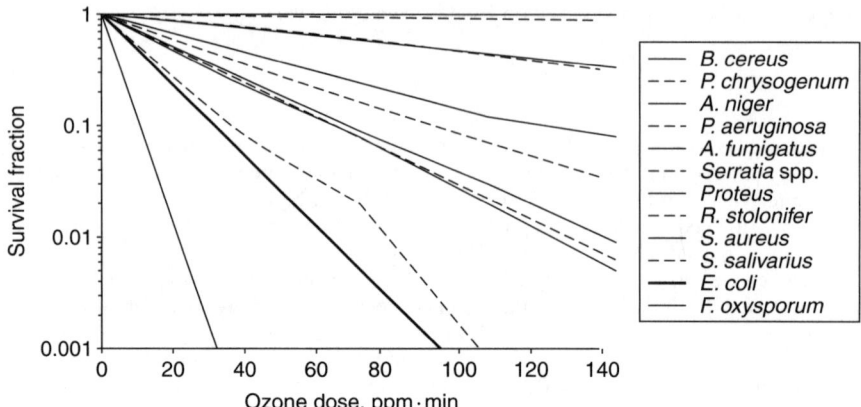

FIGURE 17.5 Comparison of decay rates vs. airborne ozone dose for several microbial species.

properties of airborne ozone for the disinfection of air or surfaces (Heindel et al., 1993; Foarde et al., 1997; Masaoka et al., 1982). The studies by Ishizaki et al. (1986) demonstrated that relative humidity has a major impact on the biocidal effect of ozone, and results suggest that humidities above 80 percent result in efficient sporocidal activity. High RH, about 90 to 95 percent, seems to maximize the disinfectant rate of ozone. Other studies report that RH below about 50 percent has negligible or limited effects (Elford and van den Eude, 1942).

Ozone does not react significantly with water or air in the absence of UV radiation over the short periods required for pathogen inactivation. These fluids merely provide the medium in which concentrations of ozone diffuse and react with organic molecules. Under ultraviolet irradiation ozone reacts with water and decomposes into various short-lived radicals, such as the highly reactive hydroxyl radical. Theoretical and empirical evidence suggests that much of the biocidal effect results from the radicals produced (Rice, 1997; Beltran, 1995). The decomposition reaction can be enhanced in air by the use of ultraviolet irradiation and through controlled humidity (NIST, 1992). The effects of ozone in air should parallel the effects of ozone in water, and the effectiveness of ozone for eliminating airborne pathogens in either medium may be comparable.

The threshold concentrations at which ozone inactivates bacteria in water are remarkably low. For example, the threshold for *Escherichia coli* lies between 0.1 and 0.2 ppm (Katzenelson and Shuval, 1973; Broadwater et al., 1973). Viruses are also sensitive to low levels of ozone in water (Harakeh and Butler, 1985). Airborne ozone studies have also found viruses to be susceptible to ozone in aerosols (de Mik and de Groot, 1977). Figure 17.5 illustrates the susceptibility of various microbes to airborne ozone dose, based on the studies in Table 17.2.

17.4 OZONE FOR SURFACE STERILIZATION

A secondary application of airborne ozone is the sterilization of surfaces or medical equipment. The use of UVGI and autoclaving typically require 20 to 30 minutes of exposure for complete sterilization of equipment. The use of ozone to sterilize equipment has the special advantage of both rapid inactivation and lower overall energy consumption. Applications could also include the sterilization of surfaces of contaminated rooms, biosafety cabinets, books, ventilation systems, or entire buildings (Elford and ven den Eude, 1942; Khurana, 2003; Masaoka et al., 1982). Ozone has also been studied as a method of decontaminating the surfaces of foodstuffs (Kells et al., 2001; Khadre et al., 2001; Kim et al., 1999). The performance of ozone in surface disinfection is similar to that of air and water systems, and some examples will be addressed in the later section on ozone system performance. Ozone can also be used for decontamination of buildings, and this subject is addressed in Chap. 26.

17.5 OZONE REMOVAL

Once ozone is used for air or surface disinfection it needs to be removed since even low levels of ozone can present a health hazard to building occupants. Several ozone removal technologies have been studied, including carbon adsorption, thermal decomposition, and catalytic destruction. Ozone can be destroyed by thermal decomposition (Zaslowsky et al., 1960). In one study, Singh et al. (1997) found that a temperature of at least 300°C would destroy ozone within about 3 seconds. The use of high temperatures is unlikely to be practical for ozone removal in ventilation airstreams due to the high energy consumption. However, the appropriate placement of ozonation systems may be able to take advantage of thermal breakdown through heating coils, especially when levels of ozone are initially low.

The half-life of ozone is approximately 15 minutes in open air, and although this is to be accounted for by the travel time in the ventilation system ductwork, this and other factors like dilution may assist in reducing levels of ozone to below hazardous levels (Khurana, 2003).

The use of carbon adsorption to remove airborne ozone has been studied by Deitz and Bitner (1972, 1973), Rakitskaya et al. (1994, 1996), Kowalski et al. (2003), and Weschler et al. (1994). The ozone removal efficiency of 10 different commercial filters was studied by Lee and Davidson (1999). Test conditions included an airstream concentration of 0.12 ppm, a relative humidity of 50 percent, and an adsorber face velocity of 2.54 m/s. The initial removal efficiency of carbon adsorbers varied from about 5 to 98 percent. Changes in relative humidity from 20 to 80 percent had no significant effect on the performance of activated carbon.

One of these filters was impregnated with metal catalysts. The removal of ozone with catalytic materials has been studied by Dhandapani and Oyama (1997), Heisig et al. (1997), and Kowalski et al. (2003). In the last study, ozone levels of about 20 ppm were reduced to nondetectable levels after passage through a metal oxide catalyst at low velocity. Figure 17.6

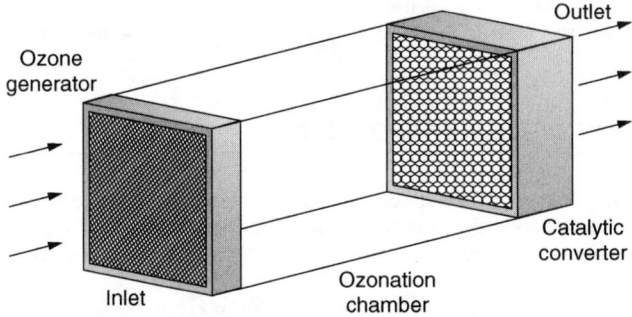

FIGURE 17.6 Schematic of an in-duct ozonation unit with integral ozone generator and catalytic converter.

shows a schematic of an air ozonation system in which the airstream is disinfected inside a portion of duct after which the air passes through a catalytic filter to remove the ozone.

In addition to creating ozone, ionizing radiation can also be used to decompose ozone (Harteck et al., 1965; Norrish and Wayne, 1965; Jones and Wayne, 1969). Photocatalysis has been used to decompose ozone over titanium dioxide (Ohtani et al., 1992).

Any system that uses ozone in a ventilation airstream must reduce levels to below the *threshold limit value* (TLV) in indoor air. Ideally, the ozone levels would be reduced to nondetectable levels in indoor air. One way to ensure this is to use the lowest levels of ozone necessary in the disinfection unit. As suggested by previous studies, very high levels of ozone (i.e., above 300 ppm) may not provide any added benefit over much lower levels (i.e., 6 to 20 ppm) due to threshold effects (Kowalski et al., 1998, 2003). The exact level of ozone necessary to make for an effective air disinfection system has not yet been established and more research is needed in this area.

17.6 OZONE HEALTH HAZARDS

The toxicity of high ozone concentrations to humans presents an obvious obstacle to the use of ozone for the disinfection of air. Residual ozone at any level is considered a hazardous pollutant. The U.S. Occupational Safety and Health Administration (OSHA) requires that the indoor TLV of an 8-hours exposure be limited to less than 0.1 ppm. The World Health Organization's air quality guidelines for Europe has set a limit of 0.075 to 0.1 ppm for 1 hour and the EPA ambient clear air standard for ozone is 0.12 ppm for 1 hour (Steiber, 1995). The use of airborne ozone for disinfection of airstreams and surfaces is limited by the fact that some means of completely removing ozone is needed before any possible applications involving indoor air disinfection can be developed. Although some ozone generators are currently being sold for the purpose of indoor air disinfection, the efficacy of using levels of ozone low enough to be harmless to human cells may have limited effect for the purpose of destroying bacterial cells. Conversely, if the ozone levels used in indoor air are sufficient to kill bacterial cells, then the question of whether damage to human cells may occur is one that needs to be more fully addressed before such applications can be sanctioned.

Methods for removing ozone from airstreams are currently available, and therefore there is no necessity to use ozone as an indoor aerial disinfectant in occupied buildings. However, the use of ozone to remediate unoccupied, contaminated buildings will be further addressed in Chap. 26.

Ozone by-products may include hazardous pollutants. Table 17.3 identifies some of the most common indoor pollutants and their associated precursors (Masschelein, 1982; Hollick and Sangiovanni, 2000). This list is not inclusive but merely representative and shows some of the precursor compounds to the VOCs, and the potential by-products of the VOCs.

17.7 OZONE SYSTEM PERFORMANCE

Few systems, if any, have yet been installed that use ozone as a surface or air disinfectant, although systems to deodorize unoccupied buildings are currently in use. The following research results are summarized as examples of how ozone performs in various tests against different microbes.

The efficacy of ozone as a terminal disinfectant for food microorganisms was evaluated by Moore et al. (2000) under laboratory conditions. Microbes tested included *E. coli, Serratia liquefaciens, Staphylococcus aureus, Listeria innocua,* and *Rhodotorula rubra*.

TABLE 17.3 Some Indoor VOCs and Ozone Precursors and By-products

Ozonation precursor	Common indoor VOC	Ozone by-product
Acetone, cresol, 4-chloro-0-cresol, ethylic alcohol, mesitylene, methylglyoxal, methylethylketone, phenol, xylenol	Acetic acid	Glyoxylic acid, oxalic acid, diacetyl
Isopropanol	Acetone	Acetic acid, formic acid, formaldehyde, oxalic acid
Albumin, 4-aminobenzoic acid, histidine, leucine	Ammonia	Nitrates
—	Benzene	Glyoxal, glyoxylic acid
Acetone, alkylbenzenesulfonic acid, decylbenzene sulfonic acid, nonylbenzenesulfonic acid, phenol, styrene	Formaldehyde	—
—	Naphthalene	Phthalic aldehyde, hydrogen peroxide, phthalic acid, phthaldehyde
Salicylic acid	Phenol	Aldehyde muconic acid, pyrocatechol, hydroquinone, glyoxylic acid, resorcinol, muconic acid, oxalic acid, formic acid, acetic acid, glyoxal, formaldehyde, muconaldehyde, hydrogen peroxide
—	Styrene	Benzaldehyde, benzoic acid, formaldehyde

Ozone was used against these microbes on stainless steel and incubated at various temperatures and relative humidities for up to 4 hours. Exposure of the samples to 2 to 5 ppm of ozone for 4 hours resulted in log reduction from 2.41 to 7.56. The loss in viability of ozonated samples was greater than the nonozonated samples for all microbes, as shown in Fig. 17.7. Gram-negative bacteria proved more sensitive to ozone than gram-positive organisms and bacteria were more sensitive than yeast. Results of this investigation suggest that if applied after adequate cleaning ozone could be used as an effective disinfectant.

Ozone has been suggested as a means of decontaminating buildings (Elford and van den Eude, 1942; Kowalski et al., 2003). Chlorine dioxide has been used for decontaminating buildings of anthrax spores, such as the Hart Senate Office Building. The effectiveness of ozone in comparison with chlorine dioxide is therefore of some interest. One such comparative test performed against *E. coli, Legionella pneumophila,* and *Bacillus subtilis* spores proved ozone to be about 10 times more efficient than chlorine dioxide (Botzenhart et al., 1993). *E. coli* was least resistant to ozone, followed by *L. pneumophila,* and *Bacillus subtilis* spores being the most resistant. These effects were somewhat dependent on temperature and pH. Whether this comparative effectiveness in water would also be true in air is not known, but since the effects of ozone increase with increasing humidity, it could be reasonably postulated that airborne ozone would be a more effective disinfectant than chlorine dioxide in high humidity.

A comparison of ozone inactivation in water of hepatitis A virus, poliovirus 1, and indicator organisms was performed by Herbold et al. (1989). Five microorganisms were found

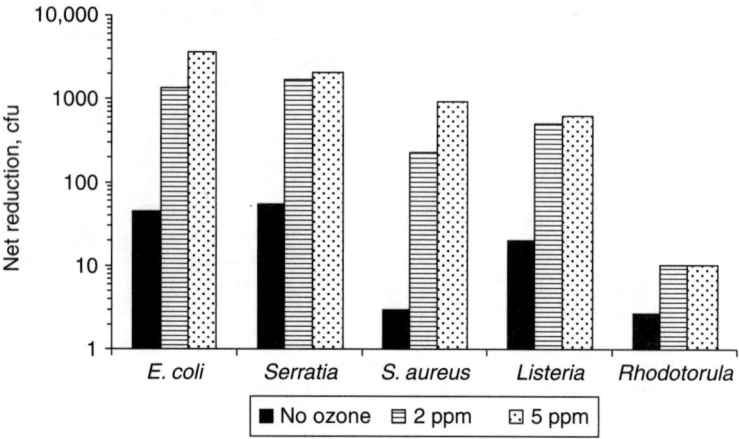

FIGURE 17.7 Reduction of microbes on steel surfaces exposed to ozone for 4 hours. [*Based on data from Moore et al. (2000).*]

to have the following order of resistance to constant ozone levels: most resistant were *Bacillus subtilis* spores, followed by *Legionella pneumophila*, then *hepatitis A virus* (HAV), then *E. coli*, and least resistant of all was *poliovirus 1* (PV1). Approximately 0.25 to 0.38 ppm (mg/L) was required for complete inactivation of HAV but only 0.13 ppm (mg/L) was required for complete inactivation of PV1.

The comparative effects of oxidizing biocides were tested on *Legionella pneumophila* by Domingue et al. (1988). The biocidal effects of ozone were compared to those of hydrogen peroxide and free chlorine in demand-free buffers. Ozone was the most potent of the three biocides, with a greater than 99 percent inactivation of *L. pneumophila* occurring during a 5-minute exposure to 0.10 to 0.30 ppm (mg/L). The bactericidal action of ozone was not markedly affected by changes in pH or temperature. Concentrations of 0.30 to 0.40 ppm (mg/L) of free chlorine killed 99 percent of the *L. pneumophila* after 30- and 5-minutes exposures, respectively. A 30-minute exposure to 1000 mg/L of hydrogen peroxide was required to effect a 99 percent reduction of the viable *L. pneumophila* population. No viable *L. pneumophila* could be detected after a 24-hour exposure to 100 or 300 mg/L of hydrogen peroxide.

Muraca et al. (1987) found that 1 to 2 ppm (mg/L) of continuous ozone over a 6-hour contact time in water produced a five-logarithm decrease of *L. pneumophila*. In this test they compared the efficacy of ozonation, UV light, hyperchlorination, and heat eradication using a model plumbing system constructed of copper piping, brass spigots, Plexiglas reservoir, electric hot water tank, and a pump. *Legionella pneumophila* was added to the system at 10^7 cfu/mL under various operating conditions. UV light and heat killed *L. pneumophila* most rapidly (producing a five-log inactivation in less than 1 hour) and required minimal maintenance. In contrast, both chlorine and ozone required 5 hours of exposure to produce a five-log decrease. Neither turbidity nor temperatures impaired the efficacy of any of the disinfectant methods.

L. pneumophila can be killed in water at less than 1 ppm (mg/L) of ozone based on a study of the efficacy of ozone for disinfecting hospital plumbing fixtures (Edelstein et al., 1982). The effect of ozonation of supply water for one wing of an unoccupied hospital building that had positive cultures for *L. pneumophila* from multiple potable water fixtures was studied in a prospective, controlled fashion. Mean ozone residual concentrations of

0.79 ppm (mg/L) eradicated *L. pneumophila* from the fixtures, but so did nonozonated water in the control wing fixtures.

Investigations into various alternative techniques for decontamination of the surfaces of artificially contaminated shell eggs were carried out by Davies and Breslin (2003). Ionized air, exposure to ozone in a dry air, and use of a commercial herbal antibacterial product were not effective. Application of ozone in a humid environment was partially effective but it was found that a commercial ionized water anolyte was much more effective in eliminating *Salmonella* from egg surfaces than ozone.

A high-speed ozone sterilizing device has been developed, using a ceramic-based ozonizer of the high-frequency surface discharge type (Masuda et al., 1990). This ozonizer can produce a high concentration of ozone of 20,000 to 30,000 ppm. Even at this concentration, complete sterilization of *Bacillus subtilis* in approximately 3 to 5 minutes is only possible when adequate temperature and humidity are maintained in the ozonized gas (50°C and 80 percent relative humidity). For reference bacteria, this device produces the standard sterilizing effect of killing one million cells in 3 to 5 minutes. One run, including processing, lasts 12 to 14 minutes as compared with approximately 30 minutes for a conventional autoclave.

In one study of the inactivation kinetics of ozone against food spoilage and pathogenic bacteria, ozone was used on *Pseudomonas fluorescens, Escherichia coli, Leuconostoc mesenteroides*, and *Listeria monocytogenes* (Kim and Yousef, 2000). Exposure of bacteria to 2.5 ppm for 40 seconds caused a five- to six-log decrease in counts. Most resistant was *E. coli*, followed by *P. fluorescens, L. mesenteroides*, with *Listeria* being least resistant. All measured decay curves showed a second stage of decay. Figure 17.8 shows the results for the survival of ozonated *Listeria* plotted as a function of dose (ppm-s). There are clearly two stages of dose-dependent decay in this graph and the same two-stage effect was evident in the other three microbes tested.

Ozone is currently being used for the reduction of microbial populations in rooms for storing fruits and vegetables, and rooms for curing cheese. Ozone has been evaluated as a means of disinfecting hatcheries (Whistler and Sheldon, 1989). In the latter study, ozone levels of about 9091 ppm (1.51 percent by weight in air) were found to result in four to seven log bacterial reductions and less than four log fungal reductions. It was also found that ozone was almost as effective as formaldehyde in reducing levels of *E. coli, Pseudomonas, Salmonella, Proteus, Aspergillus, Staphylococcus, Bacillus*, and *Streptococcus*. Davies and Breslin (2003) studied the use of ozone as a surface disinfectant for the control of *Salmonella enteritidis* on eggs but found it was ineffective.

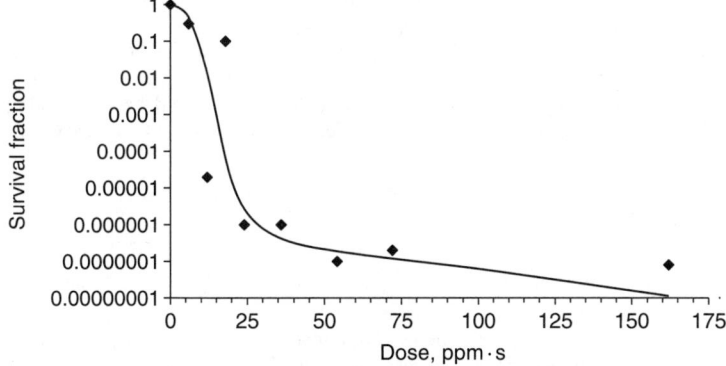

FIGURE 17.8 Graph of *Listeria* survival under ozone exposure at various doses. Line represents a two-stage shoulder model fitted to data from Kim and Yousef (2000).

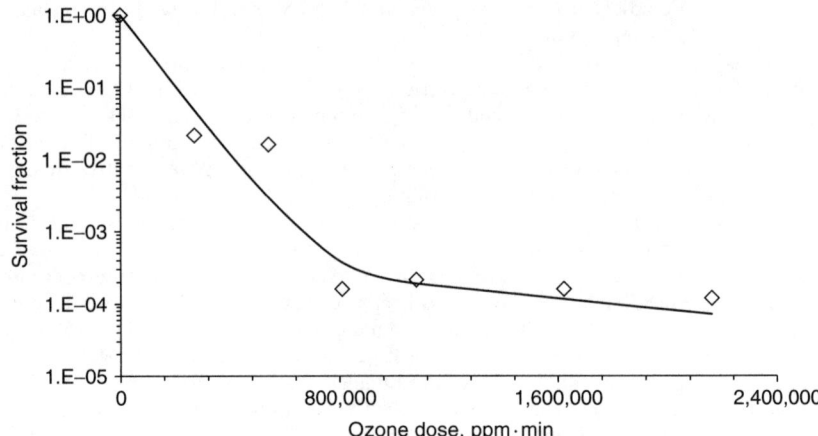

FIGURE 17.9 Survival of *Bacillus subtilis* spores under ozone exposure. Line represents a fitted two-stage curve. [*Based on data from Currier et al. (2001).*]

In a study of the use of airborne ozone to decontaminate biological pathogens, Currier et al. (2001) used humidified air with 9000 ppm of ozone against spores of *Bacillus subtilis* var. *niger* (formerly *B. globigii*). Figure 17.9 plots the data for 50 percent RH along with a fitted two-stage decay curve. With 70 percent RH, over seven logs of reduction were achieved within 60 minutes.

Khurana (2003) investigated ozone as a method of preventing microbial growth in air-conditioning systems. Findings of this study indicate that frequent treatment with ozone at 9 ppm was sufficient to prevent microbial growth. Levels as high as 45 ppm could be used for 15 minutes a day to inhibit microbial growth. Levels below 0 ppm were less effective at preventing microbial growth. Doses of 600 to 2400 ppm- minute, when the level was 11 ppm or higher, were effective in preventing microbial growth.

Fan et al. (2002) found some degree of synergy between ozone and negative air ionization and reported that the combination of these technologies could reduce airborne bacteria. Whereas the study showed no biocidal effects due to ionization alone, the combination produced a 99.3 percent reduction in viability of *Pseudomonas fluorescens* after about 6 hours of exposure to 100 nL/L of ozone (0.17 ppm) and 10^6 electron/mL ion density. Results varied with other bacterial species.

Ozone has been successfully applied as a deodorizer in unoccupied homes in a decontamination process (Schmidt, 2005). Ozone levels of about 20 ppm are maintained for 12 to 48 hours, after which the building is purged with outdoor air. Odors due to mold, bacteria, and pets have been successfully eliminated. For more on ozone decontamination processes see Chap. 26.

REFERENCES

Beltran, F. J. (1995). "Theoretical aspects of the kinetics of competitive ozone reactions in water." *Ozone Sci Eng* 17:163–181.

Botzenhart, K., Tarcson, G. M., and Ostruschka, M. (1993). "Inactivation of bacteria and coliphages by ozone and chlorine dioxide in a continuous flow reactor." *Wat Sci Tech* 27(3–4):363–370.

Broadwater, W. T., Hoehn, R. C., and King, P. H. (1973). "Sensitivity of three selected bacterial species to ozone." *Appl Microbiol* 26(3):393–393.

Chang, C. Y., Chiu, C. Y., Lee, S. J., Huang, W. H., Yu, Y. H., Liou, H. T., Ku, Y., and Chen, J. N. (1996). "Combined self-absorption and self-decomposition of ozone in aqueous solutions with interfacial resistance." *Ozone Sci Eng* 18:183–194.

Currier, R. P., Torraco, D. J., Cross, J. B., Wagner, G. L., Gladden, P. D., and Vandenberg, L. A. (2001). "Deactivation of clumped and dirty spores of *Bacillus globigii*." *Ozone Sci Engin* 23:285–294.

Davies, R. H., and Breslin, M. (2003). "Investigations into possible alternative decontamination methods for *Salmonella enteritidis* on the surface of table eggs." *J Vet Med B Infect Dis Vet Public Health* 50(1):38–41.

Deitz, V. R., and Bitner, J. L. (1972). "The reaction of ozone with adsorbent charcoal." *Carbon* 10:145–154.

Deitz, V. R., and Bitner, J. L. (1973). "Interaction of ozone with adsorbent charcoals." *Carbon* 11:393–401.

Dhandapani, B., and Oyama, S. T. (1997). "Gas phase ozone decomposition catalysts." *Appl Catal* 11:129–166.

Domingue, E. L., Tyndall, R. L., Mayberry, W. R., and Pancorbo, O. C. (1988). "Effects of three oxidizing biocides on *Legionella pneumophila* serogroup 1." *Appl Environ Microbiol* 54(3):741–747.

Dyas, A., Boughton, B. J., and Das, B. C. (1983). "Ozone killing action against bacterial and fungal species; microbiological testing of a domestic ozone generator." *J Clin Path* 36:1102–1104.

Edelstein, P. H., Whittaker, R. E., Kreiling, R. L., and Howell, C. L. (1982). "Efficacy of ozone in eradication of *Legionella pneumophila* from hospital plumbing fixtures." *Appl Environ Microbiol* 44(6):1330–1333.

Elford, W. J., and van den Eude, J. (1942). "An investigation of the merits of ozone as an aerial disinfectant." *J Hyg* 42:240–265.

Fan, L., Song, J., Hildebrand, P. D., and Forney, C. F. (2002). "Interaction of ozone and negative air ions to control microorganisms." *J Appl Microbiol* 93:144–148.

Foarde, K. K., VanOsdell, D. W., and Steiber, R. S. (1997). "Investigation of gas-phase ozone as a potential biocide." *Appl Occup Envrion Hyg* 12(8):535–541.

Harakeh, M. S., and Butler, M. (1985). "Factors influencing the ozone inactivation of enteric viruses in effluent." *Ozone Sci Eng* 6:235–243.

Hart, J., Walker, I., and Armstrong, D. C. (1995). "The use of high concentration ozone for water treatment." *Ozone Sci Eng* 17:485–497.

Harteck, P., Dondes, S., and Thompson, B. (1965). "Ozone: Decomposition by ionizing radiation." *Science* 147:393–394.

Heindel, T. H., Streib, R., and Botzenhart, K. (1993). "Effect of ozone on airborne microorganisms." *Zbl Hyg* 194:464–480.

Heisig, C., Zhang, W., and Oyama, S. T. (1997). "Decomposition of ozone using carbon-supported metal oxide catalysts." *Appl Catal* 14:117–129.

Herbold, K., Flehmig, B., and Botzenhart, K. (1989). "Comparison of ozone inactivation, in flowing water, of hepatitis A virus, poliovirus 1, and indicator organisms." *Appl Environ Microbiol* 55(11):2949–2953.

Hibben, C. R., and Stotzky, G. (1969). "Effects of ozone on the germination of fungus spores." *Canadian J Microbiol* 15:1187–1196.

Hollick, H. H., and Sangiovanni, J. J. (2000). A proposed indoor air quality metric for estimation of the combined effects of gaseous contaminants on human health and comfort. *Air Quality and Comfort in Airliner Cabins ASTM STP 1393*, E. L. Nagda, ed., American Society for Testing and Materials, West Conshohocken, PA.

Horvath, M., Bilitzky, L., and Huttnerr, J. (1985). *Ozone*. Elsevier, Amsterdam, The Netherlands.

Ishizaki, K., Shinriki, N., and Matsuyama, H. (1986). "Inactivation of *Bacillus* spores by gaseous ozone." *J Appl Bacteriol* 60:67–72.

Jones, I. T. N., and Wayne, R. P. (1969). "Photolysis of ozone by 254-, 313-, and 334-nm radiation." *J Chem Phys* 51:3617–3620.

Katzenelson, E., and Shuval, H. I. (1973). *Studies on the Disinfection of Water by Ozone: Viruses and Bacteria*, R. G. Rice and M. E. Browning, eds., Hampson Press, Washington, DC.

Kells, S. A., Mason, L. J., Maier, D. E., and Woloshuk, C. P. (2001). "Efficacy and fumigation characteristics of ozone in stored maize." *J Stored Prod Res* 37(4):371–381.

Khadre, M. A., Yousef, A. E., and Kim, J. G. (2001). "Microbiological aspects of ozone applications in food: A review." *J Food Sci* 66(9):1242–1252.

Khurana, A. (2003). *Ozone Treatment for Prevention of Microbial Growth in Air Conditioning Systems.* University of Florida, Gainesville, FL.

Kim, J. G., Yousef, A. E., and Dave, S. (1999). "Application of ozone for enhancing the microbiological safety and quality of foods: A review." *J Food Prot* 62(9):1071–1087.

Kim, J. G., and Yousef, A. E. (2000). "Inactivation kinetics of foodborne spoilage and pathogenic bacteria by ozone." *J Food Sci* 65(3):521–528.

Kowalski, W. J., Bahnfleth, W. P., and Whittam, T. S. (1998). "Bactericidal effects of high airborne ozone concentrations on *Escherichia coli* and *Staphylococcus aureus*." *Ozone Sci Eng* 20(3):205–221. http://www.bio.psu.edu/people/faculty/whittam/research/ozone.html.

Kowalski, W. J., Bahnfleth, W. P., Striebig, B. A., and Whittam, T. S. (2003). "Demonstration of a hermetic airborne ozone disinfection system: Studies on *E. coli*." *AIHA J* 64:222–227.

Langlais, B., Reckhow, D. A., and Brink, D. R. (1991). *"Ozone in Water Treatment: Application and Engineering."* C. R. Council, ed., Lewis Publishers. Chelsea, MI.

Lee, P., and Davidson, J. (1999). "Evaluation of activated carbon filters for removal of ozone at the PPB level." *Am Ind Hyg Assoc J* 60:589–600.

Masaoka, T., Kubota, Y., Namiuchi, S., Takubo, T., Ueda, T., Shibata, H., Nakamura, H., Yoshitake, J., Yamayoshi, T., Doi, H., and Kamiki, T. (1982). "Ozone decontamination of bioclean rooms." *Appl Environ Microbiol* 43(3):509–513.

Masschelein, W. J. (1982). *Ozonization Manual for Water and Wastewater Treatment.* John Wiley & Sons, New York.

Masuda, S., Kiss, E., Ishida, K., and Asai, H. (1990). "Ceramic-based ozonizer for high-speed sterilization." *IEEE T Ind Appl* 26(1):36–41.

McCarthy, J. J., and Smith, C. H. (1974). "A review of ozone and its application to domestic wastewater treatment." *J Am Wat Works Assoc* 74:718–729.

de Mik, G., and de Groot, I. (1977). "Mechanisms of inactivation of bacteriophage pX174 and its DNA in aerosols by ozone and ozonized cyclohexene." *J Hyg* 78:199–211.

Moore, G., Griffith, C., and Peters, A. (2000). "Bactericidal properties of ozone and its potential application as a terminal disinfectant." *J Food Prot* 63:1100–1106.

Mortimer, C. E. (1975). *Chemistry: A Conceptual Approach.* D. Van Nostrand Company, New York.

Muraca, P., Stout, J. E., and Yu, V. L. (1987). "Comparative assessment of chlorine, heat, ozone, and UV light for killing *Legionella pneumophila* within a model plumbing system." *Appl Environ Microbiol* 53(2):447–453.

NIST (1992). "Photoinitiated ozone-water reaction." *J Res NIST* 97(4):499.

Norrish, R. G. W., and Wayne, R. P. (1965). "The photolysis of ozone by ultraviolet radiation." *Proceedings of the Royal Society of London* 288(1413):200–211.

Oeuderni, A., Limvorapituk, Q., Bes, R., and Mora, J. C. (1996). "Ozone decomposition on glass and silica." *Ozone Sci Eng* 18:385–415.

Ohtani, B., Zhang, S., Nishimoto, S., and Kagiya, T. (1992). "Catalytic and photocatalytic decomposition of ozone at room temperature over titanium(iv) oxide." *J Chem Soc Faraday Transac* 88(7):1049–1053.

Rakitskaya, T. L., Bandurko, A. Y., Ennan, A. A., and Litvinskaya, V. V. (1994). "Kinetics of the low-temperature decomposition of ozone by carbon fiber materials." *Kinet Catal* 35(5):705–707.

Rakitskaya, T. L., Bandurko, A. Y., and Boginskaya, O. V. (1996). "Low-temperature decomposition of ozone trace concentrations by fibrous carbon materials." *Russian J Appl Chem* 69(1):148–150.

Reiger, I. H., Feucht, G., and Schonfeld, A. (1995). "Selective adsorption of noxon for the detection of ozone." *Odours VOC's J* 1(December):39–44.

Rice, R. G. (1997). "Applications of ozone for industrial wastewater treatment—A review." *Ozone Sci Eng* 18:477–515.

Rodberg, J. A., Miller, J. F., Keller, G. E., and Woods, J. E. (1991). A novel technique to permanently remove indoor air pollutants. *IAQ '91, Healthy Buildings/IAQ*, Washington, DC.

Schmidt, D. (2005). "The safe and effective use of ozone to purify/decontaminate an indoor environment." *Enviro2005*, Atlantic City, NJ.

Scott, D. B. M., and Lesher, E. C. (1962). "Effect of ozone on survival and permeability of *Escherichia coli*." *J Bacteriol* 85:567–576.

Singh, N., Pisarczyk, K. S., and Sigmund, J. J. (1997). "Catalytic destruction of ozone at room temperature." *90th Annual Meeting & Exhibition*, Toronto, Ontario.

Steiber, R. S. (1995). "Ozone Generators in Indoor Air Systems." *EPA-600/R-95-154*. Environmental Protection Agency. Washington, DC.

Takeuchi, Y., and Itoh, T. (1993). "Removal of ozone from air by activated carbon treatment." *Sep Technol* 3:168–175.

Weschler, C. J., Shields, H. C., and Naik, D. V. (1994). "Ozone-removal efficiencies of activated carbon filters after more than three years of continuous service." *ASHRAE Trans* 100:1121–1129.

Whistler, P. E., and Sheldon, B. W. (1989). "Biocidal activity of ozone versus formaldehyde against poultry pathogens inoculated in a prototype setter." *Poultry Sci* 68:1068–1073.

Zaslowsky, J. A., Urbach, H. B., Leighton, F., Wnuk, R. J., and Wojtowicz, J. A. (1960). "The kinetics of the homogeneous gas phase thermal decomposition of ozone." *J Am Chem Soc* 82:2682–2686.

CHAPTER 18

GREEN TECHNOLOGIES

18.1 INTRODUCTION

Green building design includes technologies that incorporate active, passive, or natural technologies, or sustainable materials and renewable energy resources. For aerobiological engineering, three technologies fit this category—solar exposure disinfection, vegetation air cleaning, and biofiltration of air. These are still primarily developmental technologies and may have limitations in terms of application and effectiveness, but combined with other systems they may contribute to an integrated solution to indoor air quality problems and the production of healthier buildings.

Following closely on the heels of building design for healthy living are the two additional categories of material selectivity and hygienic protocols. The appropriate selection of building materials to avoid those that may contribute to microbial growth is a developing science but one that begs attention today. Hygienic protocols refer to practices that maintain personal health, indoor cleanliness, and common sense in dealing with infected individuals. Although such protocols may not exactly comprise a technology per se, they represent what is surely the most ancient and natural means of controlling diseases and therefore may be the "greenest" technology of all.

The field of green building design bears a lot in common with aerobiological engineering since they both aim at providing healthy habitats in the most practical or feasible manner (ASHE, 2002). The U.S. Green Building Council (USGBC) has developed the Leadership in Energy & Environmental Design (LEED) Green Building Rating System for commercial interiors (USGBC, 2002). This voluntary rating system provides guidance to building owners, occupants, interior designers, architects, and others who design and install building interiors. It addresses topics related to sustainable design, such as space usage, water and energy efficiency, and *indoor environmental quality* (IEQ). There is currently little credit that can be taken in the LEED rating system for air treatment systems (other than for air filters used during and after construction) but it is likely the matter will receive increasing attention in the future. The subject of daylighting does receive LEED credit, however, and as will be seen in the following section, maximizing daylighting can provide natural disinfection to building air and surfaces.

Natural ventilation could be considered a green building technology due to the low energy consumption, but natural ventilation systems must be designed to facilitate airflow through a building. When climate permits, natural ventilation can provide an adequate number of air exchanges but cannot easily provide for filtration of outdoor spores. The use of 100 percent outdoor air with enthalpy recovery wheels is a green building option, as is the minimization of biological contamination through envelope design (Rosenbaum, 2002). Air disinfection through the use of solar exposure may provide part of the solution, and it only remains for future designers to integrate both possibilities, natural ventilation and air treatment.

Incorporating sustainable air cleaning technologies into the design of a green building brings both technologies into convergence. Although no building has yet been built that specifically incorporates green building design techniques along with the specific intention of controlling indoor aerobiology, some general suggestions are made here regarding how such a building might be designed. And, as has been pointed out before, the health impact of any building stretches far beyond its immediate community, and this is as true of the material and energy usage as it is of the health of the occupants (ASHE, 2002).

18.2 PASSIVE SOLAR EXPOSURE

Passive exposure to solar irradiation as a means of destroying airborne pathogens is based on the fact that sunlight contains some ultraviolet radiation and is lethal to airborne human pathogens (El-Adhami et al., 1994; Fernandez, 1996; Beebe, 1959). In the UVB spectrum at about 300 nm, for example, sunlight produces approximately 1 $\mu W/cm^2$ (Webb, 1991). Even visible light from fluorescent lamps has some biocidal effects (Futter and Richardson, 1967; Griego and Spence, 1978). Based on the data on *Bacillus subtilis* spores graphed in Fig. 18.1, the time to sterilize indoor surfaces, assuming six logs of reduction for sterilization, would be about 90 days when the indoor irradiance from fluorescent light is about 1/3 that of northern daylight. This assumes continuous exposure and does not account for any possible second stage or resistant fraction of the bacterial population.

In interior rooms, the total power to lighting fixtures is typically about 1 to 5 W/ft^2 (Bell, 2000). Assuming 30 percent efficiency, this converts to visible light irradiance of about 279 to 1394 $\mu W/cm^2$. Continuous lighting at this level would not be economical, but irradiance levels higher than this are likely to be produced by sunlight or daylight in rooms

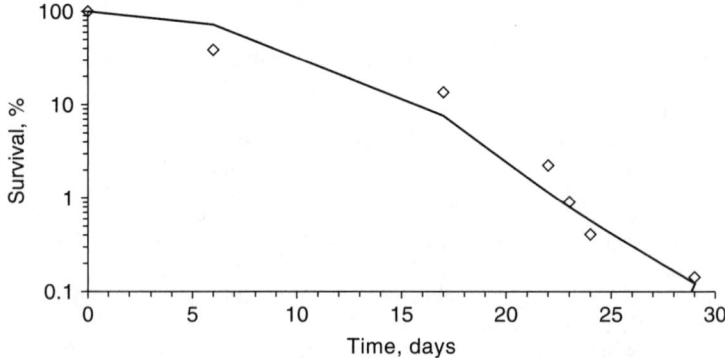

FIGURE 18.1 Survival of *Bacillus subtilis* spores in visible light from a fluorescent lamp. [*Based on data from Futter and Richardson (1967).*]

with windows. Inactivation rates for microbes can't be accurately predicted due to a lack of empirical data, and existing UV rate constants cannot be used as a substitute for visible light rate constants, however, the data presented do suggest that some advantage can be taken from the disinfecting effects of normal daylight and solar exposure through glass.

Sunlight is one of the primary reasons that most human pathogenic microorganisms die rapidly in the outdoor air (Harper, 1961; Mitscherlich and Marth, 1984). Unfiltered sunlight can destroy about 99 percent of *Staphylococcus aureus* cells within 70 minutes (El-Adhami et al., 1994). Based on the empirical data for filtered light in Fig. 18.2, it is estimated that *S. aureus* would be sterilized by Perspex filtered sunlight in about 240 minutes, while it would take about 625 minutes to sterilize with sunlight in which all the UVB was removed. Sunlight in which all UVB and UVA was removed showed no significant disinfection within the first 120 minutes. Clearly, partially filtered sunlight that passes at least some of the UVA or UVB wavelengths can have a potentially significant disinfection effect over periods of days or weeks.

Although visible light may cause photoreactivation of bacterial cells, this effect is somewhat wavelength dependent and, for extended or continuous exposure, the decay effect is dominant (Webb, 1961).

In the indoor environment, the visible light that passes through glass may be sufficient over extended periods of time to disinfect surfaces of mold spores. Maximizing the amount of light in a room by design may be an effective means of controlling indoor aerobiology. It may not be necessary to depend on direct sunlight for this effect, although it can certainly help. The main problem with using direct sunlight through glass is the heating effects and the discoloration of some materials.

Figure 18.3 shows the annual cycle of estimates of the monthly integrated surface irradiance weighted by the erythema (i.e., sunburn) action spectrum based on averaged ozone distributions in the indicated latitudes (Frederick et al., 1991). Although the integrated doses suggested by this graph, especially in the summer, would appear adequate for sterilization of most microbes, the spectra includes little UV and therefore a direct computation of sunlight inactivation rates cannot be made using UV rate constants. This chart does suggest, however, that generous quantities of sunlight and skylight are available even in the northernmost latitudes, and that careful attention to the design of indoor illumination via maximized window space may enhance the self-disinfection of buildings.

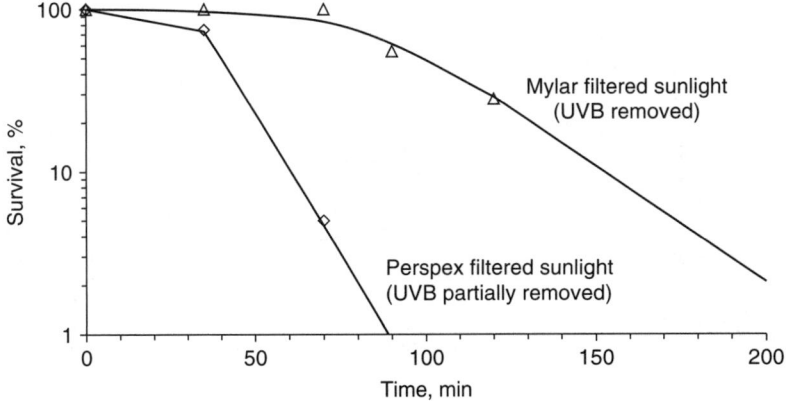

FIGURE 18.2 The effect of filtered sunlight on *Staphylococcus aureus*. [*Based on data from El-Adhami et al. (1994).*]

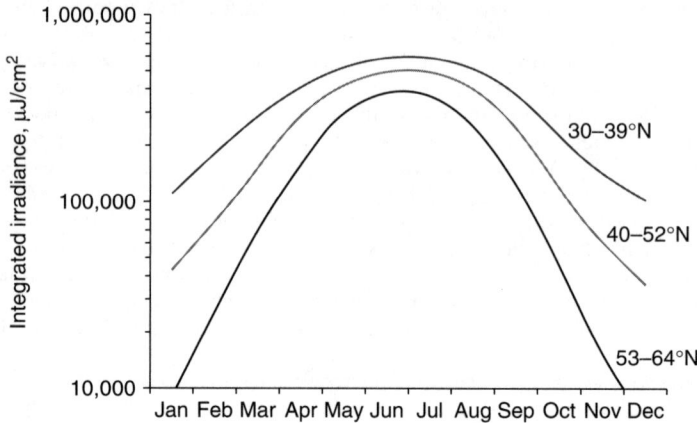

FIGURE 18.3 Daytime integrated surface irradiance weighted by the erythema action spectrum. [*Based on data from Frederick et al. (1991).*]

When window space cannot be maximized it may be possible to use UV-transmittant glass. In a hypothetical design for a building using passive solar exposure to control airborne microbes, the windows form a plenum for return air, with the outside panes being UV transmitting glass (Ehrt et al., 1994). This design requires no additional space, but may increase the cooling load.

Figure 18.4 shows the effects of sunlight on bacterial cells of *Pastuerella tularensis*. The doses shown in the figure represent simulated sunlight for 10 minutes at irradiances of 0, 30, 60, and 90 μW/cm^2 (Beebe, 1959). These results show that increasing relative humidity tends to decrease the biocidal effect of sunlight, as it tends to do with UVGI also. Although the irradiance values used in this figure may not be entirely realistic for sunlight penetrating glass, the data do indicate that disinfection is a real possibility.

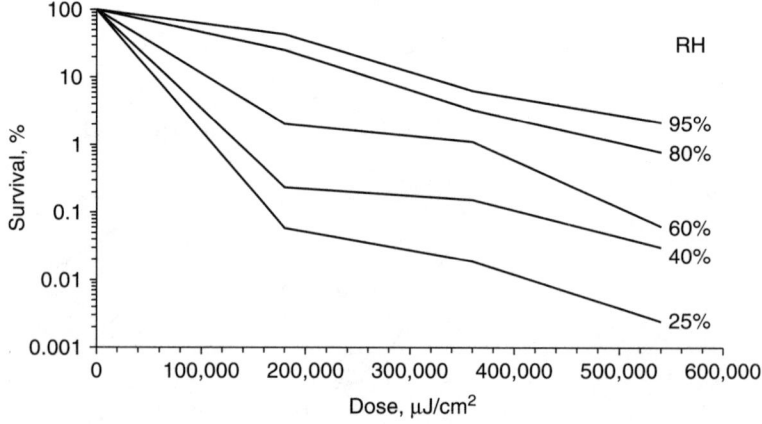

FIGURE 18.4 The effect of sunlight and relative humidity on wet-disseminated *Pastuerella tularensis*. [*Based on data from Beebe (1959).*]

The data in Fig. 18.4 could be extrapolated, for example, to determine the time to sterilization t_s from sunlight. The approximate rate constant for *P. tularensis* at 60 percent RH is 0.0088 cm²/µJ. Assuming an indoor irradiance of 5 µW/cm², the time to sterilization for *P. tularensis* would be as follows:

$$t_s = \frac{\ln S}{-kE} = \frac{\ln(1-0.999999)}{-0.000014(5)} = 197,364 \text{ s}$$

The dose for sterilization, based on these assumptions, would be 5(197364) = 986,820 µJ/cm² at 60 percent RH. Assuming 4 hours of direct sunlight a day, this would translate into about 14 days exposure.

Other potential enhancements to passive solar exposure include the incorporation of *photocatalytic oxidation* (PCO) coatings on surfaces likely to collect fomites, like doorknobs, antimicrobial coatings, and the use of light pipes to bring sunlight to internal areas of the building. Sunlight can also be used with good effect to heat sterilized water supplies through the use of solar reflectors (Safapour and Metcalf, 1999). Atriums provide an opportunity to combine vegetation air cleaning and passive solar exposure (see Fig. 18.5).

Generous amounts of sunlight will tend to destroy spores over time. Direct sunlight can be effective in destroying mold and mold spores in rugs, blankets, furniture, and the like after only a few days of exposure. The common practice of exposing mattresses, blankets, and rugs to open air in the spring could be extended to other seasons and house furnishings where mildew or water damage has occurred. However, the most practical approach is probably to maximize solar exposure indoors.

Although indoor solar exposure is limited by transmittance through glass and total window area, given sufficient time almost any material can be disinfected with proper amounts of sunlight. Books or materials can be placed in sunlight indoors to disinfect them of mold. Direct sunlight through one or more panes of glass will take much longer to disinfect materials. Although variations in cloud cover make it difficult to estimate the dose of sunlight

FIGURE 18.5 Both solar exposure and vegetation contribute to reducing the bacterial load of indoor air.

TABLE 18.1 Estimated Times for Passive Solar Disinfection

Microbe	Exposed to	Time to sterilize	
		Hours	Days
Bacteria	Direct outdoor sunlight	8	1
	Direct indoor sunlight (through glass)	12	2
	Indoor daylight (through glass)	52	5
	Indoor lighting (no daylight)	156	16
Mold and spores	Direct outdoor sunlight	240	10
	Direct indoor sunlight (through glass)	313	31
	Indoor daylight (through glass)	6912	288
	Indoor lighting (no daylight)	20736	864

that may transmit through glass, experiments by the author suggest that about 5 to 25 days are required to kill mold on books, depending on available sunlight. This time period is short enough that no significant discoloration should occur. In wet environments, like showers, it is considerably more difficult to achieve disinfection with only sunlight through bathroom windows due both to the limited amount of sunlight entering, and because the moisture levels will continuously feed growth of mold and mildew on walls, grout, curtains, and the like. It would appear that bathrooms, which typically have minimum window area, or fenestration, should instead be designed to maximize fenestration.

Table 18.1 has been prepared in summary of the above limited data on microbial disinfection with sunlight, daylight, and artificial light sources. These sterilization times are merely estimates since the data set is too limited to quantify precise rate constants and also because the exact values of indoor irradiance can vary considerably. It is assumed that 10 hours of daylight or lighting are available, and no account is taken of recovery effects, which may be significant for extended low-level cyclical exposures.

18.3 VEGETATION AIR CLEANING AND BIOFILTRATION

Large amounts of living vegetation can act as a natural biofilter, removing or reducing levels of airborne microorganisms (Darlington et al., 1998; Rautiala et al., 1999). Winter gardens may act as "buffer zones" in moderating the indoor climate (Watson and Buchanan, 1993). For example, the surface area of large amounts of vegetation may absorb or adsorb microbes or dust. The oxygen generation of the plants may have an oxidative effect on microbes. The increased humidity may have an effect on reducing some microbial species although it may favor others. The presence of symbiotic microbes such as *Streptomyces* may cause some disinfection of the air. Natural plant defenses against bacteria may operate against mammalian pathogens. Finally, gardens and vegetation may have an effect on the psyche of occupants that may stimulate a sense of well-being. This effect did not go unnoticed in hospitals through the ages and it is not uncommon for them to include "restorative gardens" as part of their architectural landscape (Gerlach-Spriggs et al., 1998). A number of specific house plants have been identified that may contribute to passive cleaning of indoor air pollutants but their effects on airborne microbes are not known (Wolverton, 1996).

Although houseplants are often considered a source of potential fungal spores, a study by Rautiala et al. (1999) indicated that no significant increase in concentrations of fungi in air or surface samples occurred when houseplants were added to an office environment.

In a vegetation air cleaning system air flows through areas filled with vegetation or through entire greenhouses (see Fig. 18.6) before entering the ventilation system. Often, such vegetation areas include water or waterfalls, which may also have an effect on local ionization levels. One downside to keeping large amounts of vegetation indoors is that the potting soil may include potentially allergenic fungi.

Vegetation has also been shown to have the capability of removing indoor pollutants and chemicals (Darlington et al., 2001). Atriums can be successful solutions to IAQ problems in commercial and institutional buildings when large spaces are provided for various purposes in multistory buildings but also present complex interfacing with ventilation and environmental control systems (Kainlauri and Vilmain, 1993). Combining a residence with a greenhouse also presents some energy performance questions and some energy advantages (Tanaka 1996; Watson and Buchanan, 1993).

Figure 18.7 illustrates a hypothetical arrangement in which greenhouse air is drawn into the supply airstream where it is filtered by the air handling unit, and then used to pressurize the house. The exfiltrated air could be used as supply for the greenhouse to save heating energy, or else air-to-air heat exchangers could be used for this purpose. This may only be economical if a greenhouse already exists or is built to serve some other primary purpose.

Biofiltration has been in successful use in the water industry for some time and offers a potential alternative method for controlling indoor air quality through the use of filtration material containing bacteria that are antagonistic to harmful pathogens. In this sense it is similar to aspects of vegetation air cleaning.

In one suggested approach, the building structure is used as the biofilter (Darlington et al., 2000). Prior to its acceptance for dealing with *volatile organic compounds* (VOCs) and CO_2, efforts were made to determine whether the incorporation of this amount of biomass into the indoor space could impact indoor air quality. A relatively large ecologically complex biofilter composed of a bioscrubber of about 10 m^2, 30 m^2 of vegetation, and a 3500-L

FIGURE 18.6 Greenhouses can be used as sources of clean air for indoor environments. (*Photo courtesy of Eric Burnett, Penn State.*)

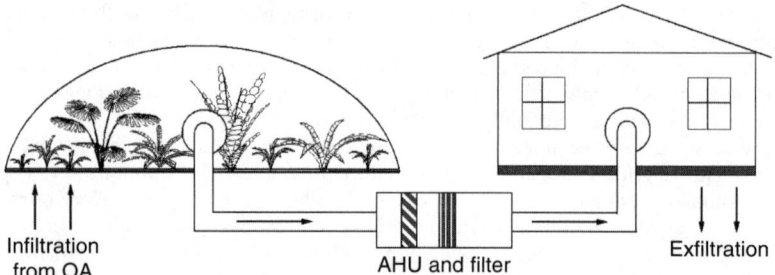

FIGURE 18.7 Depiction of a hypothetical vegetation air cleaning system in which outside air (OA) enters the greenhouse, and is supplied to the house via the air handling unit.

aquarium were established in an airtight 160 m² room in a new constructed office building in downtown Toronto. This space maintained about 0.2 *air changes per hour* (ACH) compared to the 15- to 20-ACH (with 30 percent outside air) of other spaces in the same building. *Total VOCs* (TVOCs) and formaldehyde levels in the biofilter room were the same as, or significantly less than, other spaces in the building. Aerial spore levels were slightly higher than other indoor spaces but were well within reported values for "healthy" indoor spaces. Levels appeared to be dependent on horticultural management practices within the space. Most of the fungal genera present were either common indoor spores or other genera associated with living or dead plant material or soil.

18.4 MATERIAL SELECTIVITY

The evolution of building construction toward improved comfort and functionality has been subject to a simultaneous decrease in the quality of construction due to the profit-motivated drive to produce cheaper buildings faster. This trend has resulted in the adoption of a variety of construction materials and practices that have contributed to unhealthy buildings. Among these developments are the excessive use of inexpensive wood products, other cellulose-based products, excessive airtightness, overly expedient construction, synthetic materials with hazardous potential, and fragile materials subject to damage and moisture accumulation, or which provide substrates for microbial growth. A return to the use of natural materials like stone, glass, lime or mud plasters, adobe or rammed earth, bricks, tiles, untreated wood, cork, reeds, bamboo, and all natural fibers can improve both the aesthetics and healthiness of a house (Pearson, 1998).

Apart from the fact that carpets often contain various chemical by-products that may act as indoor pollutants, carpets and rugs also provide a substrate that may both collect and grow mold and mildew under moist conditions. Since mold spores tend to settle over time, they inevitably collect in carpets and are further ground in by traffic, awaiting only a wet spill to begin germination. Once it is realized that floor coverings may do more long-term harm than good, the actual purpose of carpets and rugs may be questioned. It is considered that floor coverings serve the purpose of protecting feet from cold floors. One solution is to use radiant floor heating systems, which eliminate cold floors and might thereby eliminate the necessity for carpets and rugs. Another alternative is to use healthy materials like natural linoleum, natural cork and rubber, or even soft synthetic materials that do not absorb spores.

TABLE 18.2 Healthy Replacements for Common Building Materials

Commonly used materials	Suggested replacement materials
Wooden walls, structural wood products	Concrete, stone, brick, steel, glass
Wood lumber	Plastic lumber, steel beams
Nonsustainable wood products	Certified sustainable wood products (i.e., Forest Stewardship Council (FSC) certified wood products), recycled wood products
Arsenic-treated wood	Any safely treated wood product or substitute
PVC plastic, vinyl	Any PVC-free materials
Carpeting, rugs	Radiant floor heating
	Natural linoleum, rubber and synthetics, cork
	Bamboo flooring, wood flooring
Alkyd oil-based paint	Water-based paint
Solvent-based adhesives	NonVOC or low-VOC producing adhesives
Paints and sealants	Surfaces requiring no sealants or painting: natural stone, ceramic tile, stucco, raw plaster, waxed concrete, glass
Concrete formwork, precast concrete	High fly-ash (i.e., 15 percent) concrete
Fabrics	Natural fibers, low-VOC materials, low dust
Furnishings/furniture	Low dust-producing synthetics or leather

Table 18.2 provides a select summary of green building materials that have been proposed or used as replacements for more hazardous materials (Spiegel and Meadows, 1999). Although some of the material replacements represent choices based on VOCs rather than microbiological concerns, they all contribute to healthier buildings.

An alternative view of the building envelope is that it should not necessarily be airtight, but should breathe and even act as a natural filter for outdoor air contaminants (Pearson, 1998). An engineered approach would be to place the building under a slight negative pressure to induce such filtration effects, although the problems of biocontamination of the envelope would remain to be solved. A translucent exterior subject to solar exposure may be one option.

Various sources of information are available regarding practical and healthy building materials and construction practices and the reader should consult these for more detailed information (ASHE, 2002; Pearson, 1998; Berman, 2001; Spiegel and Meadows, 1999; Watson and Buchanan, 1993).

18.5 HYGIENIC PROTOCOLS

Perhaps the greenest technology of all is the education of people in the practice of hygiene, which costs little but pays endless dividends in health to the community. Personal hygiene and the cleanliness of indoor environments are the oldest means of controlling diseases. Many successful ancient cultures like the Greeks, the Romans, and the Egyptians were meticulous about maintaining personal cleanliness. Regular bathing was a pastime and cleaning of their clothes and households were daily priorities to a greater degree than they are today. In the modern antiseptic world, hygiene seems to be taken for granted rather than diligently practiced. Home hygiene has been a subject of past interest but it has previously focused on food preparation (IFH, 1998).

When it comes to airborne and respiratory infections there is considerable progress that can be made by merely keeping our hands, our bodies, and our living environment clean. Data provided previously in Chap. 4 suggest that perhaps 50 percent of colds may actually be transmitted by direct contact rather than by the airborne route (Buckland et al., 1965). Studies on handwashing as a preventive measure for controlling both respiratory and gastrointestinal diseases among schoolchildren suggest this infection route is responsible for a significant fraction of absenteeism. One study produced a reduction of 50 percent in school absenteeism when a mandatory handwashing program was implemented (Guinan et al., 2002). An earlier study in which hand sanitizers were installed in a school showed a reduction of approximately 20 percent in absenteeism (Hammond et al., 2000). If we make the assumption that at least 50 percent of absenteeism was due to respiratory infections, we could estimate a 10 to 25 percent reduction in respiratory infections in schools due to handwashing alone. Given that UVGI systems (see Chap. 11) can reduce respiratory infections at least 40 percent on the average, it is conceivable that a handwashing protocol combined with an air cleaning system could reduce school respiratory infections by at least 50 to 65 percent. Admittedly this is speculative arithmetic, but the potential benefits could be significant.

Isolation of individuals who are contagious with respiratory infections is a practice unheard of outside hospitals and yet it would be a most prudent practice in offices, schools, or even homes, the places where most infections are transmitted (Wells, 1955). Unfortunately, no architectural provisions are currently made that acknowledge the value of isolation or that allow for implementation of such protective measures. Protocols, however, would be simple to implement. Students with respiratory infections should remain at home during their infectious period, typically the first 3 to 5 days. Employees likewise should be ordered home during periods of contagiousness so as to protect the other employees and maintain productivity. The problem of preventing contagious infections from spreading among family members at home is not so simple, but it highlights the need for both home air cleaning systems, which are a rarity, and family education regarding hygiene.

Hygienic protocols can be fostered to some degree by the actual building design. Entryways to buildings could provide washing areas where people could clean up prior to entering other areas of the building. In residential buildings, the entryway should lead both to change areas and bathrooms where occupants could clean up or even bathe on entry into the home. The latter process would be especially effective for children, the least hygienic members of any household.

Ideally, a house designed to minimize the spread of airborne microbes or fomites would feature comfortable bathing and toilet facilities with more space and attention to design than is currently placed on these areas. In short, the paradigm of modern residential design needs to be altered to make the cleaning and bathing facilities the most, rather than the least, desirable part of modern housing. For additional information on designing natural, comfortable bathrooms that encourage rather than dissuade attention to hygiene, see Pearson (1998). The kitchen is also a major site for contamination, primarily of the foodborne kind, and protocols for kitchen hygiene have been suggested in Beumer and Kusumaningrum (2003).

Children under the age of three are often the primary and initial route by which a respiratory infection enters a household. Children under 1 year of age have an average of nine respiratory infections a year, while 4-year olds have only about 4.4 infections a year (Sauver et al., 1998). Adults get approximately two respiratory infections a year and the general trend is shown in Fig. 18.8.

The hands are important vehicles of respiratory disease transmission and intervention programs have had some success in this regard (Strazdas et al., 2003). Indoor surfaces also play a role in respiratory disease transmission since microorganisms can survive on them for long periods of time (Archer, 1989). Respiratory infections account for 75 to 90 percent

Hygienic Protocols for Home, Office, and School

1. **General**

   1.1. Wash or disinfect hands before eating, don't share drinks.

   1.2. Keep disinfectant wipes handy. Disinfect hands before rubbing eyes or touching nose, and after touching common used objects.

   1.3. Treat recovered individuals as if they are still contagious—diseases remain infectious for several days after immunity develops.

   1.4. Recognize those who cough, sneeze, or sniffle and avoid them regardless of propriety or excuses they may give.

2. **Home**

   2.1. Isolate sick family members as much as possible—use separate dining and living rooms, or keep them in separate bedrooms.

   2.2. Provide extra care to infants under 2 years of age—their immune systems are incompletely developed and they will both become infected and transmit infections (esp. to other children) more easily.

   2.3. Wash up and change clothes after handling sick infants, and disinfect the diapering area. Don't share towels.

   2.4. Don't allow children to play with other infectious children.

   2.5. Don't send children to day care centers or schools that don't enforce hygienic protocols.

   2.6. When family members have infectious diseases focus on cleaning up all touchable objects and surfaces and enforce frequent hand washing.

   2.7. Don't let infectious family members handle common objects like telephones and TV remotes without disinfecting afterward.

   2.8. Don't send infectious children to school.

   2.9. Infectious members can safely remain outdoors (in mild climates).

   2.10. Quiz relatives before they come over—don't let infectious friends or people into your home, regardless of their feelings.

   2.11. Don't enter the homes of infectious people or touch them. Explain your reasons and ask their consideration.

3. **Office and School**

   3.1. Sick employees should be ordered to work or stay at home, and/or to take cough suppressants so as not to spread bioaerosols.

   3.2. Don't remain near (i.e., 5 ft) infectious employees or students for more than 30 min. Change seats if necessary.

   3.3. Don't remain in the same room as infectious employees or students for more than 4 hours.

   3.4. Wear gloves while in transit or shopping, or avoid touching doorknobs and elevator buttons with fingertips.

   3.5. Don't tarry in places where people are coughing and sneezing.

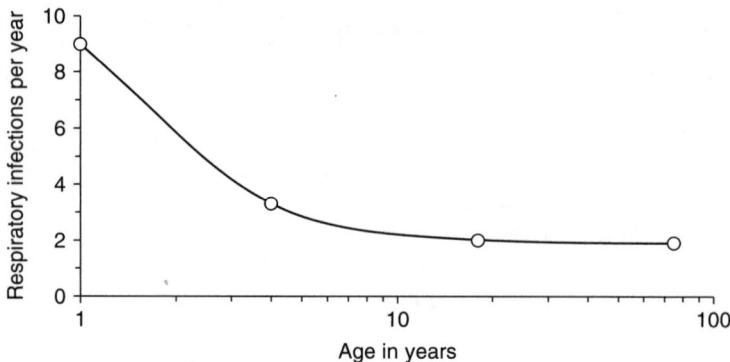

FIGURE 18.8 General trend of respiratory infections vs. age. [*Based on data from Sauver et al. (1998).*]

of infections in child care settings. Children with viral respiratory infections excrete the virus 4 to 5 days before they show symptoms (Collier and Henderson, 1997). In summary of the above respiratory disease epidemiological factors, a set of Hygienic Protocols for Home, Office, and School has been prepared to assist families and office workers in better protecting themselves against respiratory infections. These protocols do not represent any official standards or government agency guidelines but are presented for the benefit of people who wish to reduce the incidence of respiratory diseases in their households, offices, or schools. Although there is no absolute guarantee that these protocols will reduce disease incidence, readers may implement them and judge for themselves.

18.6 AEROBIOLOGICALLY GREEN BUILDINGS

To truly be a green building an indoor environment must possess an integrated design that includes green technologies and healthy building design practices. This section summarizes some fundamentals that would of necessity be both green and healthy.

It has already been shown in the previous section on passive solar exposure how a building's air quality could be enhanced by maximizing fenestration. Ideally there would be no part of a building that did not receive generous amounts of sunlight. Direct sunlight can be controlled through overhangs, automatic blinds, or low-E glass. For residential applications, it is recommended that all exterior walls include large amounts of window space and that interior walls and doors be kept to a minimum. Use of skylights can also provide options where interior walls are considered a necessity. The actual amount of sunlight or skylight, or the proper amount of fenestration that should be provided is a matter that requires further research and economic evaluation, but maximizing fenestration while keeping costs reasonable is the proper approach until guidelines can be established.

Wood is a convenient, cheap, and even aesthetic material for building construction but it has some drawbacks. It provides a substrate for moisture problems and mold growth. Ideally, wood would not be used in building construction for any but minor decorative purposes, thereby eliminating the health hazards of fires also. There are few functions performed by wood that could not be performed at least as well by stone, concrete, and steel. Therefore, it is recommended that new building construction employ stone, brick, concrete, and steel for structural components, and that such materials be favored for use in walls and floors also.

Another problem with traditional residential building design is that the inclusion of basements creates space that is often uncomfortable, accumulates storage items, and can develop moisture problems due to condensation or leakage. Basements typically have limited sunlight and condensation will often produce mold growth. It would be better overall for human comfort and health if basements were not constructed if there was no good reason to include one, and it would tend to reduce the cost of construction. Similarly, attics tend to accumulate stored items, dust, insects, and sometimes mold if the roof leaks. Attics often provide an insulating space, but this could be accomplished without turning them into partly accessible junk-collection sites. Residential homes would be healthier overall if no attics were constructed.

The huge bathhouses in Roman cities were open to the public and were places where people of all classes would congregate, bathe, exercise, or otherwise linger for hours. With steam rooms, saunas, hot baths, pools, cold dips, and massage rooms, these remarkable institutions have only a shadow of an equivalent in modern health clubs where, in spite of the plethora of exercise equipment, the bath and shower facilities are often little more than transitory points of departure. There may be value to society in returning focus to such healthy relaxations, as opposed to other more detrimental forms of diversion. The previous section on hygienic protocols suggested a reversal of the modern architectural trend in which bathrooms and toilets are relegated to the most obscure locations and given the minimum permissible floor space and window space. Perhaps instead of designing bathrooms to be the smallest and least communal rooms in our houses they should be nearer the opposite, as shown in Fig. 18.9.

In the ideal aerobiologically green building, the bathroom would be one of the largest rooms in the home and include a large bathtub and/or hot tub, an oversized shower, have seating so that it could double as a steam room, and connected changing rooms. Modern cultural mores seem to require that toilets and bidets be noncommunal, and although they should form part of an integrated bathing complex, they can be designed to provide simultaneous privacy and connectivity. Floor pedals for sinks are used in health care and might

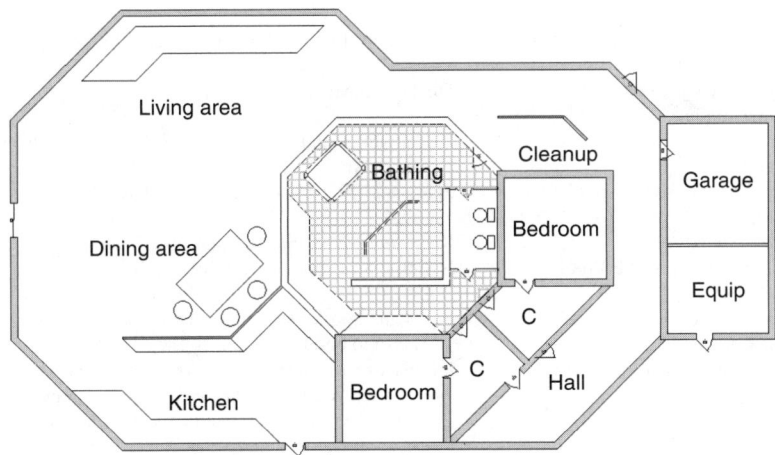

FIGURE 18.9 Floor plan for house with large centralized bathing room, window views to living area, and interconnected bedroom changing (C) rooms.

TABLE 18.3 Suggested Guidelines for Aerobiologically Green Housing

Recommendations	Notes
Avoid using wood for structure, walls, and roofs.	Wood provides a substrate and nutrients for mold.
Avoid using plasterboard walls with cellulose.	Cellulose provides a substrate and nutrients for mold.
Avoid carpeting and rugs.	Use radiant floor heating for floor warmth, use carpeting substitutes.
Avoid wallpaper.	Cellulose-based wallpaper may foster mold growth.
Minimize or avoid use of paint, especially latex paint.	Some paints can support mold growth and release VOCs. Paint is generally used to cover up cheap building materials, but the use of natural and aesthetic materials requires no expense for painting.
Avoid VOC-producing sealants and caulking.	Use healthy substitutes for sealants and caulking.
Do not create basements.	Deepen foundations if necessary, create a separate equipment room at ground level.
Do not create attics.	Expand room space to the ceiling to improve air movement, provide alternate ceiling insulation.
Avoid cloth curtains.	Use alternative nondust producing materials, blinds, and the like.
Avoid cloth furnishings.	Use nonabsorbing, nondust producing materials, such as synthetics, and treated leather.
Centralize and enlarge bathrooms.	Comfortable bathing facilities encourage hygiene.
Provide overlook from bathroom (i.e., from hot tub).	Bath, shower, and toiletries should have a central focus.
Provide bathroom access on entry from outdoors.	Entering occupants can proceed directly to bathing.
Provide bathroom exit to bedroom or living areas.	Occupants can clean up before entering main living areas.
Locate laundry between entryway and bathroom.	Clothes containing spores or allergens should be cleaned and not allowed to contaminate the living spaces.
Create entryway for removing outer clothing and shoes.	Spores cling to clothing and shoes.
Maximize window area throughout perimeter.	Solar exposure will disinfect internal surfaces.
Add skylights for interior lighting (i.e., bathroom).	Solar exposure of bathroom will control mildew.
Create overhangs to block direct summer sunlight.	Direct summer sunlight impacts cooling load.
Allow passage of direct winter sunlight.	Direct winter sunlight provides heating and disinfection.
Include air treatment in ventilation system.	Select a MERV 11–15 filter and an URV 11-15 UVGI system
Pressurize building with filtered air.	Keeps out environmental spores and bacteria.
Consider isolating one bedroom or living space.	Isolation for contagious family member(s).
Minimize use of interior walls and doors.	Walls and doors restrict free airflow and sunlight, add cost to building.
Do not create fireplaces.	Fireplaces waste energy, add first cost, and generally go unused. When used they create soot and raw wood will harbor fungi. Outdoor wood furnaces are a superior energy-saving option.
Hire builders devoted to quality instead of speed.	A house of quality worth living in for a lifetime is not built in a day.

be practical for home use also. The bathroom in this hygienically designed house would have access from the main entry and through changing rooms connected to the bedrooms. Changing areas could be linked to the laundry room such that no dirty laundry collects anywhere else. Bathroom skylights would serve the function of providing solar exposure to control mildew or mold. Surround windows in the bathing room would create openness and bring central light into the living areas. With a layout that focuses on the bathing facilities, a new paradigm may exist for healthy buildings that challenges current views on residential architecture.

Table 18.3 summarizes some design principles that would foster and sustain healthier indoor environments beyond the point in time at which typical modern residences begin to decay, mold, and otherwise disintegrate from inherent cheapness and lack of quality construction. The focus here is on residential housing since most other types of buildings use more lasting building materials and since the home generally provides a nexus of respiratory disease exchange as well as the longest exposures to allergens. These are only suggestions, some of which may seem too radical for serious consideration, and yet many of them were recommended long ago (Wright, 1954).

REFERENCES

Archer, D. (1989). "Disease in day care: A public health problem for the entire community." *J Environ Health* 51:143–147.

ASHE (2002). "Green healthcare design guidance statement." Green Building Committee of the American Society for Healthcare Engineering. American Society for Healthcare Engineering, Chicago, IL.

Beebe, J. M. (1959). "Stability of disseminated aerosols of Pastuerella tularensis subjected to simulated solar radiations at various humidities." *J Bacteriol* 78:18–24.

Bell, A. A., Jr. (2000). *HVAC Equations, Data, and Rules of Thumb*. McGraw-Hill, New York.

Berman, A. (2001). *Your Naturally Healthy Home: Stylish, Safe, Simple*. St. Martin's Press, New York.

Beumer, R. R., and Kusumaningrum, H. (2003). "Kitchen hygiene in daily life." *Intl Biodeter Biodegrad* 51:299–303.

Buckland, F. E., Bynoe, M. L., and Tyrrell, D. A. J. (1965). "Experiments on the spread of colds." *J Hyg* 63(3):327–343.

Collier, A. M., and Henderson, F. W. (1997). "Respiratory Disease in Infants and Toddlers." *Research into Practice in Infant/Toddler Care*, National Center for Early Development & Learning, Chapel Hill, NC.

Darlington, A., Dixon, M. A., and Pilger, C. (1998). "The use of biofilters to improve indoor air quality: the removal of toluene, TCE, and formaldehyde." *Life Support Biosph Sci* 5(1):63–69.

Darlington, A., Chan, M., Malloch, D., Pilger, C., and Dixon, M. A. (2000). "The biofiltration of indoor air: implications for air quality." *Indoor Air 2000* 10(1):39–46.

Darlington, A. B., Dat, J. F., and Dixon, M. A. (2001). "The biofiltration of indoor air: air flux and temperature influences the removal of toluene, ethylbenzene, and xylene." *Environ Sci Technol* 35(1):240–246.

Ehrt, D., Carl, M., Kittel, T., Muller, M., and Seeber, W. (1994). "High performance glass for the deep ultraviolet range." *J Non-Cryst Solids* 177:405–419.

El-Adhami, W., Daly, S., and Stewart, P. R. (1994). "Biochemical studies on the lethal effects of solar and artificial ultraviolet radiation on Staphylococcus aureus." *Arch Microbiol* 161:82–87.

Fernandez, R. O. (1996). "Lethal effect induced in Pseudomonas aeruginosa exposed to ultraviolet-A radiation." *Photochem Photobiol* 64(2):334–339.

Frederick, J. E., Weatherhead, E. C., and Haywood, E. K. (1991). "Long-term variations in ultraviolet sunlight reaching the biosphere: Calculations for the past three decades." *Photochem Photobiol* 54(5):781–788.

Futter, B. V., and Richardson, G. (1967). "Inactivation of bacterial spores by visible radiation." *J Appl Bacteriol* 30(2):347–353.

Gerlach-Spriggs, N., Kaufman, R. E., and Warner, S. B., Jr. (1998). *Restorative Gardens: The Healing Landscape*. Yale University Press, New Haven, CT.

Griego, V. M., and Spence, K. D. (1978). "Inactivation of Bacillus thuringiensis spores by ultraviolet and visible light." *Appl Environ Microbiol* 35(5):906–910.

Guinan, M., McGuckin, M., and Ali, Y. (2002). "The effect of a comprehensive handwashing program on absenteeism in elementary schools." *Am J Infect Control* 30(4):217–220.

Hammond, B., Ali, Y., Fendler, E., Dolan, M., and Donovan, S. (2000). "Effect of hand sanitizer use on elementary school absenteeism." *Am J Infect Control* 28:340–346.

Harper, G. J. (1961). "Airborne microorganisms: survival tests with four viruses." *J Hyg* 59:479–486.

IFH (1998). "Guidelines for prevention of infection and cross infection in the domestic environment." International Scientific Forum on Home Hygiene. Milano, Italy.

Kainlauri, E. O., and Vilmain, M. P. (1993). "Atrium design criteria resulting from comparative studies of atriums with different orientation and complex interfacing of environmental systems." *ASHRAE Trans* 99(1):1061–1069.

Mitscherlich, E., and Marth, E. H. (1984). *Microbial Survival in the Environment*. Springer-Verlag, Berlin, Germany.

Pearson, D. (1998). *The New Natural House Book: Creating a Healthy, Harmonious and Ecologically Sound Home*. Fireside, New York.

Rautiala, S., Haatainen, S., Kallunki, H., Kujanpaa, L., Laitinen, S., Miihkinen, A., Reiman, M., and Seuri, M. (1999). "Do plants in office have any effect on indoor air microorganisms?" *Indoor Air 99: Proceedings of the 8th International Conference on Indoor Air Quality and Climate*, Edinburgh, Scotland. 704–709.

Rosenbaum, M. (2002). "A green building on campus." *ASHRAE J* 44(1):41–44.

Safapour, N., and Metcalf, R. H. (1999). "Enhancement of solar water pasteurization with reflectors." *Appl Environ Microbiol* 65(2):859–861.

Sauver, J. S., Khurana, M., Kao, A., and Foxman, B. (1998). "Hygienic practices and acute respiratory illnesses in family and group day care home." *Public Health Rep* 113(6):544–551.

Spiegel, R., and Meadows, D. (1999). *Green Building Materials: A Guide to Product Selection and Specification*. John Wiley & Sons, New York.

Strazdas, L. A., Rawiel, U., Bronson, D. L., Orosz-Coghlan, P., Gerba, C. P., and Lebowitz, M. D. (2003). "Impact of a hygiene intervention program on illness in child care centers: Phase III." Epidemiology in Arizona Forum, Tucson, AZ

Tanaka, H. (1996). "Study on the energy performance of a residence with greenhouse." *Proceedings of the 7th International Conference on IAQ and Climate*, Nagoya, Japan. 235–240.

USGBC (2002). "LEED Green Building Rating System." United States Green Building Council. http://www.usgbc.org.

Watson, D., and Buchanan, G. (1993). *Designing Healthy Buildings*. American Institute of Architects, Washington, DC.

Webb, S. J. (1961). "Factors affecting the viability of airborne bacteria: IV. The inactivation and reactivation of airborne Serratia marcescens by ultraviolet and visible light." *Can J Microbiol* 7:607–619.

Webb, A. R. (1991). "Solar ultraviolet radiation in southeast England: The case for spectral measurements." *Photochem Photobiol* 54(5):789–794.

Wells, W. F. (1955). *Airborne Contagion*. New York Academy of Sciences, New York.

Wolverton, B. C. (1996). *How to Grow Fresh Air: 50 Plants That Purify Your Home or Office*. Penguin Books, Baltimore, MD.

Wright, F. L. (1954). *The Natural House*. Mentor Books, New York.

CHAPTER 19

THERMAL DISINFECTION, CRYOGENICS, AND DESICCATION

19.1 INTRODUCTION

The physical processes of heating, freezing, and dehydration can be effective means of disinfecting air and surfaces in buildings. These processes occur naturally in the outdoor air and often operate as a by-product of normal ventilation system operation. A few systems exist in which heating and freezing are used for the disinfection of air or food products but the applications of such technologies to the disinfection of building ventilation airstreams are unlikely to be economical except in special circumstances. One additional technology, however, that of cycling ventilation systems to inhibit microbial growth, has the potential to be very cost-effective since it only employs the controls, without any additional equipment.

19.2 THERMAL DISINFECTION

Heating can inactivate microorganisms, including spores and viruses, but high temperatures and long durations are typically required (Morrissey and Phillips, 1993; Woese, 1960). The food industry has studied this matter extensively and guidelines for sterilization of processed food and equipment are in common use (Heijden et al., 1999; Ray, 1996). A variety of heating systems are available for sterilizing medical and laboratory equipment and for destroying biocontaminants. Steam-heating, microwave-heating, and dry-heating ovens are commonly used for sterilizing equipment, but they typically require 20 to 30 minutes at high temperature for complete inactivation of microbes. This precludes their use for sterilizing airstreams.

Sewage and waste processing facilities sometimes incinerate the exhaust air with direct gas flames to destroy bacteria and control odors (Waid, 1969). Such systems would not be practical for use in buildings because of the high energy costs, and the fact that after heating the air would have to be cooled down again, which incurs even greater costs. Thermal incineration for the destruction of *volatile organic compounds* (VOCs) has been estimated to have a cost of about $112,000 per year (Nunez et al., 1993). This estimate is based on a flow rate of 3000 cfm. It could be expected that thermal incineration of airborne microbes would bear similar high costs.

The optimum growth temperatures for microbes have been well documented (see Mitscherlich and Marth, 1984) and temperatures above the optimum growth range would presumably result in a decrease in microbial populations. Criteria for heat sterilization of airborne microbes are included in App. A under the Inactivation category, in which both moist heat treatment and dry heat treatment are addressed where they are known. Typically, moist heat (i.e., steam) at 121°C will sterilize microbes in about 15 minutes, whereas dry heat at about 170°C will sterilize microbes in about 1 hour.

The death of microbes under heat exposure follows typical single- and two-stage decay curves as detailed in Chap. 7 (see Fig. 7.11), with shoulder curves figuring prominently as a function of decreased temperature (King et al., 1979; Moats et al., 1971). The rate constant for thermal inactivation curves is a direct function of temperature (Sapru and Labuza, 1993). This is illustrated in Fig. 19.1 in which the decay curve of *Aspergillus niger* spores are shown at two different temperatures. The lines represent two-stage shoulder curves fitted to data from Fujikawa and Itoh (1996).

The two-stage shoulder curve used in Fig. 19.1 to fit the data is as follows:

$$S(t) = (1-f)[1-(1-e^{-k_1 It})^{n_1}] + f[1-(1-e^{-k_2 It})^{n_2}] \quad (19.1)$$

Equation (19.1) was fitted to each of the data sets in Fig. 19.1 by separating the curve into two stages and adjusting the rate constants and exponents manually to obtain the best fit as described in Chap. 7. For the 58°C condition, the fitted curve that resulted is as follows:

$$S = (1-0.07)[1-(1-e^{-0.5t})^7] + 0.07e^{-0.045t} \quad (19.2)$$

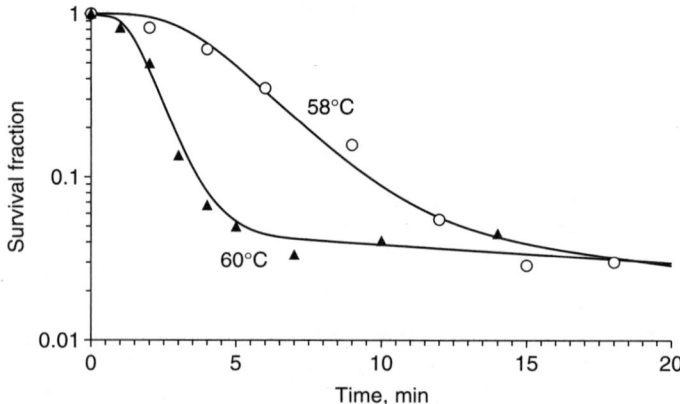

FIGURE 19.1 Thermal inactivation data for *Aspergillus niger* spores at two different temperatures. The curves fitted to the data are two-stage curves with shoulders. [*Based on data from Fujikawa and Itoh (1996).*]

In Eq. (19.2), the second stage is simplified because the exponent $n_2 = 1$, as is often the case with second stages. Similarly, the equation fitted to the data for 60°C is as follows:

$$S = (1 - 0.05)[1 - (1 - e^{-1.3t})^7] + 0.05e^{-0.026t} \qquad (19.3)$$

It may seem odd that the 58°C thermal decay curve would have a higher decay constant than the 60°C curve, but this is purely a result of the anomalous data set. Normally the higher temperature will produce a higher second-stage rate constant.

Heat resistance can be a complex function of strain variation, previous growth conditions, exposure to heat shock, environmental stresses, and composition of the heating media, and predictions using the above models should be used with caution regarding these additional factors (Doyle et al., 2001).

19.3 CRYOGENICS

The use of cryogenic freezing to inactivate pathogens on surfaces has been used with success in certain areas of the food industry (Andrews et al., 2000). The destruction of microbes by freezing is species dependent, and some species will simply become dormant (Mitscherlich and Marth, 1984). The cost of cryogenic freezing, like the cost of heat destruction, is also high and impractical for the sterilization of airstreams. However, as a means of remediating buildings, the use of freezing temperatures has the advantage of being simple to accomplish in cold climates without necessarily damaging the furnishings. That is, a building contaminated with certain pathogens that were susceptible to cold temperatures could be opened to the elements in winter.

Some variants of *Bacillus subtilis* can survive in Martian-like atmospheric conditions as well as in space (Sussman and Halvorson, 1966; La Duc et al., 2004). Some microbes, especially bacterial and fungal spores, have an extraordinary ability to survive near absolute zero temperatures and extended exposure to liquid hydrogen.

Freeze drying has a biocidal effect that varies with species. In a study by Miyamoto-Shinohara et al. (2000), the survival rates of 10 species of microorganisms were investigated after freeze drying and preservation in a vacuum at 5°C. The survival rates immediately after freeze drying of gram-positive bacteria, including *Corynebacterium acetoacidophilum* and *Streptococcus mutans*, were around 80 percent. Survival rates after the drying of gram-negative bacteria, including *Escherichia coli*, *Pseudomonas putida*, *Serratia marcescens*, and *Alcaligenes faecalis*, were around 50 percent.

Liquid nitrogen has been used for the eradication of house dust mites in carpeting as a remedial measure to assist allergic individuals, and for killing dust mites in office furniture, although some doubt exists as to the long-term effectiveness of this approach (Roys et al., 1993; Kalra et al., 1993).

19.4 DESICCATION AND DEHUMIDIFICATION

Dehydration can be effective in destroying microorganisms, but the effect is very species dependent, and may require extended durations. Spores have only limited susceptibility to dehydration and some bacterial species may be induced into dormancy by dehydration (Russell, 1982; Austin, 1991). The means of attaining dehydration is often simply heating, which results in the same basic high costs of energy discussed in the section on heating.

Desiccant dehumidification usually involves the use of a desiccant wheel that adsorbs moisture and then releases it to a secondary airstream that exhausts to the outside. Desiccant wheels can be used as part of evaporative cooling systems. Desiccant cooling has been shown to reduce levels of indoor species of microorganisms, and since this occurs as a by-product of the cooling process, the disinfection effect can be had for no additional cost. Desiccant cooling has been shown to reduce bacterial levels in the entering airstream by 10 to 80 percent or more (Kovak et al., 1997). In the same tests, airborne fungal levels were reduced 30 to 65 percent or more.

Desiccant systems have also been shown to have the capacity to remove volatile organic compounds and other pollutants from indoor air (Hines et al., 1992). Although desiccant dehumidifiers are not intended to be air cleaning systems, their presence in a ventilation system clearly offers some degree of potential protection against airborne contaminants.

19.5 THERMAL CYCLING

Variations in humidity and temperature can influence the germination and growth of fungal spores (Tang et al., 1996). An innovative application of the disinfection properties of thermal cycling has been examined by Sakuma and Abe (1996). In this study, which involved radiant ceiling panels, the system was cycled to produce several hours of warm, moist conditions that would promote the germination of fungal spores. The system would then be put into a cooling and dehydration cycle during which the germinated mold would die from dehydration. This process takes advantage of the fact that the resistance of fungal mycelia to dehydration is greatly reduced from that in the spore form. A normal cycle of daytime cooling (i.e., >90 percent RH to germinate spores) and night-time dehumidification has the potential to significantly reduce fungal growth inside AHUs without any additional equipment. The same disinfection principle, inducing germination to reduce resistance to biocidal factors, may have applications in HVAC systems that are yet to be discovered. A related approach, the use of low levels of ionizing radiation to stimulate germination, may also have applications in combination with thermal or dehydration disinfection methods (Kuzin et al, 1976, Kuzin, 1977).

REFERENCES

Andrews, L. S., Park, D. L., and Chen, Y. P. (2000). "Low temperature pasteurization to reduce the risk of vibrio infections from raw shell-stock oysters." *Food Addit Contam* 17(9):787–791.

Austin, B. (1991). *Pathogens in the Environment.* Blackwell Scientific Publications, Oxford, UK.

Doyle, M. E., Mazzotta, A. S., Wang, T., Wiseman, D. W., and Scott, V. N. (2001). "Heat resistance of *Listeria monocytogenes*." *J Food Prot* 64(3):2001.

Fujikawa, H., and Itoh, T. (1996). "Tailing of thermal inactivation curve of *Aspergillus niger* spores." *Appl Microbiol* 62(10):3745–3749.

Heijden, K. v. d., Younes, M., Fishbein, L., and Miller, S. (1999). *International Food Safety Handbook: Science, International Regulation, and Control.* Marcel Dekker. New York.

Hines, A. L., Ghosh, T. K., Loyalka, S. K., and Warder, R. C., Jr. (1992). "A summary of pollutant removal capacities of solid and liquid desiccants from indoor air." *NTIS No. PB95-104683.* Gas Research Institute, Chicago, IL.

Kalra, S., Crank, P., Hepworth, J., Pickering, C. A., and Woodcock, A. A. (1993). "Concentrations of the domestic house dust mite allergen Der p I after treatment with solidified benzyl benzoate (Acarosan) or liquid nitrogen." *Thorax* 48(5):582.

King, A. D., Bayne, H. G., and Alderton, G. (1979). "Nonlogarithmic death rate calculations for *Byssochlamys fulva* and other microorganisms." *Appl Microbiol* 37:596–600.

Kovak, B., Heiman, P. R., and Hammel, J. (1997). "The sanitizing effects of desiccant-based cooling." *ASHRAE J* 39(4):60–64.

Kuzin, A. M., Nikitina, A. N., Iurov, S. S., and Primak, V. N. (1976). "Stimulating effect of chronic low-power gamma radiation on the growth and development of Aspergillus niger." *Radiobiologiia* 16:70–72.

Kuzin, A. M. (1977). *Stimulatory Effect of Ionizing Irradiation*. Atomizdat, Moscow, USSR. 133.

La Duc, M. T., Satomi, M., and Venkateswaran, K. (2004). "*Bacillus odysseyi* sp. nov., a round-spore-forming bacillus isolated from the Mars Odyssey spacecraft." *Int J Syst Evol Microbiol* 54(2004):195–201.

Mitscherlich, E., and Marth, E. H. (1984). *Microbial Survival in the Environment*. Springer-Verlag, Berlin, Germany.

Miyamoto-Shinohara Y, I. T., Sukenobe J, Murakami Y, Kawamura S, Komatsu Y. (2000). "Survival rate of microbes after freeze-drying and long-term storage." *Cryobiology* 41(3):251–255.

Moats, W. A., Dabbah, R., and Edwards, V. M. (1971). "Interpretation of nonlogarithmic survivor curves of heated bacteria." *J Food Sci* 36:523–526.

Morrissey, R. F., and Phillips, G. B. (1993). *Sterilization Technology*. Van Nostrand Reinhold, New York.

Nunez, C. M., Ramsey, G. H., Ponder, W. H., Abbott, J. H., Hamel, L. E., and Kariher, P. H. (1993). "Corona destruction: An innovative control technology for VOCs and air toxics." *J Air Waste Manage* 43:242–247.

Ray, B. (1996). *Fundamental Food Microbiology*. CRC Press, Boca Raton, FL.

Roys, M. S., Raw, G. J., and Whitehead, C. (1993). "Sick building syndrome: Cleanliness is next to healthiness." *Indoor Air: Proceedings of the 6th International Conference on Indoor Air Quality and Climate*, Helsinki, Finland. 261–266.

Russell, A. D. (1982). *The Destruction of Bacterial Spores*. Academic Press, New York.

Sakuma, S., and Abe, K. (1996). "Prevention of fungal growth on a panel cooling system by intermittent operation." *1996 The 7th International Conference on IAQ and Climate*, Nagoya, Japan. 179–184.

Sapru, V., and Labuza, T. P. (1993). "Temperature dependence of thermal inactivation rate constants of bacterial spores in a glassy state." *J Ind Microbiol* 12:247–250.

Sussman, A. F., and Halvorson, H. O. (1966). *Spores: Their Dormancy and Germination*. Harper & Row, New York.

Tang, H., Yoshizawa, S., and Deguchi, N. (1996). "Influences of timely variations of environmental temperature and humidity on fungal growth speed." *The 7th International Conference on IAQ and Climate*, Nagoya, Japan. 191–196.

Waid, D. E. (1969). "Incineration of organic materials by direct gas flame for air pollution control." *Am Ind Hyg Assoc J* 30:291–297.

Woese, C. (1960). "Thermal inactivation of animal viruses." *Ann N Y Acad Sci* 82:741–751.

CHAPTER 20

ANTIMICROBIAL COATINGS

20.1 INTRODUCTION

Antimicrobial coatings, also sometimes called hygienic coatings, are available for a host of products, including ductwork, filters, and appliances. A wide variety of antimicrobial coatings are becoming available, including those with antibiotic properties, coatings that release antimicrobials over time, coatings that react biocidally with bacterial cells, and coatings that prevent surface adhesion of microorganisms (PRA, 2004; Engel et al., 2003; Harris and Richards, 2004).

Antimicrobial coatings have applications in health care, pharmaceuticals, the food industry, and commercial or residential applications. Antimicrobial coatings also have applications where moisture or biofouling is a concern, such as in ductwork. Structural materials are often painted or coated with organically based compounds like latex that may be susceptible to microbial growth (Klens and Yoho, 1984). The replacement of these materials or the use of antimicrobials can counteract this effect (Kennedy, 2002).

Materials like silver and copper have long been known to inhibit the growth of microorganisms and even to destroy them in time (Thurman and Gerba, 1989). Coatings made from silver ions can be bonded to metallic or other surfaces and they may have an extremely long life span, depending on wear.

Other antimicrobial materials are currently available or under study include antimicrobial fabrics, antimicrobial-coated steel duct, and antimicrobial duct fabric. This is a developmental field and research is ongoing. For CBW defense, one of the potential applications of antimicrobial duct is to facilitate remediation.

Antimicrobial filters are currently available, but their actual effectiveness is still under study (Foarde et al., 2000; Price et al., 1999). The primary purpose of most antimicrobial filters is to prevent microbial growth on the filter media if it should become moist. To this end they are effective, and public concerns about antimicrobial particulates becoming re-entrained have not been borne out by industry experience (Foarde et al., 2000). Some sources debate the efficacy of such treatments and comparisons of different filters in terms of their effectiveness in inhibiting microbial growth gave mixed results (Fellman, 1999). The primary concern with microbial growth on filters is that such growth, especially fungal growth, may penetrate the filter and release spores or other biological components downstream. Another concern is that contaminated filters may be a hazard to maintenance crews. To these ends antimicrobial filters may be effective, but further study is warranted.

The antimicrobial effect of natural Mexican zeolite supporting silver ions was investigated by Rivera-Garza et al. (2000). It was found that the Mexican silver clinoptilolite-heulandite mineral eliminated the pathogenic microorganisms *E. coli* and *S. faecalis* from water with the highest amount of silver supported on the mineral after 2 hours of contact time.

The use of nanotechnology combined with controlled morphology and functionality to create a natural antimicrobial agent has been reviewed by Ross (2004). The nano silver particles have an extremely large specific surface area, increasing the contact area with bacteria or fungi, which vastly improves its bactericidal and fungicidal effectiveness.

The antibacterial effects of addition of silver oxide to $Na_2O \cdot CaO \cdot 2SiO_2$ glass have been studied by Catauro et al. (2004). Testing of samples against *E. coli* and *Streptococcus mutans* produced significant antibacterial effects.

Other materials besides silver and copper ions can be used as antimicrobials, including immobilized enzymes, quaternary ammonium compounds, synthetic polymers, titanium dioxide (which is addressed in Chap. 14), and antibiotics (Tiller et al., 2001; Foarde and VanOsdell, 2001).

20.2 PERFORMANCE OF MATERIALS

The survival of microorganisms on surfaces is a function of the type of material, the relative humidity or moisture availability, the temperature, exposure to light, porosity of the surface, and the survival characteristics of each species. Microbes lose infectivity on drying (Buckland and Tyrrell, 1962). Microbes ultimately die out on surfaces but as fomites they may survive long enough to cause secondary infections (McEldowney and Fletcher, 1988; Wilkinson, 1966; McDade et al., 1964; Reed, 1975). Figure 20.1 shows the decay of a population of *Staphylococcus aureus* on a dry surface, based on data from Askew (2004). Obviously this microbe can survive as a fomite long enough to potentially cause secondary contamination.

Copper and silver ions have long been known to have certain biocidal effects. One of the principal mechanisms of biocidal action of these ions is thought to be cell penetration

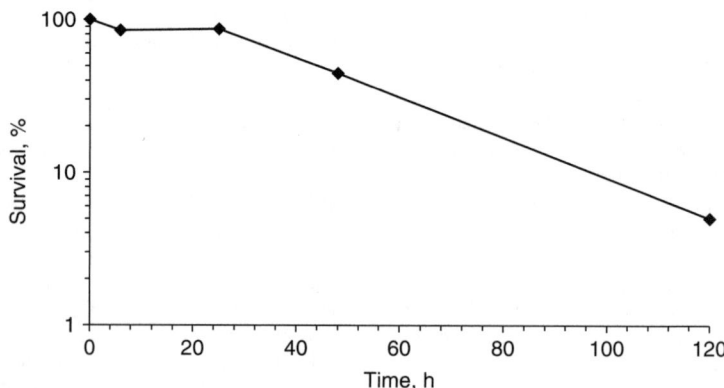

FIGURE 20.1 Survival of *Staphylococcus aureus* on a dry surface. [*Based on data from Askew (2004).*]

(Hambidge, 2001; Turpin, 2003). The positively charged copper ions form electrostatic bonds with negatively charged sites on the cell wall. The cell membrane is thus distorted, allowing ingress of silver ions that attack the cell by binding at specific sites to DNA, RNA, and other components, causing catastrophic failure of cellular life support systems. A simplified version of the process is as follows: monovalent or ionic silver has an affinity for hydrogen ions, bonding with them on the sulfhydryl groups present in microbial cells, and disrupting electron transfer and respiration. Other nonionic forms of silver employ other equally effective mechanisms, such as catalyzing the interaction of atomic hydrogen with the sulfhydryl group resulting in an OH molecule and a sulfur bond that prevents further microbial respiration (McGill, 2003).

Limited data are available on the performance of antimicrobial materials. Figure 20.2 shows results from one study in which a material called PVP outperformed a variety of common materials in terms of reducing surface survival rates.

In one process for creating antimicrobial coatings on steel, a silver ion compound, usually a zeolite matrix containing silver ions, is blended into an epoxy resin and the resin is then applied to the steel using conventional application processes (Mazurkiewicz, 2002; McGill, 2003). The coating is typically about 0.1 mil thick and is durable but still subject to wear. The expected life is said to be up to 30 years. Figure 20.3 shows an example of the suppression of microbial growth of *E. coli* on a stainless steel surface.

In a study carried out by the Centre for Applied Microbiology & Research (CAMR) researchers found that at room temperatures it takes 34 days for *E. coli* to die on stainless steel tiles, 4 days to die on brass tiles, and just 4 hours to die on copper tiles (CDA, 2000).

Figure 20.4 shows the results of a study on biocidal nanosol coating on materials used for catheters (Haufe et al., 2004). The same study demonstrated the biocidal effects of nanosol coatings on wood and textiles.

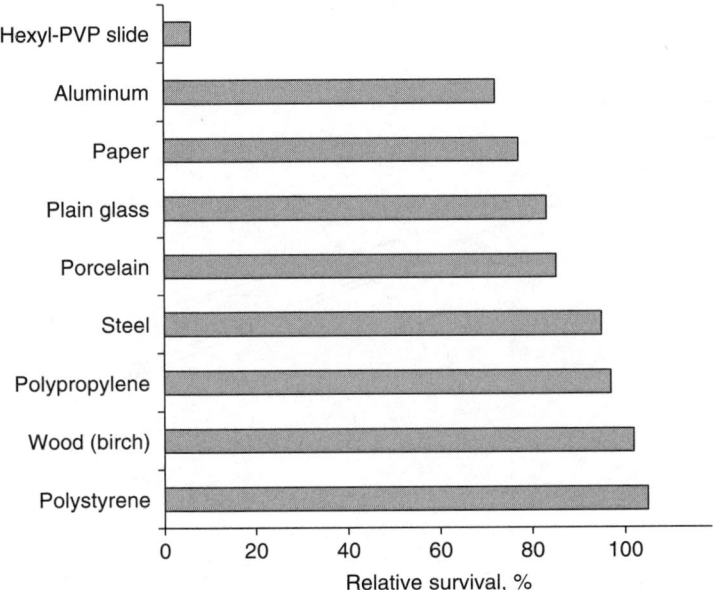

FIGURE 20.2 Relative survival of bacteria on different surface materials. [*Based on data from Tiller et al. (2001).*]

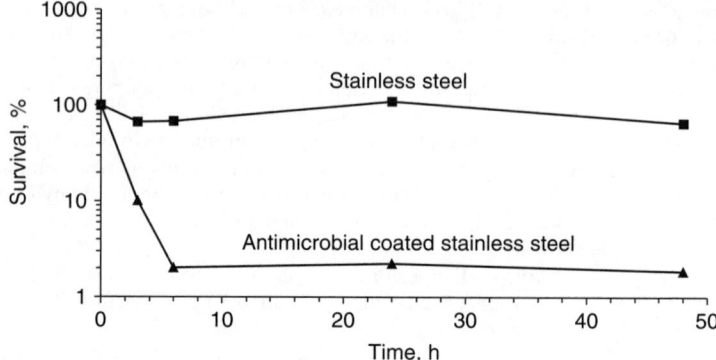

FIGURE 20.3 Survival of E. coli on stainless steel with and without an antimicrobial coating. [Based on data from McGill (2003).]

Harris and Richards (2004) studied the effect of polishing and antimicrobials to reduce surface adhesion of S. aureus on titanium. They found that polishing had no effect on preventing adhesion but that application of sodium hyaluronate decreased the density of adhering microorganisms.

Unanswered questions regarding the use of these new coatings in ventilation systems and other locations are their stability over time, their susceptibility to scratching and damage, the effects of accumulation of dust, whether parts or all of internal ductwork should be coated, and what coating thickness should be required.

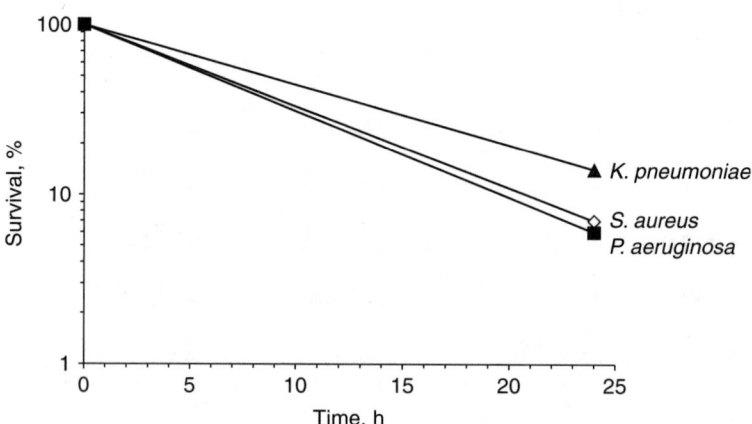

FIGURE 20.4 Survival of three species on a surface with a silica coating with embedded silver ions. [Based on data from Haufe et al. (2004).]

20.3 HEALTH CARE AND PHARMACEUTICAL APPLICATIONS

The health care industry has a vested interest in the development of antibacterial materials for use in various applications, including catheters, clothing, dressings, and virtually all indoor surfaces. Biomedical devices resistant to bacterial adherence and colonization have potentially widespread applications in health care and biotechnology. A photochemical surface modification process has been investigated as a means of applying antimicrobial coatings to biomedical devices (Dunkirk et al., 1991). The photochemical process results in covalent immobilization of coatings to all classes of medical device polymers and preliminary results indicate success in creating microbial-resistant surfaces.

The antimicrobial properties of some metals and coatings for medical products were examined in experimental studies by Studenikina et al. (1993). Results indicate that samples coated with vacuum technology-made materials, such as titanium mixtures with chromium, hafnium, zirconium, and copper, have antimicrobial properties against staphylococci, *Escherichia coli, Proteus,* and *Pseudomonas aeruginosa.*

A new thermally stable silver-based antimicrobial treatment has been developed for fabrics called Alphasan (Wright, 2002). Research addressed the effectiveness against a wide range of bacteria. The new treatment uses the element silver in its ionic form (Ag^+). The new process is being used in the production of a range of synthetic textiles in Japan, Europe, and the United States.

Modified cellulose materials (CM) and bleached cotton cloth modified by surfactant antiseptics (CA) have been prepared by the method of adsorption interaction between CM and CA (Kotelnikova et al., 2000). Microbiological tests of modified cellulose materials show that they exhibit antimicrobial activity and that they may be useful for applications in clinics as dressing materials.

A case study has been summarized by Boarini (1990) in which a new antimicrobial treated foley catheter was provided to the infection control urology market. The product was designed to help prevent catheter-associated *urinary tract infection* (UTI). The challenge involved selecting materials that were compatible with the substrate of the device in order to achieve a safe nontoxic product that achieved significant antimicrobial efficacy.

Biosurfactants are biologically produced compounds of which some possess certain antiadhesion properties. Several biosurfactants have strong antibacterial, antifungal, and antiviral activity and specific biosurfactants have been identified that inhibit the microbial adhesion or have antimicrobial effects on species of *Listeria, Bacillus, Candida, Mycobacterium, Staphylococcus,* and *Streptococcus* (Singh and Cameootra, 2004).

Researchers have suggested that switching hospital work surfaces and door handles from stainless steel to copper could help combat Methicillin-resistant *Staphylococcus aureus* (MRSA) (BBC, 2004). Copper doorknobs may also be a simple and economic solution for residential and commercial building applications.

20.4 FOOD INDUSTRY APPLICATIONS

The food industry has always had a special interest in materials and surfaces that resist biocontamination because of concerns about food pathogens. Antimicrobial surfaces may have widespread applications for appliances in both the food industries and the home environment, including possible applications for common home appliances and humidifiers (Topping, 2004). Microbes that colonize appliances may grow to hazardous concentrations and cause illness. Stainless steel is one common material that is easily cleaned with formulated disinfectants and has a natural ability to resist microbial attachment when electropolished. Surfaces treated with antimicrobials like silver ions can inhibit the growth of bacteria and prevent the formation of biofilms. Zinc oxide and other metal oxides are available that can be incorporated into plastics to prevent microbial growth. Humidifying systems, dehumidifiers, ice machines, showerheads, and other appliances are candidates for antimicrobial material applications.

In a study by Heaton et al. (1991), the efficacy of antimicrobial compounds was determined using various fungi from food industry plants. The antifungal performance of urethane oil-based paints containing C9211 was tested in laboratory experiments and in field trials at two food industry plants. The surface of hard glossy paints without the antimicrobial compound proved to be effective at preventing the germination and growth of fungi during exposure in the humidity cabinet and reasons for this are discussed. In field trials, paints without biocide developed extensive mould growth after several months of exposure while all paints containing C9211 remained mould free throughout the 18-month trial.

The transfer rates of *Salmonella enteritidis, Staphylococcus aureus*, and *Campylobacter jejuni* from kitchen sponges to stainless steel surfaces, and from these surfaces to foods, were investigated by Kusumaningrum et al. (2003). *S. aureus* was recovered from the surfaces for at least 4 days when the contamination level was high or moderate. At low levels, the surviving numbers decreased below the detection limit within 2 days. *S. enteritidis* was recovered from surfaces for at least 4 days at high contamination levels. Levels of *C. jejuni* decreased below the detection limit within 4 hours. The test microorganisms were readily transmitted from the wet sponges to the stainless steel surfaces with the transfer rates varied from 20 to 100 percent. This study has highlighted the fact that pathogens remain viable on dry stainless steel surfaces and present a contamination hazard for considerable periods of time, depending on the contamination levels and pathogen species.

20.5 RESIDENTIAL AND COMMERCIAL BUILDING APPLICATIONS

Residential applications for antimicrobial surfaces exist in abundance, especially as fomites are thought to be a primary cause of respiratory disease transmission in the home environment. Doorknobs, towels, telephones, kitchen appliances, children's toys, TV remotes, and a host of other common items could theoretically include antimicrobial coatings. Building materials, including paints, kitchen counters, furniture, ductwork, and the like, could also benefit from antimicrobial coatings. A "concept" home designed and constructed with antimicrobial materials has been developed to highlight the safe and effective use of antimicrobials (Skaer, 2001). The project involves the use of silver ion coated steel in both the structure and the ductwork, and in various locations throughout the home considered to be "high touch" zones.

Ventilation systems are a potentially significant source of microbial growth, especially in older systems that have deteriorated from wear, that have moisture problems, or that have accumulated contamination (McGill, 2003). Ventilation systems may benefit especially from the use of antimicrobial coatings. Options include steel duct impregnated or coated with silver or copper ions, antimicrobial duct fabric, and the use of antimicrobial filters. A variety of antimicrobial materials and equipment are currently available to satisfy the needs of designers seeking self-disinfecting ventilation systems.

A study by Dever et al. (1997) evaluated melt-blown filter media containing antimicrobial compounds for mechanical properties, filtration efficiency, and ability to prevent bacterial growth on the filter. Filters impregnated with three different antimicrobial compounds were challenged with gram-negative and gram-positive bacteria. Antimicrobial Compound A prevented growth of the gram-positive *Staphylococcus aureus* bacteria at pH 4.8. Compound A was ineffective against the gram-negative *Klebsiella pneumoniae* bacteria at any pH. Compound B had no effect on bacterial growth. Compound C prevented growth of gram-negative and -positive bacteria at all pH levels. Filter fiber diameter increased with increasing concentrations of antimicrobial compounds and filtration efficiency decreased with increasing fiber diameter.

Basic building materials like concrete, cement, plaster, grout, woodwork, and brick do not provide surfaces that are easily maintained in a hygienic condition (PRA, 2004). Bacteria attach most easily to porous surfaces such as wood and ceramic materials because of their accommodating microcavities. Glass is least porous surface, followed by stainless steel, then aluminum, rubber, and plastics.

REFERENCES

Askew, P. D. (2004). "Paper 6: Hygienic Control: Assessing the Role of Antimicrobial Surfaces." *The Second Global Congress Dedicated to Hygienic Coatings and Surfaces,* Orlando, FL.

BBC (2004). "Copper surfaces can kill off MRSA." BBC World News. http://news.bbc.co.uk/1/hi/health/3867781.stm.

Boarini, E. (1990). "Development of a device with an antimicrobial treatment." *Medical Plastics Conference,* Anaheim, CA. 5.

Buckland, F. E., Tyrrell, D. A. J. (1962). "Loss of infectivity on drying various viruses." *Nature* 195:1063–1064.

Catauro, M., Raucci, M. G., De Gaetano, F., and Marotta, A. (2004). "Antibacterial and bioactive silver-containing Na 2O·CaO·2SiO2 glass prepared by sol-gel method." *J Mater Sci: Mater M* 15(7):831–837.

CDA (2000). "Antibacterial properties of copper and brass demonstrate potential to combat toxic *E. coli* O157 outbreaks in the food processing industry." Copper Development Association. http://www.copper.org/about/pressreleases/2000/DemonstratePotential.html.

Dever, M., Davis, W. T., Arrage, A. A., White, D. C., and Benson, R. S. (1997). "Characterization of melt-blown filters made of polypropylene and polypropylene-antimicrobial blends." *TAPPI J* 80(3):157–168.

Dunkirk, S. G., Gregg, S. L., Duran, L. W., Monfils, J. D., Haapala, J. E., Marcy, J. A., Clapper, D. L., Amos, R. A., and Guire, P. E. (1991). "Photochemical coatings for the prevention of bacterial colonization." *J Biomater Appl* 6(2):131–156.

Engel, R., Cohen, J. I., and K. Melkonian (2003). "Antimicrobial Surfaces." *Joint Scientific Conference on Chemical & Biological Defense Research,* Bethesda, MD.

Fellman, G. (1999). "ASHRAE study shows mixed results for antimicrobial filters." Indoor Environment Connections Online. http://www.ieconnections.com/archive/nov_99/1199_article2.htm.

Foarde, K. K., Hanley, J. T., and Veeck, A. C. (2000). "Efficacy of antimicrobial filter treatments." *ASHRAE J* 42(1):52–58.

Foarde, K. K., and VanOsdell, D. W. (2001). "Investigation of the potential antimicrobial efficacy of sealants used in HVAC systems." *J Air Waste Manage Assoc* 51:1219–1226.

Hambidge, A. (2001). "Reviewing efficacy of alternative water treatment techniques." *Health Estate* 55(6):23–25.

Harris, L. G., and Richards, R. G. (2004). "*Staphylococcus aureus* adhesion to different treated titanium surfaces." *J Matl Sci: Matls in Med* 15:311–314.

Haufe, H., Mahltig, B., and Bottcher, H. (2004). "Paper 18: Biocidal nanosol coatings." *The Second Global Congress Dedicated to Hygienic Coatings and Surfaces,* Orlando, FL

Heaton, P. E., Callow, M. E., Butler, G. M., and Milne, A. (1991). "Control of mould growth by antifungal paints." *Int Biodeterior* 27(2):163–173.

Kennedy, R. (2002). "Antimicrobial control by coatings through the selection of raw materials." European Coatings Net. http://www.coatings.de/articles/ecs01papers/Kennedy/kennedy.htm.

Klens, P. F., and Yoho, J. R. (1984). "Occurrence of Alternaria Species on Latex paint." *The 6th International Biodeterioration Symposium,* Washington, DC.

Kotelnikova, N. E., Hou, Y. F., Panarin, E. F., Kudina, N. P., Li, S. X., and Zaikina, N. A. (2000). "Modification of cellulose materials by antiseptics and their antimicrobial properties (II)—Release of antiseptics from modified cellulose materials and their antimicrobial activity." *Linchan Huaxue Yu Gongye/Chemistry and Industry of Forest Products* 20(4):45–49.

Kusumaningrum, H. D., Riboldi, G., Hazeleger, W. C., and Beumer, R. R. (2003). "Survival of foodborne pathogens on stainless steel surfaces and cross-contamination to foods." *Int J Food Microbiol* 85(3):227–236.

Mazurkiewicz, G. (2002). "Coated ducts keep microbes at bay; antimicrobial coating designed to help prevent IAQ problems." *Air Conditioning, Heating & Refrigeration News* 217(8). http://www.iuoe.org/cm/iaq_mark_prod.asp?Item=331.

McDade, J. J., Hall, L. B., and Street, A. R. (1964). "Survival of *Staphylococcus aureus* in the environment." *Am J Hyg* 80:184–191.

McEldowney, S., and Fletcher, M. (1988). "The effect of temperature and relative humidity on the survival of bacteria attached to dry solid surfaces." *Lett in Appl Microbiol* 7:83–86.

McGill (2003). *The Role of Antimicrobial-Coated Ductwork in Indoor Air Quality.* McGill AirFlow Corporation, Groveport, OH. http://www.mcgillairflow.com/textDocs/silvergrd/silverGuard_WP.pdf.

PRA (2004). "Hygienic coatings and surfaces." *The Second Global Congress Dedicated to Hygienic Coatings and Surfaces,* Orlando, FL

Price, D. L., Ahearn, D. G., Ramey, D. L., and Harrison, S. L. (1999). "The role and efficacy testing of antimicrobial preservatives in air filters." *The First NSF International Conference on Indoor Air Health,* Denver, CO.

Reed, S. E. (1975). "An investigation of the possible transmission of rhinovirus colds through indirect contact." *J Hyg* 75:249–258.

Rivera-Garza, M., Olguin, M. T., Garcia-Sosa, I., Alcantara, D., and Rodriguez-Fuentes, G. (2000). "Silver supported on natural Mexican zeolite as an antibacterial material." *Micropor Mesopor Mat* 39(3):431–444.

Ross, J. (2004). "Novel hygienic solutions." *Polym Paint Colour J* 194(4479):18–20.

Singh, P., and Cameotra, S. S. (2004). "Potential applications of microbial surfactants in biomedical sciences." *Trends Biotechnol* 22(3):142–146.

Skaer, M. (2001). "Concept home billed as nation's first antimicrobial home." ACHR News. www.akconcepthome.com.

Studenikina, F. G., Mironov, M. M., Denisova, V. P., and Pokrovskaya, G. M. (1993). "Examining the antimicrobial properties of some metals and coatings for medical products." *Med Tekh* 5(Sep.–Oct.):6–8.

Thurman, R., and Gerba, C. (1989). "The molecular mechanisms of copper and silver ion disinfection of bacteria and viruses." *CRC Crit Rev Environ Control* 18:295–315.

Tiller, J. C., Liao, C.-J., Lewis, K., and Klibanov, A. M. (2001). "Designing surfaces that kill bacteria on contact." *PNAS* 98(11):5981–5985.

Topping, D. (2004). "Antimicrobial surfaces for appliances." *Appliance* 61(5):23.

Turpin, J. (2003). "New ducts have antimicrobial protection." *Air Conditioning, Heating and Refrigeration News* 9/29/03. http://www.lindab.com/usa/web_LINDAB/achrnewsarticle.htm.

Wilkinson, T. R. (1966). "Survival of bacteria on metal surfaces." *Appl Microbiol* 14:303–307.

Wright, T. (2002). "Alphasan: A thermally stable silver based inorganic antimicrobial technology." *Chem Fibers Int* 52(2):125.

CHAPTER 21

MICROWAVES

21.1 INTRODUCTION

Microwaves are a form of nonionizing electromagnetic radiation with a frequency range that extends from about 300 MHz to 300 GHz. Microwaves are a subset of the radiofrequency (RF) range which extends from about 3 kHz to 300 GHz (Shore, 1998). Microwave ovens operate at 2.45 GHz, which is the approximate resonant frequency of water molecules (Cheung and Levien, 1985; Debye, 1929). Microwave heating is a commonly used method of disinfection that is virtually identical to thermal disinfection (Chipley, 1980; Hoffman and Hanley, 1994; Welt et al., 1994). The effects of microwaves on bacterial and human cells have been the subject of intensive research since the invention of radar, but the true nature of microwave biological effects has eluded researchers. Many believe that microwaves accomplish disinfection entirely through heating effects, but evidence exists to suggest that microwaves possess an athermal effect, often called the microwave effect, that exists in addition to but separate from heating effects (Steneck, 1984; Davis et al., 1986). This chapter addresses the potential air and surface disinfection applications of microwave heating and athermal effects, and reviews the literature regarding the microwave effect as a mechanism with potential future applications for disinfection and sterilization of air and surfaces. One of the unique characteristics of microwaves that distinguishes them from other air disinfection technologies is the fact that microwave energy can be delivered directly to target molecules in air without significant waste, and therefore it holds the promise of possibly becoming the most efficient air disinfection technology by far, should it ever be successfully developed.

21.2 MICROWAVE THERMAL DISINFECTION

The effectiveness of microwaves for sterilization has been well established by numerous studies over the previous decades (Latimer and Matsen, 1977; Goldblith and Wang, 1967; Fung and Cunningham, 1980; Diaz-Cinco and Martinelli, 1991). Microwaves heat liquids and the heating effect has often been considered to be the sole reason for the resulting disinfection (Welt et al., 1994).

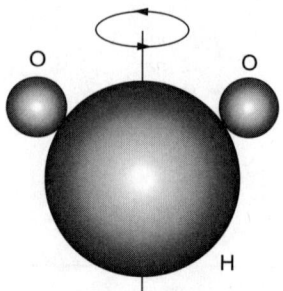

FIGURE 21.1 In a water molecule, the oxygen atoms (O) are located at 104.5° from each other relative to an axis through the center of the hydrogen (H) atom.

Microwaves consist of mutually perpendicular electric and electromagnetic waves that combine in their impact on biological systems. The primary effect of these fields is to induce rotation of polar molecules, or molecules that possess a dipole moment, such as water molecules (Pethig, 1979). Polar molecules are those that possess an uneven charge distribution and respond to an electromagnetic field by rotating. This effect is illustrated in Fig. 21.1, in which the dipole moment axis is shown through the center of the water molecule.

The dielectric effect on polar molecules has been known since 1912 (Debye, 1929). The resulting angular momentum developed by these molecules results in friction with neighboring molecules. The spinning momentum then converts to linear momentum, which is the thermodynamic definition of heat in liquids and gases.

Regardless of the biocidal mechanism, microwaves have one advantage over UVGI and ozone when it comes to the problems of disinfecting mail or packages—microwaves tend to penetrate paper and would provide internal disinfection. Studies with spores indicate that they can be sterilized in as little as 30 minutes (Cavalcante and Muchovej, 1993). Sanborn et al. (1982) found that certain common bacteria, including species of *Pseudomonas, Klebsiella, Proteus, Corynebacterium*, and *Streptococcus*, could be sterilized in as little as 3 minutes.

In an early study on microwave disinfection, Goldblith and Wang (1967) measured the degree of inactivation of *Escherichia coli* and *Bacillus subtilis* under exposure to 2450 Hz microwave radiation and found that it was identical to the inactivation by conventional heating. Similar results were found by Lechowich et al. (1969) for *Streptococcus faecalis*. Vela and Wu (1979) drew the same conclusions for various bacteria, actinomycetes, fungi, and bacteriophages. It has been suggested by other researchers that microwave heating is not identical to thermal disinfection because thermally heating bacterial cells to the same temperature as microwaves requires a longer time for inactivation (Latimer and Matsen, 1977; Dreyfuss and Chipley, 1980; Rosaspina et al., 1994). Such a result is suggested by the comparison of microwave heating with conventional heating of *E. coli* in Fig. 21.2. The differences seem slight enough that they could be neglected for most purposes.

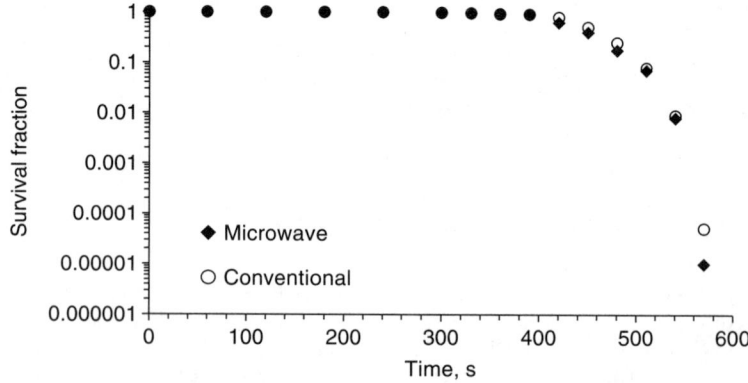

FIGURE 21.2 Comparison of microwave heating with conventional heating of *E. coli*. [*Based on data from Fujikawa et al. (1992).*]

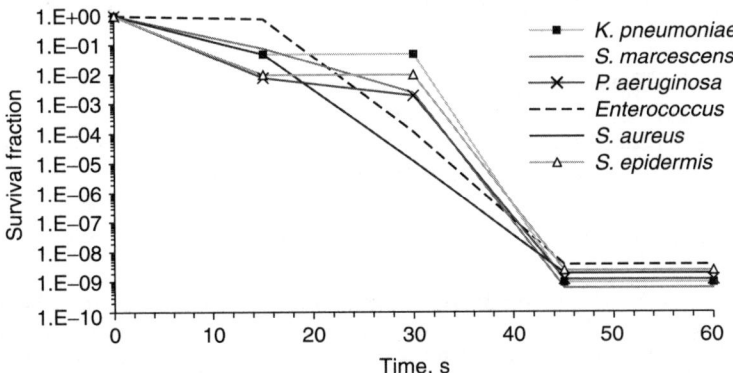

FIGURE 21.3 Survival curves for several microbes exposed to microwave radiation. [*Based on data from Latimer and Matsen (1977).*]

Most survival curves under microwave exposure will, of necessity, resemble survival curves under thermal disinfection. Figure 21.3 shows survival curves for a variety of microbes under 2450 kHz exposure. Note that these curves all tend to exhibit a shoulder and a second stage, as do the thermal disinfection curves shown in Chap. 19.

Microwave disinfection can be described by the same mathematical models used for any other disinfection process. Many of the survival curves in the literature display significant shoulder effects, such as the curves shown in the previous figures. Figure 21.4 shows a survival curve for *Bacillus subtilis* under microwave heating and in which a shoulder is clearly evident.

The shoulder curve, based on the multihit model as described in Chap. 7, is as follows:

$$S(t) = 1 - (1 - e^{-kD})^n \qquad (21.1)$$

In the case of microwave exposure, the actual dose is in terms of joules or watts delivered per unit mass or surface area, but in general it can be assumed for modeling purposes

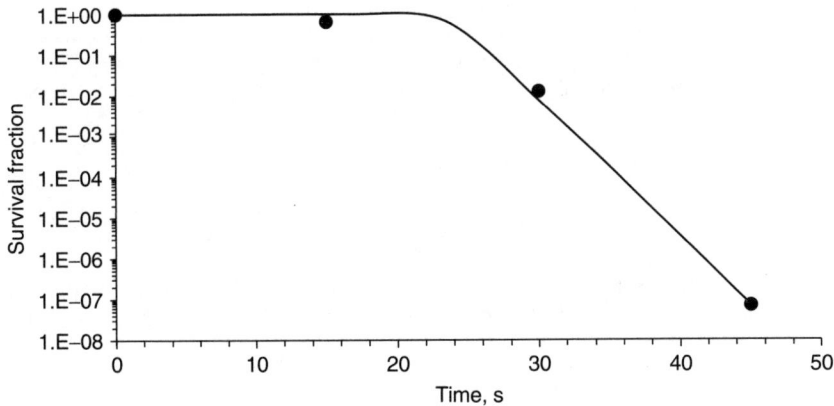

FIGURE 21.4 Microwave heating of *Bacillus subtilis* spores. [*Data from Latimer and Matsen (1977) fitted to a single-stage shoulder curve.*]

that the dose quantity D is represented by the exposure time (i.e., assuming the intensity $I = 1$). Therefore the heating model simplifies to

$$S(t) = 1 - (1 - e^{-kt})^n \tag{21.2}$$

Using Eq. (21.2), the data in Fig. 21.4 have been fitted by manually adjusting the rate constant and the exponent n until a reasonable fit has been obtained. The resulting fitted curve for Fig. 21.4 is as follows:

$$S(t) = 1 - (1 - e^{-0.77t})^{80000000} \tag{21.3}$$

Note the extreme high value of the exponent in Eq. (21.3). This results from the fact that the shoulder is extended so far out in time that it has, in effect, a delayed response. This would also be true if the classical shoulder model (see Chap. 7) were used. When the shoulder is greatly extended as in Fig. 21.4 both of the shoulder equations presented in Chap. 7 will likely provide a rather clumsy and ill-fitting curve. In such cases it is best to divide the curve into two regions, one in which the survival remains constant at unity and one in which the survival behaves as a shoulder curve set back by $(t - t_1)$, as follows:

$$\begin{array}{ll} 0 \leq t \leq t_1 & S(t) \approx 1 \\ t_1 \leq t \leq t_\infty & S(t) = f(t - t_1) \end{array} \tag{21.4}$$

In Eq. (21.4) the function $f(t - t_1)$ can be filled by either of the shoulder models presented in Chap. 7. Revisiting the curve for microwave thermal disinfection of *E. coli* shown previously in Fig. 21.2, it can be seen that the extremely delayed response in the curve will likely provide a poorly fitting curve when Eq. (21.1) is applied directly. An improved fit can be obtained by dividing the curve into two regions and applying a shoulder curve equation to the second region only. The classical shoulder equation is used because it tends to provide a smoother fit over larger time spans than the multihit model. The classical shoulder model is as follows:

$$S(t) = \exp\left(-\frac{kI}{4t_c}t^2\right) \tag{21.5}$$

The second or log-linear portion of the classical shoulder model (see Chap. 7) can be ignored here since the data in Fig. 21.2 are virtually all shoulder. The curve must be shifted ahead in time by a distance t_1 by substituting $(t - t_1)$ for t in Eq. (21.5). Rewriting Eq. (21.5) with this substitution, with unitary intensity $I = 1$, and inserting it into Eq. (21.5), we obtain the following:

$$\begin{array}{ll} 0 \leq t \leq t_1 & S(t) = 1 \\ t_1 \leq t \leq t_\infty & S(t) = \exp\left(-\frac{k(t - t_1)^2}{4t_c}\right) \end{array} \tag{21.6}$$

Figure 21.5 shows the survival curve for *E. coli* from Fig. 21.2 and the curve fitted to the data using the multihit model. The resulting equation, fitted by setting $t_1 = 410$, and by manually adjusting the exponent n and the rate constant k, is as follows:

$$\begin{array}{ll} 0 \leq t \leq 410 & S(t) = 1 \\ 410 \leq t \leq t_\infty & S(t) = \exp\left(-\frac{0.26(t - 410)^2}{4(180)}\right) \end{array} \tag{21.7}$$

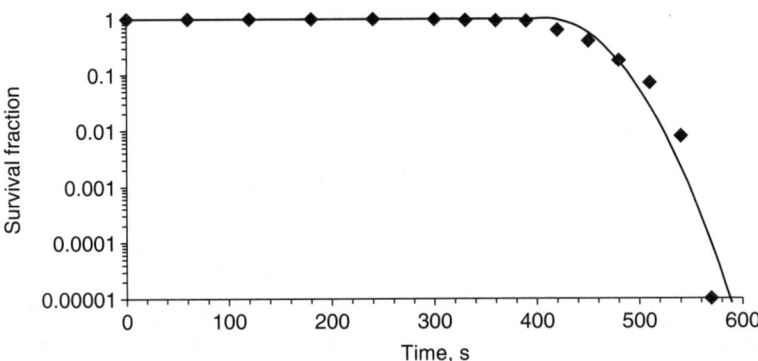

FIGURE 21.5 Microwave heating of *E. coli* fitted to a modified shoulder curve. [*Based on data from Fujikawa et al. (1992).*]

Even with these modifications, only a less than perfect fit is obtained. Still, such disinfection curves may be adequate for predictive purposes until improved predictive models can be developed.

Microwave thermal disinfection can be applied in either the dry state or wet conditions. Wet conditions invariably produce higher disinfection rates since the water molecules absorb microwaves and disperse the heat (Vela and Wu, 1979).

Another development in microwave engineering is the use of pulsed microwaves. Pulsed microwaves can generate acoustic waves in fluids and may have biocidal effects (Kiel et al., 1999). In this capacity they may be useful for sterilization of food or water. Plasmas can also be generated on surfaces by microwaves and this approach has been studied as a means of low-temperature dry sterilization (Chau et al., 1996). The removal of bacteria or viruses from material surfaces is caused by the reaction of activated oxygen species in the plasma with hydrocarbon bonds of the cell wall of the bacteria or the capsid of viruses.

21.3 MICROWAVE ATHERMAL DISINFECTION

The existence of microwave athermal effects has been hinted at in the literature for decades and sufficient evidence seems to exist that the so-called microwave effect is real, if not difficult to isolate and quantify. Heller and Texiera-Pinto (1959) noted that pulsed radio frequencies in the megacycle range could cause chromosomal aberrations. Dreyfuss and Chipley (1980) found that 2450-MHz microwave exposure affected the activity of metabolic enzymes in *S. aureus* cells. A variety of subsequent tests showed various types of microwave effects in microorganisms that could not be explained by simple heating (Ponne and Bartels, 1995). Certainly, microwave heating is not quite identical to ordinary thermal heating, or heating with external energy input, due to the fact that microwaves cause molecules to rotate first before they develop the random linear momentum which defines heat. There are some minor secondary effects of microwaves, including ionic conduction, which are negligible in external heating. Superconduction has also been postulated as a mechanism for nonthermal effects of microwaves (Cope, 1976).

Figure 21.6 shows the decay curve of *Staphylococcus aureus* under microwave irradiation. The death curves exhibit classic exponential decay with a shoulder, as well as a

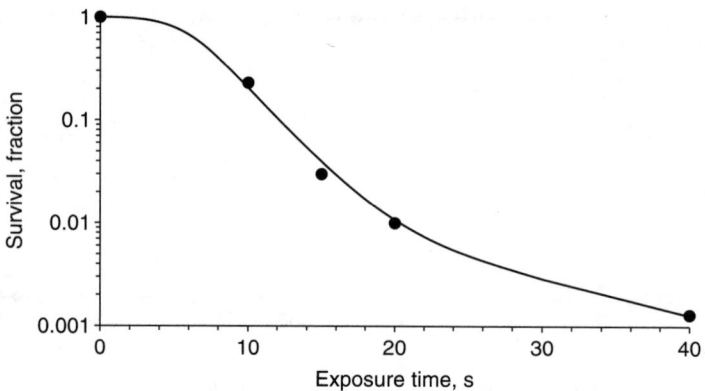

FIGURE 21.6 The results of microwave irradiation of *S. aureus*. [Based on data from Kakita et al. (1999), fitted to a two-stage curve with a shoulder.]

second stage. The fitted two-stage shoulder curve shown in Fig. 21.6, with the second stage simplified since $n = 1$, is as follows:

$$S(t) = (1 - 0.03)[1 - (1 - e^{-0.4t})^{12}] + 0.03e^{-0.079t} \quad (21.8)$$

In Fig. 21.7 is shown the decay curve of *E. coli* under microwave irradiation. The fitted two-stage shoulder curve shown in Fig. 21.7 is as follows:

$$S(t) = (1 - 0.002)[1 - (1 - e^{-0.43t})^{16}] + 0.002e^{-0.014t} \quad (21.9)$$

In the above graphic examples, the disinfection effects are essentially identical to the thermal disinfection effects that would occur from direct heating. However, an earlier test by the same researchers using the identical microwave irradiation of a bacteriophage

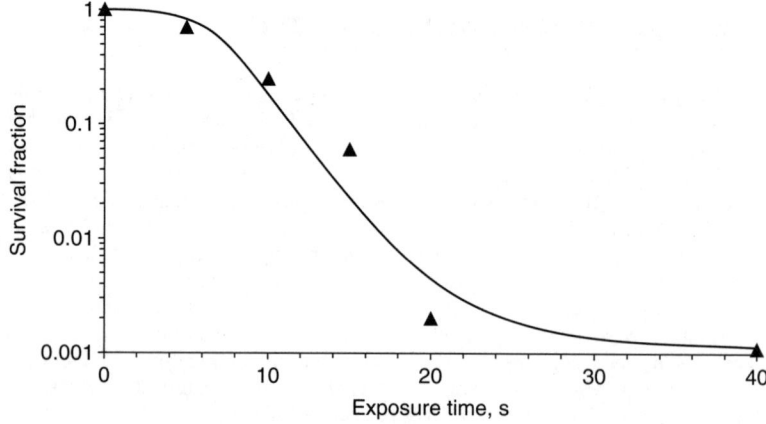

FIGURE 21.7 The results of microwave irradiation of *E. coli*. [Based on data from Kakita et al. (1999), fitted to a two-stage curve with a shoulder.]

produced DNA strand breaks (Kakita et al., 1995), suggesting that although the decay curves may look much the same, the process involves some athermal factors.

The energy level of a microwave photon is only 10^{-5} eV, whereas the energy required to break a covalent bond is 10 eV, or a million times greater. Based on this fact, it has been stated in the literature that microwaves are incapable of breaking the covalent bonds of DNA (Fujikawa et al., 1992; Jeng et al., 1987). DNA has a dielectric dispersion, in which microwaves are readily absorbed, at much lower frequencies than water (Takashima et al., 1984; Takashima, 1963), and some evidence exists to suggest that microwaves can indeed disrupt the covalent bonds of DNA. It has been demonstrated that the microwave effect is distinguishable from external heating by the fact that it is capable of extensively fragmenting viral DNA (Kakita et al., 1995). This experiment consisted of irradiating a bacteriophage PL-1 culture at 2450 MHz and comparing this with a separate culture heated to the same temperature. Figure 21.8 illustrates this striking result, in which the DNA in these phages has mostly been destroyed, a result that does not occur from heating alone.

In the Kakita experiment the survival percentage was approximately the same whether the samples were heated or irradiated with microwaves, but evaluation by electrophoresis and electron microscopy showed that the DNA of the microwaved samples had mostly disappeared. Electron microscopy has been used to study the bactericidal effects of microwaves (Rosaspina et al. 1993, 1994) and the results indicate that the effects of microwaves were distinguishable from those of external heating. There is, in fact, plenty of evidence to indicate that there are alternate mechanisms for causing DNA covalent bond breakage without invoking the energy levels of ionizing radiation (Watanabe et al., 1985, 1989; Ishibashi et al., 1982; Kashige et al., 1995). Still, no theory currently exists to explain the phenomenon of DNA fragmentation by microwaves.

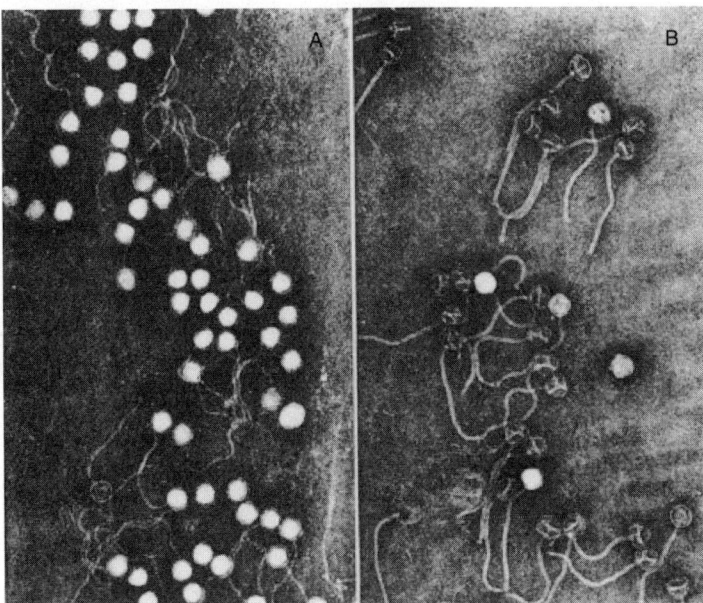

FIGURE 21.8 Phage PL-1 before (left) and after (right) irradiation with microwaves. [*Images reprinted with permission from Kakita et al. (1995).*]

A review of the data from the various referenced experiments shows a common pattern—for the first few minutes of irradiation there is no pronounced effect, and then a cascade of microbial destruction occurs. The second stage of this process may very well be the accumulation of oxygen radicals, which have a considerable propensity for dissociating the covalent bonds of DNA. Oxygen radicals can be generated by the disruption of a hydrogen bond on a water molecule. Water molecules exist alongside DNA molecules as "bound" water, two or three layers thick. These water molecules share a hydrogen bond with component atoms of the DNA backbone, including carbon, nitrogen, and other oxygen atoms. At any given point in time one of the hydrogen atoms may be primarily bonded to either an oxygen atom on the water molecule, or to an oxygen (or other) atom on the DNA backbone. The fluctuating character of these shared and exchanged bonds is enhanced by temperature and by the dynamics induced by microwaves. Although the relative quantity of oxygen radicals that may be produced by this process cannot presently be determined, the production of some number of oxygen radicals is inevitable in these circumstances. Most of the oxygen radicals produced in this manner would exist only briefly, but may result in a covalent bond break, albeit probably only a brief one. Although DNA tends to repair itself naturally, the simultaneous breakage of a sufficient number of covalent bonds could lead to a catastrophic failure of the entire DNA molecule.

There has been some suggestion in the literature that frequencies other than 2450 MHz may be better able to cause nonthermal biocidal effects (Webb and Booth, 1969; Grant et al., 1978). In fact, studies on nonthermal microwave effects have focused almost exclusively on the 2.45-GHz frequency and there seems to be no data available to analyze inactivation rates of microbes at alternate frequencies. Takashima (1963) showed that the dielectric dispersion of DNA was well below 1000 Hz. Anecdotal reports suggest the inactivation rate is more rapid at these frequencies but no confirmation could be obtained. There appears to be a need for further research in this lower frequency range that may bear useful fruit and even answer the ongoing questions about biological effects of electromagnetic radiation.

The ability of electromagnetic radiation to cause DNA damage has been a subject of considerable interest in health care fields due to the fact that both cell phones and high-voltage power lines have been suspected of causing cell damage in humans. Obviously if microwaves can damage DNA in bacterial cells, then there is a possibility that human cell damage can also be produced. In summary, the possibility that electromagnetic radiation in the nonionizing frequency range can cause genetic damage may have profound implications both for disinfection technology and for the current controversy involving EM antennae, power lines, and cell phones.

REFERENCES

Cavalcante, M. J. B., and Muchovej, J. J. (1993). "Microwave irradiation of seeds and selected fungal spores." *Seed Sci Technol* 21:247–253.

Chau, T. T., Kao, K. C., Blank, G., and Madrid, F. (1996). "Microwave plasmas for low-temperature dry sterilization." *Biomaterials* 17(13):1273–1277.

Cheung, W. S., and Levien, F. H. (1985). *Microwaves Made Simple, Principles and Applications.* Artech House, Denham, MA.

Chipley, J. R. (1980). "Effects of microwave irradiation on microorganisms." *Adv Appl Microbiol* 26:129–144.

Cope, F. W. (1976). "Superconductivity—a possible mechanism for nonthermal biological effects of microwaves." *J of Microw Power* 11:267–270.

Davis, C. C., Edwards, G. S., Swicord, M. L., Sagripanti, J., and Saffer, J. (1986). "Direct excitation of DNA internal modes by microwaves." *Bioelectrochem Bioenerg* 16:63–76.

Debye, P. (1929). *Polar Molecules*. Lancaster Press., Lancaster, PA.

Diaz-Cinco, M., and Martinelli, S. (1991). "The use of microwaves in sterilization." *Dairy Food Environ Sanit* 11(12):722–724.

Dreyfuss, M. S., and Chipley, J. R. (1980). "Comparison of effects of sublethal microwave radiation and conventional heating on the metabolic activity of *Staphylococcus aureus*." *Appl Microbiol* 39(1):13–16.

Fujikawa, H., Ushioda, H., and Kudo, Y. (1992). "Kinetics of *Escherichia coli* destruction by microwave irradiation." *Appl Environ Microbiol* 58:920–924.

Fung, D. Y. C., and Cunningham, F. E. (1980). "Effect of microwaves on microorganisms in foods." *J Food Prot* 43:641–650.

Goldblith, S. A., and Wang, D. I. C. (1967). "Effect of microwaves on *Escherichia coli* and *Bacillus subtilis*." *Appl Microbiol* 15:1371–1375.

Grant, E. H., Sheppard, R. J., and South, G. P. (1978). *Dielectric behaviour of biological molecules in solution*. Oxford University Press, Great Britain.

Heller, J. H., and Teixeira-Pinto, A. A. (1959). "A new physical method of creating chromosomal aberrations." *Nature* 183(March):905–906.

Hoffman, P. N., and Hanley, M. J. (1994). "Assessment of a microwave-based clinical waste decontamination unit." *J Appl Bacteriol* 77:607–612.

Ishibashi, K., Sasaki, T., Takesue, S., and Watanabe, K. (1982). "In vitro phage-inactivating action of d-glucosamine on Lactobacillus phage PL-1." *Agric Biol Chem* 46:1961–1962.

Jeng, D. K. H., Kaczmarek, K. A., Woodworth, A. G., and Balasky, G. (1987). "Mechanism of microwave sterilization in the dry state." *Appl Environ Microbiol* 53:2133–2137.

Kakita, Y., Kashige, N., Murata, K., Kuroiwa, A., Funatsu, M., and Watanabe, K. (1995). "Inactivation of *Lactobacillus* bacteriophage PL-1 by microwave irradiation." *Microbiol Immunol* 39:571–576.

Kakita, Y., Funatsu, M., Miake, F., and Watanabe, K. (1999). "Effects of microwave irradiation on bacteria attached to the hospital white coats." *Int J Occup Med Environ Health* 12(2):123–126.

Kashige, N., Yamaguchi, T., Ohtakara, A., Mitsutomi, M., Brimacombe, J. S., Miake, F., and Watanabe, K. (1994). "Structure-activity relationships in the induction of single-strand breakage in plasmid pBR322 DNA by amino sugars and derivatives." *Carbohydr Res* 257:285–291.

Kiel, J. L., Seaman, R. L., Marthur, S. P., Parker, J. E., Wright, J. R., Alls, J. L., and Morales, P. J. (1999). "Pulsed microwave induced light, sound, and electrical discharge enhanced by a biopolymer." *Bioelectromagnetics* 20(4):216–223.

Latimer, J. M., and Matsen, J. M. (1977). "Microwave oven irradiation as a method for bacterial decontamination in a clinical microbiology laboratory." *J Clin Microbiol* 4:340–342.

Lechowich, R. V., Beuchat, L. R., Fox, K. J., and Webster, F. H. (1969). "Procedure for evaluating the effects of 2450 MHz microwaves upon *Streptococcus faecalis* and *Saccharomyces cerevisiae*." *Appl Microbiol* 17:106–110.

Pethig, R. (1979). *Dielectric and electronic properties of biological materials*. John Wiley & Sons, Chichester, UK.

Ponne, C. T., and Bartels, P. V. (1995). "Interaction of electromagnetic energy with biological material—relation to food processing." *Radiat Phys Chem* 45(4):591–607.

Rosaspina, S., Anzanel, D. and Salvatorelli, G. (1993). "Microwave sterilization of enterobacteria." *Microbios* 76:263–270.

Rosaspina, S., Salvatorelli, G., Anazanel, D., and Bovolenta, R. (1994). "Effect of microwave radiation on *Candida albicans*." *Microbios* 78:55–59.

Sanborn, M. R., Wan, S. K., and Bulard, R. (1982). "Microwave sterilization of plastic tissue culture vessels for reuse." *Appl Environ Microbiol* 44:960–964.

Steneck, N. H. (1984). *The Microwave Debate*. The MIT Press, Cambridge, MA.

Takashima, S. (1963). "Dielectric dispersion of DNA." *J Mol Biol* 7:455–467.

Takashima, S., Gabriel, C., Sheppard, R. J., and Grant, E. H. (1984). "Dielectric behaviour of DNA solution at radio frequency and microwave frequencies." *J Biophysics* 46:29–34.

Vela, G. R., and Wu, J. F. (1979). "Mechanism of lethal action of 2450 MHz radiation on microorganisms." *Appl Environ Microbiol* 37:550–553.

Watanabe, K., Kashige, N., Kojima, M., Nakashima, Y., Hayashida, M., and Sumoto, K. (1985). "DNA strand scission by d-glucosamine and its phosphates in plasmid pBR322." *Agric Biol Chem* 50:1459–1465.

Watanabe, K., Kashige, N., Kojima, M., and Nakashima, Y. (1989). "Specificity of nucleotide sequence in DNA cleavage induced by d-glucosamine and d-glucosamine-6-phosphate in the presence of Cu2+." *Agric Biol Chem* 54:519–525.

Webb, S. J., and Booth, A. D. (1969). "Absorption of microwaves by microorganisms." *Nature* 222(June):1199–1200.

Welt, B. A., Tong, C. H., Rossen, J. L., and Lund, D. B. (1994). "Effect of microwave radiation on inactivation of *Clostridium sporogenes* spores." *Appl Environ Microbiol* 60:482–488.

CHAPTER 22

ALTERNATIVE AND DEVELOPMENTAL TECHNOLOGIES

22.1 INTRODUCTION

The basic air disinfection technologies covered in previous chapters included dilution ventilation, filtration, UVGI, and carbon adsorption. Other disinfection technologies exist that could be used in aerobiological applications but most of these are either still undergoing research, are expensive to apply for air disinfection, or are limited to highly specialized applications. For completeness, and to assist the search for new solutions, these emerging technologies are described here. Information on designing systems using the types of equipment presented in this chapter is, at present, limited or unpublished in most cases. A few of these technologies are fairly well understood but not widely used. The references cited in each section can be consulted for additional information.

The technologies presented here should not be considered exclusive, as research continues in many areas and new technologies will surely be developed in the future. Not all of these technologies necessarily have any practical application in aerobiological engineering but they may be useful in specialized applications, such as potable water or food disinfection.

22.2 IONIZING RADIATION

Various forms of ionizing radiation exist that can destroy bacterial cells and inactivate microorganisms, including gamma rays, alpha rays, beta rays, x-rays, cosmic rays, proton beams, neutron beams, particle beams, and electron beams. Electrons, alpha particles, neutrons, and other fission fragments are considered to be particles, while the remaining types of ionizing radiation are considered to be electromagnetic radiation with wavelengths mostly below the range of ultraviolet radiation (Casarett, 1968).

Gamma rays are a form of ionizing radiation and can be used to destroy microorganisms and to sterilize equipment (Smith et al., 2001). Gamma rays are particularly hazardous to all forms of life and can cause cellular and genetic damage. Gamma rays are emitted by radioactive materials and have the ability to penetrate solid and liquid matter. This gives them a unique advantage in the sterilization of packages and envelopes. However, materials can be damaged by gamma rays and because of the health hazards they must be used with caution.

The biocidal effects of gamma rays reduce microbial populations according to relations that are similar to UVGI. The decay rate is exponential and the response of any species to gamma rays can be defined with a rate constant, just as with UVGI. Gamma rays are used to sterilize some packaged foods (Alimov et al., 2000) and have been considered as one method for sterilizing mail, but the expense and hazards may limit its use as part of any immune building system. Figure 22.1 shows a gamma irradiation unit with heavy shielding.

Iordanov (1977) carried out studies to establish the sensitivity of *Bacillus anthracis* spores to gamma radiation in a suspension of saline and in pelts and wool. Five strains of *B. anthracis* spores were tested. The density of spore suspensions ranged from 1.6×10^6 to 9.0×10^9 per cm^3, and for the pelts and wool the spore count ranged from 7.8×10^8 to 5.0×10^{10} (for the material as a whole). The irradiation was carried out at a temperature of 15°C and with rates of up to 2.75 Mrad. Based on the results, the dose necessary to decontaminate anthrax infected pelts and wool is 2 Mrad. In a study of the ability of gamma rays to eliminate mold spore contamination from home furnishings, Wilson et al. (2004) found that 1 to 1.3 Mrad were sufficient to inactivate *Stachybotrys*, *Penicillium*, and *Chaetomium* spores, but did not eradicate mycotoxins.

In a study on the effects of gamma irradiation, Anellis et al. (1965) exposed *Clostridium botulinum* spores to levels of up to 9 Mrad. Figure 22.2 shows the data from this study and a two-stage survival curve fitted to the data.

FIGURE 22.1 Gamma irradiation system from J. L. Shepherd. (*Photo courtesy of Defense Microelectronics Activity, U.S. Secretary of Defense.*)

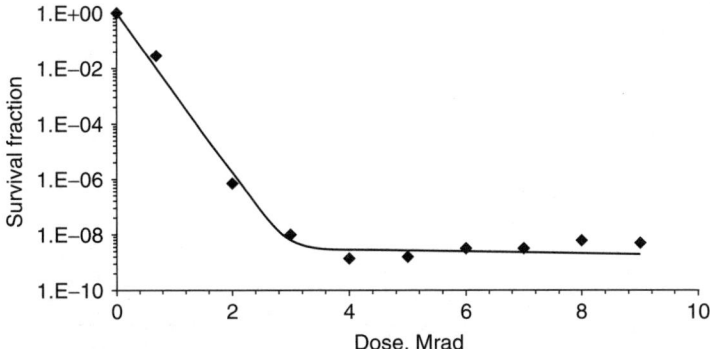

FIGURE 22.2 Survival of *Clostridium botulinum* spores under gamma irradiation. [*Two-stage curve fitted to data from Anellis et al. (1965).*]

The data in Fig. 22.2 were fitted to the following basic two-stage curve:

$$S = (1-f)e^{-k_1 D} + fe^{-k_2 D} \tag{22.1}$$

In this case the dose D is specified in Mrad. The data can be split into two sections, the first stage and the second stage, from which the rate constants and the population fraction f can be estimated. The curve in Fig. 22.2 was then fitted by adjusting the parameters manually until a reasonable fit was obtained. The resulting equation is as follows:

$$S = 0.999999996 e^{-6.6D} + 0.000000004 e^{-0.08D} \tag{22.2}$$

Bertani (1960) evaluated the effects of ionizing radiation on various bacteriophages and found that there was a species-dependent sensitivity to different frequencies. He also postulated a relationship between the size of the DNA molecule and its relative sensitivity.

Alpha rays have very limited penetrating ability and cannot pass through a sheet of paper. For this reason they are not considered to have any application as a disinfection method, but such a disadvantage may be of benefit where surface contamination is a problem and penetration is undesirable, such as in nosocomial surgical site infections. It may be possible to use alpha radiation in a limited way to disinfect the surface of open wounds during surgical procedures. Tavera et al. (2003) have reported that it takes only eight alpha particles to induce a lethal event in *E. coli*.

The use of x-rays to disinfect surfaces and foods is one of the oldest technologies but one that sees very limited use due to the associated hazards. X-rays have been tested as an alternative to thermal disinfection of meat (Curry et al., 2000). In this latter test, an accelerator was used to generate a 40-ns, 200-keV electron beam which was then converted to bremsstrahlung x-rays. A high dose rate of approximately 107 rad per x-ray pulse was used to sterilize frozen beef of *E. coli*. Figure 22.3 shows an example of the survival of bacteria exposed to x-ray irradiation, based on data from Casarett (1968). In this data a shoulder is clearly manifest.

The data in Fig. 22.3 were fitted to a single-stage curve with a shoulder. The basic equation is:

$$S(t) = 1 - (1 - e^{-kD})^n \tag{22.3}$$

The equation was fitted to the data by first estimating the single-stage rate constant from the data beyond the initial point, and then adjusting the rate constant and the exponent until

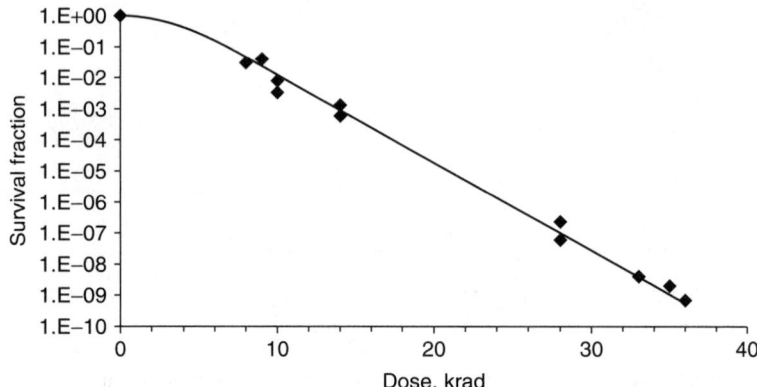

FIGURE 22.3 Survival of *Serratia marcescens* under X-ray exposure. [*Based on data from Casarett (1968).*]

a satisfactory fit was obtained. The resulting fitted equation is as follows:

$$S = 1 - (1 - e^{-0.65D})^8 \tag{22.4}$$

X-rays can also be used to sterilize air, but it is unlikely to be practical or economical. A summary of x-ray doses to kill various microbes was given by Sussman and Halvorson (1966) and these included the LD_{50} dose for *Bacillus megaterium* (200 krad), *E. coli* (5.6 krad), and *Aspergillus terreus* (30 krad).

The use of electron beams as a sterilization technology has become more common as the technology has advanced (VanLancker and Bastiaansen, 2000). Electron beams are related to gamma ray technology and to x-ray technology, but they don't depend on radioactive sources (Alimov et al., 2000). They have the same basic penetrating ability as gamma rays and have application in equipment and food sterilization. They have also been tested in applications involving the sterilization of contaminated mail.

Researchers at the Miami Electron Beam Research Facility have been testing this technology using a 1.5-MV, 50-mA electron accelerator on a 120-gpm wastewater stream and have found it to be an effective method for removing toxic organics but no data are currently available on the destruction of microorganisms (Kurucz et al., 1994).

Curiously, low levels of ionizing radiation are capable of stimulating microbiological processes and may even induce germination (Lea et al., 1941; Lea, 1955; Kuzin et al., 1976; Kuzin, 1977). Germinated spores are several orders of magnitude more susceptible to irradiation than dormant spores (Sussman and Halvorson, 1966). The fact that spores could be germinated by low-level radiation and thereby rendered vulnerable to other biocidal factors is similar to the thermal cycling method discussed in Chap. 19, and may lead to new methods of disinfection.

22.3 ULTRASONICATION

Ultrasonic waves can exhibit germicidal effects and can cause fragmentation or disintegration of bacterial cells (Hamre, 1949; Scherba et al., 1991; Gould et al., 1969). When sufficient power is applied, viruses and bacteria can be atomized like liquids.

There are two methods by which this may be accomplished, supersonic nozzles and sonic generators.

If the airstream is forced through a supersonic nozzle, a standing shock wave develops at the nozzle outlet. This shock wave dissipates energy by imparting it to the airstream, causing it to expand suddenly and rapidly. This results in the atomization, or reduction to gas, of all bioaerosols in the airstream. This technique is used effectively in ultrasonic humidifiers, but forcing high volumes of airflow through an ultrasonic nozzle would be prohibitively expensive.

A sonic generator that was tuned to resonate within a cavity would create a standing shock wave through which an airstream could pass, and in which bioaerosols may be atomized, but the power levels needed in air may be prohibitive, not to mention noisy. In liquids, a mechanism that may have biocidal effects is the generation of microscopic bubbles, which, on collapse, can damage bacterial cells (Vollmer et al., 1998).

The true value of ultrasonics may lie in its combined use with other agents, such as ozone or UVGI, in which enhancement of chemical reactions may occur. In particular, ultrasonication has been shown to reduce clumping, limiting this natural protective effect (Burleson et al., 1975; Harakeh and Butler, 1985). *Mycobacteria* tend to clump in moist environments, suggesting the possibility that ultrasonics may be used to enhance the efficacy of other technologies like filtration (Ryan, 1994; Tarleton, 1990; Tavossi, 1986). Ultrasonics may also decrease the heat resistance of spores (Sanz et al., 1985) and may also have the potential to break down some chemical compounds (Hoffman et al., 1996; Hirai et al., 1996).

22.4 PLASMA TECHNOLOGY

Plasma is considered to be an intermediate state of matter between gas and solid, consisting of ions, electrons, and neutral particles in which the behavior is dominated by the electromagnetic interactions. Plasma systems operate in air and produce uniform plasma without filamentary discharges at room temperature. Such systems can be used to disinfect or sterilize air, water, and surfaces. The typical electric field is about 8.5 kV/cm. Plasma generators create reactive oxygen species including atomic oxygen, oxygen free radicals, and hydroxyl radicals in humid airstreams. Plasma technology has been shown to be capable of destroying chemical agents (McLean and Roth, 1998). Plasma has been used to create antifouling layers on steel surfaces, to enhance deposition of silver nanoparticles onto polymer and metal surfaces to create antimicrobial characteristics, and for the disinfection of water (Jiang et al., 2004; Manolache et al., 2001; Denes et al., 2001; Denes, 1997). The use and theory of plasma technology is more complex than can be adequately covered here and readers should refer to the various citations for more technical details.

Recent experiments using plasma technology have explored the mechanism of killing of various microorganisms. Experiments on bacterial cells indicate rapid destruction of outer cell membranes. Results from electron microscopy studies show that bacteria with outer membranes are more susceptible to fragmentation than those without an outer membrane, probably due to the sensitivity of lipids to the oxygen radicals. Consistent with these findings are results that show the relative resistance of nonenveloped viruses to the plasma (Kelly-Wintenberg et al., 1999).

Montie et al. (2000) report the test results of a plasma source, called the One Atmosphere Uniform Glow Discharge Plasma (OAUGDP), which operates at atmospheric pressure in air and produces antimicrobial active species at room temperature. OAUGDP exposures have been effective at reducing microbial populations of gram-negative bacteria, gram-positive bacteria, bacterial endospores, yeast, and bacteriophages on a variety of surfaces. The nature of the surface impacted the inactivation rate, with microorganisms on polypropylene being most vulnerable, followed by glass, and cells embedded in agar. Experimental results showed at least a five-log (base 10) cfu reduction in bacteria within a range of 50 to 90 seconds of plasma exposure. After 10 to 25 seconds of exposure, macromolecular leakage and bacterial fragmentation were observed. Vulnerability of cell membranes to reactive oxygen species is hypothesized. A planar electrode configuration has been developed as a filter for use in air disinfection.

In a test of the feasibility of plasma technology to disinfect medical equipment, diffused low-power pulsed-periodical discharge (average current 1 to 10 mA) was used to form plasma in air gaps with an interelectrode gap of 10 to 15 cm for the inactivation of microbiological cultures (Karelin et al., 2001). Park et al. (2003) used a new method, microwave-induced argon plasma at atmospheric pressure, for the sterilization of microorganisms. Experimental results for six species of microorganisms, including both bacteria and fungi, indicated that all six species were fully sterilized within 20 seconds.

Roland et al. (2002) found that the performance of nonthermal plasma for the removal of volatile organic compounds is improved by the introduction of catalytically active materials into the discharge zone of a nonthermal plasma reactor. The main effects of plasma catalysis were an enhanced conversion of pollutants based on the presence of short-lived oxidizing species in the inner volume of porous catalysts.

Plasma systems can be combined with chemical disinfection technologies such as ethylene oxide. In a study by Matsumoto and Kanitani (2003), a high-speed ethylene oxide plasma sterilization system was demonstrated to be more efficient than ethylene oxide alone and produced no toxic residuals.

In a study by Nagatsu et al. (2003), spores of *Bacillus stearothermophilus* and *Bacillus subtilis* were sterilized using low-temperature plasma. Oxygen plasma discharges were generated using microwave power and exposing the spores for 3 minutes or longer. Oxygen radicals were believed to be responsible for some of the lethal cell damage.

22.5 CORONA DISCHARGE

Similar to plasma discharge systems, corona discharge involves generating a localized corona in or around electrical conductors. A corona is a relatively low-current, low-power electrical discharge that is maintained when the electrical field is high enough to induce ionization in a limited volume around the wire or other metallic conductor (Vincent, 1995). No spark is produced provided the potential is not too high. Corona destruction involves the use of a packed bed of ferroelectric pellets or wire meshes across which a high voltage alternating current is maintained (Ramsey et al., 1990; Nunez et al., 1993). Coronas form at the interstitial spaces between the pellets or wires and generate electrons. These free electrons induce chemical reactions and ionization in the passing airstream. A variety of chemical reactions are possible including the ionization of molecules and the dissociation of molecules into free radicals. These products, in turn, can induce the breaking of chemical bonds in organic molecules.

Aerosol particles receive a charge after being bombarded by the free ions produced by the corona discharge. Ions will continue to stream toward a particle until the charge builds up to the point that the local surrounding field becomes relatively neutral. At this point the particle is considered charged to saturation. During the charging process the charge q_f on a spherical particle of diameter d varies with time according to the following (Vincent, 1995):

$$q_f = \left\{\frac{3\varepsilon_r}{\varepsilon_r + 2}\right\}\left(\pi\varepsilon_0 E_c d^2\right)\left\{\frac{\pi Z_i N_i t}{1 + \pi Z_i N_i t}\right\} \tag{22.5}$$

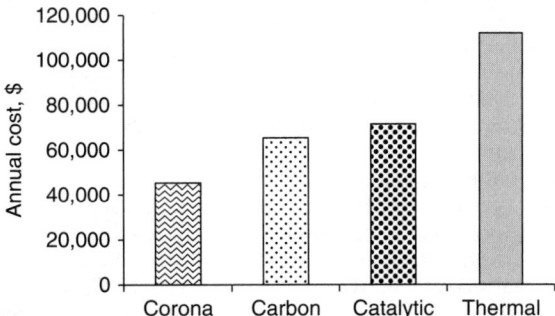

FIGURE 22.4 Annual cost of four VOC control technologies.

where q_f = charge on a spherical particle
  ε_r = dielectric constant (1 for vacuum, 1 to 10 for other materials)
  ε_0 = absolute permittivity of vacuum, 8.85×10^{-12} A·s/V·m
  E_c = electrical field strength
  d = particle diameter
  Z_i = mobility
  N_i = ion density
  t = time

The final saturation charge achieved by the particle is given by:

$$q_{fs} = \left\{\frac{3\varepsilon_r}{\varepsilon_r + 2}\right\}\left(\pi\varepsilon_0 E_c d^2\right) \tag{22.6}$$

where q_{fs} is the saturation charge on a spherical particle.

Current applications of corona destruction involve the reduction of *volatile organic compounds* (VOCs) and air toxics (Harry, 1993). No data are available to demonstrate the effectiveness of corona destruction against airborne microbes but obviously the creation of radicals, such as the hydroxyl radical OH⁻, would likely have a biocidal effect just as it does with ozone.

The operating costs of corona destruction have been reported by Nunez et al. (1993) as being less than that of carbon adsorbers. Figure 22.4 illustrates the annual cost of corona destruction compared with carbon adsorption, catalytic incineration, and thermal incineration.

Pulsed corona discharge systems are also under development that used 10 to 100 ns discharges and these also show promise as being economic alternatives for removing VOCs from airstreams (Hutcherson et al., 1995; Korzekwa et al., 1993, 1998).

Although corona discharge systems can be used to disinfect airstreams, equipment disinfection is limited by the fact that the space between the corona-producing elements is typically too small to allow for insertion of equipment. One alternative method that overcomes this obstacle is the use of pressurized gas jets to distribute the plasma (Lee et al., 2003).

22.6 CHEMICAL DISINFECTION

Chemicals were perhaps the first surface disinfection technology used when Lister developed the carbolic spray for surgical use. Since then, several attempts have been made to develop air disinfection systems using chemicals but none that appeared to be practical.

Obviously any chemical disinfectant that is biocidal to microbes is likely to be hazardous to humans as well (Popendorf and Selim, 1995). The chemical must therefore be used in concentrations that pose no threat to humans. There are an almost unlimited number of disinfectant chemicals available today and so no attempt is made here to list or identify them all although App. A lists a variety of common disinfectants that can be used effectively against specific microorganisms.

Ethylene oxide (EtO) is a commonly used disinfectant that has proven effective for sterilizing medical devices. In gaseous form it can be used inside sterilization chambers, or even to disinfect rooms or entire buildings. Gas sterilizers that use either pure ethylene oxide or an admixture with some inert diluent are often used for treating products which cannot withstand heat sterilization (Desai and Buonicore, 1990). Due to the highly toxic nature of ethylene oxide and its classification as a potential carcinogen, special precautions must be taken to minimize human exposure levels.

A variety of past field trials have provided insight into the applications and effectiveness of chemical disinfectant aerosols. Cruickshank and Muir (1940) sprayed a disinfectant aerosol throughout hospital ward to bring an outbreak of streptococcal infections under control. The spraying of a hypochlorite solution was used in army huts in an attempt to control an epidemic of respiratory infections (Middleton and Gilliland, 1941). Propylene glycol was aerosolized in the Seashore Home for Convalescent Children in Atlantic City and reportedly caused a marked decrease in respiratory infections (Harris and Stokes, 1945). Aerial use of triethylene glycol was tested at the Harriet Lane Home for Invalid Children in Baltimore but with inconclusive results (Loosli et al., 1947). Triethylene glycol was also tested as an aerosol disinfectant at Bellevue Hospital in New York City but failed to register a consistent reduction in infections rates (Krugman and Ward, 1951).

Disinfection of air with gaseous formaldehyde has been shown to inactivate *Streptococcus* and *Staphylococcus* (Casella and Schmidt-Lorenz, 1989). The aerosolization of a variety of disinfectants, including peracetic acid, hydrogen peroxide, and formaldehyde, and their effectiveness in inactivating spores were tested in a series of studies that demonstrated the best results occurred with formaldehyde at temperatures above 10°C and relative humidities within 65 to 95 percent (Theilen, 1987; Theilen et al., 1987). Other studies have shown aerosolized formaldehyde to be effective against bacteria (Fuhrmann et al., 1986). Aerial disinfectant techniques have been studied in relation to controlling diseases in poultry houses, a specialized application that may have some merit (Petkov et al., 1987).

However effective such aerial disinfectant techniques may be, it may still be necessary to remove the disinfectant from the airstream if such a system were to be used in an air handling unit. Spray washers and moisture eliminators or wet scrubbers may provide a means of removing large droplets, but some additional means like filtration or catalysis may be necessary to render the airstream safe for breathing.

The use of aerial disinfectants may have applications for buildings that are unoccupied, either temporarily or for remediation purposes. In France thermal fogging with an alkyl dimethylbenzylammonium chloride solution has been used successfully for the cleansing of fungi in indoor air (Rakotonirainy et al., 1999). Spraying of this fungicide inhibited airborne spores but had a limited effect on surface contamination. The results of this study indicate that a solution of thiabendazole [(thiazolyl-4)-2 benzimidazole] applied at 10 percent by thermal fogging, at a rate of 5 mL/m^3, disinfected the air of spores. This product deposited no films on any surfaces and did not leave any unpleasant odor in the treated buildings. At this concentration, thiabendazole did not damage paper or cause any visible degradation of painted surfaces and metal shelves.

A hydrogen peroxide gas plasma sterilization system is available that uses a low temperature, low moisture, sterilization process and leaves no toxic residuals (Spry, 2004). This sterilization device is a chamber into which hydrogen peroxide is vaporized at low pressure, and then is irradiated with RF frequencies to generate free radicals. Total cycle time is about 55 minutes.

22.7 FREE RADICALS

Free radicals are electrically neutral molecules with unpaired electrons in the outer orbit that render them extremely reactive (Casarett, 1968). Various free radicals have disinfecting properties, including OH radicals and oxygen radicals. Some of these are directly produced by ozone and indirectly produced by UV and ionizing radiation and are partially responsible for the disinfecting effects of these technologies. Because of their high reactivity with organic compounds, these radicals tend to damage bacterial cell walls and cause cell death, and because of this same high reactivity, free radicals tend to be rather short-lived. Free radicals will typically be consumed in humid air within a matter of seconds, or even microseconds. This latter characteristic can be useful for air disinfection since the radicals will tend to disappear before any disinfected air is inhaled, provided they are being generated upstream of a ventilation supply outlet, and not within the breathing space. The short decomposition time also makes leakage a minimal health hazard. The end products of free radical decomposition are primarily water and oxygen.

Although there are no current air cleaning systems that rely on free radicals for disinfecting air, there are various technologies that could be adapted to this end. Electrical plasma generating systems can be used to create airborne free radicals. In one study using atmospheric discharge plasma to generate OH radicals by decomposing hydrogen peroxide, sterilization was achieved against several microbes on plastic surfaces, including *E. coli* and *Bacillus subtilis* spores (Imaizumi et al., 2003). This study achieved sterilization within time periods that varied from 90 to 300 seconds, and demonstrated the feasibility of such an approach for applications in medical equipment decontamination and potential applications in the food industry.

REFERENCES

Alimov, A. S., Knapp, E. A., Shvedunov, V. I., and Trower, W. P. (2000). "High-power CW LINAC for food irradiation." *Appl Radiat Isot* 53(4–5): 815–820.

Anellis, A., Grecz, N., and Berkowitz, D. (1965). "Survival of *Clostridium botulinum* spores." *Appl Microbiol* 13(3):397–401.

Bertani, G. (1960). "Sensitivities of different bacteriophage species to ionizing radiations." *J Bacteriol* 79:387–393.

Burleson, G. R., Murray, T. M., and Pollard, M. (1975). "Inactivation of viruses and bacteria by ozone, with and without sonication." *Appl Microbiol* 29(3):340–344.

Casarett, A. P. (1968). *Radiation Biology*. Prentice-Hall, Englewood, CO.

Casella, M. L., and Schmidt-Lorenz, W. (1989). "Disinfection with gaseous formaldehyde. First Part: Bactericidal and sporicidal effectiveness of formaldehyde with and without formation of a condensing layer." *Zentralbl Hyg Umweltmed* 188(1–2):144–165.

Cruickshank, R., and Muir, C. (1940). "Air-borne streptococcal infection following influenza." *Lancet* 1:1155–1157.

Curry, R. D., Unklesbay, K., Unklesbay, N., Clevenger, T. E., Brazos, B. J., Mesyats, G., and Filatov, A. (2000). "Effect of high-dose-rate X-rays on E. coli O157:H7 in ground beef." *IEEE T Plasma Sci* 28(1):122–127.

Denes, F. (1997). "Synthesis and surface modification by macromolecular plasma chemistry." *Trends Polym Sci* 5(1):23–31.

Denes, A. R., Somers, E. B., Wong, A. C. L., and Denes, F. (2001). "12-crown-4-ether and tri(ethylene glycol) dimethyl-ether plasma-coated stainless steel surfaces and their ability to reduce bacterial biofilm deposition." *J Appl Polym* 81:3425–3438.

Desai, P., and Buonicore, A. J. (1990). "Engineering controls to minimize worker exposure to ethylene oxide at sterilization facilities." *Plant/Operations Progress* 9(2):103–107.

Fuhrmann, H., Floerke, I., and Bohm, K. H. (1986). "The problem of heat activation of bacterial spores after disinfection with regard to an aerosol method of decontaminating equipment and rooms." *Zentralbl Bakteriol Mikrobiol Hyg* 182(5–6):515–524.

Gould, H., Herbert, B. N., and Loviny, T. (1969). "Polysomes from *Bacillus subtilis* and *Bacillus thuringiensis.*" *Nature* 223:855–856.

Hamre, D. (1949). "The effect of ultrasonic waves upon *Klebsiella pneumoniae, Saccharomyces cerevisiae, Miyaga wanella felis,* and *Influenza virus A.*" *J Bacteriol* 57:279–295.

Harakeh, M. S., and Butler, M. (1985). "Factors influencing the ozone inactivation of enteric viruses in effluent." *Ozone Sci Eng* 6:235–243.

Harris, T. N., and Stokes, J. (1945). "Summary of a 3-year study of the clinical applications of the disinfection of air by glycol vapors." *Am J Med Sci* 209:152–156.

Harry, J. (1993). "Destruction of waste and toxic materials using electric discharges." *Engin Sci Education J* 2(4):171–176.

Hirai, K., Nagata, Y., and Maeda, Y. (1996). "Decomposition of chlorofluorocarbons and hydrofluorocarbons in water by ultrasonic irradiation." *Ultrason Sonochem* 3:S205–S207.

Hoffman, M. R., Hua, I., and Hochemer, R. (1996). "Application of ultrasonic irradiation for the degradation of chemical contaminants in water." *Ultrason Sonochem* 3:S163–S172.

Hutcherson, K., Roush, R., and Brown, R. (1995). "Chemical destruction using a pulsed corona reactor." *10th IEEE International Pulsed Power Conference,* Albuquerque, NM. 150–154.

Imaizumi, Y., Takashima, K., Katura, S., and Mizuno, A. (2003). "Sterilization using OH radical produced by atmospheric discharge plasma." *IEEE International Conference on Plasma Science,* Jeju, South Korea.

Iordanov, I. (1977). "Gamma ray decontamination of skins and wool contaminated with Bac. anthracis spores." *Vet Med Nauki* 14(8):14–19.

Jiang, H., Manolache, S., Wong, A. C. L., and Denes, F. (2004). "Plasma-enhanced deposition of silver nanoparticles onto polymer and metal surfaces for the generation of antimicrobial characteristics." *J Appl Polymer Sci* 93(3):1411–1422.

Karelin, V. I., Buranov, S. N., Voevodin, S. V., Voevodina, I. A., Matvey, T. N., Repin, P. B., and Selemir, V. D. (2001). "Study of high-voltage diffused gas discharge effect on microbiological cultures." *IEEE International Conference on Plasma Science,* Las Vegas, NV.

Kelly-Wintenberg, K., Montie, T., Hodge, A., Gaskins, J., Roth, J. R., and Chen, Z. (1999). "Mechanism of killing of microorganisms by a one atmosphere uniform glow discharge plasma." *IEEE International Conference on Plasma Science,* Monterey, CA. 202.

Korzekwa, R., Grothaus, M., Hutcherson, K., and Roush, R. (1993). "Nanosecond pulsed corona reactor for efficient destruction of hazardous gases." *Conference Record—Abstracts 1993 IEEE International Conference on Plasma Science*: 2P30, Vancouver, BC.

Korzekwa, R. A., Grothaus, M. G., Hutcherson, R. K., Roush, R. A., and Brown, R. (1998). "Destruction of hazardous air pollutants using a fast rise time pulsed corona reactor." *Rev Sci Instrum* 69(4):1886.

Krugman, S., and Ward, R. (1951). "Air sterilization in infants' ward; effect of triethylene glycol vapor and dust-suppressive measures on respiratory cross-infection rate." *JAMA* 145:775–780.

Kurucz, C. N., Waite, T. D., and Cooper, W. J. (1994). "Use of electron beam irradiation for the treatment of water and other environmental applications." *Proceedings of the IEEE International Conference on Plasma Science,* Santa Fe, NM. 141.

Kuzin, A. M., Nikitina, A. N., Iurov, S. S., and Primak, V. N. (1976). "Stimulating effect of chronic low-power gamma radiation on the growth and development of Aspergillus niger." *Radiobiologiia* 16:70–72.

Kuzin, A. M. (1977). Stimulatory effect of ionizing irradiation. Atomizdat, Moscow, USSR. 133.

Lea, D. E., Haines, R. B., and Bretscher, E. (1941). "The bactericidal action of X-rays, neutrons and radioactive radiations." *J Hyg* 41(1):1–16.

Lea, D. E. (1955). *Actions of Radiations on Living Cells.* Cambridge University Press, Cambridge, England.

Lee, Y. K., Choi, J. H., Lee, E. S., Lee, S. J., Song, K. M., and Baik, H. K. (2003). "Investigation of sterilization mechanism by atmospheric pressure plasma jet system (APPJS)." *IEEE International Conference on Plasma Science,* Jeju, South Korea.

Loosli, C. G., Smith, M. H. D., Gould, R. L., Robertson, O. H., and Puck, T. T. (1947). "Control of cross-infections in the infants' wards by the use of triethylene glycol vapor." *Am J Pub Health* 37:1385.

Manolache, S., Somers, E. B., Wong, A. C. L., Shamamian, V., and Denes, F. (2001). "Dense medium plasma environments: A new approach for the disinfection of water." *Environ Sci Technol* 18:3780–3785.

Matsumoto, K., and Kanitani, M. (2003). "High-speed ethylene-oxide plasma sterilization system without toxic residuals." *IEEE International Conference on Plasma Science,* Jeju, South Korea.

McLean, M. R., and Roth, J. R. (1998). "Utilizing a one-atmosphere uniform glow discharge plasma for chemical/biological warfare agent decontamination." *IEEE Conference on Plasma Science,* Raleigh, NC. 278.

Middleton, D. S., and Gilliland, I. C. (1941). "Prevention of droplet-borne infections by spray; a field experiment." *Lancet* 2:598–599.

Montie, T. C., Kelly-Wintenberg, K., and Roth, J. R. (2000). "Overview of research using the one atmosphere uniform glow discharge plasma (OAUGDP) for sterilization of surfaces and materials." *IEEE Transactions on Plasma Science* 28(1):41–50.

Nagatsu, M., Terashita, F., and Koide, Y. (2003). "Low-temperature sterilization with surface-wave-excited oxygen plasma." *Jpn J Appl Phys 2* 42(7B):L856–L859.

Nunez, C. M., Ramsey, G. H., Ponder, W. H., Abbott, J. H., Hamel, L. E., and Kariher, P. H. (1993). "Corona destruction: An innovative control technology for VOCs and air toxics." *J Air Waste Manage* 43:242–247.

Park, B., Lee, D. H., Park, J. -C., Lee, I. -S., Lee, K. -Y., Hyun, S. O., Chun, M.-S., and Chung, K. -H. (2003). "Sterilization using a microwave-induced argon plasma system at atmospheric pressure." *Phys Plasmas* 10(11):4539–4544.

Petkov, G., Baikov, B. D., and Russak, G. (1987). "Possibilities for decontaminating the air in commercial poultry breeding." *Vet Med Nauki* 24(3):67–72.

Popendorf, W., and Selim, M. (1995). "Exposures while applying commercial disinfectants." *Am Ind Hyg Assoc J* 56(11):1111–1120.

Rakotonirainy, M., Fohrer, F., and Flieder, F. (1999). "Research on fungicides for aerial disinfection by thermal fogging in libraries and archives." *Int Biodeter Biodegrad* 44(2–3):133–139.

Ramsey, G. H., Plaks, N., and Vogel, C. A. (1990). "The destruction of volatile organic compounds by an innovative corona technology." *The Eighth Symposium on the Transfer and Utilization of Particulate Control Technology,* San Diego, CA.

Roland, U., Holzer, F., and Kopinke, F.-D. (2002). "Improved oxidation of air pollutants in a nonthermal plasma." *Catal Today* 73(3–4):315–323.

Ryan, K. J. (1994). *Sherris Medical Microbiology.* Appleton & Lange, Norwalk, CT.

Sanz, B., Palacios, P., Lopez, P., and Ordonez, J. A. (1985). Effect of ultrasonic waves on the heat resistance of Bacillus stearothermophilus spores. *Fundamental and Applied Aspects of Bacterial Spores,* Academic Press, London. 251.

Scherba, G., Weigel, R. M., and O'Brien, W. D., Jr. (1991). "Quantitative assessment of the germicidal efficacy of ultrasonic energy." *Appl Microbiol* 57:2079–2084.

Smith, R. A., Ingels, J., Lochemes, J. J., Dutkowsky, J. P., and Pifer, L. L. (2001). "Gamma irradiation of HIV-1." *J Orthop Res* 19(5):815–819.

Spry, C. (2004). "Low-temperature sterilization." Virgo Publishing. http://www.infectioncontroltoday.com/articles/151steriliz.html.

Sussman, A. F., and Halvorson, H. O. (1966). *Spores: Their Dormancy and Germination.* Harper & Row, New York.

Tarleton (1990). "Microfiltration enhancement by electrical and ultrasonic force fields." *Filtech 89*, Karlesruhe, West Germany.

Tavera, L., Brena, M., Perez, M., Serment, J., and Balcazar, M. (2003). "Response to alpha and gamma radiations of Escherichia coli strains defective in repair or protective mechanisms." *Radiat Meas* 36(1–6 SPEC):591–595.

Tavossi, H. (1986). "Effects of an acoustic wave on the aerosol collection efficiency of a packed bed of spheres." *Langmuir* 2(6):757–760.

Theilen, U., Wilsberg, F. J., Bohm, R., and Strauch, D. (1987). "Aerosol disinfection of bacterial spores." *Zentralbl Bakteriol Mikrobiol Hyg* 184(3–4):229–252.

Theilen, U. (1987). "The mechanism of action of aerosol disinfection." *Zentralbl Bakteriol Mikrobiol Hyg* 184(2):146–159.

vanLancker, M., and Bastiaansen, L. (2000). "Electron-beam sterilization. Trends and developments." *Med Device Technol* 11(4):18–21.

Vincent, J. H. (1995). *Aerosol Science for Industrial Hygienists*. Pergamon, New York.

Vollmer, A. C., Kwakye, S., Halpern, M., and Everbach, F. C. (1998). "Bacterial stress response to 1-megahertz pulsed ultrasound in the presence of microbubbles." *Appl Environ Microbiol* 64(10):3927–391.

Wilson, S. C., Brasel, T. L., Carriker, C. G., Fortenberry, G. D., Fogle, M. R., Martin, J. M., Wu, C., Andriychuk, L. A., Karunasena, E., and Straus, D. C. (2004). "An investigation into techniques for cleaning mold-contaminated home contents." *J Occup Environ Hyg* 1:442–447.

SECTION 3

TESTING AND REMEDIATION

CHAPTER 23

AIR AND SURFACE SAMPLING

23.1 INTRODUCTION

Air sampling and surface sampling are two of the most reliable methods for determining the presence and concentration of microbial contamination in buildings, and this method has been in use for over a century. Air sampling can be used to determine the airborne concentration of bacteria and fungal spores. Surface sampling can be used to determine the presence or concentration of bacteria and fungal spores on surfaces like walls, building materials, cooling coils, and ductwork interiors.

Although some rudimentary knowledge of microbiology is useful in taking samples, it is not essential if proper procedures are observed. Equipment like air samplers, petri dishes, and sterile swabs are necessary for doing sampling in any building, but such equipment can be obtained, along with training in their use, at most microbiology laboratories. Laboratories can provide supplies and will culture and analyze the samples, freeing the engineer from anything but analysis of the test results. This chapter is intended to familiarize engineers and technicians with sampling protocols, equipment, costs, and analysis and interpretation of sampling results to the degree that they will be able to do a professional job if called upon to evaluate a building contamination problem.

Three types of microbial contamination are common in buildings—(1) obvious surface contamination on building materials that is regarded as a potential airborne problem, (2) suspected airborne quality problems with no apparent source, and (3) contamination of equipment like cooling coils whether it is obvious or otherwise. Surface contamination and equipment contamination can be investigated using surface sampling techniques. Air sampling can be used to investigate the effects of all three types of contamination. These methods are explained in the following sections.

23.2 AIR AND SURFACE MICROFLORA

Indoor and outdoor air is constantly filled with natural ambient microflora such as fungal spores, environmental bacteria, and bacteria that hail from human sources (see Fig. 23.1).

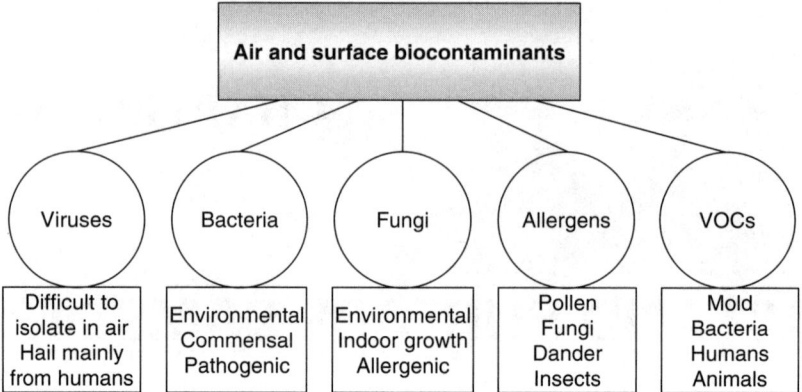

FIGURE 23.1 Breakdown of the biocontaminants that may be sampled from air and surfaces, with sources and notes as shown.

This abundance of airborne microbes allows testing to be performed without having to inject test microbes into indoor air.

Viruses are known to exist in indoor environments by inference from epidemiological studies but they have proven elusive to capture and culture from indoor air (Cox and Wathes, 1995). This characteristic makes it most unfruitful to pursue sampling them from indoor air or surfaces, although the actual culturing is not much more complicated than culturing bacteria. Typically, viruses are cultured in the laboratory by inserting them in biological host cells or bacteria (Fields and Knipe, 1991), but because of the difficulty in sampling viruses nothing further will be said on them in this chapter.

Fungi exist indoors and outdoors and are relatively easy to sample and culture. Fungal spores tend to settle by gravity over undisturbed time, like dust, and will tend to end up on the lowest flat surfaces. Sampling of fungi is best performed at floor level, at breathing height, or anywhere in-between. They are typically identified via visual examination of the spores and mycelia under a microscope (Sutton et al., 1998).

Bacteria in indoor air generally come from two primary sources, outdoor environmental bacteria and indoor generation from human and animal sources (Cox, 1987). In general the environmental bacteria are not hazardous to healthy people and it is the pathogenic bacteria, and sometimes the commensal bacteria, that are of most concern to human health (Braude et al., 1981). Bacteria, like spores, have a tendency to settle over time and air sampling may best be performed at levels between head height and the floor. Bacteria are identified by culturing and examination under a microscope, or sometimes by more advanced methods (Murray, 1999).

Pollen spores are plant seedlings that can cause allergic reactions in atopic individuals (Pope et al., 1993). They are generally not grown but merely identified via microscopy (Lewis et al., 1983).

Allergens and respiratory irritants (other than VOCs) include pollen, dust mite allergens, cockroach allergens, animal dander, and many others. These allergens cannot be cultured or grown in petri dishes but must be assayed by other methods (Lacey et al., 1996). Dust mites are often identified as allergens, but generally only where there is considerable dust disturbance (Pope et al., 1993). They can be found in high concentrations in upholstered furniture, mattresses, pillows, and carpets. Because of their size—some 300 μm—dust mites are not truly capable of airborne transport for any but the shortest distances, and therefore air sampling may not provide an accurate picture of dust mite contamination. Allergic reactions to dust mites are usually due to much smaller particulate matter that cannot be well characterized in terms of size or shape but that may become aerosolized through disturbances caused by walking over rugs, changing bed sheets, or working with stored food or grain products (Platts-Mills and Chapman, 1987). The most significant insect allergens besides dust mites in indoor environments are cockroach allergens (Rosenstreich et al., 1997).

Dander refers to particles of dead skin (squames) or hair that are normally shed by animals. Dander can come from animals and pets like dogs, cats, birds, and farm animals (Middleton et al., 1983). When these particles are microscopic in size they may become airborne and end up being inhaled. Dander will tend to accumulate in upholstered furniture, carpets, and other furnishings, and will become aerosolized on disturbance. Dander may occur in various sizes and shapes but studies suggest they will often tend to be particles of 2 μm or larger.

Endotoxins and exotoxins may be measured through the use of air sampling and analysis which generally indicate the amount of gram-negative bacteria in the air (Milton, 1996). Endotoxins are toxic lipopolysaccharides present in or released from the outer membrane of certain gram-negative bacteria (Ryan, 1994). Exotoxins are toxic proteins liberated from certain bacterial cells, usually gram-positive bacteria (Ryan, 1994). Testing for endotoxins is one method of assaying biocontamination but this may not be a critical test for most applications since, in general, bacterial air sampling will be sufficient to identify all toxin-producing bacteria.

Pollutants generated by microbial growth (i.e., VOCs or MVOCs) can be measured by traditional chemical analysis methods (Yocom and McCarthy, 1991). Such VOCs are secondary products of indoor microbial contamination and are not directly addressed here. The methods of collection generally involve the capture of gaseous samples, or depend on the reaction of VOCs with chemical or electrochemical sensors (AQS, 2003).

23.3 GROWING MICROBIAL CULTURES

Bacteria and fungi are commonly grown on petri dishes, often called plates, which are plastic dishes containing a nutrient agar. The nutrient agar provides both a substrate and food for the microbe to grow and thrive on. Older petri dishes were made of glass but these are less common today. Petri dishes for culturing microbes are often preprepared in packs of 10 to 20, either by a lab or a supplier, and kept chilled until they are needed. There are separate types of plates for bacteria and fungi that contain different types of nutrient agar. There are, in fact, numerous types of agar that are specialized for certain microbial species. Table 23.1 lists some common varieties of agar plates and their target growth microorganisms. Many of these are commercially available and can be stored under refrigerated conditions for about 3 months.

Culturing microbes on a petri dish, like the one shown in Fig. 23.2, involves placing it in an incubator that holds the temperature and humidity constant. For bacteria these conditions are typically 27°C and high humidity although the ideal growth temperature can vary widely depending on species. For fungi the conditions are typically 37°C and high humidity although, again, ideal growing conditions may vary from species to species.

Figure 23.3 shows a typical incubator, which is basically a low-temperature oven with humidity control. Incubators can range from desktop size to walk-in room size incubators.

Since microbes are invisible to the naked eye, they are grown on nutrient agar to allow them to develop into visible colonies. Theoretically, each bacterial cell will develop into a full-blown colony in the space of 8 to 24 hours under ideal conditions. If we begin with a single cell, it will double after a short period of time. The equations describing this growth pattern were introduced in Chap. 7 and the exponential growth of microbial populations is

TABLE 23.1 Typical Growth Media

Medium	Target microorganisms	Incubation temperature
Fungi		
Malt extract agar (MEA)	Saprophytic or potentially allergenic and pathogenic fungi, esp. *Penicillium* and *Aspergillus*	20–25°C/35–37°C
Inhibitory mold agar	Same as MEA but suppresses bacterial colonies	20–25°C/35–37°C
Rose bengal agar	Same as MEA but suppresses bacterial colonies	20–25°C/35–37°C
Malt extract agar with NaCl, sucrose, or dichloranglycerol	Xerophilic fungi	20–25°C/35–37°C
Benomyl agar	Basidiomycetes	20–25°C/35–37°C
Czapek-Dox agar	*Aspergillus* spp.	20–25°C/35–37°C
Mycosel/mycobiotic agar	Dermatophytes, yeast, and fungal speciation	20–25°C/35–37°C
Potato flakes agar (PFA)	Fungi (rapid identification)	20–25°C/35–37°C
Sabouraud dextrose agar (various types)	Isolation of yeast, fungi	20–25°C/35–37°C
Bacteria		
R2A with cycloheximide	Environmental bacteria with fungal suppression	20–30°C
Soybean-casein digest agar	Environmental bacteria with fungal suppression	20–30°C
Soybean-casein digest agar	Thermophilic bacteria	50–55°C
Heart infusion blood agar	Human commensal bacteria	35–37°C
Buffered charcoal yeast extract agar	*Legionella*	35–37°C
Trypticase soy agar (TSA)	Bacteria	35–37°C

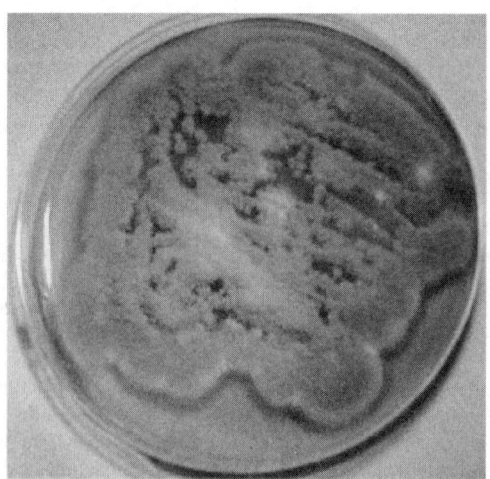

FIGURE 23.2 Petri dish on which colonies of *Penicillium* have been cultured. (*Image courtesy of Pathogenus, Inc., Canada.*)

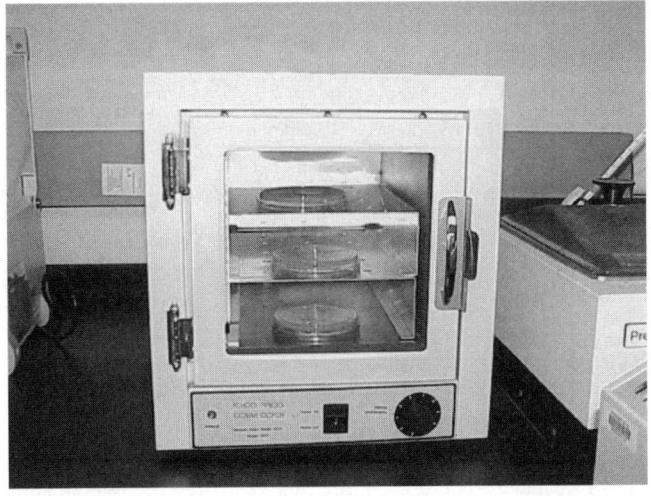

FIGURE 23.3 A small incubator for use in culturing plates of bacteria or fungi.

described as following:

$$\frac{N}{N_0} = e^{kt} \quad (23.1)$$

where N_0 = initial population
k = growth constant for the species, s^{-1}
t = time, second

Growth will rarely occur without limit. In any nutrient source such as a petri dish the population will eventually reach a steady state condition in which no further growth is possible and births equal deaths. The growth will level off creating a sigmoid (or Gompertz) style curve (Alocilja, 2001; Koch, 1969). The normalized logistic equation describing bacterial population growth in a limited environment can be written as follows (Turner et al., 1976):

$$N = \frac{1}{1 + e^{-k(t-\tau)}} \quad (23.2)$$

where k is the growth rate constant and τ is the time at which the population is one-half of maximum.

Figure 23.4 shows the growth rate of *Staphylococcus aureus* modeled with $k = 0.34$ and $\tau = 0.35$, and where the population was initially 0.476 of the maximum.

For most bacteria, the growth limit is reached in about 8 to 12 hours. Fungi take longer to grow and the incubation time may be 24 to 48 hours. Bacterial plates can therefore be cultured overnight, while fungal plates may take up to 2 days.

After growth has developed on a plate the colonies are counted, either manually or with digital imaging systems. Figure 23.5 shows a plate of *Escherichia coli* that has been digitally imaged. Each dot represents a single colony that has grown from a single bacterial cell. Digital imaging systems are capable of automatically counting colonies. Often, the species of bacteria or fungi can be identified from the appearance of the colonies (Jensen and

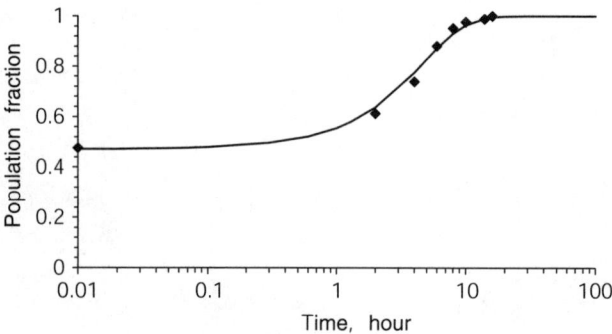

FIGURE 23.4 Growth rate of *Staphylococcus aureus* modeled with the logistic equation. [*Based on data from Kamau et al (1990).*]

Schafer, 1998; Aerotech, 2001). For verification of the species it is usually necessary to examine the colonies under a microscope. In cases where the species is still uncertain, other methods, including genetic analysis, are available.

23.4 SURFACE SAMPLING

Surface sampling involves swabbing a surface like a wall or cooling coil fins with a cotton swab that may or may not be soaked in a sterile solution. The swab must be drawn across the area, gently rolling the swab or turning it over as the surface is sampled. If the swabs are dipped in a sterile solution, the solution is cultured overnight to determine the microbial content. Alternatively, after swabbing the cotton swabs are gently brushed on a petri dish containing nutrient agar and the petri dish is cultured overnight. A third, but less common, technique is to take the petri dish itself and press it on the surface contamination. Some

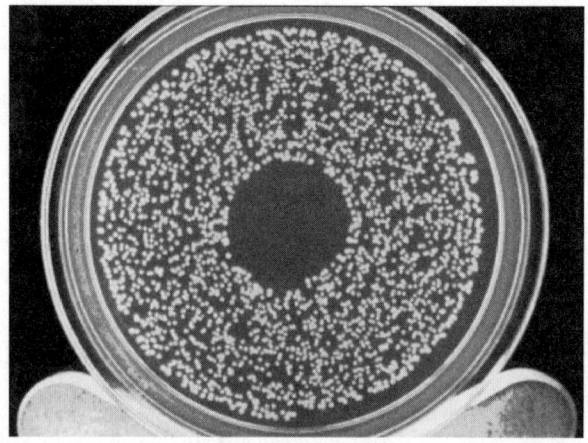

FIGURE 23.5 Digitally imaged plate of *E. coli*.

prepackaged plates are available just for this purpose and are useful for sampling the surfaces of cooling coils. They can be cultured at room temperature for 3 to 5 days or else incubated.

Typically, a sample is taken on a finite surface area of about 1 to 2 in.^{2} or approximately 2.5 to 5 cm^2. Sterile latex or other type of gloves should be used when handling sampling materials and care must be taken not to touch the sample area since it may contaminate the gloves.

About five samples should be taken inside a room, AHU, or other area. Typically this would mean four wall samples (North, South, East, and West) and one floor sample. Sterile swabs can be obtained from a microbiology laboratory. Some swabs are dry while others include a sterile solution. The choice of which type to use is arbitrary just so long as the same types of swabs are used throughout the tests.

Nonreusable sterile paper templates are available to mask the area to be sampled. These templates must be discarded after each use. Reusable templates made of metal or plastic are often used for chemical sampling. These types of templates can be used if and only if they are sterilized before each use. If they are not, then a template contaminated in the first sample will likely contaminate all subsequent samples for which it is used.

Surface sampling results, being more qualitative than quantitative, do not critically depend on the exact area being sampled since they do not tell what the airborne concentrations might be. In essence, this type of test is usually to verify sterilization or to determine the species present. If it is desired to sample a precise surface area, then templates may be used.

23.5 AIR SAMPLING

Air sampling is in common use for identifying airborne microbial contaminants such as bacteria and fungi for the purpose of remediating *indoor air quality* (IAQ) problems (Boss and Day, 2001). Surface sampling is also used to detect contaminants in buildings but this has more relevance to remediation or cleanup activities. Air sampling may take anywhere from hours to days to identify the presence of an airborne contaminant but this is not necessarily a problem because detection may still occur in sufficient time to treat any exposed occupants.

23.5.1 Settle Plates

A common method of assaying the airborne microflora of a room, settle plates are either petri dishes that can be placed around a room in several locations, such as on the floor or at breathing height while sitting or standing, and in the corners, sides, or center of a room. Settle plates are ordinary petri dishes of agar obtained from a local laboratory and are sometimes referred to as sedimentation samplers. One set of, say, three plates would be used for bacteria and another set for fungi. Settle plates can be used to sample for bacteria, fungi, or allergens.

23.5.2 Air Samplers

Air sampling normally involves drawing a sample from an airstream or room and directing it to impinge on a petri dish or sample plate such as shown in Fig. 23.6. The gelatinous medium is a growth medium that is usually selected for the particular microbes that are expected. The plate is exposed to a measured flow of air for a specific period of time, usually about 20 minutes. The plate is then incubated for a period of 12 to 48 hours during which colonies grow and become visible. The plate culture can often be visually identified, but microscopy or other tests are available to make any final determination of what microbes are present (Aerotech, 2001). A variety of air sampler types are available, including

FIGURE 23.6 Air sampler. (*Image courtesy of Aerotech Laboratories, Inc. Phoenix, Arizona.*)

cascade impactors, rotating arm samplers, liquid impingers, centrifugal or cyclone samplers, filtration samplers, and electrostatic samplers. The choice of which sampler to use is sometimes dictated by the application and how long a sampling period is desired. Multistage samplers may be used to selectively sample at different size ranges, and so provide an indication of the particle sizes.

Sampling can cause considerable stress to microorganisms and most will not survive the sampling process. Li et al. (1999) studied three types of samplers, an all-glass impinger, a nucleopore filtration and elution method, and a gelatin filter. Assessment of these types of filters showed that the hardy spore *Bacillus subtilis* had a colony survival rate of about 9.5 to 14.4 percent, while the more fragile *E. coli* had a colony survival rate of only 3.9 to 5.3 percent. The survival was found to be inversely related to the sampling time, with shorter sampling times providing the highest survival rates. Collection efficiencies of samplers can vary considerably between models, often between 50 and 90 percent for liquid impingers, and also tend to increase as the aerodynamic diameter of the sampled particle increases (Willeke et al., 1995, 1998).

Many air samplers require the use of a compressor, such as the one shown in Fig. 23.7, to draw the air sample through the sampler.

Some progress has been made in the development of automated sampling systems and samplers that are geared to identify a wide range of possible microbes (Hobson et al., 1996). The use of strips containing selective growth media, accelerated incubation times, and reduction of colonies to droplet size capable of being evaluated under a microscope are some of the methods being used to provide rapid detection for air samples.

Air samplers typically house the petri dish and direct an airstream toward the surface of the dish containing the growth media. A variety of air samplers are available today and they have diverse characteristics, applications, and varying levels of accuracy (Jensen et al., 1992). Details on the use and performance of aerosol samplers can be found in Bradley et al. (1992), Griffiths et al. (1993), Han et al. (1993), Li et al. (1999), Straja and Leonard (1996), and Aerotech (2001).

Air samplers may be driven by an air pump or an integral fan. Figure 23.8 shows a self-contained air sampler in which a small fan draws air over the surface of the plate near the entrance. After a prescribed period of time the petri dish is removed and the culture incubated.

AIR AND SURFACE SAMPLING

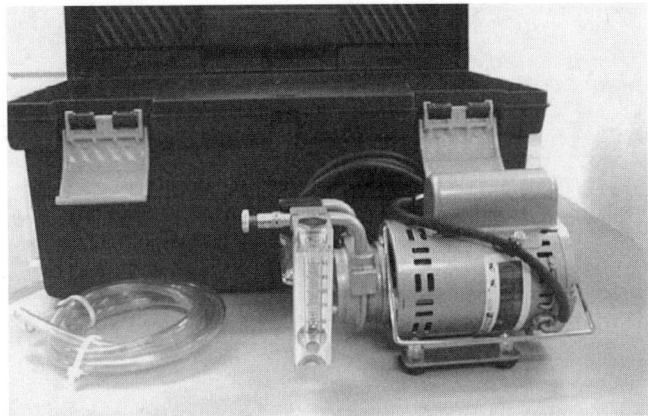

FIGURE 23.7 Air compressor for drawing air samplers. (*Image courtesy of Aerotech Laboratories, Inc. Phoenix, Arizona.*)

A great deal of research has been directed toward the development of samplers that mimic the inhalable or respirable particle size range in indoor environments (Vincent, 1995). Although this approach has considerable value in the study of industrial hygiene in the workplace, the size range of most pathogenic bacteria, fungal spores, and airborne allergens is basically encompassed by the respirable particle size range and so no special attention is warranted in this regard.

Figure 23.9 shows an example of a rotating slit sampler used to develop a dynamic record of airborne concentrations of a test microbe in a laboratory test of a room air disinfection unit.

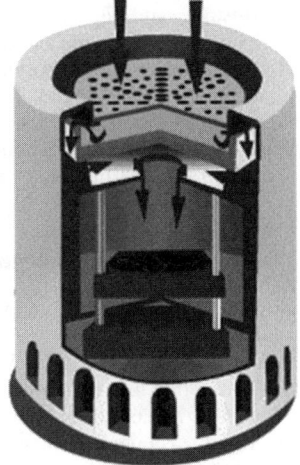

FIGURE 23.8 Air sampler with an integral fan. Cutaway diagram at left shows airflow direction. Petri dish is at the top under the grated inlet. (*Image courtesy of Pathogenus, Inc., Canada.*)

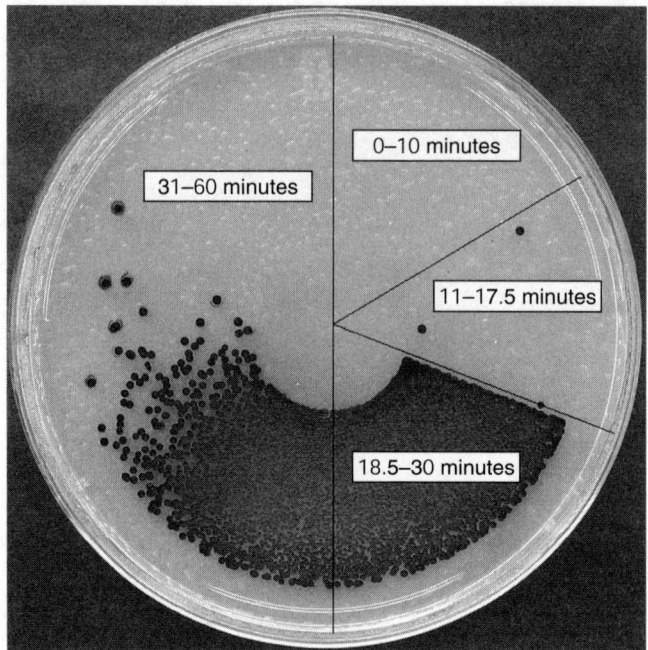

FIGURE 23.9 Example of a Reynier slit sampler used to monitor airborne concentrations in a test facility. From 0–10 min the background air was monitored, then a nebulizer was engaged, next the condition was allowed to stabilize, finally from 31–60 min an air cleaner was engaged to disinfect room air. (*Photo courtesy of Bill Carey, NQ Environmental, Inc.*)

Electrostatic sampling devices are currently available that can operate for extended periods and are sensitive enough to collect airborne microbes. In a study on experimentally infected laying hens, an electrostatic air sampling device was applied to detect *Salmonella enteritidis* in a room (Gast et al., 2004). The floor of the room was cleaned once per week but air samples were positive for *S. enteritidis* for up to 4 weeks postinoculation. In terms of positive cultures, the collection efficiency of the electrostatic device was significantly greater than that of settling plates and was similar to that of impaction samplers.

23.5.3 Air Sampler Performance

Performance may vary from sampler to sampler but the variations are generally within an acceptable range, especially when considering that exact levels of healthy or hazardous microflora have not been established.

Most investigations of mold growth in buildings employ culture-based methods. Flannigan (1997) has suggested these are inadequate for assessing health risks since culturable organisms comprise a small fraction of the total number of potentially allergenic and toxigenic particles in air. For epidemiological studies, measurement of airborne fungal biomass over extended periods may be more relevant than total counts.

The various methods used for sampling indoor bioaerosol levels typically require an incubation period of about 2 to 7 days. Moschandreas et al. (1996) undertook to develop a measuring technique using filtration fluorescence direct counting for screening purposes that would eliminate the incubation wait time. The method compared favorably with data from side-by-side sampling using conventional methods and may be an appropriate screening tool for evaluating the bioburden of indoor and outdoor environments.

It is often difficult to draw an explicit association between indoor airborne microbes and human health disorders (Morey et al., 1990; Reid et al., 1956). One of the contributing problems is the lack of standardized collection and sampling methods. In a study by Ambroise et al. (1999), quantitative assessments were made of the coherence and variability of airborne sampling results collected from different bioaerosol samplers. First, measurements were performed in a large unoccupied university amphitheatre, where air quality was supposed to be stable. Measurements were then taken in a small occupied lecture room of 150 m^3 where large variations in concentrations were expected. Four bioaerosol samplers were used in this test, a Bioimpactuer (BIM) impaction type sampler, a Merck (MAS) impaction sampler, a two-stage Anderson (IA2) sampler, and a centrifugal impaction sampler from Biotest (RCS). Overall, the bioaerosol samplers provided fairly consistent and well-correlated results as evidenced in Fig. 23.10, which represents 10 test replicants. The coefficient of variation for these samplers was within 18 to 30 percent.

Conventional sampling of bioaerosols into liquid impingers can only be performed with water or another low-viscosity liquid as the collection medium. Since these liquids evaporate quickly, sampling is generally limited to short-time periods of 15 to 30 minutes. In a study by Lin et al. (1999), a recently developed sampler, called a BioSampler, that uses a nonevaporating, higher-viscosity inert liquid has been used to sample airborne bacteria and fungi for several hours. The bioefficiency of conventional impingers decreases rapidly with sampling time until the liquid evaporates after about $1^1/_2$ hours. When the BioSampler was operated for 4 hours with nonevaporating heavy white mineral oil, the collection efficiency decreased only moderately.

23.5.4 Particle Counting

An alternative to air sampling is to use a particle counter to measure the levels of particulates in the air. Although particle counts could be expected to reflect bioaerosol concentrations, there is no absolute guarantee that the particulate levels in a building will always correspond to the bioaerosol levels in the building (Parat et al., 1999). Furthermore, occupied buildings may generate particulates internally at much higher levels than they generate fungi or bacteria. Results from testing with particle counters should therefore be used

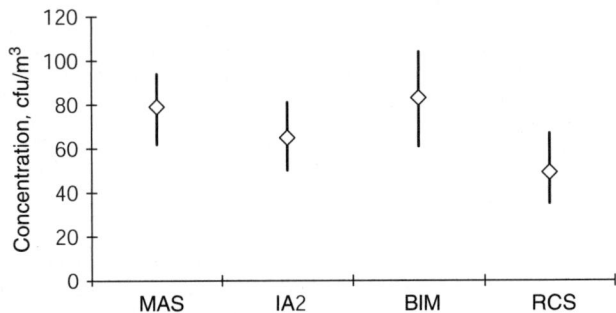

FIGURE 23.10 Comparison of 10 sampler measurements for four samplers used in same location. [*Based on data from Ambroise et al. (1999).*]

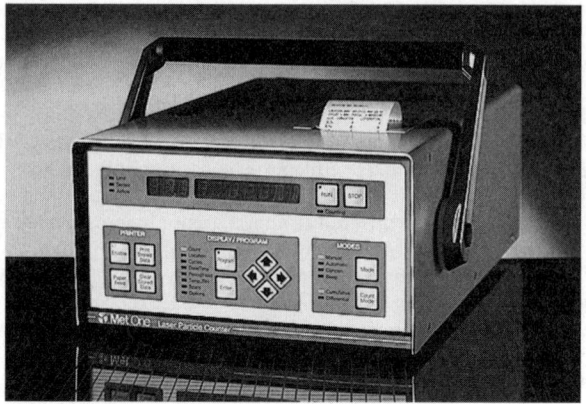

FIGURE 23.11 Laser particle counter capable of resolving micron-sized airborne pathogens and allergens. (*Photo provided courtesy of Hach Ultra Analytics, Grants Pass, Oregon.*)

with caution or as an auxiliary means of obtaining data to compare Before versus After conditions.

One type of particle counter, a particle sizer, uses laser spectrometry to measure the size distribution of particles in the 0.3 to 20 μm size range, which is the range that includes most bacteria and all spores. Particle sizers are available that can measure particles down to 0.005 μm in diameter, which is below the size range of the smallest viruses (approximately, 0.03 μm). Such discriminating information on the particle size and content of the air may be useful to some degree, but will be unable to absolutely differentiate species. Figure 23.11 shows an example of a particle counter. More information is provided on particle counters and particle sizers in the following chapter.

23.5.5 Allergen Sampling

Allergens may be present within pollen, fungal and bacterial spores, animal dander, algae, and parts of insects. Some specific allergens have been identified and categorized and these can be identified with specific immunoassays or immunochemical techniques (Cox and Wathes, 1995). Almost any type of sampler can be used to collect allergens from the air, including cascade impactors, liquid impingers, cyclones, filter-type samplers, and electrostatic samplers (Lacey et al., 1996). In general, sampling for allergens will not involve culturing bacteria or fungi, but will employ laboratory analysis or immunoassays to identify the allergens present in the collected sample.

23.6 SAMPLING PROCEDURES

Air and surface sampling procedures for aerobiological applications should be employed in a manner that is consistent with the intended purpose, practical to use, and repeatable insofar as they are necessary for comparative purposes. Most aerobiological engineering applications, other than epidemiological investigations, will involve the comparison of a Before condition and an After condition that represents either remediation of a problem or the retrofit of an air cleaning system. As a result, the following sampling procedures are geared for comparing Before versus After conditions in a modified indoor environment. Some test applications, such as laboratory testing of air cleaning equipment, involve comparing upstream versus downstream airborne concentrations to determine the once-through removal efficiency of air cleaning devices (Foarde et al., 1999), and the procedures summarized here can be adapted for such purposes. Outdoor air sampling, a subject that has received extensive attention in the field of aerobiology, is addressed here only insofar as outdoor airborne concentrations are compared with indoor concentrations.

Although no strict protocols exist for either air or surface sampling, some guidelines have been published by manufacturers and by government agencies and laboratories providing sample materials and analysis will often assist the process with basic training. The following sections provide basic guidelines for air and surface sampling and the reference documents, where they exist, should be consulted if more detailed information is required.

23.6.1 Surface Sampling Procedures

Surface sampling applications in aerobiological engineering include mold remediation, investigation of *building related illness* (BRI) and nosocomial infections, and test protocols to verify the effectiveness of installed surface or air disinfection systems. Surface sampling is qualitative only—it can establish the presence or absence of biocontamination—but the exact counts may have little quantitative significance. Surface sampling can be particularly useful for detecting the presence of mold and to verify the effectiveness of UVGI cooling coil irradiation systems. Surface sampling can also be used to check for biocontamination on filters, and to verify that UVGI air cleaning systems have disinfected or sterilized internal AHU surfaces.

Any test for surface contamination, whether inside a room or an AHU, should typically target at least three separate surfaces. Floors are ideal choices since most microbes tend to settle downward, but walls and vertical surfaces are also suitable candidates. Once test preparations are made and materials procured, follow the Test Protocol for Surface Sampling as provided in this chapter. The results can then be evaluated using the form in App. D, which may also serve as a guide for designing the test.

If room surfaces are to be sampled, then it is recommended that one or more samples be taken each from the floor and from the walls. If the interior of an AHU is being sampled, then one or more samples should also be taken from the floor and from the walls.

If cooling coils are to be sampled, the recommendation remains that an area of 1 to 2 in.^{2} be sampled, in spite of the fin spacing. Most cooling coils consist of thin aluminum fins spaced about 8 to 16 fins/in., meaning the area sampled is not a perfectly smooth surface, but this should not be a concern since the test is qualitative only and the exact counts have limited meaning. Typically, a cooling coil may be sampled before cleaning or before installation of a UVGI system. In the latter case the Before test represents the contaminated condition and the After test represents the cleaned or decontaminated condition. A cooling coil UVGI system should be allowed to operate for at least 2 days prior to taking the After samples. Ideally, samples will be taken both upstream and downstream of the cooling coils, Three samples taken upstream and downstream in the Before condition plus the same number for the After condition will mean a total of 12 surface samples. If triplicate samples are taken, a total of 36 plates will be needed.

Follow the Test Protocols for Surface Sampling provided in this chapter, or use laboratory procedures if they are given. In all surface sampling tests of bacteria and fungi the total count of cfu/in^2 is of primary interest for comparative purposes. The actual identification of bacterial or fungal species is not critical in most situations. In certain settings,

Test Protocol for Surface Sampling

The following procedure applies whether the surface to be sampled is a vertical wall, a horizontal surface, or a cooling coil.

1.1. Procure surface sampling materials as necessary. These should include sterile swabs (wet or dry), sterile gloves, and petri dishes (if necessary). The types of acceptable media have been addressed in Table 23.1.

1.2. If the sampling materials such as the swabs or gloves are not already sterile, they should be sterilized through the use of a disinfectant such as alcohol. The disinfectant can be wiped or sprayed on the materials, then wiped off or allowed to dry.

1.3. The area to be sampled can be visually estimated (unless quantified results are strictly required) or else a template can be used for precision. If sterile paper templates are used, then only one should be used per sample area, and they should not be handled prior to use. If a reusable template is used to mask off a surface area, then the template must be disinfected before the first sample and before each and every additional sample.

1.4. Shut down any operating equipment in the vicinity that may be a hazard to test personnel. If entry into an air handling unit is required, the fan should be shut down. If UV lamps are operating, these should be shut down for the duration of the test.

1.5. Identify and record the approximate location of a suitable surface sampling point. The most critical factor is that the sample location be near but not exactly the same in subsequent testing.

1.6. If using a wet sterile swab, it may be necessary to break or crush the capsule to saturate the swab with the sterile solution prior to use.

1.7. Using the wet or dry swab, draw it gently across the 1 to 2 in.^{2} area with a back and forth motion. The process may be repeated with an up and down motion to assure thorough coverage.

1.8. Insert the now-contaminated swab back into the sterile container.

1.9. If the swab is being used to transfer samples to a petri dish, then draw the swab gently across the plate, first back and forth in one direction and then again, if desired, up and down in a direction perpendicular to the first application. Close up the plates and seal them as necessary, and then store them upside down.

1.10. Using a marker, label the swabs and/or the plates with an identifying code or description of the location sampled.

1.11. Repeat the above process as necessary until all the preselected sample locations have been completed.

1.12. Deliver the samples (either swabs or plates) to a laboratory. Or else proceed with incubation and counting, placing the plates in an incubator as soon as possible, and incubate for 24 to 48 hours for bacteria or 48 to 96 hours for fungi, at the required temperature (see Table 23.1).

1.13. After incubation, and before the plates become overgrown, remove plates and count each plate. Digital images may be made of the plates and then used for counting in their place.

1.14. Tabulate and summarize the results as necessary on the form in App. D.

1.15. If identification of the bacterial or fungal species is necessary, proceed with visual or other identification, or instruct the laboratory to perform this function.

such as in health care facilities, it may be of interest to speciate or identify the actual species of microbes that were counted.

23.6.2 Air Sampling Procedures

Three basic types of air sampling procedures may be used to assess indoor bioaerosols or verify air cleaner operation, the settle plate procedure, volumetric air sampling, and particle counting. Volumetric air sampling is recommended for precision testing. The settle plate method can be a useful and inexpensive means of verifying the successful operation of an air cleaning system, especially in residential housing, and so is provided here for completeness. The particle counting method has both advantages and limitations, but is addressed here as an optional approach.

Settle Plate Procedures. Petri dishes with the appropriate nutrient (see Table 23.1) or other sedimentation equipment must first be procured from a local laboratory or by mail order. Typically at least three plates should be used for the Before condition and three more for the After condition. Normal room occupancy should be maintained in both tests. If more precise results are desired, samples should be taken in triplicate at each location (making a total of 18 plates) and averaged per the form in App. E. Settle plates should be placed around the room or zone in several locations, such as on the floor or at breathing height while sitting or standing, and in the corners, sides, or center of a room. Figure 23.12 shows examples of room placement points. Follow the Test Protocol for Settle Plate Sampling and compile the results on the form in App. E, which may also be used as a guide for laying out the test plan.

Although the settle plate method is an inexact approach, the end result of effective remediation or air treatment will probably manifest itself as a significant reduction in fungal and bacterial counts. Although not quantitative, this method is suitable for speciating bioaerosols and is a comparatively inexpensive method that might be suitable for homeowners who wish to assay the relative daily levels of airborne microbes in their homes or keep seasonal records of fungal spores.

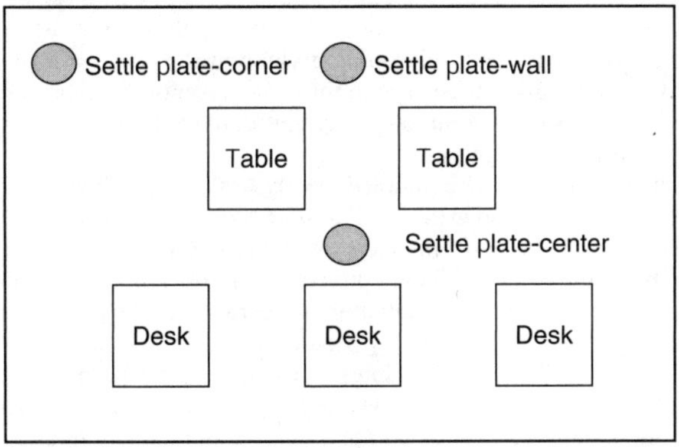

FIGURE 23.12 Schematic of the three primary settle plate locations in a typical rectangular room, classroom, or office. Plates should be located between the floor and head height.

> **Test Protocol for Settle Plate Sampling**
>
> The following test protocol is suggested for use when the settle plate method is used for assessment of indoor bioaerosols.
>
> **1.1.** Select and procure the appropriate growth media in petri dishes or other sedimentation equipment. A tabulation of some typical growth media has been provided in Table 23.1.
>
> **1.2.** Room conditions should be within normal ranges of air temperature, 21 to 24°C (70 to 75°F), and relative humidity, 50 to 70 percent, without any excessive local air currents if they can be avoided.
>
> **1.3.** Place the settle plates at the designated locations in room, preferably as shown in Fig. 23.12.
>
> **1.4.** After a period of 1 to 2 hours, or longer as required, cover and remove the settle plates, storing them upside down.
>
> **1.5.** Incubate the plates at the required temperature (see Table 23.1) or deliver them to a laboratory for incubation and counting. Plates may be incubated at room temperature but colonies may require much longer to grow and some species may be unduly favored.
>
> **1.6.** After incubation for 24 to 48 hours for bacteria or 48 to 96 hours for fungi, count the colonies on the plates and record them on the form in App. E. Make sure to remove the plates from the incubator before they become overgrown and uncountable.
>
> **1.7.** Repeat the above procedure as necessary to obtain at least three samples per location Before and three samples per location After. Use triplicate samples for enhanced precision.
>
> **1.8.** Repeat the above procedure for fungi, bacteria, or other bioaerosols as necessary.
>
> **1.9.** Record all results and complete the evaluation form in App. E.

Indoor Air Sampling Procedures. Air sampling procedures are often specific to the type of sampler used and are provided in operation manuals. Some training or instruction may be required in order to operate the air sampler correctly, although most sampler operation is self-evident. Most professional microbiology laboratories can perform air sampling or offer training and materials. Often, an engineer, microbiologist, or *certified industrial hygienist* (CIH) will perform the air sampling and deliver the samples to a microbiology laboratory for culturing and counting.

Sometimes the indoor air will be sampled merely to assess the bioaerosol levels or to speciate the bacteria, fungi, or pollen in the air. For most aerobiological engineering applications, however, air sampling will be performed to compare a Before condition with an After condition, usually as the result of a remediation project or a retrofit of an air cleaning system. The first test will be performed prior to installation or operation of the air treatment system. The second test will repeat the first test after a period of at least 2 days after the air treatment system has begun operation. For large buildings, it would be prudent to allow at least 2 weeks before performing the After test, so that the building has a fair chance to clean itself out.

Sometimes, the indoor air is to be compared to the outdoor air to assess the status of a healthy building, or as a demonstration of a newly constructed building that has a built-in air cleaning system. In such cases there is no Before condition to be tested. Each round of air sampling will typically require at least two tests, one for fungi and one for bacteria.

For outdoor air sampling, it is recommended that at least three samples be taken near the outside air intakes. This is true whether the outside air intakes are at ground level or on the roof. Air samples should be taken near the inlets in such a way as to avoid high air velocities across the sampler.

In buildings, indoor air samples should be taken in general areas such as the first floor and one or two other floors, if any. In multistory buildings the choice of which floors to sample is arbitrary, but an occupied location is preferable to an unoccupied location for bacterial sampling, and areas should be preferred that are most suspect for fungal contamination, such as basements, kitchens, and storage areas. It is recommended that at least nine air samples be taken in large buildings and with this number of samples the use of triplicate samples may not be essential except for academic precision.

The specifics of taking each air sample may vary with the type of air sampler. For air samplers using petri dishes such as Andersen air samplers, the petri dish must be briefly uncovered and placed within the sampler. The sampler is then turned on and run for a period of time, typically about 2 to 20 minutes depending on the flow rate. The calibration of the sampler may dictate the amount of time the sampler should be operated. Consult the operating manual for specifics.

Laboratory media blanks should be kept cool when possible and stored upside down to prevent possible contamination from settling inside. After sampling the plates should again be turned upside down. Laboratory media blanks can be tested for sterility although laboratories generally test their own media. If a sterility concern develops, place three blank plates in incubation and watch for any growth. Filter-type samplers usually involve the placement of a sterile coupon or filter cassette inside an air sampler. These coupons or cassettes must then be delivered to a laboratory for culturing and counting.

Samplers may also be placed directly inside return ventilation ducts to sample average room or zone air. Although this may distort the results of some air samplers due to air velocities over the sampler, this factor may not always be critical if the test objective is comparison only. If, however, it is also desired to establish indoor airborne levels to meet some criteria, then care should be taken to avoid drafts over the air sampler so as to provide the most accurate absolute readings of airborne concentrations.

If air samples are taken in occupied rooms, then care must be taken to maintain the same occupancy level in both the Before and After tests. Likewise, if air samples are taken in unoccupied rooms, then the rooms should be kept unoccupied in both tests. Personnel operating the air sampler, if they remain in the room, should keep at some distance (i.e., 6 ft or more) from the air sampler since they may release biocontaminants directly into the air sampler intake stream. This consideration applies more to bacterial contamination than to fungal spores; however, occupants may not only carry fungal spores on their clothes but they may also stir up spores from the floor depending on activity level.

Use the Test Protocol for Air Sampling to conduct the test unless another test procedure has been mandated for the application. Compile the results on the form in App. F, which may also be used for planning the test.

Particle Counting Procedure for Buildings. As an alternative to microbiological air sampling, particle counters may be used to assess the effectiveness of air disinfection systems in buildings. It should be noted, however, that particles will include dust and debris of nonmicrobiological origin and that the demonstration of the system effectiveness in Before and After testing will have a higher degree of uncertainty as airborne particulate levels may swamp the airborne bacterial and fungal spore levels. For example, results may both overestimate and underestimate performance of air disinfection systems since any filtration component will likely have a greater impact on particulates than systems UVGI component. The test results evaluation forms provided in App. E may be adapted for the purpose of particle counting.

Test Protocol for Indoor Air Sampling

The following is a suggested test protocol for air sampling indoors. This is not meant to be an exclusive test protocol and variations are possible insofar as they are in agreement with common practices.

1.1. Select and procure an appropriate air sampler and growth media in plates suitable for the selected air sampler. A tabulation of some typical growth media has been provided in Table 23.1.

1.2. Room conditions should be within normal indoor ranges for temperature, 21 to 24°C (70 to 75°F), and relative humidity, 50 to 70 percent, without any excessive local air currents if they can be avoided.

1.3. Install the petri dish, filter, or other media in the air sampler.

1.4. Place the air sampler in a preselected location and operate it for the time period specified by the manufacturer or the instructions.

1.5. After the elapsed time period, shut down the air sampler and carefully remove the plate (or filter).

1.6. Label the plate with an appropriate code or description of the location, the type of sample (bacteria or fungi), and date as necessary and record the sample information on a separate log of all samples.

1.7. If required, repeat the above procedure for outdoor air, taking at least three samples.

1.8. Either deliver the sample to a designated laboratory or begin the incubation process by placing the sample in an incubator at the appropriate temperature (see Table 23.1).

1.9. Repeat the above procedure as necessary for the remaining plates (or filters). There is no interval time necessary between multiple plates placed at the same location so these can be run in parallel if more than one air sampler is available.

1.10. If samples are being incubated (by other than a lab), they should be monitored periodically to assure they do not overgrow the plate and make counting impossible. Plates with fast growth should be removed from the incubator and all plates counted before overgrowth invalidates results.

1.11. If samples have been delivered to a laboratory, then the results provided by the lab can be directly inspected. The plate counts should be tabulated with each sample taken and averaged for each specific location.

1.12. If samples are counted (by other than a lab), they should be checked and rechecked to assure correct counts. The counts should then be tabulated for review per the form in App. F.

1.13. If plates are digitally imaged for the record, the digital images may be used for counting and the plates disposed of.

1.14. If speciation is required by the client, then the laboratory will identify the species and provide this information. The species can be reviewed to verify that only normally occurring microbes have been found. The appearance of unusual pathogens should be brought to the attention of the client or medical authorities (i.e., CDC) as necessary.

23.7 SAMPLING APPLICATIONS

Various sampling applications have been addressed in the preceding sections. Table 23.2 presents a summary of these and other possible applications for surface sampling, settle plates, volumetric air sampling, and particle counting. Some of the applications listed in Table 23.2 warrant further discussion and these are addressed in the following sections.

Whenever bioaerosol levels are of concern and air sampling or particle counting is performed to assess the health hazard, the measured levels must be compared with some standard or guideline for the particular bioaerosol. The few guidelines that have been proposed for bioaerosol levels have been summarized in Chap. 25 and these should be consulted for further information.

23.7.1 UVGI Cooling Coil Disinfection Systems

Disinfection of cooling coils with UVGI has been a topic of great interest recently because of its effectiveness and favorable energy economics, and short payback period. Testing of cooling coil disinfection systems designed to control microbial growth will generally require surface sampling, although performance testing of the AHU can also be used as an indicator of UVGI effectiveness (i.e., cooling coil performance should gradually return to the original design conditions). The procedures described in the previous section on surface sampling apply here also with some additional considerations.

The objective of this test is to demonstrate that a significant reduction of coil biocontamination has been achieved by the operation of the UVGI system. Ideally, coil surfaces would be sterilized, but this may be difficult to demonstrate in practice since the very act of sampling can introduce a few cfu of bacteria or fungi to the coil surfaces. Since sterilization is typically assumed to be six logs of reduction in the microbial population, even a single stray bacteria or spore may prevent such verification. It should suffice as proof that a significant reduction in biocontamination has been demonstrated and it is not necessary to undertake the more difficult task of demonstrating six logs of reduction.

Another question is whether to sample the upstream side of the coils, the downstream side, or the internal fin surfaces. Often the UVGI system is placed on only one side of the coils, and it may be thought that this is the best side to test to demonstrate results. All of

TABLE 23.2 Aerobiological Air Sampling Applications

Application	Type of sampling	Comparison
Investigation of indoor airborne species	SUR, SET, AIR	None
Investigation of indoor airborne bioaerosol levels	SET, AIR, PAR	With guidelines
Indoor vs. outdoor comparison	AIR	Indoor vs. outdoor
Laboratory equipment test	AIR	Upstream vs. downstream
Cooling coil disinfection/cleaning	SUR	Before vs. after
Indoor mold remediation	SUR, SET, AIR	Before vs. after
Air cleaning retrofit	SUR, SET, AIR	Before vs. after
New building air cleaning performance test	AIR	Indoor vs. outdoor
Animal houses, animal facilities	AIR, PAR	With guidelines

Abbreviations: SUR = surface sampling, SET = settle plates, AIR = air sampling, PAR = particle counts.

these concerns are irrelevant if the UVGI system is given sufficient time to disinfect the coils, since even a system on one side of the coils will eventually disinfect the internal fin surfaces and the downstream side of the coils, provided the UVGI system is adequately sized. The prudent approach is to test both the upstream and downstream sides with at least three samples each, and if disinfection seems insufficient on one side of the coil, then more time (i.e., weeks) should be allowed for the UVGI system to achieve the desired result. If a second test fails to show improvement, it may well be that the UVGI system design (wattage and lamp placement) needs to be reconsidered.

Finally, some shadow areas always exist where the levels of UV irradiance may be inadequate. Additional surface samples should be taken in areas suspected of being in the shadow of the UV lamps, with careful attention given to the fact that visible light shadows are not necessarily the same as the UV shadows.

It is common for old and obviously fouled cooling coils to be cleaned before the installation of a UVGI surface disinfection system. This may invalidate surface sampling results in the Before condition. In such a case it is sufficient to demonstrate that the surface contamination levels in the After condition are either absolutely low (i.e., single digits of cfu/m^2) or that they are significantly lower than the condition before cleaning.

It is becoming increasingly common to install an air disinfection system around the cooling coils so as to disinfect both the coils and the air with the same UVGI system. In such cases the choice of which test, surface sampling of the coils or air sampling in the building, is the owner's or engineer's prerogative.

23.7.2 UVGI Air Disinfection Systems

UVGI systems installed for the purpose of air disinfection can be tested in a variety of ways. The ultimate proof of system effectiveness is whether or not it significantly decreases indoor airborne bioaerosols. The best test is, therefore, air sampling of bacteria and fungi in the room, zone, or building general areas. In general, filtration will form an integral part of any well-designed UVGI system installed in the ductwork or air handling unit, and the benefits of filtration extend to reducing particulates and pollen. It can be inferred that if the air cleaning system has reduced bacteria or fungal spores, then it is likely also reducing pollen, particulates, and allergens, and therefore it is not essential that these additional contaminants also be sampled. If there are concerns about these bioaerosols, then particle counting may be one expedient method of addressing them. Subjective evidence from allergic or atopic building occupants may also suffice as proof of reduced allergens and this will require an epidemiological survey.

Since air cleaning systems incorporating UVGI have not, until most recently, been included in building ventilation system design, such systems must be retrofitted into existing buildings. The best test for air disinfection systems is, of course, air sampling in the rooms, zones, or general areas, in order to compare the Before condition with the After condition. It has been noted previously that at least 2 days, and preferably at least 2 weeks be allowed for a retrofitted air cleaning system to purge the building of bioaerosols. One good reason for giving such a generous amount of time is that although the air change rate in the building may be 2 to 4 ACH or more, there are always considerable quantities of spores embedded and saturated in building materials like rugs and furniture. As these spores will tend to get stirred up and reaerosolized by human activity, they will also tend to get drawn into the return airstream and filtered or disinfected over time, thereby reducing the building's net bioburden.

Sometimes building owners are not receptive to the idea of sampling indoor air as a test of a UVGI air disinfection system when there have been no previous indications of any building-related illness of health problems. Needless to say, such concerns are unwarranted since there are no legal standards for indoor bioaerosol levels, and such data would be private or proprietary anyway. In any event, it may sometimes be necessary to substitute surface sampling of the interior of the air handling unit for an actual air sampling test. Such a test will demonstrate only that the UVGI system is sterilizing or disinfecting internal surfaces, but this may be sufficient evidence for a client. Another alternative is to use particle counting instead of air sampling, or to use settle plates, since either of these methods will demonstrate air cleaning system effectiveness without any direct implications of the (prior) presence of potentially hazardous biocontaminants in a building.

In some cases two identical AHU equipment trains are present in the same building and the owner may desire a comparison of a retrofitted AHU with an unmodified AHU. Such a comparison would be a sufficient test of the efficacy of an installed UVGI air treatment system provided the initial levels of biocontamination in each unit are approximately the same. Furthermore, air sampling may not be an appropriate test if both AHUs service the same zones.

Many investigators and engineers often propose to test the upstream side of an installed air cleaning system with the downstream side in order to determine the once-through or single-pass disinfection efficiency. Although this may be a standard and desirable test of equipment in a laboratory setting, it is fraught with problems in an actual building. First of all, injecting test microbes into a ventilation system is a most undesirable approach for an occupied building regardless of how harmless the test microbes may be. Secondly, the air disinfection system will tend to gradually reduce the indoor contamination level and so the upstream airborne concentrations will continually change, making the interpretation of data less than straightforward.

If the Before samples prove to be so highly contaminated that they are uncountable, it is not necessarily a problem and the test need not be repeated. It will be sufficient in such cases if the After samples show a significant reduction. Additional information on testing UVGI air cleaning systems is available from the International Ultraviolet Association (IUVA, 2005).

23.7.3 Upper Room UVGI Systems

Upper room UVGI systems require additional testing to verify that human indoor exposure limits are not exceeded. This requires measuring UVC levels at head height at various locations around an irradiated area. Air sampling, as detailed previously, will provide adequate indication of the effectiveness of such systems in reducing indoor bioaerosols; however, an appropriate guideline has been developed by the CDC to which readers are referred for both design and testing information (NIOSH, 2005).

REFERENCES

Aerotech (2001). *IAQ Sampling Guide*. Aerotech Laboratories, Phoenix, AZ.

Alocilja, E. (2001). "Chapter 4: Growth and feedback in population biology." Michigan State University. 79–108. http://www.egr.msu.edu/classes/be230/chapter4.pdf.

Ambroise, D., Greff-Mirguet, G., Gorner, P., Fabries, J. F., and Hartemann, P. (1999). "Measurement of indoor viable airborne bacteria with different bioaerosol samplers." *J Aerosol Sci* 30(Suppl. 1): S699–S700.

AQS (2003). "Detecting mold, MVOCs in problem buildings." Air Quality Sciences Inc., Marietta, GA. http://www.snipsmag.com/CDA/ArticleInformation/features/BNP_Features_Item/0,3374,69666,.

Boss, M. J., and Day, D. W. (2001). *Air Sampling and Industrial Hygiene Engineering*. Lewis Publishers, Boca Raton, FL.

Bradley, D., Burdett, G. J., Griffiths, W. D., and Lyons, C. P. (1992). "Design and performance of size selective microbiological samplers." *J Aerosol Sci* 23(S1):s659–s662.

Braude, A. I., Davis, C. E., and Fierer, J. (1981). *Infectious Diseases and Medical Microbiology*, 2d ed., W. B. Saunders Company. Philadelphia, PA.

Cox, C. S. (1987). *The Aerobiological Pathway of Microorganisms.* John Wiley & Sons, New York.

Cox, C. S., and Wathes, C. M. (1995). *Bioaerosols Handbook.* CRC Lewis Publishers, Boca Raton, FL.

Fields, B. N., and Knipe, D. M. (1991). *Fundamental Virology.* Raven Press, New York.

Flannigan, B. (1997). "Air sampling for fungi in indoor environments." *J Aerosol Sci* 28(3):381–392.

Foarde, K. K., Hanley, J. T., Ensor, D. S., and Roessler, P. (1999). "Development of a method for measuring single-pass bioaerosol removal efficiencies of a room air cleaner." *Aerosol Sci Technol* 30:223–234.

Gast, R. K., Mitchell, B. W., and Holt, P. S. (2004). "Detection of Airborne Salmonella enteritidis in the Environment of Experimentally Infected Laying Hens by an Electrostatic Sampling Device." *Avian Dis* 48(1):148–154.

Griffiths, W. D., Upton, S. L., and Mark, D. (1993). "An investigation into the collection efficiency & bioefficiencies of a number of aerosol samplers." *J Aerosol Sci* 24(S1):s541–s542.

Han, R., Wu, J. R., and Gentry, J. W. (1993). "The development of a sampling train and test chamber for sampling biological aerosols." *J Aerosol Sci* 24(S1):s543–s544.

Hobson, N. S., Tothill, I., and Turner, A. P. F. (1996). "Microbial detection." *Biosens Bioelectron* 11(5):455–477.

IUVA (2005). *Draft Standard for the Testing and Commissioning of UVGI Air Treatment Systems.* International Ultraviolet Association, Ayr, Ontario, Canada. www.iuva.org.

Jensen, P. A., Todd, W. F., Davis, G. N., and Scarpino, P. Y. (1992). "Evaluation of eight bioaerosol samplers challenged with aerosols of free bacteria." *Am Ind Hyg Assoc J* 53(10):660–667.

Jensen, P. A., and Schafer, M. P. (1998). Chapter J: Sampling and Characterization of Bioaerosols. *NIOSH Manual of Analytical Methods*, NIOSH Publication 94-113 M. E. Cassinelli and P. F. O'Connor, eds., National Institute for Occupational Safety and Health, Atlanta, GA. 82–112.

Kamau, D. N., Doores, S., and Pruitt, K. M. (1990). "Enhanced thermal destruction of *Listeria monocytogenes* and *Staphylococcus aureus* by the lactoperoxidase system." *Appl Environ Microbiol* 56:2711–2716.

Koch, A. L. (1969). "The logarithm in biology: Distributions simulating the log-normal." *J Theoret Biol* 23:251–268.

Lacey, J., Crook, B., and Bai, A. J. (1996). Chapter 12: The detection of airborne allergens implicated in occupational asthma. *Pathogens in the Environment*, B. Austin, ed., Blackwell Scientific, Oxford, UK.

Lewis, W. H., Vinay, P., and Zenger, V. E. (1983). *Airborne and Allergenic Pollen of North America.* Johns Hopkins University Press, Baltimore, MD.

Li, C. -S., Hao, M. L., Lin, W. H., Chang, C. W., and Wang, C. S. (1999). "Evaluation of microbial samplers for bacterial microorganisms." *Aerosol Sci Technol* 30:100–108.

Lin, X., Reponen, T. A., Willeke, K., Grinshpun, S. A., Foarde, K. K., and Ensor, D. S. (1999). "Long-term sampling of airborne bacteria and fungi into a nonevaporating liquid." *Atmos Environ* 33(26):4291–4298.

Middleton, E., Reed, C. E., and Ellis, E. F. (1983). *Allergy: Principles and Practice*, Vol. 2. C.V. Mosby, St. Louis, MO.

Milton, D. K. (1996). Bacterial endotoxins: A review of health effects and potential impact in the indoor environment. *Indoor Air and Human Health*, CRC Press, Boca Raton, FL.

Morey, P. R., Feeley, J. C., and Otten, J. A. (1990). "Biological Contaminants in Indoor Environments." *STP 1071* ASTM. Philadelphia, PA.

Moschandreas, D. J., Cha, D. K., and Qian, J. (1996). "Measurement of indoor bioaerosol levels by a direct counting method." *J Environ Eng* 122(5):374–378.

Murray, P. R. (1999). *Manual of Clinical Microbiology.* ASM Press, Washington, DC.

NIOSH (2005). *Engineering Controls for Tuberculosis: Upper-Air Ultraviolet Germicidal Irradiation.* National Institute for Occupational Safety and Health, Cincinnati, OH.

Parat, S., Perdrix, A., Mann, S., and Baconnier, P. (1999). "Contribution of particle counting in assessment of exposure to airborne microorganisms." *Atmos Environ* 33(6):951–959.

Platts-Mills, T. A. E., and Chapman, M. D. (1987). "Dust mites: Immunology, allergic disease, and environmental control." *J Allergy Clin Immunol* 80:755–775.

Pope, A. M., Patterson, R., and Burge, H. (1993). *Indoor Allergens.*, I. o. Medicine, ed., National Academy Press. Washington, DC.

Reid, D. D., Lidwell, O. M., and Williams, R. E. O. (1956). "Counts of airborne bacteria as indices of air hygiene." *J Hyg* 54:524–532.

Rosenstreich, D. L., Eggleston, P., Kattan, M., Baker, D., Slavin, R. G., Gergen, P., Mitchell, H., McNiff-Mortimer, K., Lynn, H., Ownby, D., and Malveaux, F. (1997). "The role of cockroach allergy and exposure to cockroach allergen in causing morbidity among inner-city children with asthma." *N Engl J Med* 336:1356–1384.

Ryan, K. J., ed. (1994). *Sherris Medical Microbiology.* Appleton & Lange, Norwalk, CT.

Straja, S., and Leonard, R. T. (1996). "Statistical analysis of indoor bacterial air concentration and comparison of four RCS biotest samplers." *Environ Int* 22(4):389.

Sutton, D. A., Fothergill, A. W., and Rinaldi, M. G. (1998). *Guide to Clinically Significant Fungi.* Williams & Wilkins, Baltimore, MD.

Turner, M. E., Bradley, E. L., Kirk, K. A., and Pruitt, K. M. (1976). "A theory of growth." *Math Biosci* 29:367–373.

Vincent, J. H. (1995). *Aerosol Science for Industrial Hygienists.* Pergamon, New York.

Willeke, K., Grinshpun, S. A., Ulevicius, V., Terzieva, S., Donnelly, J., Stewart, S., and Jouzaitis, A. (1995). "Microbial stress, bounce and re-aerosolization in bioaerosol samples." *J Aerosol Sci* 26(S1):s883–s884.

Willeke, K., Lin, X., and Grinshpun, S. A. (1998). "Improved aerosol collection by combined impaction and centrifugal motion." *Aerosol Sci Technol* 28:439–456.

Yocom, J. E., and McCarthy, S. M. (1991). *Measuring Indoor Air Quality: A Practical Guide.* John Wiley & Sons, New York.

CHAPTER 24

BIODETECTION AND MONITORING

24.1 INTRODUCTION

In the previous chapter air and surface sampling was described as a method of identifying indoor microbial contaminants. This chapter addresses systems for automatic detection and identification of biological agents. Several types of automatic detection systems are available today but the number of microbes that can be detected are limited. Figure 24.1 shows a breakdown of some of the main automatic biodetection systems today. The costs associated with advanced detection technologies can be prohibitive. This section summarizes some of the technologies that can be used or adapted for detection of biological agents, including toxins and pathogens, in indoor environments and in ducted airstreams.

24.2 PARTICLE DETECTORS

Two types of particle detectors are currently available that have potential aerobiological detection applications—particle counters and particle sizers. Only those instruments sensitive enough to detect particles down to the range of about 0.01 to 1 μm are feasible for use in detecting bioaerosols. Such detectors may provide a feasible alternative to the use of air samplers for detecting airborne pathogens. In theory, all pathogens have a characteristic size range and shape, and it may be possible, although not necessary, to identify many pathogens on this basis alone. Particle sizers and particle counters use a variety of techniques. The types that are perhaps the most adaptable to real-time air sampling are the optical particle sizers such as the one shown in Fig. 24.2. This unit uses laser spectrometry to measure the size distribution of particles in the 0.3- to 20-μm size range, which is the range that includes most bacteria and all spores. Particle sizers are available that can measure particles down to 0.005 μm in diameter, which is below the size range of the smallest airborne viruses (approximately, 0.03 μm).

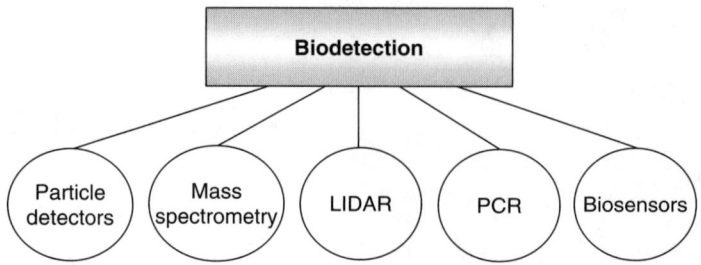

FIGURE 24.1 Breakdown of basic types of biodetection systems.

The actual identification of microbes based on particle size distribution has certain limitations. Airborne microbes do not always occur in singlets, multiple species and particles may be present in any sample, and equipment may not be sensitive enough to establish a complete size range, and so the identification of airborne microbes on the basis of size alone may be fraught with difficulties.

However, if it is not critical that airborne contaminants be specifically identified, particle detectors can be used to determine if any unusual concentrations of particles are present. For example, if particle concentrations in the microbial size range were monitored downstream of air filters, any high concentrations would likely represent either a filter breakthrough or a massive concentration of particles upstream. That is, if the filter is functioning normally, then no high concentrations of particles should exist downstream unless there was a major release of them upstream of the filter or a release downstream of the filter. In any event, detection of large concentrations of particles downstream of an air filter would warrant investigation of the cause.

Many of the particle sizers available today are relatively expensive, but in comparison with other biosensing technology they may be quite feasible for high-end immune building applications. Even basic optical sensors might be adaptable to the detection of high concentrations of particles in ductwork, especially over long spans. Table 24.1 summarizes a few of the commonly available types of particle detectors and the size ranges they cover.

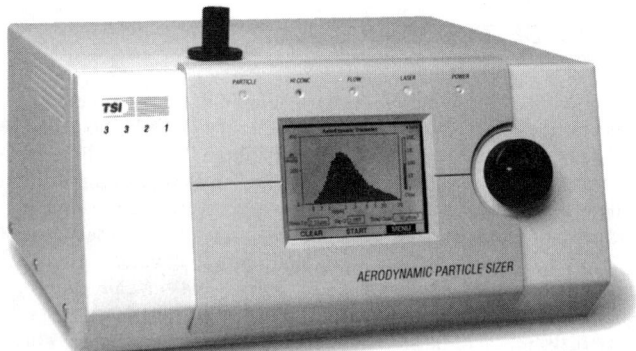

FIGURE 24.2 The aerodynamic particle sizer shown here can be used to detect particles in the bioaerosol size range of 0.5 to 20 μm. (*Photo provided courtesy of TSI Incorporated, Shoreview, MN.*)

TABLE 24.1 Characteristics of Some Direct-Reading Particle Counters

Type	Averaging time, minutes	Measured property	Size range, μm
Photometer	Continuous	Total light scattering	0.3–15
Beta-gauge	1–30	Absorption of beta radiation	1–15
Optical particle counter	0.1–10	Light scattering	0.3–15
Laser optical particle counter	0.1–10	Laser light scattering	0.08–100
Piezobalance	0.5–20	Particle size	0.02–10

Many lower-cost instruments are available that are designed to size particles in fluid suspension, like Coulter counters, and these could analyze particles after sampling air into a solution. Particle sedimentation of centrifugation can be used after the capture of an air sample to identify the presence and size range of particles. This method, however, may take 30 minutes or longer to complete.

Nephelometry is a method for measuring the light scattered by microorganisms in fluid suspension under controlled conditions. The technique produces signals that are directly proportional to the cell mass (Koch, 1961).

Dynamic light scattering is one of the primary methods of detecting or sizing particles in air. Submicron particles subject to Brownian motion are analyzed in terms of the random light intensity fluctuations of scattered light. Algorithms are then used to interpret the fluctuations to establish particle size and particle density. These methods can be accurate and can rapidly determine mean size and size distribution of particles. See the airborne particle counter in Fig. 24.3.

Particle sizers have the ability to distinguish some species based on their size distribution. The distribution of aerodynamic diameters measured by a particle sizer will be somewhat characteristic of each microbe. Although not every microbial species may be identifiable or differentiated by this method, it does offer the possibility of classifying bioaerosol type (i.e., virus, bacteria, fungal spore, or pollen).

Other methods using light scattering techniques have also been applied to the identification of bioaerosols. The use of high-resolution *two-dimensional angular optical*

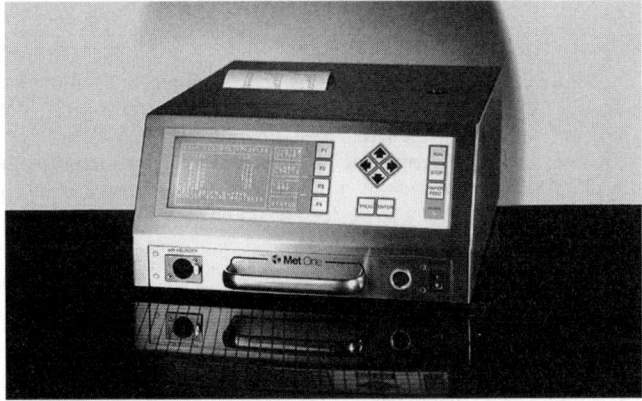

FIGURE 24.3 An airborne particle counter. (*Photo provided courtesy of Hach Ultra Analytics, Grants Pass, OR.*)

scattering (TAOS) has been shown to be capable of characterizing clusters of particles, including spores (Holler et al., 1998).

A study was performed by Parat et al. (1999) to determine the relationship between airborne bacterial concentrations and particle counts simultaneously measured at different sites. An Andersen single-stage sampler was used for microbial measurements and a laser particle counter gave the cumulative counts of particles 0.5 μm and larger in diameter. Peaks of concentrations were generated by human activity and both bacterial and particle counts were monitored over 1 hour. Measurements were also taken for several days in three occupied buildings at 10 minutes intervals. Data revealed strong correlations between bacterial and particle counts in most sites but analyses of covariance indicated that for most sites no absolute relationship could be established between bacterial and particle counts. In other words, although particle counts may often reflect airborne microbial concentrations, there is no guarantee that particle counts truly reflect microbial counts.

24.3 MASS SPECTROMETRY

Mass spectrometry uses light of varying wavelengths to determine the spectra of chemical and organic compounds. Although microbes may contain hundreds or thousands of compounds, certain compounds may be dominant, or certain groups of compounds may occur uniquely with specific microorganisms. Rapid spectroscopic analysis of the various spectra provides a possible means of identifying biological agents. This approach is complex and still under development, but it holds the promise of producing a broad-range biological detector.

Short wavelength near-infrared spectrometry can be used to rapidly measure hundreds of spectra and can estimate biomass (Singh et al., 1994). A related method, UV resonance Raman spectroscopy, has also been reported to be capable of identifying bacteria and spores (Farquharson and Smith, 1998). *Ultraviolet laser-induced fluorescence* (UV LIF) is under development as a means of monitoring the number, density, and fluorescent emissions of airborne particles (Eversole et al., 1998). Computer-assisted mass spectrometers with databases of microbial agents are currently under development or are available for a number of biological agents (Hayek et al., 1999).

24.4 LIDAR

Light detection and ranging (LIDAR) systems are developmental military technologies intended for standoff chemical and biological detection for atmospheric aerosols. These systems use lasers of various wavelengths, including ultraviolet, to detect agents at ranges of several kilometers and can discriminate between biological and nonbiological aerosols (Cannaliato et al., 2000). Such systems are designed for outdoor use and have limited effectiveness, but they may one day be perfected and become available for aerobiological applications.

24.5 POLYMERASE CHAIN REACTION

Current methods for the detection of pathogenic viruses, bacteria, and fungi tend to be inaccurate, time-consuming, and expensive. *Polymerase chain reaction* (PCR) is one of the main alternative detection methods currently under widespread development. PCR has been shown to be a rapid, highly sensitive, and accurate method. It has already been used experimentally to detect pathogenic viruses, bacteria, and protozoa in water and wastewaters. Polymerase chain reaction is a nucleic acid amplification technique that selectively amplifies specific DNA or RNA sequences and thereby allows identification of particular bacteria, viruses, or fungi. PCR is an in vitro enzymatic reaction that takes place rapidly and can be far more accurate than other identification techniques. PCR has been successfully applied to the detection of viruses in water although it cannot distinguish inactivated viruses that are no longer infective (Sobsey et al., 1998). Both DNA and RNA viruses can be detected using this technique and multiple viruses can be detected simultaneously as well as different serotypes (Alexander and Morris, 1990). PCR has certain limitations besides detecting inactivated virus. False positives can be generated through the detection of naked nucleic acids or through contamination in the laboratory. Other limitations include inhibition by environmental contaminants and difficulty in quantification (Toze, 1999). See the automated PCR system in Fig. 24.4.

Viruses that have been detected with PCR in wastewater include enteroviruses, hepatitis A virus, reoviruses, and rotavirus (Lees et al., 1995; Reynolds et al., 1997; Muscillo et al., 2001). Detection of adenoviruses and enteroviruses in tap water and river water has been accomplished using reverse transcription multiplex PCR (Cho et al., 2000).

Bacteria like *Legionella* have been proven detectable by PCR (Lye et al., 1997). A rapid multiplex PCR (m-PCR) method was used to simultaneously detect six common waterborne pathogens in a single tube, including species of *Aeromonas*, with a turnaround time of less than 12 hours (Kong et al., 2002).

Airborne PCR systems have been tested that usually involve drawing air through a filter or solution, then apply PCR techniques to the sample. An airborne species of *Mycoplasma* was detected in this fashion using a nested PCR array (Stark et al., 1998).

Automated systems using microchip-based electrochemical analysis of PCR amplified DNA molecules have proven successful (Hodko et al., 2001). *Legionella* species have been detected by microchip detection systems (Zhou et al., 1998). Rapid detection of *Bacillus cereus*,

FIGURE 24.4 Photo of a polymerase chain reaction system from the Civil & Environmental Engineering Department of Arizona State University. [*Photo courtesy of Professor Morteza Abbaszadegan (Abbaszadegan, 1999)*]

which is closely related to *Bacillus anthrax* has been demonstrated (Mantynen and Lindstrom, 1998). The development of a multiplex PCR system to diagnose tuberculosis and other bacterial infections indicates that this system could simultaneously detect tuberculosis and other common bacteria with higher specificity and sensitivity than previous methods (Fang et al., 2001). PCR has been used to identify *E. coli* in filtered water samples within 2 hours with a 100 percent agreement with standard methods (Fricker et al., 1999).

PCR systems for the identification of fungi have also been developed that can detect *Aureobasidium* (Li et al., 1996). Research continues for methods to detect *Aspergillus*, *Phialophora*, and other species (Aufavre-Brown et al., 1993; Bowman, 1993; Yan et al., 1995). *Stachybotrys* is also reported to be identifiable by PCR methods (Aerotech, 2000).

Research into automated PCR methods continues at an explosive pace and new developments may some day produce a technology capable of rapid and broad range detection of pathogens. One of the limiting factors is that it is necessary to have a specific genome sequence for each pathogen or group of pathogens identified, and this often depends on progress in genome sequencing since most pathogens have yet to have their genome fully sequenced. At present only a limited number of airborne pathogens have been sequenced to the point that they can be identified by PCR.

24.6 BIOSENSORS

Biosensors may be any type of detection system that provides reasonably rapid automatic identification of biological agents. Some biosensors are solid state devices that provide an electrical output signal on detection (Paula, 1998). Many of these devices involve the fixing of a biologically sensitive material or compound onto a hard surface (Gardner, 2000). In general, biosensors can detect only one microbial agent and can only be used once, since the process of detection consumes the sensing material (Leonelli and Althouse, 2000). Some biosensors are currently in use in the food industry for detecting food pathogens (Hobson et al., 1996). Piezoelectric biosensors have been developed that can identify *Salmonella typhi* in water (Prusak-Schoczewski et al., 1990).

Some progress is being made in the development of combined sensors that can automatically detect a wide range of microbial agents (Belgrader et al., 1998; Donlon and Jackman, 1999), but such technology is still in an incipient stage. See the experimental biosensor in Fig. 24.5. Biochemical compounds such as MVOCs can also be used as markers to identify specific fungi (Gao et al., 2002).

When a pathogen invades a host, the immune system often recognizes the invader by the presence of antigens, or surface molecules that are unique to particular species. The immune system may then attempt to release antibodies to neutralize or otherwise limit the damage done by the pathogen. These antibodies have a specific affinity for individual antigens and can be used to identify the presence of specific pathogens. Biosensors have been developed that immobilize antibodies on an inert surface and use them to generate a signal through electrochemical reactions. Biosensors using this technique have been created that can detect *Clostridium perfingens* (Cardosi et al., 1991), *Salmonella*, (Plomer et al., 1992), *Francisella tularensis* (Thompson and Lee, 1992), and *Yersinia pestis* (Cao et al., 1995).

Many other approaches are possible that can identify the reaction of an antigen with an antibody. Some of these methods use optics, such as *surface plasmon resonance* (SPR), *evanescent waves* (EW), luminescence, and fluorescence to identify an antigen-antibody reaction. Electrochemical methods can also be used to detect such reactions, such as potentiometric, amperometric, and conductimetric. Piezoelectric techniques can be used to detect viruses and bacteria (Konig and Gratzel, 1993). In addition, there are acoustic techniques

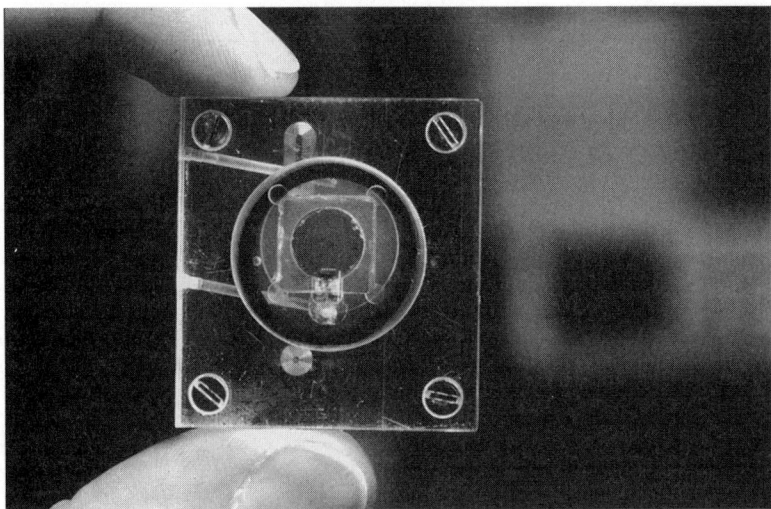

FIGURE 24.5 Biosensor experimental setup for detecting specific biological compounds (Fritz et al., 2000). (*Photo provided courtesy of IBM Research GmbH, Zurich Research Laboratory, Rüschlikon, Switzerland.*)

that use piezoelectric devices and calorimetric techniques that use thermistors or thermopiles to detect reactions (Paddle, 1996). Photoelectric detection of agents is also possible (Arrieta and Huebner, 2000). These techniques are presently under development and the number of pathogens that can be detected is small. In time, biosensors may provide an ideal and economic solution to the problem of detecting airborne pathogens but at present such technology is not quite feasible for incorporation into engineered building systems.

Biosensors depend on selective recognition of the biological molecule that is being targeted. Two distinct types of biosensors can be identified—catalytic and noncatalytic. Catalytic sensors include enzymes, bacterial cells, and tissue cells. Noncatalytic biosensors are also known as affinity class biological sensors and these include antibodies, lectins, receptors, and nucleic acids. One of the main limitations today on the development of biosensors for microbes is a lack of information on the target molecules on their surfaces. The molecules necessary for use in biosensing have either not yet been identified or can't be isolated in a stable form.

Table 24.2 lists some of the microbes for which biosensors have been developed or reported. This list is summarized from Paddle (1996), where the source documents may be found for each biosensor type. The column showing the medium refers to the media in which the biological agent was detected in the test, whether liquid solution or dry air. Although some tests were done in a medium of dry air, no results were available. The possible dependence of the biosensor on liquid media is not necessarily a problem since air can be sampled into a liquid solution.

Systems are under development for military applications that combine mass spectrometry and biosensors for both toxic gas and biological detection (Donlon and Jackman, 1999). These systems use a mass spectrometer to perform initial identification of the presence of any aerosolized agents (McLoughlin et al., 1999; Cornish and Bryden, 1999; Scholl et al., 1999).

TABLE 24.2 Biosensors for Some Microbes and Assay Times

Microbe	Biosensor type	Medium	Assay time	Reference
Bacillus anthracis spores	Optical fiber (evanescent wave-fluorescent dye)	Liquid	35 seconds	Wijesuriya et al., 1994
Yersinia pestis	Optical fiber (evanescent wave, fluorescence antibody)	Liquid	30 minutes	Cao et al., 1995
Yersinia pestis	Acoustic (QCM) (antibody capture)	Liquid	45 minutes	Konig and Gratzel, 1993
Shigella dysenteriae	Acoustic (QCM) (antibody capture)	Liquid	46 minutes	Konig and Gratzel, 1993
Francisella tularensis	LAPS (antigen-antibody, enzyme-labeled urease)	Liquid	65 minutes	Thompson and Lee, 1993
Salmonella typhimurium	Acoustic-piezoelectric (antibody-antigen)	Liquid	5 hours	Prusak-Sochaczewski et al., 1990
Salmonella typhimurium	Optical (optride-luminescence) (antibody capture membrane-enzyme label)	Liquid	30 minutes	Downs, 1991
Brucella melitensis	LAPS	Liquid	—	Lee et al., 1993
Coxiella burnetii	LAPS	Liquid	—	Menking and Goode, 1993

REFERENCES

Abbaszadegan, M., Stewart, P., and LeChevallier, M. (1999). "Strategy for detection of viruses in groundwater by PCR." *Appl Environ Microbiol* 65(2):444–449.

Aerotech (2000). "*Stachybotrys chartarum* detection technology." *Aerotech Monitor* 3(3):1–3.

Alexander, L. M., and Morris, R. (1990). "PCR and environmental monitoring. The way forward." *Water Sci Technol* 24(2):291–294.

Arrieta, R. T., and Huebner, J. S. (2000). "Photo-electric chemical and biological sensors." *2000 Chemical and Biological Sensing,* Orlando, FL. 132–142.

Aufavre-Brown, A., Tang, C. M., and Holden, D. W. (1993). "Detection of gene-disruption events in *Aspergillus* transformants by polymerase chain reaction direct from conidiospores." *Curr Genet* 24:177–178.

Belgrader, P., Benett, W., Begman, W., Langlois, R., R. Mariella, J., Milanovich, F., Miles, R., Venkateswaran, K., Long, G., and Nelson, W. (1998). Autonomous system for pathogen detection and identification." *Air Monitoring and Detection of Chemical and Biological Agents,* Boston, MA 198–206.

Bowman, B. (1993). A model PCR/probe system for the detection of fungal pathogens. *Diagnostic Medical Microbiology: Principles and Applications,* D. H. Persing, T. F. Smith, F. C. Tenover, and T. J. White, eds., ASM, Washington, DC.

Cannaliato, V. J., Jezek, B. W., Hyttinen, L., Strawbridge, J., and Ginley, W. J. (2000). "Short range biological standoff detection system (SR-BSDS)." *2000 Chemical and Biological Sensing,* Orlando, FL. 219–223.

Cao, K. L., Anderson, G. P., Ligler, F. S., and Ezzell, J. (1995). "Detection of *Yersinia pestis* Fraction 1 antigen with a fiberoptic biosensor." *J Clin Microbiol* 33:336–341.

Cardosi, M., Birch, S., Talbot, J., and Phillips, A. (1991). "An electrochemical immunoassay for *Clostridium perfringens* phospholipase C." *Electroanal* 3:1–8.

Cho, H. B., Lee, S.-H., Cho, J. C., and Kim, S. L. (2000). "Detection of adenoviruses and enteroviruses in tap water and river water by reverse transcription multiplex PCR." *Can J Microbiol* 46(5):417–424.

Cornish, T. J., and Bryden, W. A. (1999). "Miniature time-of-flight mass spectrometer for a field-portable biodetection system." *J Hopkins APL Tech D* 20(3):335–342.

Donlon, M., and Jackman, J. (1999). "DARPA integrated chemical and biological detection system." *J Hopkins APL Tech D* 20(3):320–325.

Downs, M. E. A. (1991). "Prospects for nucleic acid biosensors." *Biochem Soc Trans* 19:39–43.

Eversole, J. D., Roselle, D., and Seaver, M. E. (1998). "Monitoring biological aerosols using UV fluorescence." *1998 Air Monitoring and Detection of Chemical and Biological Agents*, Boston, MA. 34–42.

Fang, F., Xiang, Z., and Chen, R. (2001). "Establishment of a multiplex PCR system to diagnose tuberculosis and other bacterial infections." *J Tongji Medical Univ* 20(4):324–326.

Farquharson, S., and Smith, W. W. (1998). "Biological agent identification by nucleic acid base-pair analysis using surface-enhanced Raman spectroscopy." *1998 Air Monitoring and Detection of Chemical and Biological Agents*, Boston, MA. 207–214.

Fricker, E. J., Murrin, K., and Fricker, C. R. (1999). "Use of PCR for the detection of bacteria and viruses." *Water Supp* 17(2):5–16.

Fritz, J., Baller, M. K., Lang, H. P., Rothuizen, H., Vettiger, P., Meyer, E., Güntherodt, H.-J., Gerber, C., and Gimzewski, J. K. (2000). "Translating biomolecular recognition into nanomechanics." *Science* 288:316–318.

Gao, P., Korley, F., Martin, J., and Chen, B. T. (2002). "Determination of unique microbial volatile organic compounds produced by five *Aspergillus* species commonly found in problem buildings." *AIHA J* 63:135-140.

Gardner, P. J. (2000). "Chemical and biological sensing." *Proceedings of the SPIE—The International Society for Optical Engineering*. Orlando, FL.

Hayek, C. S., Pineda, F. J., Doss, O. W., III, and Lin, J. S. (1999). "Computer-assisted interpretation of mass spectra." *J Hopkins APL Tech D* 20(3):363–371.

Hobson, N. S., Tothill, I., and Turner, A. P. F. (1996). "Microbial detection." *Biosens Bioelectron* 11(5):455–477.

Hodko, D., Raymer, L., Herbst, S. M., Magnuson, J. W., and Gaskin, D. (2001). "Detection of pathogens using on-chip electrochemical analysis of PCR amplified DNA molecules." *Proceedings of SPIE—The International Society for Optical Engineering* 4265:65–74.

Holler, S., Pan, Y., Bottiger, J. R., Hill, S. C., Hillis, D. B., and Chang, R. K. (1998). "Two-dimensional angular scattering measurements of single airborne microparticles." *1998 Air Monitoring and Detection of Chemical and Biological Agents*, Boston, MA. 64–72.

Koch, A. L. (1961). "Some calculations on the turbidity of mitochondria and bacteria." *Biochim Biophys Acta* 51:429–441.

Kong, R. Y. C., Lee, S. K. Y., Law, T. W. F., Law, S. H. W., and Wu, R. S. S. (2002). "Rapid detection of six types of bacterial pathogens in marine waters by multiplex PCR." *Water Res* 36(11):2802–2812.

Konig, B., and Gratzel, M. (1993). "Detection of viruses and bacteria with piezoelectric immunosensors." *Anal Lett* 26:1567–1585.

Lee, W. E., Jacobson, T. V., and Thompson, H. G. (1993). "Characteristics of the biochemical detector sensor." *Suffield Memorandum 580*. Defence Research Establishment Suffield, Canada.

Lees, D. N., Henshilwood, K., Gallimore, C. I., and Brown, D. W. G. (1995). "Detection of small round structured viruses in shellfish by reverse transcription-PCR." *Appl Environ Microbiol* 61:4418–4424.

Leonelli, J., and Althouse, M. L. (1998). "Air Monitoring and Detection of Chemical and Biological Agents." *Proceedings of the SPIE—The International Society for Optical Engineering,* Boston, MA.

Li, S., Cullen, D., Hjort, M., Spear, R., and Andrews, J. H. (1996). "Development of an oligonucleotide probe for *Aureobasidium pullulans* based on the small-subunit rRNA gene." *Appl Environ Microbiol* 62(5):1514–1518.

Lye, D., Fout, G. S., Crout, S. R., Danielson, R., Thio, C. L., and Paszko-Kolva, C. M. (1997). "Survey of ground, surface, and potable waters for the presence of *Legionella* species by EnviroAmpR PCR *Legionella* kit, culture, and immunofluorescent staining." *Water Res* 31(2):287–293.

Mantynen, V., and Lindstrom, K. (1998). "Rapid PCR-based DNA test for enterotoxic *Bacillus cereus.*" *Appl Environ Microbiol* 64(5):1634–1639.

McLoughlin, M. P., Allmon, W. R., Anderson, C. W., Carlson, M. A., DeCicco, D. J., and Evancich, N. H. (1999). "Development of a field-portable time-of-flight mass spectrometer system." *J Hopkins APL Tech D* 20(3):326–334.

Menking, D. G., and Goode, M. T. (1993). "Evaluation of cocktailed antibodies for toxin and pathogen assays on the light addressable potentiometric sensor." *1992 ERDEC Scientific Conference on Chemical Defense Research,* Aberdeen Proving Ground, MD, *Report No. ERDEC-SP-007,* 103–109.

Muscillo, M., Rosa, G. L., Marianelli, C., Zaniratti, S., Capobianchi, M. R., Cantiani, L., and Carducci, A. (2001). "New RT-PCR method for the identification of reoviruses in seawater samples." *Water Res* 35(2):548–556.

Paddle, B. M. (1996). "Biosensors for chemical and biological agents of defence interest." *Biosens Bioelectron* 11(11):1079–1113.

Parat, S., Perdrix, A., Mann, S., and Baconnier, P. (1999). "Contribution of particle counting in assessment of exposure to airborne microorganisms." *Atmos Environ* 33(6):951–959.

Paula, G. (1998). "Crime-fighting sensors." *Mechanical Engineering* 120(1):66–72.

Plomer, M., Guilbault, G. G., and Hock, B. (1992). "Development of a piezoelectric immunosensor for the detection of enterobacteria." *Enzyme Microbiol Technol* 14:230–235.

Prusak-Sochaczewski, E., Luong, J. H. T., and Guilbault, G. G. (1990). "Development of a piezoelectric immunosensor for the detection of *Salmonella typhimurium.*" *Enzyme Microbiol Technol* 12:173–177.

Reynolds, K. S., Gerba, C. P., and Pepper, I. L. (1997). "Rapid PCR-based monitoring of infectious enteroviruses in drinking water." *Water Sci Technol* 35(11–12):423–427.

Scholl, P. F., Leonardo, M. A., Rule, A. M., Carlson, M. A., Antoine, M. D., and Buckley, T. J. (1999). "The development of a matrix-assisted laser desorption/ionization time-of-flight mass spectrometry for the detection of biological warfare agent aerosols." *J Hopkins APL Tech D* 20(3):343.

Singh, A., Kuhad, R. C., Sahai, V., and Ghosh, P. (1994). "Evaluation of biomass." *Adv Biochem Eng Biotechnol* 51:48–66.

Sobsey, M. D., Battigelli, D. A., Shin G. -A., and Newland, S. (1998). "RT-PCR amplification detects inactivated viruses in water and wastewater." *Water Sci Technol* 38(12):91–94.

Stark, K. D. C., Nicolet, J., and Frey, J. (1998). "Detection of *Mycoplasma hyopneumoniae* by air sampling with a nested PCR assay." *Appl Environ Microbiol* 64(2):543–548.

Thompson, H. G., and Lee, W. E. (1992). "Rapid immunofiltration assay of *Francisella tularensis.*" *1992 Suffield Memorandum No 1376,* Canada. 1–17.

Toze, S. (1999). "PCR and the detection of microbial pathogens in water and wastewater." *Water Res* 33(17):3545–3556.

Wijesuriya, D. C., Anderson, G. P., and Ligler, F. S. (1994). "A rapid and sensitive immunoassay for bacterial cells. Report No. ERDEC-SP-024." *ERDEC Scientific Conference on Chemical Defense Research,* Aberdeen Proving Ground, MD.

Yan, Z. H., Rogers, S. O., and Wang, C. J. K. (1995). "Assessment of *Phialophora* species based on ribosomal DNA internal transcribed spacers and morphology." *Mycologia* 87:72–83.

Zhou, S., Lu, Y., Chan, W. T., Burnett, J., and Tang, P. L. (1998). "Detection of Legionella spp. in water and air by PCR amplification of rDNA and mip gene." *Zhongshan Daxue Xuebao/Acta Scientiarum Natralium Universitatis Sunyatseni* 37(6):127–128.

CHAPTER 25

INDOOR LIMITS AND GUIDELINES

25.1 INTRODUCTION

The basis for any system designed to control the aerobiology of indoor air needs to rest on firmly established limits for airborne concentrations of the various microbial contaminants. Unfortunately, this is one of the most difficult questions to answer with any precision due to a general lack of clinical or empirical data defining the lethal and infectious doses of airborne pathogens and allergens. Further complicating the problem is the nature of the building or occupants to which the limits must apply. The season of the year also plays a part in defining what limits are reasonably achievable. This chapter explores the variety of indoor conditions where limits may be useful, the various limits and guidelines that have been suggested, and the limits in indoor air that may be reasonably achievable. These limits are addressed in a generally applicable form; for levels and limits applicable to specific types of indoor environments refer to Chaps. 28 to 40.

25.2 INDOOR BIOAEROSOL LEVELS

Indoor concentrations of environmental bacteria and spores will normally be lower than in the outdoor air, except during winter, regardless of the type of ventilation system (Woods et al., 1997). Table 25.1 provides a checklist for the types of bioaerosols for which limits are needed, their primary sources, and a description of typical indoor and outdoor levels.

There are over 200 bioaerosols represented by the categories in Table 25.1 but fewer than a dozen have actual threshold dose limits or infectious doses definitively established (see App. A). Since so few pathogens and allergens have been studied to the point that even approximate limits can be defined, only general recommendations can be made regarding acceptable levels of fungi and bacteria indoors. For dangerous pathogens, however, no level is considered acceptable.

TABLE 25.1 Checklist of Bioaerosol Sources and Typical Levels

Bioaerosol	Primary source	Indoor levels	Outdoor levels
Viruses	Humans	Vary	None
Pathogenic bacteria	Humans	Vary	None
Environmental bacteria	Environment	Low	Seasonal
Actinomycetes	Agriculture	Low	Seasonal
Fungal spores	Environment	Seasonally high	Seasonally high
Pollen	Environment	Seasonally high	Seasonally high
Animal dander	Pets	Vary	None
Dust mites	Home	Vary	None
Insect allergens	Home	Vary	None

Occasionally conditions in a building may cause indoor levels of these microbes to exceed normal outdoor levels (Rao and Burge, 1996). Such buildings are known as amplifiers, and the problem may be the result of moisture intrusion, water damage, or contamination in the ventilation system.

25.2.1 Airborne Fungal Spores

Airborne fungal spores are probably the most commonly studied health hazard in indoor environments and much is known about the levels, mixtures, and health problems associated with high concentrations. Noncommunicable respiratory diseases result primarily from fungi and the fungilike bacteria called actinomycetes. The ultimate reservoir for these microbes is outdoor soil, water, and sewage. These microorganisms grow and distribute spores seasonally. Dry, windy conditions, especially in the southwest, can enhance airborne spore levels, but outdoor spore concentrations rarely become hazardous. Upon entrance into ventilation systems, spores germinate and grow in the presence of moisture and nutrients. In general, only excessive growth, or amplification, and the generation of new spores will cause indoor airborne levels to exceed outdoor levels and endanger building occupants. The composition of fungal species indoors tends to reflect that of the outdoors (Burge, 1990). Some fungal species, most notably *Aspergillus* and *Penicillium,* are often found to account for 80 percent of indoor spores (Woods et al., 1997). The most common indoor molds are shown in Fig. 25.1, based on averages taken from several studies (Kemp et al., 1997; Li and Kendrick, 1995; Godish, 1995; Meldrum et al., 1993; Yang et al., 1993; Hyvarinen et al., 1993).

Indoor airborne spore levels can differ from outdoor air in both concentration and composition. In normal buildings spore levels tend to be anywhere from 10 to 100 percent of outdoor spore levels, and are mostly less than 200 cfu/m^3 (Godish, 1995). Problem-free multistory office buildings typically have ratios of indoor to outdoor spores of about 0.1 to 0.31.

A study on fungal allergens in indoor air estimated the threshold levels for causing symptoms were 3000 cfu/m^3 for *Cladosporium* (Gravesen, 1979). Morey et al. (1986) suggest an upper limit of 1000 cfu/m^3 for fungal spores and found levels well above this during sampling of an office building.

In one study of indoor airborne spore counts, urban residences had the lowest levels and rural residences the highest. Moisture problems in the urban housing did not increase total viable spore counts but affected the airborne composition. Spore counts were much higher in old houses than in new ones. *Penicillium* was the most common airborne fungal spore in

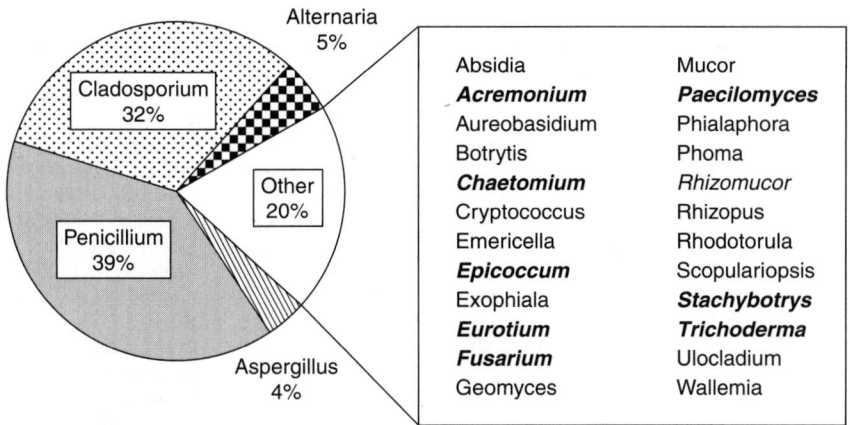

FIGURE 25.1 The most common indoor molds. Based on averages from several studies. Bold italics signify those genera that include toxigenic species.

all the residences studied. *Aspergillus* and *Cladosporium* spores were more common in damp residences and old rural houses than in the other residences (Pasanen, 1992).

Studies suggest that exposure to dampness and mold in homes is a significant risk factor for a number of respiratory symptoms (Dales et al., 1991). There is evidence that other consequences of exposure to spores, such as exposure to mycotoxins, may be an important factor associated with symptoms (Miller, 1992). Spore counts alone, therefore, may not paint a complete picture of fungal health hazards. Another potential distorting factor is that fungal fragments, which may not register as viable fungi in air samples, may contribute to human health effects to a potentially significant degree (Gorny et al., 2002).

Flood damage can result in elevated levels of airborne microorganisms within buildings and contribute to respiratory diseases (Jarvis, 2002). The results of a survey of airborne microorganisms in flooded and nonflooded houses indicated that total indoor bacterial counts in flood-damaged houses could exceed outdoor counts by two or more orders of magnitude, and fungal counts were three to four times higher than concurrent outdoor samples (Fabian et al., 2000). Theoretical research suggests that indoor sources will tend to increase indoor concentrations continuously (Kulmala et al., 1999).

The lower threshold at which fungal and bacterial spores may present a potential issue has yet to be fully explored. One study suggests that levels above 300 cfu/m^3 of nontoxigenic or nonpathogenic organisms should warrant further investigation (Robertson, 1998). This study also suggests that no organism should individually contribute more than 50 cfu/m^3 to the whole count, excepting *Cladosporium*.

The effect of indoor activities tends to increase fungal spore concentrations and size distributions. In a study by Reponen et al. (1992), the aerosolization of fungal spores by normal domestic activities, including handling of organic materials like food and firewood, cleaning activities, and normal movement through the house, had an impact on the spore counts. The concentration of the largest sized spores decreased rapidly after activity due to faster gravitational settling. A similar study by Lehtonen et al. (1993) found that resuspension of spores occurred from various domestic activities but that vacuum cleaning caused no marked changes in airborne spore concentrations.

Table 25.2 summarizes the results of a number of studies on the airborne concentrations of fungal spores in residences and office buildings, for both normal and problem buildings.

TABLE 25.2 Typical Fungal Spore Levels in Indoor Air

Concentration, cfu/m^3	Building type	Season	Condition	Reference
150	Residential	Summer	Normal	Miller et al., 1988
200	Residential	All	Normal	Godish, 1995
300	Residential	All	Normal	Robertson, 1998
500	Residential	Winter	Normal	Reynolds et al., 1990
1150	Residential	—	Normal	Kozak et al., 1985
10–100	Office buildings	Winter	Normal	Pastuszka et al., 2000
10–1000	Residential	Summer	Normal	Pastuszka et al., 2000
10–500	Residential	All	Normal	ACGIH, 1973
250–1000	Residential	All	Normal	Li and Kuo, 1992
3641	Residential	—	Problem	Kozak et al., 1985
1000–10,000	Residential	Summer	Problem	Pastuszka et al., 2000
74–1531	Residential	—	Problem	Miller, 2001
20–360	Office buildings	—	Water damaged	Sudakin, 1998

Of course, atopic, asthmatic, and hypersensitive individuals who are susceptible to allergens like pollen and fungal spores need higher levels of indoor air cleanliness, and these typical levels may pose problems for them (Cormier, 1998; Brooks, 1998).

25.2.2 Airborne Bacteria

Numerous studies have been performed on the levels of microbes found in indoor environments, including problem buildings. One of the first studies ever performed suggested an indoor limit for indoor bacteria of 700 cfu/m^3 while the threshold at which remediation was warranted is 1775 cfu/m^3 (Bourdillon et al., 1948). In a much later study suggested that levels above 1700 cfu/m^3 were seldom found indoors (Wright et al., 1969). In a study by Pastuszka et al. (2000) of buildings in Poland found that the typical level of bacterial aerosol indoors is about 1000 cfu/m^3 in homes and 100 cfu/m^3 in offices. *Micrococcus* species were found in all homes studied, constituting 36 percent of the bacterial genera with *Staphylococcus epidermidis* present in 76 percent of homes and constituting 14 percent of the total. In a study of office buildings in Hong Kong, airborne bacteria concentrations were recorded as high as 2912 cfu/m^3 during start-up of the HVAC systems in the morning (Law et al., 2001).

Bacterial levels in indoor air tend to correlate strongly with human occupation. In a study on two elementary schools in the southeastern United States, concentrations of CO_2, which relate to human occupancy and ventilation rates, correlated significantly with bacteria measurements and suggest the children and teachers are the main sources of bacterial contamination (Liu et al., 2000).

Gram-negative bacteria are possible etiological agents of respiratory diseases like hypersensitivity pneumonitis and are often singled out for study and investigation in problem homes since health problems associated with ventilation systems and humidifiers have implicated gram-negative bacteria in the past (Hood, 1989). Endotoxins can come from a variety of gram-negative bacteria. Exposure to endotoxins and to their derivative lipopolysaccharide (LPS) is related to several occupational pulmonary diseases and to asthma. Studies on the inhalation of pure LPS demonstrated responses at dose levels of 5 to 50 µg, while the threshold dose at which no response occurred was less than 0.5 µg (Michel et al., 1997). Table 25.3 summarizes a selection of studies on indoor levels of airborne bacteria in occupied normal and problem buildings.

TABLE 25.3 Typical Bacterial Levels in Indoor Air

Concentration cfu/m^3	Building type	Condition	Reference
170–2,540	Residential	Normal	Nevalainen et al., 1988
140–1,200	Residential	Normal	Nevalainen et al., 1991
564–5,360	Office	Normal	Bayer and Black, 1988
35–22,000	Residential	Normal	Flannigan et al., 1999
60–12,200	Residential	Problem	Nevalainen et al., 1988

25.2.3 Airborne Viruses

The existence of airborne viruses can be inferred from epidemiological studies on epidemics and a variety of laboratory studies. However, no viruses have ever been isolated in normal indoor environments. Viruses have been isolated in the air in several test facilities and in animal laboratories. Airborne concentrations of viruses that may occur in indoor environments can only be inferred or guessed at based on clinical studies. No data exist that would allow estimation of typical airborne levels of viruses. It can be stated categorically, of course, that no levels of pathogenic viruses should be tolerated in indoor environments and therefore there are no limits for viruses that can be proposed.

25.2.4 Airborne Pollen

The available data on indoor pollen levels are barely sufficient to establish limits but for those with allergies to pollen levels must of necessity be kept as low as possible. Spiegelman and Friedman (1968) measured the range of indoor pollen at 6 to 600 grains/m^3 in a residential home with unfiltered air during summer. With the air filtered, the pollen levels were reduced to 1 to 19 grains/m^3. It could be assumed, based on this limited data, that a mean level of less than about 10 grains/m^3 might be acceptable as a limit, although more study is needed in this area.

25.2.5 Airborne Allergens

Fungal spores have been addressed separately, but there are no specific suggestions provided in the literature for limits on dander, insect allergens, or other respiratory irritants. It could be assumed, however, that any indoor environment that meets the limits suggested herein for bacteria, pollen, and fungal spores will likely have considerably reduced levels of airborne allergens also, especially if this result is achieved through air cleaning.

25.3 INDOOR/OUTDOOR RATIOS

The ratio of indoor levels to outdoor levels provides an indicator of the aerobiological quality of the indoor air. In general, bacterial levels will be higher than outdoors due to the fact that occupants generate airborne bacteria and because outdoor bacterial levels are typically moderate except in some agricultural areas. Pollen should be kept considerably lower indoors than outdoors but there are insufficient data available on indoor pollen levels on

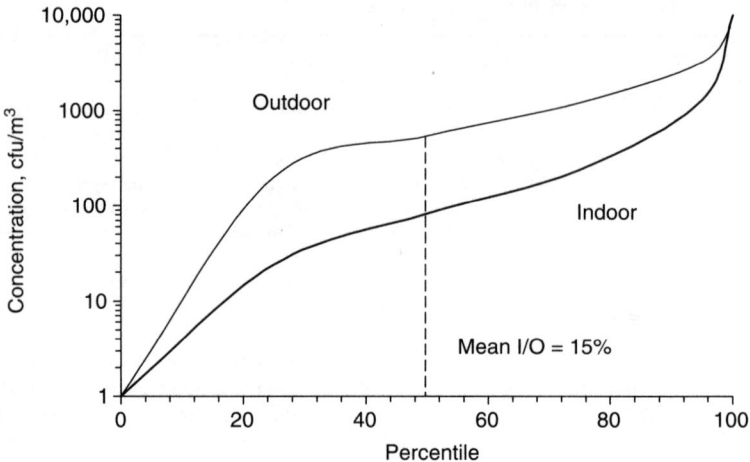

FIGURE 25.2 Levels of spores in indoor and outdoor air as a percentile of cases. [*Based on data from Shelton et al. (2002).*]

which to base an estimate of a normal *indoor/outdoor* (I/O) ratio. There is an abundance of data on indoor and outdoor fungal spore levels and these allow estimation of acceptable indoor/outdoor ratios.

Figure 25.2 shows a summary of the levels of spores in both outdoor and indoor air measured in a study that encompassed over 1700 buildings in the United States (Shelton et al., 2002). Clearly, the norm is for indoor air to have lower levels of fungi than outdoor air, and taking the mean as a basis, the normal I/O ratio is approximately 0.15, which could be used as a guideline for assessing indoor aerobiological quality in conjunction with indoor levels.

We could expect a healthy building to have indoor levels of fungi that are significantly lower when the outdoor air contains high levels of spores. In cold winter climates, however, outdoor levels may decrease to single digits (i.e., 1 to 10 cfu/m^3) and the I/O ratio may easily exceed unity. Therefore the I/O ratio is a seasonal indicator of air quality only. Figure 25.3

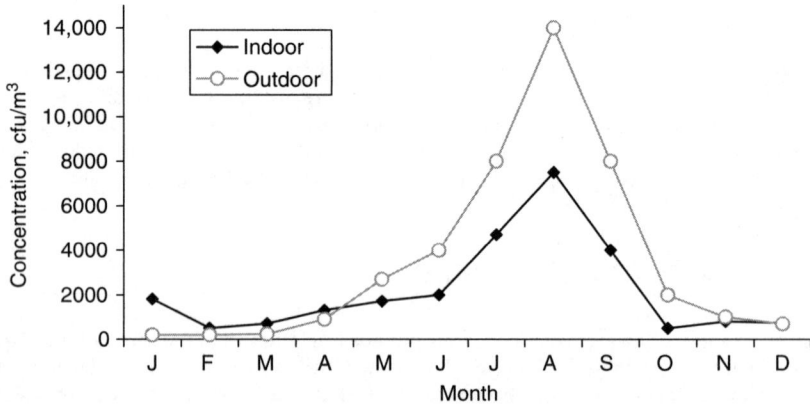

FIGURE 25.3 Monthly variation of indoor and outdoor spores. [*Based on data from Li and Kendrick (1995).*]

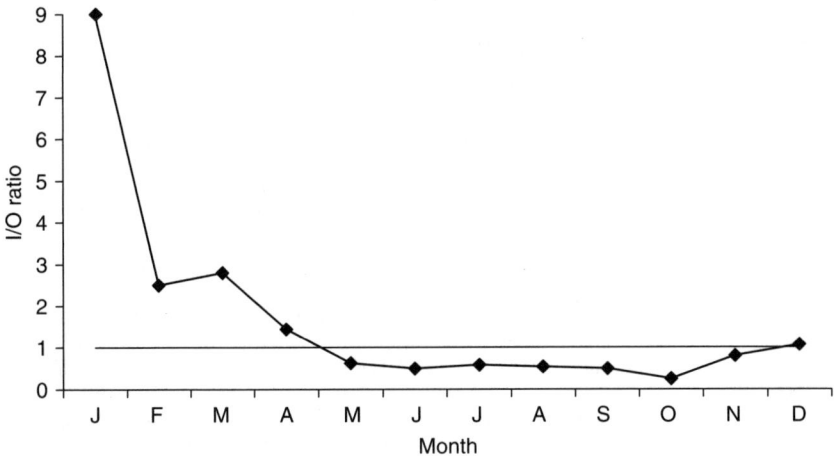

FIGURE 25.4 Monthly variation in indoor/outdoor ratio. [*Based on data from Li and Kendrick (1995).*]

illustrates the typical variation of indoor and outdoor airborne concentrations of spores over the course of the year.

Figure 25.4 shows the indoor/outdoor ratio for the data in Fig. 25.3. Note that in the winter months the indoor levels exceed the outdoor levels.

Some suggested standards for indoor/outdoor fungal spore ratios have been made predicated on the assumption that the indoor mix of fungal species does not differ significantly from the outdoor species and that no significant toxigenic species are present indoors. An I/O ratio of less than 1 was suggested as an acceptable value by ACGIH (1989). Etkin (1994) suggested that the I/O range of acceptability is 0.1 to 1.0. AIHA (1989) suggested that any high I/O ratio indicated indoor amplification.

25.4 INDOOR LIMITS

Based on the previous summary of typical indoor spore levels in Table 25.3 it could be estimated that the normal level of fungal spores indoors in healthy homes during winter is between 10 and 500 cfu/m^3 and during summer 10 and 1000 cfu/m^3. This estimate ignores, of course, the actual composition of spores, but it is assumed the mixture contains no dangerous pathogenic fungi. Various guidelines and studies suggest that upper limits for indoor environments be about 100 cfu/m^3 or some value lower than local outdoor conditions (Woods et al., 1997; Rao and Burge, 1996). Table 25.4 lists the results of various studies that include suggested limits for indoor levels in various buildings. Levels exceeding these limits do not necessarily pose a health threat.

Limited definitions exist for which pathogens constitute dangerous health hazards and which do not. This can often depend on who the victim may be, as well as the dose. The elderly may be highly susceptible to pneumonia from influenza while the majority of healthy people can withstand the flu without treatment. Tuberculosis is a danger to everyone. Any microbe, even commensals, may be dangerous to the immunocompromised. It has been suggested that no level of dangerous pathogens is acceptable in the indoor environment (AIHA, 1989; Rao and Burge, 1996), and hence no organization will sanction any

TABLE 25.4 Prior Suggested Airborne Microbial Limits in Indoor Air

Concentration, cfu/m³	Building type	Microbe	Notes	Reference
4500	Residential	Bacteria	Nonpathogenic	Flannigan et al., 1991
500–1000	Offices	Bacteria	Nonpathogenic	EPD, 1999
1775	Offices	Bacteria	Investigation threshold	Bourdillon et al., 1948
100	All	Fungi	—	ACGIH 1989
100	Schools	Fungi	—	Haverinen et al., 1999
150	Residential	Fungi	Nonpathogenic	Flannigan et al., 1991
200	Residential	Fungi	Low	CEC, 1994
200	Residential	Fungi	No investigation	CMHC, 1991
300	Residential	Fungi	Common fungi	IAQA, 1995
500	Residential	Fungi	Abnormal	Reynolds et al., 1990
1000	All	Fungi	Atypical	AIHA, 1989
1000	Residential	Fungi	Contamination	OSHA, 1992

levels above zero. Such trepidation may not be conducive to finding appropriate solutions, however. Anthrax, for example, can shut down buildings if even a few dozen spores are present, yet cattle ranchers may breathe more than this amount every day without ill effects. Even deadly pathogens must have some limit below which they are not a true health threat to healthy people.

Indoor airborne bacteria are typically either human commensals or environmental bacteria brought in with fresh air, and are, in general, not considered dangerous pathogens except to the immunocompromised. Previously, Flannigan et al. (1991) suggested 4500 cfu/m³ as a limit for bacteria in the indoors while another source recommended a limiting range of 500 to 1000 cfu/m³ (EPD, 1999). These limits do not apply to pathogens and toxic species.

Based on the preceding summaries of normal and abnormal levels of bioaerosols in indoor environments, and the summary of proposed limits, some limits can be suggested for guidance in assessing indoor bioaerosol problems and in evaluating the performance of air cleaning systems. Table 25.5 summarizes descriptive categories and general limits of bioaerosols suggested by the author. These limits do not apply to dangerous pathogens or potentially toxic allergens. These levels do not supersede other suggested limits provided in the specific applications in Chaps. 28 to 40, not do they supersede any requirements that are or may be issued by any other authorities. These limits are, furthermore, not guaranteed

TABLE 25.5 Suggested indoor Bioaerosol Assessment Categories and Guidelines

Defined ranges of total, cfu/m³	Bioclean	0–10
	Very low	10–100
	Low	100–300
	Normal	300–1,000
	High	1,000–10,000
	Extreme	10,000–100,000
Suggested limit for bacteria	Maximum, cfu/m³	500
Suggested limit for fungal spores	Maximum, cfu/m³	150
Suggested I/O ratio	Subject to above limit	0.1
Suggested limit for pollen	Maximum, g/m³	10

Note: These recommendations do not supercede those for any specific applications.

to aid those who are atopic, allergic, or hypersensitive, and for such occupants it is recommended that standards provided for health care facilities dictate the design objective.
The defined categories in Table 25.5 are provided to facilitate classification of indoor aerobiological quality. Buildings in the low to normal categories, for example, would be acceptable for ordinary residences and office buildings, whereas the categories of Bioclean and Very low might be more appropriate for hospitals and homes of the immunocompromised. The category of High may be acceptable for all but atopic individuals, but this range is a matter for further research. The category of Extreme may signal potential problems that warrant investigation.

REFERENCES

ACGIH (1973). *Threshold Limit Values of Physical Agents*. The American Conference of Government Industrial Hygienists, Cincinnati, OH.

ACGIH (1989). *Guidelines for the Assessment of Bioaerosols in the Indoor Environment*. American Conference of Government Industrial Hygienists, Cincinnati, OH.

AIHA (1989). "The practitioner's approach to indoor air quality investigations." *Indoor Air Quality International Symposium,* Fairfax, VA. 43–66.

Bayer, C. W., and Black, M. S. (1988). "IAQ evaluations of three office buildings." *ASHRAE J* 30(7):48–53.

Bourdillon, R. B., Lidwell, O. M., Lovelock, J. E., and Raymond, W. F. (1948). "Airborne bacteria found in factories and other places: Suggested limits of bacterial contamination." *Special Report Series No. 262*. Medical Research Council, London.

Brooks, S. M. (1998). Chapter 33: Occupational and environmental asthma. *Environmental & Occupational Medicine*, W. H. Rom, ed., Lippincott-Raven Publishers, Philadelphia, PA. 481–524.

Burge, H. (1990). "Bioaerosols: Prevalence and health effects in the indoor environment." *J Allerg Clin Immunol* 86(5):687–781.

CEC (1994). *Biological Particles in Indoor Environments. 12* Commission of the European Communities. Luxembourg, Belgium.

CMHC (1991). *Testing of Older Houses for Microbiological Pollutants*. I. Bowser Technical Canadian Mortgage and Housing Corp., Ottawa, Canada.

Cormier, Y. (1998). Chapter 31: Hypersensitivity pneumonitis. *Environmental & Occupational Medicine*, W. H. Rom, ed., Lippincott-Raven Publishers, Philadelphia, PA. 457–465.

Dales, R. E., Burnett, R., and Zwanenburg, H. (1991). "Adverse health effects among adults exposed to home dampness and molds." *Am Rev Resp Dis* 143:505–509.

EPD (1999). "Guidance notes for the management of indoor air quality in offices and public places." Indoor Air Quality Management Group, The Hong Kong Government of Special Administrative Region, Environmental Protection Department. Hong Kong.

Etkin, D. S. (1994). *Indoor Air Quality Update: Biocontaminants in Indoor Environments*. Cutter Information Corp., Arlington, MA.

Fabian, M. P., Reponen, T., Miller, S. L., and Hernandez, M. T. (2000). "Total and culturable airborne bacteria and fungi in arid region flood-damaged residences." *J Aerosol Sci* 31(Suppl. 1):S35–S36.

Flannigan, B., McCabe, E. M., and McGarry, F. (1991). Allergenic and toxigenic microorganisms in houses. *Pathogens in the Environment,* B. Austin, ed., Blackwell Scientific Publications, Oxford, UK.

Flannigan, B., McEvoy, E. M., and McGarry, F. (1999). "Investigation of airborne and surface bacteria in homes." *Indoor Air 99: Proceedings of the 8th International Conference on Indoor Air Quality and Climate*, Edinburgh, Scotland. 884–889.

Godish, T. (1995). *Sick Buildings : Definition, Diagnosis and Mitigation*. CRC-Lewis Publishers, Boca Raton, FL.

Gorny, R. L., Reponen, T., Willeke, K., Schmechel, D., Robine, E., Boissier, M., and Grinshpun, S. A. (2002). "Fungal fragments as indoor air biocontaminants." *Appl Environ Microbiol* 68(7):3522–3531.

Gravesen, S. (1979). "Fungi as a cause of allergenic disease." *Allergy* 34:135–154.

Haverinen, U., Husman, T., Toivola, M., Suonketo, J., Pentti, M., Lindberg, R., Leinonen, J., Hyvarinen, A., Meklin, T., and Nevalainen, A. (1999). "An Approach to management of critical indoor air problems in school buildings." *Environ Health Perspect* 107(Suppl. 3):509–513. http://ehp.niehs.nih.gov/members/1999/suppl-3/509-514haverinen/haverinen-full.html.

Hood, M. A. (1989). "Gram negative bacteria as bioaerosols." *ASTM Special Technical Publication* 1071:60–70.

Hyvarinen, A., Reponen, T., Husman, T., Ruuskanen, J., and Nevalainen, A. (1993). "Composition of fungal flora in mold problem houses determined with four different methods." *IAQ '93'*, Helsinki, Finland. 273–278.

IAQA (1995). *Indoor Air Quality Standard #95-1 recommended for Florida.* Indoor Air Quality Association, Longwood, FL.

Jarvis, B. B. (2002). Chemistry and toxicology of molds isolated from water-damaged buildings. *Mycotoxins & Food Safety,* Kluwer Academic/Plenum Publishers, Norwell, MA.

Kemp, P. C., Neumeister, H. G., Kircheis, U., Schleibinger, H., Franklin, P., and Ruden, H. (1997). "Fungal genera in an office building with a central HVAC system in an Australian mediterranean climate." *Healthy Buildings/IAQ '97',* Bethesda, MD. 257–260.

Kozak, P. P. J., Gallup, J., Cummins, L. H., and Gillman, S. A. (1985). Endogenous mold exposure: Environmental risk to atopic and nonatopic patients. *Indoor Air and Human Health,* R. B. Gammage and S. V. Kaye, eds., Lewis Publishers, Chelsea, MI.

Kulmala, M., Asmi, A., and Pirjola, L. (1999). "Indoor air aerosol model: The effect of outdoor air, filtration and ventilation on indoor concentrations." *Atmos Environ* 33(14):2133–2144.

Law, A. K. Y., Chau, C. K., and Chan, G. Y. S. (2001). "Characteristics of bioaerosol profile in office buildings in Hong Kong." *Building Environ* 36(4):527–541.

Lehtonen, M., Reponen, T., and Nevalainen, A. (1993). "Everyday activities and variation of fungal spore concentrations in indoor air." *Int Biodet Biodeg* 31(1):25–39.

Li, C., and Kuo, Y. (1992). "Airborne characterization of fungi indoors and outdoors." *J Aerosol Sci* 23(S1):s667–s670.

Li, D.-W., and Kendrick, B. (1995). "A year-round comparison of fungal spores in indoor and outdoor air." *Mycologia* 87(2):190–195.

Liu, L.-J. S., Krahmer, M., Fox, A., Feigley, C. E., Featherstone, A., Saraf, A., and Larsson, L. (2000). "Investigation of the concentration of bacteria and their cell envelope components in indoor air in two elementary schools." *J Air Waste Manage Assoc* 50(11):1957–1967.

Meldrum, J. R., O'Rourke, M. K., Boyer-Pfersdorf, P., and Stetzenbach, L. D. (1993). "Indoor residential mold concentrations as represented by spore and colony counts." *IAQ '93',* Helsinki, Finland. 189–194.

Michel, O., Nagy, A. M., Schroeven, M., Duchateau, J., Neve, J., Fondu, P., and Sergysels, R. (1997). "Dose-response relationship to inhaled endotoxin in normal subjects." *Am J Respir Crit Care Med* 156(4 Pt. 1):1157–64.

Miller, J. D., Laflamme, A. M., Sobol, Y., Lafontaine, P., and Greenlaugh, R. (1988). "Fungi and Fungal Products in some Canadian Houses." *Int Biodeterior* 24:103–120.

Miller, J. D. (1992). "Fungi as contaminants in indoor air." *Atmos Environ* 26A(12):2163–2172.

Miller, J. D. (2001). *Mycological investigations of indoor environments.* Microorganisms in Home and Work Environments, B. Flannigan, R. A. Samson, and J. D. Miller, eds., Taylor & Francis, London. 218–231.

Morey, P. R., Hodgson, M. J., Sorenson, W. G., Kullman, G. J., Rhodes, W. W., and Visvesvara, G. S. (1986). "Environmental studies in moldy office buildings." *ASHRAE Trans* 92(1B):399–416.

Nevalainen, A., Jantunen, M. J., Rytkonen, M., Niininen, M., Reponen, T., and Kalliokoski, P. (1988). The indoor air quality of Finnish homes with mold problems. *Healthy Buildings,* B. Petterson and T. Lindvall, eds., Swedish Council for Building Research, Stockholm. 319–323.

Nevalainen, A., Reponen, T., Heinonen-Tanski, H., and Raunemaa, T. (1991). Indoor air bacteria in apartment homes before and after occupancy. *Healthy Buildings/IAQ '91'*, Washington, DC.

OSHA (1992). *OSHA Technical Manual.* Occupational Health and Safety Administration, U.S. Department of Labor, Washington, DC. http://www.osha-slc.gov/dts/osta/otm/otm_toc.html.

Pasanen, A.-L. (1992). "Airborne mesophilic fungal spores in various residential environments." *Atmos Environ* 26A(16):2861–2868.

Pastuszka, J. S., Paw, U., Kyaw, T., Lis, D. O., Wlazlo, A., and Ulfig, K. (2000). "Bacterial and fungal aerosol in indoor environment in Upper Silesia, Poland." *Atmos Environ* 34(22):3833–3842.

Rao, C. Y., and Burge, H. A. (1996). "Review of quantitative standards and guidelines for fungi in indoor air." *J Air Waste Manage Assoc* 46(Sep.):899–908.

Reponen, T., Lehtonen, M., Raunemaa, T., and Nevalainen, A. (1992). "Effect of indoor sources on fungal spore concentrations and size distributions." *J Aerosol Sci* 23(Suppl. 1):S663–S666.

Reynolds, S. J., Streifel, A. J., and McJilton, C. E. (1990). "Elevated airborne concentrations of fungi in residential and office environments." *Am Ind Hyg Assoc J* 51:601–604.

Robertson, L. D. (1998). "Monitoring viable fungal and bacterial bioaerosol concentrations to identify acceptable levels for common indoor environments." *44th Annual Technical Meeting,* Phoenix, AZ.

Shelton, B. G., Kirkland, K. H., Flanders, W. D., and Morris, G. K. (2002). "Profiles of airborne fungi in buildings and outdoor environments in the United States." *Appl Environ Microbiol* 68(4):1743–1753.

Spiegelman, J., and Friedman, H. (1968). "The effect of air filtration and air conditioning on pollen and microbial contamination." *J Allergy* 42(4):193–202.

Sudakin, D. L. (1998). "Toxigenic fungi in a water-damaged building: An intervention study." *Am J Ind Med* 34:183–190.

Woods, J. E., Grimsrud, D. T., and Boschi, N. (1997). *Healthy Buildings/IAQ '97'*. ASHRAE, Washington, DC.

Wright, T. J., Greene, V. W., and Paulus, H. J. (1969). "Viable microorganism in an urban atmosphere." *J Air Pollution Contr Assoc* 19:337–341.

Yang, C. S., Hung, L.-L., Lewis, F. A., and Zampiello, F. A. (1993). "Airborne fungal populations in nonresidential buildings in the United States." *IAQ '93',* Helsinki, Finland. 219–224.

CHAPTER 26

REMEDIATION

26.1 INTRODUCTION

The discovery of a microbiological problem in a building by air sampling, surface sampling, by inspection, or following occupant complaints may warrant immediate remediation. Typically, remediation addresses mold or bacterial contamination due to water damage or excessive moisture. It can also be necessitated as the result of intentional acts of bioterrorism with agents like anthrax. Remediation may involve air cleaning, surface cleaning, physical removal of contaminated building materials, or surface decontamination of the whole building. In some cases the cost of remediating a building proves to be prohibitively expensive and the building must be torn down. Figure 26.1 illustrates the basic types of remediation strategies, excluding the building destruction option.

These remediation methods are discussed in the following sections for the purpose of familiarizing engineers with the basic techniques and their applicability. More detailed information on providing such services is available in the indicated references.

Water damage and dampness in buildings are perhaps the greatest contributors to serious mold problems in buildings. Indoor humidity is a minor or indirect contributor to the problem of mold growth and indoor air quality (Arens and Baughman, 1996). Dampness in buildings can be a major cause of mold problems and is highly correlated with health problems, especially in children (Dales et al., 1991a, 1991b; Verhoeff et al., 1992; Oliver, 1988). In one study on water damage, approximately, 90 percent of naturally ventilated apartments that had suffered various kinds of water damage had significant amounts of fungi in wall cavities (Miller et al., 2000). Readers wishing to learn more about prevention of dampness and water damage should consult EPA (2002, 2003), Oliver (1988), Rivin (2001), AIA (1993), Arens and Baughman (1996), and Nathan et al. (2002).

26.2 AIR CLEANING

The use of air cleaning technologies to remediate an aerobiological problem in a building may be an effective approach in some cases but it is not a complete solution if it does not resolve the root cause of the problem. Obviously, any of the air cleaning technologies

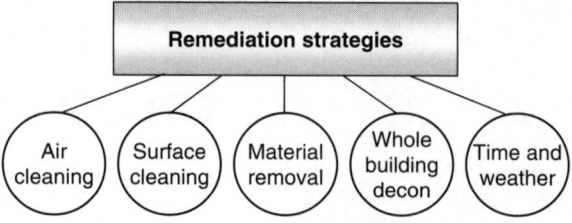

FIGURE 26.1 Breakdown of basic building remediation strategies.

presented in previous chapters, such as filtration, UVGI, or air dilution, can reduce indoor air contaminants but the achievement of acceptable indoor air quality depends on whether the removal rate is sufficient to overcome the indoor microbial generation rate that is causing the problem. The actual problem may be indoor mold growth on damaged materials, growth of mold or bacteria on air handling equipment like cooling coils, insufficient dilution ventilation or poor mixing, or even an outdoor contaminant source.

Air cleaning should never be used as an alternative to a proper remediation program if it should be warranted by the degree of the health threat. Engineers should beware, furthermore, that companies providing remediation of air quality problems without either investigating or addressing potentially serious causes of air quality problems may not be providing anything but a "band aid" which might do little more than cover up a problem. Companies that provide such air remediation services, sometimes covertly, to buildings that are involved in related litigation are, in fact, remediating the litigation and not the problem. Investigators involved in such litigation should take note of any technologies that may have been retrofit in a ventilation system to hide air quality problems after initial reports of health problems. Such actions can lower the airborne levels of spores and bacteria and shift focus from the true cause of reported health problems, thus obfuscating air quality investigations. Ozone systems, for example, might be used to purge a building of contaminants overnight, and then be removed or turned off prior to official air sampling. Such practices border on the unethical if they result in claimants being denied compensation for damages due to problem buildings.

26.3 SURFACE CLEANING

Environmental fungi and bacteria can become a potential *indoor air quality* (IAQ) problem when adequate moisture and nutrient are present in building materials. Manual cleaning of building surfaces and ventilation systems can be an effective method of remediation provided building materials are not extensively or internally damaged.

Because of their potential to rapidly spread contamination throughout a building, ventilation systems are of particular significance as potential microbial contamination sources. Fungal spore levels on HVAC duct surfaces could be substantially reduced by thorough vacuum cleaning (Foarde et al., 1997). HVAC components are often cited as bioaerosol and volatile organic compound emission sources (Batterman and Burge, 1995). A variety of methods are currently available for cleaning HVAC systems, including bleach scrubbing, vacuuming, and automatic scouring systems.

Biological contamination is often dealt with by first washing the area with water. Many agents have some degree of solubility in water but water with detergents added can be effective for removing most CW agents. Some agents are neutralized by water through hydrolysis. One of the problems with using water is that it will become a contaminated material and must be disposed of properly. Remediation can be further pursued through the use of bleaches, disinfectants, and manual scrubbing of the contaminated surfaces, including ductwork, with disinfectants (Pasanen et al., 1993; Eweis et al., 1998).

In a review of duct materials contaminated with fungi, Foarde et al. (1997) suggested that mold growth on heating, ventilating, and air-conditioning components could be substantially reduced by thorough contact vacuuming.

A study was conducted by Ahmad et al. (2001) to determine the effectiveness of three commercial HVAC duct cleaning processes in reducing airborne particulates and viable bioaerosols. The three sanitation processes included conventional vacuum cleaning of interior duct surfaces, the use of compressed air to dislodging dirt and debris, and insertion of a rotary brush into the ductwork to dislodge debris. The effectiveness of these sanitation processes was evaluated in terms of airborne particulates and viable bioaerosol concentrations in eight identical homes in the same neighborhood. It was found that both particle count readings and bioaerosol concentrations were higher during cleaning. Particle counts at a mean size of 1 µm were reduced as a result of the HVAC duct cleaning. Bioaerosol concentrations measured 2 days after cleaning were found to be lower than the prelevel concentrations. Homes cleaned with the compressed air method showed the highest reduction of bioaerosol concentrations of the three methods investigated.

26.4 REMOVAL OF BUILDING MATERIALS

Careful design, construction, and maintenance can eliminate building interior fungal growth. In the event fungal growth does occur, fixing defects in the building envelope and correction of building envelope system component problems is critical to remediating mold problems and preventing recurrence (Rivin, 2001). Concealed and unabated water damage may allow unlimited mold growth and spread in indoor environments and so any water-damaged materials ought to be removed and either discarded or replaced. Often in cases of water damage it is necessary to remove not only furnishings but also building materials such as walls and floorboards. The cost of removing and discarding building materials may often be less than attempting to clean the materials in place, an alternative that does not always succeed. Needless to say, if the expedient of cleaning heavily overgrown materials in-place fails, the long-term costs can be considerable.

Laks et al., (2002) investigated the fungal susceptibility of interior commercial building panels and found that the decay and mold resistance of wood-based building panels can impact the extent of structural damage after flooding or failure of the exterior building envelope, affecting indoor air quality in turn. Four common untreated commercial sheathing panels were evaluated for decay resistance. The most resistant was Douglas-fir plywood, followed by southern pine plywood, then southern pine *oriented strandboard* (OSB), with the least resistant being aspen OSB. A commercial paper-surfaced gypsum board was also evaluated and found to be as susceptible to surface mold growth as pine plywood. All of the tested materials were prone to fungal decay and mold growth and none could be considered resistant to fungal degradation.

26.5 WHOLE BUILDING DECONTAMINATION

Several decontamination technologies have recently developed as a result not only of mold contamination but because of bioterrorism concerns. The remediation methods discussed here are, at best, developmental technologies and only limited engineering guidelines or design information can be provided at present. Therefore, these methods are summarized only in terms of their present or hypothetical use. If new disinfectants are developed in the future for such applications, it is likely they would be applied in a manner similar to those described here, and so these techniques can be considered to be generally applicable regardless of the agent used. The use of UVGI as a surface disinfectant is not addressed here since it has been adequately covered in Chap. 11.

A number of chemical decontaminants are available that can be effective against biological agents. These chemicals are available in various forms, including liquids, powders, foams, and vapors or gases. Chemicals specifically used for disinfection of biological agents are known as disinfectants, and only the most common or representative disinfectants are covered here, since there are a large number of them. Figure 26.2 provides a breakdown of most of the basic technologies used for decontaminating chemical and biological agents on surfaces, for cleaning up chemical spills, or for remediating contaminated buildings. Each of these is addressed in the following sections.

26.5.1 Solvents and Detergents

A variety of solvents and detergents can be used to either attenuate a microbiological contamination problem or to remove the mold or contaminant (EPA, 2002). Water can be used, for example, to prevent a powdered microbiological agent from becoming aerosolized. Of course, none of these solvents will sterilize or disinfect the agent, and so ultimately these are merely expedient or temporary remedial measures that might be undertaken until a true disinfection measures can be brought to bear on the problem.

26.5.2 Thermal Removal Methods

Hot air or steam can be used to decontaminate surfaces, but the degree to which the agent is destroyed depends on its characteristics. Steam is generally considered more effective for sterilizing equipment and surfaces than hot air and is the method of choice for sterilizing most laboratory equipment (Morrissey and Philips, 1993). Steam cleaning is often used to

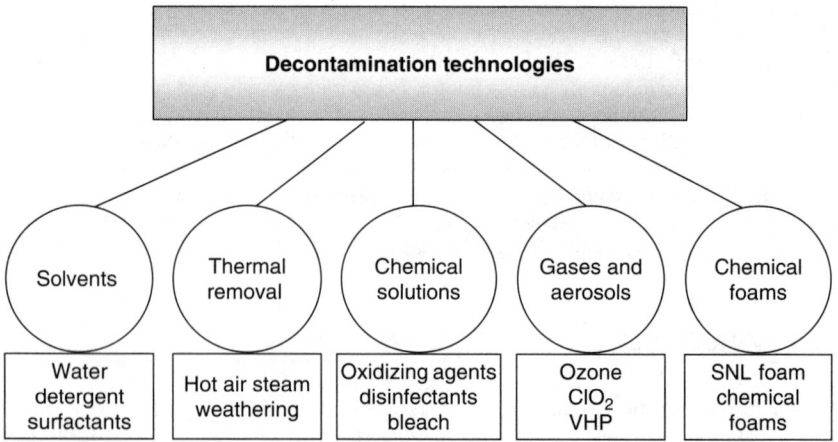

FIGURE 26.2 Breakdown of common decontamination and remediation agents and methods.

clean carpets of allergens and mold spores and studies show that it decreases spore contamination levels although not quite as effectively as bleaching (Wilson et al., 2004).

26.5.3 Chemical Disinfectants

A wide variety of chemical disinfectants are available for use in remediating mold or other biological contamination, including ethylene oxide, ethyl alcohol, formaldehyde, peracetic acid, hydrogen peroxide, formalin, sodium hypochlorite, ethanol, glutaraldehyde, and iodine (Middleton and Gilliland, 1941; Mitscherlich and Marth, 1984; Casella and Schmidt-Lorenz, 1989). Chloramine-B is an oxidant that is commonly used as an antibacterial agent (Fatah et al., 2001). Any chemical that acts as a fungicide or a biocide will also pose a hazard to humans (Popendorf and Selim, 1995).

26.5.4 Gaseous and Aerosol Fumigants

Various chemical disinfectants can be aerosolized or used in gaseous form to disinfect surfaces in rooms or entire buildings, including chlorine dioxide (ClO_2), ozone, and hydrogen peroxide, formaldehyde, ethylene oxide, hypochlorite, hydrogen peroxide, peracetic acid, and propylene glycol. Chlorine dioxide is a disinfectant that has been used to disinfect the Hart Senate Office Building in Washington, DC, of anthrax spores. This technology is also relatively new in this application and no data are available to use for prediction of disinfection rates. Chlorine dioxide is also an oxidant like ozone, but is used in both vapor and liquid form whereas ozone is always used in gaseous form. Chlorine dioxide is used for odor and taste control in water supplies. It is also used as a bleaching agent. It has a boiling point of 11°C (51.8°F) and so can be stored as a liquid when kept cold or pressurized. Toxicity to humans has not been well studied, but the U.S. *threshold limit value* (TLV) (STEL) is 0.3 ppm, or similar to that of ozone (ACGIH, 1973). In one case an exposure of a worker to 19 ppm caused fatality, but the duration of exposure was unspecified (Richardson and Gangolli, 1992).

The use of ozone as a method of disinfecting entire rooms was first suggested by Masaoka et al. (1982), but it was studied as a disinfectant of air and surfaces by several researchers much earlier (Elford and van den Eude, 1942). Ozone is a corrosive form of oxygen that is harmful to humans in concentrations above about 1 ppm. At levels of between 1 and 20 ppm it can sterilize spores in a matter of hours (Kowalski et al., 2003). It can be used at much higher levels inside enclosed rooms or buildings, with the only major side effect being the destruction of organic rubber and a few other susceptible materials. Ozone tends to break down rapidly in air, especially in sunlight or in high humidity, and so it can be vented to atmosphere after its use. It can also be removed with carbon adsorbers to some degree or it can be catalytically converted to oxygen by materials like Carulite (Carus, 2000).

Vaporized hydrogen peroxide (VHP) is a feasible method for decontamination for large areas (McDonnell, 2004). Generators are available for this disinfection scheme that operate to form a closed loop with the zone to be fumigated. Air is circulated through the generator in a four-stage fumigation process that consists of dehumidification, conditioning, decontamination, and aeration.

26.5.5 Chemical Foams

An alternative to the use of gaseous ozone or chlorine dioxide is a foam developed at Sandia National Laboratories in Albuquerque, New Mexico, under the U.S. Department of

Energy's DOE-NN-20 CBNP Program's Decontamination and Restoration Thrust Area (Modec, 2001; IITRI, 1999). The objective of this research was to develop foams, fogs, liquids, sprays, or aerosols that could be used to neutralize chemical agents and act as a microbiocide against biological agents.

26.6 TIME AND WEATHERING

Weathering is a form of passive decontamination in which natural exposure to temperature or sunlight slowly breaks down any contaminants, including biological agents. Sunlight consists of broad spectrum white light, with some UV and infrared light, and tends to break down many chemical compounds after prolonged exposure (Koller, 1952). Sunlight also has biocidal effects and will even destroy spores in time (Futter and Richardson, 1967; El-Adhami et al., 1994; Beebe 1959). In the natural environment spores usually survive in the soil or in shade, but open exposure to sunlight will eventually destroy them (Austin, 1991).

Exposing a building interior to sunlight for the purpose of destroying contaminants may be a time-consuming but nontoxic alternative to using disinfectants. One problem is that windows screen out UV and so the biocidal effect of sunlight can be enhanced by opening windows, and also by the use of reflectors to direct sunlight to shadowed areas. Open windows will also allow outside air to enter a building, and the effects of temperature extremes and variations in relative humidity also have a certain biocidal effect (Wilkinson, 1966). All pathogens have a natural decay rate that is species dependent. The natural rate of die-off can be accelerated due to exposure to the elements, and especially from temperature extremes and dehydration. Even spores will die off at some natural rate, although they may persist if they are embedded deep within building materials.

One alternative means of destroying spores is to cause them to germinate or grow and then to change the conditions. Once spores have germinated they become as vulnerable to biocidal agents as any bacteria. The way to induce germination is to provide moisture, a substrate, and warm temperatures (Sakuma and Abe, 1996). Under ideal conditions, spores can be induced to germinate within a matter of hours, after which they can be disinfected in minutes (i.e., by UVGI) when in their vulnerable growth state. They can also be destroyed by lowering or raising the temperature to destroy the mycelia, a process that is called thermal cycling.

26.7 MOLD REMEDIATION PROCEDURES

Mold remediation is a common process in buildings today that are subject to health-threatening levels of mold growth. Remediation of mold problems generally involves scrubbing with disinfectants, removal of contaminated materials, and some form of air cleaning either during remediation or after. Since the cause of the mold problem is usually water damage, parallel or follow-up work will usually include repair or improvement of the building envelope. See the images of remediation in progress in Fig. 26.3.

The following General Mold Remediation Procedures have been summarized from procedures recommended by other sources (EPA, 2001; DEHS, 2000; White, 1999; DOH, 2002). These procedures are intended to inform and provide general guidance and do not supersede any local or other applicable procedures. In some cases, permanent air treatment systems including filtration and UVGI may be implemented to assure the continued safety of the employees.

General Mold Remediation Procedures

1. Develop a plan identifying surfaces to be cleaned and contaminated materials to be removed.
2. Isolate the containment area with plastic sheeting sealed with duct tape and an exhaust fan with an integral HEPA or high-efficiency filter with integral filter to develop negative pressure in work area. Fan may vent to the local area or to outside air.
3. If space permits, create a two-stage change area and decontamination room attached to the entrance of the containment area.
4. Workers should be trained in the handling of hazardous materials.
5. Workers should be equipped with appropriate face masks (full-face negative pressure respirators with integral filters), gloves, and tyvek coveralls.
6. Provide workers with tools—cutters, pliers, crowbars, and other tools necessary to break up plasterboard walls.
7. Anyone in the surrounding areas who may be unusually susceptible to molds or allergens, such as the immunocompromised or the elderly, should be relocated while the work is in progress.
8. Apply bleach or other strong disinfectant to the contaminated areas, and wait 2 to 5 minutes to begin tearing down and removing the materials.
9. Materials in the containment that cannot be cleaned should be removed from the building in sealed plastic bags.
10. Before removing the bags from the containment area, they should be cleaned on the exterior with a detergent or disinfectant.
11. Workers exiting the containment area should wash their hands and remove their face masks and outerwear carefully so as not to recontaminate themselves.
12. Once removed from the site, bagged moldy materials can be disposed of along with normal trash.
13. Other heavy building materials may be removed in large drums, which should be sealed before taking them outside the containment.
14. After contaminated materials are removed, the area should be vacuumed and cleaned with detergent and damp cloths or mops.
15. Disengage the fan and remove the plastic sheeting in a manner that contains and encloses the remaining dust and debris.
16. The exhaust fan filter should be removed, bagged, and disposed of along with the other trash.
17. Appropriately trained personnel should then conduct visual inspection and/or air and surface sampling.

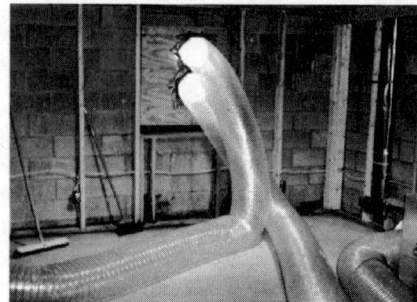

FIGURE 26.3 Mold remediation project in progress. [*Photos provided courtesy of Timothy J. Pladson, EnviroTech Remediation Services, Inc., Blaine, MN.*]

26.8 CHEMICAL DECONTAMINATION PROCEDURES

Three examples are presented here of gaseous decontamination, ClO_2, ozone, and SNL foam, although all of the previously mentioned disinfectants would probably provide the same ease of use and applicability to a wide variety of remediation problems. Disinfecting buildings with most of these other fumigants will be similar in many respects to these and so no further examples or protocols are provided. Consult remediation firms for more detailed information on specific decontamination chemicals or methods.

26.8.1 Chlorine Dioxide Procedures

The application of chlorine dioxide for building remediation is as straightforward as ozone, except that chlorine dioxide would be provided in liquid form and sprayed into the air. The evaporation in air would produce gaseous chlorine dioxide which would then be circulated throughout the building by the ventilation system. In the remediation of the Hart building, the apparatus was set up outside the building and a doorway was modified for use as a supply duct.

The actual effectiveness of the remediation of the Hart building with chlorine dioxide was less than expected, and the process had to be performed twice with the exposure period being several days. Reports suggest that chlorine dioxide experiences the same susceptibility to relative humidity that ozone does, although data are not yet available to verify this. Anecdotal reports also suggest that residue deposited on some of the furnishings, and caused some limited damage.

Prior to performing remediation of a building with chlorine dioxide, or any gaseous disinfectant, some estimation should be made of the levels needed to achieve sterilization over the specified time period. Some consideration must be given to flammability at high levels and possible damage to interior furnishings, and therefore it is necessary to use the highest level that can safely be used without damage or undue hazard. A level of 4 mg/L of ClO_2 will cause approximately a 99.9 percent reduction in *Listeria* in about 10 to 30 minutes (Steeves, 2002). In water, ClO_2 levels of about 0.25 to 0.5 g/L are able to inactivate *Bacillus subtilis spores* within about 10 minutes, based on data from Radziminski et al. (2002). Assuming that typical mold spores have a ClO_2 susceptibility that is roughly equivalent

Procedure for Building Decontamination with Chlorine Dioxide

1. Estimate the quantity of ClO_2 that will be required to fill the building volume and maintain it at the desired concentration for at least 24 hours.
2. Choose a windless day, if possible, to avoid envelope leakage.
3. Evacuate the building and remove any sensitive equipment or furnishings that may be subject to damage from ClO_2.
4. Isolate the building envelope by closing and sealing all exits, doors, shafts, windows, and other potential leak points.
5. Isolate the outdoor air intakes by sealing and/or closing the mixing dampers.
6. Locate ClO_2 sensors at appropriate locations throughout the building with external or remote readouts.
7. Locate air pressure sensors or airflow direction indicators at appropriate points across the four exterior walls and roof.
8. Configure the ventilation system to run in full recirculation mode.
9. Connect a blower to a chlorine dioxide aerosolizer/generator and run a piped or ducted connection to the ventilation return ductwork, or to a building general area.
10. Engage the ventilation system and monitor the building pressure or airflow direction indicators to verify no significant leakage of ClO_2 to the outside is occurring.
11. Engage the ClO_2 generator and begin flooding the building with ClO_2, monitoring building levels until the desired concentration is reached.
12. Operate the system for a specified time period (i.e., 24 hours).
13. Upon completion, vent the building with outdoor air for an additional day.
14. Verify that the sensors indicate residual chlorine dioxide levels are below the TLV (1 ppm or 3 mg/L) before allowing workers to enter the building.
15. Survey the interior with portable ClO_2 sensors to further verify levels are safe.
16. Conduct surface sampling of contaminated areas.
17. Once sampling indicates decontamination is complete (i.e., sterilization has been achieved) the decontamination system may be dismantled and the building restored to normal operation.
18. If contamination remains, repeat the procedure or investigate the cause of the failed decontamination and correct any problems (i.e., limited decontamination or nondecontaminated areas).

to *B. subtilis* spores, it would require a dose of about 125 mg·min/L to sterilize building surfaces. This would require maintaining an indoor concentration of ClO_2 of about 0.1 mg/L for 24 hours. Of course, the vicissitudes of airflow and distribution in a large building would probably require a higher concentration than this to be effective.

Figure 26.4 illustrates the setup of a chlorine dioxide remediation system. The outside air intakes must be closed, as well as the doors and windows. The ventilation system is then run in full recirculation mode. The ozone generator could inject ozone directly into the duct or air handling unit. It could also deliver ozone directly into the general areas if the duct is inaccessible. Ozone sensors should be distributed around the building to ensure that sufficient levels of ozone are reached everywhere.

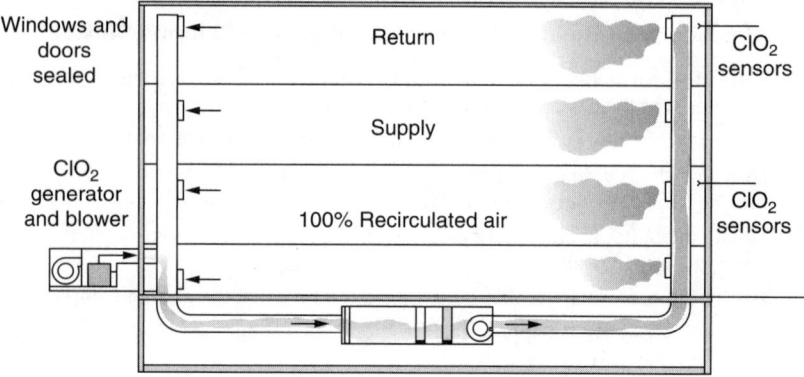

FIGURE 26.4 Schematic representation of a building in full recirculation mode, with chlorine dioxide being injected into the supply air ductwork using an external blower.

26.8.2 Ozone Procedures

The procedures for decontaminating a building with ozone are essentially identical to those given previously for ClO_2. Ozone can be generated fairly easily with various types of ozone generators like the one shown in Fig. 26.5. Some of these units, like those used for industrial water disinfection, can produce over 1500 ppm.

Sufficient information is available in the literature to approximate the disinfection rate constants for spores like *Aspergillus niger* and *Bacillus cereus*, and these can be used as models (Dyas et al., 1983; Ishizaki et al., 1986). Studies show that lower levels of ozone in the range 1 to -20 ppm may be sufficient for disinfection purposes (Kowalski et al., 1998, 2003; Lee and Deininger, 2000). An appropriate period of ozonation would be about 12 to 48 hours,

FIGURE 26.5 Ozone generator designed for remediation applications. [*Image provided courtesy of Clarence Marsden, Trio3 Industries, Inc., Fort Pierce, FL.*]

REMEDIATION

but sampling would have to be used to determine the effectiveness. Lee (2004) suggests using ozone at 0.01 to 0.025 ppm for overnight decontamination of schools and provides some procedures for this process.

As stated previously, ozonation as a remediation technology is in a developmental stage and no actual field data are presently available. The degree to which ozone may damage building materials is not completely known. Organic rubber is particularly susceptible to high levels of ozone exposure, but data on possible damage to other organic materials, or electrical and computer components, are not available at present. Ozone has been used for remediating odors in residential homes using methods, levels, and exposure times much as described above (Schmidt, 2005).

26.8.3 SNL Foam Procedures

Like ozone and ClO_2, the procedures previously given are essentially identical for disinfecting a building with foam. SNL foam consists of a combination of quaternary ammonium salts, cationic hydrotopes, and hydrogen peroxide. The hydrogen peroxide reacts with bicarbonates in the formulation to create effective oxidizing agents.

When generated as foam, the foam remains stable and in contact for extended periods. The half-life of the foam is a measure of how much time it takes for the foam to lose half of its liquid mass to drainage or evaporation. The half-life of SNL foam is several hours.

The pH of SNL foam can be adjusted to suit the particular biological agent that is to be decontaminated. Powdered formulations are used to mix the foam base just prior to use, and these packets of powder can be selected to produce the desired pH. The optimum pH is 8.0 for anthrax spores (Modec, 2001).

Figure 26.6 shows the decay rate for anthrax spores (strain ANR-1) under exposure to SNL foam. The actual survival at 1 hours was below the detection limit, but the curve has been extended, based on the data at 30 minutes, to establish the time of complete sterilization, which is approximately 78 minutes.

Deployment methods for SNL foam include delivery by equipment similar to that used with fire fighting foam agents, spraying as a liquid, or aerosolization as a vapor. Generation of the foam can be accomplished through the use of compressed air injection systems.

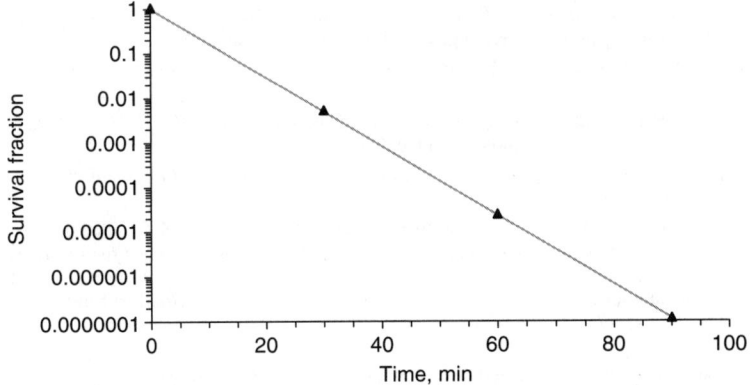

FIGURE 26.6 Survival of anthrax spores in contact with SNL foam with a pH of 8.0. [*Based on data from IITRI (1999).*]

REFERENCES

ACGIH (1973). *Threshold Limit Values of Physical Agents.* The American Conference of Government Industrial Hygienists, Cincinnati, OH.

Ahmad, I., Tansel, B., and Mitrani, J. D. (2001). "Effectiveness of HVAC duct cleaning procedures in improving indoor air quality." *Environ Monit Assess* 72(3):265–276.

AIA (1993). *Designing Healthy Buildings: Indoor Air Quality.* American Institute of Architects, Washington, DC.

Arens, E. A., and Baughman, A. V. (1996). "Indoor humidity and human health: Part II—Buildings and their systems." *ASHRAE Trans* 102(1):212–221.

Austin, B. (1991). *Pathogens in the Environment.* Blackwell Scientific Publications, Oxford, UK.

Batterman, S. A., and Burge, H. (1995). "HVAC systems as emission sources affecting indoor air quality: A critical review." *HVAC&R Res* 1(1):61–80.

Beebe, J. M. (1959). "Stability of disseminated aerosols of *Pastuerella tularensis* subjected to simulated solar radiations at various humidities." *J Bacteriol* 78:18–24.

Carus (2000). "Carulite 200 Catalyst Material Safety Data Sheet (MSDS)." Carus Chemical Corporation, LaSalle, IL.

Casella, M. L., and Schmidt-Lorenz, W. (1989). "Disinfection with gaseous formaldehyde. First Part: Bactericidal and sporicidal effectiveness of formaldehyde with and without formation of a condensing layer." *Zentralbl Hyg Umweltmed* 188(1–2):144–165.

Dales, R. E., Burnett, R., and Zwanenburg, H. (1991a). "Adverse health effects among adults exposed to home dampness and molds." *Am Rev Resp Dis* 143:505–509.

Dales, R. E., Zwanenburg, H., Burnett, R., and Franklin, C. A. (1991b). "Respiratory health effects of home dampness and molds among Canadian children." *Am J Epidem* 134(2):196–203.

DEHS (2000). "Fungal abatement safe operating procedure." University of Minnesota Department of Environmental Health and Safety. http://www.dehs.umn.edu/iaq/sop.html.

DOH (2002). "Guidelines on assessment and remediation of fungi in indoor environments." Department of Health and Mental Hygiene, Bureau of Environmental & Occupational Disease Epidemiology. New York. http://www.lchd.org/environhealth/aq/pdfs/NYC%20DOH%20Guidelines.pdf.

Dyas, A., Boughton, B. J., and Das, B. C. (1983). "Ozone killing action against bacterial and fungal species; microbiological testing of a domestic ozone generator." *J Clin Path* 36:1102–1104.

El-Adhami, W., Daly, S., and Stewart, P. R. (1994). "Biochemical studies on the lethal effects of solar and artificial ultraviolet radiation on *Staphylococcus aureus*." *Arch Microbiol* 161:82–87.

Elford, W. J., and van den Eude, J. (1942). "An investigation of the merits of ozone as an aerial disinfectant." *J Hyg* 42:240–265.

EPA (2001). "Mold Remediation in Schools and Commercial Buildings." *EPA 402-K-01-001* U.S. Environ. Prot. Agency. http://www.epa.gov/cgi-bin/epaprintonly.cgi.

EPA (2002). "Mold Resources." U.S. Environ. Prot. Agency. http://www.epa.gov/cgi-bin/epaprintonly.cgi.

EPA (2003). "A Brief Guide to Mold, Moisture, and Your Home." *EPA 402-K-02-003* U.S. Environ. Prot. Agency. www.epa.gov/iaq/molds/moldguide.html.

Eweis, J. B., Ergas, S. J., Chang, D. P. Y., and Schroeder, E. D. (1998). *Bioremediation Principles.* McGraw-Hill, Boston, MA.

Fatah, A. A., Barrett, J. A., Arcilesi, R. D., Ewing, K. J., Lattin, C. H., Helinski, M. S., and Baig, I. A. (2001). *Guide for the Selection of Chemical and Biological Decontamination Equipment for Emergency First Responders,* Vol.1. Office of Law Enforcement Standards U.S. Department of Justice, Office of Justice Program, National Institute of Justice, Washington, DC. http://www.ncjrs.org/pdffiles1/nij/189724.pdf (v. 1); http://www.ncjrs.org/pdffiles1/nij/189725.pdf (v. 2).

Foarde, K., VanOsdell, D., Meyers, E., and Chang, J. (1997). "Investigation of contact vacuuming for remediation of fungally contaminated duct materials." *Environ Int* 23(6):751–762.

Futter, B. V., and Richardson, G. (1967). "Inactivation of bacterial spores by visible radiation." *J Appl Bacteriol* 30(2):347–353.

IITRI (1999). "Laboratory Tests for Kill of Anthrax Spores in SNL Foam." *IITRI Project No. 1084, Study Nos. 1-4.* Illinois Institute of Technology Research Institute, Chicago, IL.

Ishizaki, K., Shinriki, N., and Matsuyama, H. (1986). "Inactivation of *Bacillus* spores by gaseous ozone." *J Appl Bacteriol* 60:67–72.

Koller, L. R. (1952). *Ultraviolet Radiation.* John Wiley & Sons, New York.

Kowalski, W. J., Bahnfleth, W. P., and Whittam, T. S. (1998). "Bactericidal effects of high airborne ozone concentrations on *Escherichia coli* and *Staphylococcus aureus*." *Ozone Sci Eng* 20(3):205–221. http://www.bio.psu.edu/people/faculty/whittam/research/ozone.html.

Kowalski, W. J., Bahnfleth, W. P., Striebig, B. A., and Whittam, T. S. (2003). "Demonstration of a hermetic airborne ozone disinfection system: Studies on *E. coli*." *AIHA J* 64:222–227.

Laks, P. E., Richter, D. L., and Larkin, G. M. (2002). "Fungal susceptibility of interior commercial building panels." *Forest Prod J* 52(5):41–44.

Lee, J. Y., and Deininger, R. A. (2000). "Survival of bacteria after ozonation." *Ozone Sci Eng* 22:65–75.

Lee, T. G. (2004). "Strategies for achieving a healthier environment: George Park School, Brock, Alberta." University of Calgary. http://www.ucalgary.ca/UofC/faculties/EV/people/faculty/profiles/lee/iaq/brooks.htm.

Masaoka, T., Kubota, Y., Namiuchi, S., Takubo, T., Ueda, T., Shibata, H., Nakamura, H., Yoshitake, J., Yamayoshi, T., Doi, H., and Kamiki, T. (1982). "Ozone decontamination of bioclean rooms." *Appl Environ Microbiol* 43(3):509–513.

McDonnell, G. (2004). "Large area decontamination." *Cleanroom Technol* 10(6):18–20.

Middleton, D. S., and Gilliland, I. C. (1941). "Prevention of droplet-borne infections by spray; a field experiment." *Lancet* 2:598–599.

Miller, J. D., Haisley, P. D., and Reinhardt, J. H. (2000). "Air sampling results in relation to extent of fungal colonization of building materials in some water-damaged buildings." *Indoor Air* 10(3):146–151.

Mitscherlich, E., and Marth, E. H. (1984). *Microbial Survival in the Environment.* Springer-Verlag, Berlin.

Modec, I. (2001). "Technical Report MDF2001-1002: Formulations for the decontamination and mitigation of CB warfare agents, toxic hazardous materials, viruses, bacteria and bacterial spores." Denver, CO. http://www.reevesmfg.com/Literature/Foam_Technical_Report.pdf.

Morrissey, R. F., and Phillips, G. B. (1993). *Sterilization Technology.* Van Nostrand Reinhold, New York.

Nathan, Y., Lstiburek, J., and Brennan, T. (2002). *Mold Remediation in Occupied Homes.* Buildings Science Corporation, Westford, MA, http://www.buildingscience.com/resources/mold/mold_remediation.pdf.

Oliver, A. C. (1988). *Dampness in Buildings.* Nichols Publishing, New York.

Pasanen, P., Pasanen, A.-L., Luoma, M., and Kallioski, P. (1993). Effect of duct-cleaning detergents and disinfection substances on mold growth. *IAQ 93: Operating and Maintaining Buildings for Health, Comfort, and Productivity,* K. Y. Teichman, ed., American Society of Heating, Refrigerating and Air-Conditioning Engineers, Atlanta, GA. 139–142.

Popendorf, W., and Selim, M. (1995). "Exposures while applying commercial disinfectants." *Am Ind Hyg Assoc J* 56(11):1111–1120.

Radziminski, C., Ballantyne, L., Hodson, J., Creason, R., Andrews, R. C., and Chauret, C. (2002). "Disinfection of *Bacillus subtilis* spores with chlorine dioxide: a bench-scale and pilot-scale study." *Water Res* 36:1629–1639.

Richardson, M. I., and Gangolli, S. (1992). *The Dictionary of Substances and Their Effects.* Clays Ltd., Bugbrooke, Northamptonshire.

Rivin, R. (2001). "Building envelope defects and interior mold remediation." *Construction Specifier* 54(10):57–61.

Sakuma, S., and Abe, K. (1996). "Prevention of fungal growth on a panel cooling system by intermittent operation." *1996 The Seventh International Conference on IAQ and Climate,* Nagoya, Japan. 179–184.

Schmidt, D. (2005). "The safe and effective use of ozone to purify/decontaminate an indoor environment." *Enviro2005,* Atlantic City, NJ.

Steeves, S. A. (2002). "Chlorine dioxide gas kills dangerous biological contaminants." Purdue News. http://news.uns.purdue.edu/UNS/html4ever/020912.Linton.chlorinediox.html.

Verhoeff, A. P., VanWijnen, J. H., Brunekreef, B., Fischer, P., VanReenen-Hoekstra, E. S., and Samson, R. A. (1992). "The presence of viable mould propagules in indoor air in relation to home dampness and outdoor air." *Allergy* 47:83–91.

White, T. (1999). "Toxic mold cleanup procedures." *Indoor Air 99 : Proceedings of the 8th International Conference on Indoor Air Quality and Climate,* Edinburgh, Scotland. 879–883.

Wilkinson, T. R. (1966). "Survival of bacteria on metal surfaces." *Appl Microbiol* 14:303–307.

Wilson, S. C., Brasel, T. L., Carriker, C. G., Fortenberry, G. D., Fogle, M. R., Martin, J. M., Wu, C., Andriychuk, L. A., Karunasena, E., and Straus, D. C. (2004). "An investigation into techniques for cleaning mold-contaminated home contents." *J Occup Environ Hyg* 1:442–447.

CHAPTER 27

TESTING AND COMMISSIONING

27.1 INTRODUCTION

Once air cleaning technologies have been installed in a building, it is important to verify that they perform in accordance with their specifications and that they do not adversely affect the ventilation system. *Testing and balancing* (TAB) procedures are available for ventilation system commissioning, and in-place filter testing procedures are available for filters. However, specific testing procedures for air disinfection systems have not previously been addressed. This chapter addresses those aspects of testing and commissioning that relate to air cleaning systems used for air disinfection. Ventilation system testing, filter performance, and in-place filter testing procedures are addressed by reference to existing documents. Guidelines for in-place testing of air disinfection systems like UVGI are proposed here as a precursor to the development of formal procedures and a performance index for buildings, called a building protection factor, is introduced.

27.2 VENTILATION SYSTEM TESTING

Ventilation systems provide dilution and purging of airborne contaminants in buildings and should be tested to verify that they perform in accordance with design specifications and applicable standards or guidelines. Ventilation systems are typically expected, or sometimes required, to meet the guideline recommendations of the ASHRAE Standard 62, "Ventilation for Acceptable Indoor Air Quality" (ASHRAE, 2001). The aerobiological quality of indoor air can be checked through air sampling as discussed in Chap. 24. Practical limits of indoor airborne microorganisms have been suggested in Chap. 25. These latter chapters should provide sufficient information to pursue sampling of indoor air to ascertain whether or not there is an aerobiological problem in any indoor environment.

Testing and balancing of ventilation systems is addressed by a variety of documents and guidelines including the ASHRAE *Handbook of Applications*, Chapter 36 (ASHRAE, 1999), the National Environmental Balancing Bureau *Procedural Standards for Testing, Balancing and Adjusting of Environmental System* (NEBB, 1991), the Associated Air Balance Council *Testing and Balancing Procedures* (AABC, 1997), and the Sheet Metal and Air Conditioning Contractors' National Association *HVAC Systems—Testing, Adjusting, and Balancing* (SMACNA, 1993), among others.

27.3 BUILDING LEAK AND PRESSURE TESTING

Building leakage can impact indoor contaminant levels and can bypass any filtration and air disinfection systems. Leakage can be measured empirically by pressure testing and by thermal imaging. There are currently no standards for leak testing of buildings but there have been some recent studies on the subject. A protocol for field testing of tall buildings to determine envelope air leakage rate has been proposed by Bahnfleth et al. (1999). In this project, two fan pressurization test techniques, the floor-by-floor blower door method and the air-handler method, were developed and tested on two buildings. Criteria for conducting accurate tests were developed, including limitations on outdoor air temperature and wind speed. The floor-by-floor blower door method permits isolation and measurement of the leakage flow rate of a single floor, but it is difficult and time-consuming to apply. The air-handler method uses building air distribution fans for pressurization. It is most easily applied on a system-by-system level rather than floor by floor. Fan airflow techniques including orifice plate, pitot traverse, and tracer gas dilution were considered. The tracer gas method was found to be relatively easy to apply and highly accurate. Fan airflow rate measurement uncertainty by tracer gas was estimated to be 5.4 to 8.8 percent for the cases considered, assuming a 5 percent uncertainty in interzonal leakage.

Feldman et al. (1998) developed a simple apparatus for the evaluation of air infiltration through building envelope components. The study showed that this small apparatus is capable of measuring air infiltration accurately and efficiently for engineering purposes. Experimental results of airflow through and static pressure difference across building envelope have been found to be in good agreement with the literature.

Field studies can be conducted to measure air leakage using large capacity mobile fans. Air leakage rates can be determined using inside-to-outside pressure differences of between 10 and 60 Pa (Masse et al., 1994).

Pressure, temperature, and tracer-gas measurements can be used to determine the ventilation and air leakage performance of low-rise commercial buildings (Sherman, 1990, Sherman and Dickerhoff, 1994). In this study, the ventilation rate of the entire building was measured to be about 2 air changes per hour of outdoor air. The effective leakage area of the exposed building envelope was measured as 0.5 m^2 with an additional 0.1 m^2 of leakage area (estimated) to other parts of the building. With no mechanical systems operating in the test section, this leakage induced an air infiltration rate of 0.6 air changes per hour, which alone would satisfy ASHRAE Standard 62-1989 for the normal occupancy of the building (ASHRAE, 1989).

The use of thermal infrared imaging is a valuable tool for detecting where and how leakage occurs from a building's envelope (Balaras and Argiriou, 2002). Information collected by infrared imaging can be used as part of investigative procedures to identify potential problems.

Buildings, zones, or rooms that are maintained at positive or negative pressure to control airborne contaminants need to be tested or monitored to assure that pressurization requirements are being maintained. The purpose of the pressure differential between zones is to maintain airflow in one direction. The exact value of the pressure differential is not critical as long as the airflow is going in the proper direction. Often, this can be verified

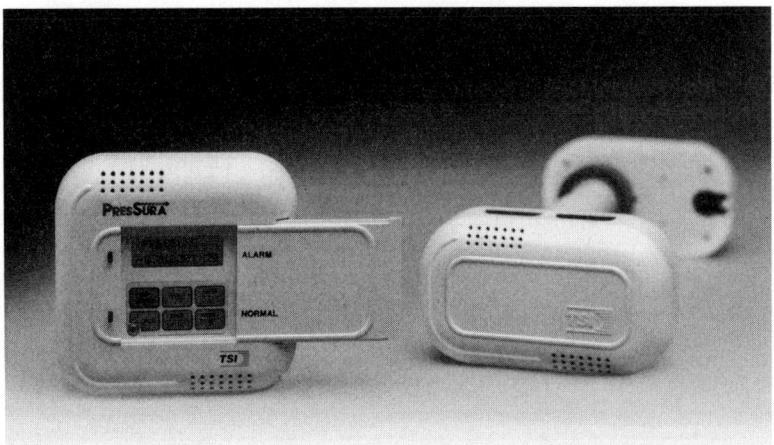

FIGURE 27.1 Commercial pressure sensor suitable for measuring or controlling pressure in positive or negative pressure zones. [*Photo provided courtesy of TSI Incorporated Shoreview, MN (PresSura is a registered TSI trademark name)*].

with a simple handheld smoke generator. In many isolation rooms today the pressure differential is monitored with a sensor that either shows the differential pressure, the airflow rate, or simply indicates the airflow direction. Figure 27.1 shows an example of a pressure sensor that is sensitive enough to detect the kinds of pressure differentials typical in modern buildings.

27.4 IN-PLACE FILTER TESTING

Although filters are individually tested and rated according to current standards like ASHRAE 52.2-1999 and ASHRAE 52.1-1992, actual installations of filters suffer from a variety of problems that will decrease filter efficiency. Poorly fitting filters or leaky frames can cause so much air to bypass the filters that they may become ineffective. This can be particularly true for the highest efficiency filters like HEPA and ULPA filters. Damaged filters that are deformed, dented, or contain pinholes may also have decreased efficiency.

Filters should be physically inspected, either after installation, or periodically during operation, to verify that they are tightly fitted in place and free from damage. They should also be replaced regularly or whenever they become loaded to the point that the pressure drop increases to the manufacturer's specified upper limit. Filters that are operated when they are overloaded may break and become completely ineffective.

In-place filter testing can be performed using most any of the methods described in the ASHRAE Standards, including arrestance, dust spot efficiency, dust-holding capacity, and DOP Penetration. The latter test might be the simplest test to use for in-place testing since it does not require penetration particle sizes. DOP stands for dioctyl phthalate, which is an oily liquid that is aerosolized to produce droplets of a specific size. The size range of the DOP particles is about 0.03 to 0.3 μm. This test is normally used for HEPA filters and testing is outlined in U.S. Military Standard MIL-STD-282 (1956) and U.S. Army document 136-300-175A (1965). When such testing is performed in support of air filtration in nuclear facilities the governing standard is ANSI/ASME N510-1989. In a DOP in-place filter test,

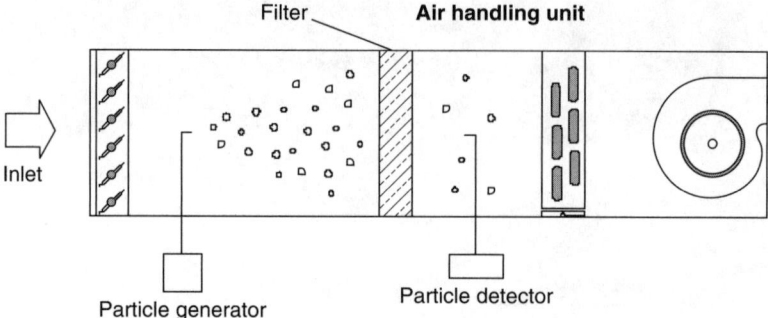

FIGURE 27.2 Test configuration for in-place filter testing using a monodisperse particle generator upstream, and a particle detector downstream, of the filter.

the aerosol is generated and injected into the duct upstream of the filters as shown in Fig. 27.2. Samples are taken upstream to determine inlet DOP concentration. Samples are then taken downstream of the filters and the concentrations are compared to the upstream samples. The reduction of particles at a particular size, or within a specific size range, should approximately match that predicted by the filter performance curve. Any major deviations from expected performance would indicate a problem such as a leaky frame or a pinhole. Samples are usually taken at a series of points that form a rectangular matrix across the duct or the face of the filters. A photometer or penetrometer is typically used to measure the upstream and downstream concentration of DOP.

It is not essential that a DOP generator be used for such nonnuclear in-place filter testing unless required by the user. Any particle generator could be used and filter performance could be established in the same way as previously described. Figure 27.3 shows one aerosol generator, or atomizer, that is capable of producing particles in the 0.2- to 1-µm size range.

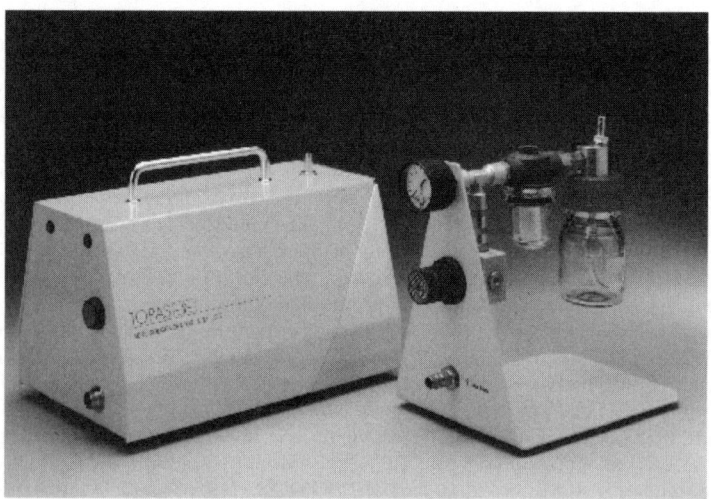

FIGURE 27.3 Aerosol Generator capable of producing liquid aerosol particles in a narrow size range for testing purposes. [*Image courtesy of Topas GmbH, Dresden, Germany.*]

TESTING AND COMMISSIONING

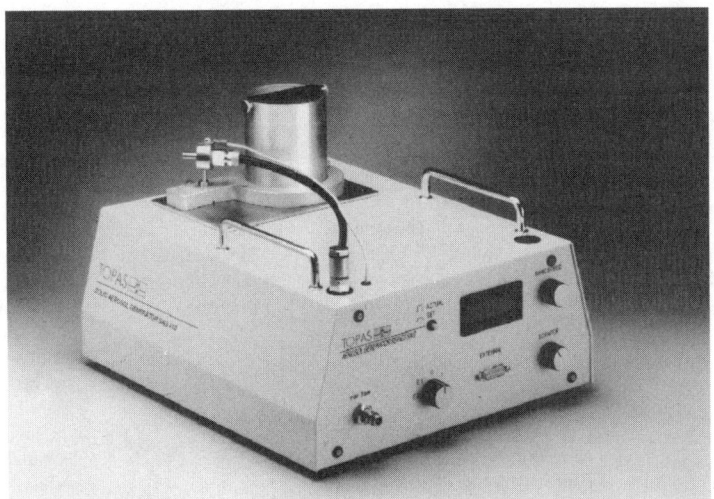

FIGURE 27.4 A powder dispersion unit capable of generating solid particles in the submicron to 100 μm particle size range. [*Image courtesy of Topas GmbH, Dresden, Germany.*]

Solid particles can be generated using powder dispersion units such as the one shown in Fig. 27.4. Such units can be used to generate solid particles over a wide size range for the purpose of filter testing.

Verification that filters perform according to specification is sufficient to ensure that they will filter out microorganisms since microbes tend to behave as particles. Only slight differences exist between microbes and particles that might potentially impact their filterability, such as the mobility of some microbes, the protective coats of lipids and water they may have, and the fact that they often exist inside water droplets (droplet nuclei) or exist as clumps rather than individual microbes. Tests that have been performed on the filtration of bacteria and viruses indicate that they are filtered out as ordinary particles (see Chap. 11).

Certain factors may require additional consideration when filters are used in aerobiological applications. In moist environments some microbes may grow on filters and even grow through to the opposite side (Kemp et al., 1995; Neumeister et al., 1997). Control of the relative humidity to prevent moisture accumulation on filters is one approach, as is the use of antimicrobial filters (Foarde et al., 2000). However, the antimicrobial approach was shown to diminish significantly on loading of the filter. Another consideration is that filters used alongside UVGI systems must be composed of materials (i.e., glass fibers) that resist breakdown under UV exposure. Some filters that are made of certain materials (i.e., polyester) or that have plastic fittings may be subject to long-term damage from UV exposure that may impact their structural integrity.

27.5 IN-PLACE UVGI SYSTEM TESTING

No standards or guidelines exist for testing UVGI air disinfection systems individually or in-place, although some are currently drafted or in development (IUVA, 2005; NIOSH 2005). Testing could be conducted in a manner similar to the in-place testing of filters. There are at least three basic methods by which in-place testing of a UVGI system could be

TABLE 27.1 Types of Tests for Air Disinfection Systems

Test	Measured quantity	Target	Advantage	Disadvantage
Injection of bioaerosols upstream of unit	Upstream concentrations vs. downstream of unit	Bacteria or fungal spores	True indicator of once-through performance	Unlikely to be permitted
Natural microflora in indoor air	Upstream concentrations vs. downstream of system	Bacteria or fungal spores	Does not introduce any new airborne microbes	Airborne levels will converge to steady state and results will not be significant.
	Before vs. After 1–2 weeks of system operation	Bacteria or fungal spores	Simple	Not easy to separate filter effects vs. UVGI effects.
Photosensor readings	Intensity field	NA	Simple, no air sampling required	Disinfection must be assumed based on irradiance readings.

accomplished, as shown in Table 27.1, which summarizes their associated advantages and disadvantages.

The use of a UV photometer is a far simpler and more economical approach to verifying the in-place performance of a UVGI system and it is recommended that such a device be procured so that maintenance personnel can both verify system operation and also check for UV leakage. Handheld devices such as the UV photometer shown in Fig. 27.5 are comparatively inexpensive and fairly easy to calibrate. Care must be exercised to duplicate system operating conditions exactly when using a sensor to measure UV irradiance. Changes in the air velocity and temperature can significantly alter the UV output and test results.

Photosensor readings can verify the lamp rating or check the in-place UV output, but they cannot provide absolute verification that the UV system is disinfecting air. For this reason some form of UV air treatment testing may be needed to establish either the once-through efficiency of the system, or the impact on building bioaerosol levels.

An air sampler (see Chap. 23) can be used to measure the bioaerosol concentrations (either fungi or bacteria) upstream and downstream of the UVGI system to determine the single-pass efficiency. At least three plates should be used for each inlet and outlet location and at least three locations should be sampled in order to average out the normal errors. Once the petri dishes have been cultured and counted, the average survival or inactivation rates will be used to compare the upstream concentrations versus the downstream concentrations. Any major deviation in the expected inactivation rates may indicate a problem with the system. It is necessary that the test microorganisms have a well-established UV rate constant in order for the test results to be verifiable in terms of existing literature. The test microbes, *Serratia marcescens* and *Bacillus subtilis* spores, are typical choices although the use of spores for air disinfection testing may produce marginal results.

An alternative to injecting test microbes into a building ventilation system is to sample the natural microflora. The normal bioaerosols present in the air of an occupied building provide an excellent basis for testing the effectiveness of any air disinfection system, as long as the levels are suitably high to begin with so that major reductions may be observed. The natural airborne microflora of any building consists mainly of bacteria and fungal spores. The spores may be a poor basis for evaluating UVGI because of their inherent resistance to UV, but any UVGI system will typically include a filter. Therefore, what should be tested is the whole air treatment system, including both the filters and the UVGI system, and for this fungal spores are appropriate.

Natural levels of bacteria in an occupied building should provide an adequate basis for evaluating the performance of a UVGI system. Bacteria are normally produced by human occupants, including *Staphylococcus* species and *Streptococcus* species, and the presence of humans inside a building may be necessary to ensure an adequate amount of microbial flora for testing purposes. Two different approaches can be taken when using the natural indoor microflora as indicators of UVGI system performance—upstream versus downstream samples or before versus after samples. If the air is sampled upstream before the air disinfection unit and compared with an air sample taken downstream of the unit, it provides direct indication of system performance. However, this may not be feasible due to the fact that most systems recirculate air and the disinfection process will cause indoor levels of microbes to converge to some minimum steady state concentration, at which point the differences may not be significant.

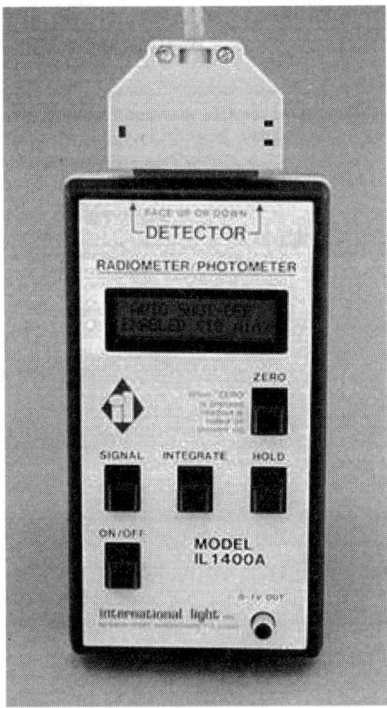

FIGURE 27.5 Handheld photometer suitable for checking the output of UV lamps. [*Image courtesy of International Light, Inc., Newburyport, MA.*]

It is, admittedly, insufficient to simply measure the once-through efficiency of an air treatment system since this efficiency value tells us nothing about how the system will perform in any given application. Therefore the second type of test, sampling indoor airborne bacteria and fungi before and after the system has been put into operation, may be a much better indicator of system performance. This type of test is actually measuring the combined performance of the air treatment system and the building or zone in which it is placed.

27.6 IN-PLACE TESTING OF SURFACE DISINFECTION SYSTEMS

Air disinfection systems like UVGI or pulsed light systems are often used for surface disinfection of cooling coils, filters, duct surfaces, walls, equipment, and the like. Table 27.2 shows the types of surface sampling tests that can be performed to verify operation of a UVGI system.

TABLE 27.2 Types of Tests for Surface Disinfection Systems

Test	Measured quantity	Target	Advantage	Disadvantage
Surface sampling	Before vs. After 1–2 weeks of system operation	Fungal spores	True indicator of performance	Requires sampling expert and lab support
Duct insulation sampling	Before vs. After 1–2 weeks of system operation	Fungal spores	True indicator of performance	Requires lab support
Photosensor readings	Irradiance field	NA	Simple, no air sampling required	Disinfection must be assumed based on irradiance readings

Indirect verification of the effectiveness of UV systems can be achieved by the use of photosensor to measure the irradiance of UV on exposed surfaces. Since the literature on UV inactivation of microbes on surfaces (i.e., petri dishes) is extensive, the dose-response relationship for many microbes is well-established. A field test of the UV irradiance at any exposed surface should include measurements of the most distant surfaces and any shaded or partially shaded surfaces.

Almost any measurable value of UV irradiance will provide long-term sterilization of exposed surfaces. If a system is designed to provide sterilization within some finite amount of time (i.e., equipment or mail disinfection systems), then the measured irradiance on the exposed surfaces should be at least whatever value would produce six logs of reduction over the exposed time period. It is assumed for computational purposes that six logs of reduction represents sterilization, although this may not always be true if initial contaminant levels are too high. It should be noted here that when sterilization is the intended goal, the standard single-stage logarithmic decay equation may not provide an accurate prediction of the inactivation rate. This is because high inactivation rates often highlight the existence of a second stage representing a resistant fraction of any microbial population, as discussed in Chaps. 7 and 12. As a result, it may be necessary to use the complete two-stage decay curve equation, if available, to predict inactivation rates in surface disinfection systems. For example, the complete two-stage decay curve for anthrax spores, based on estimates from the literature (Kowalski, 2003), is as follows:

$$S(t) = 0.9991[1-(1-e^{-0.0000424D})65 + 0.0009 e^{-0.0000042D}] \tag{27.1}$$

where D is the dose, $\mu J/cm^2$.

Figure 27.6 plots Eq. (27.1) versus a curve based on a single-stage rate constant of 0.0031 m^2/J. Although the single-stage equation may be adequate for disinfection estimations (i.e., three to four logs reduction), it would clearly deviate from expectations when applied to sterilization (six logs reduction) applications. Based on Fig. 27.6, and assuming an irradiance of 100 $\mu W/cm^2$, the time to reach sterilization would be the time at which a dose of about 1,230,000 $\mu J/cm^2$ was achieved, or

$$t = \frac{D}{I} = \frac{1230000}{100(3600)} = 3.42 \text{ hours} \tag{27.2}$$

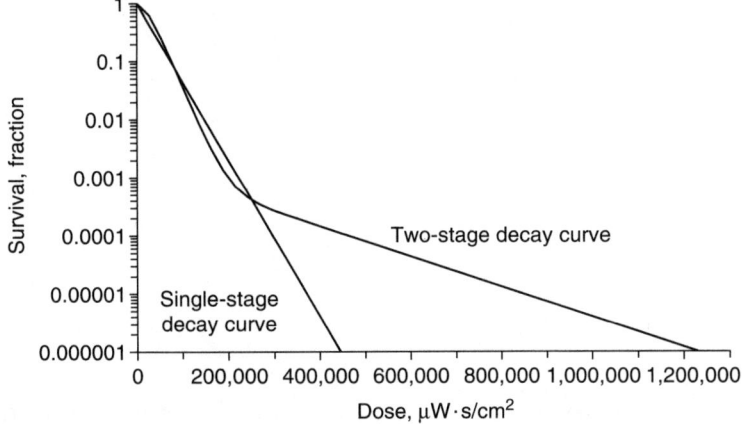

FIGURE 27.6 Comparison of two-stage and single-stage survival curves for anthrax spores under UV exposure, plotted vs. dose.

Equation (27.1) and Fig. 27.6 may be useful for biodefense applications but a more practical microbe may be necessary for applications in which surface mold is to be irradiated. Limited data are available for common molds but following is an equation for *Aspergillus niger* based on data provided by UVDI (2001).

$$S(t) = 0.593[1-(1-e^{-0.0000767D})^2 + 0.407e^{-0.0000203D}] \tag{27.3}$$

Equation (27.3) is plotted in Fig. 27.7. Based on this chart, coil surface sterilization would be achieved at a dose of about 640,000 $\mu J/cm^2$. If the surface irradiance was 100 $\mu W/cm^2$, the time to reach sterilization would be

$$t = \frac{D}{I} = \frac{640000}{100(3600)} = 1.8 \text{ hours} \tag{27.4}$$

Obviously, sterilization can be achieved with surface irradiance levels far less than 100 $\mu W/cm^2$ if continuous exposure is provided. In general, some nominal level (i.e., at least 50 $\mu W/cm^2$) should be provided for surface sterilization to assure that every crevice and corner inside an air handling unit may receive some exposure. If an irradiance in this range is measured at the face of the cooling coils or at the drain pan, sterilization would occur eventually.

A typical surface UV disinfection system for cooling coils or filters might consist of one or more lamps placed in front or above a cooling coil. The lamp is typically located anywhere from 12 to 36 inches from of the face of the coil. Consider the example shown in Fig. 27.8, where a 42-inches lamp of 48.7 UV watts power is placed centrally 15 inches from a coil with a width of 56 inches and a height of 30 inches. The average irradiance at the surface of the coil proves to be 1210 $\mu W/cm^2$ and the minimum irradiance in the corners is 460 $\mu W/cm^2$. Figure 27.9 shows the cooling coil surface irradiance contours. Although this system is considerably oversized for cooling coil irradiation, it will likely have good penetration to the opposite side of the coil, in addition to performing air disinfection.

In-place testing of a surface decontamination system for air handling units, cooling coils, drain pans, and the like can also be performed fairly easily by surface sampling. The

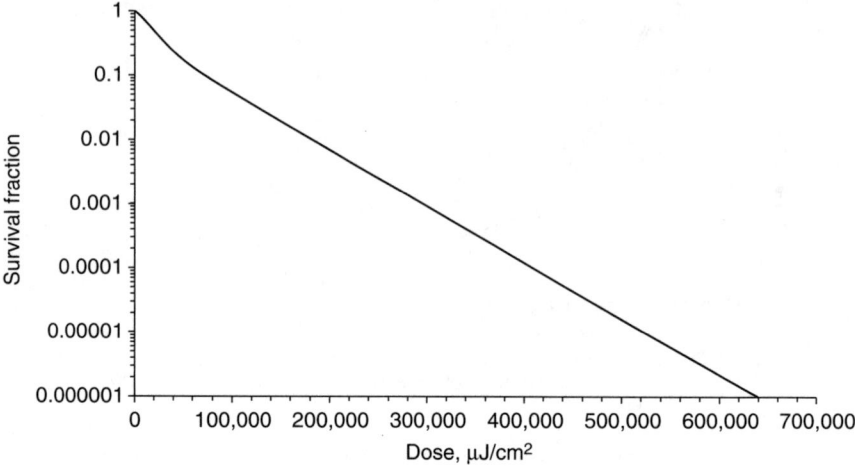

FIGURE 27.7 Two-stage survival curve for *Aspergillus niger* spores. [Based on data from UVDI (2001).]

irradiated surface can be sampled before installation or operation of the UV (or other) system for the presence of bacteria of fungi. Typically, a 1- to 2-inches section of a surface would be swabbed with sterile cotton swabs and it would then be inserted into a sterile solution and mixed. Sterile templates are available for such applications that can be used once and then disposed of. Figure 27.10 shows some examples of materials that can be produced from many microbiology laboratories for use in surface sampling. The solution would then

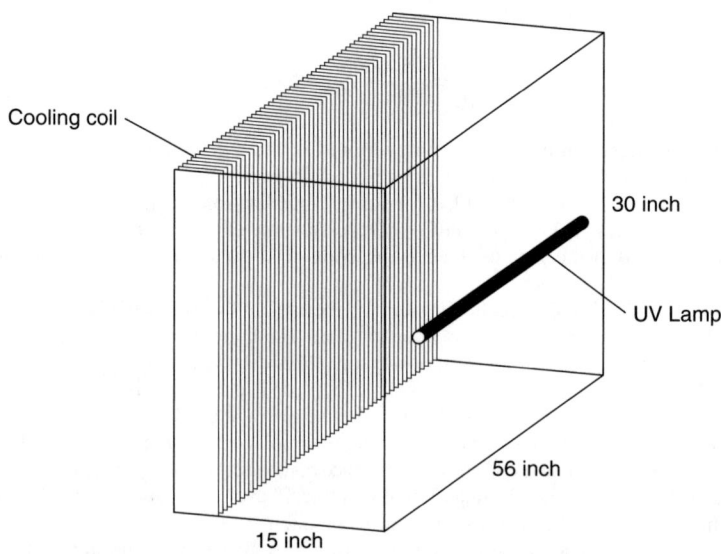

FIGURE 27.8 Layout of a typical cooling coil UV irradiation system.

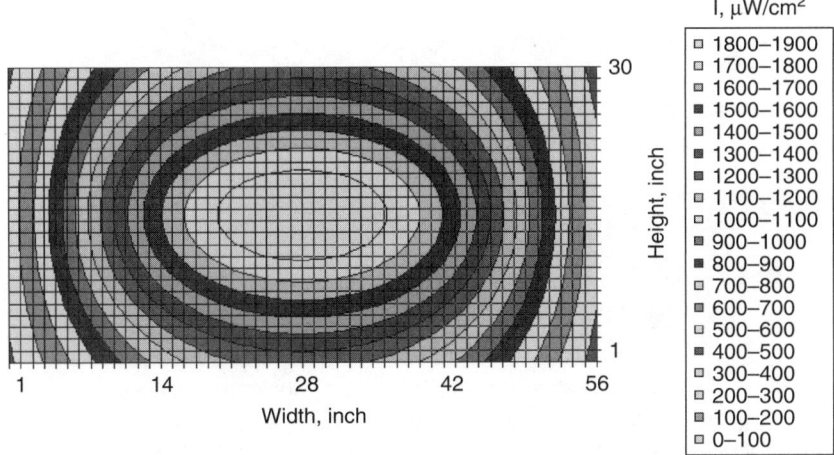

FIGURE 27.9 Irradiance field at the surface of the cooling coil shown in Fig. 27.8.

be spread on a petri dish and cultured. Two separate sets of petri dishes would be needed—one with bacterial agar and one with fungal agar.

There may or may not be any presence of microbial contamination in the "Before" samples, but there should definitely not be any growth in the "After" samples if sterilization has been achieved. The surface disinfection system should operate for some period of time before taking the second samples. Typically, 2 weeks should be sufficient, but a minimum of a day or two may be adequate.

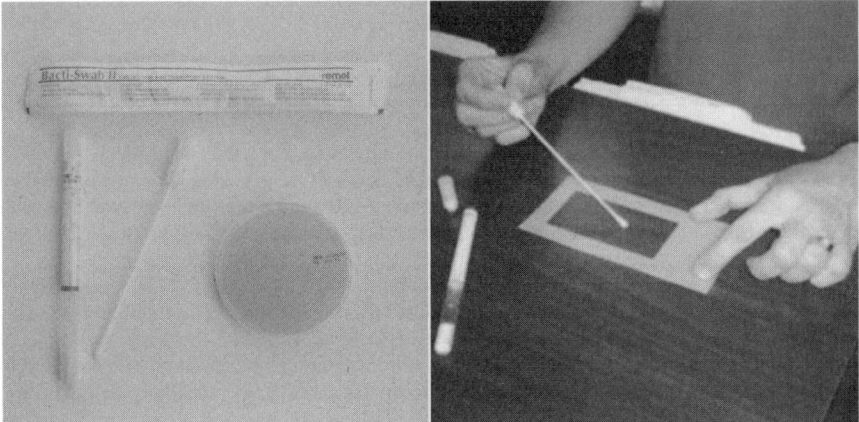

FIGURE 27.10 Examples of surface sampling materials including (left) self-contained sterile swabs in solution and petri dishes with MEA agar, and (right) sterile paper templates. [*Photo on right courtesy of SKC Inc., Eighty Four, PA.*]

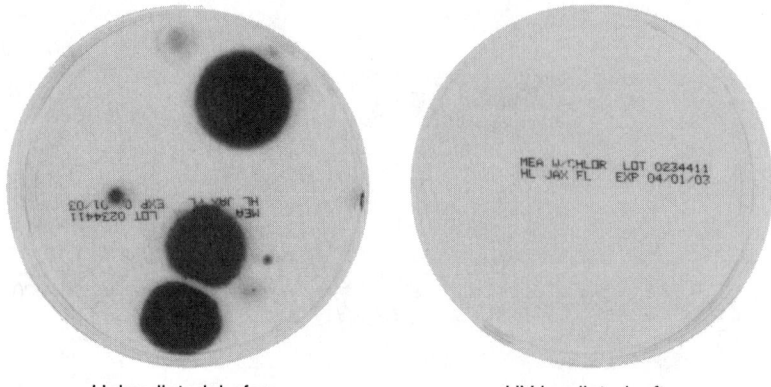

Unirradiated: before   UV irradiated: after

FIGURE 27.11 Fungal cultures grown from the insulation of an air handling unit before and after installation of a UV irradiation system. The system was operated for approximately 24 hours. No fungi grew in the After sample.

Many air handling units have internal insulation that often becomes contaminated with fungal spores during normal use. A small sample of insulation can be removed and soaked in sterile solution, and then the solution is spread on a petri dish and cultured for fungi. Figure 27.11 shows an example of fungal cultures taken in this way. The UV irradiation system in this case had been operated for only about 24 hours. The identified fungal species appeared to be *Aspergillus* and *Penicillium*.

Energy testing is an alternative means of verifying the performance of UVGI cooling coil disinfection systems. Since UVGI will gradually destroy any biofouling on the cooling coil, the heat transfer rate should eventually return to design conditions. This increase in heat transfer efficiency should register as a decrease in outlet temperatures or as a reduction in cooling water flowrates if the system is automatically controlled. Also, the pressure drop on the air side of the cooling coils should be gradually reduced. The net reduction in energy consumption may require monitoring and analysis but will suffice as proof of.

27.7 IN-PLACE CARBON ADSORBER TESTING

In-place testing of carbon adsorbers is almost identical to that used for in-place testing of filters except that a tracer gas is used instead of aerosols or particles. The gas is injected upstream and concentrations measured before and after the charcoal adsorber. The concentrations are determined with a gas detector, such as a gas chromatograph. Any major deviations from expected carbon adsorber performance would indicate leakage.

Since carbon adsorbers are used primarily for removing VOCs, they are not necessarily for aerobiological purposes. If the cause of the odors or airborne VOCs is microbiological, then the source of the problem should be pursued and remediated. There are some situations, animal facilities, for example, in which the biological sources cannot be removed but these are rarely likely to be specifically microbiological in origin. As a result, it is unlikely that a carbon adsorber would be installed to control an aerobiological problem and no type of microbiological testing can be implemented or suggested that would verify performance. Instead, readers are advised to seek additional information on carbon adsorber testing, which is available in the standard for testing nuclear air cleaning systems, ANSI/ASME N510-1989.

27.8 VENTILATION SYSTEM RETROFIT TESTING

The addition of a filter, a UVGI system, and a carbon adsorber can have a major impact on fan performance and total system airflow. These components will increase the total pressure loss through the system and will either increase fan motor energy consumption or decrease total airflow, depending on the fan type and the operating point on the fan curve (McQuiston and Parker, 1994).

The two most common types of fans used in ventilation systems are forward-tipped centrifugal and vaneaxial fans. Figures 27.12 and 27.13 show the generic performance curves for both these types of fans with typical operating points identified with circles. If an increase in pressure loss occurs in the ductwork from the addition of new components, the operating points of these fans will move back toward the left. For the vaneaxial fan the power consumption may increase, up to a point, as the pressure loss increases. For the centrifugal fan, the power consumption may actually decrease as the pressure loss increases. In both cases the total airflow decreases.

Total system airflow should be measured before and after the installation of any air cleaning components that increase the pressure loss or decrease the airflow. If airflow decreases significantly enough to impact overall system performance, measures should be taken to restore design operating conditions, or to verify the acceptability of reduced airflow.

27.9 COMMISSIONING

Commissioning of air cleaning systems is a quality assurance process by which the requirements of relevant systems are verified through a detailed review of the specifications, design documents, performance test results, and operational needs. The commissioning process assures that the system is properly designed and implemented in accordance with the owner's functional criteria and includes the preparation of manuals and training for operating and maintenance personnel. This process can be applied to air treatment systems and surface treatment systems whether they are retrofitted in existing buildings or designed and installed in new building construction. General guidelines for building commissioning are provided in the *ASHRAE Application Handbook*, Chapter 41 (ASHRAE, 1999). Aspects of commissioning as they specifically relate to air treatment systems are summarized here.

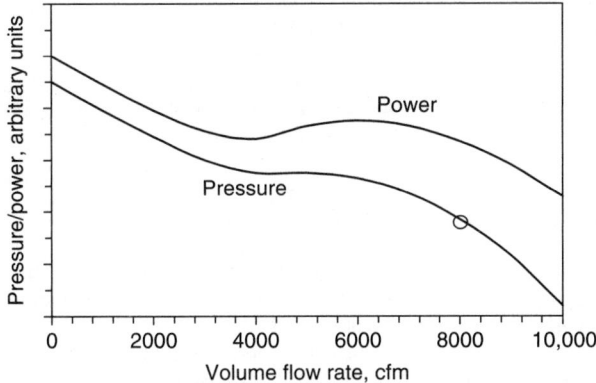

FIGURE 27.12 Fan performance curves for generic vaneaxial fan. Typical design operating point is shown by circle.

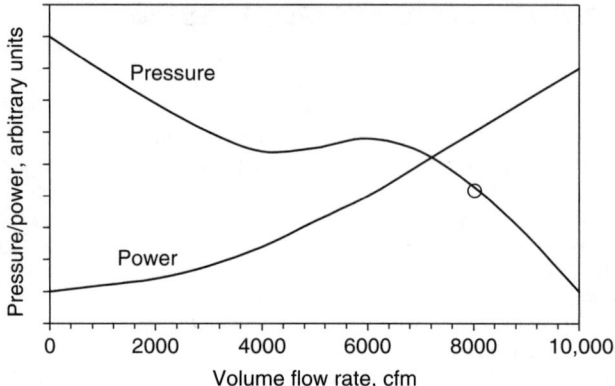

FIGURE 27.13 Fan performance curves for generic centrifugal fan. Typical design operating point is shown by circle.

27.9.1 Predesign Commissioning Phase

The predesign commissioning phase should expressly define the design intent of the air treatment system and document the design requirements, which will serve as the performance criteria to be verified during performance testing. Responsibilities should also be assigned at this stage and a plan or schedule developed so that all the design, construction, testing, training, maintenance, and operations aspects are completed in a timely and economical manner. The responsibility for testing should be assigned separately from design, since the potential exists for conflicting interests. In general, an independent testing company or microbiology laboratory should be brought in to do the testing, although the building owners may also desire to perform their own testing and develop their own ongoing survey of building air.

In general, the design intent of an air treatment system is to reduce the indoor airborne concentrations of bioaerosols, including bacteria, fungal spores, pollen, and other allergens. It may also include odor control via charcoal adsorbers, although this is not strictly an aerobiological function and should be addressed separately so as to avoid confusion. The type of air treatment used, whether filtration, UVGI, or some other technology, is not always critical since the design intent remains the same.

The design intent of cooling coil irradiation systems is to reduce the biological contamination of cooling coils and preserve or restore the original design operating conditions of the cooling coils. Often, cooling coil irradiation systems are combined with air treatment systems and the design intent becomes twofold, requiring two separate tests.

27.9.2 Design Commissioning Phase

The design commissioning phase consists of preparing the contract documents that clearly identify the design intent, and developing the design documents, including specifications and design drawings. The basis of the design should comply with the design intent and be sufficiently documented so that performance can be verified during the testing phase.

For air treatment systems, the performance criteria should either be defined in terms of the single-pass efficiency for a benchmark microorganism like *Serratia marcescens* or

Bacillus subtilis spores, or else should be defined in terms of the predicted reduction in bioaerosol levels. An alternative performance criterion for UVGI systems is the average fluence or dose produced inside the air handling unit. The fluence may, in turn, be defined in terms of the average irradiance inside the air handling unit and the travel time within the irradiance field. It may, in fact, be rather difficult to either define the requisite performance or to demonstrate it during testing, and therefore the criteria may be couched in flexible terms. For example, any properly designed and installed air treatment system should be capable of producing *a significant reduction in indoor concentrations of bacteria and fungal spores,* and it is then left to the owners or designers, or even the laboratory to interpret the data and decide if such a goal has been achieved. In time a body of data will accumulate that will aid in defining what results are realistic and achievable, but until then performance criteria may have to be kept flexible.

27.9.3 Construction Commissioning

The construction commissioning phase includes all construction activities and start-up operations of installed systems. In this phase the details of the system will be finalized and testing and balancing documentation will be generated including recorded operating conditions (i.e., airflows, temperatures, and the like) and as-built drawings. In this phase any details relevant to running the performance tests can be finalized.

27.9.4 Acceptance Commissioning

In the acceptance commissioning phase all the final performance tests that demonstrate the achievement of the performance criteria will be run. This phase will verify that the performance criteria have been met, all documentation has been finalized and updated to the as-built condition, all training has been completed, and all documents have been turned over. The performance test of an air treatment system will typically consist of air sampling inside the building to verify that significant reductions of indoor bioaerosols have been achieved. Testing may also include verifying that cooling coil surfaces have achieved high levels of disinfection. The appropriate tests as described in preceding sections of this chapter will have been performed as necessary and any deficiencies noted. Any problems encountered in this final phase, such as unexpectedly poor performance, may warrant further investigation and retesting.

Final acceptance will generally consist of a final commissioning report indicating that all systems are operating as originally specified and that all performance requirements have been met. Any recommendations for future enhancements or modifications, based on what may have been learned during testing may be included in this final report.

27.10 BUILDING PROTECTION FACTOR

The true performance of any air cleaning system is dependent on its interaction with the ventilation system and building in which it is placed. Although the various indoor limits that have been proposed (see Chap. 25) are one means of assessing performance, there is a need for a performance index for buildings equipped with air treatment. To this end has been proposed the *building protection factor* (BPF), which defines the percent of occupants theoretically protected from infection during the release of airborne pathogens (Kowalski and Bahnfleth, 2004). Methods used in Chaps. 4, 9, 10, and others can be applied here. This method has been used in a previous work to determine the protection afforded in buildings against biological weapons and are equally applicable to releases of naturally occurring viruses, bacteria, or fungal spores (Kowalski, 2003).

The method for determining the BPF for any building is straightforward if not simple. The design basis is defined as an 8-hours continuous release of a quantity of an agent that will cause 99 percent infections in a building without air cleaning but with 15 percent outside air and 8 ACH. Other scenarios are possible but this primary scenario mimics the natural release of contagious airborne diseases in buildings. Figure 27.14 illustrates the sequence of steps necessary to estimate the BPF. First the buildings parameters, including the *air change rate* (ACH) and filter *removal efficiency* (RE) are established and the design basis *release rate* (RR_T) is determined. Then, using a model of the building, the airborne concentrations are determined and the inhaled 8-hours dose computed. An epidemiological model is then used to compute the percent infections when air cleaning is applied.

Various methods can be used to create a building model, including steady state models, transient single-zone complete mixing models, and transient multizone models (i.e., CONTAMW), that have been detailed in previous chapters. It is possible to create a simplified steady state model that will provide fairly accurate predictions by using empirical results from detailed transient analysis of buildings based on complete single-zone mixing. This simplified model, called a quasi-steady state model, is summarized here for use.

The design basis release rate (in cfu/min) for a generic microbe with an $LD_{50} = 1$ that will cause approximately 99 percent infections in a building modeled as a single zone with complete mixing has been found by the author to be a function of the volume, Vol, as follows:

$$RR_T = 0.000373 \cdot \text{Vol} \qquad (27.5)$$

Equation (27.5) is a result of transient modeling but can be used in a *steady state* (SS) equation to estimate the airborne concentration based on the *clean air delivery rate* (CADR) as follows:

$$C_{ss} = \frac{RR_T}{CADR} \qquad (27.6)$$

The CADR is the total clean air delivered to the zone—the sum of the OA (assumed to be clean air) and the fraction of filtered return air delivered by the air cleaner:

$$CADR = Q_{oa} + RE(Q - Q_{oa}) \qquad (27.7)$$

where Q_{oa} is the outside airflow, cfm (or m³/min), and Q is the total airflow, cfm (or m³/min).

Assuming the sedentary breathing rate is 0.01 m³/min (0.353 cfm), the 8-hours inhaled dose D_8 from the SS condition is as follows:

$$D_8 = 8(60)B_r \cdot C_{ss} = 4.8\left(\frac{RR_T}{Q_{oa} + RE(Q - Q_{oa})}\right) \qquad (27.8)$$

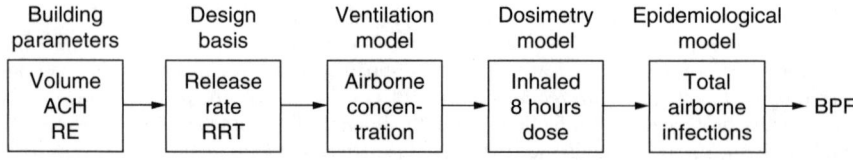

FIGURE 27.14 Sequence of steps for computing a building's BPF rating.

When no air filtration is present (i.e., the baseline condition), Eq. (27.8) reduces to

$$D_8 = 4.8\left(\frac{RR_T}{Q_{oa}}\right) \tag{27.9}$$

Since the LD_{50} has been normalized to unity, the percent of infections (see Chap. 4) can be written as follows:

$$\text{Inf} = 0.5^{0.1^{(D_8-1)}} \tag{27.10}$$

The BPF is simply the complement of the infections as follows:

$$\text{BPF} = 1 - 0.5^{0.1^{(D_8-1)}} \tag{27.11}$$

For the baseline condition with approximately 99 percent infections (with no air filtration) the BPF will be approximately 1 percent. Under the normal operating condition, where a building may have air filters installed and may operate with higher than 15 percent OA, the BPF may range from about 25 to 95 percent. The method may be used to compare the effects of increasing OA, increasing total airflow, or increasing removal efficiency, which may provide a basis for economic evaluation of options. A coded spreadsheet of the BPF model, called a BPF calculator, is available as a public domain download on the Internet (IEC, 2005). The maximum error of the quasi-SS model has been estimated by the author at about ±3 percent based on comparison with detailed multizone CONTAMW models incorporating window, wall, and door leakage, and stack effects through elevator and stairwells. This may not apply to multizone systems with multiple air handling units in which the ACH varies from zone to zone. For such complex systems it is best to use a complete multizone model, such as the CONTAMW program (Dols and Walton 2002).

This method is generally applicable to all air cleaning technologies (i.e., filtration, UVGI, ozone, electrostatic filters, cold plasma systems) provided data are available on the removal rates for the various design basis microbes. This method could be used to examine specific microbes (i.e., anthrax, influenza, TB) but the simplest and most conservative approach is to choose a design basis microbe in the *most penetrating particle* (MPP) size range. For most filters, the MPP is about 0.1 to 0.3 μm (Kowalski and Bahnfleth, 2002). This size range corresponds approximately to coronaviruses (including SARS virus) and influenza. For combinations of filtration and UVGI this is also an appropriate choice (Kowalski, 2003). For other air cleaning technologies some evaluation may be necessary to determine the MPP.

It should be noted that since the LD_{50} is assumed to be unity, the analysis is generic for all microbes regardless of whether the actual LD_{50} is known or not. The only factor that distinguishes microbes in this analysis is the *removal efficiency* (RE), which is dependent on particle size (for filters) or other factors like susceptibility (for UVGI and ozone). Since the epidemiological model is normalized, per Eq. (27.10), it should reliably predict reductions in infections regardless of the actual LD_{50}.

This method will work for any large building with balanced air distribution (i.e., relatively constant cfm/ft^2) and reasonably good air mixing. For large buildings in which the airflow distribution may vary from floor to floor (or zone to zone), the single-zone model may not be accurate and a better estimate of the overall BPF can be had by computing each *zonal protection factor* (ZPF) and an area-weighted average as follows:

$$\text{BPF} = \frac{1}{FA_{total}} \sum_{i=1}^{n} FA_i(ZPF_i) \tag{27.12}$$

where FA = floor area, ft² or m²
 n = number of zones
 ZPF_i = zonal protection factor for zone i

The BPF can be used as an index of performance to size air cleaning systems and as a method for predicting epidemic spread of diseases through entire cities. See Fig. 4.27 in Chap. 4 in which the "% buildings immunized" can be considered synonymous with a city-wide average BPF. It would appear that most buildings will have a BPF rating in the 10 to 30 percent range, while an acceptable minimum might be a BPF of at least 50 percent, although further research remains to be done on these matters. A draft standard for buildings using BPF ratings is under development (IUVA, 2005). The method is also amenable to corroboration by testing, and testing methods could also be developed to establish a BPF for any building. Refer to Kowalski (2003) and Kowalski et al. (2003) for specific examples of applications of the analytical method.

REFERENCES

AABC (1997). "Testing and balancing procedures." Associated Air Balance Council, Washington, DC.

ASHRAE (1989). "Standard 62R: Ventilation for acceptable indoor air quality." American Society of Heating, Refrigerating and Air-Conditioning Engineers, Atlanta, GA.

ASHRAE (1999). *Handbook of Applications*. American Society of Heating, Refrigerating and Air-Conditioning Engineers, Atlanta, GA.

ASHRAE (2001). "Standard 62-2001: Ventilation for acceptable indoor air quality." American Society of Heating Refrigerating and Air-Conditioning Engineers, Atlanta, GA.

Bahnfleth, W. P., Yuill, G. K., and Lee, B. W. (1999). "Protocol for field testing of tall buildings to determine envelope air leakage rate." *ASHRAE Trans* 105:27–37.

Balaras, C. A., and Argiriou, A. A. (2002). "Infrared thermography for building diagnostics." *Energ Buildings* 34(2):171–183.

Dols, W. S. and Walton, G. N. (2002). "CONTAMW 2.0 Users Manual." NIST, Springfield, VA, http://fire.nist.gov/bfrlpubs/build00/art041.html.

Feldman, D., Stathopoulos, T., Cosmulescu, C., and Wu, H. (1998). "Simple apparatus for the evaluation of air infiltration through building envelope components." *J Wind Eng Ind Aerod* 77–78:479–489.

Foarde, K. K., Hanley, J. T., and Veeck, A. C. (2000). "Efficacy of antimicrobial filter treatments." *ASHRAE J* 42(12):52–58.

IEC (2005). "Building protection factor calculator." Penn State Indoor Environment. Center. http://www.engr.psu.edu/ae/wjk/bpf.iec/publications/papers-article.htm.

IUVA (2005). "Standard for the Testing and Commissioning of UVGI In-Duct Air Treatment Systems." *IUVA-S01A-2005*. International Ultraviolet Association, Ayr, Ontario, Canada. http://www.iuva.org.

Kemp, S. J., Kuehn, T. H., Pui, D. Y. H., Vesley, D., and Streifel, A. J. (1995). "Filter collection efficiency and growth of microorganisms on filters loaded with outdoor air." *ASHRAE Trans* 101(1):228.

Kowalski, W. J., and Bahnfleth, W. P. (2002). "MERV filter models for aerobiological applications." *Air Media* Summer:13–17.

Kowalski, W. J. (2003). *Immune Building Systems Technology*. McGraw-Hill, New York.

Kowalski, W. J., Bahnfleth, W. P., and Musser, A. (2003). "Modeling immune building systems for bioterrorism defense." *J Arch Eng* 9(2):86–96.

Kowalski, W. J., and Bahnfleth, W. P. (2004). "Proposed standards and guidelines for UVGI air disinfection." *IUVA News* 6(1):20–25.

Masse, D. I., Munroe, J. A., and Jackson, H. A. (1994). "Mobile test rig for determining the air leakage characteristics of farm buildings." *Can Agric Eng* 36(3):185–188.

McQuiston, F. C., and Parker, J. D. (1994). *Heating, Ventilating, and Air Conditioning Analysis and Design*. John Wiley & Sons, New York.

NEBB (1991). *Procedural Standards for Testing, Balancing and Adjusting of Environmental System*. National Environmental Balancing Bureau, Vienna, VA.

Neumeister, H. G., Kemp, P. C., Kircheis, U., Schleibinger, H. W., and Ruden, H. (1997). "Fungal growth on air filtration media in heating ventilation and air conditioning systems." *Healthy Buildings/IAQ '97'*, Bethesda, MD. 569–574.

NIOSH (2005). *Engineering Controls for Tuberculosis: Upper-Air Ultraviolet Germicidal Irradiation*. National Institute for Occupational Safety and Health, Cincinnati, OH.

Sherman, M., and Dickerhoff, D. (1994). *Monitoring ventilation and leakage in a low-rise commercial building (LBL-34562)*. International Solar Energy Conference, ASME, ed., ASME, New York. 291–297.

Sherman, M. H. (1990). "Tracer gas techniques for measuring ventilation in a single zone, LBL-29328." *Build Environ* 25(4):365–374.

SMACNA (1993). *HVAC Systems—Testing, Adjusting, and Balancing*. Sheet Metal and Air Conditioning Contractor's National Association, Merrifield, VA.

UVDI (2001). "Report on survival data for *A. niger* and *R. nigricans* under UVGI exposure." Valencia, CA.

# SECTION · 4

APPLICATIONS

CHAPTER 28

COMMERCIAL OFFICE BUILDINGS

28.1 INTRODUCTION

Commercial buildings are the single largest type of building in the United States today, consisting of over 60 billion ft^2 of floorspace and numbering over 4.5 million buildings (EIA, 1989). In this chapter commercial office buildings are considered exclusive of hospitals, schools, laboratories, stadiums, residential housing, animal facilities, malls, airports, and other specialized structures, these being addressed in the following chapters due to their differing ventilation system requirements. Commercial office buildings typically range from 2500 to 500,000 ft^2 of floor space although the upper limit of size is constantly being challenged. Occupancies are typically between 7 and 13 people per thousand square feet, implying total occupancies between about 15 and 7000 people per building.

Due to the daily occupancy and extensive interaction of people within office buildings, many respiratory infections are regularly transmitted inside these structures. The type of ventilation system used in modern office buildings can have a major impact on the aerobiological air quality and thus influence the rate of airborne disease transmission and other respiratory problems. This chapter addresses the types of ventilation systems common in office buildings today, the typical levels of aerobiological contaminants, and various engineering solutions that may be used to control these problems.

28.2 OFFICE BUILDING VENTILATION

The specific types of ventilation systems have been addressed in Chap. 9 and include natural ventilation, constant volume, variable air volume, 100 percent OA, DOAS, and displacement ventilation systems (see Fig. 9.1). The way these ventilation systems interact with the building volume determines the rate at which airborne microbes will be purged. The rate of exchange of indoor air with outdoor air is typically stated in terms of the number of *air changes per hour* (ACH). The ACH is computed as follows:

$$\text{ACH} = \frac{\text{airflow}}{\text{volume}} \tag{28.1}$$

The units of airflow and volume are arbitrary so long as they are consistent (i.e., either cfm and ft^3 or m^3/min and m^3, respectively). The average ACH can be stated for the building as a whole, but since airflows and volumes can vary through the floors, zones, or rooms of a building, it is often necessary to compute the ACH for each individual area served.

28.3 OFFICE BUILDING EPIDEMIOLOGY

Office buildings have been identified as a contributing cause in airborne respiratory diseases. Proximity and length of exposure are major factors in the transmission of respiratory infections among office workers, with risks increasing when one worker is within about 5 feet and in front or alongside an infectious individual for 8 hours (Lidwell and Williams, 1961). MacIntyre et al. (1995) found in a study that delayed diagnosis was the major factor responsible for the spread of TB in an office in Melbourne, Victoria, Australia. The prevalence of TB infection was 24 percent among coworkers of two TB infected workers. The normal TB infection rate was 2 to 7 percent in the general community. The coworkers were exposed to infectious TB for 4 months and there was an association with sitting in proximity to the infectious workers during the period of exposure. On-site workers had a higher risk of being infected than did visiting workers.

The risk of catching the common cold is increased by shared office space. Jaakkola and Heinonen (1995) studied the impact of sharing an office with one or more colleagues in a modern, mechanically ventilated, eight-story office building in central Helsinki. The risk for more than two episodes of common cold during a 12-month period was increased in subjects with one or more office colleagues versus those working alone. Among all workers higher risk also emerged for those with young children or a history of hay fever.

Kenyon et al. (2000) studied the transmission of *Mycobacterium tuberculosis* among employees in a U.S. government office in Gaborone, Botswana. Of 79 office contacts investigated 94.7 percent born in high TB prevalence countries had a positive *tuberculin skin test* (TST) compared with 18.2 percent from low prevalence countries. Of 20 U.S.-born contacts 15 percent had documented TST conversion, two of whom were coworkers of the employee with TB. Delayed diagnosis in a setting of high TB prevalence may have contributed to transmission.

Dampness in air-conditioned office buildings has a dose-response effect on airway inflammation and systemic symptoms in workers that include eye irritation, cough, and lethargy or fatigue, according to a study by Wan and Li (1999). This study evaluated dose-response relationships from risk factors among 1237 employees in 19 air-conditioned office buildings in the Taipei area. A positive relationship existed between symptoms when mold was present along with mold odors, and when water damage or dampness was present.

Menzies et al. (1998) investigated the association between respiratory tract symptoms in office workers and immediate skin test reactions with exposure to fungal and house dust mite aeroallergens. For approximately 17 percent of workers, symptoms were associated with exposure to total concentrations of house dust mite allergens greater than 1 µg/g of floor dust or to detectable airborne *Alternaria* allergens in their offices and in the ventilation system supplying their offices. Workers with positive skin test reactions to *Alternaria* extract were exposed at their work site to airborne *Alternaria* allergen and reported significantly more respiratory symptoms. Detection of airborne *Alternaria* allergen at work sites was strongly associated with detection in the ventilation system this was in turn associated with lower efficiency filters.

The recent appearance of SARS virus highlighted the vulnerability of office workers to the spread of airborne diseases (Yu et al., 2004). In early 2003, SARS virus spread through office buildings and apartments in Beijing, China, infecting over 5000 people and causing over 500 deaths. A study of the transmission mechanisms by Jiang et al. (2003) suggested, based on epidemiological evidence and analysis of building dilution rates, that an airflow rate of at least 50 m^3/min is needed per infected individual to dilute the airborne virus below hazardous levels, and that at least 20 percent outside air is needed for such dilution. Of course, equivalent air exchanges might also be achieved through air disinfection without the need to increase outside air (First et al., 1999).

Sick building syndrome (SBS), often referred to today as *building related illness* (BRI), is a general category for a number of ailments, allergies, and complaints, all due to some physical aspect of a building. The existence of low levels of pollutants, synthetic irritants, fungi or other microorganisms, sonic or subsonic vibrations, or simply a lack of adequate fresh air, are sufficient factors to cause reactions in a certain fraction of building occupants (Godish, 1995; Lundin, 1991; Burt, 1996). Sometimes a combination of extremely low levels of several different pollutants or irritants is sufficient to induce SBS/BRI symptoms in certain sensitive individuals (Lee and Kotin, 1972). The diversity of both causes and effects of SBS/BRI seem to defy diagnosis. Often, the only common denominator is insufficient ventilation air to remove contaminant buildup, although some studies implicate ventilation systems as factors (Fisk, 1994). Sometimes the source of the problem is microbial growth inside moist ductwork or other air-handling equipment, in which case increasing the ventilation may even worsen the symptoms (Godish, 1995).

28.4 OFFICE BUILDING AEROBIOLOGY

Office buildings are subject to contaminants brought in with the outside air and contaminants from indoor sources, including the occupants. Office buildings can act like short-term incubators for airborne microorganisms. The aerobiology of office buildings depends on both the composition and number of microbes in the outdoor air as well as the indoor sources. Airborne microbes can enter via the ventilation system, occupants, and infiltration, especially from high-traffic lobbies and when buoyancy effects create high negative pressures (see Fig. 28.1).

In one study on the levels of microorganisms in a natural ventilated building versus an air-conditioned building showed the former had higher levels of airborne microbes (Parat et al., 1997). In the naturally ventilated building the indoor fungal content was strongly dependent on the outdoor content, while in the air-conditioned building fungal concentrations remained constant despite significant variations in the microbial content in the outdoor air. The differences are primarily due to the presence of high-efficiency filters and other HVAC equipment. The type of ventilation equipment can also influence airborne fungal levels (Hyvarinen et al., 1995). The presence of cooling coils and humidifiers may increase or decrease airborne fungi depending partly on operating conditions and water cleanliness. In general, air-conditioned buildings will have lower levels of bioaerosols (Fisk, 1994). Certain systems that use desiccant dehumidification, like DOAS, may also result in reduced levels of bioaerosols (Mumma, 2001; Kovak et al., 1997).

In a study of the indoor concentration of airborne bacteria and fungi in a university auditorium and other buildings, Sessa et al. (2002) measured the indoor concentrations of airborne bacteria and fungi in the presence and in absence of occupants and furnishings. In the presence of people and furnishings the average office air concentrations of bacteria were

FIGURE 28.1 Lobbies are commonly subject to an influx of unfiltered air due to periodic high negative pressures and infiltration through doors.

493 cfu/m^3 while in their absence the levels were 126 cfu/m^3, presumably due mainly to generation of bacteria by occupants. This proved to be true in other types of buildings also. The average air concentrations of fungal spores were also higher in the presence of people and furnishings at 858 cfu/m^3 versus 224 cfu/m^3 in their absence, presumably due to disturbance of dust.

In a study by Yap et al. (1996), 44 offices in 42 Singapore buildings were studied for their bioaerosol content. Bacteria included *Micrococcus, Staphylococcus, Bacillus*, and *Corynebacterium*. Fungi included *Penicillium, Cladosporium, Aspergillus*, and sterilia mycelia. Table 28.1 summarizes these and other studies on indoor bioaerosol levels.

In a study by Fouad et al. (1999), bacteria and fungi in 235 offices in 58 buildings were studied in two New Zealand cities, Auckland and Wellington, for both summer and winter. The mean indoor levels are shown in Table 28.1. In a study of indoor environmental quality in six commercial office buildings in the midwestern United States, Reynolds et al. (2001) took air samples of the indoor air from November to April. The outdoor air provided to occupants ranging from 10 to 79 cfm per person. Relative humidity ranged from 12 to 24 percent. Indoor geometric mean concentrations of microbes, endotoxins, and TVOCs are shown in Table 28.1.

In another study of office buildings the highest airborne fungal spore concentrations were found to be 3852 cfu/m^3 during weekend mornings (Law et al., 2001). This study suggests that the air change rate inside the office building had less effect on bioaerosol levels than filtration. The dominant fungal genera were *Cladosporium* and *Penicillium*. The airborne fungal spore concentrations had a strong correlation with the indoor relative humidity over extended time periods and 80 percent of the sampled bacteria were gram-positive.

Differences always seem to exist between the mixture of fungal species in indoor air and that of the outdoor air, regardless of whether the buildings are problem buildings or what climate they are in. Kemp et al. (2003) studied the changes in occurrence and distribution of airborne

TABLE 28.1 Typical Bioaerosol Levels in Office Building Air

Microbe	Concentration, cfu/m^3	Season	Condition notes	Reference
Bacteria, cfu/m^3	19–585	Summer	Normal	Yap et al., 1996
	28–611	Summer	Normal	Maroni et al., 1993
	6–266	Winter	Normal	Maroni et al., 1993
	564–5360	—	Normal	Bayer and Black, 1988
	12–1201	All	Mean of 326 offices	Gallup et al., 1993
	20–92	All	Normal	Fouad et al., 1999
	68	Winter	Normal	Ross et al., 2004
	252	Fall	Normal	EPA, 1995
	493	Summer	Normal, occupied	Sessa et al., 2002
	150	Summer/winter	Maximum	Reynolds et al., 2001
Fungal spores, cfu/m^3	3852	—	Highest level	Law et al., 2001
	20–360	—	Water damaged	Sudakin, 1998
	9–235	All	Normal	Yap et al., 1996
	34–46	All	Normal	Fouad et al., 1999
	14–5168	All	Normal	Womble et al., 1999
	260–5000	All	Water damaged	Morey et al., 1986
	44	Fall	Normal	EPA, 1995
	6–45	Summer	Normal	Maroni et al., 1993
	224–858	Summer	Normal	Sessa et al., 2002
	312	Winter	Normal	Ross et al., 2004
	150	Summer/winter	Maximum	Reynolds et al., 2001
	6–83	Winter	Normal	Maroni et al., 1993
	10–100	Winter	Normal	Pastuszka et al., 2000
Endotoxin, EU/m^3	0.5–3.0	Winter/spring	Normal	Reynolds et al., 2001
TVOCs, µg/m^3	73–235	Winter/spring	Normal	Reynolds et al., 2001
	0.2–2.2	Fall	Normal	EPA, 1995

fungi as they were transported from the outdoor air through the *heating, ventilation, and air-conditioning* (HVAC) system to the indoor air of two large office buildings in different climate zones. Air samples were taken inside the chambers of the HVAC systems of each building. Results showed that fungal species mix changed with different locations in the HVAC systems. The outdoor air intake produced the greatest filtration effect for both the counts and the species mixture. The counts and species mixture was further altered just downstream of the filters. The cooling coils also had a substantial filtration effect. However, the airborne concentrations in room air were doubled over that in the supply air in the duct and the species mixture changed considerably, suggesting indoor generation of spores.

A study on fungal spores in an office building in Brazil is shown in Fig. 28.2. In this chart the frequency of occurrence of the fungi is indicated as a percentage of samples taken (Brickus et al., 1997).

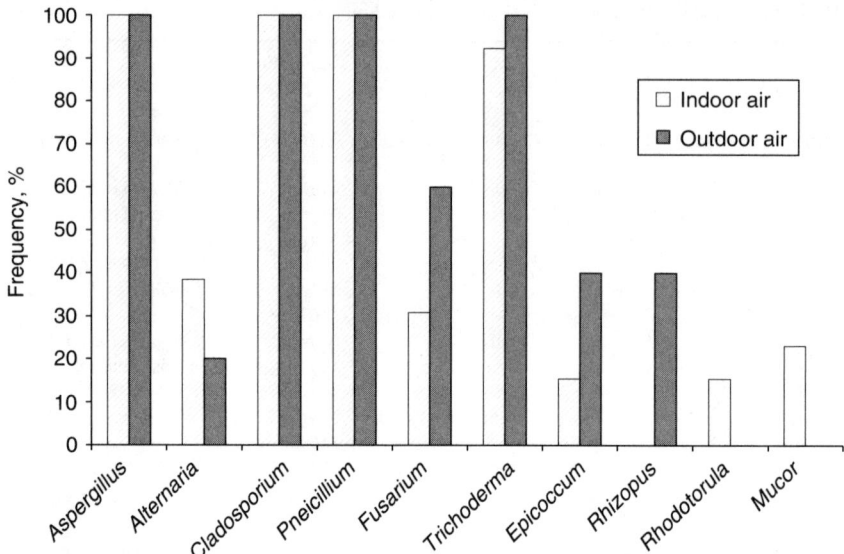

FIGURE 28.2 Frequency of occurrence of fungi in indoor and outdoor air. [*Based on data from Brickus et al. (1997).*]

In a study by Womble et al. (1999), bioaerosol samples were taken during a 4-year period in 86 randomly selected U.S. office buildings. The most commonly identified species were *Cladosporium*, sterilia mycelia, *Penicillium*, Yeast, *Aspergillus*, *Alternaria*, unknown, and *Aureobasidium*. The results of this study are summarized in Table 28.2, which shows the maximum concentrations measured and the indoor percentage frequency.

A study by the EPA (1995) found that levels of airborne bacteria averaged 226 cfu/m^3 in the HVAC system, 252 cfu/m^3 in the indoor air. The highest concentrations (848 cfu/m^3) were found in the central return at the return air fan, which was consistent with findings at return air grilles throughout the building. Airborne bacteria concentrations are more uniform throughout the building than fungi since the occupants are a major contributor.

Fungal spores enter buildings in a variety of ways, including in the outside airstream, via infiltration, and are brought in with people and objects. Once inside, some species may proliferate and the building becomes an amplifier for these species. Fungal spores often settle into carpets and furnishings and then become regularly reaerosolized and recirculate in the building until they germinate in the presence of moisture and nutrients. Various furnishings and fabrics may absorb fungal spores and maintain them almost indefinitely until reaerosolization occurs. In one study of modular fabric office panels, it was found that fungal spores existed at levels of up to 5000 cfu/g of dust vacuumed from the panels (Hung et al., 1993b). Thorough cleaning of these panels by HEPA vacuuming effectively reduced the spore loading.

Legionella is often found in the potable and nonpotable water supplies in many buildings in the United States. In a study of five office buildings by Hung et al. (1993a), *Legionella* was found in four out of five buildings. In the remaining buildings, low levels of *Legionella* were found in some potable and nonpotable water supplies including the cooling tower water, sump water, hot water tanks, and in a Kathabar system. Although no Legionnaires' disease had been reported in these buildings, the study highlighted the fact that preventive maintenance should be implemented in all buildings as a precaution.

TABLE 28.2 Summary of Airborne Allergenic Fungi in Office Buildings

Fungal taxa	Indoor concentration (maximum), cfu/m^3	Outdoor concentration (maximum), cfu/m^3	Indoor % frequency
Cladosporium	3490	5370	50–75
Penicillium	763	1130	25–50
Aspergillus	63	1130	5–25
Alternaria	21	212	5–25
Aureobasidium	21	51	5–25
Fusarium	21	42	<5%
Epicoccum	14	94	<5%
Trichoderma	14	37	<5%
Ulocladium	7	30	<5%
Botrytis	30	125	<5%
Drechslera	8	15	<5%
Paecilomyces	7	8	<5%
Curvularia	7	21	<5%
Acremonium	7	33	<5%
Verticillium	7	ND	<5%
Wallemia	7	7	<5%

28.5 STANDARDS AND GUIDELINES

Although guidelines exist for the ventilation of office buildings, such as ASHRAE Standard 62-01 (ASHRAE, 2001), "Ventilation for Acceptable Indoor Air Quality," no specific guidelines exist for the aerobiology of the indoor environment. Such guidelines are difficult to establish due to limitations of current knowledge about the dose response of exposure to airborne microorganisms. Standards such as ASHRAE 62-01 are designed to provide basic airflow rates to suit the needs of occupants for fresh air but these cannot guarantee that microbial air quality will be acceptable. See Chap. 25 for general recommendations regarding indoor limits of bioaerosols.

28.6 AEROBIOLOGICAL SOLUTIONS

In order to improve the air quality of an office building and effectively reduce indoor air microbial contaminants, solutions are required that address the building's particular concerns. These concerns may include mold spores, allergens, respiratory disease transmission, inadequate airflow or distribution, insufficient humidity control or moisture problems, MVOCs, or nonaerobiological problems like dust and VOCs. Basic solutions can include upgrading filters, addition of UVGI, duct cleaning or remediation, revamping of the ventilation system and airflow distribution, cooling coil cleaning or irradiation, and modifications to the building envelope to restrict infiltration.

Most buildings contain dust filters (in the MERV 6–8 range) but these have a very limited effect on reducing fungal spores, bacteria, and viruses from both indoor and outdoor sources. Unfortunately, simply dictating what size filter might be best for office buildings in general ignores each building's unique characteristics. MERV filters in the 9 to 13 range can be highly effective against most airborne pathogens and allergens but retrofitting such filters may not always be feasible in most buildings because of the increased pressure losses and possible reduction in airflow (Kowalski and Bahnfleth, 2002).

The use of UVGI systems can complement filtration and improve overall performance against bioaerosols (Kowalski and Bahnfleth, 2003). UVGI systems can improve indoor aerobiology when located in-duct or inside air handling units, or when recirculating units are located so as to deal with local air contamination problems. Other applications of UVGI may include overnight decontamination of kitchens, bathrooms, and storage areas with after-hours UVGI units.

The accumulation of mold growth on cooling coils can lead to odors, respiratory irritation, and even hypersensitivity pneumonitis in office workers (Fink, 1970). Cooling coils can be steam cleaned to remove biological contamination but the most effective recourse today is to add a UVGI system to continuously irradiate the coils, which not only destroys microbial growth but tends to restore cooling coils to their original design operating level of performance (ACEEE, 2004, Kelly, 2001).

In one study that compared the effectiveness of different methods to alleviate a case of SBS/BRI, Roys et al. (1993) evaluated ventilation system cleaning, air filtration, intensive office cleaning, and the use of liquid nitrogen to kill dust mites in office furniture. Only the latter two methods had any effect on this particular building. When the source of an air contamination problem is microbial growth inside ductwork or other air-handling equipment, increasing the ventilation may even worsen the symptoms (Godish, 1995). The solution is to remediate the ductwork and ventilation equipment with a cleaning and disinfecting program and adjust the system to better control humidity. In new buildings problems may result from the use of synthetic materials such as insulation or carpeting that release VOCs into the air at a very low rate (Godish, 1995). The solution in this case may be to remove the offending material and replace it with an innocuous alternative. In some instances the outside air intakes may draw in foul air from waste storage or processing areas, or from parking areas heavily laden with auto exhaust. The solution to an SBS/BRI problem may include remediation of mold problems, air treatment to control dust and bioaerosols, carbon adsorbers (for gases and VOCs), or increasing the outside airflow rate.

In an evaluation of five large office buildings in which respiratory problems or diseases were reported, the National Institute of Occupational Safety and Health developed several recommendations for reducing health hazards (Morey et al., 1986). These are summarized in Table 28.3.

A comprehensive approach to managing indoor aerobiological problems should address four primary factors: (1) the building structure and furnishings, (2) the ventilation system, (3) the aerobiology, and (4) the epidemiology or reported symptoms. The process involves evaluating the problem through inspection and resolving the problem through selection of appropriate methods. Figure 28.3 presents a generalized approach to the management of

TABLE 28.3 Preventive Measures for Reducing Microbial Contamination

1. Prevent moisture incursion into occupied space and HVAC components.
2. Remove stagnant water and slimes from building mechanical systems.
3. Use steam as a moisture source in humidifiers.
4. Eliminate the use of water sprays as components of office building HVAC systems.
5. Keep relative humidity (RH) below approximately 70 percent.
6. Use air filters with at least a 50 percent rated efficiency (i.e., DSP 50 percent or MERV 9–10).
7. Discard microbial damaged office furniture.
8. Initiate a maintenance program for HVAC system air handling and fan coil units.

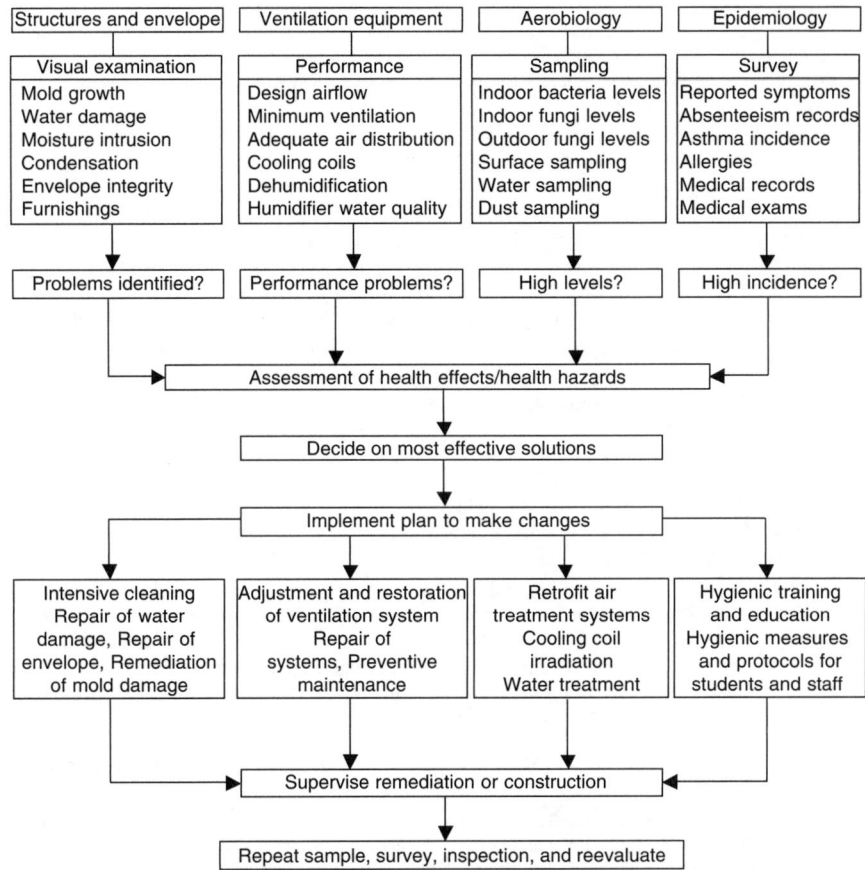

FIGURE 28.3 Flowchart for the management of aerobiological problems in buildings.

aerobiological problems in buildings in the form of a flowchart that is based partly on a similar approaches developed for addressing problems in school buildings and methods suggested by various sources (Haverinen et al.,1999; Morey et al., 1986; Roys et al. 1993; Godish, 1995). This top-down structured approach to resolving problems is generally applicable to all buildings although not every aspect will necessarily apply to every building.

New methods for designing commercial office buildings, and buildings in general, could benefit from consideration of the principles of aerobiological engineering in the design phase. Guidelines have been proposed that address the concept of designing buildings that make allowance for air cleaning systems (IUVA, 2005). In addition to creating space and operating conditions that would facilitate effective air treatment systems, more attention could be paid to developing effective air distribution systems, the use of aerobiologically green technologies (see Chap. 18), and the rating of buildings (via a building protection factor, or BPF) in terms of their overall performance at purging aerobiological contaminants (Kowalski and Bahnfleth, 2004).

REFERENCES

ACEEE (2004). "Ultraviolet germicidal irradiation (UVGI) for HVAC systems." American Council for an Energy-Efficient Economy. http://www.aceee.org/pubs/a042_h6.pdf.

ASHRAE (2001). *Standard 62-2001: Ventilation for Acceptable Indoor Air Quality.* American Society of Heating Refrigerating and Air-Conditioning Engineers. Atlanta, GA.

Bayer, C. W., and Black, M. S. (1988). "IAQ evaluations of three office buildings." *ASHRAE J* 30(7):48–53.

Brickus, L. D., Siquiera, L. F. G., Silveira, M. J., Cardaso, J. N., and Neto, F. R. de Aquino Neto (1997). "Characteristics of indoor and outdoor airborne microorganisms in southeastern Brazilian offices." *Healthy Buildings/IAQ '97'*, Bethesda, MD. 239–244.

Burt, T. S. (1996). "Sick building syndrome: acoustic aspects." *Indoor Built Environ* 5:44–59.

EIA (1989). *Commercial Buildings Characteristics.* Energy Information Administration, Washington, DC.

EPA (1995). "Inside IAQ: EPA's Indoor Air Quality Research Update." *EPA/600/N-95/007.* Environmental Protection Agency. http://www.epa.gov/appcdwww/iemb/insideiaq/fw95.pdf.

Fink, J. N. (1970). "Mold in air conditioner causes pneumonitis in office workers." *JAMA* 211(10):1627.

First, M. W., Nardell, E. A., Chaisson, W., and Riley, R. (1999). "CH-99-12-1: Guidelines for the application of upper-room ultraviolet germicidal irradiation for preventing transmission of airborne contagion." *ASHRAE Trans* 105(Pt. 1):869–876.

Fisk, W. (1994). "The California healthy buildings study." *Center for Building Science News* 12:7, 13.

Fouad, H. G., Donn, M. R., Isaacs, N. P., and Baird, G. (1999). "Results of an analysis of airborne bacterial and fungal levels in fully sealed New Zealand offices." *Indoor Air 99 : Proceedings of the 8th International Conference on Indoor Air Quality and Climate,* Edinburgh, Scotland. 246–251.

Gallup, J. M., Zanolli, J., and Olson, L. (1993). "Airborne bacterial exposure: Preliminary results of volumetric studies performed in office buildings, schools, and homes in California." *Indoor Air: Proceedings of the 6th International Conference on Indoor Air Quality and Climate,* Helsinki, Finland. 167–170.

Godish, T. (1995). *Sick Buildings : Definition, Diagnosis and Mitigation.* CRC-Lewis Publishers, Boca Raton, FL.

Haverinen, U., Husman, T., Toivola, M., Suonketo, J., Pentti, M., Lindberg, R., Leinonen, J., Hyvarinen, A., Meklin, T., and Nevalainen, A. (1999). "An Approach to Management of Critical Indoor Air Problems in School Buildings." *Environ Health Perspect* 107(Suppl. 3):509–513. http://ehp.niehs.nih.gov/members/1999/suppl-3/509-514haverinen/haverinen-full.html.

Hung, L.-L., Copperthite, D. C., Yang, C. S., Lewis, F. A., and Zampiello, F. A. (1993a). "Legionella assessment in office buildings of the continental United States." *Indoor Air: Proceedings of the 6th International Conference on Indoor Air Quality and Climate,* Helsinki, Finland. 379–384.

Hung, L.-L., Copperthite, D. C., Yang, C. S., Lewis, F. A., and Zampiello, F. A. (1993b). "Accumulation of fungal spores in fabric modular office panels." *Indoor Air: Proceedings of the 6th International Conference on Indoor Air Quality and Climate,* Helsinki, Finland. 385–389.

Hyvarinen, A., O'Rourke, M. K., Meldrum, J., Stetzenbach, L., and Reid, H. (1995). "Influence of cooling type on airborne viable fungi." *J Aerosol Sci* 26(S1):s887–s888.

IUVA (2005). *Draft Guideline for the Design and Installation UVGI Air Treatment Systems.* International Ultraviolet Association, Ayr, Ontario, Canada. http://www.iuva.org.

Jaakkola, J. J., and Heinonen, O. P. (1995). "Shared office space and the risk of the common cold." *Eur J Epidemiol* 11(2):213–216.

Jiang, Y., Li, X.-F., Zhao, B., Zhang, Z.-Q., and Zhang, Y. F. (2003). "SARS and Ventilation." *The Fourth International Symposium on HVAC,* Beijing, China. 27–36.

Kelly, T. (2001). "Illuminating your mold." Environmental Design and Construction. http://www.edc-mag.com/CDA/ArticleInformation/features/BNP_Features_Item/0,4120,21188,00.html.

Kemp, P. C., Neumeister-Kemp, H. G., Esposito, B., Lysek, G., and Murray, F. (2003). "Changes in Airborne Fungi from the Outdoors to Indoor Air; Large HVAC Systems in Nonproblem Buildings in Two Different Climates." *AIHA J* 64(2):269–75.

Kenyon, T. A., Copeland, J. E., Moeti, T., Oyewo, R., and Binkin, N. (2000). "Transmission of Mycobacterium tuberculosis among employees in a US government office, Gaborone, Botswana." *Int J Tuberc Lung Dis* 4(10):962–967.

Kovak, B., Heiman, P. R., and Hammel, J. (1997). "The sanitizing effects of desiccant-based cooling." *ASHRAE J* 39(4):60–64.

Kowalski, W. J., and Bahnfleth, W. P. (2002). "Airborne-Microbe Filtration in Indoor Environments." *HPAC Eng* 74(1):57–69. http://www.bio.psu.edu/people/faculty/whittam/research/amf.pdf.

Kowalski, W. J., and Bahnfleth, W. P. (2003). "Immune-Building Technology and Bioterrorism Defense." *HPAC Eng* 75 (Jan.)(1):57–62.

Kowalski, W. J., and Bahnfleth, W. P. (2004). "Proposed standards and guidelines for UVGI air disinfection." *IUVA News* 6(1):20–25.

Law, A. K. Y., Chau, C. K., and Chan, G. Y. S. (2001). "Characteristics of bioaerosol profile in office buildings in Hong Kong." *Build Environ* 36(4):527–541.

Lee, D. H. K., and Kotin, P. (1972). *Multiple Factors in the Causation of Environmentally Induced Disease.* Academic Press, New York.

Lidwell, O. M., and Williams, R. E. O. (1961). "The epidemiology of the common cold." *J Hyg* 59:309–334.

Lundin, L. (1991). *On Building-related Causes of the Sick Building Syndrome.* Almqvist & Wikseil Intl., Stockholm, Sweden.

MacIntyre, C. R., Plant, A. J., Hulls, J., Streeton, J. A., Graham, N. M. H., and Rouch, G. J. (1995). "High rate of transmission of tuberculosis in an office: Impact of delayed diagnosis." *Clin Infect Dis* 21(5):1170–1174.

Maroni, M., Bersani, M., Cavallo, D., Anversa, A., and Alcini, D. (1993). "Microbial contamination in buildings: Comparison between seasons and ventilation systems." *Indoor Air: Proceedings of the 6th International Conference on Indoor Air Quality and Climate,* Helsinki, Finland. 137–142.

Menzies, D., Comtois, P., Pasztor, J., Nunes, F., and Hanley, J. A. (1998). "Aeroallergens and work-related respiratory symptoms among office workers." *J Allergy Clin Immunol* 101(1 Part 1):38–44.

Morey, P. R., Hodgson, M. J., Sorenson, W. G., Kullman, G. J., Rhodes, W. W., and Visvesvara, G. S. (1986). "Environmental studies in moldy office buildings." *ASHRAE Trans* 92(1B):399–416.

Mumma, S. A. (2001). "Dedicated Outside Air Systems." *ASHRAE IAQ Applications* 2(1). http://doas-radiant.psu.ed/IAQ_Winter2001pgs20-22.pdf.

Parat, S., Perdrix, A., Fricker-Hidalgo, H., Saude, I., Grillot, R., and Baconnier, P. (1997). "Multivariate analysis comparing microbial air content of an air-conditioned building and a naturally ventilated building over one year." *Atmos Environ* 31(3):441–449.

Pastuszka, J. S., Paw, U., Kyaw, T., Lis, D. O., Wlazlo, A., and Ulfig, K. (2000). "Bacterial and fungal aerosol in indoor environment in Upper Silesia, Poland." *Atmos Environ* 34(22):3833–3842.

Reynolds, S. J., Black, D. W., Borin, S. S., Breuer, G., Burmeister, L. F., Fuortes, L. J., Smith, T. F., Stein, M. A., Subramanian, P., Thorne, P. S., and Whitten, P. (2001). "Indoor environmental quality in six commercial office buildings in the midwest United States." *Appl Occup Environ Hyg* 16(11):1065–1077.

Ross, C., deMenezes, J. R., Svidzinski, T. I. E., Albino, U., and Andrade, G. (2004). "Studies on fungal and bacterial population of air-conditioned environments." *Brazilian Arch Biology Technol* 47(5):827–835. http://www.scielo.br/pdf/babt/v47n5/a20v47n5.pdf.

Roys, M. S., Raw, G. J., and Whitehead, C. (1993). "Sick building syndrome: Cleanliness is next to healthiness." *Indoor Air: Proceedings of the 6th International Conference on Indoor Air Quality and Climate,* Helsinki, Finland. 261–266.

Sessa, R., Di, P. M., Schiavoni, G., Santino, I., Altieri, A., Pinelli, S., and Del, P. M. (2002). "Microbiological indoor air quality in healthy buildings." *New Microbiol* 25(1):51–56.

Sudakin, D. L. (1998). "Toxigenic fungi in a water-damaged building: An intervention study." *Am J Ind Med* 34:183–190.

Wan, G. H., and Li, C. S. (1999). "Dampness and airway inflammation and systemic symptoms in office building workers." *Arch Environ Health* 54(1):58–63.

Womble, S. E., Burton, L. E., Kolb, L., Girman, J. R., Hadwen, G. E., Carpenter, M., and McCarthy, J. F. (1999). "Prevalence and concentrations of culturable airborne fungal spores in 86 office buildings from the building assessment survey and evaluation (base) study." *Indoor Air 99 : Proceedings of the 8th International Conference on Indoor Air Quality and Climate,* Edinburgh, Scotland. 261–266.

Yap, H. M., Char, K. C., Chan, Y. W., Tan, T. K., and Foo, S. C. (1996). "Airborne bacteria and fungi in Singapore commercial offices." *Proceedings of the 7th International Conference on IAQ and Climate,* Nagoya, Japan 1143–1147.

Yu, I. T. S., Li, Y., Wong, T. W., Tam, W., Chan, A. T., Lee, J. H. W., Leung, D. Y. C., and Ho, T. (2004). "Evidence of airborne transmission of the Severe Acute Respiratory Syndrome virus." *N Engl J Med* 350(17):1731–1739.

CHAPTER 29

EDUCATIONAL FACILITIES

29.1 INTRODUCTION

School buildings, whether K-12 facilities, high schools, or universities, comprise buildings that resemble office buildings in structure and form except that they often have a much higher occupancy. The approach to improving the aerobiology of educational facilities is similar to that for office buildings, although the nature of the diseases and transmission mechanisms are somewhat different.

Libraries and museums at schools present their own unique aerobiological hazards and these are addressed in Chap. 35. Many schools also have biology and chemistry laboratories and for these facilities refer to Chap. 31. Likewise, for auditoriums and administrative or office areas see Chaps. 36 and 28, respectively. This chapter addresses the types of buildings and ventilation systems common in school facilities, the aerobiology of school buildings, and some of the possible solutions to the more common aerobiological problems that can occur.

29.2 EDUCATIONAL BUILDINGS

There are basically four types of schools: preschools or kindergartens, elementary schools, middle or high schools, and colleges and universities. A fifth category, day care centers, should be included because of general similarities to preschools even though education is not a primary function. Each of these five facilities will typically have certain characteristic types of buildings as shown in Fig. 29.1.

Day care and preschools have the smallest and simplest types of buildings, and often are either a single building or a zone within a larger office building. Elementary schools, high schools, colleges, and universities often have a large number and variety of buildings including many more types, such as animal facilities, theatres, and shops, than are listed in Fig. 29.1. The basic type of building addressed here is the classroom building (or just the classroom).

498 APPLICATIONS

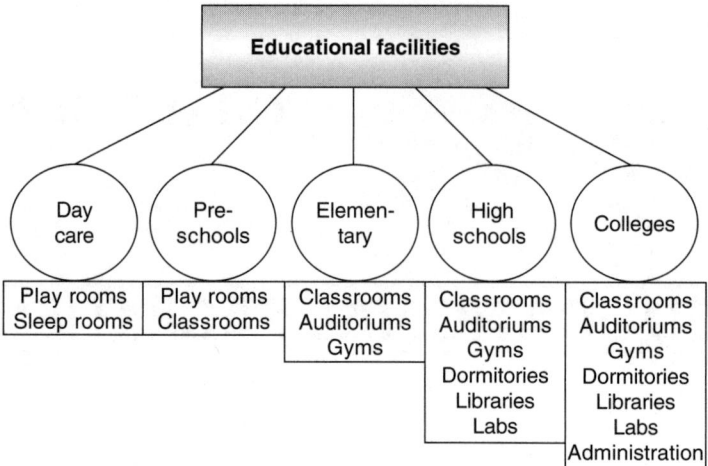

FIGURE 29.1 Breakdown of schools and associated building types.

29.3 SCHOOL BUILDING VENTILATION SYSTEMS

Like office buildings, the ventilations systems in most educational facilities will be one of the four basic types, natural ventilation, constant volume, variable air volume, or 100 percent outside air. DOAS ventilation is an uncommon choice for high occupancy schools since the outside air requirements will approach those of a constant volume or 100 percent outside air system. However, there are some unique advantages to DOAS systems for schools that will be addressed in a later section.

The ventilation requirements are based on the number of students and staff in a building. ASHRAE Standard 62-2001 provides guidelines for these ventilation requirements so as to maintain acceptable indoor air quality (ASHRAE, 2001). Many schools have ventilation systems with outdoor airflow rates that are below the minimum published standards like that of ASHRAE Standard 62-2001, either as a result of older buildings or in attempts to save energy costs. Through the use of energy-saving reductions in ventilation rate requirements, the margins may have been reduced too far in some schools. An outdoor air rate of at least 10 to 13 cfm (5 to 6 L/s) per person is recommended in order to keep indoor odors at reasonably low levels (Berglund et al., 1984). Reducing airflow rates below standards is not usually a good practice but is not necessarily hazardous either. Cost-effective systems can be designed, however, without resorting to reducing outdoor air supply below 15 cfm (7 L/s) per person (Thompson, 1998).

Maintaining relative humidity is essential to avoiding problems with indoor mold growth. ASHRAE Standard 62-2001 provides guidelines for maintaining acceptable indoor temperatures and humidity levels. Results of a simulation study indicate that conventional HVAC systems will have problems maintaining proper indoor humidity levels in some southern schools with the prescribed ventilation rates (Davanagere et al., 1997). Increases in indoor air quality problems may also be due to the accumulation of pollutants in school buildings by energy conservation measures that result in diminished outside fresh air, such as tightly sealed buildings that prevent dilution of contaminants by infiltrating airflows (Kalmaz and Cetilote, 1992).

Other engineering options exist to make school ventilation systems more economical to operate without impacting air quality. The application of desiccant-based energy recovery systems which decouple the outdoor air latent load from the heating or cooling loads can allow schools to minimize energy costs and still provide acceptable indoor air quality without the risk of mold and mildew that may result from the lack of humidity control (Smith, 1996).

Green building design techniques, such as careful sizing of the HVAC equipment, occupant control, and the use of digital control systems, have been shown to be environmentally friendly and energy-saving approaches to the design of school facilities (Rosenbaum, 2002). Other green design technologies (see Chap. 18), such as heat recovery and solar energy systems, may find an appropriate venue in school building design.

29.4 SCHOOL BUILDING EPIDEMIOLOGY

Due to the high concentration of children and adolescents in schools, a variety of pathogens pose transmission hazards that are much greater than in office buildings housing adults. Measles and mumps, for example, spread epidemically and seasonally through schools although they are almost nonexistent elsewhere. Limited information is available on the airborne composition in schools, but extensive information is available on the types of diseases that have occurred in schools. Often, these diseases occur in epidemic fashion, as singular outbreaks, or as ongoing problems.

Preschools or kindergartens are commonly subject to *upper respiratory infection* (URI) outbreaks caused by influenza, *Chlamydia*, and *Mycoplasma*, as well as exposure to allergens. Elementary schools or Primary schools are commonly subject to outbreaks caused by *Bordetella*, influenza, mumps, measles, *Mycoplasma*, varicella-zoster, and *Streptococcus*. Elementary schools have occasionally been subject to high levels of allergens and mycotoxins, usually as the result of water damage and mold growth. In some cases the damage is so severe that the schools had to be abandoned and demolished for the safety of the students (ENR, 1999). High schools and military academies are commonly subject to respiratory infections caused by adenovirus, *Chlamydia*, coxsackievirus, influenza, *Mycobacterium tuberculosis*, and *Streptococcus* species.

Colleges, universities, and medical schools have been commonly subject to respiratory infectious outbreaks caused by adenovirus, *Bordetella*, *Chlamydia*, influenza, *Mycoplasma*, parainfluenza, *Neisseria*, and *Streptococcus*. Medical schools and colleges have also been subject to outbreaks of mumps and *respiratory syncytial virus* (RSV) although these do not occur normally or regularly.

Table 29.1 presents a cross section of the most common respiratory diseases in schools and the types of children and students they impact.

Sick building syndrome or building-related illness (SBS/BRI) can occur in schools, especially when water damage has occurred. The association between allergenic components in dust, such as mites, mold, animal dander and pollen, is well established. Complaints from students concerning eye, nose and throat problems, hoarseness, headache, and fatigue are commonly associated with badly cleaned or water damaged wall-to-wall carpets (Gravesen et al., 1983). Moisture damage and exposure to molds increased the indoor air problems of schools and affected the respiratory health of children (Savilahti et al., 2000). Evidence exists of an association between moisture or mold problems in the school building and the occurrence of respiratory infections, repeated wheezing and prolonged cough in school children (Taskinen et al., 1999).

The local air quality, in terms of outdoor air pollution, has a significant effect on the prevalence of respiratory symptoms and pulmonary diseases in schoolchildren. School children growing up in a high pollution area had more respiratory problems than those in a low

TABLE 29.1 Some Common Respiratory Diseases Affecting Schools

Agent	Disease	Type of school	Reference
Adenovirus	Colds	College	Jackson et al., 2000
	Colds	Medical school	Gerth et al., 1987
	URI	Naval academy	Gray et al., 2001
	Colds	Nursery school	Blacklow et al., 1968
Airborne particulates	Respiratory disorders	Elementary	Heinrich et al., 2000
Allergens	Respiratory distress	Elementary (water damaged)	Savilahti et al., 2000
	Allergy and asthma	Preschool/elementary	Ribeiro et al., 2002
	Asthma	Elementary/high school	Gergen et al., 2002
Bordetella pertussis	Whooping cough	College	Jackson et al., 2000
	Whooping cough	Elementary	Gonzalez et al., 2002
Varicella-zoster virus	Chickenpox	Elementary	Bahlke et al., 1949
Chlamydia pneumoniae	URI	Preschool/elementary	Schmidt et al., 2002
	Pneumonia	College	Jackson et al., 2000
	URI	Naval academy	Gray et al., 2001
Coxsackievirus	Pleurodynia	High school	Ikeda et al., 1993
Echovirus	Colds	Nursery school	Hartmann, 1967
Influenza	Flu	College	Jackson et al., 2000
	Flu	Naval academy	Gray et al., 2001
	Flu	Preschool/elementary	Neuzil et al., 2002
	Flu	School dormitory	Mizuta et al., 1995
	Flu	Medical school	Gerth et al., 1987
Measles virus	Measles	Elementary	Perkins et al., 1947
Mumps virus	Mumps	Medical school	Gerth et al., 1987
Mycobacterium tuberculosis	TB	Elementary	Watson, 2001
	TB	High school	Kim et al., 2001
Mycoplasma pneumoniae	Pneumonia	College	Jackson et al., 2000
	URI	College	McMillan et al., 1993
	URI	Medical school	Gerth et al., 1987
	URI	Naval academy	Gray et al., 2001
	URI	Preschool/elementary	Bosnak et al., 2002
Neisseria meningitidis	Meningitis	University dormitory	Round et al., 2001
Parainfluenza	URI	Medical school	Gerth et al., 1987
	URI	Nursery school	Zilisteanu et al., 1966
Respiratory syncytial virus	RSV	College	Jackson et al., 2000
	RSV	Medical school	Gerth et al., 1987
Rhinovirus	Colds	Nursery school	Beem, 1969
Streptococcus pneumoniae	URI	Naval academy	Gray et al., 2001
Streptococcus pyogenes	URI	College	McMillan et al., 1993
	Rheumatic fever	Elementary	Dierksen et al., 2000
	Pharyngitis	Elementary	Hoebe et al., 2000
	Rheumatic fever	Elementary/high school	Olivier, 2000

pollution area, with over 75 percent of children with problem being sensitized to housedust and other potentially allergens (Gurzau et al., 1993). Volatile organic compounds have also been implicated in respiratory problems in schools (Johansson, 1978). These may be either man-made or produced by fungal growth.

Classrooms, as well as the family environment, are a major factor in spreading whooping cough from *Bordetella pertussis* (Gonzalez et al., 2002). The degree to which this occurs by the airborne route is unclear, as many of these infections appear to transmit by direct contact between students. Incidence rates are highest among children aged <1 year, 10 to 14 years, and 15 to 19 years and school-aged children play an important role in pertussis epidemics (Brennan et al., 2000). The risk of developing rheumatic fever following untreated tonsillopharyngitis is 1 percent in the civilian population (Olivier, 2000).

Infections and epidemics are common in the childhood populations. One study on epidemics of *Chlamydia pneumoniae* showed a prevalence of up to 24 percent in primary schools in December and April (Schmidt et al., 2002). Asymptomatic infections were common, reaching 54 percent of the students, but persistent infection in teenagers is rare. Asymptomatic infection with *C. pneumoniae* may occur in subjectively healthy adults (Miyashita et al., 2001). Of 1018 sera tested, 64 percent of men and 58 percent of women had antibody to *C. pneumoniae*. The overall prevalence of antibody was 61 percent.

Rhinovirus infections are perhaps the most common respiratory infection in schoolchildren. Rhinovirus is involved in 40 percent of asthma attacks in school children, and in about 60 percent of exacerbations in adults with chronic obstructive pulmonary diseases (Herlov-Nielsen and Permin, 2001). Both total respiratory illness and rhinovirus infection peak during the fall and spring seasons. Viral respiratory infections are common and are associated with considerable costs in terms of decreased productivity and time lost from work or school, visits to health-care providers, and the amount of drugs prescribed (Bertino, 2002).

A study on respiratory infections in midshipmen at a U.S. Naval Academy is summarized in Fig. 29.2 (Gray et al., 2001). Additionally, 81 percent of the midshipmen studied complained of having one or more respiratory symptoms during their first year of school. Results suggest that respiratory infections were frequent, had a significant adverse impact on training, and were often attributable to bacterial pathogens.

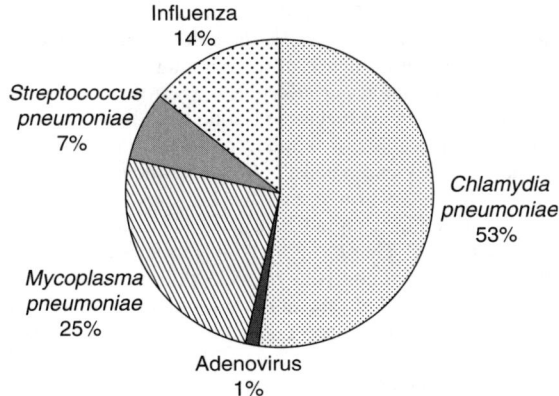

FIGURE 29.2 Breakdown of respiratory infections among midshipmen at a naval academy. [*Based on data from Gray et al. (2001).*]

Pneumonia in children may be caused by unusual organisms like *Mycoplasma*, *Chlamydia*, *Pneumocystis carinii*, and *Legionella* and may form a fairly substantial percentage of community-acquired pneumonia in children (Chugh, 1999). *Mycoplasma pneumoniae* and *Chlamydia pneumoniae* are common in school-age children while *Chlamydia trachomatis* occurs in early infancy.

Day care centers or nursery schools are commonly subject to high rates of adenovirus, echovirus, and rhinovirus. Children who attend day care have an increased risk of asthma, as reported in several studies (Nystad et al., 2001). Attendance at a large day care center was associated with more common colds during the preschool years. However, it was found to protect against the common cold during the early school years (Ball et al., 2002). Rates of illness were higher in children in child care than for children reared exclusively at home during the first 2 years of life (NICHHD, 2001). Respiratory infections were greater among children at day care centers than among children who were mostly at home and attending day care is the most important risk factor for respiratory infections in children aged 2 to 5 years (Forssell et al., 2001). Decreases in respiratory infection rates in day care centers can occur through the implementation of comprehensive educational and environmental infection control (Krilov et al., 1996).

A study of acute respiratory diseases in a student population at a university in New Orleans during a 4-year period was performed by Mogabgab (1968). Influenza and parainfluenza viruses were isolated relatively frequently, and various types of rhinoviruses were the most commonly isolated virus infections. Somewhat less common were adenovirus and coxsackievirus. The respiratory bacteria identified in the study included *Mycoplasma pneumoniae* and various Group A streptococci.

Tuberculosis outbreaks in schools have increased recently. Watson (2001) reports that the incidence of TB in British schools is about 11 per 100,000, but that in Leicester the incidence reached as high as 52 per 100,000. Washko et al. (1998) reported a case in which one infective in a high school choir caused several TB infections. Singing can increase the aerosolization of infectious bacteria and droplets can remain airborne for as long as 30 minutes.

In an earlier case of TB spread, the index patient was a 13-year-old student who attended classes, commuted on a school bus, and sang in a church choir (Sacks et al., 1985). Epidemiologic investigation involving skin testing of over 900 people revealed a 40 percent tuberculin reactor rate for persons in the junior high school she attended compared to a 2 percent rate for control schools. Schoolteachers showed a sevenfold increase in the prevalence of positive skin-test reactions following the outbreak. The more classes shared with the index patient, the higher the probability of a positive skin test. Among students who were not classmates of the index patient, the highest rates of tuberculin reactions occurred in those who had entered a classroom immediately after the index patient had left it.

Ball et al. (2002) studied the influence of the common cold on attendance at day care among children in the 0 to 13 age group. Results suggest that compared with children at home, those in large day care centers had more frequent colds at year 2, less frequent colds at years 6 to 11, and the same odds of frequent colds at year 13. It was inferred by the authors that acquired immunity explained the lower frequency of colds during the early school years, and that this protection waned by 13 years of age. Similar results were obtained in other studies (see Fig. 18.8 in Chap. 18).

A study on asthma among school children in Northern Sweden was conducted by Ronmark et al. (2002). The cohort consisted of 3525 children, 7 to 8 years old at the start, who had a cumulative incidence of physician-diagnosed asthma of 1.7 percent. The cumulative incidence of wheezing was 6.3 percent, and of frequent or daily users of asthma medicines 2.1 percent. Significant risk factors for incident asthma included allergy/rhinitis, eczema, a family history of asthma, low birth weight, respiratory infections, male gender, and a smoking mother. Nystad et al. (2001) found that children who attend day care have an increased risk of asthma, with early infections as an asthma risk factor.

Schools appear to be a major site of exposure to cat and dog allergens. In a study by Perzanowski et al. (1999), dust samples were collected from the classrooms of 22 schools in Sweden and compared with dust from 24 Swedish homes and two schools in the United States. Results of antibody assays on 165 children with respiratory symptoms confirmed that there was a high degree of sensitization to cat, dog, and birch allergens. Cat and dog allergens were present in almost all of the school samples in Sweden. By contrast, dust mites and cockroach allergens were generally immeasurable. The highest levels of cat and dog allergens were found in samples from desks and chairs. Cat and dog allergen levels in the schools were comparable with but higher than those in the homes without pets. The schools in Virginia had similar allergen levels, except that samples also had significant mite allergens. The prevalence of asthma in the northernmost region of Sweden has been estimated at 6 to 8 percent. The causes of the increase in asthma are not clear, but domestic animals are thought to be the primary source of indoor allergens.

Frequent cleaning is important in schools, as is the obvious advantages of linoleum floors over carpets. Dybendal and Elsayed (1990) investigated the prevalence of allergens from cat, dog, house dust mite, pollen, mold, and food in dust samples from schools in Norway and found the occurrence of cat and dog allergens in schools was surprisingly high, especially in classrooms with carpeted floors. Cat and dog allergens are brought into schools by children who are in contact with pets at home or elsewhere and carry animal hair and dander into the classrooms on their clothes, shoes, and bags. Mite infestation, however, was low as verified by their relative absence in both carpeted and smooth school floors.

Both total respiratory illness and rhinovirus infection peak during the fall and spring seasons, when most children attend classes (Bertino, 2002). Respiratory infections acquired by schoolchildren tend to be brought home to their families (Dick et al., 1967). Neuzil et al. (2002) attempted to quantify the effect of influenza season on illness and absenteeism among the families of children enrolled in a Seattle elementary school during the 2000 to 2001 influenza season. Total illness episodes, febrile illness episodes, analgesic use, school absenteeism, parental industrial absenteeism, and secondary illness among family members were significantly higher during influenza season compared with the noninfluenza winter season. For every 100 children studied during the influenza season, which included 37 school days, an excess 28 illness episodes and 63 missed school days occurred. For every 100 children influenza accounted for an estimated 20 days of work missed by the parents and 22 secondary illness episodes among family members.

The simple fact of crowding in school is sufficient to increase respiratory disease transmission rates. Bosnak et al. (2002) studied the prevalence of *Mycoplasma pneumoniae* among school-age children in southeast Turkey. A total of 276 blood samples were collected among two age groups, 0 to 6 and 7 to 14 years, with the latter selected from elementary schools. The overall *M. pneumoniae* seropositivity rate was 27 percent for all ages, with the highest rate occurring at 10 years of age (65 percent) and a 0 percent rate at 2 years of age. An increase was observed at 6 to 7 years of age, the age at which school began.

The incidence of whooping cough has been on the increase in recent years among school-age children. Classrooms and the family environment are a factor in spreading this disease, according to Gonzalez et al. (2002), who studied a whooping cough epidemic outbreak in Castellon, Spain, and the effectiveness of immunization. The study included students from several schools and their family members. The average age of the cases was 10.5 years, 42.6 percent being males and 84 percent school-age children, with 71.3 percent showing signs of recent infection. The effectiveness of the whooping cough booster shot was 66 percent.

Chlamydia pneumoniae is a common cause of respiratory infections but little is known about asymptomatic infection, duration of infections, and seasonal changes of prevalence in a pediatric population. Schmidt et al. (2002) investigated these factors in kindergarten and school-age children. A total of 1211 children of three school-age groups: kindergarten, first and second grades, and seventh and eighth grades. Some 5.6 percent of the children were found to be positive for the infection. Epidemics were confirmed with a prevalence up to 24 percent in a primary school. Asymptomatic infection existed at a 6 percent rate. Of the 32 asymptomatic children, 81 percent were attending seventh and eighth grade classes. The infection appears to be common in schoolchildren with seasonal epidemic-like occurrence. Asymptomatic infection occurs, especially in teenagers, but persistent infection is rare.

Gray et al. (2001) studied occurrences of respiratory infections among midshipmen at the U.S. Naval Academy, who have recently suffered epidemics of upper respiratory tract infections. Some 1249 new midshipmen were followed them during their first 11 months of training, and 85 sought medical attention for acute respiratory diseases. The causative agents were identified as *Chlamydia pneumoniae* (52.6 percent), *Mycoplasma pneumoniae* (25.3 percent), influenza (14.2 percent), *Streptococcus pneumoniae* (7.3 percent), and adenovirus (1.2 percent). Over 81 percent complained of having one or more respiratory symptoms during their first year of school.

Scarlet fever is another respiratory infection that has been increasing lately among schoolchildren. Hoebe et al. (2000) studied an epidemic of scarlet fever cases in a Dutch primary school. Within a period of 1 month, 72 percent of schoolchildren, with a mean age of 5 years, presented with symptoms caused by streptococcal infection. Eight had scarlet fever, five suffered from impetigo, and eight had pharyngitis. A further six children, outside of this class, had complaints of scarlet fever, impetigo, or pharyngitis. The pattern of the outbreak was typical of person-to-person transmission.

Meningitis, a potentially fatal disease that can result from infections with *Neisseria meningitidis* and other microbes, has been responsible for outbreaks on university campuses and dormitories around the United States and elsewhere recently (Nelson et al., 2001). Round et al. (2001) investigated outbreaks in England that involved students in university residence halls. Clinical cases were linked by a discrete network of social contacts within the residence halls. An association was also found with patronage of a campus bar.

Teachers are at higher risk for respiratory diseases and symptoms due to their exposure and contact with children in the classroom. In one study teachers were found to have an incidence of chest illness over 2.5 times that of the reference population, and had an incidence of increased colds and flu (more than five per year) some 3.5 times higher than the reference population (Whelan et al., 2003). Adults normally contract an average of just over two colds or flu per year, while this rate is higher for children, depending on age (see Chap. 4).

29.5 SCHOOL BUILDING AEROBIOLOGY

Recent studies show that indoor air in homes, schools, offices, and other nonresidential buildings contain VOCs, fungi, bacteria, viruses, and other pollutants, sometimes in high enough concentrations to pose a serious threat to human health (Kalmaz and Cetilote, 1992).

A number of studies have been performed on the microbial content of classroom air. The results of a study done in a Paris classroom are shown in Fig. 29.3. As might be expected in a room with high occupancy, the bacteria produced by the occupants are dominant in comparison with the fungi that presumably arrive from outdoors. Total bacterial levels in this classroom were 3000 cfu/m^3 (Mouilleseaux et al., 1993).

The bacterial flora of classrooms in two schools was evaluated by Williams et al. (1956). The bacteria included mostly gram-positive micrococci at 1237 cfu/m^3 and the rest, including streptococci, staphylococci, and diphtheroid microorganisms, totaled 280 cfu/m^3.

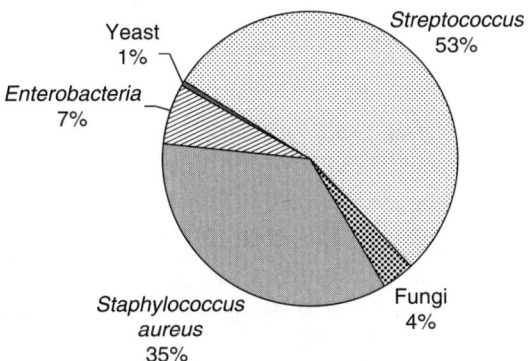

FIGURE 29.3 Breakdown of fungi and bacteria in a classroom. [*Based on data from Mouilleseaux et al. (1993).*]

The mean in the infant classrooms was 2790 cfu/m^3 as compared to 2048 cfu/m^3 in the junior classrooms. Outside air measured 1660 cfu/m^3 in the same period of testing. Table 29.2 summarizes this study and a number of other studies on the bacterial and fungal levels in school air.

The most frequently encountered fungal taxa identified in a study of dust samples from 12 schools in Spain were *Alternaria, Aspergillus*, and *Penicillium* (Angulo-Romero et al., 1996). Some 38 percent of the taxa identified are among those considered to cause diseases, including allergies, hypersensitivity, or infectious diseases.

Meklin et al. (2002) examined the impact of moisture damage in school buildings on respiratory disease symptoms in schoolchildren. Symptoms accumulated during the Spring term as the result of prolonged exposure. In concrete and brick buildings the moisture

TABLE 29.2 Typical Bioaerosol Levels in School Building Air

Microbe	Concentration, cfu/m^3	Season	Condition notes	Reference
Bacteria	100	Winter	Northern climates	Dotterud et al., 1995
	243	Summer/winter	Normal	Maroni et al., 1993
	592	Fall/spring	Water damage	Meklin et al., 2002
	919	Fall/spring	Normal, occupied	Gallup et al., 1993
	1000	Fall/spring	Natural ventilation	Dungy et al., 1986
	1000	Fall/spring	Normal	Levetin et al., 1995
	2472	Fall/spring	Normal	Williams et al., 1956
	3000	Fall/spring	Normal	Mouilleseaux et al., 1993
Fungal spores	15	Summer/winter	Normal	Maroni et al., 1993
	35	Fall/spring	Normal	Klanova and Drahonovska, 1999
	26	Fall/spring	Water damage	Meklin et al., 2002
	53	Fall/spring	Normal	Meklin et al., 1996
	150	Fall/spring	Problem	Meklin et al., 1996
	260–1297	Winter	OSHA study	Santilli and Rockwell, 2003

damage seemed to result in more elevated levels of airborne fungi than in wooden buildings. Differences in levels between moisture-damaged and normal schools were not significant.

29.6 SCHOOL BUILDING SOLUTIONS

A number of attempts have been made to control the spread of respiratory and other diseases in classrooms using approaches like upper air irradiation, chemical disinfectants, and handwashing protocols. Most of these trials experienced a moderate to high degree of success.

As in other buildings, any school with moisture problems or flood damage is likely to cause reported health effects or SBS/BRI. The resolution of such problems is remediation and moisture control. Ahman et al. (2000) reports on a Swedish school with floor moisture problems in which personnel complained of respiratory and other symptoms. Remediative measures included the installation of a ventilated floor. Symptoms virtually disappeared after the intervention.

The air-exchange rate is often low in schools, resulting in potential heath risks. Smedje and Norback (2000) studied the possible impact of improving school ventilation on health and exposure of pupils in Sweden. In 12 percent of the classrooms in 39 schools, new ventilation systems were installed between 1993 and 1995, resulting in an increase in the air-exchange rate. The relative humidity and concentration of airborne pollutants were also reduced compared with classrooms in unmodified buildings. Symptoms decreased among the pupils who attended schools with new ventilation systems.

Bayer et al. (1999) report that problems leading to high indoor levels of fungi included relative humidity control problems, the failure of school ventilation systems to filter out fungi from outdoor air, and that failure of some classrooms to meet the ASHRAE 62-1989 ventilation standard.

One option for new school building construction is *dedicated outside air systems* (DOAS) since they work well with high occupancy applications (Mumma, 2001). They have the same advantages of constant pressure relationships, meet the ventilation requirements with the least expenditure of energy, have low operating costs, and avoid sending air from classrooms with sick kids to the balance of the building.

A number of studies were undertaken in earlier decades to determine the effectiveness of upper air UVGI systems on respiratory infections in schools. See Chap. 11 for a summary of the results of these studies. Most of the studies appeared to produce a reduction of disease transmission and the net average reduction in respiratory infections was approximately 30 percent. All of these studies focused only on Upper Room systems and there are no studies that have addressed forced air UVGI systems, which would be likely to prove even more effective if adequately designed.

Aerobiological problems in school buildings are similar to those in other buildings and often relate to either water and moisture damage or poor ventilation, and the remedial measures are also often similar. Haverinen et al. (1999) suggested a comprehensive investigative approach to dealing with indoor air problems in school buildings. The program proposed includes investigation of moisture problems and the causes of water damage, review of the ventilation system and air distribution, surface sampling, air sampling, and health questionnaires for students and employees. Attention is to be given to the appearance of fungal genera typical of buildings with mold problems especially when levels that exceed those in outdoor air. This methodological approach has been generalized for all buildings and is presented in Chap. 28. For more specifics on remediation, refer to Chap. 26 and the associated references.

As a general recommendation for indoor levels in schools, and based on typical levels shown in Table 29.2, the indoor levels of bacteria should not exceed about 1000 cfu/m^3 of bacteria. In terms of fungi the recommended maximum level for a well-designed school should not exceed about 150 cfu/m^3 during all seasons. These are suggested limits and do not necessarily indicate a problem if they are exceeded, and are not supported by any current official guidelines.

Hand hygiene is a key element in preventing the transmission of cold and flu viruses. White et al. (2005) conducted a study on hand hygiene in four campus residence halls and found that handwashing decreased cold and flu illnesses and school and work absenteeism. Students who were exposed to a message campaign and provided with gel hand sanitizer experienced fewer cold and flu illnesses during the study than those in the control group and missed fewer classes. Perhaps only 25 to 50 percent of respiratory diseases are actually transmitted via the airborne route. Studies on handwashing programs have demonstrated reductions in disease transmission (Guinan et al., 2002). Since many infections are communicated either by direct contact or through fomites on building surfaces and objects handled by children, the ultimate solution may lie in a combination of technology and improved hygiene, a matter in which most children are somewhat ill-trained. For suggestions on improving the hygiene of students see "Hygienic Protocols for Home, Office, and School" in Chap. 18 or the associated references.

REFERENCES

Ahman, M., Lundin, A., Musabasic, V., and Soderman, E. (2000). "Improved health after intervention in a school with moisture problems." *Indoor Air* 10(1):57–62.

Angulo-Romero, J., and F. Infante-Garcia-Pantaleon, (1996). Chapter 5: Pathogenic and antigenic fungi in school dust of the south of Spain. *Pathogens in the Environment*, B. Austin, ed., Blackwell Scientific, Oxford, UK.

ASHRAE (2001). *Standard 62-2001: Ventilation for Acceptable Indoor Air Quality*. American Society of Heating Refrigerating and Air-Conditioning Engineers. Atlanta, GA.

Bahlke, A. M., Silverman, H. F., and Ingraham, H. S. (1949). "Effect of ultra-violet irradiation of classrooms on spread of mumps and chickenpox in large rural central schools." *Am J Pub Health* 41:1321–1330.

Ball, T. M., Holberg, C. J., Aldous, M. B., Martinez, F. D., and Wright, A. L. (2002). "Influence of attendance at day care on the common cold from birth through 13 years of age." *Arch Pediatr Adolesc Med* 156(2):121–126.

Bayer, C. W., Crow, S. A., and Fischer, J. (1999). "Causes of indoor air quality problems in schools." *ORNL/M-6633* Dept. of Energy, Washington, DC.

Beem, M. (1969). "Rhinovirus infections in nursery school children." *J Pediatr* 74(5):818.

Berglund, B., Berglund, U., and Lindvall, T. (1984). "Characterization of indoor air quality and 'Sick Buildings.'" *ASHRAE Trans* 90(1B):1045–1055.

Bertino, J. S. (2002). "Cost burden of viral respiratory infections: Issues for formulary decision makers." *Am J Med* 112(Suppl. 6A):42S–49S.

Blacklow, N. R., Hoggan, M. D., Kapikian, A. Z., Austin, J. B., and Rowe, W. P. (1968). "Epidemiology of adenovirus-associated virus infection in a nursery population." *Am J Epidemiol* 88(3):368–378.

Bosnak, M., Dikici, B., Bosnak, V., Dogru, O., Ozkan, I., Ceylan, A., and Haspolat, K. (2002). "Prevalence of *Mycoplasma pneumoniae* in children in Diyarbakir, the south-east of Turkey." *Pediatr Int* 44(5):510–512.

Brennan, M., Strebel, P., George, H., Yih, W. K., Tachdjian, R., Lett, S. M., Cassiday, P., Sanden, G., and Wharton, M. (2000). "Evidence for transmission of pertussis in schools, Massachusetts, 1996: epidemiologic data supported by pulsed-field gel electrophoresis studies." *J Infect Dis* 181(1):210–215.

Chugh, K. (1999). "Pneumonia due to unusual organisms in children." *Indian J Pediatr* 66(6):929–936.

Davanagere, B. S., Shirey, D. B., Rengarajan, K., and Colacino, F. (1997). "Mitigating the impacts of ASHRAE standard 62-1989 on Florida Schools." *ASHRAE Trans* 103(1):241–258.

Dick, E. C., Blumer, C. R., and Evans, A. S. (1967). "Epidemiology of infections with rhinovirus types 43 and 55 in a group of University of Wisconsin student families." *Am J Epid* 86:386–400.

Dierksen, K. P., Inglis, M., and Tagg, J. R. (2000). "High pharyngeal carriage rates of *Streptococcus pyogenes* in Dunedin school children with a low incidence of rheumatic fever." *N Z Med J* 113(1122):496–499.

Dotterud, L. K., Vorland, L. H., and Falk, S. (1995). "Viable fungi in indoor air in homes and schools in the Sor-Varanger community during winter." *Pediatr Allergy Immunol* 6:181–186.

Dungy, C. I., Kozak, P. P., Gallup, J., and Galant, S. P. (1986). "Aeroallergen exposure in the elementary school setting." *Ann Allergy* 56:218–221.

Dybendal, T., Vik, H., and Elsayed, S. (1990). "Dust from carpeted and smooth floors. III. Trials on denaturation of allergenic proteins by household cleaning solutions and chemical detergents." *Ann Occup Hyg* 34(2):215–229.

ENR (1999). "Indoor air quality: Health problems caused by toxic mold close two Toronto primary schools." *Eng News Rec (ENR)* 242(24):12.

Forssell, G., Hakansson, A., and Mansson, N. O. (2001). "Risk factors for respiratory tract infections in children aged 2-5 years." *Scand J Prim Health Care* 19(2):122–125.

Gallup, J. M., Zanolli, J., and Olson, L. (1993). "Airborne bacterial exposure: Preliminary results of volumetric studies performed in office buildings, schools, and homes in California." *Indoor Air: Proceedings of the 6th International Conference on Indoor Air Quality and Climate,* Helsinki, Finland. 167–170.

Gergen, P. J., Mitchell, H., and Lynn, H. (2002). "Understanding the seasonal pattern of childhood asthma: results from the National Cooperative Inner-City Asthma Study (NCICAS)." *J Pediatr* 141(5):631–636.

Gerth, H. J., Gruner, C., Muller, R., and Dietz, K. (1987). "Seroepidemiological studies on the occurrence of common respiratory infections in pediatric student nurses and medical technology students." *Epidemiol Infect* 98(1):47–63.

Gonzalez, M. F., Moreno, C. A., Amela, H. C., Pachon, A. I., Garcia, B. A., Herrero, C. C., Herrera, G. D., and Martinez, N. F. (2002). "A study of whooping cough epidemic outbreak in Castellon, Spain." *Rev Esp Salud Publica* 76(4):311–319.

Gravesen, S., Larsen, L., and Skov, P. (1983). "Aerobiology of schools and public institutions—part of a study." *Ecol Dis* 2(4):411–413.

Gray, G. C., Schultz, R. G., Gackstetter, G. D., McKeehan, J. A., Aldridge, K. V., Hudspeth, M. K., Malasig, M. D., Fuller, J. M., and McBride, W. Z. (2001). "Prospective study of respiratory infections at the U.S. Naval Academy." *Mil Med* 166(9):759–763.

Guinan, M., McGuckin, M., and Ali, Y. (2002). "The effect of a comprehensive handwashing program on absenteeism in elementary schools." *Am J Infect Control* 30(4):217–220.

Gurzau, E. S., Gurzau, A., Muresan, M., Bodor, E., Zehan, Z., and Radulescu, N. (1993). "Lung function, atopy, and chronic exposure to air pollution in schoolchildren living in two cities of different air quality." *Proc. SPIE—Int Soc Opt Eng* 1716:468–475.

Hartmann, D. (1967). "Silent passage of Echo virus type 20 together with dyspepsia coli strains in a day nursery school." *Zentralbl Bakteriol* 200(2):274–276.

Haverinen, U., Husman, T., Toivola, M., Suonketo, J., Pentti, M., Lindberg, R., Leinonen, J., Hyvarinen, A., Meklin, T., and Nevalainen, A. (1999). "An Approach to Management of Critical Indoor Air Problems in School Buildings." *Environ Health Perspect* 107(Suppl. 3):509–513. http://ehp.niehs.nih.gov/members/1999/suppl-3/509-514haverinen/haverinen-full.html.

Heinrich, J., Hoelscher, B., and Wichmann, H. E. (2000). "Decline of ambient air pollution and respiratory symptoms in children." *Am J Respir Crit Care Med* 161(6):1930–1936.

Herlov-Nielsen, H., and Permin, H. (2001). "Common cold—risk factors, transmission and treatment." *Ugeskr Laeger* 163(41):5643–5646.

Hoebe, C. J., Wagenvoort, J. H., and Schellekens, J. F. (2000). "An outbreak of scarlet fever, impetigo and pharyngitis caused by the same *Streptococcus pyogenes* type T4M4 in a primary school." *Ned Tijdschr Geneeskd* 144(45):2148–2152.

Ikeda, R. M., Kondracki, S. F., Drabkin, P. D., Birkhead, G. S., and Morse, D. L. (1993). "Pleurodynia among football players at a high school. An outbreak associated with coxsackievirus B1." *JAMA* 270(18):2205–2206.

Jackson, L. A., Cherry, J. D., Wang, S. P., and Grayston, J. T. (2000). "Frequency of serological evidence of *Bordetella* infections and mixed infections with other respiratory pathogens in university students with cough illnesses." *Clin Infect Dis* 31(1):3–6.

Johansson, I. (1978). "Determination of organic compounds in indoor air with potential reference to air quality." *Atmos Environ* 12(6–7):1371–1377.

Kalmaz, E. E., and Cetilote, M. (1992). "Methods for evaluation of mutagens in indoor environment." *Proc, Annu Tech Meet—Inst Environ Sci* 2:571–574.

Kim, S. J., Bai, G. H., Lee, H., Kim, H. J., Lew, W. J., Park, Y. K., and Kim, Y. (2001). "Transmission of *Mycobacterium tuberculosis* among high school students in Korea." *Int J Tuberc Lung Dis* 5(9):824–830.

Klanova, K., and Drahonovska, H. (1999). "The concentrations of mixed populations of fungi in indoor air: Rooms with and without mould problems; rooms with and without health complaints." *Indoor Air 99: Proceedings of the 8th International Conference on Indoor Air Quality and Climate,* Edinburgh, Scotland. 920–924.

Krilov, L. R., Barone, S. R., Mandel, F. S., Cusack, T. M., Gaber, D. J., and Rubino, J. R. (1996). "Impact of an infection control program in a specialized preschool." *Am J Infect Control* 24(3):167–173.

Levetin, E., Shaughnessey, R., Fisher, E., Ligman, B., Harrison, J., and Brennan, T. (1995). "Indoor air quality in schools." *Aerobiologia* 11:27–34.

Maroni, M., Bersani, M., Cavallo, D., Anversa, A., and Alcini, D. (1993). "Microbial contamination in buildings: Comparison between seasons and ventilation systems." *Indoor Air: Proceedings of the 6th International Conference on Indoor Air Quality and Climate,* Helsinki, Finland. 137–142.

McMillan, J. A., Weiner, L. B., Higgins, A. M., and Lamparella, V. J. (1993). "Pharyngitis associated with herpes simplex virus in college students." *Pediatr Infect Dis J* 12(4):280–284.

Meklin, T., Taskinen, T., and Nevalainen, A. (1996). "Microbial characterization of four school buildings." *Proceedings of the 7th International Conference on IAQ and Climate,* Nagoya, Japan. 1083–1094.

Meklin, T., Reponen, T., Toivola, M., Koponen, V., Husman, T., Hyvarinen, A., and Nevalainen, A. (2002). "Size distributions of airborne microbes in moisture-damaged and reference school buildings of two construction types." *Atmos Environ* 36(39–40):6031–6039.

Miyashita, N., Niki, Y., Nakajima, M., Fukano, H., and Matsushima, T. (2001). "Prevalence of asymptomatic infection with *Chlamydia pneumoniae* in subjectively healthy adults." *Chest* 119(5):1416–1419.

Mizuta, K., Oshitani, H., Mpabalwani, E. M., Kasolo, F. C., Luo, N. P., Suzuki, H., and Numazaki, Y. (1995). "An outbreak of influenza A/H3N2 in a Zambian school dormitory." *East Afr Med J* 72(3):189–190.

Mogabgab, W. J. (1968). "Acute respiratory illnesses in university (1962–1966), military and industrial (1962–1963) populations." *Am Rev Resp Dis* 98(3):359–379.

Mouilleseaux, A., Squinazi, F., and Festy, B. (1993). "Microbial characterization of air quality in classrooms." *Indoor Air 93: Proceedings of the 6th International Conference on Indoor Air Quality and Climate,* Helsinki, Finland. 195–200.

Mumma, S. A. (2001). "Designing dedicated outdoor air systems." *ASHRAE J* 43(5):28–31. http://doas-radiant.psu.ed/journal_01_doas.pdf.

Nelson, S. J., Charlett, A., Orr, H. J., Barker, R. M., Neal, K. R., Taylor, C., Monk, P. N., Evans, M. R., and Stuart, J. M. (2001). "Risk factors for meningococcal disease in university halls of residence." *Epidemiol Infect* 126(2):211–217.

Neuzil, K. M., Hohlbein, C., and Zhu, Y. (2002). "Illness among schoolchildren during influenza season: effect on school absenteeism, parental absenteeism from work, and secondary illness in families." *Arch Pediatr Adolesc Med* 156(10):986–991.

NICHHD (2001). "Child care and common communicable illnesses: results from the National Institute of Child Health and Human Development Study of Early Child Care." *Arch Pediatr Adolesc Med* 155(4):481–488.

Nystad, W., Skrondal, A., and Magnus, P. (2001). "Day care centers, infections and asthma." *Tidsskr Nor Laegeforen* 121(3):282–286.

Olivier, C. (2000). "Rheumatic fever—is it still a problem?" *J Antimicrob Chemother* 45(Suppl.):13–21.

Perkins, J. E., Bahlke, A. M., and Silverman, H. F. (1947). "Effect of ultra-violet irradiation of classrooms on the spread of measles in large rural central schools." *Am J Pub Health* 37:529–537.

Perzanowski, M. S., Ronmark, E., Nold, B., Lundback, B., and Platts-Mills, T. A. (1999). "Relevance of allergens from cats and dogs to asthma in the northernmost province of Sweden: schools as a major site of exposure." *J Allergy Clin Immunol* 103(6):1002–1004.

Ribeiro, S. A., Furuyama, T., Schenkman, S., and Jardim, J. R. (2002). "Atopy, passive smoking, respiratory infections and asthma among children from kindergarten and elementary school." *Sao Paulo Med J* 120(4):109–112.

Ronmark, E., Perzanowski, M., Platts-Mills, T., and Lundback, B. (2002). "Incidence rates and risk factors for asthma among school children: a 2-year follow-up report from the obstructive lung disease in Northern Sweden (OLIN) studies." *Respir Med* 96(12):1006–1013.

Rosenbaum, M. (2002). "A green building on campus." *ASHRAE J* 44(1):41–44.

Round, A., Evans, M. R., Salmon, R. L., Hosein, I. K., Mukerjee, A. K., Smith, R. W., and Palmer, S. R. (2001). "Public health management of an outbreak of group C meningococcal disease in university campus residents." *Eur J Public Health* 11(4):431–436.

Sacks, J. J., Brenner, E. R., Breeden, D. C., Anders, H. M., and Parker, R. L. (1985). "Epidemiology of a tuberculosis outbreak in a South Carolina junior high school." *Am J Public Health* 75(4):361–365.

Santilli, J., and Rockwell, W. (2003). "Fungal contamination of elementary schools: A new environmental hazard." *Ann Allerg Asthma Immunol* 90:203–208. http://www.moldallergy.com/presentation/New_Envior_Hazard.pdf.

Savilahti, R., Uitti, J., Laippala, P., Husman, T., and Roto, P. (2000). "Respiratory morbidity among children following renovation of a water-damaged school." *Arch Environ Health* 55(6):405–410.

Schmidt, S. M., Muller, C. E., Mahner, B., and Wiersbitzky, S. K. (2002). "Prevalence, rate of persistence and respiratory tract symptoms of Chlamydia pneumoniae infection in 1211 kindergarten and school age children." *Pediatr Infect Dis J* 21(8):758–762.

Smedje, G., and Norback, D. (2000). "New ventilation systems at select schools in Sweden—effects on asthma and exposure." *Arch Environ Health* 55(1):18–25.

Smith, J. C. (1996). "Schools resolve IAQ/humidity problems with desiccant preconditioning." *HPAC Eng* 68(4):73–78.

Taskinen, T., Hyvarinen, A., Meklin, T., Husman, T., Nevalainen, A., and Korppi, M. (1999). "Asthma and respiratory infections in school children with special reference to moisture and mold problems in the school." *Acta Paediatr* 88(12):1373–1379.

Thompson, B. (1998). "Engineers, IAQ, and schools." *ASHRAE J* 40(6):22–26.

Washko, R., Robinson, E., Fehrs, L. J., and Frieden, T. R. (1998). "Tuberculosis transmission in a high school choir." *J Sch Health* 68(6):256–259.

Watson, J. M. (2001). "TB in Leicester: out of control, or just one of those things?" *BMJ* 322:1133–1134.

Whelan, E. A., Lawson, C. C., Grajewski, B., Petersen, M. R., Pinkerton, L. E., Ward, E. M., and Schnorr, T. M. (2003). "Prevalence of respiratory symptoms among female flight attendants and teachers." *Occup Environ Med* 62:929–934.

White, C., Kolble, R., Carlson, R., and Lipson, N. (2005). "The impact of a health campaign on hand hygiene and upper respiratory illness among college students living in residence halls." *J Am Coll Health* 53(4):175–181.

Williams, R. E. O., Lidwell, O. M., and A, H. (1956). "The bacterial flora of the air of occupied rooms." *J Hyg* 54:512–523.

Zilisteanu, E., Nafta, I., Cretesco, L., Nicoulesco, I., and Focsaneanu, M. (1966). "Strains of parainfluenza virus type 4, isolated in a day nursery." *Arch Roum Pathol Exp Microbiol* 25(2):459–464.

CHAPTER 30

RESIDENTIAL HOUSING

30.1 INTRODUCTION

Residential housing consists of all those buildings used for lodging, including residential homes, apartments, dormitories, hotels, motels, and related structures. Smaller residential homes often depend on natural ventilation. Larger residential buildings like apartments and hotels differ considerably from commercial office buildings due to the fact that they usually do not depend on a single central air handling system. The characteristics of residential buildings and their associated vulnerabilities to aerobiological hazards are addressed for each type of building in the following sections.

30.2 RESIDENTIAL HOMES

Residential homes are where most individuals spend most of their time and for families the home may be the source of most of the respiratory infections they contract in their lives. Older homes often depend on natural ventilation augmented with heating and air conditioning. Newer homes often have central air handling units with refrigerant coils. Home ventilation systems rarely use filtration or other air cleaning devices, beyond simple dust filters.

The aerobiology of homes is multifaceted. In family homes the aerobiology will often be dominated by airborne bacteria and viruses produced by family members during the heating season. Dust mites are common due to the extensive use of carpeting and bedding, and the various sources of moisture available. Figure 30.1 shows an example of the most frequently identified fungi in homes. The fungal aerobiology of indoor air in homes will often resemble that of the outdoor air, except for water-damaged or problem homes. In the latter case the airborne fungal spores may reach extreme levels. Homes with pets may have high levels of animal dander.

Problem homes are those in which the levels of airborne fungal spores significantly exceed normal levels and those of the outdoor air, or otherwise have occupants who report symptoms. Figure 30.2 shows a comparison of a moisture problem home and a normal home in terms of the airborne fungi throughout the year. Such problem homes can amplify indoor levels of fungi and actinomycetes and are associated with respiratory distress and

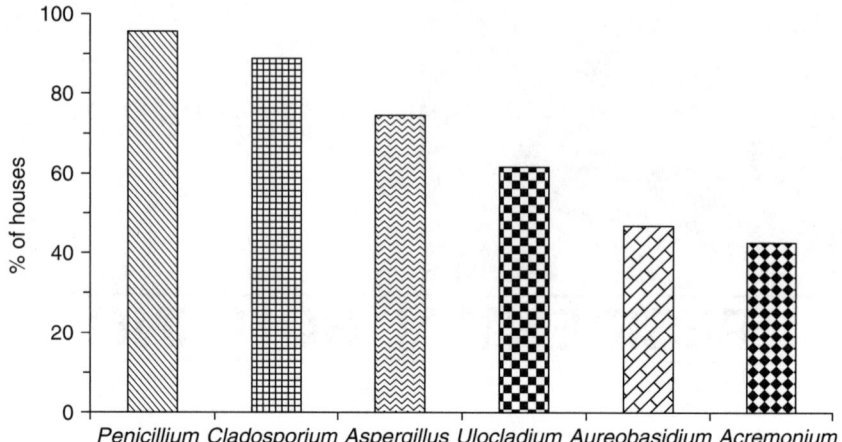

FIGURE 30.1 Comparison of the most frequently isolated fungi in homes. [*Based on data from Flannigan and Hunter (1988)*.]

allergic reactions in atopic individuals or those with asthma (Peltola et al., 1999; Lane, 1979; Brightman and Moss, 2001).

The levels of bacteria in a home will vary depending on occupancy, since most bacteria hail from human sources. Figure 30.3 illustrates the variety of bacteria that have been found in Scottish homes.

Table 30.1 lists a variety of studies and the indoor levels with each associated bioaerosol in both normal and problem homes. Reynolds et al. (1990) suggested that fungi levels above 500 cfu/m^3 correlated with increased reports of symptoms. Reponen et al. (1992) suggested the same limit for fungi, 500 cfu/m^3, for Finland, as did Jovanovic et al. (2001) as an indicator of elevated levels. For bacteria, levels exceeding 1000 cfu/m^3 have been proposed as limits for Hong Kong (Lee et al., 2002).

Stachybotrys chartarum is a fungus for which no indoor level is considered safe due to the dangerous toxins it can produce and the potentially serious health consequences of exposure. Water damage can result in growth and multiplication of this mold spore on gypsum and cellulose products. Immediate evacuation and remediation have been recommended wherever this microbe is found indoors (Scott and Yang, 1997).

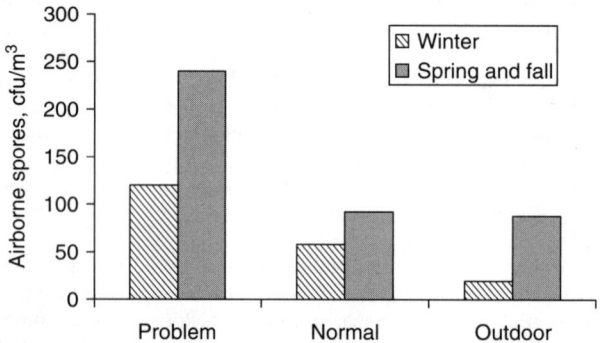

FIGURE 30.2 Comparison of airborne spores in a normal vs. a moisture problem house. [*Based on data from Hyvarinen et al. (1996a)*.]

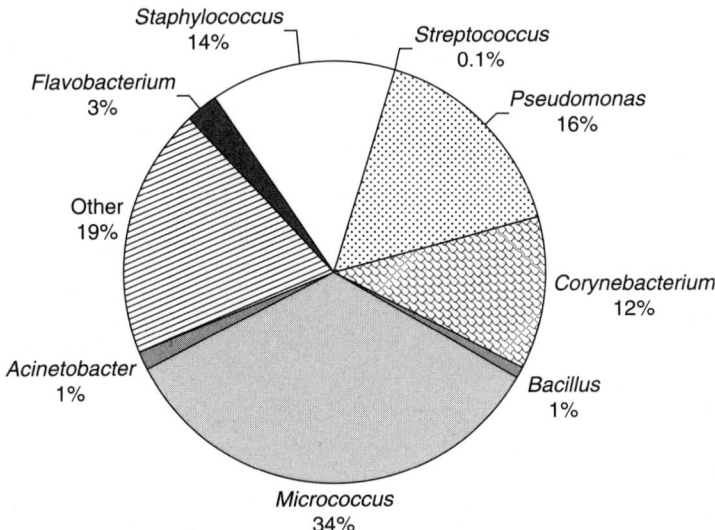

FIGURE 30.3 Breakdown of airborne bacteria found in 11 Scottish homes. [*Based on data from Flannigan et al. (1999)*.]

The species of fungi found indoors in a healthy home should bear a distinct relationship to that found in the outdoor air. Any significant deviation of the indoor air levels from the composition or the quantity in the outdoor air will suggest some degree of amplification, which may be species specific. Figure 30.4 illustrates a comparison of indoor air in a normal home with the outdoor air, based on a survey of 40 homes in Australia. It can be observed that the indoor air in these homes closely resembles the outdoor air in terms of composition.

Table 30.2 summarizes the results of a study on the types of fungi found in typical home furnishings (Kemp et al., 1999). Similar varieties of fungi may be found on other building materials like gypsum, mineral wool, base boarding, and even inside moisture damaged concrete walls (Kujanpaa et al., 1999; Pessi et al., 1999).

Hirsch and Sosman (1976) conducted a survey of fungal species in 12 homes in and around Milwaukee, Wisconsin, and found that certain species tended to favor specific parts of houses and types of ventilation systems. *Cladosporium* was the most commonly isolated airborne mold except in the basement where *Aspergillus* and *Penicillium* were more frequent. *Alternaria* was found more often in rooms with carpeting than in the basements or bathrooms where there is none. The total incidence of mold was greatest in the basement. *Cladosporium* was found more often in homes with air conditioning than in those without. *Aspergillus* was more frequent in homes with pets. *Alternaria* was found more often in wool carpets than in synthetic carpets, and more often in deeper piles.

Dust mites, agents of allergies and asthma, tend to thrive when indoor *relative humidity* (RH) is 60 to 75 percent or at temperatures of about 60 to 75°F, and tend to die out after several days when the RH is 40 to 50 percent or lower. House dust mites are rarely found in homes with low RH (i.e., less than 35 percent), according to a study by O'Rourke et al. (1996). Figure 30.5 illustrates some of the results of this study that describe locations in homes where dust mites are often found. Some 69 percent of the homes in the Sonoran desert that were studied were found to have some form of dust mites present. The mites tended to grow indoors in winter more than in other seasons, apparently due to the more favorable humidity and temperature conditions.

TABLE 30.1 Typical Levels of Contaminants in Residences

Contaminant	Normal home	Problem home	Reference
Bacteria (cfu/m^3)	140–680	—	Nevalainen et al., 1991
	380	—	Cole et al., 1993
	585	—	Godish et al., 1993
	590	960	Hyvarinen et al., 1996b
	655	—	Gallup et al., 1993
	1200	2165	Ren et al., 1993 (new vs. old homes)
	220–400	—	Lee et al., 2002
Actinomycetes (cfu/m^3)	3	60	Peltola et al., 1999
	3	3	Hyvarinen et al., 1996b
Fungi (cfu/m^3)	56	120	Hyvarinen et al., 1993
	109	—	Cole et al., 1993
	114	4013	Klanova and Drahonovska, 1999
	135	—	Jovanovic et al., 2001
	160	250	Reponen et al, 1993
	240	92	Hyvarinen et al., 1996a
	271	624	Flannigan et al., 1993
	19–889	74–1531	Miller, 2001 (apartments)
	409	—	Godish et al., 1993
	1776	—	Ren and Leaderer, 1999
Dust mite allergens (ng/m^3)	0.082	—	Deng et al., 1999
	0.12	—	Irie et al., 1993
Pollen (g/m^3)	6–79	—	Carinanos et al., 2004
	7–98	—	Sterling and Lewis, 1998 (mobile homes)
Particulates (PM$_{10}$), (mg/m^3)	0.049	—	Deng et al., 1999
	0.148	—	Lee et al., 2002
MVOCs (mg/m^3)	0.15	1.49	Hyvarinen et al., 1996a

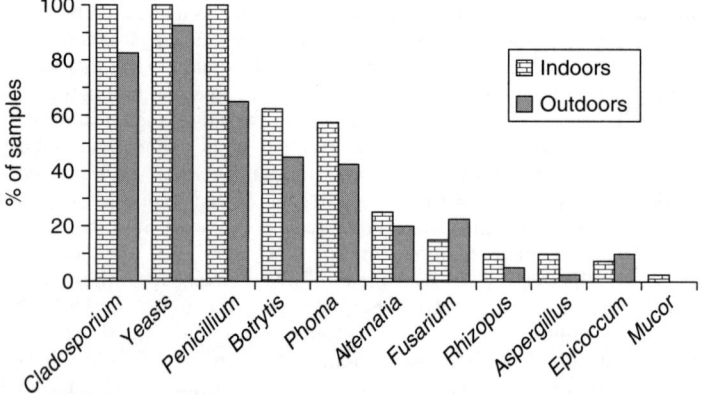

FIGURE 30.4 Comparison of indoor air fungal composition outdoor air. [*Based on data from Godish et al. (1993).*]

TABLE 30.2 Fungi Found on Typical Home Furnishings

Fungal genera	Carpet	Mats	Futon mattress	Latex mattress	Pillow	Bath mat	Sofa chair
Alternaria	X				X	X	
Acremonium	X						
Aspergillus	X	X	X	X	X		
Botrytis	X			X	X		
Cladosporium	X	X	X	X	X	X	X
Epicoccum		X	X				X
Fusarium	X			X			
Mucor	X						
Paecilomyces	X			X			
Penicillium	X	X		X		X	
Rhizopus	X	X	X	X	X	X	X
Trichoderma	X						
Other	X		X				

Animal allergens are common causes of both acute and chronic allergic diseases, and sensitization may occur at home or in the workplace (Chapman and Wood, 2001). The most important animal allergens derive principally from cats, dogs, rats, mice, horses, and cows, which release allergens into the environment. Cat and dog allergens commonly cause allergies in the home and affect the general population. Most of the mammalian allergens that have thus far been cloned belong to a single family of proteins called the lipocalins, and most animal allergens behave quite similarly with respect to their aerodynamic properties. The steps that may be necessary to control exposure to these allergens involve environmental modifications that include source control, air filtration devices, barrier devices, removal of carpeting and other reservoirs, and regular washing of the pets or animals.

Many types of allergens present in the indoor environment may lead to sensitization and respiratory allergy, including dust mites, animal dander, cockroach allergens, and molds.

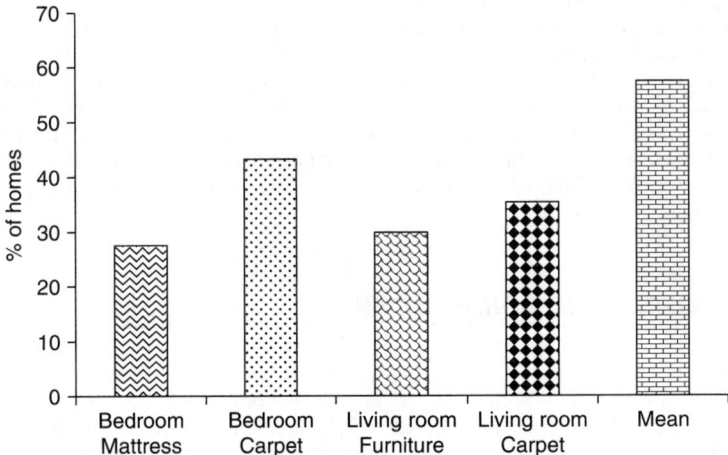

FIGURE 30.5 Dust Mites in indoor homes. [*Based on data from O'Rourke et al. (1996).*]

Research has demonstrated that asthma symptoms correlate with levels of domestic dust mite and cockroach exposure (Sarpong and Corey, 1998). In the case of dust mites, ending exposure results in symptomatic relief. Cockroach allergens have been found in house dust in 44 percent of homes that had no history of infestation, and over 8 percent of the residents were found to be sensitized to the allergen (Van Wijnen et al., 1997). Human dander in house dust has also been shown to be an allergen but its contribution to allergies is thought to be relatively minor (Voorhorst, 1977).

Cat allergens were found in all homes of asthmatic patients in one Canadian study, which may explain the high frequency of cat sensitization among patients with asthma (Quirce et al., 1995). Many patients who are allergic to cats do not own a cat and there is no obvious source of allergens in their homes. Cat allergens were found in dust from homes of 120 subjects with asthma in two Canadian cities although only 15 percent were cat owners. Rather high levels were found in homes of patients who did not have a cat but visited homes with cats.

Lind et al. (1987) examined mattress dust samples from 42 homes in Baltimore, Maryland. Dust mite allergens were detected 57 to 92 percent of the time. Cat and dog antigens were detected 63 to 77 percent of the time. Horse antigen was detected in 5 percent of homes.

In a study of home dampness and molds in Canadian homes, Dales et al. (1991) found molds were reported in 32.4 percent of homes, with flooding in 24.1 percent, and moisture in 14.1 percent. The prevalence of respiratory symptoms was consistently higher in homes with reported molds or dampness.

30.3 APARTMENT BUILDINGS

Apartment buildings vary widely in their size and occupancy, the presence of any central ventilation systems, and other factors. Most apartment buildings, at least older apartment buildings, often have individual air-conditioning units and operable windows. The aerobiology of apartment buildings is usually similar to that of residential homes.

In a study of the indoor concentration of airborne bacteria and fungi in apartments and other buildings, Sessa et al. (2002) measured the indoor concentrations of airborne bacteria and fungi in the presence and absence of occupants and furnishings. In presence of people and furnishings the average air concentrations of bacteria in an apartment, 92 to 182 cfu/m^3 were higher than in their absence, 66 to 80 cfu/m^3. The average fungal air concentrations were higher in the presence of people and furnishings, 147 to 297 cfu/m^3, than in their absence, 102 to 132 cfu/m^3.

Rautiala et al. (2002) report a case study on extensive fungal growth that occurred on building materials in an apartment 1 week after firefighting efforts. Airborne concentrations of fungal spores reached 10,000 cfu/m^3 and increased by two orders of magnitude during the removal of moldy building materials and during the cleanup operations. *Paecilomyces* was the main fungal genus in the indoor air during the removal, while *Penicillium* dominated during the reconstruction.

30.4 HOTELS AND DORMITORIES

Hotels and dormitories are unique in that the living quarters are often small and do not always have direct supply of air, only air conditioning or heaters under occupant control. Hotels will typically have one or more central air handling units providing air to the lobbies, hallways, restaurants, and other large areas. This supply of air is assumed to infiltrate

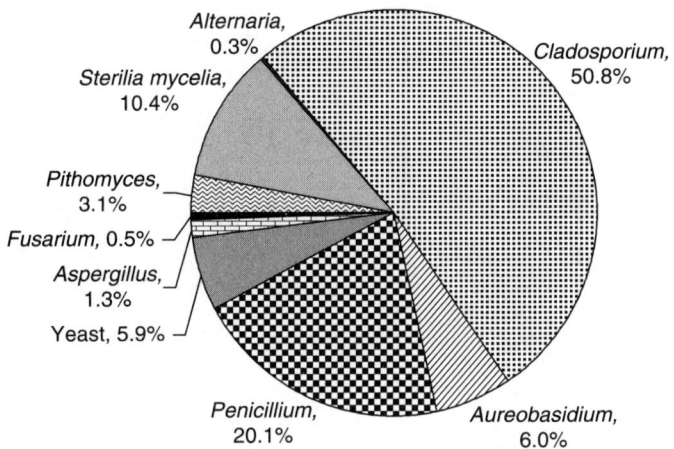

FIGURE 30.6 Airborne fungal spores in a hotel that had water damage. (*Based on average of two rooms from author's data.*)

into the individual hotel rooms. Sometimes the heaters or air-conditioning units may have individual outside air dampers.

Although the central air handling units in hotels may have medium-to-low efficiency filters, the room air conditioners rarely have anything more than a simple dust filter fabric. As a result, these air conditioners tend to accumulate spores over time. With the presence of condensation, these spores may even amplify and lead to air quality problems in rooms.

Regular maintenance of room air conditioners normally involves removing the unit and cleaning with an acid or fungicide once a year. In a 200- to 400-unit hotel where only one or two units can be cleaned per day, this means that about half the units will have months of accumulation during the times of the year they need it most—summer and fall. As a result, hotel patrons often discover that when they turn on the air conditioner it produces a somewhat unpleasant odor.

Figure 30.6 shows the results of a survey taken by the author of the air in two hotel rooms in winter, in a hotel that had experienced water damage from a leaky roof. Although outdoor spore levels were less than 10 cfu/m^3, indoor spore levels in one of the rooms exceeded a few hundred cfu/m^3.

Overcrowding has been noted as a risk factor for meningitis in university halls of residence, especially among first year students, as is bar patronage (Nelson et al., 2001). Hygiene becomes more critical under crowded conditions and the subject could certainly use more attention at all educational levels. See the suggested "Hygienic Protocols for Home, Office, and School" in Chap. 18 for guidelines on this matter.

30.5 SOLUTIONS FOR HOMES AND APARTMENTS

The solution to aerobiological problems in residential environments depends on the nature of the bioaerosols and the health concerns of the occupants. Figure 30.7 provides a figurative summary of the various bioaerosols of most concern and the various common options for dealing with them. There are two basic groups of biocontaminants represented in this

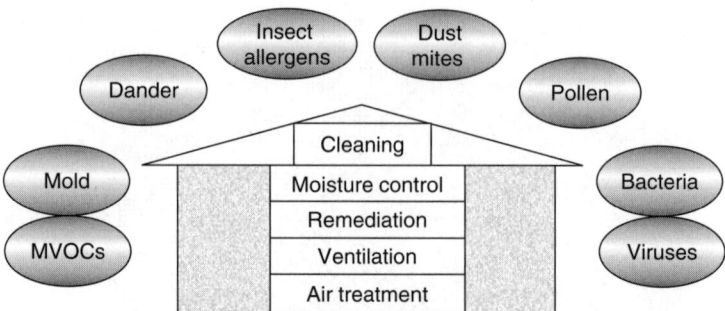

FIGURE 30.7 Illustration of the major biocontaminants in households and the basic options available for controlling indoor levels.

figure, allergens and pathogens. Allergens like dander and dust mites can be controlled to a degree by regular cleaning. Allergens from the outdoors like mold spores and pollen require some form of air cleaning and building envelope control. If, however, mold spores are being generated internally, then moisture is likely the problem and ventilation and drying up problem areas or dehumidification may be the solution. In cases of excessive mold, remediation may be the answer.

Pathogenic viruses and bacteria may be a primary concern among families, especially when children are present. In this case, improved ventilation or air treatment may be the only feasible solution. Modifying the ventilation system and incorporating filtration and/or UVGI is one alternative. Another is to add a recirculating filter/UVGI system.

Many homes and apartments are naturally ventilated and have few options for improving indoor air quality short of installing forced air systems. In such situations the use of local recirculating air treatment units are perhaps the best option. The choice of the system may depend on the nature of the problem or the microbes of concern. If the main concern is pollen, dander, and allergens, a basic recirculating filter unit may be adequate. Figure 30.8 shows an example of a home HEPA filter recirculation unit.

In homes where *volatile organic compounds* (VOCs) are a problem, then improved ventilation or recirculating carbon adsorber units may provide some relief. If, however, the problem is due to *microbial volatile organic compounds* (MVOCs), then they are being generated by fungi or bacteria, and a solution needs to be sought that addresses the source of the problem. That solution may again be cleaning, dehumidification, remediation, or air treatment.

Water damage, whether from flooding, roof leakage, excessive condensation, rising damp, or other causes, should be investigated or remediated per the methods described in Chap. 26. Any kind of major water damage or chronic moisture problem should be addressed immediately because once fungi begin to grow the problem becomes orders of magnitude more costly.

The use of air cleaning to control air quality in a house that has an ongoing mold problem may only be a superficial solution if it does not address the root cause. Although it can be effective, it may only hide the mold contamination problem, which will remain until the mold is cleaned and/or contaminated materials are removed. In a series of 11 homes in which the residents suffered from hypersensitivity pneumonitis, due apparently to *Trichosporon* contamination, the remedial measures shown in Table 30.3 were undertaken and eliminated the problem (Yoshida et al., 1989).

Atopic individuals concerned with allergens from outdoor air should shun naturally ventilated houses in favor of those with central air handling units that can be fitted with

FIGURE 30.8 HEPA filter recirculation unit for home and apartment use. (*Photo courtesy of Access Business Group, Ada, MI.*)

proper filters. Pollen are relatively easy to control with moderate efficiency filters and even with air conditioners alone (Solomon et al., 1980). Shutting windows in a naturally ventilated home during allergy season offers little relief from spores that are already present inside. In such situations, the use of recirculating filter units may provide some relief if the unit is sufficient to purge the entire house of airborne allergens. Often, this will depend on how well the air in all the rooms of the house mixes or finds its way to the filter unit. The use of portable fans or ceiling fans can assist the mixing of air in such cases.

Domestic pets are a source of allergens for which the only complete solution is to remove the pet from the home. Some remedial measures can be taken, however, to minimize the problem to allergic occupants and visitors, including frequent washing of the pet, limiting the pet's access to some areas of the house, and frequent vacuuming with a high-efficiency filtered cleaner (Short, 1999). Pet ownership is high in families where one or more members have an allergy to pet dander and air filtration has sometimes been recommended as a means of reducing pet allergen exposure. Although a properly designed and installed air filtration system should theoretically reduce airborne allergens to tolerable levels, no conclusive evidence has yet surfaced demonstrating the effectiveness of this option.

Carpeting is a collector and a sink for a variety of particulates that can become reaerosolized from ordinary traffic (Cole et al., 1993). Thick piles tend to accumulate more biocontaminants than thin piles, and organic materials sustain more fungal spores than synthetics (Hirsch and Sosman, 1976). Replacement of carpeting with hard flooring or vinyl is

TABLE 30.3 Remedial Measures for Trichosporon Mold Contamination

1. Removal of damp and decayed wood, carpets, mats, and pets
2. Thorough cleaning and disinfecting of the houses, using disinfectants such as sodium hypochlorite, ethanol, and chlorhexidine gluconate
3. Repair of water problems, improvement of ventilation, reduction of relative humidity

one solution to minimized mold spore accumulation, as is regular vacuuming and cleaning. Increasing solar exposure can also aid in limiting spore survival and mold growth. Furnishings can also absorb spores with leather and vinyl being materials considered to have much less capacity. Additional recommendations on healthy building materials and designs may be found in Chap. 18.

Although cleaning with detergents may physically remove allergens from carpeting, building materials, and furnishings, they may not destroy the allergenic activity. Dybendal et al., (1990) studied the effects of nine household cleaning solutions and five chemical detergents on dust allergens from carpeted and smooth floors, including allergens from pollen extracts, codfish, hen egg-white, cat dander, and house dust mites. Soft soap, guanidine hydrochloride, and sodium lauryl sulphate had the greatest impact, but none of the detergents totally eliminated allergenic activities, even when used in concentrations up to 10 times more than recommended. Dust mites on bedding can be controlled by washing them weekly and by exposing the sheets and pillows to light and air during the daytime, rather then covering them up.

Old ventilation systems should be regularly inspected, and cleaned or remediated as necessary. There is a strong association between respiratory problems and buildings having ventilation systems in operation for 20 years or more (Graudenz et al., 2002).

Central air handling units can themselves become a source of aerobiological hazards if the cooling coils become contaminated with spores. Since these spores may germinate and grow on the coils, or on moist filters or ductwork, the air handling unit may be serving to distribute the microbes throughout the house. One of the best solutions to this problem is to install a UVGI lamp to irradiate the coils, filters, or internal duct surfaces. For the typical house, a UVGI lamp of about 8 to 24 W of UV output (see Fig. 30.9) should be sufficient if it can be located so as to irradiate most of the surface area. Smaller lamps of 4 UV W are available and these can be placed at two or more locations (i.e., upstream and downstream of the coils) to provide good coverage. It is often necessary for a filter of at least a

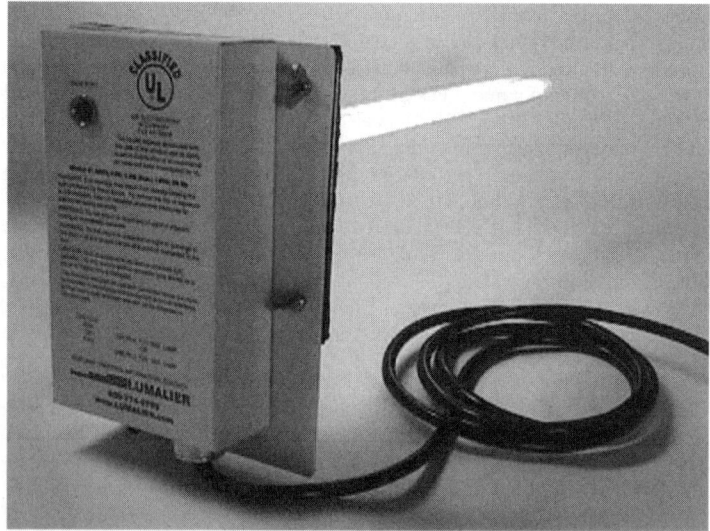

FIGURE 30.9 UL-approved UVGI system for retrofitting in-duct or microbial growth control applications. (*Photo provided courtesy of Lumalier, Memphis, TN.*)

MERV 6–8 rating (20 to 25 percent DSP) to be present to keep the lamp clean of dust. Filters should be selected for the lowest pressure drop available so as not to impact fan performance, since most home air handling units have limited capacity.

Installation of UVGI lamps on home central air handling units can be a simple matter for a homeowner but it is recommended that any such fixtures be UL rated. It is also recommended that careful consideration be given to the possible presence of water or condensation inside the air handling unit as this could conceivably short out any electrical equipment and cause a fire. If in doubt, consult an electrician or the lamp manufacturer for specific directions for such installations.

Apartment dwellers in buildings without central air are limited to the same options as homeowners in naturally ventilated buildings. Recirculating filter/UVGI systems may be the best option in such situations. Such units are widely available but performance can vary considerably. As a general rule, at least 2 air changes per hour are necessary to obtain adequate performance. Table 30.4 summarizes some suggested guidelines for recirculation units for apartments and residential homes. It is assumed in these estimates that the air velocity through the unit is 400 to 600 fpm and that there is at least a 2-feet length of clean galvanized steel that the UV lamp irradiates. It also assumes that at least a MERV 6–8 filter is included for protection of the UV lamp and that the lamp is operating at or near room temperature and humidity conditions. These wattages are suggested based on a system rating of URV 13. Lower URV ratings (i.e., MERV 9–11) may be equally capable of maintaining low levels of airborne microbes in indoor environments provided other aspects of the ventilation system are in good order (i.e., airflow distribution, no amplifiers, and like). In fact, a review of current residential home systems and installers suggests that systems of URV 15 or better are typically provided. Conservatism in the residential environment is a good thing and the differences in energy cost are not significant.

TABLE 30.4 Recirculation Air Cleaner Sizes for Apartments and Homes

Floor space			Minimum outside air		Recirculation airflow		URV 13 UV watts
ft^2	m^2	Occupancy	cfm	m^3/min	cfm	m^3/min	
500	46	1	30	0.85	550	15.57	10–15
		2	60	1.70			
1000	93	1	30	0.85	1100	31.15	20–30
		2	60	1.70			
		4	120	3.40			
2000	186	1	30	0.85	2200	62.30	40–60
		2	60	2.70			
		4	120	3.40			
3000	279	2	60	1.70	3300	93.45	60–90
		4	120	3.40			
		6	180	5.10			
4000	372	4	120	3.40	4400	124.59	80–120
		6	180	5.10			
		8	240	6.80			
5000	465	4	120	3.40	5500	155.74	100–150
		6	180	5.10			
		8	240	6.80			

30.6 SOLUTIONS FOR HOTELS

Hotels required a multifaceted approach since they typically contain both central air handling units and individual room heating and air-conditioning units. Typically, these units provide filtered outdoor air. The first step to improving or remediating indoor air is to review the filter capacity and check the filter bypass on existing air handling units. If higher efficiency filters can be installed, then the highest MERV rated filter that can be handled by the fan (from about MERV 9 up to about MERV 13–15 maximum) should be considered provided it does not significantly impact total volumetric airflow. Such upgrades will have the effect of decreasing indoor levels of fungal spores and ultimately may reduce the amount of fungal contamination that accrues on room air conditioners. It is important to keep in mind that filter bypass may be increased with higher filter efficiency, and the associated higher pressure losses. To counter this effect, filters seals should be inspected and replaced if they permit too much leakage.

Central air handling units in hotels are subject to the same fungal contamination of coils and filters as other buildings except that since they often draw large quantities of direct outside air, the coils and filters can become highly contaminated with mold spores. The best solution, short of regular cleaning of the coils with fungicides, is to install UVGI lamps to control microbial growth. Such systems have proven not only to be highly effective at suppressing all fungal and bacterial contamination, but they also pay dividends in terms of reduced energy costs. Many manufacturers now provide optional UVGI systems with package air conditioners and can provide retrofits for existing equipment. Professionals can also be consulted for design and installation assistance. In all cases, UL-approved lamps should be used and installations should be inspected to assure that no electrical problems or fire hazards have been introduced by such retrofits.

Hotel air conditioners are routinely cleaned, often with steam, bleach, or disinfectants, according to regularly scheduled maintenance programs. For a 300-room hotel, for example, some six units might have to be cleaned a week. This means that late in the year over half the units may have spring and summer mold accumulation. Some consideration should be given to adjusting procedures to provide aggressive cleaning schedules during or following the periods of peak airborne fungi, which may vary depending on climate. UVGI presents what may be a more economical option except that many window or wall air-conditioning units are so compactly constructed that UV lamps cannot be satisfactorily installed. Furthermore, these units often have little, if any, filtration and the lack of a good dust filter will cause UV lamp performance to deteriorate over time. What is needed is the redesign of window air units for hotels such that they can accept better filtration and UVGI lamps.

It is important that no stray UV light escapes from any retrofitted air-conditioning unit as this may potentially cause harm to room patrons or damage room furnishings. The maximum UV exposure level for humans is $0.2\ \mu W/cm^2$ for 8 hours. If a UV lamp cannot be installed internally in an air-conditioning unit such that the coils, filter, and any grille will reduce the external exposure to below this level, then such installations may not be viable. An exception may be where overhead cooling units exist, since the stray UV rays will tend to run parallel to the ceiling much like an upper air UVGI system. As long as UV levels at eye level are not hazardous, such modifications can be implemented.

30.7 HOMES FOR THE IMMUNOCOMPROMISED

Houses can be modified as a cost-effective alternative for providing protective isolation for immunocompromised individuals and AIDS patients (Linscomb, 1994). Immunocompromised individuals require higher standards of indoor air quality than normal healthy people

Guidelines for Controlling Opportunistic Fungi

1. **Control Moisture**
   1.1. Keep relative humidity as low as practical (i.e., 30 to 50 percent).
   1.2. Clean and dry damp areas—plumbing and flooring under sinks, toilets, and faucets; washing machines, refrigerators, drain pans, countertops; condensate on window frames, basements, drains; install exhaust fans in bathrooms and damp areas.
   1.3. Fix plumbing and roof leaks, inspect regularly.
   1.4. Dry water damaged areas immediately.
   1.5. Replace all water damaged materials, drywall, wood, carpets, and the like.
   1.6. Avoid living in below ground areas.
   1.7. Inspect air-conditioning and ventilation equipment—check for standing water and condensation; clean any biofouling on coils or in drain pans; steam clean coils if necessary.

2. **Clean Up Microbial Reservoirs**
   2.1. Floors and Countertops—use an effective cleaning agent, bleach, or disinfectant.
   2.2. Garbage Cans and Trash Containers—clean and disinfect weekly, store outdoors.
   2.3. Carpet and Upholstery—professionally steam clean twice a year and vacuum regularly; use a vacuum with a high-efficiency filter or a HEPA filter; central vacuum cleaners that exhaust to outside may be best; remove carpeting, or replace with synthetics and a thin pile.
   2.4. Furnishings—clean and disinfect periodically, avoid thick piles, select synthetics; vacuum and wipe curtains and blinds, remove curtains if possible.
   2.5. Stoves and Refrigerators—vacuum around and behind, clean with disinfectant.
   2.6. Food Storage Areas—inspect and clean regularly, at least weekly.

3. **Prevent Amplification**
   3.1. Dry Moist Areas—tubs, showers, curtains, liners, use desiccants.
   3.2. Apply Disinfectants—use after cleaning moisture-prone areas in bathrooms and kitchens; use nonaqueous disinfectant sprays, alcohol based, or bleaches.
   3.3. Maximize Interior Sunlight—open or remove blinds and curtains, clean windows.

4. **Restrict Contaminant Buildup and Entry**
   4.1. Entryway Mats—place at doorways to clean shoes on entry.
   4.2. Eliminate Stored Materials—remove to exterior storage or use plastic airtight boxes.
   4.3. Carpets—remove and replace with hard flooring or thin pile synthetics.
   4.4. Furnishings—replace with vinyl, leather, or synthetics.
   4.5. Mattresses—use foam rubber, or encase in plastic.
   4.6. Doors and Windows—tighten up and add seals or sealants.
   4.7. Houseplants and Pets—remove them.
   4.8. HVAC Equipment—inspect and clean, seal up leakpoints.

5. **Install Air Treatment**
   5.1. Air Filters—use central A/C filters MERV 11–15 (DSP 75 to 95 percent).
   5.2. UVGI—add to HVAC, irradiate recirculated air and/or cooling coils.
   5.3. Recirculation Air Cleaners—add to main rooms, living rooms, bedroom, kitchen.

and must be protected from outdoor mold spores and environmental bacteria that would not normally harm healthy individuals. When a house is converted to this end, it must have a high degree of leak-tightness to maintain the building envelope as a barrier against the outdoor air microbes. It must also be pressurized to ensure what leakage occurs is to the outside air and not vice versa. Essential to the indoor air cleanliness is an air treatment system that filters and irradiates recirculated and mixed outside air (Linscomb, 1994).

In addition to the modification of the building envelope, if necessary, and the incorporation of air treatment, is preliminary and ongoing cleaning remediation as a means of controlling opportunistic fungi. Some guidelines for controlling opportunistic fungi, adapted from Cook et al., (1999) are summarized here. These guidelines may be applied to mold problems in general are similar to other remedial approaches (i.e., see Chaps. 26 and 29) and professionals should be consulted for implementation of such programs.

REFERENCES

Brightman, H. S., and Moss, N. (2001). Chapter 3: Sick Building Syndrome Studies and the Compilation of Normative and Comparative Values. *Indoor Air Quality Handbook*, J. D. Spengler, J. M. Samet, and J. F. McCarthy, eds., McGraw-Hill, New York. 3.1–3.32.

Carinanos, P., Alcazar, P., Galan, C., Navrro, R., and Diminguez, E. (2004). "Aerobiology as a tool to help in episodes of occupational allergy in workplaces." *J Invest Allergol Clin Immunol* 14(4): 300–308.

Chapman, M. D., and Wood, R. A. (2001). "The role and remediation of animal allergens in allergic diseases." *J Allergy Clin Immunol* 107(3 Suppl.):S414–S421.

Cole, E. C., Foarde, K. K., Leese, K. E., Franke, D. L., and Berry, M. A. (1993). "Biocontaminants in carpeted environments." *Indoor Air 93*, Helsinki, Finland.

Cook, C. E., Cole, E. C., Dulaney, P. D., and Leese, K. E. (1999). "Reservoirs for opportunistic fungi in the home environment: A guide for exposure reduction in the immunocompromised." *Indoor Air 99: Proceedings of the 8th International Conference on Indoor Air Quality and Climate*, Edinburgh, Scotland. 905–910.

Dales, R. E., Zwanenburg, H., Burnett, R., and Franklin, C. A. (1991). "Respiratory health effects of home dampness and molds among Canadian children." *Am J Epidem* 134(2):196–203.

Deng, X., Irie, T., and Ohmura, M. (1999). "Field survey of living mites and airborne allergens in dwellings in Shinshu district, Japan." *Indoor Air 99: Proceedings of the 8th International Conference on Indoor Air Quality and Climate*, Edinburgh, Scotland. 935–940.

Dunn, K. (1990). *The World's First Allergy-Free House*. The South Australian Housing Trust, Adelaide, Australia.

Dybendal, T., Vik, H., and Elsayed, S. (1990). "Dust from carpeted and smooth floors. III. Trials on denaturation of allergenic proteins by household cleaning solutions and chemical detergents." *Ann Occup Hyg* 34(2):215–229.

Flannigan, B., and Hunter, C. A. (1988). Factors affecting airborne moulds in domestic dwellings. *Indoor Air and Ambient Air Quality*, R. Perry and P. W. Kirk, eds. Selper Publications, London. 461–468.

Flannigan, B., McCabe, E. M., Jupe, S. V., and Jeffrey, I. G. (1993). "Mycological and acralogical investigation of complaint and noncomplaint houses in Scotland." *Proceedings of the 6th International Conference on Indoor Air Quality and Climate*, Helsinki, Finland. 143–148.

Flannigan, B., McEvoy, E. M., and McGarry, F. (1999). "Investigation of airborne and surface bacteria in homes." *Indoor Air 99: Proceedings of the 8th International Conference on Indoor Air Quality and Climate*, Edinburgh, Scotland. 884–889.

Gallup, J. M., Zanolli, J., and Olson, L. (1993). "Airborne bacterial exposure: Preliminary results of volumetric studies performed in office buildings, schools, and homes in California." *Indoor Air:*

Proceedings of the 6th International Conference on Indoor Air Quality and Climate, Helsinki, Finland. 167–170.

Godish, D., Godish, T., Hooper, B., Panter, C., Cole, M., and Hooper, M. (1993). "Airborne mold and bacteria levels in selected houses in the Latrobe Valley, Victoria, Australia." *Indoor Air '93'*, Helsinki, Finland. 171–175.

Graudenz, G. S., Kalil, J., Saldiva, P. H., Gambale, W., Latorre, M. R., and Morato-Castro, F. F. (2002). "Upper respiratory symptoms associated with aging of the ventilation system in artificially ventilated offices in Sao Paulo, Brazil." *Chest* 122(2):729–735.

Hirsch, S. R., and Sosman, J. A. (1976). "A one-year study of mold growth inside twelve homes." *Ann Allergy* 36:30–38.

Hyvarinen, A., Reponen, T., Husman, T., Ruuskanen, J., and Nevalainen, A. (1993). "Composition of fungal flora in mold problem houses determined with four different methods." *IAQ '93'*, Helsinki, Finland. 273–278.

Hyvarinen, A., Meklin, T., Reponen, T., Kotimaa, M., and Nevalainen, A. (1996a). "Indoor air fungi during the heating season." *Proceedings of the 7th International Conference on IAQ and Climate,* Nagoya, Japan. 1089–1094.

Hyvarinen, A., Reponen, T., Husman, T., and Nevalainen, A. (1996b). "Characterizing mold problem buildings—differences in indoor air characteristics." *Proceedings of the 7th International Conference on IAQ and Climate,* Nagoya, Japan. 1125–1129.

Irie, T., Sakaguchi, M., Murakami, S., Namba, H., Fujino, S., Komine, H., Yasueda, H., and Minomiya, Y. (1993). "Study on the behavior of mite allergens in high-rise residential buildings." *Proceedings of the 6th International Conference on Indoor Air Quality and Climate,* Helsinki, Finland. 177–182.

Jovanovic, S., Piechotowski, I., Gabrio, T., Weidner, U., Zollner, I., and Schwenk, M. (2001). "Assessment of mould pollution in residences in southwest Germany." *Gesundheitswesen* 63(6):404–411.

Kemp, P. C., Neumeister-Kemp, H. G., Nickelmann, A., and Murray, F. (1999). "Fungi in the dust extracted from fabric covered furnishings: Preliminary results during method standardisation." *Indoor Air 99: Proceedings of the 8th International Conference on Indoor Air Quality and Climate,* Edinburgh, Scotland. 890–891.

Klanova, K., and Drahonovska, H. (1999). "The concentrations of mixed populations of fungi in indoor air: Rooms with and without mould problems; rooms with and without health complaints." *Indoor Air 99: Proceedings of the 8th International Conference on Indoor Air Quality and Climate,* Edinburgh, Scotland. 920–924.

Kujanpaa, L., Haatainen, S., Kujanpaa, R., Vilkki, R., and Reiman, M. (1999). "Microbes in material samples taken from base boardings, gypsum boards and mineral wool insulation." *Indoor Air 99: Proceedings of the 8th International Conference on Indoor Air Quality and Climate,* Edinburgh, Scotland. 892–896.

Lane, D. J. (1979). *Asthma: The Facts.* Oxford University Press, New York.

Lee, S. C., Li, W.-M., and Ao, C.-H. (2002). "Investigation of indoor air quality at residential homes in Hong Kong—case study." *Atmos Environ* 36:225–237.

Lind, P., Norman, P. S., Newton, M., Lowenstein, H., and Schwartz, B. (1987). "The prevalence of indoor allergens in the Baltimore area: house dust-mite and animal-dander antigens measured by immunochemical techniques." *J Allergy Clin Immunol* 80(4):541–547.

Linscomb, M. (1994). "AIDS clinic HVAC system limits spread of TB." *HPAC Eng* 66(2):55–57.

Miller, J. D. (2001). Mycological investigations of indoor environments. *Microorganisms in Home and Work Environments,* B. Flannigan, R. A. Samson, and J. D. Miller, eds., Taylor & Francis, London. 218–231.

Nelson, S. J., Charlett, A., Orr, H. J., Barker, R. M., Neal, K. R., Taylor, C., Monk, P. N., Evans, M. R., and Stuart, J. M. (2001). "Risk factors for meningococcal disease in university halls of residence." *Epidemiol Infect* 126(2):211–217.

Nevalainen, A., Reponen, T., Heinonen-Tanski, H., and Raunemaa, T. (1991). Indoor air bacteria in apartment homes before and after occupancy. *Healthy Buildings/IAQ '91'*, Washington, DC.

O'Rourke, M. K., Moore, C. L., and Arlian, L. G. (1996). Chapter 6: Prevalence of dust mites from homes in the Sonoran desert, Arizona. *Pathogens in the Environment*, B. Austin, ed., Blackwell Scientific, Oxford, UK.

Peltola, J., Andersson, M., Haahtela, T., Mussalo-Rauhamaa, H., and Salkinoja-Salonen, M. (1999). "Toxigenic indoor actinomycetes and fungi case study." *Indoor Air 99: Proceedings of the 8th International Conference on Indoor Air Quality and Climate,* Edinburgh, Scotland. 560–561.

Pessi, A. M., Helkio, K., Suonketo, J., Pentti, M., and Rantio-Lehtimaki, A. (1999). "Microbial growth inside exterior walls of precast concrete buildings as a possible risk factor for indoor air quality." *Indoor Air 99: Proceedings of the 8th International Conference on Indoor Air Quality and Climate,* Edinburgh, Scotland. 899–904.

Quirce, S., Dimich-Ward, H., Chan, H., Ferguson, A., Becker, A., Manfreda, J., Simons, E., and Chan-Yeung, M. (1995). "Major cat allergen (Fel d I) levels in the homes of patients with asthma and their relationship to sensitization to cat dander." *Ann Allergy Asthma Immunol* 75(4):325–330.

Rautiala, S. H., Nevalainen, A. I., and Kalliokoski, P. J. (2002). "Firefighting efforts may lead to massive fungal growth and exposure within one week. A case report." *Int J Occup Med Environ Health* 15(3):303–308.

Ren, J., Guo, R. R., and Wang, X. (1993). "Indoor airborne bacteria concentrations and respiratory disease in old and new rural dwelling." *Proceedings of the 6th International Conference on Indoor Air Quality and Climate,* Helsinki, Finland. 207–212.

Ren, P., and Leaderer, D. C. (1999). "The nature and concentration of fungi inside and outside homes." *Indoor Air 99 : Proceedings of the 8th International Conference on Indoor Air Quality and Climate,* Edinburgh, Scotland. 930–934.

Reponen, T., Navalainen, A., Jantunen, M., Pellikka, M., and Kalliokoski, P. (1992). "Normal range criteria for indoor air bacteria and fungal spores in a subarctic climate." *Indoor Air* 2:26–31.

Reponen, T., Hyvarinen, A., Ruuskanen, J., Raunemaa, T., and Nevalainen, A. (1993). "Size distribution of fungal spores in houses with mold problems." *Proceedings of the 6th International Conference on Indoor Air Quality and Climate,* Helsinki, Finland. 243–248.

Reynolds, S. J., Streifel, A. J., and McJilton, C. E. (1990). "Elevated airborne concentrations of fungi in residential and office environments." *Am Ind Hyg Assoc J* 51:601–604.

Sarpong, S. B., and Corey, J. P. (1998). "Assessment of the indoor environment in respiratory allergy." *Ear Nose Throat J* 77(12):960–964.

Scott, R., and Yang, C. (1997). "Comparison of successful and unsuccessful *Stachybotrys chartarum* remediation projects." *Healthy Buildings/IAQ '97',* Bethesda, MD. 269.

Sessa, R., Di, P. M., Schiavoni, G., Santino, I., Altieri, A., Pinelli, S., and Del, P. M. (2002). "Microbiological indoor air quality in healthy buildings." *New Microbiol* 25(1):51–56.

Short, S. D. (1999). "Allergens in the home: Risk factors and control measures." *Indoor Air 99: Proceedings of the 8th International Conference on Indoor Air Quality and Climate,* Edinburgh, Scotland. 925–929.

Solomon, W. R., Burge, H. A., and Boise, J. R. (1980). "Exclusion of particulate allergens by window air conditioners." *J Allergy Clin Immunol* 65(4):305–308.

Sterling, D. A., and Lewis, R. D. (1998). "Pollen and fungal spores indoor and outdoor of mobile homes." *Ann Allergy Asthma Immunol* 80:279–285.

Van Wijnen, J. H., Verhoeff, A. P., Mulder-Folkerts, D. K., Brachel, H. J., and Schou, C. (1997). "Cockroach allergen in house dust." *Allergy* 52(4):460–464.

Voorhorst, R. (1977). "The human dander atopy. I. The prototype of auto-atopy." *Ann Allergy* 39(3):205–212.

Yoshida, K., Ando, M., Sakata, T., and Araki, S. (1989). "Prevention of summer-type hypersensitivity pneumonitis: Effect of elimination of *Trichosporon cutaneum* from the patients' homes." *Arch Environ Health* 44(5):317–322.

CHAPTER 31

HEALTH CARE FACILITIES

31.1 INTRODUCTION

Health care facilities consist of all types of buildings associated with health care, such as hospitals, clinics, doctor's offices, dental offices, and homes or offices converted for health care use. These facilities are subject to airborne hazards that can threaten both patients and health care workers. Hospitals typically include a variety of buildings with various types of ventilation requirements. Hospital facilities may include general patient areas, procedure rooms, neonatal care units, operating suites, burn wards, isolation rooms for TB or AIDS patients, laboratories, and office buildings. Laboratories are covered separately in Chap. 32 while office buildings are covered in Chap. 28. Dental offices are addressed in this chapter because of their similarity to procedure rooms, although their concerns are somewhat different than health care facilities.

This chapter presents a short review of the guidelines, codes, and standards that are in current use to aid the design of health care facilities and only addresses those fundamental aspects of HVAC system design that relate to aerobiological problems and air treatment options. It seeks to provide infection control professionals with an improved perspective of how aerobiological contaminants impact nosocomial and related concerns and how they can be better controlled at all levels through the combined use of ventilation, pressurization, filtration, UVGI, and other disinfection technologies. This chapter attempts to cover the problem of nosocomial infections and related hazards from an aerobiological engineering perspective, and addresses the various applications of air and surface disinfection technologies that may aid in the reduction of nosocomial infections.

31.2 GUIDELINES, CODES, AND STANDARDS

A variety of guidelines, codes, and standards have been established that dictate specific requirements for health care facility ventilation system design, construction, and operation (AIA, 2001; ASHRAE, 1999a, 2003; CDC, 2003). In addition, there are guidelines that focus on specific problems like tuberculosis, nosocomial infections, and *surgical site infections* (SSIs) (CDC, 1994; Wenzel, 1981; Mangram et al., 1999; Kubica, 1996; Tablan et al., 1994).

In spite of the wealth of information in the codes and standards on ventilation system design, there are few guidelines on the design and installation of UVGI air disinfection systems for health care applications and the subject has, in the past, received only passing mention (Nardell, 1988; NIOSH, 2005; Luciano, 1977).

Guidelines for the design of health care ventilation systems have evolved over the years and have proven adequate in providing the highest levels of air cleanliness in hospital wards, operating rooms, and associated facilities. Much of the focus of these design guidelines is on providing high rates of air exchange using 100 percent outside air supplied through high-efficiency filters. Figure 31.1 illustrates the effectiveness of purging contaminants with various rates of filtered outside air with complete air mixing in a room. It can be observed that virtually sterility can be achieved in a room assuming complete air mixing, sterile supply air, and no source of airborne microbes in the room. Of course, there will always be some leakage and generation of microbes in an occupied building, and neither air mixing nor filtration is ever quite perfect, but the example illustrates the potential effectiveness of this approach. Treatment and operating rooms typically have an ACH of 12 to 25, whereas patient and intensive care rooms typically have an ACH of 4 to 6 (ASHRAE, 2003).

Outside air is filtered before being supplied to hospital areas with levels of filtration such as those shown in Table 31.1, based on ASHRAE (1999b, 1999c). The filters are located in sequence as indicated and often have a dust filter preceding them. For details on the performance of these filters against airborne microbes, see Chap. 10.

In addition, filters are often used when air is recirculated in isolation rooms and other areas (ASHRAE, 2003). The filters used in such applications are typically *high-efficiency particulate air* (HEPA) filters. HEPA filters are also often used to filter exhaust air from isolation rooms, laboratories, and other facilities. Figure 31.2 illustrates the effectiveness of purging contaminants at different air change rates with HEPA-filtered recirculation air with complete air mixing in a room. Note that the results are virtually identical to purging with filtered outside air as shown in Fig.31.1, which would be expected since they both purge with clean air.

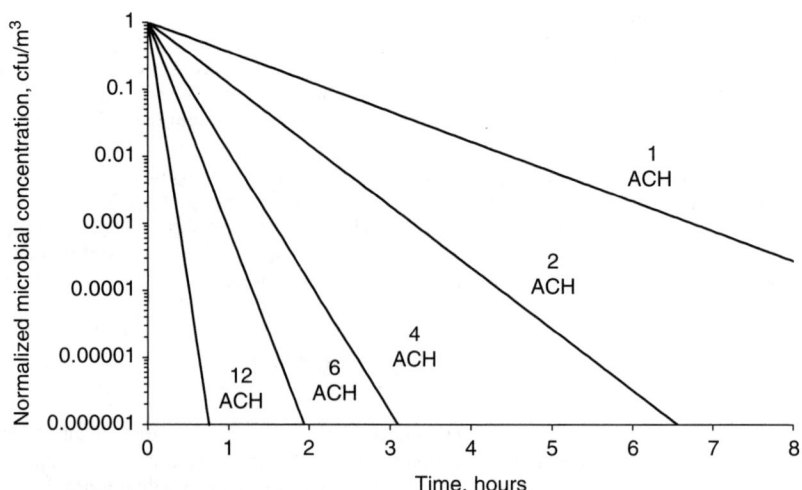

FIGURE 31.1 Effect of filtered outside air change per hour (ACH) on an initial level of airborne microbial contamination over an 8-hours period.

HEALTH CARE FACILITIES

TABLE 31.1 Filter Ratings for Health Care Facilities

Area	1st Filter	2d Filter	3d Filter
Operating rooms (OR)	MERV 7	MERV 13–14	HEPA
Procedure rooms	MERV 7	MERV 13–14	—
Treatment rooms	MERV 7	MERV 13–14	—
Intensive care units (ICUs)	MERV 7	MERV 13–14	—
Laboratories	MERV 13	—	—
Administrative areas	MERV 7	—	—

In addition to requirements for ventilation air and filtration, there are specific requirements for the pressurization of operating rooms and isolation rooms spelled out in the codes and guidelines. These systems provide barriers against the infiltration and exfiltration of airborne microorganisms and can be highly effective in protecting both patients and workers. TB isolation rooms, for example, maintain negative pressure so that no infectious TB bacilli exit the room, while operating rooms and protective isolation rooms for the immunodeficient provide positive pressure to protect the patient from ambient airborne microorganisms (ASHRAE, 2003).

All of the above approaches to infection control have proven to be reliable and effective in practice and the various referenced guidelines and documents should be consulted for more detailed information on the design and performance requirements of these fundamental infection control technologies. It must be noted here, however, that certain improvements are possible based on recent and developing research. The use of HEPA filters, which were originally designed for control of radiological particles, may represent overkill when applied to the control of airborne microbes, and in one operating room test the HEPA filter provided

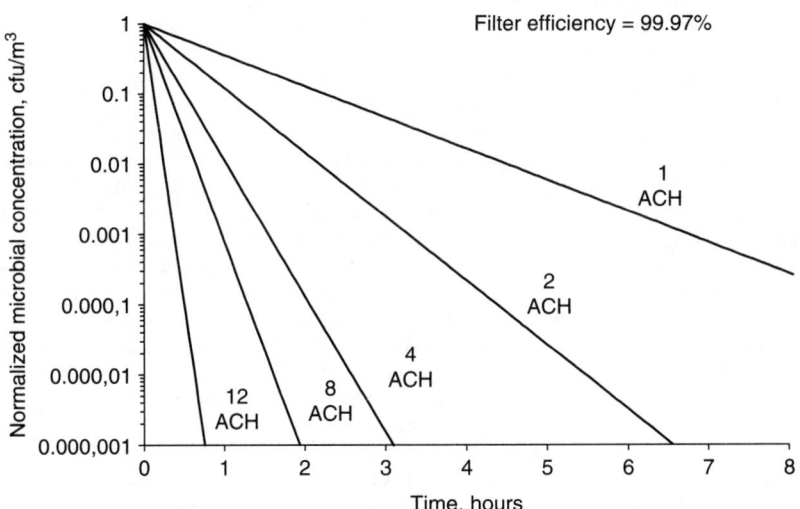

FIGURE 31.2 Effect of HEPA filtration at various air change rates (in ACH) on an initial level of airborne microbial contamination over an 8-hours period.

no better reduction in airborne bacterial load than a 95 percent DSP filter (Luciano, 1977, 1984). Furthermore, the combination of UVGI and high-efficiency filters in the MERV 13–15 range may be able to provide performance virtually equivalent to HEPA filtration, thus offering health care facilities the possibility of reducing energy costs without increasing health risks (Kowalski, 2003). Finally, the various guideline recommendations regarding the use of 100 percent outside air and the HEPA filtration of exhaust air are two areas that energy-conscious designers ought to reevaluate in terms of risk benefits.

31.3 ISOLATION ROOMS

Because of the uniqueness of isolation wards as facilities for controlling infections they are briefly reviewed here. Biological laboratories are treated separately in the following chapter. Isolation rooms incorporate pressurization control to protect those inside the room or outside, and often include HEPA filtration and UVGI systems. Since isolation room systems are essentially 100 percent outside air purge air systems, there is no difference in performance characteristics (see Fig. 31.1) and they are not analyzed further but are discussed here for completeness. Isolation rooms can be classified in three basic categories:

- Negative pressure isolation rooms
- Positive pressure isolation rooms
- Dual purpose isolation rooms (positive or negative)

The modern approach to designing isolation rooms usually includes an anteroom that separates the isolation room from the corridor of the facility, thereby maintaining pressurization integrity during access (ASHRAE, 1999a). Air is supplied to the isolation room and exhausted from both the isolation room and the anteroom. The balance of airflow, or the difference between supply and exhaust, will dictate whether the room experiences positive or negative pressure with respect to ambient. In a positive pressure isolation room, infiltrating air flows between the isolation room and the anteroom, mostly through the gaps in and around the door, and then is exhausted partly by the exhaust duct and partly by flowing out to the corridor. In a negative pressure room, air would flow from the anteroom to the isolation room. Pressure control is maintained by modulating the main supply and exhaust dampers based on a signal from a pressure transducer located inside the isolation room (Gill, 1994). This is not the only possible design—there are various configurations of supply and exhaust ductwork, dampers, and control systems that will accomplish pressurization. Figure 31.3 illustrates the basic configuration for controlling airflow in an isolation room.

Negative pressure isolation rooms keep contaminants and pathogens from reaching external areas. The most common application of these rooms in the health industry today is for isolating tuberculosis patients (CDC, 1994). The CDC recommends 6 to 12 *air changes per hour* (ACH) for TB rooms. The exhaust air is normally filtered through a HEPA filter before being exhausted to the outside. If air is recirculated within the room, it is also normally filtered. *Ultraviolet germicidal irradiation* (UVGI) may be used to augment filtration but is never recommended for use without filtration.

The exact air pressure differential that must be maintained is nominal only, as it merely indicates the airflow direction (Galson and Guisbond, 1995). It is sometimes stated as 0.001 iwg, but this is not a pressure that is practical to measure or maintain, and therefore other criteria such as maintaining an inward velocity of 100 fpm, exhausting 10 percent of the airflow, or exhausting 50 cfm more than the supply are often specified (ASHRAE, 1999a). The exact criteria will always be dependent on both the size and the airtightness of the subject facility.

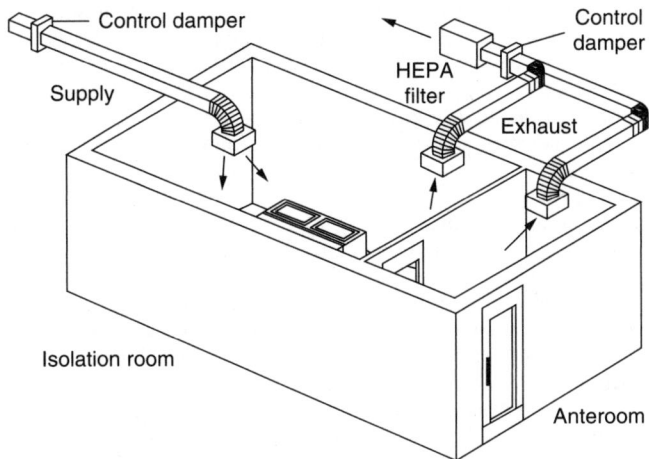

FIGURE 31.3 Isolation room with an anteroom and HEPA-filtered exhaust air.

Positive pressure isolation rooms maintain a flow of air out of the room, thus protecting the patient from possible contaminants and pathogens that might otherwise enter. The most common application today is for patients with various types of immunodeficiency, including AIDS or HIV patients (Linscomb, 1994). For such patients it is paramount to prevent the ingress of any microorganisms, including common fungi and commensal bacteria that may be harmless to healthy people. Design criteria for HIV rooms are similar to those for TB rooms. Air supplied to, or recirculated in, HIV rooms is normally filtered through HEPA filters, and UVGI systems are sometimes used in conjunction with these. The requirements for air pressure differential are the same as described for negative pressure rooms. Approximately 15 percent of AIDS patients also suffer from TB, and this presents a unique design problem (ASHRAE, 1999a). One possible solution is to nest a positive pressure (HIV) room within a negative pressure (TB) room or vice versa (Gill, 1994).

In a typical isolation ward, a floor with isolation rooms has a corridor separating it from the other areas like labs. Transfer air (exfiltration/infiltration) occurs between the corridor and the other rooms. Facilities often differ markedly in layouts, and the pressurization scheme must be adapted individually for each facility (ASHRAE, 1999a; Ruys, 1990).

31.4 AIRBORNE NOSOCOMIAL EPIDEMIOLOGY

The most serious aerobiological problems in hospitals today are nosocomial infections that are acquired incidentally in hospital environments and that were not present, symptomatically or otherwise, on admission. Airborne nosocomial infections are those that are directly or indirectly transmitted as the result of airborne microorganisms.

In intensive care units, almost a third of nosocomial infections are respiratory in nature, but not all of these are airborne since some are transmitted by contact (Wilson, 2001; Wenzel, 1981). Surgical site infections are often airborne but nonrespiratory, such as when common microbes like *Staphylococcus* settle on open wounds, burns, or medical equipment (Emmerson, 1998). These microbes may only be airborne for a fraction of a second, but if they should fall directly onto a surgical site, burn, or open wound, they do not need the ability to survive in air for very long. Patients who succumb to nosocomial infections are often those whose natural defenses have been compromised, either as a result of disease, injury, or medical treatment. Common microbes can become opportunistic pathogens when provided with such relatively defenseless victims. To those with a lower threshold of vulnerability, the endogenous flora of every healthy person may become a potential source of fatal disease. Even a patient's own endogenous microflora can become infectious if conditions permit.

Figure 31.4 illustrates the basic types of nosocomial infections. It can be observed that airborne infections comprise perhaps a third or less of all hospital-acquired infections. One source suggests that 10 percent of nosocomial infections are airborne (McDowell, 2003). Some nosocomial pneumonia and surgical site infections are due to airborne microbes. It also is possible that some urinary tract infections, and even some blood infections, may result from equipment contaminated by microbes settling on equipment.

Munzinger et al. (1983) studied the rates of nosocomial infections in a university hospital in the surgical intensive care unit (SICU), the surgical ward (SW), and in the medical ward (MW). The overall nosocomial infection rate was 14 percent and the infection rate varied greatly from ward to ward due to different patient populations, invasive procedures and severity of underlying diseases. Incidence infection rates were 42.5 percent for the SICU, 19.6 percent for the SW, and 4.1 percent for the MW. Some bacteria like *Staphylococcus aureus* and *Streptococcus pyogenes* have a greater propensity to cause surgical site infections, so extensive infection control practices are necessary to prevent or contain these pathogens (Emmerson, 1998).

Almost 30 percent of sporadic cases of nosocomial pneumonia are caused by *Legionella* (Hart and Makin, 1991). *Legionella* is a common contaminant of potable water and can proliferate in hot water systems of large buildings such as hospitals. Formation of aerosols from contaminated water is a major mode of spread of *Legionella*, but there is evidence to suggest that aspiration is also a mode of entry. Control can involve a mixture of physical and chemical methods like heating, UVGI, and ozone, combined with good plumbing practice like elimination of dead-legs where stagnant water may sit.

Operating rooms require the highest level of air cleanliness so as to avoid contaminating open wounds during surgery. Laminar flow systems with high air-exchange rates and HEPA filtration have long been relied on for control of operating room air quality and are largely

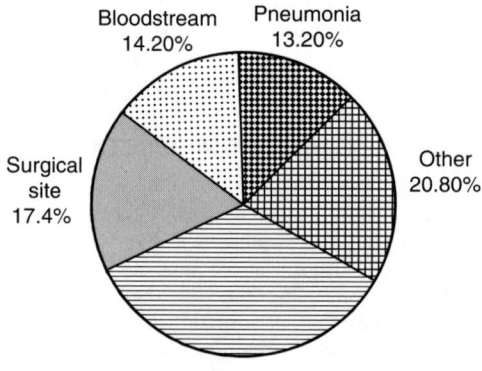

FIGURE 31.4 Types of nosocomial infection. [*Based on data from the NNIS report (CDC, 1996).*]

successful in this regard (ASHRAE, 2003). However, the rate of SSIs continues to prove an intractable barrier and they remain the most common nosocomial infection. For example, fungal surgical site infections, ranged from 0.1 to 0.3 per 1000 discharges from 1991 to 1995 (Mangram et al., 1999). Although it is difficult to estimate what portion of these SSIs are airborne infections, it is likely that at least a significant fraction of them resulted from airborne transmission, and these should be amenable to air disinfection technologies.

Hospital workers as well as patients are at risk from nosocomial infections and worker fatalities have occurred. Health care professionals routinely undergo exposure to contagious respiratory infections like TB and influenza. In most cases where medical workers have contracted respiratory infections from inhalation, the root cause has often been inadequate local ventilation, malfunctioning systems and equipment, or administrative control problems (Castle and Ajemian, 1987). Appropriate engineering design, maintenance, and procedural attentiveness can greatly reduce the risk of airborne infections for health care professionals.

Tuberculosis infections among health care workers are strongly associated with inadequate ventilation in general patient rooms and with the type and duration of work. However, no associations have been identified with ventilation of isolation rooms, based on a study of 17 acute care facilities or university hospitals (Menzies et al., 2000). Tuberculin conversion was associated with ventilation of general or patient rooms with less than 2 air exchanges per hour, in moderate to high-risk hospitals.

Although some studies have found that health care workers have an increased risk of tuberculosis, other studies have reported the opposite. Raitio and Tala (2000) report a study of Finnish health care workers in which the incidence of tuberculosis among health care workers decreased from 57.9 to 6.1 per 100,000 over a 30-year period. The overall risk in health care workers was lower than in the general population throughout the study period.

Health care workers are routinely vaccinated for potential infections but are still at risk. Among the highest risks are infections with *Bordetella pertussis*. Wright et al. (1999) investigated pertussis infections in health care workers and found the annual incidence to be 1.3 percent, finding both symptomatic and asymptomatic infections among the workers.

31.5 NOSOCOMIAL AEROBIOLOGY

Airborne nosocomial pathogens can include respiratory pathogens, opportunistic pathogens that cause surgical site or other infections, and even food pathogens if they are airborne. Although allergens are not addressed here since they are not considered nosocomial agents, ambient or environmental mold spores can cause opportunistic infections in the immunocompromised and are therefore treated as potential pathogens. All respiratory pathogens are potentially nosocomial, or hospital-acquired, infections.

Table 31.2 lists all the respiratory and other pathogens that have occurred, or are of concern, as potentially airborne nosocomial infections. This list is not exclusive as practically every agent listed in App. A, the airborne pathogen and allergen database, may potentially become a nosocomial agent, as may some food and animal pathogens and others that may have not even been addressed in this book. Table 31.2 breaks down the type of agent as a communicable (C) agent, a noncommunicable (N) agent, or an endogenous (E) agent. Endogenous microbes may exist commensally in humans but transmit only to a susceptible individual, usually meaning immunocompromised. Although these agents are potentially airborne, neither the annual cases nor the fatalities represent airborne transmission since most of these infections may occur through direct contact.

TABLE 31.2 Potential Airborne Agents of Nosocomial Infections

Airborne pathogen	Type	Primary infection caused	Annual cases	Annual fatalities
Influenza A virus	C	Flu, secondary pneumonia	2,000,000	20,000
Measles virus	C	Measles (rubeola)	500,000	Rare
Streptococcus pneumoniae	C	Lobar pneumonia, sinusitis, meningitis	500,000	50,000
Streptococcus pyogenes	C	Scarlet fever, pharyngitis	213,962	—
Mycobacterium tuberculosis	C	Tuberculosis, TB	20,000	—
Bordetella pertussis	C	Whooping cough	6,564	15
Rubella virus	C	Rubella (German measles)	3,000	None
Staphylococcus aureus	E	Staphylococcal pneumonia, opportunistic	2,750	—
Pseudomonas aeruginosa	NC	Pneumonia	2,626	—
Klebsiella pneumoniae	E	Opportunistic, pneumonia	1,488	—
Legionella pneumophila	NC	Legionnaires' disease, opportunistic	1,163	10
Haemophilus influenzae	C	Meningitis, pneumonia, endocarditis	1,162	—
Serratia marcescens	E	Bacteremia, endocarditis, pneumonia.	479	—
Acinetobacter	E	Opportunistic/septic, meningitis	147	—
Corynebacterium diphtheriae	C	Diphtheria, toxin produced	10	—
SARS virus	C	Severe acute respiratory syndrome	10	—
Chlamydia pneumoniae	C	Pneumonia, bronchitis, pharyngitis	Uncommon	—
Aspergillus	NC	Aspergillosis, alveolitis, asthma	Uncommon	—
Coccidioides immitis	NC	Coccidioidomycosis, valley fever	Uncommon	—
Nocardia asteroides	NC	Nocardiosis	Uncommon	Rare
Nocardia brasiliensis	NC	Nocardiosis	Uncommon	—
Alcaligenes	E	Opportunistic	Rare	Rare
Cardiobacterium	E	Opportunistic infections, endocarditis	Rare	—

TABLE 31.2 Potential Airborne Agents of Nosocomial Infections (*Continued*)

Airborne pathogen	Type	Primary infection caused	Annual cases	Annual fatalities
Moraxella	E	Otitis media, opportunistic	Rare	0
Burkholderia pseudomallei	NC	Melioidosis, opportunistic	Rare	Rare
Blastomyces dermatitidis	NC	Blastomycosis, Gilchrist's disease	Rare	—
Mucor plumbeus	NC	Mucormycosis, rhinitis	Rare	Rare
Pneumocystis carinii	NC	Pneumocystosis	Rare	Rare
Rhizopus stolonifer	NC	Zygomycosis, allergic reactions	Rare	—
Cryptococcus neoformans	NC	Cryptococcosis, cryptococcal meningitis	High	Rare
Parainfluenza virus	C	Flu, colds, croup, pneumonia	Common	—
Respiratory syncytial Virus	C	Pneumonia, bronchiolitis	Common	Rare
Varicella-zoster virus	C	Chickenpox	Common	250
Haemophilus parainfluenzae	E	Conjunctivitis, pneumonia, meningitis	Common	—
Histoplasma capsulatum	NC	Histoplasmosis, fever, malaise	Common	—
Burkholderia mallei	NC	Glanders, fever, opportunistic	—	None
SARS virus	C	SARS	Rare	—

In a study of airborne microbial contamination in the operating room and ICUs of a surgery clinic, Holcatova et al. (1993) measured bacterial concentrations of 150 to 250 cfu/m^3. The most frequently isolated microorganisms included *Staphylococcus epidermis, S. haemolyticus, Enterococcus* spp., *Enterobacter, Pseudomonas* spp., *Micrococcus, Corynebacteria*, and *Streptococcus faecalis*.

In one of the rare studies of pollen and spores in the air of a hospital ward, Tormo et al. (2002) conducted aerobiological studies and found 20 types of pollen grains whose concentrations ranged from 2.7 to 25.1 grains/m^3. The most frequently isolated were, in order, grasses, evergreen oak, water plantain, and olive. Twenty-two different types of mold spores were found with concentrations of 175 to 1396 spores/m^3, with the most frequent being *Cladosporium, Ustilago*, and various basidiospores. Comparison with outdoor levels showed that the three most abundant pollen types had an indoor/outdoor ratio of 30:1. For *Aspergillus-Penicillium* spores, the concentration was even higher indoors than outdoors, although for most spores, lower levels were found indoors, with a mean outdoor/indoor ratio of 4:1.

Figure 31.5 shows the variety of pathogens that are responsible for nosocomial infections. Almost all the pathogens shown are, in fact, potentially airborne, although only a fraction of them actually produce infections directly or indirectly via the airborne route.

536 APPLICATIONS

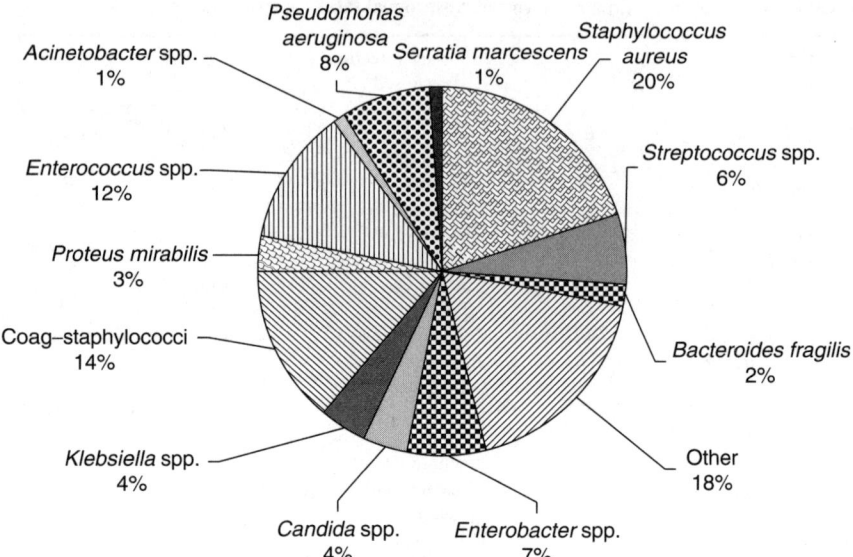

FIGURE 31.5 Primary causes of nosocomial infections. [*Based on data from NNIS (Mangram et al., 1999).*]

Patients may bring airborne infections into waiting areas. Remington et al. (1985) reported on an unusual outbreak of measles in a pediatrician's office in which three children developed measles, after arriving about an hour after an infectious child had departed. Based on an airborne transmission model, it was estimated that the index patient was producing 144 units of infection (quanta) per minute while in the office. Characteristics such as coughing, increased warm air recirculation, and low relative humidity may have increased the likelihood of transmission.

A flowchart of the major aerobiological pathways by which both patients and health care workers become infected with airborne nosocomial pathogens is shown in Fig. 31.6. Only the first- and second-order pathways are shown, although it is possible for a microbe to become reaerosolized several times or pass from person to person before causing an infection. And of course, it may be unlikely that a worker contract a respiratory infection from handling contaminated intrusive medical equipment, but this flowchart does illustrate the potential complexity of the aerobiological pathways of nosocomial infections.

Aerosolized blood pathogens, which facemasks may not protect surgeons from, can be hazardous to surgeons. Aerosols in the respirable size range (less than 5 μm) containing blood can be generated in an operating room during surgery through the use of common surgical power tools (Jewett et al., 1992).

There is a direct relationship between the number of personnel in an operating suite and the airborne bacteria (Hart, 1938; Hambraeus et al., 1977; Mangram et al., 1999; Duvlis and Drescher, 1980). Even conversation among personnel can influence the bacterial load of the air as well as causing some 9 to 10 percent of surgeons and nurses to have postoperatively contaminated facemasks (Ritter, 1984; Dubuc et al., 1973). A person may shed from 3000 to 50,000 microorganisms per minute depending on activity and the effectiveness of protective clothing. The microbes most frequently cultured in operating room air include *Staphylococcus epidermis* and *S. aureus* (Nelson et al., 1973). *Streptococcus pyogenes*

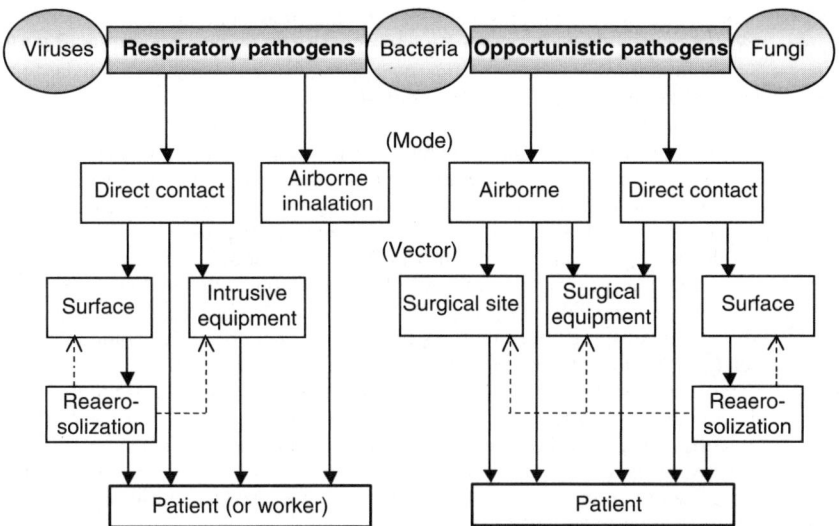

FIGURE 31.6 The major aerobiological pathways of airborne nosocomial pathogens.

has been found in about 12 to 18 percent of preoperative throat swabs from patients (Dubuc et al., 1973).

Methicillin-resistant *Staphylococcus aureus* (MRSA) has become a major nosocomial pathogen in many hospitals and are isolated with increasing frequency. The major sources of *S. aureus* in hospitals are septic lesions and carriage sites of patients and personnel (Solberg, 2000). The anterior nares are the most common carriage site, followed by the perineal area. Nasal carriage is thought to result primarily from airborne transmission. The degree of skin contamination and aerial dissemination varies significantly between carriers and are most pronounced for combined nasal and perineal carriers. The principal mode of transmission is via transiently contaminated hands of hospital personnel.

Airborne MRSA plays a role in colonization of nasal cavities and in respiratory tract MRSA infections. In a study by Shiomori et al. (2001), MRSA was found in air samples collected in single-patient rooms during both rest periods and during bed sheet changes. About 20 percent of the MRSA were less than 4 μm in size. MRSA was also isolated from inanimate environments, such as sinks, floors, and bed sheets, as well as from the patients' hands. The clinical isolates of MRSA were of one origin and were identical to the MRSA strains that infected or colonized new patients. MRSA was recirculated among the patients, the air, and the local room environments, especially during movement in the rooms.

SARS virus, one of the newest airborne nosocomial agents, is also one of the most hazardous for hospital personnel. In a study by He et al. (2003) it was found that index patients were the first generation source of transmission and they infected inpatients and medical staff, creating second generation patients. The major transmission routes were close proximity airborne droplet infection and close contact infection. There was also evidence for the likelihood of aerosol transmission of infections through the ventilation system. A similar report comes from Ho et al. (2003), who found that hospital outbreaks of SARS typically occurred within the first week of admission of the first SARS case before recognition and before isolation measures were implemented. In the majority of hospital infections, there was close contact with a SARS patient, and transmission occurred via large droplets, direct contact with infectious fluids, or by contact with fomites from infectious fluids. In some instances, potential airborne transmission was reported in association with endotracheal intubation, nebulized medications, and noninvasive positive pressure ventilation of SARS patients. Nosocomial transmission was effectively halted by enforcement of standard routines, contact and droplet precautions in all clinical areas, and additional airborne precautions in high-risk areas.

Construction-related nosocomial infections can occur from fungal spores like *Aspergillus* and *Rhizopus*. In a special care unit adjacent to an area under renovation, nosocomial fungal pulmonary infections developed in two premature infants (Krasinski et al., 1985). Inspection showed that inadequate barriers permitted the passage of airborne particles between the two areas, and a significantly higher density of mold spores were found in the SCU compared to a construction-free area. The major source of mold was dust above the hospital's suspended ceiling. Physical barriers and air filtration are essential for controlling airborne fungal spore concentrations in high-risk area during renovation projects. Mahieu et al. (2000) studied the relationship between air contamination and fungal spores, especially *Aspergillus*, in three renovation areas of a neonatal intensive care unit (NICU) and colonization and infection rates in a high care area equipped with HEPA filtration. Renovation work increased airborne concentrations of *Aspergillus* spores significantly in the medium care area and resulted in a significant increase in NICU concentrations. On the other hand, the use of a mobile HEPA air filtration system caused a significant decrease in the spore levels. There was no evidence of invasive aspergillosis during the renovation. The application of HEPA filtration in one hospital building reduced the airborne concentrations of *Aspergillus* from a maximum of 0.4 to 0.009 cfu/m^3 (Sherertz et al., 1987). In a previous hospital location the Sherertz et al. (1987) had found an airborne concentration of *Aspergillus* of approximately 47 cfu/m^3.

Table 31.3 lists the levels of bioaerosols that have been sampled in various health care facilities. In general, these levels are far lower than those that are encountered in typical office buildings and yet they appear somewhat higher than what might be expected for buildings with high levels of air filtration that requires the cleanest air supplies.

Considering that the WHO recommends not more than 50 cfu/m^3 of fungi in hospital air, it would seem over half of the facilities tested exceeded this limit (Ross et al., 2004). For bacteria, WHO recommends a limit of 100 cfu/m^3 and here we see less than 30 percent of the above results beyond this limit. There is a definite need for new guidelines and procedures for routine air sampling in health care facilities both to identify potential problems and accumulate data.

A study by Streifel (1996) on Aspergillosis in hospitals reviewed the literature on infection rates and summarized the associated levels shown in Table 31.4. This table suggests that a threshold may exist at about 0.1 cfu/m^3 below which infection rates may be negligible. Figure 31.7 graphically illustrates the possibility, based on this data, that infection rates could approach 0 percent as the contamination level in the air approaches sterility.

Dental offices have needs that are similar to operating suites except that the hazards are generally much less severe. The hazards include airborne pathogens settling or being imparted to open wounds during dental surgery or procedures, and inhalation of microbes from the patient. Facemasks are in common use by dental workers now, as shown in Fig. 31.8. The microbes of primary concern are species of *Staphylococcus* and *Streptococcus* just as in hospital operating rooms. *Legionella* also poses a potential threat to dental workers as it has been found in up to 68 percent of dental water samples, and 36 percent of samples exceeded 1000 organisms/mL (Atlas et al., 1995).

Health care workers in dentistry are at risk for infection with various airborne pathogens such as *Mycobacterium tuberculosis*, influenza, and cold viruses (Araujo and Andreana, 2002). Workers should maintain consistent adherence to recommended infection control strategies, including the use of protective barriers and appropriate methods of sterilization

TABLE 31.3 Typical Airborne Concentrations in Hospitals

Contaminant	Level	Notes	Reference
Bacteria (cfu/m^3)	5–33	OR (Joubert system)	Luciano, 1984
	7–49	OR	Nelson, 1978
	46	OR zonal ventilation	Hambraeus et al., 1977
	74	OR	Hambraeus et al., 1977
	74	OR	Tighe and Warden, 1995
	150–250	OR and ICU	Holcatova et al., 1993
	83	ICU and critical care	Tighe and Warden, 1995
	36	NICU	Kowalski and Bahnfleth, 2003
	52	Nurse's stations	Tighe and Warden, 1995
	55	General areas	Ross et al., 2004
	104	Patient rooms	Tighe and Warden, 1995
	207	General areas	Tighe and Warden, 1995
	0.8–160	General areas	Andrade and Brown, 2003
Fungi (cfu/m^3)	0–104	OR	Tighe and Warden, 1995
	9	ICU	Centeno and Machado, 2004
	23	ICU and critical care	Tighe and Warden, 1995
	15	NICU	Kowalski and Bahnfleth, 2003
	23	Nurse's stations	Tighe and Warden, 1995
	17	Wards	Streifel and Rhame, 1993
	43	Wards	Tighe and Warden, 1995
	58	General lobby	Streifel and Rhame, 1993
	84	General areas	Tighe and Warden, 1995
	194	General areas	Ross et al., 2004
	0.9–200	General areas	Andrade and Brown, 2003
	175–1396	General areas	Tormo et al., 2002
	140	Medical compr. air	Andrade and Brown, 2003
Pollen (grains/m^3)	2.7–25	General areas	Tormo et al., 2002

TABLE 31.4 Aspergillosis Infection Rates

Airborne concentration	Infection rate	Reference
cfu/m^3	%	
<1	0	Lentino et al., 1982
0.2	0.3	Arnow et al., 1991
1.65	1.2	Arnow et al., 1991
0.9	5.4	Rhame et al., 1984
2	20	Petersen et al., 1983
0.009	0	Sherertz et al., 1987

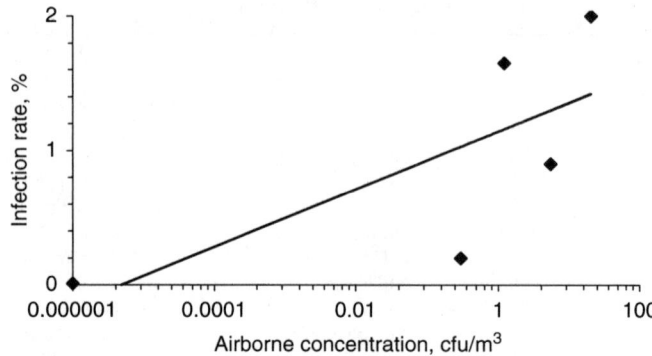

FIGURE 31.7 Plot of data in Table 31.4, assuming sterility (0.000001) for the 0 percent infection rate. [*Based on data summary from Streifel (1996).*]

or disinfection. Recirculation UVGI units are one option available to dental offices, and some benefits have even been demonstrated by ionization systems (Gabbay, 1990).

31.6 NOSOCOMIAL CONTROL OPTIONS

Current methods of controlling nosocomial infections, including isolation rooms and procedures, have proven adequate, and yet there is still some room for improvement. Areas that need further study include the establishment of specific airborne contamination limits,

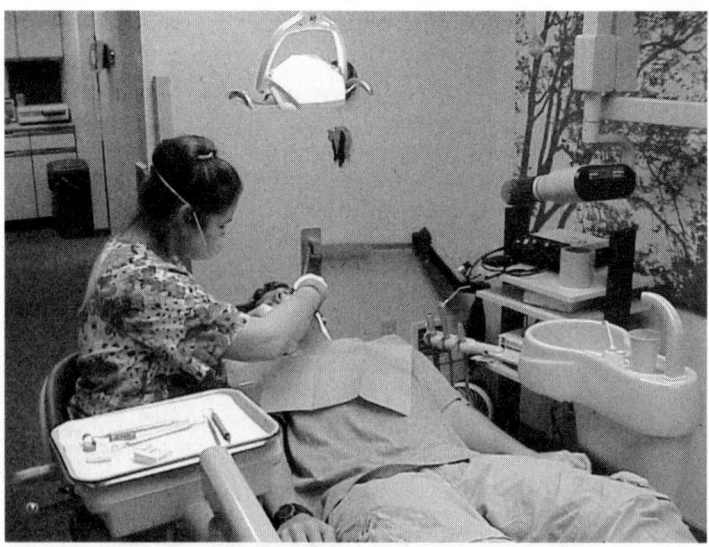

FIGURE 31.8 Dental offices also face concerns about airborne pathogens.

the use of air sampling in health care facilities to identify problems with existing control strategies, the use of UVGI on a larger scale than is currently the case, exploration of more energy-efficient methods for achieving control of airborne microorganisms, and the use of new antimicrobial materials for surfaces and equipment. Finally, further research is needed on the aerobiological pathways of nosocomial infections so that appropriate solution can be better explored.

It has been suggested in some foreign guidelines that airborne microbial levels in an operating room should be below 10 cfu/m^3 during procedures (Wilson, 2001; Gruendemann and Stonehocker, 2001). There is little evidence to indicate that any facility has attempted to conform to any such limits or even to measure levels and it is apparent from Table 31.3 that this limit is rarely achieved. Most facilities appear to have no air sampling guidelines or procedures, and it appears to be tacitly assumed that if the ventilation systems were installed per requisite codes or standards, they would perform at this or some other acceptable level. It would be prudent for all major health care facilities to implement regular air sampling in order to identify potential system performance deficiencies before they become problems (see Chap. 23). For general areas, the limit suggested by Tighe and Warden (1995), 100 cfu/m^3, of either fungi or bacteria would be a reasonable level at which investigation should be triggered. For operating rooms, a target of 10 cfu/m^3 during procedures might be a reasonable goal to strive for, if not actually achieved.

UVGI systems have been in use in some operating rooms since 1937 (Hart and Sanger, 1939). Reductions in postoperative infection rates of between 24 to 44 percent have been demonstrated (Goldner and Allen, 1973). Duke University has successfully used overhead UVGI systems to maintain an acceptably low level of orthopedic infections (Lowell et al., 1980). Upper-room UVGI systems have been used at The New England Deaconess Hospital, The Infant and Children's Hospital in Boston, The Cradle in Evanston, and St. Luke's Hospital in New York to control both surgical site infections, which decreased by a net average of 68 percent, and for the control of respiratory infections, which decreased by a net average of 50 percent (Overholt and Betts, 1940; Del Mundo and McKhann, 1941; Sauer et al., 1942; Higgons and Hyde, 1947). The Home for Hebrew Infants in New York was able to successfully bring a halt to a varicella epidemic using UVGI (Wells, 1955). In spite of these early successes, UVGI continues to be largely ignored as an option by the health care field and government agencies, although recent guidelines have begun to acknowledge its potential effectiveness (CDC, 2003; ASHRAE, 2003). A variety of mostly anecdotal reports on UVGI attest to reductions in airborne nosocomial pathogens with associated decreases in infections, and some formal studies are currently underway but no recent published data are available. In spite of past evidence of success, consistent opposition to the use of UVGI in health care facilities from some official sources, especially in England, and the tendency of guidelines to ignore UVGI as an option, has hampered its acceptance and prevented the accumulation of potentially valuable epidemiological data, but this may change in the near future.

Another application in which hospitals may find UVGI beneficial is to reduce microbial contamination of cooling coils, an approach which pays energy dividends and saves costs (Keikavousi, 2004). Combined UVGI/filtration recirculation units are often used in TB rooms to provide additional levels of air cleanliness (Kowalski and Bahnfleth, 2000). An example is shown in Fig. 31.9.

Another developing technology is the use of pulsed light systems (Wekhof et al., 2001). Pulsed light may be used to disinfect air but a variation of pulsed light with the UV component removed has the unique ability to destroy bacterial cells on the surface of skin with minimal hazard to human cells. In the absence of UV, a high intensity pulsed light flash can cause cell rupture from sudden heating, but skin cells, being locked into a matrix, may remain undamaged. This technology may be able to replace overhead UVGI systems that irradiate surgical sites and more information is provided on this application in Chap. 15.

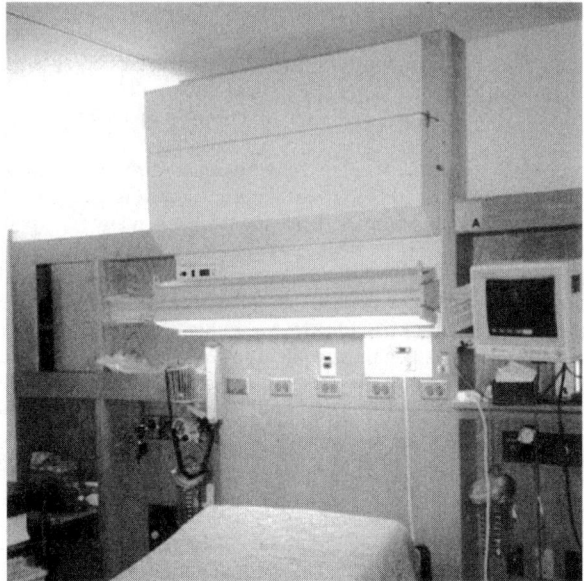

FIGURE 31.9 Overhead filter/UVGI recirculation unit for TB rooms. (*Photo courtesy of Medical Air Technology Corporation, Canton, MD.*)

Some evidence exists to suggest that air ionization can help in reducing the bacterial load of the air and control airborne nosocomial infections, completely eliminating *Acinetobacter* and associated infections in one study (McDowell, 2003). Such systems have been used in dental offices with partial success (Gabbay, 1990).

Antimicrobials are a promising development for the health care industry, and various protective clothing, dressings, surfaces, and medical equipment are becoming increasingly available as options for controlling nosocomial infections (Boarini, 1990; Studenikina et al., 1993). See Chap. 20 for more information on the various antimicrobial materials and applications.

Intrusion of airborne fungi and environmental bacteria, or contamination from visitors, can cause accumulation of microbes in areas like ICUs that should be kept at a high level of cleanliness. It would appear from Table 31.3 that much higher levels of fungi occur in general areas than in ICUs and ORs. Many hospitals have multiple wards and interconnected facilities that use separate ventilation systems designed to keep critical areas under slight positive pressure with respect to hallways and public access areas. Normally, swing doors are provided between sections to help maintain separation, but often these doors are left open as the result of heavy traffic and air flows freely between sections. Furthermore, ventilation system performance may change over time and reversals of airflow direction may occur. It would be prudent for facilities to periodically check airflow direction, which can be done with smoke testing or other means. Another source of fungal contamination in hospitals is filter bypass and maintenance problems. Since large quantities of outside air are filtered with MERV 7–13 filters, these will likely accumulate spores. Filters should be checked for bypass and maintenance procedures requiring shutdown of the fans should be followed diligently, otherwise spores may enter the ventilation system and accumulate over time in carpeting. Periodic surface sampling of the cooling coils, drain pans, and carpets can help identify potential problems.

REFERENCES

AIA (2001). *Guidelines for Design and Construction of Hospital and Health Care Facilities.* The American Institute of Architects Press, Washington, DC.

Andrade, C. M., and Brown, T. (2003). "Microbial contamination of central supply systems for medical air." *Brazilian J Microbiol* 34(Suppl. 1):29–32.

Araujo, M. W., and Andreana, S. (2002). "Risk and prevention of transmission of infectious diseases in dentistry." *Quintessence Int* 33(5):376–382.

Arnow, P. M., Sadigh, M., Costas, C., Weil, D., and Chudy, R. (1991). "Endemic and epidemic Aspergillosis associated with in-hospital replication of Aspergillus organisms." *J Infect Dis* 164:998–1002.

ASHRAE (1999a). "Standard 52.2-1999: Method of Testing General Ventilation Air-Cleaning Devices for Removal Efficiency by Particle Size." American Society of Heating, Refrigerating and Air-Conditioning Engineers. Atlanta, GA.

ASHRAE (1999b). Chapter 7: Health Care Facilities. *ASHRAE Handbook of Applications*, American Society of Heating, Refrigerating and Air Conditioning Engineers, Atlanta, GA. 7.1–7.13.

ASHRAE (1999c). *Handbook of Applications*. American Society of Heating, Refrigerating and Air-Conditioning Engineers, Atlanta, GA.

ASHRAE (2003). *HVAC Design Manual for Hospitals and Clinics*. American Society of Heating, Refrigerating and Air-Conditioning Engineers, Atlanta, GA.

Atlas, R. M., Williams, J. F., and Huntington, M. K. (1995). "Legionella contamination of dental-unit waters." *Appl Environ Microbiol* 61(4):1208–1213.

Boarini, E. (1990). "Development of a device with an antimicrobial treatment." *Medical Plastics Conference*, Anaheim, CA. 5.

Castle, M., and Ajemian, E. (1987). *Hospital Infection Control*. John Wiley & Sons, New York.

CDC (1994). *Guidelines for Preventing the Transmission of Mycobacterium Tuberculosis in Health-Care Facilities*. Federal Register, CDC, ed., US Govt. Printing Office, Washington, DC.

CDC (1996). "National Nosocomial Infections Surveillance (NNIS) Report, data summary from October 1986–April 1996, Issued May 1996." *AJIC* 24(5):380–388.

CDC (2003). "Guidelines for environmental infection control in health-care facilities." *MMWR* 52(RR-10):1–44.

Centeno, S., and Machado, S. (2004). "Assessment of airborne mycoflora in critical areas of the Principal Hospital of Cumana, state of Sucre, Venezuela." *Invest Clin* 45(2):137–144.

Del Mundo, F., and McKhann, C. F. (1941). "Effect of ultra-violet irradiation of air on incidence of infections in an infant's hospital." *Am J Dis Child* 61:213–225.

Dubuc, F., Guimont, A., Roy, L., and Ferland, J. J. (1973). "A study of some factors which contribute to surgical wound contamination." *Clin Ortho* 96:176–178.

Duvlis, Z., and Drescher, J. (1980). "Investigations on the concentration of airborne germs in conventionally air-conditioned operating theaters." *Zentralbl Bakteriol [B]* 170(1–2):185–198.

Emmerson, M. (1998). "A microbiologist's view of factors contributing to infection." *New Horiz* 6(2 Suppl.):S3–S10.

Gabbay, J. (1990). "Effect of ionization on microbial air pollution in the dental clinic." *Environ Res* 52(1):99.

Galson, E., and Guisbond, J. (1995). "Hospital sepsis control and TB transmission." *ASHRAE J* 37(5):48–52.

Gill, K. E. (1994). "HVAC design for isolation rooms." *HPAC Eng* 66(7):45–52.

Goldner, J. L., and Allen, B. L. (1973). "Ultraviolet light in orthopedic operating rooms at Duke University." *Clin Ortho* 96:195–205.

Gruendemann, B. J., and Stonehocker, S. (2001). *Infection Prevention in Surgical Settings*. W.B. Saunders, Philadelphia, PA.

Hambraeus, A., Bengtsson, S., and Laurell, G. (1977). "Bacterial contamination in a modern operating suite." *J Hyg* 79:121–132.

Hart, D. (1938). "Pathogenic bacteria in the air of operating rooms." *Arch Surg* 37:521.
Hart, D., and Sanger, P. W. (1939). "Effect on wound healing of bactericidal ultraviolet radiation from a special unit: Experimental study." *Arch Surg* 38(5):797–815.
Hart, C. A., and Makin, T. (1991). "Legionella in hospitals: a review." *J Hosp Infect* 18(June Suppl. A): 481–489.
He, Y., Jiang, Y., Xing, Y. B., Zhong, G. L., Wang, L., Sun, Z. J., Jia, H., Chang, Q., Wang, Y., Ni, B., and Chen, S. P. (2003). "Preliminary result on the nosocomial infection of severe acute respiratory syndrome in one hospital of Beijing." *Zhonghua Liu Xing Bing Xue Za Zhi* 24(7):554–556.
Higgons, R. A., and Hyde, G. M. (1947). "Effect of ultra-violet air sterilization upon incidence of respiratory infections in a children's institution." *N Y State J Med* 47(7)
Ho, P. L., Tang, X. P., and Seto, W. H. (2003). "SARS: hospital infection control and admission strategies." *Respirology* 8(Suppl.):S41–S45.
Holcatova, I., Bencko, V., and Binek, B. (1993). "Indoor air microbial contamination in the operating theatre and intensive care units of the surgery clinic." *Indoor Air 93*. Helsinki, Finland. 375–378.
Jewett, D. L., Heinsohn, P., Bennett, C., Rosen, A., and Neuilly, C. (1992). "Blood-containing aerosols generated by surgical techniques: a possible infectious hazard." *Am Ind Hyg Assoc J* 53(4):228–231.
Keikavousi, F. (2004). "UVC: Florida hospital puts HVAC maintenance under a new light." *Engin Sys* 21(3):60–66.
Kowalski, W. J., and Bahnfleth, W. P. (2000). "UVGI Design Basics for Air and Surface Disinfection." *HPAC Eng* 72(1):100–110. http://www.engr.psu.edu/ae/wjk/uvhpac.html.
Kowalski, W. J., and Bahnfleth, W. P. (2003). "Air sampling survey data collected in PA hospital NICU." *(personal communication/proprietary report)* PA public hospital.
Kowalski, W. J. (2003). *Immune Building Systems Technology*. McGraw-Hill, New York.
Krasinski, K., Holzman, R. S., Hanna, B., Greco, M. A., Graff, M., and Bhogal, M. (1985). "Nosocomial fungal infection during hospital renovation." *Infect Control* 6(7):278–282.
Kubica, G. P. (1996). Exposure risk and prevention of aerial transmission of tuberculosis in health care settings. *Indoor Air and Human Health*, CRC Press, Boca Raton, FL.
Lentino, J. R., Rosenkranz, M. A., Michaels, J. A., Kurup, V. P., Rose, H. D., and Rytel, M. W. (1982). "Nosocomial aspergillosis: A retrospective review of airborne disease secondary to road construction and contaminated air conditioners." *Am J Epidemiol* 116(8):430–437.
Linscomb, M. (1994). "AIDS clinic HVAC system limits spread of TB." *HPAC Eng* 56(2):5557.
Lowell, J. D., Kundsin, R. B., Schwartz, C. M., and Pozin, D. (1980). "Documentation of airborne infection during surgery." *Airborne Contagion, Ann N Y Acad Sci*, R. B. Kundsin, ed., NYAS, New York. 285–293.
Luciano, J. R. (1977). *Air Contamination Control in Hospitals*. Plenum Press, New York.
Luciano, J. R. (1984). "New concept in French hospital operating room HVAC systems." *ASHRAE J* 26(2).:30–34.
Mahieu, L. M., Dooy, J. J. D., Laer, F. A. V., Jansens, H., and Ieven, M. M. (2000). "A prospective study on factors influencing aspergillus spore load in the air during renovation works in a neonatal intensive care unit." *J Hosp Infect* 45(3):191–197.
Mangram, A. J., Horan, T. C., Pearson, M. L., Silver, L. C., Jarvis, W. R., and HICPAC (1999). "Guideline for prevention of surgical site infection." *Am J Infect Control* 27(2):97–134.
McDowell, N. (2003). "Air ionisers wipe out hospital infections." NewScientist.com news service. http://www.newscientist.com/article.ns?id=dn3228.
Menzies, D., Fanning, A., Yuan, L., and FitzGerald, J. M. (2000). "Hospital ventilation and risk for tuberculous infection in Canadian health care workers. Canadian Collaborative Group in Nosocomial Transmission of TB." *Ann Intern Med* 133(10):779–789.
Munzinger, J., Buhler, M., Geroulanos, S., Luthy, R., and Graevenitz, A. V. (1983). "Nosocomial infections in a university hospital. Results of a prospective study of infections in a medical and surgical ward and a surgical intensive care unit." *Schweiz Med Wochenschr* 113(48):1782–1790.
Nardell, E. A. (1988). Chapter 12: Ultraviolet air disinfection to control tuberculosis. *Architectural Design and Indoor Microbial Pollution*, R. B. Kundsin, ed., Oxford University Press, New York. 296–308.

Nelson, J. P., Glassburn, A. R., Talbott, R. D., and McElhinney, J. P. (1973). "Clean room operating rooms." *Clin Ortho* 96:179–187.

Nelson, P. J. (1978). "Clinical use of facilities with special air handling equipment." *Hosp Topics* 57(5):32–39.

NIOSH (2005). *Engineering Controls for Tuberculosis: Upper-Air Ultraviolet Germicidal Irradiation*. Department of Health and Human Services, National Institute for Occupational Safety and Health, Cincinnati, OH.

Overholt, R. H., and Betts, R. H. (1940). "A comparative report on infection of thoracoplasty wounds." *J Thoracic Surg* 9:520–529.

Petersen, P. K., McGlave, P., Ramsay, N. K., and Rhame, F. S. (1983). "A prospective study of infectious diseases following bone marrow transplantation: Emergence of Aspergillus and Cytomegalovirus as the major causes of mortality." *Infect Control* 42(2):81–89.

Raitio, M., and Tala, E. (2000). "Tuberculosis among health care workers during three recent decades." *Eur Respir J* 15(2):304–307.

Remington, P. L., Hall, W. N., Davis, I. H., Herald, A., and Gunn, R. A. (1985). "Airborne transmission of measles in a physician's office." *JAMA* 253(11):1574–1577.

Rhame, F., Streifel, A., Kersey, J., and McGlave, P. (1984). "Extrinsic risk factors for pneumonia in the patient at high risk of infection." *Am J Med*:42–52.

Ritter, M. A. (1984). "Conversation in the operating theater as a cause of airborne bacterial contamination." *J Bone Joint Surg Am* 66(3):472.

Ross, C., deMenezes, J. R., Svidzinski, T. I. E., Albino, U., and Andrade, G. (2004). "Studies on fungal and bacterial population of air-conditioned environments." *Brazilian Arch Biology Technol* 47(5):827–835. http://www.scielo.br/pdf/babt/v47n5/a20v47n5.pdf.

Ruys, T. (1990). *Handbook of Facilities Planning Volume I: Laboratory Facilities*. Van Nostrand Reinhold, New York.

Sauer, L. W., Minsk, L. D., and Rosenstern, I. (1942). "Control of cross infections of respiratory tract in nursery for young infants." *JAMA* 118:1271–1274.

Sherertz, R. J., Belani, A., Kramer, B. S., Elfenbein, G. J., Weiner, R. S., Sullivan, M. L., Thomas, R. G., and Samsa, G. P. (1987). "Impact of air filtration on nosocomial Aspergillus infections." *Amer J Med* 83:709–718.

Shiomori, T., Miyamoto, H., and Makishima, K. (2001). "Significance of airborne transmission of methicillin-resistant Staphylococcus aureus in an otolaryngology-head and neck surgery unit." *Arch Otolaryngol Head Neck Surg* 127(6):644–648.

Solberg, C. O. (2000). "Spread of Staphylococcus aureus in hospitals: causes and prevention." *Scand J Infect Dis* 32(6):587–595.

Streifel, A. J., and Rhame, F. S. (1993). "Hospital air filamentous fungal spore and particle counts in a specially designed hospital." *Indoor Air 93*, Helsinki, Finland. 161–165.

Streifel, A. J. (1996). Controlling aspergillosis and Legionella in hospitals. *Indoor Air and Human Health*, CRC Press, Boca Raton, FL.

Studenikina, F. G., Mironov, M. M., Denisova, V. P., and Pokrovskaya, G. M. (1993). "Examining the antimicrobial properties of some metals and coatings for medical products." *Med Tek* 5(Sep–Oct):6–8.

Tablan, O. C., Anderson, L. J., Arden, N. H., Beiman, R. F., Butler, J. C., MacNeil, M. M., and HICPAC (1994). "Guideline for the prevention of nosocomial pneumonia." *Am J Infect Control* 22:247–292.

Tighe, S. W., and Warden, P. S. (1995). "An investigation of microbials in hospital air environments." Indoor Air Rev May. http://www.analyticalservices.com/iaq_bioaerosols/hospital_air.html.

Tormo, M. R., Gonzalo, G. M. A., Munoz, R. A. F., and Silva, P. I. (2002). "Pollen and spores in the air of a hospital out-patient ward." *Allergol Immunopathol* 30(4):232–238.

Wekhof, A., Trompeter, I.-J., and O. Franken (2001). "Pulsed UV-Disintegration, a new sterilization mechanism for broad packaging and medical-hospital applications." *Proceedings of the First International Congress on UV-Technologies*, Washington, DC. http://www.wektec.com/.

Wells, W. F. (1955). *Airborne Contagion and Air Hygiene*. Harvard University Press, Cambridge, MA.

Wenzel, R. P. (1981). *CRC Handbook of Hospital Acquired Infections.* CRC Press, Boca Raton, FL.

WHO (1988). "Indoor air quality: biological contaminants." *European Series 31.* World Health Organization, Copenhagen, Denmark.

Wilson, J. (2001). *Infection Control in Clinical Practice.* Balliere Tindall, Edinburgh, Scotland.

Wright, S. W., Decker, M. D., and Edwards, K. M. (1999). "Incidence of pertussis infection in healthcare workers." *Infect Control Hosp Epidemiol* 20(2):120–123.

CHAPTER 32
BIOLOGICAL AND ANIMAL LABORATORIES

32.1 INTRODUCTION

Laboratories that deal with biological agents face potential inhalation hazards from handling mishaps and casual exposure. Laboratories that handle animals are subject to the risk of a wide variety of infections, especially respiratory infections, which can transmit from animal to man or from man to animal. This chapter examines the nature of these risks and the aerobiological engineering solutions that such facilities have relied on in the past and new alternatives that can be implemented to augment existing systems. The two major types of laboratories, biological and animal, are identical in many respects and are treated together here except where specified otherwise, and certain aspects of these laboratories, like the types of pathogens present, are addressed in separate sections. Clinical laboratories in hospitals are also encompassed by the information here except that Chap. 31 should be consulted for information on the specific aerobiological hazards.

32.2 LABORATORY GUIDELINES

Biological and animal laboratories may handle any number of dangerous pathogens and it is essential that the air and all accessible surfaces be maintained as free as possible from biocontamination. Because of the high concentrations of pathogenic microbes that may exist in laboratories, accidental release is a constant potential hazard. In addition, unknown hazards may exist when new or unusual microorganisms are handled. Biological laboratories normally have a variety of systems and protocols to protect workers from such laboratory hazards, including laboratory hoods, air cleaning systems, pressurization zones, sterilization equipment, biohazard-rated facilities, personnel protective suits, and strict procedures for handling hazardous agents. Each laboratory will be aware of its own unique hazards as determined by their safety officers and will use the appropriate procedures and equipment. Full coverage of the applicable guidelines, codes, and standards is not performed here since these subjects are adequately treated elsewhere, but a brief review is provided to acquaint readers with the appropriate source documents.

Biological laboratories require regulated temperature, humidity, relative static pressure, air cleanliness, and exhaust. Heating, ventilating, and air-conditioning systems serve to control these parameters and are subject to various applicable safety and environmental regulations and guidelines that may include AIA (2001), AIHA (2003), ANSI (1992), ICC (2003), DHHS (1993), NRC (1989), and ASHRAE (1999a, 2003). In addition to the HVAC system, laboratories typically employ exhaust hoods and biological safety cabinets and applicable guidelines include CDC (2000), ACGIH (2001), and ASHRAE (1995, 1999b). Additional guidance is available from Fleming et al. (1995), Richmond and McKinney (1999), Wedum (1961), Crane (1994), and Ruys (1990). Furthermore, other federal, state, and local or institutional regulations may apply. A variety of formal guidelines are available to assist designers and managers of animal research facilities in addition to the above. Perhaps the one in most common use in the United States is the Guide for the Care and Use of Laboratory Animals (NRC, 1996).

All of these guidelines offer similar guidance about the design and operation of the ventilation or air cleaning systems. Typically they recommend about 6 to 15 *air changes per hour* (ACH), where an air change refers to a complete change of the volume of air in a room. The use of filtered 100 percent outside air is generally specified as an option and this is the predominant approach taken today. Air is recommended to be exhausted to outside, and some codes may require HEPA filtration of the exhaust air (ASHRAE, 2003). For systems that recirculate air, a minimum of 50 percent outside air (or maximum 50 percent return air) is suggested by some of the guidelines. HEPA filtration is also recommended for recirculated or exhaust air from biosafety cabinets (ASHRAE, 1999a).

The adequacy of high air-exchange rates and the use of HEPA filtration in recirculation systems have been addressed in Chap. 31 and the examples of Figs. 31.1 and 31.2 can be referred to for more information. Additional information on air exchange and filtration is available in Chaps. 10 and 11. Apart from ventilation system design guidance, much of the information on controlling airborne microbial hazards provided in the various guidelines involves the design and performance of laboratory safety cabinets and containment laboratories. The major types of biological safety cabinets and their use are summarized in Table 32.1. Figure 32.1 shows some typical biological safety cabinets. Note that a Class II Type A cabinet becomes a Class II Type B3 if it is vented to the outside instead of internally.

Containment laboratories for handling infectious agents are classified according to four levels, *biosafety level* (BSL) 1, 2, 3, and 4 (DHHS, 1993; CDC, 2000). The basic characteristics of these laboratories are summarized in Table 32.2.

TABLE 32.1 Basic Characteristics of Biological Safety Cabinets

Class	Type	Inlet filter	Recirculation %	Exhaust to	Exhaust filter	Application
I	—	None	0	OA	HEPA	Low to moderate risk biological agents
II	A	None	70	Room	HEPA	Biosafety level 2 agents
	B1	None	30	OA	HEPA	Biosafety level 3 agents
	B2	None	0	OA	HEPA	Biosafety level 3 agents
	B3	None	70	OA	HEPA	Biosafety level 3 agents
III	—	HEPA	0	OA	HEPA	Biosafety level 3 agents

 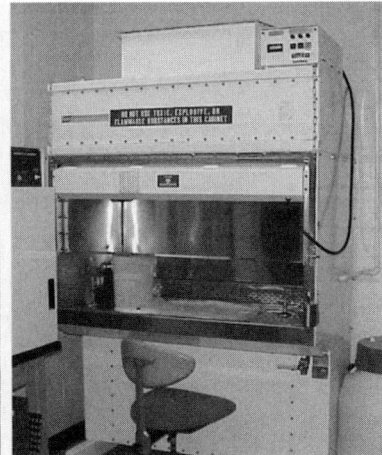

FIGURE 32.1 Typical laboratory hoods. Left is a Class II Type B3 (vented outside). Right is a Class II Type A (vented internally).

All of the relevant design guidelines cited in Table 32.2 have proven adequate for the control of aerobiological hazards in practice, although certain aspects warrant some further consideration. The use of 100 percent outside air systems consumes considerable energy in warm and cold climates, and in spite of the availability of energy recovery systems (ASHRAE, 2003), some question can be raised as to the necessity of this approach when air treatment systems are available to disinfect recirculated air. Furthermore, the reliance on HEPA filtration, as opposed to the use of high-efficiency filters, for controlling airborne microbial hazards is an approach that could bear reevaluation in terms of the energy consumption problem, and especially considering that filtration in combination with UVGI can offer comparable performance without increasing risks (Kowalski et al., 1999; Kowalski, 2003). Finally, almost none of the guidelines address the use of UVGI, in spite of the wealth

TABLE 32.2 Basic Characteristics of BSL Containment Laboratories

BSL	Requirements	Recommendations	Application
1	No specific HVAC requirements	3–4 ACH, slight negative pressure	Agents of no known or minimal hazard
2	No specific HVAC requirements	100% OA, 6–15 ACH, slight negative pressure, use of safety cabinets	Agents of moderate potential hazard
3	Physical barrier, double doors, no recirculation, maintain negative pressure	Exhaust may require HEPA filtration	Agents that pose a serious hazard via inhalation
4	Physical barrier, double doors, no recirculation, maintain negative pressure, and the like.	Requirements determined by biological safety officer	Agents that pose a high risk of lethality via inhalation

of information demonstrating its effectiveness in disinfecting air (Kowalski and Bahnfleth, 1998; Kowalski et al., 2000; VanOsdell and Foarde, 2002).

32.3 BIOLOGICAL LABORATORY PATHOGENS

The wide variety of potential airborne pathogens and allergens that may be present in laboratories would encompass the entire array of identified microbes in App. A and many more that are not. Zoonotic pathogens are addressed in the section on animal laboratories but these could also be found in biological laboratories. Some nonairborne agents may even pose previously unknown inhalation hazards. For example, several vector-borne pathogens, such as *Venezuelan equine encephalitis* (VEE), Chikungunya virus, and Dengue fever viruses, have been successfully aerosolized as bioweapon agents and the potential for accidental aerosolization of such pathogens is a danger that may not be fully appreciated (Kowalski, 2003; Kortepeter and Parker, 1999).

New pathogens, like SARS virus, may appear in medical or clinical laboratories without warning (Lim et al., 2004). Other microbes, including bioweapon agents, have been genetically engineered in laboratories and these may pose previously unknown threats to humans (Collins and Kennedy, 1993). For example, a mutant strain of the fungus *Fusarium oxysporum* has been developed for release into the environment to destroy drug crops (Belvadi, 2001). A common indoor allergen, *Fusarium oxysporum*, is known to be capable of producing the toxins moniliformin, zearalenone, isoverrucarol, T-2 toxin, wortmannin, HT-2 toxin, and diacetoxyscirpenol.

Microbiological toxins from both bacteria and fungi may be produced in laboratory settings, either intentionally or otherwise. Toxins are often produced by microbes surviving under stressful environmental conditions. The failure to carefully regulate the conditions of growth may result in unexpected production of toxins. The aerosolization of toxins is a serious inhalation threat regardless of the type of toxin. The variety of equipment used in laboratories, such as aerosolizers, mixers, and shakers, poses a constant danger of accidental release and aerosolization (Dimmick et al., 1973).

Approximately 13 percent of laboratory acquired infections result from the generation of aerosols (Collins and Kennedy, 1993). Aerosols can be generated by equipment, the bursting of bubbles, spills, blenders, centrifuges, pouring, shakers, opening containers, pipetting, and various other forms of activity. Table 32.3 lists a number of potentially airborne microbes that have caused or may cause laboratory-acquired infections, based on a review by Collins and Kennedy (1993).

Some of the fungal infections in Table 32.3 are skin infections, which can be acquired by contact or by airborne transmission (i.e., by spores settling on skin). Others are practically obsolete today, like plague and typhoid fever. Workers in autopsy rooms, including pathologists, technicians, and assistants, are also at risk for some of these infections (Collins and Kennedy, 1993). Workers in vaccine laboratories are also at risk and a number of infections by vaccine agents have occurred, although modern pharmaceutical operations are designed to protect against such hazards.

32.4 ANIMAL LABORATORY PATHOGENS

The potential for airborne disease transmission is an acute occupational hazard in laboratories where animals are used (Benirschke et al., 1978; Besch, 1980). Zoonotic diseases, diseases of the animal kingdom, may be transmitted in laboratories from animals to

TABLE 32.3 Potentially Airborne Laboratory Acquired Infections

Microbe	Infection	Microbe	Infection
Coccidioides	Coccidiomycosis	*Brucella*	Brucellosis
Lymphocytic choriomeningitis	Lymphocytic meningitis	*Coxiella burnetti*	Q Fever
Mycobacterium tuberculosis	Tuberculosis	*Francisella tularemia*	Tularaemia
Sporotrichioides	Sporotrichosis	Hantaan virus	Hantavirus
Parvovirus B19	Parvovirus (fever, fifth disease)	*Rickettsia prowazeki*	Typhus fever
Dengue fever virus	Dengue fever	*Chlamydophila psittaci*	Psittacosis
Corynebacteria diphtheria	Diphtheria	various fungal spores	Fungal infections
Neisseria meningitidis	Meningitis	various fungal spores	Dermatomycoses
VEE	Venezuelan equine encephalitis	*Yersinia pestis*	Plague

humans, from humans to animals, and between animals (Nicklas et al., 1999; Tuffery, 1995). In addition to pathogenic disease hazards, laboratory workers can develop various allergies from prolonged or chronic exposure to animals (Hunskaar and Fosse, 1993).

Many other species of animal diseases have the ability to transmit to humans or vice versa. Whenever such interspecies transmission occurs, secondary transmissions are rare, although there are some singular exceptions. Often, such interspecies transmissions occur by direct or very close contact, but the airborne transmission is an ever-present possibility. Contact transmissions may be controlled procedurally, but airborne transmission cannot be absolutely controlled without engineered systems. Many common human diseases have the potential to cause infections in animals and thus harm the animals or ruin expensive experimental trials (Hansen, 2000). The biological and physiological similarities between humans and animals are sufficient to permit such exchanges.

All airborne pathogens, allergens, and respiratory irritants that may conceivably occur in animal laboratories have been identified in Table 32.4, along with their sources, based on a review of the literature. Although the list is intended to be comprehensive, there may be some unidentified pathogens that belong on this list, as well as pathogens that have not yet emerged.

32.5 ANIMAL LABORATORY EPIDEMIOLOGY

It can be helpful to review the epidemiology of airborne hazards in animal laboratories in order to understand the nature of the airborne hazards. Most of the literature involves exposure to allergens and hypersensitivity, and there seems to be little epidemiological information available on accidental releases of bioaerosols, although such incidents certainly occur.

Laboratory animal handlers often have allergic reactions to rats and mice (Chapman and Wood, 2001). In a study of exposure sensitization relationships to rat allergens in The Netherlands, England, and Sweden, Heederik et al. (1994) reviewed data for 1062 animal

TABLE 32.4 Airborne Pathogens and Allergens in Animal Laboratories

Airborne pathogen	Group	Source or infected animal	Disease	Mean dia., μm
Avian adenovirus (FAV)	Virus	Birds	Respiratory disease, bronchitis	0.08
Bovine adenovirus	Virus	Bovines	Respiratory disease	0.08
Canine distemper virus (CDV)	Virus	Dogs	Canine distemper	0.14
Coxsackievirus	Virus	Humans, mice, rabbits, hamsters, swine, primates	Colds, acute respiratory disorder	0.03
Echovirus	Virus	Humans, mice, primates	Colds, meningitis possible	0.03
Equine rhinopneumonitis	Virus	Horses	Equine rhinopneumonitis	0.02
Feline picornavirus	Virus	Cats	Feline pneumonitis, URD	0.02
Guinea pig adenovirus	Virus	Guinea pigs	Respiratory disease	0.08
Hantaan virus	Virus	Rodents	Hem. fever, Korean hem. fever	0.10
Influenza A virus	Virus	Humans, birds, pigs, nosocomial	Flu, secondary pneumonia	0.10
Junin virus	Virus	Rodents	Hemorrhagic fever	0.12
Marburg virus	Virus	Humans, monkeys	Hemorrhagic fever	0.04
Measles virus	Virus	Humans, monkeys	Measles (rubeola)	0.16
Pneumonia virus of mice (PVM)	Virus	Mice	Pneumonia	0.20
Mumps virus	Virus	Humans, primates, rodents	Mumps, viral encephalitis	0.16
Newcastle disease virus (NDV)	Virus	Birds	NDV	0.14
Parainfluenza virus	Virus	Humans, monkeys, dogs, rodents	Flu, colds, croup, pneumonia	0.19
Paravaccinia	Virus	Cattle, humans	Pseudo cowpox	0.24
Poxviruses	Virus	Rabbits, sheep, swine, mice, horses, fowl, goats, cows	Mousepox, swinepox, rabbitpox, swinepox, monkeypox, horsepox, fowlpox, goatpox, cowpox	0.24
Reovirus	Virus	Humans, birds, mice	Colds, fever, pneumonia, rhinorrhea	0.08

TABLE 32.4 Airborne Pathogens and Allergens in Animal Laboratories (*Continued*)

Airborne pathogen	Group	Source or infected animal	Disease	Mean dia., µm
Respiratory syncytial virus	Virus	Humans, chimpanzees	Pneumonia, bronchiolitis	0.19
Reston virus	Virus	Monkeys	Reston virus	0.04
Rubella virus	Virus	Humans, monkeys	Rubella (German measles)	0.06
Sendai virus	Virus	Rodents, hamsters	Sendai virus	0.14
Sialodacryoadenitis virus (SDAV)	Virus	Rats	Rat coronavirus	0.11
Simian adenovirus	Virus	Primates	URD, enteritis	0.08
Theiler's virus	Virus	Mice	Encephalomyelitis	0.02
Vaccinia virus	Virus	Agricultural	Pox	0.22
Acinetobacter	Bacteria	Env., soil, sewage, Rats, swine bldgs	OI, septic infections, meningitis	1.22
Actinomyces bovis	Bacteria	Hamsters	Actinomycosis	0.90
Actinomyces israelii	Bacteria	Humans, cattle, rabbits, hamsters	Actinomycosis	0.90
Aerococcus viridans	Bacteria	Rodents, rabbits	Meningitis	1.41
Aeromonas spp.	Bacteria	Env., rodents, soil	Nonrespiratory OI, bacteremia	2.10
Alcaligenes	Bacteria	Humans, soil, water, swine bldgs	OI	0.78
Bacteroides fragilis	Bacteria	Humans, rodents, rabbits	OI	3.16
Bordetella bronchiseptica	Bacteria	Rabbits, cats	URD, pneumonia	0.32
Brucella	Bacteria	Goats, cattle, swine, dogs	Brucellosis, undulant fever	0.57
Burkholderia cepacia	Bacteria	Env. rabbits	OI.	0.71
Burkholderia mallei	Bacteria	Env., horses, mules, nosocomial	Glanders, fever, OI	0.67
Burkholderia pseudomallei	Bacteria	Env., rodents, soil, nosocomial	Meliodosis, OI	0.49
Chlamydophila psittaci	Bacteria	Birds, fowl	Psittacosis/ornithosis, pneumonitis	0.28
Clostridium perfringens	Bacteria	Env., humans, animals, soil	Sepsis, toxins, food poisoning	5.00
Corynebacterium bovis	Bacteria	Mice	Hyperkeratosis	0.70
Corynebacterium kutscheri	Bacteria	Mice	Pseudotuberculosis	0.70
Coxiella burnetii	Bacteria	Cattle, sheep	Q fever	0.28
Diplococcus pneumoniae	Bacteria	Monkeys	Pneumonia	0.71

(*Continued*)

TABLE 32.4 Airborne Pathogens and Allergens in Animal Laboratories (*Continued*)

Airborne pathogen	Group	Source or infected animal	Disease	Mean dia., μm
Enterobacter cloacae	Bacteria	Humans, env., rabbits	OI	1.41
Francisella tularensis	Bacteria	Animals, hamsters	Tularemia, pneumonia, fever	0.20
Haemophilus spp.	Bacteria	Rodents, guinea pigs, rabbits	Pneumonia, conjunctivitis, meningitis	1.00
Klebsiella orthinolytica	Bacteria	Rodents, rabbits	Pneumonia	0.67
Klebsiella oxytoca	Bacteria	Rodents, rabbits	Pneumonia	0.67
Klebsiella planticola	Bacteria	Rodents, rabbits	Pneumonia	0.67
Klebsiella pneumoniae	Bacteria	Env., soil, humans, monkeys, mice, swine blgs	OI, pneumonia	0.67
Mycobacterium africanum	Bacteria	Monkeys	TB	0.90
Mycobacterium avium	Bacteria	Env., water, mice	Cavitary pulmonary disease	1.12
Mycobacterium bovis	Bacteria	Monkeys	TB-like infections	0.90
Mycobacterium lepraemurium	Bacteria	Rodents	Leprosy	0.90
Mycobacterium microti	Bacteria	Rodents	Tuberculosis	0.90
Mycobacterium tuberculosis	Bacteria	Humans, sewage, monkeys	Tuberculosis, TB	0.64
Mycoplasma pulmonis	Bacteria	Rats, mice	CRD, murine mycoplasmosis	0.49
Pasteurella lepisceptica	Bacteria	Rabbits	URD, pneumonia	0.24
Pasteurella multocida	Bacteria	Rabbits, rodents	CRD, rhinitis, otitis media, pneumonia	0.24
Pasteurella pneumotropica	Bacteria	Rodents	Rhinitis, sinusitis, otitis media	0.24
Pasteurella spp.	Bacteria	Monkeys	Pneumonia	0.24
Pneumococcus Type II	Bacteria	Rats, guinea pigs	Bacterial pneumonia	0.71
Pseudomonas aeruginosa	Bacteria	Env., sewage, swine bldgs	Pneumonia, toxins	0.49
Pseudomonas diminuta	Bacteria	Rats, guinea pigs	Rhinitis, conjunctivitis	0.49
Staphylococcus aureus	Bacteria	Humans, sewage, rodents	Staphylococcal pneumonia, OI	0.87
Staphylococcus cohnii	Bacteria	Rats	Pneumonia, dermatitis	0.87
Staphylococcus haemolyticus	Bacteria	Rats	Pneumonia, dermatitis	0.87

TABLE 32.4 Airborne Pathogens and Allergens in Animal Laboratories (*Continued*)

Airborne pathogen	Group	Source or infected animal	Disease	Mean dia., μm
Staphylococcus sciuri	Bacteria	Rats	Pneumonia, dermatitis	0.87
Staphylococcus xylosus	Bacteria	Rats	Pneumonia, dermatitis	0.87
Streptobacillus moniliformis	Bacteria	Rats	Inner ear infection	0.64
Streptococcus pneumoniae	Bacteria	Humans, rats, guinea pigs, rabbits, mice.	Pneumonia, meningitis, otitis media, toxins	0.71
Streptococcus pyogenes	Bacteria	Humans, guinea pigs	Scarlet fever, pharyngitis, toxins	0.89
Yersinia pestis	Bacteria	Rodents, fleas, humans	Bubonic plague, pneumonic plague, sylvatic plague	0.71
Yersinia pseudotuberculosis	Bacteria	Rodents, rabbits, guinea pigs	Pseudotuberculosis (GP)	0.63
Bacillus anthracis	Bacterial Spore	Cattle, sheep, mice, horses	Anthrax, woolsorter's disease.	1.12
Micromonospora faeni	Bacterial Spore	Agricultural, moldy hay, indoor growth	Farmer's lung, pulmonary fibrosis, AR	0.87
Micropolyspora faeni	Bacterial Spore	Agricultural, indoor growth	Farmer's lung, alveolitis, asthma	1.34
Nocardia asteroides	Bacterial Spore	Env., sewage, rodents, rabbits	Nocardiosis, pneumonia	1.12
Coccidioides immitis	Fungal Spore	Env., soil, guinea pigs, rabbits	Coccidioidomycosis, valley fever, desert rheumatism	3.46
Mucor plumbeus	Fungal Spore	Env., sewage, guinea pigs	Mucormycosis, rhinitis	7.07
Paecilomyces variotii	Fungal Spore	Env., rats	Paecilomycosis, AA, toxins, VOC	2.83
Pneumocystis carinii	Fungal Spore	Env., monkeys, animals	Pneumocystosis	2.00
Animal dander	Allergen	Rats, dogs, cats, horses, and the like.	Allergic reactions, asthma, ALA	7.00

Abbreviations: AA = allergic alveolitis; ALA = allergy to lab animals; AR = allergic reactions; bldgs = buildings; Env. = environmental; URD = upper respiratory disease; CRD = chronic respiratory disease; GP = guinea pigs; Hem. = hemorrhagic; OI = opportunistic infections
Source: Kowalski et al., 2002.

laboratory workers. The prevalence of work-related sensitization to rat urinary allergens was 9.7 percent. Rat urinary allergen-sensitization risk increased with increasing exposure intensity. Workers who were atopic had a clearly elevated sensitization risk at low allergen exposure levels. Airborne fungal spores present potential hazards to workers and to laboratory animals but studies show that although sporadically high fungus levels have been encountered, counts of viable fungi are generally too low to be hazardous (Burge et al., 1979).

Laboratory animal allergy (LAA) is a form of occupational sensitivity affecting up to one-third or more of exposed workers and in which symptoms involve the eyes, nose, skin, and lower respiratory tract (Bush, 2001). Asthma may develop in 20 to 30 percent of sensitized individuals.

Laboratory animal workers are at high risk of developing occupational allergy. Hollander et al. (1997) studied the exposure of laboratory animal workers to airborne rat and mouse urinary allergens. Animal caretakers appeared to experience the highest exposure to aeroallergens. The highest exposure levels were found during removal of contaminated bedding from the cages. However, rat and mouse allergen exposure levels during this task varied enormously between facilities, being 1.1 to 158 ng/m^3 and 0.63 to 2000 ng/m^3, respectively.

32.6 LABORATORY AIR CLEANING

Laboratories rely on the ventilation systems, containment areas, and exhaust hoods to maintain air cleanliness. When 100 percent outside air is used for biological and clinical laboratories the supply air is typically filtered with MERV 13–14 filters (ASHRAE, 2003, 1999b). The variety of methods used to control the spread of airborne pathogens and allergens can be grouped into four categories, although they are not entirely independent from each other. Figure 32.2 provides a breakdown of these four basic methods.

Source control includes the use of safety cabinets and the local ventilation of animal cages. Good source control will greatly reduce the airborne hazards and place less dependence

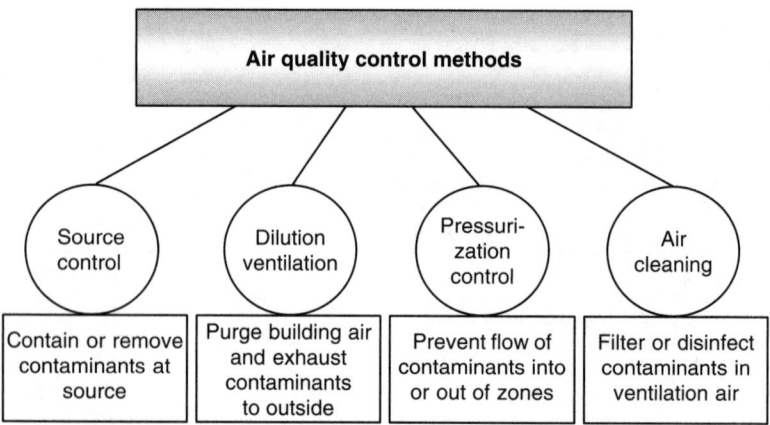

FIGURE 32.2 Breakdown of laboratory aerobiological control methods.

on the ventilation system for dilution and purging of building air, along with its inherent air mixing uncertainties. Dilution ventilation results from the outside airflow rate through the facility and is essentially a function of the air change rate, assuming that good air mixing occurs. An ACH of 6 to 12 might be a reasonable goal for any laboratory, but any increase above these levels may have limited value. If a facility has poor air mixing, however, then higher rates of air exchange may make up for such deficiencies to some degree, but a well-designed air distribution system can pay dividends in both safety and energy savings. One study on rat rooms found that 172 ACH was necessary to control rat allergens, but such a result highlights the need for source control rather than increased ventilation (Swanson et al., 1990).

The pressurization of zones isolates areas from each other and theoretically controls the flow of contaminants between zones, although it is no absolute guarantee since the simple opening of doors may temporarily disturb the airflow direction. Any deficiencies in the design, operation, or maintenance of the isolation systems can negate the biological safeguards afforded by a good air handling system (Sullivan and Songer, 1966). Figure 32.3 illustrates the basic principle of pressurization control, with the containment areas typically being held at the most negative pressure, causing the air to cascade from areas of least contamination potential to the areas of greatest contamination potential where they are exhausted to the outside.

The total airflow rate for a laboratory is determined by minimum ventilation requirements, cooling requirements, and the total exhaust from laboratory hoods and other equipment. In some high-risk laboratories the supply air may be HEPA filtered. The exhaust air may be filtered depending on the type of research performed and facility requirements. Due to the variable usage of safety cabinets and other equipment, variable air volume systems are not uncommon in laboratories as a means of saving energy. Many laboratories, especially animal laboratory facilities, will often use 100 percent outside air.

Air that is exhausted to the outdoors will be naturally disinfected by sunlight, dehydration, and temperature extremes, often quite rapidly. The dispersion of exhaust air from

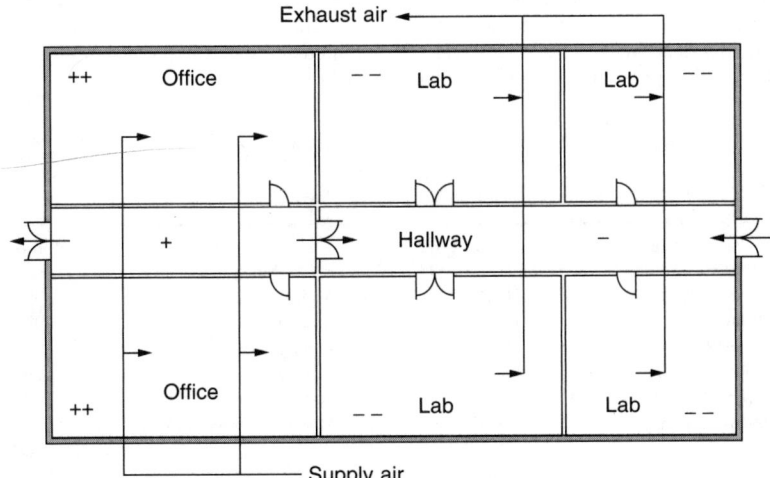

FIGURE 32.3 Schematic of pressurization control in a laboratory and adjacent areas. The "+" and "–" indicate relative positive and negative pressure, while the "++" and "– –" represent higher positive pressure and lower negative pressure, respectively.

rooftop exhaust vents is likely to be more than adequate to reduce concentrations of any but the most dangerous pathogens to harmless levels. Many facilities routinely use HEPA filters to process the exhaust air prior to release, although the practice is of questionable value and not subject to any specific regulations other than those relating to pollution. Only when exhaust air is vented directly to outdoor locations frequented by passers-by would it be prudent to filter the air, and even then a MERV 13–15 filter would provide adequate protection.

Filtration is well understood and performance is highly predictable. The most popular filter type in laboratories today is the HEPA filter, which is rated to remove at least 99.97 percent of all particles 0.3 μm in diameter or larger when operated at design air velocity. However, a MERV 14 (i.e., 95 percent DSP) will perform at almost the same level in the size range of bacteria and viruses. It is informative to review the literature on filter testing and performance as it relates to laboratories. Table 32.5 summarizes several tests on filters using viruses, which are the smallest microbes and among the most difficult to remove by filtration. It's clear that in these once-through tests the measured removal rates meet or exceed all filter performance expectations. One informative result in Table 32.5 is the last test on *minute virus of mice* (MVM), which was removed at a 100 percent rate even without filtration, probably because it didn't survive aerosolization, not being an airborne pathogen (Mrozek et al., 1994). The European "EU" rated filters are comparable to typical high-efficiency filters in the 65 to 95 percent DSP range.

Safety equipment can substantially reduce exposure to mouse allergens (Gordon et al., 2001). Allergen exposure in holding rooms will be minimized if mice are housed in individually ventilated cage systems operated at negative pressure or in airtight systems at positive pressure.

Laboratory hoods are designed to capture contaminants generated in the laboratory. Laboratory exhaust hoods attempt to maintain a certain air velocity through the front opening of the hood so that airborne agents will be prevented from escaping. The velocities through the front of the hood are typically between 50 and 200 fpm (0.25 and 0.10 m/s). Even with such velocities, there is no absolute guarantee that pathogens will not escape the hood during handling procedures. Exhaust from equipment and safety cabinets may be manifolded to a common exhaust duct, where it is vented to the outdoors. Exhaust stacks are often located on roofs to avoid recirculating potentially contaminated air back into the laboratory. Sometimes exhaust air may be incinerated at 650°F (343°C) or more to destroy hazardous biological or chemical agents (Ruys, 1990).

TABLE 32.5 Summary of Virus Filtration Test Results

Researcher	Filter type	Microbe	Removal %
Roelants et al., 1968	HEPA	Actinophage *S. virginiae* S1	99.997
Washam, 1966	HEPA	T phage *E. coli* B	99.9915
Jensen, 1967	HEPA	T phage *E. coli* B	99.99
Harstad et al., 1967	HEPA	T phage *E. coli* B	99.997
Thorne and Burrows, 1960	HEPA	Foot-and-mouth disease virus	99.998
Burmester and Witter, 1972	40–45%	Marek's disease virus (MDV)	0
	80–85%		100
	HEPA		100
Mrozek et al., 1994	EU6 and EU9	Minute virus of mice (MVM)	100
	EU6 and HEPA		100
	No filter		100

In many laboratories there may be several fan systems and the fans are often located in the penthouse. Good design of laboratory ventilation requires that the duct be negative in occupied spaces, however, it is not possible to design a fan room or penthouse with the duct negative downstream of the fan. If the system leaks after the fan, contaminated air may be released into the penthouse with potential exposure to maintenance personnel and testing has demonstrated that air may leak from the exhaust systems into the penthouse under normal operation (Knutson, 1997).

Ultraviolet germicidal irradiation (UVGI) is perhaps the most underused laboratory air cleaning technology. Few of the relevant guidelines even mention it as an option and yet it can be used in diverse applications to enhance air cleanliness and boost levels of protection for laboratory workers (Phillips and Novak, 1955). The combination of UVGI and high-efficiency filtration can result in removal efficiencies for airborne microbes that approach HEPA filter performance (Kowalski, 2003). UVGI can also be used inside laboratory hoods to disinfect surfaces but its use as such has not been popular, perhaps due to concerns about personnel exposure. Other UVGI applications include irradiation of cooling coils to save energy costs, and irradiation of supply air, exhaust air, or for sterilizing air filters (Kowalski and Bahnfleth, 1998).

32.7 ANIMAL LABORATORY PROBLEMS AND SOLUTIONS

Animal laboratories pose some special problems, not the least of which is that the concentrations of airborne dust, allergens, and bacteria in animal laboratories and animal facilities can reach high levels, as shown in the survey of results in Table 32.6. The bacterial results in this table are perhaps more reflective of historical rather than present conditions, but current data are limited.

Few upper limits for airborne particles, bacteria or allergens have been proposed, but Table 32.6 shows one suggested limit for dander and allergens, above which the incidence of LAA increases. The bacteria levels shown in Table 32.6 are in line with normal indoor levels for human dwellings, but this would be expected since most of the fungi probably enter from the outdoor air. To place these airborne concentrations in some perspective, the bacteria, levels listed in Table 32.6 are mostly far beyond any limits that have been proposed for human dwellings. Suggested indoor levels of bacteria range from 500 to 1000 cfu/m^3 (Kowalski and Bahnfleth, 1998). A well-kept laboratory, of course, would be likely to have levels far lower than even these human indoor limits.

The main problem with establishing limits for airborne microorganisms in animal facilities is the same as that for human habitats—few exposure limits are known with any

TABLE 32.6 Measured Levels of Bioaerosols in Animal Laboratories

Contaminant	Avg	Units	Location	Reference
Bacteria	2,851	cfu/m^3	Rat house unfiltered	Baskerville, 1982
	427	cfu/m^3	Rat house filtered	
Rat allergen	62	particles/m^3	Rat rooms, undisturbed	Platts-Mills et al., 1986
	158	particles/m^3	Rat rooms, disturbed	
Dander, allergens	70,629	particles/m^3	Suggested maximum for LAA	Hunskaar and Fosse, 1993

certainty. Most data available are based on specific microbes and limited epidemiological studies. As a result, few experts are willing to commit to upper limits on indoor airborne microorganisms that may cause undue concern or even legal problems. The upper limits used in the subsequent analyses here must be recognized for what they are—convenient numerical targets for use in engineering calculations that do not necessarily reflect any specific health risks and are not necessarily based on any established standards or guidelines.

Allergy and asthma caused by proteins of laboratory animals, particularly rats and mice, are the most important occupational health hazards for the scientists and technicians who work with such animals. Gordon et al. (1992) studied the influence of different cage litters, cage design, and stock density on measured *rat urinary aeroallergen* (RUA) concentrations in a room housing rats. High RUA concentrations, mean value 6.16 to 7.79 $\mu g/m^3$, occurred when the animals were housed on wood based, contact litter. The use of noncontact absorbent pads was associated with a significant decrease in RUA concentrations with a mean value of 2.47 $\mu g/m^3$. Rat urinary aeroallergen concentrations fell more than fourfold when the animals were housed on wood-based, contact litter in filter top cages (mean value 0.33 $\mu g/m^3$) rather than conventional open top cages (mean value 1.43 $\mu g/m^3$).

The source of most airborne contaminants of concern is usually the animals and animal cages. People can, themselves, be a source of contaminants, but the prime focus should always be the protection of the workers. It is implicit that a system capable of protecting lab workers from animal diseases will also protect the animals. No ventilation system or air treatment technology can make up for a source that is uncontrolled. Several approaches are in use today for controlling airborne contamination at the source (Riskowski et al., 1998). Naturally ventilated cages tend to contain much of the dander and dust. Such cages are sometimes designed using computational fluid dynamics programs to ensure that air flows through the cages and mixes in a manner sufficient to remove both contaminants and heat generated by the animals (Rivard, 2000). The room ventilation must remove any airborne contamination produced from such assemblies.

Figure 32.4 shows the approach that is increasingly more common today—cages with forced or induced ventilation. The flexible hose at the back of the assembly is connected to the exhaust ventilation duct and draws air through the cages. Air enters the cages through

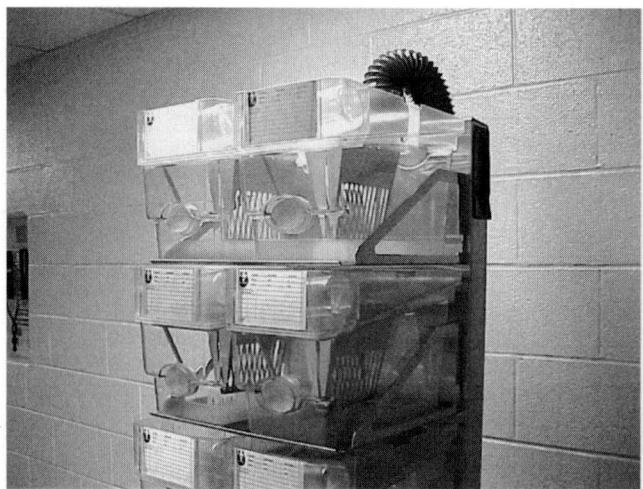

FIGURE 32.4 Animal cages with induced ventilation

slits located near the front. One of the potential problems with induced airflow is that it depends on sufficient negative pressure existing in the ductwork to which it is connected. Normally, the adequacy of the airflow will be measured and verified during installation.

Some ventilated cage racks provide supply air instead of exhausting air. Some racks also contain integral filters and fans. One of the problems with systems in which fans are mounted directly on the cage racks is the production of noise and vibrations. Noise, especially in the high-frequency range, can be rather distressing to mice, rabbits, and certain other animals (NRC, 1996). Another concern is that the filters, often HEPA filters, are sometimes operated well beyond their design face velocity, which is typically 1.27 m/s (250 fpm). Although it may be possible to operate HEPA filters at much higher than design velocity, the engineer should usually review such specifications. Doubling the air velocity through a HEPA filter, for example, can cause a tenfold increase in particle penetration.

Figure 32.5 shows a typical supply diffuser (the large ceiling outlet in the foreground) in a procedures room. An exhaust grille is also present on the ceiling near the corner of the room. This is not an uncommon configuration but is susceptible to the problem of short-circuited airflow. That is, some of the supply air may travel directly to the exhaust grille and air quality may potentially suffer.

Figure 32.6 shows a typical exhaust grille located near the floor of an animal housing room. This is the preferred placement of exhaust grilles since dust and dander tend to settle downward over time. However, when only a single exhaust grille exists in a room there is the potential for cages more distant from the exhaust grille to experience higher levels of

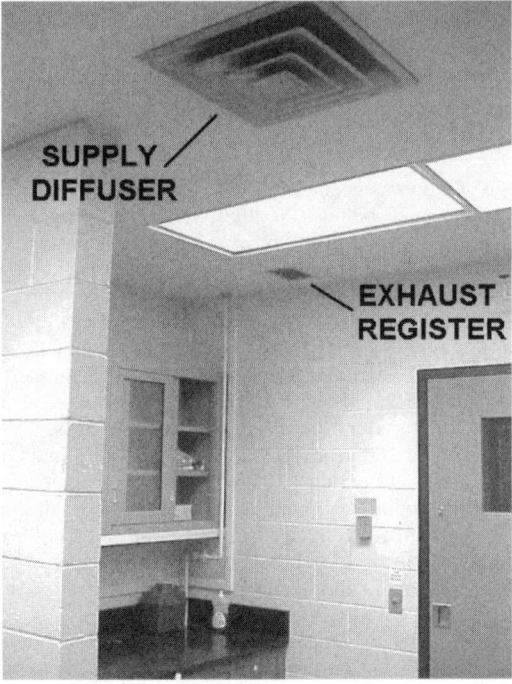

FIGURE 32.5 Ceiling supply diffuser and exhaust register in a procedures room.

FIGURE 32.6 Exhaust air grille located near floor in an animal housing room.

airborne contaminants. Ideally, air would be exhausted all around the perimeter of the room such that an equal amount of air is exhausted from each cage.

Rooms and entire building zones may be pressurized to prevent contamination from passing from one area to another. The design principle is to ensure that airflow proceeds from areas of low contamination potential to areas of high contamination potential. In other words, air must flow from clean areas to dirty areas and not vice versa. The contaminated exhaust air is directly exhausted to outdoors or else treated and then either recirculated or exhausted.

The normal range of indoor concentrations of allergens is about 100 to 1000 cfu/m^3, and so we could seek indoor levels below 100 cfu/m^3 as a rough target. For microbes pathogenic to humans, we would prefer levels of zero, although this might not be realistically achievable. Perhaps a reasonable upper limit for any type of airborne microbe that is not noted for producing human fatalities would be about 10 to 100 cfu/m^3, based on limits that are typical for general areas in hospitals (Kowalski and Bahnfleth, 1998). These are merely suggested limits for use in engineering design and should not be taken as across-the-board recommendations since some microbes are more infectious than others. Whenever particular microbes are of concern they should be evaluated in-depth.

The other major technology for controlling airborne microbes, UVGI, has been in use in various capacities for over 100 years (Kowalski et al., 2000). In a typical air UVGI disinfection system, a UV lamp of perhaps 100 W (total power) is located transverse to the airstream in a duct. Airborne microorganisms passing through this section of duct will be exposed to UV irradiation, both directly from the lamp and from UV light reflected off the duct or any reflective surfaces. Depending on the susceptibility of the specific microbe, the inactivation rate may be rapid and sufficient to sterilize the air, or it may only be sufficient to destroy a fraction of the microbes.

Viruses are the smallest and, in general, easiest microbes to destroy with UVGI. Spores, on the other hand, are the most difficult to inactivate with UVGI, but they are the easiest to filter out of the air. The bacteria are a mixture of vulnerable and resistant species and present somewhat more of a challenge. Some bacteria easily succumb to UVGI while other bacteria are extraordinarily resistant and require other means of control. However, since large microbes tend to be resistant to UVGI but are easy to filter, these two technologies are mutually complementary.

Various other technologies have been applied or are being studied as means of disinfecting air or removing dust. Included among these are *photocatalytic oxidation* (PCO), electrostatic filtration, and antimicrobial filters (Kowalski and Bahnfleth, 1998). Refer to the previous chapters for descriptions of these alternative technologies.

REFERENCES

ACGIH (2001). *Industrial Ventilation: A Manual of Recommended Practice*. ACGIH, Cincinnati, OH.

AIA (2001). *Guidelines for Design and Construction of Hospital and Health Care Facilities*. The American Institute of Architects Press, Washington, DC.

AIHA (2003). *ANSI/AIHA Z9.5-2003 Laboratory Ventilation Standard*. American Industrial Hygiene Association, Fairfax, VA.

ANSI (1992). *American National Standard for Laboratory Ventilation*. ANSI/AIHA American National Standards Institute, New York.

ASHRAE (1995). "Standard 110-1995 Method of Testing the Performance of Laboratory Hoods." American Society of Heating, Refrigerating and Air-Conditioning Engineers. Atlanta, GA.

ASHRAE (1999a). *Handbook of Applications*. American Society of Heating, Refrigerating and Air-Conditioning Engineers, Atlanta, GA.

ASHRAE (1999b). "Standard 52.2-1999: Method of Testing General Ventilation Air-Cleaning Devices for Removal Efficiency by Particle Size." American Society of Heating, Refrigerating and Air-Conditioning Engineers. Atlanta, GA.

ASHRAE (2003). *HVAC Design Manual for Hospitals and Clinics*. American Society of Heating, Refrigerating and Air-Conditioning Engineers, Atlanta, GA.

Baskerville, M., and Seamer, J. H. (1982). "Use of portable filter units to control the animal house environment." *Lab Anim* 16:356–362.

Belvadi, M. (2001). "'Agent Green'—The 'US' Leads Efforts to Use Biological Weapons in the War on Drugs." *Synthesis/Regeneration 26*. http://www.greens.org/s-r/26/26-14.html.

Benirschke, K., Garner, F. M., and Jones, T. C. (1978). *Pathology of Laboratory Animals*. Springer-Verlag. New York.

Besch, E. L. (1980). "Environmental quality within animal facilities." *Lab Animal Sci* 30(2):385–398.

Brede, H. D. (1980). "The human factor—the weakest link?" *Lab Anim Sci* 30(2 Pt. 2):451–459.

Burge, H. A., Solomon, W. R., and Williams, P. (1979). "Aerometric study of viable fungus spores in an animal care facility." *Lab Anim* 13(4):333–338.

Burmester, B. R., and Witter, R. L. (1972). "Efficiency of commercial air filters against Marek's disease virus." *Appl Microbiol* 23(3):505–508.

Bush, R. K. (2001). "Assessment and treatment of laboratory animal allergy." *ILAR J* 42(1):55–64.

CDC (2000). *Primary Containment for Biohazards: Selection, Installation and Use of Biological Safety Cabinets*. U.S. Department of Health and Human Services, Public Health Service, Centers for Disease Control and Prevention and National Institutes of Health, Atlanta, GA.

Chapman, M. D., and Wood, R. A. (2001). "The role and remediation of animal allergens in allergic diseases." *J Allergy Clin Immunol* 107(3 Suppl.):S414–S421.

Collins, C. H., and Kennedy, D. A. (1993). *Laboratory-Acquired Infections*. Butterworth Heineman, Oxford, UK.

Crane, J. T. (1994). "Biological laboratory ventilation and architectural and mechanical implications of biological safety cabinet selection, location, and venting." *ASHRAE Trans* 100:1257–1265.

DHHS (1993). *Biosafety in Microbiological and Biomedical Laboratories*. U.S. Department of Health and Human Services, Cincinnati, OH.

Dimmick, R. L., Vogl, W. F., and Chatigny, M. A. (1973). Potential for accidental microbial aerosol transmission in the biological laboratory. *Biohazards in Biological Research*, A. Hellman, M. N. Oxman, and R. Pollack, eds., Cold Spring Harbor Laboratory, Pacific Grove, CA. 246–266.

Fleming, D. O., Richardson, J. H., Tulis, J. J., and Vesley, D. (1995). *Laboratory Safety Principles and Practices*, 2d ed., ASM Press. Washington, DC.

Gordon, S., Tee, R. D., Lowson, D., Wallace, J., and Taylor, A. J. N. (1992). "Reduction of airborne allergenic urinary proteins from laboratory rats." *Br J Ind Med* 49(6):416–422.

Gordon, S., Fisher, S. W., and Raymond, R. H. (2001). "Elimination of mouse allergens in the working environment: assessment of individually ventilated cage systems and ventilated cabinets in the containment of mouse allergens." *J Allergy Clin Immunol* 108(2):288–294.

Hansen, A. K. (2000). *Handbook of Laboratory Animal Bacteriology*. CRC Press, Boca Raton, FL.

Harstad, J. B., Decker, H. M., Buchanan, L. M., and Filler, M. E. (1967). "Air filtration of submicron virus aerosols." *Am J Public Health* 57(12):2186–2193.

Heederik, D., Smid, T., Houba, R., and Quanjer, P. H. (1994). "Dust-related decline in lung function among animal feed workers." *Am J Ind Med* 25(1):117–119.

Hollander, A., Run, P. V., Spithoven, J., Heederik, D., and Doekes, G. (1997). "Exposure of laboratory animal workers to airborne rat and mouse urinary allergens." *Clin Exp Allergy* 27(6):617–626.

Hunskaar, S., and Fosse, R. T. (1993). "Allergy to laboratory mice and rats: A review of its prevention, management, and treatment." *Lab Anim* 27:206–221.

ICC (2003). *International Mechanical Code*. International Code Council, Falls Church, VA.

Jensen, M. (1967). "Bacteriophage aerosol challenge of installed air contamination control systems." *Appl Microbiol* 15(6):1447–1449.

Knutson, G. W. (1997). "Potential exposure to airborne contamination in fan penthouses." *ASHRAE Trans* 103(Pt. 2):873–878.

Kortepeter, M. G., and Parker, G. W. (1999). "Potential biological weapons threats." *Emerg Infect Dis* 5(4):523–527.

Kowalski, W., and Bahnfleth, W. P. (1998). "Airborne respiratory diseases and technologies for control of microbes." *HPAC Eng* 70(6):34–48. http://www.engr.psu.edu/ae/wjk/ardtie.html.

Kowalski, W. J., Bahnfleth, W. P., Whittam, T. S. (1999). "Filtration of Airborne Microorganisms: Modeling and prediction." *ASHRAE Trans* 105(2):4–17. http://www.engr.psu.edu/ae/wjk/fom.html.

Kowalski, W. J., Bahnfleth, W. P., Witham, D. L., Severin, B. F., and Whittam, T. S. (2000). "Mathematical modeling of UVGI for air disinfection." *Quant Microbiol* 2(3):249–270. http://www.kluweronline.com/issn/1388-3593.

Kowalski, W. J., Bahnfleth, W. P., and Carey, D. D. (2002). "Engineering control of airborne disease transmission in animal research laboratories." *Contemp Top Lab Anim Sci* 41(3):9–17.

Kowalski, W. J. (2003). *Immune Building Systems Technology*. McGraw-Hill, New York.

Lim, P. L., Kurup, A., Gopalakrishna, G., Chan, K. P., Wong, C. W., Ng, L. C., Se-Thoe, S. Y., Oon, L., Bai, X., Stanton, L. W., Ruan, Y., Miller, L. D., Vega, V. B., James, L., Ooi, P. L., Kai, C. S., Olsen, S. J., Ang, B., and Leo, Y.-S. (2004). "Laboratory-acquired Severe Acute Respiratory Syndrome." *N Engl J Med* 350(17):1740–1745.

Mrozek, M., Zillman, U., Nicklas, W., Kraft, V., Meyer, B., Sickel, E., Lehr, B., and Wetzel, A. (1994). "Efficiency of air filter sets for the prevention of airborne infections in laboratory animal houses." *Lab Anim* 28:347–354.

Nicklas, W., Homberger, F. R., Illgen-Wilcke, B., Jacobi, K., Kraft, V., Kunstyr, I., Mahler, M., Meyer, H., and Pohlmeyer-Esch, G. (1999). "Implications of infectious agents on results of laboratory animal experiments." *Lab Anim* 33(Suppl.1):S1:39–S1:87.

NRC (1989). *Biosafety in the Laboratory: Prudent Practices for Handling and Disposal of Infectious Materials*. National Research Council, National Academy Press, Washington, DC. http://bob.nap.edu/html/terrorism/index.html.

NRC (1996). *Guide for the Care and Use of Laboratory Animals*. National Research Council, National Academy Press. Washington, DC.

Phillips, G. B., and Novak, F. E. (1955). "Applications of germicidal ultraviolet in infectious disease laboratories." *Appl Microbiol* 4:95–96.

Platts-Mills, T. A. E., Heymann, P. W., Longbottom, J. L., and Wilkins, S. R. (1986). "Airborne allergens associated with asthma: Particles sizes carrying dust mite and rat allergens measured with a cascade impactor." *J Allergy Clin Immunol* 77:850–857.

Richmond, J. Y., and McKinney, R. W. (1999). *Biosafety in Microbiological and Biomedical Laboratories*, 4th ed., US Government Printing Office, Washington, DC.

Riskowski, G. L., Maghirang, R. G., Funk, T. L., Christianson, L. L., and Priest, J. B. (1998). "Environmental quality in animal production housing facilities: A review and evaluation of alternative ventilation strategies." *ASHRAE Trans* 104:104–113.

Rivard, G. F., Neff, D. E., Cullen, J. F., and Welch, S. W. J. (2000). "A novel vented microisolation container for caging animals: Microenvironmental comfort in a closed-system filter cage." *Contemp Top Lab Anim Sci* 39(1):22–27.

Roelants, P., Boon, B., and Lhoest, W. (1968). "Evaluation of a commercial air filter for removal of viruses from the air." *Appl Microbiol* 16(10):1465–1467.

Ruys, T. (1990). *Handbook of Facilities Planning Volume I: Laboratory Facilities*. Van Nostrand Reinhold, New York.

Sullivan, J. F., and Songer, J. R. (1966). "Role of differential air pressure zones in the control of aerosols in a large animal isolation facility." *Appl Microbiol* 14(4):674–678.

Swanson, M. C., Campbell, A. R., O'Hollaren, M. T., and Reed, C. E. (1990). "Role of ventilation, air filtration, and allergen production rate in determining concentrations of rat allergens in the air of animal quarters." *Am Rev Respir Dis* 141(6):1578–1581.

Thorne, H. V., and Burrows, T. M. (1960). "Aerosol sampling methods for the virus of foot-and-mouth disease and the measurement of virus penetration through aerosol filters." *J Hyg* 58:409–417.

Tuffery, A. A. (1995). *Laboratory Animals: An Introduction for Experimenters*. John Wiley & Sons. Chichester.

VanOsdell, D., and Foarde, K. (2002). "Defining the Effectiveness of UV Lamps Installed in Circulating Air Ductwork." ARTI-21CR/610-40030-01. ARTI, Arlington, VA.

Washam, C. J. (1966). "Evaluation of filters for removal of bacteriophages from air." *Appl Microbiol* 14(4):497–505.

Wedum, A. G. (1961). "Control of laboratory airborne infection." *Bacter Rev* 25:210–216.

CHAPTER 33

FOOD INDUSTRY FACILITIES

33.1 INTRODUCTION

The food industry comprises food and beverage processing facilities, food handling industries, food storage facilities other than agricultural, and kitchens in the restaurant and hospitality industry. Agricultural facilities are treated in the following chapter but the information in this chapter applies to almost any facility involved in food handling and preparation, including even the home environment. There are at least four types of potential aerobiological health hazards associated with food handling—airborne pathogens, aerosolized food pathogens, microbial allergens, and food allergens. In addition, there are spoilage microbes that are not necessarily health hazards but can be destructive to food preparation processes. Foodborne pathogens are generally transmitted by the oral route and almost exclusively cause stomach or intestinal diseases. However, some foodborne pathogens may be airborne at some stage of the processing, storage, cooking, or consumption cycle. In addition, the processing and handling of foods may create opportunities for airborne mold spores to germinate and grow, resulting in secondary airborne inhalation hazards. Surface contamination of equipment and facilities in the food industry is an ever-present problem, and surface disinfection technologies have applications here also. This chapter does not specifically address food disinfection technologies, as this subject is adequately covered elsewhere, but addresses those aspects of the food industry where aerobiology plays a part in either the inception of foodborne pathogen spread or where it is the end result of food industry processes. Waterborne pathogens are not specifically addressed here except where they also act as foodborne pathogens and/or have airborne transmission potential.

33.2 FOOD INDUSTRY CODES AND STANDARDS

The food industry is regulated via inspection from several agencies, including the Food and Drug Administration (FDA), the Environmental Protection Agency (EPA), the U.S. Department of Agriculture (USDA), and state and local health agencies. The FDA manages *good manufacturing practices* (GMPs), which deals with sanitation in manufacturing, processing, packing, and holding food. The GMP establishes basic rules for food establishment sanitation, including minimum demands on sanitary facilities for water, plumbing design, sewage disposal, toilet facilities, hand-washing facilities and supplies, and solid waste disposal. The Federal Insecticide, Fungicide, and Rodenticide Act (FIFRA) covers the use of insecticides, rodenticides, and sanitizing solutions used by food processors. The USDA applies laws to food processors offering products containing meat, poultry, and eggs and inspectors have authority over processing plants. Also, the Department of Defense sets standards for those food processors who supply products for military installations. The military standards are similar to GMPs but include certain specifics. State and local governments usually have specific laws regarding food processing and storage that are coordinated with GMPs. Finally, the Occupational Safety and Health Act (OSHA) provides for a safe working environment for employees of the food industry.

In spite of the wealth of regulations and standards regarding food processing, there is little or no regulation of ventilation system design and most of the food and beverage industry, and the hospitality industry, have adopted the ASHRAE standard written for commercial and residential buildings, Standard 62-2001: Ventilation for Acceptable Indoor Air Quality (ASHRAE, 2001). This standard provides for minimum ventilation based on occupancy but does not dictate any limits for airborne bacteria or fungi. In fact, there are no limits that have been set for indoor airborne microorganisms for any nonmedical facilities. The food industry, especially the food processing industry, can generate high levels of airborne fungi, bacteria, toxins, and various allergens, including airborne food pathogens, that may pose a health hazard to workers as well as potentially causing food contamination. There is also precious little guidance available to the food and beverage industry on air filtration or air disinfection technologies, although air cleaning systems have been in use in the industry for decades. The following sections detail the various airborne hazards that may occur in different sectors of the food and beverage industry and address some of the engineering solutions to aerobiological problems.

33.3 FOODBORNE PATHOGENS AND ALLERGENS

Foodborne pathogens are, in general, not highly transmissible by the airborne route. However, some food pathogens can settle on foodstuffs during processing and the food production process can forcibly aerosolize them and cause unusual hazards to workers. A relatively small number of foodborne pathogens exist compared to all of the potential airborne pathogens and allergens that have been identified. Part of this reason is that the stomach is well adapted to dealing with foreign bacterial matter. Stomach acids are effective at destroying or breaking down pathogens that enter by the oral route, and this includes most airborne pathogens. Inhaled pathogens, for example, that are cleared by the lungs are often swallowed and subsequently destroyed by stomach acids. Foodborne pathogens have evolved mechanisms that enable them to resist or avoid destruction in the stomach and the fewer number of foodborne pathogens that exist compared to respiratory pathogens is probably because this is a much more difficult route of entry.

The foodborne pathogens that do exist today are often uniquely virulent and hazardous. Some of these pathogens, like *Salmonella* and *Shigella*, are contagious and depend on either on excretion in feces or vomiting to facilitate their epidemic spread. Some agents of food poisoning, like *Staphylococcus* and *Clostridium*, are opportunistic or incidental contaminants of foods. Table 33.1 lists the most common foodborne pathogens. In addition, a considerable number of mold spores, including most of those identified in App. A, can also cause problems in the food industry (Samson et al., 2000).

TABLE 33.1 Major and Potential Pathogens in the Food Industry

Microbe	Potentially airborne?	Potentially respiratory?	Comment	Source	Reference
Acinetobacter	Yes	Yes	Common environmental bacteria	H, E	Ray, 1996
Aeromonas	Yes	Yes	Potential foodborne pathogen	H, E	Ray, 1996
Bacillus cereus	Yes	Yes	Foodborne pathogen	E	Heijden et al., 1999
Brucella suis	Yes	Yes	Foodborne brucellosis	A, E	Ray, 1996
Campylobacter	Yes	No	Foodborne pathogen	E	Heijden et al., 1999
Clostridium botulinum	Yes	Yes	Foodborne pathogen	E	Gorham, 1991a
Clostridium perfingens	Yes	Yes	Foodborne pathogen	E	Hui et al., 2003
Coxiella	Yes	Yes	May cause Q fever	A	Ray, 1996
Cryptosporidium	No	No	Foodborne pathogen	E	Hui et al., 2003
Cyclospora	No	No	Foodborne pathogen	E	Hui et al., 2003
Enterobacter	Yes	Yes	Fecal indicator	H, E	Ray, 1996
Escherichia coli	No	No	Some strains pathogenic	H, E	Heijden et al., 1999
Listeria	Yes	No	Foodborne pathogen	E	Heijden et al., 1999
Moraxella	Yes	Yes	Common endogenous bacteria	H, E	Ray, 1996
Norwalk virus	Yes	No	Waterborne food pathogen	E	Knowles, 2002
Salmonella	Yes	No	Foodborne pathogen	H, E	Gorham, 1991a
Shigella	No	No	Foodborne pathogen	H, E	Hui et al., 2003
Staphylococcus aureus	Yes	Yes	May cause food poisoning	H, E	Gorham, 1991a
Streptococcus	Yes	Yes	Potential foodborne pathogen	H, E	Ray, 1996
Streptococcus suis	Yes	No	Potential pathogen	A	Bartelink and vanKregten, 1995
Vibrio	No	No	Foodborne pathogen	H, E	Hui et al., 2003
Yersinia enterolitica	Unknown	No	Foodborne pathogen	E	Ray, 1996

Abbreviations: A = animals, H = humans, E = environment.

Many molds like *Aspergillus* and *Penicillium* are common contaminants of the outdoor and indoor air and can grow on microscopic particles of food, and although they are not food pathogens, they are potentially inhalation hazards. Because virtually all of the fungal spores listed in App. A are potential contaminants in the food industry, these are not listed in Table 33.1.

The spectrum of foodborne pathogens has changed dramatically over time, as well-established pathogens have been controlled or eliminated, and new ones have emerged (Tauxe, 2002). New pathogens can emerge because of changing ecology or changing technology that connects a potential pathogen with the food chain. Almost all foodborne pathogens are bacteria but one virus has recently emerged as a dangerous foodborne pathogen, Norwalk virus. Norwalk virus is also a waterborne pathogen and has caused outbreaks on cruise ships (Marks et al., 2000). This virus apparently inhabits some warm waters and can splash onto ships or even become airborne and settle on ship surfaces, but it is also highly contagious and can be brought into a facility or ship by human carriers.

TABLE 33.2 Major Spoilage Microbes in the Food Industry

Microbe	Food class	Type	Potential disease?	Potential toxin?	Potential allergen?
Alcaligenes	Protein rich	Bacteria	Yes	No	No
Aspergillus	Bakery, cereal, fruits, vegetables, meat	Fungi	Yes	Yes	Yes
Botrytis	Fruits, vegetables, meat	Fungi	No	No	Yes
Cladosporium	Dairy, fruits, vegetables, meat	Fungi	Yes	Yes	Yes
Colletotrichum	Fruits, vegetables	Fungi	No	—	No
Corynebacterium	—	Bacteria	Yes	No	No
Diplodia	Fruits, vegetables	Fungi	No	—	No
Fusarium	Cereal, fruits, vegetables	Fungi	No	Yes	Yes
Geotrichum	Dairy	Fungi	No	No	No
Monila	Bakery, dairy, fruits, vegetables, meat	Fungi	No	—	No
Mucor	Bakery, cereal, fruits, vegetables, meat	Fungi	Yes	No	Yes
Oospora	Dairy, fruits, vegetables, meat	Fungi	No	—	No
Penicillium	Bakery, cereal, dairy, fruits, V	Fungi	No	Yes	Yes
Phomopsis	Fruits, vegetables	Fungi	No	—	No
Phytophthora	Fruits, vegetables	Fungi	No	—	No
Pseudomonas	Various	Bacteria	Yes	No	No
Rhizopus	Bakery, cereal, fruits, vegetables, meat	Fungi	Yes	No	Yes
Sclerotinia	Fruits, vegetables	Fungi	No	—	No
Serratia	—	Bacteria	Yes	No	No
Sporotrichum	Bakery, meat	Fungi	No	—	Yes
Thamnidium	Meat	Fungi	No	—	Yes
Trichoderma	Fruits, vegetables	Fungi	No	Yes	Yes

Food spoilage microbes are less of a health hazard than they are a nuisance in the food industry due to the damage they can cause to processed food (Samson et al., 2000). In general, standard cooking methods and disinfection procedures are adequate to control spoilage microbes but occasionally they get out of hand (Woollen, 1970; Hui et al., 2003). Most of the common food spoilage have the capacity to circulate via the airborne route and hence should be controllable to some degree by ventilation and air cleaning systems. Table 33.2 lists the major causes of food spoilage in the food industry (Ray, 1996; Gorham, 1991a). Most of these are molds but some bacteria are included. The bacteria may come from human or environmental sources while all the molds come from the environment. All of the microbes in Table 33.2 are potentially airborne and many of them are potential allergens. Various yeasts may be causes of spoilage (intentional or otherwise) but almost none of these are potentially hazardous and even less are likely to become airborne.

33.4 FOOD INDUSTRY EPIDEMIOLOGY

There are two major epidemiological aspects of the food industry of concern to human health—foodborne pathogens, and workers' exposure to allergens that may include bacteria, bacterial endotoxins, fungi, fungal mycotoxins, and food allergens. A considerable body of literature exists in regard to foodborne pathogens and a short review of these microbes follows which highlights those microbes that are potentially airborne. The contribution of the top 10 foodborne pathogens to total cases of diseases from 1973 through 1997 is shown graphically in Fig. 33.1. Although none of these microbes, except perhaps *Clostridium botulinum*, are of any real concern as inhalation hazards, all of them have the potential to be forcibly aerosolized and spread through a facility via the airborne route.

The increasing occurrence of allergies has drawn attention to the presence of allergens in stored agricultural products. Stored food products can contain allergenic contaminants of plant, microbial, or animal origin, including allergenic insect and mite species (Korunic, 2001). The most important allergens appear to be storage mites which cause occupational asthma by inhalation and anaphylactic reactions when ingested in high numbers. The urine of rats and mice can also be a significant source of allergens that become airborne in dust particles. Storage molds are also confirmed as potential source of allergens and many of these are potential producers of mycotoxins. Careful attention to processing techniques and the use of air cleaning and surface disinfection technologies can minimize the risk.

A large number of cases of brucellosis occurred among employees at one slaughterhouse handling sheep led to the conclusion that airborne spread was the primary mode of transmission (Rodriguez et al., 2001). No significant differences were found among the attack rates by the sections of the slaughterhouse where the employees in question worked. The slaughtering analysis revealed a correlation between the volume of culled sheep and the epidemic, but specific location within the facility was not a factor. Nor were any significant differences found among the employees in terms of cuts/wounds or the use of protective measures.

Allergic reactions in food industry workers are a common but underreported occupational hazard. In 1998, a study on 107 workers at a crab processing facility showed that asthma-like symptoms developed in 26 percent and bronchitis-like symptoms in 19 percent (Ortega et al., 2001). Among the crab processing jobs, butchering and degilling workers had the highest incidence of respiratory symptoms.

Anisakis simplex, a fish and cephalopodes parasite, may cause allergic reactions in humans who eat or handle contaminated fish (Scala et al., 2001). Occupational hypersensitivity and asthmatic symptoms may be caused by exposure to this microbe and symptoms disappear immediately after exposure in the workplace ceases.

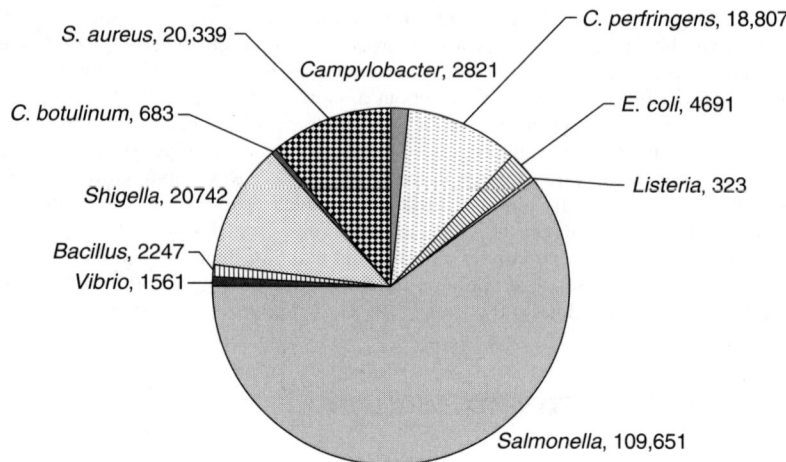

FIGURE 33.1 Cases of foodborne diseases reported to the CDC 1973–1997. [*Based on data from Hui et al. (2003).*]

The relation between bioaerosol exposure in the potato starch industry and work-related respiratory symptoms was examined in a study by Hollander et al., (1994). Groups of workers were exposed to dust concentrations of low endotoxin and antigen concentrations. Twenty of the 48 workers developed specific reactions to the dust extract. It was concluded that exposure to dust made airborne during the refining process of potato starch may cause work related respiratory symptoms.

Table 33.3 lists a variety of mostly allergenic airborne hazards in food processing facilities. Although there are more microbial hazards than just these, this summary provides an idea of the wide variety of health threats that are present in the food industry. Some examples are discussed following that provide some sense of how common such allergies, diseases, and respiratory problems are among food handlers and food processors.

In a study of a potato processing plant, workers were exposed to large concentrations of airborne microorganisms and endotoxins (Dutkiewicz, 1994). Antigenic reactions were significantly correlated with the occurrence of work-related respiratory and general symptoms in 45.9 percent of the examined workers.

A study of worker exposure to airborne fungi in a sugar beet refinery evaluates the level of exposure and whether controls could be implemented to lower these exposures (Jensen et al., 1993). A previous study at this refinery identified one worker who had reactions to *Aspergillus niger* while two other employees were diagnosed with occupational asthma. This study showed high exposure of pellet loaders and pellet silo workers to various species of *Aspergillus*. Other hazardous fungal species were detected. Exposures to fungi during the postproduction cleanup and maintenance phase were much higher than those measured during the production campaign.

Workers occupationally exposed to grain dust have a high prevalence of respiratory symptoms, but their sensitization to cereal flour cannot be demonstrated, other allergens like storage mites, tenebroids, and cockroaches are stored-grain pests frequently found in grain and cereal products that may be the real culprits. Armentia et al. (1997) performed an epidemiological analysis of sensitization of these stored-grain pests on 4379 patients residing in an area of cereal industries. The prevalence of mite sensitization in the total sample studied was 18.96 percent. The prevalence of sensitization to storage mites among

TABLE 33.3 Allergies and Hypersensitivity in Food Industry Facilities

Food industry or occupation	Allergen or source	Primary symptoms	Reference
Bakeries	Dough allergens, fungal alpha-amylase	Allergies	Houba et al., 1997
Bakeries	Alternaria, aspergillus	Baker's asthma	Klaustermeyer et al., 1977
Carob bean flour	Carob bean flour	Asthma	Scoditti et al., 1996
Cereal millers, farmers	Cereals, soya bean, storage mites	Occupational asthma	Alvarez et al., 1996
Coffee roastery	Green coffee beans, coffee leaf allergen	Allergy	Axelsson, 1994
Flour millers	Wheat flour dust, wheat flour allergens	Respiratory irritation	Smith et al., 2000
Food and spice	Mugwort, paprika, short ragweed, black pepper	Rhinitis, asthma	Schwartz et al., 1997
Ginseng processors	Brazil Pfaffia paniculata root powder	Occupational asthma	Subiza et al., 1991
Ham, sausage, cheese	Mites, fungal spores	Fungal disease	Palmas et al., 1996
Homemaker	Green beans, raw bean cooking vapors	Hypersensitivity	Igea et al., 1994
Honey packing	Honeybee body dust	Asthma	Ostrom et al., 1986
Pea flour processors	Pea flour	Occupational asthma	Bhagat et al., 1995
Popcorn plant	Volatile butter-flavoring ingredients	Bronchiolitis obliterans	Kreiss et al., 2002
Potato processors	Potato dust, fungi, bacterial endotoxins	Respiratory irritation	Dutkiewicz et al., 2002
Potato starch industry	Potato dust	Respiratory irritation	Hollander et al., 1994
Raw sugar mills	*Thermoactinomyces sacchari* spores	Bagassosis	Dawson et al., 1996
Rice granary workers	Rice dust	Respiratory irritation	Ye et al., 1998
Salmon processing	Salmon processing aerosols	Occupational asthma	Douglas et al., 1995
Sausage manufacture	Paprika, mace	Rhinitis, asthma	Sastre et al., 1996
Seafood industry	Lobster, shrimp	Occupational asthma	Lemiere et al., 1996
Seafood industry	Crustaceans, mollusks, bony fish, seafood	Allergy, dermatitis	Jeebhay et al., 2001
Seafood industry	Trout, processing water	Allergy	Tougard et al., 1997
Soybean harvesters	Soybean hull dust	Asthma, allergic rhinitis	Codina et al., 2000
Sugar beet processors	Sugar beet pollen	Allergy	Hohenleutner et al., 1996
Sugar beet processors	Moldy sugar beet pulp	Occupational asthma	Rosenman et al., 1992
Sunflower seeds	Sunflower seed dust	Occupational asthma	Vandenplas et al., 1998
Tea	Tea fluff	Respiratory irritation	Hill and Waldron, 1996

mite-sensitive patients was 11.88 percent. Among the patients studied the most frequent sensitization was that to *Dermatophagoides pteronyssinus* (58 percent), followed by *Dermatophagoides farinae* (48 percent), *Lepidoglyphus destructor*, and *Tyrophagus putrescentiae* (38 percent), *Blomia kulagini* (34 percent), and *Acarus siro* and *Chortoglyphus arcuatus* (24 percent). Fifty percent of the patients were sensitized to *Tenebrio molitor*, and 36 percent to *Blatta orientalis*.

Cooking had its associated hazards due to levels of VOCs that may be produced and some epidemiological data indicate an excess of lung and bladder cancer among cooks and kitchen workers. The process of cooking beef substances can cause formation of basic mutagens that depend on cooking temperatures. In a study by Rappaport (1979), mutagenic activity increased exponentially with cooking temperature between 137 and 252°C. This study suggests inhalation of mutagens may be a critical route of exposure.

Foodborne streptococcal pharyngitis may be caused by poor handling and preservation of cold salads, usually those which contain eggs and are prepared some hours before serving (Katzenell et al., 2001). Epidemics of streptococcal pharyngitis tend to occur in warm climates and in the hottest months of the year. *Streptococcus pyogenes* is a common human commensal and often comes from the mouth, nose, or skin. Food handlers may be the primary source of streptococcal food contamination. Food handlers working with cold salads should be supervised to ensure they comply with strict rules of preparation and storage of food and cold salads, especially those containing eggs, should not be left overnight before serving.

33.5 FOOD PLANT AEROBIOLOGY

The airflow provided by plant ventilation systems is a possible source of microbial contamination. Airborne dust particles can include microbial contaminants and these can be introduced into food products. One of the main sources of bacterial contamination is the personnel and appropriate accessories and garments should be used where necessary (Wirtanen et al., 2002). Sampling of air and dust can be used to determine the presence of bacteria and fungi.

Figure 33.2 illustrates the various pathogens and allergens that may occur in the food industry, especially in food processing plants, and the aerobiological pathways they may take to contaminate food or cause inhalation diseases in workers. The food pathogens here

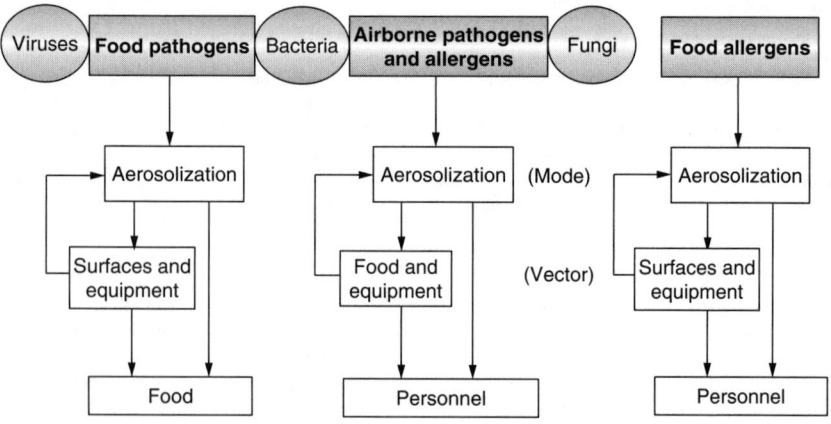

FIGURE 33.2 The primary aerobiological pathways of pathogens and allergens in food processing plants.

are those viruses and bacteria that specifically cause foodborne diseases but may transmit by the airborne route. Airborne pathogens refer to those nonfood bacterial pathogens like *Brucella* that may occur in the food industry. Food allergens are generally allergenic food particles or by-products of food manufacturing processes. Food spoilage microbes are encompassed by the food pathogen flowchart also, although they are mostly fungi.

Hot smoked fish can experience contamination with pathogenic bacteria like *Clostridium botulinum*, *Listeria monocytogenes*, *Staphylococcus aureus*, and *Vibrio parahaemolyticus*, although at very low levels (Sikorski and Kalodziejska, 2002). During processing, contamination may increase due to unsanitary procedures, rotation of assigned duties of workers, and airborne microorganisms entering and settling during packing of the product. *Listeria* has recently been a focus of attention due to several outbreaks in food processing plants and its potential lethality. Evidence exists to suggest *Listeria* can settle on foodstuffs via the airborne route (Bell and Kyriakides, 1998). See Table 33.4 for typical bioaerosol levels in food industry facilities.

Proteinaceous materials in the air can be highly allergenic and cause a variety of respiratory problems, including asthma. One study performed on airborne egg protein concentrations in an egg processing facility that had cases of occupational asthma showed the highest concentrations occurred in the egg washing room (mean exposure 644 $\mu g/m^3$) and in the egg breaking room (255 $\mu g/m^3$). These were the areas where the risk of being sensitized to the allergen was the greatest. Size-selective sampling indicated that most of the aerosol was capable of reaching the small airways (Boeniger et al., 2001).

Airborne microbial contaminants are common in dusty poultry processing plants. In a field study by Whyte et al. (2001), found that plate counts were highest in the defeathering areas of plants. *Enterobacteriaceae* counts were highest in the defeathering areas with airborne counts averaging 72 cfu/m^3. *E. coli* and *Enterobacteriaceae* were each found at approximately 78 cfu/m^3. Thermophilic campylobacters were isolated in the defeathering areas at 49 cfu/m^3.

Norwalk virus may be the only known purely intestinal pathogen that can transmit by the airborne route. An outbreak of gastroenteritis in a large hotel suggested a Norwalk-like virus infection, which was confirmed by laboratory analysis (Marks et al., 2000). The foods served could not be conclusively demonstrated to be the cause of the outbreak but a patron who vomited was believed to be responsible for producing airborne quantities of the virus. Inhalation of the virus may be followed by lung clearance and ingestion, which then produces gastroenteritis.

Slaughterhouses are susceptible to microbial airborne contamination from normal processing of carcasses. One study of a facility examined the back-splitting and weighing areas and found the air bacterial level in the back-splitting areas to be approximately 1800 cfu/m^3 was generally higher than that in the weighing areas which had 1100 cfu/m^3 (Rahkio and Korkeala, 1997). Associations were found between the contamination of air and the carcasses and the movements of workers. The layout of the slaughtering line was shown to be an important factor as separation of the clean and unclean parts of the line as well as separation of the weighing area from the other clean parts of the line decreased the contamination level. In one series of outbreaks of brucellosis in a packing house, airborne *Brucella* was implicated, but not proved, as the likely cause of the infections that afflicted almost 8 percent of employees (Hendricks et al., 1962).

Concentrations of airborne culturable fungi were measured in the kitchen of a bakery to evaluate the effects associated with common worker activities, outdoor aerosol distributions, and season (Levy et al., 1999). Activities included early morning preparation, cornmeal sifting and tossing, flour dumping and mixing, sweeping, and low-level activities. Elevated levels of total culturable fungi were found during all functions except the low-level activities. *Penicillium* was the dominant fungi in the spring, while *Cladosporium* was the dominant genus during the summer.

TABLE 33.4 Typical Bioaerosol Levels in Food Industry Facilities

Microbe	Concentration	Facility or area	Notes	Reference
Bacteria (cfu/m^3)	72–78	Poultry processing	*Enterobacteriaceae*	Whyte et al., 2001
	78	Poultry processing	*E. coli*	Whyte et al., 2001
	1,100–1,800	Slaughterhouses		Rahkio and Korkeala, 1997
	400,000–4,000,000	Slaughterhouses	Mainly *Staphylococcus*	Hagmar et al., 1990
	50,000	Poultry processing	From fecal matter	Lenhart et al., 1982
	0–8,000	Poultry processing	*Staphylococcus*	Patterson, 1973
Fungi (cfu/m^3)	119,0000,000	Salami factory	(propagules/m^3)	Jesenska et al., 1981
	500–4,000	Slaughterhouses		Hagmar et al., 1990
Total cfu/m^3	28,000–93,000	Potato processing	Potato dust	Dutkiewicz et al., 2002
Dust (mg/m^3)	6.2	Flour mills	Flour dust	Smith et al., 2000
	6.6–69.8	Rice granaries	Rice dust	Ye et al., 1998
	115–200	Potato processing	Potato dust	Dutkiewicz et al., 2002
	56	Potato processing	Potato dust	Hollander et al., 1994
Allergen (mg/m^3)	0.255–0.644	Egg processing	Egg allergen	Boeniger et al., 2001
	0–0.040	Bakeries	Dough allergen	Houba et al., 1997
	50–100	Bakeries	Wheat antigen	Burstyn et al., 1999
	46–1894	Potato processing	Postblanching	Dutkiewicz et al., 2002
Endotoxin (μg/m^3)	0.011–0.089	Potato processing	Preprocessing	Dutkiewicz et al., 2002
	0.013	Potato processing	Potato dust	Hollander et al., 1994
	0.4	Slaughterhouses	Dust	Hagmar et al., 1990

Ghosh et al. (1997) studied the problem of airborne aflatoxin generated in rice and maize processing plants. Levels of airborne aflatoxin in the rice mill always had higher numbers of respirable dust particles with airborne concentrations of 26 pg/m^3 for the workplace and 19 pg/m^3 for the storage area. Airborne aflatoxin was not detected in any of the grain processing plants. In a maize processing plant the elevator had 18 pg/m^3, the loading/unloading area had 800 pg/m^3, and the oil mill had 816 pg/m^3, and showed the presence of airborne aflatoxin in respirable dust samples.

Several studies of the baking industry have reported sensitization and respiratory disorders among bakery workers caused by enzymes in dough improvers and fungal alpha-amylase is the most frequently reported cause of allergy (Houba et al., 1997). Allergen exposure levels varied considerably among bakery workers, depending on the type of bakery and job category (range, 0 to 40 ng/m^3). In confectioneries no alpha-amylase allergens were detected. In other bakeries alpha-amylase exposure was only found for workers directly involved in dough making.

Bagassosis is a common occupational hazard among workers in various types of mills. In one study of raw sugar mills, *Thermoactinomyces sacchari* spores were measured to determine whether they were sufficient to cause acute bagassosis, and whether there was any evidence of previous exposure to sufficient airborne *T. sacchari* spores to cause the development of chronic bagassosis in any of the workforce (Dawson et al., 1996). The results showed that the total airborne bacteria spore count was lower than similar counts reported in other industries, such as cotton milling and wood chip handling. Medical data indicated that no cases of acute bagassosis and there was no evidence of the development of chronic bagassosis in any members of the workforces of either mill.

In a study in a series of poultry slaughterhouse plants, 23 dust-exposed shacklers in the hanging departments of the plants were examined immediately before and after work and the levels of both total dust and endotoxins were monitored during the work-shift (Hagmar et al., 1990). The mean level of total dust was 6.3 mg/m^3 and 0.40 µg/m^3 for endotoxins. Airborne levels of bacteria were measured at 400,000 to 4,000,000 cfu/m^3. The bacteria were mainly *Staphylococcus*. Fungi concentrations of 500 to 4000 cfu/m^3 were found in the hanging departments. Some respiratory symptoms were found but none of the workers had any symptoms indicating extrinsic allergic alveolitis or ODTS.

Epidemic and endemic brucellosis at six abattoirs implicated airflow between the kill department and other areas within an abattoir (Kaufmann et al., 1980). Air from the kill department caused abnormally high brucellosis attack rates for persons who worked in these areas at two of the abattoirs. Complete physical separation and the maintaining of negative air pressure in the kill department reduced the risk for workers in other areas at four of the abattoirs. Brucellosis is also contracted through skin contact with infectious animal tissues, but this route of transmission was less important than the airborne route.

An outbreak of psittacosis at a Nebraska poultry processing plant prompted a study of the etiology (Anderson et al., 1978). The attack rates and the nature of the spread among various processing departments suggested airborne transmission, but the investigation also suggested that workers having both frequent contact with turkey tissues and skin injuries were more likely to be infected than with other processing workers. A single turkey flock was implicated as the source of infection. This flock had been screened on a voluntary basis for evidence of infection prior to slaughter, using criteria developed by the United States Department of Agriculture. Although the preslaughter screening failed to detect psittacosis infection, two turkeys from the same flock that inadvertently had not been sent to slaughter were subsequently found to be infected.

Approximately 90,000 workers are employed in the poultry processing industries in the United States with an additional 30,000 workers employed growing chickens. Lenhart et al. (1982) sampled airborne bacteria in a poultry processing plant and found levels that averaged 50,000 cfu/m^3. Many of the microbes identified are sources of endotoxins and were believed to be of chicken fecal origin. The predominant bacterial species identified that pose potential respiratory hazards were *Corynebacterium, Staphylococcus, Acinetobacter*, and *Alcaligenes*.

Microbes can remain viable in air for various periods of time depending on species. Although respiratory pathogens often have the ability to survive for long periods in air,

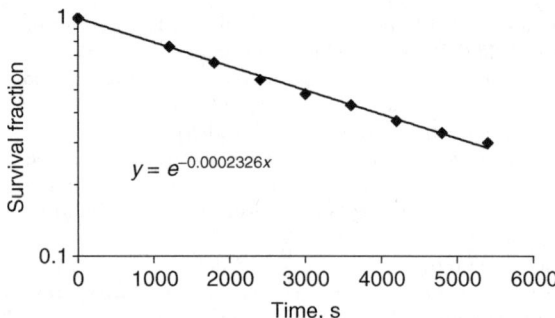

FIGURE 33.3 Survival curve of *Salmonella newbrunswick* in air. [Based on data from Stersky et al. (1972).]

little is known about the airborne survivability of foodborne pathogens. In a study of the airborne viability of the bacteria *Salmonella newbrunswick*, results indicated unusual airborne resilience. Figure 33.3 illustrates the decay curve of *Salmonella newbrunswick* in air, based on data from Stersky et al. (1972). The fitted exponential decay equation indicates a decay rate constant of 0.0002326 s^{-1}. Obviously this microbe could remain viable in air for several hours, at least. Although this may not be representative of foodborne pathogens in general, it is suggestive that such bacteria could remain suspended and viable in air long enough to transmit throughout a facility and possibly settle on foods during processing.

Volatile organic compounds (VOCs) can be released into the air as a result of the industrial manufacture and processing of food and drinks and can cause a variety of respiratory problems. However, most of these are produced by chemical agents or compounds in foods and are not the result of microbial contamination, airborne or otherwise. Odors are also generated by cooking food in kitchens and by intrusive outdoors pollutants. They are best dealt with by traditional ventilation methods and the use of exhaust hoods (ASHRAE, 1999; Turner, 2000).

Endotoxins, exotoxins, and mycotoxins are products of bacterial or fungal growth under certain environmental conditions. The key to controlling these toxins is to control microbial growth, and microbes can, in turn, be controlled by food plant sanitation, adequate ventilation, and air cleaning technologies.

33.6 FOOD PLANT SOLUTIONS

Common methods for remediating airborne problems include improved ventilation, cleaning, filtration, chemical fogging, and UVGI. HVAC systems can cause major contamination risks if they are not maintained, cleaned, and inspected regularly. If air is recirculated through plant ventilation systems, organic debris may accumulate and fungi and bacteria may grow inside ductwork and on air handling equipment (Wallin and Haycock, 1998). The filters used on such ventilation systems are typically dust filters that may be rated MERV 6–8 (DSP 25 percent) or lower, and this is inadequate to control airborne microbes effectively (see Chap. 10 for filter microbial removal efficiencies). It is recommended that recirculated air in facilities subject to aerosolization of food particles be filtered with a filter in the MERV 11–15 range to assure the highest levels of aerobiological cleanliness. Filter frames and cases are usually manufactured from galvanized mild steel or stainless steel while prefilters use card frames. The design, installation, and sealing of a suitable filter-framing system are essential to minimize filter bypass, which can seriously degrade performance. Coupling filters with UVGI systems will provide maximum benefits with minimum energy costs (Kowalski and Bahnfleth, 2000). The use of UVGI for food processing has been hampered by concerns about lamp breakage, but new lamps are available encased in plastic that are shatterproof (see Fig. 33.4).

Mold and biofilms can develop on surfaces and equipment in the food and beverage industries, including tanks and vats, cooking equipment, walls and floors, and cooling coils in breweries (Carpentier and Cerf, 1993). In general, standard cleaning and disinfection procedures are adequate to contain these problems but alternatives are available, including antimicrobial coatings like copper, UV irradiation of cooling coils and after-hours irradiation of surfaces when personnel are not present (CDA, 2000; Philips, 1985; Kowalski and Dunn, 2002). After-hours UVGI systems (see Chap. 11) have been installed in some food plants recently to control mold growth.

Ionizing radiation can be effectively applied to fruits and vegetables for disinfestation and disinfection as a quarantine treatment (Moy, 1993). Vapor heating methods can also be used in such applications but require longer treatment times, and may cause quality problems. A synergy between gamma irradiation and heating for the control of mold has been reported (Barkai-Golan et al., 1969). It is possible for low levels of radiation to germinate fungal spores, after which they may become much more vulnerable to heat and other biocidal factors. See Chaps. 19 and 22 for additional information.

Cleanroom technology is often employed in the cheese industry and other areas to ensure that such processes have a high level of hygiene. Cleanroom technology involves airtight control of areas and filtration to reduce airborne contamination to levels comparable to that in operating rooms. Such sterile processing minimizes the use of chemicals and cleaning agents, reduces the required labor, often lowers the energy requirements for heating and refrigeration, and yields financial gains through enhanced product quality (Todt, 1996).

Compressed air, which is routinely used in food processing industry, can become contaminated with bacteria and fungi (Maeda, 2001). The use of antibacterial filters can prevent this, as can regular maintenance.

Fumigation is well established in the food and beverage industry as a remediation technology for fungal and bacterial contamination (Gorham, 1991b). Fumigants may be either gases or vapors and the most commonly used fumigants include ethylene oxide, methyl bromide, sulfuryl fluoride, hydrogen cyanide, aluminum phosphide, and magnesium phosphide. These disinfectants are hazardous and must be handled according to the appropriate procedures with all precautions taken to avoid human exposure. Typically, an area is sealed off and fumigated for a period of about 24 hours. Chapter 26 may be consulted for more specific procedures for remediating rooms or buildings with fumigants. Ozone can also be used as a fumigant, as well as chlorine dioxide—see Chaps. 17 and 26 for additional information on these disinfection technologies. Ozone has also been tried as a means of decontaminating the surfaces of foods, but with mixed results (Kim and Yousef, 2000).

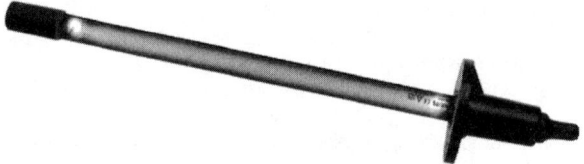

FIGURE 33.4 Waterproof, plastic-encased shatterproof UV lamp for processes where breakage may be a concern. (*Photo provided courtesy of Brad Hollander, UVCM, Inc., Minden, NV.*)

The survival of bacteria on various surfaces has been studied and the characteristics of various materials in this regard are fairly well established. One of the most commonly used materials in the food processing industry is stainless steel due to its durability and ease of cleaning. Studies have shown that foodborne bacteria like *Staphylococcus* and *Salmonella* can survive on stainless steel surfaces for up to 4 days (Kusumaningrum et al., 2003). Transfer rates of contamination from stainless steel vary from 20 to 100 percent. The use of antimicrobial coatings, such as silver-based resins, on stainless steel surfaces and appliances is a promising area of research in the HVAC industry that may have applications in the food industry (Mazurkiewicz, 2002; Topping, 2004). Antimicrobial food preservatives may also be of some benefit in terms of controlling microbial contamination (Luck and Jager, 1997). Filters used in the food industry are subject to accumulation of food particles and can become a source of microbial growth, and antimicrobial filter media may be able to control such problems (Maeda, 2001).

Fungal and other microorganisms may degrade food in shipment and storage and certain of these microorganisms may be inhalation or ingestion hazards. Surface applications of waxes and water-loss barriers are often applied to fruits and vegetables to maintain quality and sensory attributes between harvest and market (Law and Cooper, 2001). However, damage in transit can still allow for microbial growth on foodstuffs and such protective sprays are not an absolute guarantee against spoilage microorganisms.

Poultry hanging workers commonly show symptoms of bronchitis and respiratory problems due to high aerosol particle concentrations, and suggestion for their protection included the use of air cleaners, improved ventilation, and protective masks (VanWicklen et al., 1997).

Negative ionization has been used for the reduction of airborne bacteria and dust in poultry facilities and may have applications in the food industry (Estola et al., 1979). In one laboratory study, aerosolized *Salmonella* was exposed to ionization in a sealed plastic chamber. An average of greater than 1000 cfu/plate was observed on plates exposed to the aerosol without ionization compared with an average of 53 cfu/plate on the ionizer-treated plates (Seo et al., 2001). The authors of the latter study suggest that high levels of negative air ions were responsible for direct killing of the microorganisms.

REFERENCES

Alvarez, M. J., Tabar, A. I., Quirce, S., Olaguibel, J. M., Lizaso, M. T., Echechipia, S., Rodriguez, A., and Garcia, B. E. (1996). "Diversity of allergens causing occupational asthma among cereal workers as demonstrated by exposure procedures." *Clin Exp Allergy* 26(2):147–153.

Anderson, D. C., Stoesz, P. A., and Kaufmann, A. F. (1978). "Psittacosis outbreak in employees of a turkey-processing plant." *Am J Epidemiol* 107(2):140–148.

Armentia, A., Martinez, A., Castrodeza, R., Martinez, J., Jimeno, A., Mendez, J., and Stolle, R. (1997). "Occupational allergic disease in cereal workers by stored grain pests." *J Asthma* 34(5):369–378.

ASHRAE (1999). Chapter 30: Kitchen Ventilation. *HVAC Applications*, American Society of Heating, Refrigerating and Air-Conditioning Engineers, Atlanta, GA.

ASHRAE (2001). "Standard 62-2001: Ventilation for acceptable indoor air quality." American Society of Heating Refrigerating and Air-Conditioning Engineers. Atlanta, GA.

Axelsson, I. G. (1994). "Allergy to the coffee plant." *Allergy* 49(10):885–887.

Barkai-Golan, R., Kaham, R. S., and Padova, R. (1969). "Synergistic effects of gamma irradiation and heat on the development of *Penicillium digitatum* in vitro and in stored citrus fruits." *Phytopathology* 59:922–924.

Bartelink, A. K. M., and vanKregten, E. (1995). "*Streptococcus suis* as threat to pig-farmers and abattoir workers." *Lancet* 346:1707.

Bell, C., and Kyriakides, A. (1998). *Listeria: A Practical Approach to the Organism and its Control in Foods.* Blackie Academic & Professional, London.

Bhagat, R., Swystun, V. A., and Cockcroft, D. W. (1995). "Occupational asthma caused by pea flour." *Chest* 107(6):1772.

Boeniger, M. F., Lummus, Z. L., Biagini, R. E., Bernstein, D. I., Swanson, M. C., Reed, C., and Massoudi, M. (2001). "Exposure to protein aeroallergens in egg processing facilities." *Appl Occup Environ Hyg* 16(6):660–670.

Burstyn, I., Heederik, D., Bartlett, K., Doerkes, G., Houba, R., Teschke, K., and Kennedy, S. M. (1999). "Wheat antigen content of inhalable dust in bakeries: Modeling and an inter-study comparison." *Appl Occup Environ Hyg* 14(11):791–798.

Carpentier, B., and Cerf, O. (1993). "Biofilms and their consequences, with particular reference to hygiene in the food industry." *J Appl Bacteriol* 75:499–511.

CDA (2000). "Antibacterial properties of copper and brass demonstrate potential to combat toxic *E. coli* O157 outbreaks in the food processing industry." Copper Development Association. http://www.copper.org/about/pressreleases/2000/DemonstratePotential.html.

Codina, R., Ardusso, L., Lockey, R. F., Crisci, C., and Bertoya, N. (2000). "Sensitization to soybean hull allergens in subjects exposed to different levels of soybean dust inhalation in Argentina." *J Allergy Clin Immunol* 105(3):570–576.

Dawson, M. W., Scott, J. G., and Cox, L. M. (1996). "The medical and epidemiologic effects on workers of the levels of airborne *Thermoactinomyces* Spp. spores present in Australian raw sugar mills." *Am Ind Hyg Assoc J* 57(11):1002–1012.

Douglas, J. D., McSharry, C., Blaikie, L., Morrow, T., Miles, S., and Franklin, D. (1995). "Occupational asthma caused by automated salmon processing." *Lancet* 346(8977):737–740.

Dutkiewicz, J. (1994). "Bacteria, fungi, and endotoxin as potential agents of occupational hazard in a potato processing plant." *Am J Ind Med* 25(1):43–46.

Dutkiewicz, J., Krysinska-Traczyk, E., Skorska, C., Cholewa, G., and Sitkowska, J. (2002). "Exposure to airborne microorganisms and endotoxin in a potato processing plant." *Ann Agric Environ Med* 9(2):225–235.

Estola, T., Makela, P., and Hovi, T. (1979). "The effect of air ionization on the airborne transmission of experimental Newcastle disease virus infections in chickens." *J Hyg* 83:59–67.

Ghosh, S. K., Desai, M. R., Pandya, G. L., and Venkaiah, K. (1997). "Airborne aflatoxin in the grain processing industries in India." *Am Ind Hyg Assoc J* 58(8):583–588.

Gorham, J. R. (1991a). Food pests as disease vectors. *Ecology and Management of Food-Industry Pests,* J. R. Gorham, ed., Association of Official Analytical Chemists, Arlington, VA.

Gorham, J. R. (1991b). "Ecology and management of food-industry pests." *FDA Technical Bulletin* 4.

Hagmar, L., Schutz, A., Hallberg, T., and Sjoholm, A. (1990). "Health effects of exposure to endotoxins and organic dust in poultry slaughterhouse workers." *Int Arch Occup Environ Health* 62(2):159–164.

Heijden, K. v. d., Younes, M., Fishbein, L., and Miller, S. (1999). *International Food Safety Handbook: Science, International Regulation, and Control.* Marcel Dekker. New York.

Hendricks, S. L., Borts, I. H., Heeren, R. H., Hausler, W. J., and Held, J. R. (1962). "Brucellosis outbreak in an Iowa packing house." *J Pub Health* 52(7):1166–1178.

Hill, B., and Waldron, H. A. (1996). "Respiratory symptoms and respiratory function in workers exposed to tea fluff." *Ann Occup Hyg* 40(5):491–497.

Hohenleutner, S., Pfau, A., Hohenleutner, U., and Landthaler, M. (1996). "Sugar beet pollen allergy as a rare occupational disease." *Hautarzt* 47(6):462–464.

Hollander, A., Heederik, D., and Kauffman, H. (1994). "Acute respiratory effects in the potato processing industry due to a bioaerosol exposure." *Occup Environ Med* 51(2):73–78.

Houba, R., vanRun, P., Doekes, G., Heederik, D., and Spithoven, J. (1997). "Airborne levels of alpha-amylase allergens in bakeries." *J Allergy Clin Immunol* 99(3):286–292.

Hui, Y. H., Bruinsma, B. L., Gorham, J. R., Nip, W.-K., Tong, P. S., and Ventresca, P. (2003). *Food Plant Sanitation.* Marcel Dekker. New York.

Igea, J. M., Fernandez, M., Quirce, S., Hoz, B. de la, and Gomez, M. L. D. (1994). "Green bean hypersensitivity: an occupational allergy in a homemaker." *J Allergy Clin Immunol* 94(1):33–35.

Jeebhay, M. F., Robins, T. G., Lehrer, S. B., and Lopata, A. L. (2001). "Occupational seafood allergy: a review." *Occup Environ Med* 58(9):553–562.

Jensen, P. A., Todd, W. F., Hart, M. E., Mickelsen, R. L., and O'Brien, D. M. (1993). "Evaluation and control of worker exposure to fungi in a beet sugar refinery." *Am Ind Hyg Assoc J* 54(12):742–748.

Jesenska, Z., Polakova, O., Polster, M., Koskova, L., and Vaszilkova, A. (1981). "Potential producers of aflatoxin in working environment." *Zentralbl Bakteriol Mikrobiol Hyg [B]* 172(4–5):382–389.

Katzenell, U., Shemer, J., and Bar-Dayan, Y. (2001). "Streptococcal contamination of food: an unusual cause of epidemic pharyngitis." *Epidemiol Infect* 127(2):179–184.

Kaufmann, A. F., Fox, M. D., Boyce, J. M., Anderson, D. C., Potter, M. E., Martone, W. J., and Patton, C. M. (1980). "Airborne spread of brucellosis." *Ann N Y Acad Sci* 353:105–114.

Kim, J. G., and Yousef, A. E. (2000). "Inactivation kinetics of foodborne spoilage and pathogenic bacteria by ozone." *J Food Sci* 65(3):521–528.

Klaustermeyer, W. B., Bardana, E. J., and Hale, F. C. (1977). "Pulmonary hypersensitivity to *Alternaria* and *Aspergillus* in baker's asthma." *Clin Allergy* 7(3):227–233.

Knowles, T. (2002). *Food Safety in the Hospitality Industry.* Butterworth Heineman, Oxford, UK.

Korunic, Z. (2001). "Allergenic components of stored agro products." *Arh Hig Rada Toksikol* 52(1):43–48.

Kowalski, W. J., and Bahnfleth, W. P. (2000). "UVGI Design Basics for Air and Surface Disinfection." *HPAC* 72(1):100–110. http://www.engr.psu.edu/ae/wjk/uvhpac.html.

Kowalski, W. J., and Dunn, C. E. (2002). "Current Trends in UVGI Air and Surface Disinfection." *INvironment Professional* 8(6):4–6.

Kreiss, K., Gomaa, A., Kullman, G., Fedan, K., Simoes, E. J., and Enright, P. L. (2002). "Clinical bronchiolitis obliterans in workers at a microwave-popcorn plant." *N Engl J Med* 347(5):330–338.

Kusumaningrum, H. D., Riboldi, G., Hazeleger, W. C., and Beumer, R. R. (2003). "Survival of foodborne pathogens on stainless steel surfaces and cross-contamination to foods." *Int J Food Microbiol* 85(3):227–236.

Law, S. E., and Cooper, S. C. (2001). "Air-assisted electrostatic sprays for postharvest control of fruit and vegetable spoilage microorganisms." *IEEE Trans Ind Appl* 37(6):1597–1602.

Lemiere, C., Desjardins, A., Lehrer, S., and Malo, J. L. (1996). "Occupational asthma to lobster and shrimp." *Allergy* 51(4):272–273.

Lenhart, S. W., Olenchock, S. A., and Cole, E. C. (1982). "Viable sampling for airborne bacteria in a poultry processing plant." *J Toxicol Environ Health* 10(4–5):613–619.

Levy, J. I., Nishioka, Y., Gilbert, K., Cheng, C. H., and Burge, H. A. (1999). "Variabilities in aerosolizing activities and airborne fungal concentrations in a bakery." *Am Ind Hyg Assoc J* 60(3):317–325.

Luck, E., and Jager, M. (1997). *Antimicrobial Food Preservatives.* Springer, Berlin, Germany.

Maeda, S. (2001). "Antibacterial air filter's role in the food processing industry." *Filtr Separat* 38(7):38–40.

Marks, P. J., Vipond, I. B., Carlisle, D., Deakin, D., Fey, R. E., and Caul, E. O. (2000). "Evidence for airborne transmission of Norwalk-like virus (NLV) in a hotel restaurant." *Epidemiol Infect* 124(3):481–487.

Mazurkiewicz, G. (2002). "Coated ducts keep microbes at bay; antimicrobial coating designed to help prevent IAQ problems." *Air Conditioning, Heating & Refrigeration News* 217(8). http://www.iuoe.org/cm/iaq_mark_prod.asp?Item=331.

Moy, J. H. (1993). "Efficacy of irradiation vs thermal methods as quarantine treatments for tropical fruits." *Rad Physics and Chem* 42(1–3 Pt. 1):269–272.

Ortega, H. G., Daroowalla, F., Petsonk, E. L., Lewis, D., Berardinelli, S., Jones, W., Kreiss, K., and Weissman, D. N. (2001). "Respiratory symptoms among crab processing workers in Alaska: epidemiological and environmental assessment." *Am J Ind Med* 39(6):598–607.

Ostrom, N. K., Swanson, M. C., Agarwal, M. K., and Yunginger, J. W. (1986). "Occupational allergy to honeybee-body dust in a honey-processing plant." *J Allergy Clin Immunol* 77(5):736–740.

Palmas, F., Meloni, V., Tinti, M., Deplano, M., and Fadda, M. E. (1996). "Pollution by mites and/or mold spores in work environments and the risk of occupational respiratory pathology." *Med Lav* 87(5):411–422.

Patterson, J. T. (1973). "Airborne microorganisms in poultry processing plants." *Br Poult Sci* 14(2):161–165.

Philips (1985). *UVGI Catalog and Design Guide*. Catalog No. U.D.C. 628.9, Philips Co., The Netherlands.

Rahkio, T. M., and Korkeala, H. J. (1997). "Airborne bacteria and carcass contamination in slaughterhouses." *J Food Prot* 60(1):38–42.

Rappaport, S. M., McCartney, M. C., and Wei, E. T. (1979). "Volatilization of mutagens from beef during cooking." *Cancer Lett* 8(2):139–145.

Ray, B. (1996). *Fundamental Food Microbiology*. CRC Press, Boca Raton, FL.

Rodriguez, V. M. E., Pousa, O. A., Pons, S. C., Larrosa, M. A., Sanchez, S. L. P., and Martinez, N. F. (2001). "Brucellosis as occupational disease: study of an outbreak of airborn transmission at a slaughter house." *Rev Esp Salud Publica* 75(2):159–169.

Rosenman, K. D., Hart, M., and Ownby, D. R. (1992). "Occupational asthma in a beet sugar processing plant." *Chest* 101(6):1720–1722.

Samson, R. A., Hoekstra, E. S., Frisvad, J. C., and Filtenborg, O. (2000). *Introduction to Food and Airborne Fungi*. Ponsen Looyen, Waganingen, The Netherlands.

Sastre, J., Olmo, M., Novalvos, A., Ibanez, D., and Lahoz, C. (1996). "Occupational asthma due to different spices." *Allergy* 51(2):117–120.

Scala, E., Giani, M., Pirrotta, L., Guerra, E. C., Cadoni, S., Girardelli, C. R., Pita, O. D., and Puddu, P. (2001). "Occupational generalised urticaria and allergic airborne asthma due to anisakis simplex." *Eur J Dermatol* 11(3):249–250.

Schwartz, H. J., Jones, R. T., Rojas, A. R., Squillace, D. L., and Yunginger, J. W. (1997). "Occupational allergic rhinoconjunctivitis and asthma due to fennel seed." *Ann Allergy Asthma Immunol* 78(1):37–40.

Scoditti, A., Peluso, P., Pezzuto, R., Giordano, T., and Melica, A. (1996). "Asthma to carob bean flour." *Ann Allergy Asthma Immunol* 77(1):81.

Seo, K. H., Mitchell, B. W., Holt, P. S., and Gast, R. K. (2001). "Bactericidal effects of negative air ions on airborne and surface *Salmonella enteritidis* from an artificially generated aerosol." *J Food Prot* 64(1):113–116.

Sikorski, Z. E., and Kalodziejska, I. (2002). "Microbial risks in mild hot smoking of fish." *Crit Rev Food Sci Nutr* 42(1):35–51.

Smith, T. A., Parker, G., and Hussain, T. (2000). "Respiratory symptoms and wheat flour exposure: a study of flour millers." *Occup Med (Lond)* 50(1):25–29.

Stersky, A. K., Heldman, D. R., and Hedrick, T. I. (1972). "Viability of airborne *Salmonella newbrunswick* under various conditions." *J Dairy Sci* 55(1):14–18.

Subiza, J., Subiza, J. L., Escribano, P. M., Hinojosa, M., Garcia, R., Jerez, M., and Subiza, E. (1991). "Occupational asthma caused by Brazil ginseng dust." *J Allergy Clin Immunol* 88(5):731–736.

Tauxe, R. V. (2002). "Emerging foodborne pathogens." *Int J Food Microbiol* 78(1–2):31–41.

Todt, W. (1996). "Clean rooms for select moulds." *Sulzer Technical Rev* 78(3):28–31.

Topping, D. (2004). "Antimicrobial surfaces for appliances." *Appliance* 61(5):23.

Tougard, A. B., Bach, B., Taudorf, E., and Stouby, V. L. (1997). "Occupational respiratory tract allergy in trout processing workers." *Ugeskr Laeger* 159(39):5800–5804.

Turner, W. A. (2000). "Ventilation IAQ for the hospitality industry." *HPAC Eng* 72(7):37–44.

Vandenplas, O., Borght, T. V., and Delwiche, J. P. (1998). "Occupational asthma caused by sunflower-seed dust." *Allergy* 53(9):907–908.

VanWicklen, G. L., Czarick, M., and Lacy, M. P. (1997). "Protecting worker respiratory health in poultry hanging areas." Rep. No. 975011, ASAE, St. Joseph, MI.

Wallin, P., and Haycock, P. (1998). *Foreign Body Prevention, Detection and Control*. Blackie Academic & Professional, London.

Whyte, P., Collins, J. D., McGill, K., Monahan, C., and O'Mahony, H. (2001). "Distribution and prevalence of airborne microorganisms in three commercial poultry processing plants." *Food Prot* 64(3):388–391.

Wirtanen, G., Miettinen, H., Pahkala, S., Enbom, S., and Vanne, L. (2002). *Clean Air Solutions in Food Processing*. VTT Publications, Espoo, Finland.

Woollen, A. (1970). "*Food Industries Manual.*" Chemical Publishing, New York.

Ye, T. T., Huang, J. X., Shen, Y. E., Lu, P. L., and Christiani, D. C. (1998). "Respiratory symptoms and pulmonary function among Chinese rice-granary workers." *Int J Occup Environ Health* 4(3):155–159.

CHAPTER 34

AGRICULTURAL AND ANIMAL FACILITIES

34.1 INTRODUCTION

Agricultural facilities range from small country farms run by families to the large industrialized megafarms run by corporations. Agricultural facilities pose a variety of airborne occupational hazards including diseases from farm animals, allergies from animal dander and foodstuffs, health threats from mold and actinomycetes, and gases produced by animals, mold, bacteria, and waste or sewage. Animal facilities like barns, poultry houses, swine houses, and kennels pose unique hazards to both man and farm animals due to high levels of bioaerosols. Levels of airborne microbes in agricultural facilities are among the highest found in any indoor environments. Composting and agricultural waste problems are partly addressed in this chapter but Chap. 39 treats sewage and waste facilities directly and should be referred to for more details.

This chapter is intended to address those aerobiological concerns resulting from farm-related activities and the health impacts on both farm workers and farm animals, which are often closely related. Animal facilities may be any building or structure used to house or service animals, whether they be farm animals, pets, or zoos. Indoor facilities for veterinary care could be considered more similar to animal laboratories and Chap. 32 should be consulted for more specific information.

34.2 AGRICULTURAL PATHOGENS AND ALLERGENS

Airborne microorganisms of concern to human health can hail from the farm environment, farm animals, cattle feed or fodder, or mold that may grow on fodder or various materials. In addition, farm produce itself may be allergenic to some workers and the high levels of dust encountered may cause various forms of respiratory distress or diseases (Rangaswami, 1966). Many animal pathogens can transmit to humans either by direct contact, via the airborne

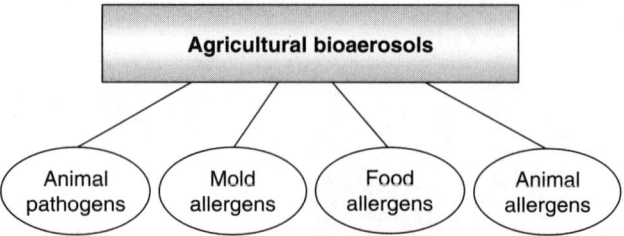

FIGURE 34.1 Breakdown of bioaerosol health threats in agriculture.

route, or by other mechanisms. The microorganisms of primary concern are those that can become airborne in indoor facilities, and these are mainly respiratory pathogens and allergens.

Allergens may be produced as a by-product of animal husbanding or from animal waste, animal feeds, or farm produce. The bacteria classified as actinomycetes are particularly common in agriculture and can grow on materials like moldy hay. Farmers may be routinely exposed to actinomycetes in very high concentrations and may inhale as many as 750,000 spores per minute of which 98 percent may be actinomycetes (Lacey and Crook, 1988). Farmer's lung is a name given to a group of respiratory problems experienced by farm workers chronically exposed to excessive concentrations of actinomycetes (Pepys et al., 1963). Figure 34.1 shows a breakdown of aerobiological health threats in agriculture. Not shown are toxins and *microbial volatile organic compounds* (MVOCs) but these come from microbial sources and so are part of the same hazard. VOCs such as ammonia from waste and dung are not specifically the product of airborne microorganisms but can be considered bioaerosols and are treated here for completeness.

Table 34.1 lists many of the pathogens and allergens that have occurred or may occur in agriculture, based on a review of the literature (see References section). In addition to these, many of the food pathogens and food allergens addressed in the previous chapter may occur in agricultural facilities where the associated foods are handled, such as grain elevators for wheat, potatoes, or rice. Pollen may also act as allergens to susceptible individuals, but these have also been addressed in previous chapters, including Chaps. 28 and 30, and these chapters should be referred to for additional information. More details on allergens from plants and foodstuffs are provided in the following sections.

34.3 AGRICULTURAL EPIDEMIOLOGY

Farmers may suffer from a variety of allergies, infections, and respiratory ailments associated with breathing the air in agricultural facilities. The nature of the respiratory problems may depend on the agricultural product being handled. Many insecticides and chemicals used on farms may also cause respiratory problems but these are not the subjects here as they are not aerobiological contaminants. Such pollutants, however, may contribute to rendering farm workers susceptible to microbiological agents.

A number of studies are available on the epidemiology of farmworker diseases (McDuffie et al., 1995). Various studies on farmers have reviewed the epidemiological data collected from clinics and hospitals and found a high prevalence among farmworkers of allergic asthma, allergic rhinitis, hypersensitivity pneumonitis, *organic toxic dust syndrome* (ODTS), pulmonary tuberculosis, pneumonia, upper respiratory tract infections, chronic

TABLE 34.1 Airborne Pathogens and Allergens in Agricultural Environments

Pathogen or allergen	Type	Disease or infection	Natural source	Toxins
Acinetobacter	Bacteria	Opportunistic/septic infections, meningitis	Environmental, soil, sewage	None
Actinomyces israelii	Bacteria	Actinomycosis	Humans, cattle	None
Aeromonas	Bacteria	Nonrespiratory opportunistic infections, gastroenteritis	Environmental, water, soil	None
Alcaligenes	Bacteria	Opportunistic infections	Humans, soil, water, nosocomial	None
Alternaria alternata	Fungi	Allergic alveolitis, rhinitis, irritation, asthma, toxic reactions	Environmental, indoor growth on paint, dust	Alternariol, tenuazonic acid
Aspergillus	Fungi	Aspergillosis, alveolitis, asthma, allergic fungal sinusitis, ODTS, toxic reactions, VOCs	Environmental, indoor growth on insulation and coils	Aflatoxin, fumigaclavines, gliotoxin, fumigatoxin, helvolic acid, etc.
Bacillus anthracis	Bacteria	Anthrax, woolsorter's disease	Cattle, sheep, other animals, soil	None
Brucella	Bacteria	Brucellosis, undulant fever	Goats, cattle, swine, dogs, sheep, caribou, elk, coyotes, camels	None
Cladosporium	Fungi	Chromoblastomycosis, allergic reactions, rhinitis, asthma	Environmental, indoor growth on dust	Cladosporin, emodin, epicladosporic acid
Coxiella burnetii	Bacteria	Q fever	Cattle, sheep, goats	None
Cryptostroma corticale	Fungi	Alveolitis, asthma, maple bark pneumonitis, maple bark disease	Environmental, found on maple and sycamore bark	None
Dander, hair	Allergen	Cows, horses, farm animals	Agricultural	None
Foodstuffs, fodder	Allergen	Allergies, hypersensitivity	Agricultural	None
Influenza A virus	Virus	Flu, secondary pneumonia	Humans, birds, pigs, nosocomial	None
Lymphocytic choriomeningitis	Virus	LCM, lymphocytic meningitis	House mouse, swine, dogs, hamsters, guinea pigs	None
Micromonospora faeni	Bacteria	Farmer's lung, pulmonary fibrosis, allergic reactions, UR irritation	Agricultural, moldy hay, indoor growth	None
Mycobacterium kansasii	Bacteria	Cavitary pulmonary disease	Water, cattle, swine	None

(*Continued*)

TABLE 34.1 Airborne Pathogens and Allergens in Agricultural Environments (*Continued*)

Pathogen or allergen	Type	Disease or infection	Natural source	Toxins
Penicillium	Fungi	Alveolitis, rhinitis, asthma, allergic reactions, irritation, ODTS, toxic reactions, VOCs produced	Environmental, indoor growth on paint, filters, coils, and humidifiers	Penicillic acid, peptide nephrotoxin, viomellein, etc.
Saccharopolyspora rectivirgula	Bacteria	Farmer's lung, alveolitis, asthma	Agricultural	None
Salmonella	Bacteria	Poultry, eggs	Agricultural.	None
Sporothrix schenckii	Fungi	Sporotrichosis, rose gardeners' disease	Environmental, plant material	None
Staphylococcus aureus	Bacteria	Staphylococcal pneumonia	Humans, sewage	None
Thermoactinomyces sacchari	Bacteria	Bagassosis, alveolitis, HP	Agricultural, bagasse	None
Thermoactinomyces vulgaris	Bacteria	Farmer's lung, pulmonary fibrosis, allergic reactions, asthma, HP	Agricultural, indoor growth in air conditioners	None
Thermomonospora viridis	Bacteria	Farmer's lung, HP	Agricultural	None
Vaccinia virus	Virus	Cowpox	Agricultural	None

bronchitis, and other respiratory problems (Kirkhorn and Garry, 2000; Strauch 1987a). There is often a strong correlation in these studies between the diseases and the typical levels of airborne dust, allergens, endotoxins, bacteria, and fungi. Table 34.2 lists a nonexclusive variety of these occupations and associated respiratory hazards.

Agriculture is among the most hazardous of occupations and one in which organic dusts and toxic gases constitute some of the most common and potentially disabling occupational hazards (Kirkhorn and Garry, 2000). Animal confinement operations with increasing animal density, particularly swine confinement, have contributed significantly to increased bioaerosol levels and exposures. Current research implicates bacterial endotoxins, fungal spores, and toxic grain dusts in grain and animal production as primary causes of respiratory problems and diseases.

Farmworkers are constantly subject to the occupational hazard of contracting infectious diseases from farm animals, including forms of tuberculosis, cowpox, brucellosis, influenza, and other potentially fatal diseases. Animal dander is a source of allergens. Cow dander allergy is usually caused by occupational exposure and occurs in farmers and farm workers (Chapman and Wood, 2001). Horse allergy occurs among people who regularly handle horses, either professionally or for recreational purposes (Mackiewicz et al., 1996). Foodstuffs and agricultural plants are a source of allergens to susceptible individuals, as are the mold spores and actinomycetes that may grow on them and on animal waste (Chan-Yeung, 1979). Veterinarians are also subject to an increased risk of zoonotic diseases like swine influenza and bacterial infections (Nowotny et al., 1997).

Avian influenza virus (AIV) is an emerging pathogen that has recently been transmitting from chickens and ducks to humans, is caused by type A influenza orthomyxovirus (del Rey Calero, 2004). Migration birds and wild ducks are the main reservoir. In 1997 it was observed

TABLE 34.2 Respiratory Disease and Allergies in Agricultural Workers

Farm industry or occupation	Allergen, pathogen, or source	Primary symptom/disease	Reference
Carnation winter quarters	Carnation, *Dianthus caryophyllus*	Occupational allergy	Sanchez-Guerrero et al., 1999
Cattle farmers	Cow hair allergen	Occupational asthma	Hinze et al., 1997
Cattle farmers	*Streptococcus suis*	Infection	Bartelink and vanKregten 1995
Compost production	Contaminated dust, fungi, bacteria	Respiratory disorders	Giubileo et al., 1998
Dairy farmer	*Saccharomyces cerevisiae*	Hypersensitivity pneumonitis	Yamamoto et al., 2002
Farmers	Oilseed rape flour in animal fodder	Occupational asthma	Alvarez et al., 2001
Farming	*T. vulgaris, S. rectivirgula*	Farmer's lung	Reijula, 1993
Farming	Dust, mold spores	Allergic alveolitis, ODTS	Malmberg et al., 1993
Farming	Dust mites	Allergies	vanHage-Hamsten et al., 1994
Farming	Endotoxins, *Alcaligenes faecalis, Erwinia herbicola*	Respiratory disease	Skorska et al., 1996
Flower caretaker	Spathe flower, *Spathiphyllum wallisii*	Occupational asthma	Kanerva et al., 1995
Flower growers	Spider mites	Allergy, asthma	Kim and Kim, 2002
Flower Supplier	*Dianthus caryophyllus, Gypsophila paniculata, Lilium longiflorum.*	Occupational allergy	Vidal and Polo, 1998
Fur industry	Mink urine	Occupational asthma	Jimenez et al., 1996
Grain elevator workers	Grain dust	Respiratory irritation	Chan-Yeung et al., 1979
Greenhouse Flowers	Workplace flowers and molds	Occupational asthma	Monso et al., 2002
Mushroom workers	Oyster mushroom, *Pleutrotus* Florida	Mushroom worker's lung	Noster et al., 1976
Poultry workers	Influenza A (H5N1), avian influenza	Influenza	Bridges et al., 2002
Purebred horse farms	*S. rectivirgula, Acinetobacter calcoaceticus*	ODTS	Mackiewicz et al., 1996
Rose cultivators	*Rosa domescena*	Allergic rhinitis	Unlu et al., 2001
Saffron workers	Saffron flower	Occupational allergy	Feo et al., 1997
Sunflower workers	Sunflower pollen	Allergic rhinitis	Atis et al., 2002
Tomato greenhouse	Insects	Occupational asthma	Erlam et al., 1996

that one strain jumped the interspecies barrier and affected 18 humans, and 6 of them died. Molecular characterization of the ssRNA sequence indicates it has taken an evolutionary path from a new genotype of AIV that was present during the 2003 to 2004 Asian bird flu outbreaks (Wan et al., 2005). Thai authorities believe a recent case of avian flu in a human patient was transmitted directly to other family members (Williams, 2005). There is, as yet, no evidence of airborne transmission, but influenza viruses have evolved a unique propensity for such adaptation.

The epidemiology of airborne zoonotic diseases that transmit only among farm animals is too extensive to cover here, and readers interested in this matter should see the summary of studies and the bibliography in Wathes (1995).

34.4 AGRICULTURAL AEROBIOLOGY

The aerobiology of agricultural environments presents some of the highest levels of dust and airborne microbes of almost any indoor facilities. The extremely high levels of bioaerosols that can occur may cause acute short-term doses due to inhalation or exposure. Much of the airborne contamination in animal facilities may be due to dust stirred up by the animals or activities. Dust deposits in barns and animal housing can contain high concentrations of bacteria and fungi and increased incidence of chronic respiratory symptoms in pig workers has been well documented. Table 34.3 summarizes the results of a number of studies on airborne levels of biocontaminants and dust in agricultural facilities. Many of the levels found are extraordinary by any standards.

In one study of the bacterial and fungal flora of dust deposits in a newly built pig grower finisher building, viable bacterial counts and microbial species found in a barn which had never housed pigs were compared with those in a barn housing 144 pigs (Martin et al., 1996). The lowest viable bacterial count was found in the barn with no pigs, with a concentration of 48,000 cfu/mg of dust. In the barn with pigs the highest viable bacterial count was in dust from the top of a partition close to pig activity, with a value of 2,100,000 cfu/mg of dust. The study concludes that pigs are not only a source but also a disperser of airborne bacteria in pig buildings.

Cattle behavior and feeding patterns can affect the amount of dust generated. In a study of the quantity and diversity of microorganisms in cattle feedlot air, cattle held in eight commercial feedlot pens were studied during the dust peak for 4 days at sites both upwind and downwind of the feedlot pens (Wilson et al., 2002). Results showed that the only viable bacteria recovered were nonpathogenic gram-positive organisms, although gram-negative bacteria may have been present in a viable but nonculturable state. Fungi were recovered in smaller numbers than bacteria. Results also suggested that some feeding patterns produced cattle behavior that generated higher levels of downwind dust.

The dispersion of airborne viruses in outdoor animal facilities has been suggested to play an important role in the spread of animal disease. In such cases animals may be protected to some degree by sheltering in indoor facilities (Casal et al., 1999).

Facilities rearing hens in egg production plants produce VOCs like ammonia and carbon dioxide, and also produce large quantities of dust in the respirable range that include endotoxins and airborne microorganisms. In a study by Martensson and Pehrson (1997) multiple tier systems for hens produce significantly higher levels of total dust and endotoxins than in battery type systems although airborne microorganisms were found in considerable concentrations in both. Unloading activities tended to exacerbate levels of airborne contaminants due both to raised activity levels of the animals and activities of the workers.

TABLE 34.3 Typical Airborne Concentrations in Agricultural Facilities

Contaminant	Level	Area	Reference
Bacteria (cfu/m^3)	384,000–528,000	Dairy farms	Duchaine et al., 1999
	7,000,000	Poultry houses	Clark et al., 1983
	79,000,000	Poultry houses	Radon et al., 2002
	87,589–477,540	Poultry houses	Carpenter et al., 1986
	100,000–10,000,000	Swine houses	Crook et al., 1991
	75,000–237,000	Swine houses	Carpenter et al., 1986
	100,000	Swine houses	Clark et al., 1983
	5,800,000	Swine houses	Radon et al., 2002
Fungi (cfu/m^3)	3,180,000–4,500,000	Dairy farms	Duchaine et al., 1999
	400,000–46,000,000	Farm buildings	Lacey and Lacey, 1964
	13,000,000	Farms (outdoor air)	Malmberg et al., 1993
	647,000	Grain elev. (*Aspergillus*)	Krysinska-Traczyk, 2000
	500	Poultry houses	Clark et al., 1983
	440,000	Poultry houses	Radon et al., 2002
	300	Swine houses	Clark et al., 1983
	380,000	Swine houses	Radon et al., 2002
Actinomycetes (cfu/m^3)	4,000,000–46,000,000	Farm buildings	Lacey and Lacey, 1964
Fungi + bacteria (cfu/m^3)	231,000	Herb processing plant	Dutkiewicz et al., 2001
	930,600	Pig farms	Mackiewicz, 1998
Dust (mg/m^3)	2,600,000	Farms (outdoor air)	Malmberg et al., 1993
	2.34	Poultry houses	Clark et al., 1983
	7.01	Poultry houses	Radon et al., 2002
	4–7.01	Poultry houses	Carpenter et al. 1986
	3.95–5	Swine houses	Radon et al., 2002
	3	Swine houses	Clark et al., 1983
	1.67–21.0	Swine houses	Crook et al., 1991
	2.8	Swine production facility	Donham et al., 1995
Endotoxin (ng/m^3)	136–174*	Dairy farms	Duchaine et al., 1999
	0.44–74.4*	Wheat farms	Viet et al., 2001
	0.36	Greenhouses	Radon et al., 2002
	58–76	Swine houses	Radon et al., 2002
	257.6	Poultry farms	Radon et al., 2002

*Assuming 1 endotoxin unit (EU) = 0.1 ng.

Volatile organic compounds (VOCs) discharged from industrial and agricultural sources consist of many odor-producing components. Some odor control techniques, like isolating compost and the use of odor suppressants, were found to have limited success during small-scale field tests (Botsford et al., 1997). The only VOCs that are relevant to aerobiology are those produced by bacteria and fungi, or microbial volatile organic compounds (MVOCs), and the control of these will generally result in reductions in MVOCs.

34.5 VENTILATION AND SOLUTIONS

The housing of animals has developed along with animal husbandry over the ages and the result is that today each type of farm animal has its own unique form of housing (Slack and Gerencser, 1985). These various facilities require appropriate levels of ventilation, heating and cooling, but limited focus has been placed on air cleaning until recent decades when the problems of airborne diseases in farm workers became apparent. Appropriate design of both buildings and mechanical ventilation systems can be used to minimize airborne bacteria, dust, humidity, and ammonia and help control the problems. Most indoor animal facilities, however, are not designed for high levels of indoor air quality and are primarily designed to remove the heat produced by large numbers of animals kept in close quarters (MWPS, 1989, 1990).

The high concentration of particulate matter is often cited as the most prevalent indoor air quality problem in animal housing facilities. Conditions can vary from summer to winter. Results of a survey indicated that for larger, more mature animals, properly designed and controlled natural ventilation systems are effective in providing good environments (Riskowski et al., 1998). For colder climates and more sensitive animals, a combined system with mechanical ventilation for cold weather and natural ventilation for warm weather works well. The researchers suggested that poor design and management of existing technologies are responsible for most problems with poor indoor air quality.

One common method for reducing airborne concentrations or dust and microbes in animal facilities is the use of recirculating air filtration systems, but experience suggests such air systems must be appropriately adapted to the existing structures and ventilation systems to ensure performance. Simply placing a recirculation unit inside at an arbitrary location may not be sufficient to control the air quality adequately. Attention must be paid to both the sources of contaminants and the air currents or airflow directions in the barns. Filters load up rapidly in dusty environments and the question of economics comes into play.

In one test of the reduction of dust and bacteria levels inside feeder barns of a hog farm, multiple fabric filters and a high-voltage, plate-type electrostatic precipitator were used as dust collection devices for grower and finisher barns in a hog farm (Lau et al., 1996). Each barn was partitioned into halves and one-half was equipped with a 20-ACH recirculating air filtration system. Respirable and inhalable dust levels, bacteria counts, and ammonia levels were monitored for 18 months. The air filtration systems effectively reduced dust levels and aerial bacteria counts inside the pig barns although their efficiency was dependent on air recirculation flow rate, location, size, and number of inlets and outlets, and humidity. Results showed a reduced prevalence of enzootic pneumonia and atrophic rhinitis among the pigs. It also accelerated animal growth and overall productivity. The electrostatic precipitator exhibited higher dust removal efficiency during winter and spring time and was almost as effective as the fabric filter during the other seasons of the year. The electrostatic precipitator was consistently more effective than the fabric filters at reducing airborne bacteria levels, probably due to the fact that in such environments the dust can be a major carrier of bacteria.

Principles of source control can be applied by exhausting and supplying air at ground level in barns where airborne contaminants tend to be produced. In one test of a pig barn, air quality was shown to be improved by placing the ventilation air inlet low on the floor of the feeding passage and placing the outlet just above the floor and slurry pit, instead of at higher locations (Aarnink and Wagemans, 1997). The study evaluated ammonia volatilization and dust concentration inside houses for fattening pigs. Ammonia concentrations and dust concentrations were significantly reduced by this type of system in comparison with the use of inlets and outlets placed at higher elevations.

TABLE 34.4 Natural versus Mechanical Ventilation in Barns

Airborne particles	Natural ventilation	Mechanical ventilation	Change %
Inhalable dust, mg/m^3	2.19	2.13	−3
Respirable dust, mg/m^3	0.1	0.11	10
Total viable particles, cfu/m^3	60,000	17,000	−72
Respirable viable particles, cfu/m^3	9,800	4,500	−54

Natural ventilation is common in agricultural facilities but mechanical ventilation will generally provide superior control of air quality. In a test of air quality in two commercial swine-finishing barns, one with natural ventilation and one with mechanical ventilation, concentrations of inhalable dust, respirable dust, bioaerosols, carbon dioxide, and ammonia were monitored for 41 weeks (Predicala et al., 2001). The mean levels of these contaminants are summarized in Table 34.4. The changes in total airborne dust were not significant but the airborne bacteria appeared to be significantly lower with mechanical ventilation.

Two problems confront designers of ventilation systems for animal houses—(1) noxious odors and gases tend to rise to the ceiling areas with thermal currents from the animals, and (2) dander, dust, spores, and bacteria tend to settle downward over time due to gravity. Various types of ventilation systems have been designed and installed in animal houses, including naturally ventilated buildings venting from side to side, forced roof ventilation with ceiling exhaust, forced ventilation with exhaust from the side walls, and forced ventilation exhausting from the floor through a plenum (Maton et al., 1985; MWPS, 1989, 1990). In addition, various types of internal recirculating filters have been used in an attempt to reduce indoor contaminant levels, but with limited success (Carpenter et al., 1986; Wathes et al., 1991). Perhaps what is needed is a combination of roof exhaust, floor exhaust, and recirculation, as shown in Fig. 34.2, coupled with a well-designed air distribution system.

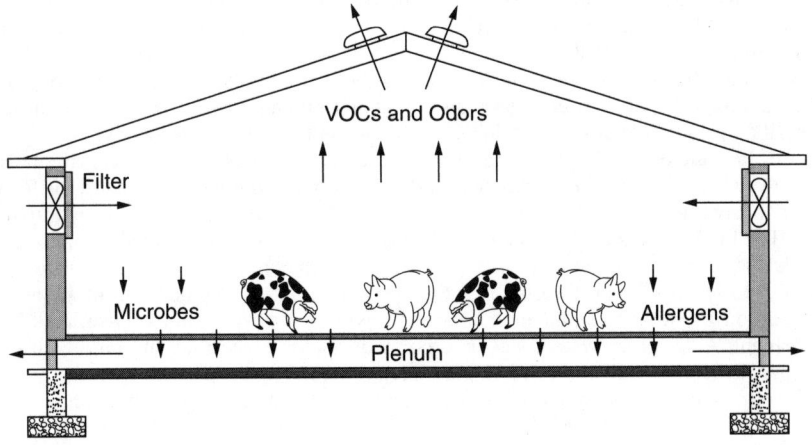

FIGURE 34.2 A split exhaust ventilation system in which lighter bioaerosols and gases are exhausted from the roof and heavier microbes and allergens are exhausted through a floor plenum.

Various methods of controlling VOCs and odors from agricultural animal facilities have been investigated and applied (Strauch, 1987b). The North Carolina State University Animal and Poultry Waste Management Center program identified several such technologies that are either in developmental stages or application (Williams, 2001). Performance data have shown that swine manure treatment systems including a covered in-ground anaerobic digester, a sequencing batch reactor, and an upflow biological aerated filter system can significantly improve odor emissions.

Hillman et al. (1992) installed a ventilation system that minimized airborne contaminants in a calf nursery. This system prevented cross transfer of airborne pathogens between neighboring calves by providing uniform air distribution throughout the calf nursery and greatly reduced respiratory problems among the calves.

Electrostatic precipitators, ionization, and other types of electrostatic filters can be effective and economical in controlling air quality in animal facilities due to the excessive levels of airborne dust that may be produced. In a study of ionization in a poultry hatching cabinet, two configurations of negative air ionizers were tested for their ability to remove inhalable airborne particulates and to determine their potential for reducing airborne disease transmission (Mitchell, 1998). The treatments were applied to ambient dust where particle counts in the inhalable range were high. The overall efficiency of the room ionizer system was approximately 92 percent.

According to Ellen et al. (2000), dust concentrations in poultry houses vary from 0.02 to 81.33 mg/m^3 for inhalable dust and from 0.01 to 6.5 mg/m^3 for respirable dust. Houses with caged laying hens showed the lowest dust concentrations at less than 2 mg/m^3, while the dust concentrations in the other housing systems, like the perchery and aviary systems, were often four to five times higher. Factors that influenced dust concentrations include type of animal, amount of animal activity, type of bedding materials, and season. The most important sources of dust are generally the animals and waste products. It has been found that maintaining the relative humidity of the air in a broiler house at 75 percent will tend to control inhalable dust but not respirable dust. Some reduction of respirable dust is possible by fogging with pure water or water with rapeseed oil. In an aviary system, a 50 to 65 percent reduction of the inhalable dust concentration was found after spraying water with 10 percent of oil and pure water, respectively.

Spraying with a mixture of water and rapeseed oil was also found to reduce airborne dust concentrations in pig houses (Takai et al., 1995). No detrimental effects on the pigs were observed. The daily dose of oil used was 5 to 64 mL per pig per day. Oil concentrations of 5 to 20 percent were used for spraying durations of 5 to 90 s per day. Long-term observations in houses for piglets, young pigs, and fattening pigs at a commercial farm showed that respirable dust was reduced by 76, 54, and 52 percent, respectively.

Biofilters are one option for controlling air quality problems at the source. Nicolai and Janni (1997) described an experimental biofilter that was built to treat exhaust air from a continuously running ventilation fan that exhausted air from a manure pit under a farrowing barn. The biofilter was constructed out of wooden poles laid on the ground and perforated plastic slats covered with a mixture of compost and kidney bean straw. Performance of the biofilter in terms of odor threshold, hydrogen sulfide and ammonia concentrations was monitored for 9 months and average odor and hydrogen sulfide removal averaged 75 to 90 percent. A sprinkler system was incorporated to add moisture during mild and warm weather. Cost was estimated at $0.28 per piglet.

Electrostatic charging of particles in an enclosed hatching cabinet has been shown to be effective at reducing airborne levels of *Salmonella* (Mitchell et al., 2002). Dust generated during the hatching process has been strongly implicated in *Salmonella* transmission. Four trials of an *electrostatic space charge system* (ESCS) on the levels of total aerobic bacteria, *Enterobacteria*, and *Salmonella*, within an experimental hatching cabinet showed that the ESCS reduced the total airborne bacteria and the *Enterobacteria* by 85 to 93 percent. Dust concentrations were reduced by an average of 93.6 percent.

The production and use of animal wastes involve certain hazards due to the pathogenic and allergenic agents that may be present, and biotechnological methods may be used to disinfect manure and livestock slurry (Strauch and Ballarini, 1994). Additives to manure and waste products are one means of source control for odor and air quality problems in animal facilities. Although the efficiency of additives have been questioned in the literature, field measurements by Hendriks et al. (1998) showed that the application of additives to the manure in commercial pig house could significantly reduce ammonia emissions in the exhaust air. In measurements taken over a 1-year period, the mean emission rate for untreated manure was 7.2 mg/(kg·h) per kg of animal weight. For the treated manure, the mean emission rate was 5.8 mg/(kg·h) per kg of animal weight, for a net reduction of about 20 percent. Subsequent testing showed a mean reduction of 51 percent during the winter months. The reduction efficiency of the additive is a function of climatic conditions, number of animals, and animal weight.

An aerobic biofilter system was studied to assess its effectiveness for reducing enteric microbial indicators in flushed swine wastewater under different seasonal conditions by (Hill et al., 2002). Fecal coliforms, *Escherichia coli*, enterococci, somatic coliphages, and male-specific coliphages were reduced by 97 to 99 percent in the biofilter system.

Nonthermal plasma technology has been studied as a means of controlling odors in agricultural settings (Ruan et al., 1999). Nonthermal plasma is capable of decomposing dilute, complex polluting gases, and has the potential for animal house and waste odor reduction. In this study silent discharge, nonthermal plasma reactors were studied in terms of ammonia reduction efficiency and energy efficiency under different reactor and operational conditions. Results showed that ammonia reduction was possible and that efficiencies depended on the operational conditions.

Another study on plasma technology for the control of noxious gases in an animal facility used an ac-powered ferroelectric packed-bed plasma reactor to decompose ammonia and odors in the air in a swine house (Zhang et al., 1996). Ammonia decomposition efficiencies of up to 95 percent were obtained under a combination of test conditions and the unit was found to be effective at decomposing odors in the swine house air.

Davies and Breslin (2003) carried out investigations into various alternative techniques for decontamination of the surfaces of shell eggs artificially contaminated with *Salmonella*. Ionized air, exposure to ozone in dry air, and use of a commercial herbal antibacterial product were not effective. Application of ozone in a humid environment was partially effective but not as effective as a commercial ionized water anolyte in eliminating *Salmonella* from egg surfaces. The efficacy and fumigation characteristics of ozone used for stored maize were studied by Kells et al. (2001). Treatment of maize with 50 ppm ozone for 3 days resulted in 92 to 100 percent mortality of adult insects, and reduced by 63 percent the contamination level of the fungus *Aspergillus parasiticus* on the kernel surface. Ozone fumigation of maize had two distinct phases lasting several days in which the first phase was characterized by rapid degradation of the ozone and slow movement through the grain. In the second phase the ozone flowed freely through the grain with little degradation and occurred once the molecular sites responsible for ozone degradation became saturated. The rate of saturation depended on the velocity of the ozone/air stream. The optimum velocity for deep penetration of ozone into the grain mass was 0.03 m/s.

Disinfection methods are commonly used to deal with viral and bacterial contamination problems and various disinfectants have been used, including phenols, cresols, alcohols, formaldehyde, glutaraldehyde, phenyl mercurial compounds, quaternary ammonium compounds, chlorine, iodine, sodium carbonate, and ammonia (Sainsbury and Sainsbury, 1988). Commercial hatcheries typically infuse hydrogen peroxide or formaldehyde gas into hatching cabinets to reduce airborne pathogens that may lead to disease transmission during the hatch.

Other common fumigants include formalin vapor and mixtures of chloroxylenol and triethylene glycol. Sander and Wilson (1999) achieved a significant reduction in airborne bacterial counts in a chicken hatchery through the use of hydrogen peroxide fogging without any impact on egg productivity. Ozone has been used as a fumigant in poultry facilities (Whistler and Sheldon, 1989).

Ionization has been used in poultry houses as a means of controlling dust, which may carry disease agents. The generation of negative ions tends to cause dust particles to agglomerate and settle out of the air, and ions may also have bactericidal effects. They may also improve the efficiency of filter units. Published data suggest that the incidence of disease in poultry houses is reduced when negative ionization is used to charge the air (Estola et al., 1979).

An ESCS was tested extensively in broiler hatcheries by Mitchell and Waltman (2003). The ESCS cleans air by transferring a strong negative electrostatic charge to dust and microorganisms that are aerosolized during the hatch and collecting the charged particles on grounded plates or surfaces. In studies with three poultry companies, the ESCS resulted in 77 to 79 percent reduction of airborne dust, 93 to 96 percent reduction in *Enterobacteriaceae*, and reduced airborne *Salmonella* by 33 to 83 percent. Richardson et al. (2003) used an electrostatic space charging system in a treatment room with a grounded floor to achieve a 61 percent reduction in dust concentrations and a 76 percent reduction in airborne bacteria in a broiler breeder. See Chap. 16 for more information on ionization technology.

Technological developments in animal husbandry have been increasing the number of animals raised per unit area, decreasing labor costs by automated animal feeding, watering, and housing (Cole et al., 1999), and increasing animal growth rates through the use of antibiotics. These changes in animal production practices have resulted in reduced disease risks in some cases, but also have introduced new risks to farm workers. Exposure pathways of infectious microbes include animal excreta, birthing wastes, carcasses, and airborne hazards from VOCs, dust particles, aerosols, and odors. Measures for reducing risks to animal handlers include the use of waste management and treatment techniques.

Biofilters have been used to reduce specific microbial bioaerosol consisting of bacteria, fungi, airborne endotoxins, microbial volatile organic compounds (MVOCs), including odor and ammonia emissions from a pig facility (Martens et al., 2001). Five biofilter units filled with different filter materials such as biochips, coconut-peat, wood-bark, pellets, bark, and compost were used to reduce airborne bacteria by 70 to 95 percent. Total fungi were reduced by over 60 percent. Airborne endotoxins and MVOCs were effectively reduced by at least 90 percent. Odors were reduced, on the average, by 40 to 83 percent, although not all the filters were effective against the ammonia emissions.

In a study conducted on a swine facility, Dosman et al. (2000) evaluated the benefits of wearing an N-95 disposable respirator in a swine confinement facility and determined that the device significantly reduced hazardous health effects in subjects not previously exposed to a swine barn environment.

Farmers routinely carry contaminants from their work areas into their homes, and regular cleaning or avoidance may provide them relief from respiratory problems. Hinze et al. (1997) studied farmers who suffered from occupational cow hair asthma and found cow hair allergens in their house dust. Dust samples of carpets often contained high concentrations of the allergen while dust from tiles and linoleum was difficult to detect. Cessation of cattle farming and avoiding the barns resulted in a marked reduction of allergen concentrations in living quarters and reduced symptoms. It is recommend the farmers with cow hair asthma should avoid cattle and thoroughly clean all carpets in the living quarters. Radon et al. (2000) investigated the distribution of mite allergens in pig-farming environments in comparison to urban homes and found that allergen concentrations in the mattresses of farmers were much higher than in those of urban environments.

TABLE 34.5 Proposed Bioaerosol Limits for Agricultural Facilities

Contaminant	Level	Area	Reference
Fungi (cfu/m^3)	50,000	Grain farming, threshing	Krysinska-Traczyk, 2000
Total Microbes (cfu/m^3)	430,000	Swine buildings	McDuffie et al., 1995
Endotoxin (ng/m^3)	100	Animal confinement buildings	Iversen et al., 2000
	100	Agricultural facilities	Dutkiewicz et al., 2001
Endotoxin (EU/m^3)	100	Agricultural facilities	Fairfax et al., 1999
Dust (mg/m^3)	4	Agricultural facilities (TLV)	ACGIH, 1991
	2.4–3.7	Swine buildings	McDuffie et al., 1995

Some limits have been established in foreign countries or proposed by various researchers that may be used to assess aerobiological problems via air sampling (see Chap. 23) and these are shown in Table 34.5.

REFERENCES

Aarnink, A. J. A., and Wagemans, M. J. M. (1997). "Ammonia volatilization and dust concentration as affected by ventilation systems in houses for fattening pigs." *Trans ASAE* 40(4):1161–1170.

ACGIH (1991). "Threshold Limit Values and Biological Exposure Indices for 1991-1992." American Conference of Governmental Industrial Hygienists. Cincinnati, OH.

Alvarez, M. J., Estrada, J. L., Gozalo, F., Fernandez-Rojo, F., and Barber, D. (2001). "Oilseed rape flour: another allergen causing occupational asthma among farmers." *Allergy* 56(2):185–188.

Atis, S., Tutluoglu, B., Sahin, K., Yaman, M., Kucukusta, A. R., and Oktay, I. (2002). "Sensitization to sunflower pollen and lung functions in sunflower processing workers." *Allergy* 57(1):35–39.

Bartelink, A. K. M., and vanKregten, E. (1995). "*Streptococcus suis* as threat to pig-farmers and abattoir workers." *Lancet* 346:1707.

Botsford, C. W., Huang, E., and Magee, R. (1997). "Control of odor from composting sources." *Proceedings of the Air & Waste Management Association's Annual Meeting & Exhibition* 97-FA159.01:12, Toronto, Ontario, Canada.

Bridges, C. B., Lim, W., Hu-Primmer, J., Sims, L., Fukuda, K., Mak, K. H., Rowe, T., Thompson, W. W., Conn, L., Lu, X., Cox, N. J., and Katz, J. M. (2002). "Risk of influenza A (H5N1) infection among poultry workers, Hong Kong, 1997-1998." *J Infect Dis* 185(8):1005–1010.

Carpenter, G. A., Smith, W. K., MacLaren, A. P. C., and Spackman, D. (1986). "Effect of internal air filtration on the performance of broilers and the aerial concentrations of dust and bacteria." *Br Poult Sci* 27:471–480.

Casal, J., Planas-Cuchi, E., and Casal, J. (1999). "Sheltering as a protective measure against airborne virus spread." *J Hazard Mater* 68(3):179–189.

Chan-Yeung, M., Wong, R., and MacLean, L. (1979). "Respiratory abnormalities among grain elevator workers." *Chest* 75(4):461–467.

Chapman, M. D., and Wood, R. A. (2001). "The role and remediation of animal allergens in allergic diseases." *J Allergy Clin Immunol* 107(3 Suppl):S414–S421.

Clark, S., Rylander, R., and Larsson, L. (1983). "Airborne bacteria, endotoxin and fungi in dust in poultry and swine confinement buildings." *Amer Ind Hyg Assoc J* 44(7):537–541.

Cole, D. J., Hill, V. R., Humenik, F. J., and Sobsey, M. D. (1999). "Health, safety, and environmental concerns of farm animal waste." *Occup Med* 14(2):423–448.

Crook, B., Robertson, J. F., Glass, S. A., Botheroyd, E. M., Lacey, J., and Topping, M. D. (1991). "Airborne dust, ammonia, microorganisms, and antigens in pig confinement houses and the respiratory health of exposed farm workers." *Am Ind Hyg Assoc J* 52(7):271–279.

Davies, R. H., and Breslin, M. (2003). "Investigations into possible alternative decontamination methods for *Salmonella enteritidis* on the surface of table eggs." *J Vet Med B Infect Dis Vet Public Health* 50(1):38–41.

del Rey Calero, J. (2004). "Epidemiological perspectives on SARS and avian influenza." *An R Acad Nac Med (Madr)* 121(2):289–304.

Donham, K. J., Reynolds, S. J., Whitten, P., Merchant, J. A., Burmeister, L., and Popendorf, W. J. (1995). "Respiratory dysfunction in swine production facility workers: dose-response relationships of environmental exposures and pulmonary function." *Am J Ind Med* 27(3):405–418.

Dosman, J. A., Senthilselvan, A., Kirychuk, S. P., Lemay, S., Barber, E. M., Willson, P., Cormier, Y., and Hurst, T. S. (2000). "Positive human health effects of wearing a respirator in a swine barn." *Chest* 118(3):852–860.

Duchaine, C., Meriaux, A., Brochu, G., and Cormier, Y. (1999). "Airborne microflora in Quebec dairy farms: lack of effect of bacterial hay preservatives." *Am Ind Hyg Assoc J* 60(1):89–95.

Dutkiewicz, J., Krysinska-Traczyk, E., Skorska, C., Sitkowska, J., Prazmo, Z., and Golec, M. (2001). "Exposure to airborne microorganisms and endotoxin in herb processing plants." *Ann Agric Environ Med* 8(2):201–211.

Ellen, H. H., Bottcher, R. W., vonWachenfelt, E., and Takai, H. (2000). "Dust levels and control methods in poultry houses." *J Agric Safety and Health* 6(4):275–282.

Erlam, A. R., Johnson, A. J., and Wiley, K. N. (1996). "Occupational asthma in greenhouse tomato growing." *Occup Med (Lond)* 46(2):163–164.

Estola, T., Makela, P., and Hovi, T. (1979). "The effect of air ionization on the airborne transmission of experimental Newcastle disease virus infections in chickens." *J Hyg* 83:59–67.

Fairfax, R., Eberts, H., and Wilson, D. (1999). "Health hazards at an egg farm." *Appl Occup Environ Hyg* 14(11):728–731.

Feo, F., Martinez, J., Martinez, A., Galindo, P. A., Cruz, A., Garcia, R., Guerra, F., and Palacios, R. (1997). "Occupational allergy in saffron workers." *Allergy* 52(6):633–641.

Giubileo, L., Sarti, A. M., Bianchi, L. A., Calcaterra, E., and Colombi, A. (1998). "Review of risks of biological agents and preventive measures to safeguard the health of compost production workers." *Med Lav* 89(4):301–315.

Hendriks, J., Berckmans, D., and Vinckier, C. (1998). "Field tests of bio-additives to reduce ammonia emission from and ammonia concentration in pig houses." *ASHRAE Trans* 104(Pt. 1B):1699–1705.

Hill, V. R., Kantardjieff, A., Sobsey, M. D., and Westerman, P. W. (2002). "Reduction of enteric microbes in flushed swine wastewater treated by a biological aerated filter and UV irradiation." *Water Environ Res* 74(1):91–99.

Hillman, P., Gebremedhin, K., and Warner, R. (1992). "Ventilation system to minimize airborne bacteria, dust, humidity, and ammonia in calf nurseries." *J Dairy Sci* 75(5):1305–1312.

Hinze, S., Bergmann, K. C., Lowenstein, H., and Hansen, G. N. (1997). "Cow hair allergen (Bos d 2) content in house dust: correlation with sensitization in farmers with cow hair asthma." *Int Arch Allergy Immunol* 112(3):231–237.

Iversen, M., Kirychuk, S., Drost, H., and Jacobson, L. (2000). "Human health effects of dust exposure in animal confinement buildings." *J Agric Saf Health* 6(4):283–288.

Jimenez, G. I., Anton, E., Picans, I., Jerez, J., and Obispo, T. (1996). "Occupational asthma caused by mink urine." *Allergy* 51(5):364–365.

Kanerva, L., Makinen-Kiljunen, S., Kiistala, R., and Granlund, H. (1995). "Occupational allergy caused by spathe flower (Spathiphyllum wallisii)." *Allergy* 50(2):174–178.

Kells, S. A., Mason, L. J., Maier, D. E., and Woloshuk, C. P. (2001). "Efficacy and fumigation characteristics of ozone in stored maize." *J Stored Prod Res* 37(4):371–381.

Kim, Y. K., and Kim, Y. Y. (2002). "Spider-mite allergy and asthma in fruit growers." *Curr Opin Allergy Clin Immunol* 2(2):103–107.

Kirkhorn, S. R., and Garry, V. F. (2000). "Agricultural lung diseases." *Environ Health Perspect* 108(Suppl. 4):705–712.

Krysinska-Traczyk, E. (2000). "Microflora of the farming work environment as an occupational risk factor." *Med Pr* 51(4):351–355.

Lacey, J., and Lacey, M. E. (1964). "Spore concentrations in the air of farm buildings." *Trans Brit Mycol Soc* 47(4):547–552.

Lacey, J., and Crook, B. (1988). "Fungal and actinomycete spores as pollutants of the workplace and occupational illness." 32:515–533.

Lau, A. K., Vizcarra, A. T., Lo, K. V., and Luymes, J. (1996). "Recirculation of filtered air in pig barns." *Canadian Agric Eng* 38(4):297–304.

Mackiewicz, B., Prazmo, Z., Milanowski, J., Dutkiewicz, J., and Fafrowicz, B. (1996). "Exposure to organic dust and microorganisms as a factor affecting respiratory function of workers of purebred horse farms." *Pneumonol Alergol Pol* 64(Suppl. 1):19–24.

Mackiewicz, B. (1998). "Study on exposure of pig farm workers to bioaerosols, immunologic reactivity and health effects." *Ann Agric Environ Med* 5(2):169–175.

Malmberg, P., Rask-Andersen, A., and Rosenhall, L. (1993). "Exposure to microorganisms associated with allergic alveolitis and febrile reactions to mold dust in farmers." *Chest* 103(4):1202–1209.

Martens, W., Martinec, M., Zapirain, R., Stark, M., Hartung, E., and Palmgren, U. (2001). "Reduction potential of microbial, odour and ammonia emissions from a pig facility by biofilters." *Int J Hyg Environ Health* 203(4):335–345.

Martensson, L., and Pehrson, C. (1997). "Air quality in a multiple tier rearing system for layer type pullets." *J of Agric Safety and Health* 3(4):217–228.

Martin, W. T., Zhang, Y., Willson, P., Archer, T. P., Kinahan, C., and Barber, E. M. (1996). "Bacterial and fungal flora of dust deposits in a pig building." *Occup Environ Med* 53(7):484–487.

Maton, A., Daelemans, J., and Lambrecht, J. (1985). *Housing of Animals: Construction and Equipment of Animal Houses.* Elsevier, Amsterdam.

McDuffie, H. H., Dosman, J. A., Semchuk, K. M., Olenchock, S. A., and Senthilselvan, A. (1995). *Agricultural Health and Safety: Workplace Environment Sustainability.* CRC Press. Boca Raton, FL.

Mitchell, B. W. (1998). "Effect of negative air ionization on ambient particulates in a hatching cabinet." *Appl Eng Agric* 14(5):551–555.

Mitchell, B. W., Buhr, R. J., Berrang, M. E., Bailey, J. S., and Cox, N. A. (2002). "Reducing airborne pathogens, dust and *Salmonella* transmission in experimental hatching cabinets using an electrostatic space charge system." *Poult Sci* 81(1):49–55.

Mitchell, B. W., and Waltman, W. D. (2003). "Reducing airborne pathogens and dust in commercial hatching cabinets with an electrostatic space charge system." *Avian Dis* 47(2):247–253.

Monso, E., Magarolas, R., Badorrey, I., Radon, K., Nowak, D., and Morera, J. (2002). "Occupational asthma in greenhouse flower and ornamental plant growers." *Am J Respir Crit Care Med* 165(7):954–960.

MWPS (1989). *Natural Ventilating Systems for Livestock Housing.* M. P. S. Committee, Iowa State University, Ames, IA.

MWPS (1990). *Mechanical Ventilating Systems for Livestock Housing.* M. P. S. Committee, Iowa State University, Ames, IA.

Nicolai, R. E., and Janni, K. A. (1997). "Development of a low-cost biofilter for swine production facilities." *Paper: Am Soc Agric Eng* 3(974040):9.

Noster, U., Hausen, B. M., Felten, G., and Schulz, K. H. (1976). "Mushroom worker's lung caused by inhalation of spores of the edible fungus *Pleurotus florida* ('oyster mushroom')." *Dtsch Med Wochenschr* 101(34):1241–1245.

Nowotny, N., Deutz, A., Fuchs, K., Schuller, W., Hinterdorfer, F., Auer, H., and Aspock, H. (1997). "Prevalence of swine influenza and other viral, bacterial, and parasitic zoonoses in veterinarians." *J Inf Dis* 176:1414–1415.

Pepys, J., Jenkins, P., Festenstein, G., Gregory, P. H., Lacey, M. E., and Skinner, F. (1963). "Farmer's Lung: thermophilic actinomycetes as a source." *Lancet* 2:607.

Predicala, B. Z., Maghirang, R. G., Jerez, S. B., Urban, J. E., and Goodband, R. D. (2001). "Dust and bioaerosol concentrations in two swine-finishing buildings in Kansas." *Trans Am Soc Agric Eng* 44(5):1291–1298.

Radon, K., Schottky, A., Garz, S., Koops, F., Szadkowski, D., Radon, K., Nowak, D., and Luczynska, C. (2000). "Distribution of dust-mite allergens (Lep d 2, Der p 1, Der f 1, Der 2) in pig-farming environments and sensitization of the respective farmers." *Allergy* 55(3):219–225.

Radon, K., Danuser, B., Iversen, M., Monso, E., Weber, C., Hartung, J., Donham, K. J., Palmgren, U., and Nowak, D. (2002). "Air contaminants in different European farming environments." *Ann Agric Environ Med* 9:41–48.

Rangaswami, G. (1966). *Agricultural Microbiology*. Asia Publishing House, New York.

Reijula, K. E. (1993). "Two bacteria causing farmer's lung: fine structure of *Thermoactinomyces vulgaris* and *Saccharopolyspora rectivirgula*." *Mycopathologia* 121(3):143–147.

Richardson, L. J., Hofacre, C. L., Mitchell, B. W., and Wilson, J. L. (2003). "Effect of electrostatic space charge on reduction of airborne transmission of Salmonella and other bacteria in broiler breeders in production and their progeny." *Avian Dis* 47(4):1352–1361.

Riskowski, G. L., Maghirang, R. G., Funk, T. L., Christianson, L. L., and Priest, J. B. (1998). "Environmental quality in animal production housing facilities: A review and evaluation of alternative ventilation strategies." *ASHRAE Trans* 104:104–113.

Ruan, R., Han, W., Ning, A., Deng, S., Chen, P. L., and Goodrich, P. (1999). "Effects of design parameters of planar, silent discharge, plasma reactors on gaseous ammonia reduction." *Trans ASAE* 42(6):1841–1846.

Sainsbury, D., and Sainsbury, P. (1988). *Livestock Health and Housing*. Bailliere Tindall, London.

Sanchez-Guerrero, I. M., Escudero, A. I., Bartolom, B., and Palacios, R. (1999). "Occupational allergy caused by carnation (*Dianthus caryophyllus*)." *J Allergy Clin Immunol* 104(1):181–185.

Sander, J. E., and Wilson, J. L. (1999). "Effect of hydrogen peroxide disinfection during incubation of chicken eggs on microbial levels and productivity." *Avian Dis* 43(2):227–233.

Skorska, C., Milanowski, J., Dutkiewicz, J., and Fafrowicz, B. (1996). "Bacterial endotoxins produced by *Alcaligenes faecalis* and *Erwinia herbicola* as potential occupational hazards for agricultural workers." *Pneumonol Alergol Pol* 64(Suppl. 1):9–18.

Slack, J. M., and Gerencser, M. A. (1975). *Actinomycetes, Filamentous Bacteria: Biology and Pathogenicity*. Burgess Publishing, Minneapolis MN.

Strauch, D. (1987a). Animal production and environmental health. *World Animal Science*. Elsevier, Amsterdam, The Netherlands.

Strauch, D. (1987b). Chapter 5. Hygiene of animal waste management. *Animal Production and Environmental Health*, D. Strauch, ed., Elsevier, Amsterdam, The Netherlands.

Strauch, D., and Ballarini, G. (1994). "Hygienic aspects of the production and agricultural use of animal wastes." *Zentralbl Veterinarmed B* 41(3):176–228.

Takai, H., Moller, F., Iversen, M., Jorsal, S. E., and Bille-Hansen, V. (1995). "Dust control in pig houses by spraying rapeseed oil." *Trans ASAE* 38(5):1513–1518.

Unlu, M., Sahin, U., Yariktas, M., Demirci, M., Akkaya, A., Ozturk, M., and Orman, A. (2001). "Allergic rhinitis in *Rosa domescena* cultivators: a novel type of occupational allergy?" *Asian Pac J Allergy Immunol* 19(4):231–235.

vanHage-Hamsten, M., Harfast, B., and Johansson, S. G. (1994). "Dust mite allergy: an important cause of respiratory disease in farmers." *Am J Ind Med* 25(1):47–48.

Vidal, C., and Polo, F. (1998). "Occupational allergy caused by *Dianthus caryophillus, Gypsophila paniculata*, and *Lilium longiflorum*." *Allergy* 53(10):995–998.

Viet, S. M., Buchan, R., and Stallones, L. (2001). "Acute respiratory effects and endotoxin exposure during wheat harvest in Northeastern Colorado." *Appl Occup Environ Hyg* 16(6):685–697.

Wan, X. F., Ren, T., Luo, K. J., Liao, M., Zhang, G. H., Chen, J. D., Cao, W. S., Li, Y., Jin, N. Y., Xu, D., and Xin, C. A. (2005)" Genetic characterization of H5N1 avian influenza viruses isolated in southern China during the 2003-04 avian influenza outbreaks." *Arch Virol*150(6):1257–1266.

Wathes, C. M., Johnson, H. E., and Carpenter, G. A. (1991). "Air hygiene in a pullet house: effects of air filtration on aerial pollutants measured in vivo and in vitro." *Br Poult Sci* 32:31–46.

Wathes, C. M. (1995). Bioaerosols in animal houses. *Bioaerosols Handbook*, C. S. Cox and C. M. Wathes, eds., CRC Lewis Publishers, Boca Raton, FL.

Whistler, P. E., and Sheldon, B. W. (1989). "Biocidal activity of ozone versus formaldehyde against poultry pathogens inoculated in a prototype setter." *Poultry Sci* 68:1068–1073.

Williams, C. M. (2001). "Technologies to address air quality issues impacting animal agriculture." *Wat Sci Technol* 44(9):233–236.

Williams, N. (2005). "Alarm bells ring over bird flu threat." *Curr Biol* 15(4):R107–R108.

Wilson, S. C., Morrow-Tesch, J., Straus, D. C., Cooley, J. D., Wong, W. C., Mitlohner, F. M., and McGlone, J. J. (2002). "Airborne microbial flora in a cattle feedlot." *Appl Environ Microbiol* 68(7):3238–3242.

Yamamoto, Y., Osanai, S., Fujiuchi, S., Akiba, Y., Honda, H., Nakano, H., Ohsaki, Y., and Kikuchi, K. (2002). "Saccharomyces-induced hypersensitivity pneumonitis in a dairy farmer: a case report." *Nihon Kokyuki Gakkai Zasshi* 40(6):484–488.

Zhang, R., Yamamoto, T., and Bundy, D. S. (1996). "Control of ammonia and odors in animal houses by a ferroelectric plasma reactor." *IEEE Trans Ind Applic* 32(1):113–117.

CHAPTER 35

LIBRARIES AND MUSEUMS

35.1 INTRODUCTION

Museums, libraries, archives, and other storage facilities generally require stringent control over indoor air conditions. Temperature and humidity are maintained within a narrow range so as to minimize the deterioration of stored artifacts, books, and other materials. In facilities frequented by visitors conditions are maintained for human comfort as well. In archives and storage facilities temperatures may be maintained at lower than required for human comfort so as to reduce deterioration and preserve sensitive materials for extended periods of time. The materials that are kept in museums, libraries, and archives may be a source of allergens and may also be a nutrient source for microorganisms, especially fungi. The growth of fungi and some bacteria on stored materials contributes to biodeterioration and this can be a major concern, especially where water damage has occurred. The aerobiology in such facilities may therefore impact both the occupants as well as the stored materials themselves. This chapter addresses aspects of aerobiological quality in museums, libraries, and other types of storage facilities that may be amenable to both air cleaning technologies and surface disinfection technologies. The specific design of environmental control systems for such facilities is not addressed in detail here since these have been adequately addressed in other sources (ASHRAE, 1999a; NBS, 1983).

35.2 LIBRARIES

Libraries contain books and other materials that may be subject to biodeterioration from microorganisms and insects, and that may be sources of allergens. Most books are made from cellulose, which can be degraded by a variety of microbes and insects. These microbes include allergens and may include a few potential pathogens. Insects and animals that may feed off book materials may also be sources of allergens.

Guidelines for the design of air conditioning systems and the operating conditions of libraries are provided in *ASHRAE Applications Handbook* (ASHRAE, 1999b). Guidelines for the control of indoor air quality for human occupation are addressed in ASHRAE Standard 62-2001 and these apply generally to libraries (ASHRAE, 2001). Within the

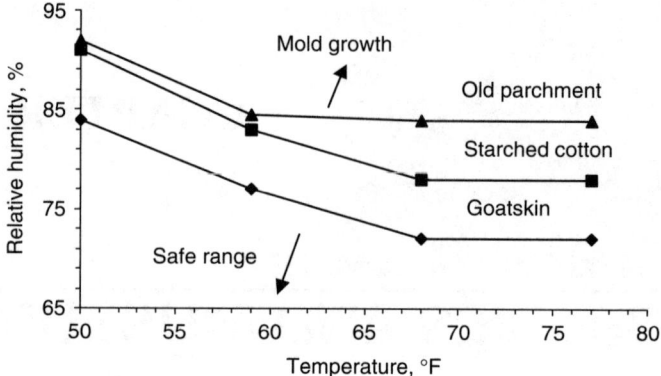

FIGURE 35.1 Range of mold growth on sensitive materials. [*Based on data from Groom and Panisset (1933).*]

range of human comfort, typically 68 to 79°F (20 to 26°C) and 30 to 60 percent RH, mold growth may occur when nutrients are present. Figure 35.1 illustrates the range of conditions in which mold growth may occur after 100 to 200 days, based on data from Groom and Panisset (1933) for sensitive book materials.

Mold damage can pose a serious threat to library and archival collections, especially when humidity is high (Ogden, 1992). Figure 35.2 illustrates the growth rate of mold on susceptible materials under various relative humidities at approximately 77°F (25°C) based on data from Snow et al. (1944). The fitted curve in Fig. 35.2 is defined as follows:

$$\ln(t) = e^{\left(\frac{RH-93.2}{14.9}\right)} \qquad (35.1)$$

where t is the time for visible growth, days, and RH is the relative humidity, %.

The main contributing factors to mold growth include incorrect or fluctuating humidity and temperature levels, high airborne spore counts, and lack of adequate airflow in the stacks.

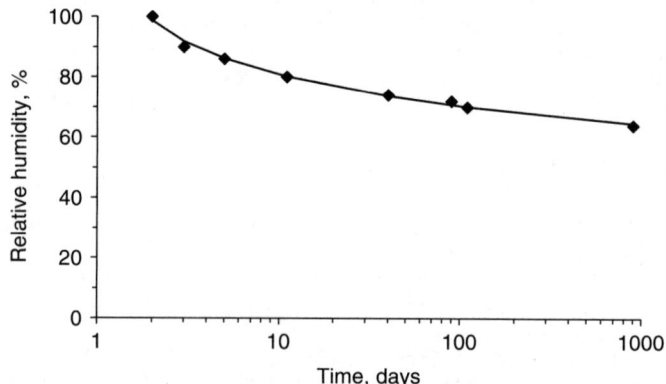

FIGURE 35.2 Time for visible growth of mold on sensitive materials at 77°F (25°C). [*Based on data from Snow et al. (1944).*]

Even with a high spore count in a room, mold may not germinate and grow if the temperature and humidity are kept steady and at appropriate levels. Environmental controls are key to controlling mold. The importance of keeping clean stacks is probably not fully appreciated. This is often a problem in special collections and archives, where the paper is very old and deteriorating. Procedures for regularly cleaning shelves, HEPA-filter vacuuming of books, and cleaning the stacks in general should be made a standard part of library maintenance if they are not already.

Libraries may contain microenvironments where isolated areas of books stacks or storage have poor air circulation. Older libraries in particular often have such microenvironments that are prone to mold blooms or continuous mold growth. Preventing mold in these areas is achieved by improving air circulation and controlling humidity and temperature.

Over 84 genera, representing 234 species, have been isolated from such library materials as books, paper, parchment, feather, textiles, animal and vegetable glues, inks, wax seals, moving pictures, magnetic tapes, microfilms, black and white photographs, papyrus, wood, and synthetic materials in books (Zyska, 1997). Thirty-four genera of fungi have been isolated in the air of three archives in Warsaw, Poland. Most of these have also been found in library materials. Among the fungal species identified some 19 percent may be a source of various diseases caused by mycotoxins.

A survey conducted in Uganda found libraries to be one of the indoor environments with the highest levels of airborne fungi, alongside bathrooms (Ismail et al., 1999). A total of 39 genera and 52 species and some unidentified fungi were sampled from the outdoors while the indoor environments were found to have 35 genera and 49 species. The most prevalent fungi from both outdoor and indoor air were species of *Mycosphaerella*, Yeasts, *Penicillium, Fusarium, Aspergillus, Cochliobolus*, and *Alternaria*.

Spores that enter library facilities will gradually settle out of the air and onto books, carpets, or other surfaces. A study of the airborne fungal spores in the rooms of a university library in China using settle plates found almost 12 spores per plate would precipitate in 5 minutes (Huang et al., 1997). The book storage rooms and reading rooms had an average precipitation of 15 and 8 spores per plate in 5 minutes, respectively, at a humidity of 41.7 and 11.4 percent, respectively.

Gift collections of books are also a prime source for introducing bacteria and mold spores to libraries. Some libraries have the facilities to treat these additions before they enter the library, but most do not. Libraries should have a procedure for quarantining and/or treating gift collections to ensure they do not introduce microbial problems into the library.

Microorganisms that thrive on paper materials and other components of books like starch-based glues may enter the materials at any point, including the production process. Microbiological surveys by Raaska et al. (2002) verified that the production and use of pasteurized starch-based glue was the most important factor threatening process hygiene and product safety. A total of 33 spore-forming bacterial and 15 enterobacterial isolates were identified. The most common spore-forming bacteria (55 percent of the isolates) were *Paenibacillus* species. The most common enterobacteria (87 percent) were *Enterobacter cloacae* and *Citrobacter freundii* or species closely related to them. Table 35.1 summarizes many of the microbes that can be found in libraries.

The fungi most frequently found in the air and the books of 28 libraries in Brazil were studied by Gambale et al. (1993). This study also evaluated librarians regarding asthmatic symptoms or rhinitis, and the relationship with the site of work. Forty-nine percent of them reported such symptoms and 80 percent related them to their place of work. Almost 6 percent of librarians tested positive for reactivity to the 20 fungi most frequently isolated in libraries. The airborne fungi isolated in libraries are common in the outdoor air but were present in higher concentrations in libraries, and were associated with respiratory allergies.

Libraries generally prefer to use carpeting, like that shown in Fig. 35.3, to quiet the noise of traffic, but carpets tend to accumulate fungi and bacteria due to settling in the air

TABLE 35.1 Air and Surface Microbes in Libraries

Microbe	Type	Location	Source	Hazard	Reference
Alternaria	Fungi	Library air, paper	Environment	Allergen	Ismail et al., 1999, Burge et al., 1978
Arthrinium	Fungi	Library air	Environment	Allergen	Burge et al., 1978
Aspergillus	Fungi	Books, library air	Humans	Allergen	Florian and Manning, 2000, Burge et al., 1978
Aureobasidium	Fungi	Library air	Environment	Allergen	Burge et al., 1978
Citrobacter	Bacteria	Books	—	–	Raaska et al., 2002
Cladosporium	Fungi	Library air, paper	Environment	Allergen	Burge et al., 1978
C. diphtheria	Bacteria	Books	Humans	Pathogen	Ferson, 2001
Enterobacter	Bacteria	Books	Humans	Pathogen	Raaska et al., 2002
Epicoccum	Fungi	Library air, paper	Environment	Allergen	Burge et al., 1978
Eurotium	Fungi	Books	Humans	Allergen	Florian and Manning, 2000
Fusarium	Fungi	Library air, paper	Environment	Allergen	Ismail et al., 1999, Burge et al., 1978
M. tuberculosis	Bacteria	Books	Humans	Pathogen	Kenwood and Dove, 1915
Paecilomyces	Fungi	Library air	Environment	Allergen	Burge et al., 1978
Paenibacillus	Bacteria	Books	–	–	Raaska et al., 2002
Penicillium	Fungi	Library air	Environment	Allergen	Ismail et al., 1999, Burge et al., 1978
Pithomyces	Fungi	Library air, paper	Environment	Allergen	Burge et al., 1978
Smallpox	Virus	Books	Humans	Pathogen	Reinick, 1914
Staphylococcus	Bacteria	Books	Humans	Pathogen	Ferson, 2001
Streptococcus	Bacteria	Books	Humans	Pathogen	Balmain, 1927
Trichoderma	Fungi	Library air, paper	Environment	Allergen	Burge et al., 1978
Tritirachium	Fungi	Library air	Environment	Allergen	Burge et al., 1978

or from being brought in on shoes and clothes. In a study of bacteria and fungi in carpeted buildings, Cole et al. (1993) found carpet dust contained 85,000 cfu/g of fungi, 12,000,000 cfu/g of mesophilic bacteria, and 4500 cfu/g of thermophilic bacteria. With the controlled environment inside libraries, these levels are likely to be less than in the buildings tested, but obviously the potential of carpets to harbor microorganisms is considerable. Camuffo et al. (1999) found that airborne bacteria levels were higher in carpeted rooms than in rooms with stone floors, and noted that people walking on carpets stir up the dust and microbial flora. In addition, carpets have a high equilibrium moisture content that favors microbial growth (IEA, 1996).

Foxing is the spotting or staining of books caused by fungi. Arai (2000) has investigated foxing and has developed techniques that enable the isolation and culturing of fungi that produce the foxing effect, and has identified the conditions that favor the appearance of foxing. Foxing also occurs on other materials like silk. Florian and Manning (2000) examined 111 fungal fox spots in a book published in 1854 and examined the morphology of the fungal structures. The results showed that there were two fungal species randomly distributed on pages throughout the book, suggesting that the pages were contaminated during the paper making process or book preparation and the fox spots were considered not to have been caused by airborne contamination during use. The two species identified were of the *Aspergillus glauca* group and two different *Eurotium* species. One of the species was found beneath the ink between the paper fibers. Comparison with another book published in 1785

FIGURE 35.3 Library books and carpets provide niches for a variety of microbes that may be brought in by patrons or enter via the air.

showed similar results but different fungal species. The presence of a minute mitelike animal's eggshells, silk threads, and fecal pellets suggests a complex ecosystem is involved in the deterioration of the materials.

Books may even function as carriers of diseases when they become contaminated with fomites. The issue was first raised over a century ago and caused sufficient concern that a number of researchers pursued investigations of the matter in the early 1900s (McLary, 1985). In 1890, the Victorian Health Act of Australia suggested removing books to the open air and sunshine to prevent the spread of contagious diseases (Ferson, 2001). Some of the early results of these studies produced varying conclusions as to the possibility that books could transmit diseases like smallpox, tuberculosis, scarlet fever, and diphtheria (Reinick, 1914; Kenwood and Dove 1915; McCartney, 1925). The first methods and guidelines for the disinfection of books were developed in about 1912, although antiseptic cleaning methods had been in use since about 1896 (Beebe, 1911; Nice, 1912). Some evidence was found that books could indeed preserve certain bacteria like *Streptococcus* and TB bacilli in a viable state for limited periods of time (Balmain 1927; Smith, 1942). These and other studies suggest that storing books for 1 month in a warm, dry room minimizes the risks of infection from tuberculosis, *Streptococcus*, and staphylococci (Smiley, 1933).

In a study by Burge et al. (1978), the role of fungi as allergic contaminants in book collections was investigated in 11 University of Michigan libraries. Air sampling was used during periods of book handling and periods of inactivity. Library spore levels were generally low (see Table 35.4) while outdoor levels consistently exceeded indoor levels. Air-conditioned libraries had lower spore levels and indoor/outdoor ratios than conventionally ventilated libraries. Activity in the libraries always increased spore counts. Fungal species recovered were similar to those encountered in domestic interiors and outside locations. Figure 35.4 illustrates the proportions of the various fungi found in these libraries

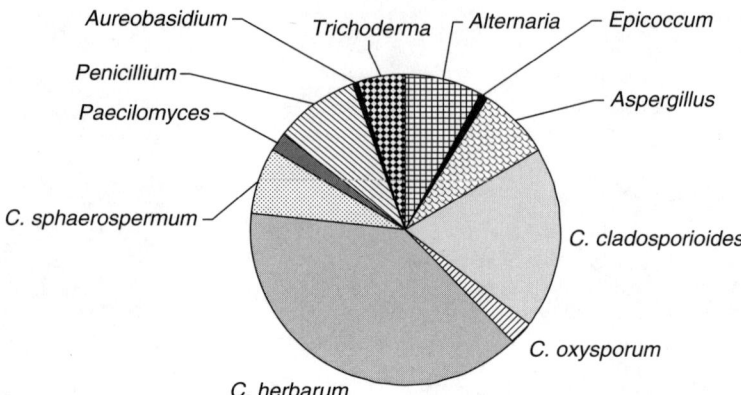

FIGURE 35.4 Breakdown of fungi found in eleven university libraries in summer. [*Based on data from Burge et al. (1978)*.]

during summer. All of these fungi are potential allergens (see App. A for more information), however, no specific etiology tied any of the fungal species to respiratory symptoms experienced by library employees. It is possible that multispecies exposure, combined with exposure to other pollutants in the library, may account for some of the symptoms.

In addition to microbial contamination, a variety of other pollutants may be present in libraries, including chemicals and *volatile organic compounds* (VOCs) that may be of microbiological origin (i.e., MVOCs). In one study, four libraries in an Italian university were found to have formaldehyde and benzene, toluene, and xylenes in the air, although not at hazardous levels (Righi et al., 2002). Total airborne dust in these libraries was measured at 40 to 350 $\mu g/m^3$ and total VOCs were 203 to 749 $\mu g/m^3$.

Symptoms of *sick building syndrome* (SBS) have been observed in recent decades among people spending most of their time in various indoor environments, including libraries, which have sometimes been cited as the primary cause of self-reported symptoms (Righi et al., 2002). This has sometimes been referred to as sick library syndrome (Hay, 1995). Yeung et al. (1991) performed a survey on a case of SBS at a library building in Hong Kong in which the environmental conditions, indoor air quality, and species of common bacteria were measured for a period of 5 months. Results of this case study indicated that indoor environmental conditions could generally be maintained within acceptable standards but could not assure the absence of symptoms.

Cleaning contaminated books is a form of source control for air quality in libraries. Some equipment is available for the cleaning of delicate archival materials. Procedures for using a mobile clean-air workstation to clean dust and other contaminants from fragile books and other materials have been described in the literature (Koreck, 1999). Once mold growth appears on a book, the affected materials should be isolated from the collections and removed by personnel wearing protective gloves and respirators (Ogden, 1992). The affected materials should be thoroughly dried and then cleaned. Dried mold and spores can be brushed off with a sable-tipped paintbrush and covers can be wiped off with a damp cloth (Swartzburg, 1980). A fumigation chamber can be constructed and used to disinfect the materials using various fumigants (Cunha and Tucker, 1972). Modern methods for sterilizing books generally involve vacuuming them clean and wiping with a disinfectant. Ethylene oxide vapor has been used for this purpose but is not in use today because of concerns about damage. Wiping with a dilute solution of ethyl alcohol is a common practice today. Gamma radiation is an expensive alternative that can penetrate books but may cause damage to paper products. A dose of 800 krad was found to be effective at destroying molds but also causes degradation of treated paper, but a dose of 50 krad combined with a temperature of 60°C (140°F) reportedly disinfected books without doing damage (Hanus, 1985). Low levels of radiation and heat can act synergistically to destroy fungi, possibly by inducing germination of spores and desiccation of mycelia.

Ionizing radiation has been investigated and applied as an effective means of disinfecting archival and manuscript materials of insects and fungi by Chappas and McCall (1984). Both cobalt-60 and electron accelerators can be used to irradiate materials. Insects can be disinfected with a dose of 0.03 to 0.2 kGy. Sterilizing materials of bacteria and molds requires 1 to 10 kGy. Chappas and McCall (1984) used a dose of 4.5 kGy to disinfect a few hundred boxes of papers and books that were sealed in plastic bags and only trace levels were found after irradiation. Gamma irradiation can also be used to disinfect ancient documents and books but attention must be paid to the type of printing ink as well as the type or condition of the paper, as some inks are more susceptible to discoloration under gamma irradiation (Rocchetti et al., 2002).

Other alternatives for mold contaminated books include the same methods used for surface disinfection such as ozone and SNL foam (see Chap. 26). Although these disinfectants are well known to destroy spores, limited information is available at this time on whether or not they may do damage to paper or other materials. Ozone systems for disinfecting letters and books are, in fact, currently available from some manufacturers and are typically small ovenlike enclosures in which books can be ozonated for extended periods. Presumably the ozone will be filtered or vented safely.

Chlorine dioxide, released from small self-activating packets, has been used to disinfect small rooms with library collections (Weaver-Meyers et al., 1998; Southwell, 2003). In the libraries of the University of Oklahoma, chlorine dioxide has been used in solution for wet wiping of moldy books and also as a fumigant (Weaver-Meyers et al., 2000). Results suggest that chlorine dioxide is relative safe to use for fumigation and health risks of long-term exposure are low. For large-scale decontamination it is recommended that firms who specialize in such remediation be contacted.

Ultraviolet light is another option for disinfecting surfaces, but it can cause damage and discoloration to pigments and paint. UVGI is best used for disinfecting the air and especially for removing mold from air conditioning systems (see Chap. 11). Filtered pulsed light may have applications in this regard also (see Chap. 15).

Microwaves are another promising alternative since they are capable of destroying spores after about 30 minutes or less (in a typical microwave oven), but the effects on book materials, binding glues, and ink are not known. Books should be dry when attempting microwave disinfection procedures since the generation of steam could cause damage. If microwave disinfection proves to be safe and effective, cooking the books could take on new meaning.

Aerial disinfection of libraries and archives by thermal fogging was researched by Rakotonirainy et al. (1999) as a means of supplementing the normal cleaning of contaminated documents by disinfection of the storerooms once they are emptied of their contents. Thermal fogging with an alkyl dimethylbenzyl ammonium chloride solution has been employed successfully in France for the cleaning of library atmospheres contaminated by fungi. This fungicide is effective in inhibiting the spores suspended in the air but only has a weak action on those deposited on surfaces. The results of this study show that a solution of thiabendazole applied at 10 percent by thermal fogging, at a rate of 5 mL/m^3, produces effective disinfection of library air while also acting on spores deposited on surfaces. Unlike other methods such as chlorine dioxide, this product deposits no films on surfaces and does not leave unpleasant odors in buildings. Thiabendazole has not been found to damage paper and does not cause any visible degradation of painted surfaces and metal shelves.

Fungicides can be used to control fungal growth on paper materials but some evidence exists that fungi may adapt to their prolonged use. Strzelczyk (2001) found that these compounds are not sufficiently stable and that experimentally induced adaptation of fungi to phenolic and organomercuric compounds causes them to develop increasing resistance.

The maintenance of library environmental conditions requires special care to equipment sizing and operation. Libraries may be subject to severe budget constraints, but many engineering options are available today that can produce acceptable environments without undue operating expense or first costs. Advances in the design of systems with part-load operation can offer even older libraries cost-effective retrofits that will provide both adequate air quality and effective environmental control for preservation of collections (Hartman, 1996).

Chemically treated adsorbent air cleaners have been developed and tested for use specifically in libraries, archives, and museums (Kelly and Kinkead, 1993). This testing demonstrated the effectiveness of the adsorbers at low contaminant levels in the parts-per-billion range.

Air sampling for fungi that are damaging to books or to human health is one way of identifying and preempting potential problems and readers should consult Chap. 23 for air sampling test procedures. Also see Table 35.3 at the end of this chapter for typical bioaerosol levels in libraries and museums.

35.3 MUSEUMS

Museums may store and display any variety of artifacts, art, photographs, and printed materials. Some of these stored materials may themselves be causes of indoor air quality problems via their microbiological contents or due to the fact that they provide a nutrient source for normal environmental microbes and insects. The microbes that may grow on stored materials may be allergenic or even pathogenic. The insects, and animals, that may live off stored materials may produce allergens also.

Dust in museums is a problem that can pose a hazard to the cleanliness of museum collections. Soiling of museum artifacts is related to visitor flow and distance between displays and visitors, with the degree of soiling dropping off at a distance of about 0.5 m (Yoon and Brimblecombe, 2001). Since dust can also contain bacteria and fungi, it is likely that biological contaminants are regularly brought in by museum patrons.

Heating, ventilating, and air conditioning systems designed specifically for human comfort may not be suitable for preservation of museum collections if they are prone to fluctuations in environmental conditions. A survey of a Venice museum that monitored the seasonal climate variations, air pollution, deposition and origin of suspended particulate matter and microorganisms was performed by Camuffo et al. (1999). This study focused on the problems caused by the heating and air conditioning system and the effects due to carpets. The results suggest that the carpet has a negative influence on the museum environment as it retains particles and bacteria that are reaerosolized under traffic. The ratio of indoor to outdoor pollutants was greater in the summertime, when doors and windows are opened more frequently. Bacteria are potentially damaging to paintings, especially lipase-producing bacteria which have been found in rooms where paintings are stored (Camuffo et al., 1999).

Most museums, especially art museums, may restrict the amount of light that can impinge on displays so as to minimize light-induced discoloration and deterioration. This absence of light may inadvertently favor the survival of microorganisms. Bright skylit rooms like the one shown in Fig. 35.5 may discourage the growth of fungi (see Chap. 18).

A study by Brimblecombe et al. (1999) examined the indoor climate, gaseous and particulate pollution, and the concentrations of bacteria in summer and winter. Results suggest that the most serious problems resulted from the large temperature and humidity fluctuations that occurred in the summer.

FIGURE 35.5 Maximizing skylight and avoiding the use of carpets can aid in preventing the growth and accumulation of fungi.

In a case study, Kolmodin-Hedman et al. (1986) investigated a staff member of a local arts and crafts museum who had reported symptoms after handling books and bookkeeping registers in the basement archives. Symptoms included fever, chills, nausea, and cough over a period of 1 year. The symptoms appeared at the end of the working day and disappeared after a few days at home. No symptoms occurred during the summer holidays. The books stored in the basement archives had visible mold growth from previous water damage from a leaky roof. Measurements of airborne spore levels in the basement were between 1,000,000 and 100,000,000 cfu/m^3.

Lichens, bacteria, and fungi can all play a role in the biodeterioration of historical glass. In a study on the fungal diversity of two historical church window glasses, Schabereiter-Gurtner et al. (2001) found complex fungal communities consisting of members and relatives of the genera *Aspergillus, Aureobasidium, Geomyces, Rhodotorula, Ustilago, Verticillium,* and others that inhabited the glass surfaces.

Even stone monuments are subject to damage from biological contamination. An evaluation of field data from historical buildings in Germany showed that bacteria were more numerous than fungi in building stones (Mansch and Bock, 1998). The highest cell counts were usually found near the surface and the colonization of natural stone by the bacteria took several years. The highest cell numbers were in some cases found underneath the surface. Data strongly suggested that microbial colonization of historical buildings was enhanced by air pollution. Samples taken from stone material with a pore radius approximately 1 μm had significantly higher cell numbers when they were covered with black crusts. In a study by Gorbushina et al. (2002) outdoor monuments in St. Petersburg and Moscow made of marble and limestone were investigated for microbial contamination. The monuments in St. Petersburg were found to be severely contaminated with micromycetes belonging to 24 genera, primarily of the class Deuteromycetes. The monuments in Moscow were contaminated with bacteria at a density of about 200,000 cfu/g. A total of 12 bacterial genera were identified. Table 35.2 summarizes the various fungi found in museums and the materials they may be found in or grow on.

Some museum objects may be more susceptible to microbial contamination than others and may contain unusual species. In samples collected from feather and leather objects in museums in India, and also from their surroundings and deposited dusts, were found to

TABLE 35.2 Air and Surface Fungi in Museums

Microbe	Location	Hazard	Reference
Acremonium	Feathers, leather	Allergen	Nigam et al., 1994
Alternaria	Paintings	Allergen	Inoue and Koyano, 1991
Aspergillus	Stained glass	Allergen	Schabereiter-Gurtner et al., 2001
Aspergillus	Paintings, feathers, leather, ancient documents, limestone, textiles	Allergen	Nigam et al.,1994; Inoue and Koyano, 1991; Szczepanowska and Cavaliere, 2000; Leznicka et al., 1991; Montegut et al., 1991
Aureobasidium	Paintings, stained glass, limestone	Allergen	Schabereiter-Gurtner et al., 2001; Inoue and Koyano, 1991; Leznicka et al., 1991
Chaetomium	Ancient documents, textiles	Allergen	Szczepanowska and Cavaliere, 2000; Montegut et al., 1991
Chrysosporium	Feathers, leather	Allergen	Nigam et al.,1994
Cladosporium	Paintings, limestone, frescoes, cotton	Allergen	Inoue and Koyano, 1991; Leznicka et al., 1991; Karpovich-Tate and Rebrikova, 1990; Siu, 1951
Curvularia	Cotton textiles	Allergen	Siu, 1951
Exophiala	Limestone	Allergen	Leznicka et al., 1991
Fusarium	Paintings, textiles	Allergen	Nigam et al.,1994; Inoue and Koyano, 1991; Szczepanowska and Cavaliere, 2000; Montegut et al., 1991
Geomyces	Stained glass	Allergen	Schabereiter-Gurtner et al., 2001
Mucor	Cotton textiles	Allergen	Siu, 1951
Paecilomyces	Limestone, cotton textiles	Allergen	Leznicka et al., 1991; Siu, 1951
Penicillium	Paintings, feathers, leather, ancient documents, limestone, textiles	Allergen	Nigam et al.,1994; Inoue and Koyano, 1991; Szczepanowska and Cavaliere, 2000; Leznicka et al., 1991; Montegut et al., 1991
Phialaphora	Frescoes, building materials	Deteriogen	Karpovich-Tate and Rebrikova, 1990
Phoma	Limestone, cotton textiles	Allergen	Leznicka et al., 1991; Siu, 1951
Rhizopus	Paintings, textiles	Allergen	Inoue and Koyano, 1991; Montegut et al., 1991
Rhodotorula	Stained glass	Allergen	Schabereiter-Gurtner et al., 2001
Stachybotrys	Textiles	Allergen	Montegut et al., 1991
Trichoderma	Paintings, limestone, textiles	Allergen	Inoue and Koyano, 1991; Leznicka et al., 1991; Montegut et al., 1991
Tritirachium	Frescoes, building materials	Deteriogen	Karpovich-Tate and Rebrikova, 1990
Ulocladium	Paintings	Allergen	Inoue and Koyano, 1991
Ustilago	Stained glass	Allergen	Schabereiter-Gurtner et al., 2001
Verticillium	Stained glass, limestone, frescoes	Deteriogen	Schabereiter-Gurtner et al., 2001; Leznicka et al., 1991; Karpovich-Tate and Rebrikova, 1990

TABLE 35.3 Air and Surface Bacteria in Museums

Microbe	Location	Hazard	Reference
Actinomyces	Cotton textiles	Allergen	Siu, 1951
Arthrobacter	Museum air, paintings	Deteriogen	Camuffo et al., 1999
Arthrobacter	Frescoes, building materials	Deteriogen	Karpovich-Tate and Rebrikova, 1990
Aureobacterium	Museum air, paintings	Deteriogen	Camuffo et al., 1999
Bacillus	Paintings, frescoes, building materials, cotton textiles	Deteriogen	Inoue and Koyano, 1991; Karpovich-Tate and Rebrikova, 1990; Siu, 1951
Corynebacterium	Cotton textiles	Pathogen	Siu, 1951
Cyanobacteria	Building materials	Deteriogen	Ortega-Calvo et al., 1991
Microbacterium	Museum air, paintings	Deteriogen	Camuffo et al., 1999
Micrococcus	Museum air, paintings, cotton textiles	Deteriogen	Camuffo et al., 1999; Inoue and Koyano, 1991; Siu, 1951
Micromonospora	Cotton textiles	Allergen	Siu, 1951
Proteus vulgaris	Paintings	Deteriogen	Inoue and Koyano, 1991
Streptomyces	Frescoes, building materials	Deteriogen	Karpovich-Tate and Rebrikova, 1990

contain five species of *Chrysosporium*, four of *Aspergillus*, two of *Penicillium*, and two each of *Acremonium* and *Fusarium* (Nigam et al., 1994).

Table 35.3 summarizes the various bacteria found in museums and the materials they may be found in or grow on.

Although outdoor gaseous pollutants are mostly not of microbiological origin and not the subject of this book, they can contribute both to damage of museum collections and to exacerbating the susceptibility of those exposed to airborne allergens. A study by Druzik et al. (1990) was performed at 11 museums, art galleries, historical houses, and a museum library during successive summers to determine whether high outdoor ozone concentrations are transferred to the indoor atmosphere of museums. Museums having conventional air conditioning systems showed peak indoor ozone concentrations about 30 to 40 percent of outdoor levels. Buildings without air conditioning but which had a high rate of air exchange with the outdoors showed peak indoor ozone levels 69 to 84 percent of the outdoor concentrations, while other buildings where slow air infiltration provides the only means of air exchange have indoor ozone levels typically 10 to 20 percent of those outdoors. For purposes of comparison, the recommended ozone level in places where works of art and historical documents are stored ranges from 0.013 to 0.001 ppm depending on the authority cited.

The use of appropriate levels of filtration can limit soiling of museum collections, as well as reduce indoor levels of airborne microorganisms. In one study by Nazaroff and Cass (1991), it was found that objects in Southern California museums may become visibly soiled within as little as 1 year due to the deposition of airborne contaminants on their surfaces. Methods proposed for reducing the soiling rate include reducing the building ventilation rate, increasing the effectiveness of particle filtration, reducing the particle deposition velocity onto surfaces of concern, placing objects within display cases or glass frames, managing a site to achieve low outdoor aerosol concentrations, and eliminating indoor particle sources. According to results of a theoretical model, the soiling rate can be reduced by at least two orders of magnitude through practical application of these control measures and combining improved filtration with either a reduced ventilation rate for the entire building or low-air-exchange display cases can be an effective approach to reducing the soiling hazard.

The question of to what degree patrons contribute to indoor dust and airborne microbial pollution has never been fully resolved. Experiments conducted inside the Spencer Museum of Art at the University of Kansas identified the sources and concentrations of indoor generated particulate matter and assessed the soiling impacts of these sources on works of art (Roshanaei and Braaten, 1996). Results of this study showed no significant relationship between the high particle concentrations and the presence of the public in the gallery. However, the changing of the filters in the ventilation system contributed 12 percent to the total mass deposited to a gallery painting. Cleaning activities contributed some 5 to 15 percent to the total mass flux.

A study of an embalmed elephant trunk stored in a museum turned up the mold species *Penicillium notatum*, *P. variabilis*, *P. purpurogenum*, and *Aspergillus niger* (Riha et al., 1991). The trunk was then preserved for approximately 1 year in a 50 percent ethylene glycol solution with 0.9 percent glutaraldehyde, after which no viable organisms were found. Subsequently, the fluid was removed and thymol and a desiccant were placed in the sealed display case. One year later, the trunk was examined for fungi, with no growth detected.

Fabrics in museums, especially ancient cloth, are subject to biodeterioration from fungi and bacteria. Fungi grow extensively on cotton fibers and growth rates are highest on 100 percent cotton (Goynes et al., 1995). The control of environmental conditions can be an effective means of static control of fungal biodegradative agents (Montegut et al., 1991). The various types of materials in museums often include wood, textiles, and paintings, which are all subject to biodeterioration by fungi. Studies of deterioration caused by the growth of fungi on paintings have isolated over 100 culprit species of fungi (Inoue and Koyano, 1991). The most typical species are *Alternaria*, *Cladosporium*, *Fusarium*, *Aspergillus*, *Trichoderma*, and *Penicillium*. A varnish made from synthetic resin containing Vinyzene, a fungicide, was sprayed on about 100 paint samples and annual observations by the researchers have confirmed its effectiveness as a fungicide.

Cellulose devouring fungi pose one of the greatest hazards to ancient books and documents. In a case study by Szczepanowska and Cavaliere (2000) iron chests containing eighteenth and nineteenth century documents that had been stored in a cellar in Maryland had been periodically submerged in water during seasonal floods over a period of 80 years were compacted into brick-shaped forms as a result of biological deterioration by fungi. The fungi were identified as four species representing members in the genus *Chaetomium*, including *Chaetomium globosum*. In addition to these, several species of *Penicillium* and *Aspergillus* were also found.

Damaged frescoes and building materials in the Cathedral of the Nativity of the Virgin at the Pafnutii-Borovskii Monastery were contaminated with various fungi and bacteria, including streptomycetes (Karpovich-Tate and Rebrikova, 1991). Growth of microorganisms was associated with the formation of a compact yellowish-white coating on the cleared and restored frescoes, plaster and brick, and with the powdery destruction of building stone and brickwork at the lower parts of the walls. Evidence suggests that although the microorganisms could use organic substances in dust, admixtures in plaster, binding medium of paints and restoration material as nutrient sources for growth on inorganic building materials and frescoes, that in fact complex microbiological interrelationships played the main role in such biodeterioration.

Biocides are a common approach to cleaning and disinfecting museum artifacts. In a study on the efficiency and threshold concentration of biocides on fungi and algae responsible for attacking museum exhibits inside and in the open air, Mamonova et al. (1992) found that AB catamin, which is used in museums, was effective, and that the use of catapol had advantages over traditional cleaning methods.

Gamma irradiation is a novel alternative technology for the prevention of decay of waterlogged archaeological wood that offers potentially greater effectiveness than traditional methods (Pointing et al., 1998). A dose of 15 kGy was found to be sufficient to inactivate a

TABLE 35.4 Bioaerosol Levels in Libraries and Museums

Contaminant	Level	Area/notes	Reference
Fungi (cfu/m^3)	6–38	4th and 5th floor libraries	MUN, 2000
	96	Library, air conditioned, winter	Burge et al., 1978
	409	Library, air conditioned, summer	Burge et al., 1978
	100	Library, nonair conditioned, winter	Burge et al., 1978
	671	Library, nonair conditioned, summer	Burge et al., 1978
	205	Museum	Krake et al., 1999
	200	Museum library	Krake et al., 1999
	320	Painting storage	Krake et al., 1999
	1,000,000	Water damaged book storage	Kolmodin-Hedman et al., 1986
Dust (µg/m^3)	190	16 libraries, Univ. of Modena	Fantuzzi et al., 1996

large number of wood biodeteriogens, including fungi, bacteria, and invertebrates. For timbers excavated from polluted sites, a dose of 25 kGy is suggested to inactivate human pathogens. No adverse effects on the physical properties of degraded archaeological wood were detected at doses of up to 100 kGy. This is the maximum recommended single or cumulative lifetime dose for any timber.

Table 35.4 provides a summary of some air sampling results in libraries and museums that can be used as a guide for determining whether measured levels are typical or not. See Chap. 23 for information on air and surface sampling.

REFERENCES

Arai, H. (2000). "Foxing caused by Fungi: Twenty-five years of study." *Int Biodeter Biodegrad* 46(3):181–188.

ASHRAE (1999a). *Handbook of Applications.* American Society of Heating, Refrigerating and Air-Conditioning Engineers, Atlanta, GA.

ASHRAE (1999b). Chapter 20: Museums, Libraries, and Archives. *HVAC Applications,* American Society of Heating, Refrigerating and Air-Conditioning Engineers, Atlanta, GA.

ASHRAE (2001). "Standard 62-2001: Ventilation for acceptable indoor air quality." American Society of Heating Refrigerating and Air-Conditioning Engineers. Atlanta, GA.

Balmain, A. R. (1927). "Recovery of *Streptococcus scarlatinae* from experimentally infected books." *Lancet* 2:1128.

Beebe, W. L. (1911). "Carbo gasoline method for the disinfection of books." *J Am Public Health Assoc* 1:54–60.

Brimblecombe, P., Blades, N., Camuffo, D., Sturaro, G., Valentino, A., Gysels, K., VanGrieken, R., Busse, H. J., Kim, O., Ulrych, U., and Wieser, M. (1999). "The indoor environment of a modern museum building, the Sainsbury Centre for Visual Arts, Norwich, UK." *Indoor Air* 9(3):146–164.

Burge, H. P., Boise, J. R., Solomon, W. R., and Bandera, E. (1978). "Fungi in libraries: an aerometric survey." *Mycopathologia* 64(2):67–72.

Camuffo, D., Brimblecombe, P., VanGrieken, R., Busse, H. J., Sturaro, G., Valentino, A., Bernardi, A., Blades, N., Shooter, D., DeBock, L., Gysels, K., Wieser, M., and Kim, O. (1999). "Indoor air quality at the Correr Museum, Venice, Italy." *Sci Total Environ* 236(1–3):135–152.

Chappas, W. J., and McCall, N. (1984). "The use of ionizing radiation in disinfestation of archival and manuscript materials." *Biodeterioration 6: The 6th International Biodeterioration Symposium,* Washington, DC. 370–373.

Cole, E. C., Foarde, K. K., Leese, K. E., Franke, D. L., and Berry, M. A. (1993). "Biocontaminants in carpeted environments." *Indoor Air 93,* Helsinki, Finland.

Cunha, G. M., and Tucker, N. P. (1972). *Libraries and Archives Conservation.* Boston Athenaeum, Boston, MA.

Druzik, J. R., Adams, M. S., Tiller, C., and Cass, G. R. (1990). "Measurement and model predictions of indoor ozone concentrations in museums." *Atmos Environ* 24A(7):1813–1823.

Fantuzzi, G., Aggazzotti, G., Righi, E., Cavazzuti, L., Predieri, G., and Franceschelli, A. (1996). "Indoor air quality in the university libraries of Modena." *Science of the Total Environ* 193(1):49–56.

Ferson, M. J. (2001). "Books as carriers of disease." *Med J Australia* 175:663–664. http://www.mja.com.au/public/issues/175_12_171201/ferson/ferson.html.

Florian, M.-L. E., and Manning, L. (2000). "SEM analysis of irregular fungal fox spots in an 1854 book: Population dynamics and species identification." *Int Biodeter Biodegrad* 46(3):205–220.

Gambale, W., Croce, J., Costa-Manso, E., Croce, M., and Sales, M. (1993). "Library fungi at the University of Sao Paulo and their relationship with respiratory allergy." *J Investig Allergol Clin Immunol* 3(1):45–50.

Gorbushina, A. A., Lialikova, N. N., Vlasov, D., and Khizhniak, T. V. (2002). "Microbial communities on the monuments of Moscow and St. Petersburg: biodiversity and trophic relations." *Mikrobiologiia* 71(3):409–417.

Goynes, W. R., Moreau, J. P., DeLucca, A. J., and Ingber, B. F. (1995). "Biodeterioration of nonwoven fabrics." *Textile Res J* 65(8):489–494.

Groom, P., and Panisset, T. (1933). "Studies in *Penicillium chrysogenum thom* in relation to temperature and relative humidity of the air." *Ann Appl Biol* 20:633–660.

Hanus, J. (1985). "Gamma radiation for use in archives and libraries." *Abbey Newsletter* 9(2):1–2. http://palimpsest.stanford.edu/byorg/abbey/an/an09/an09-02/an09-209.html.

Hartman, T. (1996). "Library and museum HVAC: new technologies/new opportunities—Part 1." *HPAC Eng* 68(4):57–60.

Hay, R. J. (1995). "Sick library syndrome." *Lancet* 346(8990):1573–1574.

Huang, M., Guan, L., Zhou, C., Li, D., and Ling, D. (1997). "An investigation of precipitation of airborne fungi particles in the rooms of university library." *Hunan Yi Ke Da Xue Xue Bao* 22(6): 494–496.

IEA (1996). *IEA Annex 24—Volume 3: Heat and Moisture Transfer*, International Energy Agency, Paris, France.

Inoue, M., and Koyano, M. (1991). "Fungal contamination of oil paintings in Japan." *Int Biodeter* 28(1–4):23–35.

Ismail, M. A., Chebon, S. K., and Nakamya, R. (1999). "Preliminary surveys of outdoor and indoor aeromycobiota in Uganda." *Mycopathologia* 148(1):41–51.

Karpovich-Tate, N., and Rebrikova, N. L. (1991). "Microbial communities on damaged frescoes and building materials in the Cathedral of the Nativity of the Virgin in the Pafnutii-Borovskii Monastery, Russia." *Int Biodeter* 27(3):281–296.

Kelly, T. J., and Kinkead, D. A. (1993). "Testing of chemically treated adsorbent air purifiers." *ASHRAE J* 35(7):7.

Kenwood, H., and Dove, E. L. (1915). "The risks from tuberculosis infection retained in books." *Lancet* 2:66–68.

Kolmodin-Hedman, B., Blomquist, G., and Sikstrom, E. (1986). "Mould exposure in museum personnel." *Int Arch Occup Environ Health* 57(4):321–323.

Koreck, S. (1999). "Cleaning books in the Bodleian Library." *Paper Conservation News* 90:1.

Krake, A. M., Worthington, K. A., Wallingford, K. M., and Martinez, K. F. (1999). "Evaluation of microbiological contamination in a museum." *Appl Occup Environ Hyg* 14(8):499–509.

Leznicka, S., Kuroczkin, J., Krumbein, W. E., Strelczyk, A. B., and Petersen, K. (1991). "Studies on the growth of selected fungal strains on limestones impregnated with silicone resins (Steinfestiger H and Elastosil E-41)." *Int Biodeter* 28:91–111.

Mamonova, I. V., Kurochkin, V. E., Paramonov, G. A., Petrjakov, A. O., and Panina, L. K. (1992). "Determination of threshold concentrations of biocides for museum exhibits by a photometric method." *Int Biodeter Biodegrad* 30(4):303–312.

Mansch, R., and Bock, E. (1998). "Biodeterioration of natural stone with special reference to nitrifying bacteria." *Biodegradation* 9(1):47–64.
McCartney, J. E. (1925). "Infection by books." *Lancet* 2:212.
McLary, A. (1985). "Beware the deadly books: A forgotten episode in library history." *J Libr Hist* 20:427–433.
Montegut, D., Indictor, N., and Koestler, R. J. (1991). "Fungal deterioration of cellulosic textiles: a review." *Int Biodeter* 28(1–4):209–226.
MUN (2000). "Memorial University of Newfoundland Environmental Assessment: Project No. 86519." Memorial University of Newfoundland Facilities Management. St. John's. http://www.mun.ca/facman/enviroaudit/chap8.html.
Nazaroff, W. W., and Cass, G. R. (1991). "Protecting museum collections from soiling due to the deposition of airborne particles." *Atmos Environ* 25A(5–6):841–852.
NBS (1983). "Air quality criteria for storage of paper-based archival records." *NBSIR 83-2795*. National Institute of Standards and Technology. Gaithersburg, MD.
Nice, L. B. (1912). "The disinfection of books." *Bull Med Library Assoc* 1:61–66.
Nigam, N., Dhawan, S., and Nair, M. V. (1994). "Deterioration of feather and leather objects of some Indian museums by keratinophilic and nonkeratinophilic fungi." *Int Biodeter Biodegrad* 33(2):145–152.
Ogden, S. (1992). *Preservation of Library and Archival Materials: A Manual*. Northeast Document Conservation Center. Andover, MA.
Ortega-Calvo, J. J., Hernandez-Marine, M., and Saiz-Jimenez, C. (1991). "Biodeterioration of building materials by cyanobacteria and algae." *Int Biodeter* 28:165–185.
Pointing, S. B., Jones, E. B. G., and Jones, A. M. (1998). "Decay prevention in waterlogged archaeological wood using gamma irradiation." *Int Biodeter Biodegrad* 42(1):17–24.
Raaska, L., Sillanpaa, J., Sjoberg, A. M., and Suihko, M. L. (2002). "Potential microbiological hazards in the production of refined paper products for food applications." *J Ind Microbiol Biotechnol* 4:225–231.
Rakotonirainy, M., Fohrer, F., and Flieder, F. (1999). "Research on fungicides for aerial disinfection by thermal fogging in libraries and archives." *Int Biodeter Biodegrad* 44(2–3):133–139.
Reinick, W. R. (1914). "Books as a source of disease." *Am J Pharm* 86:13–25.
Righi, E., Aggazzotti, G., Fantuzzi, G., Ciccarese, V., and Predieri, G. (2002). "Air quality and well-being perception in subjects attending university libraries in Modena (Italy)." *Sci Total Environ* 286(1–3):41–50.
Riha, V. F., Peterson, L. J., Rossmoore, H. W., and Shoshani, J. (1991). "Detection and inhibition of fungi from the excised trunk of an Asian elephant." *Int Biodeter* 28(1–4):113–124.
Rocchetti, F., Adamo, M., and Magaudda, G. (2002). "Fastness of printing inks subjected to gamma-ray irradiation and accelerated aging." *Restaurator* 23(1):15–26.
Roshanaei, H., and Braaten, D. A. (1996). "Indoor sources of airborne particulate matter in a museum and its impact on works of art." *J Aerosol Sci* 27(Suppl. 1):S443–S444.
Schabereiter-Gurtner, C., Pinar, G., Lubitz, W., and Rolleke, S. (2001). "Analysis of fungal communities on historical church window glass by denaturing gradient gel electrophoresis and phylogenetic 18S rDNA sequence analysis." *J Microbiol Methods* 47(3):345–354.
Siu, R. G. H. (1951). *Microbial Decomposition of Cellulose*. Reinhold Publishing Corp., New York.
Smiley, H. E. (1933). "Books—shall they be sterilized?" *Rhode Island Med J* 16:5–6.
Smith, C. R. (1942). "Survival of tubercle bacilli in books." *Am Rev Tuberculosis* 46:549–559.
Snow, D., Crichton, M. H. G., and Wright, N. C. (1944). "Mould deterioration of feeding stuffs in relation to humidity of storage." *Ann Appl Biol* 31:102–110.
Southwell, K. L. (2003). "Chlorine Dioxide: A Treatment for Mold in Libraries." *Abbey Newsletter* 26(6). http://palimpsest.stanford.edu/byorg/abbey/an/an26/an26-6/an26-618.html.
Strzelczyk, A. B. (2001). "Adaptation to fungicides of fungi damaging paper 1. Influence of adaptation on morphology and anatomy of colonies and hyphae." *Int Biodeter and Biodegrad* 48(1–4):255–262.
Swartzburg, S. G. (1980). *Preserving Library Materials*. The Scarecrow Press, London.

Szczepanowska, H., and Cavaliere, A. R. (2000). "Fungal deterioration of 18th and 19th century documents: A case study of the Tilghman Family Collection, Wye House, Easton, Maryland." *Int Biodeter Biodegrad* 46(3):245–249.

Weaver-Meyers, P., Stolt, W. A., and 455-458., B. K. (1998). "Controlling Mold on Library Materials with Chlorine Dioxide: An Eight-Year Case Study." *J Acad Librariansh* 24:455–458.

Weaver-Meyers, P. L., Stolt, W. A., and Kowaleski, B. (2000). "Controlling mold on library materials with chlorine dioxide: An eight-year case study." *Abbey Newsletter* 24(4):5. http://palimpsest.stanford.edu/byorg/abbey/an/an24/an24-4/an24-402.html.

Yeung, Y. N. A., Chow, W. K., and Lam, V. Y. K. (1991). "Sick building syndrome. A case study." *Building and Environ* 26(4):319–330.

Yoon, Y. H., and Brimblecombe, P. (2001). "The distribution of soiling by coarse particulate matter in the museum environment." *Indoor Air* 11(4):232–240.

Zyska, B. (1997). "Fungi isolated from library materials: A review of the literature." *Int Biodeter Biodegrad* 40(1):43–51.

CHAPTER 36

PLACES OF ASSEMBLY

36.1 INTRODUCTION

Places of assembly may include all structures enclosing large indoor spaces, such as auditoriums, stadiums, theaters, malls, gymnasiums, natatoriums, arenas, town halls, churches, cathedrals, temples, mosques, industrial halls, convention centers, atriums, malls, shopping centers, airports, and other places where large public gatherings may occur indoors or through which heavy cyclical occupancy may occur. In such places of assembly any number of airborne diseases can be exchanged and large numbers of people may be exposed to airborne hazards. The large volumes of air enclosed in such facilities often guarantee good diffusion and mixing of air, and as a result the air quality is often acceptable for extended periods of occupancy. At the same time, ventilation systems installed in such facilities may be unable to guarantee delivery of a constant rate of outdoor air to all corners. The result is a great deal of unpredictability in the performance and airflow distribution inside large enclosed spaces.

The natural airflows and forced air currents inside large ventilated structures make it difficult to predict performance of air cleaning systems, as well as to design ventilation systems to begin with. Air cleaning systems can be sized for these facilities, but designing a system that can efficiently clean the air and deliver it to each occupant is a challenging problem that may require additional distribution equipment and special analytical software. These topics and the aerobiology of indoor auditoriums, shopping malls, and other facilities are addressed in the following sections.

36.2 VENTILATION SYSTEMS

Many places of public assembly have intermittent occupancy and the ventilation system may operate in two basic modes—low occupancy and high occupancy. The more critical design condition, high occupancy, will typically require huge volumes of outside air that are cooled and dehumidified, or sometimes heated. Guidelines for outdoor air ventilation rates and air conditioning for human comfort are provided in ASHRAE Standard 62-2001 and this and related references can be consulted for specific design information (ASHRAE, 1999, 2001). Only aspects of ventilation system design that relate to aerobiological concerns are addressed here.

Some indoor stadiums are naturally ventilated, but it is more common, even in older stadiums, for large fans to be used to supply or exhaust air without ductwork. These fans may be located on the roof or others areas. In such systems, with large vaneaxial fans, there will be little or no filtration of outside air. The level of filtration used in places of public assembly is typically minimal and mainly for dust control, or about 30 to 35 percent DSP (i.e., MERV 7–8). Places requiring higher levels of air cleanliness may use 80 percent DSP filters (MERV 13) or better, which will considerably reduce intake of fungal spores from the outdoors. Due to intermittent occupancy, the lifetime of high-efficiency filters may be prolonged, compared to systems that operate at full capacity every day. Even so, it is common to use prefilters in these situations to extend the life of high-efficiency filters. Off-load conditions are not completely without aerobiological concerns. When large facilities are unoccupied, it is often necessary to control temperature and humidity so as to preclude mold growth.

In mild climates, very large stadiums may use fans located near the roof to supply direct outside air without filtration, heating, or cooling. In other climates, stadiums and other large facilities will typically have a full complement of mechanical HVAC equipment in a separate equipment room. Due to the intermittent nature of the cooling loads in stadiums and other large facilities, thermal storage systems are often economically viable. Systems without thermal storage often precool the stadiums prior to events, a practice than can produce localized condensation, even if only for a short while. Condensation in general areas or in equipment rooms around cooling or thermal storage equipment may cause microbial growth and efforts should be made to ensure adequate drainage is provided.

Computational fluid dynamics (CFD) can be a very useful tool to help designers evaluate the indoor air quality and thermal comfort provided by HVAC systems in domed stadiums. CFD programs can be used to guide the design process and optimize ventilation systems. In one design of an indoor autoracing complex, Zhai et al. (2002) describe the facility as primarily a single space building with a floor area of over 2,000,000 ft^2 (200,000 m^2) with a ceiling height of 150 ft (46 m). The occupancy is estimated at 60,000 spectators in the grandstands and/or 60,000 spectators in the infield. The worst-case occupancy condition was used for the design basis. The CFD results were used to improve the conventional ventilation system design by incorporating a combination of displacement ventilation and overhead ducting.

The spatial distribution of contaminants in large semiopen building areas such as sports stadiums, warehouses, malls, and other industrial halls is critical to estimating exposure, health risks, and building energy performance. Nonuniformity of airflows presents a challenge to experimental measurements of how effectively the ventilation system removes or dilutes air pollutants in such facilities. Tracer gas testing is one method for evaluating airflow distributions or verifying CFD models. An experimental method developed by Demokritou et al. (2002) has been described that can be used for both the experimental evaluation of ventilation effectiveness as well for CFD modeling validation in large open space applications. The developed method is based on using a passive tracer gas and has been successfully used for the experimental evaluation of ventilation effectiveness in an ice skating arena and for the validation of a CFD model for the analysis of IAQ in ice rink facilities.

Domed stadiums with retractable roofs provide one option for effectively switching to natural ventilation when outdoor conditions are favorable and saving ventilation energy costs. In the design of such a stadium in Phoenix, wind tunnel tests and a CFD model were used to predict performance of the air distribution system and fine tune its effectiveness in cooling the facility after closing the roof (Gamble et al., 1996).

Heating and ventilation system design for gymnasiums and other large volume buildings can be accomplished without any ducts. In one such design by Blank (2001), the doors to the rooms remote from the heat source are opened to allow natural convection and uniform temperatures were obtained by proper selection of the throw from the supply grilles.

Large industrial halls used for welding and other types of work often have ceiling air outlets which blow air from the ceiling down to the floor level exhaust grilles. This method of ventilation may not be suitable for welding halls, according to a study by Pfeiffer and Brunk (1990), because the welding process produces thermal currents that tend to rise. In an alternative system, supply air is discharged in the hall floor area so that any pollutant released during welding is drawn up into the exhaust ducts. The former type of ventilation, with air currents traveling downward into floor grilles, is probably preferable from the point of view of microbial contaminants, which will tend to settle downward.

Industrial halls are more likely to contain gaseous pollution than high levels of airborne microbes, but the distribution of pollutants in such large open buildings provides some insight into the distribution of airborne microbes in similar facilities. An investigation of the effect of displacement flow on the distribution of aerosol concentrations in an industrial hall was performed by Hameri et al. (2003). According to the displacement ventilation principle, vertical upflow is accomplished by introducing fresh air, cooler than room air, into the occupied zone near floor level. The fresh air is introduced from low-velocity devices and heated by warm processes. This technique allows warm air contaminants to rise to the ceiling, and the rising plume is then exhausted close to the ceiling. The aerosol properties and behavior, especially the vertical gradients, are characterized by a displacement flow field. The results indicate that fine particles less than 1 μm in diameter are transported away from the breathing zone by this displacement ventilation process. However, the air quality is significantly influenced by the emission source, and therefore the concentration of fine and ultrafine aerosol particles smaller than 0.1 μm in the breathing zone increased compared to that of the incoming clean air. The vertical gradients displayed clear particle size dependence with the strongest gradients between 0.003 and 0.015 μm in diameter. The problem with this approach is that microbiological contaminants, especially bacteria and fungal spores, will tend to settle over time and accumulate on floors and carpets. As suggested in Chap. 9, a solution that addresses both VOCs and aerobiological concerns might be a split displacement ventilation system in which air is withdrawn from both the floor and the upper areas, with fresh air supplied at breathing level.

Underfloor air supply systems may be one of the most viable approaches to guaranteeing air quality for the individual occupants in a stadium or theater. Measurements and evaluation of indoor thermal environments is another means of assessing air distribution in large domed stadiums. In a study of the thermal environment within a domed stadium, Nishioka et al. (2000) evaluated a multizonal air-conditioning system that used under-seat supply of conditioned air and displacement ventilation. Measurements, including temperature distribution, humidity, airflow, and outdoor weather conditions, were carried out for three seasons, with each measurement period lasting for a week. The horizontal temperature distribution indicated that the air-conditioning system effectively separated the space and that each zone operated adequately. The vertical temperature distribution indicated that the under-seat supply air-conditioning system provided adequate air conditioning to a limited area or to the occupant zone only.

Due to the complexity of air currents in occupied stadiums, CFD provides one of the best alternatives for creating an efficient and functional design to remove both heat and biocontaminants. Baker et al. (2002) used CFD methods to design a ventilation system for an indoor auto-racing complex. The facility is primarily a single space building with a floor area of over 2×10^6 ft^2 (0.2×10^6 m^2) and a ceiling height of 150 ft (46 m). The facility is being designed to accommodate a variety of future possible occupancy conditions for a wide variety of events—60,000 spectators in the grandstands and/or 60,000 spectators using a combination of displacement ventilation and overhead duct system, and partial air curtains.

36.3 EPIDEMIOLOGY

Little is known about the epidemiology of indoor malls, shopping centers, airports, and other facilities with large transient occupancies. A handful of outbreaks have occurred in such settings as stadiums, convention centers, tent fairs, stadiums, and churches, and these are reviewed here.

One of the most famous outbreaks of a respiratory disease in a heavily occupied building was the eponymous Legionnaires' convention in Philadelphia in 1976 where 182 attendees came down with a mysterious and intractable lung infection that caused 29 deaths (Barry, 2001). It was traced to *Legionella* contamination in a cooling tower that apparently wafted from outdoors into a crowded entryway through open doors. Subsequent outbreaks at other locations continued until intense focus on the problem resulted in standard procedures for monitoring and remediating water-based contamination of cooling towers and similar reservoirs. *Legionella* is noncontagious and a curious example of how engineered systems can create new niches for adaptation of pathogenic microorganisms.

Other technological niches for *Legionella* that have resulted in outbreaks include saunas and sprinklers. Den Boer et al. (2002) report on an outbreak of Legionnaires' disease that affected over 100 visitors to a flower show in the Netherlands. The suspected sources of the outbreak included two whirlpool spas and a sprinkler in three separate halls of the exhibition. Persons who paused at the whirlpool spa in one hall were put at a definitive risk of legionellosis. De Schrijver et al. (2003) report on an outbreak of Legionnaires' disease in Belgium among visitors of the annual fair who entered a large tent. Stand employees, technical staff, and visitors at the fair were affected with 41 cases confirmed and five deaths. The source of the epidemic was an aerosol-producing whirlpool near the center of the tent.

A prime example of how a contagious airborne disease can be transmitted inside a domed stadium comes from a study by Ehresmann et al. (1995). In this epidemiological investigation, an outbreak of measles that occurred in a stadium in the Minneapolis-St. Paul metropolitan area during July 1991 resulted in 16 outbreak-associated cases of measles among U.S. residents from seven states, with nine additional cases resulting from subsequent transmission. Patient zero was a track and field athlete from Argentina and transmission occurred in three settings inside the stadium—the opening ceremonies, track and field events, and first-aid stations. Eight secondary cases had their exposure at the opening ceremonies, including two spectators sitting in the same section of the upper deck about 30 minutes above the athlete's entrance. Although it is surprising that such outbreaks don't occur more often, they may, in fact, occur all the time with more common respiratory infections at a level that escapes general notice.

Tuberculosis is one of the highly infectious agents that can transmit easily in crowded places of assembly where people remain for long periods. Outbreaks of TB usually arise within households or extended families where extended close contact provides more opportunity for exposure, but community outbreaks of TB can occur after only casual contact. Several outbreaks have occurred in churches and mosques due to a single infectious person. Dutt et al. (1995) reports on an outbreak of tuberculosis in a church where one man who was apparently unaware his infection was TB exposed 42 percent of his congregation, as well as 26 percent of his coworkers. Some 4.3 percent of exposed parishioners developed active TB. Cook et al. (2000) reports on a similar outbreak of tuberculosis in a local church that spread among an extended family network and other church contacts. One of the infectious cases visited a local church on two occasions, resulting in a further 16 cases.

Singing in crowded halls has been identified as a possible cause of outbreaks, including one classic case of a rock singer with the flu who infected a number of fans who were in the front of the audience. Choir singing can put the singers themselves at risk if one individual is infected. A case of a tuberculosis outbreak in a choir occurred among members of a famous gospel choir in Newark. Five cases of TB in the choir were identified including the index case. Of the 306 choir members who had tuberculin tests, 19 percent were positive for exposure. Four of the five infected individuals were tenors, and one was an alto. An air ventilation outlet was directly in front of the tenor section and was suspected of being a contributing factor. It would seem prudent that since singing promotes aerosolization, any member of a singing group with a respiratory infection should be denied participation in practices or performances.

In Japan, where public saunas remain popular, several outbreaks of tuberculosis in public saunas occurred that were associated with homeless people in the metropolitan area (Nakanishi et al., 1997). Pulmonary tuberculosis occurred among employees and customers of public saunas in Shijuku, a large office, shopping, and amusement quarter in Tokyo. Twenty-four individuals became infected from what were apparently multiple index cases among the homeless. Japanese saunas have been used as hotels and homeless people have traditionally been permitted to use the saunas as their home.

An outbreak of meningococcal disease occurred in the Islamic holy city of Mecca in 1992 (al-Gahtani et al., 1995). Over 100 cases were confirmed and *Neisseria meningitidis* was identified as the cause. Almost three times as many males as females were affected. The case-fatality ratio was 14.7 percent among confirmed cases. Visitors comprised 59 percent of cases and the rest were residents. The carriage rate was 86 percent among residents living near the Holy Mosque.

Evans et al. (2002) reviewed an outbreak of viral gastroenteritis following environmental contamination at a concert hall. Illnesses occurred over a 5-day period and were identified as being due to *Norwalk-like virus* (NLV). The index case was a concert attendee who vomited in the auditorium and adjacent male toilet. Gastrointestinal illness occurred among members of school children who attended the following day. Transmission was most likely not airborne but occurred through direct contact with contaminated fomites. The disinfection procedures were inadequate and the disinfectant used contained no sodium hypochlorite.

36.4 AEROBIOLOGY

The greatest aerobiological hazard in auditoriums, stadiums, and other places of assembly is the exchange of human pathogens. The presence of allergens that may enter from the environment is unlikely to be any worse a hazard than it is in other buildings. Furthermore, due to the wide-open nature of most modern facilities and the fact that they are often constructed from steel and concrete without a great deal of carpeting or other internal furnishings that may support mold growth, these facilities are probably unlikely to become amplifiers of indoor mold spores. Still, under extreme conditions, like those of water damage or flood damage, such facilities could become unhealthy environments in terms of air quality.

The available data on the aerobiology of large indoor facilities are quite limited, but a summary of some available studies is shown in Table 36.1. Only some of the types of facilities are represented in this table but results could be expected to be similar for any place of assembly.

Shopping malls and airports resemble stadiums in their size, general ventilation characteristics, and occupancies. The transmission of airborne diseases is perhaps less likely in shopping malls due to the fact that proximity to contagious individuals plays a major role (Lidwell and Williams, 1961). In a survey of the indoor air quality in shopping malls in Hong Kong, Li et al. (2001) measured various gaseous pollutants as well as indoor levels of bacteria. More than 40 percent of the shopping malls had 1-hours average CO_2 levels above the 1000 ppm limit of ASHRAE standards on both weekdays and weekends. The highest indoor airborne respirable particle level found was 380 $\mu g/m^3$. Of the malls

TABLE 36.1 Typical Airborne Concentrations in Places of Assembly

Contaminant	Level	Area	Reference
Bacteria (cfu/m^3)	16	Public auditorium	Ross et al., 2004
	84	Shopping center	Ross et al., 2004
	252	University auditorium (unoccupied)	Sessa et al., 2002
	1,075	University auditorium (occupied)	Sessa et al., 2002
	>1,000	Shopping Malls	Li et al., 2001
Fungi (cfu/m^3)	178	Public auditorium	Ross et al., 2004
	338	Shopping Center	Ross et al., 2004
	353	Lecture hall (winter)	Medrela-Kuder, 2003
	366	University auditorium (unoccupied)	Sessa et al., 2002
	939	Lecture hall (summer)	Medrela-Kuder, 2003
	1,512	University auditorium (occupied)	Sessa et al., 2002
	>14,661	High School and college gyms	Dacarro et al., 2003
Dust (μg/m^3)	380	Shopping malls (highest level)	Li et al., 2001

surveyed, 30 percent had indoor airborne bacteria levels above 1000 cfu/m^3, which is a limit set by the Hong Kong Indoor Air Quality Objective. The elevated indoor CO_2 and bacteria levels resulted from a combination of high occupancy and insufficient ventilation. The increased respirable particle levels were attributed to illegal smoking, pollution from public transports, and cooking in food courts inside these establishments.

In a study of the indoor concentration of airborne bacteria and fungi in a university auditorium and other buildings, Sessa et al. (2002) measured the indoor concentrations of airborne bacteria and fungi in the presence and in absence of occupants and furnishings. In the presence of people and furnishings the average air concentration of bacteria in a university auditorium was 1075 cfu/m^3. In the absence of people and furnishings the average was 252 cfu/m^3. The average airborne concentrations of fungi were higher in presence of people and furnishings in the university auditorium at 1512 cfu/m^3 while in their absence they were only 366 cfu/m^3. These results are shown graphically in Fig. 36.1.

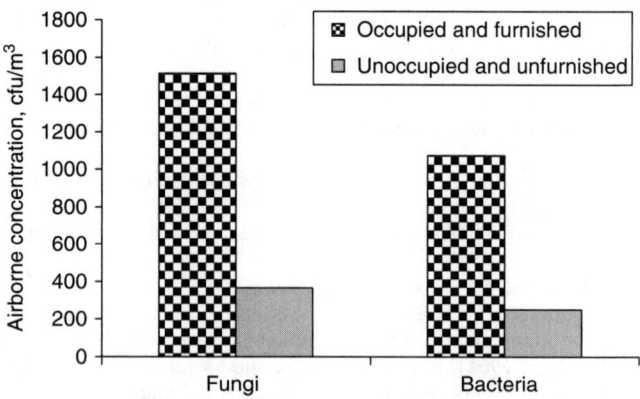

FIGURE 36.1 Comparison of airborne microbes in occupied vs. unoccupied stadium. [*Based on data from Sessa et al. (2000)*.]

Dacarro et al. (2003) evaluated the microbiological *indoor air quality* (IAQ) of high school and college gyms during physical training lessons using air samples from 11 high school and college gyms of Pavia, Italy. Using the global index of microbiological contamination (GIMC per m^3), it was found that the highest values of GIMC were greater than 14,661 per m^3, and that these levels were observed between April and October when the central heating systems were switched off. The lowest fungal counts were detected in modern buildings equipped with forced ventilation systems. Some 45 fungal taxa were identified.

Dust is another indicator of air quality since dust may often contain allergens like fungal spores or dust mite dander. In old theaters, singers and actors may be exposed to dusty and dry on-stage environments, which may induce respiratory tract indispositions and decrements in vocal performance, according to a study by Richter et al. (2000). This study evaluated the effects of humidification units installed in a previously dusty and dry theater where performers had experienced respiratory symptoms. Following the activation of the humidification units, humidities and fine dust concentrations were improved and generally within acceptable ranges for the generic workplace based on state guidelines.

Animal exhibits in indoor arenas can generate high levels of dander and allergens. Mussalo-Rauhamaa et al. (2001) studied the allergen content in indoor air and dust during dog shows in an indoor arena. Dog dander is one of the most important indoor allergens in Nordic countries and can be found practically everywhere indoors. In facilities used for dog shows, the dog allergen content was exceptionally high but can be reduced by proper cleaning. Cleaning the furniture and textiles is critical.

Air quality measured by *volatile organic compounds* (VOCs) is one indicator that may provide an idea of how well air is exchanged or distributed inside large open indoor spaces. In a study of four town halls in Copenhagen, Denmark, three buildings with two different levels of respiratory complaints were examined (Zweers et al., 1990). The total concentration of VOCs and the air temperature correlated significantly with odor intensity in the rooms. The concentrations of VOCs also correlated strongly with the air-exchange rate. Based on the results, the threshold for reduced indoor air quality caused by VOCs in these four buildings is estimated to be between 0.19 and 0.66 mg/m^3.

REFERENCES

al-Gahtani, Y. M., el Bushra, H. E., al-Qarawi, S. M., al-Zubaidi, A. A., and Fontaine, R. E. (1995). "Epidemiological investigation of an outbreak of meningococcal meningitis in Makkah (Mecca), Saudi Arabia, 1992." *Epidemiol Infect* 115(3):399–409.

ASHRAE (1999). Chapter 4: Places of Assembly. *HVAC Applications,* American Society of Heating, Refrigerating and Air-Conditioning Engineers, Atlanta, GA. 4.1–4.8.

ASHRAE (2001). "Standard 62-2001: Ventilation for acceptable indoor air quality." American Society of Heating Refrigerating and Air-Conditioning Engineers. Atlanta, GA.

Baker, A. J., Zhai, Z., Chen, Q., and Scanlon, P. W. (2002). "Design of a ventilation system for an indoor auto racing complex." *ASHRAE Trans* 108(Pt. 2):989–998.

Barry, B. E. (2001). Chapter 48: Legionella. *Indoor Air Quality Handbook,* J. D. Spengler, J. M. Samet, and J. F. McCarthy, eds., McGraw-Hill, New York. 48.1–48.15.

Blank, J. J. (2001). "Where are the ducts?" *ASHRAE J* 43(9):14.

Cook, S. A., Blair, I., Tyers, M. (2000). "Outbreak of tuberculosis associated with a church." *Commun Dis Public Health* 3(3):181–183.

Dacarro, C., Picco, A. M., Grisoli, P., and Rodolfi, M. (2003) "Determination of aerial microbiological contamination in scholastic sports environments." *J Appl Microbiol* 95(5):904–912.

De Schrijver K, D. K., Van Bouwel K, Mortelmans L, Van Rossom P, De Beukelaar T, Vael C, Fajo M, Ronveaux O, Peeters MF, Van der Zee A, Bergmans A, Ieven M, Goossens H. (2003). "An outbreak of Legionnaire's disease among visitors to a fair in Belgium in 1999." *Public Health* 117(2):117–124.

Demokritou, P., Yang, C., Chen, Q., and Spengler, J. D. (2002). "An experimental method for contaminant dispersal characterization in large industrial buildings for indoor air quality (IAQ) applications." *Build Environ* 37(3):305–312.

Den Boer, J. W. Y. E., Schellekens J, Lettinga KD, Boshuizen HC, Van Steenbergen, J. E., Bosman A, Van den Hof S, Van Vliet, H. A., Peeters, M. F., Van Ketel, R. J., Speelman, P., Kool, J. L., Conyn-Van Spaendonck, M. A. (2002). "A large outbreak of Legionnaires' disease at a flower show, the Netherlands, 1999." *Emerg Infect Dis* 8(1):37–43.

Dutt, A. K., Mehta, J. B., Whitaker, B. J., Westmoreland, H. (1995). "Outbreak of tuberculosis in a church." *Chest* 107(2):447–452.

Ehresmann, K. R., Hedberg, C. W., Grimm, M. B., Norton, C. A., MacDonald, K. L., and Osterholm, M. T. (1995). "An outbreak of measles at an international sporting event with airborne transmission in a domed stadium." *J Infect Dis* 171(3):679–683.

Evans, M. R., Meldrum, R, Lane, W., Gardner, D., Ribeiro, C. D., Gallimore, C. I., Westmoreland, D. (2002). "An outbreak of viral gastroenteritis following environmental contamination at a concert hall." *Epidemiol Infect* 129(2):355–360.

Gamble, S. L., Sinclair, R. J., Matsui, K. M., and Barrett, M. R. D. (1996). "Ventilation requirements for new Phoenix ballpark." *AMSE FED Publication* 238(3):519–527.

Hameri, K., Gaman, A., Hussein, T., Raisanen, J., Niemela, R., Aalto, P. P., and Kulmala, M. (2003). "Particle concentration profile in a vertical displacement flow: a study in an industrial hall." *Appl Occup Environ Hyg* 18(3):183–192.

Li, W. M., Lee, S. C., and Chan, L. Y. (2001). "Indoor air quality at nine shopping malls in Hong Kong." *Sci Total Environ* 273(1–3):27–40.

Lidwell, O. M., and Williams, R. E. O. (1961). "The epidemiology of the common cold." *J Hyg* 59:309–334.

Medrela-Kuder, E. (2003). "Seasonal variations in the occurrence of culturable airborne fungi in outdoor and indoor air in Cracow." *Int Biodeterior Biodegr* 52:203–205.

Mussalo-Rauhamaa, H., Reijula, K., Malmberg, M., Makinen-Kiljunen, S., and Lapinlampi, T. (2001). "Dog allergen in indoor air and dust during dog shows." *Allergy* 56(9):878–882.

Nakanishi, Y., Oyama, Y., Takahashi, M., Mori, T. (1997). "A molecular epidemiological analysis of several outbreaks of tuberculosis in public saunas. A problem of tuberculosis among homeless people in the metropolitan area." *Nippon Koshu Eisei Zasshi* 44(10):769–768.

Nishioka, T., Ohtaka, K., Hashimoto, N., and Onojima, H. (2000). "Measurement and evaluation of the indoor thermal environment in a large domed stadium." *Energ Buildings* 32(2):217–223.

Pfeiffer, W., and Brunk, M. F. (1990). "Ventilation of welding halls." *Energy and Buildings* 14(3):215–219.

Richter, B., Lohle, E., Maier, W., Kliemann, B., and Verdolini, K. (2000). "Working conditions on stage: climatic considerations." *Logoped Phoniatr Vocol* 25(2):80–86.

Ross, C., deMenezes, J. R., Svidzinski, T. I. E., Albino, U., and Andrade, G. (2004). "Studies on fungal and bacterial population of air-conditioned environments." *Brazilian Arch Biology Technol* 47(5):827–835. http://www.scielo.br/pdf/babt/v47n5/a20v47n5.pdf.

Sessa, R., Di, P. M., Schiavoni, G., Santino, I., Altieri, A., Pinelli, S., and Del, P. M. (2002). "Microbiological indoor air quality in healthy buildings." *New Microbiol* 25(1):51–56.

Zhai, Z., Chen, Q., Scanlon, P. W., and Baker, A. J. (2002). "Design of a ventilation system for an indoor auto racing complex." *ASHRAE Trans* 108(Pt. 1):989–998.

Zweers, T., Skov, P., Valbjorn, O., and Molhave, L. (1990). "Effect of ventilation and air pollution on perceived indoor air quality in five town halls." *Energy and Buildings* 14(3):175–181.

CHAPTER 37

AIRCRAFT AND TRANSPORTATION

37.1 INTRODUCTION

Aircraft, trains, ships, cars, and other compact enclosed environments can pose aerobiological hazards similar to those of other indoor environments except that the typically smaller total volumes often increase the risk due to proximity and make air distribution and air cleanliness a more critical factor. Although large cruise ships may resemble hotels and apartment buildings in terms of their ventilation systems, smaller craft like cars and planes offer airborne microbes extended opportunities to be exchanged between hosts due to the close quarters, shared breathing air, potentially extended periods of occupancy, and the limited degree of outside air that may be brought in, especially in cold climates. Other microenvironments like elevators and city buses probably don't allow enough exposure time for occupants to become infected with contagious airborne diseases. This chapter addresses the unique aerobiological problems associated with enclosed forms of transportation insofar as they have been studied. Except for airlines and cruise ships, there has been limited study on airborne diseases exchanged inside transportation vehicles, although it probably happens somewhere every day.

37.2 AIRCRAFT

Aircraft present one of the most crowded environments in which people remain for extended periods of time (see Fig. 37.1) and the potential for airborne disease transmission, as well as allergen exposure, is fairly obvious. Airplanes are potential vectors for the transmission of airborne diseases between countries and the inherent risk of spreading epidemics internationally should warrant special attention to cabin aerobiology (Hunt et al., 1995; Masterson and Green, 1991). This section examines the problem of the airline cabin environment from an aerobiological perspective. The topic of VOCs is only incidentally addressed here and the references should be consulted for further information on this subject.

FIGURE 37.1 Airline cabins are one of the most crowded indoor environments in which people remain for extended periods of time. (*Photo provided courtesy of Franz Zwart and Airliners.Net.*)

The question of airborne disease transmission aboard aircraft has been a matter of debate for many years. On the one hand there have been various studies claiming there is no evidence of disease transmission, that air-exchange rates exceed standards for buildings, and that proximity to infectious travelers is the primary determining factor for secondary infections [Zitter et al., 2002; Wenzel, 1996, AMA(CSA), 1998]. On the other hand, an increasing variety of evidence suggests the contrary conclusion—that airline crews and passengers have a higher risk of contracting infections on long flights (Klontz et al., 1989; Mixeu et al., 2002; NRC, 2002; AFA, 2003; Ungs and Sangal, 1990).

Complaints of air quality among airline passengers in commercial aircraft are accentuated when they come from crew members. The various symptoms reported, include headache, fatigue, fever, and respiratory difficulties, particularly by flight attendants on long-haul routes (Rayman, 2002). Although poor cabin air quality, including buildup of VOCs and carbon dioxide, is the primary suspect, other factors inherent in-flight, such as lowered barometric pressure, hypoxia, low humidity, circadian dysynchrony, and vibration, may also be contributing causes. Respiratory diseases are the most frequent cause of absenteeism among flight crews in the airline industry and airline crews run a high risk of contracting viruses because they daily come in contact with hundreds of potentially infected individuals. Vaccination of crews can result in 33 percent fewer episodes of any severe flulike illness but does not eliminate the problem (Mixeu et al., 2002).

Transmission and transport of influenza aboard a military aircraft was implied as part of an epidemiological study of an outbreak in the Naval Air Station in Key West, Florida in 1986 (Klontz et al., 1989). Active duty personnel reported experiencing a respiratory illness characterized by fever, cough, sore throat, and myalgia. Influenza virus was recovered from three symptomatic patients. Some 68 percent of case-patients belonged to a squadron that had traveled to Puerto Rico for a temporary assignment and among them the attack rate was 37 percent. Transmission of infection among squadron personnel appeared to have commenced in Key West and continued in a barracks in Puerto Rico and aboard two DC-9 aircraft that transported the squadron back to Key West. This was the first reported outbreak of this strain of influenza in the continental United States in the 1986 to 1987 influenza season.

One of the most well known outbreaks of influenza aboard an airliner apparently occurred on the ground during a delayed takeoff (Moser et al., 1979). Some 72 percent of the passengers became ill with many testing positive for influenza. A single person with the infection was identified as the index case. The ventilation system was inoperative during the delay, presumably accounting for the high attack rate.

In a survey of influenza viruses entering Japan via airline travelers Sato et al. (2000) collected samples from 504 passengers who complained of respiratory symptoms. From these samples several influenza virus strains were isolated. Twenty-eight of the isolates were influenza A viruses and two were influenza B viruses. The results of this study suggested that imported virus by travelers may have influenced the domestic influenza epidemics. Table 37.1 summarizes a number of airborne respiratory allergens and pathogens that have been identified onboard airlines or are suspected to be commonly airborne.

In-flight disease transmission of bactericidal meningitis has been defined as a patient who meets the case definition of meningococcal disease within 14 days of travel on a flight of at least 8 hours duration (MMWR, 2001a). *Neisseria meningitidis* is a leading cause of bacterial meningitis and sepsis in children and young adults in the United States and is spread through either direct contact with respiratory secretions or via droplet inhalation. Persons in close contact with patients who have meningococcal disease are at increased risk for contracting the disease. Because of concerns about disease transmission aboard aircraft, CDC has developed recommendations to ensure a standard approach to management of airline contacts (Navigant, 2003).

In a study conducted among 1100 passengers traveling in the western U.S. during January through early April 1999, no significant increase in upper respiratory infections was found among passengers in planes with recirculated air, as opposed to passengers that had 100 percent outside air (Zitter et al., 2002). This study showed that passengers on airplanes with and without air recirculation had similar rates of postflight respiratory symptoms, or about 19 to 21 percent. Although no evidence was found in this study that aircraft cabin air recirculation increases the risk for *upper respiratory infection* (URI) symptoms in passengers, the duration

TABLE 37.1 Potential Airborne Pathogens and Allergens in Airline Cabins

Contaminant	Type	Reference
Adenovirus	Virus, pathogen	Ungs and Sangal, 1990 (URIs)*
Aspergillus niger	Allergen	Pierce et al., 1999
Cat dander	Allergen	Various
Chickenpox	Virus, pathogen	Hocking, 2001
Coronavirus	Virus, pathogen	Ungs and Sangal, 1990 (URIs)*
Dog dander	Allergen	(various)
Dust mites	Allergen	AT, 2000
Influenza	Virus, pathogen	Moser et al., 1979
Measles	Virus, pathogen	Hocking, 2001
Mumps	Virus, pathogen	Hocking, 2001
Mycobacterium tuberculosis	Bacterial pathogen	Parmet, 1999; Kenyon et al., 1996
Neisseria meningitidis	Bacterial pathogen	MMWR, 2001a
Paecilomyces variotti	Allergen	Pierce et al., 1999
SARS virus	Virus, pathogen	Wilder-Smith et al., 2003

*Upper respiratory infections (URIs) were transmitted but not specifically identified.

of the flights were only 2 hours and apparently of insufficient duration to induce infections. The study itself acknowledged this defect. Studies by Lidwell and Williams (1961) suggest that 8 hours of exposure might be a mean time within which the transmission of colds could be expected. Domestic flights would rarely meet this criterion and transatlantic flights would be a more appropriate test of the possibility of in-flight disease transmission.

Mucosal irritation is a common complaint among travelers and flight attendants in aircraft cabins that may be related more to low relative humidity than to microbial contamination (Nagda and Hodgson, 2001). Volatile organic compounds can occur in airplane compartments and may increase sensitivity to respiratory infections. Aircraft disinfection is commonly implemented and is required by some countries. Some of the disinfectants used include aerosols of 2 percent phenothrin and permethrin emulsions. Passengers and crew exposed to such disinfectants often complain of skin rashes, respiratory problems, tingling and numbness in fingertips and lips, and burning eyes (vanNetten, 2002). The occasional occurrence of certain types of aircraft malfunction can introduce more serious contaminants to the aircraft cabin.

One of the strongest indicators that airline cabins create a higher risk of respiratory disease transmission is the higher incidence of such symptoms among flight attendants, who typically spend extended periods on board planes. Flight attendants contract significantly more colds and flus than the normal population of adults, and have over quadruple the rate of chest illnesses (Whelan et al., 2003). This rate exceeds that of teachers, who are themselves subject to higher rates of infections and symptoms, as illustrated in Fig. 37.2.

Controlling levels of airborne microbes is not a primary function of cabin ventilation (NRC, 2002). It has been shown that the lower the air-exchange rate, the higher the levels of bacteria (Nagda et al., 1992). A limited number of studies have been conducted on airline cabin aerobiology, and some of these results are shown in Table 37.2. These levels of bacteria and fungi look normal in comparison with buildings (see Chap. 28), which has led some to state that the levels of microbial contamination in airline cabins is "unlikely to cause adverse health effects" [AMA(CSA), 1998]. This perspective ignores the fact that people in airplanes have much higher proximity to each other and may require air quality levels superior to those in ordinary buildings.

Although the reported levels of bacteria in cabin air are low, the exposure to airborne pathogens is a dose-dependent process, and the longer the flight, the higher the risk of

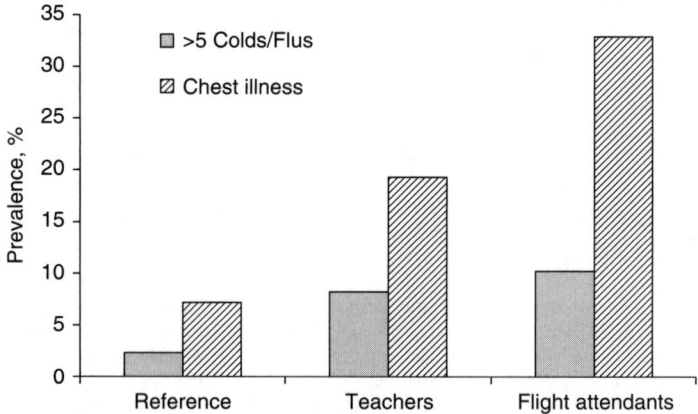

FIGURE 37.2 Comparison of prevalence of respiratory illness among female flight attendants compared with teachers and a normal reference population. [*Based on data from Whelan et al. (2003).*]

TABLE 37.2 Typical Levels of Contaminants in Airline Cabins

Contaminant	Level	Reference
Bacteria (cfu/m^3)	44–93	Nagda et al., 1992
	40–76	Lee et al., 1998, 1999
	50–244	Pierce et al., 1999 (domestic)
	39–133	Pierce et al., 1999 (international)
	201	Weisel, 2001
	0–1763	NRC, 2002
Fungi (cfu/m^3)	17–107	Nagda et al., 1992
	1–37	Pierce et al., 1999 (domestic)
	1–13	Pierce et al., 1999 (international)
	0–450	NRC, 2002
Particulates (PM$_{2.5}$), (μg/m^3)	3–10	NRC, 2002
	<10	Pierce et al., 1999
	<10	Nagda et al., 1992

infection from a fellow passenger (Zitter et al., 2002). In studies by Lidwell and Williams (1961) on the risk of infection as a function of proximity to an infected individual, it was shown that the risk increased considerably within distances of about 5 feet (see Chap. 4).

Human bioeffluents can become a primary contaminant of airplane air during long flights, as countless passengers have noted. The small air space available per person in a fully occupied aircraft passenger cabin increases the exposure to bioaerosols. The outside air ventilation required to minimize carbon dioxide concentration in the air of an occupied enclosed space remains the same regardless of the volume of that space. A study by Hocking (2002) suggests improved air quality benefits from returning to the higher outside air ventilation rates of the 1960s. Until the 1980s, airlines provided generous amounts of ventilation, often using 100 percent outside air, and there were no reports of problems. The current ASHRAE Standard 62 recommend 20 cfm of outdoor air be supplied for each and every occupant of a building (ASHRAE, 2001). New commercial aircraft are typically designed to recirculate approximately 50 percent or more of cabin air and provide about 8 to 15 cfm per passenger (ASHRAE, 1999a; Moore, 2000).

ASHRAE has proposed standards to ensure that cabin air quality is safe for flight occupants, minimizes the potential for adverse health effects, and is comfortable to occupants (Pierce et al., 1999). The standard includes development of a testing protocol to measure aircraft contaminants, the conduction of air quality monitoring, and testing for compliance (Miro and Cox, 2000).

The system of air recirculation in airplanes has undergone improvements in recent years but questions remain about the air quality in the cabin (Barthorpe, 1999). During flight, the outside air is heated through the engine intakes and then filtered to remove ozone. Typically, the outside air is mixed with return air in a plenum, as shown in Fig. 37.3, and drawn through a filter before being supplied to the overhead supply ducts. The return air is removed at floor level and drawn into a common duct before entering the mixing plenum.

Although the net air-exchange rate is high in airline cabins and 8 to 15 cfm is supplied per passenger (Pierce et al., 1999; Hunt et al., 1995), the actual amount of fresh air supplied decreases from the front to the back of the plane, as illustrated in Fig. 37.4. While the cockpit, of necessity for cooling, receives preferential volumes of fresh air, the cfm per passenger decreases from first class to economy class (AT, 2000). Higher levels of airborne microorganisms have been measured in economy class relative to first class (NRC, 2002). In addition, the nature of the typical supply and return duct designs in planes will likely draw more air from the front portion of the plane and less from the rear, leaving a dead spot at the back of the plane

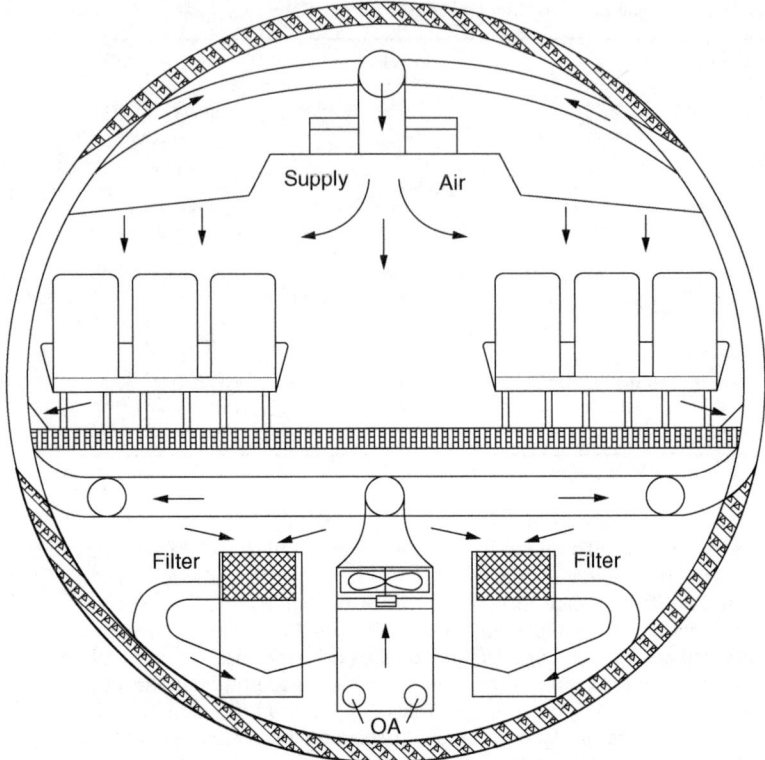

FIGURE 37.3 Schematic of typical ventilation system in large aircraft. Return air from floor level enters a plenum and is filtered and then mixed with processed outside air (OA) from the engines, and supplied through overhead registers.

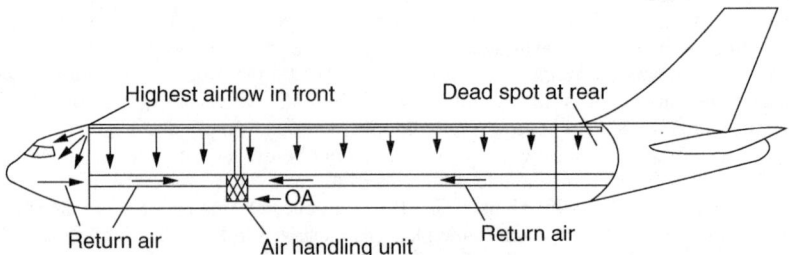

FIGURE 37.4 The nature of the supply and return air system on typical airliners may create increased air-exchange rates in the cockpit and up front, and decreased air exchange toward the rear, including a dead spot in the back.

where contamination may remain or accumulate (Clayton et al., 1976). It has also been noted that airline ventilation systems do not always function as designed (Olander and Westlin, 1991).

Recirculated building air has been identified as a risk factor for transmission of airborne diseases (Zeterberg, 1973). In a smaller, enclosed environment such as an airplane, it might be expected that the risk of disease transmission would be higher but data are inconclusive at present. Bergau (1999) states that the overall risk of getting an infectious disease is significantly lower than in other ground operated public means of transportation. Due to the low doses but long exposure possible in aircraft, the danger to passengers may be limited to infectious airborne microorganisms that are uniquely adapted to airborne transport, such as *Mycobacterium tuberculosis*, and influenza, and for which the infectious doses are low. Such microbes are transported by droplet nuclei and can remain viable for extended periods.

Smoke patterns have been used to assess airflow inside aircraft cabins. In a study on the problem of carrying patients with highly virulent communicable diseases Clayton et al. (1976) examined transport aircraft of the Canadian Forces. Smoke patterns suggested that an infected patient should be placed at the rear of the aircraft to minimize dissemination of the infection to other passengers. This study was followed by the evaluation of the spread on nonpathogenic organisms disseminated within Boeing 707 and C13OE (Hercules) aircraft by attempting to recover respiratory tract viruses during transatlantic flights. The spread of the nonpathogenic organisms in a 707 indicated that contamination was largely confined to the rear, except when the aircraft was in an unpressurized mode. In the C13OE, contamination was shown to occur throughout the whole aircraft. No respiratory tract viruses were recovered during the transatlantic flights.

In a study by Olander and Westlin (1991), measured values for different air contaminants from smoking were used to evaluate the air distribution systems in various aircraft. The study concluded that the air distribution systems in the aircraft studied did not function as intended. In another study on tobacco smoke inside airplanes, Nagda et al. (1992) evaluated 92 randomly selected flights that were monitored to determine prevailing levels of pollutants. Levels of microbial aerosols were below those in residential environments that have been characterized through cross-sectional studies.

In a study by Dechow et al. (1997) the concentrations of selected air quality parameters in aircraft cabins were investigated, including particle numbers in cabin air compared to fresh air and recirculation air and microbiological contamination. Airborne particles were found to be mainly emitted by the passengers, especially by smokers. The particle content of recirculation air was lower than or equal to that of fresh air depending on recirculation filter efficiency. The concentration of *volatile organic compounds* (VOCs) was well below threshold values. The amount of bacteria exceeded reported concentrations within other indoor spaces. The detected species were mainly nonpathogenic.

Proper distribution of conditioned air is an important factor in the air quality of aircraft. Experimental investigations of airflow and temperature distributions for indoor environmental conditions can be expensive and numerical simulation of such flow conditions is a feasible alternative today. Blockage and heat loads from occupants in an enclosure could have a significant effect on air distribution. One *computational fluid dynamics* (CFD) model was developed for an aircraft cabin section by Singh et al. (2002) to determine the effects of occupant heat loads. In this model, conditioned air entered the cabin through a three-slot diffuser located on the ceiling, flowed along the aisle, and exited through a return located on the floor on both sides along the cabin. Good agreement between the numerical simulation results and experimental data were achieved.

Reportedly, 85 percent of newer airplanes have HEPA filters (GAO, 2004). Some documentation suggests that airline filters referred to as *HEPA-type filters* are actually 95 percent DSP (i.e., MERV 14) filters (Hunt et al., 1995; Pierce et al., 1999). MERV 14 filters, while being excellent filters for controlling fungal spores and most bacteria, cannot guarantee complete protection against viruses and smaller bacteria (Kowalski and Bahnfleth, 2002). Coupling a MERV 14 filter with a UVGI system would provide superior overall performance (Kowalski, 2003).

All of the debate on quantities of fresh air supplied to the cabin and VOCs or CO_2 levels seems to obscure the problem of aerobiological quality. Simply increasing outside air-exchange rates and adding HEPA filters may not be the best or most economical solution. The energy costs of using HEPA filters may not be justified when a simple combination of a MERV 13–15 filter and an URV 13-15 UVGI system will provide comparable results (Kowalski, 2003). Furthermore, applying building air quality criteria to a crowded enclosed environment like an airplane may be inappropriate and new standards for air quality are needed that account for both proximity and duration of exposure. If the problem is really one of providing clean air to each passenger, then the air distribution system needs to be reevaluated to ensure that each passenger actually receives a proper amount of fresh air. If satisfactory air distribution cannot be obtained, then perhaps localized air treatment, supplying disinfected air directly from the overhead gaspers, would be the next best alternative.

37.3 SPACECRAFT

Although spacecraft are among the cleanest indoor man-made environments, they are not free from microbial contamination. Microbes in spacecraft may come from the astronauts and from the materials they bring on board, including experimental apparatus. Spores have even been isolated from the exterior of spacecraft, including the Mars Odyssey, where they have survived in space (La Duc et al., 2004). Air cleaning inside spacecraft is usually accomplished with the use of HEPA filters and carbon adsorbers. Scrubbers for removing excess carbon dioxide are also used. The International Space Station (see Fig. 37.5) already

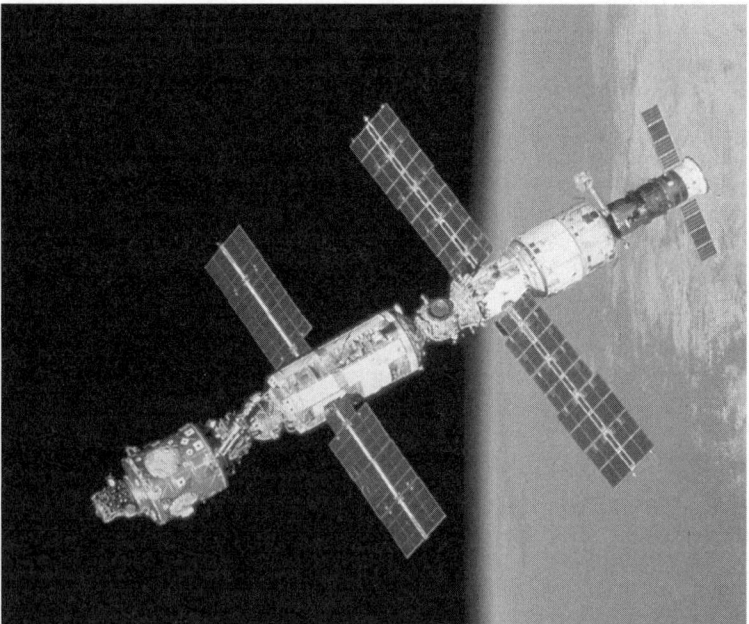

FIGURE 37.5 The International Space Station hosts a variety of inadvertent space-traveling microorganisms. (*Image provided courtesy of NASA.*)

hosts a variety of microorganisms that have been brought aboard from the astronauts and equipment, including experiments.

Table 37.3 lists the variety of airborne microorganisms that have been isolated on American and Soviet spacecraft and the number of times they have been detected (Horneck et al., 1985; Nicogossian et al., 1993, 1994; Johnson and Dietlin, 1977; Decelle and Taylor, 1976; Johnston et al., 1975; Gurovsky et al., 1973). The last column identifies the probable originating source, although some of the fungi may have come from experimental apparatus.

In a seminal incident during an Apollo mission, one crew member developed symptoms of the flu after launch. A few days later, a second member of the crew developed the same symptoms, but the third crewmember never developed the infection at all, despite breathing the same cabin air for weeks. NASA subsequently began quarantining crew members prior to spaceflights to ensure they would not carry any infections into space asymptomatically. This incident also highlights the fact that in spite of extended exposure, not everyone will succumb to the same infection.

Spacecraft typically use monitoring or quarantine to prevent astronauts from going into space with infections and use filtration to cleanse the air of allergens. Although air disinfection systems like UVGI could be used in combination with filtration aboard spacecraft, present methods are probably more than adequate. Curiously, UV radiation from sunlight is abundant in free space, as is cosmic radiation, and both are highly biocidal. Space disinfection systems could be developed for sterilizing internal air and compartments without having to import anything more than UV-transmittant quartz glass windows.

TABLE 37.3 Airborne Microorganisms Detected in Spacecraft

	Type	Apollo	Soviet	Shuttle	Source
Acinetobacter	Bacteria	>1	1	>1	Humans
Acremonium	Fungi			>1	Environmental
Alkaligenes	Bacteria		1		Humans
Alternaria	Fungi			>1	Environmental
Aspergillus	Fungi		11	>1	Environmental
Bacillus	Bacteria		9	>1	Environmental
Cladosporium	Fungi		2	>1	Environmental
Corynebacterium	Bacteria		7	>1	Humans
Curvularia	Fungi			>1	Environmental
Fusarium	Fungi		2		Environmental
Haemophilus influenzae	Bacteria	1			Humans
Haemophilus parainfluenzae	Bacteria	1			Humans
Klebsiella pneumoniae	Bacteria	>1	4		Humans
Moraxella	Bacteria	>1	2		Humans
Mucor	Fungi		1		Environmental
Mycoplasma	Bacteria	1			Humans
Neisseria	Bacteria		5		Humans
Penicillium	Fungi		13	>1	Environmental
Pseudomonas	Bacteria		5		Humans
Staphylococcus	Bacteria	>1	3	>1	Humans
Streptococcus	Bacteria		6	>1	Humans

Note: ">1" signifies the microbe was detected at least once.

37.4 SHIPS

Ships that include enclosed compartments range in size from small boats to cruise ships and ocean liners that are virtually the size of commercial office buildings. Larger ships are in many ways similar to large buildings except that they tend to be airtight. Ships are expected to meet the requirements of ASHRAE Standard 62 and these and other guidelines are available to assist ventilation system design and operation (ASHRAE, 2001, 1999b). The transmission of airborne diseases aboard ships is a recurring phenomenon, with some diseases appearing to be favorably transmitted onboard ships, such as Norwalk virus and Legionnaires' disease. In addition, industrial fishing ships may have their own unique occupational hazards.

The incidence of respiratory illness aboard ships in the military increases as ship size (see Fig. 37.6) decreased (Blood and Griffith, 1990). Outpatient data from ships of three different sizes (destroyers/frigates, cruisers, aircraft carriers) were surveyed and it was found that overall rates of illness were lower for the largest ships when contrasted with the smallest vessels for all operational theaters. These rate differences were particularly significant for the East Asia and Indian Ocean regions. Among major categories of disease, significantly higher rates aboard the small vessels were seen in at least two of the geographic regions for respiratory disorders and digestive diseases.

The cramped quarters and closed nature of military vessels apparently amplifies natural transmission rates for airborne and other diseases. Attack rates for upper respiratory diseases on troopships of World War II exceeded those attained experimentally by inoculation of volunteers (Gordon and Ingalls, 1957). In a study by Houk (1980) on a tuberculosis outbreak aboard a navy ship it was concluded that although proximity and direct contact played a role in transmission of the infection, airborne transmission of droplet nuclei, including via the ventilation system, was responsible for most of the secondary infections.

Diarrhea and respiratory diseases were common problems among ground troops deployed to the Middle East during Operation Desert Shield. An epidemiologic survey was conducted on a hospital ship by Paparello et al. (1993). An episode of acute diarrhea was reported by 46 percent of the surveyed population and 79 percent reported upper respiratory symptoms. Six percent of personnel were temporarily incapacitated due to gastrointestinal

FIGURE 37.6 In the U.S. Navy, higher rates of respiratory diseases occur onboard smaller ships than on larger ones. (*U.S. Navy photo by Photographer's Mate 1st Class David C. Lloyd.*)

symptoms and 7 percent due to respiratory symptoms. Officers were at increased risk of experiencing an episode of diarrhea, and female crew members more often reported respiratory complaints.

A study was conducted by Cross et al. (1992) to determine the risk of upper respiratory disease among U.S. Navy shipboard personnel on 10 ships between January and June 1989. Of 967 cases of URI, 64.4 percent occurred while at sea, with an average daily rate of 0.5 per 1000 crew members. Some 35.4 percent of the cases occurred while in port, with an average daily rate of 0.4 per 1000 crew members. There was an increase in URI rates after 9 days at sea or in port. The results of this study suggest that there are defined periods of increased transmission of upper respiratory infections aboard ships, both at sea and in port. Outbreaks of diseases like influenza can be explosive in the close quarters (Olson et al., 1979; Ksiazek et al., 1980).

In an extensive study of shipboard diseases, Bohnker et al. (2003) identified respiratory conditions as accounting for 0.65 visits per 100 person-weeks aboard ship. The mean rate of all medical problems for cruisers/destroyers/ frigates was not significantly different from aircraft carriers but was statistically different from supply ships and amphibious support ships. In a study by Blood (2000), hospitalizations aboard aircraft carriers were examined to ascertain differences in illness type attributable to theater of operations and combat deployment status. It was found that respiratory diseases were significantly higher for Vietnam combat support deployments than during subsequent peacetime deployments. During peacetime deployments, the Western Pacific deployments had higher percentages of respiratory disorders.

Aerosolized crab allergens are suspected causative agents for asthma among crab-processing workers. In a study by Beaudet et al. (2002) crab allergen concentrations and respiratory symptom prevalence among processing workers aboard crab-processing vessels were examined. Air samples were collected and analyzed along with questionnaires used to assess respiratory symptoms. Aerosolized crab allergen concentrations ranged from 79 to 21,093 ng/m^3. The highest concentrations were measured at workstations. A significant percentage of workers reported developing respiratory symptoms during the crab-processing season. Cough developed in 28 percent of workers, phlegm in 11 percent of workers, and wheeze and other asthma-like symptoms developed in 4 percent of workers. Despite variations in crab allergen levels, respiratory symptom prevalence was similar across all job categories. The high prevalence of reported respiratory symptoms across all job categories suggests that potential adverse respiratory effects result from long-term exposure to crab allergens.

A retrospective epidemiologic study was carried out by Peake et al. (1999) to evaluate diseases occurring on cruises originating in a calendar-year period from the United States. Approximately 3.6 percent of passengers reported illnesses during the cruises with females experiencing more problems than males. The most common diagnosis was respiratory tract infection, which occurred in 29.1 percent of ship patients. Of these, 11 percent of patients had a serious or potentially life-threatening diagnosis. The spectrum of diseases is similar in many respects to patients presenting to emergency departments.

Outbreaks of Legionnaires' disease aboard cruise ships and ocean liners present a public-health challenge since rapid detection of clusters of disease among travelers is difficult. Sometimes outbreaks are not recognized for months, since Legionnaires' disease can resemble other infections or can resemble pneumonia from other causes. In a series of outbreaks in which 50 passengers with Legionnaires' disease were identified from nine cruises embarking in 1994, exposure to whirlpool spas was strongly associated with disease and the risk of acquiring Legionnaires' disease increased by 64 percent for every hour spent in the spa water (Jernigan et al., 1996). Passengers spending time around the whirlpool spas, but not in the water, were also significantly more likely to have acquired infection. *Legionella pneumophila* was isolated from the sand filter in the ship's whirlpool spa. New strategies for whirlpool spa maintenance and decontamination, such as ozonation, chemical treatment, and UV disinfection, have been used successfully to minimize transmission of legionellae from these aerosol-producing devices.

In a study by Jensen (1996) on mortality patterns in Danish commercial fishermen between 1970 and 1985, the *standardized mortality ratio* (SMR) for all causes among crew members was found to be 1.50 and highest in the age group 20 to 34 years who had an SMR of 2.09. The SMR is the ratio of observed death rate to the normally expected death rate. The increased SMR among fishermen was primarily due to deaths by accident, ischemic heart disease, and unknown causes. The SMR due to bronchitis and emphysema among 35- to 64-year-old crew members was 1.96. This study confirms earlier findings of a high mortality among fishermen with a slightly increased risk of dying from cancer, respiratory and cardiovascular diseases.

A study by Medvedev et al. (1996) determined that tuberculosis morbidity among fishery workers in the Russian Far East remains high and that it is higher in fishermen working in the sea than in those who are engaged in fish factories. Pulmonary tuberculosis was typical for both groups, but the fishery workers developed destruction of pulmonary tissue and discharged bacteria significantly more frequently due to impaired cellular immunity, apparently due to some occupational hazard.

Seamen tend to experience higher death rates from cancer than land-based workers. Among 1922 deaths in the American merchant marine population who were patients in 1973 to 1978, 46 percent were cancer associated, a pattern which was contrary to that found among patients of acute general care hospitals nationwide for the same period (Kelman and Kavaler, 1990). Respiratory cancer amounted to 19.3 percent of the total, more than twice the number of such deaths among nonseamen patients. These excess cancer-associated deaths, particularly deaths from respiratory cancer, may indicate an occupational hazard.

It is not uncommon for outbreaks aboard ships to consist of multiple viral and bacterial infections, including diarrheal illness and influenza (Ruben and Ehreth, 2002). A large outbreak of respiratory and diarrheal illness took place over a period of 21 days in 1984 on board a cruise ship in southern Europe (Christenson et al., 1987). Some 86 percent of the 391 passengers interviewed were affected. Of the ill passengers, 88 percent had an influenza-like illness, including 20 with signs of lower respiratory tract infection. In 24 passengers viral infections with influenza B, influenza A, parainfluenza, respiratory syncytial virus, and Epstein-Barr virus infections were diagnosed. In two patients *Legionella* antibodies were discovered. The outbreak was thus evidently caused by multiple pathogens mainly affecting the respiratory tract. Although most of the passengers acquired their infections on board the ship, a common source was not discovered. A steep rise in the epidemic curve the day after the air conditioning was switched on, however, was mentioned as a possible precipitating factor of this incident of "sick boat syndrome."

Outbreaks of respiratory and gastrointestinal illnesses have been a recurring problem on cruise ships as the industry has expanded in recent years. During the summer of 2000 an outbreak of respiratory illnesses occurred on one such ship during a 12-day Baltic cruise from the United Kingdom to Germany via Russia (MMWR, 2001b). The ship carried 1311 passengers, primarily from the United States, and 506 crew members from many countries. Testing implicated an influenza B virus infection as the cause of the outbreak.

Norwalk viruses have recently caused numerous outbreaks of gastroenteritis on cruise ships. Also called *Norwalk-like viruses* (NLV) and noroviruses, these pathogens are responsible for 23 million cases of illness each year. Although no other gastrointestinal viruses are known to transport by the airborne route, there is reason to believe that airborne transport or transmission is a factor in Norwalk virus outbreaks. Fomites on ship surfaces are one etiologic pathway, and they can arrive via droplets. During 2002, cruise ships with foreign itineraries sailing into U.S. ports reported 21 outbreaks on 17 cruise ships (Cramer et al., 2003). Of the 21 outbreaks, nine were confirmed by laboratory analysis of stool specimens from affected persons to be associated with noroviruses, three were attributable to bacterial agents, and nine were of unknown etiology. Seven outbreaks were reported in 2001, and of these, four were confirmed to be associated with norovirus.

Norwalk virus and the related calicivirus called *snow mountain agent* (SMA) are major waterborne pathogens. Evidence from outbreaks suggests that these viruses have a very low infectious dose, perhaps only 10 virions. One such virus was implicated as the cause of recurrent outbreaks of gastroenteritis on a cruise ship (Ho et al., 1989). There was no identifiable relation to food or water consumption, but the risk of gastroenteritis among passengers who had shared toilet facilities was twice that of those who had a private bathroom. Contaminated bathrooms were suggested as an important vehicle for person-to-person spread. Patients who had vomited in their cabins were more likely to have infected cabinmates than were patients who had not vomited. These epidemiological findings implicate vomitus in the transmission of viral gastroenteritis and they are consistent with the transmission of viral agents by both airborne droplets and person-to-person contact.

In an explosive outbreak of gastroenteritis aboard a cruise ship in 1977 caused by a Norwalk-like virus some 64 percent of cruise ship passengers were infected (Gunn et al., 1980). Principal symptoms experienced by ill passengers were nausea, vomiting, and diarrhea. Fever was reported for approximately 25 percent of the patients. The outbreak was compatible with a common-source exposure, but no source was found, and person-to-person transmission was suspected. Passengers on the next four cruises suffered a similar gastrointestinal illness but the outbreaks affected fewer persons, and symptoms appeared milder on each subsequent cruise.

Infected crew members may also serve as a reservoir for secondary infections among passengers (Cramer et al., 2003). The incubation period of Norwalk virus is 12 to 48 hours and it lasts 12 to 60 hours. It is believed that Norwalk viruses can contaminate shellfish from raw sewage dumped overboard from recreational and commercial boats and that contaminated tropical waters may become aerosolized with the passage of other cruise ships. In theory, the virus may travel from a cruise ship to warm waters, and then to other cruise ships.

In an outbreak of influenza aboard a cruise ship from Hong Kong in 1987, 38 percent of passengers came down with acute respiratory illnesses (Berlingberg et al., 1988). Some 2.5 percent of passengers were diagnosed with pneumonia but no fatalities occurred. Some secondary infections were reported to occur in homes of infected travelers. Although a number of the infected patients had been vaccinated, the vaccine appeared to provide little protection. This study suggested that prolonged exposure in a partially closed setting overcame the level of immunity that would have been expected under normal conditions. It was also suggested that other respiratory pathogens circulating in the ship cabins might have caused some of the illnesses. Miller et al. (1998) reported on a similar outbreak of influenza A aboard a cruise ship.

Illnesses, allergenic reaction, discomfort, and nuisance may result from poor air quality and badly maintained ventilation systems aboard passenger ships. Research by Lloyd's Register on air quality and ventilation systems indicates a number of problems on board passenger ships and a poor appreciation of maintenance requirements and contamination control options (Webster, 1997). Operators need to be aware of the importance of proper filtration and ductwork cleanliness, as well as the benefits of regular indoor air quality surveys in order to assist early identification of potential problems.

Ship owners, operators, and captains need to be diligent in maintaining the biological cleanliness of their water and air-conditioning systems, and pay particular attention to whirlpool spas. In the period between 1977 and 1997, over 100 cases of Legionnaires' disease have been linked to ships, including 10 fatalities (Rowbotham, 1998). Most of the cases were associated with cruise ships, but a variety of other vessels were involved. Cases of infection appear to be less common among crew members than among passengers.

37.5 SUBMARINES

Submarines are essentially ships that have completely enclosed systems and that may run on recirculated and processed air for indefinite periods (Shadle and Daley, 1991). The close quarters, (see Fig. 37.7) necessitate the use of air cleaning systems. As a result of air filtration

FIGURE 37.7 Submarine crews have normal rates of onboard respiratory diseases, possibly because of advanced air cleaning systems. (*U.S. Navy Photo by Chief Photographer's Mate John E. Gay.*)

practices and the use of air cleaning systems like carbon adsorption and carbon dioxide scrubbers, respiratory problems aboard submarines appear to be less serious than those aboard open vessels.

The primary purpose of the ventilation system onboard US Navy submarines is to condition air and provide ventilation to maintain design conditions in all compartments. A second and equally important function of the system is to provide emergency ventilation of any compartment within the vessel (Shadle, 1992). The emergency ventilation mode is the means used for evacuation of specific compartments of atmospheric contaminants such as smoke, toxic gas, or other objectionable or noxious gases. In this mode, a supply of fresh air is provided to the affected compartment while exhaust air is routed through a second path.

In a study by Burr and Palinkas (1987) health risks associated with the U.S. Navy submarine duty were evaluated by comparing hospitalization rates of submariners with surface ship personnel. Submarine personnel did not have significantly higher hospitalization rates for any diagnostic categories nor for any of the submarine-associated illnesses. Submariners' relative risk of hospital admissions was slightly greater for a few selected diagnoses but differences were not statistically significant. Submarine personnel had lower hospitalization rates for nearly all of the diagnostic categories examined. Some of the reasons for these lower rates were believed to include the stringent screening of submariners and higher levels of education among submariners.

37.6 CARS, BUSES, AND TRAINS

Smaller enclosed vehicles like cars, trucks, buses, and trains are potential microenvironments within which airborne diseases can be transmitted and which may also support microbial growth. Design guidelines for ventilation of surface transportation vehicles are available from ASHRAE (1999c). Information of the aerobiology of vehicles is not available but it could be assumed that the primary health threat is exposure to other infected passengers.

The duration of time people spend inside such vehicles is often too short to foster airborne disease transmission, and yet because of close contact and surfaces that may become contaminated with fomites, such transmissions probably occur frequently. Such may be the case in elevators, where the time spent is too short to inhale an infectious dose of airborne microbes but where the elevator buttons may become contaminated with fomites.

Studies on disease transmission inside cars and trains are almost nonexistent but a few have been summarized here. Additional aerobiological problems exist due to biocontamination of such vehicles, especially the air conditioners, and the confounding factors of pollution that may exacerbate respiratory problems or even reduce immunity to infections.

In a study by Kumar et al. (1990), eight cars belonging to patients with symptoms of allergic rhinitis and bronchial asthma were examined. With the car air conditioners off, mold concentrations inside the passenger compartment were lower than concentrations in the outside air. After turning on the air conditioners, cultures revealed that mold concentrations in the passenger compartment decreased over time. It was found that placement of a filter at the portal of entry of outside air significantly reduced the mold concentration in the passenger compartment. Car air conditioners, especially in older cars, should be cleaned periodically to remove mold.

The effects of pollution may exacerbate respiratory problems and infections in drivers that are constantly exposed to auto exhaust and fuel vapors as an occupational hazard. In a study of the disease patterns among male professional drivers in Denmark, Hannerz and Tuchsen (2001) found that *standardized hospital admission ratios* (SHRs) for diseases in practically all systems and organs of the body were higher among professional drivers than they were in the male working population at large. Also drivers of passenger transport, compared with drivers of goods vehicles, had significantly high SHRs due to various diseases including respiratory infections.

Even though relationships between asthma and atmospheric pollution have been well documented, the relationship between pollution and allergic diseases has not yet been confirmed by epidemiological studies (Dutau and Charpin, 1998). Comparison of the frequency of asthma in urban and rural surroundings in general shows, in the urban environment, higher irritation symptoms but not asthma. Several recent investigations showed that respiratory symptoms related to the intensity of automobile traffic, but no relationship was shown with allergies. In the same way, no epidemiological investigation has shown a relationship between automobile pollution and prevalence of allergy.

Transmission of respiratory infection between passengers on long train or bus trips could be expected to be similar to those that occur on airplanes and ships. In 1996 a man who had traveled on two U.S. passenger trains for 29.1 hours and a bus for 5.5 hours was suspected of causing secondary infections in other passengers (Moore et al., 1999). Of the 240 persons who completed screening, 2 percent had a documented conversion. For two of the passengers no other risk factors for a TB conversion were identified other than exposure to the ill passenger during the trip. These findings support the idea of limited transmission of *Mycobacterium tuberculosis* from a potentially highly infectious passenger to other persons during extended train and bus travel.

An outbreak of tuberculosis on a commuter bus was reported by Yagi et al. (1999). The index patient was a 22-year-old woman who was an employee of an electronics company in which five more employees tested positive for exposure. These five employees worked separately from the index patient but rode the same commuter bus. The air conditioning on the bus used a closed recirculation system and it was believed inadequate ventilation in the bus contributed to the secondary tuberculosis infections.

Filtration has been suggested as one means of cleaning the air inside cars, trains, and other transportation vehicles but it is not commonly used. At least one UVGI system has been developed specifically to combat biological terrorist attacks that could be installed in trains and aircraft and that could help control common infections (Lee, 2002).

REFERENCES

AFA (2003). "Flight attendants demand protection from SARS." *Press Release, Association of Flight Attendants* 3(April):3.

AMA(CSA) (1998). "Airborne infections on commercial flights." *Report 10 of the Council on Scientific Affairs (A-98)* American Medical Association Council on Scientific Affairs, Washington, DC.

ASHRAE (1999a). Chapter 9: Aircraft. *HVAC Applications*. American Society of Heating, Refrigerating and Air-Conditioning Engineers, Atlanta, GA.

ASHRAE (1999b). Chapter 10: Ships. *HVAC Applications*. American Society of Heating, Refrigerating and Air-Conditioning Engineers, Atlanta, GA.

ASHRAE (1999c). Chapter 8: Surface transportation. *HVAC Applications*. American Society of Heating, Refrigerating and Air-Conditioning Engineers, Atlanta, GA.

ASHRAE (2001). "Standard 62-2001: Ventilation for acceptable indoor air quality." American Society of Heating Refrigerating and Air-Conditioning Engineers, Atlanta, GA.

AT (2000). "Cabin safety: Contaminated cabins can spread disease." Aviation Today. http://www.aviationtoday.com/reports/disease.htm.

Barthorpe, F. (1999). "Air circulation in aeroplanes." *Environ Eng* 12(3):12–14.

Beaudet, N., Brodkin, C. A., Stover, B., Daroowalla, F., Flack, J., and Doherty, D. (2002). "Crab allergen exposures aboard five crab-processing vessels." *AIHA J* 63(5):605–609.

Bergau, L. (1999). "Radiation exposure and air quality aboard commercial airplanes." *Z Arztl Fortbild Qualitatssich* 93(7):491–494.

Berlingberg, C. D., Kahn, F. H., Chun, L. Y., Cruz, J. M., Gellert, G., Giles, M. P., Mascola, L., Sorvillo, F. J., Tormey, M., Waterman, S. H., Murray, R. A., and Acree, K. H. (1988). "Acute respiratory illness among cruise ship passengers—Asia." *MMWR* 259(9):13051306.

Blood, C. G., and Griffith, D. K. (1990). "Ship size as a factor in illness incidence among U.S. Navy vessels." *Mil Med* 155(7):310–314.

Blood, C. G. (2000). "Shipboard medical admissions during peacetime and combat support deployments." *Mil Med* 165(3):228–236.

Bohnker, B. K., Sherman, S. S., and McGinnis, J. A. (2003). "Disease and nonbattle injury patterns: afloat data from the U.S. Fifth Fleet (2000–2001)." *Mil Med* 168(2):131–134.

Burr, R. G., and Palinkas, L. A. (1987). "Health risks among submarine personnel in the U.S. Navy, 1974–1979." *Undersea Biomed Res* 14(6):535–544.

Christenson, B., Lidin-Janson, G., and Kallings, I. (1987). "Outbreak of respiratory illness on board a ship cruising to ports in southern Europe and northern Africa." *J Infect* 14(3):247–254.

Clayton, A. J., O'Connell, D. C., Gaunt, R. A., and Clarke, R. E. (1976). "Study of the microbiological environment within long- and medium-range Canadian Forces aircraft." *Aviat Space Environ Med* 47(5):471–482.

Cramer, E. H., Forney, D., Dannenberg, A. L., Widdowson, M. A., Bresee, J. S., Monroe, S., Beard, R. S., White, H., Bulens, S., Mintz, E., Stover, C., Mullins, J., Wright, J., Hsu, V., Chege, W., and Varma, J. (2003). "Outbreaks of Gastroenteritis associated with Noroviruses on cruise ships—United States, 2002." *MMWR* 289(2):167–169.

Cross, E. R., Hermansen, L. A., Pugh, W. M., White, M. R., Hayes, C., and Hyams, K. C. (1992). "Upper respiratory disease in deployed U.S. Navy shipboard personnel." *Mil Med* 157(12):649–651.

Decelle, J. G., and Taylor, G. R. (1976). "Autoflora in the upper respiratory tract of Apollo astronauts." *Appl Environ Microbiol* 32(5):659–665.

Dechow, M., Sohn, H., and Steinhanses, J. (1997). "Concentrations of selected contaminants in cabin air of airbus aircrafts." *Chemosphere* 35(1–2):21–31.

Dutau, H., and Charpin, D. (1998). "Pollution and allergy: the epidemiological data." *Allerg Immunol (Paris)* 30(10):329–336.

GAO (2004). "More Research Needed on the Effects of Air Quality on Airliner Cabin Occupants." *GAO-05-54*. General Accounting Office. Washington, DC.

Gordon, J. E., and Ingalls, T. H. (1957). "Preventive medicine and epidemiology." *Prog Med Science* 233:334–357.

Gunn, R. A., Terranova, W. A., Greenberg, H. B., Yashuk, J., Gary, G. W., Wells, J. G., Taylor, P. R., and Feldman, R. A. (1980). "Norwalk virus gastroenteritis aboard a cruise ship: an outbreak on five consecutive cruises." *Am J Epidemiol* 112(6):820–827.

Gurovsky, N. N., Gazenko, O. G., Rudnyi, N. M., Lebedev, A. A., and Egorov, A. D. (1973). "Some results of medical investigations performed during the flight of the research orbital station Salyut." *Life Sci Space Res* 11:77–88.

Hannerz, H., and Tuchsen, F. (2001). "Hospital admissions among male drivers in Denmark." *Occup Environ Med* 58(4):253–260.

Ho, M. S., Glass, R. I., Monroe, S. S., Madore, H. P., Stine, S., Pinsky, P. F., Cubitt, D., Ashley, C., and Caul, E. O. (1989). "Viral gastroenteritis aboard a cruise ship." *Lancet* 2(8669):961–965.

Hocking, M. (2001). *Air Quality of Aircraft Passenger Cabins*. The Aviation Health Institute, Oxford, UK.

Hocking, M. B. (2002). "Trends in cabin air quality of commercial aircraft: industry and passenger perspectives." *Rev Environ Health* 17(3):248.

Horneck, G., Bucker, H., and Reitz, G. (1985). Bacillus subtilis spores on Spacelab I: Response to solar UV radiation in free space. *Fundamental and Applied Aspects of Bacterial Spores*, G. J. Dring, D. J. Ellar, and G. W. Gould, eds., Academic Press, London. 241–249.

Houk, V. N. (1980). Spread of tuberculosis via recirculated air in a naval vessel: The Byrd study. *Airborne Contagion, Ann N Y Acad Sci,* R. B. Kundsin, ed., NYAS, New York. 10–24.

Hunt, E. H., Reid, D. H., Space, D. R., and Tilton, F. E. (1995). "Commercial Airliner Environmental Control System." *Aerospace Medical Association Annual Meeting,* Anaheim, CA

Jensen, O. C. (1996). "Mortality in Danish fishermen." *Bull Inst Marit Trop Med Gdynia* 47(1–4):5–10.

Jernigan, D. B., Hofmann, J., Cetron, M. S., Genese, C. A., Nuorti, J. P., Fields, B. S., Benson, R. F., Carter, R. J., Edelstein, P. H., Guerrero, I. C., Paul, S. M., Lipman, H. B., and Breiman, R. (1996). "Outbreak of Legionnaires' disease among cruise ship passengers exposed to a contaminated whirlpool spa." *Lancet* 347(9000):494–499.

Johnson, R. S., and Dietlin, L. F. (1977). *Biomedical Results from Skylab*. NASA Scientific and Technical Information Office, Washington, DC.

Johnston, R. S., Dietlin, L. F., and Berry, C. A. (1975). *Biomedical Results of Apollo*. US Government Printing Office, Washington, DC.

Kelman, H. R., and Kavaler, F. (1990). "Mortality patterns of American merchant seamen 1973-1978." *Am J Ind Med* 17(4):423–433.

Kenyon, T. A., Valway, S. E., Ihle, W. W., Onorato, I. M., and Castro, K. G. (1996). "Transmission of multidrug-resistant *Mycobacterium tuberculosis* during a long airplane flight." *New Engl J Med* 334(15):933–945.

Klontz, K. C., Hynes, N. A., Gunn, R. A., Wilder, M. H., Harmon, M. W., and Kendal, A. P. (1989). "An outbreak of influenza A/Taiwan/1/86 (H1N1) infections at a naval base and its association with airplane travel." *Am J Epidemiol* 129(2):341–348.

Kowalski, W. J., and Bahnfleth, W. P. (2002). "MERV filter models for aerobiological applications." *Air Media* Summer:13–17.

Kowalski, W. J. (2003). *Immune Building Systems Technology*. McGraw-Hill, New York.

Ksiazek, T. G., Olson, J. G., Irving, G. S., Settle, C. S., White, R., and Petrusso, R. (1980). "An influenza outbreak due to a/USSR/77-like virus aboard a US Navy ship." *Am J Epidem* 112(4):487–494.

Kumar, P., Lopez, M., Fan, W., Cambre, K., and Elston, R. C. (1990). "Mold contamination of automobile air conditioner systems." *Ann Allergy* 64(2 Pt. 1):174–177.

La Duc, M. T., Satomi, M., and Venkateswaran, K. (2004). "*Bacillus odysseyi* sp. nov., a round-spore-forming bacillus isolated from the Mars Odyssey spacecraft." *Int J Syst Evol Microbiol* 54(2004):195–201.

Lee, S.-C., Poon, C.-S., Li, X.-D., and Luk, F. (1998). "Indoor air quality investigation on commercial aircraft." *Proceedings of the Air & Waste Management Association's Annual Meeting & Exhibition* 98-TP48.05:15, San Diego, CA.

Lee, S.-C., Poon, C.-S., Li, X.-D., and Luk, F. (1999). "Indoor air quality investigation on commercial aircraft." *Indoor Air* 9(3):180.

Lee, A. (2002). "Protecting commuters." *Engineer* 291(7616):8.

Lidwell, O. M., and Williams, R. E. O. (1961). "The epidemiology of the common cold." *J Hyg* 59:309–334.

Masterson, R. G., and Green, A. D. (1991). Dissemination of human pathogens by airline travel. *Pathogens in the Environment*, B. Austin, ed., Blackwell Scientific Publications, Oxford, UK.

Medvedev, V. I., Volobueva, E. M., Paukova, T. A., and Gordievskaia, N. V. (1996). "Role of occupational factors in the incidence and course of respiratory system tuberculosis in fishery workers in the South Far East." *Probl Tuberk* 2:6–7.

Miller, J., Tam, T., Afif, C., Maloney, S., Cetron, M., Fukata, K., Klimov, A., Hall, H., Kertesz, D., and Hockin, J. (1998). "Influenza A outbreak on a cruise ship." *Can Commun Dis Rep* 24(2):9–11.

Miro, C. R., and Cox, J. E. (2000). "Evaluating Aircraft Air Cabin Quality." *ASHRAE J* 42(9):12.

Mixeu, M. A., Vespa, G. N., Forleo-Neto, E., Toniolo-Neto, J., and Alves, P. M. (2002). "Impact of influenza vaccination on civilian aircrew illness and absenteeism." *Aviat Space Environ Med* 73(9):876–880.

MMWR (2001a). "Exposure to patients with meningococcal disease on aircrafts—United States, 1999–2001." *Morb Mortal Wkly* 50(23):485–489.

MMWR (2001b). "Influenza B virus outbreak on a cruise ship—Northern Europe, 2000." *Morb Mortal Wkly Rep* 50(8):137–140.

Moore, M., Valway, S. E., Ihle, W., and Onorato, I. M. (1999). "A train passenger with pulmonary tuberculosis: evidence of limited transmission during travel." *Clin Infect Dis* 28(1):57–58.

Moore, T. (2000). "Cabin safety: Contaminated cabins can spread disease." *Washingtonian* 35:5.

Moser, M. R., Bender, T. R., Margolis, H. S., Noble, G. R., Kendal, A. P., and Ritter, D. G. (1979). "An outbreak of influenza aboard a commercial airliner." *Am J Epidemiol* 110(1):1–6.

Nagda, N. L., Koontz, M. D., Konheim, A. G., and Hammond, S. K. (1992). "Measurement of cabin air quality aboard commercial airliners." *Atmos Environ* 26A(12):2203–2210.

Nagda, N. L., and Hodgson, M. (2001). "Low relative humidity and aircraft cabin air quality." *Indoor Air* 11(3):200–214.

Navigant (2003). "Meningitis risk on aircraft." http://www.navigant.com/travelResources/health/Menigitis.asp.

Nicogossian, A. E., Mohler, S. R., Gazenko, O. G., and Grigoryev, A. I. (1993). *Space Biology and Medicine*. AIAA, Washington, DC.

Nicogossian, A. E., Huntoon, C. L., and Pool, S. L. (1994). *Space Biology and Medicine*. Lea & Febiger, Philadelphia, PA.

NRC (2002). *The Airliner Cabin Environment and the Health of Passengers and Crew*. N. R. Council National Academy Press, Washington, DC.

Olander, L., and Westlin, A. (1991). "Air flow in aircraft cabins." *Staub—Reinhaltung der Luft* 51(7–8):283–288.

Olson, J. G., Irving, G. S., Ksiazek, T. G., and Rendin, R. W. (1979). "An explosive outbreak of influenza caused by A/USSR/77-like virus on a United States naval ship." *Mil Med* 144(11):743–745.

Paparello, S. F., Garst, P., Bourgeois, A. L., and Hyams, K. C. (1993). "Diarrheal and respiratory disease aboard the hospital ship, USNS-Mercy T-AH 19, during Operation Desert Shield." *Mil Med* 158(6):392–395.

Parmet, A. J. (1999). "Tuberculosis on the flight deck." *Aviat Space Environ Med* 70(8):817–818.

Peake, D. E., Gray, C. L., Ludwig, M. R., and Hill, C. D. (1999). "Descriptive epidemiology of injury and illness among cruise ship passengers." *Ann Emerg Med* 33(1):67–72.

Pierce, W. M., Janczewski, J. N., Roethlisberger, B., and Janczewski, M. G. (1999). "Air quality on commercial aircraft." *ASHRAE J* 41(9):8.

Rayman, R. B. (2002). "Cabin air quality: an overview." *Aviat Space Environ Med* 73(3):211–215.

Rowbotham, T. J. (1998). "Legionellosis associated with ships: 1977 to 1997." *Commun Dis Public Health* 1(3):146–151.

Ruben, F. L., and Ehreth, J. (2002). "Maritime health: a case for preventing influenza on the high seas." *Int Marit Health* 53(1–4):36–42.

Sato, K., Morishita, T., Nobusawa, E., Suzuki, Y., Miyazaki, Y., Fukui, Y., Suzuki, S., and Nakajima, K. (2000). "Surveillance of influenza viruses isolated from travellers at Nagoya International Airport." *Epidemiol Infect* 124(3):507–514.

Shadle, T., and Daley, T. (1991). "U.S. navy submarine life support systems." *SAE Technical Paper Series* 911329:1–10.

Shadle, T. (1992). "U.S. navy submarine normal and emergency ventilation systems." *SAE Technical Paper Series* 921413:1–8.

Singh, A., Hosni, M. H., Horstman, R. H., VanGilder, J., and May, R. (2002). "Numerical simulation of airflow in an aircraft cabin section." *ASHRAE Trans* 108(Pt. 1):1005–1013.

Ungs, T. J., and Sangal, S. P. (1990). "Flight crews with upper respiratory tract infections: epidemiology and failure to seek aeromedical attention." *Aviat Space Environ Med* 61:938–941.

vanNetten, C. (2002). "Analysis and implications of aircraft disinfectants." *Sci Total Environ* 293(1–3):257–262.

Webster, A. D. (1997). "Contribution of ventilation system design and maintenance to air quality on passenger ships." *Trans—Ins Marine Eng* 109(Pt. 2):145–159.

Weisel, C. P. (2001). Chapter 68: Transportation. *Indoor Air Quality Handbook,* J. D. Spengler, J. M. Samet, and J. F. McCarthy, eds., McGraw-Hill, New York. 68.1–68.2.

Wenzel, R. P. (1996). "Editorial: Airline travel and infection." *New Engl J Med* 334(15):981–982.

Whelan, E. A., Lawson, C. C., Grajewski, B., Petersen, M. R., Pinkerton, L. E., Ward, E. M., and Schnorr, T. M. (2003). "Prevalence of respiratory symptoms among female flight attendants and teachers." *Occup Environ Med* 62:929–934.

Wilder-Smith, A., Paton, N. I., and Goh, K. T. (2003). "Low risk of transmission of severe acute respiratory syndrome on airplanes: the Singapore experience." *Trop Med Int Health* 8(11):1035–1037.

Yagi, T., Sasaki, Y., Yamagishi, F., Mizutani, F., Wada, A., and Kuroda, F. (1999). "Tuberculosis microepidemic in a commuter bus." *Kekkaku* 74(6):507–511.

Zeterberg, J. M. (1973). "A review of respiratory virology and the spread of virulent and possibly antigenic viruses via air conditioning systems." *Ann Allergy* 31:228–299.

Zitter, J. N., Mazonson, P. D., Miller, D. P., Hulley, S. B., and Balmes, J. R. (2002). "Aircraft cabin air recirculation and symptoms of the common cold." *JAMA* 288(23):2972–2973.

CHAPTER 38
INDUSTRIAL FACILITIES

38.1 INTRODUCTION

The wide variety of industrial facilities precludes any generalization of the types of respiratory diseases to which workers may be subject but the engineering solutions are, in general, the same as for other facilities. Industrial facilities that deal with microbes or other organic materials are associated with greater microbiological hazards. Industrial facilities that process or produce chemicals, minerals, equipment, and other nonorganic products usually have much greater occupational hazards due to pollution and exposure to chemical agents than to pathogens or allergens, but there are strong associations between exposure to pollutants and respiratory infections. This chapter reviews the occupational aerobiological hazards of various industrial environments. The effects of pollutants are addressed only insofar as they relate to exacerbating microbial respiratory illnesses, but the etiologies are complex and often there is difficulty in distinguishing respiratory problems due to pollutants and those due to bioaerosols. Agricultural industries, food industries, and sewage and waste treatment facilities have been addressed in other chapters and are specifically excluded from the following sections except as they relate to the subject matter.

38.2 OCCUPATIONAL AIRBORNE DISEASES

There are many types of occupational diseases but the ones of primary concern here are those with an aerobiological etiology. These include mainly respiratory diseases but can also include skin, eye, ear, and other infections that result from airborne pathogens or allergens. The chain of causation of occupational respiratory diseases is complicated by the fact that pollution appears to play a major role in both asthma and allergic reactions and respiratory irritation. It is not always possible to separate the effects of chemical pollutants from those of aerobiological pollutants when it comes to respiratory symptoms. Asthma is itself a complex disease that is not completely understood at present. The evidence suggests that a predisposition to asthma can become manifest under exposure to either chemical or biological pollutants of the air. Once manifest, asthma may be further triggered by a variety of causes, not all of which are aerobiological. The following review of occupational airborne diseases describes the complex interrelationships between asthma and other pollutants as well as the basic types of respiratory diseases that can occur.

Figure 38.1 illustrates the complex relationship between the basic types of diseases that may occur in occupational settings. Pollution from nonmicrobial sources may cause all sorts of respiratory problems, including cancer, emphysema, black lung disease, silicosis, asbestosis, and the like. Infections due to airborne pathogens are a distinct class of diseases except that they may trigger asthma and the existence of pollution (or pollution-induced respiratory disease) may exacerbate pathogenic infections. Asthma may be induced by pollution, may be preexisting in workers, or may be induced by allergies or other factors. Allergies may be due to biological sources or pollution. Once allergies have developed, they may be exacerbated by pollution, or may trigger asthmatic reactions. Allergic reactions are not necessarily allergies, but may include all types of respiratory irritation, including chronic respiratory problems. Allergic reactions may occur in individuals who are not atopic due to prolonged or heavy exposure to almost any kind of biological or chemical pollutant. Chronic exposure to almost any such pollutants may ultimately produce an allergy, or even asthma.

The actual interactions between the types of diseases and the specific breakdown of diseases, as shown in Fig. 38.1, may be much more complex than described here, but this basic framework should assist in the understanding of the subsequent descriptions of specific occupational diseases.

Occupational airway diseases include occupational rhinitis, industrial bronchitis, reactive airway dysfunction syndrome, bronchiolitis obliterans, and occupational asthma (Schachter et al., 2001). Rhinitis is the condition often associated with hay fever, which is characterized by inflammation of the nasal mucous, congestion, sneezing, and watery discharge from the nose. There are two types of occupational rhinitis, allergic rhinitis or nonallergic rhinitis, due to irritation (Drake-Lee et al., 2002). Occupational rhinitis can be likened to occupational asthma. Irritation causes symptoms during exposure that cease afterward unless clinically obvious damage has occurred. Occupational rhinitis has been a recognized industrial disease since 1907 but it has only relatively recently received significant attention by occupational and industrial health physicians (Welch et al., 1995). At the present time the precise mechanisms of pathogenesis are unclear and would appear to be multiple. There is difficulty in differentiating between occupational and nonoccupational rhinitis.

Occupational rhinitis is a common but generally underreported entity. It is frequently but not always associated with occupational asthma. Occupational asthma may have several

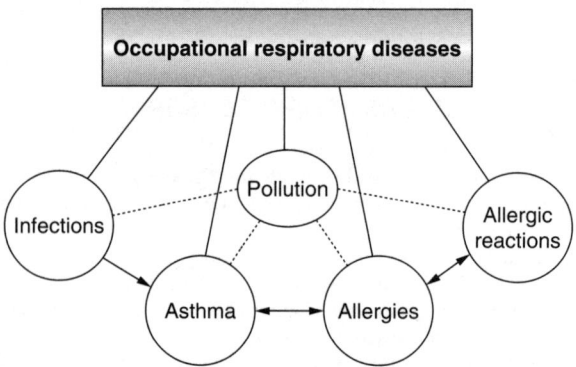

FIGURE 38.1 Types of occupational respiratory diseases and their relative interactions with each other and with pollution.

different presentations and it is difficult to distinguish it from asthma induced at home. The respiratory tract acts as a common pathway for all inhaled environmental pollutants, whether encountered in the home or at work. More than 200 chemicals have been incriminated as a cause of work-related asthma and about 2 percent of the 10 million Americans who have asthma acquired it as a result of some chemical irritant or immunogen in their work environment (Bardana, 1995). A number of predisposing factors facilitate the development of work-related asthma that include industrial conditions, climatic factors, atopic predisposition, smoking, viral infection, nonspecific bronchial hyperreactivity, and other factors. Occupational asthma may be immunologic or nonimmunologic in nature where the immunologic variants involve sensitization to a variety of large-molecular-weight constituents. The nonimmunologic variant is referred to as inflammatory bronchoconstriction or reactive airways dysfunction syndrome. The most common presentation of occupational asthma involves workers with preexisting asthma who have been adversely affected by work exposures. Occupational asthma is clearly not a single, simple homogeneous entity, even when a single specific cause can be identified in the workplace.

Zock et al. (2001) studied the relationship between occupational exposures, chronic bronchitis, and lung function in a general population survey in 14 industrialized countries. An increased risk for chronic bronchitis was found in agricultural, textile, paper, wood, chemical, and food processing workers, being more pronounced in smokers. Findings were similar for asthmatic and nonasthmatic subjects. Occupational exposures contributed to the occurrence of chronic (industrial) bronchitis in young adults.

Zhestkov (2000) examined patients with different forms of dust-induced lung disease. Impaired immune system function and nonspecific resistance in dust-induced lung disease were found to depend on the type of disease and the subject's predisposition to infections. The workers' risk for any infection appears to be largely defined by their inherent immunity and their natural resistance under exposure to high concentrations of industrial aerosols.

Workers often live in industrial environments that may be polluted. An epidemiological study by Heinrich et al. (1999) found that children residing in a polluted environment have about a 50 percent increased lifetime prevalence for physician-diagnosed allergies, eczema, and bronchitis compared to children who do not. Children in polluted environments have about twice the number of respiratory symptoms such as wheeze, shortness of breath, and cough without cold, as well as increased rates of asthma. Industrial pollution is associated with a higher lifetime prevalence of respiratory disorders and an increased rate of allergic sensitization in children.

Airborne particulate dust in industrial environments may contain chemicals, microbes, biological compounds, or inert materials. Particulate agents may be sensitizers—agents that sensitize individuals to other disease-causing factors via their allergen content. Allergen-bearing airborne particulates are causative agents of lung inflammation via their immunotoxic properties and as inducers of inflammatory alveolitis (Salvaggio, 1994).

The aerobiological conditions inside modern industrial environments are in a state of constant flux due to both improvements in air quality and the introduction of new threats. The risk of humans acquiring pneumonia as a result of their occupation appears to have declined during the twentieth century in developed countries (Esposito, 1992). Some conditions that are traditionally associated with the workplace, such as woolsorters' disease, now have historical interest only. The problem of occupation-associated pneumonia remains substantial. Large outbreaks of zoonotic infections continue to occur, especially psittacosis among poultry farmers and abattoir workers. Clusters of illnesses caused by recently recognized pathogens are now being reported, including Legionnaires' disease in industrial workers and *Chlamydia pneumoniae* infections in military personnel. Epidemics of old diseases are also appearing in new settings, such as tuberculosis among nursing home workers.

Industrial bronchitis is defined as a chronic bronchitis due to prolonged exposure to polluted air at the professional workplace (Thorens, 1991). Up until recently its existence was open to controversy because outside factors, such as tobacco, were considered to be predominant causes. However, recent epidemiological studies have shown that exposure to dust, in conjunction with fumes and gases, is responsible for chronic obstructive syndromes.

Data from a French study were used to assess the effect of moderate occupational exposure to dust, gases, or chemical fumes on the prevalence of respiratory symptoms and ventilatory function (Krzyzanowski and Kauffmann, 1988). In this community-based population, without houscholds headed by manual workers, 34 percent of the men and 23 percent of the women, 25 to 59 years of age, occupationally active, reported some exposure. For men and women, some 50 percent increase in chronic cough, chronic bronchitis, dyspnoea, and wheezing prevalence was observed in the exposed group compared to the never exposed, with the strongest association for wheezing. Results suggest that occupational exposures of relatively low intensity, encountered in nonindustrial workplaces, may constitute a risk for respiratory health.

Although occupational asthma is the most commonly recognized occupational lung disease, the condition remains underreported. As many as 20 percent of adult asthma cases are related to factors at the place of employment (Arnaiz and Kaufman, 2002). Occupational asthma is the most prevalent occupational lung disease in industrialized countries, accounting for approximately 5 percent of asthma in adults (Galdi and Moscato, 2002). The most important remedial measure is prevention at the workplace by completely eliminating the causative substances, or, failing that, reducing exposure to the lowest practicable levels. This goal may be achieved in a variety of ways, including automation of processes, improvement of ventilation, modification of the process or agents, and use of personal protection devices. However, the control of environmental exposure presents several difficulties, including the fact that exposure levels that induce respiratory sensitization to other agents are not completely defined.

A diversity of airborne dusts, gases, fumes, and vapors can cause dose-related symptoms in individuals exposed in the workplace. Occupational asthma occurs in 2 to 6 percent of the asthmatic population (Bardana, 2003). Respiratory infection can be a predisposing factor for occupational asthma. New-onset occupational asthma may be immunologic or nonimmunologic in origin and symptoms may ensue after a latent period of months to years. Nonimmunologic occupational asthma can be precipitated by a brief, high-level exposure to a potent irritant. Symptoms occur immediately or within a few hours of the exposure. Prevention is the best therapeutic intervention.

Data collected on occupational illnesses from annual labor statistics, workers' compensation reports, and physician reports, were studied by Morse et al. (2002). Manufacturing, state employment, and municipalities tended to have the highest number of cases and rates. Acute respiratory conditions were the most common lung condition, followed by occupational asthma and asbestos-related conditions. Occupational illnesses appear to be increasing.

Workers who spend time in cold workrooms or outdoors in cold weather may be subject to higher rates of respiratory problems. Breathing cold air can cause bronchospasm and increases respiratory tract secretions in asthmatic patients as well as in normal individuals. In a 1-year study on breathing in cold environments, Jammes et al. (2002) examined individuals who spent 6 hours a day in cold stores (3°C to 10°C) and spent approximately 25 percent of that time at 3°C. Compared with a control group of six subjects engaged at the same time but who did not work in cold stores, six of 11 individuals who worked 12 months in a cold environment experienced increased symptoms of rhinitis, sore throat, and cough. The report concluded that 1 year of daily exposure to a cold occupational environment elicits a modest but significant airflow limitation, accompanied by bronchial hyperresponsiveness, with the effects beginning within 6 months of exposure.

About one quarter of working adults with asthma often find their work environment makes their asthma worse and experienced workplace exacerbation of asthma (Henneberger et al., 2002). Percentages of workers with workplace exacerbation of asthma were highest for mining and construction (36 percent), wholesale and retail trade (33 percent), and public administration (33 percent), and lowest for educational services (22 percent), finance, insurance, and real estate (22 percent), and nonmedical and noneducational services (18 percent). Results are summarized graphically in Fig. 38.2.

Animal laboratory workers (treated in Chap. 32) are not the only people who may be exposed to animals in the course of their work. Workers who have frequent or occasional contact with rodents as part of their occupation, such as trash collectors, sewer workers, and workers who clean or enter rarely used rodent infested structures, may be at increased risk of exposure to rodent-borne viruses such as hantavirus (Armstrong et al., 1995). One employee of a California utility company who was apparently occupationally exposed to Sin Nombre virus was presented by Jay et al. (1996). Employees of utility companies and similar industries may be an important risk group in areas where hantavirus is endemic and emphasizes the need for occupational safety protocols for new diseases. In a survey of 81 persons with possible occupational exposure to rodents, 72 reported handling rodents as part of their job (Fritz et al., 2002).

In a study by Hnizdo et al. (2002), it was found that occupations associated with increased risk for chronic obstructive pulmonary disease (COPD) included freight, stock, and material handlers, records processing and distribution clerks, sales, transportation-related occupations, machine operators, construction trades, and waitresses. The fraction of COPD attributable to work was estimated as 19.2 percent overall and 31.1 percent among nonsmokers.

Although studies suggest that allergic rhinitis impairs workers, national estimates of the magnitude of this impairment vary widely. A study by Kessler et al. (2001) indicated that pollen and mold counts are significantly related to work impairments only among respondents with allergic rhinitis, with annual work-impairment costs of $5.4 to $7.7 billion. Successful treatment of allergies or other remedial measures to improve the work environment can obviously result in considerable savings on a national scale.

In certain occupations respiratory diseases can have a more significant impact on lost work time. Based on a study of the national sick-leave register in Sweden, respiratory diseases accounted for 4.4 percent of the total number of sick leaves (Nathell et al., 2000). The incidence of long-term sick leave due to respiratory diseases was three times higher in occupations with a

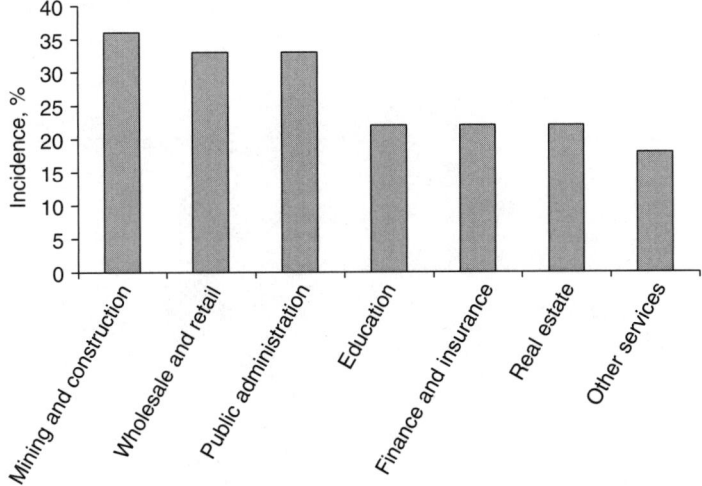

FIGURE 38.2 Incidence of exacerbated asthma in the workplace by profession. [*Based on data from Henneberger et al. (2002).*]

high incidence of respiratory diseases than in those with a low incidence. Agricultural workers had a 46 percent higher proportion of long-term respiratory diseases than metal workers. Industrial workers, food industry workers, and painters had an increased risk.

Karjalainen et al. (2000) found that occupational asthma in Finland had a mean annual incidence rate of 17.4 cases per 100,000 workers. Incidence was highest in bakers, painters and lacquerers, veterinary surgeons, chemical workers, farmers, animal husbandry workers, other food industry workers, welders, plastic product workers, butchers, sausage makers, and floor layers. Cases caused by animal epithelia, hairs, and secretions or flours, grains, and fodders accounted for 60 percent of the total.

Meyer et al. (1999) surveyed reports from physicians and found that occupational asthma was the most-reported respiratory condition, with 27 percent of total cases with mesothelioma at 23 percent, benign pleural disease at 21 percent, pneumoconiosis at 7 percent, and inhalation injuries at 6 percent. Of these diseases, only occupational asthma and pleural disease may have aerobiological etiology. Figure 38.3 illustrates these proportions. The most commonly identified agents causing asthma were enzymes, isocyanates, laboratory animals and insects, colophony and fluxes, flour, latex, and glutaraldehyde.

Only a few threshold limit values (TLVs) and occupational exposure limits (OELs) exist at present for allergens in the workplace known to cause bronchial asthma. Recently published studies provide evidence of dose-response relationships for occupational allergens of plant, microbiological, animal, or man-made origin. Allergen exposure levels below some TLV are not associated with an increased risk of occupational asthma. Corresponding TLV and OEL data suggested by various sources are summarized in Table 38.1. Such limits may be useful for establishing the performance of air cleaning systems. If no reliable data on the health risk of an occupational airborne hazard exist, the lowest reasonably practicable exposure level should be sought. Up to 15 percent of all asthma cases are of occupational origin or have at least a significant causal occupational factor. The main asthma-inducing agents in the workplace are flour, grain and feed dust, animal dander/urinary proteins, and isocyanates. In addition to chronic exposures, singular accidental exposure to high concentrations of irritative or toxic airborne substances can cause occupational asthma. This condition is frequently called reactive airways dysfunction.

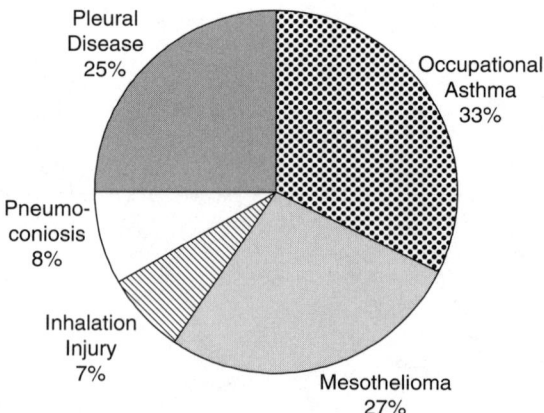

FIGURE 38.3 Breakdown of occupational respiratory diseases in England. [*Based on data from Meyer et al. (1999).*]

TABLE 38. 1 Some Suggested Limits for Bioaerosols

Allergen	TLV/OEL	Reference
Endotoxin	20 ng/m^3	Alwis et al., 1999
Endotoxin	50 EU/m^3	Dutkiewicz et al., 2001
Fungal alpha-amylase	0.25 ng/m^3	Baur et al., 1998
Gram-bacteria	1000 cfu/m^3	Malmros et al., 1992
Hardwood dust	0.5 mg/m^3	Carton et al., 2002
Natural latex	0.6 ng/m^3	Baur et al., 1998
Rat allergen	0.7 µg/m^3	Baur et al., 1998
Textile dust	0.5–1 mg/m^3	Da Costa et al., 1998
Total microbes	10,000 cfu/m^3	Malmros et al., 1992
Western red cedar	0.4 mg/m^3	Baur et al., 1998
Wheat flour	1 mg/m^3	Baur et al., 1998

38.3 WOOD AND PAPER INDUSTRIES

Industries that process wood or paper products, or that make heavy use of paper products have long been known to have airborne hazards, especially in relation to allergens. Cellulose provides a nutrient source for a variety of fungi, including several that are known allergens. The dust from wood or paper processing may itself cause respiratory problems.

In a study in a newspaper company, workers were identified who had been admitted to hospital during their employment (Liu et al., 2002). Some 37 percent of the people who had worked in this company had been admitted to a hospital at least once. Chronic pharyngitis or sinusitis showed a significant relationship with printing activities, as did other disorders. Such problems, however, may be largely related to the use of chemicals and printing inks. Papa et al. (1996) studied workers employed in a paper-making/printing factory and a control population of office workers from the same urban area. The percentage of subjects with allergies in the factory-worker group (24.4 percent) was significantly higher than that observed among the office workers (12.5 percent). Of the factory workers with allergies, 94 percent had histories of daily exposure to aliphatic hydrocarbons.

In a study of small-scale wood industry workers in Africa, Rongo et al. (2002) found that prevalence of respiratory symptoms was significantly higher in an exposed group compared with nonexposed office workers. Inhalable dust measurements indicated mean exposure levels of 2.9 to 22.8 mg/m^3. Allergy and sensitivity symptoms were reported regularly in the exposed group as opposed to the controls.

According to Haug et al. (2002), previous studies of paper machine operators have mainly focused on endotoxins as a possible health hazard, but not on culturable microorganisms. Based on exposure assessment in 11 paper mills, workers exposed to bioaerosols were compared with unexposed workers in terms of infections and associated symptoms. Results showed that concentrations of culturable bacteria in process waters varied in the range 10^4 to 10^6 cfu/mL. Bioaerosols levels are shown in Table 38.2. Operators exposed to bioaerosols reported higher cumulative incidence of symptoms associated with infections compared to the reference population.

In 1993, four cases of histoplasmosis were reported among employees in a Michigan pulp paper factory (Stibierski et al., 1996). Among the 96 employees surveyed, 18 persons had the illness and the attack rate among maintenance employees (30 percent) was much higher than among other employees (5 percent). On October 22, a dry, windy day, one

TABLE 38. 2 Airborne Contaminant Levels in Various Industries

Contaminant	Level	Industry	Reference
Bacteria (cfu/m^3)	24,357	Cotton spinning mill	Cinkotai et al., 1977
	2,725	Cotton waste mill	Cinkotai et al., 1977
	7,700	Cotton willowing mill	Cinkotai et al., 1977
	10,700	Furniture factory (Beechwood)	Krysinska-Traczyk et al., 2002
	3,600	Furniture factory (Chipboards)	Krysinska-Traczyk et al., 2002
	10,000–100,000	Paper industry	Haug et al., 2002
	2,100	Pipe tobacco factory	Cinkotai et al., 1977
	800	Tea packing factory	Cinkotai et al., 1977
	14,800	Wool spinning mill	Cinkotai et al., 1977
Total microbes (cfu/m^3)	10,000–100,000	Cigar factory	Reiman and Uitti, 2000
	1,000–10,000	Sawmills	Dutkiewicz et al., 2001
	14,000	Wood debarking site	Sarantila et al., 2001
Fungi (cfu/m^3)	6,214	Cotton spinning mill	Cinkotai et al., 1977
	1,595	Cotton waste mill	Cinkotai et al., 1977
	22,500	Cotton willowing mill	Cinkotai et al., 1977
	950	Pipe tobacco factory	Cinkotai et al., 1977
	1,000–10,000	Sawmills	Alwis et al., 1999
	5,200	Tea packing factory	Cinkotai et al., 1977
	660	Wood debarking site	Sarantila et al., 2001
	100–10,000	Wood joineries	Alwis et al., 1999
	1,000–100,000	Woodchipping mills	Alwis et al., 1999
	280–791	Wool mill	Sigsgaard et al., 1992
	2,500	Wool spinning mill	Cinkotai et al., 1977
Dust (mg/m^3)	3.3	Cigar factory (wick dept.)	Reiman and Uitti, 2000
	1.8	Cotton spinning mill	Cinkotai et al., 1977
	0.77	Cotton waste mill	Cinkotai et al., 1977
	0.28–7.8	Cotton waste mill	Engelberg et al., 1985
	15	Cotton willowing mill	Cinkotai et al., 1977
	1.19	Furniture factories	Schlunssen et al., 2002 (Inhalable)
	0.35	Pipe tobacco factory	Cinkotai et al., 1977
	4.5	Raw tobacco handling	Reiman and Uitti, 2000
	10.3	Soft tissue industry	Kraus et al., 2002 (Inhalable)
	2.27	Tea packing factory	Cinkotai et al., 1977
	0.1–1.25	Textile factories	Da Costa et al., 1998
	2.9–22.8	Wood industry	Rongo et al., 2002 (Inhalable)
	1.6–20	Wool carding factory	Moscato et al., 2000 (Inhalable)
	0.3–0.7	Wool combing factory	Moscato et al., 2000 (Inhalable)
	1.89	Wool spinning mill	Cinkotai et al., 1977
Endotoxin (ng/m^3)	38	Cigar factory (wick dept.)	Reiman and Uitti, 2000
	9–126	Cotton mills	Sigsgaard et al., 1992
	106	Raw tobacco handling	Reiman and Uitti, 2000
	375	Wood debarking site	Sarantila et al., 2001 (Maximum)

maintenance worker swept bird guano from an adjacent roof within 20 meters from the maintenance building. Methods should be used to discourage birds from roosting at facilities and safe procedures should be used for cleanup and disposal of bird droppings.

Microbiological air sampling was performed in two furniture factories located in eastern Poland by Krysinska-Traczyk et al. (2002). In one factory furniture was made from fiberboard and chipboards while in the other from beech wood. Results are shown in Table 38.2. On the average, the most common microorganisms in the air of the furniture factories were *Corynebacterium* spp., *Arthrobacter* spp., and *Brevibacterium* spp. which formed 18 to 50 percent of the total airborne microflora, and fungi (mostly *Aspergillus* spp., *Penicillium* spp., *Absidia* spp., and yeasts) which formed 6 to 54 percent of the total count. Altogether, 28 species or genera of bacteria and 12 species or genera of fungi were identified in the air of examined factories, of which seven to eight species or genera were reported as having allergenic and/or immunotoxic properties.

Microbiological studies of the air and allergological examinations of the workers performed in two sawmills processing deciduous wood (mainly oak) and in one sawmill processing coniferous wood (mainly pine) were conducted by Dutkiewicz et al. (1996). The most common organisms were *Arthrobacter, Corynebacterium, Brevibacterium, Microbacterium, Bacillus spp.*, gram-negative bacteria (*Rahnella*), and filamentous fungi (*Aspergillus, Penicillium*). The workers reacted to the extract of pine dust with much higher frequenc y than to the extract of oak dust. The workers processing pine were often sensitized to *Rahnella* while those processing oak were commonly sensitized to *Penicillium*. Levels of dust, bacteria, fungi, and endotoxins were high in the initial stages of production and decreased sharply during the final stages (Dutkiewicz et al., 2001).

The causative role of wood dust in the onset of sinonasal cancers is solidly established by numerous epidemiological studies, and the magnitude of the risk is particularly high for adenocarcinoma induced by exposure to hardwood dust. Various impairments of lung function have frequently been associated with exposure to both allergenic and nonallergenic wood dusts and may occur at very low concentrations. Measurements performed in France show that exposure levels are high with about 50 percent of the samples being over 1 mg/m^3. A limit value of 0.5 mg/m^3 would credibly allow for protection of exposed workers from most of the risks of nonmalignant pulmonary effects (Carton et al., 2002).

A cross-sectional study including 54 furniture factories and three control factories was conducted by Schlunssen et al. (2002) to survey lung function and prevalence of respiratory symptoms among woodworkers. The arithmetic mean was slightly above some suggested TLVs (see Tables 38.1 and 38.2). Woodworkers had increased frequency of coughing. A dose-response relationship was seen between dust exposure and asthma symptoms, and a positive interaction for asthma was seen between female gender and dust exposure.

Microbes previously considered innocuous to healthy people could present as allergens or pathogens due to chronic occupational exposures to high concentrations. An occurrence of multiple chronic lung abscesses was described in a nonimmunocompromised wood pulp worker (Odell et al., 2000). The organism isolated was *Ochroconis gallopavum*, a dematiaceous fungus known to cause diseases in immunocompromised patients and epidemic encephalitis in poultry. The fungus is typically found in warm environments and in decaying compost and is not otherwise known as an airborne allergen or pathogen. *Chrysonilia sitophila* has been found to induce occupational asthma in the logging industry (Tarlo et al., 1996).

Working in sawmills is associated with bioaerosol exposure and respiratory health problems due to the high concentrations of dust, and the possible use of chemicals. A study by Duchaine and Meriaux (2000) analyzed the mycoflora of eastern Canadian sawmills and 50 work sites (debarking, sawing, planing, and sorting) within 17 sawmills were sampled for airborne fungi. One thousand seven hundred strains were isolated. In eastern Canadian sawmills, *Penicillium* species are the most frequently isolated fungi.

In a study by Alwis et al. (1999), exposure levels to fungi at logging sites were measured at sawmills, woodchipping mills, and joineries (see Table 38.2). Although mean endotoxin levels were lower than the suggested TLV, some personal exposures exceeded the standard. Highly significant associations were found between mean personal inhalable endotoxin exposures and gram-negative bacteria levels and fungi levels. The prevalence of cough, phlegm, chronic bronchitis, nasal symptoms, frequent headaches, and eye and throat irritations was significantly higher among woodworkers than among controls. Dose-response relationships were found between personal exposures and work-related symptoms among the workers.

A study was undertaken by Rylander et al. (1999) on workers in a paper industry to determine the presence of airway inflammation using clinical diagnostic procedures. There was a relation between exposure to endotoxins and airway responsiveness when controlling for age, sex, smoking habits, atopy, and asthma. The results suggest an increased prevalence of subjective respiratory symptoms, and an increased airway responsiveness among exposed workers.

Industrial wastewater may also pose certain airborne hazards to workers. In a study on wastewater from the industrial debarking of wood by Sarantila et al. (2001), airborne levels of endotoxins were found as high as 375 ng/m^3 in some areas. See Table 38.2 for measured levels. Table 38.3 summarizes the potentially pathogenic or allergenic fungi identified in various areas of the facility including the outdoors.

A study conducted by Osim and Esin (1996) found that chronic exposure to large quantities of old and dirty currency notes withdrawn from circulation may impair the lung function of treasury workers. The mean measured lung function values of the female workers were not significantly different from their control values. However, the mean measured values of lung function values of the males were significantly lower than their control values, which indicated general restrictive ventilatory effects.

Kraus et al. (2002) performed a study to correlate the prevalence of respiratory tract symptoms and diseases with dust and fiber exposure in the soft tissue industry in Germany. The mean concentrations for inhalable, respirable, and fibrous dusts were 10.3, 0.22 mg/m^3, and 415,000 fibers/m^3, respectively. Correlations were significant for symptoms of blocked nose, mucosal irritation, dry nose, cough, phlegm, exercise induced dyspnea, hoarseness, and sneezing attacks. Dose-response relationships for intensity and duration of exposure were found for all these symptoms.

TABLE 38.3 Airborne Fungi in a Debarking Facility

Fungi	Debarking	Sludge treatment	Basins	Outdoor
Penicillium	X	X	X	X
Aspergillus	X	X	X	X
Rhodotorula	X	X	X	X
Cladosporium	X	X	X	X
Aureobasidium	X	X		X
Paecilomyces	X	X	X	X
Mucor	X	X	X	
Trichoderma	X	X	X	X
Rhizopus	X			
Alternaria	X	X	X	
Botrytis	X	X	X	X
Phialophora	X	X	X	
Uloclodium	X			ˇ

38.4 TEXTILE, LEATHER, AND FUR INDUSTRIES

Textile industry workers have long been known to suffer respiratory problems due to high levels of dust and fibers in the air of facilities. Similar problems have not been as common in the leather and fur industries but these facilities seem to have their own unique occupational hazards. Anthrax infections among workers handling goat hair and wool were once common but such problems are rare today.

Moscato et al. (2000) studied the frequency and types of respiratory and allergic symptoms in textile workers employed in the wool processing industry. This cross-sectional study was carried out in four wool textile mills and included subjects employed in the early stages of wool processing. An environmental survey showed higher air dust concentrations during carding operations and lower concentrations during combing (see Table 38.2). Almost 35 percent of the subjects reported work-related symptoms. Overall, 13.4 percent of the subjects had occupational respiratory symptoms. Some subjects had only seasonal respiratory symptoms unrelated to work. Within the group of symptomatic subjects, 75.6 percent had serum specific IgE for common pneumoallergens.

Workers in the nylon flocking industry have been found to be at increased risk of chronic nongranulomatous interstitial lung disease due to inhalation of respirable nylon particulates. Although a spectrum of cytologic and histopathologic abnormalities has been observed, nonspecific interstitial pneumonia, lymphoid nodules, and lymphocytic bronchiolitis predominated in the reported cases of flock worker's lung (Kern et al., 2000).

In a study of 11 textile industries of the north of Portugal by Da Costa et al. (1998), the airborne dust levels were found in many cases to exceed the Portuguese standard levels of exposure. For these dust levels they found 23 percent of workers had respiratory symptoms with occupational characteristics in 10.8, and 5.7 percent presenting byssinosis. Workers exposed to cotton fibers in spinning areas had the highest prevalence of symptoms. These characteristics were related to dust levels and were higher in the initial phases of the spinning processes.

Although uncommon today, inhalation anthrax and cutaneous anthrax can still occur in the textile industry. An industrial epidemic of 24 cases of cutaneous anthrax included an apparently minor infection of the upper respiratory tract in a 37-year-old textile worker (Winter and Pfisterer, 1991). The subject developed dyspnea, hematemesis, melaena, and symptoms of shock following the onset of acute illness. Outbreaks of anthrax have occurred in goat-hair mills in Europe. Studies in a mill in the United States where outbreaks had occurred measured anthrax levels at up to 300 cfu/m^3 (Crook and Swan, 2001).

Outbreaks of Q fever have been reported among workers in a Philadelphia wool and hair processing plant (Sigel et al., 1950). Q fever can result from inhalation or exposure to *Coxiella burnetii*. Workers typically mistook the infection for a cold or flu.

There are both acute and chronic effects due to cotton exposure in the cotton waste utilization industry. The overall prevalence of symptoms suggesting byssinosis was 5.9 percent in cotton workers and 4.7 percent in control subjects (Engelberg et al., 1985). Cotton workers with less than 2 years of employment had a significantly greater prevalence of bronchitis than the controls. The cotton workers with 2 years or more of employment had significantly greater prevalences of bronchitis.

A 31-year-old male wool worker presenting rhinoconjunctivitis and asthma episodes linked to exposure to *Dermestidae*-infected wool was investigated by Brito et al. (2002). *Dermestidae* beetle larvae feed on dead, dry animal matter such as furs, skins, cured meat, and decaying carcasses. *Dermestidae* exposure in wool workers when handling parasitized wool can be a cause of IgE-mediated rhinoconjunctivitis and asthma.

Workers in industries that process fur or feathers may be at some risk of allergy and occupational asthma. The prevalence of respiratory symptoms and hypersensitivity to feather and fur allergens is increased among zoological garden workers (Krakowiak et al., 2002).

Nose or eye symptoms were reported most frequently while rhinitis and asthma were reported by atopic subjects more often than by nonatopic subjects. Occupational asthma due to feathers was diagnosed in 2 percent of zoo workers, and to fur in 10 percent of subjects working in contact with birds and furred animals.

Kapok is a natural fiber often used for pillows. It is suspected of having allergenic potential. In a study of a patient believed to be suffering from occupational exposure to kapok, Kern and Kohn (1994) evaluated the patient and performed an environmental assessment. The patient's asthma disappeared after cessation of kapok exposure and, with no other incriminating agent, the test results suggest that kapok fiber itself is allergenic.

Workers in the shoe manufacturing industry may be subject to acute and chronic respiratory problems. A study of respiratory problems in female workers employed in a shoe manufacturing plant showed significantly higher prevalences of all chronic respiratory symptoms compared to control workers (Zuskin et al., 1997). Chest tightness symptoms (exposed: 44.7 percent; control: 0 percent), dyspnea (exposed: 42.6 percent; control: 2 percent), and rhinitis (exposed: 46.3 percent; control: 2.4 percent) were far more prevalent in workers than in controls. There was also a high prevalence of acute symptoms among shoe workers during the work shift, being most pronounced for nose and throat irritation (61.4 percent). The prevalence of acute and chronic respiratory symptoms increased with duration of employment. Environmental measurements demonstrated that benzene, fur, and synthetic fibers were found at higher than maximal concentrations allowed by Croatian standards.

A study of respiratory disorders and atopy in Danish textile industry workers, including cotton mills, a wool mill, and a synthetic fiber mill, was conducted by Sigsgaard et al. (1992). The concentrations of airborne respirable dust and endotoxins were highest in the cotton industry. Mold spores were highest in the wool mill (see Table 38.2). Only low concentrations of microorganisms were found in the synthetic fiber mill. Reduced lung capacity in the cotton workers was strongly associated with cumulative exposure to respirable endotoxin. Byssinosis was diagnosed only in the cotton industry. A dose-response relationship was found between endotoxin exposure and byssinosis.

Esparto is a plant in the grass family that is used in the production of ropes, canvas, sandals, mats, baskets, and for the production of paper paste (Moreno-Ancillo et al 1997). Inhalation of esparto dust has been reported as a cause of hypersensitivity pneumonitis. During the esparto fiber manufacturing process, esparto grass can be contaminated by molds and thermophilic actinomycetes, which have been described as causative agents of hypersensitivity pneumonitis in plaster workers. In at least one case of hypersensitivity pneumonitis linked to esparto, a species of *Aspergillus* contaminating esparto was identified as the likely cause of the disease.

38.5 METAL INDUSTRIES

Although metal and metalworking industries seem unlikely places for airborne microorganisms to proliferate, some niches exist that may produce unique hazards to workers. In addition, the pollution and dust levels that may exist in such industries may contribute to rendering workers more susceptible to infections and allergies.

Chronic bronchitis and decreased respiratory function could be expected to occur in any metalworking industry without high levels of air quality as a result of pollutant levels. A study was carried out on a group of workers of a primary aluminum production plant located in Italy by Alessandri et al. (1992). The results showed a higher prevalence of chronic bronchitis (5.6 versus 2.6 percent) and a greater reduction in vital capacity but only in the nonsmokers (5.4 versus 1.7 percent) among the workers as opposed to control subjects. The highest prevalence of chronic bronchitis (20 percent) and reduced vital capacity (27 percent) was observed in the rodding section. The frequency of bronchial obstruction was similar among potroom and casting workers.

In a study on nonbiological pollutants by Rastogi et al. (1992), a random sample of workers representing each suboccupation in the brassware industry was studied for the prevalence of chronic bronchitis in relation to occupational and environmental factors. Although the exposed group showed higher prevalence of chronic bronchitis in comparison to that observed in the controls (10.5 versus 5.3 percent), the difference was not considered significant. It was observed that the workers engaged in nondusty occupations, such as brass sheet cutting (see Fig. 38.4) and engraving, showed the lowest prevalence of the disease (5 percent) while those engaged in the dusty occupations, such as casting, soldering, electroplating, and polishing, showed the highest prevalence (12 percent). The study thus showed a dose-response relationship between length of exposure and chronic bronchitis.

Allergic reactions can be caused by nonorganic materials as well as by organic or microbiological sources, a factor that may confuse investigations where both sources are present. In a study of occupational allergic diseases of the respiratory system and skin by Obtulowicz et al. (2000), a population of 17,600 employees of a steel factory was examined who were suspected of having occupational allergy. In 15 cases, some 48 percent suffered from upper respiratory tract diseases or bronchial asthma and 52 percent suffered from allergic skin disease (dermatitis). Industrial dust containing metals (nickel, chrome, iron, copper) turned out to be the main allergic factor.

Inhalation of ferrous and other metal fumes in metalworking industries may increase susceptibility of workers to infections like pneumonia. In a study of men admitted to 11 hospitals in England with community-acquired pneumonia Palmer et al. (2003) found that pneumonia was associated with reported occupational exposure to metal fumes in the previous year. The risk was highest for lobar pneumonia and recent exposure to ferrous fumes. Various microorganisms can cause pneumonia but the association was not specific to any one microorganism.

FIGURE 38.4 Metalworking industries like sheet cutting produce metallic dust that may cause allergic reactions and increase susceptibility to infections. (*Photo courtesy of Eric Burnett, Penn State.*)

Hypersensitivity pneumonitis is an acute or chronic respiratory problem also known as extrinsic allergic alveolitis, is distinct from asthma, and is characterized by cough, dyspnea, fever, chills, myalgias, and malaise. Type II hypersensitivity reactions, exemplified by Goodpasture's syndrome, have been associated with exposure to certain metals (Kirchner, 2002). In 2001, three machinists at an automobile brake manufacturing facility in Ohio were hospitalized with respiratory illness characterized by dyspnea, cough, fatigue, weight loss, hypoxia, and pulmonary infiltrates and hypersensitivity pneumonitis was diagnosed in all three workers (MMWR, 2002). It was found that exposure to aerosolized nontuberculous mycobacteria might be contributing to the observed respiratory illnesses in this manufacturing facility.

Fluids used in some metal industries can become contaminated with certain types of microorganisms and may subsequently become aerosolized. Contamination of air and metalworking fluid systems with a rapidly growing mycobacterium was detected in 1995 in a single manufacturing plant in which there had been recent cases of hypersensitivity pneumonitis (Moore et al., 2000). Extensive environmental sampling showed that mycobacteria were present in many indoor air samples, 100 percent of the main metalworking fluid storage tanks, and 75 percent of the freestanding cutting, drilling, and grinding machines. Contamination was essentially limited to a recently introduced brand of semisynthetic metalworking fluid used in almost all of the facility's machining operations. The degree of contamination ranges from 100 to 10^7 cfu/mL. The mycobacterial isolates consisted of a single strain and represented a previously undescribed taxon closely related to *Mycobacterium chelonae/abscessus*.

In another case of contaminated metalworking fluid causing respiratory infection, a worker without immunodeficiency acquired pneumonia complicated by pulmonary abscess (Zell et al., 1999). The source of infection was identified as aerosolized metalworking fluid at his workplace contaminated with *Pseudomonas aeruginosa*. The patient had high levels of IgG antibodies against *Pseudomonas aeruginosa* in his serum, indicating longstanding occupational airborne exposure.

In a study of the health problems in lock factory workers by Hassan et al. (2002) some 45.7 percent of workers were found to be suffering from one or more diseases. Results showed that 73 percent were suffering from respiratory diseases with a majority of them having upper respiratory tract infection. Diseases included chronic bronchitis, emphysema, upper respiratory tract infection, bronchial asthma, and respiratory injuries that varied according to the age and duration of work. Polishing & filing appeared to be the most hazardous type of the work as shown in the proportion of chest diseases with rates of 56.6 and 38.3 percent, respectively.

Tuberculosis continues to be an occupational hazard in other parts of the world. In a study at Russian plants by Khudushina et al. (1991) the incidence of tuberculosis was found to be highest at a foundry plant at 39.9 per 100,000 and in an automatic-assembly plant at 64.4 per 100,000 with 73.9 percent of the patients being males of all occupations. The proportion of patients was highest among the adjusting fitters aged 30 to 39 years (40.4 percent) and founders (36.3 percent).

38.6 BIOTECHNOLOGY INDUSTRIES

Occupational hazards in the biotechnology sector could be expected to be similar to those of laboratory workers as addressed in Chap. 32, but the potential for high levels of airborne pollutants and microbiological hazards may be greater due to the industrial scale of the processes. Previously unknown hazards, including environmental hazards, may even be created as new technologies develop.

The manufacture of citric acid by fermentation of molasses with *Aspergillus niger* was considered the cause of occupational asthma in factory workers, for which a survey of the workforce was carried out (Seaton and Wales, 1994). Airborne spore counts of *A. niger* in the factory averaged about 100 times those in the outside air. Remedial measures included partial enclosure of the process and installation of exhaust ventilation that were effective in preventing any new cases of occupational asthma over the 8-year period.

Equipment used in biotechnology industries for mixing and pressurized devices may inadvertently aerosolize microbes or other products. A research microbiologist developed rhinoconjunctivitis and asthma after release of *D. discoideum* from a pressurized canister (Gottlieb et al., 1999). *Dictyostelium discoideum* is a slime mold that can produce unique lysosomal enzymes but is not known as an airborne allergen or pathogen. Tests revealed IgE antibody in the patient against *D. discoideum* but with the strongest response being directed toward lysosomal enzymes. This was the first report of occupational rhinoconjunctivitis and asthma from a slime mold.

Artamonova et al. (1989) studied the results of immunologic studies of different groups of workers engaged in microbiologic synthesis production. The results suggest individual immunity plays an important role in the development of various forms of diseases. Differentiated shifts in cellular and humoral immunity were identified in healthy people and in patients with occupational diseases. The results suggest that the selection of personnel with stable, adequate immune systems may enable them to go on working in biotechnology facilities without any significant pathologic changes in the body.

The efficacy of environmental protection in the areas surrounding protein synthesis microbiological plants was analyzed by Pogorel'skaia et al. (1989). It was found that when proper operational procedures were observed and the gas phase filtration systems for industrial effluents were properly maintained no air pollution of residential districts resulted from the production of fungi and protein dust of the finished product.

38.7 CHEMICAL AND MINERAL INDUSTRIES

Chemical and mineral industries are unlikely candidates for occupational respiratory allergies or infections due to microorganisms, but, like other industries that generate high levels of dust and pollutants, the workers may be subject to decreased resistance to respiratory infections.

The major hazardous substances in cobalt acetate, chloride, nitrate and sulphate processing facilities are dusts (disintegration aerosols) that occur in concentrations from 0.05 to 53 mg/m^3 depending on the stage of the process (Talakin et al., 1991). An analysis of morbidity and the medical records of workers in these industries showed particular prevalence of upper respiratory tract diseases that included rhinitis, pharyngitis, and laryngitis.

A random sample of workers engaged in chipping and grinding of agate stones were surveyed in a cross-sectional study to assess the prevalence of respiratory illness in the agate industry (Rastogi et al., 1991). Results of the study showed a significantly higher prevalence of lung diseases among agate workers than among controls at 63.4 versus 35.5 percent. The prevalence of pneumoconiosis in agate workers at 18.4 percent was highly significant as compared with controls, in whom not a single case was found. Among the cases of pulmonary diseases in agate workers, pneumoconiosis formed the largest group at 18.4 percent, whereas among controls it was tuberculosis at 12.1 percent. The prevalence of chronic bronchitis was found to be higher among the control population as compared with the exposed group suggesting that agate dust had no role in precipitating chronic bronchitis. However, bronchial asthma appeared to have been aggravated due to agate dust, as the risk among agate workers was sevenfold that found among the controls.

Moiseseu et al. (1979) studied the dynamics of endemic tuberculosis in a district in Romania. An industrial environment was compared with a nonindustrial one and the results show a marked decrease in the incidence of tuberculosis in the industrial environment over 11 years, but, at the same time, it was demonstrated that morbidity due to tuberculosis was over six times higher in the industrial area as compared with the nonindustrial area. Certain factors of the industrial environment, such as pneumoconioses—of which the most important are those due to silica—are significant risk factors for tuberculosis.

Although allergic reactions are more frequently attributed to protein exposure, there is increasing evidence that certain chemicals can produce allergic diseases, including hypersensitivity reactions. Kirchner (2002) studied the types of allergic responses that can be seen among employees working in the chemical industry. Type I hypersensitivity reactions are seen with certain low-molecular-weight chemicals. Type II hypersensitivity reactions, exemplified by Goodpasture's syndrome, have been associated with certain metal exposures. Low-molecular-weight chemicals have been reported to cause type III hypersensitivity reactions such as those seen in hypersensitivity pneumonitis.

The detergent industry has experienced enzyme-induced occupational asthma and allergy, including rhinitis and conjunctivitis but has had success in controlling the problem through monitoring of worker sensitization to enzymes. Guidelines have been developed for the safe handling of enzymes in order to reduce the risk of occupational allergy and asthma (Sarlo and Kirchner, 2002). Manufacturing facilities that follow all of the guidelines enjoy little or no incidence of asthma and allergy among workers exposed to enzymes. The key to the success of the management of enzyme-induced allergy and asthma is surveillance for the development of antibody before the onset of allergic symptoms, which allows for continuing intervention to reduce exposures.

Occupational sensitization to senna is uncommon. Helin and Makinen-Kiljunen (1996) describe a 21-year-old male atopic factory worker who developed asthma and rhinoconjunctivitis 5 months after exposure to senna while he was working for a company manufacturing hair dyes. In the bronchial challenge test with senna, he exhibited a strong reaction. The patient became free of symptoms after he was transferred to a different position within the same company.

38.8 OCCUPATIONAL LEGIONNAIRES' DISEASE

Occasional outbreaks of Legionnaire's disease have been reported in the work environment. In a study by Muraca et al. (1988) two industrial plants were found to contain high numbers of *Legionella pneumophila* in the potable water after several employees became infected. *L. pneumophila* was eradicated from this plant using acidic and caustic scale removers, calcium hypochlorite, and a biocide.

After a worker from a plastics factory was diagnosed with *Legionella* pneumonia it was discovered that a retired factory employee had also been in the hospital with a serious chest infection 6 months earlier (Allen et al., 1999). Investigation revealed that the likely source was a machine cooling system that took water from an uncovered water tank outdoors from which *Legionella pneumophila* was isolated. The tank generated an aerosol through a crack in the flow meter sight glass.

In 1981 six cases of pneumonia occurred among men working in a power station in England that was under construction of which at least three were positively identified as cases of *Legionella* pneumonia (Morton et al., 1986). Cases of pneumonia were associated with four small capacity cooling towers on the site. *Legionella pneumophila* serogroup 1 was isolated from the water systems of these four towers but was not found anywhere else on the site. It appeared that airborne spread of the organism occurred from these cooling water systems that had not received conventional treatment with biocides.

Two mechanics working on a cargo ship under repair in Barcelona had died from a fever that was diagnosed as influenza (Cayla et al., 2001). Both men had been working with the pump of the ship's water system. Various serogroups of *L. pneumophila* were isolated from the ship's water pump and distribution system. The contaminated water system was treated with sodium hypochlorite. The first postintervention water samples showed no further growth of *Legionella*, but serogroups 4 and 8 were identified 8 months later. This legionellosis outbreak, although small, was highly lethal, probably due to the high levels of bacteria to which the patients were exposed and also because of the failure of correct diagnosis.

A large outbreak of Legionnaires' disease occurred among exhibitors at a 1999 floral trade show where a whirlpool spa was on display (Boshuizen et al., 2001). The exhibitors had higher average antibody levels than did the general population and the closer to the whirlpool that the exhibitors worked, the higher were their antibody levels. Whirlpool spas have been associated with a number of Legionnaires' disease outbreaks recently, most notably on cruise ships (see Chap. 37).

Workers on sea drilling platforms may be subject to increased risk of Legionnaires' disease. Based on a study of workers on sea drilling platforms, Lapinski and Kruminis-Lozowski (1997) found that antibody to *Legionella pneumophila* was detected in 25 percent of the workers. Workers had frequently been diagnosed with bronchitis, suggesting a possible misdiagnosis of legionellosis.

The drilling of artesian water wells may present a risk for Legionnaires' disease. In 1990, pneumonia due to *Legionella pneumophila* was diagnosed in two employees working in the area of Apulia, southern Italy, where artesian wells were under construction (Miragliotta et al., 1992). The illness occurred only in those employees who were present when the water emerged from the ground under high pressure, in which legionellae were presumably aerosolized and subsequently inhaled.

38.9 OTHER INDUSTRIES AND HAZARDS

Considering the myriad of industries and situations that may produce occupational exposure to allergens and pathogens, known or otherwise, it is not surprising that many singular examples of respiratory diseases occur occupationally that defy generalization. Following are some examples of occupational diseases that do not fit the previous categories.

A work-related case of inhalational fever associated with an aerosol generated by a pool contaminated with *Pseudomonas*, in a building used for testing scientific equipment, is described by Anderson et al. (1996). Of the 20 subjects who reported symptoms, fever was most common in those with the highest exposure in the sump bay of the pool when the water pumps torrentially recirculated the water. Symptoms occurred late in the working day only on days when the water pumps were operated. The bacterial content of the aerosol rose from 6 to over 10,000 cfu/m^3 when the pumps were operating and was primarily composed of environmental *Pseudomonas*. High endotoxin concentrations were measured in the waters and oil sumps in the pumps. The fever symptoms disappeared completely after the water was changed and the pumps were cleaned.

Like health care workers, prison guards as well as inmates may be at higher risk for tuberculosis. A 1-year prospective study of inmates in Geneva showed that the prevalence of active and residual tuberculosis is five to ten times higher among prisoners than in the general population (Chevallay et al., 1983). Many prisons in the United States incorporate filtration and UVGI systems to help control such problems.

Jones et al. (1999) report that bridge workers have been subject to outbreaks of histoplasmosis, an infection usually associated with exposure to soil contaminated with *Histoplasma capsulatum* in bird droppings or bat guano. The infected workers reported exposure to dust from disturbed bat guano during their normal jobs.

The concentrations of airborne microbes, endotoxins, and total dust in one cigar and two cigarette factories were measured in order to evaluate the risk of respiratory symptoms (Reiman and Uitti, 2000). Air samples produced gram-negative bacteria, mesophilic fungi, thermotolerant fungi, and thermophilic actinomycetes that were found in higher concentrations in the cigar factory than in the cigarette factories. Measurements are shown in Table 38.2. The spray humidifiers in the cigar factory were determined to be a more important source of microbes than was the raw tobacco. In the cigarette factories, where steam humidifiers were used, the humidified air was free of microbes.

Industrial use of induction heaters generating electromagnetic energy at frequencies below 9 kHz exposes operators to magnetic fields of high intensity and to electric fields of relatively low intensity. The operators of such heaters have suffered repeatedly from infections in the upper respiratory tract and in the skin of the forearm (Mikolajczyk et al., 1993). Biological reactions to low frequency electromagnetic fields have long been suspected to include reduced immune system function, and may represent an occupational risk for increased respiratory diseases due to microbial sources (see Chap. 21).

In a survey of 1850 persons working in air-conditioned facilities, persons were examined who reported respiratory disorders as compared with a control group (Molina et al., 1982). Regular checkups during the last 2 years have failed to reveal any serious disease but the most frequent complaints were rhinitis and tracheitis. No alveolitis was observed. The finding of *Bacillus subtilis* in samples of ambient air and air-conditioner filters in conjunction with the presence of precipitating antibodies against crude extracts from these samples, suggested that the respiratory disorders might have been due to this microorganism. *Bacillus subtilis* is ubiquitous in the environment, however, and cases of respiratory problems associated with this microbe are almost nonexistent.

Workers in water-damaged buildings are at increased risk of exposure to allergens if mold growth has occurred. Trout et al. (2001) report the case of a worker with a respiratory illness related to bioaerosol exposure in a water-damaged building with extensive fungal contamination. Extensive fungal contamination was documented in many areas of the building. *Penicillium, Aspergillus*, and *Stachybotrys* species were the most predominant fungi found in air sampling. Proper remediation procedures include the use of respiratory protection (see Chap. 26 for more information). Workers renovating water-damaged buildings were exposed to 590,000 cfu/m^3 of fungi before demolition and 1,300,000 during demolition, according to Rautiala et al. (1996).

38.10 INDUSTRIAL SOLUTIONS

The diversity of workplace ventilation systems, facilities, and associated aerobiological problems precludes any simple general solution, but control measures for aerobiological contamination in the workplace are likely to be similar to those for buildings in general. Improved air filtration, improved air distribution, the increase of outside air if necessary, combined with energy recovery systems, and the use of UVGI for air disinfection can all have applications in various industries that depend on the nature of the problem and the associated economics. The usefulness of personnel protective devices such as facemasks cannot be underestimated in areas of heavy bioaerosol contamination, and the importance of personal hygiene in preventing the transmission of infectious diseases cannot be overemphasized. And, above all, source control in areas that generate high levels of airborne microorganisms will perhaps be the most effective measure in reducing allergies, occupational asthma, and infectious diseases. See Chaps. 10, 11, and 12 for information on basic air cleaning technologies and see Table 28.3 for information on managing aerobiological problems.

REFERENCES

Alessandri, M. V., Baretta, L., and Magarotto, G. (1992). "Chronic bronchitis and respiratory function in those employed in primary aluminum production." *Med Lav* 83(5):445–450.

Allen, K. W., Prempeh, H., and Osman, M. S. (1999). "Legionella pneumonia from a novel industrial aerosol." *Commun Dis Public Health* 2(4):294–296.

Alwis, K. U., Mandryk, J., and Hocking, A. D. (1999). "Exposure to biohazards in wood dust: bacteria, fungi, endotoxins, and (13)-beta-D-glucans." *Appl Occup Environ Hyg* 14(9):598–608.

Anderson, K., McSharry, C. P., Clark, C., Clark, C. J., Barclay, G. R., and Morris, G. P. (1996). "Sump bay fever: inhalational fever associated with a biologically contaminated water aerosol." *Occup Environ Med* 53(2):106–111.

Armstrong, L. R., Zaki, S. R., Goldoft, M. J., Todd, R. L., Khan, A. S., Khabbaz, R. F., Ksiazek, T. G., and Peters, C. J. (1995). "Hantavirus pulmonary syndrome associated with entering or cleaning rarely used, rodent-infested structures." *J Infect Dis* 172(4):1166.

Arnaiz, N. O., and Kaufman, J. D. (2002). "New developments in work-related asthma." *Clin Chest Med* 23(4):737–747.

Artamonova, V. G., Dzhaginian, A. I., Andreeva, L. N., Cherednik, A. N., and Svitina, N. N. (1989). "Status of immunologic reactivity and prospects of the implementation of occupational selection of workers at a microbiological plant." *Gig Tr Prof Zabol* 1:5–8.

Bardana, E. J. J. (1995). "Occupational asthma and related respiratory disorders." *Dis Mon* 41(3):143–199.

Bardana, E. J. J. (2003). "8. Occupational asthma and allergies." *J Allergy Clin Immunol* 111(2 Suppl.): S530–S539.

Baur, X., Chen, Z., and Liebers, V. (1998). "Exposure-response relationships of occupational inhalative allergens." *Clin Exp Allergy* 28(5):537–544.

Boshuizen, H. C., Neppelenbroek, S. E., vanVliet, H., Schellekens, J. F., denBoer, J. W., Peeters, M. F., and Spaendonck, M. A. C.-v. (2001). "Subclinical Legionella infection in workers near the source of a large outbreak of Legionnaires disease." *J Infect Dis* 184(4):515–518.

Brito, F. F., Mur, P., Barber, D., Lombardero, M., Galindo, P. A., Gomez, E., and Borja, J. (2002). "Occupational rhinoconjunctivitis and asthma in a wool worker caused by Dermestidae spp." *Allergy* 57(12):1191–1194.

Carton, M., Goldberg, M., and Luce, D. (2002). "Occupational exposure to wood dust. Health effects and exposure limit values." *Rev Epidemiol Sante Publique* 50(2):159–178.

Cayla, J. A., Maldonado, R., Gonzalez, J., Pellicer, T., Ferrer, D., Pelaz, C., Gracia, J., Baladron, B., Plasencia, A., and group., L. s. (2001). "A small outbreak of Legionnaires' disease in a cargo ship under repair." 17(6):1322–1327.

Chevallay, B., Haller, R. d., and Bernheim, J. (1983). "Epidemiology of pulmonary tuberculosis in the prison environment." *Schweiz Med Wochenschr* 113(7):261–265.

Cinkotai, F. F., Lockwood, M. G., and Rylander, R. (1977). "Airborne microorganisms and prevalence of byssinotic symptoms in cotton mills." *AIHA J* 38:554–559.

Crook, B., and Swan, J. R. M. (2001). Bacteria and other bioaerosols in industrial workplaces. *Microorganisms in Home and Indoor Work Environments,* B. Flannigan, R. A. Samson, and J. D. Miller, eds., Taylor and Francis, New York.

Da Costa, J. T., Barros, H., Macedo, J. A., Ribeiro, H., Mayan, O., and Pinto, A. S. (1998). "Prevalence of respiratory diseases in the textile industry. Relation with dust levels." *Acta Med Port* 11(4):301–309.

Drake-Lee, A., Ruckley, R., and Parker, A. (2002). "Occupational rhinitis: a poorly diagnosed condition." *J Laryngol Otol* 116(8):580–585.

Duchaine, C., and Meriaux, A. (2000). "Airborne microfungi from eastern Canadian sawmills." *Can J Microbiol* 46(7):612–617.

Dutkiewicz, J., Krysinska-Traczyk, E., Skorska, C., Milanowski, J., Sitkowska, J., Dutkiewicz, E., Matuszyk, A., and Fafrowicz, B. (1996). "Microflora of the air in sawmills as a potential occupational hazard: concentration and composition of microflora and immunologic reactivity of workers to microbial aeroallergens." *Pneumonol Alergol Pol* 64(Suppl. 1):25–31.

Dutkiewicz, J., Olenchock, S. A., Krysinska-Traczyk, E., Skorska, C., Sitkowska, J., and Prazmo, Z. (2001). "Exposure to airborne microorganisms in fiberboard and chipboard factories." *Ann Agric Environ Med* 8:191–199.

Engelberg, A. L., Piacitelli, G. M., Petersen, M., Zey, J., Piccirillo, R., Morey, P. R., Carlson, M. L., and Merchant, J. A. (1985). "Medical and industrial hygiene characterization of the cotton waste utilization industry." *Am J Ind Med* 7(2):93–108.

Esposito, A. L. (1992). "Pulmonary infections acquired in the workplace. A review of occupation-associated pneumonia." *Clin Chest Med* 13(2):355–365.

Fritz, C. L., Fulhorst, C. F., Enge, B., Winthrop, K. L., Glaser, C. A., and Vugia, D. J. (2002). "Exposure to rodents and rodent-borne viruses among persons with elevated occupational risk." *J Occup Environ Med* 44(10):962–967.

Galdi, E., and Moscato, G. (2002). "Prevention of occupational asthma." *Monaldi Arch Chest Dis* 57(3–4):211–212.

Gottlieb, S. J., Garibaldi, E., Hutcheson, P. S., and Slavin, R. G. (1999). "Occupational asthma to the slime mold *Dictyostelium discoideum*." *J Occup Med* 35(12):1231–1235.

Hassan, M. A., Khan, Z., Yunus, M., and Bhargava, R. (2002). "Health profile of lock factory workers in Aligarh." *Indian J Public Health* 46(2):39–45.

Haug, T., Sostrand, P., and Langard, S. (2002). "Exposure to culturable microorganisms in paper mills and presence of symptoms associated with infections." *Am J Ind Med* 41(6):498–505.

Heinrich, J., Hoelscher, B., Wjst, M., Ritz, B., Cyrys, J., and Wichmann, H. (1999). "Respiratory diseases and allergies in two polluted areas in East Germany." *Environ Health Perspect* 107(1):53–62.

Helin, T., and Makinen-Kiljunen, S. (1996). "Occupational asthma and rhinoconjunctivitis caused by senna." *Allergy* 51(3):181–184.

Henneberger, P. K., Hoffman, C. D., Magid, D. J., and Lyons, E. E. (2002). "Work-related exacerbation of asthma." *Int J Occup Environ Health* 8(4):291–296.

Hnizdo, E., Sullivan, P. A., Bang, K. M., and Wagner, G. (2002). "Association between chronic obstructive pulmonary disease and employment by industry and occupation in the US population: a study of data from the Third National Health and Nutrition Examination Survey." *Am J Epidemiol* 156(8):738–746.

Jammes, Y., Delvolgo-Gori, M. J., Badier, M., Guillot, C., Gazazian, G., and Parlenti, L. (2002). "One-year occupational exposure to a cold environment alters lung function." *Arch Environ Health* 57(4):360–365.

Jay, M., Hjelle, B., Davis, R., Ascher, M., Baylies, H. N., Reilly, K., and Vugia, D. (1996). "Occupational exposure leading to hantavirus pulmonary syndrome in a utility company employee." *Clin Infect Dis* 22(5):841–844.

Jones, T. F., Swinger, G. L., Craig, A. S., McNeil, M. M., Kaufman, L., and Schaffner, W. (1999). "Acute pulmonary histoplasmosis in bridge workers: A persistent problem." *Am J Med* 106:480–481.

Karjalainen, A., Kurppa, K., Virtanen, S., Keskinen, H., and Nordman, H. (2000). "Incidence of occupational asthma by occupation and industry in Finland." *Am J Ind Med* 37(5):451–458.

Kern, D. G., and Kohn, R. (1994). "Occupational asthma following kapok exposure." *J Asthma* 31(4):243–250.

Kern, D. G., Ely, C. K., Pransky, E. W., Mello, G. S., Fraire, C. J., and Muller, J. (2000). "Flock worker's lung: broadening the spectrum of clinicopathology, narrowing the spectrum of suspected etiologies." *Chest* 117(1):251–259.

Kessler, R. C., Almeida, D. M., Berglund, P., and Stang, P. (2001). "Pollen and mold exposure impairs the work performance of employees with allergic rhinitis." *Ann Allergy Asthma Immunol* 87(4):289–295.

Khudushina, T. A., Altynova, M. P., and Sagirova, G. M. (1991). "The optimal system of ambulatory examination of groups with high risk of tuberculosis at a large industrial plant." *Probl Tuberk* 5:10–13.

Kirchner, D. B. (2002). "The spectrum of allergic disease in the chemical industry." *Int Arch Occup Environ Health* 75(Suppl.):S107–112.

Krakowiak, A., Palczynski, C., Walusiak, J., Wittczak, T., Ruta, U., Dudek, W., and Szulc, B. (2002). "Allergy to animal fur and feathers among zoo workers." *Int Arch Occup Environ Health* 75(Suppl.):S113–S116.

Kraus, T., Pfahlberg, A., Gefeller, O., and Raithel, H. J. (2002). "Respiratory symptoms and diseases among workers in the soft tissue producing industry." *Occup Environ Med* 59(12): 830–835.

Krysinska-Traczyk, E., Skorska, C., Cholewa, G., Sitkowska, J., Milanowski, J., and Dutkiewicz, J. (2002). "Exposure to airborne microorganisms in furniture factories." *Ann Agric Environ Med* 9(1):85–90.

Krzyzanowski, M., and Kauffmann, F. (1988). "The relation of respiratory symptoms and ventilatory function to moderate occupational exposure in a general population. Results from the French PAARC study of 16,000 adults." *Int J Epidemiol* 17(2):397–406.

Lapinski, T. W., and Kruminis-Lozowski, J. (1997). "Infection with Legionella pneumophila among workers of Polish sea drilling platforms." *Wiad Lek* 50(1–3):11–15.

Liu, Y. H., Du, C. L., Lin, C. T., Chan, C. C., Chen, C. J., and Wang, J. D. (2002). "Increased morbidity from nasopharyngeal carcinoma and chronic pharyngitis or sinusitis among workers at a newspaper printing company." *Occup Environ Med* 59(1):18–22.

Malmros, P., Sigsgaard, T., and Bach, B. (1992). "Occupational health problems due to garbage sorting." *Waste Manag Res* 10:227–234.

Meyer, J. D., Holt, D. L., Cherry, N. M., and McDonald, J. C. (1999). "SWORD '98': surveillance of work-related and occupational respiratory disease in the UK." *Occup Med* 49(8):485–489.

Mikolajczyk, H. J., Kamedula, M., and Kamedula, T. (1993). "Biological accounts emerging from some kinds of electromagnetic waves in the environment." *Pol J Occup Med Environ Health* 6(3):263–271.

Miragliotta, G., DelPrete, R., Sabato, R., Cassano, A., and Carnimeo, N. (1992). "Legionellosis associated with artesian well excavation." *Eur J Epidemiol* 8(5):748–749.

MMWR (2002). "Respiratory illness in workers exposed to metalworking fluid contaminated with nontuberculous mycobacteria—Ohio, 2001." *MMWR* 51(16):349–352.

Moiseseu, V., Papahagi, C., and Popescu, M. (1979). "Aspects of tuberculosis prevention in the industrial environment." *Rev Ig Bacteriol Virusol Parazitol Epidemiol Pneumoftiziol* 28(2):67–73.

Molina, C., Aiache, J. M., Bedu, M., Menaut, P., Wahl, D., Brestowski, J., and Grall, Y. (1982). "Air-conditioner disease: Results of an industrial medicine survey." *Nouv Presse Med* 11(31):2325–2329.

Moore, J. S., Christensen, M., Wilson, R. W., Jr., R. J. W., Zhang, Y., Nash, D. R., and B., B. S. (2000). "Mycobacterial contamination of metalworking fluids: involvement of a possible new tax on of rapidly growing mycobacteria." *AIHAJ* 61(2):205–213.

Moreno-Ancillo, A., Padial, M. A., Lopez-Serrano, M. C., and Granado, S. (1997). "Hypersensitivity pneumonitis due to inhalation of fungi-contaminated esparto dust in a plaster worker." *Allergy Asthma Proc* 18(6):355–357.

Morse, T., Grey, M., Storey, E., and Kenta-Bibi, E. (2002). "Occupational disease in Connecticut, 2000." *Conn Med* 66(12):723–730.

Morton, S., Bartlett, C. L., Bibby, L. F., Hutchinson, D. N., Dyer, J. V., and Dennis, P. J. (1986). "Outbreak of Legionnaires' disease from a cooling water system in a power station." *Br J Ind Med* 43(9):630–635.

Moscato, G., Catenacci, G., Dellabianca, A., Lecchi, A., Omodeo, P., Manfredi, S., and Tonin, C. (2000). "A respiratory and allergy survey in textile workers employed in early stages of wool processing." *G Ital Med Lav Ergon* 22(3):236–240.

Muraca, P. W., Stout, J. E., Yu, V. L., and Yee, Y. C. (1988). "Legionnaires' disease in the work environment: implications for environmental health." *Am Ind Hyg Assoc J* 49(11):584–590.

Nathell, L., Malmberg, P., Lundback, B., and Nygren, A. (2000). "Impact of occupation on respiratory disease." *Scand J Work Environ Health* 26(5):382–389.

Obtulowicz, K., Kolarzyk, E., Laczkowska, T., Porebski, G., Zapolska, I., and Hudzik, A. (2000). "Occupational allergic diseases in the steel industry. Population studies." *Przegl Lek* 57(9):446–450.

Odell, J. A., Alvarez, S., Cvitkovich, D. G., Cortese, D. A., and McComb, B. L. (2000). "Multiple lung abscesses due to *Ochroconis gallopavum*, a dematiaceous fungus, in a nonimmunocompromised wood pulp worker." *Chest* 118(5):1503–1505.

Osim, E. E., and Esin, R. A. (1996). "Lung function studies in some Nigerian bank workers." *Cent Afr J Med* 42(2):43–46.

Palmer, K. T., Poole, J., Ayres, J. G., Mann, J., Burge, P. S., and Coggon, D. (2003). "Exposure to metal fume and infectious pneumonia." *Am J Epidemiol* 157(3):227–233.

Papa, G., Quaratino, D., DiFonso, M., Guiffreda, F., Romano, A., and Venuti, A. (1996). "Allergic respiratory diseases and environmental pollution: experience in the printing/paper-manufacturing industry." *Allergy* 51(11):833–836.

Pogorel'skaia, S. A., Mokeeva, N. V., Litovskaia, A. V., Chebotarev, P. A., and Bundakov, A. (1989). "Medico-biological criteria of the evaluation of air quality at the site of microbiological synthesis plants." *Gig Sanit* 3(Mar.):53–54.

Rastogi, S. K., Gupta, B. N., Chandra, H., Mathur, N., Mahendra, P. N., and Husain, T. (1991). "A study of the prevalence of respiratory morbidity among agate workers." *Int Arch Occup Environ Health* 63(1):21–26.

Rastogi, S. K., Gupta, B. N., Mathur, N., Husain, T., Mahendra, P. N., Pangtey, B. S., and Srivastava, S. (1992). "A survey of chronic bronchitis among brassware workers." *Ann Occup Hyg* 36(3):283–294.

Rautiala, S., Reponen, T., Hybarinen, A., Nevalainen, A., Husman, T., Vehvilainen, A., and Kalliokoski, P. (1996). "Exposure to airborne microbes during the repair of moldy buildings." *Am Indust Hyg J* 57:279–284.

Reiman, M., and Uitti, J. (2000). "Exposure to microbes, endotoxins and total dust in cigarette and cigar manufacturing: an evaluation of health hazards." *Ann Occup Hyg* 44(6):467–473.

Rongo, L. M., Besselink, A., Douwes, J., Barten, F., Msamanga, G. I., Dolmans, W. M., Demers, P. A., and Heederik, D. (2002). "Respiratory symptoms and dust exposure among male workers in small-scale wood industries in Tanzania." *J Occup Environ Med* 44(12):1153–1160.

Rylander, R., Thorn, J., and Attefors, R. (1999). "Airways inflammation among workers in a paper industry." *Eur Respir J* 13(5):1151–1157.

Salvaggio, J. E. (1994). "Inhaled particles and respiratory disease." *J Allergy Clin Immunol* 94(2 Pt. 2):304–309.

Sarantila, R., Reiman, M., Kangas, J., Husman, K., and Savolainen, H. (2001). "Exposure to endotoxins and microbes in the treatment of wastewater and in the industrial debarking of wood." *Bull Environ Contam Toxicol* 67:171–178.

Sarlo, K., and Kirchner, D. B. (2002). "Occupational asthma and allergy in the detergent industry: new developments." *Curr Opin Allergy Clin Immunol* 2(2):97–101.

Schachter, E. N., Zuskin, E., and Saric, M. (2001). "Occupational airway diseases." *Rev Environ Health* 16(2):87–95.

Schlunssen, V., Schaumburg, I., Taudorf, E., Mikkelsen, A. B., and Sigsgaard, T. (2002). "Respiratory symptoms and lung function among Danish woodworkers." *J Occup Environ Med* 44(1):82–98.

Seaton, A., and Wales, D. (1994). "Clinical reactions to *Aspergillus niger* in a biotechnology plant: an eight-year follow up." *Occup Environ Med* 51(1):54–56.

Sigel, M. M., MvNair, T. F., and Henle, W. (1950). "Q Fever in a wool and hair processing plant." *Am J Pub Health* 40:524–532.

Sigsgaard, T., Pedersen, O. F., Juul, S., and Gravesen, S. (1992). "Respiratory disorders and atopy in cotton, wool, and other textile mill workers in Denmark." *Am J Ind Med* 22(2):163–184.

Talakin, I. N., Ivanova, L. A., Kostetskaia, N. I., Komissarov, V. N., and Beliaeva, I. V. (1991). "Hygienic characteristics of work conditions and health status of workers engaged in cobalt salt production." *Gig Tr Prof Zabol* 1:10–11.

Tarlo, S. M., Wai, Y., Dolovich, J., and Summerbell, R. (1996). "Occupational asthma induced by *Chrysonilia sitophila* in the logging industry." *J Allergy Clin Immunol* 97(6):1409–1413.

Thorens, B. (1991). "Occupational bronchitis." *Schweiz Rundsch Med Prax* 80(18):496–501.

Trout, D., Bernstein, J., Martinez, K., Biagini, R., and Wallingford, K. (2001). "Bioaerosol lung damage in a worker with repeated exposure to fungi in a water-damaged building." *Environ Health Perspect* 109(6):641–644.

Welch, A. R., Birchall, J. P., and Stafford, F. W. (1995). "Occupational rhinitis—possible mechanisms of pathogenesis." *J Laryngol Otol* 109(2):104–107.

Winter, H., and Pfisterer, R. M. (1991). "Inhalation anthrax in a textile worker: nonfatal course." *Schweiz Med Wochenschr* 121(22):832–835.

Zell, L., Mack, U., Sommerfeld, A., Buchter, A., and Sybrecht, G. W. (1999). "Abscessed pneumonia caused by *Pseudomonas aeruginosa* as an occupational disease in a metal driller." *Pneumologie* 53(12):620–625.

Zhestkov, A. V. (2000). "Immunological changes in dust-induced lung diseases." *Gig Sanit* 6:30–33.

Zock, J. P., Sunyer, J., Kogevinas, M., Kromhout, H., Burney, P., and Anto, J. M. (2001). "Occupation, chronic bronchitis, and lung function in young adults. An international study." *Am J Respir Crit Care Med* 163(7):1572–1577.

Zuskin, E., Mustajbegovic, J., Schachter, E. N., Doko-Jelinic, J., and Bradic, V. (1997). "Respiratory function in shoe manufacturing workers." *Am J Ind Med* 31(1):50–55.

CHAPTER 39

SEWAGE AND WASTE PROCESSING

39.1 INTRODUCTION

The processing of sewage and waste involves the handling and disinfection of biocontaminants and organic materials that can be breeding grounds for bacteria, viruses, and fungi. Most of the microorganisms associated with waste are not respiratory pathogens or normally airborne but workers in sewage, wastewater, and waste processing industries are subject to occupational hazards for which they must take precautions. Global population increases, coupled with intensive animal and livestock production practices, have resulted in the generation, accumulation, and disposal of large amounts of wastes around the world (Pillai and Ricke, 2002). Aerosolization of microbial pathogens, endotoxins, allergens, odors, and dust particles is an inevitable consequence of the generation and handling of waste material. There are also concerns associated with the proximity of human populations to municipal waste treatment facilities in many parts of the world and their exposure to bioaerosols.

This chapter addresses the types of microbes that may be found in facilities that handle, process, or recycle waste, sewage, compost, or wastewater and the aerobiological hazards to which workers in these industries, and local populations, may be exposed. Agricultural facilities that treat compost locally are addressed in Chap. 34, which can be consulted for further information. Waterborne pathogens are not specifically addressed here unless they are found in wastewater and have airborne transmission potential.

39.2 SEWAGE AND WASTE AEROBIOLOGY

Sewage and waste facilities, whether they are part of an agricultural facility or otherwise, are subject to a wide variety of bioaerosols, only a few of which are potential airborne respiratory pathogens and allergens. Table 39.1 lists some of the respiratory hazards that have been isolated in sewage.

TABLE 39.1 Airborne Microbes That May Grow or Occur in Sewage

Airborne Pathogen	Type	Group	Reference
Acinetobacter	Bacteria	Pathogen	Rosas et al., 1996
Adenovirus	Virus	Pathogen	Hers and Winkler, 1973
Adenovirus	Virus	Pathogen	Rao and Melnick, 1987
Aeromonas	Bacteria	Pathogen	Dumontet et al., 2001
Alcaligenes	Bacteria	Pathogen	Rosas et al., 1996
Alternaria	Fungi	Allergen	Rosas et al., 1996
Aspergillus	Fungi	Allergen	Rosas et al., 1996
Bacillus anthracis	Bacteria	Pathogen	Dumontet et al., 2001
Brucella spp.	Bacteria	Pathogen	Dumontet et al., 2001
Candida spp.	Bacteria	Pathogen	Dumontet et al., 2001
Cladosporium	Fungi	Allergen	Rosas et al., 1996
Clostridium botulinum	Bacteria	Pathogen	Dumontet et al., 2001
Clostridium perfringens	Bacteria	Pathogen	Dumontet et al., 2001
Corynebacterium spp.	Bacteria	Pathogen	Higgins and Burns, 1975
Coxsackievirus	Virus	Pathogen	Dumontet et al., 2001
Cryptococcus neoformans	Bacteria	Pathogen	Dumontet et al., 2001
Echovirus	Virus	Pathogen	Dumontet et al., 2001
Exophiala	Fungi	Allergen	Sutton et al., 1998
Fusarium	Fungi	Allergen	Lavoie et al., 1996
Klebsiella spp.	Bacteria	Pathogen	Rao and Melnick, 1987
Listeria monocytogenes	Bacteria	Pathogen	Pillai and Ricke, 2002
Mucor	Fungi	Allergen	Higgins and Burns, 1975
Mycobacterium	Bacteria	Pathogen	Higgins and Burns, 1975
Nocardia	Bacteria	Pathogen	Higgins and Burns, 1975
Norwalk virus	Virus	Pathogen	Rao and Melnick, 1987
Parvovirus	Virus	Pathogen	Rao and Melnick, 1987
Penicillium	Fungi	Allergen	Lavoie et al., 1996
Phialaphora	Fungi	Allergen	Dumontet et al., 2001
Pseudomonas aeruginosa	Bacteria	Pathogen	Higgins and Burns, 1975
Reovirus	Virus	Pathogen	Rao and Melnick, 1987
Rhizopus	Fungi	Allergen	Rosas et al., 1996
Serratia	Bacteria	Pathogen	Rosas et al., 1996
Staphylococcus spp.	Bacteria	Pathogen	Higgins and Burns, 1975
Streptococcus faecalis	Bacteria	Pathogen	Rao and Melnick, 1987

Sewage sludge is a by-product of the processing and treatment of municipal wastewater. It is primarily composed of sediments of organic matter of wastewater. The sediments precipitate out of the wastewater in settling basins. The organic matter in sewage sludge has a complex and unpredictable mixture that may vary seasonally and geographically. The chief component is the remnants of human feces after chemical or other treatments that biodegrade the microbiota. Human feces are originally about 30 percent bacteria when excreted. The remaining organic matter in sewage is easily fermented and must undergo stabilization before it can be used for agricultural or other uses (Dumontet et al., 2001). The stabilization process consists of drying with air or heat, chemical treatments, aerobic stabilization in the liquid state, anaerobic stabilization, and composting. The sanitized organic matter from sewage sludge is commonly used in agriculture to enrich the soil and the bioresidues of sewage are useful as a source of nutrients for plants. Figure 39.1 illustrates a typical mixture of airborne fungi that may be found airborne in or around sewage treatment plants.

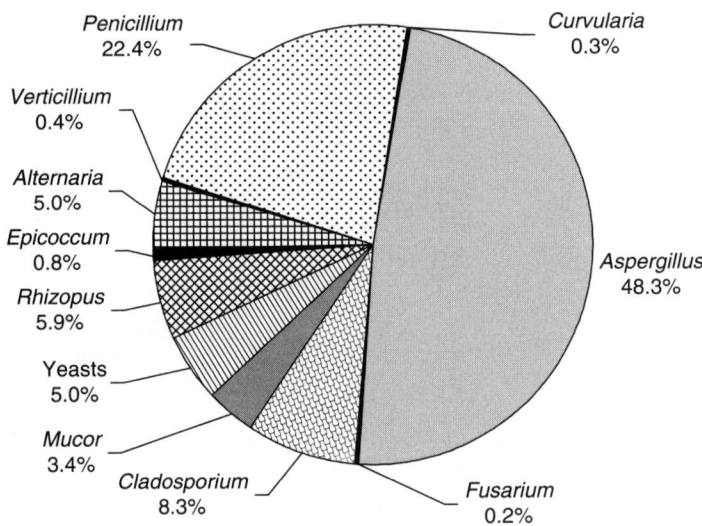

FIGURE 39.1 Typical mixture of airborne fungi in a waste composting facility. [*Based on data from Hryhorczuk et al. (2001).*]

Viruses have occasionally been isolated in sewage but no data on their frequency of occurrence are available (Bitton, 1980; Hers and Winkler, 1973).

Airborne microbes in the vicinity of sewage treatment plants are potential hazards to the surrounding community, although little evidence exists to support this view. The primary hazard from sewage treatment plants may be annoyance due to odors. Steuer (1986) took measurements of airborne microbes over a 1-year period at several sewage treatment plants using sewage treatment methods. Results indicate that neighboring residential areas or factories in the vicinity of sewage treatment plants are not exposed to airborne hazards.

Wastewater processes include various types of bubble ventilation and sprayers that may result in aerosolization of biocontaminants. Measurements of the microbial emissions in various activated sludge units with different ventilations systems have shown that the lowest airborne levels occur with bubble ventilation and that the spray devices used for foam elimination caused particularly high airborne levels of microbes (Wanner, 1975). Airborne microbial levels above sewage sludge pools measured 50 to 100,000 cfu/m^3 depending on the weather. Levels between 500 and 1500 cfu/m^3 were measured at a distance of 50 and 100 meters from the sewage facility. At distances of 200 to 400 m the airborne levels of microbes were about the same as normally measured in the outside air, or about 100 to 500 cfu/m^3. In a closed sewage sludge pool considerably higher levels were measured. In the immediate vicinity of the activated sludge unit airborne levels were found to be about 10,000 to 25,000 cfu/m^3. In the hall of the enclosed facility levels of 3000 to 4000 cfu/m^3 were measured. The composition included coliform bacteria at 1 to 2 percent and enterococci at 2 to 4 percent. In addition, it was noted that besides the intestinal flora other unidentified pathogenic agents were present in the aerosols. Enclosed sewage sludge processing units, therefore, harbor a greater risk of airborne infections, while the hazard in open-air pools is much lower. Figure 39.2 illustrates a typical mixture of airborne bacteria in a waste composting facility.

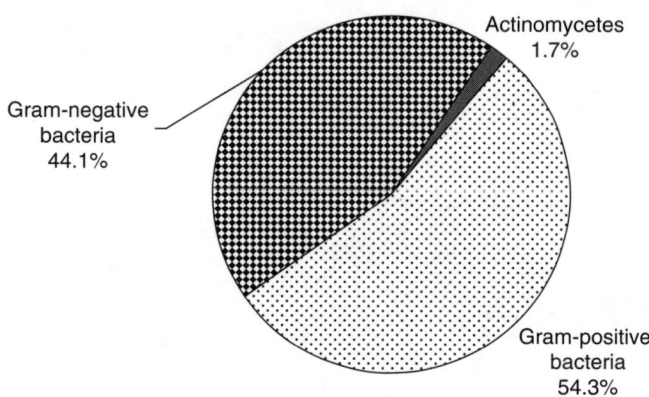

FIGURE 39.2 Typical mixture of airborne bacteria in a waste composting facility. [*Based on data from Hryhorczuk et al. (2001).*]

Many companies produce pellets from paper and plastic waste for burning as fuel. During 1998 and 1999, several measurements were made to determine the dust, particle, microbe, and endotoxin concentrations in a Finnish plant by Tolvanen (2001). The dust and the particle concentrations were low, but the microbe concentrations, especially in the summer and in the autumn, were at a level that may present occupational health hazards. The total concentration of airborne microbes was approximately 4.8×10^6 cfu/m^3. The concentrations of endotoxins were high in summer and in autumn, being 340 to 1000 ng/m^3 and exceeded recommended values, considered to be 30 ng/m^3 (Lavoie et al., 1996). In the winter, the concentration of the endotoxin was lower, being 4.7 to 33 ng/m^3. Table 39.2 summarizes various studies showing typical levels at these and various other various facilities.

The diversity and distribution of microbes within brewery-degrading anaerobic sludge granules were studied by Liu et al. (2002). Microbes found included *Methanosaeta concilii, Clostridium, Xanthomonas* species, and *Desulfovibrio* species, but only the *Clostridium* species are a potential health threat to humans.

Airborne fungal spores and their metabolites present health hazards to workers in composting facilities. Fischer et al. (1999) investigated worker exposures to airborne fungal spores at various working places in compost plants. The results indicate that the spectrum of *microbial volatile organic compounds* (MVOCs) and mycotoxins produced can be specific for certain species. In addition to the pathogenic and allergenic hazards, fungi may have different toxicological health impacts. An evaluation of health effects caused by the great variety of MVOCs remains difficult, since information on their health effects impact is lacking.

39.3 SEWAGE AND WASTE EPIDEMIOLOGY

The nature of waste and sewage incurs an ever-present potential for exposure to hazardous bioaerosols. A number of studies have been performed on sewage pathogens and the occupational hazards of workers in waste processing plants or trash heaps and several of these are

TABLE 39.2 Typical Airborne Concentrations in Waste Facilities

Contaminant	Level	Facility	Reference
Bacteria (cfu/m^3)	11,879	Waste composting	Hryhorczuk et al., 2001
	21,201–84,806	Waste composting (shedding)	Jager et al., 1994
	972	Waste composting (maximum)	Folmsbee and Stewart, 1999
	1,091	Wastewater treatment (cyclone)	Lavoie et al., 1996
	662	Wastewater treatment (sludge)	Lavoie et al., 1996
	6,700	Waste transfer station	Rosas et al., 1996
Fungi (cfu/m^3)	13,451	Waste composting	Hryhorczuk et al., 2001
	1,000,000–10,000,000	Waste composting	Fischer et al., 1999
	78,000	Waste composting	Lacey and Crook, 1988
	5,151–19,064	Waste composting (shedding)	Jager et al., 1994
	5,059	Waste composting (maximum)	Folmsbee and Stewart, 1999
	181	Wastewater treatment (cyclone)	Lavoie et al., 1996
	1,317	Wastewater treatment (sludge)	Lavoie et al., 1996
	4,900	Waste transfer station	Rosas et al., 1996
Total microbes (cfu/m^3)	700	Waste composting (shipping)	Van Der Werf, 1996
	137,400	Waste composting (curing)	Van Der Werf, 1996
	50–100,000	Sewage sludge pools	Wanner, 1975
	3,000,000,000	Trash recycling (maximum)	Sigsgaard et al., 1990
	46,000	Garbage handling	Sigsgaard et al., 1994
	54,000	Composting areas	Sigsgaard et al., 1994
	8,000	Water purification	Sigsgaard et al., 1994
	4,700	Paper sorting	Sigsgaard et al., 1994
	4,800,000	Paper and plastic recycling	Tolvanen, 2001
Endotoxins (ng/m^3)	1.94	Waste composting	Hryhorczuk et al., 2001
	1.9–47	Waste composting	Van Der Werf, 1996
	3–39	Sewage treatment (water basins)	Rylander, 1999
	3.8–32,170	Sewage treatment (sludge handling)	Rylander, 1999
	340–1,000	Paper and plastic recycling (summer)	Tolvanen, 2001

reviewed here to illustrate the epidemiology and aerobiology of such facilities. Some outdoor facilities are considered here since they may be subject to the same aerobiological problems.

The risks of airborne hazards associated with wastewater treatment plants were reviewed by Seidel (1983), who found the risks to be controllable. Risks are considerably lower when wastewater is handled directly than when water contaminated by wastewater is handled. The sewage treatment technologies presently in use reduce the concentrations of viruses and bacteria to varying extents. The various processes of sludge stabilization influence their number and degree of infectiousness to a varying extent. The most important risk of infection from wastewater originates from contamination of raw water and drinking water.

A survey of work related symptoms among sewage workers was performed in Sweden by Thorn et al. (2002a) to assess the risks for workers. Results show significantly increased risks for respiratory symptoms, including chronic bronchitis and toxic pneumonitis, as well as central nervous system problems such as headache, fatigue, and concentration difficulties over workers in nonsewage industries. An increased risk for nonspecific work-related gastrointestinal symptoms was found among the sewage workers. Causal relationships were not specifically identified.

Aspergillosis is a lung infection that may be caused by various species of *Aspergillus* in different settings. As an occupational disease it most often occurs in relation to sewage or composting (Joseph, 1983). In the latter environments the causative species is most often *Aspergillus fumigatus*.

Wastewater treatment workers are potentially exposed to a variety of infectious agents and toxic materials. Khuder et al. (1998) examined the prevalence of infectious diseases and associated symptoms in 242 wastewater treatment workers over a 12-month period. Data were compared with 54 college maintenance and oil refinery workers as a control group. The wastewater workers exhibited a significantly higher prevalence of gastroenteritis, gastrointestinal symptoms, and headaches but no significant differences were found with regard to respiratory and other symptoms.

Although the multistage treatment and processing of sewage sludge greatly reduces the microbial content, complete sterilization may not be an achievable goal. In addition to low levels of microbes remaining in processing sewage, the possibility of processing mistakes and cross-contamination of potable water with contaminated water in sewage treatment plants exists. The occurrence of viruses, bacteria, yeasts, fungi, and zooparasites in sewage sludge has epidemiologic concerns for sewage sludge recycling facilities and their employees (Dumontet et al., 2001).

Perhaps the greatest potential hazard from sewage is the potential for human pathogenic viruses and bacteria to cause infections among workers. According to Seidel (1983) the epidemiological hygienic risk of handling community wastewater can be controlled and sewage treatment technologies currently in use effectively reduce the levels of viruses, bacteria, and protozoa to varying extents. The most serious risks of infection from wastewater involve the potential contamination of raw water and treated drinking water. Although few cases of employee infections have occurred, U.S. statistics indicate that bacteria, protozoa, and some viruses have periodically caused drinking water epidemics.

The occupational hazard of working in sewage plants can involve exposure to different types of bacteria, fungi, viruses, and chemicals. In a study of sewage treatment plants in three municipalities in Sweden by Thorn et al. (2002b), the amounts of airborne endotoxin were found to be generally low. At certain worksites, however, higher endotoxin values were found, with the highest airborne levels being at worksites located indoors. The results suggest that the exposure levels are relatively stable over short time periods but that higher levels can be recorded during work processes involving agitation of wastewater.

Rylander (1999) studied the health effects among workers in sewage treatment plants. Employees in sewage treatment plants were compared with a control group. Workers in sewage plants reported significantly higher nose irritation, fatigue, and diarrhea. The results confirmed previous studies on the presence of airways and intestinal inflammation among workers in sewage treatment plants. The most likely causative agent is endotoxin, and at 14 of 23 workplaces, concentrations exceeded recommended guidelines.

Pontiac fever is a less serious form of Legionnaires' disease that can occur from the inhalation of contaminated water droplets. In one outbreak of Pontiac fever at a sewage treatment plant in the food industry during the summer months, workers became ill with fever and flulike symptoms after repairing a decanter for sludge concentration (Gregersen et al., 1999). The work took place over a period of 10 days in a small closed room, while another decanter was in operation and was consequently spraying an aerosol into the environment and to which the workers were exposed. Samples of the sludge were collected and *L. pneumophila* serogroup 1 was cultured in high amounts from sludge from the decanter. It was concluded that the fever was caused by *L. pneumophila* emitted to the environment by the uncovered decanter.

Solid waste carries with it the hazard of exposure to various microorganisms, even though the processing of such waste rarely involves spraying or other types of wastewater processing techniques. The health hazards to which waste management workers in Denmark are exposed were studied by Sigsgaard (1999). Investigation of garbage recycling workers began in the early 1990s. The results showed a wide range of symptoms and diseases from occupational asthma to gastrointestinal and skin symptoms. From the Danish experience it would seem that under normal circumstances with good hygienic practices and use of the proper protective equipment by an educated workforce, garbage handling induces a small but significant risk of occupational asthma. The majority of the asthma cases identified are linked to a poor perception of the risks related to organic dust exposure. However, there is an increased prevalence of respiratory symptoms such as chest tightness and toxic alveolitis among workers. In addition, gastrointestinal and skin symptoms are more frequent among waste recycling workers compared to other blue-collar workers.

During the last two decades, a growing interest in recycling of domestic waste has developed, and plans to increase the recycling of domestic waste have been adopted by many municipal governments. A common feature of these plans is the implementation of new systems and equipment for the collection of domestic waste that has been separated at the pick-up sites. A review of occupational health problems associated with collection of domestic waste was undertaken by Poulsen et al. (1995) to assess their significance and possible causes. An excess risk for chronic bronchitis was reported for waste collectors in Geneva and data indicate an excess risk for pulmonary problems among waste collectors compared with the total workforce. Recent studies have indicated that implementation of some new waste collection and recycling systems may result in an increased risk of occupational health problems. High incidence rates of gastrointestinal problems, irritation of the eye and skin, and perhaps symptoms of organic dust toxic syndrome (flulike symptoms, cough, muscle aches, fever, fatigue, headache) have been reported among workers collecting the biodegradable fraction of domestic waste. The limited data available on exposure to bioaerosols and volatile organic compounds indicate that these waste collectors may be simultaneously exposed to multiple agents such as dust containing bacteria, endotoxin, mold spores, glucans, and volatile organic compounds. Diesel exhaust pollution may also play a role in causing or exacerbating respiratory problems.

In another study on the recycling of trash in Denmark, Sigsgaard et al. (1990) studied a plant that opened in 1986, where household trash and industrial refuse was sorted by machine in a large partially open plant and was converted into fuel pellets. During a period of 8 months, eight out of 15 employees developed respiratory symptoms. In seven employees bronchial asthma was diagnosed and chronic bronchitis in one person. Four had initial symptoms of the organic dust toxic syndrome. After further 6 months, another case of occupational asthma occurred in the plant. Only two out of nine of the employees had previously had asthma or atopic disease. In spring of 1989, from six to 18 months after the onset of the symptoms, six still had dyspnea on exertion. See Table 39.2 for the measurements taken. The plant design had required considerable contact with the refuse in open systems and it was reconstructed to lower employee exposures. The redesign and reconstruction resulted in hygienic conditions that are now acceptable.

Another survey of respiratory disorders and atopy in Danish refuse workers was conducted by Sigsgaard et al. (1994). Respiratory and mucosal symptoms of paper-sorting workers, compost workers, and garbage-handling workers were compared with a control group of workers from water purification plants. Garbage-handling workers were exposed to a significantly higher mean concentration of total dust (0.74 mg/m^3) than were water supply workers (0.42 mg/m^3). Levels of airborne microorganisms are shown in Table 39.2. Significantly higher amounts of gram-negative bacteria were found in composting and garbage-handling plants than in water-supply plants. In garbage-handling plants only, there were significantly higher amounts of endotoxins than in paper-sorting plants. Higher prevalence of chest tightness (14 percent), flulike symptoms (14 percent), itching eyes (27 percent), itching nose (14), and sore or itching throat (21 percent) were found among garbage-handling workers, compared with water-supply workers (1, 1, 11, and 0 percent, respectively).

The urban poor in Third World countries often eke out a living scavenging through garbage dumps and are exposed to a variety of health hazards (Fry et al., 2002). This activity is often performed by children who reportedly suffer from various diseases including TB, acute respiratory infections, and measles, but no epidemiological data are available to assess the actual risks.

39.4 AIR TREATMENT SOLUTIONS

Air treatment solutions for sewage and waste treatment plants include all those technologies that have previously been evaluated, but special focus needs to be placed on source control and disinfection due to the concentrated microbial hazards that may be present. Certain technologies, like incineration, may reduce the microbial hazard but simultaneously create a pollution hazard. In the processing of wastewater and sewage, technologies for disinfecting water, like biocides and UVGI, will simultaneously promote the reduction of airborne microbes. Water disinfection technologies are not specifically addressed here but sufficient information on this subject is available elsewhere (AWWA, 1971).

Some new gas phase filtration technologies for controlling odors may have useful applications in sewage treatment facilities. The possibility of using nonthermal plasma destruction of dilute, complex polluting gases has been investigated by Ruan et al. (1999) for its potential use in animal houses for waste odor reduction. Silent discharge, nonthermal plasma reactors were developed for energy efficient decomposition of ammonia. Results of testing showed that these systems could achieve a wide range of ammonia reduction efficiencies depending on the operating conditions.

Another study on plasma technology for the control of noxious gases produced by waste in a swine house used an ac-powered ferroelectric packed-bed plasma reactor to decompose ammonia and odors (Zhang et al., 1996). Ammonia decomposition efficiencies of up to 95 percent were obtained and the unit was found to be effective at removing odors in the swine house air.

Reduction of airborne microbes and pollutants is probably best approached using source control methods. In modern compact sewage treatment plants, the removal of nitrogen and phosphorus often requires additional stages of processing which both complicates matters and provide opportunities. Rogalla et al. (1992) studied the minimization of escaping gases and aerosols by covering sewage treatment plants. They describe large-scale examples of compact technology and the additional upgrading flexibility they provided. Some new facilities are implemented in sensitive neighborhoods by creative siting under sports stadiums, parks, or buildings. In covered plants, odor emission control becomes of primary importance. Some plants are sited in inner-city locations to avoid the costs of pumping sewage to remote sites. Most of these plants incorporate space-saving settling facilities and high-rate biological reactors to reduce overall size and thus allow for coverage. Parallel plates in primary settlers reduce the surface to about one-tenth of conventional systems. Biocarbone aerated filters combine biodegradation with very high removal rates and retention of particles in one reactor, without additional clarification or filtration. Air treatment for large plants is mostly performed by chemical scrubbing, completely eliminating environmental nuisances. A large treatment plant has operated since 1987 under a stadium in Marseille while in Monaco the sewage treatment plant is located in the city center underneath a large building. Primary settlers are followed by biological treatment on biocarbone aerated filters and air is chemically deodorized.

Biofilters are a promising technology for the reduction of microbial composition of waste and the corresponding reduction of any airborne hazards. Hill et al. (2002) studied the use of an aerobic biofilter system for reducing enteric microbial indicators in flushed swine wastewater under different seasonal conditions. This study also investigated the effectiveness of UV irradiation for inactivating enteric bacteria, coliphages, and bacterial spores in treated and untreated swine wastewater. Fecal coliforms, *Escherichia coli*, enterococci, and coliphages were reduced by 97 to 99 percent in the biofilter system. *Salmonella* were reduced by 95 to 97 percent. Of the six microbial indicators studied, *Clostridium perfringens* spores were typically reduced the least by the biofilter system. At an average absorbed UV irradiation dose of 13 mJ/cm^2, maximum reductions of fecal coliforms, *E. coli*, enterococci, *C. perfringens* spores, and somatic coliphages in biofilter system effluent were 99.4, 99.2, 94.6, 26.7, and 99.5 percent, respectively. The results of this study show that the aerobic biofilter system can be an effective alternative for treatment of flushed swine waste. Ultraviolet irradiation can be effective for further reducing enteric microbe concentrations in biologically treated swine waste, as well as in lower quality wastewaters. Biotechnological methods may be used to disinfect manure and livestock slurry of the hazards due to pathogenic and allergenic agents (Strauch and Ballarini, 1994).

Collection, separation, and composting of household waste generates organic dusts that may contain the endotoxin and (13)-beta-D-glucan, a cell wall component of fungi, plants and certain bacteria, as well as other bioaerosols. A study performed by Thorn (2001) found that the amount of airborne endotoxin was higher during the warm season, and there was a relationship between exposure levels and outdoor temperature. Household waste collectors handling compostable waste can be exposed to airborne (13)-beta-D-glucan, especially during the warm season, when more symptoms have been reported among waste collectors, according to previous studies.

Koe and Tan (1990) investigated the field performance of activated carbon adsorption for sewage air by routing sewage air through parallel trains of carbon material with and without alkali impregnation. The odor and H$_2$S concentrations of the influent sewage air varied significantly throughout the study period, averaging 120 SOU/m^3 and 5.3 ppm, respectively. Although the alkali-impregnated carbon was capable of removing a much larger quantity of H$_2$S than the nonalkali-impregnated grade, it was less effective at removing other odorous but non-H$_2$S gaseous compounds.

Source control is one of the most fruitful means of reducing aerosol exposure and there are many methods, including disinfection of the water or sewage. Cassells et al. (1995) described a combined system of electronically generated copper and silver ions and free chlorine that could be used for inactivation of *Naegleria fowleri* amoebas in water. Copper and silver alone caused no significant inactivation of *N. fowleri* even after extended exposure but addition of 1.0 mg/L free chlorine to water resulted in 99 percent inactivation rates compared to either chlorine alone or the metals alone. Enhanced inactivation of *N. fowleri* by a combined system of free chlorine and copper and silver suggests synergistic effects. Similar results were obtained by Landeen et al. (1989) against *Legionella pneumophila, Staphylococcus aureus, Pseudomonas aeruginosa, Escherichia coli,* and *Streptococcus faecalis.* In general, methods used to disinfect water, such as the use of chlorine and ozone, may have similarly effective applications for source control in wastewater processing (Craun, 1986).

REFERENCES

AWWA (1971). *Water Quality and Treatment.* The American Water Works Association, McGraw-Hill, New York.

Bitton, G. (1980). *Introduction to Environmental Virology.* John Wiley & Sons, New York.

Cassells, J. M., Yahya, M. T., Gerba, C. P., and Rose, J. B. (1995). "Efficacy of a combined system of copper and silver and free chlorine for inactivation of Naegleria fowleri amoebas in water." *Water Sci Technol* 31(5–6):119–122.

Craun, G. F. (1986). *Waterborne Diseases in the United States.* CRC Press. Boca Raton, FL.

Dumontet, S., Scopa, A., Kerje, S., and Krovacek, K. (2001). "The importance of pathogenic organisms in sewage and sewage sludge." *J Air Waste Manage Assoc* 51(6):848–860.

Fischer, G., Schwalbe, R., Moeller, M., Ostrowski, R., and Dott, W. (1999). "Species-specific production of microbial volatile organic compounds (MVOC) by airborne fungi from a compost facility." *Chemosphere* 39(5):796–810.

Folmsbee, M., and Stewart, K. (1999). "Bioaerosol concentrations at an outdoor composting center." *J Air Waste Manage Assoc* 49:554–561.

Fry, S., Cousins, B., and Olivola, K. (2002). "Health of children living in urban slums in Asia and the Near East: Review of existing literature and data." *Environmental Health Project Report 109.* U.S. Agency for International Development. Washington, DC. http://www.ehproject.org/PDF/Activity_Reports/AR109ANEUrbHlthweb.pdf.

Gregersen, P., Grunnet, K., Uldum, S. A., Andersen, B. H., and Madsen, H. (1999). "Pontiac fever at a sewage treatment plant in the food industry." *Scand J Work Environ Health* 25(3):291–295.

Hers, J. F. P., and Winkler, K. C. (1973). *Airborne Transmission and Airborne Infection.* Enschede Oosthoek Publishing Company, Utrecht, The Netherlands.

Higgins, I. J., and Burns, R. G. (1975). *The Chemistry and Microbiology of Pollution.* Academic Press, London.

Hill, V. R., Kantardjieff, A., Sobsey, M. D., and Westerman, P. W. (2002). "Reduction of enteric microbes in flushed swine wastewater treated by a biological aerated filter and UV irradiation." *Water Environ Res* 74(1):91–99.

Hryhorczuk, D., Curtis, L., Scheff, P., Chung, J., Rizzo, M., Lewis, C., Keys, N., and Mooney, M. (2001). "Bioaerosol emissions from a suburban yard waste composting facility." *Ann Agric Environ Med* 8:177–185.

Jager, E., Ruden, H., and Zeschmar-Lahl, B. (1994). "Kompostierunglangen 2.Mitteilung: Aerogene Keimbelastung an verscheidenen Arbeitsbereichen von Kompostierungsanlagen." *Zentralblatt Hyg Unweltmed* 196:367–379.

Joseph, J. M. (1983). Aspergillosis. *Occupational Mycoses,* A. F. DiSalvo, ed., Lea & Febiger, Philadelphia, PA.

Khuder, S. A., Arthur, T., Bisesi, M. S., and Schaub, E. A. (1998). "Prevalence of infectious diseases and associated symptoms in wastewater treatment workers." *Am J Ind Med* 33(6):571–577.

Koe, L. C. C., and Tan, N. C. (1990). "Field performance of activated carbon adsorption for sewage air." *J Environ Eng* 116(4):721–734.

Lacey, J., and Crook, B. (1988). "Fungal and actinomycete spores as pollutants of the workplace and occupational illness." *Ann Occup Hyg* 32:515–533.

Landeen, L. K., Yahya, M. T., Kutz, S. M., and Gerba, C. P. (1989). "Microbiological evaluation of copper: silver disinfection units for use in swimming pools." *Water Sci Technol* 21(3):267–270.

Lavoie, J., Pineau, S., and Marchland, G. (1996). Chapter 7: Aeromicrobial analyses in a wastewater treatment plant. *Pathogens in the Environment,* B. Austin, ed., Blackwell Scientific, Oxford, UK.

Liu, W. T., Chan, O. C., and Fang, H. H. (2002). "Characterization of microbial community in granular sludge treating brewery wastewater." *Water Res* 36(7):1767–1775.

Pillai, S. D., and Ricke, S. C. (2002). "Bioaerosols from municipal and animal wastes: background and contemporary issues." *Can J Microbiol* 48(8):681–696.

Poulsen, O. M., Breum, N. O., Ebbehoj, N., Hansen, A. M., Ivens, U. I., vanLelieveld, D., Malmros, P., Matthiasen, L., Nielsen, B. H., Nielsen, E. M. et al. (1995). "Collection of domestic waste. Review of occupational health problems and their possible causes." *Sci Total Environ* 170(1–2):1–19.

Rao, V. C., and Melnick, J. L. (1987). *Human Viruses in Sediments, Sludges, and Soils.* CRC Press, Boca Raton, FL.

Rogalla, F., Roudon, G., Sibony, J., and Blondeau, F. (1992). "Minimizing nuisances by covering compact sewage treatment plants." *Water Sci Tech* 25(4–5):363–374.

Rosas, I., Calderon, C., Salinas, E., and Lacey, J. (1996). Chapter 8: Airborne microorganisms in a domestic waste transfer station. *Pathogens in the Environment,* B. Austin, ed., Blackwell Scientific, Oxford, UK.

Ruan, R., Han, W., Ning, A., Deng, S., Chen, P. L., and Goodrich, P. (1999). "Effects of design parameters of planar, silent discharge, plasma reactors on gaseous ammonia reduction." *Trans ASAE* 42(6):1841–1846.

Rylander, R. (1999). "Health effects among workers in sewage treatment plants." *Occup Environ Med* 56(5):354–357.

Seidel, K. (1983). "Communicable disease problems of sewage with special reference to human pathogenic viruses." *Zentralbl Bakteriol Mikrobiol Hyg [B]* 178(1–2):98–110.

Sigsgaard, T. I., Bach, B., Taudorf, E., Malmros, P., and Gravesen, S. (1990). "Accumulation of respiratory diseases among employees at a recently established refuse sorting plant." *Ugeskr Laeger* 152(35):2485–2488.

Sigsgaard, T., Malmros, P., Nersting, L., and Petersen, C. (1994). "Respiratory disorders and atopy in Danish refuse workers." *Am J Respir Crit Care Med* 149(6):1407–1412.

Sigsgaard, T. (1999). "Health hazards to waste management workers in Denmark." *Schriftenr Ver Wasser Boden Lufthyg* 104:563–568.

Steuer, W. (1986). "Airborne microbes in the vicinity of sewage treatment plants." *Zentralbl Bakteriol Mikrobiol Hyg [B]* 182(2):202–214.

Strauch, D., and Ballarini, G. (1994). "Hygienic aspects of the production and agricultural use of animal wastes." *Zentralbl Veterinarmed B* 41(3):176–228.

Sutton, D. A., Fothergill, A. W., and Rinaldi, M. G. (1998). *Guide to Clinically Significant Fungi.* Williams & Wilkins, Baltimore, MD.

Thorn, J. (2001). "Seasonal variations in exposure to microbial cell wall components among household waste collectors." *Ann Occup Hyg* 45(2):153–156.

Thorn, J., Beijer, L., and Rylander, R. (2002a). "Work related symptoms among sewage workers: a nationwide survey in Sweden." *Occup Environ Med* 59(8):562–566.

Thorn, J., Beijer, L., Jonsson, T., and Rylander, R. (2002b). "Measurement strategies for the determination of airborne bacterial endotoxin in sewage treatment plants." *Ann Occup Hyg* 46(6):549–554.

Tolvanen, O. K. (2001). "Airborne bioaerosols and noise in a dry waste treatment plant in Pietarsaari, Finland." *Waste Manag Res* 19(2):108–114.

Van Der Werf, P. (1996). "Bioaerosols at a Canadian composting facility." *Biocycle* 37:78–82.

Wanner, H. U. (1975). "Microbial contamination of air by activated sludge units." *Zentralbl Bakteriol [Orig B]* 161(1):46–53.

Zhang, R., Yamamoto, T., and Bundy, D. S. (1996). "Control of ammonia and odors in animal houses by a ferroelectric plasma reactor." *IEEE Trans Ind Applic* 32(1):113–117.

APPENDIX A
AIRBORNE PATHOGENS AND ALLERGENS

Absidia

GROUP	Fungal Spore
CLASS	Zygomycetes
ORDER	Mucorales
FAMILY	Mucoraceae
DISEASE GROUP	Non-communicable
BIOSAFETY LEVEL	Risk Group 2
Infectious Dose (ID_{50})	NA
Lethal Dose (LD_{50})	NA
Infection Rate	none
Incubation Period	-
Peak Infection	NA
Annual Cases	rare
Annual Fatalities	-

Absidia represents a genera of fungi that are opportunistic pathogens for man. They can infect the lungs and other locations, where spores will germinate and mycelia will grow. It may cause zygomycosis and other opportunistic infections. This type of infection can be fatal to those with immunodeficiency. Other than fever and dyspnea, there are rarely any other symptoms. With continued tissue necrosis, hemoptysis may develop, and the end result may be fatal pulmonary hemorrhage. Water Activity Aw = 0.98.

Disease or Infection	zygomycosis, opportunistic infections, pneumonia				
Natural Source	Environmental				
Toxins	none				
Point of Infection	Upper Respiratory Tract				
Symptoms	The only clinical symptoms usually manifest are fever and dyspnea. Tissue necrosis may result in hemoptysis.				
Treatment	Treatment with amphotericin B remains the only reliable therapy.				
Untreated Fatality Rate	-	Prophylaxis: none		Vaccine:	none
Shape	ovoid spore				
Mean Diameter, µm	3.536	Size Range:	2.5–5 microns		
Growth Temperature	37°C	Survival Outside Host:		survives outdoors	
Inactivation	Moist heat: 121°C for 30 min.				
Disinfectants	1% sodium hypochlorite, phenolics, formaldehyde, glutaraldehyde.				
Filter Nominal Rating	MERV 6	MERV 8	MERV 11	MERV 13	MERV 14
Removal Efficiency, %	59.1	72.7	92.2	99.0	99.0
UVGI Rate Constant	(unknown)	Media		-	
Dose for 90% Inactivation	-	Ref.		-	
Suggested Indoor Limit	150–500	cfu/m^3	(in mix with other nonpathogenic fungi)		
Genome Size (bp)	unknown				
Related Species	A. corymbifera, A. ramosa				
Notes	CDC Reportable.				
Photo Credit	Image courtesy of Neil Carlson, University of Minnesota.				
References	Freeman 1985, Howard 1983, Lacey 1988, Murray 1999, Rao 1996, Ryan 1994, Smith 1989, Mandell 2000				

Acinetobacter

GROUP	Bacteria
TYPE	Gram-
GENUS	Acinetobacter
FAMILY	Moraxellaceae
DISEASE GROUP	Endogenous
BIOSAFETY LEVEL	Risk Group 2
Infectious Dose (ID_{50})	unknown
Lethal Dose (LD_{50})	NA
Infection Rate	-
Incubation Period	-
Peak Infection	NA
Annual Cases	147
Annual Fatalities	-

Occurs mainly in the immunocompromised. Acinetobacter calcoaceticus commonly inhabit soil, sewage, and water, and are frequently found in humans, apparently as biological contaminants. Evidence suggests they are opportunistic pathogens that can cause meningitis and septic infections, mostly in immunocompromised hosts. Transient colonization of the pharynx occurs in 7% of healthy people, while cutaneous colonization occurs in 25%. Infection can occur at any body site.

Disease or Infection	opportunistic/septic infections, nosocomial infections, meningitis				
Natural Source	Environmental, soil, sewage, indoor growth in potable water, nosocomial				
Toxins	none				
Point of Infection	Upper Respiratory Tract, skin				
Symptoms	Causes suppurative infections of every organ system. Respiratory infections include bronchiolitis and tracheobronchitis in children.				
Treatment	Combined use of beta-lactams and aminoglycosides.				
Untreated Fatality Rate	-	Prophylaxis:		Vaccine:	none
Shape	cocci, rods				
Mean Diameter, μm	1.225	Size Range:	1–1.5 × 1.5–2.5 microns		
Growth Temperature	30–32°C	Survival Outside Host:	survives outdoors		
Inactivation	Moist heat: 121°C for 30 min.				
Disinfectants	1% sodium hypochlorite, phenolics, formaldehyde, glutaraldehyde.				
Filter Nominal Rating	MERV 6	MERV 8	MERV 11	MERV 13	MERV 14
Removal Efficiency, %	27.2	38.9	61.3	94.4	98.8
UVGI Rate Constant	0.00021	m^2/J	Media	water	
Dose for 90% Inactivation	10,965	J/m^2	Ref.	Keller 1982	
Suggested Indoor Limit	50	cfu/m^3			
Genome Size (bp)	3,598,621		40.3% G+C	59.7% T+A	
Related Species	A. baumanii (formerly A. calcoaceticus), Moraxella				
Notes	Nosocomial infections related to contaminated inhalation equipment.				
Photo Credit	Centers for Disease Control PHIL# 201, Janice Carr.				
References	Braude 1981, Castle 1987, Mitscherlich 1984, Murray 1999, Prescott 1996, Ryan 1994, Weinstein 1991, Woods 1997, Mandell 2000				

Acremonium

GROUP	Fungal Spore
PHYLUM	Ascomycota
ORDER	Hypocreales
FAMILY	Hypocreaceae
DISEASE GROUP	Non-communicable
BIOSAFETY LEVEL	Risk Group 1-2
Infectious Dose (ID_{50})	NA
Lethal Dose (LD_{50})	-
Infection Rate	none
Incubation Period	-
Peak Infection	NA
Annual Cases	-
Annual Fatalities	-

Reportedly allergenic, it can produce toxins. Has been found growing indoors. Can cause cutaneous mycetoma, or eumycetoma, of the foot and in the nails, onychomycosis, corneal ulcers, endophthalmitis, meningitis, and endocarditis. Occurs worldwide between the tropics of capricorn and cancer. Occurs most often in India, Mexico, Niger, Senegal, and other countries. Can produce a trichothecene toxin which is toxic if ingested.

Disease or Infection	allergic alveolitis, mycetoma, toxic reactions, keratitis, lesions of hard palate.				
Natural Source	Environmental, indoor growth in humidifiers and insulation.				
Toxins	crotocin, trichothecine				
Point of Infection	Upper Respiratory Tract, skin, foot, eyes				
Symptoms	Mycetoma appears as swellings and deep infections of the skin.				
Treatment	Eumycetoma is treated with ketoconazole.				
Untreated Fatality Rate	-	Prophylaxis: none		Vaccine:	none
Shape	ovoid spore				
Mean Diameter, μm	2.449	Size Range:	2–3 × 4–6 microns		
Growth Temperature	37°C	Survival Outside Host:		survives outdoors	
Inactivation	Moist heat: 121°C for 30 min.				
Disinfectants	1% sodium hypochlorite, phenolics, formaldehyde, glutaraldehyde.				
Filter Nominal Rating	MERV 6	MERV 8	MERV 11	MERV 13	MERV 14
Removal Efficiency, %	50.1	65.5	86.8	98.9	99.0
UVGI Rate Constant	(unknown)	Media		-	
Dose for 90% Inactivation	-	Ref.		-	
Suggested Indoor Limit	150–500	cfu/m^3	(in mix with other nonpathogenic fungi)		
Genome Size (bp)					
Related Species	various				
Notes	CDC Reportable.				
Photo Credit	Image courtesy of Neil Carlson, University of Minnesota.				
References	Freeman 1985, Howard 1983, Lacey 1988, Murray 1999, Pope 1993, Rao 1996, Ryan 1994, Smith 1989, Woods 1997				

Actinomyces israelii

GROUP	Bacteria
TYPE	Gram+
GENUS	Actinomyces
FAMILY	Actinobacteria
DISEASE GROUP	Endogenous
BIOSAFETY LEVEL	Risk Group 2
Infectious Dose (ID_{50})	unknown
Lethal Dose (LD_{50})	-
Infection Rate	
Incubation Period	-
Peak Infection	NA
Annual Cases	uncommon
Annual Fatalities	-

An opportunistic pathogen. Occurs as normal flora of the gastrointestinal and genitourinary tracts but can cause cervicofacial infections. Often chronic or benign. Cervicofacial infections may follow dental infections. Thoracic infection may involve lungs, pleura, mediastinum, or chest wall. Infection may follow oropharyngeal aspiration or esophageal penetration.

Disease or Infection	actinomycosis				
Natural Source	Humans, cattle				
Toxins	none				
Point of Infection	Upper Respiratory Tract				
Symptoms	Bronchitis, emphysema, chronic pneumonitis may be features. Infection may mimic tuberculosis or malignancy.				
Treatment	Penicillin G or any of the penicilin alternatives.				
Untreated Fatality Rate	-	Prophylaxis: none		Vaccine: none	
Shape	rods				
Mean Diameter, μm	0.901	Size Range:	0.2–1 × 2–5 microns		
Growth Temperature	35–37°C	Survival Outside Host:	survives outdoors		
Inactivation	Moist heat: 121°C for 30 min.				
Disinfectants	1% sodium hypochlorite, phenolics, formaldehyde, glutaraldehyde.				
Filter Nominal Rating	MERV 6	MERV 8	MERV 11	MERV 13	MERV 14
Removal Efficiency, %	18.8	27.4	47.2	86.3	97.6
UVGI Rate Constant	(unknown)	Media		-	
Dose for 90% Inactivation	-	Ref.		-	
Suggested Indoor Limit	-				
Genome Size (bp)					
Related Species	Type Species: A. bovis				
Notes	CDC Reportable.				
Photo Credit	Image of actinomyces provided courtesy of Dick Eikelboom, Activated Sludge Information Systems (ASIS), The Netherlands.				
References	Braude 1981, Freeman 1985, Mitscherlich 1984, Murray 1999, Prescott 1996, Ryan 1994, Mandell 2000				

Adenovirus

GROUP	Virus
TYPE	dsDNA, no RNA
GENUS	Mastadenovirus
FAMILY	Adenoviridae
DISEASE GROUP	Communicable
BIOSAFETY LEVEL	Risk Group 2
Infectious Dose (ID_{50})	32–150
Lethal Dose (LD_{50})	none
Infection Rate	0.51–0.75
Incubation Period	1–10 days
Peak Infection	3–4 days
Annual Cases	common
Annual Fatalities	-

Common in the adult population as mild colds. Adenovirus causes acute respiratory infections of the lungs and sometimes the eyes. Mild respiratory infections resemble the common cold. It can occur in epidemic form. Some types of this virus occur primarily in infants. Commonly spreads by fomites as well as by airborne route. Possible spread by the fecal-oral route, and in swimming pools. Prolonged survival outside hosts—Type 2 lasted 3–8 weeks on environmental surfaces.

Disease or Infection	colds, fever, pharyngitis, Acute Respiratory Disorder, pneumonia				
Natural Source	Humans, sewage				
Toxins	none				
Point of Infection	Upper Respiratory Tract, eyes				
Symptoms	Varies in clinical manifestation and severity. Symptoms may include fever, pharyngitis, rhinitis, tonsilitis, cough, and conjunctivitis.				
Treatment	Supportive therapy is the only treatment. No prophylaxis available.				
Untreated Fatality Rate	NA	Prophylaxis: None		Vaccine:	Type 4 & 7
Shape	naked icosahedral				
Mean Diameter, μm	0.079	Size Range:	0.07–0.09 microns		
Growth Temperature	na	Survival Outside Host:		10 days on paper (T3)	
Inactivation	30 minutes contact with chemical disinfectants.				
Disinfectants	1% sodium hypochlorite				
Filter Nominal Rating	MERV 6	MERV 8	MERV 11	MERV 13	MERV 14
Removal Efficiency, %	11.0	13.0	22.9	53.9	74.8
UVGI Rate Constant	0.055	m^2/J	Media	Air	
Dose for 90% Inactivation	41.87	J/m^2	Ref.	Jensen 1964	
Suggested Indoor Limit	-	cfu/m^3			
Genome Size (bp)	34,794		48.9 % G+C	51.1 % T+A	
Related Species	Adenovirus Types 1, 2, 3, 4, 5, and 7				
Notes	CDC Reportable. Can survive 3–8 weeks on glass, steel, and tile.				
Photo Credit	Centers for Disease Control, PHIL# 237, Dr. G. WilliamGary, Jr.				
References	Dalton 1973, Fraenkel-Conrat 1985, Freeman 1985, Mahy 1975, Murray 1999, Ryan 1994, Canada 2001, Collins & Kennedy 1993				

Aeromonas

GROUP	Bacteria
TYPE	Gram-
GENUS	Aeromonas
FAMILY	Aeromonadaceae
DISEASE GROUP	Non-communicable
BIOSAFETY LEVEL	Risk Group 2
Infectious Dose (ID_{50})	unknown
Lethal Dose (LD_{50})	-
Infection Rate	none
Incubation Period	unknown
Peak Infection	NA
Annual Cases	-
Annual Fatalities	-

Can be found in natural waters and soil, and may contaminate medical equipment or wounds as a nosocomial pathogen. At least four species are recognized, of which A. hydrophila, A. sobris, and A. caviae are associated with disease. Have been reported as opportunistic pathogens in immunocompromised hosts. Evidence suggests an occasional role in gastroenteritis through production of a toxin with enterotoxic and cytotoxic properties, but the association is not definitive.

Disease or Infection	Non-respiratory opportunistic infections, gastroenteritis, bacteremia.				
Natural Source	Environmental, water, soil.				
Toxins	possible				
Point of Infection	burns, wounds				
Symptoms	Acute diarrhea is the most common symptom. Wound infections and cellulitis are possible, especially in the immunocompromised.				
Treatment	Chloramphenicol, trimethoprim-sulfamethoxazole, quinolones.				
Untreated Fatality Rate	-	Prophylaxis:		Vaccine: none	
Shape	rods				
Mean Diameter, μm	2.098	Size Range:	1–4.4 microns		
Growth Temperature	10–41°C	Survival Outside Host:		survives outdoors	
Inactivation	Moist heat: 121°C for 30 min.				
Disinfectants	1% sodium hypochlorite, phenolics, formaldehyde, glutaraldehyde.				
Filter Nominal Rating	MERV 6	MERV 8	MERV 11	MERV 13	MERV 14
Removal Efficiency, %	45.2	60.5	82.7	98.8	99.0
UVGI Rate Constant	0.2031	m^2/J	Media	-	
Dose for 90% Inactivation	11.34	J/m^2	Ref.	Hyllseth 1998	
Suggested Indoor Limit	0	cfu/m^3			
Genome Size (bp)	4,700,000				
Related Species	Aeromonas salmonicida (not pathogenic), Plesiomonas.				
Notes	Shows resistance to penicillins and cephalosporins.				
Photo Credit	Centers for Disease Control, PHIL# 1255, Dr. William A. Clark.				
References	Braude 1981, Freeman 1985, Mitscherlich 1984, Murray 1999, Prescott 1996, Ryan 1994, Castle 1987, Weinstein 1991, Mandell 2000				

Alcaligenes

	GROUP	Bacteria
	TYPE	Gram-
	GENUS	Alcaligenes
	FAMILY	(glucose nonoxidizer)
	DISEASE GROUP	Endogenous
	BIOSAFETY LEVEL	Risk Group 2
	Infectious Dose (ID_{50})	unknown
	Lethal Dose (LD_{50})	-
	Infection Rate	-
	Incubation Period	-
	Peak Infection	NA
	Annual Cases	rare
	Annual Fatalities	rare

Can be found in soil and water and may be an innocuous inhabitant of man, especially in the respiratory tract and gastrointestinal tract in hospitalized patients. It can infect immunocompromised hosts but is uncommon. Infection results when microbes are introduced into wounds or colonize immunosuppressed hosts. Most isolates of A. faecalis from blood or respiratory secretions are related to contamination of hospital equipment or fluids. Some species affect other areas of the body.

Disease or Infection	opportunistic infections				
Natural Source	Humans, soil, water, nosocomial				
Toxins	none				
Point of Infection	Upper Respiratory Tract, blood, urine, wounds				
Symptoms	Opportunistic infections of wounds or other parts of the body.				
Treatment	Aminoglycosides, beta-lactams, fluoroquinolones. Nosocomial strains may be resistant to common antibiotics.				
Untreated Fatality Rate	-	Prophylaxis: -		Vaccine:	none
Shape	rods				
Mean Diameter, μm	0.775	Size Range:	0.5–1.2 × 0.5–2.6 microns		
Growth Temperature	25–37°C	Survival Outside Host:		unknown	
Inactivation	Moist heat: 121°C for 30 min.				
Disinfectants	1% sodium hypochlorite, phenolics, formaldehyde, glutaraldehyde.				
Filter Nominal Rating	MERV 6	MERV 8	MERV 11	MERV 13	MERV 14
Removal Efficiency, %	15.5	22.6	40.7	80.5	95.8
UVGI Rate Constant	(unknown)		Media	-	
Dose for 90% Inactivation	-		Ref.	-	
Suggested Indoor Limit	-	cfu/m^3			
Genome Size (bp)					
Related Species	aka Alkaligenes, Type Species: A. faecalis, A. odorans, A. piechaudii				
Notes	Usual habitat is the respiratory tract or the intestinal tract.				
Photo Credit	Centers for Disease Control, PHIL# 1038, Dr. William A. Clark.				
References	Braude 1981, Freeman 1985, Mitscherlich 1984, Murray 1999, Prescott 1996, Ryan 1994, Castle 1987, Weinstein 1991, Mandell 2000				

Alternaria alternata

GROUP	Fungal Spore
CLASS	Euascomycetes
ORDER	Pleiosporales
FAMILY	Pleiosporaceae
DISEASE GROUP	Non-communicable
BIOSAFETY LEVEL	Risk Group 1
Infectious Dose (ID_{50})	NA
Lethal Dose (LD_{50})	-
Infection Rate	none
Incubation Period	-
Peak Infection	NA
Annual Cases	-
Annual Fatalities	-

Non-pathogenic but can cause allergic reactions. A common indoor air contaminant. Indoor levels can exceed outdoor levels. Can produce toxins. Alternaria is considered a nonpathogenic fungi, but it can cause considerable distress to allergic patients. As a fungal allergen, it can be a contributing factor in Sick Building Syndrome. It can produce opportunistic skin infections in the immunocompromised. A common tomato pathogen. Can occur on various foodstuffs and textiles.

Disease or Infection	allergic alveolitis, rhinitis, sinusitis, asthma, toxic reactions				
Natural Source	Environmental, indoor growth on paint, dust, filters, & cooling coils.				
Toxins	alternariol, tenuazonic acid, altenuisol				
Point of Infection	Upper Respiratory Tract				
Symptoms	Respiratory irritation, rhinitis, toxic reactions, skin infections in the immunosuppressed.				
Treatment	antifungal therapy, amphotericin B				
Untreated Fatality Rate	-	Prophylaxis: none		Vaccine: none	
Shape	ovoid spore				
Mean Diameter, μm	11.225	Size Range:	7–18 × 18–83 microns		
Growth Temperature	25–37°C	Survival Outside Host:	survives outdoors		
Inactivation	Moist heat: 121°C for 30 min.				
Disinfectants	1% sodium hypochlorite, phenolics, formaldehyde, glutaraldehyde.				
Filter Nominal Rating	MERV 6	MERV 8	MERV 11	MERV 13	MERV 14
Removal Efficiency, %	65.0	75.2	94.0	99.0	99.0
UVGI Rate Constant	(unknown)	Media		-	
Dose for 90% Inactivation	-	Ref.		-	
Suggested Indoor Limit	150–500 cfu/m^3	(in mix with other nonpathogenic fungi)			
Genome Size (bp)					
Related Species	Formerly A. tenuis, A. tomato. Ascomycetes teleomorph.				
Notes	Dermatiaceous hyphomycetes				
Photo Credit	Image courtesy of Neil Carlson, University of Minnesota.				
References	Freeman 1985, Howard 1983, Lacey 1988, Murray 1999, Pope 1993, Rao 1996, Ryan 1994, Smith 1989, Woods 1997, Mandell 2000				

Arthrinium phaeospermum

GROUP	Fungal Spore
CLASS	Euascomycetes
ORDER	Sordariales
FAMILY	Lasiosphaeriaceae
DISEASE GROUP	Non-communicable
BIOSAFETY LEVEL	Risk Group 2
Infectious Dose (ID_{50})	NA
Lethal Dose (LD_{50})	NA
Infection Rate	none
Incubation Period	-
Peak Infection	NA
Annual Cases	-
Annual Fatalities	-

Non-pathogenic but can cause allergic reactions. A suspected fungal allergen, it can be a contributing factor in Sick Building Syndrome/Building Related Illnes (SBS/BRI). Often identified as a black yeast. Widespread saprophyte on dead plant material, particularly swampy grasses. Less frequently recovered from soil. This fungus has also been identified in various subcutaneous infections. Decomposes cellulose. Susceptible to cycloheximide. No toxic related diseases are of record to date.

Disease or Infection	rhinitis, asthma				
Natural Source	Environmental				
Toxins	none				
Point of Infection	Upper Respiratory Tract				
Symptoms	Respiratory irritation, allergic reactions, cutaneous infections.				
Treatment	NA				
Untreated Fatality Rate	NA	**Prophylaxis:** none		**Vaccine:** none	
Shape	ovoid spore				
Mean Diameter, μm	5	**Size Range:**	5×5–7.5 microns		
Growth Temperature	38°C	**Survival Outside Host:**		survives outdoors	
Inactivation	Moist heat: 121°C for 30 min.				
Disinfectants	1% sodium hypochlorite, phenolics, formaldehyde, glutaraldehyde.				
Filter Nominal Rating	MERV 6	MERV 8	MERV 11	MERV 13	MERV 14
Removal Efficiency, %	63.5	75.0	93.8	99.0	99.0
UVGI Rate Constant	(unknown)	**Media**		-	
Dose for 90% Inactivation	-	**Ref.**		-	
Suggested Indoor Limit	150–500 cfu/m^3	(in mix with other nonpathogenic fungi)			
Genome Size (bp)					
Related Species	Ascomycetes teleomorph				
Notes	Dermatiaceous or moniliaceous hyphomycetes				
Photo Credit	Image courtesy of Neil Carlson, University of Minnesota.				
References	Freeman 1985, Howard 1983, Lacey 1988, Murray 1999, Pope 1993, Rao 1996, Ryan 1994, Smith 1989, Woods 1997, Sutton 1998				

Aspergillus

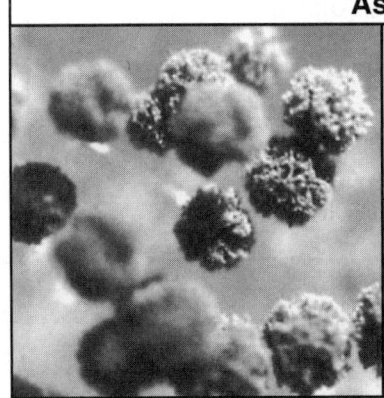

GROUP	Fungal Spore
CLASS	Hyphomycetes
ORDER	Eurotiales
FAMILY	Trichocomaceae
DISEASE GROUP	Non-communicable
BIOSAFETY LEVEL	Risk Group 2
Infectious Dose (ID_{50})	9643 - 58154
Lethal Dose (LD_{50})	-
Infection Rate	none
Incubation Period	3–30 days
Peak Infection	NA
Annual Cases	uncommon
Annual Fatalities	-

Aspergillus represents several related fungi which cause aspergillosis. This disease most often affects the external ear, but also the lungs. Aspergillus species are common in the soil, and spores become airborne in dry windy weather. Spores can germinate in moist areas of buildings and ventilation systems. Sometimes associated with Sick Building Syndrome. Can be fatal to those with immunodeficiency. Sometimes found on human body surfaces. May produces MVOCs.

Disease or Infection	aspergillosis, alveolitis, asthma, allergic fungal sinusitis, ODTS, toxic reactions, pneumonia possible				
Natural Source	Environmental, nosocomial, indoor growth on insulation & coils.				
Toxins	aflatoxin, fumigaclavines, gliotoxin, fumigatoxin, helvolic acid, etc.				
Point of Infection	Upper Respiratory Tract				
Symptoms	Clinical manifestation of symptoms is largely determined by the immunological state of the patient.				
Treatment	Infection is susceptible to amphotecirin B, itraconazole, or voriconazole.				
Untreated Fatality Rate	NA	Prophylaxis: none		Vaccine: none	
Shape	spherical spore				
Mean Diameter, μm	3.354	Size Range:	2.5–4.5 microns		
Growth Temperature	37°C	Survival Outside Host:	survives outdoors		
Inactivation	Moist heat: 121°C for 30 min.				
Disinfectants	1% sodium hypochlorite, 2% glutaraldehyde.				
Filter Nominal Rating	MERV 6	MERV 8	MERV 11	MERV 13	MERV 14
Removal Efficiency, %	58.2	72.1	91.8	99.0	99.0
UVGI Rate Constant	0.0007	m^2/J	Media	Air	
Dose for 90% Inactivation	3289	J/m^2	Ref.	Luckiesh 1946	
Suggested Indoor Limit	150–500	cfu/m^3	(in mix with other nonpathogenic fungi)		
Genome Size (bp)	35,900,000 (A. niger)				
Related Species	A. niger, A. flavus, A. terreus, A. versicolor, A. ustus, A. sydowi, etc.				
Notes	CDC Reportable. Water activity Aw = 0.78–0.82				
Photo Credit	Image courtesy of Neil Carlson, University of Minnesota.				
References	Joseph 1983, Lacey 1988, Murray 1999, Rao 1996, Ryan 1994, Smith 1989, Weinstein 1991, Arnow 1991, Canada 2001, Sutton 1998				

Aureobasidium pullulans

GROUP	Fungal Spore
CLASS	Euascomycetes
ORDER	Dothideales
FAMILY	Dothioraceae
DISEASE GROUP	Non-communicable
BIOSAFETY LEVEL	Risk Group 1
Infectious Dose (ID_{50})	NA
Lethal Dose (LD_{50})	NA
Infection Rate	none
Incubation Period	-
Peak Infection	NA
Annual Cases	-
Annual Fatalities	-

Reportedly allergenic. Can grow indoors on gypsum. Frequently found in moist environments and main habitat appears to be on the aerial parts of plants. This species has been associated with deratitis, peritonitis, pulmonary infection, and invasive disease in AIDS patients. Sometimes recovered as a skin contaminant from humans. Does not cause any toxic diseases or reactions. May be recovered as a contaminant from human cutaneous sites.

Disease or Infection	allergic alveolitis, sequoiosis, keratitis, peritonitis, pulmonary infection, invasive disease in AIDS patients				
Natural Source	Environmental, indoor growth on building materials, gypsum, paint, and filters.				
Toxins	none				
Point of Infection	Upper Respiratory Tract				
Symptoms	Allergic reactions.				
Treatment	NA				
Untreated Fatality Rate	NA	Prophylaxis: none		Vaccine:	none
Shape	ovoid spore				
Mean Diameter, μm	4.899	Size Range:	4–6 × 8–12 microns		
Growth Temperature	25°C	Survival Outside Host:		survives 2 years outdoors	
Inactivation	Moist heat: 121°C for 30 min.				
Disinfectants	1% sodium hypochlorite, phenolics, formaldehyde, glutaraldehyde.				
Filter Nominal Rating	MERV 6	MERV 8	MERV 11	MERV 13	MERV 14
Removal Efficiency, %	63.3	75.0	93.7	99.0	99.0
UVGI Rate Constant	(unknown)		Media	-	
Dose for 90% Inactivation	-		Ref.	-	
Suggested Indoor Limit	150–500	cfu/m^3	(in mix with other nonpathogenic fungi)		
Genome Size (bp)					
Related Species	Formerly Pullularia pullulans				
Notes	Ascomycetes teleomorph, max temp for growth 35°C, CDC Reportable				
Photo Credit	Image courtesy of Neil Carlson, University of Minnesota.				
References	Freeman 1985, Howard 1983, Lacey 1988, Murray 1999, Rao 1996, Ryan 1994, Smith 1989, Woods 1997, Sutton 1998				

Bacillus anthracis		
GROUP	Bacterial Spore	
TYPE	Gram+	
GENUS	Bacillus	
FAMILY	Bacillaceae	
DISEASE GROUP	Non-communicable	
BIOSAFETY LEVEL	Risk Group 2	
Infectious Dose (ID_{50})	1300–10000	
Lethal Dose (LD_{50})	28000	
Infection Rate	none	
Incubation Period	2–3 days	
Peak Infection	NA	
Annual Cases	rare	
Annual Fatalities	rare	

Causes anthrax, which is primarily a disease of lower animals that can be transmitted to man by contact and by inhalation of spores. Pulmonary anthrax was commonly known as "woolsorter's disease." It occasionally causes minor outbreaks in agriculture and related industries. Anthrax rapidly spreads from the lungs to other organs, and is deadly if untreated. The United States, Japan, Iraq, the Soviet Union, and other countries have researched and developed anthrax as a biological weapon.

Disease or Infection	anthrax, woolsorter's disease				
Natural Source	Cattle, sheep, other animals, soil.				
Toxins	none				
Point of Infection	Upper Respiratory Tract				
Symptoms	Inhalation anthrax: respiratory distress, fever and shock with death shortly thereafter.				
Treatment	Ciprofloxacin, doxycycline, tetracyclines, erythromycin, chloramphenicol				
Untreated Fatality Rate	5–20%	**Prophylaxis:** Antibiotics		**Vaccine:** Available	
Shape	spherical spore				
Mean Diameter, μm	1.118	**Size Range:**	1–1.25 microns		
Growth Temperature	0–45°C	**Survival Outside Host:**	years		
Inactivation	Moist heat: 121°C for 30 minutes.				
Disinfectants	5% formalin, 2% glutaraldehyde formaldehyde				
Filter Nominal Rating	MERV 6	MERV 8	MERV 11	MERV 13	MERV 14
Removal Efficiency, %	24.4	35.3	57.1	92.6	98.6
UVGI Rate Constant	0.0031	m^2/J	**Media**	Plates (spores)	
Dose for 90% Inactivation	743	J/m^2	**Ref.**	Knudson 1986	
Suggested Indoor Limit	0	cfu/m^3			
Genome Size (bp)	5,227,293		35.3 % G+C	64.7 % T+A	
Related Species	B. cereus, Type Species: B. subtilis (B. globigii)				
Notes	CDC Reportable.				
Photo Credit	Centers for Disease Control, PHIL# 1064, Dr. William A. Clark.				
References	Brachman 1966, Braude 1981, Inglesby 1999, Murray 1999, Ryan 1994, Canada 2001, Collins & Kennedy 1993				

Bacteroides fragilis

GROUP	Bacteria
TYPE	Gram-
GENUS	Bacteroides
FAMILY	Bacteroidaceae
DISEASE GROUP	Endogenous
BIOSAFETY LEVEL	Risk Group 2
Infectious Dose (ID_{50})	unknown
Lethal Dose (LD_{50})	-
Infection Rate	none
Incubation Period	unknown
Peak Infection	-
Annual Cases	uncommon
Annual Fatalities	rare

Nonrespiratory but can become airborne. Predominantly intestinal tract flora. Rarely causes infections in the absence of other microorganisms. Can cause severe infection in proximity to mucosal surfaces where they exist normally. Infection can be prolonged and sometimes fatal. Sometimes found on human body surfaces.

Disease or Infection	opportunistic infections.				
Natural Source	Humans				
Toxins	none				
Point of Infection	burns, wounds				
Symptoms	Bacteremia, abcesses and lesions in lung, abdomen, brain. Periodontis, endocarditis, wound infections.				
Treatment	Susceptible to metrinidazole. May be resistant to penicillins.				
Untreated Fatality Rate	-	**Prophylaxis:** Vancomycin **Vaccine:** none			
Shape	rods, nonmotile or motile with perritrichous flagella				
Mean Diameter, μm	3.162	**Size Range:** 1–10 microns			
Growth Temperature	25–45°C	**Survival Outside Host:** 24–72 hours in air			
Inactivation	Moist heat, 121°C for 15 minutes. Dry heat, 160–170°C, for 1 hour.				
Disinfectants	1% sodium hypochlorite, 70% ethanol, 2% glutaraldehyde				
Filter Nominal Rating	MERV 6	MERV 8	MERV 11	MERV 13	MERV 14
Removal Efficiency, %	56.9	71.2	91.1	99.0	99.0
UVGI Rate Constant	(unknown)	**Media**	na		
Dose for 90% Inactivation	-	**Ref.**	-		
Suggested Indoor Limit	0 cfu/m³				
Genome Size (bp)	5,205,140	43.19 % G+C 56.81 % T+A			
Related Species	B. ureolyticus, B. capillosis, B. bivius, B. disiens, B. vulgatus				
Photo Credit	Photo reprinted from D.B. de Tesoro (1975), Rev. Fac. Agron. (Maracay), IX (3) ; 5–53 with permission from E.I. Chaparro-Martìnez, Revista de la Facultad de Agronomìa, Universidad Central de Venezuela.				
References	Braude 1981, Freeman 1985, Mitscherlich 1984, Murray 1999, Prescott 1996, Ryan 1994				

Blastomyces dermatitidis

GROUP	Fungal Spore
CLASS	Euascomycetes
ORDER	Onygenales
FAMILY	Onygenaceae
DISEASE GROUP	Non-communicable
BIOSAFETY LEVEL	Risk Group 2
Infectious Dose (ID_{50})	11000
Lethal Dose (LD_{50})	unknown
Infection Rate	none
Incubation Period	weeks
Peak Infection	NA
Annual Cases	rare
Annual Fatalities	-

Can resemble TB and spread beyond the lungs. Found mainly in north central and eastern United States. Causes blastomycosis. This malady is also known as Gilchrist's disease and Chicago disease. Entry is through the upper respiratory tract, but can spread to other locations. Males are more susceptible to this progressive disease than females. Like other fungi pathogenic for man, this fungi exhibits dimorphism, existing in one form in nature and another when causing infection.

Disease or Infection	blastomycosis, Gilchrist's Disease, Chicago disease, pneumonia possible				
Natural Source	Environmental, nosocomial.				
Toxins	none				
Point of Infection	Upper Respiratory Tract, skin				
Symptoms	Acute or chronic granulomatous mycosis of the lungs or skin. Indolent onset becoming chronic pulmonary infection.				
Treatment	Amphotericin B, itraconazole, ketoconazole, hydroxystilbamidine				
Untreated Fatality Rate	-	Prophylaxis:	none	Vaccine:	none
Shape	oval to spherical spore, dumb-bell shaped				
Mean Diameter, μm	12.649	Size Range:	8–20 microns		
Growth Temperature	37°C	Survival Outside Host:		survives outdoors	
Inactivation	Moist heat: 121°C for 30 min.				
Disinfectants	1% sodium hypochlorite, phenolics, formaldehyde, 10% formalin.				
Filter Nominal Rating	MERV 6	MERV 8	MERV 11	MERV 13	MERV 14
Removal Efficiency, %	65.0	75.2	94.0	99.0	99.0
UVGI Rate Constant	0.0247	m^2/J	Media	Plates (estimated)	
Dose for 90% Inactivation	93.22	J/m^2	Ref.	Chick 1963	
Suggested Indoor Limit	0	cfu/m^3			
Genome Size (bp)					
Related Species	Ascomycetes teleomorph, Formerly B. tulanensis				
Notes	CDC Reportable.				
Photo Credit	Centers for Disease Control, PHIL# 494, Dr. Libero Ajello.				
References	Collins 1993, Freeman 1985, Howard 1983, Lacey 1988, Murray 1999, Ryan 1994, Smith 1989, Miyaji 1987, Sorensen 1999, DiSalvo 1983, Canada 2001				

Bordetella pertussis

GROUP	Bacteria
TYPE	Gram-
GENUS	Bordetella
FAMILY	Alcaligenaceae
DISEASE GROUP	Communicable
BIOSAFETY LEVEL	Risk Group 2
Infectious Dose (ID_{50})	(4)
Lethal Dose (LD_{50})	(1314)
Infection Rate	high
Incubation Period	7–10 days
Peak Infection	7–14 days
Annual Cases	6,564
Annual Fatalities	15

Bordetella pertussis is the cause of whooping cough. It produces microbial toxins, which are primarily responsible for the disease symptoms. Occurring worldwide, this infection almost exclusively affects children. Almost two-thirds of cases are under 1 year of age. Asymptomatic cases, however, are more frequent. It is highly contagious and transmits via airborne bioaerosols, by fomites, and by direct contact.

Disease or Infection	whooping cough, toxic reactions				
Natural Source	Humans, nosocomial				
Toxins	none				
Point of Infection	Upper Respiratory Tract, trachea				
Symptoms	Three stages occur: coughing (1–2 weeks), violent whooping cough (2–6 weeks), recovery for several weeks.				
Treatment	14 days treatment with erythromycin or TMP-SMX, oxygenation, hydration & electrolyte balance				
Untreated Fatality Rate	-	**Prophylaxis:** Antibiotics		**Vaccine:**	Available
Shape	coccobacilli				
Mean Diameter, μm	0.245	**Size Range:**	0.2–0.3 × 0.5–1 microns		
Growth Temperature	35–37°C	**Survival Outside Host:**	1 hrs–7 days		
Inactivation	Moist heat: 121°C for 15 minutes. Dry heat: 170°C for 1 hour.				
Disinfectants	1% sodium hypochlorite, 70% ethanol, iodines, phenolics, glutaraldehyde, formaldehyde.				
Filter Nominal Rating	MERV 6	MERV 8	MERV 11	MERV 13	MERV 14
Removal Efficiency, %	5.6	7.3	14.5	38.5	60.7
UVGI Rate Constant	(unknown)		**Media**	-	
Dose for 90% Inactivation	-		**Ref.**	-	
Suggested Indoor Limit	0	cfu/m³			
Genome Size (bp)	4,086,189		61.58 % G+C	38.42 % T+A	
Notes	CDC Reportable.				
Photo Credit	Centers for Disease Control, PHIL# 254, Janice Carr.				
References	Braude 1981, Freeman 1985, Mitscherlich 1984, Murray 1999, Prescott 1996, Ryan 1994, Castle 1987, Weinstein 1991, Canada 2001				

Brucella

GROUP	Bacteria
TYPE	Gram-
GENUS	Brucella
FAMILY	Brucellaceae
DISEASE GROUP	Non-communicable
BIOSAFETY LEVEL	Risk Group 3
Infectious Dose (ID_{50})	1300
Lethal Dose (LD_{50})	-
Infection Rate	-
Incubation Period	5–60 days
Peak Infection	-
Annual Cases	98
Annual Fatalities	rare

Zoonotic. Contact with animals is usually required for infection but airborne infection has occurred. May become chronic. Predominantly an occupational disease of those who work with infected animals. Route of infection may be via ingestion, abraded skin, or through the mucous membranes. A genitourinary infection of sheep, cattle, pigs, and other animals.

Disease or Infection	Brucellosis, undulant fever, Malta fever.				
Natural Source	Goats, cattle, swine, dogs, sheep, caribou, elk, coyotes, camels.				
Toxins	none				
Point of Infection	Upper Respiratory Tract, skin				
Symptoms	Acute onset with intermittent fever, fatigue, headache, profuse sweating, chills, arthralgia, local infections, long recovery period.				
Treatment	Tetracyclines, streptomycin usually combined with doxycycline, TMP-SMX, combined with aminoglycoside				
Untreated Fatality Rate	<2%	Prophylaxis: none		Vaccine: none	
Shape	short rods				
Mean Diameter, μm	0.566	Size Range:	0.4–0.8 × 0.4–1.5 microns		
Growth Temperature	37.5°C	Survival Outside Host:	32–135 days		
Inactivation	Moist heat: 121°C for 15 minutes. Dry heat: 170°C for 1 hour.				
Disinfectants	1% sodium hypochlorite, 70% ethanol, glutaraldehyde, formaldehyde.				
Filter Nominal Rating	MERV 6	MERV 8	MERV 11	MERV 13	MERV 14
Removal Efficiency, %	10.4	15.1	29.2	66.3	88.6
UVGI Rate Constant	(unknown)		Media	-	
Dose for 90% Inactivation	-		Ref.	-	
Suggested Indoor Limit	0				
Genome Size (bp)	3,315,173		57.2 % G+C	42.8 % T+A	
Related Species	B. melitensis, B.suis, B.abortus, B.canis.				
Notes	CDC Reportable. Open Air Factor rate constant k = 1.5–10 per minute.				
Photo Credit	Centers for Disease Control, PHIL# 734, Dr. Marshall Fox.				
References	Braude 1981, Murray 1999, Mitscherlich 1984, Prescott 1996, Ryan 1994, Canada 2001, Mandell 2000, Ellison 2000.				

Burkholderia cenocepacia

GROUP	Bacteria
TYPE	Gram-
GENUS	Burkholderia
FAMILY	Burkholderiaceae
DISEASE GROUP	Non-communicable
BIOSAFETY LEVEL	Risk Group 1
Infectious Dose (ID_{50})	unknown
Lethal Dose (LD_{50})	unknown
Infection Rate	-
Incubation Period	unknown
Peak Infection	-
Annual Cases	-
Annual Fatalities	-

Non-respiratory. Formerly Pseudomonas cepacia. Can contaminate medical devices, reagents, and disinfectants in much the same manner as P. aeruginosa. Causes only opportunistic infections in hospital settings. Often associated with cases of P. aeruginosa infections. Has complicated the course of cystic fibrosis patients. Reports vary regarding frequency of isolation.

Disease or Infection	Opportunistic infections.				
Natural Source	Environmental				
Toxins	none				
Point of Infection	Upper Respiratory Tract, burns, wounds				
Symptoms	Opportunistic infections.				
Treatment	Ceftazidime, imipenem, doxycycline, minocycline, ciprofloxacin, gentamicin				
Untreated Fatality Rate	-	Prophylaxis: none		Vaccine: none	
Shape	rods				
Mean Diameter, μm	0.707	Size Range: 0.5–1 × 1.5–4 microns			
Growth Temperature	4–43°C	Survival Outside Host: -			
Inactivation	Moist heat: 55°C for 10 minutes.				
Disinfectants	1% sodium hypochlorite, 70% ethanol, 2% glutaraldehyde.				
Filter Nominal Rating	MERV 6	MERV 8	MERV 11	MERV 13	MERV 14
Removal Efficiency, %	13.7	20.1	37.1	76.6	94.3
UVGI Rate Constant	0.0396	m^2/J	Media	Water	
Dose for 90% Inactivation	58	J/m^2	Ref.	Abshire 1981	
Suggested Indoor Limit	0	cfu/m^3			
Genome Size (bp)	8,056,000		66.9 % G+C	33.1 % T+A	
Related Species	Formerly B. cepacia, and Pseudomonas cepacia				
Notes	Often found with Pseudomonas aeruginosa in nosocomial infections.				
Photo Credit	Image shows B. cepacia. Centers for Disease Control, PHIL# 255, Janice Carr.				
References	Braude 1981, Freeman 1985, Mitscherlich 1984, Murray 1999, Prescott 1996, Ryan 1994				

Burkholderia mallei

GROUP	Bacteria
TYPE	Gram-
GENUS	Burkholderia
FAMILY	Burkholderiaceae
DISEASE GROUP	Non-communicable
BIOSAFETY LEVEL	Risk Group 3
Infectious Dose (ID_{50})	3200
Lethal Dose (LD_{50})	-
Infection Rate	none
Incubation Period	1–14 days
Peak Infection	NA
Annual Cases	-
Annual Fatalities	none

Formerly Pseudomonas mallei, B. mallei causes a disease called Glanders, which is a disease of horses and mules. It has, on rare occasions been transmitted to humans, but has been eradicated in the West. The infection route is most probably through skin abrasions. The disease manifests itself as local suppurative or acute pulmonary infection.

Disease or Infection	Glanders, fever, opportunistic infections				
Natural Source	Environmental, horses, mules, nosocomial.				
Toxins	none				
Point of Infection	skin				
Symptoms	Chronic form with cough and mucous discharge. Septicemic form with fever, chills, death within 7–10 days.				
Treatment	Ceftazidime, imipenem, doxycycline, minocycline, ciprofloxacin, gentamicin				
Untreated Fatality Rate	-	Prophylaxis: none		Vaccine:	none
Shape	rods				
Mean Diameter, µm	0.674	Size Range:	0.3–0.8 × 1.4–4 microns		
Growth Temperature	20–42°C	Survival Outside Host:	30 days in water		
Inactivation	Moist heat: 55°C for 10 minutes.				
Disinfectants	1% sodium hypochlorite, 70% ethanol, 2% glutaraldehyde.				
Filter Nominal Rating	MERV 6	MERV 8	MERV 11	MERV 13	MERV 14
Removal Efficiency, %	12.9	18.9	35.3	74.4	93.3
UVGI Rate Constant	(unknown)		Media	-	
Dose for 90% Inactivation	-		Ref.	-	
Suggested Indoor Limit	50	cfu/m^3			
Genome Size (bp)	5,835,527		68.4% G+C	31.6% T+A	
Related Species	Formerly Pseudomonas mallei. Related to B. pseudomallei.				
Notes	Occupational risk in agriculture, veterinarians				
Photo Credit	Image shows a glanders infection, photo courtesy of Dr. Dave Fritz of USAMRIID.				
References	Braude 1981, Freeman 1985, Mitscherlich 1984, Murray 1999, Prescott 1996, Ryan 1994, Castle 1987, Weinstein 1991, Canada 2001				

Burkholderia pseudomallei

GROUP	Bacteria
TYPE	Gram-
GENUS	Burkholderia
FAMILY	Burkholderiaceae
DISEASE GROUP	Non-communicable
BIOSAFETY LEVEL	Risk Group 3
Infectious Dose (ID_{50})	unknown
Lethal Dose (LD_{50})	-
Infection Rate	none
Incubation Period	2 days min.
Peak Infection	NA
Annual Cases	rare
Annual Fatalities	rare

Formerly Pseudomonas pseudomallei, B. pseudomallei occurs primarily in rodents and other animals in the South Pacific. In these areas, it is endemic and can be acquired by humans, but is very rare in the West. Being an animal respiratory disease, like brucellosis, it can be transmitted to humans outdoors via direct contact or inhalation. It can also be found in moist soil or warm water, from which it may contaminate open wounds.

Disease or Infection	meliodosis, opportunistic infections				
Natural Source	Environmental, rodents, soil, water, nosocomial.				
Toxins	none				
Point of Infection	Upper Respiratory Tract				
Symptoms	May resemble typhoid fever or tuberculosis. Can vary from chronic infection to rapidly fatal septicemia.				
Treatment	TMP-SMX, ceftazidime, imipenem, doxycycline, ciprofoxacinsulphas, tetracycline, chloramphenicol				
Untreated Fatality Rate	-	Prophylaxis:	none	Vaccine:	none
Shape	rods				
Mean Diameter, μm	0.494	Size Range:	0.3–0.8 × 1–3 microns		
Growth Temperature	5–42°C	Survival Outside Host:	years in soil & water		
Inactivation	Moist heat: 121°C for 15 min. Dry heat: 170°C for 1 hr.				
Disinfectants	1% sodium hypochlorite, 70% ethanol, glutaraldehyde.				
Filter Nominal Rating	MERV 6	MERV 8	MERV 11	MERV 13	MERV 14
Removal Efficiency, %	8.8	12.8	25.2	60.0	84.0
UVGI Rate Constant	(unknown)	Media		-	
Dose for 90% Inactivation	-	Ref.		-	
Suggested Indoor Limit	50	cfu/m^3			
Genome Size (bp)	7,247,547		68.06 % G+C	31.94 % T+A	
Related Species	(previously Pseudomonas pseudomallei)				
Photo Credit	Donald Woods of the Canadian Bacterial Diseases Network, University of Calgary, Alberta.				
References	Braude 1981, Freeman 1985, Mitscherlich 1984, Murray 1999, Prescott 1996, Ryan 1994, Canada 2001				

Candida albicans

GROUP	Fungal Spore
CLASS	Ascomycetes
ORDER	Saccharomycetales
FAMILY	Saccharomycetaceae
DISEASE GROUP	Endogenous
BIOSAFETY LEVEL	Risk Group 1
Infectious Dose (ID_{50})	unknown
Lethal Dose (LD_{50})	NA
Infection Rate	-
Incubation Period	variable
Peak Infection	-
Annual Cases	uncommon
Annual Fatalities	-

Opportunistic and non-respiratory. Common in the skin, mouth, vagina, and intestines. Requires compromised health to become a problem. Spreads by contact with excretions of mouth, skin, and feces from carriers. May become airborne and settle on equipment. May originate from mucosal lesions. Common cause of superficial skin infection, oral and vaginal infection, sepsis and disseminated disease. Related species can cause some infections.

Disease or Infection	candidamycosis, candidiasis, bronchitis, pneumonitis, onychomycosis, pneumonia possible in the immunocompromised				
Natural Source	Humans, leaves and flowers, water and soil.				
Toxins	none				
Point of Infection	Upper Respiratory Tract, mouth, skin				
Symptoms	Mycosis of superficial layers of skin or mucous membranes, ulcers in esophagus, gastrointestinal tract or bladder.				
Treatment	Nystatin, clotrimazole, ketoconazole, fluconazole, amphotecerin B.				
Untreated Fatality Rate	NA	Prophylaxis: none		Vaccine: none	
Shape	spherical spore				
Mean Diameter, µm	4.899	Size Range:	4–6 microns		
Growth Temperature	37°C	Survival Outside Host:		can survive outside host	
Inactivation	Moist heat: 121°C for 15 min. Dry heat: 170°C for 1 hr.				
Disinfectants	1% sodium hypochlorite, formaldehyde, 2% glutaraldehyde.				
Filter Nominal Rating	MERV 6	MERV 8	MERV 11	MERV 13	MERV 14
Removal Efficiency, %	63.3	75.0	93.7	99.0	99.0
UVGI Rate Constant	0.0100	m^2/J	Media	-	
Dose for 90% Inactivation	23,026	J/m^2	Ref.	Dolman 1989 (Yeast)	
Suggested Indoor Limit	0	cfu/m^3			
Genome Size (bp)	20,000,000				
Related Species	C. krusei, C. glabrata, C. lipolytica, C. tropicalis				
Notes	Formerly Monilia albicans				
Photo Credit	Centers for Disease Control, PHIL# 291, Maxine Jalbert, Dr. Leo Kaufman.				
References	Freeman 1985, Howard 1983, Lacey 1988, Murray 1999, Rao 1996, Ryan 1994, Smith 1989, Canada 2001				

Cardiobacterium

GROUP	Bacteria
TYPE	Gram-
GENUS	Cardiobacterium
FAMILY	Cardiobacteriaceae
DISEASE GROUP	Endogenous
BIOSAFETY LEVEL	Risk Group 2
Infectious Dose (ID_{50})	unknown
Lethal Dose (LD_{50})	-
Infection Rate	-
Incubation Period	-
Peak Infection	NA
Annual Cases	rare
Annual Fatalities	-

This bacteria normally inhabits the respiratory tract, nose, throat, mouth, or intestines. Generally requires compromised health to be a problem and so is basically a nosocomial infection or a concern to the immunodeficient. This organism requires incubation in an atmosphere enhanced with carbon dioxide and has a predilection for the heart valve when causing disease.

Disease or Infection	opportunistic infections, endocarditis				
Natural Source	Humans, nosocomial.				
Toxins	none				
Point of Infection	Upper Respiratory Tract				
Symptoms	Subacute presentation with insidious onset.				
Treatment	Chloramphenicol, beta-lactams, and tetracycline. Variable susceptibility to vancomycin, aminoglycosides, erythromycin, and clindamycin.				
Untreated Fatality Rate	-	**Prophylaxis:** none		**Vaccine:** none	
Shape	rods				
Mean Diameter, μm	0.612	**Size Range:**	0.5–0.75 × 1–3 microns		
Growth Temperature	30–37°C	**Survival Outside Host:**	does not survive well		
Inactivation	Moist heat: 121°C for 30 min.				
Disinfectants	1% sodium hypochlorite, phenolics, formaldehyde, glutaraldehyde.				
Filter Nominal Rating	MERV 6	MERV 8	MERV 11	MERV 13	MERV 14
Removal Efficiency, %	11.4	16.7	31.8	70.0	90.9
UVGI Rate Constant	(unknown)		**Media**	-	
Dose for 90% Inactivation	-		**Ref.**	-	
Suggested Indoor Limit	-				
Genome Size (bp)					
Related Species	C. hominis				
Notes					
Photo Credit	Centers for Disease Control, PHIL# 761.				
References	Braude 1981, Freeman 1985, Mitscherlich 1984, Murray 1999, Prescott 1996, Ryan 1994, Castle 1987, Weinstein 1991, Mandell 2000				

Chaetomium globosum

GROUP	Fungal Spore
CLASS	Euascomycetes
ORDER	Sordariales
FAMILY	Chaetomiaceae
DISEASE GROUP	Non-communicable
BIOSAFETY LEVEL	Risk Group 1
Infectious Dose (ID_{50})	unknown
Lethal Dose (LD_{50})	-
Infection Rate	none
Incubation Period	-
Peak Infection	NA
Annual Cases	-
Annual Fatalities	-

Reportedly allergenic, it can be found growing on water-damaged cellulose products. Found in the soil, air and on plant debris. Is considered an agent of onychomycosis, peritonitis, and cutaneous lesions. C. globosum and C. atrobrunneum have been implicated in fatal systemic mycoses. C. atrobrunneum is found in the soil, air, and on plant debris and should be considered as allergenic. Rarely, if ever, causes any toxic diseases or reactions. C. strumarium should be considered allergenic. Has been implicated in fatal brain abscesses in drug abusers. May produce MVOCs. Will grow indoors on dust, filters, and insulation. C. globosum is the most widely distributed species.

Disease or Infection	allergic alveolitis, onychomycosis, peritonitis, cutaneous lesions				
Natural Source	Environmental, soil, plant debris, compost, indoor growth				
Toxins	chaetoglobosins				
Point of Infection	Upper Respiratory Tract				
Symptoms	Allergic reactions.				
Treatment	Amphotericin B				
Untreated Fatality Rate	NA	Prophylaxis:	none	Vaccine:	none
Shape	ovoid spore				
Mean Diameter, μm	5.455	Size Range:	4.8–6.2 × 5.9–6.8 microns		
Growth Temperature	25°C	Survival Outside Host:	survives outdoors		
Inactivation	Moist heat: 121°C for 30 min.				
Disinfectants	1% sodium hypochlorite, phenolics, formaldehyde, glutaraldehyde.				
Filter Nominal Rating	MERV 6	MERV 8	MERV 11	MERV 13	MERV 14
Removal Efficiency, %	64.0	75.2	93.9	99.0	99.0
UVGI Rate Constant	(unknown)	Media	-		
Dose for 90% Inactivation	-	Ref.	-		
Suggested Indoor Limit	150–500 cfu/m³	(in mix with other nonpathogenic fungi)			
Genome Size (bp)					
Related Species	C. atrobrunneum, C. strumarium. Formerly C. cinnamomeum				
Notes	CDC Reportable.				
Photo Credit	Image courtesy of Neil Carlson, University of Minnesota.				
References	Freeman 1985, Howard 1983, Lacey 1988, Murray 1999, Pope 1993, Rao 1996, Ryan 1994, Smith 1989, Woods 1997, Sutton 1998				

Chlamydia pneumoniae

GROUP	Bacteria
TYPE	Gram-
GENUS	Chlamydia
FAMILY	Chlamydiaceae
DISEASE GROUP	Communicable
BIOSAFETY LEVEL	Risk Group 2
Infectious Dose (ID_{50})	unknown
Lethal Dose (LD_{50})	-
Infection Rate	0.5
Incubation Period	7 days
Peak Infection	-
Annual Cases	uncommon
Annual Fatalities	-

Common cause of "walking pneumonia" in young adults. Causes both upper and lower respiratory tract infections, including pneumonia. Non-seasonal outbreaks occur worldwide. This microbe is distinct from both Chlamydia (Chlamydophila) psittaci and Chlamydia trachomatis. Spread appears to be from person to person. May be capable of producing infections with a wide range of symptoms.

Disease or Infection	pneumonia, bronchitis, pharyngitis.				
Natural Source	Humans, nosocomial				
Toxins	none				
Point of Infection	Upper Respiratory Tract				
Symptoms	Mild respiratory tract infections, pneumonia.				
Treatment	Tetracycline may be effective.				
Untreated Fatality Rate	-	Prophylaxis: none		Vaccine:	none
Shape	spherical				
Mean Diameter, μm	0.548	Size Range:	0.2–1.5 microns		
Growth Temperature	33–41°C	Survival Outside Host:		unknown	
Inactivation	Moist heat: 121°C for 30 min.				
Disinfectants	1% sodium hypochlorite, 70% ethanol, glutaraldehyde.				
Filter Nominal Rating	MERV 6	MERV 8	MERV 11	MERV 13	MERV 14
Removal Efficiency, %	10.0	14.5	28.2	64.7	87.6
UVGI Rate Constant	(unknown)	Media		-	
Dose for 90% Inactivation	-	Ref.		-	
Suggested Indoor Limit	0	cfu/m^3			
Genome Size (bp)	1,230,230		59.4% G+C	40.6% T+A	
Related Species	C. psittaci. Type Species: C. trachomatis				
Notes	aka TWAR. CDC Reportable.				
Photo Credit	Centers for Disease Control, PHIL# 1349, Dr. Martin Hicklin.				
References	Braude 1981, Freeman 1985, Mitscherlich 1984, Murray 1999, Prescott 1996, Ryan 1994, Storz 1971, Castle 1987, Weinstein 1991, Mandell 2000.				

Chlamydophila psittaci

GROUP	Bacteria
TYPE	Gram-
GENUS	Chlamydia
FAMILY	Chlamydiaceae
DISEASE GROUP	Non-communicable
BIOSAFETY LEVEL	Risk Group 2
Infectious Dose (ID_{50})	unknown
Lethal Dose (LD_{50})	-
Infection Rate	none
Incubation Period	5–15 days
Peak Infection	NA
Annual Cases	33
Annual Fatalities	rare

Generally parasitizes animals and birds in the wild. Man is but an incidental host who becomes infected as the result of contact with domesticated birds and fowl. Human infections are rare, and often mild. These bacteria are obligate intracellular parasites that use ATP produced by the host cell, hence, they are termed "energy parasites." Formerly known as Chlamydia psittaci.

Disease or Infection	psittacosis/ornithosis (parrot fever), pneumonitis				
Natural Source	Birds, fowl				
Toxins	none				
Point of Infection	Upper Respiratory Tract				
Symptoms	Fever, myalgia, headache, chills, respiratory distress, pneumonia, fatigue, anorexia, encephalitis.				
Treatment	Penicillin, tetracyclines, erythromycin				
Untreated Fatality Rate	<6%	Prophylaxis: Anitbiotics		Vaccine:	none
Shape	spherical				
Mean Diameter, μm	0.283	Size Range:	0.2–0.4 × 0.4–1 microns		
Growth Temperature	33–41°C	Survival Outside Host:		2-20 days	
Inactivation	Moist heat: 121°C for 15 min. Dry heat: 170°C for 1 hr.				
Disinfectants	1% sodium hypochlorite, 70% ethanol, glutaraldehyde.				
Filter Nominal Rating	MERV 6	MERV 8	MERV 11	MERV 13	MERV 14
Removal Efficiency, %	5.7	7.7	15.5	40.8	64.0
UVGI Rate Constant	(unknown)		Media	-	
Dose for 90% Inactivation	-		Ref.	-	
Suggested Indoor Limit	0	cfu/m³			
Genome Size (bp)	1,144,377	(C. abortus)	39.9 % G+C	60.1 % T+A	
Related Species	Formerly Chlamydia psittaci, Type Species: C. trachomatis				
Notes	CDC Reportable.				
Photo Credit	Centers for Disease Control, PHIL# 1351, Dr. Martin Hicklin.				
References	Braude 1981, Freeman 1985, Mitscherlich 1984, Murray 1999, Prescott 1996, Ryan 1994, Storz 1971, Canada 2001, Mandell 2000				

Cladosporium

GROUP	Fungal Spore
PHYLUM	Ascomycota
SUBPHYLUM	Ascomycotina
GENUS	Moniliales
DISEASE GROUP	Non-communicable
BIOSAFETY LEVEL	Risk Group 2
Infectious Dose (ID_{50})	unknown
Lethal Dose (LD_{50})	NA
Infection Rate	none
Incubation Period	-
Peak Infection	NA
Annual Cases	-
Annual Fatalities	-

Considered nonpathogenic but can cause distress to allergic patients. It can be a contributing factor in SBS/BRI. A common indoor contaminant, it can grow indoors on dust, filters, ducts, and coils. This dematiaceous fungi may produce a brain abscess known as cerebral cladosporiosis, usually as the result of a lung lesion or history of penetrating trauma. C. cladosporioides and C. herbarum are the most common species found on dead organic matter and in air.

Disease or Infection	chromoblastomycosis, allergic reactions, rhinitis, asthma				
Natural Source	Environmental, widespread distribution.				
Toxins	cladosporin, emodin, epicladosporic acid				
Point of Infection	Upper Respiratory Tract				
Symptoms	Allergic reactions, respiratory irritation, possible sinus infections, pulmonary and cutaneous infections.				
Treatment	Intravenous amphotericin B produces a marginal response. Combinations of flucytosine and amphotericin B may be beneficial.				
Untreated Fatality Rate	-	Prophylaxis: none		Vaccine: none	
Shape	spherical spore or ellipsoid to lemon-shaped.				
Mean Diameter, μm	8.062	Size Range:	5–13 microns		
Growth Temperature	37°C	Survival Outside Host:	survives outdoors		
Inactivation	Moist heat: 121°C for 30 min.				
Disinfectants	1% sodium hypochlorite, phenolics, formaldehyde, glutaraldehyde.				
Filter Nominal Rating	MERV 6	MERV 8	MERV 11	MERV 13	MERV 14
Removal Efficiency, %	64.9	75.3	94.0	99.0	99.0
UVGI Rate Constant	0.00384	m^2/J	Media	Air	
Dose for 90% Inactivation	600	J/m^2	Ref.	Luckiesh 1946	
Suggested Indoor Limit	150–500	cfu/m^3	(in mix with other nonpathogenic fungi)		
Genome Size (bp)					
Related Species	C. cladosporioides, C. herbarum, C. spaerospermum				
Notes	CDC Reportable. Water activity Aw = 0.84–0.88.				
Photo Credit	Image courtesy of Neil Carlson, University of Minnesota.				
References	Freeman 1985, Howard 1983, Lacey 1988, Murray 1999, Ryan 1994, Smith 1989, Godish 1995, Braude 1981, Woods 1997, Sutton 1998				

Clostridium botulinum

	GROUP	Bacteria
	TYPE	Gram+
	GENUS	Clostridium
	FAMILY	Clostridiaceae
	DISEASE GROUP	Non-communicable
	BIOSAFETY LEVEL	Risk Group 2
	Infectious Dose (ID_{50})	unknown
	Lethal Dose (LD_{50})	-
	Infection Rate	-
	Incubation Period	12–36 hours
	Peak Infection	-
	Annual Cases	-
	Annual Fatalities	-

A bacterial spore that can cause food poisoning when conditions are right for toxin production. May settle on food. May also settle on open wounds causing wound botulism. Infant botulism occurs when infants under one year of age ingest spores. Produces a neurotoxin under anaerobic conditions and in low acid foods.

Disease or Infection	Botulism, toxic poisoning.				
Natural Source	Environmental				
Toxins	Botulinum toxin				
Point of Infection	Ingested.				
Symptoms	Acute flaccid paralysis involving the muscles of the face, head, and pharynx, down to the thorax and extremities, respiratory failure.				
Treatment	Antibiotic treatment generally not effective, trivalent (types A, B, & E) equine antitoxin, Penicillin G sometimes used				
Untreated Fatality Rate	-	Prophylaxis: Antitoxin		Vaccine:	B. Toxoid
Shape	Rods				
Mean Diameter, µm	1.975	Size Range:	1.3–3 microns		
Growth Temperature	10–50°C	Survival Outside Host:		indefinitely in soil, water	
Inactivation	Boiling 10 mins. Moist heat: 120°C for 15 min.				
Disinfectants	1% sodium hypochlorite, 70% ethanol.				
Filter Nominal Rating	MERV 6	MERV 8	MERV 11	MERV 13	MERV 14
Removal Efficiency, %	43.2	58.4	80.8	98.7	99.0
UVGI Rate Constant	0.0170	m^2/J	Media	Plates	
Dose for 90% Inactivation	135	J/m^2	Ref.	Jepson 1973	
Suggested Indoor Limit	-				
Genome Size (bp)	3,886,916		28.2 % G+C	71.8 % T+A	
Related Species	C. perfringens (used for UVGI rate constant), C. butyricum				
Notes	Not necessarily toxic if inhaled.				
Photo Credit	Centers for Disease Control, PHIL# 1932, Larry Stauffer, Oregon State Public Health Laboratory				
References	Braude 1981, Freeman 1985, Mitscherlich 1984, Murray 1999, Prescott 1996, Ryan 1994, Canada 2001, Mandell 2000				

Clostridium perfringens

	GROUP	Bacteria
	TYPE	Gram+
	GENUS	Clostridium
	FAMILY	Clostridiaceae
	DISEASE GROUP	Non-communicable
	BIOSAFETY LEVEL	Risk Group 2
	Infectious Dose (ID_{50})	10 per g of food
	Lethal Dose (LD_{50})	-
	Infection Rate	none
	Incubation Period	6–24 hrs.
	Peak Infection	NA
	Annual Cases	10,000
	Annual Fatalities	10

Non-respiratory but may settle on exposed foods. A food-borne pathogen that produces four major enterotoxins and nine minor toxins. Classified by the type of toxin produced, with Type A being the most important and common in the human colon and the soil. May grow on foods like meat. Often found in the intestines and in feces. Can cause wound contamination. Forms spores that are resistant to heat. May cause gas gangrene in infections.

Disease or Infection	sepsis, toxic reactions, food poisoning.				
Natural Source	Environmental, Humans, animals, soil. Spp: C. novyi, septicum.				
Toxins	Exotoxins, several types.				
Point of Infection	intestines				
Symptoms	Sudden onset of colic, diarrhea, nausea, short duration, rarely fatal.				
Treatment	Penicillin				
Untreated Fatality Rate	-	Prophylaxis: none		Vaccine:	none
Shape	spherical spore, "box-car" vegetative cell				
Mean Diameter, μm	5	Size Range:	2.5–10 microns		
Growth Temperature	6–52°C	Survival Outside Host:		years	
Inactivation	Moist heat: 121°C for >15 min.				
Disinfectants	1% sodium hypochlorite, prolonged contact with glutaraldehyde.				
Filter Nominal Rating	MERV 6	MERV 8	MERV 11	MERV 13	MERV 14
Removal Efficiency, %	63.5	75.0	93.8	99.0	99.0
UVGI Rate Constant	0.017	m^2/J	Media	Plates	
Dose for 90% Inactivation	135	J/m^2	Ref.	Jepson 1973	
Suggested Indoor Limit	0	cfu/m^3			
Genome Size (bp)	3,031,430		28.6 % G+C	71.4 % T+A	
Related Species	Type Species: C. butyricum. C. botulinum.				
Notes	Not necessarily toxic if inhaled.				
Photo Credit	Centers for Disease Control (Clostridium novyi shown), PHIL #1033, Dr. William A. Clark.				
References	Braude 1981, Freeman 1985, Mitscherlich 1984, Murray 1999, Prescott 1996, Ryan 1994, Canada 2001, Mandell 2000				

Coccidioides immitis

GROUP	Fungal Spore
CLASS	Euascomycetes
ORDER	Onygenales
FAMILY	Onygenaceae
DISEASE GROUP	Non-communicable
BIOSAFETY LEVEL	Risk Group 3
Infectious Dose (ID_{50})	100–1350
Lethal Dose (LD_{50})	-
Infection Rate	none
Incubation Period	1–4 weeks
Peak Infection	13 days
Annual Cases	uncommon
Annual Fatalities	-

The most dangerous fungal infection. Causes the disease coccidioidomycosis. Estimates are that 20–40 million people in the Southwest have had coccidioidomycosis. Only about 40% of infections are symptomatic, and only 5% are clinically diagnosed. A self-limiting, non-progressive form of the infection is commonly known as valley fever or desert rheumatism. Transmission occurs by inhalation and no direct transmission from animals occurs. The natural reservoir is the soil.

Disease or Infection	coccidioidomycosis, valley fever, desert rheumatism, chronic pneumonia possible, potential nosocomial infection.				
Natural Source	Environmental, found in alkali soil in warm, dry regions, SW USA, etc.				
Toxins	none				
Point of Infection	Upper Respiratory Tract				
Symptoms	Respiratory influenza-like infection, erythema nodosum in 20% cases, may progress to lung lesions, abcesses. Some 60% of infections are normally asymptomatic.				
Treatment	Amphotericin B, ketoconazole, itraconazole, fluoconazole for meningeal infections				
Untreated Fatality Rate	0.9	**Prophylaxis:** none		**Vaccine:** none	
Shape	barrel-shaped arthroconidia				
Mean Diameter, μm	3.464	**Size Range:** 2–6 microns			
Growth Temperature	25°C	**Survival Outside Host:** years			
Inactivation	Moist heat: 121°C for >15 min.				
Disinfectants	1% sodium hypochlorite, phenolics, glutaraldehyde, formaldehyde.				
Filter Nominal Rating	MERV 6	MERV 8	MERV 11	MERV 13	MERV 14
Removal Efficiency, %	58.8	72.5	92.1	99.0	99.0
UVGI Rate Constant	(unknown)	**Media**		-	
Dose for 90% Inactivation	-	**Ref.**		-	
Suggested Indoor Limit	0	cfu/m^3			
Genome Size (bp)					
Notes	CDC Reportable. Grows well at 20–40°C, pH 3.5–9.0.				
Photo Credit	Centers for Disease Control, PHIL #481, Mercy Hospital Toledo, OH, Brian J. Harrington.				
References	Sarosi 1970, Pappagianis 1983, Murray 1999, Ryan 1994, Smith 1989, Sorensen 1999, Miyaji 1987, Canada 2001, DiSalvo 1983, Sutton 1998				

Coronavirus

	GROUP	Virus
	TYPE	ssRNA, positive
	GENUS	Coronavirus
	FAMILY	Coronaviridae
	DISEASE GROUP	Communicable
	BIOSAFETY LEVEL	Risk Group 2
	Infectious Dose (ID_{50})	unknown
	Lethal Dose (LD_{50})	-
	Infection Rate	0.34–0.5
	Incubation Period	2–5 days
	Peak Infection	3–4 days
	Annual Cases	1,700,000
	Annual Fatalities	0

Accounts for about 10–30% of all colds with ages 14–24 most affected. Can affect other animals. Coronaviruses are one of the causes of the common cold. They account for about 15% of cases of the common cold. Coronaviruses can infect other animals besides humans but strains are general specific to one host. Occurs worldwide with predominance in late fall and early winter.

Disease or Infection	colds, croup				
Natural Source	Humans				
Toxins	none				
Point of Infection	Upper Respiratory Tract				
Symptoms	Afebrile cold in adults, nasal discharge, malaise, may exacerbate asthma or chronic respiratory disease.				
Treatment	No specific treatment, therapy, or antivirals.				
Untreated Fatality Rate	none **Prophylaxis:** none **Vaccine:** none				
Shape	enveloped helical				
Mean Diameter, μm	0.110 **Size Range:** 0.08–0.15 microns				
Growth Temperature	na **Survival Outside Host:** up to 24 hours on metal				
Inactivation	heat will inactivate				
Disinfectants	1% sodium hypochlorite, 2% glutaraldehyde.				
Filter Nominal Rating	MERV 6	MERV 8	MERV 11	MERV 13	MERV 14
Removal Efficiency, %	8.4	10.0	18.0	44.7	65.7
UVGI Rate Constant	(unknown)		**Media**	na	
Dose for 90% Inactivation	-		**Ref.**	-	
Suggested Indoor Limit	-				
Genome Size (bp)	27,000–32,000		34. % G+C	66. % T+A	
Related Species	SARS virus				
Notes	CDC Reportable.				
Photo Credit	Centers for Disease Control, PHIL# 189, Dr. Erskine Palmer.				
References	Dalton 1973, Fraenkel-Conrat 1985, Freeman 1985, Mahy 1975, Murray 1999, Ryan 1994, Canada 2001, Myint 1995				

Corynebacterium diphtheriae

GROUP	Bacteria
TYPE	Gram+
GENUS	Corynebacterium
FAMILY	Mycobacteriaceae
DISEASE GROUP	Communicable
BIOSAFETY LEVEL	Risk Group 2
Infectious Dose (ID_{50})	unknown
Lethal Dose (LD_{50})	-
Infection Rate	varies
Incubation Period	2–5 days
Peak Infection	-
Annual Cases	10
Annual Fatalities	-

Corynebacterium diphtheria is the causative agent of diphtheria, which was historically a disease of children. In modern times this disease is less prevalent, but increasingly afflicts those in older age groups. Healthy carriers may harbor the bacteria in their throats and upper respiratory tracts asymptomatically for a lifetime. Transmitted by droplet spread, by direct contact, and via fomites. Small outbreaks occur periodically.

Disease or Infection	diphtheria, toxin produced, opportunistic.				
Natural Source	Humans (only known reservoir), nosocomial				
Toxins	diphtheria toxin				
Point of Infection	Upper Respiratory Tract				
Symptoms	Pharyngitis, fever, malaise, swelling of the neck, headache, hypoxia, toxic effects on nervous system.				
Treatment	Administer antitoxin in conjunction with erythromycin, penicillin				
Untreated Fatality Rate	5–10%	Prophylaxis: DTP		Vaccine:	Available
Shape	rods				
Mean Diameter, μm	0.6981	Size Range:	0.3–0.8 × 1–6 microns		
Growth Temperature	15–40°C	Survival Outside Host:	2.5 hrs in air, <1 year soil		
Inactivation	Moist heat: 121°C for 15 minutes. Dry heat: 170°C for 1 hour.				
Disinfectants	1% sodium hypochlorite, phenolics, glutaraldehyde, formaldehyde.				
Filter Nominal Rating	MERV 6	MERV 8	MERV 11	MERV 13	MERV 14
Removal Efficiency, %	13.5	19.8	36.6	76.0	94.0
UVGI Rate Constant	0.0701	m²/J	Media	Plates	
Dose for 90% Inactivation	32.85	J/m²	Ref.	Sharp 1939	
Suggested Indoor Limit	0	cfu/m³			
Genome Size (bp)	2,488,635		53.5 % G+C	46.5 % T+A	
Related Species	C. ulcerans, C. jeikeium.				
Notes	CDC Reportable.				
Photo Credit	Centers for Disease Control PHIL# 1943.				
References	Braude 1981, Freeman 1985, Mitscherlich 1984, Murray 1999, Prescott 1996, Ryan 1994, Canada 2001, Mandell 2000				

Coxiella burnetii

GROUP	Bacteria / Rickettsiae
TYPE	Gram-
GENUS	Coxiella
FAMILY	Rickettsiae
DISEASE GROUP	Non-communicable
BIOSAFETY LEVEL	Risk Group 3
Infectious Dose (ID_{50})	10
Lethal Dose (LD_{50})	-
Infection Rate	none
Incubation Period	9–18 days
Peak Infection	NA
Annual Cases	rare
Annual Fatalities	-

Rickettsiae are bacteria-like microorganisms. Coxiella burnetii are transmitted from animals to humans by inhalation. Coxiella is more resistant to drying than other Rickettsiae and can survive in fomites for years. A high risk for slaughterhouses and animal research. Endemic in many areas. Resistant to heat and dessication. Q fever is primarily a zoonotic disease that is widely distributed among many mammals. Infection can also occur from ingestion of animal products such as unpasteurized milk.

Disease or Infection	Q fever				
Natural Source	Cattle, sheep, goats.				
Toxins	none				
Point of Infection	Upper Respiratory Tract				
Symptoms	Sudden onset of acute febrile disease, chills, headache, fatigue, sweats, pneumonitis, pericarditis, hepatitis.				
Treatment	Tetracycline, chloramphenicol, rifampin				
Untreated Fatality Rate	<1%	Prophylaxis: ineffective		Vaccine:	Available
Shape	short rods, highly pleomorphic				
Mean Diameter, μm	0.283	Size Range:	0.2–0.4 × 0.4–1 microns		
Growth Temperature	32–35°C	Survival Outside Host:		years	
Inactivation	Moist heat: 130°C for 1 hour.				
Disinfectants	Ethanol, glutaraldehyde, gaseous formaldehyde.				
Filter Nominal Rating	MERV 6	MERV 8	MERV 11	MERV 13	MERV 14
Removal Efficiency, %	5.7	7.7	15.5	40.8	64.0
UVGI Rate Constant	0.1535	m^2/J	Media	Water	
Dose for 90% Inactivation	15.00	J/m^2	Ref.	Little 1980	
Suggested Indoor Limit	0	cfu/m^3			
Genome Size (bp)	1,995,275		42.7 % G+C	57.3 % T+A	
Related Species	Rickettsia prowazekii, R. typhi, Rochalimaea quintana.				
Notes	CDC Reportable.				
Photo Credit	Image shows rickettsia in an endothelial cell. Centers for Disease Control, PHIL# 931, Edwin P. Ewing.				
References	Braude 1981, McCaul 1981, Murray 1999, Prescott 1996, Ryan 1994, Walker 1988, Canada 2001, NATO 1996, Mandell 2000				

Coxsackievirus

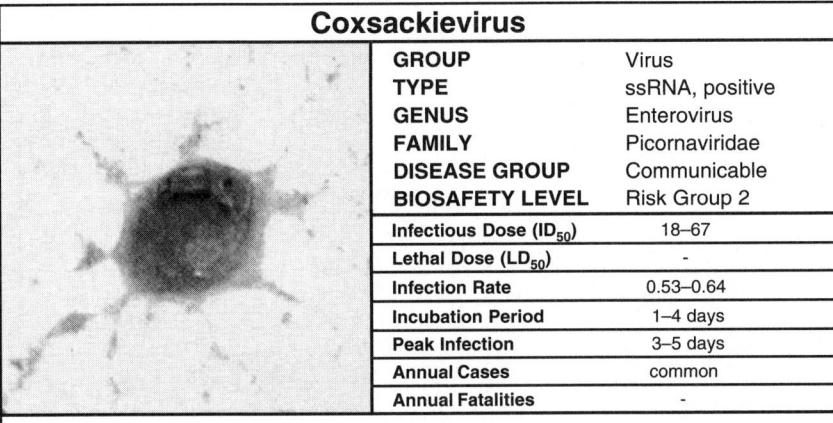

GROUP	Virus
TYPE	ssRNA, positive
GENUS	Enterovirus
FAMILY	Picornaviridae
DISEASE GROUP	Communicable
BIOSAFETY LEVEL	Risk Group 2
Infectious Dose (ID_{50})	18–67
Lethal Dose (LD_{50})	-
Infection Rate	0.53–0.64
Incubation Period	1–4 days
Peak Infection	3–5 days
Annual Cases	common
Annual Fatalities	-

A common cause of colds. Often subclinical. It can sometimes be found in feces and sewage, but is commonly isolated from the throat. This particular cold virus is prevalent in the summer and fall. Human infections, mostly in children unders 10 years, are common and can be subclinical. Typically self-limiting. Occurs worldwide.

Disease or Infection	colds, Acute Respiratory Disorder.				
Natural Source	Humans, feces, sewage.				
Toxins	none				
Point of Infection	Upper Respiratory Tract				
Symptoms	Abrupt onset of fever, sore throat, anorexia, disphagia, vomiting, small lesions in the oral region. Most frequent in children.				
Treatment	No specific treatment. No antivirals available.				
Untreated Fatality Rate	none **Prophylaxis:** none **Vaccine:** none				
Shape	spherical				
Mean Diameter, μm	0.027 **Size Range:** 0.024–0.03 microns				
Growth Temperature	na **Survival Outside Host:** 2 weeks on glass, steel				
Inactivation	56–60°C for 30 minutes, pH 2.3–9.4 for 1 day				
Disinfectants	70% ethanol, 5% lysol, 1% sodium hypochlorite				
Filter Nominal Rating	MERV 6	MERV 8	MERV 11	MERV 13	MERV 14
Removal Efficiency, %	24.6	28.7	47.3	85.1	96.0
UVGI Rate Constant	0.111	m^2/J	**Media**	Air	
Dose for 90% Inactivation	20.74	J/m^2	**Ref.**	Jensen 1964	
Suggested Indoor Limit	-	cfu/m^3			
Genome Size (bp)	7,401				
Related Species	Human Coronavirus OC43.				
Notes	CDC Reportable. Can survive 24 hours on paper & plastic.				
Photo Credit	Reprinted w/ permission from Malherbe and Strickland 1980, Viral Cytopathology, Copyright CRC Press, Boca Raton, FL				
References	Bendinelli 1988, Dalton 1973, Collins & Kennedy 1993, Mahy 1975, Freeman 1985, Murray 1999, Ryan 1994, Canada 2001				

Cryptococcus neoformans

GROUP	Fungal Yeast
CLASS	Hyphomycetes
ORDER	Sporidiales
FAMILY	Sporidiobolaceae
DISEASE GROUP	Non-communicable
BIOSAFETY LEVEL	Risk Group 2
Infectious Dose (ID_{50})	1000
Lethal Dose (LD_{50})	NA
Infection Rate	none
Incubation Period	unknown
Peak Infection	NA
Annual Cases	high
Annual Fatalities	rare

The cause of cryptococcosis, and can result in cryptococus meningitis also. It is an opportunistic pathogen that can fatally infect those with impaired immune systems. Cells can enter the lungs, germinate, and produce mycelial growth. This is always in yeast form and can occur in dried pigeon excrement, from which it can become airborne. It is a contributing factor in some cases of SBS/BRI. Most infections are subclinical and asymptomatic. Sometimes found on human body surfaces.

Disease or Infection	cryptococcosis, cryptococcal meningitis, pneumonia possible				
Natural Source	Environmental, nosocomial, indoor growth on floor dust, SBS/BRI.				
Toxins	none				
Point of Infection	Upper Respiratory Tract, cutaneous infections possible.				
Symptoms	Mycosis appearing as subacute or chronic meningitis, infection of lung or kidney, skin lesions, ulcers, subcutaneous tumor-like masses.				
Treatment	Combination therapy with amphotericin B and 5-fluorocytosine or ketoconazole.				
Untreated Fatality Rate	-	Prophylaxis: none		Vaccine:	none
Shape	spherical cell				
Mean Diameter, μm	4.899	Size Range:	4–6 microns		
Growth Temperature	37°C	Survival Outside Host:		survives in pigeon feces	
Inactivation	Moist heat: 121°C for 15 minutes.				
Disinfectants	1% sodium hypochlorite, iodine, phenolics, glutaraldehyde				
Filter Nominal Rating	MERV 6	MERV 8	MERV 11	MERV 13	MERV 14
Removal Efficiency, %	63.3	75.0	93.7	99.0	99.0
UVGI Rate Constant	0.0167	m^2/J	Media	Plates	
Dose for 90% Inactivation	138	J/m^2	Ref.	Wang 1994	
Suggested Indoor Limit	0	cfu/m^3			
Genome Size (bp)					
Related Species	C. albidus, C.laurentii, C. neoformans var. gatti				
Notes	Cryptococcus neoformans var, neoformans. CDC Reportable.				
Photo Credit	Centers for Disease Control, PHIL# 827, Dr. Edwin P. Ewing, Jr.				
References	Gordon 1983, Freeman 1985, Howard 1983, Murray 1999, Ryan 1994, Smith 1989, Wang 1994, Godish 1995, Braude 1981, Woods 1997				

Cryptostroma corticale

GROUP	Fungal Spore	
TYPE	Mitosporic fungi	
CLASS	Hyphomycetes	
FAMILY	Cryptococcaceae	
DISEASE GROUP	Non-communicable	
BIOSAFETY LEVEL	Risk Group 2	
Infectious Dose (ID_{50})	NA	
Lethal Dose (LD_{50})	-	
Infection Rate	-	
Incubation Period	-	
Peak Infection	-	
Annual Cases	0	
Annual Fatalities	-	

Sometimes found growing on logs and bark, especially of maple trees, sycamore trees, and on stored logs. An occupational hazard in wood mills and agriculture. This tree pathogen may survive endophytically within host tissues for many years, only developing to cause visible symptoms when the host is under stress. Cryptostroma corticale is the cause of Sooty bark disease of sycamore (Acer pseudoplatanus) in north west Europe, which only causes damage following hot dry summers.

Disease or Infection	alveolitis, asthma, maple bark pneumonitis, maple bark disease				
Natural Source	Environmental, found on maple & sycamore bark.				
Toxins	none				
Point of Infection	Upper Respiratory Tract				
Symptoms	Pneumonitis.				
Treatment					
Untreated Fatality Rate	-	Prophylaxis: none		Vaccine: none	
Shape	ovoid spore				
Mean Diameter, μm	3.742	Size Range:	3.5–4 × 4–6.5 microns		
Growth Temperature	38°C	Survival Outside Host:	survives outdoors		
Inactivation	Moist heat: 121°C for 30 min.				
Disinfectants	1% sodium hypochlorite, phenolics, formaldehyde, glutaraldehyde.				
Filter Nominal Rating	MERV 6	MERV 8	MERV 11	MERV 13	MERV 14
Removal Efficiency, %	60.2	73.4	92.7	99.0	99.0
UVGI Rate Constant	(unknown)		Media	-	
Dose for 90% Inactivation	-		Ref.	-	
Suggested Indoor Limit	150–500	cfu/m³	(in mix with other nonpathogenic fungi)		
Genome Size (bp)					
Related Species	-				
Notes	CDC Reportable.				
Photo Credit	Penn State AE Dept.				
References	Freeman 1985, Howard 1983, Lacey 1988, Murray 1999, Rao 1996, Ryan 1994, Shapton 1972, Smith 1989				

Curvularia lunata

GROUP	Fungal Spore
CLASS	Euascomycetes
ORDER	Pleosporales
FAMILY	Pleosporaceae
DISEASE GROUP	Non-communicable
BIOSAFETY LEVEL	Risk Group 1
Infectious Dose (ID_{50})	unknown
Lethal Dose (LD_{50})	NA
Infection Rate	-
Incubation Period	-
Peak Infection	-
Annual Cases	rare
Annual Fatalities	-

Reported to be allergenic. May cause sinusitis, mycetoma, corneal infections, and opportunistic infections in the immunocompromised. Curvularia is a demitiaceous fungus that may cause brain abscesses either by spreading from a lung infection or from a contiguous paranasal sinus. A widely distributed contaminant of seed crops, it can be isolated from cereals, floor dust, mattress dust, wallpaper, and painted wood surfaces. Curvularia lunata is the species of most concern.

Disease or Infection	mycetoma, opportunistic infections, allergic fungal sinusitis				
Natural Source	Environmental, wide distribution, plant pathogen.				
Toxins	none.				
Point of Infection	Eyes, upper respiratory tract, skin.				
Symptoms	Allergic reactions, respiratory irritation, agent of phaeohyphomycosis causing sinusitis, keratitis, subcutaneous and pulmonary disease.				
Treatment	Intravenous amphotericin B produces a marginal response. Combinations of flucytosine and amphotericin B may be beneficial.				
Untreated Fatality Rate	-	Prophylaxis:	none	Vaccine:	none
Shape	elongated spore				
Mean Diameter, μm	11.619	Size Range:	9–15 × 20–32 microns		
Growth Temperature	39°C	Survival Outside Host:	survives outdoors		
Inactivation	Moist heat: 121°C for 30 min.				
Disinfectants	1% sodium hypochlorite, phenolics, formaldehyde, glutaraldehyde.				
Filter Nominal Rating	MERV 6	MERV 8	MERV 11	MERV 13	MERV 14
Removal Efficiency, %	65.0	75.2	94.0	99.0	99.0
UVGI Rate Constant	(unknown)	Media	-		
Dose for 90% Inactivation	-	Ref.	-		
Suggested Indoor Limit	150–500	cfu/m^3	(in mix with other nonpathogenic fungi)		
Genome Size (bp)					
Related Species	Curvularia lunata var. aeria, C. geniculata				
Notes	Several other species of Curvularia are less common agents of disease.				
Photo Credit	Image courtesy of Neil Carlson, University of Minnesota.				
References	Freeman 1985, Howard 1983, Lacey 1988, Murray 1999, Pope 1993, Rao 1996, Ryan 1994, Slack 1975, Sutton 1998				

Echovirus

GROUP	Virus
TYPE	ssRNA, positive
GENUS	Parechovirus
FAMILY	Picornaviridae
DISEASE GROUP	Communicable
BIOSAFETY LEVEL	Risk Group 2
Infectious Dose (ID_{50})	unknown
Lethal Dose (LD_{50})	-
Infection Rate	0.43–0.80
Incubation Period	2–14 days
Peak Infection	3–4 days
Annual Cases	common
Annual Fatalities	-

A uncommon cause of colds. Often subclinical. Related to viruses found in cattle, pigs, and monkeys. Infections are most common in children, but are often subclinical. The most serious disease that echoviruses can induce is meningitis.

Disease or Infection	colds, meningitis possible				
Natural Source	Humans				
Toxins	none				
Point of Infection	Upper Respiratory Tract				
Symptoms	Most infections are asymptomatic, but may vary from mild to acute, chronic, or lethal. May be associated with aseptic meningitis, muscle weakness, paralysis, common cold, acute febrile respiratory illness.				
Treatment	No specific treatment or antivirals are available.				
Untreated Fatality Rate	-	Prophylaxis: none		Vaccine: none	
Shape	icosahedral				
Mean Diameter, μm	0.024	Size Range:	0.020–0.03 microns		
Growth Temperature	na	Survival Outside Host:		up to 3 weeks dry	
Inactivation	heat at 50°C for 2 hours,				
Disinfectants	70% ethanol, 5% lysol, 11% quaternary ammonium compounds				
Filter Nominal Rating	MERV 6	MERV 8	MERV 11	MERV 13	MERV 14
Removal Efficiency, %	25.9	30.1	49.5	86.8	96.6
UVGI Rate Constant	0.217	m^2/J	Media	Water	
Dose for 90% Inactivation	10.61	J/m^2	Ref.	Hill 1970	
Suggested Indoor Limit	-				
Genome Size (bp)	7,450				
Related Species	Human Parechovirus 1 and 2, Enterovirus				
Notes	CDC Reportable.				
Photo Credit	Reprinted w/ permission from Malherbe and Strickland 1980, Viral Cytopathology, Copyright CRC Press, Boca Raton, FL				
References	Dalton 1973, Fraenkel-Conrat 1985, Freeman 1985, Mahy 1975, Murray 1999, Ryan 1994, Canada 2001				

Emericella nidulans

GROUP	Fungal Spore
CLASS	Ascomycetes
ORDER	Eurotiales
FAMILY	Trichocomaceae
DISEASE GROUP	Non-communicable
BIOSAFETY LEVEL	Risk Group 1
Infectious Dose (ID_{50})	NA
Lethal Dose (LD_{50})	-
Infection Rate	none
Incubation Period	-
Peak Infection	NA
Annual Cases	-
Annual Fatalities	-

A reportedly allergenic fungus that hails from environmental sources, it has occasionally been found indoors gowing in house dust and a variety of building materials. A ubiquitous soil fungus. Most often isolated from tropical and subtropical climates. Emericella can cause allergenic and asthmatic reactions in sensitive individuals. Emericella and its anamorph state Aspergillus can cause infection in humans and animals including aspergilloma and haemorrhage particularly in immunocompromised individuals. No toxic or invasive diseases have been reported but are possible.

Disease or Infection	allergic alveolitis, toxic reactions possible				
Natural Source	Environmental.				
Toxins	sterigmatocystin, asperthecin, penicillin, corycepin, pentostatin, asperugin, emerin, emericellin, nidurufin, aspergiline, echinocandin B				
Point of Infection	Upper Respiratory Tract				
Symptoms	Allergic reactions, respiratory irritation.				
Treatment	NA				
Untreated Fatality Rate	-	Prophylaxis: none		Vaccine: none	
Shape	spherical spore				
Mean Diameter, μm	3.240	Size Range:	3–3.5 microns		
Growth Temperature	25°C	Survival Outside Host:		survives outdoors	
Inactivation	Moist heat: 121°C for 30 min.				
Disinfectants	1% sodium hypochlorite, phenolics, formaldehyde, glutaraldehyde.				
Filter Nominal Rating	MERV 6	MERV 8	MERV 11	MERV 13	MERV 14
Removal Efficiency, %	57.5	71.6	91.4	99.0	99.0
UVGI Rate Constant	(unknown)	Media		-	
Dose for 90% Inactivation	-	Ref.		-	
Suggested Indoor Limit	150–500 cfu/m^3	(in mix with other nonpathogenic fungi)			
Genome Size (bp)					
Related Species	E. quadrillineata, E. rugulosa				
Notes	CDC Reportable.				
Photo Credit	Image courtesy of Neil Carlson, University of Minnesota.				
References	Freeman 1985, Howard 1983, Lacey 1988, Murray 1999, Pope 1993, Rao 1996, Ryan 1994, Smith 1989				

Enterobacter cloacae

GROUP	Bacteria
TYPE	Gram-
GENUS	Enterobacter
FAMILY	Enterobacteriaceae
DISEASE GROUP	Endogenous
BIOSAFETY LEVEL	Risk Group 1
Infectious Dose (ID_{50})	unknown
Lethal Dose (LD_{50})	-
Infection Rate	-
Incubation Period	unknown
Peak Infection	-
Annual Cases	uncommon
Annual Fatalities	-

Often found as commensals in the intestines. Associated with a variety of infections, especially nosocomial including: urinary, pulmonary, wound, bloodstream, and other opportunistic infections. Occurs worldwide in hospital settings. Has caused a septicemia epidemic. Equipment contamination and the fecal oral route are possible transmission mechanisms. Can resist some antibiotics.

Disease or Infection	Opportunistic infections, pneumonia possible from some Enterobacter species				
Natural Source	Humans, environmental, soil, and water.				
Toxins	Endotoxins				
Point of Infection	wounds				
Symptoms	Opportunistic infections of the lungs, blood, and urinary tract.				
Treatment	Aminoglycosides, chloramphenicol, tetracyclines, TMP-SMX, nalidixic acid, nitrofurantoin.				
Untreated Fatality Rate	-	Prophylaxis: possible		Vaccine: none	
Shape	rods				
Mean Diameter, μm	1.414	Size Range:	1–2 microns		
Growth Temperature	37°C	Survival Outside Host:	7–21 days in food		
Inactivation	Moist heat: 121°C for 15 minutes. Dry heat: 170°C for 1 hour.				
Disinfectants	1% sodium hypochlorite, phenolics, glutaraldehyde, formaldehyde.				
Filter Nominal Rating	MERV 6	MERV 8	MERV 11	MERV 13	MERV 14
Removal Efficiency, %	31.8	44.9	67.8	96.6	99.0
UVGI Rate Constant	0.03598	m^2/J	Media	Water	
Dose for 90% Inactivation	64.00	J/m^2	Ref.	Zemke 1990	
Suggested Indoor Limit	0	cfu/m^3			
Genome Size (bp)					
Related Species	Related to Klebsiella, Serratia, and E. aerogenes				
Notes	Usually found in mixed infections.				
Photo Credit	Photo reprinted from Farmer et al 1980, Intl. J. System. Bact. 30(3):569–584. by permission of the Intl. Union of Microbiol. Societies				
References	Braude 1981, Farmer 1980, Freeman 1985, Mitscherlich 1984, Murray 1999, Prescott 1996, Ryan 1994				

Enterococcus

GROUP	Bacteria
TYPE	Gram+
GENUS	Enterococcus
FAMILY	Enterococcaceae
DISEASE GROUP	Endogenous
BIOSAFETY LEVEL	Risk Group 1-2
Infectious Dose (ID_{50})	unknown
Lethal Dose (LD_{50})	-
Infection Rate	-
Incubation Period	-
Peak Infection	-
Annual Cases	rare
Annual Fatalities	-

Includes Enterococcus species other than E. faecalis. Enterococcus faecium occasionally causes human disease. Enterococcus durans accounts for less than 2% of enterococci isolates. Non-respiratory but may be airborne in nosocomial settings. Can cause opportunistic infections of the urinary tract and wounds. Normally resident in the intestines and other areas of the body as commensals. E. faecium is found in the feces of about 25% of normal adults. These species are related to Group D streptococci and pneumococci, and cause similar clinical infections.

Disease or Infection	Opportunistic infections, endocarditis, bacteremia				
Natural Source	Humans.				
Toxins	none				
Point of Infection	wounds				
Symptoms	Urinary tract and soft tissue infections, bacteremia.				
Treatment	Ampicillin, combinations of penicillin and aminoglycoside.				
Untreated Fatality Rate	-	Prophylaxis:		Vaccine:	none
Shape	coccoid				
Mean Diameter, μm	1.414	Size Range:	1–2 microns		
Growth Temperature	37°C	Survival Outside Host:			
Inactivation	Moist heat: 121°C for 30 min.				
Disinfectants	1% sodium hypochlorite, 2% glutaraldehyde, formaldehyde, iodines.				
Filter Nominal Rating	MERV 6	MERV 8	MERV 11	MERV 13	MERV 14
Removal Efficiency, %	31.8	44.9	67.8	96.6	99.0
UVGI Rate Constant	(unknown)	Media			
Dose for 90% Inactivation	-	Ref.		-	
Suggested Indoor Limit	0	cfu/m^3			
Genome Size (bp)	3,218,031	(E. faecalis)	37.5% G+C	62.5% T+A	
Related Species	E. faecalis				
Notes	Often resistant to antibiotics, especially sulfanilomides.				
Photo Credit	Centers for Disease Control, PHIL# 209.				
References	Braude 1981, Freeman 1985, Mitscherlich 1984, Murray 1999, Prescott 1996, Ryan 1994, Mandell 2000				

Enterococcus faecalis

GROUP	Bacteria
TYPE	Gram+
GENUS	Enterococcus
FAMILY	Enterococcaceae
DISEASE GROUP	Endogenous
BIOSAFETY LEVEL	Risk Group 1
Infectious Dose (ID_{50})	unknown
Lethal Dose (LD_{50})	-
Infection Rate	-
Incubation Period	-
Peak Infection	-
Annual Cases	-
Annual Fatalities	-

Non-respiratory but may become airborne in nosocomial settings. An opportunistic pathogen. Causes urinary tract and other infections. E. faecalis causes more disease than other Group D streptococci. Is commonly present in the mouth of normal adults, and in small numbers throughout the intestines. Can be isolated in the feces of normal adults. Commonly implicated in endocarditis and is the most frequent isolate in polymicrobial bacteremia. Infection at the site of intravenous lines is possible.

Disease or Infection	Endocarditis, neonatal septicemia, meningitis				
Natural Source	feces.				
Toxins	none				
Point of Infection	Upper Respiratory Tract, urinary tract				
Symptoms	Opportunistic infections in urinary tract and soft tissues.				
Treatment	Ampicillin, combinations of penicillin and aminoglycoside.				
Untreated Fatality Rate	-	Prophylaxis:		Vaccine:	none
Shape	coccoid				
Mean Diameter, μm	0.707	Size Range:		0.5–1 microns	
Growth Temperature	10–45°C	Survival Outside Host:			
Inactivation	Moist heat: 121°C for 30 min.				
Disinfectants	1% sodium hypochlorite, 2% glutaraldehyde, formaldehyde, iodines.				
Filter Nominal Rating	MERV 6	MERV 8	MERV 11	MERV 13	MERV 14
Removal Efficiency, %	13.7	20.1	37.1	76.6	94.3
UVGI Rate Constant	(unknown)		Media		-
Dose for 90% Inactivation	-		Ref.		-
Suggested Indoor Limit	0	cfu/m^3			
Genome Size (bp)	3,218,031		37.5% G+C	62.5% T+A	
Related Species	(previously Streptococcus faecalis)				
Notes	Often resistant to antibiotics, especially sulfanilomides.				
Photo Credit	Centers for Disease Control, PHIL# 258, Pete Wardell.				
References	Braude 1981, Freeman 1985, Mitscherlich 1984, Murray 1999, Prescott 1996, Ryan 1994, Mandell 2000				

Epicoccum purpurascens

GROUP	Fungal Spore
TYPE	Mitosporic fungi
CLASS	Ascomycetes
FAMILY	Damataiaceae
DISEASE GROUP	Non-communicable
BIOSAFETY LEVEL	Risk Group 1
Infectious Dose (ID_{50})	NA
Lethal Dose (LD_{50})	NA
Infection Rate	none
Incubation Period	NA
Peak Infection	NA
Annual Cases	-
Annual Fatalities	-

Considered a common allergen and widely distributed. Can grow on grains, plants, soil, cellulose, paper products, and textiles. Colonizes dead plants and occurs in soil. Can be isolated from air, animals, foodstuffs, floor, carpet, and mattress dust and exposed acrylic paint. Indoor levels can exceed outdoor levels. A common early secondary invader of numerous plants and may cause leaf spots. Not documented as an etiologic agent in human or animal disease.

Disease or Infection	possible allergic alveolitis, rhinitis				
Natural Source	Environmental, indoor growth on fiberglass insulation.				
Toxins	flavipin, epicorazine A, epicorazine B, indole-3-acetonitrile				
Point of Infection	Upper Respiratory Tract				
Symptoms	Allergic reactions and respiratory irritation presumed.				
Treatment	NA				
Untreated Fatality Rate	-	**Prophylaxis:** none		**Vaccine:** none	
Shape	spherical spore				
Mean Diameter, μm	17.321	**Size Range:**	12–25 microns		
Growth Temperature	25°C	**Survival Outside Host:**	survives outdoors		
Inactivation	Moist heat: 121°C for 30 min.				
Disinfectants	1% sodium hypochlorite, phenolics, formaldehyde, glutaraldehyde.				
Filter Nominal Rating	MERV 6	MERV 8	MERV 11	MERV 13	MERV 14
Removal Efficiency, %	65.0	75.2	94.0	99.0	99.0
UVGI Rate Constant	(unknown)	**Media**		-	
Dose for 90% Inactivation	-	**Ref.**		-	
Suggested Indoor Limit	150–500 cfu/m^3	(in mix with other nonpathogenic fungi)			
Genome Size (bp)					
Related Species	Formerly Epicoccum nigrum.				
Notes					
Photo Credit	Image courtesy of Neil Carlson, University of Minnesota.				
References	Freeman 1985, Howard 1983, Lacey 1988, Murray 1999, Pope 1993, Rao 1996, Ryan 1994, Smith 1989, Woods 1997, Sutton 1998				

Eurotium

GROUP	Fungal Spore
CLASS	Ascomycetes
ORDER	Ascomycetes
FAMILY	Trichocomaceae
DISEASE GROUP	Non-communicable
BIOSAFETY LEVEL	Risk Group 1
Infectious Dose (ID_{50})	NA
Lethal Dose (LD_{50})	-
Infection Rate	none
Incubation Period	-
Peak Infection	NA
Annual Cases	-
Annual Fatalities	-

A reportedly allergenic fungus that hails from environmental sources. Frequently encountered in tropical and subtropical climates. Can grow indoors on gypsum. E. rubrum, E. chevalieri, and E. amstelodami have been reported to be allergenic. It is frequently reported from soils and dried or concentrated food products, leather goods, cotton, seeds, and other dried products. E. chevalieri and E. rubrum are considered to be xerophiles while E. amstelodami is not. No toxic or invasive diseases have been reported for these fungi. Water Activity Aw = 0.8–0.9.

Disease or Infection	allergic alveolitis				
Natural Source	Environmental, indoor growth on gypsum.				
Toxins	xanthocillin				
Point of Infection	Upper Respiratory Tract				
Symptoms	Allergic reactions, respiratory irritation.				
Treatment	NA				
Untreated Fatality Rate	-	Prophylaxis: none		Vaccine: none	
Shape	spherical spore				
Mean Diameter, μm	5.612	Size Range:	4.5–7 microns		
Growth Temperature	25°C	Survival Outside Host:	survives outdoors		
Inactivation	Moist heat: 121°C for 30 min.				
Disinfectants	1% sodium hypochlorite, phenolics, formaldehyde, glutaraldehyde.				
Filter Nominal Rating	MERV 6	MERV 8	MERV 11	MERV 13	MERV 14
Removal Efficiency, %	64.1	75.2	93.9	99.0	99.0
UVGI Rate Constant	(unknown)		Media	-	
Dose for 90% Inactivation	-		Ref.	-	
Suggested Indoor Limit	150–500 cfu/m^3	(in mix with other nonpathogenic fungi)			
Genome Size (bp)					
Related Species	E. herbariorum, E. amstelodami, E. rubrum, E. chevalieri				
Notes	CDC Reportable.				
Photo Credit	Image courtesy of Neil Carlson, University of Minnesota.				
References	Freeman 1985, Howard 1983, Lacey 1988, Murray 1999, Pope 1993, Rao 1996, Ryan 1994, Smith 1989, Woods 1997				

Exophiala

GROUP	Fungal Spore
PHYLUM	Ascomycota
ORDER	Chaetothyriales
GENUS	Exophiala
DISEASE GROUP	Non-communicable
BIOSAFETY LEVEL	Risk Group 2
Infectious Dose (ID_{50})	NA
Lethal Dose (LD_{50})	NA
Infection Rate	none
Incubation Period	-
Peak Infection	NA
Annual Cases	-
Annual Fatalities	-

Reportedly allergenic. Hails from environmental sources. Can grow in humidifier water. This dematiaceous fungus has been known to cause nodular and macropapular skin lesions. These fungi live as saprophytes in nature, such as on rotted wood, water, and may enter the skin via minor trauma. Several of the related species and variants can cause subcutaneous infections and disseminated phaeohyphomycosis and nosocomial infections. May be neurotrophic in otherwise healthy hosts.

Disease or Infection	allergic reactions				
Natural Source	Environmental, soil, water, rotten wood, sewage sludge, pulp.				
Toxins	none				
Point of Infection	Upper Respiratory Tract, skin				
Symptoms	Allergic reactions, respiratory irritation.				
Treatment	Intravenous amphotericin B produces a marginal response. Combinations of flucytosine and amphotericin B may be beneficial.				
Untreated Fatality Rate	-	Prophylaxis: none		Vaccine:	none
Shape	ovoid spore, ellipsoid conidia				
Mean Diameter, μm	2.121	Size Range:	1.5–3 × 2.5–4 microns		
Growth Temperature	37°C	Survival Outside Host:	survives outdoors		
Inactivation	Moist heat: 121°C for 30 min.				
Disinfectants	1% sodium hypochlorite, phenolics, formaldehyde, glutaraldehyde.				
Filter Nominal Rating	MERV 6	MERV 8	MERV 11	MERV 13	MERV 14
Removal Efficiency, %	45.6	60.9	83.1	98.8	99.0
UVGI Rate Constant	(unknown)		Media	-	
Dose for 90% Inactivation	-		Ref.	-	
Suggested Indoor Limit	150–500 cfu/m³		(in mix with other nonpathogenic fungi)		
Genome Size (bp)	19,000,000				
Related Species	E. jeanselmei var. jeanselmei, E. jeanselmei var. lecanii-corni, E. dermatitidis, E. castellanii, E. dermatitidis, E. moniliae, E. pisciphila				
Notes	No growth at 40°C. CDC Reportable.				
Photo Credit	Image courtesy of Doctor Fungus, by permission of John H. Rex.				
References	Freeman 1985, Howard 1983, Lacey 1988, Murray 1999, Pope 1993, Rao 1996, Ryan 1994, Smith 1989, Woods 1997, Sutton 1998				

Francisella tularensis

GROUP	Bacteria
TYPE	Gram-
FAMILY	Francisellaceae
GENUS	Francisella
DISEASE GROUP	Non-communicable
BIOSAFETY LEVEL	Risk Group 3
Infectious Dose (ID_{50})	10–100
Lethal Dose (LD_{50})	-
Infection Rate	none
Incubation Period	1–14 days
Peak Infection	NA
Annual Cases	rare
Annual Fatalities	-

A zoonotic microbe that primarily causes a blood infection (tularemia) contracted from rabbits, squirrels, muskrats, beavers, deer, or other wild animals, but sometimes from insect vectors like ticks. It can be transmitted via the airborne route, but cases are rare. Infections are more commonly due to bites or scratches. Infected animals may not show signs of infection. Common throughout North America, Europe (except UK), and Asia throughout the year.

Disease or Infection	tularemia, pneumonia, fever				
Natural Source	wild animals, natural waters				
Toxins	none				
Point of Infection	Upper Respiratory Tract, skin				
Symptoms	Indolent ulcer at site of infection, with swelling of local lymph nodes, pain, fever, pneumonic disease may follow.				
Treatment	Gentamicin, aminoglycosides, streptomycin, tobramicin, kanamycin, tetracyclines, chloramphenicol				
Untreated Fatality Rate	5–15%	**Prophylaxis:** Antibiotics		**Vaccine:** Available	
Shape	pleomorphic coccobacillus				
Mean Diameter, μm	0.2	**Size Range:**	0.2 × 0.2–0.7 microns		
Growth Temperature	37°C	**Survival Outside Host:**	31–133 days		
Inactivation	Moist heat: 121°C for 15 min. Dry heat: 170°C for 1 hr.				
Disinfectants	1% sodium hypochlorite, 70% ethanol, glutaraldehyde, formaldehyde.				
Filter Nominal Rating	MERV 6	MERV 8	MERV 11	MERV 13	MERV 14
Removal Efficiency, %	5.8	7.3	14.0	37.1	58.2
UVGI Rate Constant	0.01474	m^2/J	**Media**	Air	
Dose for 90% Inactivation	156	J/m^2	**Ref.**	Beebe 1959	
Suggested Indoor Limit	0	cfu/m^3			
Genome Size (bp)	1,892,819		32.3 % G+C	67.7 % T+A	
Related Species	Brucella has similar morphology.				
Notes	CDC Reportable. Open Air Factor rate constant k = 1.5–10 per minute.				
Photo Credit	Centers for Disease Control, PHIL# 2985, Dr. P. B. Smith.				
References	Braude 1981, Freeman 1985, Mitscherlich 1984, Murray 1999, Prescott 1996, Ryan 1994, Canada 2001, Mandell 2000				

Fusarium

GROUP	Fungal Spore
PHYLUM	Ascomycota
ORDER	Hypocreales
FAMILY	Hypocreaceaa
DISEASE GROUP	Non-communicable
BIOSAFETY LEVEL	Risk Group 1
Infectious Dose (ID_{50})	NA
Lethal Dose (LD_{50})	-
Infection Rate	none
Incubation Period	-
Peak Infection	NA
Annual Cases	-
Annual Fatalities	-

Reportedly allergenic. Can grow on damp grains and a wide range of plants. Has been found growing in humidifiers. A common soil fungus. Several species in this genus can produce potent toxins. Produces vomitoxin on grains during damp growing conditions. Symptoms may occur either through ingestion of contaminated grains or possibly inhalation of spores. The genera can produce hemorrhagic syndrome in humans. Frequently involved in eye, skin, and nail infections.

Disease or Infection	allergic alveolitis, allergic fungal sinusitis, toxic reactions, MVOCs				
Natural Source	Environmental, indoor growth on floor dust filters, & in humidifiers.				
Toxins	trichothecenes (Type B), T-2 toxin, zearalenone (F-2 toxin), vomitoxin, deoxynivalenol, fumonisin, and others				
Point of Infection	Upper Respiratory Tract, skin, eyes				
Symptoms	Fusarium infections can have a wide variety of clinical manifestations.				
Treatment	NA				
Untreated Fatality Rate	-	Prophylaxis: none		Vaccine: none	
Shape	ovoid spore				
Mean Diameter, μm	11.225	Size Range:	9–14 microns		
Growth Temperature	25°C	Survival Outside Host:		survives outdoors	
Inactivation	Moist heat: 121°C for 30 min.				
Disinfectants	1% sodium hypochlorite, phenolics, formaldehyde, glutaraldehyde.				
Filter Nominal Rating	MERV 6	MERV 8	MERV 11	MERV 13	MERV 14
Removal Efficiency, %	65.0	75.2	94.0	99.0	99.0
UVGI Rate Constant	0.0071	m^2/J	Media	Plates	
Dose for 90% Inactivation	324	J/m^2	Ref.	Asthana 1992	
Suggested Indoor Limit	150–500	cfu/m^3	(in mix with other nonpathogenic fungi)		
Genome Size (bp)	33,000,000–50,000,000				
Related Species	F. oxysporum, F. solani, F. sporotrichioides, F. culmorum, F. chlamydosporum, F. dimerum, F. moniliforme, F. napiforme, F. proliferatum, F. semitectum.				
Notes	Water activity Aw = 0.90.				
Photo Credit	Image courtesy of Neil Carlson, University of Minnesota.				
References	Freeman 1985, Howard 1983, Lacey 1988, Murray 1999, Pope 1993, Rao 1996, Ryan 1994, Smith 1989, Woods 1997, Sutton 1998				

Haemophilus influenzae

GROUP	Bacteria
TYPE	Gram-
GENUS	Haemophilus
FAMILY	Pasteurellaceae
DISEASE GROUP	Communicable
BIOSAFETY LEVEL	Risk Group 2
Infectious Dose (ID$_{50}$)	unknown
Lethal Dose (LD$_{50}$)	-
Infection Rate	0.2–0.5
Incubation Period	2–4 days
Peak Infection	3–4 days
Annual Cases	1,162
Annual Fatalities	-

A leading cause of meningitis before vaccine development. Infants are main victims and it can be fatal under age 2. Can be pleomorphic in shape. In spite of the name, this microbe is the cause of meningitis, but not a major cause of the flu. It can occur as a secondary invader when Influenza virus is present. Some species occur naturally as human oral flora. Sometimes found on human body surfaces.

Disease or Infection	meningitis, pneumonia, endocarditis, otitis media, and flu. Opportunistic.				
Natural Source	Humans, Nosocomial				
Toxins	none				
Point of Infection	nasopharyngeal				
Symptoms	Bacterial meningitis, otitis media, sinusitis, sudden onset of fever, vomiting, lethargy, stiff neck, progressive stupor.				
Treatment	Antibiotic therapy for 10–14 days, chloramphenicol, or cephalosporins.				
Untreated Fatality Rate	-	Prophylaxis: Rifampin		Vaccine:	available
Shape	coccobacilli				
Mean Diameter, μm	0.285	Size Range:	0.2–0.3 × 0.5–2 microns		
Growth Temperature	22–45°C	Survival Outside Host:	12 days in sputum		
Inactivation	Moist heat: 121°C for 15 min. Dry heat: 170°C for 1 hr.				
Disinfectants	1% sodium hypochlorite, 70% ethanol, glutaraldehyde, formaldehyde.				
Filter Nominal Rating	MERV 6	MERV 8	MERV 11	MERV 13	MERV 14
Removal Efficiency, %	5.8	7.7	15.5	41.0	64.2
UVGI Rate Constant	0.0656	m^2/J	Media	Plates	
Dose for 90% Inactivation	35.10	J/m^2	Ref.	Mongold 1992	
Suggested Indoor Limit	0	cfu/m^3			
Genome Size (bp)	1830138		38.2 % G+C	61.8 % T+A	
Related Species	H. parainfluenzae, H. aphrophilus, H. paraphrophilus, H. hemolyticus.				
Notes	CDC Reportable.				
Photo Credit	Centers for Disease Control, PHIL# 1617, Dr. Wiliam A. Clark.				
References	Braude 1981, Freeman 1985, Mitscherlich 1984, Murray 1999, Prescott 1996, Ryan 1994, Castle 1987, Weinstein 1991				

Haemophilus parainfluenzae

GROUP	Bacteria
TYPE	Gram-
GENUS	Haemophilus
FAMILY	Pasteurellaceae
DISEASE GROUP	Endogenous
BIOSAFETY LEVEL	Risk Group 2
Infectious Dose (ID_{50})	unknown
Lethal Dose (LD_{50})	-
Infection Rate	-
Incubation Period	-
Peak Infection	NA
Annual Cases	common
Annual Fatalities	-

Causes infections similar to or associated with H. influenzae, but is more common. A member of the normal flora in the upper respiratory tract (oral cavity and pharynx). Can be recovered in the throat of 10–25% of children. May cause pharyngitis, epiglottis, otitis media, conjunctivitis, pneumonia, meningitis bacteremia, endocarditis, and other infections. Other respiratory tract infections may predispose patients to infections with H. parainfluenzae.

Disease or Infection	opportunistic infections, conjunctivitis, pneumonia, meningitis				
Natural Source	Humans, nosocomial.				
Toxins	none				
Point of Infection	Upper Respiratory Tract				
Symptoms	Opportunistic infections of the respiratory tract.				
Treatment	Erythromycin, trimethoprim-sulfamethoxazole.				
Untreated Fatality Rate	-	Prophylaxis:		Vaccine:	none
Shape	rods				
Mean Diameter, μm	1.732	Size Range:	1–3 microns		
Growth Temperature	30–45°C	Survival Outside Host:		-	
Inactivation	Moist heat: 121°C for 15 min. Dry heat: 170°C for 1 hr.				
Disinfectants	1% sodium hypochlorite, 70% ethanol, glutaraldehyde, formaldehyde.				
Filter Nominal Rating	MERV 6	MERV 8	MERV 11	MERV 13	MERV 14
Removal Efficiency, %	38.7	53.3	76.2	98.2	99.0
UVGI Rate Constant	(unknown)		Media	-	
Dose for 90% Inactivation	-		Ref.	-	
Suggested Indoor Limit	-				
Genome Size (bp)	1,830,138 (H. influenzae) 38.2% G+C 61.8% T+A				
Related Species	H. aphrophilus, H. paraphrophilus, H. hemolyticus, H. parahemolyticu, H. ducreyi. Type Species: H. influenzae				
Notes	H. ducrei causes chancroid, a common venereal disease.				
Photo Credit	Centers for Disease Control, PHIL# 236, Dr. Erskine Palmer.				
References	Braude 1981, Castle 1987, Freeman 1985, Mitscherlich 1984, Murray 1999, Prescott 1996, Ryan 1994, Weinstein 1991, Mandell 2000				

Hantaan virus

GROUP	Virus
TYPE	RNA
GENUS	Hantavirus
FAMILY	Bunyaviridae
DISEASE GROUP	Non-communicable
BIOSAFETY LEVEL	Risk Group 3
Infectious Dose (ID_{50})	unknown
Lethal Dose (LD_{50})	-
Infection Rate	none
Incubation Period	14–30 days
Peak Infection	NA
Annual Cases	44
Annual Fatalities	22

Occurs from inhalation of infected rodent feces. This unusually deadly pathogen emerged in the 1980s where it killed a number of people in the Southwest. It was subsequently identified as existing in rodent populations across the country. It is not contagious. In dry climates, the feces of infected mice living indoors become airborne. Inhalation can then rapidly incapacitate and prove fatal without treatment. Has potential for use as a biological weapon.

Disease or Infection	Korean hemorrhagic fever, Hantavirus, HFRS, HPS				
Natural Source	Field rodents, Deer mouse, Rattus spp.				
Toxins	none				
Point of Infection	Upper Respiratory Tract				
Symptoms	Abrupt onset of fever lasting 3–8 days, conjunctival infection, prostration, lower back pain, headache, abdominal pain, anorexia, vomiting, respiratory distress.				
Treatment	Ribavarin (IV) during early phase if HFRS				
Untreated Fatality Rate	5–15%	**Prophylaxis:** none		**Vaccine:** none	
Shape	spherical				
Mean Diameter, μm	0.096	**Size Range:**	0.08–0.115 microns		
Growth Temperature	na	**Survival Outside Host:**	2–8 years		
Inactivation	Moist heat: 60°C for 1 hr.				
Disinfectants	1% sodium hypochlorite, 70% ethanol, 2% glutaraldehyde.				
Filter Nominal Rating	MERV 6	MERV 8	MERV 11	MERV 13	MERV 14
Removal Efficiency, %	9.4	11.1	19.8	48.2	69.3
UVGI Rate Constant	(unknown)		Media	-	
Dose for 90% Inactivation	-		Ref.	-	
Suggested Indoor Limit	0 cfu/m³				
Genome Size (bp)	10,500–22,700				
Related Species	Dobrova-Belgrade virus, Sin Nombre virus, Seoul virus				
Notes	CDC Reportable.				
Photo Credit	Centers for Disease Control, PHIL# 1139, Sherif A. Zaki.				
References	Dalton 1973, Fraenkel-Conrat 1985, Freeman 1985, Mahy 1975, Murray 1999, Ryan 1994, Canada 2001				

Histoplasma capsulatum

GROUP	Fungal Spore
CLASS	Ascomycetes
ORDER	Onygenales
FAMILY	Onygenaceae
DISEASE GROUP	Non-communicable
BIOSAFETY LEVEL	Risk Group 3
Infectious Dose (ID_{50})	10
Lethal Dose (LD_{50})	40000
Infection Rate	none
Incubation Period	4–22 days
Peak Infection	NA
Annual Cases	common
Annual Fatalities	-

Histoplasma capsulatum causes histoplasmosis, an infection estimated to have afflicted 40 million Americans, mostly in the Southeast. It most often causes mild fever and malaise, but in 0.1–0.2 % of cases the disease becomes progressive. The infection is inevitably airborne and enters through the lungs, from where it may spread to other areas. In the environment, it is most often found in pigeon roosts, bat caves, or old buildings. This infection can become fatal in some cases.

Disease or Infection	histoplasmosis, fever, malaise, pneumonia possible				
Natural Source	Environmental, nosocomial, pigeon roosts, bat caves, old buildings.				
Toxins	none				
Point of Infection	Upper Respiratory Tract				
Symptoms	Respiratory infection, mild coldlike symptoms, mild fever or cough. Severe cases may have chills, chest pain, malaise.				
Treatment	Amphotericin B.				
Untreated Fatality Rate	-	Prophylaxis: none		Vaccine: none	
Shape	spherical spore				
Mean Diameter, μm	2.236	Size Range:	1–5 microns		
Growth Temperature	37°C	Survival Outside Host:	indefinite		
Inactivation	Moist heat: 121°C for 15 min.				
Disinfectants	1% sodium hypochlorite, phenolics, formaldehyde, glutaraldehyde.				
Filter Nominal Rating	MERV 6	MERV 8	MERV 11	MERV 13	MERV 14
Removal Efficiency, %	47.3	62.7	84.5	98.9	99.0
UVGI Rate Constant	0.0247	m^2/J	Media	Plates (estimated)	
Dose for 90% Inactivation	93.22	J/m^2	Ref.	Chick 1963	
Suggested Indoor Limit	150–500	cfu/m^3	(in mix with other nonpathogenic fungi)		
Genome Size (bp)					
Related Species	H. capsulatum var. capsulatum, H. capsulatum var.duboisii				
Notes	CDC Reportable.				
Photo Credit	Centers for Disease Control, PHIL# 867, Edwin P. Ewing, Jr.				
References	Larsh 1983, Howard 1983, Lacey 1988, Murray 1999, Ryan 1994, Smith 1989, Ashford 1999, Fuortes 1988, Sorensen 1999, Miyaji 1987, Canada 2001, DiSalvo 1983, Sutton 1998				

Influenza A virus

GROUP	Virus
TYPE	ssRNA, negative
GENUS	Influenza virus A, B
FAMILY	Orthomyxoviridae
DISEASE GROUP	Communicable
BIOSAFETY LEVEL	Risk Group 2
Infectious Dose (ID_{50})	20–790
Lethal Dose (LD_{50})	-
Infection Rate	0.2–0.83
Incubation Period	2–3 days
Peak Infection	3–4 days
Annual Cases	2,000,000
Annual Fatalities	20,000

Causes periodic flu pandemics and can cause widespread fatalities, and sometimes many millions, dead. Constant antigenic variations among the main types of Influenza, Type A and Type B, ensure little chance of immunity developing. Pneumonia can result from secondary bacterial infections, usually staphylococcus or streptococcus. Current theory suggests that the virus passes to and from humans, pigs, and birds, in agricultural areas of Asia where their close association is common.

Disease or Infection	flu, secondary pneumonia				
Natural Source	Humans, birds, pigs, nosocomial				
Toxins	none				
Point of Infection	Upper Respiratory Tract				
Symptoms	Acute fever, chills, headache, myalgia, weakness, sore throat, cough, runny nose.				
Treatment	No antibiotic treatments, fluids and rest				
Untreated Fatality Rate	low	Prophylaxis: possible		Vaccine: Available	
Shape	enveloped helical				
Mean Diameter, μm	0.098	Size Range:	0.08–0.12 microns		
Growth Temperature	na	Survival Outside Host:		2–4 days on cloth, steel	
Inactivation	Moist heat: 56°C for 30 min.				
Disinfectants	1% sodium hypochlorite, 70% ethanol, glutaraldehyde, formaldehyde.				
Filter Nominal Rating	MERV 6	MERV 8	MERV 11	MERV 13	MERV 14
Removal Efficiency, %	9.2	10.9	19.5	47.6	68.7
UVGI Rate Constant	0.119	m^2/J	Media	Air	
Dose for 90% Inactivation	19.35	J/m^2	Ref.	Jensen 1964	
Suggested Indoor Limit	0	cfu/m^3			
Genome Size (bp)	13,588				
Related Species	Influenza Type A, B.				
Notes	CDC Reportable. Avian influenza (Bird Flu)				
Photo Credit	Centers for Disease Control, PHIL# 1841, C. Goldsmith, J. Katz, S. Zaki.				
References	Dalton 1973, Collins & Kennedy 1993, Freeman 1985, Mahy 1975, Murray 1999, Ryan 1994, Malherbe 1980, Canada 2001				

Junin virus

	GROUP	Virus
	TYPE	ssRNA, negative
	GENUS	Arenavirus
	FAMILY	Arenaviridae
	DISEASE GROUP	Non-communicable
	BIOSAFETY LEVEL	Risk Group 4
	Infectious Dose (ID_{50})	unknown
	Lethal Dose (LD_{50})	10–100000
	Infection Rate	low
	Incubation Period	2–14 days
	Peak Infection	7 days
	Annual Cases	2,000
	Annual Fatalities	rare

Occurs primarily in South American farm workers, in late summer or fall. It is inhaled from aerosolized rodent feces. Fatality rate 3–15%. Like several of the other arenaviruses, it causes hemorrhagic fever. It is relatively uncommon outside of South America. It occurs seasonally in concert with an increase in the rodent population. Rodent excreta contain the virus and inhalation or ingestion may cause the disease.

Disease or Infection	Argentinean hemorrhagic fever				
Natural Source	Rodents				
Toxins	none				
Point of Infection	Upper Respiratory Tract				
Symptoms	Slow onset of fever, fatigue, headache, muscular pain, bleeding may occur from nose, gums, intestines.				
Treatment	Ribavarin, human plasma treatment				
Untreated Fatality Rate	10–50%	**Prophylaxis:** none		**Vaccine:**	Available
Shape	enveloped helical				
Mean Diameter, μm	0.122	**Size Range:**	0.05–0.3 microns		
Growth Temperature	na	**Survival Outside Host:**		-	
Inactivation	Moist heat: 56°C for 30 min.				
Disinfectants	1% sodium hypochlorite, 2% glutaraldehyde				
Filter Nominal Rating	MERV 6	MERV 8	MERV 11	MERV 13	MERV 14
Removal Efficiency, %	7.7	9.2	16.7	42.2	63.0
UVGI Rate Constant	(unknown)		Media	-	
Dose for 90% Inactivation	-		Ref.	-	
Suggested Indoor Limit	0	cfu/m^3			
Genome Size (bp)	10,000–14,000				
Related Species	Machupo, Lassa				
Notes	CDC Reportable.				
Photo Credit	Photo shows a cell infected with Cupixi virus, an arenavirus. Cynthia Goldsmith & Michael Bowen, Centers for Disease Control, Atlanta.				
References	Dalton 1973, Fraenkel-Conrat 1985, Freeman 1985, Mahy 1975, Murray 1999, Ryan 1994, Salvato 1993, Kenyon 1988, Malherbe 1980				

Klebsiella pneumoniae

GROUP	Bacteria
TYPE	Gram-
GENUS	Klebsiella
FAMILY	Enterobacteriaceae
DISEASE GROUP	Endogenous
BIOSAFETY LEVEL	Risk Group 2
Infectious Dose (ID_{50})	unknown
Lethal Dose (LD_{50})	-
Infection Rate	-
Incubation Period	-
Peak Infection	NA
Annual Cases	1,488
Annual Fatalities	-

Klebsiella pneumoniae exist in the soil and in water as free-living microorganisms. They are also found in man's intestines as commensal flora. It is only when they occur in the upper respiratory tract that they become an infectious problem. Worldwide, 2/3 of infections are nosocomial, this bacteria causes 3% of cases of acute bacterial pneumonia, but the fatality rate is as high as 90% in untreated cases.

Disease or Infection	opportunistic infections, pneumonia, ozena, rhinoscleroma				
Natural Source	Environmental, soil, Humans, indoor Growth in water, nosocomial.				
Toxins	Endotoxins				
Point of Infection	Upper Respiratory Tract				
Symptoms	Nosocomial urinary and and pulmonary infections, wound infections, secondary infections of lungs in cases of chronic pulmonary disease.				
Treatment	Aminoglycosides, cephalosporins, resists some other antibiotics.				
Untreated Fatality Rate	-	Prophylaxis: possible		Vaccine: none	
Shape	rods				
Mean Diameter, μm	0.671	Size Range:	0.3–1.5 × 0.6–6 microns		
Growth Temperature	35–37°C	Survival Outside Host:		4 hours to several days	
Inactivation	Moist heat: 121°C for 15 min. Dry heat: 160–170°C for 1 hr.				
Disinfectants	1% sodium hypochlorite, 70% ethanol, 2% glutaraldehyde, iodines.				
Filter Nominal Rating	MERV 6	MERV 8	MERV 11	MERV 13	MERV 14
Removal Efficiency, %	12.8	18.8	35.1	74.2	93.2
UVGI Rate Constant	0.0548	m^2/J	Media	Water	
Dose for 90% Inactivation	42	J/m^2	Ref.	Zemke 1990	
Suggested Indoor Limit	-				
Genome Size (bp)					
Related Species	Similar to Enterobacter, Serratia, and Citrobacter.				
Notes	CDC Reportable.				
Photo Credit	Centers for Disease Control, PHIL# 2849, Dr. Thomas F. Sellers. Image shows granuloma inguinale infection.				
References	Braude 1981, Castle 1987, Freeman 1985, Mitscherlich 1984, Murray 1999, Prescott 1996, Ryan 1994, Weinstein 1991, Canada 2001				

Lassa virus

GROUP	Virus
TYPE	ssRNA, negative
GENUS	Arbovirus
FAMILY	Arenaviridae
DISEASE GROUP	Communicable
BIOSAFETY LEVEL	Risk Group 4
Infectious Dose (ID_{50})	15
Lethal Dose (LD_{50})	2–200,000
Infection Rate	high
Incubation Period	7–14 days
Peak Infection	-
Annual Cases	-
Annual Fatalities	-

This virus causes a hemorrhagic fever endemic to West Africa and which is similar to Junin (Argentinean hemorrhagic fever) and Machupo (Bolivian henorrhagic fever). Highly dangerous in terms of infectivity. Person-to-person spread occurs by contact with body fluids, but evidence exists for airborne spread. Forcible aerosolization is possible, making this a laboratory hazard also.

Disease or Infection	Lassa fever				
Natural Source	Rodents				
Toxins	none				
Point of Infection	Upper Respiratory Tract				
Symptoms	Fever accompanied by hemorrhagic manifestations, shock, neurologic disturbances, and bradycardia.				
Treatment	Intravenous ribavarin within 6 days can help				
Untreated Fatality Rate	10–50% **Prophylaxis:** none **Vaccine:** none				
Shape	enveloped helical				
Mean Diameter, μm	0.122 **Size Range:** 0.05–0.3 microns				
Growth Temperature	na **Survival Outside Host:** -				
Inactivation	Moist heat: 56°C for 30 min.				
Disinfectants	1% sodium hypochlorite, 70% ethanol, glutaraldehyde, formaldehyde.				
Filter Nominal Rating	MERV 6	MERV 8	MERV 11	MERV 13	MERV 14
Removal Efficiency, %	7.7	9.2	16.7	42.2	63.0
UVGI Rate Constant	(unknown)		**Media**	-	
Dose for 90% Inactivation	-		**Ref.**	-	
Suggested Indoor Limit	0 cfu/m^3				
Genome Size (bp)	10,000–11,000				
Related Species	Machupo, Marburg				
Notes	CDC Reportable.				
Photo Credit	Lassa virus in Vero cells (800x) Reprinted w/ permission from Malherbe and Strickland 1980, Viral Cytopathology, Copyright CRC Press, Boca Raton, FL				
References	Dalton 1973, Fraenkel-Conrat 1985, Freeman 1985, Mahy 1975, Schaal 1981, Peters 1987, Malherbe 1980, Oldstone 1987				

Legionella pneumophila

GROUP	Bacteria
TYPE	Gram-
GENUS	Legionella
FAMILY	Legionellaceae
DISEASE GROUP	Non-communicable
BIOSAFETY LEVEL	Risk Group 2
Infectious Dose (ID_{50})	<129
Lethal Dose (LD_{50})	140,000
Infection Rate	<0.01
Incubation Period	2–10 days
Peak Infection	NA
Annual Cases	1,163
Annual Fatalities	10

The well known cause of Legionnaire's Disease, Legionella pneumophila exists in warm outdoor ponds naturally and in indoor water supplies unnaturally only. It becomes a problem when amplified by air-conditioning equipment and aerosolized in ventilation systems. Extremely high concentrations of the bacteria can result in aerosolization by various means, including shower heads and sauna baths.

Disease or Infection	Legionnaire's Disease, Pontiac fever, opportunistic infections, pneumonia				
Natural Source	Environmental, Growth in cooling tower water, spas, potable water, nosocomial.				
Toxins	none				
Point of Infection	Upper Respiratory Tract				
Symptoms	Acute pneumonitis with malaise, myalgia, anorexia, headache, fever, chills, nonproductive cough, abdominal pain and diarrhea.				
Treatment	Erythromycin, rifampin, ciprofloxacin, oxygen and fluid replacement				
Untreated Fatality Rate	39–50%	**Prophylaxis:** Antibiotics		**Vaccine:** none	
Shape	rods				
Mean Diameter, μm	0.520	**Size Range:**	0.3–0.9 × 0.6–2 microns		
Growth Temperature	29–35°C	**Survival Outside Host:**	months in water		
Inactivation	Moist heat: 121°C for 15 min. Dry heat: 170°C for 1 hr.				
Disinfectants	1% sodium hypochlorite, 70% ethanol, glutaraldehyde, formaldehyde.				
Filter Nominal Rating	MERV 6	MERV 8	MERV 11	MERV 13	MERV 14
Removal Efficiency, %	9.4	13.6	26.6	62.3	85.8
UVGI Rate Constant	0.182	m^2/J	**Media**	Plates	
Dose for 90% Inactivation	12.65	J/m^2	**Ref.**	Antopol 1979	
Suggested Indoor Limit	0	cfu/m^3			
Genome Size (bp)	3,503,610		38.4 % G+C	61.6 % T+A	
Related Species	L. parisiensis (suspected)				
Notes	CDC Reportable.				
Photo Credit	Centers for Disease Control, PHIL# 1187.				
References	Braude 1981, Freeman 1985, Gilpin 1984, Murray 1999, Ryan 1994, Berendt 1980, Canada 2001, Mandell 2000				

Listeria monocytogenes

GROUP	Bacteria
TYPE	Gram+
ORDER	Listeriaceae
GENUS	Listeria
DISEASE GROUP	Non-communicable
BIOSAFETY LEVEL	Risk Group 2
Infectious Dose (ID_{50})	unknown
Lethal Dose (LD_{50})	
Infection Rate	
Incubation Period	
Peak Infection	
Annual Cases	363
Annual Fatalities	70

A food-borne pathogen that may transport and settle on foods by the airborne route, especially in the food processing industry. Grows at normal temperatures but may grow slowly below 8°C. Does not normally afflict healthy adults but pregnant women and the elderly are at risk. Causes flu-like symptoms with diarrhea possible. Has caused periodic outbreaks as a result of contamination of food processing facilities. Listeria is widespread among animals in nature, including fowl and ungulates.

Disease or Infection	Food poisoning.				
Natural Source	Environmental, humans (1–10% may carry it)				
Toxins	none				
Point of Infection	Gastrointestinal				
Symptoms	Severe flu-like symptoms that may include diarrhea, meningitis, and septicemia.				
Treatment	Combination antibiotics				
Untreated Fatality Rate	-	Prophylaxis:		Vaccine:	
Shape	short rods				
Mean Diameter, μm	0.707	Size Range:	0.5–1 microns		
Growth Temperature	2–45°C	Survival Outside Host:			
Inactivation	Moist heat: 121°C for 15 min. Dry heat: 170°C for 1 hr.				
Disinfectants	ethanol, sodium hypochlorite, glutaraldehyde				
Filter Nominal Rating	MERV 6	MERV 8	MERV 11	MERV 13	MERV 14
Removal Efficiency, %	13.7	20.1	37.1	76.6	94.3
UVGI Rate Constant	0.2303	m^2/J	Media	Air	
Dose for 90% Inactivation	10.00	J/m^2	Ref.	Collins 1971	
Suggested Indoor Limit	-	cfu/m^3			
Genome Size (bp)	2,905,310		38. % G+C	62. % T+A	
Related Species	L. ivanovii, L. seeligeri				
Notes	motile, can grow at refrigerator temperatures				
Photo Credit	Centers for Disease Control, PHIL# 2286, Dr. B. Swaminathan and P. Hayes.				
References	Knowles 2002, Bell 1998, Heijden 1999, Welshimer 1960, Collins 1971, Mitscherlich 1984, Ryan 1994, Ray 1996				

Lymphocytic choriomeningitis

	GROUP	Virus
	TYPE	ssRNA, negative
	ORDER	Nidovirales
	FAMILY	Arenaviridae
	DISEASE GROUP	Non-communicable
	BIOSAFETY LEVEL	Risk Group 2–3
	Infectious Dose (ID_{50})	unknown
	Lethal Dose (LD_{50})	<1000
	Infection Rate	na
	Incubation Period	8–13 days
	Peak Infection	3–7 days
	Annual Cases	rare
	Annual Fatalities	-

LCM virus can be inhaled or ingested from contaminated rodent feces. Prevalence in humans is 2–10%. Occasionally causes outbreaks in Europe, America, Australia, and Japan, sometimes by pet hamsters, laboratory animals, or contact with mice. Rodents may harbor the virus for life and transmit to progeny. Evidence exists for transplacental infection in humans.

Disease or Infection	LCM, lymphocytic meningitis				
Natural Source	House mouse, swine, dogs, hamsters, guinea pigs.				
Toxins	none				
Point of Infection	Inhaled or oral.				
Symptoms	Mild influenza-like symptoms, asymptomatic in 1/3 of cases, may progress to meningitis, with transverse myelitis, orchitis or protitis.				
Treatment	No treatment, ribavarin and anti-inflammatories helpful.				
Untreated Fatality Rate	<1%	Prophylaxis: none		Vaccine: none	
Shape	enveloped helical				
Mean Diameter, µm	0.087	Size Range:	0.05–15 microns		
Growth Temperature	na	Survival Outside Host:		in mouse droppings	
Inactivation	Moist heat: 121°C for 30 min.				
Disinfectants	1% sodium hypochlor., 70% ethanol, 2% glutaraldehyde, formaldehyde				
Filter Nominal Rating	MERV 6	MERV 8	MERV 11	MERV 13	MERV 14
Removal Efficiency, %	10.2	12.1	21.4	51.2	72.3
UVGI Rate Constant	(unknown)		Media	-	
Dose for 90% Inactivation	-		Ref.	-	
Suggested Indoor Limit	0				
Genome Size (bp)	10,600 (Machupo)				
Related Species	Arenaviruses				
Notes	No person-to-person transmission has been documented.				
Photo Credit	Photo of LCV budding from infected cells. Reprinted with permission from M.S.Salvato, 1993, The Arenaviridae, Plenum Press, New York.				
References	Dalton 1973, Fraenkel-Conrat 1985, Freeman 1985, Mahy 1975, Murray 1999, Schaal 1979, Peters 1987, Malherbe 1980, Canada 2001				

Machupo

GROUP	Virus
TYPE	ssRNA, negative
GENUS	Arenavirus
FAMILY	Arenaviridae
DISEASE GROUP	Non-communicable
BIOSAFETY LEVEL	Risk Group 4
Infectious Dose (ID_{50})	unknown
Lethal Dose (LD_{50})	<1000
Infection Rate	-
Incubation Period	7–16 days
Peak Infection	-
Annual Cases	-
Annual Fatalities	-

Similar to Junin virus. Inhaled from rodent feces and contaminated floor dust. Fatality rate 3–15%. This virus causes hemmorhagic fever and is relatively uncommon outside of South America. It occurs during late summer and fall, in concert with an increase in the rodent population. Rodent excreta contain the virus and are inhaled when disturbed or ingested.

Disease or Infection	Bolivian hemorrhagic fever				
Natural Source	Rodents				
Toxins	none				
Point of Infection	Upper Respiratory Tract				
Symptoms	Slow onset of fever, fatigue, headache, muscular pain, bleeding may occur from nose, gums, intestines.				
Treatment	Ribavarin, human plasma treatment				
Untreated Fatality Rate	5–30%	**Prophylaxis:** none		**Vaccine:**	none
Shape	pleomorphic				
Mean Diameter, μm	0.120	**Size Range:**	0.110–0.13 microns		
Growth Temperature	na	**Survival Outside Host:**	-		
Inactivation	Moist heat: 121°C for 30 min.				
Disinfectants	1% sodium hypochlorite, 2% glutaraldehyde				
Filter Nominal Rating	MERV 6	MERV 8	MERV 11	MERV 13	MERV 14
Removal Efficiency, %	7.8	9.3	16.9	42.7	63.5
UVGI Rate Constant	(unknown)		**Media**	-	
Dose for 90% Inactivation	-		**Ref.**	-	
Suggested Indoor Limit	0	cfu/m^3			
Genome Size (bp)	10,600				
Related Species	Junin, Lassa				
Notes	CDC Reportable.				
Photo Credit	Photo shows the arenavirus Tacaribe. Reprinted with permission from M.S.Salvato, 1993, The Arenaviridae, Plenum Press, New York.				
References	Dalton 1973, Fraenkel-Conrat 1985, Freeman 1985, Mahy 1975, Murray 1999, Ryan 1994, Salvato 1993, Wagner 1977, Franz 1997				

Marburg virus

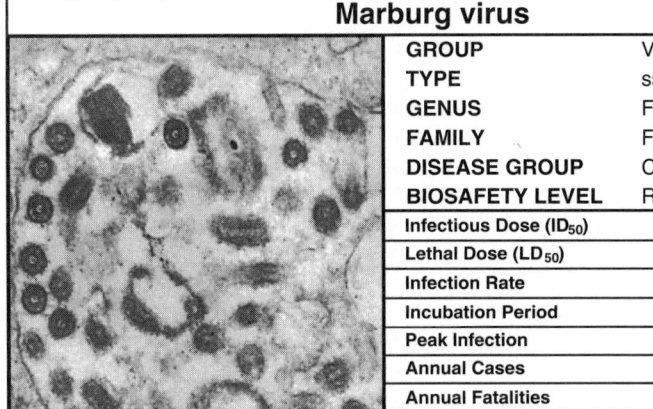

GROUP	Virus
TYPE	ssRNA, negative
GENUS	Filovirus
FAMILY	Filoviridae
DISEASE GROUP	Communicable
BIOSAFETY LEVEL	Risk Group 4
Infectious Dose (ID_{50})	unknown
Lethal Dose (LD_{50})	-
Infection Rate	-
Incubation Period	7 days
Peak Infection	-
Annual Cases	rare
Annual Fatalities	rare

Resembles Lassa fever epidemiologically. Identical to Ebola morphologically. Primarily transmitted by contact. Marburg virus was originally identified in outbreaks traced to contact with infected green monkeys from Uganda. It causes hemorrhagic fever and has a case fatality rate between 22% and 88%. Infections, though ultimately deriving from rodents or other animals, have mainly been transmitted between humans by direct contact, airborne inhalation, or contact with blood.

Disease or Infection	hemorrhagic fever				
Natural Source	Humans, monkeys				
Toxins	none				
Point of Infection	Upper Respiratory Tract				
Symptoms	Sudden onset of high fever, weakness, myalgia, vomiting, diarrhea, maculopapular rash, hemorrhagic diathesis, leukopenia.				
Treatment	No effective treatment. Maintain renal function, electrolyte balance, transfusions				
Untreated Fatality Rate	0.25 **Prophylaxis:** none **Vaccine:** none				
Shape	enveloped helical				
Mean Diameter, μm	0.039 **Size Range:** 0.03–0.05 microns				
Growth Temperature	na **Survival Outside Host:** 2 weeks in warm blood				
Inactivation	Moist heat: 121°C for 30 min.				
Disinfectants	1% sodium hypochlorite, 2% glutaraldehyde, formaldehyde.				
Filter Nominal Rating	MERV 6	MERV 8	MERV 11	MERV 13	MERV 14
Removal Efficiency, %	19.5	22.8	38.5	76.4	91.9
UVGI Rate Constant	(unknown)		Media	-	
Dose for 90% Inactivation	-		Ref.	-	
Suggested Indoor Limit	0 cfu/m^3				
Genome Size (bp)	19,112				
Related Species	related to Ebola (non-airborne)				
Notes	CDC Reportable.				
Photo Credit	CDC, PHIL# 1872, Fred Murphy, Sylvia Whitfield.				
References	Dalton 1973, Fraenkel-Conrat 1985, Freeman 1985, Johnson 1995, Mahy 1975, Murray 1999, Canada 2001, Franz 1997, Martini 1971				

Measles virus

GROUP	Virus
TYPE	ssRNA, negative
GENUS	Morbillivirus
FAMILY	Paramyxoviridae
DISEASE GROUP	Communicable
BIOSAFETY LEVEL	Risk Group 2
Infectious Dose (ID_{50})	0.2 units
Lethal Dose (LD_{50})	-
Infection Rate	0.85
Incubation Period	7–18 days
Peak Infection	9–11 days
Annual Cases	500,000
Annual Fatalities	rare

Mainly affects children in 2–3 year epidemic cycles. Morbillivirus causes the well-known children's disease measles, also called rubeola. Some 90% of adults carry immunity to this virus. It is contracted via the airborne route, and most often in schools, where recirculated air from ventilation systems has been directly implicated by epidemiology studies. It is closely related to canine distemper virus. Occurs primarily in winter and spring.

Disease or Infection	measles (rubeola), Hard measles, Red measles, Morbilli				
Natural Source	Humans, nosocomial				
Toxins	none				
Point of Infection	Upper Respiratory Tract				
Symptoms	Prodromal fever, conjunctivitis, coryza, cough, Koplik spots on buccal mucosa, red blotchy rash in 3–7 days, leukopenia, anorexia, diarrhea.				
Treatment	No antibiotic treatment.				
Untreated Fatality Rate	0.25 **Prophylaxis:** live vaccine **Vaccine:** available				
Shape	enveloped helical				
Mean Diameter, μm	0.158 **Size Range:** 0.1–0.25 microns				
Growth Temperature	na **Survival Outside Host:** 30 minutes as aerosol				
Inactivation	Moist heat: 121°C for 30 min.				
Disinfectants	1% sodium hypochlorite, 70% ethanol, glutaraldehyde, formaldehyde.				
Filter Nominal Rating	MERV 6	MERV 8	MERV 11	MERV 13	MERV 14
Removal Efficiency, %	6.4	7.8	14.6	38.2	58.8
UVGI Rate Constant	(unknown)		Media	-	
Dose for 90% Inactivation	-		Ref.	-	
Suggested Indoor Limit	0 cfu/m^3				
Genome Size (bp)	15,894				
Related Species	Mumps virus, Rubella				
Notes	CDC Reportable.				
Photo Credit	Centers for Disease Control, PHIL# 859, Dr. Edwin P. Ewing, Jr.				
References	Dalton 1973, Collins & Kennedy 1993, Freeman 1985, Mahy 1975, Murray 1999, Ryan 1994, Castle 1987, Weinstein 1991, Canada 2001				

Micromonospora faeni

GROUP	Bacterial Spore
TYPE	Micromonosporaceae
GENUS	Micromonospora
FAMILY	Actinomycetes
DISEASE GROUP	Non-communicable
BIOSAFETY LEVEL	Risk Group 1
Infectious Dose (ID_{50})	unknown
Lethal Dose (LD_{50})	unknown
Infection Rate	none
Incubation Period	-
Peak Infection	NA
Annual Cases	uncommon
Annual Fatalities	-

One of the thermophilic actinomycetes, the fungi-like bacteria that grow mycelia and produce spores. Inhalation of the spores can produce pulmonary fibrosis, which can have sudden symptoms and progress rapidly to death. It is a cause of Farmer's Lung (hypersensitivity pneumonitis), especially when it results from exposure to moldy hay that may produce chronic or acute and extraordinary airborne inhalation doses. It is primarily an agricultural hazard and flourishes in high humidity. In urban areas, mold growth on air conditioners has been tied to the disease.

Disease or Infection	Farmers Lung, pulmonary fibrosis, allergic reactions, UR irritation				
Natural Source	Agricultural, moldy hay, indoor growth, air conditioners				
Toxins	none				
Point of Infection	Upper Respiratory Tract				
Symptoms	Allergic reactions, respiratory irritation.				
Treatment	Systemic corticosteroid treatment (I.e. prednisone for 4–8 weeks)				
Untreated Fatality Rate	0–20% **Prophylaxis:** none **Vaccine:** none				
Shape	spherical spore, grows in chains of 5–15 spores				
Mean Diameter, μm	0.866 **Size Range:** 0.5–1.5 microns				
Growth Temperature	46–52°C **Survival Outside Host:** survives outdoors				
Inactivation	Moist heat: 121°C for 30 min.				
Disinfectants	various common disinfectants should be effective				
Filter Nominal Rating	MERV 6	MERV 8	MERV 11	MERV 13	MERV 14
Removal Efficiency, %	17.8	26.1	45.5	84.9	97.2
UVGI Rate Constant	(unknown)		Media	-	
Dose for 90% Inactivation	-		Ref.	-	
Suggested Indoor Limit	150–240 cfu/m³				
Genome Size (bp)					
Related Species	Type Species: M. chalcea				
Notes	Primarily an agricultural occupational hazard.				
Photo Credit	Penn State AE Dept. (image of Micropolyspora faeni)				
References	Freeman 1985, Mitscherlich 1984, Murray 1999, Sikes 1973, Schaal 1979, Woods 1997, Ortiz-Ortiz 1984, Katila 1978				

Moraxella

GROUP	Bacteria
TYPE	Gram-
GENUS	Moraxella
FAMILY	Moraxellaceae
DISEASE GROUP	Endogenous
BIOSAFETY LEVEL	Risk Group 2
Infectious Dose (ID_{50})	unknown
Lethal Dose (LD_{50})	-
Infection Rate	-
Incubation Period	-
Peak Infection	NA
Annual Cases	rare
Annual Fatalities	none

This disease is extremely rare but can infect the upper respiratory tract. It is considered to be an innocuous inhabitant of the nasopharynx in healthy humans. It can cause otitis media in children and can infect immunocompromised adults. Moraxella lacunata is extremely rare but only affects humans and can infect the eyes and occasionally the respiratory tract. Fourth most common cause of corneal infections. Occurs worldwide, especially in conditions of poor hygiene.

Disease or Infection	otitis media, opportunistic infections				
Natural Source	Humans, nosocomial.				
Toxins	none				
Point of Infection	Upper Respiratory Tract, eyes				
Symptoms	Parasitic on skin and mucous membranes, corneal infections, otitis media, endophthalmitis, sepic arthritis, pneumonia, meningitis.				
Treatment	Penicillins, cephalosporins, tetracyclines, quinolones, aminoglycosides.				
Untreated Fatality Rate	-	**Prophylaxis:** possible		**Vaccine:**	none
Shape	coccobacilli, short rods, non-motile, occur in pairs and short chains				
Mean Diameter, µm	1.225	**Size Range:**	$1-1.5 \times 1.5-2.5$ microns		
Growth Temperature	2–37°C	**Survival Outside Host:**		3–4 days	
Inactivation	Moist heat: 121°C for 15 min. Dry heat: 160–170°C for 1 hr.				
Disinfectants	1% sodium hypochlorite, 70% ethanol, 2% glutaraldehyde, formaldehyde.				
Filter Nominal Rating	MERV 6	MERV 8	MERV 11	MERV 13	MERV 14
Removal Efficiency, %	27.2	38.9	61.3	94.4	98.8
UVGI Rate Constant	0.0002	m^2/J	**Media**	Air	
Dose for 90% Inactivation	11,513	J/m^2	**Ref.**	Keller 1982	
Suggested Indoor Limit	-				
Genome Size (bp)					
Related Species	M. catarrhalis, lacunata, prev. Branhamella				
Notes	Often confused with Neisseria.				
Photo Credit	Image provided courtesy of Dr. J. Michael Miller, CDC, Atlanta.				
References	Braude 1981, Freeman 1985, Murray 1999, Prescott 1996, Ryan 1994, Castle 1987, Weinstein 1991, Canada 2001				

Mucor

GROUP	Fungal Spore
PHYLUM	Zygomycota
ORDER	Mucorales
FAMILY	Mucoraceae
DISEASE GROUP	Non-communicable
BIOSAFETY LEVEL	Risk Group 1
Infectious Dose (ID_{50})	unknown
Lethal Dose (LD_{50})	-
Infection Rate	none
Incubation Period	-
Peak Infection	NA
Annual Cases	rare
Annual Fatalities	rare

An opportunistic pathogen that can infect the lungs or other locations. It can be fatal to those with impaired immune systems. Spores will enter and germinate to produce mycelial growth. Indoor levels can exceed outdoor levels and it can grow on dust and filters. The majority of patients with mucormycosis are seriously immunocompromised. Other than fever and dyspnea, there are rarely any other symptoms. With continued tissue necrosis, hemoptysis may develop, and the end result may be fatal pulmonary hemorrhage. Sometimes found in leather, meat, animal hair, and dairy products.

Disease or Infection	mucormycosis, rhinitis, pneumonia				
Natural Source	Environmental, sewage, dead plant material, horse dung, fruits.				
Toxins	none				
Point of Infection	Upper Respiratory Tract				
Symptoms	The only clinical symptoms usually manifest are fever and dyspnea. Tissue necrosis may result in hemoptysis.				
Treatment	Treatment with amphotericin B remains the only reliable therapy.				
Untreated Fatality Rate	-	Prophylaxis: none		Vaccine: none	
Shape	spherical spore				
Mean Diameter, μm	7.071	Size Range:	5–10 microns		
Growth Temperature	25°C	Survival Outside Host:		survives outdoors	
Inactivation	Moist heat: 121°C for 30 min.				
Disinfectants	1% sodium hypochlorite, phenolics, formaldehyde, glutaraldehyde.				
Filter Nominal Rating	MERV 6	MERV 8	MERV 11	MERV 13	MERV 14
Removal Efficiency, %	64.8	75.3	94.0	99.0	99.0
UVGI Rate Constant	0.0135	m^2/J	Media	Air	
Dose for 90% Inactivation	171	J/m^2	Ref.	Luckiesh 1946	
Suggested Indoor Limit	150–500	cfu/m^2	(in mix with other nonpathogenic fungi)		
Genome Size (bp)	39,000,000	(M. racemosus)			
Related Species	M. circinelloides, M. ramosissimus, M. plumbeus, M. pusillus				
Notes	Mucor mucedo used for UVGI rate constant.				
Photo Credit	Image courtesy of Neil Carlson, University of Minnesota.				
References	Freeman 1985, Howard 1983, Lacey 1988, Murray 1999, Ryan 1994, Smith 1989, Castle 1987, Weinstein 1991, Woods 1997, Sutton 1998				

Mumps virus

GROUP	Virus
TYPE	ssRNA, negative
GENUS	Rubulavirus
FAMILY	Paramyxoviridae
DISEASE GROUP	Communicable
BIOSAFETY LEVEL	Risk Group 2
Infectious Dose (ID_{50})	unknown
Lethal Dose (LD_{50})	-
Infection Rate	0.6–0.85
Incubation Period	14–28 days
Peak Infection	3–10 days
Annual Cases	10,000
Annual Fatalities	rare

Causes mumps in about 60% of children in spring and winter. Some 70% of infections are asymptomatic. Paramyxovirus, or Mumps virus, causes the common childhood disease. It only affects humans and is seldom life-threatening. Immunity runs at 60% in the adult population. Tends to be benign and self-limiting. Normally transmitted by direct contact, droplet nuclei, or fomites and enters through the nose or mouth.

Disease or Infection	mumps, viral encephalitis				
Natural Source	Humans				
Toxins	none				
Point of Infection	Upper Respiratory Tract				
Symptoms	Nonsuppurative swelling and tenderness of the salivary glands.				
Treatment	Symptomatic and supportive treatment only.				
Untreated Fatality Rate	- **Prophylaxis:** possible **Vaccine:** available				
Shape	enveloped helical				
Mean Diameter, μm	0.164 **Size Range:** 0.09–0.30 microns				
Growth Temperature	na **Survival Outside Host:** -				
Inactivation	Moist heat: 121°C for 30 min.				
Disinfectants	1% sodium hypochlorite, 70% ethanol, glutaraldehyde, formaldehyde.				
Filter Nominal Rating	MERV 6	MERV 8	MERV 11	MERV 13	MERV 14
Removal Efficiency, %	6.3	7.7	14.5	37.8	58.5
UVGI Rate Constant	(unknown)		**Media**	-	
Dose for 90% Inactivation	-		**Ref.**	-	
Suggested Indoor Limit	-				
Genome Size (bp)	15,000–16,000				
Related Species	Parainfluenza, Newcastle Disease virus				
Notes	CDC Reportable.				
Photo Credit	Centers for Disease Control, PHIL# 1874, Dr. Ersline Palmer, B. G. Partin.				
References	Dalton 1973, Fraenkel-Conrat 1985, Freeman 1985, Mahy 1975, Murray 1999, Ryan 1994, Mandell 2000				

Mycobacterium avium

GROUP	Bacteria
TYPE	Gram+
GENUS	Mycobacterium
FAMILY	Mycobacteriaceae
DISEASE GROUP	Non-communicable
BIOSAFETY LEVEL	Risk Group 2
Infectious Dose (ID_{50})	unknown
Lethal Dose (LD_{50})	-
Infection Rate	none
Incubation Period	-
Peak Infection	NA
Annual Cases	uncommon
Annual Fatalities	rare

Has TB-like symptoms. Mycobacterium avium and Mycobacterium intracellulare are nearly identical and are members of the atypical mycobacteria. They are non-tubercle forming. They can be asymptomatic. They are often found in association with the tuberculosis bacilli. Has been isolated from soil and water. Inhalation is believed to be the common route of infection. Distributed worldwide. Nearly all cases of pulmonary infection occur in adults.

Disease or Infection	cavitary pulmonary disease, opportunistic				
Natural Source	Environmental, water, dust, plants.				
Toxins	none				
Point of Infection	Upper Respiratory Tract				
Symptoms	Mild symptoms, pulmonary disease may be prolonged.				
Treatment	Treatment with INH, RIF, and EMB.				
Untreated Fatality Rate	none	**Prophylaxis:** none		**Vaccine:** none	
Shape	rods				
Mean Diameter, μm	1.118	**Size Range:**	$1-1.25 \times 3-5$ microns		
Growth Temperature	22–45°C	**Survival Outside Host:**	survives outdoors		
Inactivation	Moist heat: 121°C for 15 min.				
Disinfectants	5% phenol, 1% sodium hypochlorite, iodine solutions, glutaraldehyde, formaldehyde.				
Filter Nominal Rating	MERV 6	MERV 8	MERV 11	MERV 13	MERV 14
Removal Efficiency, %	24.4	35.3	57.1	92.6	98.6
UVGI Rate Constant	3.8841	m^2/J	**Media**	Air	
Dose for 90% Inactivation	0.59	J/m^2	**Ref.**	David 1973	
Suggested Indoor Limit	0	cfu/m^3			
Genome Size (bp)	4,829,781		69.3% G+C	30.7% T+A	
Related Species	M. tuberculosis, M. kansasii				
Notes	CDC Reportable.				
Photo Credit	Centers for Disease Control, PHIL# 965, Edwin P. Ewing..				
References	Braude 1981, Collins 1971, Freeman 1985, Mitscherlich 1984, Murray 1999, Prescott 1996, Ryan 1994, Woods 1997, Mandell 2000				

Mycobacterium kansasii

	GROUP	Bacteria
	TYPE	Gram+
	GENUS	Mycobacterium
	FAMILY	Mycobacteriaceae
	DISEASE GROUP	Non-communicable
	BIOSAFETY LEVEL	Risk Group 2
	Infectious Dose (ID_{50})	unknown
	Lethal Dose (LD_{50})	-
	Infection Rate	none
	Incubation Period	-
	Peak Infection	NA
	Annual Cases	rare
	Annual Fatalities	rare

Non-tubercular bacteria often associated with TB, that has TB-like symptoms. Can be asymptomatic. Mycobacterium kansasii are an atypical mycobacteria. They are very uncommon. Infections are more severe than M. avium but milder than TB. Occurs worldwide. Infects males 3 times more than females. Chronic obstructive pulmonary disease predisposes patients to infection.

Disease or Infection	cavitary pulmonary disease				
Natural Source	water, cattle, swine				
Toxins	none				
Point of Infection	Upper Respiratory Tract				
Symptoms	Symptoms resemble TB but are milder.				
Treatment	Isoniazid (INH), streptomycin, ethambutol in combination.				
Untreated Fatality Rate	-	**Prophylaxis:** none		**Vaccine:** none	
Shape	rods				
Mean Diameter, µm	0.637	**Size Range:**	0.2–0.6 × 1–5 microns		
Growth Temperature	34–41°C	**Survival Outside Host:**	survives outdoors		
Inactivation	Moist heat: 121°C for 15 min.				
Disinfectants	5% phenol, 1% sodium hypochlorite, iodine solutions, glutaraldehyde, formaldehyde.				
Filter Nominal Rating	MERV 6	MERV 8	MERV 11	MERV 13	MERV 14
Removal Efficiency, %	12.0	17.6	33.2	71.8	92.0
UVGI Rate Constant	3.5829	m^2/J	**Media**	Air	
Dose for 90% Inactivation	0.64	J/m^2	**Ref.**	David 1973	
Suggested Indoor Limit	0	cfu/m^3			
Genome Size (bp)	4,345,492	(M. bovis)	65.6% G+C	34.4% T+A	
Related Species	M. avium, M. intracellulare, M. tuberculosis				
Notes	CDC Reportable.				
Photo Credit	Image of Mycobacterium bovis courtesy of Dr. Glenn Songer, University of Arizona.				
References	Braude 1981, Freeman 1985, Mitscherlich 1984, Murray 1999, Prescott 1996, Ryan 1994, Mandell 2000				

Mycobacterium tuberculosis

GROUP	Bacteria
TYPE	Gram+ (acid fast)
GENUS	Mycobacterium
FAMILY	Mycobacteriaceae
DISEASE GROUP	Communicable
BIOSAFETY LEVEL	Risk Group 3
Infectious Dose (ID_{50})	1–10
Lethal Dose (LD_{50})	-
Infection Rate	0.33
Incubation Period	4–12 weeks
Peak Infection	varies
Annual Cases	20,000
Annual Fatalities	-

Tuberculosis infects over 1/3 of the world's population. This bacteria causes TB, once called consumption because of the way it seemed to deplete a person till death, and was an ancient disease even to the Egyptians. Estimated to be at least 15,000 years old, this parasite poses one of the greatest modern health hazards due to the recent emergence of drug-resistant strains. It is highly contagious and a single bacilli is capable of causing an infection in lab animals.

Disease or Infection	tuberculosis, TB, pneumonia possible				
Natural Source	Humans, sewage (potential), nosocomial.				
Toxins	none				
Point of Infection	Upper Respiratory Tract				
Symptoms	Slow progress to pulmonary infection, fatigue, fever, cough, chest pain, hemoptysis fibrosis, cavitation.				
Treatment	Isoniazid, rifampin, streptomycin, ethambutol, pyrazinamide				
Untreated Fatality Rate	-	**Prophylaxis:** possible		**Vaccine:**	Available
Shape	rods				
Mean Diameter, μm	0.637	**Size Range:**	0.2–0.6 × 1–5 microns		
Growth Temperature	30–38°C	**Survival Outside Host:**	40–100 days		
Inactivation	Moist heat: 121°C for 15 min.				
Disinfectants	5% phenol, 1% sodium hypochlorite, iodine solutions, glutaraldehyde, formaldehyde.				
Filter Nominal Rating	MERV 6	MERV 8	MERV 11	MERV 13	MERV 14
Removal Efficiency, %	12.0	17.6	33.2	71.8	92.0
UVGI Rate Constant	0.2132	m^2/J	**Media**	Plates	
Dose for 90% Inactivation	10.80	J/m^2	**Ref.**	David 1973	
Suggested Indoor Limit	0	cfu/m^3			
Genome Size (bp)	4,411,529		65.6 % G+C	34.4 % T+A	
Related Species	M. avium, M. intracellulare, M. kansasii				
Notes	CDC Reportable.				
Photo Credit	Centers for Disease Control, PHIL# 647, Dr. Shirley E. Maddison.				
References	Braude 1981, David 1973, Freeman 1985, Higgins 1975, Murray 1999, Ryan 1994, Youmans 1979, Canada 2001, Mandell 2000				

Mycoplasma pneumoniae

GROUP	Bacteria
TYPE	no wall
GENUS	Mycoplasma
FAMILY	Mycoplasmataceae
DISEASE GROUP	Endogenous
BIOSAFETY LEVEL	Risk Group 2
Infectious Dose (ID_{50})	100
Lethal Dose (LD_{50})	-
Infection Rate	-
Incubation Period	6–23 days
Peak Infection	NA
Annual Cases	uncommon
Annual Fatalities	rare

Weakly pathogenic for man and often found as commensals. Mycoplasma pneumoniae is a member of a class called Mollicutes, which are considered to be different from bacteria since they contain no cell wall. Immune system disruption, usually by another disease, is required to produce an infection. Accounts for approximately 20% of all cases of pneumonia. Only about 3–10 % of infections result in apparent pneumonia. Endemic infections occur worldwide. Mainly affects age group 5–15 years old.

Disease or Infection	pneumonia, PPLO, walking pneumonia, (Gulf War Syndrome?)				
Natural Source	Humans				
Toxins	none				
Point of Infection	Upper Respiratory Tract				
Symptoms	Slow onset with malaise, headache, paraxysmal cough, substernal pain, leukocytosis possible, pneumonia.				
Treatment	Tetracyclines, gentamicin, doxycycline, macrolides				
Untreated Fatality Rate	-	**Prophylaxis:** Antibiotics		**Vaccine:**	none
Shape	pleomorphic				
Mean Diameter, μm	0.177	**Size Range:**	0.125–0.25 microns		
Growth Temperature	30–39°C	**Survival Outside Host:**		10–50 hours in air	
Inactivation	Moist heat: 121°C for 15 min. Dry heat: 170°C for 1 hr.				
Disinfectants	1% sodium hypochlorite, 70% ethanol, glutaraldehyde, formaldehyde.				
Filter Nominal Rating	MERV 6	MERV 8	MERV 11	MERV 13	MERV 14
Removal Efficiency, %	6.0	7.5	14.2	37.3	58.1
UVGI Rate Constant	(unknown)		**Media**	-	
Dose for 90% Inactivation	Low		**Ref.**	Kundsin 1968	
Suggested Indoor Limit	0	cfu/m^3			
Genome Size (bp)	816,394		40. % G+C	60. % T+A	
Related Species	M. hominis, Ureaplasma urealyticum				
Notes	CDC Reportable.				
Photo Credit	Centers for Disease Control, PHIL# 1351, Dr. Martin Hicklin.				
References	Braude 1981, Freeman 1985, Maniloff 1992, Murray 1999, Prescott 1996, Ryan 1994, Canada 2001, Madoff 1971, Mandell 2000				

Neisseria meningitidis

GROUP	Bacteria
TYPE	Gram-
GENUS	Neisseria
FAMILY	Neisseriaceae
DISEASE GROUP	Endogenous
BIOSAFETY LEVEL	Risk Group 2
Infectious Dose (ID_{50})	1–10
Lethal Dose (LD_{50})	-
Infection Rate	varies
Incubation Period	2–10 days
Peak Infection	2–4 days
Annual Cases	3,308
Annual Fatalities	rare

Exclusively a human parasite, it is considered commensal, and only becomes a problem when host resistance becomes reduced. Some 20–40 % of young adults can carry this microorganism asymptomatically. Less healthy hosts can exhibit a range of maladies, especially if it enters the blood. It is rarely fatal and epidemics sweep through children and adolescents at irregular intervals. Direct transmission is rare. It is the second leading cause of meningitis, after Haemophilus influenzae.

Disease or Infection	meningitis, pharyngitis				
Natural Source	Humans				
Toxins	none				
Point of Infection	Upper Respiratory Tract				
Symptoms	Sudden onset of fever, intense headache, nausea, vomiting, stiff neck, frequently a pink rash, delerium, and coma.				
Treatment	Rifampin, ceftraixone, ciprofloxacin, penicillin.				
Untreated Fatality Rate	0.5	**Prophylaxis:** Rifampin		**Vaccine:**	available
Shape	spherical, chains				
Mean Diameter, µm	0.775	**Size Range:**	0.6–1 microns		
Growth Temperature	25–42°C	**Survival Outside Host:**		does not survive well	
Inactivation	Moist heat: 121°C for 15 min. Dry heat: 160–170°C for 1 hr.				
Disinfectants	1% sodium hypochlorite, 70% ethanol, glutaraldehyde, iodines.				
Filter Nominal Rating	MERV 6	MERV 8	MERV 11	MERV 13	MERV 14
Removal Efficiency, %	15.5	22.6	40.7	80.5	95.8
UVGI Rate Constant	0.0523	m²/J	**Media**	-	
Dose for 90% Inactivation	44.00	J/m²	**Ref.**	Nagy 1964	
Suggested Indoor Limit	-				
Genome Size (bp)	2,184,406		51.8 % G+C	48.2 % T+A	
Related Species	N. gonorrhoeae, N. lactamica, N. meningitidis, N. catarrhalis				
Notes	CDC Reportable. N. catarrhalis used for UVGI rate constant.				
Photo Credit	Centers for Disease Control, PHIL# 1006, Dr. M. S. Mitchell.				
References	Braude 1981, Freeman 1985, Mitscherlich 1984, Murray 1999, Prescott 1996, Ryan 1994, Canada 2001				

Nocardia asteroides

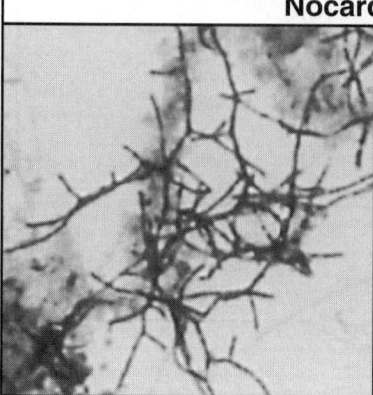

GROUP	Bacterial Spore
TYPE	Nocardiaceae
GENUS	Nocardia
FAMILY	Actinomycetes
DISEASE GROUP	Non-communicable
BIOSAFETY LEVEL	Risk Group 2
Infectious Dose (ID_{50})	unknown
Lethal Dose (LD_{50})	-
Infection Rate	none
Incubation Period	-
Peak Infection	NA
Annual Cases	uncommon
Annual Fatalities	rare

Considered pathogenic, this Gram+ bacteria is classified as a pathogenic actinomycetes. This microorganism is a bacterium barely distinguishable from fungi. It can be found in some soils. It is an opportunistic pathogen and primarily affects patients who have been rendered susceptible by other diseases, especially those involving immunodeficiency.

Disease or Infection	nocardiosis, pneumonia
Natural Source	Environmental, soils, sewage, nosocomial.
Toxins	none
Point of Infection	Upper Respiratory Tract
Symptoms	Fever, cough, chest pain, CNS disease, headache, lethargy, confusion, seizures.
Treatment	Surgical drainage, sulfonilamides (TMP-SMX, sulfisoxazole, sulfadiazine)
Untreated Fatality Rate	0.1 **Prophylaxis:** none **Vaccine:** none
Shape	ovoid spore
Mean Diameter, μm	1.118 **Size Range:** 1–1.25 × 3–5 microns
Growth Temperature	0–50°C **Survival Outside Host:** indefinitely in soil, water
Inactivation	Moist heat: 121°C for 15 min. Dry heat: 170°C for 1 hr.
Disinfectants	1% sodium hypochlorite, 2% glutaraldehyde, formaldehyde.

Filter Nominal Rating	MERV 6	MERV 8	MERV 11	MERV 13	MERV 14
Removal Efficiency, %	24.4	35.3	57.1	92.6	98.6

UVGI Rate Constant	0.0123	m^2/J	**Media**	Plates (estimated)
Dose for 90% Inactivation	187	J/m^2	**Ref.**	Chick 1963
Suggested Indoor Limit	0	cfu/m^3		
Genome Size (bp)	6,021,225 (N. farcinica) 70.8 % G+C 29.2 % T+A			
Related Species	Gram+ bacteria. N. caviae., N. brasiliensis			
Notes	CDC Reportable.			
Photo Credit	Photo courtesy of Phil Russell of the Littleton/Englewood Wastewater Treatment Plant.			
References	Austin 1991, Freeman 1985, Lacey 1988, Murray 1999, Slack 1975, Sikes 1973, Grigoriu 1987, Al-Doory 1987, Miyaji 1987, Canada 2001			

Nocardia brasiliensis

GROUP	Bacterial Spore
TYPE	Nocardiaceae
GENUS	Nocardia
FAMILY	Actinomycetes
DISEASE GROUP	Non-communicable
BIOSAFETY LEVEL	Risk Group 2
Infectious Dose (ID_{50})	unknown
Lethal Dose (LD_{50})	-
Infection Rate	none
Incubation Period	-
Peak Infection	NA
Annual Cases	uncommon
Annual Fatalities	-

Considered pathogenic, this Gram$^+$ bacteria can be found in some soils. Primarily affects those with compromised immunity. Causes chronic lung disease, 80% of cases are invasive pulmonary infections, 20% of patients develop granulomous skin lesions, 10% of pulmonary infections are fatal. Abnormalities can spread to other areas. Causes occasional sporadic disease worldwide. Male to female infection rate is 2:1.

Disease or Infection	nocardiosis, pneumonia				
Natural Source	Environmental, soils, sewage, nosocomial.				
Toxins	none				
Point of Infection	Upper Respiratory Tract				
Symptoms	Fever, cough, chest pain, CNS disease, headache, lethargy, confusion, seizures.				
Treatment	Sulfanilomides, TMP-SMX, sulfisoxazole, sulfadiazine				
Untreated Fatality Rate	0.1	**Prophylaxis:** none		**Vaccine:** none	
Shape	spherical spore				
Mean Diameter, µm	1.414	**Size Range:** 1–2 microns			
Growth Temperature	10–45°C	**Survival Outside Host:** survives outside host			
Inactivation	Moist heat: 121°C for 15 min. Dry heat: 160–170°C for 1 hr.				
Disinfectants	1% sodium hypochlorite, 2% glutaraldehyde, formaldehyde.				
Filter Nominal Rating	MERV 6	MERV 8	MERV 11	MERV 13	MERV 14
Removal Efficiency, %	31.8	44.9	67.8	96.6	99.0
UVGI Rate Constant	(unknown)		**Media**	-	
Dose for 90% Inactivation	-		**Ref.**	-	
Suggested Indoor Limit	150–240 cfu/m^3		(in mix with other nonpathogenic fungi)		
Genome Size (bp)	6,021,225 (N. farcinica) 70.8 % G+C 29.2 % T+A				
Related Species	Type Species: N. farcinica				
Notes	CDC Reportable.				
Photo Credit	Centers for Disease Control, PHIL #3038, Dr. Lucille K. Georg.				
References	Austin 1991, Freeman 1985, Lacey 1988, Mitscherlich 1984, Schaal 1981, Murray 1999, Slack 1975, Ryan 1994, Sikes 1973, Weinstein 1991, Canada 2001				

Norwalk virus

GROUP	Virus
TYPE	ssRNA, positive
GENUS	Norovirus
FAMILY	Caliciviridae
DISEASE GROUP	Communicable
BIOSAFETY LEVEL	Risk Group 2
Infectious Dose (ID_{50})	unknown
Lethal Dose (LD_{50})	NA
Infection Rate	high
Incubation Period	10–60 hours
Peak Infection	1–2 days
Annual Cases	181,000
Annual Fatalities	0

Strictly an intestinal pathogen, this is the only such known pathogen that can transmit by the airborne route. Aerosolization can occur during vomiting, or from outdoor air sprays of warm, contaminated seawater, such as may happen on cruise ships. Has caused repeated outbreaks on some cruise ships in tropical waters. Occasionally causes food-borne outbreaks on land. The viruses are passed in the stool of infected persons. People get infected by swallowing stool-contaminated food or water. Outbreaks in the United States are often linked to raw oysters.

Disease or Infection	Gastroenteritis				
Natural Source	Environmental, warm ocean waters				
Toxins	none				
Point of Infection	Gastrointestinal				
Symptoms	Diarrhea, vomiting, gastroenteritis lasting up to 4 days.				
Treatment	None, rehydration therapy.				
Untreated Fatality Rate	0%	**Prophylaxis:** none		**Vaccine:** none	
Shape	small round naked RNA				
Mean Diameter, μm	0.029	**Size Range:**	0.27–0.38 microns		
Growth Temperature	NA	**Survival Outside Host:**		survives outside host	
Inactivation	Moist heat: 121°C for 30 min.				
Disinfectants	bleach, chlorine (I.e. 0.1% hypochlorite)				
Filter Nominal Rating	MERV 6	MERV 8	MERV 11	MERV 13	MERV 14
Removal Efficiency, %	23.3	27.2	45.1	83.2	95.3
UVGI Rate Constant	(unknown)	**Media**		-	
Dose for 90% Inactivation	-	**Ref.**		-	
Suggested Indoor Limit	0	cfu/m^3			
Genome Size (bp)	8,000	(PEC virus)	53.32% G+C	46.68% T+A	
Related Species	noroviruses				
Notes	CDC Reportable.				
Photo Credit	Centers for Disease Control, PHIL #2172.				
References	Ryan 1994, Wilson et al 1982, Goodman et al 1982, Gunn et al 1980, Marks et al 2000				

Paecilomyces variotii

GROUP	Fungal Spore
CLASS	Euascomycetes
ORDER	Eurotiales
FAMILY	Trichomaceae
DISEASE GROUP	Non-communicable
BIOSAFETY LEVEL	Risk Group 1
Infectious Dose (ID_{50})	unknown
Lethal Dose (LD_{50})	NA
Infection Rate	none
Incubation Period	-
Peak Infection	NA
Annual Cases	-
Annual Fatalities	-

Common worldwide in soil, dust, and decaying vegetation. May grow indoors on cellulose or in humidifiers. Can cause paecilomycosis. Associated with wood-trimmers disease and humidifier associated illnesses. Reported to be allergenic. Some members of this genus may cause pneumonia. It may produce arsine gas if growing on arsenic substrate, which can occur on wallpapers covered with paris green. Possible nosocomial agent in the immunocompromised.

Disease or Infection	paecilomycosis, allergic alveolitis, toxic reactions, MVOCs				
Natural Source	Environmental, decaying vegetation, indoor growth in humidifier water.				
Toxins	paecilotoxins, byssochlamic acid, variotin, ferrirubin, viriditoxin, patulin, indole-3-acetic acid, fusigen				
Point of Infection	Upper Respiratory Tract, eyes.				
Symptoms	Has caused endocarditis on rare occasions, prosthetic valve endocarditis, and ocular infections.				
Treatment	NA				
Untreated Fatality Rate	-	Prophylaxis: none		Vaccine: none	
Shape	ovoid spore				
Mean Diameter, μm	2.828	Size Range:	2–4 × 3–5 microns		
Growth Temperature	25°C	Survival Outside Host:	survives outdoors		
Inactivation	Moist heat: 121°C for 30 min.				
Disinfectants	1% sodium hypochlorite, phenolics, formaldehyde, glutaraldehyde.				
Filter Nominal Rating	MERV 6	MERV 8	MERV 11	MERV 13	MERV 14
Removal Efficiency, %	54.2	69.1	89.6	99.0	99.0
UVGI Rate Constant	(unknown)		Media	-	
Dose for 90% Inactivation	-		Ref.	-	
Suggested Indoor Limit	150–500	cfu/m^3	(in mix with other nonpathogenic fungi)		
Genome Size (bp)					
Related Species	P. lilacinus				
Notes	Thermophilic, grows up to 50°C, at least. CDC Reportable.				
Photo Credit	Centers for Disease Control, PHIL# 234.				
References	Freeman 1985, Howard 1983, Lacey 1988, Murray 1999, Pope 1993, Rao 1996, Ryan 1994, Smith 1989, Woods 1997, Sutton 1998				

Paracoccidioides brasiliensis

GROUP	Fungal Spore
PHYLUM	Ascomycota
SUBPHYLUM	Ascomycotina
FAMILY	Paracoccidioides
DISEASE GROUP	Non-communicable
BIOSAFETY LEVEL	Risk Group 2
Infectious Dose (ID$_{50}$)	8,000,000.00
Lethal Dose (LD$_{50}$)	unknown
Infection Rate	none
Incubation Period	-
Peak Infection	NA
Annual Cases	rare
Annual Fatalities	-

Most common in South America, where an estimated 10 million people have been infected. Males are definitively more susceptible than females, representing 90% of all cases. May take years to manifest infection. Subclinical infection possible. One of only four fungi that cause true systemic infections. Can spread to the lymphatics and the lips and nose.

Disease or Infection	paracoccidioidomycosis, paracoccidioidal granuloma, Lutz's disease, South American blastomycosis				
Natural Source	Environmental, soil, decaying vegetation in Central & S. America.				
Toxins	none				
Point of Infection	Upper Respiratory Tract, skin, mucous membranes.				
Symptoms	Chronic mucocutaneous or cutaneous ulcers. Lymph nodes may become infected.				
Treatment	Sulfonilamides, amphotericin B, azole compounds				
Untreated Fatality Rate	-	**Prophylaxis:** none		**Vaccine:** none	
Shape	spherical spore, lemon shaped				
Mean Diameter, μm	4.472	**Size Range:** 2–10 microns			
Growth Temperature	25–37°C	**Survival Outside Host:** indefinitely in soil, water			
Inactivation	Moist heat: 121°C for 30 min.				
Disinfectants	1% sodium hypochlorite, phenolics, formaldehyde, glutaraldehyde.				
Filter Nominal Rating	MERV 6	MERV 8	MERV 11	MERV 13	MERV 14
Removal Efficiency, %	62.5	74.6	93.5	99.0	99.0
UVGI Rate Constant	(unknown)	**Media**		-	
Dose for 90% Inactivation	-	**Ref.**		-	
Suggested Indoor Limit	0 cfu/m^3				
Genome Size (bp)	25,000,000				
Related Species	P. brasiliensis				
Notes	CDC Reportable.				
Photo Credit	Centers for Disease Control, PHIL 520, Dr. Lucille K. Georg.				
References	Collins 1993, Freeman 1985, Howard 1983, Lacey 1988, Murray 1999, Ryan 1994, Sutton 1998, Tuder 1985, Kashino 1985, Miyaji 1987				

Parainfluenza virus

GROUP	Virus
TYPE	ssRNA, negative
GENUS	Rubulavirus
FAMILY	Paramyxoviridae
DISEASE GROUP	Communicable
BIOSAFETY LEVEL	Risk Group 2
Infectious Dose (ID_{50})	unknown
Lethal Dose (LD_{50})	>1.5
Infection Rate	0.2–0.75
Incubation Period	1–3 days
Peak Infection	3–4 days
Annual Cases	common
Annual Fatalities	-

Occurs worldwide and primarily affects children. Parainfluenza occurs worldwide and infects children (at a rate of 75–80%) more than adults. This virus is very contagious and also causes croup. Immunity is temporary and adults can be reinfected. This virus can infect swine and rodents. Infections occur throughout the year but outbreaks usually occur in the fall. Many infections are asymptomatic but progressive disease can be fatal. There are four types of parainfluenza viruses, Parainfluenza 1, 2, 3, and 4. Parainfluenza 1 and 2 can cause croup. Parainfluenza 3 is a major cause of respiratory disease in infants and children. Parainfluenza is the least common and causes mild symptoms.

Disease or Infection	flu, colds, croup, pneumonia				
Natural Source	Humans, nosocomial				
Toxins	none				
Point of Infection	Upper Respiratory Tract, Lower Respiratory Tract				
Symptoms	Cold-like symptoms, croup, bronchiolitis.				
Treatment	No specific treatment available.				
Untreated Fatality Rate	low	Prophylaxis: none		Vaccine: none	
Shape	enveloped helical				
Mean Diameter, μm	0.194	Size Range: 0.15–0.25 microns			
Growth Temperature	na	Survival Outside Host: 4–10 hours on steel, cloth			
Inactivation	Moist heat: 121°C for 30 min.				
Disinfectants	1% sodium hypochlorite, 70% ethanol, glutaraldehyde, formaldehyde.				
Filter Nominal Rating	MERV 6	MERV 8	MERV 11	MERV 13	MERV 14
Removal Efficiency, %	5.8	7.3	14.0	37.0	58.1
UVGI Rate Constant	(unknown)		Media	-	
Dose for 90% Inactivation	-		Ref.	-	
Suggested Indoor Limit	0 cfu/m^3				
Genome Size (bp)	15,654				
Related Species	RSV, Adenovirus, Sendai virus				
Notes	CDC Reportable.				
Photo Credit	Centers for Disease Control, PHIL# 236, Dr. Erskine Palmer.				
References	Dalton 1973, Collins & Kennedy 1993, Freeman 1985, Mahy 1975, Murray 1999, Ryan 1994, Castle 1987, Weinstein 1991, Mandell 2000				

Parvovirus B19

GROUP	Virus
TYPE	ssDNA
GENUS	Erythrovirus
FAMILY	Parvoviridae
DISEASE GROUP	Communicable
BIOSAFETY LEVEL	Risk Group 2
Infectious Dose (ID_{50})	0.5 ml of serum
Lethal Dose (LD_{50})	-
Infection Rate	0.3–0.8
Incubation Period	4–20 days
Peak Infection	-
Annual Cases	uncommon
Annual Fatalities	rare

An uncommon cause of fever. Parvovirus B19 is similar to the adenoviruses and can cause fever. Mostly occurs in children. Some 25% of infections are asymptomatic. Symptoms resolve in 7–10 days. Severe complications are unusual but anaemic patients may develop transient aplastic crisis. Severe anaemia in the immunosuppressed. Worldwide outbreaks occur, mainly in schoolchildren in winter and spring. Some 60% of adults have been exposed.

Disease or Infection	fifth disease, anemia, fever.				
Natural Source	Humans				
Toxins	none				
Point of Infection	Upper Respiratory Tract				
Symptoms	Mild, usually nonfebrile, viral disease with erythematous eruption on the cheeks and red rash on trunk and limbs.				
Treatment	Treatment of symptoms only.				
Untreated Fatality Rate	-	Prophylaxis: none		Vaccine: none	
Shape	naked icosahedral				
Mean Diameter, μm	0.022	Size Range:	0.018–0.026 microns		
Growth Temperature	na	Survival Outside Host:	survives frozen for years		
Inactivation	resistant to heat				
Disinfectants	1% sodium hypochlorite, aldehydes				
Filter Nominal Rating	MERV 6	MERV 8	MERV 11	MERV 13	MERV 14
Removal Efficiency, %	27.6	32.1	52.3	88.9	97.3
UVGI Rate Constant	0.0658	m^2/J	Media	-	
Dose for 90% Inactivation	35.00	J/m^2	Ref.	vonBrodorotti 1982 (Bovine parvovirus)	
Suggested Indoor Limit	0	cfu/m^3			
Genome Size (bp)	5,176				
Related Species	various other parvoviruses exist				
Notes	CDC Reportable.				
Photo Credit	Image courtesy of Dr. M. S. McNulty, Queen's University of Belfast.				
References	Dalton 1973, Braude 1981, Fraenkel-Conrat 1985, Freeman 1985, Mahy 1975, Murray 1999, Pattison 1988, Ryan 1994, Zerbini 1995, Canada 2001				

Penicillium

GROUP	Fungal Spore
CLASS	Euascomycetes
ORDER	Eurotiales
FAMILY	Trichomaceae
DISEASE GROUP	Non-communicable
BIOSAFETY LEVEL	Risk Group 2
Infectious Dose (ID_{50})	unknown
Lethal Dose (LD_{50})	NA
Infection Rate	none
Incubation Period	-
Peak Infection	NA
Annual Cases	rare
Annual Fatalities	-

Penicillium notatum and some closely related species of penicillium are occasional causes of infections in man. Pulmonary, or lung, infections are rare, but it can infect the ear and cornea. Penicillium notatum can produce penicillin, and some people are highly allergic to this antibiotic. Indoor levels can exceed outdoor levels. Sometimes found on human body surfaces.

Disease or Infection	alveolitis, rhinitis, asthma, allergic reactions, irritation, ODTS, toxic reactions, MVOCs produced
Natural Source	Environmental, indoor growth on paint, filters, coils, & humidifiers.
Toxins	penicillic acid, peptide nephrotoxin, viomellein, xanthomegin, xanthocillin X, mycophenolic acid, rocquefortine C, etc.
Point of Infection	Upper Respiratory Tract, ear, eyes.
Symptoms	Allergic reactions, respiratory irritation, keratitis.
Treatment	Amphotericin B, itraconazole
Untreated Fatality Rate	NA **Prophylaxis:** none **Vaccine:** none
Shape	spherical spore
Mean Diameter, µm	3.262 **Size Range:** 2.8–3.8 × 3–4 microns
Growth Temperature	25–37°C **Survival Outside Host:** survives outdoors
Inactivation	Moist heat: 121°C for 30 min.
Disinfectants	1% sodium hypochlorite, phenolics, formaldehyde, glutaraldehyde.

Filter Nominal Rating	MERV 6	MERV 8	MERV 11	MERV 13	MERV 14
Removal Efficiency, %	57.6	71.7	91.5	99.0	99.0

UVGI Rate Constant	0.00718	m^2/J	**Media**	Air
Dose for 90% Inactivation	321	J/m^2	**Ref.**	Asthana 1992
Suggested Indoor Limit	150–500	cfu/m^3	(in mix with other nonpathogenic fungi)	
Genome Size (bp)	34,000,000	(P. chrysogenum)	51.5 % G+C	48.5 % T+A
Related Species	P. notatum, P. glabrum, P. chrysogenum, P. citrinum, P. janthinellum, P. marneffi, P. purpurogenum			
Notes	P. digitatum used for UV rate constant. Water activity Aw = 0.78–0.88.			
Photo Credit	Image courtesy of Neil Carlson, University of Minnesota.			
References	Freeman 1985, Howard 1983, Lacey 1988, Murray 1999, Pope 1993, Ryan 1994, Smith 1989, Su 1992, Woods 1997, Sutton 1998			

Phialophora

	GROUP	Fungal Spore
	CLASS	Euascomycetes
	ORDER	Chaetothyriales
	FAMILY	Herpotrichiellaceae
	DISEASE GROUP	Non-communicable
	BIOSAFETY LEVEL	Risk Group 2
	Infectious Dose (ID_{50})	unknown
	Lethal Dose (LD_{50})	-
	Infection Rate	none
	Incubation Period	-
	Peak Infection	NA
	Annual Cases	-
	Annual Fatalities	-

Some species have been associated with SBS/BRI. P. verrucosa may cause skin infections. A major agent of chromoblastomycosis in tropical and subtropical regions, including Japan and South America. This dematiaceous fungus has been known to cause nodular and macropapular skin lesions. These fungi live as saprophytes in nature, such as on rotted wood, and may enter the skin via minor trauma. Phaeomycotic cysts may be caused by these fungi.

Disease or Infection	allergic alveolitis, chromomycosis				
Natural Source	Environmental, decaying wood, soil, indoor growth in humidifier water.				
Toxins	none				
Point of Infection	Upper Respiratory Tract, skin				
Symptoms	Allergic reactions, respiratory irritation, infections possible.				
Treatment	Intravenous amphotericin B produces a marginal response. Combinations of flucytosine and amphotericin B may be beneficial.				
Untreated Fatality Rate	-	Prophylaxis: none		Vaccine: none	
Shape	ovoid spore				
Mean Diameter, μm	1.470	Size Range:	1.2–1.8 × 3–4 microns		
Growth Temperature	25–30	Survival Outside Host:	survives outdoors		
Inactivation	Moist heat: 121°C for 30 min.				
Disinfectants	1% sodium hypochlorite, phenolics, formaldehyde, glutaraldehyde.				
Filter Nominal Rating	MERV 6	MERV 8	MERV 11	MERV 13	MERV 14
Removal Efficiency, %	33.0	46.5	69.5	97.0	99.0
UVGI Rate Constant	(unknown)	Media		-	
Dose for 90% Inactivation	-	Ref.		-	
Suggested Indoor Limit	150–500	cfu/m^3	(in mix with other nonpathogenic fungi)		
Genome Size (bp)					
Related Species	P. repens, P. verrucosa, P. hoffmanii, P. mutabilis, P. richardsiae.				
Notes	CDC Reportable.				
Photo Credit	Image courtesy of Dr. Sonia Rozental of Laboratório de Imagem Biológica Universidade Federal do Rio de Janeiro				
References	Freeman 1985, Howard 1983, Lacey 1988, Murray 1999, Pope 1993, Ryan 1994, Smith 1989, Sutton 1998, Braude 1981, Woods 1997				

Phoma

GROUP	Fungal Spore
CLASS	Euascomycetes
ORDER	Pleosporales
FAMILY	Pleosporaceae
DISEASE GROUP	Non-communicable
BIOSAFETY LEVEL	Risk Group 1
Infectious Dose (ID_{50})	NA
Lethal Dose (LD_{50})	NA
Infection Rate	none
Incubation Period	-
Peak Infection	NA
Annual Cases	-
Annual Fatalities	-

A common indoor allergen and widely distributed in nature. Often found in soil and potatoes and may grow on butter, paint, cement, dust, filters, humidifiers, and rubber. It is similar to the early stages of growth of Chaetomium species. Produces pink and purple spots on painted walls. It may have antigens which cross-react with those of Alternaria species. It may cause phaeohyphomycotic skin lesions.

Disease or Infection	allergic alveolitis				
Natural Source	Environmental, soil, common plant pathogen, indoor growth				
Toxins	phomenone				
Point of Infection	Upper Respiratory Tract, skin				
Symptoms	Allergic reactions, respiratory irritation.				
Treatment	NA				
Untreated Fatality Rate	NA	Prophylaxis: none		Vaccine: none	
Shape	ovoid spore				
Mean Diameter, μm	3.162	Size Range:	2.5–4 × 6–10 microns		
Growth Temperature	25°C	Survival Outside Host:	survives outdoors		
Inactivation	Moist heat: 121°C for 30 min.				
Disinfectants	1% sodium hypochlorite, phenolics, formaldehyde, glutaraldehyde.				
Filter Nominal Rating	MERV 6	MERV 8	MERV 11	MERV 13	MERV 14
Removal Efficiency, %	56.9	71.2	91.1	99.0	99.0
UVGI Rate Constant	(unknown)		Media	-	
Dose for 90% Inactivation	-		Ref.	-	
Suggested Indoor Limit	150–500	cfu/m³	(in mix with other nonpathogenic fungi)		
Genome Size (bp)					
Related Species	various				
Notes	CDC Reportable.				
Photo Credit	Image courtesy of Neil Carlson, University of Minnesota.				
References	Freeman 1985, Howard 1983, Lacey 1988, Murray 1999, Pope 1993, Rao 1996, Ryan 1994, Smith 1989, Woods 1997, Sutton 1998				

Pneumocystis carinii

GROUP	Fungal Spore
TYPE	Protozoal
ORDER	Pneumocystidales
FAMILY	Pneumocystidaceae
DISEASE GROUP	Communicable
BIOSAFETY LEVEL	Risk Group 1
Infectious Dose (ID_{50})	unknown
Lethal Dose (LD_{50})	-
Infection Rate	none
Incubation Period	4–8 weeks
Peak Infection	NA
Annual Cases	rare
Annual Fatalities	rare

Pneumocystis carinii was previously classified as a protozoa, but is now recognized as a complex fungi. It is opportunistic and dangerous mainly to those with forms of immunodeficiency. Occurs worldwide and is reportedly airborne. It is an important cause of pneumonia, especially in AIDS patients. P. carinii exists as a saprophyte in the lungs of humans and a variety of animal species. Outbreaks of pneumocystis have occurred in hospitals and orphanages. Most healthy children have been exposed at an early age.

Disease or Infection	pneumocystosis, pneumonia possible				
Natural Source	Environmental, humans, nosocomial.				
Toxins	none				
Point of Infection	Upper Respiratory Tract				
Symptoms	Nomally asymptomatic in healthy hosts. In infants, poor feeding, respiratory distress and cyanosis.				
Treatment	Trimethoprim-sulfamethoxazole (TMP-SMX).				
Untreated Fatality Rate	-	**Prophylaxis:** none		**Vaccine:**	none
Shape	spherical spore				
Mean Diameter, μm	2	**Size Range:**	1–4 microns		
Growth Temperature	37°C	**Survival Outside Host:**		survives outdoors	
Inactivation	Moist heat: 121°C for 30 min.				
Disinfectants	1% sodium hypochlorite, phenolics, formaldehyde, glutaraldehyde.				
Filter Nominal Rating	MERV 6	MERV 8	MERV 11	MERV 13	MERV 14
Removal Efficiency, %	43.6	58.8	81.3	98.7	99.0
UVGI Rate Constant	(unknown)		**Media**	-	
Dose for 90% Inactivation	-		**Ref.**	-	
Suggested Indoor Limit	150–500	cfu/m³	(in mix with other nonpathogenic fungi)		
Genome Size (bp)					
Related Species	none				
Notes	CDC Reportable.				
Photo Credit	Centers for Disease Control, PHIL# 554, Dr. Russell K. Brynes.				
References	Freeman 1985, Howard 1983, Lacey 1988, Murray 1999, Rao 1996, Ryan 1994, Smith 1989, Mandell 2000				

Proteus mirabilis

GROUP	Bacteria
TYPE	Gram-
GENUS	Enterobacteria
FAMILY	Enterobacteriaceae
DISEASE GROUP	Endogenous
BIOSAFETY LEVEL	Risk Group 2
Infectious Dose (ID_{50})	unknown
Lethal Dose (LD_{50})	-
Infection Rate	-
Incubation Period	-
Peak Infection	-
Annual Cases	-
Annual Fatalities	-

A non-respiratory opportunistic pathogen. Often found as intestinal flora and has a tendency to colonize the urinary tract. Up to 10% of urinary tract infections are caused by P. mirabilis. Can cause problems when health is compromised. Second only to E. coli in the percentage of Enterobacteriaceae encountered in clinical laboratories. Proteus cultures tend to swarm over the surface of media rather than remain confined to colonies (see image).

Disease or Infection	Opportunistic infections, pneumonia possible				
Natural Source	Humans				
Toxins	Endotoxins				
Point of Infection	Upper Respiratory Tract, burns, wounds				
Symptoms	Opportunistic infections.				
Treatment	Beta-lactams, quinolones.				
Untreated Fatality Rate	-	Prophylaxis: none		Vaccine: none	
Shape	rods				
Mean Diameter, μm	0.494	Size Range:	0.4–0.6 × 1–3 microns		
Growth Temperature	10–43°C	Survival Outside Host:	-		
Inactivation	Moist heat: 121°C for 30 min.				
Disinfectants	1% sodium hypochlorite, 70% ethanol, glutaraldehyde, formaldehyde.				
Filter Nominal Rating	MERV 6	MERV 8	MERV 11	MERV 13	MERV 14
Removal Efficiency, %	8.8	12.8	25.2	60.0	84.0
UVGI Rate Constant	0.07675		Media	-	
Dose for 90% Inactivation	30.0		Ref.	Nagy 1964 (P. vulgaris)	
Suggested Indoor Limit	-				
Genome Size (bp)					
Related Species	Morganella, Providencia, P. vulgaris, P. mirabilis				
Notes	May resist ampicillin and cephalosporin.				
Photo Credit	Centers for Disease Control, PHIL# 1046, Dr. John J. Farmer.				
References	Braude 1981, Freeman 1985, Mitscherlich 1984, Murray 1999, Prescott 1996, Ryan 1994, Mandell 2000				

Pseudomonas aeruginosa

GROUP	Bacteria
TYPE	Gram-
GENUS	Pseudomonas
FAMILY	Pseudomonadaceae
DISEASE GROUP	Non-communicable
BIOSAFETY LEVEL	Risk Group 1
Infectious Dose (ID_{50})	unknown
Lethal Dose (LD_{50})	-
Infection Rate	none
Incubation Period	2–3 days
Peak Infection	2–4 days
Annual Cases	2,626
Annual Fatalities	-

The primary cause of nosocomial pseudomonal infections. Its infectivity is limited mostly to immunosuppressed patients or those who have their health compromised by other illnesses. It can be considered communicable, but without secondary infections. It is considered to exist ubiquitously in the environment, but is amplified in hospitals. Infections sites include the lungs, burn wounds and open wounds. Can become fatal in 80% of cases. It produces some minor toxins.

Disease or Infection	pneumonia, toxic reactions				
Natural Source	Environmental, sewage, nosocomial, indoor growth in dust, water, humidifiers.				
Toxins	minor toxins				
Point of Infection	Upper Respiratory Tract, burns, wounds				
Symptoms	Primary pneumonia may occur in those with chronic lung disease or congestive heart failure.				
Treatment	Antibiotic treatment.				
Untreated Fatality Rate	-	Prophylaxis: possible		Vaccine: none	
Shape	rods				
Mean Diameter, μm	0.494	Size Range:	0.3–0.8 × 1–3 microns		
Growth Temperature	5–42°C	Survival Outside Host:	survives outdoors		
Inactivation	Moist heat: 121°C for 30 min.				
Disinfectants	1% sodium hypochlorite, 70% ethanol, 2% glutaraldehyde.				
Filter Nominal Rating	MERV 6	MERV 8	MERV 11	MERV 13	MERV 14
Removal Efficiency, %	8.8	12.8	25.2	60.0	84.0
UVGI Rate Constant	0.2375	m^2/J	Media	Air	
Dose for 90% Inactivation	9.70	J/m^2	Ref.	Collins 1971	
Suggested Indoor Limit	50	cfu/m^3			
Genome Size (bp)	6,264,403		66.6 % G+C	33.4 % T+A	
Related Species	P. pseudomallei, P. mallei				
Notes	Regularly resistant to penicillin, ampicillin, and other antibiotics.				
Photo Credit	Centers for Disease Control, PHIL# 231, Janice Carr.				
References	Braude 1981, Freeman 1985, Mitscherlich 1984, Murray 1999, Prescott 1996, Ryan 1994, Weinstein 1991, Woods 1997, Mandell 2000				

Reovirus

	GROUP	Virus
	TYPE	dsRNA
	GENUS	Reovirus
	FAMILY	Reoviridae
	DISEASE GROUP	Communicable
	BIOSAFETY LEVEL	Risk Group 2
	Infectious Dose (ID_{50})	unknown
	Lethal Dose (LD_{50})	-
	Infection Rate	-
	Incubation Period	3–4 days
	Peak Infection	-
	Annual Cases	-
	Annual Fatalities	-

A comparatively rare cause of fever and colds. Reovirus can cause mild forms of fever in infants and children. Several Types exist that have variable symptoms. Adults can be infected but symptoms are often mild. Upper respiratory infections may cause fever, pharyngitis, rhinitis, and sometimes rashes. Isolated cases of encephalitis, pneumonia, and renal disease have been reported.

Disease or Infection	colds, fever, pneumonia, rhinorrhea				
Natural Source	Humans				
Toxins	none				
Point of Infection	Upper Respiratory Tract				
Symptoms	Fever, rhinitis, headache, rashes, rhinorrhea.				
Treatment	No treatment available.				
Untreated Fatality Rate	-	Prophylaxis: none		Vaccine: none	
Shape	enveloped icosahedral				
Mean Diameter, μm	0.075	Size Range:	0.074–0.076 microns		
Growth Temperature	na	Survival Outside Host:	-		
Inactivation	Moist heat: 121°C for 30 min.				
Disinfectants	1% sodium hypochlorite, 70% ethanol, glutaraldehyde, formaldehyde.				
Filter Nominal Rating	MERV 6	MERV 8	MERV 11	MERV 13	MERV 14
Removal Efficiency, %	11.6	13.6	23.9	55.7	76.5
UVGI Rate Constant	0.0132	m^2/J	Media	Water	
Dose for 90% Inactivation	174	J/m^2	Ref.	Hill 1970	
Suggested Indoor Limit	-				
Genome Size (bp)	10,000–12,000				
Related Species	Orthoreoviruses				
Notes	CDC Reportable.				
Photo Credit	Reprinted w/ permission from Malherbe and Strickland 1980, Viral Cytopathology, Copyright CRC Press, Boca Raton, FL				
References	Dalton 1973, Fraenkel-Conrat 1985, Freeman 1985, Mahy 1975, Murray 1999, Ryan 1994, Mandell 2000				

Respiratory Syncytial Virus

GROUP	Virus
TYPE	ssRNA, negative
GENUS	Pneumovirus
FAMILY	Paramyxoviridae
DISEASE GROUP	Communicable
BIOSAFETY LEVEL	Risk Group 2
Infectious Dose (ID_{50})	160–640
Lethal Dose (LD_{50})	-
Infection Rate	0.5–0.9
Incubation Period	4–5 days
Peak Infection	1–3 days
Annual Cases	common
Annual Fatalities	0

A common cause of pneumonia (40%) and bronchiolitis (90%) in infants. Occurs within a few months of birth. About one-half of infants are affected but very few infections become life-threatening. This virus is unaffected by maternal antibody. Occurs worldwide. Most common cause of viral pneumonia in children under 5 years. Causes bronchiolitis in babies. Reinfection is common. Outbreaks peak in March and February. Can cause severe illness in the elderly and the immunocompromised.

Disease or Infection	pneumonia, bronchiolitis				
Natural Source	Humans, nosocomial				
Toxins	none				
Point of Infection	Lower Respiratory Tract				
Symptoms	Cold-like lower respiratory tract illness in infants, common colds in adults, febrile bronchitis in infants, pneumonia in infants.				
Treatment	Ribavirin is beneficial when delivered as a nasal spray.				
Untreated Fatality Rate	-	Prophylaxis: possible		Vaccine:	none
Shape	enveloped helical				
Mean Diameter, μm	0.190	Size Range:	0.12–0.3 microns		
Growth Temperature	na	Survival Outside Host:		up to 8 hours	
Inactivation	Heat: 55°C, freezing & thawing, pH < 5.				
Disinfectants	1% sodium hypochlorite, 70% ethanol, 2% glutaraldehyde.				
Filter Nominal Rating	MERV 6	MERV 8	MERV 11	MERV 13	MERV 14
Removal Efficiency, %	5.9	7.3	14.0	37.1	58.0
UVGI Rate Constant	(unknown)		Media	-	
Dose for 90% Inactivation	-		Ref.	-	
Suggested Indoor Limit	0	cfu/m^3			
Genome Size (bp)	15,222				
Related Species	Parainfluenza, Adenovirus, Sendai virus				
Notes	CDC Reportable. aka. RSV				
Photo Credit	Centers for Disease Control, PHIL# 276, Dr. Erskine Palmer.				
References	Dalton 1973, Collins & Kennedy 1993, Freeman 1985, Mahy 1975, Murray 1999, Ryan 1994, Castle 1987, Weinstein 1991, Canada 2001				

Rhinovirus

GROUP	Virus
TYPE	ssRNA, positive
GENUS	Rhinovirus
FAMILY	Picornaviridae
DISEASE GROUP	Communicable
BIOSAFETY LEVEL	Risk Group 2
Infectious Dose (ID_{50})	1–5
Lethal Dose (LD_{50})	-
Infection Rate	0.38–0.89
Incubation Period	2–4 days
Peak Infection	3–4 days
Annual Cases	common
Annual Fatalities	-

One of the causes of the common cold. Man is the only host for the human strains, but related rhinoviruses exist in horses and cows. Rhinovirus colds are predominant in adults, while other cold viruses may predominantly afflict children. Brief hand contact is thought to be a primary mechanism of transmission, but clinical studies have failed to prove this. Extended indoor exposure to airborne concentrations of virus is most likely the primary transmission mechanism in adults.

Disease or Infection	colds				
Natural Source	Humans				
Toxins	none				
Point of Infection	Upper Respiratory Tract				
Symptoms	Acute infection of the upper respiratory tract, coryza, sneezing, lacrimation, irritated nasopharynx, headache, sore throat, malaise.				
Treatment	No specific antivirals, but sensitive to alpha-2 interferon.				
Untreated Fatality Rate	none	**Prophylaxis:** none		**Vaccine:** none	
Shape	naked icosahedral				
Mean Diameter, µm	0.023	**Size Range:**	0.02–0.027 microns		
Growth Temperature	na	**Survival Outside Host:**		1–7 days on surfaces	
Inactivation	inactivated at pH < 3.6				
Disinfectants	1% sodium hypochlorite, iodine, phenol-alcohol, 2% glutaraldehyde				
Filter Nominal Rating	MERV 6	MERV 8	MERV 11	MERV 13	MERV 14
Removal Efficiency, %	26.6	31.0	50.7	87.7	96.9
UVGI Rate Constant	(unknown)		Media	-	
Dose for 90% Inactivation	-		Ref.	-	
Suggested Indoor Limit	0	cfu/m^3			
Genome Size (bp)	7,212				
Related Species	Similar to Adenoviruses, Reoviruses, RSV				
Notes	CDC Reportable.				
Photo Credit	Image provided courtesy of Dr. M. S. McNulty, Queen's University of Belfast.				
References	Dalton 1973, Collins & Kennedy 1993, Fraenkel-Conrat 1985, Freeman 1985, Mahy 1975, Murray 1999, Ryan 1994, Canada 2001				

Rhizomucor pusillus

GROUP	Fungal Spore
PHYLUM	Zygomycota
ORDER	Mucorales
FAMILY	Mucoraceae
DISEASE GROUP	Non-communicable
BIOSAFETY LEVEL	Risk Group 1
Infectious Dose (ID_{50})	unknown
Lethal Dose (LD_{50})	NA
Infection Rate	none
Incubation Period	-
Peak Infection	NA
Annual Cases	rare
Annual Fatalities	-

Reportedly allergenic, especially in occupational settings. Hazardous to the immunocompromised. The majority of patients with mucormycosis are seriously immunocompromised. Other than fever and dyspnea, there are rarely any other symptoms. With continued tissue necrosis, hemoptysis may develop, and the end result may be fatal pulmonary hemorrhage. Water Activity Aw = 1.0.

Disease or Infection	allergic alveolitis, mucormycosis, nosocomial infections.				
Natural Source	Environmental, soil, decaying fruits and vegetables.				
Toxins	rhizonin A				
Point of Infection	Upper Respiratory Tract, sinus, skin eyes				
Symptoms	The only clinical symptoms usually manifest are fever and dyspnea. Tissue necrosis may result in hemoptysis.				
Treatment	Treatment with amphotericin B remains the only reliable therapy.				
Untreated Fatality Rate	NA	Prophylaxis: none		Vaccine:	none
Shape	spherical spore				
Mean Diameter, μm	4.183	Size Range:	3.5–5 microns		
Growth Temperature	37–60°C	Survival Outside Host:		survives outdoors	
Inactivation	Moist heat: 121°C for 30 min.				
Disinfectants	1% sodium hypochlorite, phenolics, formaldehyde, glutaraldehyde.				
Filter Nominal Rating	MERV 6	MERV 8	MERV 11	MERV 13	MERV 14
Removal Efficiency, %	61.8	74.3	93.3	99.0	99.0
UVGI Rate Constant	(unknown)		Media	-	
Dose for 90% Inactivation	-		Ref.	-	
Suggested Indoor Limit	150–500 cfu/m^3		(in mix with other nonpathogenic fungi)		
Genome Size (bp)					
Related Species	R. variabilis				
Notes	Thermophilic, can grow between 20°C and up to 54–58°C.				
Photo Credit	Image provided courtesy of Doctor Fungus, by permission of John H. Rex.				
References	Freeman 1985, Howard 1983, Lacey 1988, Murray 1999, Pope 1993, Rao 1996, Ryan 1994, Smith 1989, Sutton 1998				

Rhizopus

GROUP	Fungal Spore
PHYLUM	Zygomycota
ORDER	Mucorales
FAMILY	Mucoraceae
DISEASE GROUP	Non-communicable
BIOSAFETY LEVEL	Risk Group 2
Infectious Dose (ID_{50})	unknown
Lethal Dose (LD_{50})	NA
Infection Rate	none
Incubation Period	-
Peak Infection	NA
Annual Cases	rare
Annual Fatalities	-

Can infect the lungs and other locations. Can be fatal to those with impaired immune systems. An opportunistic pathogen. Spores will germinate and mycelial growth will result. R. arrhizus is the most common agent of zygomycosis, with R. microsporus the second most common. The majority of patients with mucormycosis are seriously immunocompromised. Other than fever and dyspnea, there are rarely any other symptoms. With continued tissue necrosis, hemoptysis may develop, and the end result may be fatal pulmonary hemorrhage. Indoor growth on dust, filters, & ductwork.

Disease or Infection	zygomycosis, allergic reactions, pneumonia, mucormycosis.				
Natural Source	Environmental, decaying fruit and vegetables, compost.				
Toxins	rhizonin A				
Point of Infection	Upper Respiratory Tract, sinus, skin eyes				
Symptoms	The only clinical symptoms usually manifest are fever and dyspnea. Tissue necrosis may result in hemoptysis.				
Treatment	Treatment with amphotericin B remains the only reliable therapy.				
Untreated Fatality Rate	NA **Prophylaxis:** none **Vaccine:** none				
Shape	spherical spore				
Mean Diameter, μm	6.928 **Size Range:** 4–12 microns				
Growth Temperature	25–45°C **Survival Outside Host:** survives outdoors				
Inactivation	Moist heat: 121°C for 30 min.				
Disinfectants	1% sodium hypochlorite, phenolics, formaldehyde, glutaraldehyde.				
Filter Nominal Rating	MERV 6	MERV 8	MERV 11	MERV 13	MERV 14
Removal Efficiency, %	64.8	75.3	94.0	99.0	99.0
UVGI Rate Constant	0.00861 m^2/J **Media** Air				
Dose for 90% Inactivation	267 J/m^2 **Ref.** Luckiesh 1946				
Suggested Indoor Limit	150–500 cfu/m^2 (in mix with other nonpathogenic fungi)				
Genome Size (bp)					
Related Species	R. arrhizus, R. microsporus var. rhizopodiformis, R. schipperae, R. stolonifer				
Notes	CDC Reportable. Rhizopus nigricans used for UVGI k.				
Photo Credit	Image courtesy of Neil Carlson, University of Minnesota.				
References	Freeman 1985, Howard 1983, Lacey 1988, Pope 1993, Rao 1996, Ryan 1994, Smith 1989, Castle 1987, Weinstein 1991, Woods 1997				

Rhodotorula

GROUP	Fungal Spore
CLASS	Urediniomycetes
ORDER	Sporidiales
FAMILY	Sporidiobolaceae
DISEASE GROUP	Non-communicable
BIOSAFETY LEVEL	Risk Group 1
Infectious Dose (ID_{50})	NA
Lethal Dose (LD_{50})	NA
Infection Rate	none
Incubation Period	-
Peak Infection	NA
Annual Cases	-
Annual Fatalities	-

Reportedly allergenic. Often found growing indoors in drains, cooling coils, and damp carpets. Indoor levels can exceed outdoor levels. Sometimes found on human body surfaces. Can cause keratitis. A reddish yeast typically found in moist environments such as carpeting, cooling coils, and drain pans. In some countries it is the most common yeast genus identified in indoor air. It may colonize in terminally ill patients. Human infections are rare but generally nosocomial. Indoor growth can occur on wet carpet, wet walls, and moist building materials.

Disease or Infection	allergic alveolitis, keratitis				
Natural Source	Environmental, soil, water, air, dairy products, skin, urine, feces.				
Toxins	none				
Point of Infection	Upper Respiratory Tract, skin				
Symptoms	Allergic reactions, respiratory irritation.				
Treatment	NA				
Untreated Fatality Rate	NA	Prophylaxis: none		Vaccine:	none
Shape	spherical spore				
Mean Diameter, μm	13.856	Size Range:	12–16 microns		
Growth Temperature	25°C	Survival Outside Host:	survives outdoors		
Inactivation	Moist heat: 121°C for 30 min.				
Disinfectants	1% sodium hypochlorite, phenolics, formaldehyde, glutaraldehyde.				
Filter Nominal Rating	MERV 6	MERV 8	MERV 11	MERV 13	MERV 14
Removal Efficiency, %	65.0	75.2	94.0	99.0	99.0
UVGI Rate Constant	(unknown)		Media	-	
Dose for 90% Inactivation	-		Ref.	-	
Suggested Indoor Limit	150–500	cfu/m^3	(in mix with other nonpathogenic fungi)		
Genome Size (bp)					
Related Species	R. rubra, R. glutinus, R. minuta, R. mucilaginosa				
Notes	Upper growth limit at 28–38°C.				
Photo Credit	Image courtesy of Neil Carlson, University of Minnesota.				
References	Freeman 1985, Howard 1983, Lacey 1988, Murray 1999, Pope 1993, Rao 1996, Ryan 1994, Smith 1989, Woods 1997, Sutton 1998				

Rickettsia prowazeki

GROUP	Bacteria
TYPE	Gram-
GENUS	Rickettsiae
FAMILY	Rickettsiaceae
DISEASE GROUP	Vector-borne
BIOSAFETY LEVEL	Risk Group 3
Infectious Dose (ID_{50})	10
Lethal Dose (LD_{50})	-
Infection Rate	-
Incubation Period	1–2 weeks
Peak Infection	-
Annual Cases	-
Annual Fatalities	-

Obligate intracellular bacterium that occurs in louse-infested areas. Infected lice excrete rickettsiae in their feces and this can infect bites or abrasions. Also caused by inhalation of infective louse feces or the bites of squirrel flea. The cause of epidemic typhus. Endemic in mountainous regions of Central and South America, Africa, and Asia.

Disease or Infection	Epidemic typhus, Louse-borne typhus fever, Brill-Zinsser disease				
Natural Source	Body louse, humans, squirrels, squirrel flea.				
Toxins	none				
Point of Infection	skin				
Symptoms	Onset can be sudden with fever, chills, headache, prostration, general pain, macular eruption, toxemia.				
Treatment	Tetracycline, chloramphenicol, doxycycline				
Untreated Fatality Rate	10–40% **Prophylaxis:** none **Vaccine:** Possible				
Shape	pleomorphic				
Mean Diameter, µm	0.6 **Size Range:** 0.3–1.2 microns				
Growth Temperature	32–35°C **Survival Outside Host:** weeks				
Inactivation	Moist heat: 121°C for 15 min. Dry heat: 170 for 1 hr.				
Disinfectants	1% sodium hypochlorite, 70% ethanol, glutaraldehyde, formaldehyde.				
Filter Nominal Rating	MERV 6	MERV 8	MERV 11	MERV 13	MERV 14
Removal Efficiency, %	11.2	16.3	31.1	69.0	90.4
UVGI Rate Constant	0.0292	m^2/J	**Media**	Plates	
Dose for 90% Inactivation	78.86	J/m^2	**Ref.**	Allen 1954	
Suggested Indoor Limit	na				
Genome Size (bp)	1,111,523		29. % G+C	71. % T+A	
Related Species	R. canadensis, R. rickettsi, R. tsutsugamushi				
Notes	CDC Reportable.				
Photo Credit	Centers for Disease Control (R. tsutsugamushi shown), PHIL# 929, Dr. Edwin P. Ewing.				
References	Braude 1981, Freeman 1985, McCaul 1981, Mitscherlich 1984, Murray 1999, Prescott 1996, Ryan 1994, Canada 2001, Mandell 2000				

Rubella virus

GROUP	Virus
TYPE	ssRNA, positive
GENUS	Rubivirus
FAMILY	Togaviridae
DISEASE GROUP	Communicable
BIOSAFETY LEVEL	Risk Group 2
Infectious Dose (ID_{50})	10–60
Lethal Dose (LD_{50})	-
Infection Rate	0.3–0.8
Incubation Period	12–23 days
Peak Infection	10–14 days
Annual Cases	3,000
Annual Fatalities	none

The common cause of German measles in children. Up to 80% of adults have immunity. It is a mild disease and those infected develop life-long immunity. Occurs worldwide with prevalence in winter and spring. Endemic in most communities. Congenital rubella syndrome may occur in infants born to women with rubella in first trimester. Virus is shed in oropharyngeal secretions and is highly transmissible. In communities where vaccination is rare, spring outbreaks typically occur every few years. Children represent the largest number of cases.

Disease or Infection	rubella (German measles)				
Natural Source	Humans, nosocomial				
Toxins	none				
Point of Infection	Upper Respiratory Tract				
Symptoms	Mild febrile infection with a diffuse punctate and macular rash resembling measles. Leukopenia, arthalgia and arthritis may arise. Some 30–50% of infections are asymptomatic.				
Treatment	No antibiotic treatment, no specific treatment.				
Untreated Fatality Rate	-	**Prophylaxis:** none		**Vaccine:** none	
Shape	enveloped icosahedral				
Mean Diameter, μm	0.061	**Size Range:** 0.05–0.075 microns			
Growth Temperature	na	**Survival Outside Host:** for short periods			
Inactivation	Heat 56°C for 30 mins, 70°C for 4 mins, 100°C for 2 mins				
Disinfectants	1% sodium hypochlorite, 70% ethanol, glutaraldehyde, formaldehyde.				
Filter Nominal Rating	MERV 6	MERV 8	MERV 11	MERV 13	MERV 14
Removal Efficiency, %	13.8	16.1	27.9	62.3	82.2
UVGI Rate Constant	(unknown)		**Media** -		
Dose for 90% Inactivation	-		**Ref.** -		
Suggested Indoor Limit	0 cfu/m^3				
Genome Size (bp)	9,755				
Related Species	Mumps virus, measles				
Notes	CDC Reportable.				
Photo Credit	Centers for Disease Control, PHIL# 269, Dr. Erskine Palmer.				
References	Dalton 1973, Fraenkel-Conrat 1985, Freeman 1985, Mahy 1975, Murray 1999, Ryan 1994, Castle 1987, Weinstein 1991, Canada 2001				

Saccharopolyspora rectivirgula

GROUP	Bacterial Spore
TYPE	Micromonosporaceae
GENUS	Micropolyspora
FAMILY	Actinomycetes
DISEASE GROUP	Non-communicable
BIOSAFETY LEVEL	Risk Group 2
Infectious Dose (ID_{50})	unknown
Lethal Dose (LD_{50})	-
Infection Rate	none
Incubation Period	-
Peak Infection	NA
Annual Cases	-
Annual Fatalities	-

One of the thermophilic actinomycetes, fungi-like bacteria that grow mycelia and produce spores. Long-term inhalation of the spores can produce alveolitis and asthma. Occurs primarily in agricultural environments. They are capable of growing at elevated temperatures in moldy hay, mushroom compost, and other vegetable matter, and can also grow in the heating systems of buildings.

Disease or Infection	Farmers Lung, alveolitis, asthma				
Natural Source	Agricultural, indoor growth in humidifiers.				
Toxins	none				
Point of Infection	Upper Respiratory Tract				
Symptoms	Respiratory irritation, asthmatic reactions.				
Treatment	Systemic corticosteroid treatment (I.e. prednisone for 4-8 weeks)				
Untreated Fatality Rate	0–20% **Prophylaxis:** none **Vaccine:** available				
Shape	spherical spore				
Mean Diameter, μm	1.342 **Size Range:** 1.2–1.5 microns				
Growth Temperature	30–45°C **Survival Outside Host:** survives outdoors				
Inactivation	Moist heat: 121°C for 30 min.				
Disinfectants	various common disinfectants should be effective				
Filter Nominal Rating	MERV 6	MERV 8	MERV 11	MERV 13	MERV 14
Removal Efficiency, %	30.0	42.7	65.5	95.9	98.9
UVGI Rate Constant	(unknown)		Media	-	
Dose for 90% Inactivation	-		Ref.	-	
Suggested Indoor Limit	150–240 cfu/m^3				
Genome Size (bp)					
Related Species	aka: Micropolyspora faeni, Thermopolyspora polyspora, Faenia rectivirgula. Type Species: M. bravicatena. S. hisuta				
Notes	CDC Reportable. Water Activity Aw = 1.0.				
Photo Credit	Image courtesy of Yuumi Kobayashi, Dow Agrosciences, Dow Chemical Japan, Ltd.				
References	Austin 1991, Freeman 1985, Lacey 1988, Mitscherlich 1984, Murray 1999, Slack 1975, Ryan 1994, Schaal 1981, Sikes 1973, Woods 1997, Ortiz-Ortiz 1982.				

SARS virus

GROUP	Virus
TYPE	ssRNA, positive
GENUS	Coronavirus
FAMILY	Coronaviridae
DISEASE GROUP	Communicable
BIOSAFETY LEVEL	Risk Group 4
Infectious Dose (ID_{50})	low
Lethal Dose (LD_{50})	-
Infection Rate	high
Incubation Period	2–7 days
Peak Infection	3–4 days
Annual Cases	8,000 (in 2003)
Annual Fatalities	774 (in 2003)

SARS virus is a variant of coronavirus and causes Severe Acute Respiratory Syndrome (SARS). It is highly contagious and spreads both by direct contact, through fomites, and by the airborne route. Can cause pneumonia and can be fatal if untreated. Can survive in sewage and on cold surfaces, including plastic and toilet seats, for days. Three strains circulate - Group 1, 2, & 3. Strains from Group 1 & 2 infect humans. Outbreaks can be explosive. Has apparently transmitted aboard aircraft.

Disease or Infection	severe acute respiratory syndrome, pneumonia				
Natural Source	Humans, birds, pigs, dogs, cats, bovines, rodents.				
Toxins	none				
Point of Infection	Upper Respiratory Tract				
Symptoms	Fever of 101.5 F, more severe symptoms than ordinary coronavirus colds, nasal discharge, malaise				
Treatment	No specific treatment, therapy, or antivirals.				
Untreated Fatality Rate	2–15% **Prophylaxis:** none **Vaccine:** possible				
Shape	enveloped helical				
Mean Diameter, μm	0.110 **Size Range:** 0.08–0.15 microns				
Growth Temperature	na **Survival Outside Host:** up to days or more				
Inactivation	heat will inactivate				
Disinfectants	1% sodium hypochlorite, 2% glutaraldehyde.				
Filter Nominal Rating	MERV 6	MERV 8	MERV 11	MERV 13	MERV 14
Removal Efficiency, %	8.4	10.0	18.0	44.7	65.7
UVGI Rate Constant	(unknown)		**Media**	na	
Dose for 90% Inactivation	-		**Ref.**	-	
Suggested Indoor Limit	0	cfu/m^3			
Genome Size (bp)	27,000–32,000		34. % G+C	66. % T+A	
Related Species	Coronavirus (used for genome size)				
Notes	CDC Reportable. Vaccine under development.				
Photo Credit	Centers for Disease Control, PHIL#3492, Dr. Fred Murphy.				
References	Dalton 1973, Fraenkel-Conrat 1985, Freeman 1985, Mahy 1975, Murray 1999, Ryan 1994, Snijder 1991, Canada 2001, Myint 1995, Jiang 2003, Razum 2003, Zhang 2003, del Rey Carero 2004				

Scopulariopsis

GROUP	Fungal Spore
CLASS	Euascomycetes
ORDER	Microascales
FAMILY	Microacaceae
DISEASE GROUP	Non-communicable
BIOSAFETY LEVEL	Risk Group 2
Infectious Dose (ID_{50})	unknown
Lethal Dose (LD_{50})	NA
Infection Rate	none
Incubation Period	-
Peak Infection	NA
Annual Cases	-
Annual Fatalities	-

Allergenic. Worldwide distribution in the environment. Can grow indoors on paper, floor dust, and filters. Can produce microbial volatile organic compounds (MVOCs), including arsine gas on arsenic substrates, under some conditions that may result in irritation or allergic reactions. Most common species is Scopulariopsis brevicaulis. May cause onychomycosis, or infection of the toenails, skin lesions, and opportunistic disease in immunocompromised hosts. More common after trauma. May cause primary infection. S. brumptii may cause pulmonary hypersensitivity. Aw = 0.9–0.95.

Disease or Infection	allergic reactions, onychomycosis, skin lesions.				
Natural Source	Environmental, found in soils, plants, feathers, and insects.				
Toxins	none				
Point of Infection	Upper Respiratory Tract				
Symptoms	Allergic reactions, respiratory irritation.				
Treatment	NA				
Untreated Fatality Rate	NA	Prophylaxis: none		Vaccine:	none
Shape	ovoid spore				
Mean Diameter, μm	5.916	Size Range:	5–7 × 5–9 microns		
Growth Temperature	24–30°C	Survival Outside Host:	survives outdoors		
Inactivation	Moist heat: 121°C for 30 min.				
Disinfectants	1% sodium hypochlorite, phenolics, formaldehyde, glutaraldehyde.				
Filter Nominal Rating	MERV 6	MERV 8	MERV 11	MERV 13	MERV 14
Removal Efficiency, %	64.4	75.3	93.9	99.0	99.0
UVGI Rate Constant	0.00344	m^2/J	Media	Air	
Dose for 90% Inactivation	669	J/m^2	Ref.	Luckiesh 1946	
Suggested Indoor Limit	150–500	cfu/m^3	(in mix with other nonpathogenic fungi)		
Genome Size (bp)					
Related Species	S. brevicaulis, S. fusca, S. brumptii				
Notes	Scopulariopsis brevicaulis used for UVGI k. Upper growth limit 37°C.				
Photo Credit	Image courtesy of Neil Carlson, University of Minnesota.				
References	Freeman 1985, Howard 1983, Lacey 1988, Murray 1999, Pope 1993, Rao 1996, Ryan 1994, Smith 1989, Woods 1997, Sutton 1998				

Serratia marcescens

GROUP	Bacteria
TYPE	Gram-
GENUS	Serratia
FAMILY	Enterobacteriaceae
DISEASE GROUP	Endogenous
BIOSAFETY LEVEL	Risk Group 1
Infectious Dose (ID_{50})	unknown
Lethal Dose (LD_{50})	-
Infection Rate	-
Incubation Period	-
Peak Infection	NA
Annual Cases	479
Annual Fatalities	-

Normally benign but capable of causing serious infections in some cases, especially as nosocomial infections. Causes opportunistic infections of the eyes, blood, wounds, urinary tract, and respiratory tract. Important—causes nosocomial outbreaks in nurseries, intensive-care wards, and renal dialysis units. Responsible for 4% of nosocomial pneumonias. Usually transmitted by contaminated medical equipment.

Disease or Infection	opportunistic infections, bacteremia, endocarditis, pneumonia.				
Natural Source	Environmental, indoor growth in potable water, nosocomial.				
Toxins	Endotoxins				
Point of Infection	Upper Respiratory Tract, wounds, eyes, urinary tract				
Symptoms	Opportunistic infections of the lungs, eyes, and urinary tract.				
Treatment	Aminoglycosides, amikacin, resistant to penicillins.				
Untreated Fatality Rate	-	**Prophylaxis:** possible		**Vaccine:** none	
Shape	rods				
Mean Diameter, μm	0.632	**Size Range:**	0.5–0.8 × 0.9–2 microns		
Growth Temperature	21–37°C	**Survival Outside Host:**	35 days or more		
Inactivation	Moist heat: 121°C for 15 min. Dry heat: 160–170°C for 1 hr.				
Disinfectants	1% sodium hypochlorite, 70% ethanol, glutaraldehyde, formaldehyde.				
Filter Nominal Rating	MERV 6	MERV 8	MERV 11	MERV 13	MERV 14
Removal Efficiency, %	11.9	17.4	32.9	71.5	91.8
UVGI Rate Constant	0.2208	m^2/J	**Media**	Air	
Dose for 90% Inactivation	10	J/m^2	**Ref.**	Collins 1971	
Suggested Indoor Limit	-				
Genome Size (bp)	5,113,802		59.5 % G+C	40.5 % T+A	
Related Species	Serratia liquefaciens (causes similar nosocomial infections)				
Notes	Decay constant in air k = 0.0829, at 37% RH.				
Photo Credit	Image courtesy of the Tokyo Metropolitan Government Infectious Disease Surveillance Center				
References	Braude 1981, Freeman 1985, Mitscherlich 1984, Murray 1999, Prescott 1996, Ryan 1994, Castle 1987, Weinstein 1991, Woods 1997, Canada 2001				

Sporothrix schenckii

	GROUP	Fungal Spore
	CLASS	Euascomycetes
	ORDER	Ophiostomatales
	FAMILY	Ophiostomataceae
	DISEASE GROUP	Non-communicable
	BIOSAFETY LEVEL	Risk Group 2
	Infectious Dose (ID_{50})	unknown
	Lethal Dose (LD_{50})	NA
	Infection Rate	none
	Incubation Period	1–12 weeks
	Peak Infection	NA
	Annual Cases	rare
	Annual Fatalities	-

A hazard to the immunocompromised, but uncommon. Pulmonary sporotrichosis probably develops as the result of inhalation of spores. Skin infections may develop as the result of contamination of scratches and cuts. Outbreaks have occurred among children playing or working with hay. Occurs worldwide. Often an occupational disease in agriculture and sporadic in nature. Epidemics have occurred. Worldwide distribution in environment. Laboratory-acquired infections have occurred. S. cyanescens may cause nosocomial pneumonia in the immunocompromised.

Disease or Infection	pulmonary sporotrichosis, rose gardeners disease				
Natural Source	Environmental, soil, decaying plant material.				
Toxins	none				
Point of Infection	Upper Respiratory Tract, skin				
Symptoms	Skin infection may begin as a nodule, lymphatics draining the area become firm, form a series of nodules which may ulcerate. Arthritis, meningitis, pneumonitis and other infections may result.				
Treatment	Oral iodides or itraconazole, amphotericin B.				
Untreated Fatality Rate	uncommon **Prophylaxis:** none **Vaccine:** none				
Shape	ovoid spore				
Mean Diameter, μm	6.325 **Size Range:** 5–8 × 10–20 microns				
Growth Temperature	25–30°C **Survival Outside Host:** months in vegetation				
Inactivation	Moist heat: 121°C for 15 min.				
Disinfectants	1% sodium hypochlorite, iodine, glutaraldehyde, formaldehyde.				
Filter Nominal Rating	MERV 6	MERV 8	MERV 11	MERV 13	MERV 14
Removal Efficiency, %	64.6	75.3	94.0	99.0	99.0
UVGI Rate Constant	(unknown)		Media	-	
Dose for 90% Inactivation	-		Ref.	-	
Suggested Indoor Limit	150–500 cfu/m^3 (in mix with other nonpathogenic fungi)				
Genome Size (bp)					
Related Species	S. cyanescens				
Photo Credit	Image provided courtesy of Doctor Fungus, by permission of John H. Rex.				
References	Goodman 1983, Howard 1983, Lacey 1988, Murray 1999, Rao 1996, Ryan 1994, Smith 1989, Canada 2001, Sutton 1998				

Stachybotrys chartarum

GROUP	Fungal Spore
CLASS	Hyphomycetes
ORDER	Stachybotrys
FAMILY	Hyphomycetes
DISEASE GROUP	Non-communicable
BIOSAFETY LEVEL	Risk Group 1-2
Infectious Dose (ID_{50})	unknown
Lethal Dose (LD_{50})	-
Infection Rate	none
Incubation Period	-
Peak Infection	NA
Annual Cases	-
Annual Fatalities	-

Produces toxins that can be hazardous and poisonous by inhalation. Will grow indoors. A recently identified fungal allergen, it has been identified as a contributing factor in some cases of SBS/BRI. Possible cause of bleeding lung disease in infants. Grows on cellulose materials with low nitrogen.

Disease or Infection	allergic reactions, stachybotritoxicosis, irritation, toxic reactions				
Natural Source	Environmental, indoor growth on building materials & in humidifiers.				
Toxins	trichothecenes, verrucarin J, roridin E, satratoxin F, satratoxin G, satratoxin H, sporidesmin G, trichoverrol, cyclosporins, stachybotryolactone				
Point of Infection	Upper Respiratory Tract				
Symptoms	Lung mycotoxicosis, cough, rhinitis, buring sensation in mouth and nasal passages.				
Treatment	Antifungal therapy				
Untreated Fatality Rate	-	Prophylaxis: none		Vaccine: none	
Shape	spherical spore				
Mean Diameter, μm	5.623	Size Range:	5.1–6.2 microns		
Growth Temperature	37°C	Survival Outside Host:		indefinitely	
Inactivation	Moist heat: 121°C for 30 min.				
Disinfectants	alcohol, glutaraldehyde, peroxacetic acid				
Filter Nominal Rating	MERV 6	MERV 8	MERV 11	MERV 13	MERV 14
Removal Efficiency, %	64.2	75.2	93.9	99.0	99.0
UVGI Rate Constant	(unknown)		Media	-	
Dose for 90% Inactivation	-		Ref.	-	
Suggested Indoor Limit	0	cfu/m^3			
Genome Size (bp)					
Related Species	Previously S. atra and S. alterans.				
Notes	Water activity = 0.94–0.98.				
Photo Credit	Image courtesy of Neil Carlson, University of Minnesota, Dept. of Environmental Health & Safety.				
References	Freeman 1985, Howard 1983, Lacey 1988, Montana 1988, Murray 1999, Nikulin 1996, Ryan 1994, Smith 1989				

Staphylococcus aureus

	GROUP	Bacteria
	TYPE	Gram+
	GENUS	Staphylococcus
	FAMILY	Micrococcaceae
	DISEASE GROUP	Endogenous
	BIOSAFETY LEVEL	Risk Group 2
	Infectious Dose (ID_{50})	varies
	Lethal Dose (LD_{50})	-
	Infection Rate	-
	Incubation Period	4–10 days
	Peak Infection	NA
	Annual Cases	2,750
	Annual Fatalities	-

Generally a commensal microorganism. Can cause opportunistic infections when host resistance is compromised, especially when a primary infection such as influenza is present. Since it exists ubiquitously and commensally it can be considered Communicable, but without a secondary infection rate. The case mortality rate is high. Also causes food intoxication and toxic shock syndrome.

Disease or Infection	staphylococcal pneumonia, opportunistic infections (esp. MRSA)					
Natural Source	Humans, sewage, nosocomial.					
Toxins	none					
Point of Infection	Upper Respiratory Tract					
Symptoms	Surface infections include impetigo, folliculitis, abcesses, boils. Deep infections include endocarditis, meningitis, pneumonia. Systemic infection may cause fever, headache, malaise, myalgia.					
Treatment	Treatment depends on strains due to varied antibiotic resistance.					
Untreated Fatality Rate	-	Prophylaxis:	none	Vaccine:	none	
Shape	coccoid, usually occur in clusters, non-motile					
Mean Diameter, μm	0.866	Size Range:	0.5–1.5 microns			
Growth Temperature	7–45°C	Survival Outside Host:	7–60 days, 72 hrs on steel			
Inactivation	Moist heat: 121°C for 15 min. Dry heat: 170°C for 1 hr.					
Disinfectants	1% sodium hypochlorite, iodine/alcohol solutions, glutaraldehyde.					
Filter Nominal Rating	MERV 6	MERV 8	MERV 11	MERV 13	MERV 14	
Removal Efficiency, %	17.8	26.1	45.5	84.9	97.2	
UVGI Rate Constant	0.3476	m^2/J	Media	Air		
Dose for 90% Inactivation	6.62	J/m^2	Ref.	Sharp 1940		
Suggested Indoor Limit	-					
Genome Size (bp)	2,799,802		32.8 % G+C	67.2 % T+A		
Related Species	S. epidermis, MRSA has become common in nosocomial infections					
Notes	CDC Reportable. Decay constant in air k = 0.0276 at 37% RH,					
Photo Credit	Centers for Disease Control, PHIL# 617, Jim Biddle.					
References	Braude 1981, Castle 1987, Freeman 1985, Braude 1981, Mitscherlich 1984, Murray 1999, Ryan 1994, Weinstein 1991, Canada 2001					

Staphylococcus epidermis

GROUP	Bacteria
TYPE	Gram+
GENUS	Staphylococcus
FAMILY	Micrococcaceae
DISEASE GROUP	Endogenous
BIOSAFETY LEVEL	Risk Group 1
Infectious Dose (ID_{50})	unknown
Lethal Dose (LD_{50})	-
Infection Rate	-
Incubation Period	-
Peak Infection	-
Annual Cases	common
Annual Fatalities	-

Non-respiratory. A normal commensal of the skin and the most frequently isolated clinically. May represent up to 90% of all isolates from skin. Can contaminate medical equipment via contact or settling in air. A common cause of nosocomial urinary tract infections. Virtually all S. epidermis infections are hospital-acquired. Some people, particularly some males, shed much more S. epidermis than others.

Disease or Infection	Opportunistic infections, bacteremia				
Natural Source	Humans, sewage, nosocomial.				
Toxins	none				
Point of Infection	skin				
Symptoms	Opportunistic infections.				
Treatment	Vancomycin, rifampin, ciprofloxacin.				
Untreated Fatality Rate	-	**Prophylaxis:** none		**Vaccine:** none	
Shape	coccoid				
Mean Diameter, μm	0.866	**Size Range:**	0.5–1.5 microns		
Growth Temperature	10–45°C	**Survival Outside Host:**	-		
Inactivation	Moist heat: 121°C for 15 min. Dry heat: 170°C for 1 hr.				
Disinfectants	1% sodium hypochlorite, iodine/alcohol solutions, glutaraldehyde.				
Filter Nominal Rating	MERV 6	MERV 8	MERV 11	MERV 13	MERV 14
Removal Efficiency, %	17.8	26.1	45.5	84.9	97.2
UVGI Rate Constant	0.02093	m^2/J	**Media**	water	
Dose for 90% Inactivation	110	J/m^2	**Ref.**	Harris 1993	
Suggested Indoor Limit	0	cfu/m^3			
Genome Size (bp)	2,499,279		32.1 % G+C	67.9 % T+A	
Related Species	S. saphrophyticus, Type Species: S.aureus				
Notes	Open Air Factor rate constant k = 1.5–10 per minute.				
Photo Credit	Centers for Disease Control, PHIL# 259, Segrid McAllister.				
References	Braude 1981, Freeman 1985, Austin 1991, Braude 1981, Mitscherlich 1984, Murray 1999, Prescott 1996, Ryan 1994, Castle 1987, Weinstein 1991, Mandell 2000				

Streptococcus pneumoniae

GROUP	Bacteria
TYPE	Gram+
ORDER	Lactobacillales
GENUS	Streptococcus
DISEASE GROUP	Communicable
BIOSAFETY LEVEL	Risk Group 2
Infectious Dose (ID_{50})	unknown
Lethal Dose (LD_{50})	-
Infection Rate	0.1–0.3
Incubation Period	1–5 days
Peak Infection	2–10 days
Annual Cases	500,000
Annual Fatalities	50,000

The leading cause of death in the world. This microorganism is commonly known as pneumococcus, and is the prime agent of lobar pneumonia, which predominantly affects children. It is commonly carried asymptomatically in healthy individuals. Carriage rates among children are high -- about 30% for children and 10% for adolescents.

Disease or Infection	lobar pneumonia, sinusitis, meningitis, otitis media, toxic reactions				
Natural Source	Humans, nosocomial.				
Toxins	none				
Point of Infection	Upper Respiratory Tract				
Symptoms	Sudden onset with fever, shaking chills, pleural pain, dyspnea, coughing, leukocytosis, pneumonia, bacteremia, meningitis.				
Treatment	Penicillin, erythromicin				
Untreated Fatality Rate	5–40%	**Prophylaxis:** Antibiotics		**Vaccine:** Available	
Shape	coccoid				
Mean Diameter, μm	0.707	**Size Range:** 0.5–1 microns			
Growth Temperature	25–42°C	**Survival Outside Host:** 1–25 days			
Inactivation	Moist heat: 121°C for 15 min. Dry heat: 170°C for 1 hr.				
Disinfectants	1% sodium hypochlorite, 2% glutaraldehyde, formaldehyde, iodines, 70% ethanol.				
Filter Nominal Rating	MERV 6	MERV 8	MERV 11	MERV 13	MERV 14
Removal Efficiency, %	13.7	20.1	37.1	76.6	94.3
UVGI Rate Constant	0.055	m^2/J	Media	-	
Dose for 90% Inactivation	1.82	J/m^2	Ref.	Nagy 1964	
Suggested Indoor Limit	0	cfu/m^3			
Genome Size (bp)	2,038,615		39.7 % G+C	60.3 % T+A	
Related Species	aka pnemococcus				
Notes	CDC Reportable.				
Photo Credit	Centers for Disease Control, PHIL# 265, Dr. Richard Facklam.				
References	Braude 1981, Freeman 1985, Austin 1991, Mitscherlich 1984, Murray 1999, Prescott 1996, Ryan 1994, Canada 2001, Mandell 2000				

Streptococcus pyogenes

GROUP	Bacteria
TYPE	Gram+
GENUS	Streptococcus
FAMILY	Streptococcaceae
DISEASE GROUP	Communicable
BIOSAFETY LEVEL	Risk Group 2
Infectious Dose (ID_{50})	unknown
Lethal Dose (LD_{50})	-
Infection Rate	-
Incubation Period	1–5 days
Peak Infection	2–10 days
Annual Cases	213,962
Annual Fatalities	-

Belongs to the normal flora of the human body and only results in disease when host immunity is compromised. Often a nosocomial infection in wounds, lung infections can also result. Epidemics once occurred periodically in Europe and the United States. Infections are most common in the 5–15 age group, and during the months from December to May. Airborne spread is predominant, but evidence of direct contact spread exists.

Disease or Infection	scarlet fever, pharyngitis, toxic reactions				
Natural Source	Humans, nosocomial.				
Toxins	none				
Point of Infection	Upper Respiratory Tract, burns, wounds				
Symptoms	Sore throat, fever, tonsillitis, pharyngitis, scarlet fever, nausea, septicemia, pneumonia.				
Treatment	Penicilin, clindamycin, cephalosporin.				
Untreated Fatality Rate	0.03	**Prophylaxis:** Penicillin		**Vaccine:**	none
Shape	coccoid, usually occurring in pairs, nonmotile				
Mean Diameter, μm	0.894	**Size Range:**	0.8–1 microns		
Growth Temperature	37°C	**Survival Outside Host:**		up to 195 days in dust	
Inactivation	Moist heat: 121°C for 15 min. Dry heat: 160–170°C for 1 hr.				
Disinfectants	1% sodium hypochlorite, glutaraldehyde, formaldehyde, 70% ethanol.				
Filter Nominal Rating	MERV 6	MERV 8	MERV 11	MERV 13	MERV 14
Removal Efficiency, %	18.6	27.1	46.9	86.0	97.5
UVGI Rate Constant	0.6161	m^2/J	**Media**	Plates	
Dose for 90% Inactivation	3.74	J/m^2	**Ref.**	Lidwell 1950	
Suggested Indoor Limit	0	cfu/m^3			
Genome Size (bp)	1,899,877		38.7 % G+C	61.3 % T+A	
Related Species	Streptococcus pneumoniae				
Notes	CDC Reportable.				
Photo Credit	Image courtesy of Dr. Fusao Ota, Department of Food Microbiology, The University of Tokushima				
References	Braude 1981, Freeman 1985, Austin 1991, Murray 1999, Prescott 1996, Ryan 1994, Castle 1987, Weinstein 1991, Canada 2001				

Thermoactinomyces sacchari

GROUP	Bacterial Spore
TYPE	Micromonosporaceae
GENUS	Thermoactinomyces
FAMILY	Actinomycetes
DISEASE GROUP	Non-communicable
BIOSAFETY LEVEL	Risk Group 2
Infectious Dose (ID_{50})	unknown
Lethal Dose (LD_{50})	unknown
Infection Rate	none
Incubation Period	-
Peak Infection	NA
Annual Cases	-
Annual Fatalities	-

One of the thermophilic actinomycetes that occurs primarily in agricultural environments. Actinomycetes are higher bacteria that resemble fungi in their colonial morphology. They produce spores that may be inhaled. They are capable of growing at elevated temperatures in moldy hay, mushroom compost, and other vegetable matter, and can also grow in the heating systems of buildings. It is a cause of Farmer's Lung (hypersensitivity pneumonitis), especially when it results from exposure to moldy hay that may produce chronic or acute and extraordinary airborne inhalation doses. Incidence of Farmers Lung among farmers is high, up to 0.5%. Water Activity Aw = 1.0.

Disease or Infection	bagassosis, alveolitis, hypersensitivity pneumonitis				
Natural Source	Agricultural, bagasse, moldy hay				
Toxins	none				
Point of Infection	Upper Respiratory Tract				
Symptoms	Allergic reactions, respiratory irritation.				
Treatment	Systemic corticosteroid treatment (I.e. prednisone for 4–8 weeks)				
Untreated Fatality Rate	0–20% **Prophylaxis:** none **Vaccine:** none				
Shape	spherical spore				
Mean Diameter, μm	0.855 **Size Range:** 0.6–0.8 × 3 microns				
Growth Temperature	32–65°C **Survival Outside Host:** survives outdoors				
Inactivation	Moist heat: 121°C for 30 min.				
Disinfectants	various common disinfectants should be effective				
Filter Nominal Rating	MERV 6	MERV 8	MERV 11	MERV 13	MERV 14
Removal Efficiency, %	17.5	25.7	44.9	84.4	97.1
UVGI Rate Constant	(unknown)		Media	-	
Dose for 90% Inactivation	-		Ref.	-	
Suggested Indoor Limit	150–240 cfu/m³				
Genome Size (bp)					
Related Species	Type Species: T. vulgaris				
Notes	Farmers Lung infection may be subacute or misinterpreted.				
Photo Credit	Photo shows Micromonospora spp., Penn State AE Dept.				
References	Freeman 1985, Lacey 1988, Mitscherlich 1984, Murray 1999, Slack 1975, Ryan 1994, Schaal 1979, Sikes 1973, Ortiz-Ortiz 1982				

Thermoactinomyces vulgaris

GROUP	Bacterial Spore
TYPE	Micromonosporaceae
GENUS	Thermoactinomyces
FAMILY	Actinomycetes
DISEASE GROUP	Non-communicable
BIOSAFETY LEVEL	Risk Group 1
Infectious Dose (ID_{50})	unknown
Lethal Dose (LD_{50})	unknown
Infection Rate	none
Incubation Period	-
Peak Infection	NA
Annual Cases	uncommon
Annual Fatalities	rare

Common in agriculture and can be found in moldy hay. One of the thermophilic actinomycetes, the fungilike bacteria that grow mycelia and produce spores. Inhalation of the spores can produce pulmonary fibrosis, which can have sudden symptoms and progress rapidly to death. It is also known as Farmer's Lung. They are capable of growing at elevated temperatures in moldy hay, mushroom compost, and other vegetable matter, and can also grow in the heating systems of buildings. Growth on air conditioners has been linked to disease in urban areas. Water Activity Aw = 1.0.

Disease or Infection	Farmer's Lung, pulmonary fibrosis, allergic reactions, asthma, HP				
Natural Source	Agricultural, indoor growth in air conditioners and humidifiers.				
Toxins	none				
Point of Infection	Upper Respiratory Tract				
Symptoms	Allergic reactions, respiratory irritation.				
Treatment	Systemic corticosteroid treatment (I.e. prednisone for 4–8 weeks)				
Untreated Fatality Rate	0–20% **Prophylaxis:** none **Vaccine:** none				
Shape	spherical spore				
Mean Diameter, μm	0.866 **Size Range:** 0.5–1.5 microns				
Growth Temperature	28–69°C **Survival Outside Host:** survives outdoors				
Inactivation	Moist heat: 121°C for 30 min.				
Disinfectants	various common disinfectants should be effective				
Filter Nominal Rating	MERV 6	MERV 8	MERV 11	MERV 13	MERV 14
Removal Efficiency, %	17.8	26.1	45.5	84.9	97.2
UVGI Rate Constant	(unknown)		Media	-	
Dose for 90% Inactivation	-		Ref.	-	
Suggested Indoor Limit	150–240 cfu/m³				
Genome Size (bp)					
Related Species	T. candidus				
Notes	Farmer's Lung infection may be subacute or misinterpreted.				
Photo Credit	Photo shows spores of Actinomyces vulgaris, Penn State AE Dept.				
References	Austin 1991, Freeman 1985, Lacey 1988, Mitscherlich 1984, Murray 1999, Slack 1975, Ryan 1994, Schaal 1979, Sikes 1973, Woods 1997, Ortiz-Ortiz 1982.				

Thermomonospora viridis

GROUP	Bacterial Spore
TYPE	Micromonosporaceae
GENUS	Thermomonospora
FAMILY	Actinomycetes
DISEASE GROUP	Non-communicable
BIOSAFETY LEVEL	Risk Group 1
Infectious Dose (ID_{50})	unknown
Lethal Dose (LD_{50})	unknown
Infection Rate	none
Incubation Period	-
Peak Infection	NA
Annual Cases	-
Annual Fatalities	-

One of the thermophilic actinomycetes that occurs primarily in agricultural environments. Actinomycetes are higher bacteria that resemble fungi in their colonial morphology. They produce spores that may be inhaled. They are capable of growing at elevated in moldy hay, mushroom compost, and other vegetable matter, and can also grow in the heating systems of buildings. It is a cause of Farmer's Lung (hypersensitivity pneumonitis), especially when it results from exposure to moldy hay that may produce chronic or acute and extraordinary airborne inhalation doses.

Disease or Infection	Farmer's Lung, hypersensitivity pneumonitis				
Natural Source	Agricultural, moldy hay, indoor growth, air conditioners				
Toxins	none				
Point of Infection	Upper Respiratory Tract				
Symptoms	Allergic reactions, respiratory irritation.				
Treatment	Systemic corticosteroid treatment (I.e., prednisone for 4–8 weeks)				
Untreated Fatality Rate	0–20% **Prophylaxis:** none **Vaccine:** none				
Shape	ovoid endospores				
Mean Diameter, μm	0.520 **Size Range:** 0.3–0.9 × 0.6–1.5 microns				
Growth Temperature	25–60°C **Survival Outside Host:** survives outdoors				
Inactivation	Moist heat: 121°C for 30 min.				
Disinfectants	various common disinfectants should be effective				
Filter Nominal Rating	MERV 6	MERV 8	MERV 11	MERV 13	MERV 14
Removal Efficiency, %	9.4	13.6	26.6	62.3	85.8
UVGI Rate Constant	(unknown)		**Media**	-	
Dose for 90% Inactivation	-		**Ref.**	-	
Suggested Indoor Limit	150–240 cfu/m^3				
Genome Size (bp)					
Related Species	Type Species: T. curvata				
Notes	Farmer's Lung infection may be subacute or misinterpreted.				
Photo Credit	Photo shows Micromonospora spp., Penn State AE Dept.				
References	Austin 1991, Freeman 1985, Lacey 1988, Mitscherlich 1984, Murray 1999, Slack 1975, Ryan 1994, Schaal 1979, Sikes 1973, Ortiz-Ortiz 1982				

Trichoderma

GROUP	Fungal Spore
CLASS	Euascomycetes
ORDER	Hypocreales
FAMILY	Hypocreaceae
DISEASE GROUP	Non-communicable
BIOSAFETY LEVEL	Risk Group 1
Infectious Dose (ID_{50})	NA
Lethal Dose (LD_{50})	NA
Infection Rate	none
Incubation Period	-
Peak Infection	NA
Annual Cases	-
Annual Fatalities	-

Reportedly allergenic, this fungus can grow on other fungi. Can grow on and degrade cellulose. Widespread in the environment. Will grow indoors on moist building materials and filters. Can produce toxins. It is commonly found in soil, dead trees, pine needles, paper, and unglazed ceramics. An emerging pathogen in immunocompromised hosts. Can produce MVOCs and antibiotics that are toxic to humans.

Disease or Infection	allergic alveolitis, toxic reactions, MVOCs				
Natural Source	Environmental, soil, wood, decaying vegetation.				
Toxins	trichodermin, isocyanides				
Point of Infection	Upper Respiratory Tract				
Symptoms	Allergic reactions, respiratory irritation.				
Treatment	NA				
Untreated Fatality Rate	NA	Prophylaxis: none		Vaccine: none	
Shape	spherical spore				
Mean Diameter, μm	4.025	Size Range:	3.6–4.5 microns		
Growth Temperature	25°C	Survival Outside Host:	outdoors >= 120 days		
Inactivation	Moist heat: 121°C for 30 min.				
Disinfectants	1% sodium hypochlorite, phenolics, formaldehyde, glutaraldehyde.				
Filter Nominal Rating	MERV 6	MERV 8	MERV 11	MERV 13	MERV 14
Removal Efficiency, %	61.3	74.0	93.1	99.0	99.0
UVGI Rate Constant	(unknown)	Media	-		
Dose for 90% Inactivation	-	Ref.	-		
Suggested Indoor Limit	150–500 cfu/m^3 (in mix with other nonpathogenic fungi)				
Genome Size (bp)					
Related Species	T. polysporum, T. viride, T. harzianum				
Notes	Formerly Aleurisma, Pachybasium				
Photo Credit	Image courtesy of Neil Carlson, University of Minnesota.				
References	Freeman 1985, Howard 1983, Lacey 1988, Murray 1999, Pope 1993, Smith 1989, Woods 1997, Flannigan 2001, Sutton 1998				

Trichosporon

GROUP	Fungi/Yeast
PHYLUM	Basidiomycota
ORDER	Sporidiales
FAMILY	Sporidiobolaceae
DISEASE GROUP	Non-communicable
BIOSAFETY LEVEL	Risk Group 3
Infectious Dose (ID_{50})	unknown
Lethal Dose (LD_{50})	NA
Infection Rate	none
Incubation Period	-
Peak Infection	NA
Annual Cases	-
Annual Fatalities	0

Reportedly a cause of summertime hypersensitivity pneumonitis as a result of growth on damp wood and matting material. Although it is a yeast and does not occur in the spore form, hyphae and fragments may be released and become airborne. A cause of white piedra and onychomycosis in man. Localized infections with T. beiglii may include endocarditis, meningitis, pneumonia, and ocular infections. An agent of bronchial and pulmonary infections in immunocompromised hosts.

Disease or Infection	Hypersensitivity pneumonitis, white piedra, onychomycosis, opportunistic infections				
Natural Source	Environmental, soil, water, vegetation.				
Toxins	none				
Point of Infection	Upper Respiratory Tract, skin, hair shafts.				
Symptoms	Allergic reactions, respiratory irritation.				
Treatment	NA				
Untreated Fatality Rate	none **Prophylaxis:** none **Vaccine:** none				
Shape	irregular hyphae				
Mean Diameter, μm	8.775 **Size Range:** 7–11 microns				
Growth Temperature	25°C **Survival Outside Host:** survives outdoors				
Inactivation	Moist heat: 121°C for 30 min.				
Disinfectants	ethanol, sodium hypochlorite, chlorhexidine gluconate				
Filter Nominal Rating	MERV 6	MERV 8	MERV 11	MERV 13	MERV 14
Removal Efficiency, %	65.0	75.3	94.0	99.0	99.0
UVGI Rate Constant	(unknown)		Media	-	
Dose for 90% Inactivation	-		Ref.	-	
Suggested Indoor Limit	150–500 cfu/m^3 (in mix with other nonpathogenic fungi)				
Genome Size (bp)					
Related Species	T. cutaneum, T. beiglii, T. asahii, T. mucoides, T. asteroides, T. inkin, T. ovoides				
Notes	Upper limit for growth 29–42°C.				
Photo Credit	Centers for Disease Control, PHIL# 3064, Dr. Lucille K. Georg.				
References	Flannigan et al 2001, Sutton et al 1998, Hawksworth et al 1983, Ryan 1994, Yoshida et al 1989				

Ulocladium

GROUP	Fungal Spore
CLASS	Euascomycetes
ORDER	Pleosporales
FAMILY	Pleosporaceae
DISEASE GROUP	Non-communicable
BIOSAFETY LEVEL	Risk Group 1
Infectious Dose (ID_{50})	NA
Lethal Dose (LD_{50})	NA
Infection Rate	none
Incubation Period	-
Peak Infection	NA
Annual Cases	-
Annual Fatalities	-

Reportedly allergenic. Widely distributed in nature and common in the soil and on decaying herbaceous plants and dung. Can grow indoors on cellulose, wood, painted walls, water-based emulsion paint, and textiles. Found in floor dust, filters, humidifiers, and mattress dust. Has a water activity of Aw = 0.89. Reported as an agent of subcutaneous infection but vary rarely causes disease. Has only two active species, U. chartarum and U. botrytis.

Disease or Infection	allergic alveolitis				
Natural Source	Environmental, soil, wood, decaying plants.				
Toxins	none				
Point of Infection	Upper Respiratory Tract				
Symptoms	Allergic reactions, respiratory irritation.				
Treatment	NA				
Untreated Fatality Rate	NA	Prophylaxis: none		Vaccine:	none
Shape	spherical spore				
Mean Diameter, μm	14.142	Size Range:	10–20 microns		
Growth Temperature	25°C	Survival Outside Host:		survives outdoors	
Inactivation	Moist heat: 121°C for 30 min.				
Disinfectants	1% sodium hypochlorite, phenolics, formaldehyde, glutaraldehyde.				
Filter Nominal Rating	MERV 6	MERV 8	MERV 11	MERV 13	MERV 14
Removal Efficiency, %	65.0	75.2	94.0	99.0	99.0
UVGI Rate Constant	(unknown)	Media		-	
Dose for 90% Inactivation	-	Ref.		-	
Suggested Indoor Limit	150–500 cfu/m^3	(in mix with other nonpathogenic fungi)			
Genome Size (bp)					
Related Species	U. chartarum, U. botrytis				
Notes	Formerly Pseudostemphylium.				
Photo Credit	Image courtesy of Neil Carlson, University of Minnesota.				
References	Howard 1983, Lacey 1988, Murray 1999, Pope 1993, Rao 1996, Ryan 1994, Smith 1989, Woods 1997, Flannigan 2001, Sutton 1998				

Ustilago

GROUP	Fungal Spore
CLASS	Basidiomycetes
ORDER	Ustilaginales
FAMILY	Ustilaginaceae
DISEASE GROUP	Non-communicable
BIOSAFETY LEVEL	Risk Group 1
Infectious Dose (ID_{50})	unknown
Lethal Dose (LD_{50})	NA
Infection Rate	none
Incubation Period	unknown
Peak Infection	NA
Annual Cases	-
Annual Fatalities	-

Reportedly allergenic. Ustilago is a yeast that inhabits soil and plant material. It is a pathogen of seeds and flowers of wheat, corn, and grasses. Has been isolated from sputum and body fluids but disease associations are unclear. This fungus hails primarily from the environment and is a major pathogen of plants. Has been isolated from bodily fluids and sputum. Reports of infection are rare. Only active species is U. maydis

Disease or Infection	rhinitis, asthma, allergic reactions				
Natural Source	Environmental, plants, air.				
Toxins	none				
Point of Infection	Upper Respiratory Tract				
Symptoms	Allergic reactions, respiratory irritation.				
Treatment	NA				
Untreated Fatality Rate	NA	Prophylaxis: none		Vaccine:	none
Shape	rodlike spore				
Mean Diameter, μm	5.916	Size Range:	5–7 × 6–9 microns		
Growth Temperature	22–30°C	Survival Outside Host:		survives outdoors	
Inactivation	Moist heat: 121°C for 30 min.				
Disinfectants	1% sodium hypochlorite, phenolics, formaldehyde, glutaraldehyde.				
Filter Nominal Rating	MERV 6	MERV 8	MERV 11	MERV 13	MERV 14
Removal Efficiency, %	64.4	75.3	93.9	99.0	99.0
UVGI Rate Constant	0.0658	m^2/J	Media	-	
Dose for 90% Inactivation	35.00	J/m^2	Ref.	Sussman 1966	
Suggested Indoor Limit	150–500	cfu/m^3	(in mix with other nonpathogenic fungi)		
Genome Size (bp)					
Related Species	U. avenae, U. hordei, U. nuda, U. zeae (used for UV rate constant.)				
Notes	Previous reports of U. maydis may have been misidentifications.				
Photo Credit	Image provided courtesy of Doctor Fungus, by permission of John H. Rex.				
References	Freeman 1985, Howard 1983, Lacey 1988, Murray 1999, Rao 1996, Ryan 1994, Smith 1989, Sutton et al 1998, Hawksworth et al 1983				

Vaccinia virus

GROUP	Virus
TYPE	dsDNA
GENUS	Orthopoxvirus
FAMILY	Poxviridae
DISEASE GROUP	Non-communicable
BIOSAFETY LEVEL	Risk Group 2
Infectious Dose (ID_{50})	unknown
Lethal Dose (LD_{50})	-
Infection Rate	none
Incubation Period	-
Peak Infection	NA
Annual Cases	rare
Annual Fatalities	none

Vaccinia virus is serologically related to smallpox and it is believed to have derived from smallpox or cowpox, or both. It causes a cow disease that can occasionally be acquired by humans. Not life-threatening except to the immunodeficient. Confers immunity to smallpox. This virus, however, has the unique distinction of having led to the first vaccine since people who had cowpox were observed to be immune to smallpox. Some serious complications can arise from vaccinia infections, including encephalitis, progressive vaccinia, and allergic reactions.

Disease or Infection	cowpox				
Natural Source	Agricultural.				
Toxins	none				
Point of Infection	Upper Respiratory Tract				
Symptoms	Mild dermal reactions.				
Treatment	No treatment necessary.				
Untreated Fatality Rate	none	Prophylaxis: none		Vaccine:	none
Shape	complex				
Mean Diameter, µm	0.224	Size Range:	$0.2–0.25 \times 0.25–0.3$ microns		
Growth Temperature	na	Survival Outside Host:	2–4 weeks on cloth		
Inactivation	Heat: 60°C for 10 minutes.				
Disinfectants	Chlorine, formaldehyde, iodophores.				
Filter Nominal Rating	MERV 6	MERV 8	MERV 11	MERV 13	MERV 14
Removal Efficiency, %	5.6	7.2	14.1	37.5	59.2
UVGI Rate Constant	0.153	m^2/J	Media	Air	
Dose for 90% Inactivation	15.05	J/m^2	Ref.	Jensen 1964	
Suggested Indoor Limit	-				
Genome Size (bp)	191,737				
Related Species	Smallpox, Monkeypox, Camelpox, Mousepox.				
Notes	CDC Reportable.				
Photo Credit	Centers for Disease Control, PHIL# 1877, Dr. Fred Murphy, Sylvia Whitfield.				
References	Dalton 1973, Fraenkel-Conrat 1985, Freeman 1985, Mahy 1975, Murray 1999, Ryan 1994, Mandell 2000				

Varicella-zoster virus

GROUP	Virus
TYPE	dsDNA
GENUS	Varicellavirus
FAMILY	Herpesviridae
DISEASE GROUP	Communicable
BIOSAFETY LEVEL	Risk Group 2
Infectious Dose (ID_{50})	unknown
Lethal Dose (LD_{50})	-
Infection Rate	0.75–0.96
Incubation Period	11–21 days
Peak Infection	2–4 days
Annual Cases	common
Annual Fatalities	250

Varicella-zoster virus (VZV) causes varicella (chickenpox) in almost everyone by the age of 10 and is highly contagious. Infections can recur for those who are immunodeficient, especially bone marrow transplant patients. Occurs worldwide chiefly as a disease of children (75% of population by age 15, 90% of young adults have had disease). More frequent in winter and early spring in temperate zones. It also causes herpes zoster, which is more common in adults. Major transmission mode is respiratory, but direct contact with pustules can also produce infection.

Disease or Infection	chickenpox				
Natural Source	Humans, nosocomial				
Toxins	none				
Point of Infection	Upper Respiratory Tract				
Symptoms	Acute generalized disease with sudden onset of fever and vesicular eruption, of the skin and mucous membranes.				
Treatment	Vidarabine and acyclovir.				
Untreated Fatality Rate	rare	**Prophylaxis:** possible		**Vaccine:** available	
Shape	enveloped icosahedral				
Mean Diameter, μm	0.173	**Size Range:**	0.15–0.20 microns		
Growth Temperature	na	**Survival Outside Host:**	for short periods		
Inactivation	heat inactivates				
Disinfectants	1% sodium hypochlorite, 2% glutaraldehyde, formaldehyde.				
Filter Nominal Rating	MERV 6	MERV 8	MERV 11	MERV 13	MERV 14
Removal Efficiency, %	6.1	7.5	14.3	37.4	58.2
UVGI Rate Constant	5.86	m^2/J	**Media**	-	
Dose for 90% Inactivation	39.28	J/m^2	**Ref.**	Lytle 1971	
Suggested Indoor Limit	0	cfu/m^3			
Genome Size (bp)	-				
Related Species	Cytomegalovirus, Herpes Simplex Virus (HSV).				
Notes	CDC Reportable. (Herpes used for UVGI rate constant)				
Photo Credit	Centers for Disease Control, PHIL # 199, Dr. Erskine Palmer.				
References	Dalton 1973, Fraenkel-Conrat 1985, Freeman 1985, Mahy 1975, Murray 1999, Ryan 1994, Castle 1987, Weinstein 1991, Canada 2001				

Variola (smallpox)

GROUP	Virus
TYPE	dsDNA
GENUS	Variola
FAMILY	Poxviridae
DISEASE GROUP	Communicable
BIOSAFETY LEVEL	Risk Group 4
Infectious Dose (ID_{50})	10–100 (?)
Lethal Dose (LD_{50})	-
Infection Rate	0.3–0.9
Incubation Period	12–14 days
Peak Infection	14–17 days
Annual Cases	0
Annual Fatalities	0

The cause of smallpox, previously epidemic worldwide until eradicated in 1977. Exists in laboratory samples only, but has been developed as a biological weapon. The Greeks in Constantinople had apparently been practicing forms of variolation since before the 1400s, and Europeans adopted the technique beginning in 1718. Fatality rate is about 30% in Europeans but among Native Americans, against whom it was used as a biological weapon (1753 to late 1800s), the fatality rate was over 90%.

Disease or Infection	Smallpox				
Natural Source	Humans				
Toxins	none				
Point of Infection	Upper Respiratory Tract				
Symptoms	Pustular skin rash.				
Treatment	Rifamycin, Isatin B-thiosemicarbazone				
Untreated Fatality Rate	10–40% **Prophylaxis:** - **Vaccine:** Available				
Shape	complex capsid				
Mean Diameter, μm	0.173 **Size Range:** 0.1–0.3 microns				
Growth Temperature	na **Survival Outside Host:** (18 months at 5°C)				
Inactivation	Moist heat: 121°C for 15 min.				
Disinfectants	1% sodium hypochlorite, 2% glutaraldehyde, formaldehyde.				
Filter Nominal Rating	MERV 6	MERV 8	MERV 11	MERV 13	MERV 14
Removal Efficiency, %	6.1	7.5	14.3	37.4	58.2
UVGI Rate Constant	0.1528	m^2/J	**Media**	Air	
Dose for 90% Inactivation	15.07	J/m^2	**Ref.**	Collier 1955 (Vaccinia)	
Suggested Indoor Limit	0	cfu/m^3			
Genome Size (bp)	186,000		36.5 % G+C	63.5 % T+A	
Related Species	Monkeypox, Camelpox, Mousepox, Vaccinia (cowpox).				
Notes	CDC Reportable.				
Photo Credit	(Variola minor in monkey cells) Reprinted w/ permission from Malherbe and Strickland 1980, Viral Cytopathology, Copyright CRC Press, Boca Raton, FL				
References	Dalton 1973, Freeman 1985, Mahy 1975, Murray 1999, Henderson 1999, Malherbe 1980, Kowalski 2003, Franz 1997				

Verticillium

	GROUP	Fungal Spore
	TYPE	Mitosporic fungi
	CLASS	Hyphomycetes
	GENUS	Verticillium
	DISEASE GROUP	Non-communicable
	BIOSAFETY LEVEL	Risk Group 1
	Infectious Dose (ID_{50})	NA
	Lethal Dose (LD_{50})	NA
	Infection Rate	-
	Incubation Period	-
	Peak Infection	-
	Annual Cases	-
	Annual Fatalities	NA

Potentially allergenic. A possible cause of human keratitis. Has been isolated from cornea, lungs, bronchial fluid, and soil. A rare cause of corneal infections. A filamentous fungus that is widespread in decaying vegetation and the soil. A pathogen of various plant species and some species are parasitic on other fungi, arthropods, and plants. May very rarely cause human disease.

Disease or Infection	Allergic reactions, possible keratitis.				
Natural Source	Environmental, soil, decaying vegetation.				
Toxins	none				
Point of Infection	Eyes.				
Symptoms	Allergic reactions, respiratory irritation.				
Treatment	NA				
Untreated Fatality Rate	NA	Prophylaxis: -		Vaccine:	none
Shape	Ovoid to pyriform.				
Mean Diameter, μm	4.796	Size Range:	2–13 microns		
Growth Temperature	25°C	Survival Outside Host:		survives 14 years outdoors	
Inactivation	Moist heat: 121°C for 30 min.				
Disinfectants	1% sodium hypochlorite, 2% glutaraldehyde, formaldehyde.				
Filter Nominal Rating	MERV 6	MERV 8	MERV 11	MERV 13	MERV 14
Removal Efficiency, %	63.2	74.9	93.7	99.0	99.0
UVGI Rate Constant	(unknown)		Media	-	
Dose for 90% Inactivation	-		Ref.	-	
Suggested Indoor Limit	150–500 cfu/m^3		(in mix with other nonpathogenic fungi)		
Genome Size (bp)					
Related Species	V. affinae, V. albo-atrum, V. fusisporum, V. luteoalbum				
Notes	Classification of other Verticillium species remains uncertain.				
Photo Credit	Image courtesy of American Phytopathological Society and Fred Crow, Central Oregon Agricultural Research Center.				
References	Sutton et al 1998, Hawksworth et al 1983, Henis 1987				

Wallemia sebi

GROUP	Fungal Spore
TYPE	Mitosporic fungi
CLASS	Hyphomycetes
FAMILY	Wallemia
DISEASE GROUP	Non-communicable
BIOSAFETY LEVEL	Risk Group 1
Infectious Dose (ID_{50})	NA
Lethal Dose (LD_{50})	NA
Infection Rate	none
Incubation Period	-
Peak Infection	NA
Annual Cases	-
Annual Fatalities	-

An allergenic fungus. Can be found growing on sugars, sugary foods, salted meat, bacon, salted fish, dairy products, soil, hay, fruits, jams and jellies, bread, nuts, stored hay, and textiles. Has a water activity Aw = 0.75. Commonly present in air. Especially associated with dried foodstuffs. Has been found growing in floor dust, carpet dust, mattress dust, radiator dust, and filters. May cause subcutaneous infections.

Disease or Infection	allergic alveolitis				
Natural Source	Environmental, indoor growth on floor dust & filters.				
Toxins	none				
Point of Infection	Upper Respiratory Tract				
Symptoms	Allergic reactions, respiratory irritation.				
Treatment	NA				
Untreated Fatality Rate	NA	Prophylaxis:	none	Vaccine:	none
Shape	spherical spore				
Mean Diameter, μm	2.958	Size Range:	2.5–3.5 microns		
Growth Temperature	25°C	Survival Outside Host:	survives outdoors		
Inactivation	Moist heat: 121°C for 30 min.				
Disinfectants	1% sodium hypochlorite, phenolics, formaldehyde, glutaraldehyde.				
Filter Nominal Rating	MERV 6	MERV 8	MERV 11	MERV 13	MERV 14
Removal Efficiency, %	55.4	70.0	90.3	99.0	99.0
UVGI Rate Constant	(unknown)		Media	-	
Dose for 90% Inactivation	-		Ref.	-	
Suggested Indoor Limit	150–500	cfu/m^3	(in mix with other nonpathogenic fungi)		
Genome Size (bp)					
Related Species	W. ichthyophaga				
Notes	Formerly Wallemia ichthyophaga				
Photo Credit	Image courtesy of Neil Carlson, University of Minnesota.				
References	Freeman 1985, Howard 1983, Lacey 1988, Murray 1999, Pope 1993, Ryan 1994, Smith 1989, Woods 1997, Flannigan 2001, Sutton 1998				

Yersinia pestis

GROUP	Bacteria
TYPE	Gram-
GENUS	Yersinia
FAMILY	Enterobacteriaceae
DISEASE GROUP	Communicable
BIOSAFETY LEVEL	Risk Group 3
Infectious Dose (ID_{50})	3000
Lethal Dose (LD_{50})	-
Infection Rate	varies
Incubation Period	2–6 days
Peak Infection	-
Annual Cases	4
Annual Fatalities	-

The ancient cause of plague. Normally transmitted by flea bites. Aerosol transmission can become epidemic and is generally known as pneumonic plague. Primarily a zoonotic disease with rodents and fleas as the natural reservoir. Occasionally causes infections and localized outbreaks in the Americas, Africa, the Near East and Middle East, Asia, and Indonesia. Common in Burma and Vietnam.

Disease or Infection	Plague, bubonic plague, pneumonic plague, sylvatic plague.				
Natural Source	Rodents, fleas, humans				
Toxins	Endotoxin				
Point of Infection	Upper Respiratory Tract, skin via flea bites				
Symptoms	Bubonic plague with lymphadenitis around area of flea bite, fever. Pneumonic plague results in pneumonia, mediastinitis, pleural effusion.				
Treatment	Streptomycin, tetracycline, chloramphenicol, kanamycin 8–24 hours after onset				
Untreated Fatality Rate	0.5	**Prophylaxis:** Antibiotics		**Vaccine:** Possible	
Shape	rods				
Mean Diameter, μm	0.707	**Size Range:**	0.5–1 × 1–2 microns		
Growth Temperature	27–28°C	**Survival Outside Host:**		100-270 days in bodies	
Inactivation	Moist heat: 121°C for 15 min. Dry heat: 170 for 1 hr.				
Disinfectants	1% sodium hypochlor., 2% glutaraldehyde, formaldehyde, 70% ethanol.				
Filter Nominal Rating	MERV 6	MERV 8	MERV 11	MERV 13	MERV 14
Removal Efficiency, %	13.7	20.1	37.1	76.6	94.3
UVGI Rate Constant	0.154	m^2/J	**Media**	-	
Dose for 90% Inactivation	15.00	J/m^2	**Ref.**	Hyllseth 1998	
Suggested Indoor Limit	0	cfu/m^3			
Genome Size (bp)	4,653,728		47.6 % G+C	52.4 % T+A	
Related Species	Y. enterolitica, previously Pasteurella pestis				
Notes	CDC Reportable. (Y. enterolitica used for UVGI rate constant)				
Photo Credit	Centers for Disease Control, PHIL# 741, Dr. Marshall Fox.				
References	Braude 1981, Freeman 1985, Linton 1982, Mitscherlich 1984, Murray 1999, Ryan 1994, Canada 2001, NATO 1996, Mandell 2000				

Additional Potential Airborne Pathogens and Allergens

Fungi	Fungi	Fungi
Acrodontium salmoneum	Humicola	Pyrenochaeta
Agaricus bisporus	Hyphopichia burtonii	Pythium insidiosum
Aphanoascus fulvescens	Lasiodiplodia theobromae	Ramichloridium mackenziei
Aphanocladium album	Lechythophora	Rhinocladiella spp.
Apophysomyces elegans	Lentinus edulis	Saccharomyces cerevisiae
Basidiobolus ranarum	Leptodontium	Saksenaea vasiformis
Beauveria	Malassezia	Scedosporium prolificans
Bipolaris australensis	Malbranchea	Schizophyllum commune
Blastoschizomyces capitatus	Metarrhizium	Schizophyllum commune
Boletus edulis	Microascus ruber	Scolecobasidium terreum
Botrytis cinerea	Microsporum	Scytalidium hyalinum
Cephalosporium	Monascus	Serpula spp.
Chrysonillia sitophila	Monilia sitophila	Spegazzinia
Chrysosporium	Monocillium	Sporidiobolus salmonicolor
Cladophialophora	Mortierella	Sporobolomyces roseus
Cokeromyces recurvatus	Myceliophthora thermophila	Sporobolomyces roseus
Colletotrichum	Myrothecium	Sporotrichum pruinosum
Conidobolus	Myxotrichum deflexum	Stemphylium
Coniothyrium fuckelii	Nattrassia mangiferae	Syncephalastrum racemosum
Coremiella cubispora	Neosartorya fischeri	Tilletia caries
Cunninghamella bertholletiae	Nigrospora	Torula
Cylindrocarpon	Nodulisporium	Torulomyces lagena
Didymella	Ochroconis	Trichophyton
Doratomyces	Oidiodendron	Trichothecium
Drechslera	Pestalotiopsis	Tritirachium
Engyodontium album	Phaeoacremonium parasiticum	Wardomyces inflatus
Epidermophyton floccosum	Phaeoannellomyces werneckii	Zygosporium masonii
Erwinia	Phanerochaete chrysosporium	**Bacteria**
Fonsecaea pedrosi	Phialemonium	Achromobacter
Geomyces pannorus	Pholiota nameko	Cytophaga allerginea
Geotrichum candidum	Pichia anomala	Erwinia
Gliocladium	Pithomyces	Flavobacterium
Graphium	Pleurotus ostreatus	Nocardia caviae
Gymnoascus dankaliensis	Prototheca wickerhamii	Saccharomonospora viridis
Helminthosporium	Pseudallescheria boydii	Streptomyces spp.
Hormonema dematioides	Puccinia graminis	Thermoactinomyces dichotomicus
Virus	Avian influenza (Type A Influenza Orthomixovirus in birds)	

References: Sutton et al 1998, Sussman and Halvorson 1966, Kemp et al 1997, Kotimaa 1990, Lacey and Crook 1988, Husman 1996, Murray 1999, Godish 1995, Howard and Howard 1983, Pope 1993, Woods 1997, Morey 1990, del Rey Calero 2004.

Note: These microbes have been cited as actual or potential airborne pathogens or allergens in a limited number of cases, including immunocompromised infections, and sometimes the disease association is doubtful or unproven. Consult the References for further information.

APPENDIX B
TOXINS AND ASSOCIATED FUNGAL SPECIES

APPENDIX B

Toxin	Species	LD_{50} mg/kg (avg.)	LD_{50} Animal	Spore Content ng/spore	Reference
aflatoxin	Aspergillus fumigatus	-	-	22.5	
aflatoxin	Aspergillus parasiticus	7.56	rat	-	CAST 1989
aflatoxin B1	Aspergillus flavus	9	mice	-	Smith 1985
aflatoxin B2	Aspergillus flavus	-	-	-	
aflatrem	Aspergillus flavus	-	-	-	
altenuene	Alternaria spp.	-	-	-	
altenusin	Alternaria spp.	-	-	-	
alternariol	Alternaria alternata	-	-	-	
alternariol monomethylether	Alternaria spp.	-	-	-	
altertoxin I	Alternaria spp.	-	-	-	
altertoxin II	Alternaria spp.	-	-	-	
aspercolorin	Aspergillus versicolor	-	-	-	
austamide	Aspergillus ustus	-	-	-	
austdiol	Aspergillus ustus	-	-	-	
austocystins	Aspergillus ustus	-	-	-	
averufin	Aspergillus versicolor	-	-	-	
brevianamide	Aspergillus ustus	-	-	-	
butenolide	Fusarium nivale	-	-	-	
byssochlamic acid	Paecilomyces spp.	-	-	-	
chaetoglobosins	Chaetomium globosum	-	-	-	
chaetomin	Chaetomium spp.	-	-	-	
chrysog	Penicillium spp.	-	-	-	
citreoviridin	Penicillium spp.	15.7	mice	-	CAST 1989
citrinin	Penicillium citrinum	35	mice	-	CAST 1989
citrinin	Penicillium expansum	35	mice	-	CAST 1989
citrinin	Penicillium viridicatum	35	mice	-	CAST 1989
cladosporin	Cladosporium spp.	-	-	-	
crotocin	Acremonium spp.	-	-	-	
cyclochlorotine	Penicillium islandicum	-	-	-	
cyclopiazonic acid	Aspergillus flavus	33.77	rat	-	CAST 1989
cyclopiazonic acid	Aspergillus versicolor	33.77	rat	-	CAST 1989
cyclopiazonic acid	Penicillium spp.	33.77	rat	-	CAST 1989
cyclopiazonic acid	Penicillium cyclopium	33.77	rat	-	CAST 1989
cyclosporins	Stachybotrys chartarum	-	-	-	
cytochalasin E	Aspergillus clavatus	-	-	-	
decumbin	Penicillium spp.	-	-	-	
deoxynivalenol	Fusarium spp.	70	mice	-	Smith 1985
destruxin B	Aspergillus ochraceus	-	-	-	
emodin	Cladosporium spp.	-	-	-	
epicladosporic acid	Cladosporium spp.	-	-	-	
epicorazine A	Epicoccum spp.	-	-	-	
epicorazine B	Epicoccum spp.	-	-	-	
ergot alkaloids	Aspergillus fumigatus	-	-	-	
ferrirubin	Paecilomyces spp.	-	-	-	
flavipin	Epicoccum spp.	-	-	-	
fumigaclavines	Aspergillus fumigatus	-	-	-	
fumigatoxin	Aspergillus fumigatus	-	-	-	
fumagillin	Aspergillus fumigatus	-	-	-	
fumitremorgens	Aspergillus fumigatus	-	-	-	
fumonisin	Fusarium spp.	-	-	-	

TOXINS AND ASSOCIATED FUNGAL SPECIES

Toxin	Species	LD_{50} mg/kg (avg.)	LD_{50} Animal	Spore Content ng/spore	Reference
fusarin	Fusarium moniliforme	-	-	-	
fusigen	Paecilomyces spp.	-	-	-	
gliotoxin	Aspergillus fumigatus	-	-	-	
griseofulvin	Penicillium spp.	-	-	-	
helvolic acid	Aspergillus fumigatus	-	-	-	
indole-3-acetic acid	Paecilomyces spp.	-	-	-	
indole-3-acetonitrile	Epicoccum spp.	-	-	-	
islanditoxin	Penicillium islandicum	-	-	-	
isofumigaclavine A	Penicillium spp.	-	-	-	
kojic acid	Aspergillus flavus	-	-	-	
luteoskyrin	Penicillium islandicum	-	-	-	
malformin C	Aspergillus niger	-	-	-	
mycophenolic acid	Penicillium spp.	-	-	-	
ochratoxin	Penicillium viridicatum	27.25	mice	-	Smith 1985
ochratoxin	Penicillium cyclopium	-	-	-	
ochratoxin	Aspergillus ochraceus	-	-	-	
oxalic acid	Aspergillus niger	-	-	-	
paecilotoxins	Paecilomyces spp.	-	-	-	
patulin	Penicillium expansum	-	-	-	
patulin	Paecilomyces	-	-	-	
patulin	Aspergillus clavatus	-	-	-	
penicillic acid	Penicillium spp.	-	-	-	
penicillic acid	Penicillium cyclopium	-	-	-	
penicillic acid	Aspergillus ochraceus	-	-	-	
penicillin	Penicillium spp.	-	-	-	
penitrem A	Penicillium spp.	1.05	mice	-	CAST 1989
penitrem A	Penicillium cyclopium	1.05	mice	-	CAST 1989
penitrem A	Penicillium crustosum	1.05	mice	-	CAST 1989
peptide nephrotoxin	Penicillium spp.	-	-	-	
P.R. toxin	Penicillium roqueforti	5.8	mice	-	CAST 1989
roquefortine	Penicillium roqueforti	-	-	-	
roquefortine C	Penicillium spp.	-	-	-	
roquefortine D	Penicillium spp.	-	-	-	
roridin E	Stachybotris chartarum	-	-	-	
rubratoxin	Penicillium purpurogenum	130	mice	-	Smith 1985
satratoxin F	Stachybotrys chartarum	-	-	-	
satratoxin G	Stachybotrys chartarum	-	-	-	
satratoxin H	Stachybotrys chartarum	-	-	11.5	Burge 1996
sporidesmin G	Stachybotrys chartarum	-	-	-	
stachybotryolactone	Stachybotrys chartarum	-	-	-	
sterigmatocystin	Aspergillus versicolor	113	mice	-	Smith 1985
T-2 toxin	Fusarium spp.	5.2	mice	-	Smith 1985
tenuazonic acid	Alternaria spp.	-	-	-	
tenuazonic acid	Alternaria alternata	-	-	-	
trichodermin	Trichoderma spp.	-	-	-	
trichothecenes	Stachybotrys chartarum	-	-	-	
trichothecenes	Fusarium nivale	-	-	-	
trichothecenes	Fusarium graminearum	-	-	-	
trichothecenes	Fusarium solani	-	-	-	
trichothecenes (Type B)	Fusarium spp.	-	-	-	

Toxin	Species	LD_{50} mg/kg (avg.)	LD_{50} Animal	Spore Content ng/spore	Reference
trichoverrol	Stachybotrys chartarum	-	-	-	
tryptoquivaline	Aspergillus clavatus	-	-	-	
tryptoquivaline tremorgens	Aspergillus fumigatus	-	-	-	
variotin	Paecilomyces spp.	-	-	-	
verrucarin J	Stachybotrys chartarum	-	-	-	
verruculogen	Aspergillus fumigatus	64.55		-	
verruculogen	Penicillium spp.	-	-	-	
versicolorin	Aspergillus versicolor	-	-	-	
viomellein	Penicillium spp.	-	-	-	
viomellein	Penicillium viridicatum	-	-	-	
viridicatin	Penicillium viridicatum	-	-	-	
viriditoxin	Paecilomyces spp.	-	-	-	
viriditoxin	Aspergillus fumigatus	-	-	-	
vomitoxin	Fusarium spp.	-	-	-	
xanthocillin	Eurotium	-	-	-	
xanthocillin	Aspergillus chevalieri	-	-	-	
xanthocillin X	Penicillium spp.	-	-	-	
xanthomegnin	Penicillium spp.	-	-	-	
xanthomegnin	Penicillium viridicatin	-	-	-	
zearalenone	Fusarium graminearum	1	mice	-	Smith 1985
zearalenone (F-2 toxin)	Fusarium spp.	1	mice	-	Smith 1985
zearalenone (F-2 toxin)	Acremonium spp.	1	mice	-	Smith 1985

Note: LD_{50} are averaged for more than one value, regardless of intake route.

APPENDIX C
MICROBIAL VOLATILE ORGANIC COMPOUNDS

APPENDIX C

MVOC	CAS	Source	Notes	Reference
1-butanol	71-36-3	-	-	Kreja and Seidel 2002
1-decanol	112-30-1	-	highly toxic	Kreja and Seidel 2002
1-hexanol	111-27-3	-	-	Kreja and Seidel 2002
1-octen-3-ol	3391-86-4	inactive fungi	contamination indicator	Kreja and Seidel 2002
1-pentanol	71-41-0	-	-	Kreja and Seidel 2002
2-butanone	78-93-3	-	-	Kreja and Seidel 2002
2-ethyl-1-hexanol	104-76-7	Aspergillus	-	Pasanen et al 1997
2-heptanol	-	-	contamination indicator	Sandstrom 2003
2-heptanone	110-43-0	fungi, Aspergillus	contamination indicator	Kreja and Seidel 2002, Pasanen et al 1997, Fischer and Dott 2003
2-hexanone	591-78-6	Aspergillus	contamination indicator	Kreja and Seidel 2002, Pasanen et al 1997
2-isopropryl-3-methoxypyrazine	-	-	contamination indicator	Cochrane 2001
2-methyl-1-butanol	137-32-6	fungi	-	Kreja and Seidel 2002
2-methyl-1-propanol (isobutanol)	78-83-1	fungi	-	Kreja and Seidel 2002
2-methyl-2-butanol	598-75-4	-	contamination indicator	Kreja and Seidel 2002
2-methylfuran	-	fungi, Aspergillus	-	Fischer and Dott 2003, Pasanen et al 1997
2-methyl-isoborneol	-	-	contamination indicator	Cochrane 2001
2-methylisopentylether	-	fungi	-	Fischer and Dott 2003
2-nonanon	821-55-6	-	-	Kreja and Seidel 2002
2-octen-1-ol	-	-	contamination indicator	Cochrane 2001
2-pentanol	6032-29-7	fungi	contamination indicator	Fischer and Dott 2003
2-pentanone	107-87-9	Aspergillus	-	Pasanen et al 1997
2-pentylfuran	-	fungi	-	Fischer and Dott 2003
2,4-dinitriophenyl-hydrazone	-	-	contamination indicator	Schleibinger et al 1997
2,6-di-tert-butyl-p-benzoquinone	-	-	contamination indicator	Sandstrom 2003
3-hexanone	-	fungi	-	Fischer and Dott 2003
3-methyl-1-butanol	123-51-3	fungi	contamination indicator	Kreja and Seidel 2002
3-methyl-2-butanol	-	-	contamination indicator	Cochrane 2001
3-methylbutan-1-ol	-	fungi, Eurotium	mold indicator	Elke et al 1999
3-methylbutan-2-ol	-	fungi, Eurotium	-	Elke et al 1999
3-methylfuran	-	growing fungi	contamination indicator	Fischer and Dott 2003
3-octanol	589-98-0	fungi	contamination indicator	Kreja and Seidel 2002
3-octanone	106-68-3	fungi	contamination indicator	Kreja and Seidel 2002
4-methyl-2-hexanone	-	-	contamination indicator	Sandstrom 2003

MVOC	CAS	Source	Notes	Reference
5-methyl-2-heptanone	-	-	contamination indicator	Sandstrom 2003
acetaldehyde	75-07-0	-	contamination indicator	Schleibinger et al 1997
acetone	-	-	contamination indicator	Schleibinger et al 1997
alpha-terpineol	-	fungi, Aspergillus	-	Elke et al 1999
β-farnesene	-	fungi	-	Fischer and Dott 2003
borneol	-	fungi	-	Fischer and Dott 2003
dimethyl sulfoxide (DMSO)	67-68-5	-	-	Kreja and Seidel 2002
dimethyldisulfide	-	fungi	contamination indicator	Fischer and Dott 2003
dimethyltrisulfide	-	fungi	-	Fischer and Dott 2003
ethyl-2-methylbutyrate	-	fungi	-	Fischer and Dott 2003
ethylisobutyrate	-	fungi	-	Fischer and Dott 2003
fentone	-	fungi	-	Elke et al 1999
formaldehyde	50-00-0	-	contamination indicator	Schleibinger et al 1997
geosmin	-	fungi	contamination indicator	Fischer and Dott 2003
heptan-2-one	-	fungi, Aspergillus, Eurotium	mold indicator	Elke et al 1999
hexan-2-one	-	fungi, Aspergillus, Eurotium	mold indicator	Elke et al 1999
methyl methanesulfonate (MMS)	-	fungi	-	Kreja and Seidel 2002
methylisoborneol	-	-	contamination indicator	AQS 2003
methylisobutylether	-	fungi	-	Fischer and Dott 2003
octan-3-one	-	fungi, Aspergillus	-	Elke et al 1999
octan-3-ol	-	fungi, Aspergillus, Eurotium	mold indicator	Elke et al 1999
pentan-2-ol	-	fungi	-	Elke et al 1999
terpineol	-	-	contamination indicator	AQS 2003
thujopsene	-	fungi, Eurotium	-	Elke et al 1999
verbenone	-	-	contamination indicator	Sandstrom 2003

APPENDIX D
SURFACE SAMPLING TEST RESULTS EVALUATION FORM

Test ID Number	
Test Date	
Company	
Laboratory	
Surface Sample Area	in^2 ____ cm^2 ____
Type of Surface Sampled	
Type of Swab	☐ Dry ☐ Wet
Type of Growth Media	
Purpose of Test	☐ Survey Only ☐ Air Cleaner Performance Comparison
Test Comparison	☐ Before vs. After ☐ Upstream vs. Down ☐ Parallel Systems
Sampled Biocontaminant	☐ Pollen ☐ Bacteria ☐ Allergen ☐ Particulate ☐ Fungi ☐ Virus ☐ Endotoxin ☐ _____
Sampling Units	☐ cfu/in^2 ☐ cfu/cm^2 ☐ _____

TEST RESULTS

Location/Condition	Trial	Test Code	Before cfu/in^2	Trial	Test Code	After cfu/in^2
1.	1			1		
	2			2		
	3			3		
	Before Average cfu/in^2			After Average cfu/in^2		
2.	1			1		
	2			2		
	3			3		
	Before Average cfu/in^2			After Average cfu/in^2		
3.	1			1		
	2			2		
	3			3		
	Before Average cfu/in^2			After Average cfu/in^2		
4.	1			1		
	2			2		
	3			3		
	Before Average cfu/in^2			After Average cfu/in^2		
5.	1			1		
	2			2		
	3			3		
	Before Average cfu/in^2			After Average cfu/in^2		
6.	1			1		
	2			2		
	3			3		
	Before Average cfu/in^2			After Average cfu/in^2		

RESULTS SUMMARY	Overall Average cfu/in^2 ____ Overall Average cfu/in^2 ____ Overall % Change Overall Result: ☐ Decrease ☐ Increase ☐ No Change

APPROVAL SECTION

Responsible Individual Signature & Date:
Approval Signature & Date:
NOTES:

APPENDIX E
SETTLE PLATE TEST RESULTS EVALUATION FORM

Test ID Number	
Date	
Company	
Laboratory	
Settle Plate Type	☐ Petri Dish ☐ Liquid ☐ _____
Sample Time	seconds minutes hours days
Type of Media	
Purpose of Test	☐ Survey Only ☐ Air Cleaner Performance Comparison
Test Comparison	☐ Before vs. After ☐ Upstream vs. Down ☐ Parallel Systems
Sampled Bioaerosol	☐ Pollen ☐ Bacteria ☐ Allergen ☐ Particulate ☐ Fungi ☐ Virus ☐ Endotoxin ☐ _____
Sampling Units	☐ cfu/in^2 ☐ cfu/cm^2 ☐ _____

TEST RESULTS

Location	Trial	Test Code	Before cfu/in^2	Trial	Test Code	After cfu/in^2
1.	1			1		
	2			2		
	3			3		
	Before Average cfu/in^2			After Average cfu/in^2		
2.	1			1		
	2			2		
	3			3		
	Before Average cfu/in^2			After Average cfu/in^2		
3.	1			1		
	2			2		
	3			3		
	Before Average cfu/in^2			After Average cfu/in^2		
4.	1			1		
	2			2		
	3			3		
	Before Average cfu/in^2			After Average cfu/in^2		
5.	1			1		
	2			2		
	3			3		
	Before Average cfu/in^2			After Average cfu/in^2		
6.	1			1		
	2			2		
	3			3		
	Before Average cfu/in^2			After Average cfu/in^2		

RESULTS SUMMARY	Overall Average cfu/in^2	Overall Average cfu/in^2
	Overall % Change	
	Overall Result: ☐ Decrease ☐ Increase ☐ No Change	

APPROVAL SECTION

Responsible Individual Signature & Date:
Approval Signature & Date:
NOTES:

APPENDIX F
AIR SAMPLING TEST RESULTS EVALUATION FORM

APPENDIX F

Test ID Number							
Date							
Company							
Laboratory							
Type of Volume Sampled	Room		Zone		Building		Airstream
Air Sampler Model							
Air Volume Sampled	ft^3		in^3		L	m^3	cm^3
Type of Media							
Purpose of Test	Survey Only			Air Cleaner Performance Comparison			
Test Comparison	Before vs. After			Upstream vs. Down		Parallel Systems	
Sampled Bioaerosol	Pollen		Bacteria		Allergen	Particulate	
	Fungi		Virus		VOC	Endotoxin	
Sampling Units	cfu/m^3		cfu/L		cfu/ft^3		

TEST RESULTS

Location	Trial	Test Code	Before cfu/m^3	Trial	Test Code	After cfu/m^3
1.	1			1		
	2			2		
	3			3		
	Before Average cfu/m^3			After Average cfu/m^3		
2.	1			1		
	2			2		
	3			3		
	Before Average cfu/m^3			After Average cfu/m^3		
3.	1			1		
	2			2		
	3			3		
	Before Average cfu/m^3			After Average cfu/m^3		
4.	1			1		
	2			2		
	3			3		
	Before Average cfu/m^3			After Average cfu/m^3		
5.	1			1		
	2			2		
	3			3		
	Before Average cfu/m^3			After Average cfu/m^3		
6.	1			1		
	2			2		
	3			3		
	Before Average cfu/m^3			After Average cfu/m^3		

RESULTS SUMMARY	Overall Average cfu/m^3		Overall Average cfu/m^3	
	Overall % Change			
	Overall Result:	Decrease	Increase	No Change

APPROVAL SECTION

Responsible Individual Signature & Date:

Approval Signature & Date:

NOTES:

Glossary

ACH Air changes per hour. The number of times per hour the room volume is exchanged.

Actinomycetes A group of gram-positive bacteria that grow in filaments.

Adsorption The process by which molecules will adhere to surfaces as the result of van der Waals forces, as in carbon adsorption.

Aerobiology The study of the biology of air.

Aerobiological engineering The engineering of the indoor air environment to control aerobiological contaminants through the use of active or passive air cleaning technologies or systems.

Aerosol Fine liquid or solid particles suspended in a gas, for example, fog or smoke.

Aerosolizer A device that can generate a fine suspension of liquid or solid particles in air.

Aflatoxin Fungal metabolite with toxic properties.

After-hours UVGI system UVGI disinfection systems that operate only when personnel are not present.

Agglomeration The accumulation of particles in air. Usually the result of static charge or molecular forces.

AHU Air handling unit, usually including fans, filters, cooling coils, heating coils, and control dampers. May also include air disinfection equipment.

Air sampling The process of drawing an air sample for identification of airborne chemicals or microorganisms.

Amplifier A building that amplifies the microbial content of the air, due to moisture, mold growth, or other problems.

Antibiotic A substance that inhibits the growth of or kills microorganisms.

Antibody An immunoglobulin released by the immune system in response to the presence of an antigen.

Antimicrobial coatings Any one of a number of compounds that can be used to inhibit microbial growth on surfaces or in food.

Antiseptic A chemical used for disinfection.

Aerosol A cloud of microscopic and submicroscopic particles or droplets suspended in air.

Arclength The length of the filament of a tubular or cylindrical lamp, as opposed to the full physical lamp length.

Aspiration   To breathe.

Asymptomatic   The condition when an infection is present without apparent symptoms.

Atomization   Reducing a liquid or solid to an aerosol.

Axial flow   The UVGI configuration in which the lamp is oriented with its axis parallel to the direction of airflow.

Bacteria   Single-celled organisms that multiply by cell division and that can cause disease in humans, plants, or animals.

Bactericidal   Having the ability to kill bacteria.

Bacteriostatic   Having the ability to prevent bacterial growth.

Bell curve   A normal, or Gaussian, distribution.

Biaxial   Having an axis in two separate locations, as in biaxial lamps that have two conjoined cylindrical sections, each with its own axis.

Bioaerosol   Any airborne microbe, or airborne liquid or particle of biologic origin.

Biocidal   Having the ability to kill living organisms.

Biodetection   The process of detecting biological agents such as microbes or toxins.

Biodeterioration   Degradation of materials, especially building materials like wood or stone, due to the action of microorganisms.

Bioremediation   The removal or degradation of biological agents, especially mold, that may exist as contaminants.

Biosafety level   Four biosafety levels have been established ranked from the least dangerous (I) to the most dangerous (IV). They carry specific requirements for handling microorganisms, protective clothing and equipment or facilities.

Biosensor   Sensor for biological agents in which a sensing agent is fixed to a solid substrate and monitored for changes that indicate a reaction has taken place.

BPF   Building protection factor.

BRI   Building-related illness.

Brownian motion   The random motion of airborne particles subject to buffeting by air molecules.

BSL   Biosafety level.

Capsule   A layer of organized material that surrounds many bacteria or viruses.

Carbon adsorber   Granulated activated carbon (GAC) used as a filter for gases and vapors.

Catalyst   A substance that accelerates a chemical reaction without being consumed.

CDC   Abbreviation for the Centers for Disease Control, Atlanta.

CFD   Computational fluid dynamics.

CNS   Central nervous system.

cfu   Colony-forming unit, usually from the culturing of a plate of microbe.

Clumping   The tendency of micron-sized particles to agglomerate.

Cocci Oblong or oval bacterial cells.

Coccoid Oblong or oval in shape.

Coefficient of variation (COV) A constant that defines the distribution of lognormally distributed sizes of particles. The standard deviation divided by the mean.

Commensal Any microorganism that coexists with another organism, to the mutual benefit of both or to the harm of neither. Bacteria in the human stomach that protect us against pathogens and aid in digestion are one example.

Communicable A disease that can transmit from one infected host to another, usually by contact, airborne spread, or other mechanism.

Constant volume system Ventilation at a constant airflow rate.

Contagious Communicable. When a disease is capable of being transmitted from one person to another.

Control dampers Mechanical dampers in a ventilation system used for isolation or control of airflow.

Cross flow the UVGI system configuration in which the airflow direction is perpendicular to the axis of the UV lamp.

Culture A population of microorganisms grown in a medium such as a petri dish.

Cutaneous Of or pertaining to the skin.

Cyanobacteria A large group of photosynthetic bacteria.

Cytotoxic Causing damage to or death of a cell.

Cytotoxin A toxin that has a specific action upon cells, whether human or bacterial.

Dalton Atomic mass unit that gives the same number as atomic weight.

Death curve Decay curve or survival curve of a microbe exposed to biocidal factors.

Decay curve The decay of a microbial population under any biocidal factor.

Decontamination The process of removing harmful agents by absorbing, destroying, neutralizing, making harmless, or removing the hazardous material from the area or people who have been contaminated.

Dedicated outdoor air systems (DOAS) Ventilation systems that use 100 percent outside air to meet the minimum outside air requirements.

Dehumidification Drying the air or reducing the relative humidity by various means.

Desiccant Materials that absorb moisture from the air, lowering humidity.

Desiccant cooling The combination of a desiccant and an evaporative cooler to provide cooling or reduce enthalpy in an airstream.

Diffusion The motion of molecules through air or solid materials.

Dilution ventilation Purging of the air by mixing it with clean air or outdoor air. The normal action of any ventilation system using outside air for dilution. Also called purge ventilation.

Dioctyl phthalate (DOP) An oily liquid aerosolized to test filter penetration.

Disease progression curve Graph of the severity of disease symptoms over time.

Disinfection Having a bactericidal or viricidal effect. Total disinfection is sterilization.

DOAS Dedicated outdoor air systems.

DOP Dioctyl phthalate, a test aerosol or simulant.

Dose-response curve The relationship between the dose and the fatalities or casualties, or between the dose and the infections, of any particular agent.

Dosimetry The science of measuring dose.

Droplet nuclei Small particles in the 1 to 4 µm size range that remain after evaporation of droplets.

DSP Dust spot efficiency. A rating system for filters.

Dyspnea Shortness of breath.

Dysentery Pain and frequent defecation resulting from inflammation of the colon or other intestines.

Edema The accumulation of an excessive amount of fluid in cells or tissues.

Electret A filter using static or electrical charge to enhance efficiency.

Electrostatic filter A filter that maintains a static charge to enhance filtration effficiency.

Endemic disease A disease that is common in a particular population or geographic area.

Endogenous Native to an individual's own body.

Endogenous infection An infection caused by an individual own microbiota.

Endotoxin Toxins produced by certain gram-negative bacteria.

Energy recovery The use of heat exchangers to transfer heat energy.

Enterotoxin A toxin that specifically affects the cells of the intestinal mucosa.

Epidemic An outbreak of disease.

Epidemiology The study of the factors influencing the spread of disease in a population.

Epizootic Sudden outbreak of a disease in an animal population.

Erythema Redness of skin due to cell damage.

Exotoxin A toxin produced by certain bacteria and released into the surroundings.

Exposure time The time of exposure of a microbe to a biocidal factor like UVGI or ozone.

Febrile Fever causing.

Fluence The total amount of radiant energy from all directions incident on an infinitesimally small sphere.

Fomites Particles or droplets left on surfaces or parts of the body that contain viable infectious microorganisms and can transmit infections.

Foodborne pathogens Pathogenic microorganisms that may incubate or survive in foods.

Fungi Any of a group of plants mainly characterized by the absence of chlorophyll, the green colored compound found in other plants.

GAC Granulated activated carbon. Used in carbon adsorbers.

Gamma irradiation The use of ionizing gamma radiation to sterilize materials.

Gas phase filtration Equipment, such as carbon adsorbers, used to remove gases from the air.

Gastroenteritis An acute inflammation of the lining of the stomach.

Gastritis Inflammation of the stomach.

Genome The full set of genes present in a cell or virus.

Genus A well-defined group of one or more species.

Germination Growth of a fungus from spores.

Gompertz curve An S-shaped curve common in natural and electronic processes.

Gram stain A differential staining procedure that divides bacteria into gram-positive or gram-negative groups.

HEPA filter High-efficiency particulate air filters that have removal efficiencies of 99.97 percent or higher at 0.3 µm. May be rated MERV 17 or higher.

Herd immunity The resistance of a population to epidemic spread due to the immunity of a high percentage of the population.

High-efficiency filters Filters with rated efficiencies of 25 percent DSP or higher. This category usually excludes dust filters, prefilters, HEPA filters, and ULPA filters. May be rated MERV 7–15.

Host An animal or plant that is used by another organism, typically a microorganism, for shelter and sustenance. Sometimes this may be a commensal relationship, but often it is a parasitical.

Hydrophilic Microbial preference for wet conditions.

Hypersensitivity Exaggerated and harmful immune system response.

IAQ Indoor air quality.

ID50 Mean infectious dose. The dose or number of microorganisms that will cause infections in 50 percent of an exposed population. Units are always in terms of number of microorganisms, or, more correctly, the number of colony forming units per cubic meter (cfu/m^3).

IEQ Indoor environmental quality.

Immune building A building that has the engineered ability to prevent or protect against the spread of disease or the dissemination of chemical agents.

Immunocompromised The condition in which a person's immunity to disease has been reduced, whether due to health or injury.

Immunodeficiency The state of being unable to protect against disease. The inability to produce normal antibodies in response to antigens.

Immunodeficient Being unable to protect against disease. Those who are immunocompromised.

Immunosuppression The reduction of immune system response that leaves a person susceptible to diseases.

Impaction The collision of a particle with a filter fiber that results in attachment.

Impregnated carbon   Carbon adsorber material that has compounds added to increase adsorption efficiency for specific chemicals.

Incubation   The time during which a pathogen grows and multiplies to cause an infection. The process of maintaining a bacterial culture in ideal conditions to induce growth of colonies.

Infectious agents   Biological agents capable of causing disease in a susceptible host.

Infectious period   The period during which an infected person is contagious.

Infectivity   (1) The ability of an organism to spread. (2) The number of organisms required to cause an infection to secondary hosts. (3) The capability of an organism to spread out from the site of infection and cause disease in the host organism.

Inoculation   The use of subcutaneous vaccines. Also the process of initiating a culture on a petri dish.

Intensity   Degree of radiation exposure similar to the term irradiance but without a precise definition.

Interception   The process by which particles attach to filter fibers when they pass close enough to come into contact.

Inverse square law   A description of the decrease of light intensity as a function of the inverse of the square of the distance from the source.

Ionization   The process of stripping electrons from atoms to produce ions.

Ionizing radiation   Radiation of short wavelengths or high energy that causes atoms to lose their electrons, or become ionized.

Irradiance   The density of radiation incident upon a flat surface. Common units include $\mu W/cm^2$ or J/m^2.

Isolation room   Rooms in which the pressure is controlled to prevent microbial pathogens from entering or exiting. May be positively or negatively pressurized, depending on the function.

LD$_{50}$ Mean lethal dose. The dose or number of microorganisms that will cause fatalities in 50 percent of an exposed population. The units are number of microbes (or cfu/m3).

Logarithmic decay   The classical shape of the decay curve of a population exposed to biocidal factors.

Logmean   The average of the logarithms of any quantity, such as microbe diameters.

Lognormal   The distribution in which the logarithm of some factor such as size is distributed normally (i.e., as a bell curve).

Mean diameter   The actual median diameter for a population of microbes or particles.

Mean disease period   The mean period for any disease during which symptoms are manifest.

Mean infectious period   The mean period for any contagious disease during which the disease is communicable.

Media velocity   The actual velocity through pleated filter media, as opposed to the face velocity through the filter cartridge.

MERV   Minimum efficiency reporting value. Rating system for filters based on ASHRAE Standard 52.2-1999.

Microbe   Synonym for microorganism.

Microflora   Microbial populations that may be of mixed species.

Micron   One-millionth of a meter, also µm.

Microorganism   Any organism, such as bacteria, viruses, fungi, and protozoa, that can be seen only with a microscope.

Mildew   Common description of mold when it grows on fabrics or textiles.

Mist   A suspension of liquid droplets. Usually created by atomization or vapor condensation.

Mold   The common term describing visible growth of fungi.

Morbidity   Reduced health or lethargy due to disease.

Mucosal   Related to mucous.

Multihit model   A model of the decay of microbial populations that accounts for shoulder curves.

Multizone   Being divided into separate ventilation zones.

MVOC   Microbial volatile organic compound.

Mycetoma   A localized lesion caused by fungal infection.

Mycoplasma   Bacteria of the class Mollicutes that lack a cell wall.

Mycosis   Any disease caused by fungi.

Mycotoxin   A toxin produced by fungi, such as T-2 mycotoxin, aflatoxin, or ochratoxin.

Natural ventilation   Ventilation without the use of powered equipment.

Nebulizer   A device for producing a fine spray or aerosol.

Normal distribution   The Gaussian distribution of data in a bell curve.

Nosocomial   Refers to infections that occur in hospitals.

Organism   Any individual living thing, whether animal or plant.

Ozonation   The use of ozone to disinfect air or water.

Ozone   A corrosive form of oxygen consisting of three bound oxygen molecules.

Pandemic   An epidemic that occurs around the world.

Parasite   Any organism that lives in or exploits another organism without providing benefit in return.

Passive solar exposure   The use of solar exposure to disinfect surfaces or materials.

Pathogen   Any organism capable of producing disease or infection.

Pathogenic agents   Biological agents capable of causing serious disease.

Pathogenicity   The disease-causing ability of a pathogen. The degree of health hazard posed by such microbes.

Pathology   The study of disease or the course of a particular disease.

PCO   Photocatalytic oxidation.

Peak infection period   The mean period for any contagious disease during which the host is infectious.

Penetration   The degree to which particles or microbes penetrates a filter. The complement of removal efficiency. %Penetration = 100 − %efficiency.

Petri dish   A plastic or glass plate used for culturing microbes.

Photocatalytic oxidation (PCO)   A technology in which material coated with titanium dioxide is irradiated to produce an oxidative effect.

Pleomorphic   Capable of changing shape.

Plug flow   The type of airflow pattern in a room in which air moves from one side to the other in piston fashion.

Pneumonia   The presence of excessive fluid in the lungs. May be caused by any one of a number of pathogens.

Pollen   Seed spores of plants, grasses, and trees.

Potable water   Clean drinking water.

Prefilters   Filters used before other equipment, including UVGI systems, carbon adsorbers, and high-efficiency filters.

Protective clothing   Respiratory and physical protection.

Protozoa   A multi-celled microscopic animal. These microorganisms may sometimes be involved in human disease.

Pulmonary edema   The accumulation of excessive fluid in the lungs.

Pulsed filtered light   Pulsed white light with the UVC spectrum removed.

Pulsed light disintegration   The disintegrating effect on bacterial cells from excess power levels of pulsed light. A non-UV dependent phenomenon.

PUV   Pulsed ultraviolet light.

PWL   Pulsed white light.

Radiation view factor   A quantity that defines the amount of radiation transmitted from a radiating surface and absorbed by another surface.

Recirculation   The return of airflow, processed or otherwise, to the building volume by the ventilation system.

Remediation   The cleanup or decontamination of an area that has been exposed to CBW agents, molds, or other hazards.

Retrofit   To add a new component to an existing system.

Route of exposure   The path by which a person comes into contact with an agent or organism, for example, through breathing, digestion, or skin contact.

Second stage   The portion of a microbial population decay curve after the first stage. Usually results from a fraction of the population being resistant to some biocidal factor.

Simulant   A normally harmless microbe used to simulate a pathogenic microbe for testing.

Spore   A seed-like dormant form that some microorganisms adopt as a reproductive mechanism, or as a hedge against environmental conditions.

GLOSSARY

Standard deviation A parameter of a normal or bell curve that defines how far out the curve spreads.

Total arrestance A rating system for filters which measures the total removal efficiency over a broad range of particle sizes.

Toxicity A measure of the harmful effect produced by a given amount of a toxin on a living organism.

Toxigenic Containing or capable of producing toxins.

Toxins Biologically generated poisonous substances created by plants, insects, animals, fish, bacteria, or fungi.

Tracer gas A harmless gas used for testing ventilation systems and equipment.

Trichothecenes A class of compounds produced by certain fungi that includes some toxins.

Tuberculosis An infectious disease caused by *Mycobacterium tuberculosis*.

TVOCs Total volatile organic compounds.

Type species A representative species for the genera.

ULPA Ultra low penetration air filters. Filters that have efficiencies exceeding those of HEPA filters. May be 99.999 percent or higher at 0.3 µm.

Ultrasonic atomizers Aerosolizers that use ultrasonics to generate vapors or mists.

Ultraviolet light Light in the range of approximately 200 to 400 nm.

Upper air irradiation irradiation The use of UVGI to irradiate the air in rooms above the level of occupancy so as to disinfect the air and limit human exposure.

URI Upper respiratory illness or infection.

URD Upper respiratory disease.

URV UVGI rating value.

UV Common abbreviation for ultraviolet.

UVA Ultraviolet light in the A band region, approximately 315 to 400 nm. Has minor biocidal effects.

UVB Ultraviolet light in the B band region, approximately 280 to 315 nm. Has minor biocidal effects.

UVC Ultraviolet light in the C band region, approximately 200 to 280 nm. Responsible for the greatest part of the biocidal effects of UV light.

UVGI Ultraviolet germicidal irradiation. The term defined by the CDC to describe the technology of using ultraviolet light to disinfect air and surfaces.

Vaccine A preparation of killed or weakened microorganism products used to artificially induce immunity against a disease.

Vapor Aerosolized form of a chemical agent or liquid that exists in airborne droplet form.

Vapor pressure The pressure at which a liquid and its vapor are in equilibrium at a given temperature.

Variable air volume Ventilation systems that adjust the amount of outside air based on outside air conditions to save energy.

VAV   Variable air volume.

Vector   An agent, such as a mosquito, tick, or rat, serving to transfer a pathogen from one organism to another.

Ventilation effectiveness   A measure of how well the air is purged from a room.

View factor   Radiation view factor.

Virion   A single virus particle.

Virulence   The quality that defines how pathogenic a microbe is, or how deadly it is, or how rapidly it causes an infection.

Virulent   More pathogenic, or having greater disease-causing capacity.

Virus   An infectious microorganism that exists as a particle rather than as a complete cell.

VOCs   Volatile organic compounds.

Volatile organic compounds   Aerosols and gases that come from organic or hydrocarbon sources.

Weathering   The use of the natural elements of sunlight and temperature extremes to decontaminate or disinfect surfaces or materials.

Xerophile   Microbes that prefer dry conditions.

Zoonosis   Disease of animals that can be transmitted to humans.

INDEX

AC (alternating current) ionizers, 325
ACH (see Air changes per hour)
Adsorption media, 275
Aerial disinfection, 336, 337, 396, 609
Aerobiological engineering:
  agricultural facilities, applied to, 590–591
  airports, applied to, 623, 624
  apartment buildings, applied to, 516
  assembly places, applied to, 623–625
  background of, 1–179
  defined, 3
  dormitories, applied to, 516–517
  educational facilities, applied to, 504–506
  food plants, applied to, 574–578
  health care facilities, applied to, 527–542
  hotels, applied to, 516–517
  laboratories, applied to, 547–563
  nosocomial, 533–540
  office buildings, applied to, 487–491
  outdoor environments, applied to, 99–114
  residential homes, applied to, 511–516
  sewage processing facilities, applied to, 671–674
  shopping malls, applied to, 623, 624
Aerodynamic diameter, 123–125
Aerosol fumigants, 453
Aerosol science, 119–140 (*See also* Bioaerosol dynamics; Bioaerosols)
Aerosolization, 161
Aerosolized blood pathogens, 536
Aerosolized virus, 328
Aerosols, 550
Agricultural facilities, 585–597
  aerobiology of, 590–591
  agricultural pathogens, 585–586
  airborne viruses in, 590
  animal dander, 588
  avian influenza virus, 588, 590
  breathing in, 586

Agricultural facilities (*Cont.*):
  cattle behavior, 590
  chronic respiratory symptoms, 590
  cow dander allergy, 588
  dust, 590
  epidemiology in, 586–590
  feeding patterns, 590
  organic toxic dust syndrome, 586, 588
  pollen, 586
  toxic gases, 588
  VOCs, 591
  zoonotic diseases, in veterinarians, 588
Agricultural pathogens, 585–586
AHU (air handling unit), 232
AIDS patients, 522, 524, 531
Air changes per hour (ACH), 186
  building protection factor (BPF), 478
  laboratories, 548, 557
  for TB rooms, 530
Air cleaning, 350, 449–450, 475, 532
  laboratories, 556–559
  remediation *vs.*, 450
Air conditioning systems, 522, 603
Air disinfection systems, 389, 469–474
  in buildings, 464–465
  photocatalytic oxidation, 297–301
  (*See also specific types, e.g.:* Ultraviolet germicidal irradiation)
Air filters:
  bacteria in, 223
  ionization of, 325
  mold spores in, 223
Air handling unit (AHU), 232, 516–517
  (*See also* Central air handling units)
Air ionization, 324–325, 542
  electrostatic precipitators *vs.*, 324
  ozone generators *vs.*, 324
Air ions, 324
Air microflora, 403–405

INDEX

Air mixing effects, 248–249
Air pollution, 106, 110
Air purifiers, 330
Air quality, in aircraft, 628, 631
Air recirculation units, 234–235
Air sampling, 403–423
  and air microflora, 403–405
  air samplers, 409–412
  allergen sampling, 414
  automated sampling systems, 410
  for bioaerosol concentration measurement, 468
  in buildings, 419
  electrostatic sampling devices, 412
  indoor air quality, 409, 418–419
  microorganism stress, 410
  outdoor air sampling, 419
  particle counting, 413–414, 419
  performance, 412–413
  procedures, 417–420
  rotating slit sampler, 411
  settle plates, 409, 417
  test protocol, 420
Air temperature effects, 251–252
Air treatment systems:
  bacteria production, 477
  design intent of, 476
  fungal spore production, 477
  performance criteria for, 476–477
  sewage processing facilities, 678–680
Airborne aflatoxin, 576
Airborne allergens:
  and HEPA filters, 225
  indoor bioaerosol levels, 441
Airborne animal diseases, 111–112
Airborne bacteria:
  in auditoriums, 624
  indoor bioaerosol levels, 440–441
  in office buildings, 490
  in university auditoriums, 624
Airborne diseases, 3–13
  as biological weapons, 6
  building related illness (BRI), 5
  and building sciences, 4–5
  coevolution with habitats, 166
  control technologies, 183–397
  costs to society, 67–69
  current environment of, 3–6
  dosimetry of, 90–94
  and engineering technologies, 6
  epidemiology of, 82–87
  food poisoning, 62
  history of, 6–12

Airborne diseases (*Cont.*):
  and indoor environments, 3–13
  nonrespiratory, 62
  occupational, 647–653
  pathology of, 41–63
  sick building syndrome, 5
  skin infections, 62
  Soper equation, 82, 84
  statistics of, 67–69
  transmission, 550–555
  ventilation system model equation, 84–85
  (*See also specific types, e.g.:* Viruses)
Airborne fungi, 572
  indoor bioaerosol levels, 438–440
  in office buildings, 488
Airborne hazards, 551, 572
Airborne infections:
  noncommunicable, 46–47
  spread by patients, 536
Airborne microbes, 291
  ICUs, 535
  ionization of, 326–329
  operating rooms, 535
  in sewage, 673
  surgical site infections (SSI), 532
Airborne microflora, 469
Airborne microorganisms, 559–560
  filtration of, 207–227
  flood damage, 439
  identification of, 428
  in naturally ventilated building, 487
  in office buildings, 487
Airborne MRSA, 537
Airborne nosocomial epidemiology, 531–533
Airborne ozone disinfection, 336–338
Airborne ozone removal, 339
Airborne particulate dust, 649
Airborne pathogens, 15–32
  allergen database, 30–32
  building protection factor of, 477
  coevolution of habitats and, 12
  coevolution of man and, 10–12
  in dental offices, 538
  development of, 167
  epidemiological nature of, 4–5
  evolution of, 7–10
  and indoor environments, 168
  in lungs, 45
  passive solar exposure, 350–354
  pathogenic nature of, 4–5
  potential, 29
Airborne PCR systems, 431

INDEX

Airborne pollen, 441
Airborne respiratory infections, 48–62
  lower respiratory tract infections, 50–51
  middle respiratory tract infections, 49–50
  upper respiratory tract infections, 48–49
Airborne sampling, 291
Airborne spores, 108
Airborne viruses:
  in agricultural facilities, 590
  indoor bioaerosol levels, 441
Aircraft, aerobiological concerns with, 627–634
  air quality, 628, 631
  ASHRAE, 631
  bactericidal meningitis, 629
  bioaerosols, 631
  disease transmission, 628, 630
  HEPA filters, 633–634
  human bioeffluents, 631
  influenza, 628
  mucosal irritation, 630
  smoking, 633
  upper respiratory infection, 629
Air-exchange rates, 548
Airflow, in laboratories, 557, 562
Airports, 623, 624
Airtightness, 166, 167, 169–170, 357
Airway diseases, occupational, 648
Airway inflammation, 486
AIV (*see* Avian influenza virus)
Algae, 20–21
Allergen sampling, 414
Allergenic airborne hazards, 572
Allergenic fungi, 4
Allergens, 15–32, 54–55
  algae, 20–21
  animal dander, 25–26
  from animals, 515–516
  apartment buildings, 518
  bacteria, 18–19
  cockroach, 24, 25
  detergent removal of, 520
  domestic pets, 519
  dust mites, 23–24
  food industry, 571
  fungal spores, 19–20
  indoor bioaerosol levels of, 441
  in laboratories, 560
  MVOCs, 27, 29
  in office buildings, 486
  outdoor, 99–104
  prions, 16
  protozoa, 20

Allergens (*Cont.*):
  rat urinary allergens, 556
  residential homes, 518
  sampling, 404–405
  storage mites, 23–24
  toxins, 26–29
  viruses, 16–17
  VOCs, 27
  (*See also specific types, e.g.:* Pollen allergens)
Allergic reactions:
  caused by nonorganic materials, 659
  food industry, 571
  metal industry, 659
  from outdoor air, 106
Allergic rhinitis, nonseasonal, 56
Alpha rays, 391
Alternaria, 103
Alternating current (AC) ionizers, 325
Anabaena flosquae, 21
Animal allergens, 515–516
Animal dander, 25–26
  agricultural facilities, 588
  in indoor arenas, 625
  residential homes, 511
Animal diseases, 111–112
Animal facilities, ventilation in, 592–597
  (*See also* Agricultural facilities)
Animal handlers, 551, 556
Animal husbanding, 586
Animal laboratories:
  airborne disease transmission, 550–555
  airborne hazards in, 551
  airborne microorganism limits in, 559–560
  airflow in, 562
  animal handlers, 551, 556
  asthma, 556
  epidemiology in, 551, 556
  exhaust grille placement, 561–562
  HEPA filters, 561
  laboratory animal allergy, 556
  minute virus of mice, 558
  naturally ventilated cages, 560
  pathogens, 550–555
  photocatalytic oxidation, 562
  problems and solutions, 559–563
  rat urinary aeroallergen, 556, 560
  UVGI, 562
  workers in, 651
  work-related sensitization, 556
  zoonotic diseases, 550, 551
  (*See also* Laboratories, aerobiological
    concerns in)

Animal pathogens, 585–586
Anisakis simplex, 571
Anthrax:
  buildings with, 444
  and Gruinard Island, 106
  textile industry, 657
Antimicrobial coatings, 371–377
  and antimicrobial filters, 371
  applications of, 371
  building applications of, 376–377
  Centre for Applied Microbiology &
    Research, 373
  copper, 371
  food industry applications of, 375–376
  health care applications of, 375
  material performance, 372–374
  pharmaceutical applications of, 375
  silver, 371
Antimicrobial filters, 371
Antimicrobial food preservatives, 580
Apartment buildings, aerobiological
    concerns with, 516
  allergens, 518
  carpeting, 519
  central air handling units, 520
  pathogenic viruses, 518
  solutions for, 517–521
  ventilation, 518–520
  water damage, 518
  without central air, 521
Aspergillosis, 538, 676
Aspergillus niger, 148, 223
Assembly places, aerobiological concerns with,
    619–625
  aerobiology of, 623–625
  epidemiology, 622–623
  tuberculosis, 622
  ventilation systems, 619–622
  (*See also specific places, e.g.:* Stadiums)
Asthma, 52–53
  animal laboratories, 556
  causes of increase in, 503
  children, 502
  food industry, 571, 575
  industrial facilities, 647–648
  laboratories, 560
  occupational, 53, 648–650
  and outdoor air, 110–111
  residential homes, 516
  risk factors for, 108
Atopy, 55
Auditoriums, aerobiological concerns with, 624

Automated sampling systems, 410
Avian influenza virus (AIV), 29,
    588, 590
Aw (water activity), 174

Bacillus cereus, 336, 337, 343
Bacillus species, 103
Bacillus subtilis, 223, 341, 350
Bacteria, 18–19
  in air filter media, 223
  air treatment systems, 477
  on building materials, 172, 174
  damage via PWL, 315
  environmental, 4, 47, 175
  food industry, 575
  gram-negative, 440
  inactivation in ozone, 338
  indoor air levels, 440
  indoor bioaerosol levels, 440–441
  in indoor environments, 404
  levels in residential homes, 512
  in libraries, 605
  microbial cultures, 405
  microbial density of, 126
  in occupied buildings, 469
  in office buildings, 490
  outdoor levels of, 103
  pathogenic, 575
  pulsed light, 309
  resistance to ozone, 342
  respiratory infections caused by, 47
  source of, 4, 469
  surface characteristics of, 120–122
  survival with filters, 222
  UVGI, 562
Bacterial spores, 126
Bactericidal meningitis, 629
Bagassosis, 56–57, 577
Baking industry, 577
Beggs, Clive, 329
BIM (bioimpactor), 413
Bioaerosol dynamics, 130, 134–137
  air sampler measurement, 468
  Brownian motion, 134
  electrical forces, 134
  electromagnetic radiation, 134–135
  gravitation, 134
  inertial impaction, 136–137
  thermal gradients, 134–135
  turbulent diffusion, 135–136
Bioaerosol exposure, in food industry,
    572

Bioaerosols:
  and aerodynamic diameter, 123–125
  aircraft, 631
  concentrations of, 451
  evaporation of, 137–139
  sampling, 413
  settling of, 137–139
  size and shape of, 119–120
  viability of, 139–140
  (*See also* Indoor bioaerosol levels)
Biocontamination, in ventilation, 189–190
Biodetection, 427–434
  biosensors, 432–434
  catalytic sensors, 433
  light detection and ranging (LIDAR), 430
  mass spectrometry, 430
  noncatalytic biosensors, 433
  particle detectors, 427–430
  photoelectric detection, 433
  polymerase chain reaction, 430–432
Biofilms, 175, 177
  environmental bacteria on, 175
  food industry, 579
Biofilters (sewage processing facilities), 679
Biofiltration, 354–356
Bioimpactor (BIM), 413
Biological contamination, 450, 611
Biological laboratories, 550, 551
  aerosols, 550
  guidelines for, 548
  microbiological toxins, 550
  SARS virus, 550
  Venezuelan equine encephalitis (VEE), 550
  zoonotic pathogens, 550
  (*See also* Laboratories, aerobiological concerns in)
Biological weapons:
  airborne, 29–30
  airborne diseases, 6
Biosafety level (BSL), 548
Biosensing technology, 428
Biosensors, 432–434
Biotechnology industry, aerobiological concerns in, 660–661
Biotest (RCS), 413
Birds, and outdoor air, 106
Blastomycosis, 105
Book contamination, 608
Bordetella pertussis, 45
BPF (*see* Building protection factor)

Breathing, 42–43
  in agricultural facilities, 586
  ozone in air spaces, 337
BRI (*see* Building related illness)
Bronchitis, 50
  chronic, 658
  industrial, 649–650
  pneumonia, 50–51
Brownian motion, 134
Brucellosis, 571
BSL (biosafety level), 548
Building leak and pressure testing, 464–465
Building materials, 172–179
  fungi in, 175
  growth of bacteria on, 172, 174
  growth of fungi on, 172, 174
  removal remediation, 451
Building protection factor (BPF), 477–480
  ACH, 478
  of airborne pathogens, 477
  calculating, 479
  determining, 478
  release rate (RR_T), 478
  removal efficiency (RE), 478
Building related illness (BRI):
  airborne diseases, 5
  in office buildings, 487
  SBS *vs.*, 5
  and surface sampling, 415
Buildings, 165–179
  air disinfection systems, 464–465
  air sampling, 419
  and airborne diseases, 4–5
  airborne microflora, 469
  airtightness of, 166, 167, 169–170, 357
  with anthrax, 444
  bacteria in, 469
  BPF rating in, 480
  dampness, 459
  decontamination with ozone, 341
  development of airborne pathogens, 167
  forced ventilation, 168, 170
  fungal spores in, 439–440
  incubator quality of, 85
  indoor airborne spore levels, 438
  indoor contaminant levels, 464–465
  indoor humidity, 459
  leakage, 464–465
  mold growth, 459
  naturally ventilated, 170, 487
  ozone damage, 459
  particle counting, 419

Buildings (*Cont.*):
  pressure for contaminant control, 464
  self-disinfection of, 351
  spores, in old houses, 438
  types of, 168–169
  water damage, 459
  (*See also* Green building design)
Burkholderia cepacia, 249
Buses, aerobiological concerns with, 640–641
Byssinosis, 57

CADR (clean air delivery rate), 478
CAMR (Centre for Applied Microbiology & Research), 373
Cancers, in wood industry, 655
Carbolic spray, 395
Carbon adsorber testing, 474
Carbon adsorption, 267, 270–274
  for airborne ozone removal, 339
  applications of, 276–278
  cartridge-type, 276
  effect on microbiological particles, 275
  impregnated carbon, 274–275
  lifespan, 273
  operation and maintenance, 275–277
  regeneration, 272
  sizing, 271
  stand-alone recirculation units, 278
  testing of, 474
Carpeting:
  apartment buildings, 519
  as indoor pollutant, 356
  libraries, 605
  residential homes, 519
Cars, aerobiological concerns with, 640–641
Cartridge-type carbon adsorption, 276
Cat allergens:
  educational facilities, 503
  residential homes, 516
Catalytic incineration, 301
Catalytic sensors, 433
Cattle behavior, 590
Ceiling mounted ion generators, 330
Central air handling units:
  in apartment buildings, 520, 521
  in hotels, 522
  residential homes, 520
Centre for Applied Microbiology & Research (CAMR), 373
Centrifugal fans, 475
Certified industrial hygienist (CIH), 418

CFD (computational fluid dynamics) modeling, 204
  for stadiums, 620, 621
cfu (colony forming units), 326
Chemical decontamination procedures:
  chlorine dioxide procedures, 456–458
  ozone, 458–459
  ozone procedures, 458–459
  remediation, 456–459
  SNL foam procedures, 459
Chemical disinfection, 394, 395–396, 453
Chemical foams, 453–454
Chemical industry, aerobiological concerns in, 661–662
Chemically treated filters, 227
Children:
  asthma in, 502
  and disease transmission, 78–81
  epidemics, 501
  infections in, 501
  pneumonia in, 502
  respiratory allergy effects, 108
  respiratory infections in, 358, 502
  scarlet fever in, 504
  tuberculosis in, 502
  whooping cough in, 503
Chlamydia pneumoniae, 8, 50
Chlorine dioxide (ClO_2), 453, 456–458
  building decontamination procedures, 457
  libraries, 609
  whole building decontamination, 457
Choirs, aerobiological concerns with, 622–623
Chronic bronchitis, 658
Chronic obstructive pulmonary disease (COPD), 651
Chronic respiratory symptoms, 590
CIH (certified industrial hygienist), 418
ClO_2 (*see* Chlorine dioxide)
Cladosporium, 103, 225
Clean air delivery rate (CADR), 478
Cleanroom technology, 579
Cluster ions, 324
Cockroach allergens, 24, 25
Coefficient of variation (COV), 128
College campuses:
  crowding, 517
  meningitis, 504, 517
  (*See also* Dormitories)

Colony forming units (cfu), 326
Commercial air cleaners, 224–225
Commercial buildings, 168, 191 (*See also* Office buildings)
Commissioning, 475–477
　acceptance, 477
　of air cleaning systems, 475
　construction, 477
　design phase, 476–477
　predesign phase, 476
Common cold, 486
Communicable diseases, 47
　environmental bacteria, 47
　epidemiology of, 82–87
　fungal spores, 47
Communicable respiratory viruses, 46
Composting plants, 105
Compressed air, 579
Compressed air method, 451
Computational fluid dynamics (CFD) modeling, 204
Condensation, 171–172, 190
Constant volume (CV) ventilation, 187
Contact transmission, 551
Contamination, 450
Control options, nosocomial, 540–542
Cooking:
　food industry, 572, 574
　and VOCs, 574
Cooling coil irradiation systems:
　design intent of, 476
　ultraviolet germicidal irradiation (UVGI), 255–256
Cooling coils:
　microbial growth, 189
　mold growth on, 492
　and spores, 190
　UV decontamination of, 255–256
COPD (chronic obstructive pulmonary disease), 651
Copper:
　antimicrobial coatings, 371
　biocidal effects of, 372
Corona discharge, 394–395
　ionizer, 325
　pulsed corona discharge systems, 395
Corona precharger, 227
Corona wind, 289
Corona wind air cleaners:
　electrostatic filtration, 289–291
　ionizing air cleaners, 289
　ozone generation, 290

Coronaviruses, 11
Correction factor, 216
Coughing:
　disease transmission, 71, 75
　respiratory system, 42
Coulter counters, 429
Counter-diffusion factor, 216
COV (coefficient of variation), 128
Cow dander allergy, 588
Crab allergens, 637
Crowding, 517
Cruise ships, aerobiological concerns with, 637
Cryogenics, 367
Cunningham slip factor, 215
Currency notes, 656
CV (constant volume) ventilation, 187

Dampness:
　in buildings, 459
　in office buildings, 486
　respiratory allergy effects, 439
Dander:
　sampling, 405
　source of, 4
　(*See also* Animal dander)
DC (direct current) ionizers, 325
Decay curve, 317–318
Decay model:
　exponential, 145–148
　two-stage, 148
Decontamination:
　libraries, 609
　ozone for, 344
　surface, 473
　surface decontamination system, testing, 471–472
　(*See also* Whole building decontamination)
Dedicated outdoor air systems (DOAS), 6, 188, 498, 506
Dehumidification, 367
Dehydration, of microorganisms, 113–114
Dental offices, 538
Dermatophagoides farinae, 23, 110
Dermatophagoides pteronyssinus, 23, 110
Desicant dehumidification, 368
Desiccation, 156
　and dehumidification, 367
　of microorganisms, 113–114
Detergents:
　allergen removal with, 520
　whole building decontamination, 452
Deutsch equation, 284

828 INDEX

Developmental technologies, 389–397
  air disinfection, 389
  chemical disinfection, 394, 395–396
  corona discharge, 394–395
  free radicals, 397
  ionizing radiation, 389–392
  plasma technology, 394
  pulsed corona discharge systems, 395
Diarrhea, on ships, 636
Diffuse reflectivity, 244–245
Diffusion, 213
  counter-diffusion factor, 216
  single-fiber efficiencies for, 213
  turbulent, 135–136
Diffusion charging, 134
Diffusion efficiency correction factor, 216
Dilution, 185–204
Dilution modeling, 195–197
Dilution ventilation, 191–192, 557
Direct current (DC) ionizers, 325
Direct exposure, 234
Discharge lamp, surface, 309
Disease control, 12–13
Disease progression curves, 87–90
Disease transmission:
  airborne, 550–555
  aircraft, 628, 630
  and children, 78–81
  contact transmission, 551
  and droplet nuclei, 75
  foodborne pathogens, 71
  of Hantavirus, 72
  interspecies transmission, 551
  of noncommunicable diseases, 72–73
  of nonrespiratory diseases, 71
  nose blowing, 75
  respiratory diseases, 358, 360
  of respiratory syncytial virus (RSV), 73
  of rhinovirus, 73, 79
  sneezing, 71, 75
  stadiums, 622
  on surfaces, 78
  waterborne pathogens, 71
Diseases, 46–48
  communicable, 47
  noncommunicable airborne infections, 46–47
  noninfectious respiratory, 51–61
  nonrespiratory airborne, 62
  nosocomial (hospital-acquired) infections, 47
  outdoor airborne animal diseases, 111–112
  outdoor pathogenic disease, 104–106

Diseases (*Cont.*):
  respiratory viruses, 46
  zoonotic, 46
Disinfection:
  aerial disinfectants, 396
  air, 297–301
  air disinfection, 389
  air disinfection systems, 464–465
  airborne ozone, 336–338
  chemical disinfectants, 453
  chemical disinfection, 394, 395–396
  of food, 571
  food industry, 579
  of fruits and vegetables, 579
  in-place testing of surface disinfection systems, 469–474
  microbial, 143–161 (*See also* Microbial disinfection)
  microwave thermal disinfection, 383
  microwaves, 379–383
  pulsed electric fields, 317
  surface, 302
  surface UV disinfection system, 471
  thermal, 153–156
  ultraviolet germicidal irradiation (UVGI), 235–237
  visible light, 351
Disinfection models, 153–161
  aerosolization, 161
  desiccation, 156
  disinfectants, 160
  freezing, 159–160
  ionizing radiation, 157–158
  nonionizing radiation, 158–159
  nutrient deprivation, 159
  open air factor, 157
  oxygenation, 157
  pressure, 160
  thermal disinfection, 153–156
  ultraviolet germicidal irradiation (UVGI), 237–240
Disintegration, pulsed light, 316
Displacement ventilation, 188–189
DNA, radiation damage to, 386
DOAS (*see* Dedicated outdoor air systems)
Dog allergens, 503
Domestic pets, 519
Domestic waste (sewage processing facilities), 678
DOP Penetration, 465–466
Dormitories, aerobiological concerns with, 516–517

INDEX   **829**

Dosimetry, 90–95
  of airborne diseases, 90–94
  defined, 90
  of microbial agent ingestion, 91–92
  of ozone, 336
  of toxins, 91–92
Droplet nuclei:
  and disease transmission, 75
  surface characteristics of, 122
DSP-rated filters, 209–210
Dual purpose isolation rooms, 530–531
Dust:
  and adsorption media, 275
  agricultural facilities, 590
  and fungi, 172
  museums, 610
  paper industry, 653
  sewage processing facilities, 679
  theaters, 625
  and UVGI, 190
  in ventilation systems, 190
  wood industry, 653, 655
Dust mites, 23–24
  allergens, 486
  in India, 110
  residential homes, 511, 513, 516
Dust spot efficiency, 207–208
Dust-induced lung disease, 649
Dynamic light scattering

E. coli (see Escherichia coli)
E (overall filter efficiency), 212–213
EAC (see Electronic air cleaners)
Educational facilities, aerobiological concerns with, 497–507
  aerobiology of, 504–506
  building types, 497–498
  cat allergens, 503
  cleaning/hygiene, 503
  crowding, 503
  DOAS ventilation, 498, 506
  dog allergens, 503
  epidemiology, 499–504
  green building design, 499
  hand hygiene, 507
  humidity, 498
  moisture damage, 505–506
  mold growth, 498, 499
  pollutant accumulation, 498
  problems, 506
  respiratory disease transmission, 503
  respiratory syncytial virus (RSV), 499

Educational facilities, aerobiological concerns with (*Cont.*):
  rhinovirus, 501
  SBS/BRI, 499
  solutions for, 506–507
  upper respiratory infection (UFI), 499
  ventilation, 498–499, 506
  VOCs, 501
  water damage, 499
EEFs (*see* Electrically enhanced filters)
Efficiency filters, 465
Electret filters, 287
Electrical filtration, 283
Electrical forces, 134
Electrical wind, 289
Electrically charged filters, 227
Electrically enhanced filters (EEFs), 287–289
Electromagnetic radiation, 134–135, 386
Electron beams, 392
Electron volts (eV), 330
Electronic air cleaners (EACs), 285–286
Electronic ionizer, 325
Electrostatic filtration, 281–292
  aerobiological performance of ESPs, 291
  against airborne microbial contaminants, 291
  corona wind air cleaners, 289–291
  defined, 286
  economics of, 292
  electret filters, 287
  electrically enhanced filters, 287–289
  electronic air cleaners, 285–286
  electrostatic precipitator, 281–285
Electrostatic precipitators (ESP), 281–285
  air ionizers *vs.*, 324
  airborne sampling, 291
  droplet removal, 285
  as electronic air cleaners, 285
  models of, 283
  for pollen removal, 291
  two-stage, 281–285
  use of, 281–282
Electrostatic sampling devices, 412
Enclosure reflectivity, 243–247
Enclosures (*see* Buildings)
Endogenous microbes, 533
Endotoxins, 26–27
  and LPS, 440
  sampling, 405
Engineering technologies, 6
Environmental bacteria, 47
  on biofilms, 175
  source of, 4

Enzootic pneumonia, 112
Epidemics:
  children, 501
  statistics of, 82
Epidemiology, 67–90
  in agricultural facilities, 586–590
  of airborne diseases, 82–87
  airborne nosocomial, 531–533
  in animal laboratories, 551, 556
  in assembly places, 622–623
  of communicable diseases, 82–87
  defined, 67
  in educational facilities, 499–504
  in food industry, 571–574
  of noncommunicable diseases, 82–87
  in office buildings, 486–487
  in sewage processing facilities, 674–678
  in stadiums, 622
Epiglottis, 49–50
ePTFE filters, 227
Erythema action spectrum, 351
Escherichia coli, 222, 223, 339, 341, 432
ESP (*see* Electrostatic precipitators)
Ethylene oxide (EtO), 396
EtO (ethylene oxide), 396
eV (electron volts), 330
Evanescent waves (EW), 432
EW (evanescent waves), 432
Exhaust grilles, 561–562
Exotoxins, 26–27, 405
Exponential decay model, 145–148, 237
Exponential growth model, 143–145
Exponential phase, 144

Fabrics, aerobiologically unsafe, 613
Fans:
  forward-tipped centrifugal fans, 475
  laboratories, 559
  used in ventilation systems, 475
  veneaxial, 475
Farmer's lung, 56
FDA (Food and Drug Administration), 567–568
Feces, human, 672
Federal Insecticide, Fungicide, and Rodenticide Act (FIFRA), 568
Fevers, respiratory, 57
Fibrous filters, 269
FIFRA (Federal Insecticide, Fungicide, and Rodenticide Act), 568
Filter performance curves, 207–210
Filter testing, 465–467
Filtered pulsed light, 316–317

Filters:
  antimicrobial, 371
  chemically treated, 227
  DSP-rated filters, 209–210
  electrically charged, 227
  ePTFE filters, 227
  fibrous, 269
  filter application test results, 224–227
  high efficiency particulate air (HEPA) filters, 208
  high-efficiency filters, 224
  ionization of, 325
  medium efficiency air filters, 225
  membrane filters, 227
  modeling performance of, 212
  recirculating air filtration systems, 224
  survival of bacteria in, 222
  ultra-low penetration air (ULPA) filters, 208
  (*See also* Ventilation)
Filtration, 207–227
  of airborne microorganisms, 207–227
  ASHRAE performance scale, 210–211
  and commercial air cleaners, 224–225
  corona precharger, 227
  filter performance curves, 207–210
  fitting models to MERV data, 217–221
  of indoor air, 207
  ion generator, 227
  laboratories, 558
  mathematical model of, 212–215
  microbial filtration test results, 221–224
  microorganisms, 467
  minimum efficiency reporting value (MERV) filter ratings, 210–211
  multifiber filtration model, 215–217
  museums, 613
  and ultraviolet germicidal irradiation (UVGI), 260–262
  with UVGI, 530
  and UVGI systems, 467
  vegetation air cleaning, 354–356
  of viruses, 222
  (*See also* Electrostatic filtration)
Fishery workers, 638
Flood damage, 439
Fomites, 607
Food and Drug Administration (FDA), 567–568
Food disinfection, 571
Food industry, aerobiological concerns in, 567–580
  airborne aflatoxin, 576
  airborne fungi, 572

INDEX

Food industry, aerobiological concerns in (*Cont.*):
  allergenic airborne hazards, 572
  allergens, 571
  allergic reactions, 571
  Anisakis simplex, 571
  antimicrobial food preservatives, 580
  asthma, 575
  bagassosis, 577
  baking industry, 577
  bioaerosol exposure, 572
  biofilms, 579
  brucellosis, 571
  cleanroom technology, 579
  codes and standards, 567–568
  compressed air, 579
  cooking, 572, 574
  disinfection, 579
  epidemiology, 571–574
  Federal Insecticide, Fungicide, and Rodenticide Act (FIFRA), 568
  Food and Drug Administration (FDA), 567–568
  food plant aerobiology, 574–578
  food spoilage microbes, 571
  foodborne pathogens, 568–571, 578
  fumigation, 579
  fungal microorganisms, 580
  good manufacturing practices (GMPs), 567–568
  ionizing radiation, 579
  microbes, 577
  mite sensitization, 572, 574
  mold spores, 568, 570, 579
  negative ionization, 580
  Norwalk virus, 570, 575
  occupational asthma, 571
  Occupational Safety and Health Act (OSHA), 568
  pathogenic bacteria, 575
  poultry plants, 575, 577, 580
  psittacosis, 577
  rat urinary aeroallergen (RUA), 571
  remediation, 578–580
  respiratory symptoms, 572
  slaughterhouses, 575
  storage mites, 571
  storage molds, 571
  UVGI systems, 579
  Ventilation for Acceptable Indoor Air Quality, 568
  VOCs, 574, 578
Food plant aerobiology, 574–578

Food poisoning, 62
Food spoilage microbes, 571
Foodborne pathogens, 568–571, 578
  defined, 568–571
  disease transmission, 71
  survivability of, 578
Foodborne streptococcal pharyngitis, 574
Forced ventilation, 168, 170, 187
Formaldehyde, gaseous, 396
Forward-tipped centrifugal fans, 475
Fourier analysis, 122
Foxing, 606
Fractal dimension, 122
Franscisella tularensis, 156
Free electrons, 324, 335
Free radicals, 397
Freeze drying, 367
Freezing, 159–160
Fuller, Buckminster, 4
Fumigants, 453
  aerosol, 453
  gaseous, 453
Fumigation, 579
Fungal allergens, 106–108
  indoor environments, 438
  in office buildings, 486
Fungal microorganisms, 580
Fungal species:
  I/O ratio in offices, 488–489
  in office buildings, 488–489
  outdoor, 100
Fungal spores, 19–20, 47
  air treatment systems, 477
  in buildings, 439–440
  humidity and temperature variations and, 368
  indoor bioaerosol levels, 438–440
  insulation contamination by, 474
  microbial density of, 126
  office buildings, 490
  in office buildings, 488, 489
  outdoor levels of, 103
  pulsed light eradication, 309
  sewage processing facilities, 674
  source of, 4
Fungi:
  airborne, 572
  allergenic, 4
  on building materials, 175
  and dust, 172
  growth on building materials, 172, 174
  guidelines for controlling, 523
  indoor fungal species, 100

Fungi (Cont.):
  indoors, 513
  libraries, 605–608
  mattresses causing, 516
  microbial cultures, 405
  outdoor levels of, 100
  residential homes, 511, 513
  respiratory infections caused by, 47
  sampling, 404
  on walls in homes, 177
Fungicides, 610
Fur industry, aerobiological concerns in, 657–658

GAC (granulated activated carbon), 269
Gamma rays, 390
Gas phase filtration, 267–278
  aerobiological applications of, 268–269
  carbon adsorption, 270–274
  filtration technologies, 269–270
  impregnated carbon, 274–275
  and MVOCs, 268
  and VOCs, 268
  (See also Carbon adsorption)
Gas plasma sterilization system, 396
Gaseous formaldehyde, 396
Gaseous fumigants, 453
General Mold Remediation Procedures, 454–456
Generation, defined, 82
Geometric growth, Malthus law of, 144–145
Germicidal radiation, 6
Germinated spores, 392, 454
GMPs (good manufacturing practices), 567–568
Good manufacturing practices (GMPs), 567–568
Government buildings, 168
Gradients, thermal, 134–135
Grain fever, 58
Gram-negative bacteria, 440
Granulated activated carbon (GAC), 269
Gravitation, 134
Green building design, 349–363
  aerobiological standards for, 360–363
  air cleaning technology, 350
  airborne microbe minimization, 358
  biofiltration, 354–356
  educational facilities, 499
  hygienic protocols, 357–360
  indoor solar exposure, 353
  ionization levels, 355
  material selectivity, 356–357
  natural ventilation, 350

Green building design (Cont.):
  passive solar exposure, 350–354
  sunlight, 351–354
  vegetation air cleaning, 354–356
  ventilation and, 186
  VOCs, 355
Green Buildings Rating System, 349
Growth:
  defined, 143
  and ionization, 328
  Malthus law of geometric growth, 144–145
  microbial cultures, 407
  mold, 498, 499
  of mold on cooling coils, 492
  temperatures for microbial, 366
Growth model, exponential, 143–145
Gruinard Island, 106
Gymnasiums:
  HVAC system, 620
  IAQ, 625
Gypsum board, 175, 176

Habitats:
  coevolution of airborne diseases, 166
  coevolution with airborne pathogens, 12
  evolution of, 166–168
Haemophilus influenzae, 45, 50
Handheld smoke generator, 465
Handwashing, 358
Hantavirus, 46, 72
Hay fever, 55–56
Health care facilities, aerobiological concerns with, 527–542
  air ionization, 542
  airborne nosocomial epidemiology, 531–533
  dental offices, 538
  filtration with UVGI, 530
  guidelines, codes, and standards, 527–530
  health care workers, 538, 540
  HEPA filters, 528, 529–530
  HIV rooms, 531
  infection rates, 538
  isolation rooms, 529, 530–531
  medical ward (MW), 532
  neonatal intensive care unit (NICU), 538
  nosocomial aerobiology, 533–540
  nosocomial control options, 540–542
  operating rooms, 529, 532, 541
  outside air, 528
  pressurization requirements, 529, 531
  SARS virus, 537–538
  surgical intensive care unit (SICU), 532

Health care facilities, aerobiological concerns with (*Cont.*):
  surgical site infections (SSI), 527, 532
  surgical ward (SW), 532
  TB rooms, 530
  UVGI systems, 541
  ventilation systems, 528
  worker infection, 533
Health care workers, 533, 538, 540
Heating, ventilation, and air-conditioning (HVAC), 187
  and airborne allergens, 225
  aircraft, 633–634
  animal laboratories, 561
  gymnasiums, 620
  health care facilities, 528, 529–530
  laboratories, 549
  museums, 610
  in office buildings, 489
  and photocatalytic reactors, 300
HEPA filters (*see* High efficiency particulate air filters)
Herd immunity, 86
High efficiency particulate air (HEPA) filters, 208
  animal laboratories, 561
  health care facilities, 528, 529–530
  laboratories, 549
High-efficiency filters, 224
High-speed ozone sterilizing device, 343
Histoplasma capsulatum, 105, 106, 126
Histoplasmosis, 653
HIV patients (*see* AIDS patients)
HIV rooms, 531
Hospital-acquired infections (*see* Nosocomial infections)
Hospitals, 538
Hotels, aerobiological concerns with, 516–517
  air handling units, 516–517
  solutions for, 522
House mites:
  dust mite allergens, 486
  eradication with liquid nitrogen, 367
Houseplants, 354–355
Human bioeffluents, 631
Human feces, 672
Humidifier fever, 57
Humidifiers, 191
Humidity:
  educational facilities, 498
  indoor air, 459
  and viruses, 113

HVAC (*see* heating, ventilation, and air-conditioning (HVAC))
Hydrodynamic factor, Kuwabara, 214
Hydrogen peroxide gas plasma sterilization system, 396
Hydroxide radicals, 296
Hygienic coatings (*see* Antimicrobial coatings)
Hygienic protocols, 357–360
Hypersensitivity pneumonitis, 58–59
  airborne microbial causes of, 59
  metal industry, 660

IA2 (two-stage Anderson) sampler, 413
IAQ (*see* Indoor air quality)
ICU, airborne microbes in, 535
ID_{50} (Mean Infectious Dose), 90
Ideal gas law, 267
IE (ionization energy), 330
IEQ (indoor environmental air quality), 349
Immunity, herd, 86
Immunocompromised, 522, 524
Impaction, inertial, 136–137
Impregnated carbon, 274–275
Inactivation rates, 261
Incubators, 405, 407
India:
  dust mites in, 110
  pollen in, 109
Indoor air quality (IAQ):
  and air sampling, 409
  and airborne diseases, 5
  gymnasiums, 625
  remediation, 450
Indoor air sampling, 418–419
Indoor bioaerosol levels, 437–441
  airborne allergens, 441
  airborne bacteria, 440–441
  of airborne fungal spores, 438–440
  airborne pollen, 441
  airborne viruses, 441
Indoor environmental air quality (IEQ), 349
Indoor environments:
  air sampling, 418–419
  air sampling test protocol for, 420
  airborne diseases in, 3–13
  and airborne pathogens, 168
  bacteria, 440
  bacteria sampling, 404
  filtration of, 207
  fungal allergens, 438
  indoor humidity, 459
  mold growth in, 174

Indoor environments (*Cont.*):
  outdoor environments *vs.*, 441
  remediation, 522
  spore levels, 443
  virus sampling, 404
Indoor fungal species, 100
Indoor fungi, 513
Indoor solar exposure, 353
Indoor/outdoor (I/O) ratios (*see* I/O ratios)
In-duct air disinfection systems, 232–233
Industrial bronchitis, 649–650
Industrial facilities, aerobiological concerns with, 647–664
  airborne particulate dust, 649
  asthma, 647–648
  chronic obstructive pulmonary disease (COPD), 651
  dust-induced lung disease, 649
  industrial bronchitis, 649–650
  occupational airborne diseases, 647–653
  occupational asthma, 648–650
  occupational Legionnaires' disease, 662–663
  occupational rhinitis, 648–649
  respiratory diseases, 651–652
  solutions, 664–665
  workers, 649–651
  (*See also specific industries, e.g.:* Metal industry)
Industrial halls, ventilation in, 621
Industrial wastewater, 656
Inertial impaction, 136–137
Infection rates, health care, 538
Infections:
  airborne noncommunicable, 46–47
  children, 501
  of health care workers, 533
  hospital-acquired, 47
  from inhalation, 73
  lower respiratory tract, 50–51
  in lungs, 44–46
  middle respiratory tract, 49–50
  nosocomial, 47, 61
  respiratory, 358
  skin, 62
  spread by health care patients, 536
  upper respiratory tract, 48–49
Influenza, 10, 12, 45
  aircraft, 628
Inhalation, infections from, 73
In-place carbon adsorber testing, 474
In-place filter testing, 465–467

In-place testing of surface disinfection systems, 469–474
In-place UVGI system testing, 467–469
Insulation:
  fungal spore contamination of, 474
  and microbial growth, 177
Interception, 213
Interspecies transmission, 551
I/O (indoor/outdoor) ratios, 441–443
  of fungal species, 488–489
  of pollen, 441–442
  of spores, 442
Ion generators, 227, 324–325, 330
  (*See also* Air ionizers)
Ionic wind, 289
Ionization, 323–330
  of aerosolized virus, 328
  of air filters, 325
  of airborne microbes, 326–329
  alternating current (AC) ionizers, 325
  biocidal effects, 328
  corona-discharge ionizer, 325
  direct current (DC) ionizers, 325
  electronic ionizer, 325
  equipment, 324–326
  food industry, 580
  in hospital wards, 329
  ionization theory, 323–324
  and microbial growth, 328
  of MVOCs, 330
  negative, 580
  negative air, 326
  for odor removal, 330
  of ozonation, 328
  Salmonella enteritidis, 329
  steady-state DC ionizers, 325
  with waterfalls, 355
Ionization energy (IE), 330
Ionization theory, 323–324
Ionizers (*see specific types, e.g.:* Air ionizers)
Ionizing air cleaners, 289
Ionizing radiation, 157–158, 389–392
  alpha rays, 391
  to disinfect surfaces, 391–392
  food industry, 579
  gamma rays, 390
  germinated spores, 392
  libraries, 609
  on various bacteriophages, 391
  x-rays, 391
Ions, superoxide, 296

INDEX

Irradiance, surface, 471
Irradiation, with UV, 243–247, 289
Isolation rooms, 529, 530–531
  dual purpose, 530–531
  negative pressure, 530
  positive pressure, 530

Kakita experiment, 385
Klebsiella, 77
Kuwabara hydrodynamic factor, 214

LAA (laboratory animal allergy), 556
Laboratories, aerobiological concerns with:
  100 percent outside air systems, 549
  ACH, 557
  air changes per hour, 548
  air cleaning, 556–559
  air-exchange rates, 548
  airflow rate, 557
  allergies, 560
  asthma, 560
  biosafety level (BSL), 548
  contaminant, 548
  fan systems, 559
  filtration, 558
  guidelines for, 547–550
  for handling infectious agents, 548
  HEPA filtration, 549
  safety equipment, 558
  UVGI, 549–550, 559
  ventilation, 556–557
  (*See also specific laboratories, e.g.:* Animal laboratories)
Laboratory animal allergy (LAA), 556
Laboratory hoods, 558
Laryngitis, 50
LD_{50} (Mean Lethal Dose), 91–94
Leadership in Energy and Environmental Design (LEED), 349
Leakage:
  buildings, 464–465
  thermal infrared imaging for, 464
Leather industry, aerobiological concerns in, 657–658
LEED (Leadership in Energy and Environmental Design), 349
Legionella, 60–61, 104, 105, 172, 191
Legionella pneumophila, 18, 105, 341, 342–343
Legionnaires' disease (legionellosis), 60
  occupational, 662–663
  pontiac fever *vs.*, 57

Libraries, aerobiological concerns with, 603–610
  aerial disinfection, 609
  air conditioning systems, 603
  bacteria introduction, 605
  book contamination, 608
  carpeting, 605
  chlorine dioxide, 609
  decontamination, 609
  disease carriers, 607
  fomites, 607
  foxing, 606
  fungi, 605–608
  fungicides, 610
  ionizing radiation, 609
  microwaves, 609
  mold, 604
  SBS, 608
  thiabendazole, 609
  ultraviolet light, 609
LIDAR (light detection and ranging), 430
Light detection and ranging (LIDAR), 430
Lipopolysaccharide, 440
Lipopolysaccharide (LPS), 440
Liquid nitrogen, 367
Listeria, 152
Lodging facilities, 168
Log phase, 144
Lower respiratory tract infections, 50–51
LPS (*see* Lipopolysaccharide)
Lung disease, dust-induced, 649
Lungs:
  airborne pathogens in, 45
  diseases in (*see* Respiratory diseases)
  infection of, 44–46
  phagocytosis in, 45
  respiratory system, 42

Malthus law of geometric growth, 144–145
Mankind, and airborne pathogens, 10–12
MAS (Merck) impaction sampler, 413
Mass spectrometry, 430
Mathematical model of filtration, 212–215
Mattresses, 516
Mean disease period (MDP), 89
Mean infectious dose (ID_{50}), 90
Mean infectious period (MIP), 89
Mean lethal dose (LD_{50}), 91–94
Medical ward (MW), 532
Medium efficiency air filters, 225
Membrane filters, 227

Meningitis:
  bactericidal, 629
  college campuses, 517
  on college campuses, 504
Merck (MAS) impaction sampler, 413
MERV (minimum efficiency reporting value)
  filter ratings, 210–211
Metal industry, aerobiological concerns in,
  658–660
  allergic reactions, 659
  chronic bronchitis, 658
  fume inhalation, 659
  hypersensitivity pneumonitis, 660
  nonbiological pollutants, 659
  respiratory ailments, 658
  tuberculosis, 660
Methicillin-resistant *Staphylococcus aureus*
  (MRSA), 537
Miami Electron Beam Research Facility, 392
Microbes:
  aspect ratio of, 123
  endogenous, 533
  food industry, 577
  food spoilage, 571
  under heat exposure, 366
  ionization of, 326–329
  optimum growth temperatures for, 366
  in sewage, 673
  size distribution of, 126–131
  thermal death of, 112
  ventilation systems for growth of, 376
Microbial agent ingestion, 91–92
Microbial contamination:
  museums, 612, 613
  spacecraft, 634
Microbial cultures:
  bacteria, 405
  fungi, 405
  growth, 407
  incubators, 405, 407
  on a petri dish, 405, 406
  sampling, 405–408
Microbial density, 126
  of airborne microbe, 126
  of bacteria, 126
  of fungal spores, 126
  of pollen, 126
  of spores, 126
Microbial disinfection, 143–161
  exponential decay model, 145–148
  exponential growth model, 143–145
  multihit target model, 151–153

Microbial disinfection (*Cont.*):
  photoreactivation, 153
  recovery, 153
  shoulder model, 148–151
  two-stage decay model, 148
Microbial filtration, 221–224
Microbial growth:
  and condensation, 190
  cooling coils, 189
  on filters, 234
  and humidifiers, 191
  in insulation, 177
  and ionization, 328
  ultraviolet germicidal irradiation (UVGI) for,
    233–234
Microbial levels, in operating rooms, 541
Microbial volatile organic compounds
  (MVOCs), 3, 27, 29
  causing respiratory irritation, 44
  destruction, 301–302
  and gas phase filtration, 268
  ionization of, 330
  and photocatalytic oxidation, 295
Microbiological air sampling, 655
Microbiological toxins, 550
Micrococcus species, 103
Microflora:
  airborne, 469
  buildings, 469
Microorganisms:
  airborne, 487
  dehydration of, 113–114
  desiccation of, 113–114
  filterability, 467
  identification of, 428
  in office buildings, 487
  photocatalysis of, 300
  sampling stress, 410
  survival of, 112–114
Microwave athermal disinfection, 383–386
Microwave thermal disinfection, 383
Microwaves, 379–386
  disinfection, 380
  libraries, 609
  polar molecules, 380
  for sterilization, 379–380
  thermal disinfection, 379–383
  and UVGI, 380
Middle respiratory tract infections, 49–50
Mill fever, 57
Mineral industry, aerobiological concerns in,
  661–662

Minimum efficiency reporting value (MERV) filter ratings, 210–211
Minute virus of mice (MVM), 558
MIP (mean infectious period), 89
Mite sensitization, 572, 574
Mites, 23–24
  eradication with liquid nitrogen, 367
  food industry, 571
  office building allergens, 486
  residential homes, 511, 513
  storage, 571
  (*See also* Dust mites; Storage mites)
Moisture damage, 505–506
Molds:
  in air filter media, 223
  educational facilities, 498, 499
  food industry, 568, 570, 571, 579
  growth in buildings, 459
  growth in indoor environments, 174
  growth on cooling coils, 492
  libraries, 604
  remediation procedures, 454–456
  respiratory allergy effects, 439
  storage, 571
  visible light disinfection, 351
Monitoring (*see* Biodetection)
Most penetrating particle (MPP), 479
MRSA (methicillin-resistant *Staphylococcus aureus*), 537
Mucosal irritation, 630
Multifiber filtration model, 215–217
Multihit target model, 151–153
Multizone modeling, 201–204
Museums, aerobiological concerns with, 610–615
  biological contamination, 611
  decontamination/disinfecting, 614–615
  dust, 610
  fabrics, 613
  filtration, 613
  HVAC systems, 610
  lack of light, 610
  microbial contamination, 612, 613
MVM (minute virus of mice), 558
MVOC (*see* Microbial volatile organic compounds)
MW (medical ward), 532
Mycobacterium Leprae, 12
Mycobacterium luteus, 222, 223
Mycobacterium parafortuitum, 250
Mycobacterium tuberculosis, 8, 12, 14, 42, 239
Mycoplasma hypopneumoniae, 112

Mycoplasma pneumoniae, 45
Mycoses, occupational, 60
Mycotoxicosis, pulmonary, 60
Mycotoxins, 27–29
Myxosporea spores, 126

Natural indoor decay rate, 140
Natural ventilation, 170, 186–187, 350, 487
Naturally ventilated cages, 560
Negative air ionization, 326
Negative ionization, 580
Negative pressure isolation rooms, 530
Neisseria meningitidis, 11
Neonatal intensive care unite (NICU), 538
Nephelometry, 429
Newcastle disease, 326
NICU (neonatal intensive care unit), 538
NLV (Norwalk-like viruses), 638–639
Nonbiological pollutants, 659
Noncatalytic biosensors, 433
Noncommunicable airborne infections, 46–47
Noncommunicable diseases:
  epidemiology of, 82–87
  transmission of, 72–73
Noninfectious respiratory diseases, 51–61
  allergies, 54–55
  asthma, 52–53
  atopy, 55
  bagassosis, 56–57
  byssinosis, 57
  farmer's lung, 56
  grain fever, 58
  hay fever, 55–56
  humidifier fever, 57
  hypersensitivity pneumonitis, 58–59
  Legionella, 60–61
  Legionnaires' disease (legionellosis), 60
  mill fever, 57
  nonseasonal allergic rhinitis, 56
  occupational asthma, 53
  occupational mycoses, 60
  organic dust toxic syndrome (ODTS), 57
  pigeon breeder's disease, 57
  pontiac fever, 57–58
  pulmonary mycotoxicosis, 60
  reactive airways dysfunction syndrome (RADS), 53
  sump fever, 58
  swine fever, 58
  toxic reactions, 58, 60
  wood-trimmer's disease, 58

Nonionizing electromagnetic radiation, 379
Nonionizing radiation, 158–159
Nonrespiratory airborne diseases, 62
  food poisoning, 62
  skin infections, 62
Nonrespiratory diseases, 71
Nonseasonal allergic rhinitis, 56
Nonuniformity of airflows, 620
Norwalk virus, 570, 575
Norwalk-like viruses (NLV), 638–639
Nose blowing, 75
Nosocomial aerobiology, 533–540
Nosocomial control options, 540–542
Nosocomial (hospital-acquired) infections, 47
  airborne nosocomial epidemiology, 531–533
  construction-related nosocomial infections, 538
  nosocomial control options, 540–542
  pneumonia, 61
  and worker fatalities, 533
  (*See also specific types, e.g.:* SARS)
Nosocomial infections, construction-related, 538
Nosocomial respiratory diseases, 61–62
Nutrient deprivation, 159
Nylon flocking industry, aerobiological concerns in, 657–658

OAF (open air factor), 139
OAUGDP (One Atmosphere Uniform Glow Discharge Plasma), 393
Occupational airborne diseases, 647–653
Occupational airway diseases, 648
Occupational asthma, 53
  food industry, 571
  industrial facilities, 648–650
Occupational Legionnaires' disease, 662–663
Occupational mycoses, 60
Occupational rhinitis, 648–649
Occupational Safety and Health Act (OSHA), 568
Odor removal, 330
ODTS (*see* Organic dust toxic syndrome)
ODTS (organic toxic dust syndrome), 586, 588
Office buildings, aerobiological concerns with, 485–493
  aerobiology of, 487–491
  airborne bacteria, 490
  airborne fungal spores, 488
  airborne microorganisms, 487
  airway inflammation, 486
  building related illness, 487

Office buildings, aerobiological concerns with (*Cont.*):
  common cold, 486
  dampness, 486
  design of, 493
  epidemiology, 486–487
  fungal allergens, 486
  fungal species, 488–489
  fungal spores, 489, 490
  house dust mite allergens, 486
  HVAC systems, 489
  and reducing health hazards, 492
  respiratory tract symptoms, 486
  SARS, 487
  SBS/BRI, 487
  sick building syndrome, 487
  solutions, 491–493
  standards and guidelines, 491
  tuberculin skin test (TST), 486
  UVGI, 492
  ventilation, 485–486
One Atmosphere Uniform Glow Discharge Plasma (OAUGDP), 393
100 percent outside air systems, 188
100 percent outside air systems, 549
Open air factor (OAF), 139, 157
Operating rooms, 529, 532, 541
  air cleanliness, 532
  airborne microbes in, 535
  microbial levels in, 541
  UVGI systems in, 541
Opportunistic respiratory diseases, 61–62
Organic compounds:
  microbial volatile organic compounds (MVOCs), 3
  volatile organic compounds (VOCs), 5
Organic dust toxic syndrome (ODTS), 57
Organic toxic dust syndrome:
  sewage processing facilities, 678
Oriented strandboard (OSB), 451
OSB (oriented strandboard), 451
OSHA (Occupational Safety and Health Act), 568
Outdoor air:
  dilution with, 201
  and health care facilities, 528
Outdoor airborne animal diseases, 111–112
Outdoor allergens, 99–104
Outdoor environments:
  aerobiological engineering in, 99–114
  air sampling, 419
  allergic reactions, 106

Outdoor environments (*Cont.*):
  asthma, 110–111
  bacteria levels, 103
  birds effect on, 106
  fungal allergies, 106–108
  fungal spores, 103
  fungus, 100
  indoor environments *vs.*, 441
  and pollen, 100
  pollen allergies, 108–110
Outdoor fungal species, 100
Outdoor pathogenic disease, 104–106
Outdoor pathogens, 99–104
Overall filter efficiency (E), 212–213
Oxidation, 6
Oxygen plasma discharges, 394
Oxygen radicals, 394
Oxygenation, 157
Ozonation, 328
Ozone, 333–344
  as aerial disinfectant, 336, 337
  bacteria inactivation in, 338
  bacterial resistance, 342
  biocidal effects of, 342
  for biological pathogen decontamination, 344
  in breathing air spaces, 337
  building damage, 459
  for building decontamination, 341
  by-products, 340
  chemical decontamination procedures, 458–459
  chemistry, 333–336
  as a deodorizer, 344
  dosimetry, 336
  and *E. coli*, 341
  free electrons, 335
  half-life, 339
  health hazards, 340
  high-speed sterilizing device, 343
  for microbial population reduction, 343
  reactivity of, 335
  remediation, 450
  removal, 339–340
  for surface sterilization, 338–339
  system performance, 340–344
  as terminal disinfectant, 340–341
  ultraviolet lamp production of, 335
  for water disinfection, 333
Ozone disinfection, airborne, 336–338
Ozone generators, 324

Panonychus citri, 110
Paper industry, aerobiological concerns in, 653–656
  dust, 653
  histoplasmosis, 653
Parainfluenza, 45
Particle counting, 413–414, 419
Particle detectors, 427–430
Particle dynamics, 119–140
Particle sizers, 414, 428
Parvoviruses, 11
Passive solar exposure, 350–354
Pasteurella tularensis, 352
Pathogenic bacteria, 575
Pathogenic disease, outdoor, 104–106
Pathogenic viruses:
  apartment buildings, 518
  residential homes, 518
  source of, 4
Pathogens:
  agricultural, 585–586
  animal, 585–586
  animal husbanding, 586
  animal laboratories, 550–555
  foodborne, 568–571, 578
  outdoor, 99–104
  (*See also* Airborne pathogens)
PCO (photocatalytic oxidation), 267
Peak expiratory flow rate (PEFR), 225
Peclet number, 214
PEF (*see* Pulsed electric fields)
PEFR (peak expiratory flow rate), 225
Penetrating dampness, 171
Penetrometer, 466
Penicillium, 225
Penicillium brevicompactum, 223
Penicillium melinii, 223
Petri dish:
  microbial cultures, 405, 406
  for surface decontamination, 473
Pets allergens, 519
Phagocytosis:
  in lungs, 45
  respiratory system, 42
Pharyngitis, 49
Photocatalysis:
  of microorganisms, 300
  VOC decomposition, 296
Photocatalytic oxidation (PCO), 6, 267, 295–303, 353
  air disinfection, 297–301
  animal laboratories, 562

Photocatalytic oxidation (PCO) (*Cont.*):
  applications, 302–303
  bactericidal effect of, 298
  catalytic incineration *vs.*, 301
  defined, 295
  filters, 301
  MVOC destruction, 301–302
  and MVOCs, 295
  reaction rate, 296
  surface disinfection, 302
  theory and operation, 295–297
  and titanium dioxide (TiO_2), 295–297
  transparent PCO films, 302
  VOC conversion, 301
  for wastewater treatment, 296
Photocatalytic reactors, 300
Photoelectric detection, 433
Photometer, 466
Photophoresis, 135
Photoreactivation, 153, 250–251
Pigeon breeder's disease, 57
Plague, 7
Plant aerobiology, 574–578
Plasma technology, 393–394
  and chemical disinfection technologies, 394
  to disinfect medical equipment, 394
  hydrogen peroxide gas plasma sterilization system, 396
  oxygen plasma discharges, 394
Pneumocystis carinii, 20
Pneumonia, 50–51, 61, 502
Poliovirus (PV1), 342
Pollen allergens, 21–23
  agricultural facilities, 586
  and air pollution, 110
  climate changes effecting, 109
  in India, 109
  indoor bioaerosol levels, 441
  I/O ratio, 441–442
  microbial density of, 126
  in outdoor air, 108–110
  outdoor studies of, 100
  removal with ESP, 291
  seasonal effects of, 109
Pollen spores, 404
Pollutant accumulation, 498
Pollutants:
  nonbiological, 659
  sampling, 405
Pollution (*see* Air pollution)

Polymerase chain reaction:
  airborne PCR systems, 431
  E. coli identification, 432
  virus detection, 431
Polymerase chain reaction (PCR), 430–432
Pontiac fever, 57–58
Positive pressure isolation rooms, 530
Potable water supply, 191
Poultry houses, 326
Poultry plants, 575, 577, 580
Pressurization, 160, 191, 464, 529, 531
Prions, 16
Protozoa, 20
Psittacosis, 577
Pseudomonas fluorescens, 344
Pulmonary mycotoxicosis, 60
Pulmonary region, 42
Pulsed corona discharge systems, 395
Pulsed electric fields (PEF):
  decay curve, 317–318
  disinfection, 317
  pulsed light, 317–318
Pulsed lamp, 309
Pulsed lasers, 318–319
Pulsed light, 307–319
  bacteria eradication, 309
  disintegration, 316
  filtered, 316–317
  fungal spore eradication, 309
  number of pulses in system, 311
  pulsed electric fields, 317–318
  pulsed lasers, 318–319
  pulsed UV, 307–315
  pulsed white light, 307–315
  UVGI *vs.*, 309
Pulsed microwaves, 383
Pulsed UV (PUV), 307–315
Pulsed white light, 307–315
Pulsed white light (PWL):
  bacteria damage, 315
  dosage, 312–313

Q fever, 657

Radiation:
  electromagnetic, 134–135
  in food industry, 579
  ionizing, 157–158
  ionizing radiation, 389–392
  nonionizing, 158–159
  ultraviolet germicidal irradiation (UVGI), 6
  (*See also* Ionizing radiation)

Radiative irradiance, 240
RADS (*see* Reactive airways dysfunction syndrome)
Rat urinary aeroallergen (RUA), 556, 560, 571
RCS (Biotest), 413
RE (*see* Removal efficiency)
RE (removal efficiency), 479
Reactive airways dysfunction syndrome (RADS), 53
Recirculating air filtration systems, 224
Recirculation, 224
Recovery, 153
 of spores, 153
 during UV exposure, 153
Reflectivity:
 diffuse, 244–245
 enclosure, 243–247
 specular, 245–247
Reflectivity of enclosure, 243–247
Relative humidity (RH), 233, 249–250, 513
Release rate (RR_T), 478
Remediation, 449–459
 air cleaning, 449–450
 building material removal, 451
 chemical decontamination procedures, 456–459
 food industry, 578–580
 General Mold Remediation Procedures, 454–456
 indoor air, 522
 indoor air quality, 450
 mold remediation procedures, 454–456
 ozone systems, 450
 sunlight for, 454
 surface cleaning, 450–451
 time and weathering, 454
 whole building decontamination, 451–454
Removal efficiency (RE), 479
Residential homes, aerobiological concerns with, 511–516
 aerobiology of, 511
 allergens, 518
 animal allergens, 515–516
 animal dander, 511
 asthma symptoms, 516
 bacteria levels, 512
 carpeting, 519
 cat allergens, 516
 central air handling units, 520
 dust mites, 511, 513, 516
 fungal aerobiology, 511
 for the immunocompromised, 522, 524

Residential homes, aerobiological concerns with (*Cont.*):
 indoor fungi, 513
 mattresses, 516
 pathogenic viruses, 518
 relative humidity (RH), 513
 respiratory infections, 511
 solutions for, 517–521
 UVGI lamps, 521
 ventilation, 518–520
 water damage, 512, 518
Respiratory allergy effects:
 and children, 108
 and dampness, 439
 and mold exposure, 439
Respiratory disease transmission, 69–82, 358, 360
 coughing, 71, 75
 educational facilities, 503
Respiratory diseases:
 gram-negative bacteria, 440
 industrial facilities. aerobiological concerns with, 651–652
 medical costs of, 68–69
 noninfectious, 51–61
 nosocomial, 61–62
 opportunistic, 61–62
 and teachers, 504
 (*See also specific diseases, e.g.:* Asthma)
Respiratory infections:
 airborne, 48–62
 caused by bacteria, 47
 caused by fungus, 47
 children, 502
 and children, 358
 isolation of individuals, 358
 lower tract, 50–51
 in metal industry, 658
 middle tract, 49–50
 residential homes, 511
 on ships, 636
 textile industry, 658
 upper tract, 48–49
Respiratory irritation:
 in MVOCs, 44
 sampling, 404–405
Respiratory symptoms:
 food industry, 572
 in office building workers, 486
Respiratory syncytial virus (RSV):
 educational facilities, 499
 transmission of, 73

Respiratory system, 41–44
  breathing, 42–43
  coughing, 42
  lungs, 42
  phagocytosis, 42
  pulmonary region, 42
  sneezing, 42
  tracheobronchial region, 42
  (*See also specific parts, e.g.:* Lungs)
Respiratory viruses, communicable, 46
RH (*see* Relative humidity)
Rhinitis, 49
  allergic nonseasonal, 56
  occupational, 648–649
Rhinovirus:
  educational facilities, 501
  transmission of, 73, 79
Rhizopus nigricans, 150
Rising dampness, 171
Rotating slit sampler, 411
RR_T (release rate), 478
RSV (respiratory syncytial virus), 73, 499
RUA (rat urinary aeroallergen), 560

Safety equipment, 558
Salmonella, 343
Sampling, 403–423
  allergens, 414
  of allergens, 404–405
  applications, 421–423
  automated sampling systems, 410
  of bacteria in indoor environments, 404
  bioaerosols, 413
  of dander, 405
  electrostatic sampling devices, 412
  of endotoxins, 405
  of exotoxins, 405
  of fungi, 404
  of microbial cultures, 405–408
  of pollen spores, 404
  of pollutants, 405
  procedures, 414–420
  of respiratory irritants, 404–405
  rotating slit sampler, 411
  upper room UVGI systems, 423
  UVGI air disinfection systems, 422–423
  UVGI cooling coil disinfection systems, 421–422
  of viruses in indoor environments, 404
SARS virus, 85
  in biological laboratories, 550
  in health care facilities, 537–538
  and office buildings, 487

SBS (*see* Sick building syndrome)
SBS/BRI:
  educational facilities, 499
  in office buildings, 487
  (*See also* Sick building syndrome; Building related illness)
Scarlet fever, 504
Self-disinfection, of buildings, 351
Serratia marcescens, 249
Settle plates, 409, 417, 418
Sewage processing facilities, 365–367, 671–680
  aerobiology of, 671–674
  air treatment solutions for, 678–680
  aspergillosis, 676
  biofilters, 679
  domestic waste, 678
  dust, 679
  epidemiology in, 674–678
  fungal spores, 674
  occupational hazards, 674, 676–678
  organic toxic dust syndrome, 678
  solid waste, 677
  source control, 679–680
  viruses, 673
  waste collection, 678
Sewage sludge, 672
Shipboard diseases, 637
Ships, aerobiological concerns with, 636–639
  crab allergens, 637
  cruises, 637
  diarrhea, 636
  fishery workers, 638
  Norwalk-like viruses (NLV), 638–639
  respiratory illnesses, 636
  shipboard diseases, 637
  snow mountain agent (SMA), 638
  standardized mortality ratio (SMR), 638
  tuberculosis, 638
  waterborne pathogens, 638
Shoe industry, aerobiological concerns in, 658
Shopping malls, 623, 624
Shoulder model, 148–151
Sick building syndrome (SBS), 165
  amd volatile organic compounds, 5
  BRI *vs.*, 5
  libraries, 608
  in office buildings, 487
SICU (surgical intensive care unit), 532
Silver:
  antimicrobial coatings, 371
  biocidal effects of, 372

Single-fiber efficiencies:
  for diffusion, 213
  for interception, 213
Single-zone steady-state model, 192–195
Single-zone transient modeling, 197–201
Sinonasal cancers, in wood industry, 655
Sinusitis, 49
Size distribution curve, 128
Skin infections, 62
Slaughterhouses, 575
Slip factor, Cunningham, 215
Slit sampler, rotating, 411
Sludge, 672, 673 (*See also* Sewage processing facilities)
SMA (snow mountain agent), 638
Smoke generator, 465
Smoking, on aircraft, 633
SMR (standardized mortality ratio), 638
Sneezing:
  respiratory disease transmission, 71, 75
  respiratory system, 42
SNL foam procedures, 459
Snow mountain agent (SMA), 638
Solar exposure, 350–354
  indoor, 353
  passive, 350–354
Solid waste:
  sewage processing facilities, 677
Solvents, 452
Sonic generator, 393
Sonication, 301
Soper equation, 82, 84
Space charge, 324
Spacecraft, aerobiological concerns with, 634–635
  microbial contamination, 634
  preventative measures, 635
Spectrometry, mass, 430
Specular reflectivity, 245–247
Spoilage microbes, 571
Spores:
  and cooling coils, 190
  food industry, 568, 570, 579
  fungal, 4, 47, 674
  germinated, 454
  I/O ratio, 442
  levels for indoor airborne spores, 438
  levels in indoor air, 443
  microbial density of, 126
  office buildings, 490
  in office buildings, 488, 489
  in old houses, 438
  recovery of, 153

Spores (*Cont.*):
  sewage processing facilities, 674
  UVGI, 562
SPR (surface plamon resonance), 432
SS (steady state) equation, 478
SSI (*see* Surgical site infections)
Stachybotrys, 175, 179
Stachybotrys chartarum, 100
Stadiums:
  CFD modeling for, 620, 621
  disease transmission, 622
  epidemiology in, 622
  nonuniformity of airflows, 620
  underfloor air supply systems, 621
  ventilation, 620
Stand-alone recirculation units, 278
Standardized mortality ratio (SMR), 638
Staphylococcus, 77
Staphylococcus aureus, 151, 351
Steady state (SS) equation, 478
Steady-state DC ionizers, 325
Steady-state model, single-zone, 192–195
Sterile templates, 472
Sterilization, 161
  high-speed ozone sterilizing device, 343
  hydrogen peroxide gas plasma sterilization system, 396
  microwaves for, 379–380
  with surface irradiance, 471
  UV irradiance field, 470
  x-rays for air sterilization, 392
Stokes flow, 284
Stomatitis, 49
Storage death, 160
Storage mites, 23–24, 571
Storage molds, 571
Streptococcal pharyngitis, foodborne, 574
Streptococcus pneumoniae, 50, 51
Streptococcus pyogenes, 11, 249, 574
Streptomyces albus, 12
Submarines, aerobiological concerns with, 639–640
Sump fever, 58
Sunlight:
  green building design, 351–354
  for remediation, 454
Superoxide ions, 296
Surface cleaning, 450–451
Surface decontamination, 471–473
Surface discharge lamp, 309
Surface disinfection:
  photocatalytic oxidation, 302
  ultraviolet germicidal irradiation (UVGI), 235

Surface disinfection systems, 469–474
Surface irradiance, 471
Surface microflora, 403–405
Surface plamon resonance (SPR), 432
Surface sampling, 403–423
   building related illness, 415
   procedures, 415–417
   and surface microflora, 403–405
   test protocol for, 416
Surface sterilization, 338–339
Surface UV disinfection system, 471
Surgeons, 536
Surgical intensive care unit (SICU), 532
Surgical site infections (SSI), 527, 532
   airborne microbes, 532
   rate of, 533
Surgical ward (SW), 532
SW (surgical ward), 532
Swine fever, 58
Symbiotic microbes, 354

TAB (testing and balancing), 463
TAOS (two-dimensional angular optical scattering), 429–430
Target model, multihit, 151–153
TB rooms, 530
Teachers, 504
Templates, sterile, 472
Testing:
   building leak and pressure testing, 464–465
   and efficiency filters, 465
   guidelines for UVGI, 467
   in-place carbon adsorber testing, 474
   in-place filter testing, 465–467
   in-place testing of surface disinfection systems, 469–474
   in-place UVGI system testing, 467–469
   of surface decontamination system, 471–472
   ventilation system retrofit testing, 475
   ventilation system testing, 463–464
Testing and balancing (TAB), 463
Textile industry, aerobiological concerns in, 657–658
   anthrax, 657
   Q fever, 657
   respiratory disorders, 658
Theaters:
   dust, 625
   underfloor air supply systems, 621
Thermal cycling, 368
Thermal death:
   of microbes, 112
   time curve, 155

Thermal disinfection, 153–156, 365–367, 379–383
Thermal gradients, 134–135
Thermal infrared imaging, 464
Thermal removal methods, 452–453
Thermophoresis, 135
Thiabendazole, 609
Three-dimensional (3D) irradiance field, 240–243
Threshold limit value (TLV), 340, 453
Time curve thermal death, 155
Titanium dioxide (TiO_2):
   photocatalytic oxidation (PCO), 295–297
   sonication of, 301
TLV (*see* Threshold limit value)
TLV (threshold limit value), 340
To disinfect surfaces, 391–392
Tonsillitis, 49
Total VOCs (TVOCs), 356
Toxic gases, 588
Toxic reactions, 58, 60
Toxins, 26–29, 91–92
Tracheobronchial region, 42
Trains, aerobiological concerns with, 640–641
Transient modeling, single-zone, 197–201
Transparent PCO films, 302
TST (tuberculin skin test), 486
Tuberculin skin test (TST), 486
Tuberculosis (TB), 6–8, 12–14, 49, 50
   assembly places, 622
   in British schools, 502
   children, 502
   exhaust air elimination of, 257
   and indoor limiting levels, 443
   infections among health care workers, 533
   metal industry, 660
   Mycobacterium, 8, 12, 14, 42
   ships, 638
   TB rooms, 530
   ventilation, 533
   without treatment, 88
Turbulent diffusion, 135–136
TVOCs (total VOCs), 356
Two-dimensional angular optical scattering (TAOS), 429–430
Two-stage Anderson (IA2) sampler, 413
Two-stage decay model, 148
Two-stage electrostatic precipitator, 281–285

URI (upper respiratory infection), 499
Uganda, airborne spores in, 108
ULPA (ultra-low penetration air) filters, 208
Ultra-low penetration air (ULPA) filters, 208

INDEX **845**

Ultrasonication, 392–393
Ultraviolet (UV) exposure, 233
Ultraviolet germicidal irradiation (UVGI), 6, 231–262
   air disinfection systems, 467
   air mixing effects, 248–249
   air recirculation units, 234–235
   air temperature effects, 251–252
   animal laboratories, 562
   bacteria, 562
   cooling coil irradiation, 255–256
   defined, 231
   design guides for, 256–257
   direct exposure, 234
   disinfection modeling, 237–240
   disinfection theory, 235–237
   and dust, 190
   exponential decay model, 237
   in field trials, 257–260
   and filtration, 467
   filtration and, 260–262
   filtration with, 530
   food industry, 579
   in health care facilities, 541
   in-duct air disinfection systems, 232–233
   in laboratories, 549–550, 559
   for microbial growth control, 233–234
   microbial growth on filters, 234
   microwaves, 380
   in office buildings, 492
   in operating rooms, 541
   performance optimization, 252–255
   photoreactivation, 250–251
   pulsed light *vs.,* 309
   reflectivity of enclosure, 243–247
   relative humidity (RH), 233, 249–250
   spores, 562
   surface disinfection, 235
   testing, 467–469
   testing guidelines, 467
   three-dimensional (3D) irradiance field, 240–243
   ultraviolet (UV) exposure, 233
   UVGI rating value (URV), 243
   upper room UVGI, 233
   upper room UVGI systems, 423
   UV photometer, 467
   UVGI air disinfection systems, 422–423
   UVGI cooling coil disinfection systems, 421–422
   viruses, 562
Ultraviolet (UV) light, 158–159, 289
   for decontamination, 255–256

Ultraviolet (UV) light (*Cont.*):
   libraries, 609
   recovery during exposure to, 153
   UVGI rating value (URV), 243
Underfloor air supply systems:
   stadiums, 621
   theaters, 621
University auditoriums, 624
University of Leeds, 329
Upper respiratory infection (URI):
   aircraft, 629
   educational facilities, 499
Upper respiratory tract infections (URI), 48–49, 69, 170
Upper room UVGI, 233
URI (*see* Upper respiratory tract infections)
URV (UVGI rating value), 243
U.S. Environmental Protection Agency (USEPA), 43
U.S. Green Building Council (USGBC), 349
USEPA (*see* U.S. Environmental Protection Agency)
USGBC (U.S. Green Building Council), 349
UV (*see* Ultraviolet light)
UV irradiance field, 243–247, 470
UV lamps, 233–234
   ozone production, 335
   ratings of, 243
UV light, 522
UV photometer, 467
UV systems:
   effectiveness of, 470
   surface UV disinfection system, 471
UVGI (*see* Ultraviolet germicidal irradiation)
UVGI lamps, 521

Vaccination programs, 86
Van der Waals force, 270
Vaporized hydrogen peroxide (VHP), 453
Variable air volume (VAV) systems, 188
VEE (Venezuelan equine encephalitis), 550
Vegetation air cleaning, 354–356
Veneaxial fans, 475
Venezuelan equine encephalitis (VEE), 550
Ventilation, 168, 170, 185–204
   100 percent outside air systems, 188
   in animal facilities, 592–597
   apartment buildings, 518–520
   assembly places, 619–622
   biocontamination in, 189–190
   computational fluid dynamics (CFD) modeling, 204
   constant volume (CV) ventilation, 187

846 INDEX

Ventilation (*Cont.*):
  dedicated outside air systems (DOAS), 188
  dilution modeling, 195–197
  dilution ventilation, 191–192, 557
  displacement ventilation, 188–189
  dust in, 190
  educational facilities, 498–499, 506
  fans used in, 475
  forced, 168, 170, 187
  forced ventilation, 187
  and green building design, 186
  health care facilities, 528
  for human comfort, 619
  in industrial halls, 621
  injecting test microbes into system, 469
  laboratories, 556–557
  as microbial growth source, 376
  multizone modeling, 201–204
  natural, 170, 186–187, 350, 487
  naturally ventilated cages, 560
  in office buildings, 485–486
  pressurization control, 191
  residential homes, 518–520
  single-zone steady-state model, 192–195
  single-zone transient modeling, 197–201
  in stadiums, 620
  surface cleaning, 450
  tuberculosis, 533
  variable air volume (VAV) systems, 188
  (*See also* Filters)
Ventilation for Acceptable Indoor Air Quality, 568
Ventilation system model equation, 84–85
Ventilation system retrofit testing, 475
Ventilation system testing, 463–464
Veterinarians, 588
VHP (vaporized hydrogen peroxide), 453
Viruses, 16–17
  aerosolized, 328
  filtration of, 222
  and humidity, 113
  indoor bioaerosol levels, 441
  pathogenic, 4, 518
  PCR detection, 431
  residential homes, 518
  respiratory, 46
  in sewage, 673
  source of, 4
  UVGI, 562
Vitruvius, 167
Volatile organic compounds (VOCs), 27
  agricultural facilities, 591

Volatile organic compounds (VOCs) (*Cont.*):
  conversion, 301
  and cooking, 574
  decomposition, 296
  decomposition of, 296
  educational facilities, 501
  food industry, 574, 578
  and gas phase filtration, 268
  and SBS, 5
  and thermal disinfection, 366

Waste processing facilities (*see* Sewage processing facilities)
Wastewater treatment, 296
Water activity (Aw), 174
Water damage:
  apartment buildings, 518
  buildings, 459
  educational facilities, 499
  residential homes, 512, 518
Waterborne pathogens:
  disease transmission, 71
  ships, 638
Weathering, 454
Whole building decontamination:
  aerosol fumigants, 453
  chemical disinfectants, 453
  chemical foams, 453–454
  chlorine dioxide (ClO_2), 457
  gaseous fumigants, 453
  remediation, 451–454
  solvents and detergents, 452
  thermal removal methods, 452–453
Whooping cough, 503
Wood industry, aerobiological concerns in, 653–656
  currency notes, 656
  dust, 653, 655
  industrial wastewater, 656
  microbiological air sampling, 655
  sinonasal cancers, 655
Wood-trimmer's disease, 58
Work-related sensitization, 556

X-rays:
  electron beams, 392
  ionizing radiation, 391
  to sterilize air, 392

Zonal protection factor (ZPF), 479
Zoonotic diseases, 46, 550, 551, 588
Zoonotic pathogens, 550
ZPF (zonal protection factor), 479